UNDERSTANDING

Social Problems

11e

Linda A. Mooney
EAST CAROLINA UNIVERSITY

Molly Clever
WEST VIRGINIA WESLEYAN COLLEGE

Marieke Van Willigen
GEORGIA SOUTHERN UNIVERSITY

Australia • Brazil • Canada • Mexico • Singapore • United Kingdom • United States

Understanding Social Problems, Eleventh Edition
Linda A. Mooney, Molly Clever, and Marieke Van Willigen

SVP, Higher Education & Skills Product: Erin Joyner

VP, Higher Education & Skills Product: Thais Alencar

Product Director: Laura Ross

Product Manager: Kori Alexander

Product Assistant: Shelby Blakey

Learning Designer: Emma Guiton

Digital Delivery Lead: Matt Altieri

Senior Content Manager: Diane Bowdler

Senior Marketing Manager: Tricia Salata

Intellectual Property Analyst: Deanna Ettinger

Intellectual Property Project Manager: Kelli Besse

Production Service: MPS Limited

Text and Cover Designer: Nadine Ballard

Cover Images:
Protest Group: Jacob Lund/Shutterstock.com
Factories: TR STOK/Shutterstock.com
Medical: kali9/iStock/Getty Images Plus/Getty Images
Wallet: perfectlab/Shutterstock.com
iPhone: BLACKDAY/Shutterstock.com

Library of Congress Control Number: 2020914277

ISBN: 978-0-357-50742-1
Loose-leaf Edition ISBN: 978-0-357-50743-8

Cengage
200 Pier 4 Boulevard
Boston, MA 02210
USA

Printed at CLDPC, USA, 04-21

IN MEMORIAM

This book is dedicated to the memory of Supreme Court Justice Ruth Bader Ginsberg (1933–2020), Congressman John Lewis (1940–2020), and Senator John McCain (1936–2018), who lived their lives with integrity, humility, and righteousness.

IN HONOR

Of the lesser known heroes of the world—the doctors, nurses, and other health care workers, teachers, agricultural laborers, bus and truck drivers, first responders, retail workers, and all the other essential workers who risk their lives daily for others.

Brief Contents

Contents

PART 1 SOCIOLOGY AND THE STUDY OF SOCIAL PROBLEMS

PART 2 PROBLEMS OF WELL-BEING

3 Alcohol and Other Drugs 73

4 Crime and Social Control 115

5 Family Problems 161

PART 3 PROBLEMS OF INEQUALITY

6 Economic Inequality, Wealth, and Poverty 201

Features

Self and Society

Social Problems Research Up Close

The Human Side

Chapter 1 ("Thinking about Social Problems"), in response to the increasing politicization of social problems, now begins with new sections on "The Social Context: A Divided Nation, Politics in America" (with subsections on "The Roots of Political Partisanship," "The Growth of Political Partisanship," and "State of the Union"). All *What Do You Think?* features are new and address such topics as the meaning of democracy, the role of social sciences in fighting the pandemic, former President Trump's alleged culpability in the January 2021 attempted coup, and the impact of political partisanship on family relations. There are three new features in this heavily revised chapter, including *The Human Side,* which offers advice from student activists on getting involved in "good trouble," and a *Social Problems Research Up Close,* which examines generational variations in following the news on the 2020 election, COVID-19, and Black Lives Matter protests. New topics include globalization, the American political spectrum, political polarization, populist movements, and the media's role in defining social problems.

Chapter 2 ("Physical and Mental Health and Health Care") begins with the global impact of the COVID-19 pandemic, from contrasting country-level response strategies to the impact of testing initiatives, and from factors that contributed to failures within the U.S. public health system to the politicization of mask wearing. The new opening vignette focuses on the demanding role that health care workers played in providing emotional support to dying COVID-19 patients in the absence of families. A newly titled section on health disparities integrates the impact of inequality on COVID-19 patient outcomes as well as on other health conditions. The new *Social Problems Research Up Close* feature examines the increasing importance of education in relationship to life expectancy by race. An updated section on mental illness includes a new *The Human Side* feature that describes what it is like to live with mental illness during a pandemic, and a new *Self and Society* that asks students to examine their own mental health risks.

The chapter integrates updated information on the Affordable Care Act, Medicaid expansion, trends in health insurance coverage, and factors affecting the cost of health care in the United States compared to other countries. A revised "Strategies for Action" section examines policy initiatives to expand health care coverage, increased services to the mentally ill, and preparation for the next pandemic. New *What Do You Think?* questions throughout the chapter prompt students to consider such issues as what cultural values would promote universal mask wearing and social distancing, whether U.S. women should have access to over-the-counter birth control pills as in other countries, and what factors might explain why education is a stronger contributor to life expectancy in some U.S. states than in others. New key terms include health disparities, pandemic, contact tracing, positivity rate, death rate, and the criminalization of mental illness.

Chapter 3 ("Alcohol and Other Drugs") is thoroughly revised with all new features and *What Do You Think?* questions. Changes to the organization of the chapter reflect the emerging patterns in state-level decriminalization and prescription drug abuse that are blurring the lines between legal and illegal drug use. The chapter has an increased emphasis on the impact of the opioid addiction crisis, including a new opening vignette and *The Human Side* that focus on families mourning losses from drug overdose as well as coverage of recent lawsuits against pharmaceutical companies. The chapter also offers expanded and updated coverage on the impact of the War on Drugs, the growing shift toward a medical model of drug and alcohol abuse in public policy, the relationship between poverty and substance abuse, and the growing trend of vaporizers and e-cigarettes. Other updates and new topics include polling data on Americans' views about marijuana decriminalization and drug abuse as a social problem, updated data on drug use and abuse patterns globally and within the United States, and the complex relationships between poverty, mental health, and substance abuse.

Chapter 4 ("Crime and Social Control") has been thoroughly revised and begins with a new opening vignette. There is also new content in this chapter's three features. The *Self and Society* feature allows students to assess their fear of various crimes, the *Social Problems Research Up Close* feature examines the media's portrayal of serial killers, and in *The Human Side* a victim of a campus rape emotionally describes its impact on every facet of her life. New topics include the BLM protests and responses to them, police reform initiatives, the militarization of police, political crime, prisoners and COVID-19, the cost to families of having an incarcerated relative, several new pieces of legislation including the *George Floyd Justice in Policing Act of 2020*, and the Biden administration's proposed reform of the criminal justice system.

All of the *What Do You Think?* questions in this chapter are new and address such topics as the impact of COVID-19 on property and violent crime rates, whether sitting or former presidents should have criminal charges levied against them, the disconnect between actual crime statistics and the public's perception of crime, and accountability for lethal police violence.

Chapter 5 ("Family Problems") begins with a new opening vignette spotlighting the problem of domestic violence, as exemplified by the Turpins who were convicted of imprisoning and torturing their 13 children. Domestic violence is also addressed in the new *The World in Quarantine* feature documenting the victimization of women during the pandemic and in a revised section examining the types of domestic violence as well as factors associated with it. The chapter includes updated data on family trends, expanded coverage of foster and blended families, and a fuller integration of same-sex couples/families. A new section examines unplanned pregnancies, as well as access to contraception and abortion.

The new *Social Problems Research Up Close* feature examines how racialized immigration policies force women in mixed-status families into the role of single parents. A new *The Human Side* features a woman who was forced into an arranged marriage as a teenager, and a new *Self and Society* feature prompts students to consider their own views on abortion in a variety of circumstances. The "Strategies for Action" section has been fully revised, including strategies for improving access to contraception. New *What Do You Think?* questions prompt students to consider issues such as what age is too young to get married, why arranged marriages result in fewer divorces, and why more women over 30 are planning to be single parents.

Chapter 6 ("Economic Inequality, Wealth, and Poverty") opens with a new vignette about a family living in their car after losing their jobs during the COVID-19 pandemic. This fully revised chapter includes all new tables and figures featuring the most up-to-date data available on poverty rates and wealth inequality. The chapter focuses extensively on the growing levels of inequality globally and within the United States, with special attention given to the wealth accumulated by the world's billionaires. The chapter also includes new topics on social mobility trends, the impact of climate change on low-income Americans, anti-poverty social movements, and the impact of the COVID-19 pandemic on poverty, housing instability, and food insecurity. A new *Social Problems Research Up Close* feature examines the effectiveness of Housing First programs to address homelessness, and a new *The Human Side* feature explores how Americans with college debt increasingly see the American Dream as out of reach. New *What Do You Think?* questions ask students to think about how much CEOs should be paid, what the new minimum wage should be, what types of limits should be placed on welfare usage, and how much wealth should be considered "extreme."

Chapter 7 ("Work and Unemployment") contains extensive coverage of the impact of the COVID-19 pandemic on the lives of working Americans, including workplace health and safety, unemployment, the employment prospects for the "unlucky cohort" of 2020 graduates, work/life balance and stress, and family and medical leave policies. The chapter's new *The World in Quarantine* feature delves into the unique plight facing working mothers during the 2020 pandemic.

This chapter provides updated data on Americans' shifting attitudes about capitalism and socialism, while a new *Self and Society* feature asks students to assess their own attitudes toward capitalism and socialism. The chapter also examines work from the perspective of the global supply chain, with a new *The Human Side* feature in which child garment factory workers tell their stories, expanded coverage of the ongoing global impact of the Great Recession, new coverage of changes to international free trade agreements under the Trump administration, and a new discussion on the pattern of policy drift as it relates to globalization and outdated labor laws. Other new and updated topics include the generational divide in perceptions about capitalism and socialism, the questions include whether politicians should be allowed to buy and sell stock while in office and whether social media sites should be able to block posts that promote climate denial.

A new *Social Problems Research Up Close* examines why property owners allow fracking on their land and what happens when they do. The new *The Human Side* documents the retaliation and violence Kenyan environmental activist and whistleblower Phyllis Omido experienced in calling out lead pollution by a company in her village. A new *Self and Society* allows students to compare their opinions on the environment and climate change with those of the U.S. public. Key terms include mountaintop removal mining, strip mining, community solar gardens, solar farms, and coral bleaching, among others.

Chapter 14 ("Science and Technology") contains a new opening vignette and all new *What Do You Think?* questions on such topics as artificial intelligence, the use of algorithms in corroborating jury sentences, workplace surveillance technology, the banning and/or labeling of social media posts, and deepfake videos. Given the events of the last three years, this chapter has been reorganized and thoroughly revised to include five new subsections under the heading "Technology and the Workplace": "Robotics, Software Robotics," "Worker Error and Technological Failure," "Telecommute and Telepresence," and "Technology and Social Control"; new subsections on "Algorithms" and "Computers as Big Business" under the heading "The Computer Revolution"; a new subsection on "Smart Technologies" under "The Digital Divide"; and, under the heading Malicious Use of the Internet, four new subsections including "The Deep and Dark Web," "Malware and Hacking," "Disinformation, Deepfakes, and Conspiracy Theories," and "Politics and Election Tampering."

New topics and key terms include new social media (e.g., Triller, Tik-Tok, Parler), the anti-science administration, the science administration, Internet censorship, the use of predictive algorithms, artificial intelligence, election tampering, the use of social media for activism (e.g., #BLACKLIVESMATTER), CRISPR, three types of genetic cloning, replacement therapy, heritable genome editing, problems associated with facial recognition technology, QAnon, and national and international efforts to fight disinformation. Finally, *The World in Quarantine: The Other Virus That Kills* feature in this chapter is on the harmful effects of disinformation campaigns, along with new content for each of the other three features including a *Self and Society* that allows students to evaluate their "science and technology IQ."

Chapter 15 ("Conflict, War, and Terrorism") begins with a vignette telling the story of Nobel Peace Prize winner Nadia Murad, a former ISIS sex slave and human rights activist. New features in the chapter include *The Human Side* and *Social Problems Research Up Close,* which examine the long-term impact of experiencing war on U.S. veterans and Syrian refugees. The chapter provides extensive coverage of the impact of President Trump's America First agenda on the U.S. role in international relations, as well as a discussion of likely foreign policy changes to occur during the Biden administration. Throughout this thoroughly revised chapter, students are encouraged to critically examine their own perspectives on the role of the United States in global affairs. A new *Self and Society* feature asks students to assess the extent to which they think the United States should be involved in global affairs. New *What Do You Think?* questions ask, for example, (1) whether the United States should withdraw from NATO, (2) whether the abrupt withdrawal from Iraq and Afghanistan is worth the potential long-term risks, (3) the conditions under which the United States is justified in using force against foreign governments, and (4) whether policies limiting refugee admission is an appropriate response to the threat of terrorism.

The chapter includes updated data on military spending, the arms trade, and the costs of war as well as war trends and conflicts in Iraq, Afghanistan, Syria, and Yemen. New and updated topics also include the establishment of a new Space Force branch of the military, coverage of women's combat roles and transgender policies in the military, updated coverage of the ongoing global refugee crisis, and new coverage of the legal and ethical issues associated with private military and security contractors. Coverage of terrorism is expanded and reorganized, reflecting the blurring boundaries between domestic and international terrorism, and includes a new section on white supremacist terrorism.

Features and Pedagogical Aids

We have integrated a number of features and pedagogical aids into the text to help students learn to think about social problems from a sociological perspective. *Understanding Social Problems* was designed to actively engage students in examining social issues from a variety of perspectives. Through content that is visually appealing, connected to current events, and relevant to their everyday lives, *Understanding Social Problems* provides students with the tools to sharpen their sociological imaginations.

Boxed Features

Self and Society. Each chapter includes a *Self and Society* feature designed to help students assess their own attitudes, beliefs, knowledge, or behaviors regarding some aspect of the social problem under discussion. In Chapter 4 ("Crime and Social Control"), for example, the "Fear of Crime Assessment" invites students to evaluate their own fear of criminal victimization. The *Self and Society* feature in Chapter 11 ("Sexual Orientation and the Struggle for Equality") allows students to assess their attitudes toward gay and transgender men and women and compare their responses to a sample of respondents from all over the world.

The Human Side. Each chapter includes a boxed feature that describes personal experiences and views of individuals who have been directly affected by social problems. *The Human Side* feature in Chapter 10 ("Gender Inequality"), for example, describes U.S. House of Representatives member Alexandria Ocasio-Cortez's response to an unprovoked verbal attack by a congressman on the U.S. Capitol steps, and *The Human Side* feature in Chapter 7 ("Working and Unemployment") poignantly describes "life as a child worker in a garment factory." *The Human Side* in Chapter 14, sadly, describes the tensions between family members and friends as a result of the unfounded conspiracy theories put forth by QAnon.

Social Problems Research Up Close. This feature, found in every chapter, presents examples of social science research, summarizing the sampling and methods involved in data collection and presenting the findings and conclusions of the research study. Examples of *Social Problems Research Up Close* topics include opposition to needle exchange programs, media portrayals of serial killers, implicit bias training to reduce racial disparities, the relationship between mother's education and decreased child mortality rates, and variables that predict belief in scientific conspiracies.

In-Text Learning Aids

Learning Objectives. We have developed a set of learning objectives that are presented at the beginning of each chapter. The learning objectives are designed to help students focus on key concepts, theories, and terms as they read each chapter.

Vignettes. Each chapter begins with a vignette designed to engage students and draw them into the chapter by illustrating the current relevance of the topic under discussion. For example, Chapter 8 ("Problems in Education") begins with a teacher describing her frustration with the profession and her decision to leave it, and Chapter 10 ("Gender Inequality") introduces students to Max, a non-binary youth who must decide whether to go through female or male puberty. The opening vignette in Chapter 14 ("Science and Technology") documents the miracle of medical technology that is helping young boys with Duchenne Muscular Dystrophy to run, jump, and play.

Key Terms and Glossary. Important terms and concepts are highlighted in the text where they first appear. To reemphasize the importance of these words, they are listed at the end of every chapter and are included in the glossary at the end of the text.

Running Glossary. This eleventh edition continues the running glossary that highlights the key terms in every chapter by putting the key terms and their definitions in the text margins.

What Do You Think? Feature. Each chapter contains multiple feature boxes called *What Do You Think?* These features invite students to use critical thinking skills to answer questions about issues related to the chapter content. For example, one *What Do You Think?* in Chapter 4 ("Crime and Social Control") asks students to consider whether or not the "law should require the use of algorithms in murder cases," and a *What Do You Think?* question in Chapter 8 ("Problems in Education") asks students, "Do you think standardized tests should be eliminated entirely and, if not, how long after the end of the pandemic should they be reinstated?"

Margin Quotes and Margin Tweets. New to this edition, margin quotes and margin tweets connect with students through their interest in social media, while introducing students to alternative points of view, perhaps from someone they are "following." They also encourage students to apply sociology to everyday life as they see celebrities, politicians, authors, and the like, doing so. Margin quotes and margin tweets come from a diverse array of commentators and organizations, including, for example, former Presidents Obama and Trump, Lady Gaga, Pharrell Williams, and Taylor Swift, as well as the LGBT Foundation, the Center for Disease Control and Prevention, and the United Nations.

Understanding [Specific Social Problem] Sections. All too often, students, faced with contradictory theories and research results, walk away from social problems courses without any real understanding of their causes and consequences. To address this problem, chapter sections titled "Understanding [a specific social problem]" cap the body of each chapter just before the chapter summaries. Unlike the chapter summaries, these sections sum up the present state of knowledge and theory on the chapter topic and convey the urgency for rectifying the problems discussed in the chapter.

Supplements

The eleventh edition of *Understanding Social Problems* comes with a full complement of supplements designed for both faculty and students.

Supplements for Instructors

Online Instructor's Resource Manual. This supplement offers instructors learning objectives, key terms, lecture outlines, student projects, classroom activities, exercises, and video suggestions.

Online Test Bank. Test items include multiple-choice and true-false questions with answers and text references, as well as short-answer and essay questions for each chapter.

Cengage Learning Testing Powered by Cognero. The Test Bank is also available through Cognero, a flexible, online system that allows instructors to author, edit, and manage test bank content as well as create multiple test versions in an instant. Instructors can deliver tests from their school's learning management system, classroom, office, or home.

Online PowerPoints. These vibrant, Microsoft® PowerPoint® lecture slides for each chapter assist instructors with lectures by providing concept coverage using images, figures, and tables directly from the textbook.

MindTap™: The Personal Learning Experience

MindTap for *Understanding Social Problems* represents a highly personalized, online learning platform. A fully online learning solution, MindTap combines all of a student's learning tools, readings, and multimedia activities—into a Learning Path that guides the student through the social problems course. Highly interactive activities challenge students to think critically by exploring, analyzing, and creating content, while developing their sociological lenses through personal, local, and global issues.

MindTap Understanding Social Problems is easy to use and saves instructors time by allowing you to:

- Break course content down into manageable modules to promote personalization, encourage interactivity, and ensure student engagement.
- Bring interactivity into learning through the integration of multimedia assets (apps from Cengage and other providers) and numerous in-context exercises and supplements; student engagement will increase, leading to better student outcomes.
- Track students' use, activities, and comprehension in real-time, which provides opportunities for early intervention to influence progress and outcomes. Grades are visible and archived so students and instructors always have access to current standings in the class.
- Assess knowledge throughout each section: after readings and in automatically graded activities and assignments.
- A digital implementation guide will help you integrate the new MindTap Learning Path into your course.

Learn more at **www.cengage.com/mindtap.**

Acknowledgments

We would like to acknowledge the support and assistance of Carol L. Jenkins, Leslie Carter, Ila T. Logan, Crosby Hipes, Mary Clever, Larry Clever, Doug Mace, and Sabina Mace. To each we send our heartfelt thanks.

Additionally, we are interested in ways to improve the text and invite your feedback and suggestions for new ideas and material to be included in subsequent editions. You can contact us at mooneyl@ecu.edu, clever_m@wvwc.edu, and mvanwilligen@georgiasouthern.edu.

Ira L. Black - Corbis/Corbis News/Getty Images

"Unless someone like you cares a whole awful lot, nothing is going to get better. It's not."

DR. SEUSS
The Lorax

1

Thinking about Social Problems

Chapter Outline

Learning Objectives

After studying this chapter, you will be able to . . .

1. Describe the American political party system.
2. Discuss the causes of political partisanship in the United States.
3. Define a social problem.
4. Discuss the elements of the social structure and culture of society.
5. Explain the connections between private troubles and public issues, as well as how they relate to the sociological imagination.
6. Summarize structural functionalism, conflict theory, and symbolic interactionism and their respective theories of social problems.
7. Describe the stages in conducting a research study.
8. Distinguish between the four methods of data collection used by sociologists.

AP Images/David Zalubowski

The year 2020 was a year of activism as Black Lives Matters and supporters protested police violence, student strikes marked Sweden's day of climate action, lockdown opponents rallied against government mandates, and election results in the United States were met with both celebrations and demonstrations questioning the results.

IN AN OCTOBER 2020 survey, respondents were asked, "What is the most important problem facing this country today?" Only 14 percent of respondents reported economic problems such as the economy in general, unemployment, and the gap between the rich and poor. Eighty-seven percent of respondents reported noneconomic social problems, from the most to the least frequent, coronavirus/diseases, poor government leadership, race relations/racism, unifying the country, crime/violence, health care, the judicial system/courts, and the environment (Gallup Poll 2020a). Moreover, a 2020 survey indicates that just 14 percent of Americans are satisfied "with the way things are going in the United States"—a decrease from 33 percent in the previous year (Gallup Poll 2020b). The increase in dissatisfaction is likely, among other things, a result of the COVID-19 pandemic, racial unrest, and a contentious political environment.

Problems related to government leadership, COVID-19, race, crime and violence, divisiveness, health care, and environmental destruction, as well as many other social issues, are both national and international concerns. Because of **globalization**, i.e., the growing economic, cultural, and technological interdependence between countries and regions, some social problems are clearly universal such as climate change, while others *appear* to only impact the nation in which they occur. The economy, for example, is often discussed in terms of the U.S. gross domestic product (GDP), the U.S. inflation rate, or Americans' consumer confidence. And yet U.S. economic indicators don't operate in a vacuum. Even before COVID-19 was considered a significant threat to the United States, as it spread from Asia to Europe, U.S. financial markets fell to their lowest point in years as a result of what was happening overseas (Imbert and Huang 2020).

> "We reject globalism and embrace the doctrine of patriotism."
>
> **–DONALD TRUMP, FORMER PRESIDENT OF THE UNITED STATES**

Globalization was championed by the United States and other Western nations after World War II as a way to deter future international conflict (Goodman 2019; Posen 2018). Facilitated by advances in technology and transportation, population growth and geographic mobility, and the expansion of multinational corporations, countries became reliant on one another for the production and consumption of goods and services. Raw materials and labor, rather than coming from a single country, were drawn from all over the world leading to a global marketplace. Free trade zones were established, tariffs eliminated, trade agreements forged, and dispute resolution processes put into place.

globalization The growing economic, cultural, and technological interdependence between countries and regions.

populist movements Emphasize "the people" rather than the "government elite" and their political parties, tend to be conservative, right to far-right leaning, anti-immigrant, nationalistic, and anti-globalist.

However, fears that globalization would reduce the importance of nation-states and lead to cultural homogenization wherein the lynchpins of American society—individual achievement, self-determination, hard work, and national unity—would be lost continue today. In fact, some research suggests that Brexit (i.e., the exit of Great Britain from the European Union, framed by the slogan "take back control"), the election of Donald Trump ("America First"), and other populist movements in Europe are a direct response to such fears (Adnane 2019; Silver, Schumacher, and Mordecai 2020). **Populist movements**, which claim to represent "the people" rather than government elites and their political parties, tend to be conservative, right to far-right leaning, anti-immigrant, nationalistic, and anti-globalist (Ruzza and Salgado 2020).

Given globalization and its inevitable continuation (Goodman 2019), it is important that America maintain its standing as a world leader. And yet, in a 2020 survey of 14 countries, only respondents from South Korea and Japan named the United States as the world's leading economic power with, for example, the United Kingdom, Canada, France, Australia, and Germany naming China (Poushter and Moncus 2020).

In response to an anti-globalist stance, the Biden administration has made it clear that they reject the former president's "America First" policy and has assured foreign leaders that the United States has returned to the world stage (Ordonez 2020). President Biden is working with prominent allies to fight the COVID-19 pandemic, and rejoined the World Health Organization (WHO) (see Chapter 2). His administration is also focused on the environmental crisis by, for example, rejoining the Paris climate accord (see Chapter 13), and on restructuring foreign policy, including the Iran nuclear deal that former President Trump withdrew from in 2018 (see Chapter 15). After his election, President Biden also announced a "Democracy Summit" of heads of states from leading democratic countries to be held in the United States in 2021 (Holpuch et al. 2020)

What do you think?

Krupnikov and Ryan (2020) argue that there is an "attention divide" in the United States between those who follow politics closely, about 15 to 20 percent of the population, and the remainder who follow it casually or not all. Democrats and Republicans who don't follow politics closely are much more likely to agree on the most important problems facing America than Democrats and Republicans who do follow politics closely. Why do you think there is much more disagreement about the importance of social problems between Democrats and Republicans who follow the news closely when compared to those who don't?

Figure 1.1 graphically portrays the thoughts of American and United Kingdom focus group participants who were tasked with discussing "how people in the U.K. and the U.S. feel about globalization and how this relates to their views about their communities and their country" (Silver, Shoemaker, and Mordecai 2020, p. 1). Most participants, American and British, had difficulty defining globalization but were able to voice concerns (e.g., "diluting our culture") as well as elements of cooperation (e.g., "learning from other cultures") leading to the emergence of five key themes as indicated in Figure 1.1.

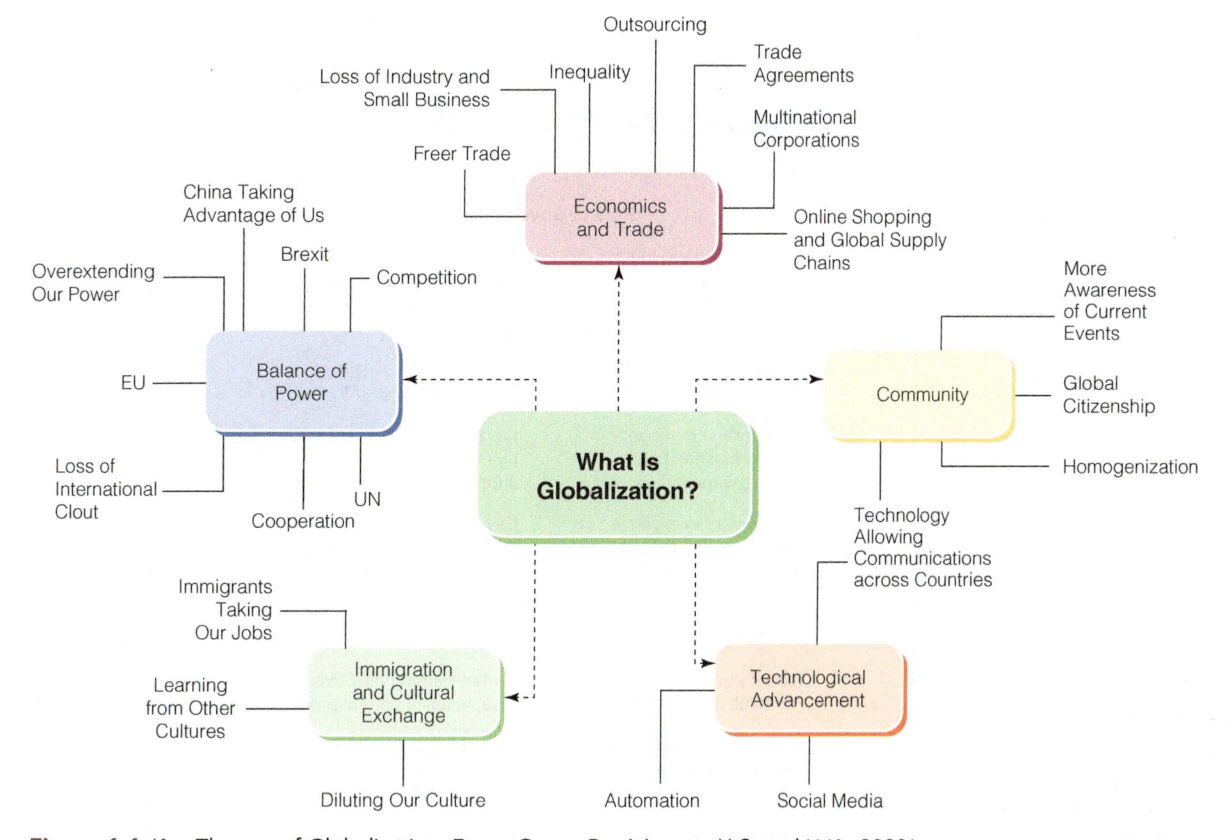

Figure 1.1 Key Themes of Globalization, Focus Group Participants, U.S. and U.K., 2020*
*When asked to define globalization, focus group participants found it easier to illustrate than to define.
SOURCE: Silver, Schumacher, and Mordecai 2020.

Because many Americans are often unfamiliar with world events, with the exception of this chapter, every subsequent chapter begins with a section on the global context of the social problem under discussion; at the end of each chapter, policy initiatives from the United States and, where appropriate, from around the world are highlighted.

The topics covered in this book vary widely; however, all chapters share common objectives: to explain how social problems are created and maintained; to indicate how they affect individuals, social groups, and societies as a whole; and to examine programs and policies for change. We begin by looking at the sociopolitical climate surrounding social problems in America.

The Social Context: A Divided Nation

In the United States, social problems are often framed within the context of *culture wars* whereby various groups, often based on political party affiliation, disagree as to what constitutes a social problem and/or how it should be addressed. In the following section, the American political spectrum, political polarization, and the state of the union are examined.

Politics in America

Although there are smaller and lesser known political parties such as the Libertarian Party, the Green Party, and the Constitutional Party, historically the United States has been characterized by a two-party system with either Democrats or Republicans winning the White House since the 1860s. Democratic presidents include Franklin D. Roosevelt, John Kennedy, Bill Clinton, Barack Obama, and presently Joe Biden. Republican presidents include Abraham Lincoln, Richard Nixon, Ronald Reagan, George W. Bush, and, most recently, Donald Trump.

The two parties differ in their philosophy of the role of government in society and on social and economic policies (see Figure 1.2). Democrats and progressives are often referred to as being on the left, while Republicans and reactionaries are often referred to as being on the right

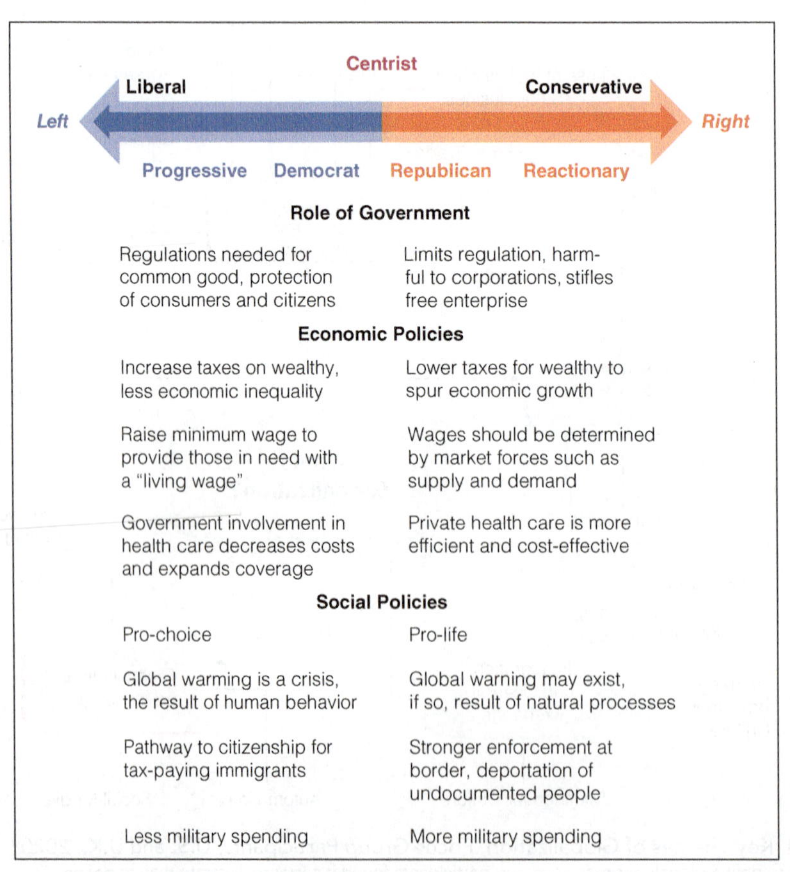

Figure 1.2 The American Political Spectrum

political partisanship Supporters of a political party are entrenched in their party's policies, with little to no motivation to compromise with opposing political views.

(Tanenhaus 2016). In general, the left is considered liberal, and the right is considered conservative. However, as political scientists note, "the very meaning of 'liberalism' and 'conservatism' changes" over time (Lewis 2019, p. 1).

AP Images/Sipa USA

The 2020 presidential election was one of the most contentious in American history. Here supporters of President-Elect Joe Biden drive by a group of supporters of the former president following the announcement of the outcome of the election.

The Roots of Political Partisanship. When supporters of a political party are entrenched in their party's policies, with little to no motivation to compromise with opposing political views, it is called **political partisanship**. Like today, the 1960s and 1970s were characterized by political and social divisiveness as those opposed to America's involvement in the Viet Nam war (i.e., doves) and those in favor it (i.e., hawks), Democrats and Republicans, battled in Congress and on the streets.

After the killing of four students by the Ohio National Guard at an antiwar protest at Kent State University in 1970, student demonstrations in support of the protesters erupted across the nation. The majority of Americans, however, supported the National Guard and when antiwar demonstrations broke out in New York City, four days after the Kent State killings, groups of construction workers, defining the students as "un-American," attacked them with crowbars, resulting in several serious injuries. Thus, as explained by Paul Kuhn (2020), author of *The Hardhat Riot: Nixon, New York City, and the Dawn of the White Working-Class Revolution*, "[I]f there's an era when tribalization … began, it's this time … between Kent State and the hardhat riot [where] you have the best microcosm that there is of the beginning of the polarization that haunts America today" (McGreal 2020).

The Growth of Political Partisanship. Political partisanship has increased dramatically over the last several decades. In 1960, just 4 percent of Republicans and 4 percent of Democrats said they would be "somewhat or very unhappy" if their son or daughter married someone from the opposite political party. In 2019, however, 45 percent of Democrats and 35 percent of Republicans said they would be "somewhat or very unhappy" if their son or daughter married someone from the opposite political party (Najle and Jones 2019). Interestingly, research indicates that ideological position as either a liberal or a conservative is a better predictor of partisan dislike of ideological opponents than positions on social issues; i.e., political party and its accompanying ideology has become a social identity in and of itself (Mason 2018). Given the significance of social identity, it is not surprising that the term *political tribalism* is sometimes used to describe unquestioning loyalty to a political belief or party.

After months of claiming that the 2020 presidential election had been stolen, on January 6, 2021, President Trump encouraged a crowd of his supporters to go to the U.S. Capitol and "take back our country." Thousands stormed the Capitol, breaking windows, assaulting Capitol police, and ransacking lawmakers' offices. As a result of the insurrection, six people died, hundreds were arrested, and Donald Trump was impeached for a second time, charged with incitement of insurrection. Do you think Donald Trump should have been impeached so close to leaving office?

What *do you* think?

Political partisanship is thought to be the result of several interacting social forces in the United States (Mansbridge 2016; Bail et al. 2018; Blankenhorn 2018; Carothers and O'Donohue 2019). These social forces include:

- movement from the center of the political spectrum;
- greater racial, religious, and ethnic diversity;
- increased division between socioeconomic classes;

@Mitt Romney

The President is within his rights to request recounts, to call for investigation of alleged voting irregularities where evidence exists, and to exhaust legal remedies—doing these things is consistent with our election process. He is wrong to say that the election was rigged, corrupt and stolen—doing so damages the cause of freedom here and around the world, weakens the institutions that lie at the foundation of the Republic, and recklessly inflames destructive and dangerous passions.

-Mitt Romney

- polarizing leaders who demonize opponents;
- residential and geographical homogeneity;
- "media ghettos" segregated by political party (e.g., MSNBC, *Huffington Post* vs. Fox News, *Breitbart*);
- viral misinformation and disinformation;
- exposure to "news" consistent with existing beliefs, i.e., social media as an echo chamber.

Note, however, it is difficult to establish causality. For example, do people with polarized beliefs seek news outlets that are consistent with those beliefs, or does consuming ideologically slanted media create polarized beliefs? The answer is probably both.

Figure 1.3 displays the differences between Republicans and Democrats who, when asked about a particular social problem, reported they believed it was a "very big problem in the country today" (Dunn 2020, p. 1). With the exception of the federal budget deficit, violent crime, terrorism, and illegal immigration, Democrats were more likely to report each of the social problems listed as a "very big problem" compared to Republicans. When political party is held constant, more Americans report that ethics in government, COVID-19, and the affordability of health care are significant problems today than, for example, those who cite illegal immigration or terrorism.

Unfortunately, beliefs about political polarization in America, whether accurate or not, increase the likelihood of further polarization. Fewer than 10 percent of Americans define themselves as at the extremes of the political spectrum. Yet extreme views, whether far right or far left, are more likely to be popularized in the news, posted on social media, and shared with others (Heltzel and Laurin 2020). As a result, Americans see their political opponents as extremists, which reinforces and hardens their own partisan resolve. In a 2020 survey, 81 percent of Republicans said that "the Democratic Party has been taken over by socialists," and 78 percent of Democrats said that "the Republican Party has been taken over by racists" (Public Religion Research Institute 2020).

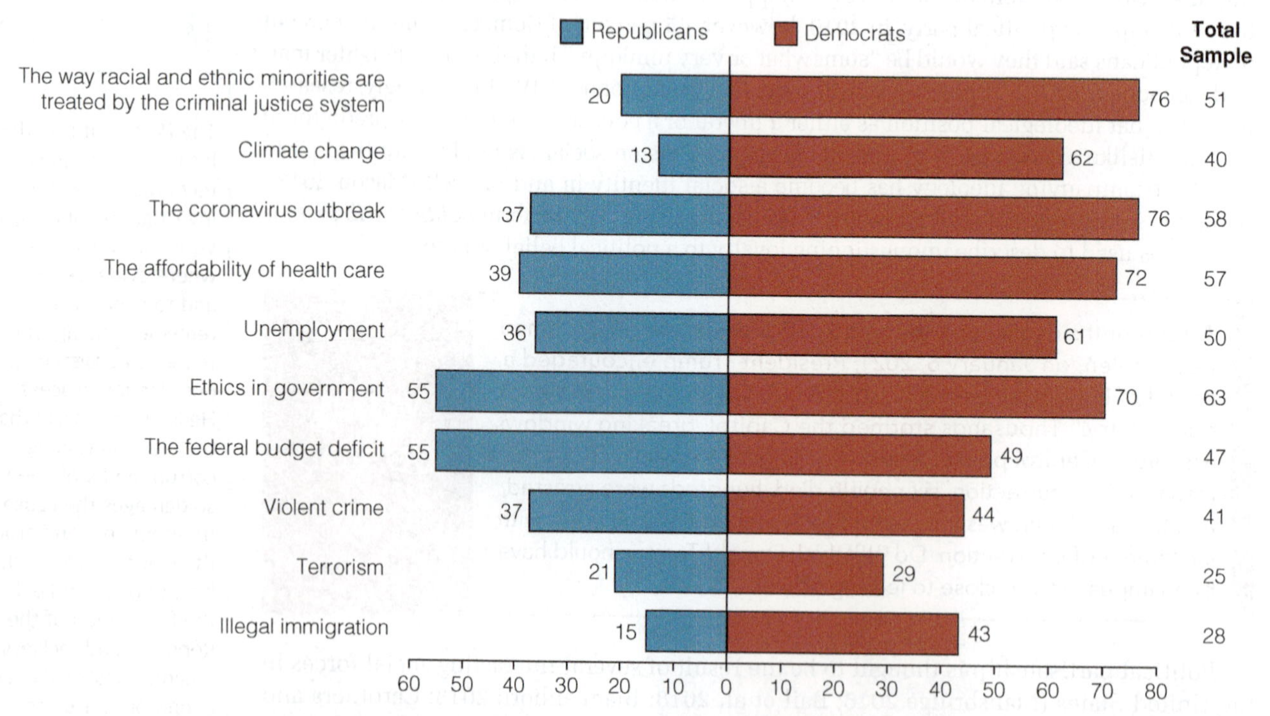

Figure 1.3 Percentage Who Said _______ Is a **Very Big Problem** in the Country Today, by Political Party, 2020*

*Survey of U.S. adults conducted June 16–22, 2020; Republicans include Republicans and those leaning toward Republican, and Democrats include Democrats and those leaning toward Democratic.
SOURCE: Dunn 2020.

Although few Americans adhere to extreme far left or right ideologies:

> political differences are ripping our country apart ... Political scientists find that our nation is more polarized than it has been at any time since the Civil War. This is especially true among partisan elites – leaders who, instead of bringing us together, depict our differences in unbridgeable, apocalyptic terms. (Brooks 2019 p. 2)

Former President Trump, for example, accused then Vice President Biden as "running on the most extreme far-left platform of any nominee in American history" and called Black Lives Matters protesters "thugs" (quoted in Wise 2020, p. 1). In the first debate, candidate Biden, after being bullied and repeatedly interrupted by former President Trump, called him a "liar" and a "clown" (quoted in Manchester 2020, p. 1). It is thus not surprising that the political divide between right- and left-leaning Americans, often seems insurmountable.

@ladygaga

🙌 Queen Kamala

-Lady Gaga

ReTweeting @KamalaHarris

I hope every little girl watching tonight sees that this is a country of possibilities.

-Kamala Harris

State of the Union

The results of the 2016 election and, to a lesser extent, the 2020 election, signaled Americans'—and particularly White working-class Americans'—dissatisfaction with the status quo. Although Democrats Joe Biden and Kamala Harris won, the results documented just how polarized the United States was at the time of the election. A record number of Americans voted, with over 72 million, 47.5 percent of the electorate, voting for Republican incumbents Donald Trump and Mike Pence (Fox News 2020). As Deane and Gramlich (2020, p. 1) note, one takeaway from the election is the:

> continuing political polarization that has come to define the United States. Democrats and Republicans could both walk away from the election with cause for disappointment, and [a] divided government in Washington. ... The elected officials who take the oath of office in January [2021] will be representing two broad coalitions of voters who are deeply distrustful of one another and who fundamentally disagree over policies, plans and even the very problems that face the country today.

Thus, one of the most daunting tasks of the Biden administration, as President Biden noted in his post-election speech, is to heal America, "to put away the harsh rhetoric, lower the temperature, see each other again, listen to each other again, [in order] to make progress" (Biden 2020, p. 1).

Although the Trump administration had some notable successes, President Biden, Vice President Harris, and Democrats in general are "eager to systematically erase what they view as the destructive policies that the president pursued on the environment, immigration, healthcare, gay rights, trade, tax cuts, civil rights, abortion, race relations, military spending, and more" (Shear and Friedman 2020, p. 1). For example, the Trump administration was responsible for the largest tax reform in 30 years. Critics, however, were quick to note that while reducing the corporate tax rate from 35 percent to 21 percent, providing a corporate windfall, it did so at the expense of middle-class Americans. Since 2016, the income gaps between upper-income, middle-income, and lower-income households increased, and the percentage of income held by middle-income households decreased (Horowitz, Igielnik, and Kochhar 2020) (see Chapter 6 and Chapter 7).

AP Images/Andrew Harnik

On January 6, 2021, in an attempted coup, rioters attacked the U.S. Capitol in the hopes of stopping lawmakers from tallying the electoral votes for the 2020 presidential election. With some of the extremists calling for the execution of then Vice President Mike Pence (R) and Speaker of the House Nancy Pelosi (D), here Capitol police, with guns drawn and furniture stacked to prevent entry, defend the integrity of the "peoples' house."

We may be opponents—
but we are not enemies.

We are Americans.

-Joe Biden

President Biden has proposed raising the corporate tax rate from the prior administration's preferred rate of 21 percent to 28 percent (Ember 2020).

The Trump administration also lobbied for the repeal of the *Affordable Care Act* (ACA), an Obama-era legislative initiative supported by then Vice President Biden. With the repeal of the individual mandate and the end to subsidies to insurance companies in the exchanges after President Trump took office, the number of *uninsured* Americans increased by 2.3 million, including over 725,000 children (Gee 2020) (see Chapter 2). Further, the Trump administration removed health care and health insurance non-discrimination protections for LGBTQ citizens (Simmons-Duffin 2020) (see Chapter 11). There were also concerns that the repeal of the ACA would make substance abuse services financially out of reach for many in need (Firozi 2019) (see Chapter 3). Not surprisingly, one of President Biden's first official actions was to sign an executive order strengthening the ACA by opening enrollment thereby allowing more Americans to sign up for health care during the pandemic (Deliso 2021) (see Chapter 2).

Concerns, primarily by Republicans, over immigration from Mexico and Central America led to a national policy of family separation in 2017 and 2018 that became a significant point of contention in the 2020 election. As of this writing, 545 children remain separated from their parents or guardians as the Trump administration has been unable to track down their families after detaining them at the U.S.–Mexico border (Lantry 2020) (see Chapter 9). Further, the former administration's delays in approving student visas has resulted in fewer international students at American colleges and universities (see Chapter 8). Some analysists believe that the Biden administration would be wise to "make a clean break from the Trump era by undoing all executive orders and proclamations on immigration that are not directly tied to health concerns related to COVID-19" (Anderson 2020, p. 1). To that end, President Biden has signed several immigration-related orders including one terminating the construction of and funding for the border wall between the United States and Mexico (Deliso 2021) (see Chapter 9).

Existing social problems, of course, have been exasperated by the onset of the COVID-19 pandemic in 2020 and by the former administration's anti-science stance (see Chapter 2 and Chapter 14). Just prior to the election, the unemployment rate hovered around 8 percent nationally (U.S. Bureau of Labor Statistics 2020), an increase of nearly 5 percent from the same time period in the previous year (see Chapter 7). Moreover, by the fall of 2020, economic growth had dropped by over 30 percent (BBC 2020) and student repayment of loans, with debt at an all-time high, had to be deferred as less educated workers, including college students, were the most likely to lose their jobs (Baum and Looney 2020) (see Chapter 8).

Acknowledging the devastating impact of the pandemic on the economy, the Biden administration believes that it is time to "build back better" and to address the "old economy's structural weaknesses and inequalities" (Economic Recovery 2020, p. 1). More specifically, for example, the new administration supports student loan forgiveness for low-income families, increasing Social Security payments, federal grants for small businesses, and creation of a Public Health Job Corps to help fight the pandemic and reduce unemployment (Economic Recovery 2020; Sherman 2020) (see Chapter 2, Chapter 7, Chapter 8, and Chapter 12).

Not surprisingly, given the trajectory of the country over the last several years, many Americans have questioned the future of the country, and political polarization has led to a lack of confidence in traditional institutions (Gallup 2020c). For example, a 2020 poll of Americans between the ages of 18 and 29 found that, at the time, fewer than 10 percent of respondents believed that the government was working as it should. Fifty-one percent of Democrats surveyed believed, "Our government has problems, and in order for them to be solved, we need to replace and create new institutions to address those challenges" compared to 38 percent of Independents and just 19 percent of Republicans (Harvard Kennedy School 2020).

Finally, in 2020, Americans reported being unhappier than they've been since 1972 (see Chapter 2), as well as more pessimistic about the future of their children with only 42 percent responding that their children will have a higher standard of living than they have, the lowest recorded level since 1994 (Lush 2020) (see Chapter 6). Although there is little doubt that the COVID-19 pandemic has contributed to the higher rates of personal unhappiness and pessimism about the future, the increases for both began in 2018, pre-dating the pandemic.

We now turn our attention to the objective and subjective components of social problems and the role of the media in defining them. We also examine the variability of social problems, i.e., how definitions of social problems change over time, both within and between societies.

What Is a Social Problem?

There is no universal, constant, or absolute definition of what constitutes a social problem. Rather, social problems are defined by a combination of objective and subjective criteria that vary across societies, among individuals and groups within a society, and across historical time periods.

Objective and Subjective Elements of Social Problems

Although social problems take many forms, they all share two important elements: an objective social condition and a subjective interpretation of that social condition. The **objective element of a social problem** refers to the existence of a social condition. We become aware of social conditions through our own life experiences, through the media, and through education. We see homelessness, hear gunfire in the streets, and see battered women in hospital emergency rooms. We read about employees losing their jobs and businesses shutting down as a result of the COVID-19 pandemic. In television news reports, we see the anguished faces of parents whose children have been killed in Afghanistan.

The **subjective element of a social problem** refers to the belief that a particular social condition is harmful to society or to a segment of society and that it should and can be changed. We know that crime, drug addiction, poverty, racism, and global warming exist. These social conditions are not considered social problems sociologically, unless at least a segment of society believes that these conditions diminish the quality of human life.

By combining these objective and subjective elements, we arrive at the following definition: A **social problem** is a social condition that a segment of society views as harmful to members of society and is in need of remedy.

Media and Social Problems

Media, including social media, print media, and television and radio, increasingly play a critical role in how social problems are defined. Ironically, a majority of Americans see intentionally misleading information in the media as *the* "major problem" in the United States, of greater concern than illegal drugs, crime, gun violence, or political partisanship (McCorkindale 2020) (see Chapter 14).

One reason for the variation between Democrats' and Republicans' rankings of social problems, with discrepancies as high as 50 percentage points in some cases (see Figure 1.3), may be attributed to differences in media consumption. When Democrats and Republicans were asked the source(s) they trust for political news, 65 percent of Republicans responded Fox News and 33 percent ABC News. Democrats, on the other hand, were much more likely to report getting their news from a variety of sources, the top five varying in frequency by just over 10 percent from a high of 67 percent (CNN) to a low of 56 percent (PBS) (Gramlich 2020). In 2021, after the assault on the U.S. Capitol, some Republicans abandoned fact-checking social media such as Twitter and Facebook in favor of, for example, Gab, which describes itself as a social network that "champions free speech, individual liberty and the free flow of information online" (Gab 2021, p.1).

Information about COVID-19, like information about climate change, on conservative-leaning media versus liberal-leaning media is very different. Calvillo et al. (2020), in an analysis of COVID-19 and media consumption, report that respondents with higher Fox News consumption were less likely to perceive themselves as personally vulnerable or to be knowledgeable about the disease, and more likely to believe that the threat is exaggerated (see Chapter 2 and Chapter 14).

objective element of a social problem Awareness of social conditions through one's own life experiences, through the media, and through education.

subjective element of a social problem The belief that a particular social condition is harmful to society, or to a segment of society, and that it should and can be changed.

social problem A social condition that a segment of society views as harmful to members of society and in need of remedy.

@TIMESUP-NOW

For just the third time in U.S. history, a woman will be a major party's #VP nominee. We won't let the media's sexist political attacks tear down @KamalaHarris or any other women candidates. Speak out & add your name to the #WeHaveHerBack open letter:

-TIME'S UP

Not surprisingly, research also indicates those who feel less vulnerable and are less well-informed about the risks of the disease are more likely to oppose government interventions such as lockdowns, school closures, and mask and social distancing mandates (Jorgenson et al. 2020). Given the foregoing, it is not unreasonable to hypothesize that watching Fox News may be linked to opposition to state lockdown orders. Indeed, research from several countries documents the relationship between right-leaning political beliefs and anti-lockdown protests (Vieten 2020). In fact, former President Trump, a Republican, called for the "liberation" of U.S. states with lockdown orders (Katsambekis and Stavrakakis 2020), and Fox News "covered the [protests] favourably while criticising the governors who implemented the lockdowns" (Ananyev, Poyker, and Tian 2020, p. 1). Thus, variability in what is defined as the *real* social problem, in this example, lockdowns or the pandemic, can be traced, at least in part, to variations in media presentations.

Variability in Definitions of Social Problems

Individuals and groups, often on the basis of demographic variables such as age, race, gender, and political party, frequently disagree about what constitutes a social problem. For example, some Americans view gun control as a necessary means of reducing gun violence, whereas others believe that gun control is a threat to civil rights and individual liberties. Similarly, some Americans view the availability of abortion as a social problem, whereas others view restrictions on abortion as a social problem.

Definitions of social problems, and their importance, vary not only within societies but also *across societies and geographic regions.* Just 3 percent of Americans listed health care as an important problem facing the country today compared to 21 percent of a sample of 16- to 64-year-olds from 27 countries. Similarly, 30 percent of the global respondents identified crime and violence as a top concern compared to just 8 percent of Americans (Gallup 2020a; Ipsos 2020). Country-specific rankings of COVID-19 also vary dramatically with 72 percent of South Koreans compared to 27 percent of Swedes ranking the virus as their country's top concern.

Win McNamee/Getty Images News/Getty Images

What constitutes a social problem varies by individuals, groups, time, and place. On October 22, 2020, the Handmaid's Brigade demonstrated against the Senate Judiciary Committee's vote to confirm Amy Coney Barrett's nomination to the U.S. Supreme Court. Justice Barrett, who was eventually approved by the full U.S. Senate, has been a vocal opponent of the pro-choice movement.

What constitutes a social problem also *varies over time.* For example, before the 19th century, a husband's legal right and marital obligation was to discipline and control his wife through the use of physical force. Today, the use of physical force is regarded as a social problem rather than a marital right. Even a matter of months can make a significant difference. In February 2020, just 3 percent of Americans thought race relations and racism were the most important problems facing the country; five months later, 19 percent thought race relations and racism were the most important problems facing the United States (Gallup 2020a).

Lastly, social problems change over time not only because definitions of conditions change, as in the example of the use of force in marriage but also because the *conditions themselves change.* The use of cell phones while driving was not considered a social problem in the 1990s, as cell phone technology was just beginning to become popular. Now, with most U.S. adults having a cell phone, the issue of "distracted driving" has become a national problem. According to the Centers for Disease Control and Prevention (CDC), every day, approximately nine people are killed and over 1,000 injured in crashes involving a distracted driver. The majority of those injured or killed are between the ages of 20 and 29. The average time a distracted driver is not paying attention is five seconds. At 55 mph, it's like driving the length of a football field blindfolded (CDC 2020a).

Because social problems can be highly complex, it is helpful to have a framework within which to view them. Sociology provides such a framework. Using a sociological perspective to examine social problems requires knowledge of the basic concepts and tools of sociology. In the remainder of this chapter, we discuss some of these concepts and tools: social structure, culture, the "sociological imagination," major theoretical perspectives, and types of research methods.

Elements of Social Structure and Culture

Although society surrounds us and permeates our lives, it is difficult to "see" society. By thinking of society in terms of a picture or image, however, we can visualize society and therefore better understand it. Imagine that society is a coin with two sides: On one side is the structure of society, and on the other is the culture of society. Although each side is distinct, both are inseparable from the whole. By looking at the various elements of social structure and culture, we can better understand the root causes of social problems.

Elements of Social Structure

The **structure** of a society refers to the way society is organized. Society is organized into different parts: institutions, social groups, statuses, and roles.

Institutions. An **institution** is an established and enduring pattern of social relationships. The five traditional institutions are family, religion, politics, economics, and education, but some sociologists argue that other social institutions—such as science and technology, mass media, medicine, sports, and the military—also play important roles in modern society. Many social problems are generated by inadequacies in various institutions. For example, unemployment may be influenced by the educational institution's failure to prepare individuals for the job market and by alterations in the structure of the economic institution.

Social Groups. Institutions are made up of social groups. A **social group** is defined as two or more people who have a common identity, interact, and form a social relationship. For example, the family in which you were reared is a social group that is part of the family institution. The religious association to which you may belong is a social group that is part of the religious institution.

Social groups can be categorized as primary or secondary. **Primary groups**, which tend to involve small numbers of individuals, are characterized by intimate and informal interaction. Families and friends are examples of primary groups. **Secondary groups**, which may involve small or large numbers of individuals, are task oriented and characterized by impersonal and formal interaction. Examples of secondary groups include employers and their employees and clerks and their customers.

Statuses. Just as institutions consist of social groups, social groups consist of statuses. A **status** is a position that a person occupies within a social group. The statuses we occupy largely define our social identity. The statuses in a family may consist of mother, father, stepmother, stepfather, wife, husband, partner, child, and so on. Statuses can be either ascribed or achieved. An **ascribed status** is one that society assigns to an individual on the basis of factors over which the individual has no control. For example, we have no control over the sex, race, ethnic background, and socioeconomic status into which we are born. Similarly, we are assigned the status of child, teenager, adult, or senior citizen on the basis of our age—something we do not choose or control.

structure The way society is organized including institutions, social groups, statuses, and roles.

institution An established and enduring pattern of social relationships.

social group Two or more people who have a common identity, interact, and form a social relationship.

primary groups Usually small numbers of individuals characterized by intimate and informal interaction.

secondary groups Involving small or large numbers of individuals, groups that are task oriented and are characterized by impersonal and formal interaction.

status A position that a person occupies within a social group.

ascribed status A status that society assigns to an individual on the basis of factors over which the individual has no control.

An **achieved status** is assigned on the basis of some characteristic or behavior over which the individual has some control. Whether you achieve the status of college graduate, spouse, parent, bank president, or prison inmate depends largely on your own efforts, behavior, and choices. One's ascribed statuses may affect the likelihood of achieving other statuses, however. For example, if you are born into a poor socioeconomic status, you may find it more difficult to achieve the status of college graduate because of the high cost of a college education.

Every individual has numerous statuses simultaneously. You may be a student, parent, tutor, volunteer fund-raiser, female, and Hispanic. A person's *master status* is the status that is considered the most significant in a person's social identity. In the United States, a person's occupational status is typically regarded as a master status. If you are a full-time student, your master status is likely to be student.

Roles. Every status is associated with many **roles**, or the set of rights, obligations, and expectations associated with a status. Roles guide our behavior and allow us to predict the behavior of others. As students, you are expected to attend class, listen and take notes, study for tests, and complete assignments. Because you know what the role of teacher involves, you can predict that your teachers will lecture, give exams, and assign grades based on your performance on tests.

A single status involves more than one role. The status of prison inmate includes one role for interacting with prison guards and another role for interacting with other prison inmates. Similarly, the status of nurse involves different roles for interacting with physicians and with patients.

> "Our shared values define us more than our differences. And acknowledging those shared values can see us through our challenges today if we have the wisdom to trust in them again."
>
> –JOHN McCAIN, U.S. SENATOR FROM ARIZONA

Elements of Culture

Whereas the social structure refers to the organization of society, the **culture** refers to the meanings and ways of life that characterize a society. The elements of culture include beliefs, values, norms, sanctions, and symbols.

Beliefs. Beliefs refer to definitions and explanations about what is assumed to be true. The beliefs of an individual or group influence whether that individual or group views a particular social condition as a social problem. Does secondhand smoke harm nonsmokers? Does wearing a mask protect others from COVID-19? Does violence in movies and on television lead to increased aggression in children? Our beliefs regarding these issues influence whether we view the issues as social problems. Beliefs influence not only how a social condition is interpreted but also the existence of the condition itself.

What do you think?

Beliefs often determine values. For example, if I believe in democracy, I value voting, free speech, and freedom. One common element of a shared culture is agreement about beliefs and values, and yet in a recent poll there was only one value Republicans and Democrats agreed on—"freedom" (Luntz 2018). While Democrats thought of freedom as **freedom from** (e.g., discrimination, poverty), Republicans thought of it as **freedom to** (e.g., own a gun, practice your religion). What do you think is the meaning of the fundamental American value freedom?

Values. Values are social agreements about what is considered good and bad, right and wrong, desirable and undesirable. Frequently, social conditions are viewed as social problems when the conditions are incompatible with or contradict closely held values. For example, poverty and homelessness violate the value of human welfare; crime contradicts the values of honesty, private property, and nonviolence; racism, sexism, and heterosexism violate the values of equality and fairness. Often responses to opinion surveys (see this chapter's *Self and Society* feature) reveal an individual's values. For example, agreeing with the statement that "a chief benefit of a college

achieved status A status that society assigns to an individual on the basis of factors over which the individual has some control.

roles The set of rights, obligations, and expectations associated with a status.

Social Problems Student Survey

Indicate with a check mark whether you agree (either somewhat agree or strongly agree) or disagree (either somewhat disagree or strongly disagree) with the following statements. When you are done, compare your responses to those that follow.

	Agree	Disagree
1. Racial discrimination is no longer a major problem in America.	_____	_____
2. Abortion should be legal.	_____	_____
3. Colleges have the right to ban extreme speakers from campus.	_____	_____
4. Wealthy people should pay a larger share of taxes than they do now.	_____	_____
5. Addressing global climate change should be a federal priority.	_____	_____
6. The federal government should have stricter gun control laws.	_____	_____
7. Affirmative action in college admissions should be abolished.	_____	_____
8. The federal government should raise taxes to reduce the deficit.	_____	_____
9. Gay men and lesbians should have the legal right to adopt a child.	_____	_____
10. The U.S. government should create a clear path to citizenship for undocumented immigrants.	_____	_____
11. My political views closely resemble those of my parent(s)/guardian(s).	_____	_____

The following percentages are from a national sample of first-semester, first-year college students, at bachelor-granting institutions in the United States who "somewhat agree" or "strongly agree" with the following statements.*

	Percentage Agreeing
1. Racial discrimination is no longer a major problem in America.	17.8
2. Abortion should be legal.	73.1
3. Colleges have the right to ban extreme speakers from campus.	51.0
4. Wealthy people should pay a larger share of taxes than they do now.	67.9
5. Addressing global climate change should be a federal priority.	85.8
6. The federal government should have stricter gun control laws.	76.3
7. Affirmative action in college admissions should be abolished.	50.2
8. The federal government should raise taxes to reduce the deficit.	36.2
9. Gay men and lesbians should have the legal right to adopt a child.	90.5
10. The U.S. government should create a clear path to citizenship for undocumented immigrants.	85.9
11. My political views closely resemble those of my parent(s)/guardian(s).	65.6

*Percentages are rounded.

SOURCE: Stolzenberg, E.B., M.C. Aragon, E. Romo, V. Couch, D. McLennan, M.K. Eagan, and N. Kang. 2020. *The American Freshman: National Norms Fall 2019*. Los Angeles: Higher Education Research Institute, UCLA.

education is that it increases one's earning power" reflects the American value of economic well-being.

Values play an important role not only in the interpretation of a condition as a social problem but also in the development of the social condition itself. For example, most Americans view capitalism, characterized by free enterprise and the private accumulation of wealth, positively. Nonetheless, a capitalist system, in part, is responsible for the inequality in American society as people compete for limited resources.

Norms and Sanctions. **Norms** are socially defined rules of behavior. Norms serve as guidelines for our behavior and for our expectations of the behavior of others.

There are three types of norms: folkways, laws, and mores. *Folkways* refer to the customs, habits, and manners of society—the ways of life that characterize a group or society. In many segments of our society, it is customary to shake hands when being introduced to a new acquaintance, to say "excuse me" after sneezing, and to give presents to family and friends on their birthdays. Although no laws require us to do these things,

culture The meanings and ways of life that characterize a society, including beliefs, values, norms, sanctions, and symbols.

beliefs Definitions and explanations about what is assumed to be true.

values Social agreements about what is considered good and bad, right and wrong, desirable and undesirable.

norms Socially defined rules of behavior, including folkways, laws, and mores.

TABLE 1.1 Types and Examples of Sanctions

	Positive	Negative
Informal	Being praised by one's neighbors for organizing a neighborhood recycling program	Being criticized by one's neighbors for refusing to participate in the neighborhood recycling program
Formal	Being granted a citizen's award for organizing a neighborhood recycling program	Being fined by the city for failing to dispose of trash properly

we are expected to do them because they are part of the cultural tradition, or folkways, of the society in which we live.

Laws are norms that are formalized and backed by political authority. It is normative for a Sikh to wear a turban and to have long hair and a beard. However, when Kanwar Singh requested a religious exemption to the Army's grooming regulations in 2014, he was denied a commission, beginning a four-year quest to join the National Guard. In 2016, Mr. Singh was granted a temporary religious accommodation and in 2018, after the Army passed a directive making the wearing of religious articles permissible, Kanwar Singh was sworn in as a commissioned officer in the U.S. Army (Lacdan 2020).

Mores are norms with a moral basis. Both littering and child sexual abuse are violations of law, but child sexual abuse is also a violation of our mores because we view such behavior as immoral.

All norms are associated with **sanctions**, or social consequences for conforming to or violating norms. When we conform to a social norm, we may be rewarded by a positive sanction. These may range from an approving smile to a public ceremony in our honor. When we violate a social norm, we may be punished by a negative sanction, which may range from a disapproving look to the death penalty or life in prison. Most sanctions are spontaneous expressions of approval or disapproval by groups or individuals—these are referred to as informal sanctions. Sanctions that are carried out according to some recognized or formal procedure are referred to as formal sanctions. Types of sanctions, then, include positive informal sanctions, positive formal sanctions, negative informal sanctions, and negative formal sanctions (see Table 1.1).

sanctions Social consequences for conforming to or violating norms.

symbol Something that represents something else.

JEFF KOWALSKY/AFP/Getty Images

Symbolic interactionists emphasize the significance of language, gestures, and objects and their social meaning in determining human behavior. One of the most universal symbols in the United States is the American flag. Here supporters of Senator Bernie Sanders, a 2020 presidential hopeful, rally in Grand Rapids, Michigan.

Symbols. A **symbol** is something that represents something else. Without symbols, we could not communicate with one another or live as social beings.

The symbols of a culture include language, gestures, and objects whose meanings the members of a society commonly understand. In our society, Uncle Sam has come to symbolize the government of the United States, a peace sign symbolizes the value of non-violence, and a white-hooded robe symbolizes the Ku Klux Klan. Sometimes people attach different meanings to the same symbol. The Confederate flag is a symbol of Southern pride to some and a symbol of racial bigotry to others.

The elements of the social structure and culture just discussed play a central role in the creation, maintenance, and social responses to various social problems. One of the goals of taking a course in social problems is to develop an awareness of how the elements of social structure and culture contribute to social problems. Sociologists refer to this awareness as the "sociological imagination."

The Sociological Imagination

The **sociological imagination**, a term C. Wright Mills (1959) coined, refers to the ability to see the connections between our personal lives and the social world in which we live. When we use our sociological imagination, we are able to distinguish between "private troubles" and "public issues" and to see connections between the events and conditions of our lives and the social and historical context in which we live.

For example, that one person is unemployed constitutes a private trouble. That millions of people are unemployed in the United States constitutes a public issue. Once we understand that other segments of society share personal troubles such as intimate partner abuse, drug addiction, criminal victimization, poverty, and racism, we can look for the elements of the social structure and culture that contribute to these public issues and private troubles. If the various elements of the social structure and culture contribute to private troubles and public issues, then society's social structure and culture must be changed if these concerns are to be resolved.

Rather than viewing the private trouble of obesity and all of its attending health concerns as a result of an individual's faulty character, lack of self-discipline, or poor choices regarding food and exercise, we may understand the obesity epidemic as a public issue that results from various social and cultural forces, including government policies that make high-calorie foods more affordable than healthier, fresh produce; powerful food lobbies that fight against proposals to restrict food advertising to children; and technological developments that have eliminated many types of manual labor and replaced them with sedentary "desk jobs."

What do you think?

Although being unable to talk about politics to family and/or friends may feel like a private trouble, of late, it is actually a public issue. In a 2019 survey of 12,043 U.S. adults, nearly half reported that they had stopped talking about politics with someone as a result of something someone said either online or in person (Jurkowitz and Mitchell 2020). There are, however, demographic differences. For example, White Americans were more likely than Black or Hispanic Americans to stop talking with someone about politics. Why do you think that is true?

@Dave Ashelman

We are all a product of our social history. Nearly everyone (who isn't a Sociologist) forgets that.

-Dave Ashelman

ReTweeting @Sociology Theory

"You can never really understand an individual unless you also understand the society, historical time period in which they live, personal troubles, and social issues." - C. Wright Mills

-Sociology Theory

Theoretical Perspectives

Theories in sociology provide us with different perspectives with which to view our social world. A perspective is simply a way of looking at the world. A **theory** is a set of interrelated propositions or principles designed to answer a question or explain a particular phenomenon; it provides us with a perspective. Sociological theories help us to explain and predict the social world in which we live.

Sociology includes three major theoretical perspectives: the structural-functionalist perspective, the conflict perspective, and the symbolic interactionist perspective. Each perspective offers a variety of explanations about the causes of and possible solutions to social problems.

Structural-Functionalist Perspective

The structural-functionalist perspective is based largely on the works of Herbert Spencer, Emile Durkheim, Talcott Parsons, and Robert Merton. According to structural functionalism, society is a system of interconnected parts that work together in harmony to maintain a state of balance and social equilibrium for the whole. For example, each of the social institutions contributes important functions for society: Family provides a context for reproducing, nurturing, and socializing children; education offers a way to transmit a society's

sociological imagination The ability to see the connections between our personal lives and the social world in which we live.

theory A set of interrelated propositions or principles designed to answer a question or explain a particular phenomenon.

skills, knowledge, and culture to its youth; politics provides a means of governing members of society; economics provides for the production, distribution, and consumption of goods and services; and religion provides moral guidance and an outlet for worship of a higher power.

The structural-functionalist perspective emphasizes the interconnectedness of society by focusing on how each part influences and is influenced by other parts. For example, the increase in dual-earner families has contributed to the increase in day cares and after-school programs. As a result of changes in technology, colleges are offering more technical programs, and many adults are returning to school to learn new skills that are required in the workplace. The increasing number of women in the workforce has contributed to the formulation of policies against sexual harassment and job discrimination.

Structural functionalists use the terms *functional* and *dysfunctional* to describe the effects of social elements on society. Elements of society are functional if they contribute to social stability and dysfunctional if they disrupt social stability. Some aspects of society can be both functional and dysfunctional. For example, crime is dysfunctional in that it is associated with physical violence, loss of property, and fear. But according to Durkheim and other functionalists, crime is also functional for society because it leads to heightened awareness of shared moral bonds and increased social cohesion.

Sociologists have identified two types of functions: manifest and latent (Merton 1968). **Manifest functions** are consequences that are intended and commonly recognized. **Latent functions** are consequences that are unintended and often hidden. For example, the manifest function of education is to transmit knowledge and skills to society's youth. But public elementary schools also serve as babysitters for employed parents, and colleges offer a place for young adults to meet potential mates. The babysitting and mate selection functions are not the intended or commonly recognized functions of education; hence, they are latent functions.

Structural-Functionalist Theories of Social Problems

Two dominant theories of social problems grew out of the structural-functionalist perspective: social pathology and social disorganization.

Social Pathology. According to the social pathology model, social problems result from some "sickness" in society. Just as the human body becomes ill when our systems, organs, and cells do not function normally, society becomes "ill" when its parts (i.e., elements of the structure and culture) no longer perform properly. For example, problems such as crime, violence, poverty, and juvenile delinquency are often attributed to the breakdown of the family institution; the decline of the religious institution; and inadequacies in our economic, educational, and political institutions.

Social "illness" also results when members of a society are not adequately socialized to adopt its norms and values. People who do not value honesty, for example, are prone to dishonesties of all sorts. Early theorists attributed the failure in socialization to "sick" people who could not be socialized. Later theorists recognized that failure in the socialization process stemmed from "sick" social conditions, not "sick" people. To prevent or solve social problems, members of society must receive proper socialization and moral education, which may be accomplished in the family, schools, places of worship, and/or through the media.

Social Disorganization. According to the social disorganization view of social problems, rapid social change (e.g., the cultural revolution of the 1960s) disrupts the norms in a society. When norms become weak or are in conflict with one another, society is in a state of **anomie**, or *normlessness.* Hence, people may steal, physically abuse their spouses or children, abuse drugs, commit rape, or engage in other deviant behavior because the norms regarding these behaviors are weak or conflicting.

According to this view, the solution to social problems lies in slowing the pace of social change and strengthening social norms. For example, although the use of alcohol by teenagers is considered a violation of a social norm in our society, this norm is weak. The media portray young people drinking alcohol, teenagers tell each other where to buy fake

manifest functions Consequences that are intended and commonly recognized.

latent functions Consequences that are unintended and often hidden.

anomie A state of normlessness in which norms and values are weak or unclear.

identification cards (IDs) to purchase alcohol, and parents model drinking behavior by having a few drinks after work or at a social event. Solutions to teenage drinking may involve strengthening norms against it through public education, restricting media depictions of youth and alcohol, imposing stronger sanctions against the use of fake IDs to purchase alcohol, and educating parents to model moderate and responsible drinking behavior.

Conflict Perspective

Contrary to the structural-functionalist perspective, the conflict perspective views society as composed of different groups and interests competing for power and resources. The conflict perspective explains various aspects of our social world by looking at which groups have power and benefit from a particular social arrangement. For example, feminist theory argues that we live in a patriarchal society—a hierarchical system of organization controlled by men. Although there are many varieties of feminist theory, most would hold that feminism "demands that existing economic, political, and social structures be changed" (Weir and Faulkner 2004, p. xii).

The origins of the conflict perspective can be traced to the classic works of Karl Marx. Marx suggested that all societies go through stages of economic development. As societies evolve from agricultural to industrial, concern over meeting survival needs is replaced by concern over making a profit, the hallmark of a capitalist system. Industrialization leads to the development of two classes of people: the bourgeoisie, or the owners of the means of production (e.g., factories, farms, businesses), and the proletariat, or the workers who earn wages.

The division of society into two broad classes of people—the "haves" and the "have-nots"—is beneficial to the owners of the means of production. The workers, who may earn only subsistence wages, are denied access to the many resources available to the wealthy owners. According to Marx, the bourgeoisie use their power to control the institutions of society to their advantage. For example, Marx suggested that religion serves as an "opiate of the masses" in that it soothes the distress and suffering associated with the working-class lifestyle and focuses the workers' attention on spirituality, God, and the afterlife rather than on worldly concerns such as living conditions. In essence, religion

THE HAVES AND THE HAVE NOTS

Mike Luckovich/THE CARTOONIST GROUP

diverts the workers so that they concentrate on being rewarded in heaven for living a moral life rather than on questioning their exploitation.

Conflict Theories of Social Problems

There are two general types of conflict theories of social problems: Marxist and non-Marxist. Marxist theories focus on social conflict that results from economic inequalities; non-Marxist theories focus on social conflict that results from competing values and interests among social groups.

Marxist Conflict Theories. According to contemporary Marxist theorists, social problems result from class inequality inherent in a capitalistic system. A system of haves and have-nots may be beneficial to the haves but often translates into poverty for the have-nots. For example, in 2019, the average annual pay for chief executive officers (CEOs) of the top 350 U.S. corporations was $21.3 million, an increase of 105 percent over the last 10 years. Alternatively, during the same time period, the average worker at one of these large corporations saw their average annual compensation, including wages and benefits, grow by just 7.6 percent (Mishel and Kandra 2020). As we will explore later in this book, many social problems, including physical and mental illness, low educational achievement, and substandard housing and homelessness, are linked to poverty.

In addition to creating an impoverished class of people, capitalism also encourages "corporate violence." *Corporate violence* can be defined as actual harm and/or risk of harm inflicted on consumers, workers, and the general public as a result of decisions by corporate executives or managers. Corporate violence can also result from corporate negligence; the quest for profits at any cost; and willful violations of health, safety, and environmental laws (Reiman and Leighton 2020). Our profit-motivated economy encourages individuals, some of whom are otherwise good, kind, and law abiding, to knowingly participate in the manufacturing and marketing of defective products.

Take, for example, Boeing's 737 Max jetliner. The jetliner crashed twice within five months in 2018 and 2019. The first crash occurred off the coast of Jakarta, Indonesia, killing 189 people, and the second took place when an Ethiopian Airline, like the Indonesian flight, crashed shortly after takeoff, killing all 157 passengers. Boeing defended its safety record insisting that both crashes were the result of a "chain of events" rather than "any single item" (Chicago Tribune Wire 2019). Nonetheless, there is evidence that both crashes were the result a financial decision (i.e., to save money) by Boeing to exclude two sensors that would have displayed the angle of the nose of the jetliner and would have allowed the pilots to override the misfunctioning MAX software system (Shin 2019). In 2021, Boeing agreed to pay $2.5 billion in a settlement with the U.S. Department of Justice after it brought criminal charges against the airline manufacturer (Schaper 2021).

Marxist conflict theories also focus on the problem of **alienation**, or powerlessness and meaninglessness in people's lives. In industrialized societies, workers often have little power or control over their jobs, a condition that fosters in them a sense of powerlessness. The specialized nature of work requires employees to perform limited and repetitive tasks; as a result, workers may come to feel that their lives are meaningless.

Alienation is bred not only in the workplace but also in the classroom. Students have little power over their education and often find that the curriculum is not meaningful to their lives. Like poverty, alienation is linked to other social problems, such as low educational achievement, violence, and suicide.

Marxist explanations of social problems imply that the solution lies in eliminating inequality among classes of people by creating a classless society. The nature of work must also change to avoid alienation. Finally, stronger controls must be applied to corporations to ensure that corporate decisions and practices are based on safety rather than on profit considerations.

Non-Marxist Conflict Theories. Non-Marxist conflict theorists, such as Ralf Dahrendorf, are concerned with conflict that arises when groups have opposing values and interests. For example, anti-abortion activists value the life of unborn embryos and fetuses; pro-choice activists value the right of women to control their own bodies and reproductive decisions. These different value positions reflect different subjective interpretations of

alienation A sense of powerlessness and meaninglessness in people's lives.

what constitutes a social problem. For anti-abortionists, the availability of abortion is the social problem; for pro-choice advocates, the restrictions on abortion are the social problem. Sometimes the social problem is not the conflict itself but rather the way that conflict is expressed. Even most pro-life advocates agree that shooting doctors who perform abortions and blowing up abortion clinics constitute unnecessary violence and lack of respect for life. Value conflicts may occur between diverse categories of people, including non-White versus White, gay versus straight, young versus old, Democrats versus Republicans, and environmentalists versus industrialists.

Solving the problems that are generated by competing values may involve ensuring that conflicting groups understand one another's views, resolving differences through negotiation or mediation or agreeing to disagree. Ideally, solutions should be win-win, with both conflicting groups satisfied with the solution. However, outcomes of value conflicts are often influenced by power; the group with the most power may use its position to influence the outcome of value conflicts. For example, when Congress could not get all states to voluntarily increase the legal drinking age to 21, it threatened to withdraw federal highway funds from those that would not comply.

Symbolic Interactionist Perspective

Both the structural-functionalist and the conflict perspectives are concerned with how broad aspects of society, such as institutions and large social groups, influence the social world. This level of sociological analysis is called *macro-sociology*: It looks at the big picture of society and suggests how social problems are affected at the institutional level.

Micro-sociology, another level of sociological analysis, is concerned with the social-psychological dynamics of individuals interacting in small groups. Symbolic interactionism reflects the micro-sociological perspective and was largely influenced by the work of early sociologists and philosophers such as Max Weber, Georg Simmel, Charles Horton Cooley, G. H. Mead, W. I. Thomas, Erving Goffman, and Howard Becker. Symbolic interactionism emphasizes that human behavior is influenced by definitions and meanings that are created and maintained through symbolic interaction with others.

Sociologist W. I. Thomas (1931/1966) emphasized the importance of definitions and meanings in social behavior and its consequences. He suggested that humans respond to their definition of a situation rather than to the objective situation itself. Hence, Thomas noted that situations that we define as real become real in their consequences.

Symbolic interactionism also suggests that social interaction shapes our identity or sense of self. We develop our self-concept by observing how others interact with us and label us. By observing how others view us, we see a reflection of ourselves, what Cooley called the "looking glass self."

Last, the symbolic interactionist perspective has important implications for how social scientists conduct research. German sociologist Max Weber argued that, to understand individual and group behavior, social scientists must see the world through the eyes of that individual or group. Weber called this approach *verstehen*, which in German means "to understand." *Verstehen* implies that, in conducting research, social scientists must try to understand others' views of reality and the subjective aspects of their experiences, including their symbols, values, attitudes, and beliefs.

@Sociology Dictionary

Thomas theorem definition: The theory that if we define something as real, or believe that something is real, it is real in its consequences.

-Sociology Dictionary

Symbolic Interactionist Theories of Social Problems

A basic premise of symbolic interactionist theories of social problems is that a condition must be *defined or recognized* as a social problem for it to *be* a social problem. Three symbolic interactionist theories of social problems are based on this general premise.

Blumer's Stages of a Social Problem. Herbert Blumer (1971) suggested that social problems develop in stages. First, social problems pass through the stage of *societal recognition*—the process by which a social problem, for example, drunk driving, is "born." Drunk driving wasn't illegal until 1939, when Indiana passed the first state law regulating alcohol consumption and driving (Indiana State Government 2013). Second, *social legitimation* takes place when the social problem achieves recognition by the larger community, including the media, schools, and churches. As the visibility of traffic fatalities associated with alcohol increased, so did the legitimation of drunk driving as a social problem. The next stage in the development of a social problem involves *mobilization for action*, which occurs when individuals and groups, such as Mothers against Drunk Driving, become concerned about how to respond to the social condition. This mobilization leads to the *development and implementation of an official plan* for dealing with the problem, involving, for example, highway checkpoints, lower legal blood-alcohol levels, and tougher regulations for driving drunk.

Blumer's stage development view of social problems is helpful in tracing the development of social problems. For example, although sexual harassment and date rape occurred throughout the 20th century, these issues did not begin to receive recognition as social problems until the 1970s. Social legitimation of these problems was achieved when high schools, colleges, churches, employers, and the media recognized their existence. Organized social groups mobilized to develop and implement plans to deal with these problems. Groups successfully lobbied for the enactment of laws against sexual harassment and the enforcement of sanctions against violators of these laws. Groups also mobilized to provide educational seminars on date rape for high school and college students and to offer support services to victims of date rape.

Some disagree with the symbolic interactionist view that social problems exist only if they are recognized. According to this view, individuals who were victims of date rape in the 1960s should be considered victims of a problem, even though date rape was not recognized as a social problem at that time.

Labeling Theory. Labeling theory, a major symbolic interactionist theory of social problems, suggests that a social condition or group is viewed as problematic if it is labeled as such. According to labeling theory, resolving social problems sometimes involves changing the meanings and definitions that are attributed to people and situations. For example, so long as teenagers define drinking alcohol in a positive way, they will continue to abuse alcohol. So long as our society defines providing sex education and contraceptives to teenagers as inappropriate or immoral, the teenage pregnancy rate in the United States will continue to be higher than that in other industrialized nations. Individuals who label their own cell phone use while driving as safe will continue to use their cell phones as they drive, endangering their own lives and the lives of others.

Social Constructionism. Social constructionism is another symbolic interactionist theory of social problems. Similar to labeling theorists and symbolic interactionism in general, social constructionists argue that individuals who interpret the social world around them socially construct reality. Society, therefore, is a social creation rather than an objective given. As such, social constructionists often question the origin and evolution of social problems. For example, social constructionist theory has been used to:

> analyze the history of the temperance and prohibition movements[,] . . . the rise of alcoholism as a disease movement in the post-prohibition era[,] . . . and the crusade against drinking and driving in the 1980s in the United States. . . . These studies [each] analyzed the shifts in social meanings attributed to alcohol beverage use and to problems within the changing landscapes of social, economic, and political power relationships in American society. (Herd 2011, p. 7)

Central to this idea of the social construction of social problems are the media, universities, research institutes, and government agencies, which are often responsible for the public's initial "take" on the problem under discussion.

Table 1.2 summarizes and compares the major theoretical perspectives, their criticisms, and social policy recommendations as they relate to social problems. The study of social problems is based on research as well as on theory, however. Indeed, research and theory are intricately related. As Wilson (1983, p. 1) stated:

> Most of us think of theorizing as quite divorced from the business of gathering facts. It seems to require an abstractness of thought remote from the practical activity of

TABLE 1.2 Comparison of Theoretical Perspectives

	Structural Functionalism	Conflict Theory	Symbolic Interactionism
Representative theorists	Emile Durkheim, Talcott Parsons, Robert Merton	Karl Marx, Ralf Dahrendorf	George H. Mead, Charles Cooley, Erving Goffman
Society	Society is a set of interrelated parts; cultural consensus exists and leads to social order; natural state of society—balance and harmony.	Society is marked by power struggles over scarce resources; inequities result in conflict; social change is inevitable; natural state of society—imbalance.	Society is a network of interlocking roles; social order is constructed through interaction as individuals, through shared meaning, make sense of their social world.
Individuals	Individuals are socialized by society's institutions; socialization is the process by which social control is exerted; people need society and its institutions.	People are inherently good but are corrupted by society and its economic structure; institutions are controlled by groups with power; "order" is part of the illusion.	Humans are interpretive and interactive; they are constantly changing as their "social beings" emerge and are molded by changing circumstances.
Cause of social problems?	Rapid social change; social disorganization that disrupts the harmony and balance; inadequate socialization and/or weak institutions.	Inequality; the dominance of groups of people over other groups of people; oppression and exploitation; competition between groups.	Different interpretations of roles; labeling of individuals, groups, or behaviors as deviant; definition of an objective condition as a social problem.
Social policy/ solutions	Repair weak institutions; assure proper socialization; cultivate a strong collective sense of right and wrong.	Minimize competition; create an equitable system for the distribution of resources.	Reduce impact of labeling and associated stigmatization; alter definitions of what is defined as a social problem.
Criticisms	Called "sunshine sociology"; supports the maintenance of the status quo; needs to ask "functional for whom?"; does not deal with issues of power and conflict; incorrectly assumes a consensus.	Utopian model; Marxist states have failed; denies existence of cooperation and equitable exchange; cannot explain cohesion and harmony.	Concentrates on micro-issues only; fails to link micro-issues to macro-level concerns; too psychological in its approach; assumes label amplifies problem.

empirical research. But theory building is not a separate activity within sociology. Without theory, the empirical researcher would find it impossible to decide what to observe, how to observe it, or what to make of the observations.

What do you think?

For the first time in 208 years, the editors of the *New England Journal of Medicine* endorsed a presidential candidate stating that the former administration, rather than relying on science, "turned to uninformed 'opinion leaders' and charlatans who obscure[d] the truth and facilitate[d] the promulgation of outright lies" (The Editors 2020, p. 1480). Alternatively, President Biden has repeatedly said that his administration will "follow the science." How do you think sociology, as one of the four social sciences, can help in the fight against pandemics?

Social Problems Research

Most students taking a course in social problems will not become researchers or conduct research on social problems. Nevertheless, we are all consumers of research that is reported in the media. Politicians, social activist groups, and organizations attempt to justify their decisions, actions, and positions by citing research results. As consumers of research, we need to understand that our personal experiences and casual observations are less reliable than generalizations based on systematic research. One strength of scientific research is that it is subjected to critical examination by other researchers (see this chapter's *Social Problems Research Up Close* feature). The more you understand how research is done, the

social problems
RESEARCH
UP CLOSE

The Sociological Enterprise, Media, and COVID-19

Each chapter in this book contains a *Social Problems Research Up Close* box describing a research study that examines some aspect of a social problem and is presented in a report, book, or journal. Academic sociologists, those teaching at community colleges, colleges, or universities, as well as other social scientists, primarily rely on journal articles as the means to exchange ideas and information. Some examples of the more prestigious journals in sociology are the *American Sociological Review*, the *American Journal of Sociology*, and *Social Forces*. Most journal articles begin with *an introduction and review of the literature*. Here, the investigator examines previous research on the topic, identifies specific research areas, and otherwise "sets the stage" for the reader. Often in this section, research hypotheses are set forth, if applicable. A researcher, for example, might want to investigate the media habits of U.S. adults in reference to three of the biggest news stories of 2020—the presidential election, George Floyd's killing, and the COVID-19 pandemic. Given the lower likelihood of getting seriously ill from COVID-19 (Centers for Disease Control and Prevention [CDC] 2020b) or of voting in an election (Misra 2019), along with greater involvement in demonstrations for racial equality (Barroso and Minkin 2020), a researcher might hypothesize that younger Americans were less likely to follow news about COVID-19 or the 2020 election and more likely to follow news about George Floyd's death when compared to older Americans.

The next major section of a journal article is *sample and methods*. In this section, an investigator describes how the research sample was selected, the characteristics of the research sample, the details of how the research was conducted, and how the data were analyzed (see Appendix A). Given the proposed research topic, a sociologist might obtain data from the American Trends Panel survey conducted by the Pew Research Center. The American Trends Panel survey is a nationally representative survey of randomly selected U.S. adults who complete a self-administered web-based questionnaire.

The final section of a journal article includes the *findings and conclusions*. The findings of a study describe the results, that is, what the researcher found as a result of the investigation. Findings are then discussed within the context of the hypotheses and the conclusions that can be drawn. Often, research results are presented in tabular form. Reading tables carefully is an important part of drawing accurate conclusions about the research hypotheses. Using the table on the next page, follow these steps to assess the association between news topic and readers' ages:

1. *Read the title of the table and make sure that you understand what the table contains.* The title of the table indicates the unit of analysis (U.S. adults), the dependent variable (media viewing habits), the independent variables (age and news topic), and what the numbers represent (percentages).
2. *Read the information contained at the bottom of the table, including the source and any other explanatory information.* For example, the information at the bottom of this table indicates that there were 9,654 respondents and how the column variables were constructed. For example, following a topic "closely" combined the responses of those who answered that they followed the news item either "very closely" or "fairly closely." Alternatively, "not closely" sums the responses by those who reported following a topic "not too closely" or "not at all closely." Finally, the exact wording of the questions used to ask about the three news items is also listed under the table.
3. *Examine the row and column headings.* This table looks at the percentage of U.S. adults in four age groups who reported closely or not closely following three 2020 news stories. The three news stories were the 2020 presidential election candidates, the demonstrations protesting the death of George Floyd, and the outbreak of COVID-19. The age groups are young adults (18–29 years old), younger middle-aged adults (30–49 years old), older middle-aged adults (50–64 years old), and the elderly (65 years old or older).
4. *Thoroughly and carefully examine the data in the table, looking for patterns between variables.* As indicated in the table, young adults, those between the ages of 18 and 29, were the least likely to closely follow news about the 2020 presidential election candidates. Alternatively, those 65 and over were the most likely to closely follow news about the 2020 presidential candidates. In fact, looking at the first column, the older respondents are, the more likely they were to report following the presidential election candidates. Note that as the numbers in the first column increase with age, the numbers in the second column decrease with age since, when looking at an age group, 18- to 29-year-olds for example, the sum of the two numbers must equal 100 percent (35 + 65 = 100 percent), the total number of respondents for that age group.

The age pattern detected for following news about the presidential election candidates is not replicated for following news about the George Floyd demonstrations. The numbers vary little by age, differing at most by just 10 percent between 30- to 49-year-old respondents (80 percent) and those 65 and older (90 percent) in terms of closely following news about the demonstrations. Although there is more variation in the percentages of those closely following news about the outbreak of COVID-19 than those following news about the demonstrations, the range is still fairly small, 75 percent compared to 94 percent. Note that as age increases, so does the percentage of respondents who reported closely following news about COVID-19, the same pattern that existed for respondents closely following the 2020 election.

Now compare percentages across rows. Young adults were most likely to closely follow news related to George Floyd's killing, followed by COVID-19 and the 2020 presidential candidates. Respondents in each of the remaining three age groups were most likely to closely follow news related to COVID-19, followed by the George Floyd demonstrations, and, lastly, the 2020 presidential election. As indicated by the last row in the table where it reads "Total," regardless of age, respondents were the most likely to follow the George Floyd demonstrations, followed by COVID-19, and the 2020 election candidates.

5. *Use the information you have gathered in step 4 to address the hypotheses.* Clearly, as hypothesized, young adults were the most likely to closely follow news about the demonstrations around the country to protest the death of George Floyd. And, although no specific hypothesis was made about the relative differences between young adults following news about COVID-19 versus news about the 2020 presidential election candidates, it is interesting to note that 75 percent of 18- to 29-year-olds closely followed news about COVID-19 compared to just 35 percent who closely followed news about the 2020 presidential election candidates. Not surprisingly, given the increased risk of contracting COVID-19, respondents over the age of 65 were not only the most likely to follow news about the pandemic, they had the highest rate of any age group following any of the three news topics.
6. *Draw conclusions consistent with the information presented.* From the results of the study, we can conclude that younger adults, that is, those between the ages of 18 and 29, were more likely to follow news reports of the demonstrations protesting the death of George Floyd when compared to the two other biggest new stories of 2020—the presidential election candidates and the COVID-19 pandemic. We can also conclude that as age increases, the likelihood of following news about the pandemic closely also increases, as does the likelihood of following news about the 2020 presidential election candidates. We cannot, however, determine the reason for these results without more information. For example, although it has been suggested that respondents 65 years of age and older were more likely to follow COVID-19 news because of being at a higher risk to contract the disease, it may simply be that those over the age of 65 have more time to follow *all* news stories. Supporting this contention, respondents over the age of 65 were more likely than each of the other age groups to follow news stories about the presidential election, the demonstrations over George Floyd's death, and the pandemic.

SOURCE: Jurkowitz 2020.

Percentage of U.S. Adults Following Each Topic "Closely" or "Not Closely" by Age Group, 2020*

	2020 NEWS TOPICS:					
	2020 Election[1]		George Floyd[2]		COVID-19[3]	
AGE GROUPS:	Closely	Not Closely	Closely	Not Closely	Closely	Not Closely
Ages 18–29	35	65	83	17	75	25
Ages 30–49	46	46	80	20	83	16
Ages 50–64	60	40	84	16	91	9
Ages 65 and over	74	26	90	10	94	6
Total	65	34	83	17	81	18

[1]Percentage of each age group who have been following news about the 2020 election candidates.
[2]Pecentage of each age group who have been following news about the demonstrations around the country to protest the death of George Floyd.
[3]Percentage of each age group who have been following news about the outbreak of the coronavirus strain known as COVID-19.
*N = 9,654; numbers may not sum to 100 due to rounding error; survey conducted between June 4, 2020, and June 10, 2020.

> "The function of sociology, as of every science, is to reveal that which is hidden."
>
> –PIERRE BOURDIEU, FRENCH SOCIOLOGIST

better able you will be to critically examine and question research rather than to passively consume research findings. In the remainder of this section, we discuss the stages of conducting a research study and the various methods of research that sociologists use.

Stages of Conducting a Research Study

Sociologists progress through various stages in conducting research on a social problem. In this section, we describe the first four stages: (1) formulating a research question, (2) reviewing the literature, (3) defining variables, and (4) formulating a hypothesis.

Formulating a Research Question. A research study usually begins with a research question. Where do research questions originate? How does a particular researcher come to ask a particular research question? In some cases, researchers have a personal interest in a specific topic because of their own life experiences. For example, a researcher who has experienced spouse abuse may wish to do research on such questions as, "What factors are associated with domestic violence?" and "How helpful are battered women's shelters in helping abused women break the cycle of abuse in their lives?" Other researchers may ask a particular research question because of their personal values—their concern for humanity and the desire to improve human life. Researchers may also want to test a particular sociological theory or some aspect of it in order to establish its validity or conduct studies to evaluate the effect of a social policy or program. Research questions may also be formulated by the concerns of community groups and social activist organizations in collaboration with academic researchers. Government and industry also hire researchers

The American Journal of Sociology

A Bimonthly edited by the Sociological Faculty of the University of Chicago, with the advice of leading sociologists in America and Europe. ALBION W. SMALL, Editor-in-Chief

THE ONLY JOURNAL IN THE ENGLISH LANGUAGE DEVOTED PRIMARILY TO PURE SOCIOLOGY

$2.00 a year; single copies, 50 cents

THE sociologists are working on the clue that human association—or " the stream of life," as it was called a generation ago—is a process, made up of lesser processes, down to the vanishing of social relations in movements within the individual consciousness which make the problems of psychology.

The goal of the sociologists is a statement of life in terms of the ultimate processes which are working out through the different incidents of human experience.

Some of the sociologists prefer to describe their work as a return to the ideal of social study proposed by Adam Smith, but developed by him only in the economic division of human activities. In the philosophy of the author of *The Wealth of Nations* the activities prompted by the wealth interests were merely one of several departments of human pursuits. In his scheme, accordingly, economic science was only one of an indefinite number of social sciences which must be worked out and correlated in order to furnish an adequate chart of actual social processes. For nearly a century the economic fraction of social science was cultivated as though it were the whole. Sociology is not a rival of economics. It is essentially a method of investigation, with the aim of making the other social processes as intelligible as the economists have made the processes which terminate in the production of wealth.

This **Journal** is a medium of publication for both general and special studies of social relations, as they appear from this point of view.

Subscriptions filed immediately to begin January, 1907, will include the November, 1906, number free.

The University of Chicago Press (Dept. 16), CHICAGO and NEW YORK

Jay Paull/Archive Photos/ Getty Images

Sociology is the scientific study of society. The first academic department of sociology was established in 1892 at the University of Chicago, and the discipline's leading journal, the *American Journal of Sociology*, was founded in 1895. Pictured is an advertisement for a subscription to the Journal, which, in 1907, was $2.00 a year. Journals are the primary means by which research is communicated among professional sociologists.

to answer questions such as, "How many vehicle crashes are caused by 'distracted driving' involving the use of cell phones?" and "What types of cell phone technologies can prevent the use of cell phones while driving?"

Reviewing the Literature. After a research question is formulated, researchers review the published material on the topic to find out what is already known about it. Reviewing the literature also provides researchers with ideas about how to conduct their research and helps them formulate new research questions. A literature review serves as an evaluation tool, allowing a comparison of research findings and other sources of information, such as expert opinions, political claims, and journalistic reports.

Defining Variables. A **variable** is any measurable event, characteristic, or property that varies or is subject to change. Researchers must operationally define the variables they study. An *operational definition* specifies how a variable is to be measured. For example, an operational definition of the variable "religiosity" might be the number of times the respondent reports going to church, temple, or other religious gatherings. Another operational definition of "religiosity" might be the respondent's answer to the question, "How important is religion in your life?" (For example, 1 is not important; 2 is somewhat important; 3 is very important.)

Operational definitions are particularly important for defining variables that cannot be directly observed. For example, researchers cannot directly observe concepts such as "mental illness," "sexual harassment," "child neglect," "job satisfaction," and "drug abuse." Nor can researchers directly observe perceptions, values, and attitudes.

Formulating a Hypothesis. After defining the research variables, researchers may formulate a **hypothesis**, which is a prediction or educated guess about how one variable is related to another variable. The **dependent variable** is the variable that researchers want to explain; that is, it is the variable of interest. The **independent variable** is the variable that is expected to explain change in the dependent variable. In formulating a hypothesis, researchers predict how the independent variable affects the dependent variable. For example, Kmec (2003) investigated the impact of segregated work environments on minority wages, concluding that "minority concentration in different jobs, occupations, and establishments is a considerable social problem because it perpetuates racial wage inequality" (p. 55). In this example, the independent variable is workplace segregation, and the dependent variable is wages.

Methods of Data Collection

After identifying a research topic, reviewing the literature, defining the variables, and developing hypotheses, researchers decide which method of data collection to use. Alternatives include experiments, surveys, field research, and secondary data.

Experiments. **Experiments** involve manipulating the independent variable to determine how it affects the dependent variable. Experiments require one or more experimental groups that are exposed to the experimental treatment(s) and a control group that is not exposed. After a researcher randomly assigns participants to either an experimental group or a control group, the researcher measures the dependent variable. After the experimental groups are exposed to the treatment, the researcher measures the dependent variable again. If participants have been randomly assigned to the different groups, the researcher may conclude that any difference in the dependent variable among the groups is due to the effect of the independent variable.

An example of a "social problems" experiment on crime would be to randomly assign men on parole to a group who receive counseling (experimental group) or a group that does not receive counseling (control group). The independent variable would be counseling and the dependent variable would be whether the parolee re-offended within six months of being released from prison. The researcher might hypothesize that men who receive counseling will be less likely to re-offend than those who do not.

variable Any measurable event, characteristic, or property that varies or is subject to change.

hypothesis A prediction or educated guess about how one variable is related to another variable.

dependent variable The variable that researchers want to explain; that is, it is the variable of interest.

independent variable The variable that is expected to explain change in the dependent variable.

experiments Manipulating the independent variable to determine how it affects the dependent variable. Experiments require one or more experimental groups that are exposed to the experimental treatment(s) and a control group that is not exposed.

@cdixon

Poll: retweet/favorite polls are an effective and unbiased polling mechanism.
retweet = agree
favorite = disagree

-Chris Dixon

The major strength of the experimental method is that it provides evidence for causal relationships, that is, how one variable affects another. A primary weakness is that experiments are often conducted on small samples, often in artificial laboratory settings; thus, the findings may not be generalized to other people in natural settings.

Surveys. **Survey research** involves eliciting information from respondents through questions. An important part of survey research is selecting a sample of those to be questioned. A **sample** is a portion of the population, selected to be representative so that the information from the sample can be generalized to a larger population. For example, instead of asking all middle school children about their delinquent activity, the researcher would ask a representative sample of them and assume that those who were not questioned would give similar responses. After selecting a representative sample, survey researchers either interview people, ask them to complete written questionnaires, or elicit responses to research questions through web-based surveys. Some surveys are conducted annually or every other year so that researchers can observe changes in responses over time. This chapter's *Self and Society* feature allows you to voice your opinion on various social issues through the use of a written questionnaire. After completing the survey you can compare your responses to a national sample of first-year college students.

survey research Eliciting information from respondents through questions.

sample A portion of the population, selected to be representative so that the information from the sample can be generalized to a larger population.

Interviews. In interview survey research, trained interviewers ask respondents a series of questions and make written notes about or tape-record the respondents' answers. Interviews may be conducted over the phone or face-to-face.

One advantage of interview research is that researchers are able to clarify questions for respondents and follow up on answers to particular questions. Researchers often conduct face-to-face interviews with groups of individuals who might otherwise be inaccessible. For example, some AIDS-related research attempts to assess the degree to which individuals engage in behavior that places them at high risk for transmitting or contracting HIV. Street youth and intravenous drug users, both high-risk groups for HIV infection, may not have a telephone or address because of their transient lifestyle. These groups may be accessible, however, if the researcher locates their hangouts and conducts face-to-face interviews.

The most serious disadvantages of interview research are cost and the lack of privacy and anonymity. Respondents may feel embarrassed or threatened when asked questions that relate to personal issues such as drug use, domestic violence, and sexual behavior. As a result, some respondents may choose not to participate in interview research on sensitive topics. Those who do participate may conceal or alter information or give socially desirable answers to the interviewer's questions (e.g., "No, I do not use drugs" or "No, I do not text while driving").

Reading Eagle/MediaNews Group/Getty Images

A sign in Pennsylvania reads in both English and Spanish: "You Count! Census 2020." The U.S. Census is the largest and most comprehensive survey in the United States. In 2020, as result of soaring costs and the pandemic, households were able to respond to the questionnaire online for the first time. Concerns over a citizenship question and mistrust of the government contributed to concerns that the results may not be reliable. The Census is conducted every 10 years as mandated by the U.S. Constitution.

Questionnaires. Instead of conducting personal or phone interviews, researchers may develop questionnaires that they either mail, post online, or give to a sample of respondents. Questionnaire research offers the advantages of being less expensive and less time-consuming than face-to-face or telephone surveys. Questionnaire research also provides privacy and anonymity to the research participants, thus increasing the likelihood that respondents will provide truthful answers.

The major disadvantage of mail or online questionnaires is that it is difficult to obtain an adequate response rate. Many people do not want to take the time or make the effort to complete a questionnaire. Others may be unable to read and understand the questionnaire.

Web-based surveys. In recent years, technological know-how and the expansion of the Internet have facilitated the use of online surveys. Web-based surveys, although still less common than interviews and questionnaires, are growing in popularity and are thought by some to reduce many of the problems associated with traditional survey research. For example, the response rate of telephone surveys has been declining as potential respondents have caller ID, unlisted telephone numbers, answering machines, or no home (i.e., landline) telephone (Farrell and Petersen 2010). On the other hand, the use of and access to the Internet continue to grow. In 2020, the number of Americans connected to the Internet was higher than in any other single year (Internet Live 2020).

Field Research. **Field research** involves observing and studying social behavior in settings in which it occurs naturally. Two types of field research are participant observation and nonparticipant observation.

In *participant observation research*, researchers participate in the phenomenon being studied so as to obtain an insider's perspective on the people and/or behavior being observed. Palacios and Fenwick (2003), two criminologists, attended dozens of raves over a 15-month period to investigate the South Florida drug culture. In *nonparticipant observation research*, researchers observe the phenomenon being studied without actively participating in the group or the activity. For example, Simi and Futrell (2009) studied White power activists by observing and talking to organizational members but did not participate in any of their organized activities.

Sometimes sociologists conduct in-depth detailed analyses or case studies of an individual, group, or event. For example, Fleming (2003) conducted a case study of young auto thieves in British Columbia. He found that, unlike professional thieves, the teenagers' behavior was primarily motivated by thrill seeking—driving fast, the rush of a possible police pursuit, and the prospect of getting caught.

The main advantage of field research on social problems is that it provides detailed information about the values, rituals, norms, behaviors, symbols, beliefs, and emotions of those being studied. A potential problem with field research is that the researcher's observations may be biased (e.g., the researcher becomes too involved in the group to be objective). In addition, because field research is usually based on small samples, the findings may not be generalizable.

Secondary Data Research. Sometimes researchers analyze secondary data, which are data that other researchers or government agencies have already collected or that exist in forms such as historical documents, police reports, school records, and official records of marriages, births, and deaths. A major advantage of using secondary data in studying social problems is that the data are readily accessible, so researchers avoid the time and expense of collecting their own data. Secondary data are also often based on large representative samples. The disadvantage of secondary data is that researchers are limited to the data already collected.

Ten Good Reasons to Read This Book

Most students reading this book are not majoring in sociology and do not plan to pursue sociology as a profession. So why should students take a course on social problems? How can reading this textbook about social problems benefit you?

1. *Understanding that the social world is too complex to be explained by just one theory will expand your thinking about how the world operates.* For example, juvenile delinquency doesn't have just one cause—it is linked to (1) an increased number of youths living in inner-city neighborhoods with little or no parental supervision (social disorganization theory); (2) young people having no legitimate means of acquiring material wealth (anomie theory); (3) youths being angry and frustrated at the inequality and racism in our society (conflict theory); and (4) teachers regarding youths as "no good" and treating them accordingly (labeling theory).

field research Observing and studying social behavior in settings in which it occurs naturally.

2. *Developing a sociological imagination will help you see the link between your personal life and the social world in which you live.* In a society that values personal responsibility, there is a tendency to define failure and success as consequences of individual free will. The sociological imagination enables us to understand how social forces influence our personal misfortunes and failures and contribute to personal successes and achievements.
3. *Understanding globalization can help you become a safe, successful, and productive world citizen.* Social problems cross national boundaries. Problems such as COVID-19, war, climate change, human trafficking, and overpopulation are global problems. Problems that originate in one part of the world may affect other parts of the world and may be caused by social policies in other nations. Thus, understanding social problems requires consideration of the global interconnectedness of the world. And solving today's social problems requires collective action among citizens across the globe.
4. *Understanding the difficulty involved in "fixing" social problems will help you make decisions about your own actions, for example, whom you vote for or what charity you donate money to.* It is important to recognize that "fixing" social problems is a very difficult and complex enterprise. One source of this difficulty is that we don't all agree on what the problems are. We also don't agree on what the root causes are of social problems. Is the problem of gun violence in the United States a problem caused by gun availability? Violence in the media? A broken mental health care system? Masculine gender norms? If we socialized boys to be more nurturing and gentler, rather than aggressive and competitive, we might reduce gun violence, but we would also potentially create a generation of boys who would not want to sign up for combat duties in the military, and our armed forces would not have enough recruits. Thus solving one social problem (gun violence) may create another social problem (too few military recruits). It should also be noted that although some would see low military recruitment as a problem, others would see it as a positive step toward a less militaristic society.
5. *Although this is a social problems book, it may actually make you more rather than less optimistic.* Yes, all the problems discussed in the book are real, and they may seem insurmountable, but they aren't. You'll read about positive social change; for example, the number of people who smoke cigarettes in the United States has declined dramatically in recent years, as has the crime rate. Life expectancy has increased, and more people go to college than ever before. Change for the better can and does happen.
6. *Knowledge is empowering.* Social problems can be frightening, in part, because most people know very little about them beyond what they see on the news, read on social media, or hear from their friends. Thus, a new feature has been added called *The World in Quarantine*, which includes such topics as the impact of the pandemic on minority communities, the rise of domestic violence as a result of lockdowns, the effect of COVID-19 on population growth, and online disinformation campaigns that question the severity of the virus.
7. *The* Self and Society *exercises increase self-awareness and allow you to position yourself within the social landscape.* For example, earlier in this chapter, you had the opportunity to assess your opinions on a variety of social problems and to compare your responses to a national sample of first-year college students' attitudes toward the same issues.
8. The Human Side *features make you a more empathetic and compassionate human being by personalizing the topic at hand.* The study of social problems is always about the quality of life of individuals. By conveying the private pain and personal triumphs associated with social problems, we hope to elicit a level of understanding that may not be attained through the academic study of social problems alone. *The Human Side* in this chapter highlights college students' advice on how to be an activist, regardless of what cause you want to champion. Other *The Human Side* features in the book include the cost of the opiate epidemic; college, debt, and the American dream; experiencing racism at school; being elderly and homeless; and the devastating effect of having a friend or loved one in QAnon.

Getting in Good Trouble

Representative John Lewis, longtime congressman from Georgia, devoted his life to racial justice as an organizer and as an activist. In 2018 he tweeted, "Be hopeful, be optimistic. ... Never, ever be afraid to make some noise and get in good trouble, necessary trouble." Congressman Lewis, who died in 2020, was referring to actions that lead to positive social change through individual and collective action. Here, student activists give their advice on how to get into some "good trouble."

> They must first have a topic or issue that they are passionate about. They have to know that there is work involved. They have to know that they need to put everything they have and everything they have got into growing their movement and making a change. A successful person is not defined as someone who never fails. A successful person is defined as someone who fails ... but never gives up.
>
> —Zuriel Oduwole, founder of Dream Up, Speak Up, Stand Up, an organization that advocates for quality education for all children

> Begin with a problem that is close to home, both physically and emotionally. What has been bothering you, what issue can you not get out of your heart and mind? Start there. From there it takes a simple Google search to find a local organization that tackles that issue. Just sit in on a meeting and take it from there!
>
> —Jamie Margolin, author of *Youth to Power: Your Voice and How to Use It*, a guide book for young activists

Some advice for youth who want to get more involved is to reach out to your local community and start there. Reach out to whatever cause or movement is in your area and get involved. It's super fun and easy. We are standing to keep our earth pretty, clean, and green. Listen to each other, learn from each other, and fight [alongside] each other.

—Quannah Chasinghorse, climate justice advocate whose work helped pass the Arctic Cultural and Coastal Plain Protection Act

If you have an idea, or if you want to start a campaign, it's achievable through the power of social media and getting together with your friends and just talking openly and often about quite difficult and often taboo subjects, you can make change.

—Amika George, founder of Free Periods, a global organization that helps girls get personal hygiene products so they don't miss school during their menstrual cycles

Taking the first step is the hardest part but once you do it and focus on the activism that matters to you, you will find you are not alone.

—Mari Copeny, an activist in Flint, Michigan, helping to make sure children have access to clean water

Start small. Look within—and towards the communities—you come from, and ask yourself about all the stories you and people know. Is there a common thread connecting all these stories? For me, I thought a lot about the floods and hurricanes, and the effects they have on the different communities I come from. As a young person who could not even vote to give these issues the limelight, activism looked a lot like me educating myself, and all those around me. The goal is not to change the world alone but to do something—anything—that can create a ripple. So, start small.

—Aryaana Khan, climate advocate dedicated to ensuring that "the voices that don't get to negotiate with global leaders" are heard

To me, youth activism means that young people are not just limited to knowledge and awareness on socially relevant issues, but are actually taking initiatives to bring about a positive change. My advice for young activist hoping to make a difference would be:

Be bold: Young activists need to be bold, to voice their opinion and make their voices heard. We should not be demotivated by the negativity that surrounds us.

Be patient: We need to learn to be patient, because we often might not get to our intended outcome when we want.

Synergize: It is very important that young activists find like-minded people and synergize their efforts in achieving their common goal.

—Franklin Gnanammuthu, contraception and sexual health activist in India who is passionate about helping young people through the use of digital technology

SOURCE: silvia li sam. (October 11, 2019).

9. *The* Social Problems Research Up Close *features teach you the basics of scientific inquiry, making you a smarter consumer of "pop" sociology, psychology, anthropology, and the like.* These boxes demonstrate the scientific enterprise, from theory and data collection to findings and conclusions. Examples of research topics featured in later chapters of this book include the portrayal of serial killers in popular films, micro-aggressions toward LGBTQ families, the relationship between education and life expectancy among Black and White Americans, attitudes toward

> "Fight for the things that you care about, but do it in a way that will lead others to join you."
>
> –RUTH BADER GINSBURG, JUSTICE, U.S. SUPREME COURT

needle exchange programs, and why some people allow fracking in their backyards and what happens when they do.

10. *Learning about social problems and their structural and cultural origins helps you—individually or collectively—make a difference in the world.* Individuals can make a difference in society through the choices they make. You may choose to vote for one candidate over another, demand the right to reproductive choice or protest government policies that permit it, drive drunk or stop a friend from driving drunk, repeat a homophobic or racist joke or chastise the person who tells it, and practice safe sex or risk the transmission of sexually transmitted diseases.

Collective social action is another, often more powerful way to make a difference. You may choose to create change by participating in a **social movement**—an organized group of individuals with a common purpose of promoting or resisting social change through collective action. Some people believe that, to promote social change, one must be in a position of political power and/or have large financial resources. However, the most important prerequisite for becoming actively involved in improving levels of social well-being may be genuine concern and dedication to a social "cause" (see this chapter's *The Human Side*).

Understanding Social Problems

At the end of each chapter, we offer a section with a title that begins with "Understanding . . .," in which we reemphasize the social origin of the problem being discussed, the consequences, and the alternative social solutions. Our hope is that readers will end each chapter with a "sociological imagination" concerning the problem and with an idea of how, as a society, we might approach a solution.

Sociologists have been studying social problems since the Industrial Revolution. Industrialization brought about massive social changes: The influence of religion declined, and families became smaller and moved from traditional, rural communities to urban settings. These and other changes have been associated with increases in crime, pollution, divorce, and juvenile delinquency. As these social problems became more widespread, the need to understand their origins and possible solutions became more urgent. The field of sociology developed in response to this urgency. Social problems provided the initial impetus for the development of the field of sociology and continue to be a major focus of sociology.

social movement An organized group of individuals with a common purpose of promoting or resisting social change through collective action.

There is no single agreed-on definition of what constitutes a social problem. Most sociologists agree, however, that all social problems share two important elements: an objective social condition and a subjective interpretation of that condition. Each of the three major theoretical perspectives in sociology—structural-functionalist, conflict, and symbolic interactionist—has its own notion of the causes, consequences, and solutions of social problems.

Chapter Review

- **What is globalization?**
Globalization is the growing economic, cultural, and technological interdependence among countries and regions. While some social problems are clearly universal, such as climate change, others *appear* to impact only the nation in which they occur. However, social problems are impacted by and have consequences for multiple nations when they occur.

- **What are some of the causes of political partisanship in the United States?**
Variables linked to political partisanship in the United States include political extremism; greater racial, religious, and ethnic diversity; leaders who demonize opponents; an increase in the division between classes; and disinformation campaigns.

- **What is a social problem?**
Social problems are defined by a combination of objective and subjective criteria. The objective element of a social problem refers to the existence of a social condition; the subjective element of a social problem refers to the belief that a particular social condition is harmful to society or to a segment of society and that it should and can be changed. By combining these objective and subjective elements, we arrive at the following definition: A social problem is a social condition that a segment of society views as harmful to members of society and in need of remedy.

- **What is meant by the structure of society?**
The structure of a society refers to the way society is organized.

- **What are the components of the structure of society?**
The components are institutions, social groups, statuses, and roles. Institutions are an established and enduring pattern of social relationships and include family, religion, politics, economics, and education. Social groups are defined as two or more people who have a common identity, interact, and form a social relationship. A status is a position that a person occupies within a social group and that can be achieved or ascribed. Every status is associated with many roles, or the set of rights, obligations, and expectations associated with a status.

- **What is meant by the culture of society?**
Whereas *social structure* refers to the organization of society, *culture* refers to the meanings and ways of life that characterize a society.

- **What are the components of the culture of society?**
The components are beliefs, values, norms, and symbols. *Beliefs* refer to definitions and explanations about what is assumed to be true. *Values* are social agreements about what is considered good and bad, right and wrong, desirable and undesirable. *Norms* are socially defined rules of behavior. Norms serve as guidelines for our behavior and for our expectations of the behavior of others. Finally, a *symbol* is something that represents something else.

- **What is the sociological imagination, and why is it important?**
The *sociological imagination*, a concept that C. Wright Mills (1959) developed, refers to the ability to see the connections between our personal lives and the social world in which we live. It is important because, when we use our sociological imagination, we are able to distinguish between "private troubles" and "public issues" and to see connections between the events and conditions of our lives and the social and historical context in which we live.

- **What are the differences among the three sociological perspectives?**
According to structural functionalism, society is a system of interconnected parts that work together in harmony to maintain a state of balance and social equilibrium for the whole. The conflict perspective views society as composed of different groups and interests competing for power and resources. Symbolic interactionism reflects the micro-sociological perspective and emphasizes that human behavior is influenced by definitions and meanings that are created and maintained through symbolic interaction with others.

- **What are the first four stages of a research study?**
The first four stages of a research study are formulating a research question, reviewing the literature, defining variables, and formulating a hypothesis.

- **How do the various research methods differ from one another?**
Experiments involve manipulating the independent variable to determine how it affects the dependent variable. Survey research involves eliciting information from respondents through questions. Field research involves observing and studying social behavior in settings in which it occurs naturally. Secondary data are data that other researchers or government agencies have already collected or that exist in forms such as historical documents, police reports, school records, and official records of marriages, births, and deaths.

- **What is a social movement?**
Social movements are one means by which social change is realized. A social movement is an organized group of individuals with a common purpose to either promote or resist social change through collective action.

Test Yourself

1. On the U.S. political spectrum, Democrats are left-leaning and are called liberals or progressives; Republicans are right-leaning and are called independents or conservatives.
   a. True
   b. False
2. The social structure of society contains
   a. statuses and roles.
   b. institutions and norms.
   c. sanctions and social groups.
   d. values and beliefs.
3. The culture of society refers to its meaning and the ways of life of its members.
   a. True
   b. False
4. Alienation
   a. refers to a sense of normlessness.
   b. is focused on by symbolic interactionists.
   c. can be defined as the powerlessness and meaninglessness in people's lives.
   d. is a manifest function of society.
5. Blumer's stages of social problems begin with
   a. mobilization for action.
   b. societal recognition.
   c. social legitimation.
   d. development and implementation of a plan.
6. The independent variable comes first in time (i.e., it precedes the dependent variable).
   a. True
   b. False

7. The third stage in defining a research study is
   a. formulating a hypothesis.
   b. reviewing the literature.
   c. defining the variables.
   d. formulating a research question.
8. A sample is a subgroup of the population—the group to whom you actually give the questionnaire.
   a. True
   b. False
9. Studying police behavior by riding along with patrol officers would be an example of
   a. participant observation.
   b. nonparticipant observation.
   c. field research.
   d. both a and c.
10. Students benefit from reading this book because it
   a. provides global coverage of social problems.
   b. highlights social problems research.
   c. encourages students to take pro-social action.
   d. all of the above

Answers: 1. B; 2. A; 3. A; 4. C; 5. B; 6. A; 7. C; 8. A; 9. D; 10. D.

Key Terms

achieved status 14
alienation 20
anomie 18
ascribed status 13
beliefs 14
culture 14
dependent variable 27
experiments 27
field research 29
globalization 4
hypothesis 27
independent variable 27
institution 13
latent functions 18
manifest functions 18
norms 15
objective element of a social problem 11
primary groups 13
political partisanship 7
populist movements 4
roles 14
sample 28
sanctions 16
secondary groups 13
social group 13
social movement 32
social problem 11
sociological imagination 17
status 13
structure 13
subjective element of a social problem 11
survey research 28
symbol 16
theory 17
values 14
variable 27

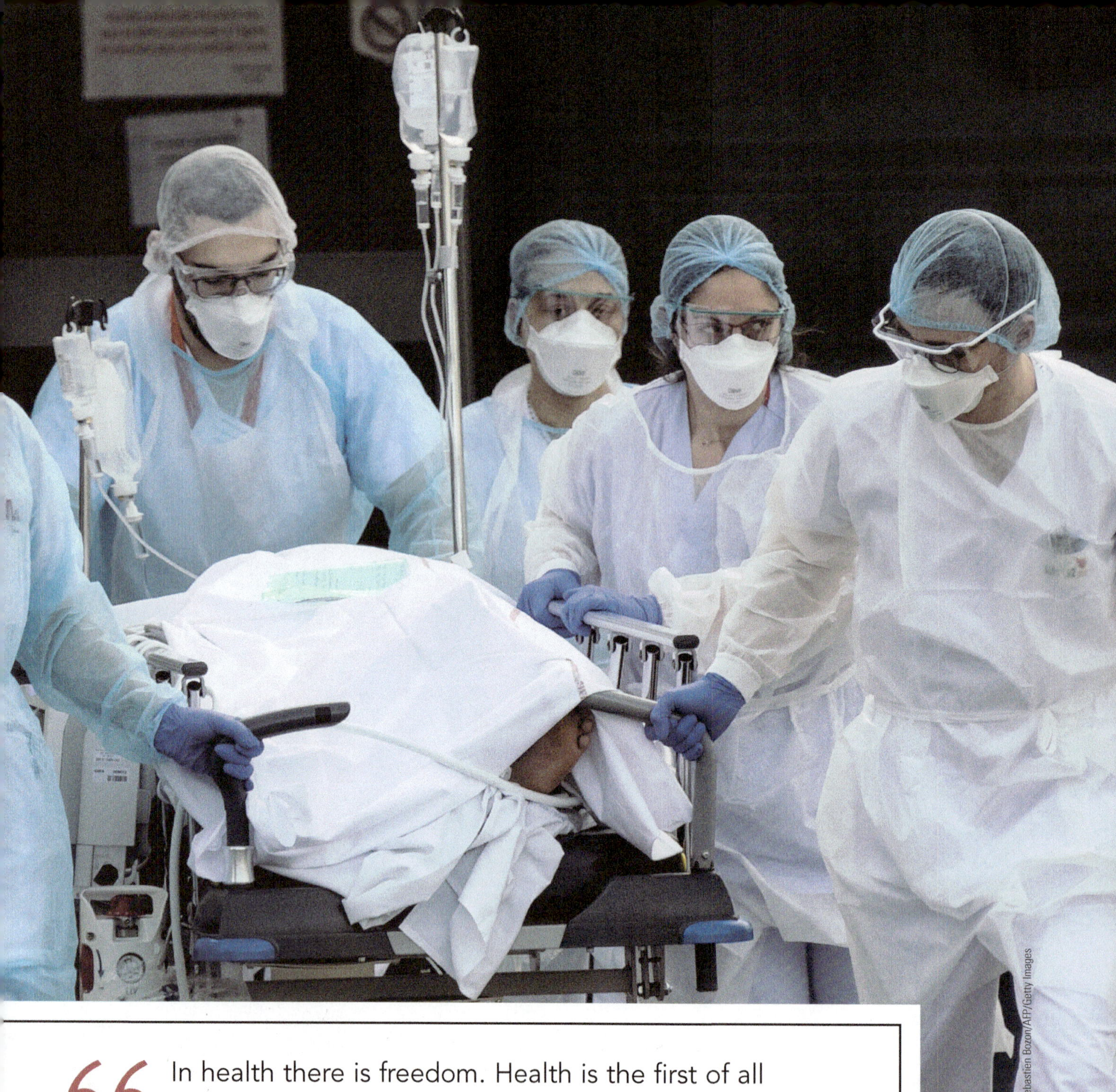
Sebastien Bozon/AFP/Getty Images

> In health there is freedom. Health is the first of all liberties."
>
> **HENRI AMIEL**
> Philosopher

2

Physical and Mental Health and Health Care

Chapter Outline

Learning Objectives

After studying this chapter, you will be able to …

1. Identify factors which contributed to the differential impact of the COVID-19 pandemic in the United States.
2. Compare life expectancy and mortality in low-, middle-, and high-income countries.
3. Identify ways in which globalization affects health and health care.
4. Explain how conflict theory, structural-functionalism, and symbolic interactionism help us understand illness and health care.
5. Describe the prevalence, impact, and causes of mental illness.
6. Give examples of how socioeconomic status, gender, and race and ethnicity affect health.
7. Describe the various types of private insurance plans and public health care insurance programs in the United States.
8. Identify factors that contribute to the high costs of health care in the United States.
9. Describe efforts to improve health in low- and middle-income countries, improve mental health care, increase access to affordable health care, and prepare for future pandemics in the United States.
10. Discuss the complexity of factors that affect health and that must be addressed in order to improve the health of a society.

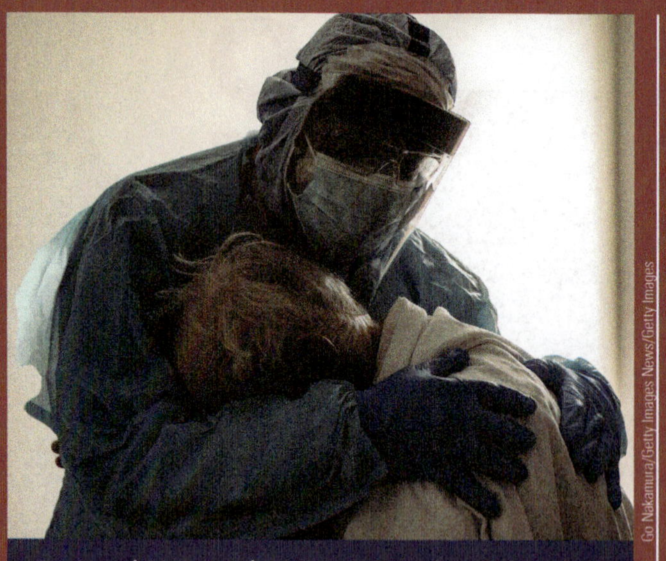

Dr. Joseph Varon, a physician at a Texas hospital, worked over 252 consecutive days treating COVID-19 patients. When he was walking through the intensive care unit (ICU) on Thanksgiving Day, he saw a distraught elderly man trying to get out of bed. Dr. Varon immediately embraced him.

IN EARLY AUGUST 2020, North Carolina couple Johnny Lee and Cathy "Darlene" Peoples both contracted COVID-19. The two, in their mid-sixties, had recently celebrated their 50th wedding anniversary. On August 11, they were both admitted to the hospital. On September 1, the couple's children were told that their parents would not survive. After weeks of caring for the couple, hospital staff put their beds side by side and clasped their hands together. Then the couple was removed from ventilators and died within minutes of each other, surrounded by their nurses (Johnson 2020). Day after day, week after week, month after month, health care professionals have worked the front lines of the COVID-19 pandemic, providing not only expert and often life-saving care but also serving far too many times as the last faces patients saw before death.

Over 1,445 U.S. health care workers had died from COVID-19 by the end of 2020 (Kaiser Health News 2020), with the global total death toll in the millions. The pandemic has been a stark reminder of the importance of health, the medical establishment and health care workers, and their impact on society as a whole. The World Health Organization (WHO 1946) defines **health** as "a state of complete physical, mental, and social well-being" (p. 3). One could easily argue that the study of social problems is, essentially, the study of health problems, as each social problem affects the physical, mental, and social well-being of humans and the social groups of which they are a part.

In this chapter, we use a sociological approach to examine physical and mental health and why some social groups experience more health problems than others, and how social forces affect and are affected by health and illness. We also address problems in health care focusing on issues related to access, cost, and quality of care.

The Global Context: The COVID-19 Pandemic

The first reported cases of what has come to be known as COVID-19 were identified in the Wuhan, Hubie province of China in late December 2019. At first thought to be a new strain of SARS (i.e., severe acute respiratory syndrome) by local whistleblowers (Smith and Liu 2020), the cases were officially identified as "viral pneumonia of unknown cause." The World Health Organization (WHO) China Country Office was informed of the cases on December 31, 2019 and immediately offered assistance to the Chinese government (WHO 2020a). The WHO reported the outbreak to their member nations including the United States via their Disease Outbreak News report on January 5, 2020. By then it was known that 59 cases of viral pneumonia of unknown origins had been identified in China.

health According to the World Health Organization, "a state of complete physical, mental, and social well-being."

By January 10th, the Chinese had identified the outbreak as being linked to a novel ("new") coronavirus, and released the genetic sequences for the virus the next day. This information was reported by the WHO which held its first of two global teleconferences on COVID-19 on the 10th, as well as publishing safety protocols related to managing the outbreak of a new disease. The first death in China from the novel virus was announced on January 11, 2020. For only the sixth time in its history, the World Health Organization declared a "public health emergency of international concern" on January 30, 2020 (WHO 2020a).

By December 2020, the **pandemic**, a worldwide disease outbreak, had infected over 69 million people worldwide resulting in 1.6 million deaths (WHO 2020b). Four days after the Chinese identified the virus, a case was confirmed in Thailand; two days later, the Japanese had identified a case and, nine days later, the first case in the United States appeared. By the end of January 2020, nearly 10,000 cases had been reported in over 21 countries (Johns Hopkins University 2020a). Retrospective studies indicate the virus was circulating in Europe as early as November 2019 (Horsley 2020).

> "The Pandemic has finally opened our eyes to the fact that health is not driven just by biology, but by the social environment in which we all find ourselves."
>
> **–SARAH HAWKES, CODIRECTOR OF GLOBAL HEALTH 5050**

The Evolution of COVID-19 in the United States

In the United States, the first 14 people who had been identified as having the virus had each traveled to China prior to the outbreak (Centers for Disease Control and Prevention [CDC] 2020). Subsequently, early cases of the virus were linked to travel in Europe. The Centers for Disease Control and Prevention (CDC) identified the first *non-travel-related* cases in California where two people had died from the disease, suggesting that the virus was present in the United States by late January or early February (CDC 2020). As Figure 2.1 illustrates, by December, 2020, the United States had surpassed several other countries in the number of new cases.

By mid-December, 2020 over 16.5 million U.S. cases of the virus had been recorded and more than 300,000 people had died. New cases in the United States exceeded 220,000 a day and the daily U.S. death toll rose to more than 3,000. In the same month, COVID-19 was the leading cause of death in the United States (Institute for Health Metrics and Evaluation 2020), and was projected to be the second leading cause of death nationally for 2020, surpassed only by heart disease. New mutations of the virus posed further challenges in early 2021, with some proving to be more contagious than the original strain (Reardon 2021).

The U.S. Centers for Disease Control and Prevention used an "excess death" analysis to determine the accuracy of reported COVID-19 death rates. **Death rates** refer to the number of people per 100,000 in a population who die in a given period of time. The CDC concluded that there were 299,028 more deaths than would ordinarily be expected between late January and early October, 2020. Of that number, 66 percent were officially caused by COVID-19 (Rossen et al. 2020), and the remaining 34 percent, although likely COVID related, had not been identified as such. If COVID-19 deaths had been so

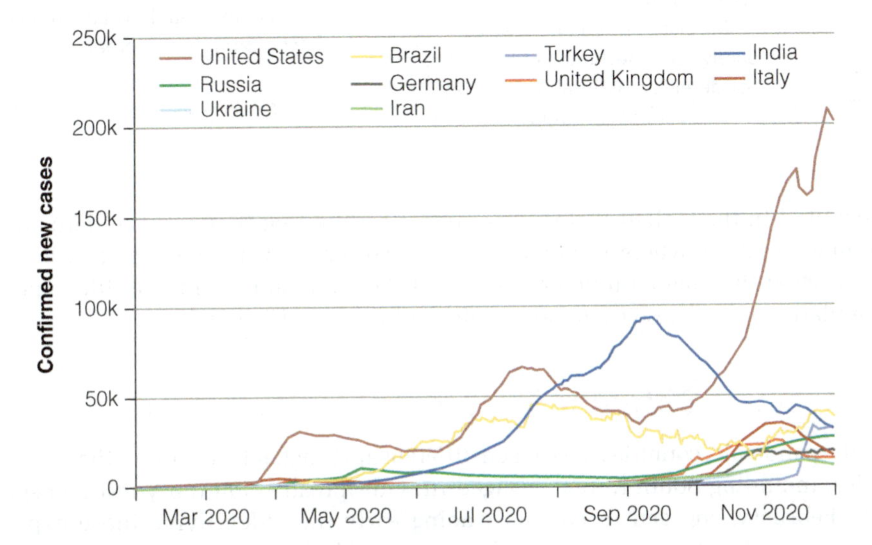

Figure 2.1 7-Day Average of New COVID-19 Cases in the Top 10 Most Affected Countries (December 10, 2020)
SOURCE: Johns Hopkins University 2020a.

pandemic A worldwide disease outbreak.

death rates The number of people per 100,000 in a population who die in a given period of time.

Coping with Mental Illness during a Pandemic

Dr. Stacey Torres is an assistant professor of sociology at the University of California–San Francisco. She was born and raised in New York, and her 77-year-old father and 37-year-old sister Erica still live in the city while she provides support from across the country.

My sister doesn't yet have the virus, but COVID-19 is already killing her.

Before the pandemic, Erica struggled. The isolation of New York City's spring lockdown pushed her off a proverbial cliff from which I'm not sure she can return.

From birth my sister has struggled. Born with a dislocated hip, at 3 years old she spent months in a full body cast. ... Unable to walk, run, and play like other children, she gained weight as a toddler and has faced lifelong mobility challenges. Erica's hard start continued with intellectual disabilities and a diagnosis of schizophrenia in her 20s. ... Her fiercest advocate—our mother—died when Erica was 13.

When I last saw her in February, Erica wasn't doing great but not awful. ... My sister had climbed out of extreme isolation last summer. It took persistent urging to get her to leave the house and return to mental health treatment. Just as she'd reestablished her routines, she had to isolate again for her safety and our father's.

She had phone therapy sessions but no longer had in-person appointments to compel her to walk to the bus and run errands. She couldn't use the internet anymore with public library closures or attend a psychosocial clubhouse where she participated in social activities. Her mobility plummeted and depression soared.

I found a disaster when I returned. ... Erica stayed in bed most days. Empty soda bottles and candy wrappers piled beneath, and a mountain of dirty clothes that no longer fit grew in the corner. I discovered her ankle monstrously swollen. ... Erica's used to being in poor health, never wanting to cause trouble. "I'm good," she answers whenever I ask how she is.

Getting her to the clinic was no small feat. The doctor warned of what he called "expiration of life" in the next 10 years at Erica's BMI of 84. "4-7-5 pounds." I had to ask him to repeat this number several times. ... Forget 10 years. I worry about her surviving 2020.

With her poor balance, I dread the fall that lands her in a nursing home or the COVID that robs her breath. We're waiting for Medicaid to approve a homecare aide and medical transportation to appointments because she cannot walk to the bus and can barely get into a taxi with my assistance. It feels too little, too late.

My sister isn't alone. We've barely scratched the surface of the pandemic's collateral damage as reports surface about increases in drug overdoses, anxiety, depression. ... With perpetual surges on the horizon and the threat of necessary future shutdowns, our failure to control the virus condemns vulnerable people like my sister to isolation, delayed medical care, and premature deaths of despair.

Before leaving I cleaned and stocked the house with healthy food and 100 rolls of toilet paper. We organized Erica's room, but there's much more to do when I return to help prepare her space for the long winter. ... I couldn't hug her before I left, having strictly masked and physically distanced from my family. I patted her arm. I didn't have time to say this before dashing to the airport:

> The world doesn't love you, Erica. The pandemic has exposed how we devalue the lives of those we deem "throwaway people," the poor, sick, disabled, old, immigrant, and minority. If our leadership cared we wouldn't allow more than 176,000 deaths and counting. But we've met many caring people, such as the patient immigrant cabdrivers that risked their own health by physically coming close to us to assist as you struggled into the taxi. We all need a little help sometimes, said the kind man who brought us home.

SOURCE: Stacy Torres - USA TODAY NETWORK

undercounted in the United States, it is almost certain that they were undercounted in developing countries where tests and medical resources were even scarcer. It remains unknown what the long-term global impact of the pandemic will be on life expectancy and population size. These issues are explored further in Chapter 12.

The Impact of Different Strategies for Prevention

contact tracing Identifying and contacting those exposed to people with positive tests.

positivity rate Percentage of positive tests out of every hundred tests conducted.

During the pandemic, countries have taken different approaches to testing for the virus. Some countries, including South Korea, engaged in widespread community testing beginning in early February, engaged in **contact tracing** early (i.e., identifying those exposed to people with positive tests), and quarantined exposed individuals to stop the virus (Hasell 2020). By the middle of March, South Korea had 600 new confirmed cases a day. However, even at that time, South Korea's **positivity rate**, the percentage of positive tests out of every one hundred tests conducted, was under 2 percent as a result of testing people with no symptoms or known exposure. Widespread testing helped with estimating the extent of the virus' spread and enabled some countries to get it under control quickly.

In the early months of the pandemic, tests were in short supply in the United States as a result of the bureaucracy of the Food and Drug Administration (FDA), the distribution of faulty test kits, and a lack of basic supplies (e.g., swabs, personal protective equipment [PPE]) (Sommer 2020; Yong 2020). Consequently, until early March, the CDC limited testing to people in the hospital or those with direct exposure to confirmed cases (Holmes, Devine, and Herb 2020).

By the third week in February, testing in the United States had been confined to just a few hundred people (The COVID Tracking Project 2020) while millions had been tested in China via a widespread community testing program similar to South Korea's (Burki 2020). Although Congress eventually appropriated funds to expand testing (Snell 2020), former President Trump was not a proponent of community-wide testing, commenting that if we stopped testing people, the positivity rate would go down.

In the United States, a lack of testing in combination with delayed shelter-in-place orders contributed to the spread of the virus by asymptomatic carriers (Schnieder 2020; Sommer 2020; Woolf et al. 2020). In March 2020, the World Health Organization (2020c) recommended that 10 to 30 tests be conducted for each positive case in order to identify infected individuals and thwart further spread of the disease.

At that time the United States was conducting approximately six tests per identified case (COVID Tracking Project 2020) with a positivity rate of over 20 percent (Our World in Data 2020). A lack of widespread testing meant health professionals had inaccurate estimates of the incidence of the virus across the United States, hindering their ability to put in place effective measures to control it. Most importantly, it diminished the impact of contact tracing and quarantine efforts, which worked effectively in other countries, including Taiwan and New Zealand (Summers et al. 2020). By late fall 2020, the most frequent situations leading to transmission in the United States were small social gatherings of family and friends who were unaware they were infected. (Brulliard 2020).

@Heather R411

450,088
Not a number. Each is a life lost.

<2% of American adults vaccinated.

#WearAMask 😷
#SociallyDistance
#COVID19

-Heather

What do you think?

In South Korea all residents and visitors are required to have an app installed on their smart phones that notifies them if they have been within 6 feet of someone who has tested positive for COVID-19. They are then required by law to quarantine. How do you think this mandate would be received in the United States and why?

An Overview of Global Health

In making international comparisons of health outcomes, social scientists commonly classify countries according to their level of economic development: (1) *high-income countries*, sometimes called *developed countries*, have relatively high gross national income per capita; (2) *middle-income countries*, also known as *less developed* or *developing countries*, have relatively low gross national income per capita; and (3) *low-income countries*, or *least developed countries*, which are the poorest countries of the world. Globally, as discussed in the following section, how long people live and what causes their deaths vary considerably.

Life Expectancy and Mortality in Low-, Middle-, and High-Income Countries

Life expectancy—the average number of years that individuals born in a given year can expect to live—is significantly greater in high-income countries than in low-income countries (see Table 2.1). Life

life expectancy The average number of years that individuals born in a given year can expect to live.

TABLE 2.1 Life Expectancy by Country Income Level, 2018

Country Income Level	Life Expectancy*
High	81
Upper middle	75
Middle	72
Lower middle	68
Low	63
WORLD	73

*Average life expectancy for countries in each income category.
SOURCE: World Bank 2019.

TABLE 2.2 Leading Causes of Death by Country Income Level, 2019

Low Income	High Income
1. Respiratory infections	Ischemic heart disease
2. Diarrheal diseases	Stroke and other cerebrovascular disease
3. Ischemic heart disease	Alzheimer's and other dementias
4. HIV/AIDS	Trachea, bronchus, lung cancers
5. Stroke	Chronic obstructive pulmonary disease
6. Malaria	Respiratory infections
7. Tuberculosis	Colon and rectum cancer
8. Preterm birth complications	Diabetes

Note: the information in this table was compiled before the COVID-19 pandemic.
SOURCE: World Health Organization 2020.

expectancy ranges from a high of 84 (in Japan) to a low of 53 (in the Central African Republic) (World Bank 2020a).

As shown in Table 2.2, the leading causes of death, or **mortality**, also vary around the world (note the data in Table 2.2 was compiled before the COVID-19 pandemic). Deaths caused by parasitic and infectious diseases, such as HIV/AIDS, tuberculosis, diarrheal diseases, and malaria, are much more common in less developed countries compared to more developed countries, with the exception of COVID-19. Parasitic and infectious diseases spread more easily in poor and overcrowded housing conditions, and in areas with lack of clean water and sanitation (see also Chapter 6).

In wealthy countries such as the United States, noninfectious diseases, also called "noncommunicable" diseases, are typically the leading causes of death, Not factoring in COVID-19 deaths during the pandemic, worldwide, nearly 70 percent of deaths are due to noninfectious diseases, primarily heart disease, stroke, cancer, and respiratory diseases (WHO 2020c). These diseases are related to tobacco use, physical inactivity, unhealthy diet, and alcohol abuse. Some are related to environmental exposures, as discussed in Chapter 13.

In recent decades, noncommunicable diseases—particularly heart disease—have also become leading causes of death in low- and middle-income countries, as rising incomes and emerging middle classes in countries such as China and India have led to (1) increased use of tobacco which is linked to cancer and respiratory diseases, (2) increased access to automobiles, televisions, and other technologies that contribute to a sedentary lifestyle, and (3) increased consumption of processed foods, high in sugar and fat, which are linked to obesity (WHO 2020d). Eighty-two percent of deaths from noncommunicable diseases now occur in low- and middle-income countries.

By mid-December 2020, globally, the number of reported deaths due to coronavirus exceeded 1.5 million (WHO 2020b). Twenty-seven percent of the total number of deaths were reported in North America, 27 percent in Europe, 21.9 percent in South America, 19.7 percent in Asia, 3.5 percent in Africa, and less than 0.1 percent in the Oceanic region (Our World in Data 2020). However, it is widely believed that these numbers underrepresent the actual number of deaths due to lack of sufficient testing. Figure 2.2 presents COVID-19 death rates (per 100,000 people) in the countries most affected.

mortality Death.

infant mortality rate The number of deaths of infants under 1 year of age per 1,000 live births.

Mortality among Infants and Children

The rates of infant mortality and under-5 mortality provide powerful indicators of the health of a population. The **infant mortality rate** is the number of deaths of infants under

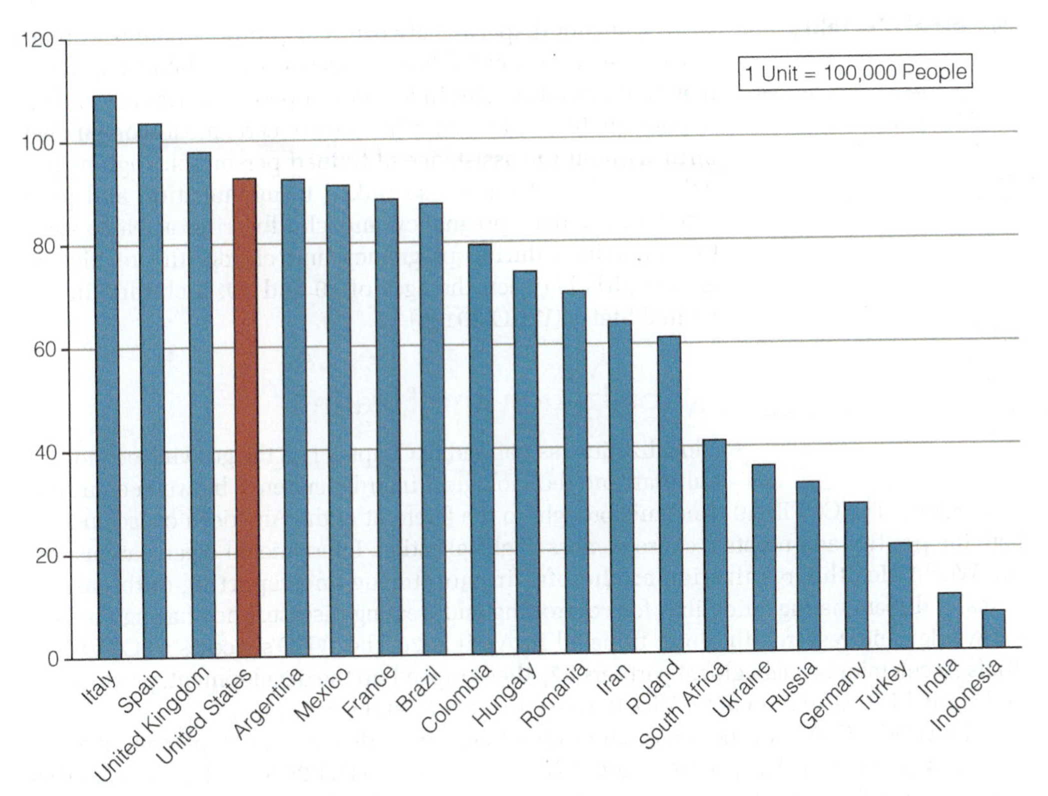

Figure 2.2 COVID-19 Deaths per 100,000 Residents, for Top 20 Most Affected Countries (December 2020)
SOURCE: Johns Hopkins University 2020b.

> "Let us be the ones who say we do not accept that a child dies every three seconds simply because he does not have the drugs you and I have. Let us be the ones to say we are not satisfied that your place of birth determines your right for life. Let us be outraged, let us be loud, let us be bold."
>
> —BRAD PITT, ACTOR

1 year of age per 1,000 live births and ranges from an average of 4 in high-income nations to 48 in low-income nations (World Bank 2020b). The **under-5 mortality rate**, or the number of deaths of children under age 5 per 1,000 live births, similarly is much lower in high-income countries than in low-income countries. Forty-five percent of all child deaths are due to infectious diseases like malaria; another 15 percent are due to pneumonia and other respiratory diseases, and 10 percent are due to diarrheal diseases (Roser, Ritchie, and Dadonaite 2019). These are treatable diseases, and most are related to lack of sanitation.

under-5 mortality rate The number of deaths of children under age 5 per 1,000 live births.

Half of the world's population does not have access to adequate sanitation facilities, one in nine people across the globe are hungry, and one in three people on the planet don't have access to safe drinking water (WHO 2019a) (see also Chapter 6). All of these factors contribute to high rates of infant and child mortality in the developing world.

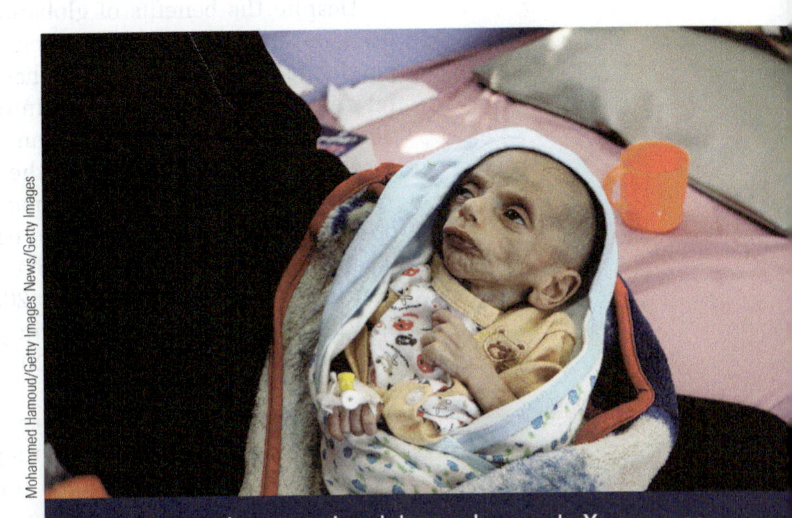

Mohammed Hamoud/Getty Images News/Getty Images

One in 9 people across the globe are hungry. In Yemen, one of the poorest countries in the world, about 70 percent of the country's population don't have enough food, including 2 million children who suffer acute malnutrition.

Maternal Mortality

Women in the United States and other developed countries generally do not experience pregnancy and childbirth as life-threatening. But for women ages 15 to 49 in developing countries, **maternal mortality**—death that results from complications associated with pregnancy and childbirth—is a leading cause of death. More than 94 percent of maternal deaths occur in low-income countries (WHO 2019b). Rates of maternal mortality

TABLE 2.3 Lifetime Risk of Maternal Mortality by Country Income

	Lifetime Risk of Maternal Mortality
Low income	1 in 45
Lower-middle income	1 in 140
Upper-middle income	1 in 1,200
High income	1 in 5,400
United States	1 in 3,000

SOURCE: World Health Organization 2019b.

show a greater disparity between rich and poor countries than any of the other societal health measures (see Table 2.3). High maternal mortality rates in less developed countries are related to poor-quality and inaccessible health care; most women give birth without the assistance of trained personnel. High maternal mortality rates are also linked to malnutrition and poor sanitation and to pregnancy and childbearing at early ages. Complications during pregnancy and childbirth are highest among girls between the ages of 10 and 19, including in the United States (WHO 2019b).

Globalization and Health

Globalization, as defined in Chapter 1, is the growing economic, cultural, and technological interdependence between countries and regions. The COVID-19 pandemic brought to the forefront of the American consciousness both the positive and negative consequences of globalization. International organizations like the World Health Organization are helpful in monitoring and reporting outbreaks of disease, disseminating guidelines for controlling and treating disease, and sharing medical knowledge and research findings. Initiated in April 2020, the WHO's Access to COVID-19 Tools Accelerator brought global partners together to speed up the development, production, and equitable distribution of COVID-19 tests, treatments, and vaccines.

> "Rich countries have the majority of the supply [of vaccines]. No country is exceptional and should cut the queue and vaccinate all of their population while some remain with no supply of the vaccine."
>
> –WHO DIRECTOR-GENERAL TEDROS ADHANOM GHEBREYESUS

The WHO's COVAX program is committed to ensuring that at least 20 percent of every country's population has access to a COVID-19 vaccine (WHO 2020e), thereby ensuring that even members of the poorest countries are protected. As of mid-January 2021, a small group of high-income countries representing only 16 percent of the world's population had purchased 60 percent of the global supply of COVID-19 vaccines (Marcus 2021). Without a global initiative to assist low-income countries, their citizens will remain unprotected, making it impossible to achieve worldwide herd immunity. Unfortunately, as travel bans are lifted, international travelers, even those who have been vaccinated, may unknowingly continue to spread the disease. Rather than work to support the World Health Organization, former President Trump announced that he was withdrawing the United States from the organization effective July 2021, and withdrew financial support prior to the end of the pandemic (Ortagus 2020). President Biden rejoined the World Health Organization, restored the White House National Security Council Directorate for Global Health Security and Biodefense, and committed $4 billion to the COVAX program (Deliso 2021).

Despite the benefits of globalization, global travel is the primary means by which illnesses are transmitted between countries. This has been clear during the COVID-19 pandemic. International travel has become commonplace with more than 241 million international passengers flying into the United States in 2019 (Bureau of Transportation Statistics 2020). Global travel can result in the spread of infectious diseases as was the case with the first 14 people in the United States identified as having COVID-19.

In the first six months of the pandemic, former President Trump signed nine executive orders restricting entry of foreign nationals traveling to the United States from a range of locations including China, Europe, Scandinavia, the United Kingdom and Ireland, and Iran (Bureau of Consular Affairs 2020). These bans, however, did not stop the spread of COVID-19 in the United States in part because Americans were allowed to return from overseas (Bollyky and Nuzzo 2020).

Effects of Global Trade Agreements on Health. Global trade agreements (see also Chapter 7) have expanded the range of goods available to consumers but at a cost to global health. The international trade of tobacco, alcohol, and sugary drinks and high-calorie processed foods, and the expansion of fast-food chains across the globe, is associated with a worldwide rise in cancer, heart disease, stroke, obesity, and diabetes (WHO 2020d). Globalization has resulted in rising incomes in the developing world, and although it has improved the quality of life for many people, it has also increased access to unhealthy foods and beverages and decreased levels of physical activity.

maternal mortality Deaths that result from complications associated with pregnancy, childbirth, and unsafe abortion.

As poor populations move toward the middle class, they can afford to buy televisions, computers, automobiles, and processed foods—products that increase caloric intake and decrease physical activity, leading to increased rates of obesity around the world. Until recently, obesity was a public health problem only in Western industrialized countries. But since 1975, obesity around the world has tripled (WHO 2020f). Indeed, a new word has emerged to refer to the high prevalence of obesity around the world: **globesity**.

> In general, as countries move up the income scale, rates of obesity increase as well. But in low-income countries, wealthier people are more likely to be overweight, whereas rates of obesity in high-income countries are higher among the poor (Templin et al. 2019). Why do you think this is so?
>
> **What *do you* think?**

Medical Tourism. The globalization of medical care involves increased international trade in health products and services. **Medical tourism**—a growing multibillion-dollar global industry (Dalen and Alpert 2018)—involves traveling across international borders to obtain medical care. Health care consumers travel to other countries for medical care for three primary reasons: (1) to obtain medical treatment that is not available in their home country, (2) to avoid waiting periods for treatment, and/or (3) to save money on the cost of medical treatment. It is estimated that more than 1.4 million Americans seek health care in another country each year (Dalen and Alpert 2018). Senator Rand Paul traveled to Canada for a hernia surgery in 2019 at a clinic offering a resort-like setting for his recovery (Healthline 2019).

Some major companies like Walmart, Boeing, and JetBlue help their employees to connect with out-of-state or out-of-country health care providers to reduce their corporate medical costs (Hyland 2019). Popular medical tourism destinations that lure health care consumers with competitively priced medical care include Mexico, Singapore, Thailand, South Korea, and India. The United States also draws medical tourists, with the U.S. medical tourism market generating 6.7 billion in 2018 (Grandview Research 2019).

globesity The high prevalence of obesity around the world.

medical tourism A global industry that involves traveling, primarily across international borders, for the purpose of obtaining medical care.

> Two million people in the United States purchase prescription drugs from outside the United States to save money (Hong et al. 2020). Most do it because they have inadequate health insurance coverage. Pharmacists warn this can be dangerous as there is no guarantee of what one might receive. Should it be legal for people to purchase drugs from outside the country? Why or why not?
>
> **What *do you* think?**

Although medical tourism can benefit some patients in providing timely, reduced-cost, quality medical care, a number of risks and problems are involved. Unlike the highly regulated health care industry in the United States, medical services, products, and facilities in other countries may not be regulated, so quality control and risk of infection are concerns (Healthline 2019). Medical tourism raises ethical concerns about health equity, as health services in popular medical tourism destinations flow not to the local population but to foreigners. And while medical tourism can promote economic development within destination countries, evidence suggests those benefits are overestimated and typically do not flow to the economically disadvantaged (Beladi et al. 2017).

During the COVID-19 pandemic, some families have traveled to get away from highly infectious areas. In New York City, there was a 256 percent increase in change of address forms submitted with the U.S. Postal Service in March 2020 alone, with people renting AirBnBs or going to stay with family (Krauth 2020). Further, results from a survey in Los Angeles indicate that 70 percent more people moved out of the city than moved into the city with many specifically mentioning COVID as their reason for leaving (Eng 2020).

Oscar Avila/Chicago Tribune/MCT via Getty Images

Thousands of Americans travel to Mexico for less expensive dental care. The small town of Los Algodones has 350 dentists who all cater to American "medical tourists," who walk over the border from California.

Sociological Theories of Illness and Health Care

The three major sociological theories—structural functionalism, conflict theory, and symbolic interactionism—each contribute to our understanding of illness and health care.

Structural-Functionalist Perspective

According to the structural-functionalist perspective, health care is a social institution that functions to maintain the well-being of societal members and, consequently, of the social system as a whole. Thus, failures in the health care system are viewed as dysfunctions of the system that affect not only the well-being of individuals but also the health of other social institutions, such as the economy, education, and the family. For example, the lack of widespread community testing early on during the COVID-19 pandemic was a significant factor in the need for virtual schooling. Families had to find ways to manage schooling and supervision of children, often restricting physical access to grandparents to protect them but sometimes utilizing them as virtual tutors (Jargon 2020). Many women and some men, faced with the choice of leaving children alone or quitting their jobs, have chosen to stay home with their children (Cleo 2020; Kochhar 2020). Many sectors of the economy have suffered as workplaces closed during shelter-in-place ordinances, and consumers have avoided public places where the virus might be spread (Bauer et al. 2020).

The structural-functionalist perspective also examines how changes in society affect health. As societies develop and provide better living conditions, life expectancy increases and birthrates decrease (Roser, Ritchie, and Ortiz-Ospina 2019). At the same time, the main causes of death and disability shift from infectious disease, and infant, child, and maternal mortality to chronic, noninfectious illness and disease such as cancer, heart disease, Alzheimer's disease, and arthritis.

Just as social change affects health, health concerns may lead to social change. The emergence of HIV and AIDS in the U.S. gay male population helped unite and mobilize gay rights activists. Concern over the effects of exposure to tobacco smoke—the greatest cause of disease and death in the United States and other developed countries—led to legislation banning smoking in public places.

Finally, the structural-functionalist perspective draws attention to latent dysfunctions, or unintended and often unrecognized negative consequences of social patterns or behavior. For example, the pervasive use of antibiotics in factory farm animal feed has led to antimicrobial resistance among human beings exposed through the food chain or through contaminated water, air, or soil (Ma et al. 2020). The impact of environmental exposures to pesticides and other dangerous chemicals is discussed further in Chapter 13.

@DrTom Frieden

There's no friction between public health and the economy. If you look around the country and around the world, the places that have done the best job protecting people from Covid have also seen the fastest economic rebound.

-Dr. Tom Frieden

Conflict Perspective

The conflict perspective focuses on how socioeconomic status, power, and the profit motive influence illness and health care. As we discuss later in this chapter, socioeconomic status affects access to quality health care and influences living and working conditions that affect our health.

The conflict perspective points to ways in which powerful groups and wealthy corporations influence health-related policies and laws through lobbying and financial contributions to politicians and political candidates. The "health care industrial complex," which includes pharmaceutical and health care product industries, and organizations representing doctors, hospitals, nursing homes, and other health services industries, spent almost $600 million lobbying Congress on a variety of issues in 2019, more than any other industry (Evers-Hillstrom 2020). In 2018, a coalition of hospitals, pharmaceutical companies, and professional organizations called the Partnership for America's Health Care organized members to lobby against a series of

Congressional bills seeking to expand Medicare; one bill would have allowed 50- to 64-year-olds to opt into Medicare, and another would have allowed any American to buy into the program (Pear 2019). The coalition flooded the media with negative advertising, engaged in public education efforts, and encouraged its members' lobbyists to express their views to Congress. Pharmaceutical and health product industry lobbyists spent $295 million in 2019 alone lobbying against proposals to regulate or put a cap on drug prices (Evers-Hillstrom 2020).

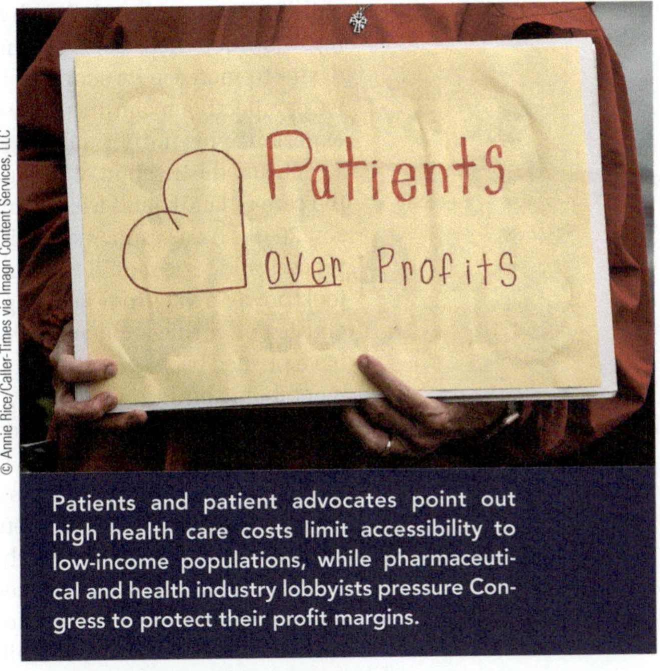

© Annie Rice/Caller-Times via Imagn Content Services, LLC

Patients and patient advocates point out high health care costs limit accessibility to low-income populations, while pharmaceutical and health industry lobbyists pressure Congress to protect their profit margins.

The conflict perspective argues that the pharmaceutical and health care industry is motivated by profits over people. The price of one common insulin medication, Humalog, increased from $21 in 1999 to $332 in 2019, an increase of more than 1,000 percent (Rajkumar 2020). Similarly, a common blood pressure medication, Benicar, increased in price by 114 percent between 2012 and 2017 (Bunis 2019), and a common ADHD medication Adsenys XR increased by 10 percent in one month in the midst of the COVID-19 pandemic bringing its cost to $439 per month (Owermohle 2020). Pharmaceutical companies implemented more than 800 price increases in the first half of 2020, during the COVID-19 pandemic. While the pharmaceutical industry argues its prices are needed to fund research and development costs, the U.S. Government Accountability Office (2017) found that the profit margins of pharmaceutical companies in 2015 exceeded that of both software companies and the largest Fortune 500 companies.

The pharmaceutical and health product industries make some decisions about which drugs and products to develop based on profit. For example, the Centers for Disease Control rank antibiotic-resistant infections as among the nation's top health threats; an estimated 23,000 people in the United States die of such infections each year, and the threat of a superbug that is resistant to all antibiotics is a recognized threat (Emanuel 2019). But only about 40 new antibiotics were in development in 2018, as compared with over 600 drugs for cancer. Why? Because antibiotics are relatively cheap and taken for only a short period of time so the profit margin is small (Emanuel 2019), whereas drugs for chronic diseases have a high profit margin since they are taken over a long period of time.

Many industries place profit above the health considerations of workers and consumers. Chapter 7, "Work and Unemployment," discusses how employers often cut costs by neglecting to provide adequate safety measures for their employees. Chapter 13, "Environmental Problems," looks at how corporations often ignore environmental laws and policies, exposing the public to harmful pollution.

Symbolic Interactionist Perspective

Symbolic interactionists focus on (1) how meanings, definitions, and labels influence health, illness, and health care and (2) how such meanings are learned through interaction with others and through media messages and portrayals. According to the symbolic interactionist perspective of illness, "[T]here are no illnesses or diseases in nature. There are only conditions that society, or groups within it, has come to define as illness or disease" (Goldstein 1999, p. 31). Psychiatrist Thomas Szasz (1961/1970) argued that what we call "mental illness" is no more than a label conferred on those individuals who are "different," that is, those who do not conform to society's definitions of appropriate behavior.

Defining or labeling behaviors and conditions as medical problems that fall within the responsibility and expertise of the medical profession is a process known as **medicalization** (Correia 2017). Behaviors and conditions that have undergone medicalization include post-traumatic stress disorder, premenstrual syndrome, menopause,

medicalization Defining or labeling behaviors and conditions as medical problems.

childbirth, attention deficit/hyperactivity disorder, and even the natural process of dying. Conflict theorists view medicalization as resulting from the medical profession's desire to increase its scope of influence in the pursuit of profits. A symbolic interactionist perspective examines how the medicalization of a condition or process impacts the experience for individuals.

The medicalization of dying took death out of the home and into the hospital, away from the care of family and friends into the hands of health care professionals (Schwarz and Benson 2018). As a result, symbolic interactionists argue, we became estranged from the death process and fearful of it. Similarly, the "medicalized birth" of the late 1800s and early 1900s took power away from laboring women through the use of anesthesia, surgical techniques, and hospital rules and procedures (Martucci 2018). Both pregnant women and the dying have pushed back against the medical model, with a resurgence of interest in family involvement, home births and deaths, minimal medication, and birth and death plan documents.

> "As twisted as it sounds, I was so happy that I had received a diagnosis."
>
> **–BETHANY STAHL, AUTHOR OF THE BOOK *ENDOMETRIOSIS: IT'S NOT IN YOUR HEAD***

Symbolic interactionists recognize that the lay public may also drive the medicalization process. Medicalization allows sufferers to "translate their individual experiences of distress into shared experiences of illness" (Barker 2002, p. 295). In the early 1990s sociologist Verta Taylor (1995) found that women suffering from depression and anxiety after childbirth took an active role in constructing the Postpartum Depression diagnosis in order to find support and legitimization from the medical profession. The diagnosis destigmatized their experiences. Similarly, the development of the ADHD diagnosis gave parents of children who had been labeled "deviant" the ability to share their experiences with others (Novotni 2017).

Symbolic interactionism draws attention to the effects that meanings and labels have on health and health risk behaviors and how such meanings are learned through interaction with others and through media messages. The meaning of mask wearing became politicized early on in the COVID-19 pandemic when then President Trump refused to wear a mask until six months into the pandemic (Associated Press 2020). He also publicly stated that some people wear masks just to show they don't like him, explained that he didn't "want to give the press the pleasure of seeing it" when he refused to wear a mask in a Ford motor plant, and regularly ridiculed Joe Biden for consistently wearing a mask (North 2020).

Not surprisingly, a poll in June 2020 found a significant partisan divide between Democrats and Republicans in mask wearing, with 63 percent of Democrats reporting they always wear a mask in public places when they might be near others as compared to 29 percent of Republicans (Pew Research Center 2020). By fall 2020, a majority favored a national mask mandate (Zoellner 2020), but a vocal minority viewed mask mandates, even in public schools, as infringements on personal freedom and a violation of the U.S. Constitution (Karalis 2020; Robinson 2020). President Joe Biden encouraged Americans to wear masks for the first 100 days of his administration in an effort to get the pandemic under control (Biden 2020).

What *do you* think?

In some countries few members of the population objected to mask ordinances and widespread mask wearing occurred quickly during the COVID-19 pandemic. The population of South Africa embraced mask wearing quickly after President Cyril Ramaphosa and his ministers began wearing them regularly (Alberga et al. 2020). In Vietnam, 91 percent of the population were wearing masks in April 2020 despite the fact that no cases of COVID-19 were reported in the country until July 2020. What types of cultural values do you think would support such a requirement by the national government, and which would not?

Health Disparities in the United States

health disparities Preventable differences in exposure to disease or injury or in opportunities to achieve optimal health across social groups.

Individual health outcomes are determined by a number of factors including genetics, individual behavior, access to resources that support health, and exposure to factors that undermine health. Differences in health are found not only between individuals but also between social groups. **Health disparities** are preventable differences in exposure to disease or injury or in opportunities to achieve optimal health across social groups. Health disparities result from **social stratification**—systems of social inequality by which a society divides people into groups with unequal access to wealth, material and social resources, and power. After reading the following sections on how socioeconomic status, gender, and race/ethnicity affect health, you may appreciate the words of sociologist Patricia Homan (2019), who wrote: "Social inequality in the United States is sickening. Literally" (p. 486).

Socioeconomic Status and Health

Socioeconomic status, or **social class**, refers to a person's position in society based on that person's level of educational attainment, occupation, and household income (see also Chapter 6). One's socioeconomic status—one's education, occupation, and income level—greatly influences one's health. Throughout the world, people living in poverty are more likely to suffer from malnutrition; hazardous environments, housing and working conditions; a lack of clean water and sanitation; and inadequate medical care (see also Chapter 6).

Low socioeconomic status is associated with lower life expectancy and is a leading causal factor in poor health. Low socioeconomic status is associated with less access to quality health care; increased likelihood of having an unhealthy lifestyle; and higher exposure to adverse living conditions, injury, and disease (Cockerham 2019).

COVID-19 deaths in the United States have been higher among residents in poor counties than in middle- and high-income counties (Finch and Finch 2020). Evidence suggests that individuals in poor counties had less access to remote work, testing, particularly early on, and quality medical care. When poor people seek medical treatment, hospitals in poor communities are more likely to be understaffed and do not have the same life-saving equipment as hospitals in wealthier communities (Rosenthal et al. 2020). During his own hospital stay for COVID-19, former President Trump's personal attorney's comments underscored the class divide in U.S. medical care: "Sometimes when you're a celebrity, they're worried if something happens to you. ... They're going to examine it more carefully and do everything right." (Quoted in Sepkowitz 2020.)

In the United States, rates of being overweight or obese—which are both linked to health problems such as heart disease, cancer, and diabetes—are higher among people living in poverty. This is, in part, because high-calorie processed foods tend to be more affordable than fresh vegetables, fruits, and lean meats or fish. Further, residents of low-income areas often live in **food deserts**—areas where residents lack access to grocery stores that sell fresh fruits and vegetables and instead must rely on convenience stores and fast-food chains that sell mostly high-calorie processed food. Prior to the COVID-19 pandemic, over 10 percent of U.S. households experienced food insecurity, i.e., they didn't know where their next meal would come from, and millions of American children relied on free-meal programs in public schools (Silva 2020). The COVID-19 pandemic exacerbated this problem as a result of school closures and unemployment.

In addition, members of the lower class are subjected to the greatest stress and have the fewest resources to cope (Cockerham 2019). U.S. adults living below the poverty threshold are nearly eight times more likely to report experiencing serious psychological distress than adults in families with an income at least four times the poverty level (National Center for Health Statistics 2016). Stress has been linked to a variety of physical and mental health problems, including high blood pressure, cancer, chronic fatigue, and substance abuse.

Educational attainment, a component of socioeconomic status, also affects health and longevity. Higher levels of education are associated with lower mortality rates, and U.S. adults with more years of education have better health than those with less education (Kemp and Montez 2020; Osario, Prisinzano, and Paulson 2020). A study of Black and White individuals in four U.S. cities found that educational attainment was a stronger predictor of life expectancy than race (Hathaway 2020). This chapter's *Social Problems Research Up Close* reviews changes in life expectancy by race and education, as well as factors contributing to these changes.

@Rep DianaDeGette

We are now one year into this pandemic.

Millions of people are out of work & millions of families are struggling to pay their bills.

We need a bold, comprehensive relief plan to provide EVERYONE the help they need.

-Rep. Diana DeGette

social stratification Systems of social inequality by which a society divides people into groups with unequal access to wealth, material and social resources, and power.

socioeconomic status (social class) A person's position in society based on the level of educational attainment, occupation, and income of that person or that person's household.

food deserts Areas where residents lack access to grocery stores that sell fresh fruits and vegetables and instead rely on convenience stores and fast-food chains that sell mostly high-calorie processed food.

social problems
RESEARCH
UP CLOSE

Education and Life Expectancy among Non-Hispanic White and Black Americans

Since the late 1980s, differences in mortality rates and adult life expectancy have increased significantly between education and income groups suggesting that socioeconomic inequality is having an increased impact on how long people live. By 2010, the gap in life expectancy between 25-year olds who had not completed high school and those who had completed college ranged from 4.6 years to 11.9 years across race-sex groups. Additionally, life expectancy at birth decreased between 2014 and 2017, the largest decrease since 1993. In this feature, Sasson and Hayward (2019) examine differences in life expectancy and major causes of death across educational groups between 2010 and 2017.

Sample and Methods

Since most people have completed their education by age 25 and cause of death is less reliably reported after age 85, Sasson and Hayword (2019) focused on U.S. adults between the ages of 25 and 84. The researchers utilized publicly available data collected by the federal government, including the U.S. National Vital Statistics System and the American Community Survey (ACS). The variables of interest were race/ethnicity, sex, and educational attainment. These measures were self-reported on the ACS and reported by funeral directors on death certificates recorded by the U.S. National Vital Statistics System. Outcome measures include life expectancy and years of life lost (YLL) which were calculated by subtracting the age at which the individual died from how long they were expected to live. The researchers chose to focus on non-Hispanic White and Black adults because other racial/ethnic groups were not sufficiently represented in the ACS sample.

Findings and Conclusions

Life Expectancy: From 2010 to 2017, changes in life expectancy varied by educational attainment. Among White men and women, life expectancy decreased between 2010 and 2017 for those with less than a four-year college degree but increased for those with a college degree. For White men and women with a high school diploma or less, life expectancy decreased by more than one year. Similarly, among Black men, life expectancy decreased for those with a high school diploma or less, although by only a fraction of a year. However, for college-educated Black men, life expectancy increased. Black women saw the greatest advantages of education with life expectancy increasing by an estimated 1.7 years among college-educated Black women. Black women experienced no decline in life expectancy across any of the educational levels.

Cause of Death: Circulatory diseases, cancer, and smoking-related diseases were the greatest contributors to years of life lost (YLL) or early death for White and Black men and women. By 2017, circulatory diseases and smoking-related diseases had a slightly lower impact than in 2010. However, the impact of drug use, alcohol use, and suicide on YLL increased significantly across all four groups during the time period. The impact of drug use increased the most of the three, especially among White and Black men. As with life expectancy, educational level also factors into cause of death. The increasing impact of drug use on early death between 2010 and 2017 is most pronounced on men and women of both races who have a high school diploma or less, followed by men and women with some college education but no college degree. Among college-educated White and Black women, drug use had no impact on changes in YLL between 2010 and 2017, while among White and Black men it continued to have a small impact. Among Black men with a high school diploma or less, firearm deaths had a significant impact on increasing YLL between 2010 and 2017.

In conclusion, although adult life expectancy increased among college-educated adults between 2010 and 2017, the increases were off-set by decreases in life expectancy among individuals without a college degree. Much of the decrease in life expectancies are attributable to the increasing impact of drug-related deaths, particularly among White men and women. Alcohol use and suicide also contributed to decreased life expectancy among Black and White women and men. These findings support other research that point to the health benefits of completing a college education.

SOURCE: Sasson and Hayward 2019.

What do you think? The association between education and health is stronger in some U.S. states than in others and, therefore, having a lower level of education is riskier for one's health in some states than in others (Kemp and Montez 2020; Montez, Hayward, and Zajacova 2019). What kinds of state characteristics and state policies do you think help protect the health of less educated residents?

Why is higher educational attainment linked to better health and lower mortality? Higher education can lead to better paying jobs that provide more social and economic resources, healthier lifestyles, and better health insurance (Berchick, Hood, and Barnett 2018).

Higher education also provides greater exposure to knowledge about health issues and encourages the development of cognitive skills that enable individuals to make better health-related choices such as the choice to exercise, avoid smoking and heavy drinking, use contraceptives and condoms to avoid sexually transmitted infections, seek prenatal care, and follow doctors' recommendations for managing health problems.

Just as socioeconomic status affects health, health affects socioeconomic status. Physical and mental health problems can limit one's ability to pursue education or vocational training and to find or keep employment. And the costs of managing health problems can impact one's economic position. Globally, nearly 90 million people in 2015 were pushed into extreme poverty by out-of-pocket health care costs (WHO 2020d). Later in this chapter, we look more closely at the high cost of health care and its consequences for individuals and families.

Gender and Health

In many societies, men have more access to social power, privileges, resources, and opportunities. Yet these advantages do not translate into living longer lives. Throughout the world, women live longer than men. Globally, the life expectancy gender gap was 4.4 years in 2018 (WHO 2020d). In the United States, average life expectancy in 2018 was 81 for women and 76 for men (Arias and Xu 2020).

Lower life expectancy in males is related to males' greater exposure to occupational hazards (see also Chapter 7), and social norms that encourage male risk-taking behavior, such as alcohol and drug use, dangerous sports, violence, and fast driving (see Chapter 10). Men are also less likely than women to seek health care when they are sick, and when they see a doctor, less likely to disclose any symptoms of disease or illness they may be experiencing (Baker et al. 2014).

Research suggests that, across eight developed countries surveyed, women have taken the COVID-19 pandemic more seriously and are more likely to take precautions, including mask wearing and social distancing (Gerdeman 2020). This is consistent with other research that suggests women tend to adhere to medical advice at higher rates than men.

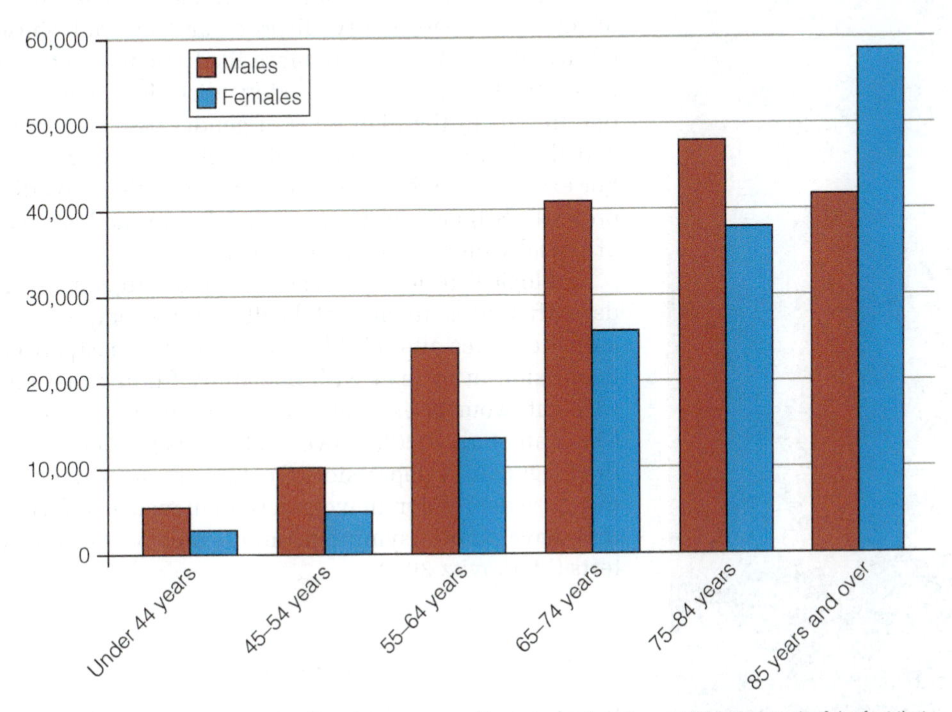

Note: The greater number of female as compared to male deaths after age 85 is the result of the fact that females are almost twice as likely to live to age 85 as are males.

Figure 2.3 U.S. COVID-19 Deaths by Age and Gender, February 1, 2020–May 2, 2020
SOURCE: Reeves and Ford 2020.

Males with COVID-19 are more likely to die (Peckham, et al. 2020), especially when adjusting for age (see Figure 2.3). For example, in the early stages of the pandemic in the United States, males between the ages of 45 and 54 were more than 2.5 times more likely to die than were women. Note as age increases, the differences between male and female deaths narrow. This pattern seems to hold true across virtually all countries and states in the United States (Peckham et al. 2020). More research is required to determine why this is the case.

Although women tend to have longer life expectancies than men, they suffer other health-related disadvantages. In many societies, women and girls are viewed and treated as socially inferior and are denied equal access to nutrition and health care, particularly reproductive health care. Traditional family responsibilities also affect women's health. Women do most of the food preparation and, in many areas of the world where solid fuels are used for cooking indoors, women are more likely than men to suffer from respiratory problems due to exposure to cooking fumes. Gender inequality also exposes women to domestic and sexual exploitation, increasing women's risk of physical injury and of acquiring HIV and other sexually transmitted infections. Globally, 30 percent of women aged 15 and over have experienced physical and/or sexual violence perpetrated by an intimate partner (WHO 2020g).

What *do you* think?

In most countries of the world, women can buy birth control pills without a prescription. But women in the United States, Canada, and much of Europe cannot purchase birth control pills unless they have a prescription. Should the pill be available in the United States without a prescription? Why or why not?

Gender inequality in power and resources exists at both the state level (inequality in economic, legal, and political institutions) and at the marital level (inequality in spousal relationships). One study found that while state-level sexism is associated with increased chronic health conditions and lower self-rated health among both women and men, gender inequality in marital relationships is associated with negative health outcomes for women but better health outcomes for men (Homan 2019).

Regarding mental health, men have higher rates of antisocial personality disorder and alcohol abuse (WHO 2020h). Women are more likely than men to experience depression and anxiety disorders, which is due in part to the high rates of gender-based violence and the lower social status of women (WHO 2020h). For example, the high rate of post-traumatic stress disorder (PTSD) among women is related to the high rate of sexual violence they experience.

Biological differences can contribute to some gender differences in mental health. For example, hormonal changes after childbirth can lead to postpartum depression in women with one study finding that one in eight women experienced postpartum depression (Bauman et al. 2020). However, there is gender bias in the diagnosis of depression. Doctors are more likely to diagnose depression in women than in men, even when they have similar symptoms and scores on diagnostic tests (Markovitz 2020).

MARK RALSTON/AFP/Getty Images

A sign warns against the COVID-19 virus near the Navajo Indian nation town of Tuba City, Arizona. Following these precautions is difficult in a territory in which an estimated 30 to 40 percent of the residents don't have access to running water or sanitation. This is a major reason why the United States' Native American population had high rates of COVID-19.

Race, Ethnicity, and Health

Racial and ethnic health disparities have received national attention during the 2020 COVID-19 pandemic, as communities of color are disproportionately affected by the novel coronavirus. Compared with White U.S. residents,

Black and Latino U.S. residents have been three times more likely to contract COVID-19 and nearly twice as likely to die from it. In New Mexico, Native Americans comprise 11 percent of the population yet accounted for more than half of COVID-19 cases (Wen and Sadeghi 2020). Minority group members are more likely to contract the virus in part because they are more likely to live in overcrowded and collective-living arrangements where it is difficult to social distance.

Members of minority groups are also more likely to work in "essential" jobs, where there is higher risk of exposure to the virus. In meat packing plants, for example, where the rate of COVID-19 has been high, nearly half of workers are Hispanic and a quarter are Black (Wen and Sadeghi 2020). Black and Hispanic Americans also have higher rates of chronic health conditions, including obesity, diabetes, and kidney disease, which place them at higher risk for severe illness and death from COVID-19.

Racial disparities in health have been well documented, with Black people in particular having poorer health and shorter lives than White people (see Table 2.4). Black males are significantly more likely to die from homicide (see Chapter 4). Black individuals suffer from disproportionately high rates of overweight and obesity, high blood pressure, and infant mortality. Compared with non-Hispanic White people, Native Americans and Alaskan Natives have high rates of HIV infection, cigarette smoking, obesity, and diabetes (National Center for Health Statistics 2019). Black women are 3 times more likely to die as a result of pregnancy-related complications than are White women (CDC 2019). These are but a few examples of the health disparities among U.S. racial/ethnic groups.

> "Every mother, everywhere, regardless of race or background deserves to have a healthy pregnancy and birth."
>
> **–SERENA WILLIAMS, TENNIS CHAMPION**

As shown in Table 2.4, Hispanic women and men live longer than either Black or White women and men. The Hispanic health advantage—known as the *Hispanic paradox*—is interesting because Hispanics share some of the same risk factors that contribute to poorer health among Black people—higher rates of poverty and obesity, and lower educational attainment compared with non-Hispanic White people (Office of Minority Health 2020). One theory for the Hispanic paradox is that Hispanic cultural values promote close and supportive family and community relationships and build strong social support, which is associated with better health outcomes. Traditional Hispanic diets tend also to include more healthy food items (de Gispert 2015). The Hispanic paradox seems to hold primarily for Hispanic immigrants, however; second- and third-generation Hispanics have health patterns that more closely match non-Hispanic White people.

Racial/ethnic differences in income, education, housing, and access to health care contribute to health disparities. Racial and ethnic minority group members are less likely than White people to have health insurance (Berchik, Hood, and Barnett 2018) and so are less likely to receive preventive services such as colon cancer screening, medical treatment for chronic conditions, or prenatal care. Minorities are also more likely than White people to live in environments where they are exposed to toxic chemicals and other environmental hazards (see also Chapter 13). One explanation for the effect of race on health, independent of socioeconomic status, is that health is affected not only by one's current socioeconomic status but also by social and economic circumstances experienced over the life course (Goldstein et al. 2020).

Health disparities are sometimes explained by differences in lifestyle behaviors, such as diet, physical exercise, and use of tobacco, alcohol, and other drugs. But lifestyle behaviors are affected by social factors. "Health behaviors are not simply a matter of individual choices. They are influenced by the social, cultural, and economic circumstances

TABLE 2.4 Life Expectancy at Birth by Race/Hispanic Origin and Sex: United States, 2018

	Non-Hispanic White	Black	Hispanic
Female	81.1	78.0	84.3
Male	76.2	71.3	79.1

SOURCE: National Center for Health Statistics 2019.

that frame and constrain them" (Cook et al. 2020, p. e21). For example, the dietary choices of minorities are influenced by their income, which, on average, is lower than the average income of White people. As noted earlier, high-calorie processed foods are more affordable and are sometimes all that is available when access to fresh and affordable produce and lean proteins is limited.

@Jerry Kovacs

Exactly one year ago, the Chinese Government locked down Wuhan, the source of Covid-19 @visit_wuhan

Forty-one years ago Jimmy Carter argued for national health care in the USA with emphasis on the prevention of disease @CarterCenter @caroloffcbc @washingtonpost @nytimes @CTVNews

-Jerry Kovacs

Another social factor that contributes to racial/ethnic health disparities is prejudice and discrimination. When Black Americans experience racism, they have an increase in inflammation, which raises the risk of chronic illness, including heart attack and cancer (Thames et al. 2019). Stress from racism is also linked to depression and to cellular changes related to accelerated aging, suggesting that racism literally ages Black Americans faster (Carter 2020). Prejudice and discrimination can also adversely affect health by reducing minorities' access to jobs and safe housing (see also Chapter 9) (Taylor 2019). Finally, racial bias can result in poorer quality health care for minorities. For example, Black patients are less likely than White patients to be treated for pain management (Hoffman et al. 2016).

Regarding mental health, data from the National Institute of Mental Health (2020) indicate that in 2019 the prevalence of serious mental illness was highest among mixed-race adults, followed by American Indian/Alaska Natives who have higher rates of substance abuse and post-traumatic stress disorder. Some studies suggest that minorities have a higher risk for mental disorders, such as anxiety and depression, in part because of racism and discrimination (Williams 2018), but this is not reflected in official diagnosis patterns. Members of racial/ethnic minority groups are less likely to seek out mental health care, and some evidence suggests that lack of cultural understanding may contribute to underdiagnosed mental illnesses in these populations (American Psychiatric Association 2017).

Problems in U.S. Health Care

The United States is the only developed country in the world that does not have a **universal health care system**, a health care system in which all residents of the country have access to health care without financial hardship. Universal health care systems vary from country to country. In the United Kingdom, health care providers are employees of the government's National Health Service and clinics and hospitals are administered by government employees.

This type of government-owned and administered health care system is unusual. In Germany all citizens are required to have health insurance which they can receive through the government's statutory health insurance plan or through a private insurance plan. Eighty-eight percent of the German population relies on the statutory health insurance plan. Some countries like Canada have universal health insurance through the government but also have private insurance options that residents can purchase for services not covered through the public plan. These variations of universal health care have two things in common: (1) all citizens have access to a standard level of care, and (2) the costs are paid out of tax dollars.

In the United States, as of this writing, there is no system by which basic medical care is guaranteed to all citizens and, even with health insurance coverage, many people incur significant costs as a result of health care needs. In this section we discuss problems of the uninsured in America and the high costs of U.S. health care.

U.S. Health Insurance Options

universal health care system System of health care, typically financed by the government, that ensures health care coverage for all citizens.

A little over half of Americans have health insurance through their employers, which typically means they pay a portion of the health insurance premiums while their employer pays the remainder (see Figure 2.4). Another 10 percent purchase private health insurance, many through the *Affordable Care Act* (ACA) "exchanges" discussed later in this chapter. Monthly health insurance premiums can range from an average of $100 for employer-provided plans to over $1,000 per month for privately purchased insurance (Fontinelle 2019).

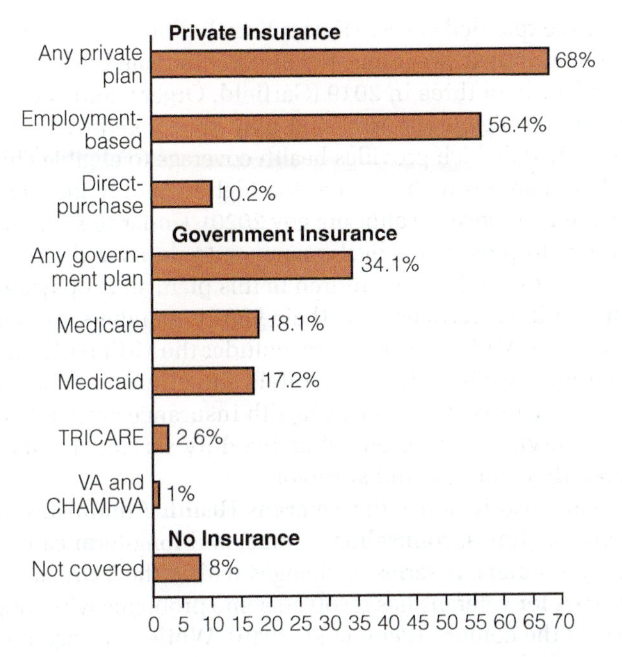

Figure 2.4 Coverage by Type of Health Insurance, 2019
SOURCE: Starkey and Bunch 2020.

In traditional health insurance plans, the insured choose a private health care provider whose fees are reimbursed by the insurance company. Insured individuals typically pay an out-of-pocket "deductible" usually ranging from a few hundred to a thousand dollars or more per year per person, as well as a "copay" or a percentage of medical expenses (e.g., 20 percent) until a maximum out-of-pocket amount is reached. After the maximum out-of-pocket amount is reached, insurance covers 100 percent of the medical costs. For individuals with chronic health conditions that require regular doctor's visits, testing, and prescriptions or for someone who is injured in an accident, these costs can quickly add up.

A third of Americans are covered by a government plan, either through one of the government health insurance plans or through government-provided health care. **Medicare**, which covers people 65 years and older, people with disabilities, and those with end-stage kidney disease, is funded by federal government general revenues, payroll taxes, and beneficiary premiums. In 2019, 18 percent of the U.S. population was covered by Medicare (see Figure 2.4). Medicare Part A provides hospital insurance for in-patient care and is premium-free for most people but enrollees may pay a deductible and a copayment as with traditional insurance plans. In order to have coverage for physician and outpatient services, Medicare recipients must enroll in a Part B insurance plan which requires they pay a monthly premium as well as copayments for services. In 2021, the standard Medicare premium was $148.50 (or higher depending on income) ("Medicare Costs at a Glance" undated).

Medicare supplementary plans can be purchased to cover long-term nursing home care, dental care, eyeglasses, and outpatient drug benefit. However, these require additional monthly premiums. Thus, while Medicare is an important support for seniors seeking medical care, it does not fully protect them from health care costs. Over 7 million seniors covered by Medicare report that they are unable to pay for medication prescribed by their doctors (Stevens and Mallory 2019). In 2019, seniors in the United States withdrew an estimated $22 billion from their long-term savings to cover health care costs.

Medicaid, which provides health care coverage for the poor, is jointly funded by the federal and state governments and covered 17 percent of Americans in 2019 (see Figure 2.4). Each state determines who is eligible for the program, and states with fewer financial resources tend to have stricter eligibility requirements. Individuals covered by Medicaid seek out their own providers, but physicians are less likely to accept Medicaid patients than patients on private insurance, as reimbursements are lower and slower to come in (Caffrey 2017).

The *Affordable Care Act* (ACA) of 2010 allows states to expand their Medicaid eligibility to all individuals who were at or below 133 percent of the federal poverty thresholds.

@Rep KatiePorter

Under current law, seniors are subject to a 10% penalty for each year they delay enrolling in Medicare. This fee can quickly skyrocket—handcuffing older Americans to higher bills for life. Today, I'm introducing a bipartisan bill to protect seniors from this unfair charge.

-Rep. Katie Porter

Medicare Federally funded program that provides health insurance benefits to the elderly, disabled, and those with advanced kidney disease.

Medicaid Public health insurance program, jointly funded by the federal and state governments, that provides health insurance coverage for the poor who meet eligibility requirements.

By 2020, 39 states had expanded access (Kaiser Family Foundation 2020), but among the states that did not expand their programs, the median income limit for Medicaid eligibility was $8,532 for a family of three in 2019 (Garfield, Orgera, and Damico 2020).

Children who are not eligible for Medicaid may be covered by the **Children's Health Insurance Program (CHIP)**, which provides health coverage to eligible children without insurance, typically from families with incomes too high to qualify for Medicaid but too low to afford private health insurance (HealthCare.gov 2020). Under this initiative, states receive matching federal funds to provide medical insurance to uninsured children. Parents must pay a monthly premium to enroll their children in this plan, and copayments are required.

Members of the military, retirees, and their family members can access health care through various programs. Military health care includes the **Military Health System (MHS)**, which provides medical care in military hospitals and clinics, in combat zones, at bases overseas, and on ships. The MHS also has a health insurance system known as TRICARE that provides health services to millions of active-duty service members, military retirees, their eligible family members, and survivors.

Military health care also includes the **Veterans Health Administration (VHA)**, which is a system of hospitals, clinics, counseling centers, and long-term care facilities that provides care to military veterans. A series of changes within the VHA including an increase in access to private care for veterans has resolved some problems with long delays for treatment in many areas of the country (Penn et al. 2019). While coverage under these various military health services does not require a monthly premium, increasingly copayments are required, particularly for family members and for conditions not related to military service.

The **Indian Health Service** is a federal agency that provides health services to the approximately 2.6 million members of 574 federally recognized American Indian and Alaska Native tribes and their descendants (Indian Health Service 2020). Health care falls under the provisions of various treaties between the federal government and Indian tribes. Most IHS funds go to providing services to American Indians and Alaska Natives who live on or near reservations or Alaskan villages, but some funding supports health care programs for American Indians and Alaska Natives who live in urban areas. Large tribes like the Cherokee Nation run their own hospitals and clinics, while smaller tribes contract with outside facilities. IHS funding, which is appropriated by Congress each year, is significantly lower per person than Medicare, Medicaid, and the VHA (see Figure 2.5), leaving many American Indians and Alaska Natives without access to needed care (King 2019).

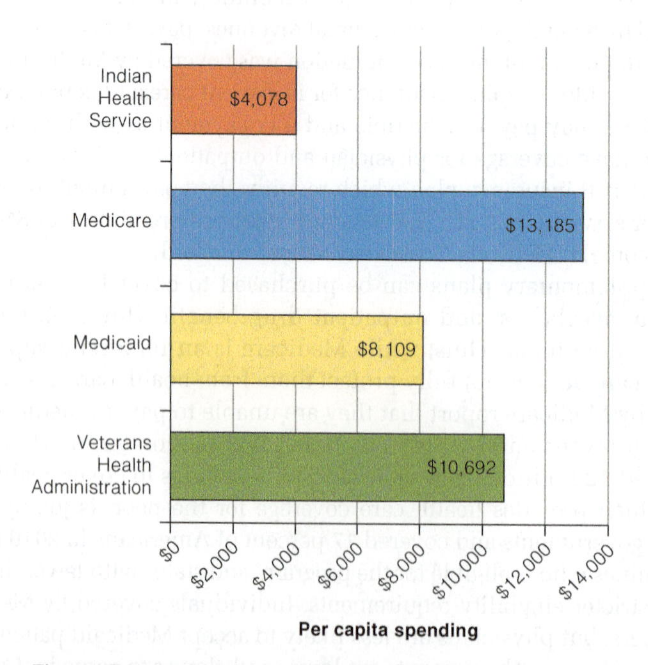

Figure 2.5 Per Capita Spending on Health Care for Four Federal Programs, 2019
SOURCE: King 2019.

Children's Health Insurance Program (CHIP) A public health insurance program, jointly funded by the federal and state governments, that provides health insurance coverage for children whose families meet income eligibility standards.

The Uninsured

In 2019, 8 percent of Americans or 29.3 million people did not have health insurance coverage for the entire year (Keisler-Starkey and Bunch 2020). In 2018, non-Hispanic White people were more likely than people of color to have health insurance. Nonelderly adults, those between the ages of 19 and 64, comprised the age group most likely to be uninsured (12.9 percent) followed by children (5.6 percent). Virtually all U.S. senior citizens who are permanent legal residents are covered by Medicare. During the COVID-19 pandemic 3.1 million people lost employer-sponsored insurance due to job losses between March and late September 2020. However, most were able to regain insurance through Medicaid/CHIP or by purchasing private insurance on the ACA exchanges which were reopened in several states (Karpman and Zuckerman 2020).

Employed individuals and individuals with higher incomes are more likely to have health insurance. However, employment is no guarantee of health care coverage; in 2019, 73 percent of families with no health insurance had at least one full-time worker (Tolbert, Orgera, and Damico 2020). Some businesses do not offer health benefits to their employees, some employees are not eligible for health benefits because of waiting periods or part-time status, and some employees who are eligible may not enroll in employer-provided health insurance because they cannot afford their share of the premiums.

AP Images/Gabriele Holtermann/Sipa USA

In June 2020, in the middle of the COVID-19 pandemic when health care systems were being overwhelmed and thousands of Americans were dying of the disease, former President Trump directed U.S. Park Police and National Guard Troops to use tear gas to move protestors from Lafayette Park. Trump then walked across to St. John's Church to pose for photos as he held a bible and declared, "We have the greatest country in the world. Keep it nice and safe" (Gjelten 2020).

High Health Care Costs in the United States

The United States spends more per person on health care than any other country in the world both in terms of actual dollars ($10,966 per person in 2019) and in terms of percentage of GDP (17 percent in 2019) (Kamal, Ramirez, and Cox 2020). The high cost of health care burdens governments, communities, individuals, and families. After K-12 education, Medicaid is the states' largest expense (Rosewicz, Theal, and Ascanio 2020). Two-thirds of all bankruptcies in the United States, involving about 530,000 families each year, are related to health care debt (Himmelstein et al. 2019), and one-fourth of U.S. adults say that in the past year, they or a family member postponed treatment for a serious medical condition due to cost (Saad 2019).

Despite the high personal and public expenditures on health care, U.S. life expectancy is lower than that of other industrialized countries, and U.S. maternal and infant mortality rates are higher. Why is it that we are spending so much and seemingly getting so little in return?

High Costs of Administration. In the United States, we spend more than three times as much on health administrative costs than other countries (Nunn, Parsons, and Shambaugh 2020). Most insurance companies and the federal government control costs through **managed care**, which involves monitoring and controlling the decisions of health care providers. The insurance company may, for example, require doctors to receive approval before they can hospitalize a patient, perform surgery, or order an expensive diagnostic test. While other countries manage care in this way too, in a universal health care system, these processes are streamlined. In the United States, each insurance policy requires different procedures and different claim forms and has different restrictions and reimbursement rates. And in a U.S. hospital, the physicians and other professionals are not necessarily employees of the hospital, so patients get different

managed care Any medical insurance plan that controls costs through monitoring and controlling the decisions of health care providers.

TABLE 2.5 Percentage of College Students Experiencing Selected Mental Health Difficulties Anytime in the Past 12 Months: 2015 and 2019

Mental Health Difficulty	Percentage	
	2015	2019
Felt so depressed it was difficult to function	35	45
Felt overwhelming anxiety	57	66
Felt very lonely	59	66
Felt things were hopeless	49	56
Seriously considered suicide	9	13
Intentionally cut, burned, bruised, or otherwise hurt yourself	6	9

Note: Percentages are rounded.
SOURCE: Adapted from American College Health Association National College Health Assessment II: Reference Groups Executive Summary Spring 2015 and 2019. Hanover, MD: American College Health Association.

Mental Illness among College Students

Mental health problems are common among college students and have been increasing in recent years (see Table 2.5). In 2019, one in three college students had been diagnosed or treated by a professional for a mental health problem within the past year—24 percent of college students had been diagnosed or treated for anxiety, 22 percent for depression, and 12 percent for panic attacks (American College Health Association [ACHA] 2019). More than one in four college students in a 2019 survey reported that anxiety affected their academic performance; one in five reported that depression affected their performance (ACHA 2019). Only 15 percent of students reported seeking help from their college counseling or health services center. A spring 2020 survey of more than 18,000 college students found that depression and anxiety had increased even further during the pandemic, although substance use had decreased (Healthy Minds Network 2020). Students also reported that access to mental health services was more difficult during the pandemic.

Seven percent of college students have been diagnosed with attention deficit and hyperactivity disorder (ADHD) (ACHA 2019). Seventy percent of college students with ADHD report that their academics are not affected by their diagnosis (ACHA 2019). Students with ADHD who do have academic difficulties report getting low test and course grades, receiving an incomplete or dropping a course, and having difficulty completing a thesis project. Academic problems among students with ADHD are not surprising: compared with students who don't have ADHD, college students with ADHD have more difficulty with organizing and planning daily activities, impulse control and delayed satisfaction, working memory (e.g., following multistep instructions), and sustained attention to tasks (Weyandt et al. 2017).

Treatment of Mental Illness

In the 1960s, the U.S. model for psychiatric care shifted from long-term inpatient care in institutions to drug therapy and community-based mental health centers (Jaffe 2017). This transition, known as **deinstitutionalization**, has resulted in a significant decrease in the number of mental health facilities with 24-hour or residential treatment and the number of psychiatric treatment beds available. Deinstitutionalization removed patients from facilities where they were sometimes treated in a neglectful or inhumane manner and restored freedom of choice to mental health consumers, including the right to refuse treatment.

deinstitutionalization The removal of individuals with psychiatric disorders from mental hospitals and large residential institutions to outpatient community mental health centers.

During the deinstitutionalization era, a variety of laws were passed making it illegal to commit psychiatric patients against their will unless they posed an immediate threat to themselves or to others. However, community mental health programs have not adequately met the need for care, and millions of Americans with mental disabilities go without care or rely on hospital emergency room care when their condition deteriorates into a major mental health breakdown (Jaffe 2017).

The most frequently used source of care for mental health problems has become primary care, and doctors and nurses. Other "nonspecialty" care providers include community health centers, schools, nursing homes, correctional institutions, and emergency rooms. This fragmented system leaves many people with mental health problems to fall through the cracks. More than 390,000 Americans with serious mental illnesses are incarcerated; 14 percent of prisoners in state and federal facilities and 26 percent of those in jails meet the criteria for serious mental illness compared to 5 percent of the general population (Lyon 2019). The largest mental health institution in the United States is the Los Angeles County Jail (Rosenberg 2019). Many argue America's correctional facilities have become the new mental health "asylums," a phenomenon called the **criminalization of mental illness** (see Chapter 4).

What do you think?

In a mental health crisis, people are more likely to encounter the police than to get medical help (National Alliance on Mental Illness 2020a). As a result, about 2 million people with serious mental illness are booked into jails each year. Once in jail most people don't get the treatment they need and get worse rather than better. What do you think needs to be done to better support people with serious mental illnesses in our society?

Strategies for Action: Improving Health and Health Care

Making meaningful change in public health requires interventions that target specific health problems such as promoting condom use to prevent HIV infections, promoting vaccines to prevent the spread of COVID-19 and the flu, and encouraging physical activity to reduce obesity. However, promoting health also requires developing and maintaining social structures and policies that support healthy lifestyles and that assist people when they face injury, illness, or disease.

Ultimately, health is impacted by social inequalities, such as poverty and economic inequality, gender inequality, and racial/ethnic discrimination. Therefore, in addition to the strategies discussed in this section, efforts to alleviate social problems discussed in other chapters of this book are also essential elements to improving health.

Students: if you're going home to see loved ones over the holidays, take precautions to reduce the spread of respiratory viruses like flu and #COVID19.
There's still time to roll your #SleeveUp and get your #flushot. #FightFlu bit.ly/2yGuyOz

-CDC

Improving Health in Low- and Middle-Income Countries

The most fundamental needs to improve health in low- and middle-income countries are access to adequate nutrition, clean water, and sanitation. A nongovernmental organization (NGO) called Water.org started by activist Gary White and actor Matt Damon assists people in building water containment systems and sanitation systems through low-interest micro-loans (Water.org 2020). The organization has already spent $1 billion working to end the water crisis in the developing world.

Other targeted interventions include increasing immunizations for diseases such as measles and distributing mosquito nets to prevent malaria. Just prior to the COVID-19 pandemic, the World Health Organization began a pilot test of a new malaria vaccine in Kenya, Ghana, and Malawi (Kelland 2020). The Global Fund, an NGO committed to ending malaria, tuberculosis, and AIDS, raises more than $4 billion per year for programs across the world that educate, distribute prevention measures, and provide testing and treatment (Global Fund 2020). In 2020, The Global Fund also began supporting COVID-19 initiatives.

criminalization of mental illness Refers to the high number of mentally ill people being held in America's correctional facilities.

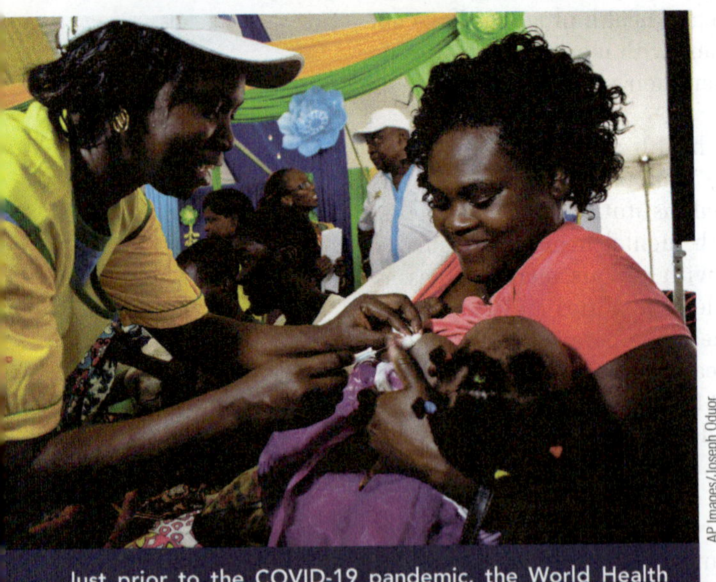

Just prior to the COVID-19 pandemic, the World Health Organization began a pilot test of a new malaria vaccine in Kenya, Ghana, and Malawi. More than 409,000 people globally—most of them babies—died of malaria in 2019 (Kelland 2020).

Efforts to reduce maternal mortality focus on providing access to quality reproductive care and family planning services (see also Chapter 12). Because maternal and infant mortality rates are highest among very young women, health advocates are fighting to pass legislation aimed at preventing child marriage including in the United States where, in the last two decades, girls as young as 12 have been married to adult men (Unchained at Last 2020).

Another strategy to improve the health of women and children in low-income countries is to provide women with education and income-producing opportunities. Promoting women's education increases the status and power of women to control their reproductive lives, exposes women to information about health issues, and also delays marriage and childbearing. One way to promote education among women and girls is to provide access to clean water so that women and girls are freed from their work of collecting water for the household (see also Chapter 12).

These and other efforts have had some success as child and maternal death rates have declined over the last decade. However, a number of countries with the highest rates of maternal mortality have made little to no progress. If all countries had the maternal outcomes of the healthiest countries in the world, the lives of nearly 300,000 women would be saved each year (Ritchie 2019).

> "Educating girls is a surefire way to raise economic productivity, lower infant and maternal mortality, improve nutritional status and health, reduce poverty, and wipe out HIV/AIDS and other diseases."
>
> —MARITZA ASCENCIOS, UN CHILDREN'S FUND (UNICEF)

Expanding U.S. Health Care Coverage

Expanding health insurance coverage in the United States has been the goal of numerous presidential administrations. Medicaid and Medicare were signed into law with the authorization of Title XIX of the *Social Security Act* in 1965 under then President Lyndon B. Johnson. The Children's Health Insurance Program (CHIP) was signed into law in 1997 by President Bill Clinton, after an unsuccessful effort to develop a universal health care program and, in 2010, the *Patient Protection and Affordable Care Act*, commonly referred to as the ***Affordable Care Act* (ACA)** or Obamacare, was signed into law by President Obama.

The *Affordable Care Act* (ACA) of 2010 was designed to expand insurance coverage by (1) allowing states to expand access to Medicaid, (2) providing subsidies so that low-middle income individuals can purchase health insurance, (3) creating health insurance exchanges—online marketplaces where consumers could shop for, compare, and enroll in insurance plans, and (4) requiring health insurance plans to provide dependent coverage for children up to age 26. The ACA also stipulated that individuals with **preexisting conditions**, illnesses or injuries that pre-date their current insurance plan, could no longer be told that problems related to those conditions would not be covered by their insurance. As Table 2.6 demonstrates, the changes implemented by the ACA have been more popular with Democrats and Independents than with Republicans.

Affordable Care Act (ACA) Health care reform legislation that President Obama signed into law in 2010, with the goal of expanding health insurance coverage to more Americans; also known as the *Patient Protection and Affordable Care Act*, or Obamacare.

preexisting conditions Illnesses or injuries that occurred before a person begins coverage under a new health insurance plan.

In the six years since the ACA was adopted, the uninsured rate declined from nearly 18 percent to 10 percent (Tolbert, Orgera, and Damico 2020). When former President Trump campaigned for the 2016 election, and after he took office, he vowed to replace "Obamacare." Rates of the uninsured increased between 2016 and 2020 as the Trump administration decreased funding for initiatives to promote the marketplace, and decreased the length of the enrollment period through the federal site (Gee 2020). States that opted to run their own marketplaces saw no decline in enrollment. During the COVID-19 pandemic, a dozen of those states opened enrollment periods through state sites for people who had lost employment-sponsored health insurance; the federal government did not.

TABLE 2.6 Opinions on the Provisions Put in Place by the *Affordable Care Act*, 2019

Percent who say it is "very important" that each of these parts of the ACA are kept in place:	Total	Democrats	Independents	Republicans
Prohibits health insurance companies from denying coverage for people with preexisting conditions	72%	88%	73%	62%
Prohibits health insurance companies from denying coverage to pregnant women	71	89	73	49
Prohibits health insurance companies from charging sick people more	64	76	64	55
Requires health insurance companies to cover the cost for most preventive services	62	80	58	49
Prohibits health insurance companies from setting a lifetime limit	62	72	65	48
Gives states the option of expanding their Medicaid programs	57	84	55	36
Provides financial help to low- and moderate-income Americans to help them purchase coverage	57	82	54	31
Prohibits private health insurance companies from setting an annual limit	51	67	46	38
Allows young adults to stay on their parents' insurance plans until age 26	51	68	50	36

SOURCE: Hamel et al. 2020a.

The former Trump administration supported a legal challenge to the individual mandate for health insurance provision under the ACA which was heard by the Supreme Court in late 2020. A decision is expected in the spring of 2021. Those who support the challenge hope that the Court will dismantle the entire program. If this were to happen, more than 20 million people could lose their health coverage and over 135 million people could lose protections for preexisting conditions, including the long-term effects of COVID-19 (Gee 2020; Rapfogel, Calsyn, and Seeberger 2020).

President Biden wants to "build upon" the ACA by restoring and expanding premium subsidies that were cut under the Trump administration (Biden 2020). President Biden has also expressed support for a "public option" which could involve expanding Medicare to people under the age of 65. Such a plan would provide competition between private insurers and thus, likely, reduce the cost of premiums.

What do you think?

A growing majority of Americans believe the federal government has a responsibility to ensure that all Americans have health care coverage (Jones 2020). Just over a third believe that this should be accomplished through a single-payer government program, while about a quarter believe that a mix of private insurance and government programs is preferable. Do you think the federal government has a responsibility to provide universal health care for its citizens?

Some worry that Biden's "public option" would result in a **single-payer health care** system, a single tax-financed public insurance program that replaces private insurance companies. Opponents argue that a single-payer national health insurance program would amount to a "government takeover" of health care and would result in higher costs, less choice, rationing, and excessive bureaucracy—the very outcomes that resulted from corporatized medicine. Not surprisingly, the insurance industry opposes the adoption of a single-payer health care system and spends a great deal of money on lobbying to influence the health reform debate.

single-payer health care Health care system in which a single tax-financed public insurance program replaces private insurance companies.

Supporters of a single-payer health care system argue that access to health care is a human right—a public responsibility that should be paid for with public monies rather than provided through a market-based system. They further point to the health disparities noted earlier in this chapter as evidence that the market-based system of health care is failing to meet the American value of justice for all. There is also an economic argument for a single-payer health care system: Replacing private insurance companies with single-payer health care would save more than $400 billion in administrative costs per year—enough to provide universal coverage to every U.S. resident without copayments, deductibles, or increases in health expenditures (Himmelstein and Woolhandler 2014).

Steve Pfost/Newsday LLC/Getty Images

The COVID-19 pandemic has taken a toll on the mental health of frontline health care workers, many of whom suffer increased stress, anxiety, exhaustion and burnout, depression, and worry over being infected by the virus or exposing their loved ones to the disease (Cabarkapa et al. 2020).

Strategies to Improve Mental Health Care

Some of the strategies for improving mental health care in the United States include eliminating the stigma associated with mental illness, improving access to mental health services, and supporting the mental health needs of college students.

Eliminating the Stigma of Mental Illness. Eliminating the stigma associated with mental illness would likely increase the number of people seeking mental health treatment. The nonprofit organization *Make It OK* focuses on raising awareness and providing examples of open discussions about mental illness (Make It OK 2020). One of their podcasts, "The Hilarious World of Depression," features humorous conversations with people who have dealt with depression. Another podcast, "Tremendous Upside," features athletes speaking about their mental health. Members of the public can add their own stories to the site. Effective anti-stigma campaigns not only focus on eradicating negative stereotypes of people with mental illness but also emphasize the positive accomplishments and contributions of people with mental illness.

Efforts to reduce the stigma of mental illness on campuses include providing information to the campus community on how to recognize common mental health problems and how to get help (see this chapter's *Self and Society* feature). Recognizing the important role that students can play in supporting one another, many colleges and universities have peer-to-peer intervention programs that train students on how to recognize and respond to individuals experiencing distress and how to refer these individuals, when appropriate, to professional resources. The University of Pennsylvania has a program called I CARE that provides students, faculty, and staff with the skills and resources needed to intervene with student stress, distress, and crisis (University of Pennsylvania 2020). On college campuses throughout the United States, students are getting involved in student organizations aimed at raising awareness of mental health issues, educating the campus community, and supporting students. These include organizations like NAMI on Campus, Active Minds, and To Write Love on Her Arms (O'Donnell 2019). Research documents that student involvement in these types of groups has an impact on campus culture and mental health (Mitchell and Ortega 2019).

Most colleges and universities offer mental health services to students and provide accommodations for students with documented mental health conditions (e.g., adjustments in test setting and times and excused absences for treatment). However, with a doubling of students on college campuses with diagnosed mental illness (Mitchell and Ortega 2019) and an extended period of budget cuts made much worse by the COVID-19 pandemic, counseling centers are struggling to meet students' needs. In the absence of a significant influx of funds and positions, most Counseling Centers are eager to partner with faculty and staff to increase awareness and support for students with mental illness.

Warning Signs for Mental Illness

Could you or someone you know, such as a roommate, friend, or family member, have a mental illness and not realize it? Read each of the following warning signs for mental illness and put a check mark next to each one that applies to yourself or to someone you are concerned about.* If you or someone you know shows any of these signs, consider seeking help from a qualified health professional. To find out what mental health help is in your area, you can call the NAMI hotline (1-800-950-6264; Monday through Friday, 10 a.m. to 6 p.m.) or e-mail info@nami.org. If there is a risk of suicide, the National Suicide Prevention Lifeline (1-800-273-8255) is available 24 hours a day.

Warning Signs:	You	Someone You Are Concerned About
1. Excessive worrying or fear	____	____
2. Feeling excessively sad or low	____	____
3. Confused thinking or problems concentrating or learning	____	____
4. Extreme mood changes, including uncontrollable "highs" or feelings of euphoria	____	____
5. Prolonged or strong feelings of irritability or anger	____	____
6. Avoiding friends and social activities	____	____
7. Difficulties understanding or relating to other people	____	____
8. Changes in sleeping habits or feeling tired and low energy	____	____
9. Changes in eating habits such as increased hunger or lack of appetite	____	____
10. Overuse of substances like alcohol or drugs	____	____
11. Changes in sex drive	____	____
12. Difficulty perceiving reality (delusions or hallucinations)	____	____
13. Inability to perceive changes in one's own feelings, behavior, or personality ("lack of insight")	____	____
14. Multiple physical ailments without obvious causes (such as headaches, stomachaches)	____	____
15. Inability to carry out daily activities or handle daily stress or problems	____	____
16. An intense fear of weight gain or concern with appearance	____	____
17. Thinking about suicide	____	____

*Please note that this tool is intended to identify potential warning signs for mental illness. It is not intended to be a diagnostic to determine whether an individual is experiencing mental illness, which is a determination that should only be made by a qualified mental health professional.
SOURCE: Know the Warning Signs (NAMI 2020b).

Over the last 10 years, the incidence of mental health disorders among college students has increased, as has their use of mental health services (Lipson et al. 2019). Yet, a survey of university administrators found that 43 percent indicated that mental health was part of their institution's strategic plan (Mitchell and Ortega 2019). Do you think mental health services should be part of the college or university mission?

What *do you* think?

Improving Access to Mental Health Care. Improving access to mental health services involves (1) recruiting more mental health professionals, especially those who are willing to serve in rural and impoverished communities and who have cultural competency to work with clients from diverse cultural backgrounds; (2) improving health insurance coverage for mental health problems; and (3) expanding mental health screening.

The 2010 *Affordable Care Act* (ACA) included a new program—the Mental and Behavioral Health Education and Training Grant program—that provides funds to institutions of higher education to recruit and train students pursuing graduate degrees in clinical mental and behavioral health. The program now includes an Opioid Workforce Expansion Program (OWEP) to increase the number of professionals trained in the treatment of opioid addiction. An estimated 2 million people in the United States are addicted to opioids (U.S. Department of Health and Human Services 2020) (see Chapter 3).

The *Affordable Care Act* also improved mental health and substance use disorder insurance coverage by requiring insurance plans to cover mental health and substance use disorder services, with the same benefits as medical and surgical benefits—a concept known as **parity**. If the former Trump administration's Supreme Court challenge of the ACA is successful, health insurance companies may return to providing little to no coverage for mental illness in their plans, which would impact millions of Americans.

Another strategy to improve access to mental health care involves making mental health screening a standard practice reimbursed by insurance companies, just like mammograms and other screening tests that are reimbursed. Only seven states require public schools to screen all students for mental health problems such as depression (Gracy et al. 2018).

Preparing for the Next Pandemic

The COVID-19 pandemic was not unexpected. National and international leaders had been discussing the potential for a global pandemic for decades. In 2013, members of the National Security Council convened a meeting with representatives of various federal departments to discuss the growing threat of infectious diseases (Jenkins 2020). That meeting led to a series of meetings with national and international partners that eventually resulted in the Global Health Security Agenda (GHSA), a global initiative led by the United States and signed onto by 67 countries. The GHSA's focus was on building countries' capacities to predict, detect, and respond to infectious disease threats, natural or human-made.

Committed to addressing these threats, former President Obama created the Office of Global Health Security and Biodefense commenting that, "No single nation can be prepared if other nations remain unprepared to counter biological threats" (Obama 2016). The United States and most other G-7 nations made financial commitments to support the efforts of the GHSA (Nuzzo and Inglesby 2018). Since the advent of the GHSA, the initiative has helped with the containment of two outbreaks of Ebola and one of Zika.

The GHSA has also conducted "peer reviews" of nations' readiness to tackle a serious outbreak. Approximately two months before the COVID-19 pandemic appeared in the United States, the Global Health Security Index ranked 195 nations in their level of preparedness to tackle a pandemic. The United States received the highest ranking (Yamey and Wenham 2020). However, the Index failed to take into account the impact that a populist, America First agenda would have over strategies to contain the virus.

not sure why we need to be the ones to tell you this, but don't drink bleach.

-Burger King

Although the United States received the high marks for its collaborative role in global health initiatives, the former administration rolled back many of these commitments. For example, in 2018, the Centers for Disease Control and Prevention staff in 39 countries was called back from their posts due to cuts in funding for the GHSA initiatives (Nuzzo 2018). Thus, at the time COVID-19 was discovered in China, the United States had no CDC representatives there to advise their government or to keep the United States fully informed. The former president withdrew further from the global community by announcing the United States' withdrawal from the World Health Organization. This action was preceded by President Trump's refusal to follow WHO recommendations to develop a robust "track and trace" program to contain the outbreak.

The GHS Index also gave high marks to the United States' Centers for Disease Control and Prevention and its strong health care system. Historically, the CDC has been an international leader in addressing outbreaks across the globe. Yet, both the United States and the United Kingdom, which had also received a high ranking, resisted the advice of their own health experts and scientists (Yamey and Wenham 2020). Further, the Trump administration limited the power of the CDC to disseminate information to the public, and ordered the CDC to remove guidelines on reopening schools that contradicted the public statements of former Secretary of Education Betsy Devos. The former president also gave unsupported medical advice (e.g., hydroxychloroquine as a cure for COVID-19) and mocked the effectiveness of wearing masks even as his own experts argued it was critical to slowing transmission of the virus (Holmes 2020).

The favorable GHS Index ranking was predicated on the assumption that the United States would act in a coordinated fashion with other countries, which it did not. In late January, former President Trump instituted selective travel bans but did little else to prevent spread of the virus, telling the American people "the coronavirus is very much under control" and "like a miracle it will disappear" (Holmes 2020). He later stated to journalist Bob Woodward: "I wanted to always play it down. I still like playing it down because I don't want to create a panic."

parity In health care, a concept requiring equality between mental health care insurance coverage and other health care coverage.

Tests were in such short supply in the United States that just hundreds of Americans had been tested by the end of February while millions had been tested in China (Yong 2020). Faced with shortages in personal protective equipment (PPE), in part, because much of it was manufactured in China, Trump told America's governors to "try getting it themselves." Without a coordinated approach, state and local leaders were forced to make their own decisions about shelter-in-place orders, mask ordinances, and managing local hospitals often in the face of White House opposition. As a result, the virus continued to spread, catapulting the United States to first place in number of COVID-19 deaths and new cases by early April 2020 (Johns Hopkins University 2020a).

The rapid development of COVID-19 vaccines was an important step in moving the country toward achieving **herd immunity**—the level at which a sufficient percentage of the population is immune to the virus either through exposure or immunization, thus stopping its spread. Herd immunity for COVID-19 is estimated to be about 70 percent (Mayo Clinic 2020). However, by the end of January 2021 the state-by-state approach established to distribute the vaccine in the United States had proven to be agonizingly slow and often disorganized (Huang and Carlsen 2021). Vaccine reserves promised by the former Trump administration did not exist, leaving some state programs empty-handed. In his first week in office President Biden signed executive orders to require masks and social distancing in federal buildings and for all interstate travel on trains, planes, and buses (Deliso 2021). He also created COVID-19 Response Coordinator and Deputy Coordinator positions within the Executive Branch to work with other federal offices to support and expand efforts to end the pandemic. And, he established a COVID-19 Health Equity Task Force charged with identifying strategies to mitigate health inequities caused by the pandemic. During the first week of the Biden administration Dr. Deborah Birx, former White House COVID-response coordinator, revealed there had been no full-time team dedicated to the COVID-19 pandemic in the previous administration (Quinn and Brennan 2021).

As the country and the world eventually move forward from the COVID-19 pandemic, experts agree we must prepare for another infectious disease outbreak in the future. While the GHS Index did not prove to be predictive in this case, its assessment of the strengths that the United States could have brought to bear on a pandemic was accurate. Our strong scientific, health, and public health communities still exist, although the pandemic did expose the challenges many underfunded state and local-level public health departments face across the country. Our close relationships with international organizations and countries across the globe can be rebuilt. To that end, in his first week in office President Biden rejoined the World Health Organization and restored the White House National Security Council Directorate for Global Health Security and Biodefense (Deliso 2021). However, we face an uphill battle in regaining public trust in science, as that trust has been declining for decades, particularly among Republicans (Rosenbaum 2020) (see Chapter 14). The task of re-establishing trust in science falls not only to government leaders but also to our educational system. Trust in government must also be restored. The majority of Americans rated the response to the COVID-19 pandemic under the former Trump Administration as less effective than that of other wealthy countries (see Figure 2.6).

Public mistrust in science and government affects people's willingness to be vaccinated against disease. Shortly after the first COVID-19 vaccinations were being given to U.S. health care workers, a survey found that about a quarter of the U.S. public said they "probably or definitely would not get a COVID-19 vaccine even if it were available for free and deemed safe by scientists" (Hamel et al. 2020b). The main reasons given for not taking the vaccine include (1) concerns about possible side effects, and (2) a lack of trust in the government to ensure the vaccine's safety and effectiveness.

Finally, the COVID-19 pandemic highlighted the need to invest in strong public health and medical systems that meet the health needs of *all* our residents.

> [The pandemic] has also brought home the importance of public health, and strong health systems and emergency preparedness, as well as the resilience of a population in the face of a new virus or pandemic, lending ever greater urgency to the quest for universal health coverage (UHC). (United Nations 2020, p. 2)

herd immunity The point at which enough people in a population have been exposed to or immunized from an infectious agent to stop its spread.

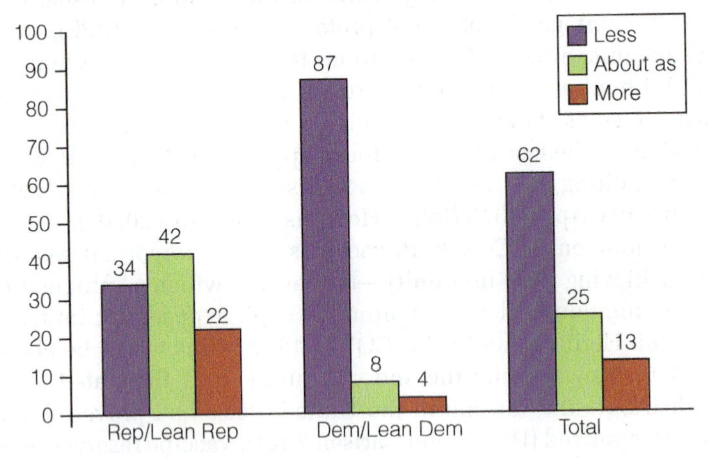

Figure 2.6 American's Views on How the United States Has Handled the COVID-19 Pandemic
SOURCE: Gramlich 2020.

> "We feel there has been over a 20–1 return. Helping young people live, get the right nutrition, contribute to their countries—that has a payback that goes beyond any typical financial return."
>
> **–BILL GATES, BILL AND MELINDA GATES FOUNDATION, ON WHY THEY CONTRIBUTE TO THE GLOBAL FUND AND OTHER INTERNATIONAL HEALTH INITIATIVES**

Understanding Problems of Illness and Health Care

In a Policy Brief on COVID-19, the United Nations (2020) argued that "The pandemic has laid bare long-ignored risks, including inadequate health systems, gaps in social protection and structural inequalities" (United Nations 2020). This assessment deftly describes the U.S. experience. As discussed earlier, structural inequalities have contributed to some Americans having more exposure to the coronavirus than others. And among those infected, unequal access to testing, hospital beds, staff, and advanced equipment have left some more vulnerable to death. The underlying health conditions that put some people at a greater risk of death from COVID-19 are not equally distributed. They are entangled with long-standing structural inequalities in access to healthy lifestyles and medical resources.

Across the globe, preventable health problems affect the quality of life for individuals, families, communities, and states. Poor countries are grappling with problems such as high maternal and infant mortality rates, malaria, Ebola, and HIV/AIDS. Cancer, once viewed as a disease that affects primarily wealthy countries, has now become prevalent in low-income countries where treatment is either not available or not affordable (Ruff et al. 2016). Obesity and its associated health problems have also spread throughout the world, adding to the burden of infectious diseases that already plague low-income countries.

Although individuals make choices that affect their health—choices such as whether to exercise, eat a healthy diet, wear a seat belt, and so on—those choices are affected by social, economic, and environmental conditions. Poverty, the stress of discrimination, unequal access to education, unequal exposure to pollution, the gendered division of labor in families, and exposure to violence are some of the many social factors that impact health disparities in the United States and around the world. A sociological approach to illness and health care looks at social solutions that address these root causes of health problems.

A sociological view of illness and health care looks not only at the social causes but also at the social *consequences* of health problems—consequences that potentially affect us all. While poverty is one of the strongest predictors of health, individuals' health can also affect their socioeconomic status by impacting their ability to work for pay or increasing their **medical debt**. When individuals cannot support themselves—due to either physical or mental illness—it impacts families, communities, and society at large. The COVID-19 pandemic illustrates how illness can affect virtually all aspects of social life.

medical debt Debt that results when people cannot afford to pay their medical bills.

While increasing access to health care is critically important for improving health, it is just one piece of the puzzle. A comprehensive approach to improving the health of a society requires a society-wide commitment to addressing diverse issues such as poverty and economic inequality, education, housing, education, crime and violence, drug and alcohol use, gender inequality, racial discrimination, workplace safety, population growth, and energy and environmental issues—topics discussed in subsequent chapters of this text.

Chapter Review

- **Why do the authors argue that the study of social problems is essentially the study of health problems?**
  According to the World Health Organization, health is "a state of complete physical, mental, and social well-being." Based on this definition, the authors suggest that the study of social problems is essentially the study of health problems because each social problem affects the physical, mental, and social well-being of humans and the social groups of which they are a part.

- **What factors led the United States to become the country with the highest number of COVID-19 cases and deaths?**
  The federal government did not respond quickly and in a coordinated way with testing and quarantining those exposed. Equipment such as personal protective equipment and COVID-19 tests were in short supply early in the pandemic. Messages about masks and social distancing were inconsistent, as former President Trump contradicted health professionals and minimized the pandemic's impact.

- **What are some major differences in the health of populations living in high-income countries compared with the health of populations living in low-income countries?**
  Life expectancy is significantly greater in high-income countries compared with low-income countries. Although the majority of deaths worldwide are caused by noncommunicable diseases such as heart disease, stroke, cancer, and respiratory disease, low-income countries have a comparatively higher rate of infectious and parasitic diseases, infant and child deaths, and maternal mortality.

- **How has globalization affected health worldwide?**
  Increased global transportation and travel contribute to the spread of infectious disease. Globalization is linked to the rise in obesity worldwide due to increased access to unhealthy foods and beverages and to televisions, computers, and motor vehicles, which are associated with increased sedentary behavior. On the positive side, globalized communications technology is helpful in monitoring and reporting on outbreaks of disease, disseminating guidelines for controlling and treating disease, and sharing medical knowledge and research findings. Another aspect of globalization and health is the growth of medical tourism—a multibillion-dollar global industry that involves traveling, primarily across international borders, for the purpose of obtaining medical care.

- **How do structural-functionalism, conflict theory, and symbolic interactionism help us understand illness and health care?**
  Structural-functionalism examines (1) how failures in the health care system affect not only the well-being of individuals but also the health of other social institutions, such as the economy and the family; (2) how changes in society affect health and how health concerns may lead to social change; and (3) latent dysfunctions or the unintended and often unrecognized negative consequences of health-related social patterns or behavior. The conflict perspective (1) focuses on how socioeconomic status, power, and the profit motive influence illness and health care; (2) points to ways in which powerful groups and wealthy corporations influence health-related policies and laws through lobbying and financial contributions to politicians and political candidates; and (3) criticizes the pharmaceutical and health care industry for placing profits above people. Symbolic interactionism focuses on (1) how meanings, definitions, and labels influence health, illness, and health care; (2) the process of medicalization whereby behaviors and conditions come to be labeled as medical problems; (3) how conceptions of health and illness are socially constructed and vary over time and across societies; and (4) the stigmas associated with certain health conditions.

- **What are three main social factors that are associated with health disparities?**
  Three main social factors associated with health disparities are socioeconomic status, race/ethnicity, and gender.

- **How does health care in the United States compare with that of many other high-income nations?**
  Many other advanced countries have national health insurance systems—typically administered and paid for by government—that provide universal health care (health care for all citizens). The United States does not have a health care system per se but rather has a patchwork that includes both private insurance (purchased individually or through employers or other groups) and public insurance plans such as Medicare and Medicaid. As a result, many face crippling medical debt and make decisions to avoid medical care or do not comply with their prescription regimens due to cost.

- **Why is mental illness referred to as a hidden epidemic?**
  Mental illness is a hidden epidemic because the shame and embarrassment associated with mental problems discourage people from acknowledging and talking about them. One of the most common stereotypes of people with mental illness is that they are dangerous and violent. In fact, people with mental illness are much more likely to be victims of violence than members of the general population.

- **How can we prepare for a future pandemic?**
  We must support and sustain our strong scientific, health, and public health communities and maintain close relationships with international health organizations and with countries across the globe. We must regain public trust in

science through strong government messaging and through education. We need to develop a medical system that meets the health needs of *all* our residents.

- **What is single-payer health care?**
  In a single-payer health care system, a single tax-financed public insurance program replaces private insurance companies. Advocates of single-payer health care financing point out that replacing private insurance companies with single-payer health care would save more than $400 billion per year—enough to provide universal coverage to every U.S. resident without copayments, deductibles, or increases in health expenditures.

Test Yourself

1. Deaths due to ________ are much more common in less developed countries compared with more developed countries.
   a. heart disease
   b. cancer
   c. stroke
   d. infectious diseases
2. The United States has the highest life expectancy of any country in the world.
   a. True
   b. False
3. Obesity is a problem only in wealthy, developed countries.
   a. True
   b. False
4. How many Americans will experience some form of mental disorder in their lifetime?
   a. One in 20
   b. One in 10
   c. One in 5
   d. About half
5. Which of the following factors helps to explain why residents of poor counties were more likely to die of COVID-19 than residents of middle- and high-income counties?
   a. less access to remote work
   b. less access to testing early on
   c. hospitals being more likely to be understaffed and less likely to have life-saving equipment
   d. all of the above
6. In the United States, which group has the lowest life expectancy?
   a. White women
   b. Black women
   c. White men
   d. Black men
7. In the United States, the most frequently used source of care for mental health problems is
   a. psychologists.
   b. psychiatrists.
   c. primary care physicians.
   d. social workers.
8. The *Affordable Care Act* required health insurance plans to provide dependent coverage for children up to age
   a. 18.
   b. 20.
   c. 24.
   d. 26.
9. In most countries of the world, women can buy birth control pills without a prescription.
   a. True
   b. False
10. Replacing U.S. private insurance companies with single-payer health care would do which of the following?
    a. save more than $400 billion in administrative costs per year
    b. save enough to provide universal coverage to every U.S. resident without copayments, deductibles, or increases in health expenditures
    c. both a and b
    d. neither a nor b

Answers: 1. D; 2. B; 3. B; 4. C; 5. D; 6. D; 7. C; 8. D; 9. A; 10. C.

Key Terms

Affordable Care Act (ACA) 62
Children's Health Insurance Program (CHIP) 56
contact tracing 40
criminalization of mental illness 61
death rate 39
deinstitutionalization 60
food deserts 49
globesity 45
health 38
health disparities 48
herd immunity 67
Indian Health Service 56
infant mortality rate 43
life expectancy 41
managed care 57
maternal mortality 44
Medicaid 55
medicalization 47
medical debt 68
medical tourism 45
Medicare 55
mental health 59
mental illness 59
Military Health System (MHS) 56
mortality 41
parity 66
pandemic 39
positivity rate 40
preexisting conditions 62
single-payer health care 63
social stratification 49
socioeconomic status (social class) 49
stigma 59
under-5 mortality rate 43
universal health care system 54
Veterans Health Administration (VHA) 56

Liz Dufour/The Enquirer

Substance abuse, the nation's number one preventable health problem, places an enormous burden on American society, harming health, family life, the economy, and public safety, and threatening many other aspects of life."

THE ROBERT WOOD JOHNSON FOUNDATION
Institute for Health Policy, Brandeis University

3

Alcohol and Other Drugs

Chapter Outline

Learning Objectives

1. Compare drug use trends globally and within the United States.
2. Explain how the different sociological theories characterize the causes of drug abuse.
3. Propose recommendations to reduce drug abuse based on each of the three sociological theories.
4. Describe patterns in the use and abuse of legal and illegal drugs.
5. Explain the relationship between legal and illegal drug use.
6. Explain how the consequences of drug abuse are harmful to society.
7. Identify viable substance abuse-fighting strategies.

UNTIL THE NIGHT THEY DIED, brothers Jack and Nick Savage had never been in trouble with drugs or alcohol. The two high schoolers had gone to a graduation party that evening. They arrived home at 12:30 p.m., said goodnight to their mom Becky who was waiting up for them, and went to bed. They did not wake up the next morning. The boys had accidentally taken a lethal combination of alcohol and prescription hydrocodone pills that a friend had passed out at the party. "We've talked to our kids about drinking," said mom Becky, "but we had never talked to them about prescription drugs, because it wasn't even on our radar" (Bergeron 2018, p. 1). With their organization the 525 Foundation, the Savage family is determined to prevent other families from experiencing a similar tragedy. Through speaking engagements, political lobbying, and partnerships with local law enforcement agencies, the 525 Foundation is organizing pill drop-offs and advocating for stricter laws around prescription drugs. "In different communities, there are still people who are unaware of the dangers," says Becky. "After I get done talking to them, the first thing they say is they're going to go home and clean out their medicine cabinets."

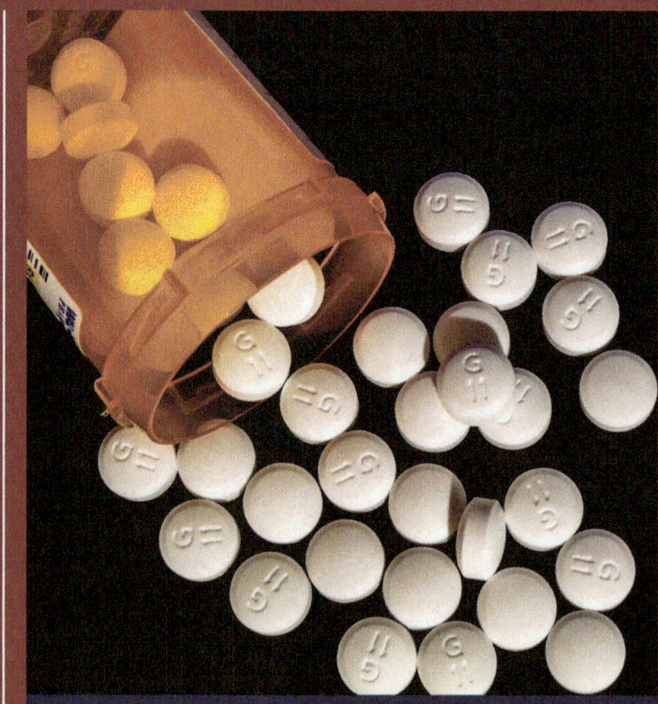

Justin Brodt/EyeEm/Getty Images

The misuse of prescription painkillers is associated with nearly two-thirds of all drug overdose deaths in the United States (CDC 2020d).

Drug-related deaths are just one of the many negative consequences of alcohol and drug abuse. The abuse of alcohol and other drugs is a social problem when it interferes with the well-being of individuals and/or the societies in which they live—when it jeopardizes health, safety, work, and academic success, family, and friends. But managing the drug problem is a difficult undertaking, particularly when attitudes vary dramatically by demographic groups (see Table 3.1). In dealing with drugs, a society must balance individual rights and civil liberties against the personal and social harm that drugs promote—fetal alcohol syndrome, suicide, drunk driving, industrial accidents, mental illness, unemployment, and teenage addiction. When to regulate, what to regulate, and whether to regulate are complex social issues. Our discussion begins by looking at how drugs are used and regulated in other societies.

The Global Context: Drug Use and Abuse

Pharmacologically, a **drug** is any substance other than food that alters the structure or functioning of a living organism when it enters the bloodstream. Using this definition, everything from vitamins to aspirin is a drug. Sociologically, the term *drug* refers to any chemical substance that (1) has a direct effect on users' physical, psychological, and/or intellectual functioning; (2) has the potential to be abused; and (3) has adverse consequences for individuals and/or society. Societies vary in how they define and respond to drug use. Thus, drug use is influenced by the social context of the particular society in which it occurs.

drug Any substance other than food that alters the structure or functioning of a living organism when it enters the bloodstream.

Drug Use and Abuse around the World

Globally, 1 in 19 adults between the ages of 15 and 64 used at least one illicit drug in 2018—269 million drug users (World Drug Report [WDR] 2020). Of those who use drugs, nearly

TABLE 3.1 Views of Drug Problem in the United States by Subgroup and Year, 2018 (N = 2,657)

	Percent who view drugs in the United States as an extremely/very serious problem			
	2000	2007	2016	2018
Men	77	66	59	57
Women	86	80	70	72
White	80	71	45	64
Black	90	80	57	73
18–29	71	71	45	63
30–49	80	63	63	62
50–64	88	77	72	65
65+	91	89	77	72
College graduate	74	65	59	60
Not college graduate	85	77	67	71
Republican	84	69	81	72
Democrat	84	75	57	63
Independent	79	74	58	62

SOURCE: Pew Research Center 2016, 2018.

36 million are estimated to suffer from drug use disorders; of these, only about one in eight will receive treatment. According to the most World Drug Report, opioid use causes the most harm, accounting for two-thirds of the deaths attributed to drug use disorders. Drug use by injection contributes to these death rates by increasing the likelihood that the user will contract illnesses such as HIV and hepatitis; about half of all drug-related deaths were caused by untreated hepatitis C. Increasingly, however, drug use deaths are tied to the rising nonmedical use of pharmaceutical opioids. Methamphetamine use is also leading to increased concern among public health professionals, with evidence indicating a more than eightfold increase in seizures of illegal methamphetamines in some regions over the past decade.

The *Global Status Report on Alcohol and Health* (World Health Organization [WHO] 2018) indicates that, in the year prior to data collection, 43 percent of respondents had consumed alcohol. Furthermore, 18.2 percent of the world's 15 years and older population were "heavy episodic" drinkers, more commonly called binge drinking. In 2016, the most recent year for which data are available, over 3 million deaths were attributable to alcohol (WHO 2018).

Harm from alcohol consumption is connected to socioeconomic status. In general, alcohol use is tied to income; those who are poorer are less likely to consume alcohol than those who are wealthier, and alcohol consumption is lower in poorer societies than in rich ones. However, at the individual level, the "harm per liter" metric shows a negative correlation between socioeconomic status and alcohol consumption (WHO 2018). This means that, among those who drink, those with low incomes are more likely than those with high incomes to experience negative effects from alcohol consumption. This occurs for a variety of contextual reasons, for example, those who are poor are more likely to be socially stigmatized, more likely to live in crowded living arrangements, to have less access to sanitation, less access to health care, and to be more likely to have other health problems that can be exacerbated by alcohol use. All of these factors increase the likelihood that alcohol consumption will lead to adverse outcomes.

Alcohol consumption is also tied to cultural and societal factors. For example, Russians' high alcohol consumption, although descending recently, reflects access and low cost, high unemployment and the resulting boredom, and peer pressure (Jargin 2012). Conversely, consumption rates are lower in the Middle East and Asia where religious dictates discourage alcohol use, availability is difficult, and punishment is severe.

Globally, nearly 17 percent of the adult population smokes cigarettes, and 80 percent of the people who smoke cigarettes are from low- and middle-income countries (WHO 2020). Prevalence of smoking, however, varies dramatically by country. For example, the WHO estimates that at least 76 percent of Indonesian men and 59 percent of Russian men smoke, but fewer than 10 percent of Ethiopian men smoke (WHO 2016). Globally, men are estimated to smoke at least five times as much as women, but these rates vary by country. In high-income countries, such as the United States and Canada, women and men smoke at almost the same rate. However, women in low- and middle-income countries smoke at much lower rates than men; in China, men are approximately 15 times more likely to smoke than women (Hitchman and Fong 2011).

As a result of robust public health efforts over the past two decades, tobacco use through cigarettes has been declining globally. However, tobacco use through vaporizers and e-cigarettes is increasing. These products are often advertised as reduced harm products, a claim which has not been scientifically verified. Instead, the World Health Organization warns that these products may lead to other types of harm, including increased cancer and heart disease risks due to the chemicals contained in flavored additives and the aerosols produced through the vaporizing process. Although there is not yet enough research to determine how the health impacts of these new forms of tobacco consumption compare to those of traditional cigarettes and cigars, emerging evidence from the Centers for Disease Control and Prevention (CDC) indicates that e-cigarette use among young adults was associated with increased risk of hospitalization, lung damage, and death from COVID-19 (Armatas et al. 2020).

Finally, drug use varies over time. Figure 3.1, based on data from Monitoring the Future (MTF), a national in-school survey of secondary school students' drug use, indicates that alcohol and illicit drug use, in general, has decreased over the last two decades. Marijuana use has mostly held steady. It should also be noted that Figure 3.1 compares 8th, 10th, and 12th graders reported drug use in the preceding 12 months. In each case, drug use increases with age (National Institute on Drug Abuse [NIDA] 2020b). This survey also indicates a dramatic shift in teen tobacco use. From 2017 to 2019, the percentage of 12th graders who reported smoking cigarettes in the past 30 days dropped from 10 percent to less than 6 percent. In that same time period, the percentage of 12th graders who reported vaping nicotine in the past 30 days increased from 11 percent to nearly 26 percent.

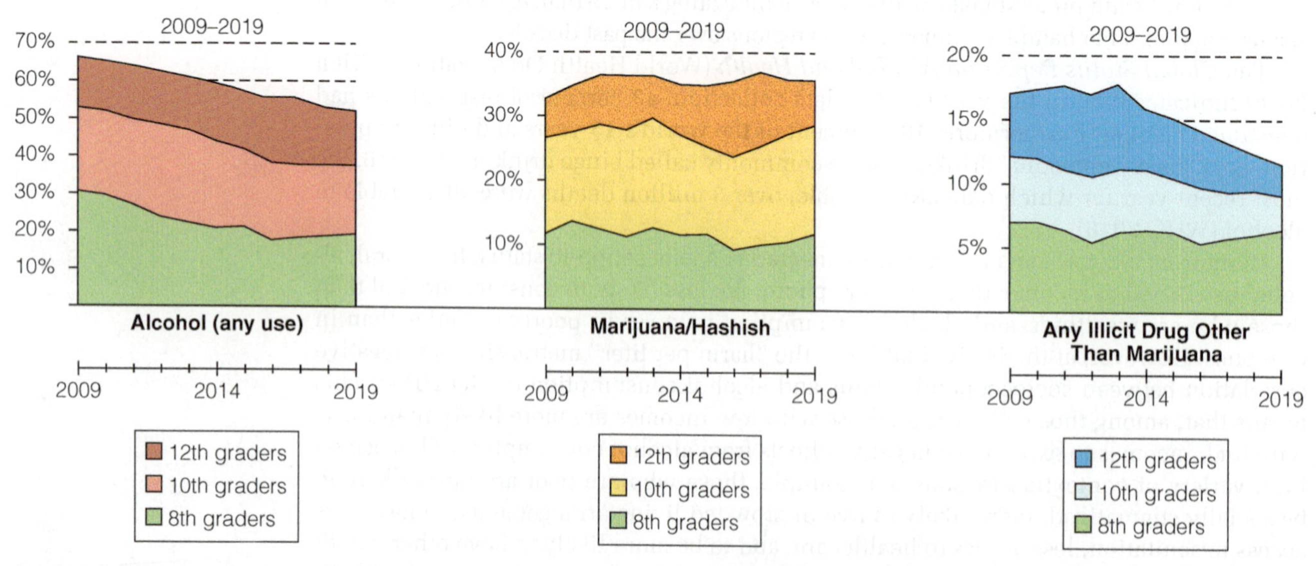

Figure 3.1 Alcohol, Marijuana, and Illicit Drug Use for 8th, 10th, and 12th graders, 2009–2019
SOURCE: National Institute on Drug Abuse (NIDA) 2020.

Some have argued that differences in drug use can be attributed to variations in drug policies. The Netherlands, for example, has had an official government policy of treating the use of "soft" drugs such as marijuana and hashish as a public health issue rather than a criminal justice issue since the mid-1970s. Despite the proliferation of so-called cannabis cafés, sales of marijuana for personal use in the Netherlands do not significantly differ from that in other European countries (Korf 2019).

Historically, Great Britain has also adopted a "medical model," particularly in regard to heroin and cocaine. As recently as the 1960s, English doctors prescribed opiates and cocaine for their drug-addicted patients who were unlikely to quit using drugs on their own and for the treatment of withdrawal symptoms. By the 1970s, however, the emphasis was on prohibition, criminalization, and incarceration and, as in the United States, it became known as the **war on drugs**. Presently, British officials are weighing the relative merits of returning to a medical model that embraces drug addiction as a disorder (British Medical Association [BMA] 2019). Similarly, in 2014 the European Union adopted a drug policy that for the first time included the objective of reducing "the health and social risks and harms caused by drugs" (The European Monitoring Centre for Drugs and Drug Addiction [EMCDDA] 2014, p. 2). After five years, this approach has shown some success. The EMCDDA reports that in the five years since implementing this new strategy of drug policy, HIV infections in EU nations caused by drug injections have declined by 40 percent (EMCDDA 2019).

war on drugs A public policy approach to the illicit drug trade in the United States, initially implemented by the Nixon administration in the 1970s, which focused on the widespread prohibition and criminalization of drug use and distribution.

What *do you* think?

Alcohol use policies vary widely by country. In the United States, it is mostly illegal for young adults under the age of 21 to purchase or consume alcohol, while many European countries have drinking ages of 16 or 18 or may have no legal drinking age at all. Scientific research has firmly established that alcohol consumption in adolescence leads to long-term health problems and increased likelihood of substance abuse as an adult. However, some studies indicate that while U.S. teens are less likely to drink alcohol overall compared to their European counterparts, they are more likely to binge drink and use other illicit drugs (Gilligan et al. 2012; Knopf 2016). Do you think public health outcomes would be improved if the United States adopted a European-style policy to adolescent alcohol use?

In stark contrast to such health-based policies, many other countries execute drug offenders or subject them to corporal punishment that may include whipping, stoning, and torture. Thirty-five countries and territories have the death penalty for drug violations, and globally, at least 4,000 people have been executed for drug violations from 2008 to 2018 (Harm Reduction International [HRI] 2019). In the Philippines, the war on drugs has taken on a literal meaning, with President Rodrigo Duterte urging police and private citizens to kill people suspected of being drug users or drug dealers. Human Rights Watch (HRW) estimates that since Duterte took office in June 2016, at least 12,000 Filipinos, mostly urban poor, have been killed as part of this war on drugs (HRW 2020).

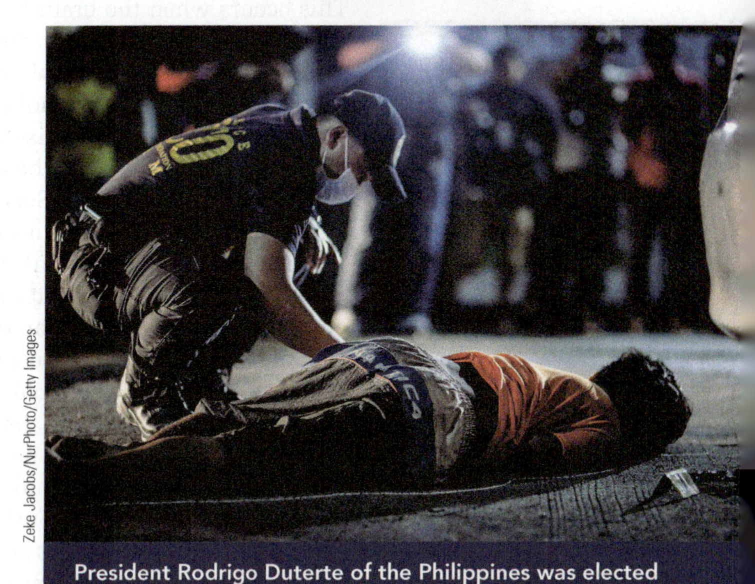

Zeke Jacobs/NurPhoto/Getty Images

President Rodrigo Duterte of the Philippines was elected in 2016, after making a campaign pledge to declare an all-out war against drugs. After his election, he urged the police and vigilantes to hunt down and kill suspected drug users and dealers. Here, a police investigator marks the corpse of a suspected drug pusher who was shot during a police roundup. Duterte's drug war has led to at least 12,000 extra-judicial killings since 2016.

Drug Use and Abuse in the United States

In the United States, cultural definitions of drug use are contradictory—condemning it, on the one hand (e.g., heroin), yet encouraging and tolerating it on the other (e.g., alcohol). At various times in U.S. history, many drugs that are illegal today were legal and readily available. In the 1800s and the early 1900s, opium was routinely used in medicines as a pain reliever, and morphine was taken as a treatment for dysentery and fatigue. Amphetamine-based inhalers were legally available until 1949, and cocaine was an ingredient in Coca-Cola until 1906, when it was replaced with another drug—caffeine (Abadinsky 2013; Witters et al. 1992). Not surprisingly, Americans' concerns with drugs have varied over the years. In 2000, 31 percent of Americans believed that marijuana should be legal; in 2019, 67 percent thought that marijuana should be legal (Daniller 2019).

Use of illegal drugs in the United States is a fairly common phenomenon. According to the most recent National Survey on Drug Use and Health (NSDUH 2019) available, in 2018, nearly one out of every five Americans aged 12 and older had used an illicit drug in the month prior to the survey year. For purposes of the survey, illicit drugs included "marijuana/hashish, cocaine (including **crack**), heroin, hallucinogens, inhalants, or prescription-type psychotherapeutics (e.g., pain relievers, tranquilizers, stimulants, and sedatives) used non-medically" (NSDUH 2019, p. 1). Furthermore, 51.1 percent of Americans age 12 and older reported alcohol consumption in the previous month, and 21.5 percent reported tobacco use in the previous month.

Sociological Theories of Drug Use and Abuse

Drug abuse occurs when acceptable social standards of drug use are violated, resulting in adverse physiological, psychological, and/or social consequences. When an individual's drug use leads to hospitalization, arrest, or divorce, such use is usually considered abusive. Drug abuse, however, does not always entail drug addiction. Drug addiction, or **chemical dependency**, refers to a condition in which drug use is compulsive—users are unable to stop because of their dependency. The dependency may be psychological (the individual needs the drug to achieve a feeling of well-being) and/or physical (withdrawal symptoms occur when the individual stops taking the drug). For example, withdrawal from alcohol can cause trembling, insomnia, anxiety, hallucinations, and seizures. Approximately 1 in 20 people who have alcohol withdrawal symptoms will experience a condition known as *delirium tremens*, which occurs when the brain's chemistry has been significantly altered due to heavy, long-term alcohol use (Harvard Health 2020). This occurs when the brain becomes unable to effectively regulate breathing and blood circulation without alcohol, and withdrawal can become life-threatening. In these cases, medical supervision is required to stop using alcohol.

If recurrent use of alcohol and/or drugs causes a person to experience clinically significant health problems, disability, and inability to meet major responsibilities at work, school, or home, they may be diagnosed with **substance use disorder** (SUD) (Substance Abuse and Mental Health Services Administration [SAMHSA] 2020). In 2018, approximately 20.3 million people aged 12 and older had SUD. Alcohol use disorder is the most common SUD (14.8 million), followed by illicit drug use disorder (8.1 million). More than half of those with illicit drug use disorder had marijuana use disorder (4.4 million), and about one-quarter had an opioid use disorder (2.0 million) (NSDUH 2019a).

Substance use disorder occurs disproportionately among some racial, ethnic, and sexual minority groups. For example, among those aged 26 or older, 5.1 percent of the overall population had alcohol use disorder, but this number is 11.2 percent among lesbian, gay, or bisexual adults, and 7.2 percent among indigenous Americans (NSDUH 2019b). Men are more likely than women to be heavy drinkers, to binge drink, and to use marijuana and other illicit drugs (Centers for Disease Control [CDC] 2018). Why are some individuals more likely to use and abuse drugs and alcohol compared to others? Drug use is not simply a matter of individual choice. Theories of drug use explain how structural and cultural forces, as well as biological and psychological factors, influence drug use and society's responses to it.

Structural-Functionalist Perspective

Structural functionalists argue that drug abuse is a response to weakening societal norms. As society becomes more complex and as rapid social change occurs, norms and values become unclear and ambiguous, resulting in anomie—a state of normlessness. Anomie

crack A crystallized illegal drug product produced by boiling a mixture of baking soda, water, and cocaine.

drug abuse The violation of social standards of acceptable drug use, resulting in adverse physiological, psychological, and/or social consequences.

chemical dependency A condition in which drug use is compulsive and users are unable to stop because of physical and/or psychological dependency.

substance use disorder A medical diagnosis used when recurrent use of alcohol and/or drugs causes clinically significant health problems, disabilities, and inability to meet major responsibilities at work, school, or home.

may exist at the societal level, resulting in social strains and inconsistencies that lead to drug use. For example, research indicates that increased alcohol consumption in the 1830s and the 1960s was a response to rapid social change and the resulting stress (Rorabaugh 1979). Similarly, the disruption to daily routines and stress produced by stay-at-home orders in many states during the 2020 COVID-19 pandemic led to a more than 32 percent increase in alcohol consumption, compared to the same period of time the year before (MacMillan 2020). Similarly, the economic disruption and social isolation that occurred after COVID-19 prompted state-wide shutdown orders are blamed for a resurgence in opioid overdose deaths in West Virginia, after two years of decline, referred to by one drug policy expert as "deaths of despair" (McGreal 2020). Anomie produces inconsistencies in cultural norms regarding drug use. For example, although public health officials and health care professionals warn of the dangers of alcohol and tobacco use, advertisers glorify the use of alcohol and tobacco, and the U.S. government subsidizes the alcohol and tobacco industries. Furthermore, cultural traditions persist, such as giving away cigars to celebrate the birth of a child and toasting a bride and groom with champagne.

Anomie may also exist at the individual level, as when a person suffers feelings of estrangement, isolation, and turmoil over appropriate and inappropriate behavior. An adolescent whose parents are experiencing a divorce, who is separated from friends and family as a consequence of moving, or who lacks parental supervision and discipline may be more vulnerable to drug use because of such conditions. Thus, from a structural-functionalist perspective, drug use is a response to the absence of a perceived bond between the individual and society and to the weakening of a consensus regarding what is considered acceptable.

We know, for example, that high levels of parental conflict in the home are associated with an increased likelihood of adolescent cigarette and alcohol use (Fosco and Feinberg 2018). Furthermore, there is evidence that clear communication between parents and their adolescent children about expectations regarding drug use can override the effects of negative societal messages and peer pressure. A meta-analysis of 57 studies on parental communication strategies and adolescent behavior found that teens were less likely to use drugs and alcohol when parents used an "active mediation" approach to media use (Collier et al. 2016). In this approach, parents encourage critical thinking about media representations of drug and alcohol use by talking directly with their children about the messaging and asking questions about characters' behaviors, motivations, and consequences of their actions.

Conflict Perspective

The conflict perspective emphasizes the importance of power differentials in influencing drug use behavior and societal values concerning drug use. From a conflict perspective, drug use occurs as a response to the inequality perpetuated by a capitalist system. Societal members, alienated from work, friends, and family as well as from society and its institutions, turn to drugs as a means of escaping the oppression and frustration caused by the inequality they experience. Furthermore, conflict theorists emphasize that the most powerful members of society influence the definitions of which drugs are illegal and the penalties associated with illegal drug production, sales, and use.

For example, alcohol is legal because it is often consumed by those who have the power and influence to define its acceptability—White males (NSDUH 2019b). This group also disproportionately profits from the sale and distribution of alcohol and can afford powerful lobbying groups in Washington, D.C., to guard the alcohol industry's interests. Because this group also commonly uses tobacco and caffeine, societal definitions of these substances are also relatively accepting. Conversely, minority group members disproportionately use crack cocaine rather than powder cocaine (Fryer et al. 2014). Although the pharmacological properties of the two drugs are the same, possession of 5 grams of crack cocaine carried the same penalty under federal law as possession of 500 grams of powdered cocaine (Taifia 2006). In 2010, Congress voted to change the 1986 law that established the 100-to-1 ratio sentencing disparity and also eliminated the five-year mandatory minimum for first-time possession of crack cocaine. Although racial inequities in drug sentencing are beginning to show decline, significant racial disparities in drug arrests and incarceration persist (Beckett and Brydolf-Horwitz 2020).

What *do you* think?

As more states move to decriminalize and in some cases fully legalize marijuana use, many criminal justice reform activists have pushed to release from prison and expunge the criminal records of those with low-level marijuana convictions (Zimmerman 2020). However, marijuana remains illegal at the federal level and is classified as a Schedule I narcotic. What do you think should be done with the marijuana conviction records of those living in states where marijuana has recently become decriminalized or legalized?

@Cory Booker

It's not enough to just legalize marijuana at the federal level—we should also expunge records of those who have served their time, and reinvest in communities hardest hit by the failed War on Drugs—which has really been a war on people.

-Cory Booker

The criminalization of drugs has followed a strongly racialized pattern. In the 1800s, Chinese immigrants who had been brought to the United States to work on the railroads also brought with them a cultural tradition of smoking opium. As unemployment among White workers increased, however, so did resentment of Chinese laborers. Attacking the use of opium became a convenient means of attacking the Chinese, and in 1877, Nevada became the first of many states to prohibit opium use. Similarly, cocaine, which is made from the coca plant, has been used for thousands of years. Coca leaves were used in the original formula for Coca-Cola, but in the early 1900s, anti-cocaine sentiment emerged as a response to the heavy use of cocaine among urban Black and poor White individuals, and criminals (Friedman-Rudovsky 2009; Thio 2007; Witters et al. 1992). Cocaine was outlawed in 1914. The criminalization of other drugs, including marijuana, follows similar patterns of social control of the powerless, political opponents, and/or minorities.

In the 1940s, marijuana was used primarily by those in the Black and Hispanic working class, and users faced severe criminal penalties. However, after White, middle-class college students began to use marijuana in the 1970s, the government reduced the penalties associated with its use. As of January 2021, marijuana is fully illegal in only six states, with other states having a mix of policies ranging from full recreational legalization to restricted medical use legalization. Although the nature and pharmacological properties of the drug have not changed, the population of users is now connected to power and influence.

Similarly, analyses of the public attention given to the ongoing prescription opioid addiction crisis compared to previous addiction crises (such as heroin and crack cocaine) have found a stark racial difference (Netherland and Hansen 2017). The opioid crisis is largely associated with White and rural populations, and the public attention to it has been generally sympathetic and focused on a medical model to understand its causes and solutions. In contrast, public attention to the crack cocaine and heroin crises of the 1980s and 1990s—which were predominantly associated with Black and Hispanic populations—was generally critical and focused on a crime and punishment model. Thus, conflict theorists regard the regulation of certain drugs, as well as drug use itself, as a reflection of differences in the political, economic, and social power of various interest groups.

Symbolic Interactionist Perspective

Symbolic interactionism, which emphasizes the importance of definitions and labeling, concentrates on the social meanings associated with drug use. If the initial drug use experience is defined as pleasurable, it is likely to recur, and the individual may earn the label of "drug user" over time. If this definition is internalized so that the individual assumes an identity of a drug user, the behavior will probably continue and may even escalate. Conversely, Copes, Hochstetler, and Williams (2008) observed that respondents who self-identified as "hustlers" rather than "crack-heads" were less likely to fall prey to the debilitating effects of the drug, for "[s]lipping into uncontrollable addiction is antithetical to the hustler identity" (p. 256).

Drug use is also learned through symbolic interaction in small groups. High school students who want to be perceived of as popular by their peer groups are more likely to engage in heavy drinking (Dumas et al. 2017). Peer influence and social media use are strong predictors of teen substance use. In a survey of over 1,000 12- to 17-year-olds, the strongest factors predicting a teen's willingness to try drugs, alcohol, or nicotine in the future were if they had a friend who used substances, if they hung out with friends unsupervised by an adult at least once a week, and if their parents did not (National Center on Addiction and Substance Abuse [CASA] 2019). Research tracking high school students' social media use and drinking behavior over the course of a year found that higher levels of exposure to friends' social media posts depicting alcohol use predicted an individual's likelihood of drinking for the first time and binge drinking one year later, even when controlling for other substance use risk factors (Nesi et al. 2017).

In another study, researchers tracked teens who had never used tobacco or marijuana and found that over a six-month period, those who checked social media platforms and text messages frequently had a higher likelihood of having started to use tobacco or marijuana during that time period (Kelleghan 2020). Interactions between teens and their parents or caregivers can also provide a strong source of drug abuse prevention. Teens who described their relationship with their parents as "excellent" were 1.8 times more likely to report that they did not intend to try substances in the future than those who did not report excellent relationships with their parents (CASA 2019). Teens were also less likely to use substances if their parents had conversations with them about the risks of drug use. Compared to teens who reported they were likely to try substances in the future, those who did not intend to use drugs were significantly more likely to report that they felt their parents knew their friends, knew what they were up to, trusted them, really listened to them, were honest with them, and spent free time with them.

Interactionists also emphasize that symbols, such as those in advertising and anti-drug campaigns, can be manipulated and used for political and economic agendas. Advertising for alcohol and tobacco products has a long history of targeting minorities, often drawing on cultural anxieties about fitting in with mainstream American culture. Anti-drug campaigns also draw on cultural symbols to target their message. First Lady Nancy Reagan's Just Say No campaign in the 1980s and the popular DARE (Drug Abuse Resistance Education) program, which was commonly used in public schools from 1983 to 2009, were strongly individualistic in their anti-drug messages. Individualism is tightly tied to American cultural values and carries powerful symbolic value among politicians. However, these messages failed to account for the social and contextual factors that contribute to adolescent drug use. Studies of DARE in the early 2000s indicated that the program was largely ineffective at curbing adolescent substance use, and the program was mostly discontinued (West and O'Neal 2004, Weiss et al. 2017). Recent efforts to revise the

The Center for the Study of Tobacco and Society

Advertising for alcohol products has a long history of targeting minorities. Here, a cigarette ad from 1988 appeals to traditional female concerns about femininity and body image, while its tagline invokes a cultural reference to gender and racial progress. Although tobacco advertising is now banned in print and television media, storefront and billboard advertising for tobacco products is disproportionately concentrated in low-income, Black, and Hispanic neighborhoods (Ribisl et al. 2017).

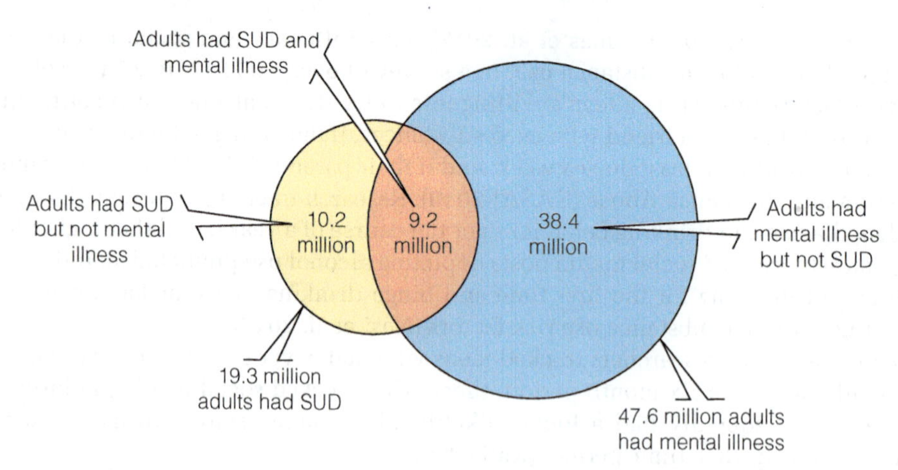

Figure 3.2 Co-occurrence of Substance Use Disorder (SUD) and Mental Illness among Adults Aged 18 or Older, 2018
SOURCE: National Survey on Drug Use and Health (NSDUH) 2019.

DARE curriculum have focused more on evidence-based practices of promoting healthy decision-making behavior among students within their social contexts rather than on lectures about the dangers of drug use (DARE 2018).

@NIH

"Thinking of #addiction as a brain disease is great to reduce #stigma & help people get treatment, but we need a more holistic view that goes beyond a person's biology; we must integrate culture" said @UofNM Dr. Venner. go.usa.gov/xvMH2 @NIDAnews #nihHEALinitiative

-NIH

Biological and Psychological Theories

Drug use and addiction are the result of a complex interplay of social, psychological, and biological forces. For example, some researchers suggest that drug use and addiction are caused by a "bio-behavioral disorder," which combines biological and psychological factors (Margolis and Zweben 2011). Biological research has primarily concentrated on the role of genetics in predisposing an individual to drug use. For example, several genes determine how quickly alcohol is metabolized and the extent to which acetaldehyde, a toxic byproduct of alcohol, is absorbed. If a person has high levels of acetaldehyde in their body, alcohol consumption will lead to negative side effects, thus reducing the likelihood of alcohol abuse (National Institute on Alcohol Abuse and Alcoholism [NIAAA] 2013).

Research also indicates that severe, early-onset alcoholism may be genetically predisposed, with some men having 10 times the risk of addiction as those without a genetic predisposition. Interestingly, other problems such as depression, chronic anxiety, and attention deficit disorder are also linked to the likelihood of addiction. Nonetheless, researchers warn "that genes alone do not determine our destiny—lifestyle choices and other environmental factors have a substantial impact" (NIAAA 2013, p. 4).

Psychological explanations focus on the tendency of certain personality types to be more susceptible to drug use. Individuals who are particularly prone to anxiety may be more likely to use drugs as a way to "self-medicate." Research indicates that child abuse or neglect, particularly among females, contributes to alcohol and drug abuse that extends into adulthood (Gilbert et al. 2009). Substance use disorder is disproportionately high among people with mental illness (see Figure 3.2). Researchers explain this pattern using the concept of "cumulative burden," in which mental illness, substance use disorder, chronic medical conditions, and poverty tend to co-occur due to social stigma and lack of access to resources (Walker and Druss 2017).

Patterns of Drug Use in the United States

Social definitions regarding which drugs are legal or illegal vary over time, circumstance, and societal forces. In the United States, two of the most dangerous and widely abused drugs, alcohol and tobacco, are legal. Compare, for example, the percentage of Americans who report past-month use (i.e., current use) of various illicit drugs, 19 percent, to the 21.5 percent who are current tobacco users and the 51.1 percent who are current alcohol drinkers (NSDUH 2019).

The lines between legal and illegal drugs are complex. In addition to being restricted by age, alcohol sales are illegal in some "dry counties" and in many parts of the United States, sales are banned on Sundays and certain holidays. Although prescription painkillers are legal (albeit highly regulated), the high demand for these drugs to treat chronic pain has driven a surge in the production of illegal counterfeit pills (Drug Enforcement Administration [DEA] 2019). The legality of marijuana is even more complex. According to the federal government, marijuana is illegal and classified as a schedule 1 controlled substance, meaning that it has no known acceptable medical use and high potential for abuse (DEA 2020). Despite this, there are currently 36 states, along with the District of Colombia and several U.S. territories that have approved some version of legal marijuana use for medical purposes; of these, 15 states also allow recreational marijuana use among adults (National Conference of State Legislators [NCSL] 2020; see Figure 3.3). Because federal law overrides state and local laws, the Department of Justice has the option to pursue or not pursue criminal prosecution of marijuana use in those states where it is currently permitted at the local level. This decision is shaped by a complex combination of social, cultural, and political pressures. In this section, we will review that most commonly used drugs in the United States and discuss the social and cultural factors that contribute to their use and abuse.

In November 2020, voters in Oregon passed a ballot initiative to approve the decriminalization of all drugs, including cocaine, methamphetamine, and heroin. Although still considered illegal, offenders in possession of small amounts of these drugs would face a citation and fine similar to a traffic ticket rather than a criminal charge and potential jail time. Proponents of the initiative say this measure is one small step toward shifting public attitudes about drug use and toward a medical and addiction treatment model and ultimately ending the war on drugs (Fuller 2020). What do you think the effects of widespread drug decriminalization would be in your community?

What *do you* think?

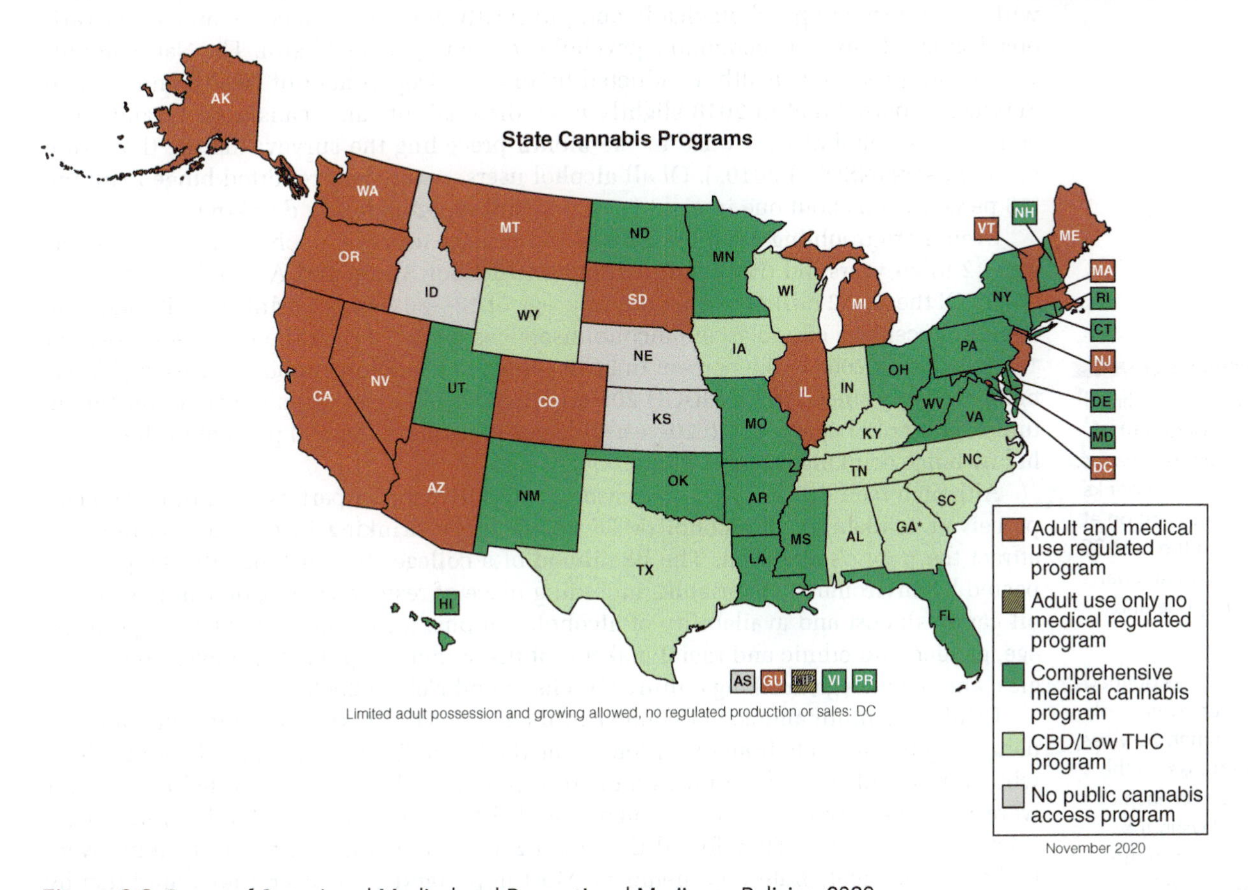

Figure 3.3 Status of State-Level Medical and Recreational Marijuana Policies, 2020
SOURCE: National Conference of State Legislators (NCSL) 2020.

Alcohol: The Drug of Choice

Americans' relationship with alcohol has a long and varied history, much longer than once thought. New research indicates that humans have been consuming alcohol, although often in the form of fermented fruit, for well over 10 million years and not the earlier estimate of 10,000 years (Carrigan et al. 2014). Throughout much of this time, "attitudes toward drinking were characterized by a continued recognition of the positive nature of moderate consumption" (Hanson 1997, n.p.). However, in the 18th century, concerns about excessive alcohol consumption and the negative effects of drinking gave rise to the emergence of a temperance movement (Hanson 2013).

By 1920, the federal government in the United States had prohibited the manufacture, sale, and distribution of alcohol through the passage of the Eighteenth Amendment to the U.S. Constitution. Many have argued that Prohibition, like the opium regulations of the late 1800s, was in fact a "moral crusade" (Gusfield 1963) against immigrant groups who were more likely to use alcohol. The amendment had little popular support and was repealed in 1933. Today, the U.S. population is experiencing a resurgence of concern about alcohol. What has been called a "new temperance" has manifested itself in federally mandated 21-year-old drinking age laws, warning labels on alcohol bottles, increased concern over fetal alcohol syndrome and underage drinking, stricter enforcement of drinking and driving regulations (e.g., checkpoint traffic stops), and zero-tolerance policies. Such practices may have had an effect on drinking norms, particularly for young people. Between 1996 and 2019, the rate of current alcohol use by 12- to 17-year-olds steadily declined, as did the number of 8th, 10th, and 12th graders who report ever being drunk (MTF 2019; NSDUH 2019).

Despite such restrictive policies, alcohol remains the most widely used and abused drug in the United States. According to a survey, 65 percent of U.S. adults drink alcohol, while 34 percent refer to themselves as "total abstainers" (Saad 2019). Although most people who drink alcohol do so moderately and experience few negative effects, people with alcoholism are psychologically and physically addicted to alcohol and suffer various degrees of physical, economic, psychological, and personal harm. The National Survey on Drug Use and Health, conducted by the U.S. Department of Health and Human Services, reported that in 2018 slightly more than half of Americans aged 12 and older consumed alcohol at least once in the month preceding the survey; that is, they were *current users* (NSDUH 2019a). Of all alcohol users, nearly half reported **binge drinking** (48 percent) and about one in eight (11.8 percent) reported **heavy drinking**.

Even more troubling were the 7.1 million current users of alcohol who were underage: 12 to 20 years old (National Institute on Alcohol Abuse and Alcoholism [NIAAA] 2020). Of these, 2.2 million were under the age of 18—adolescent drinkers. Although the overall percentage of adolescent alcohol users has declined over the past decade—from 14.7 percent in 2008 to 9.0 percent in 2018—the rate of adolescent binge drinking has remained steady since 2015 (NSDUH 2019a). Binge drinking is extremely common among underage alcohol users; 12- to 20-year-olds consume more than 90 percent of their alcohol by binge drinking (NIAAA 2020).

binge drinking As defined by the U.S. Department of Health and Human Services, drinking five or more drinks on the same occasion on at least one day in the past 30 days prior to the National Survey on Drug Use and Health.

heavy drinking As defined by the U.S. Department of Health and Human Services, five or more drinks on the same occasion on each of five or more days in the past 30 days prior to the National Survey on Drug Use and Health.

Although teen drinking has decreased in recent years, in part as a result of reduced perceived availability of alcohol (MTF 2019), binge drinking in college continues to attract the public's attention. The likelihood of a college student binge drinking is impacted by environmental variables including place of residence (e.g., on campus versus off campus); cost and availability of alcohol; campus, local, and state alcohol policies; age, gender, and ethnic and racial makeup of the student population; prevention strategies; and the college drinking culture (Wechsler and Nelson 2008).

Drinking culture and social context influence binge drinking behavior. Researchers using longitudinal data from the "Monitoring the Future" survey compared young adults who were members of fraternities and sororities with those who attended college but were not fraternity or sorority members, and those who did not attend college at all (McCabe et al. 2018). They found that among all groups, rates of binge drinking were highest among male fraternity members. Most importantly, however, they found that by age 35, 45 percent of those men who had lived in a fraternity had alcohol use disorder symptoms, compared with 33 percent of men who were members of fraternities but did not live in a fraternity house, with 30 percent of men who attended college but were not members of a fraternity, and with 33 percent of men who did not attend college. Binge drinking rates among women were lower than among men, but followed a similar pattern, with long-term alcohol use disorder highest among those women who lived in a sorority house.

Many binge drinkers began drinking in high school, with almost one-third having their first drink before age 13. Research indicates that the younger the age of onset, the higher the probability that an individual will develop a drinking disorder at some time in his or her life (Behrendt et al. 2009; Blomeyer et al. 2013; Hingson et al. 2006). For example, an individual's chance of becoming dependent on alcohol is 40 percent if the person's drinking began before the age of 13. As drinking levels increase from abstinence to heavy alcohol use, the likelihood of using illegal drugs or tobacco products also increases.

> "For me, drugs and alcohol were a solution to an emotional, and perhaps even spiritual problem. A feeling literally of disease, unhappiness, an inability to cope with life."
>
> **–RUSSELL BRAND, ACTOR AND ACTIVIST**

More males than females aged 12 to 20 report binge drinking, heavy drinking, and current alcohol use. Researchers have long questioned the relationship between gender and drinking behavior, including binge drinking. After administering two measures of gender identity to a sample of college students, Peralta et al. (2010) concluded, "[M]ales, who are socialized to be masculine, may rely on heavy alcohol use to coincide with other forms of male-associated behaviors (e.g., sports, risk-taking). Women, who are socialized to be feminine, on the other hand, may not engage in heavy drinking practices because this is not a part of normative femininity expression rituals" (p. 377).

What *do you* think?

In 2014, 19-year-old college freshman Nolan Burch died of alcohol poisoning during a hazing ritual at a West Virginia University fraternity. In 2017, 19-year-old college freshman Tim Piazza died after binge drinking and falling down stairs at the Penn State University fraternity house where he was pledging. Four fraternity brothers were sentenced to jail time for failing to seek help. High-profile incidents like these have pushed many colleges and universities to take steps to change the drinking culture at fraternities and sororities. WVU, for example, has barred fraternities from recruiting freshman (WVU Today 2018). Several universities have banned hard liquor from fraternity houses (Tsubasa Field 2018). What do you think colleges should do to address problematic drinking cultures within fraternities?

Tobacco and Nicotine

Native Americans first cultivated tobacco and introduced it to the European settlers in the 1500s. The Europeans believed that tobacco had medicinal properties, and its use spread throughout Europe, ensuring the economic success of the colonies in the New World. Tobacco was initially used primarily through chewing and snuffing, but smoking became more popular in time, even though scientific evidence that linked tobacco smoking to lung cancer existed as early as 1859 (Feagin and Feagin 1994). However, the U.S. surgeon general did not conclude that tobacco products are addictive and that nicotine causes dependency until 1989.

Tobacco has historically been one of the most widely used drugs in the United States, but its use has been steadily declining in recent decades, in large part due to widespread anti-smoking campaigns. In 2002, about one in four Americans age 12 and older had smoked cigarettes in the past 30 days; in 2018, fewer than one in six Americans were current smokers (NSDUH 2019a). However, there is wide variation in smoking habits by demographic group. According to the Centers for Disease Control (2020b), men are more likely to smoke than women, and Indigenous Americans and Black Americans are more likely to smoke than White and Asian Americans. Adults with a GED were approximately 10 times more likely to be smokers than those with a graduate degree. Smoking rates were also highest among people who identify as LGBT and among those with significant psychological distress.

AP Images/Seth Wenig

Public health experts have warned that flavored e-cigarettes increase the likelihood that teens will become addicted to nicotine and increase the risk of switching to conventional cigarettes. In response, the FDA placed restrictions on the sale of most flavored e-cigarettes in 2020. Opponents of the flavor ban protested, however, arguing that e-cigarettes are an effective harm reduction strategy that encourage long-time smokers to reduce or eventually quit smoking.

While cigarette use has declined, especially among young people, other forms of nicotine use have become more popular. The biggest surge in youth nicotine use in recent years has occurred through **e-cigarettes**, battery-operated devices that produce a vapor containing nicotine that can be inhaled. In 2019, more than one-third of 8th, 10th, and 12th grades reported using e-cigarettes in the past month, while fewer than 4 percent reported smoking conventional cigarettes (MTF 2020). Although the safety of e-cigarettes remains unknown, only 22 percent of 8th-grade students believe they are harmful.

Research evidence suggests that youth develop attitudes and beliefs about tobacco products at an early age (Freeman et al. 2005). Advertising of tobacco products continues to have an influence on youth despite the 2009 Family Smoking Prevention and Tobacco Control Act, which, among other provisions, outlawed flavored cigarettes most often marketed to children (see "Government Regulation" in the "Strategies for Action" section). Tobacco company executives, however, have argued that the 2009 act only covers "cigarettes, cigarette tobacco, roll-your-own tobacco, and smokeless tobacco" and that cigars are excluded from the control of the Food and Drug Administration (Myers 2011, p. 1). Several tobacco companies are now selling small, inexpensive, sweet-flavored cigars, and, between 2000 and 2018, cigar use more than doubled (Campaign for Tobacco Free Kids 2020). There are also concerns about e-cigarette advertising and its impact on young people. Research shows that young adults exposed to e-cigarette advertising are significantly more likely to start vaping (Loukas et al. 2019) and that teens and young adults who start using nicotine through e-cigarettes are at increased risk of later switching to conventional cigarettes (Soneji et al. 2017). Critics also argue that the flavored products offered by e-cigarette manufacturers are designed to promote nicotine addiction among teens and young adults.

> "We have been asked by our client to come up with a package design ... that is attractive to kids. While this cigarette is geared toward the youth market, no attempt (obvious) can be made to encourage persons under twenty-one to smoke. The package design should be geared to attract the youthful eye ... not the ever watchful eye of the Federal Government."
>
> **–1970 LETTER FROM LORILLARD (MANUFACTURER OF NEWPORT CIGARETTES) ADVERTISING EXECUTIVE TO A MARKETING PROFESSOR, SOLICITING HELP FOR ADVERTISING DESIGN**

What *do you* think?

The explosion of e-cigarette popularity among teens prompted a push for increased government regulation. In particular, critics argued that e-cigarette products with flavors like mint, mango, peach, and sour apple are designed to appeal to kids. In 2019, the Trump administration issued several new policy orders that increased the legal age to purchase e-cigarettes from 18 to 21 and banned the sale of e-cigarettes with fruit, mint, or dessert-style flavors. However, the policy only applied to refillable e-cigarette cartridges, allowing for disposable e-cigarette brands to continue selling flavored products (Kaplan 2020). Do you think the federal government should do more to restrict e-cigarette use among teens?

The tobacco industry has a long history of targeting women and racial minorities (Schmidt 2014; Henrisken et al. 2011). Primack et al. (2007) found that tobacco advertisements in Black communities were 2.6 times higher per person than in White communities. In addition, the likelihood of tobacco-related billboards was 70 percent higher in Black than in White communities. In an analysis of more than 1,500 cigarette advertisements from historical and contemporary magazines and newspapers, researchers at Stanford University concluded that marketing cigarettes to women and girls has always been tied to body image and the evolving role of women. For example, during the second wave of feminism in the 1960s, Philip Morris's "You've Come a Long Way, Baby" campaign was developed for Virginia Slims. One researcher noted that, even today, "women-targeted cigarette brands are almost universally promoted as slender, thin, slim, lean, or light. Some brands have even gone so far as to recommend 'cigarette diets'" (quoted in Marine-Street 2012, p. 1). Women in developing countries are also being targeted.

@FLOTUS

I am deeply concerned about the growing epidemic of e-cigarette use in our children. We need to do all we can to protect the public from tobacco-related disease and death, and prevent e-cigarettes from becoming an on-ramp to nicotine addiction for a generation of youth. @HHSGov

-Melania Trump

Marijuana

Marijuana use has had a long and legally complicated history in the United States. Globally, its use dates back to 2737 BCE in China, and marijuana has a long tradition of use in India, the Middle East, and Europe. In North America, hemp, as it was then called, was used to make rope and as a treatment for various ailments. Marijuana's active ingredient is THC (Δ^9-tetrahydrocannabinol), which in varying amounts can act as a sedative or a hallucinogen. When just the top of the marijuana plant is sold, it is called hashish. Hashish is much more potent than marijuana, which comes from the entire plant.

Concerns about the potential misuse of marijuana in the United States grew in the early part of the 20th century, in large part driven by racial tensions between White and Hispanic laborers during the Great Depression. In 1937, Congress passed the Marijuana Tax Act that restricted the use of marijuana; the law was passed as a result of a racially charged media campaign that portrayed marijuana users as "dope fiends." In the only congressional hearing on the act, the then commissioner of narcotics testified:

> There are 100,000 total marijuana users in the U.S. and most are Negroes, Hispanics, Filipinos, and entertainers. Their Satanic music, jazz and swing, result from marijuana use. This marijuana causes white women to seek sexual relations with Negroes, entertainers, and any others. The primary reason to outlaw marijuana is its effect on degenerative races. Marijuana is an addictive drug which produces in its users insanity, criminality, and death. You smoke a joint and you're likely to kill your brother. Marijuana is the most violence-causing drug in the history of mankind. (Quoted in Rabinowitz and Lurigio 2009)

Despite widespread global legal restrictions on its production and distribution, marijuana continues to be the most widely used illicit drug in the world. Globally, it is estimated that 3.9 percent of the world's population age 15–64 are marijuana users, and the overall number of annual users has increased by roughly 30 percent in the past two decades (WDR 2020). An estimated 43.5 million current marijuana users live in the United States, representing 15.9 percent of the U.S. population aged 12 and older (NSDUH 2019).

In the past decade, states have been gradually lifting marijuana restrictions at the local level, with wide variations in policies ranging from partial decriminalization, restricted medical use, and full legalization (see Figure 3.3 and this chapter's section on "Strategies for Action"). In the debate over the legalization of marijuana, as with cigarettes and alcohol, many express fears that it is a **gateway drug**, the use of which causes progression to other drugs. Indeed, marijuana, alcohol, and nicotine are the most commonly used drugs among adolescents and teenagers (see Figure 3.4). Most research, however, suggests that people who experiment with one drug are more likely to experiment with another. Indeed, most drug users use several drugs concurrently, most commonly cigarettes, alcohol, marijuana, and cocaine (Lee and Abdel-Ghany 2004). Other arguments against easing marijuana restrictions are focused on the potential harms of marijuana itself. As demand for the drug has increased, producers have focused on producing strains with more concentrated levels of THC (WDR 2020). Most medical research has concluded that marijuana use is associated with increased risk of psychosis, severe mental health problems, diminished cognitive abilities, drug dependency, and testicular cancer, among other health problems (Memedovich et al. 2018).

Still, there is widespread support for some form of marijuana decriminalization. Supporters argue that, although there may be some increased health risks for heavy users of the drug, these risks are not more serious than those associated with legal drugs, such as

e-cigarettes Battery-operated devices that produce a vapor that contains nicotine, which can then be inhaled.

gateway drug A drug (e.g., marijuana) that is believed to lead to the use of other drugs (e.g., cocaine).

Attitudes about Marijuana Legalization

Americans' views about marijuana (cannabis) have evolved rapidly over the past decade. Compare your views to those from a representative sample of 1,000 U.S. adults, age 21 and older, surveyed in April 2019 (PSB Research 2019).

	Yes	No	Not Sure
1. Do you consider cannabis a drug?	____	____	____
2. Do you support legalizing cannabis for:			
• Medical purposes	____		
• Recreational purposes	____		
• Both medical and recreational purposes	____		
• I do not support legalizing cannabis for any purposes.	____		
• Other	____		
3. Do you support legalizing "harder" drugs, such as cocaine, heroin, and methamphetamine?	____	____	____

	More Likely	About as Likely	Less Likely
4. Would you be more likely, less likely, or about as likely to vote for a presidential candidate who supports legalizing cannabis?	____	____	____

	Would Support	Would Oppose
5. There are currently several proposals being considered for ways recreational cannabis can be legalized. Please indicate how much you support or oppose each of the following potential measures:		
• Concept of a "taxation and regulation": A law that regulates and taxes cannabis like alcohol and allows adults to use it.	____	____
• Marijuana Justice Act: A law that legalizes cannabis on the national level and also punishes states that keep it illegal. The law would also get rid of criminal records for past cannabis convictions.	____	____
• Concept of "decriminalization": A law that removes penalties for cannabis use and possession but maintains that growing and selling cannabis are crimes.	____	____
• STATES Act: A law that keeps cannabis illegal at the national level but bans the federal government arresting those who comply with state legalization laws.	____	____
• Concept of "unregulated free-for-all": A law legalizing cannabis consumption and cultivation but without any government oversight (such as taxes or regulation).	____	____

Results: Percent (%) of American adults answering "yes":

	Yes	No	Not Sure
1. Do you consider cannabis a drug?	59	31	19
2. Do you support legalizing cannabis for:			
• Medical purposes	35		
• Recreational purposes	4		
• Both medical and recreational purposes	45		
• I do not support legalizing cannabis for any purposes.	16		
• Other	1		
3. Do you support legalizing "harder" drugs, such as cocaine, heroin, and methamphetamine?	7	87	6

	More Likely	About as Likely	Less Likely
4. Would you be more likely, less likely, or about as likely to vote for a presidential candidate who supports legalizing cannabis?	41	36	23

	Would Support	Would Oppose
5. There are currently several proposals being considered for ways recreational cannabis can be legalized. Please indicate how much you support or oppose each of the following potential measures:		
• Concept of a "taxation and regulation": A law that regulates and taxes cannabis like alcohol and allows adults to use it.	60	40
• Marijuana Justice Act: A law that legalizes cannabis on the national level and also punishes states that keep it illegal. The law would also get rid of criminal records for past cannabis convictions.	45	55
• Concept of "decriminalization": A law that removes penalties for cannabis use and possession but maintains that growing and selling cannabis are crimes.	37	63
• STATES Act: A law that keeps cannabis illegal at the national level but bans the federal government arresting those who comply with state legalization laws.	37	63
• Concept of "unregulated free-for-all": A law legalizing cannabis consumption and cultivation but without any government oversight (such as taxes or regulation).	29	71

alcohol and tobacco. Additionally, there are documented therapeutic benefits to marijuana use, including for conditions such as chronic pain, anxiety, multiple sclerosis, and the nausea associated with cancer treatments (National Academies of Sciences, Engineering, and Medicine [NASEM] 2017). Because federal legal restrictions limit government-funded research on its effects, little is known about the long-term effects of marijuana use, particularly in regard to the brain development of children and adolescents. Medical experts warn policy makers to exercise caution as they introduce deregulation policies in their communities. Finally, proponents of legalization argue that the criminalization

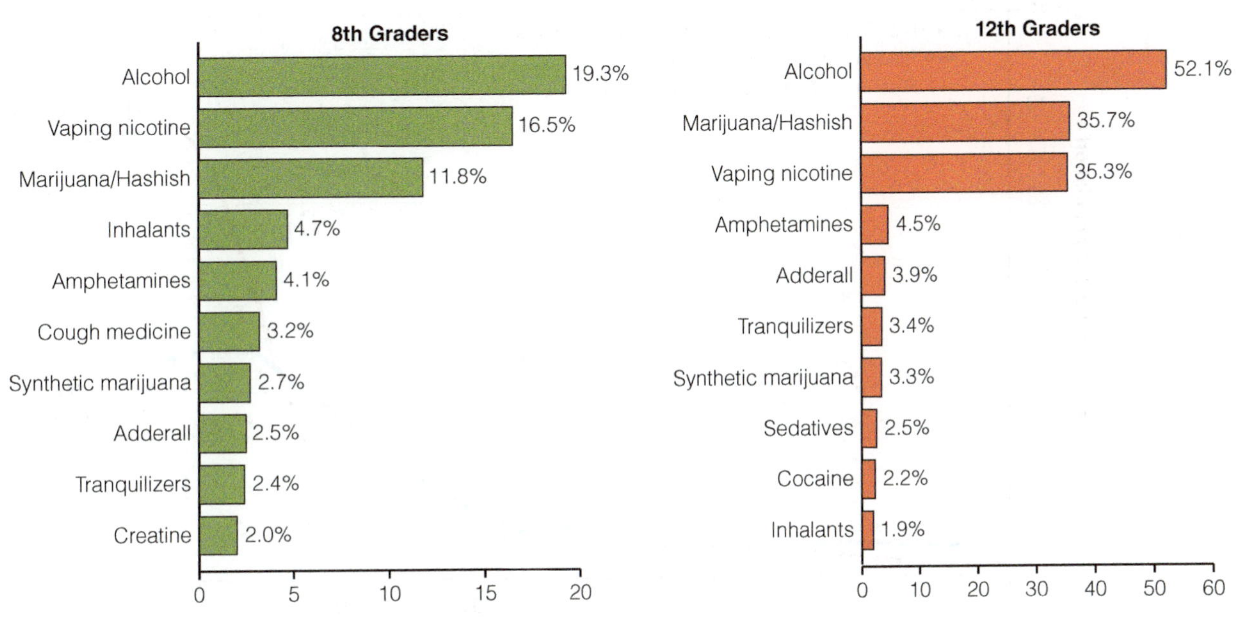

Figure 3.4 Top Drugs among 8th and 12th Graders, Past Year Use, 2019
SOURCE: Monitoring the Future, 2019.

of marijuana has had devastating effects on the lives of many recreational users, particularly those in heavily policed Black and Hispanic neighborhoods (Todd 2018; also see this chapter's section on the "War on Drugs").

Prescription Drugs and the Opioid Crisis

The most rapidly growing public health problem connected to substance use over the past decade has been the misuse of pharmaceutical, or prescription, drugs. When used for nonmedical purposes, pharmaceuticals are referred to as **psychotherapeutic drugs**. Although marijuana and alcohol are much more commonly used than psychotherapeutic drugs in the United States, nearly two-thirds of all drug overdose deaths in the United States in 2017 were associated with the misuse of prescription painkillers (CDC 2020d). The Centers for Disease Control describes the current prescription opioid crisis as occurring in three waves (see Figure 3.5). The pattern of these waves highlights the complex relationship between legal and illegal drug use.

Originally developed to treat pain in terminal cancer patients, opioid-derived pharmaceuticals gained traction in the late 1980s as early research indicated high effectiveness and low addiction potential (Quinones 2019). Institutional pressures and societal conditions exacerbated the shift toward increased opioid prescribing:

> By the early 2000s, doctors were being urged to prescribe the drugs after almost any routine surgery. ... They also prescribed them for chronic conditions such as arthritis and back pain. Chronic pain had once been treated with a combination of strategies that only sometimes involved narcotics; now ... insurance companies cut back on reimbursing patients for long-term pain therapies that did not call on the drugs. The U.S. drug industry, meanwhile, was investing heavily in marketing, hiring legions of young salespeople to convince doctors of their drugs' various miracles. (Quinones 2019, p. 1)

psychotherapeutic drug The nonmedical use of any prescription pain reliever, stimulant, sedative, or tranquilizer.

Meanwhile, ongoing research demonstrated that the early findings were misleading because these had been based on a small group of terminal cancer patients. It soon became apparent that prescription pain killers were highly addictive and increased risks of medical complications and death, especially when combined with other drugs such as alcohol (see this chapter's opening feature). The high degree of addiction potential prompted many who started using the drugs for pain management to shift to illegal opioids, such as

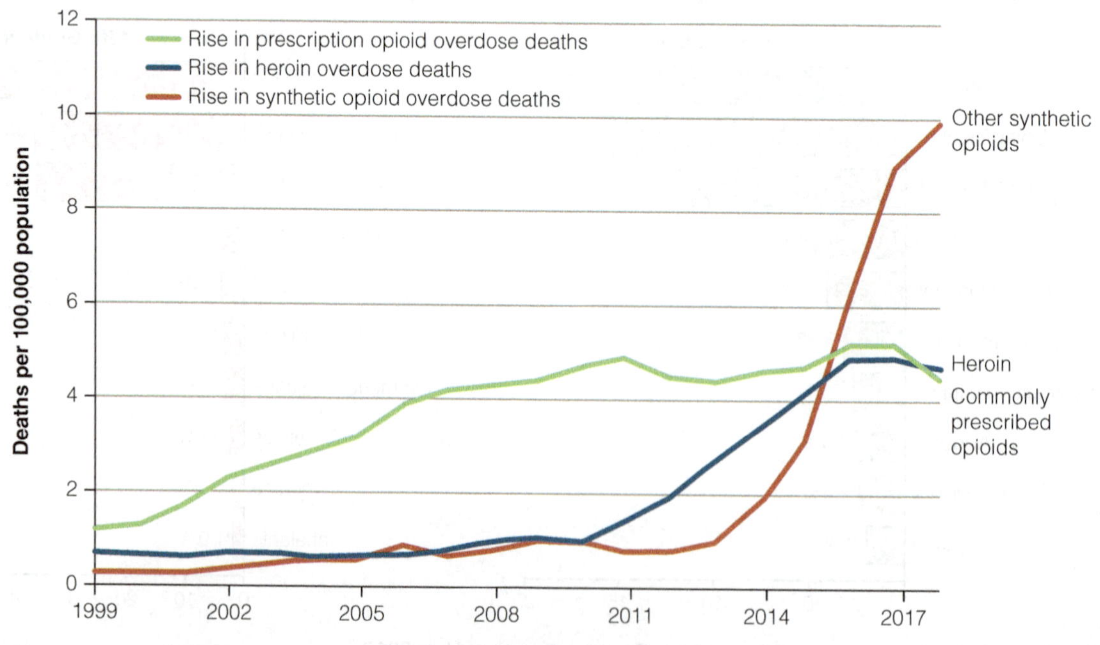

Figure 3.5 Three Waves of the Rise in Opioid Overdose Deaths
SOURCE: National Vital Statistics System (NVSS) Mortality file.

heroin (see this chapter's *Human Side*). Soon, regulatory pushback made it more difficult to access legal prescription painkillers, and opioid production globally shifted toward fentanyl, a powerful opioid synthetic. The World Drug Report (2019, p. 24) referred to this pattern as a "global paradox of too much and not enough" in which the widespread availability and misuse of pharmaceutical opioids has contributed to a dire public health emergency, while at the same time, there are severe limitations in access and availability for those who need these drugs for essential medical treatment.

As OxyContin and other painkillers become more difficult to obtain, users often turn to the less expensive and more available heroin—and not by accident. Rather, it is the plan of drug lords from Mexico and Colombia, who strategically market the drug to Middle America with new, sophisticated techniques. Packets of heroin are now stamped with popular brand names like Chevrolet or Prada or marketed using blockbuster movie names aimed at young people, like the *Twilight* series.

> "Pharmaceutical opioids are essential medicines for the management of pain and other conditions … . However, irrational prescription practices, unjustified promotion and uncontrolled availability of prescription drugs result in negative consequences and their non-medical use has become detrimental to public and individual health in many subregions worldwide."
>
> **–WORLD DRUG REPORT 2019, P. 24**

Meth: The Resurging Epidemic

Methamphetamine is a central nervous system stimulant that is highly addictive. In 2018, the proportion of persons 12 and older who used methamphetamine in the previous year was less than 1 percent (NSDUH 2019). Although its usage is less common relative to other drugs, methamphetamine use has devastating consequences on individuals and communities. The availability of meth is highest in the western part of the United States; more than 70 percent of local law enforcement officers in the pacific and west central regions of the United States reported that methamphetamine was the greatest drug threat in their area (NIH 2019). Although the drug has only become popular in the past three decades, it is not new:

> During the Second World War, soldiers on both sides used it to reduce fatigue and enhance performance. Hitler was widely believed to be a meth addict. Later, in the 1960s, President John Kennedy also used the drug and soon it caught on among so-called "speed freaks." But, because it was extremely expensive as well as difficult to obtain, meth was never close to being as widely used as cocaine. (Thio 2007, p. 276)

Methamphetamine production grew in the 1980s, although at that time its use was primarily limited to truck drivers, construction workers, and other blue-collar workers who primarily snorted the powdered form of the drug (Gonzalez et al. 2010). In the 1990s, a new crystallized form of methamphetamine, referred to as ice, began to be imported from Asia. This form of crystal meth could be heated and then smoked, which produced a more intense high. At this time, production of methamphetamine exploded in the United States, and it became more affordable and widely available.

Because methamphetamine could be made from cold medications such as Sudafed, the U.S. Congress passed the Comprehensive Methamphetamine Control Act of 1996 that made obtaining the chemicals needed to make methamphetamine more difficult (Office of National Drug Control Policy [ONDCP] 2006; Thio 2007). In 2006, the Combat Methamphetamine Epidemic Act, which further articulated standards for selling over-the-counter medications used in methamphetamine production, went into effect. Since that time, several states have passed regulations that require a prescription to obtain ephedrine and pseudoephedrine, the key ingredients in methamphetamine, and

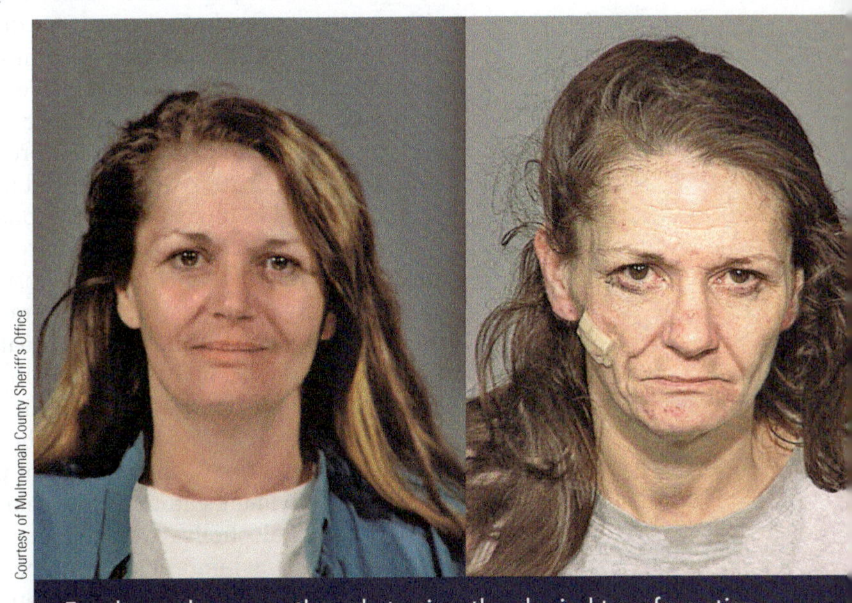
Courtesy of Multnomah County Sheriff's Office

For those who use methamphetamine, the physical transformation is remarkable. The time lapse between the before (left) and after (right) pictures of this methamphetamine user is only three years, five months.

with some success. Past-year usage rates and admissions to treatment program for methamphetamine abuse showed a steady decline in the past decade after a peak in 2005 (Gonzalez et al. 2010; NSDUH 2019).

There is some evidence, however, that meth use is resurging to its previous epidemic levels. One longitudinal study examined amphetamine-related hospitalizations across the United States from 2003 to 2015. Hospitalizations declined from 2003 to 2008 but then, between 2008 and 2015, showed a 245 percent increase—far surpassing the 45 percent increase in opioid-related hospitalizations during the same period (Winkelman et al. 2018). Increasingly, those seeking help for substance abuse disorders are reporting **polydrug abuse** disorders—a substance abuse pattern that occurs when a user becomes dependent on two or more drugs simultaneously, typically having one dominant drug addiction and then a secondary addiction to a drug that helps to counteract the negative effects of the primary drug. Some researchers believe that the resurgence in methamphetamine use is part of a polydrug abuse pattern connected to the opioid epidemic (Dembosky 2019). As new regulations decreased the availability of opioid prescriptions, users increasingly turned to street drugs such as heroin and methamphetamine to stave off withdrawal symptoms. The combination of the two types of drugs creates a "synergistic high, and [meth] balanced out the effects of opioids so one could function 'normally'" (quoted in Dembosky 2019, p. 1).

Societal Consequences of Drug Use and Abuse

Drugs are a social problem not only because of their adverse effects on individuals but also because of the negative consequences their use has for society as a whole. Everyone is a victim of drug abuse. Drugs contribute to problems within the family and to escalating crime rates, are tremendously costly, and place a heavy strain on the environment. Drug abuse also has serious consequences for health at both the individual and the societal level.

The Cost to Children and Family

The cost of drug abuse to families is incalculable. In 2018, it was estimated that more than 20 million people were classified as substance abusers in the United States—7.4 percent of the 12 and older population—resulting in one in ten children under the age of 18 living with a parent in need of treatment for drug or alcohol dependency (NSDUH 2019; Lipari and Van Horn 2017). Children raised in such homes are more likely to (1) live in an environment riddled with conflict, (2) have a higher probability of physical illness including injuries or death from an automobile accident, and (3) be victims of child abuse and neglect (Lipari and Van Horn 2017). Approximately one out of every three children who entered foster care in 2017 were removed from their families due to parental drug or alcohol abuse; this rate has increased for six consecutive years (Sepulveda and Williams 2019).

For every one child placed in foster care, 19 children are being raised by grandparents or other relatives outside of the foster care system (Generations United 2018). These households are often referred to as "grandfamilies," and their increase over the past decade is attributed to the high rates of death, incarceration, and parental neglect associated with the opioid crisis. While research indicates that children who are removed from their parents care do better when placed in grandfamilies (as opposed to foster care), these families often have limited access to the resources available to traditional foster families, low financial resources, and significant physical and mental health challenges due to the age and employment status of older relatives.

polydrug abuse A substance abuse pattern that occurs when a user becomes dependent on two or more drugs simultaneously, typically having one dominant drug addiction and then a secondary addiction to a drug that helps to counteract the negative effects of the primary drug.

Children of alcoholics—the clear majority of children who live in homes with a drug-dependent parent(s)—are at risk for anxiety disorders, depression, and problems with cognitive and verbal skills (Lipari and Van Horn 2017). Children of alcoholics, particularly female children of alcoholics, suffer from "significant mental health consequences . . . that persist far into adulthood" (Balsa et al. 2009, p. 55). Additional harms as a result of excessive drinking in the home were identified by Berends et al. (2014) in a telephone survey of 2,649 randomly selected adults in Australia. Results indicate that about one-third of those reporting a problem drinker in the family were significantly impacted by physical (e.g., sexual coercion), social (e.g., negatively affected a social occasion), psychological (e.g., feeling threatened), and/or practical (e.g., being put at risk in a car) harms. Females were more likely to report being impacted than males, physical injury was more likely to be reported between close rather than extended family members, and "significant harm" was twice as likely to be reported by those living together in the 12 months prior to the survey than those living apart (Berends et al. 2014).

> **What do you think?**
>
> In Alabama, a woman can be charged with "chemical endangerment" if she uses a controlled substance while pregnant (Alabama Code 2015). Do you think all states should have such a law? Should a pregnant woman be arrested for smoking? For drinking a glass of red wine? For refusing to take prenatal vitamins?

Crime and Drugs

Those who are incarcerated appear to have a higher rate of drug use and dependence than the nonoffender population. At least 65 percent of the U.S. prison population is estimated to have an active substance use disorder, while another 20 percent did not meet official SUD criteria but were under the influence of drugs or alcohol at the time of their crime (NIDA 2020c).

The relationship between crime and drug use, however, is complex. Sociologists disagree as to whether drugs actually "cause" crime or whether, instead, criminal activity leads to drug involvement. Alternatively, criminal involvement and drug use can occur at the same time; that is, someone can take drugs and commit crimes out of the desire to engage in risk-taking behaviors. Furthermore, because both crime and drug use are associated with low socioeconomic status, poverty may actually be the more powerful explanatory variable.

In addition to the hypothesized crime–drug use link, some criminal offenses are defined by the use of drugs: possession, cultivation, production, and sale of controlled substances; public intoxication; drunk and disorderly conduct; and driving while intoxicated. Driving while intoxicated is one of the most common drug-related crimes. In 2018, approximately 20.5 million people aged 16 or older drove under the influence of alcohol, 11.8 million drove under the influence of marijuana, and 2.4 million drove under the influence of other illicit drugs at least once in the previous year (NSDUH 2019). In 2018, at least 10,511 Americans died in alcohol-related crashes—one-third of all traffic-related deaths (National Highway Traffic Safety Administration [NHTSA] 2019).

The rapidly changing policies surrounding marijuana use have made it difficult for law enforcement agencies to address the problems associated with driving under the influence of marijuana. Unlike alcohol impairment, which can be rapidly tested on site using a blood alcohol content (BAC or breathalyzer) test, no technology is currently available to do a similar test for marijuana impairment (Azofeifa et al. 2019). Marijuana is documented as causing driving impairment through delayed response time and decreased cognitive functioning (NIDA 2018a). However, research on the effects of marijuana remains limited and inconclusive. The relationship between alcohol impairment and blood alcohol level has been well studied, allowing for policy makers to set legal standards to prosecute driving under the influence. However, similar findings connecting THC levels to impairment "have been unable to link blood tetrahydrocannabinol [THC] levels to driving impairment, and the effects of marijuana in drivers likely varies by dose, potency of the product consumed, means of consumption (e.g., smoking, eating, or vaping), length of use, and co-use of other substances, including alcohol" (Azofeifa et al. 2019).

The High Price of Alcohol and Other Drug Use

While the total costs of substance abuse is difficult to estimate, a 2009 comprehensive report by the National Center on Addiction and Substance Abuse at Columbia University

(CASA 2009) set the total annual cost of substance abuse and addiction in the United States at $467.7 billion. More importantly, the analysis revealed that more than 95 percent of those costs were spent on "shoveling up the wreckage," with less than 5 percent devoted to treatment, prevention, or regulation combined. The report's authors conclude that "[u]nder any circumstances, spending more than 95 percent of taxpayer dollars on the consequences of tobacco, alcohol, and other drug abuse and addiction and less than 2 percent to relieve individuals and taxpayers of this burden would be considered a reckless misallocation of public funds. In these economic times, such upside-down-cake public policy is unconscionable" (p. i).

It is estimated that alcohol abuse alone costs the United States $249 billion annually, or $2.05 per drink (CDC 2015). Much of that cost is passed to the public, with $179 billion associated with lost workplace productivity, $28 billion to treat people for health problems caused by excessive drinking, $25 billion connected with alcohol-related crime, and $13 billion in car crashes caused by alcohol impairment.

Nicotine use carries similar societal costs. In 2018, the global cigarette industry was worth more than $713 billion, and tobacco companies spent $8.4 billion on cigarette advertising and promotions in the United States alone (Tobacco Free Kids 2019; CDC 2020a). The CDC estimates that the annual costs of smoking-related illnesses in the United States is at least $300 billion, much of which is absorbed by the taxpayer. The high public costs associated with smoking prompted a wave of both individual and state-level lawsuits against major cigarette manufacturers. In 1998, the four largest tobacco companies settled the majority of these lawsuits with the attorneys general of 46 states for $246 billion, to be paid out over 25 years (see this chapter's discussion of "Legal Action" in the "Strategies for Action" section).

Similarly, the economic devastation wrought by the opioid epidemic over the past two decades, particularly in low socioeconomic rural communities, has prompted thousands of lawsuits against the large pharmaceutical companies who manufactured and distributed the drugs. In 2019, an Oklahoma judge ruled that pharmaceutical giant Johnson & Johnson had created a "public nuisance" in the state by promoting the use of a drug they knew to be harmful and ordered the company to pay the state $572 million in restitution (Hoffman 2019). In his ruling, the judge wrote that the company had "promulgated 'false, misleading, and dangerous marketing campaigns' that had 'caused exponentially increasing rates of addiction, overdose deaths' and babies born exposed to opioids" (quoted in Hoffman 2019, p. 1). Rather than go to trial, two other large pharmaceutical companies, including Purdue, which manufactures OxyContin, settled with Oklahoma privately for approximately $350 million.

The landmark Oklahoma ruling prompted a national push toward a comprehensive settlement. As of 2020, the attorneys general of 31 states and the District of Columbia are in negotiations with three major pharmaceutical companies to collectively settle thousands of lawsuits from individuals and municipalities for somewhere between $19 and $50 billion (Hoffman 2020). Nineteen other states—including Florida and West Virginia—have declined to participate in the negotiations. They argue that the proposed settlement deal will be insufficient to combat the financial devastation the opioid epidemic continues to cause in their communities.

fetal alcohol syndrome (FAS) A syndrome characterized by serious physical and mental handicaps as a result of maternal drinking during pregnancy.

What do you think?

The tiny town of Kermit, West Virginia—population 393—became one of the most prominent symbols of the opioid epidemic. Between 2007 and 2012, pharmaceutical companies distributed 12 million hydrocodone pills to one pharmacy in the former coal mining town, or approximately 30,000 pills for each resident (Lilly 2018). The Kermit pharmacy served as a "pill mill" hub, which distributed the highly addictive drugs throughout the Appalachian region, which experienced some of the highest rates of overdose deaths in the United States (see Figure 3.6). Do you think pharmaceutical companies should be held financially responsible for the devastation caused by the distribution of highly addicted drugs in these communities?

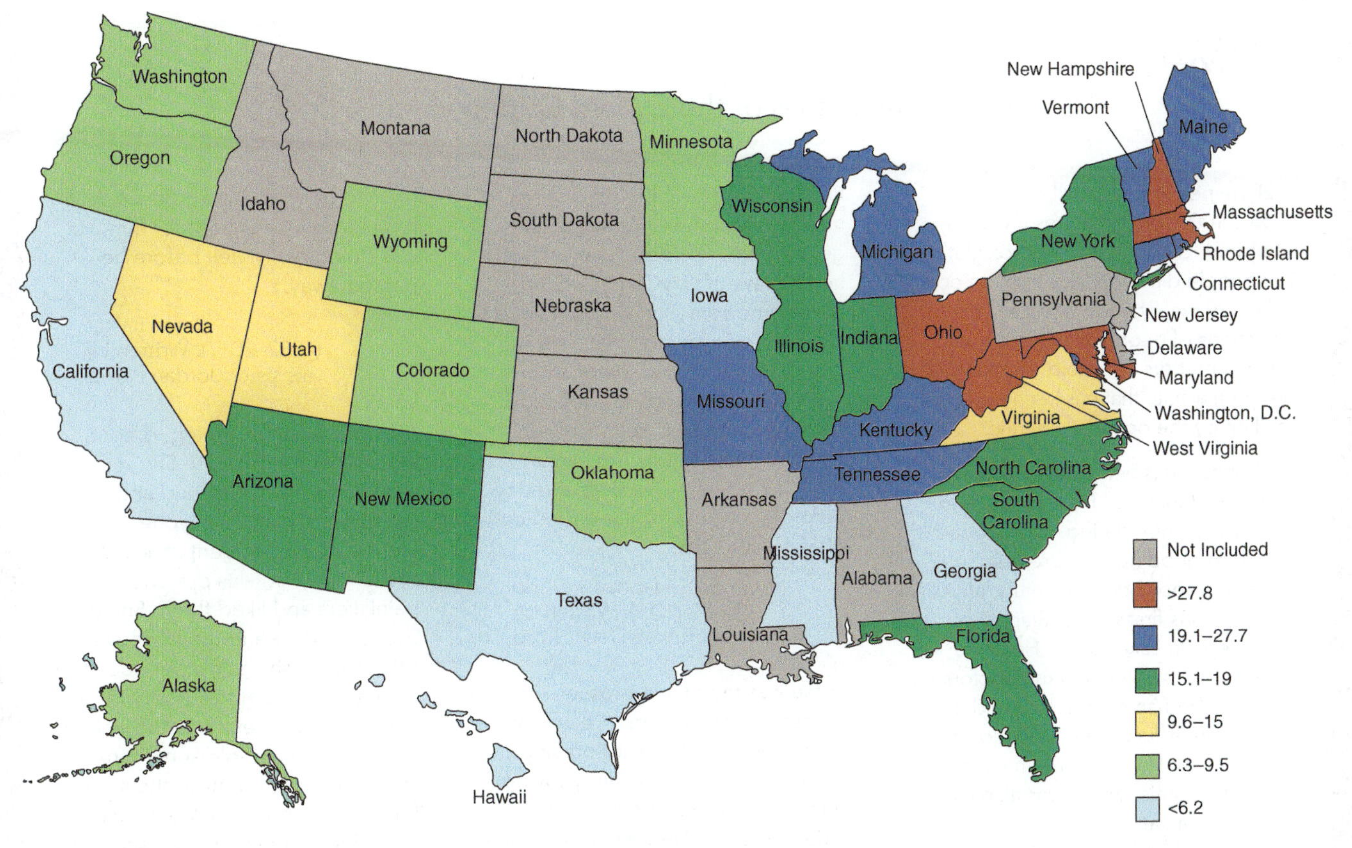

Figure 3.6 Opioid-Involved Overdose Death Rates (per 100,000 People), 2018–2019

Physical and Mental Health Costs

The physical and mental health costs of alcohol and other drugs, unlike the economic costs, are incalculable—death, disease, and injury.

Death and Disease. Annually, alcohol abuse is responsible for over 3 million deaths, more than 5 percent of all deaths worldwide and 13.5 percent of the total deaths of those aged 20 to 39 (WHO 2018). In the United States, alcohol abuse is responsible for at least 88,000 deaths per year, or about one in ten deaths among adults aged 20 to 64 (CDC 2019). Excessive alcohol use is associated with both short-term health effects such as an increased risk for violence, suicide, risky sexual behaviors, and accidents, as well as long-term chronic health risks such as cirrhosis of the liver, cancer, hypertension, dementia, and psychological disorders.

Maternal prenatal alcohol use is associated with fetal alcohol spectrum disorders (FASDs) in children, a broad term used for a variety of preventable birth defects and developmental disabilities. **Fetal alcohol syndrome (FAS)** is the most serious of the FASDs and is characterized by serious mental and physical handicaps

AP Images/Josh Reynolds

Mothers who lost children to opioid overdoses protest outside the Arthur M. Sackler Museum at Harvard University in April 2019. The Sackler family, which owns the OxyContin maker Purdue Pharma, is accused of attempting to hide more than $1 billion in assets after agreeing to a financial settlement with states in response to the opioid crisis (Hakim 2019). The Sackler family, who controlled the board of Purdue for decades, is estimated to be worth at least $13 billion. Protestors argue that this wealth was accrued due to the suffering of their loved ones as a result of opioid addiction.

Mourning the Losses of the Opioid Epidemic

In 2018, nearly 70% of all overdose deaths in the United States involved an opioid (CDC 2020d). Approximately 128 people die every day from an opioid overdose. The following are the reflections of people who have lost family members to opioid addiction in Vermont, a state that has been especially hard hit by the opioid epidemic.

> Brennan was beautiful, kind, smart and so funny ... . He was so warm and accommodating, such a good son. A good human. Brennan suffered a snowboarding injury, an ACL tear, in his first year at Sierra Nevada College in Lake Tahoe. He came home to recover, and the doctors prescribed an opiate, Dilaudid, for his pain. It turned into full-blown addiction. He eventually moved on to OxyContin, which was all over the streets at the time ... . Brennan struggled and suffered like no young man should. He was a victim of the opioid epidemic, which is taking our kids from us like something out of a science fiction film. They are just disappearing right before our eyes.
>
> —Brennan Joseph Dekeersgieter (1986–2013). Written by his mother, Margery Keasler

> From the get-go, Jenna was a helper and a lover of everyone ... . She was in her first semester of college when her boyfriend beat her up on Christmas Eve, and she went to the emergency room. The doctors prescribed her 30 days of OxyContin. Jenna turned into someone we didn't even know. I remember the first time we found out Jenna had tried heroin. I just started crying on the floor and knew we had to go to a recovery meeting immediately ... . Jenna died of a fentanyl overdose the day before she would have received her 60-day sober chip.
>
> —Jenna Rae Taro (1992–2019). Written by her mother, Dawn Tatro

> Jesse was my oldest brother—the oldest of nine siblings. He was an artist, a joker, a licensed nursing assistant ... . He was a very loving human ... . Jesse was in a car accident in 2005. He was in the intensive-care unit for two months and almost died several times. After the accident, he had chronic pain. One of his doctors prescribed opiates. When the opioid crisis became a big issue in Vermont, doctors stopped prescribing to him, and he started using off the street. Jesse became very angry and avoided our family a lot. He stole from us. He became very distant at times. He went to rehab several times in Vermont and Florida. He really tried to get past his addiction, but it was a very hard fight for him. He was doing really well before he passed away.
>
> —Jesse Palker (1982–2017). Written by his sister, Jordan Palker

> Alexa was very outgoing, gregarious and fun to be with. She was a good friend ... and kind and compassionate to everyone ... . She had a car accident at 17. She broke her wrist, was prescribed painkillers and liked them. She'd use for a while and then not use. I don't think she got super addicted until her mid-twenties ... . On the day she died, Alexa was too sick to watch [her 2-year-old son] Frankie, so Frankie and I went to the store for diapers and baby needs. We came back within an hour, and the house was quiet. I thought Alexa was still sleeping. I made Frankie a snack and changed him before I went in to find Alexa lying sideways on the bed. I screamed her name. She didn't move. So I shook her. She didn't move. I think that Frankie heard me scream.
>
> —Alexa Rose Cioffi (1985–2016) Written by her father, Frank Cioffi

SOURCE: Resmer 2019.

including facial deformities, growth problems, difficulty communicating, short attention spans, and hearing and vision problems, and speech and language delays (CDC 2020b). In an investigation of the prevalence of FASD among a representative sample of first graders, May et al. (2014) conclude that the best predictors of FAS are (1) late recognition of pregnancy by the mother, (2) amount of alcohol consumed by the mother three months prior to pregnancy, and (3) the quantity of alcohol consumed by the father.

Tobacco use is the leading preventable cause of disease and death in the world. Globally, more than 8 million people die from tobacco use each year, at least 1.2 million of these deaths are among nonsmokers with high exposure to secondhand smoke (WHO 2020a). The World Health Organization warns that tobacco use dramatically increases the risk of serious illness and death from COVID-19 (WHO 2020b).

Americans who smoke cigarettes are more than twice as likely to develop coronary heart disease and/or to have a stroke, and both male and female smokers are 25 times more likely to develop lung cancer than nonsmokers (CDC 2020e). It should be noted that the health impact of smoking goes beyond consumption and the effects of secondhand smoke. For example, children who work in tobacco fields are at risk for "green tobacco sickness," a disease caused by the absorption of nicotine through the skin from handling wet tobacco leaves (WHO 2020a). Additionally, smoking before and/or during pregnancy is linked to sudden infant death syndrome, pre-term delivery, congenital abnormalities, stillbirth, and spontaneous abortions (CDC 2020b). Much is still unknown about the health risks of e-cigarettes. Emerging research indicates that while e-cigarettes are likely less harmful overall than traditional smoking, they do increase the risk of nicotine addiction and future smoking among adolescents, and that the chemical additives in e-cigarette flavorings increase toxicity levels in the body (NASEM 2018). The long-term impact of that toxicity is still being studied.

AP Images/Dan Gleiter

Between 1999 and 2014, the number of women who had opioid use disorder while pregnant more than quadrupled (Haight et al. 2018). Because the severe nature of opioid chemical dependency and withdrawal symptoms is likely to harm both mother and fetus, doctors often do not recommend that women entirely stop using opioids during their pregnancy. Instead, as in the case of 23-year-old Logan and her newborn daughter pictured here, doctors managed her heroin addiction with the drug methadone during her pregnancy. After birth, doctors provide the baby with a morphine treatment to help ease her through her own withdrawal symptoms.

Prescription drug abuse also takes its toll. Drug overdoses have steadily increased over the last three decades and abuse of prescription drugs, specifically painkillers such as OxyContin, is in part responsible (CDC 2020d). Of overdose deaths in which a specific drug was identified, opioids were responsible for twice as many deaths in 2010 than in 1980. While pharmaceutical drug misuse is still responsible for the majority of drug overdose deaths, emerging data indicate a growing shift toward illegal drugs. From 2012 through 2018, the rate of overdose deaths from cocaine tripled, and the rate of deaths from psychostimulant drugs, such as methamphetamine, increased by nearly fivefold (Hedegaard et al. 2020).

Globally, the use of illicit drugs (excluding the misuse of prescription drugs) was associated with an estimated 167,000 deaths in 2018 (WDR 2020). Illicit drug use is also associated with a variety of diseases. For example, injecting drug users are at a higher risk for hepatitis B, hepatitis C, and HIV. Marijuana is often considered by the public to be the least dangerous illicit drug. However, its growing rate of use and abuse globally is highlighting important health concerns. Approximately 9 percent of marijuana users become dependent, but most research suggests it is not as addicting as most commonly used drugs (WDR 2020; Christensen and Wilson 2014). That is not to say that marijuana use is without risks. Because the tar in marijuana has a higher concentration of the chemicals linked to lung cancer, smoking marijuana is more dangerous than smoking an equivalent amount of tobacco. In addition, there has been an increase in the number of studies documenting the relationship between habitual marijuana use and structural changes of the brain (Filbey et al. 2014), psychosis in vulnerable populations (Pierre 2011), and reduced intellectual functioning (Nordqvist 2012).

Poverty and Substance Abuse Risks. The negative health consequences of drug use and abuse disproportionately affect the most vulnerable in society, namely, those who are poor, isolated, and suffer from mental illness. The United Nations World Drug Report (2020) argues that socioeconomic inequalities and substance abuse exist in a "vicious cycle" in which "poverty, limited education and marginalization, may increase the risk of developing drug use disorders and exacerbate their consequences" (WDR 2020, p. 9).

The factors associated with the initiation of drug use, such as unemployment, instability in housing and family life, and psychological distress, are also common outcomes of substance abuse disorder. Although drug and alcohol use is more common among men and those with higher incomes, the populations who are most adversely affected by substance abuse are those who are the most socially marginalized, namely, women, sexual minorities, indigenous and ethnic minority groups, immigrants and refugees, and rural populations. The report also warns that the global economic disruption caused by the spread of COVID-19 is likely to lead to increased drug and alcohol use among the world's most vulnerable populations, further exacerbating the health risks of both the global pandemic and substance use.

The Cost of Drug Use on the Environment

Although not something usually considered, the production of illegal drugs—as well as efforts to eradicate them—has a tremendous impact on the environment. The cultivation, production, and trafficking of illicit drugs lead to the clear-cutting of forests, nonsustainable water irrigation practices, and the irresponsible disposal of toxic byproducts, among other environmental harms. Researchers at the University of Oregon used the term *narco-deforestation* to describe the unprecedented rate of tropical forest loss in many Central and South American countries (Sesnie et al. 2017). Their research found a highly significant correlation between clusters of forest loss and the timing of increased cocaine trafficking.

Deforestation connected to illicit drug production creates long-term environmental harm to global ecosystems, as well as devastating surrounding communities. When unprecedented floods and mudslides began destroying farms in one village in Honduras, researchers went into the forest and found that drug traffickers had clear-cut trees for drug farming as well as to create cattle ranches where illegal profits could be laundered. Not only had the drug traffickers removed the natural shield of the forest that had absorbed rainfall and protected nearby communities, the local population also found themselves unable to sustain their livelihoods and unable to relocate to safer areas. According to one researcher: "Their land is taken by drug traffickers. They have no room to grow. All their activities, their houses, their roads and their use of the land are confined to a small space. If there is a flood event, they have no resilience" (Rodriguez 2018).

In the United States and Mexico, outdoor cannabis cultivation leads to contaminated water, clear-cutting of natural vegetation, the disposal of nonbiodegradable materials, and the diversion of natural waterways often polluted with toxic chemicals, which endanger fish and other wildlife (DEA 2014; U.S. Department of Justice 2019). Often, these environmental harms are an inadvertent outcome of national anti-drug efforts. This occurs as a result of **crime displacement**, in which producers of illicit drugs shift into remote areas and public lands in order to avoid detection.

Klassen and Anthony (2019) compared how different types of changes in marijuana laws affected the crime displacement of marijuana production in national forests in Oregon and Washington. They found that Oregon, which had implemented broad legalization of recreational marijuana use, production, and sales, saw a significantly greater decrease in illicit production on public lands when compared with the state of Washington, which had implemented a decriminalization of marijuana use but had more restrictive policies regarding marijuana production and sales. In this case, Oregon's more liberal legalization policy led to a more significant reduction in the environmental harm caused by marijuana production compared with its neighboring state because producers were less incentivized to take risks that shifted the potential harms of production to public lands.

crime displacement Pattern of environmental and social harm that occurs when producers of illicit drugs shift into remote areas and public lands in order to avoid detection from law enforcement.

Attempts to eradicate drug production and distribution can also take an environmental toll. Monsanto's Roundup, often used in aerial eradication of Colombia's coca fields, leads to "deforestation, contamination of water and water systems, eradication of non-coca crops and natural vegetation, and a generally negative impact on the biodiversity of the region" (Smith et al. 2014, p. 195). Government efforts to destroy production sites and seize illicit drug shipments often mean that the illicit drugs which remain on the market are more profitable. As a result, drug traffickers are incentivized to take greater production risks (McSweeney 2015). Typically, this involves carving out land, drug production sites, and trafficking routes through increasingly remote and biodiverse regions. In many areas, this leads to conflict and harm among indigenous groups who primarily occupy the most isolated areas of the country.

Strategies for Action: America Responds

Drug use is a complex social issue that is exacerbated by the structural and cultural forces of society that contribute to its existence. Although the structure of society perpetuates a system of inequality, creating in some the need to escape, the culture of society, through the media and normative contradictions, sends mixed messages about the acceptability of drug use. Thus, trying to end drug use by developing programs, laws, or initiatives may be unrealistic. Historically, Americans have held a punitive rather than rehabilitative attitude on how to deal with America's drug problem, particularly in reference to "hard drugs" such as heroin and cocaine. Some polling data indicate that attitudes are shifting, as a majority of Americans now believe that the government should focus more on providing treatment for drug users than prosecution (Pew Research Center 2014). Still, a slight majority of Americans indicate that the government does not do enough to "crack down" on drug users (see Figure 3.7). In this section, we review the numerous social policies and strategies that have been implemented or proposed to help control drug use and its negative consequences.

Alcohol, Tobacco, and Prescription Drugs

Although there may be some overlap (e.g., education), strategies to deal with alcohol, tobacco, and prescription drug abuse are often different from those initiated to deal with illegal drugs. Prohibition, the largest social policy attempt to control a drug in the United States, was a failure, and criminalizing tobacco is likely to be just as successful. By definition prescription drugs are legal—it is access to and misuse of that needs to be controlled. Research has identified several promising strategies in reducing alcohol, tobacco, and/or prescription drug misuse including economic incentives, government regulations, legal sanctions, and prevention initiatives.

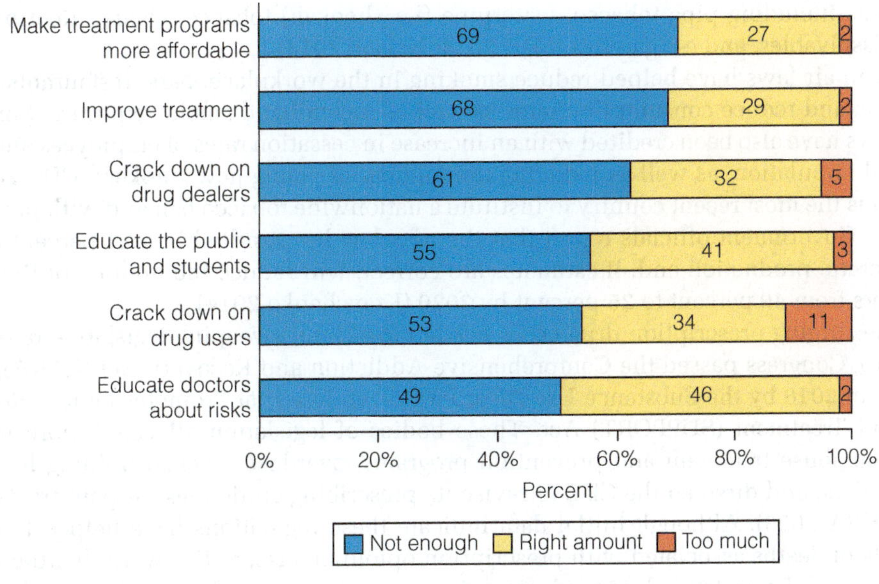

Figure 3.7 American Attitudes about Their Community's Response to the Problem of Substance Use, 2016
SOURCE: Associated Press NORC, 2016.

Economic Incentives. One method of reducing alcohol and tobacco use is to increase the cost of the product. The World Health Organization estimates that for each 10 percent increase on the price of a pack of cigarettes, demand will reduce by 4 percent in high income countries and 5 percent in low- and middle-income countries (WHO 2020a). Globally, taxes amount to 41 percent of the average price of a pack of cigarettes.

Other incentives include covering the cost of treatment and rewarding success. Behavioral economists have clearly established that even small financial incentives can effectively promote healthy behaviors and discourage unhealthy ones (Vlaev et al. 2019). However, the evidence of the long-term effectiveness of these strategies is unclear; people with substance use disorders may stop using in the short term, but start again when the financial incentives are removed. Other examples of economic incentives include reduced health insurance premiums for nonsmokers, decreased car insurance premiums for nondrinkers, advertising and promotion bans, and membership fee discounts at fitness centers.

Government Regulation. Federal, state, and local governmental regulations have each had some success in reducing tobacco and alcohol use and the problems associated with them. In 1984, Congress passed legislation that required states to raise the legal drinking age to 21 under threat of losing federal highway funds; and by 1987, all states had done so. Despite some controversy over the effectiveness of the law, a review of the research indicates that the 21 minimum drinking age requirement "has served the nation well by reducing alcohol-related traffic crashes and alcohol consumption among youths while also protecting drinkers from long-term negative outcomes they might experience in adulthood, including alcohol and other drug dependence, adverse birth outcomes, and suicide and homicide" (DeJong and Blanchette 2014, p. 113). The success of the alcohol age policy led the Food and Drug Administration (FDA) to revise its tobacco sales policy in 2019, increasing the legal age of purchase for any tobacco product—including e-cigarettes—to 21.

The 1998 tobacco settlement agreement also opened the door for increased FDA regulation of tobacco marketing, leading to the Family Smoking Prevention and Tobacco Control Act of 2009. The law gives authority to the FDA to regulate the manufacturing (e.g., tobacco companies must now disclose ingredients in their products), marketing (e.g., tobacco names or logos may no longer be used to sponsor sporting or entertainment events), and sale of tobacco products (e.g., terms like *light, mild,* and *low tar* may no longer be used). As a result of the FDA's regulations, the tobacco industry has shifted its attention to products not previously covered by the 2009 act (Sifferlin 2013). In reaction to such posturing, the FDA has proposed extending their authority to other tobacco products including pipe tobacco, waterpipe (i.e., hookah) tobacco, cigars, nicotine gel and dissolvables, and e-cigarettes (FDA 2014; Nelson 2014).

Clean air laws have helped reduce smoking in the workplace, bars, restaurants, and the like, and reduce consumption rates as well as secondhand smoke exposure. Smoke-free laws have also been credited with an increase in cessation rates of employees and the general population, as well as reducing the number of young new smokers (CDC 2014). Russia is the most recent country to institute a nationwide tobacco ban and with positive results. Government officials report that the new law has resulted in a 21 percent drop in cigarette production and, if estimates are correct, will reduce the number of Russian smokers from 40 percent to 25 percent by 2020 (Kravchenko 2014).

The surging prescription drug crisis has led to several sweeping legislative reforms. In 2016, Congress passed the Comprehensive Addiction and Recovery Act (CARA), followed in 2018 by the Substance Use-Disorder Prevention That Promotes Opioid Recovery and Treatment (SUPPORT) Act. These bodies of legislation allocated more funds for drug abuse treatment and prevention programs, overdose reversal training for first responders, and directed the CDC to revise its prescribing guidelines for pain treatment (SAMHSA 2020). Although initial data indicate these regulations have helped to curb the rate of deaths associated with prescription opioid misuse, critics warn that the policies have had unintended negative consequences. As doctors became more reluctant to prescribe opioid painkillers, those with chronic pain conditions became more likely to seek out more dangerous and illegal forms of pain management:

With no alternatives for pain control, long waiting lists for substance abuse treatment programs, and the physical and mental pressure of unremitting pain, many patients turned to illicit drugs, especially heroin. The result has been greater addiction, more deaths from overdoses, and an increase in cases of HIV/AIDS and hepatitis from contaminated syringes. (Rothstein 2017, p. 1253)

Thus, government regulation is often highly effective at modifying individual behavior, but may result in creating new and unanticipated problems.

Legal Action. Suits against retailers, distributors, and manufacturers of alcohol are more recent than and often modeled after tobacco litigation. These suits primarily concern accusations of unlawful marketing, sales to underage drinkers, and failure to adequately warn of the risks of alcohol. In 2014, suits were filed against, among others, Coors, Heineken, Bacardi, the makers of Zima, and the makers of Mike's Hard Lemonade, accusing the companies of using a "long-running, sophisticated and deceptive scheme... to market alcoholic beverages to children and other underage consumers" (Willing 2014, p. 1).

The landmark lawsuit settlements against tobacco in 1998 and pharmaceutical companies in 2020 have helped to bring much-needed funds to communities harmed by the health problems and deaths caused by addiction. However, many critics have argued that the manner in which these funds are distributed and used by state governments are ineffective at remedying the harms these settlements were intended to address. For example, the 1998 tobacco settlement agreement specified that funds should be used to cover smoking-related health care costs, tobacco cessation and smoking prevention programs, along with directing policy changes that limited cigarette advertising and sales to minors. However, 20 years after the settlement, less than 3 percent of settlement funds distributed to states were spent on tobacco-related programming (Berman 2019). Instead,

much of the money, which the industry continues to pay out, is plowed into state slush funds and used to patch budget shortfalls. In the most extreme cases, states sacrificed future payments for much smaller, but immediate infusions of cash to pay workers, or build schools and roads. (Demko 2018, p. 1)

Experts warn that policy makers should heed the lessons of the 1998 tobacco settlement when it comes to current and future lawsuit settlements from drug manufacturers, particularly those connected with the opioid epidemic.

Legal action can also be effective in shifting public opinion and the perception of harm caused by particular drugs. For example, a 2012 federal ruling required tobacco companies were ordered to publically admit their deception. Among other required statements to be placed in advertisements, some tobacco companies must state that "[a] Federal Court has ruled that the Defendant tobacco companies deliberately deceived the American public about designing cigarettes to enhance the delivery of nicotine, and has ordered those companies to make this statement. Here is the truth: Smoking kills, on average, 1,200 Americans. Every day" (Mears 2012).

Criminalization Strategies

The War on Drugs. Social, political, and economic conditions drove the pattern of increasing criminalization of drug use in the latter half of the twentieth century. While the 1960s are often associated with free love and widespread drug use, it was also a time of political and economic uncertainty. The rapid pace of change spurred many Americans to seek a perceived return to "law and order." In a June 1971 press conference, President Richard Nixon declared that "America's public enemy number one ... is drug abuse. In order to fight and defeat this enemy, it is necessary to wage a new, all-out offensive" (quoted in Friedersdorf 2011, p. 1). Two years later, Nixon established the Drug Enforcement Agency (DEA) and began a pattern of dramatically increasing funding for policing and tracking of drug-related criminal offenses. The pace of the war on drugs accelerated through the 1980s under the Reagan administration, driven by a social panic about the

> "The Nixon campaign in 1968, and the Nixon White House after that, had two enemies: the antiwar left and black people. ... We knew we couldn't make it illegal to be either against the war or blacks. But by getting the public to associate the hippies with marijuana and blacks with heroin, and then criminalizing both heavily, we could disrupt those communities."
>
> **–JOHN EHRLICHMANN, WHITE HOUSE COUNSEL AND CHIEF ADVISOR ON DOMESTIC AFFAIRS TO PRESIDENT NIXON (QUOTED IN BAUM 2016)**

drug epidemic. The proportion of Americans who saw drug abuse as the most pressing problem in America grew from under 6 percent in 1980 to 64 percent in 1989 (Drug Policy Alliance 2020).

Race, Gender, and Social Class Inequalities in the War on Drugs. As was the case with earlier drug criminalization efforts—such as with opium in the 1870s, alcohol in the 1920s, and marijuana in the 1930s—patterns of racial and socioeconomic hostility are evident in the policies of the war on drugs. The 1986 Anti-Drug Abuse Act established mandatory minimum sentencing for many nonviolent drug possession offences. Due to sentencing disparities between crack and powder cocaine, as well as the fact that Black and Hispanic communities are more heavily policed than White communities, incarceration rates among minority populations, especially Black men, skyrocketed through the 1980s and 1990s (see Figure 3.8). In 1980, approximately one-fifth of all people in federal prison were sentenced for drug offenses; by 2017, nearly half of all federal sentences were for drug offences (Sentencing Project 2018). During the same period, the total annual federal prison population increased by more than 700 percent.

The intersection between race, gender, and social class inequalities in the broader American society has meant that the war on drugs has disproportionately impacted some groups. For example, even though Black and White populations use drugs at similar rates, the rate of imprisonment on drug charges is nearly six times higher for African American people (National Association for the Advancement of Colored People [NAACP] 2020). Although men are more likely to be incarcerated than women in general, women are more likely than men to be convicted of drug offenses. In 2017, 25 percent of women in prison were convicted of a drug offense, compared with 14 percent of men (Bronson and Carson 2019). Finally, social class—particularly as measured by education and neighborhood poverty—is one of the strongest predictors of drug-related incarceration. High-poverty neighborhoods are more likely to be patrolled by police, and therefore drug users are more likely to be monitored and arrested. Once charged, those with a college education tend to receive shorter sentences—approximately 5 to 8 percent shorter—than those without a college education (O'Neill Hayes and Barnhorst 2020).

The number of women in prison has increased dramatically over the past two decades. Merolla (2008), using conflict theory, argues that the increase, to a large extent, is a function of the war on drugs rather than behavioral changes in women—that is, their increased use of drugs. Using data from government sources, Merolla concludes that the war on drugs, in and of itself, increased arrest rates of women in two ways. First, the war on drugs redefined through the media and in many other ways who a criminal is in nongendered terms (i.e., men *and* women). Second, as a consequence of this redefinition, law enforcement practices changed and more aggressively target female drug users. This targeting disproportionately impacts women of color. For example, even though Black and White women are equally likely to use drugs during pregnancy, Black women are approximately ten times more likely to be reported to child welfare services for drug use than White women (DPA 2018).

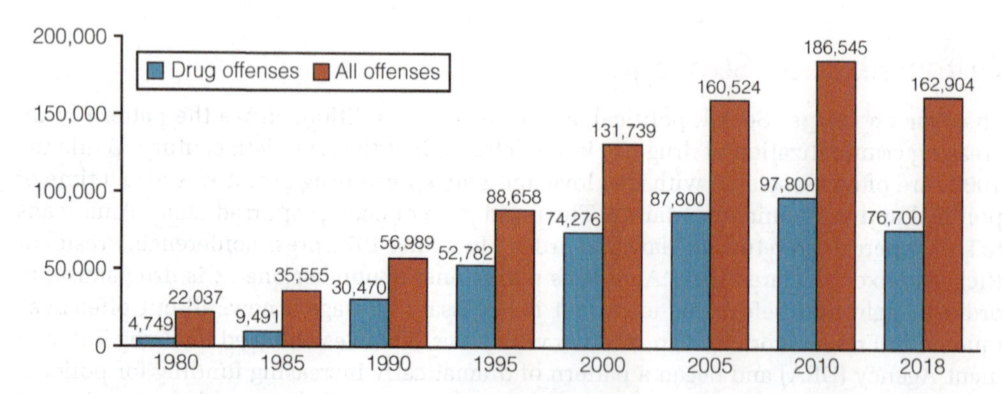

Figure 3.8 Number of Sentenced People in Federal Prisons for Drug Offenses, 1980–2018
SOURCES: Prisoners Series. Washington, DC: Bureau of Justice Statistics.

Feminist theorists have also pointed out that women's drug-related incarceration is often connected to unbalanced social power dynamics. For example, women and girls, especially in high poverty areas, are likely to be used by male partners and family members as drug mules and lower level distributors, and they often do not have access to the insider information that could be used as part of plea bargain to reduce their sentencing (Malinowska-Sempruch and Rychova 2016). Women's proximity to men who sell drugs may also make them vulnerable to drug charges. Conspiracy laws, which are designed to capture collaborators in organized crime networks, often function to ensnare women who may have had no role in drug trafficking but who lived with or rode in a car with an intimate partner who did. Criminal prosecutors rarely take into account the contextual factors associated with women's involvement in drug-related crimes, "which include pressure from a sexual partner, histories of domestic violence or other abuse, lack of mainstream livelihood opportunities, and lack of accessible treatment programs and related social support. Prison sentences are likely to exacerbate most of these factors" (p. 8).

The harsher penalties enacted as part of the "war on drugs" required prison sentences for almost all drug offenders—first-time or repeat—and limited judicial discretion in deciding what best served the public's interest. Taken together, the pattern of policies enacted through the war on drugs has meant that the risks of being incarcerated on drug-related charges are highest among people of color (especially men) who did not go to college and who live in high poverty communities. In response to public outcries and accusations of institutional discrimination (see Chapter 9), reform of such laws began (Peters 2009). As a result, the racial gap in incarceration has narrowed—although not disappeared—over the past decade (Gramlich 2020). Nonetheless, discriminatory practices continue. Alexander (2010) argues that the war on drugs simply replicated Jim Crow laws by disproportionately labeling African Americans as felons, thereby justifying discrimination in all its forms, as well as denying the right to vote and the right to serve on a jury.

Cost of the War on Drugs. Although the focus here is on the war in drugs in the United States, there remains a global initiative to control drug production, trafficking, and use. One collaborative project, *Count the Costs*, is dedicated to changing the punitive approach to drug control policy and, with it, the reduction of costs associated with such an emphasis (Count the Costs 2019). Count the Costs has identified seven categories of costs associated with the global war on drugs:

- Undermining development and creating conflict (e.g., use of military to fight drug cartels)
- Threatening public health, spreading disease and death (e.g., scarce resources are channeled toward law enforcement rather than treatment)
- Undermining human rights (e.g., execution of drug users in some countries)
- Promoting stigma and discrimination (e.g., some drug use is culturally embedded)
- Creating crime, enriching criminals (e.g., prohibiting illicit drugs may reduce supply, but with constant demand, prices increase)
- Deforestation and pollution (e.g., aerial spraying of drug crops)
- Wasting billions of dollars (e.g., spending on law enforcement)

The United States spends more than $47 billion a year on the war on drugs (DPA 2020).

Losing the War on Drugs: Drug Policy Reform. There is little debate "that the global prohibition of certain drugs and the war on drugs have largely failed to reach their stated goals" (Chouvy 2013, p. 216). Consequently, beginning in 2010, the federal government began to reform national drug policy. In general, the U.S. policy on fighting drugs is two-pronged. First is **demand reduction**, which entails reducing the demand for drugs through treatment and prevention. The second strategy is **supply reduction**. A much more punitive strategy, supply reduction relies on international efforts, interdiction, and domestic law enforcement to reduce the supply of illegal drugs.

The 2020 *National Drug Control Strategy* contains a blueprint for addressing illegal drugs in the United States through three simultaneous approaches: (1) to prevent drug

demand reduction One of two strategies in the U.S. war on drugs (the other is supply reduction); focuses on reducing the demand for drugs through treatment, prevention, and research.

supply reduction One of two strategies in the U.S. war on drugs (the other is demand reduction), supply reduction concentrates on reducing the supply of drugs available on the streets through international efforts, interdiction, and domestic law enforcement.

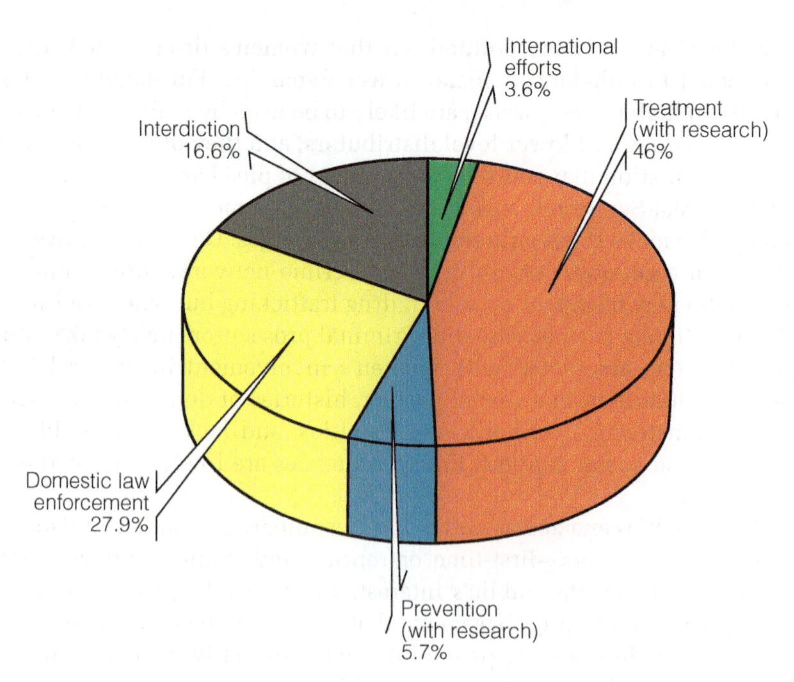

Figure 3.9 Federal Drug Control Spending by Function, FY 2021
SOURCE: White House 2020.

use initiation through education and evidence-based prevention programs; (2) to reduce barriers to treatment services; and (3) to reduce the availability of drugs in the United States through law enforcement and international cooperation (White House 2020). The allocation of funds for each of these strategies has changed over time as a more balanced and less punitive approach has been adopted (see Figure 3.9). For example, from 2019 to 2021 the budget allocation for drug interdiction (e.g. prohibition) through the Department of Defense decreased by 51 percent, while the allocation for treatment programs increased by nearly three percent (White House 2020).

The Push for Reform: Drug Courts. Concern over the punitive treatment of drug offenders and the failure of the criminal justice system to reduce recidivism rates led to the development of **drug courts**—nearly 3,000 of them in the United States alone (U.S. Department of Justice 2020). In general, after meeting eligibility requirements, defendants must complete a specialized program designed to divert nonviolent offenders with substance abuse problems from incarceration. Emphasizing collaboration between service providers (e.g., physicians, social workers) and court personnel (e.g., judges, prosecutors), drug courts emphasize counseling, treatment, education, family reunification, and self-sufficiency.

A cost–benefit analysis sponsored by the National Institute of Justice examined 23 drug courts and 1,787 drug offenders. Offenders were interviewed at the time they enrolled in a drug court program (treatment group) or when they would have enrolled in a drug court program (control group), and again 18 months later. Results of the study indicate that drug court participants, when compared to control group members, (1) had improved employment outcomes; (2) had lower rearrest and incarceration rates, thereby saving the criminal justice system money; and (3) were less likely to be involved in criminal activities. The authors conclude that "drug courts are doing exactly what they are supposed to do" (Downey and Roman 2014, p. 24).

drug courts Special courts that divert drug offenders to treatment programs in lieu of probation or incarceration.

Legal Alternatives: Deregulation and Legalization. Increasingly, the American public favors reducing or eliminating criminal penalties for nonviolent drug users, especially marijuana. Globally, there are a range of different approaches to drug legalization, with varying degrees of effectiveness. One model is decriminalization without legalization, an approach that was adopted by Portugal in 2000. While drugs such as heroin, marijuana, and cocaine remain illegal in Portugal, no one is incarcerated for drug use or possession. Instead, users are connected with social services to identify resource options such as housing, employment, or mental health care (Nadelmann 2019). Some states and municipalities in the United States have similarly moved to decriminalize marijuana. In 2014, residents of the District of Columbia voted to approve a marijuana decriminalization bill that allows residents over the age of 21 to possess, use, and grow small amounts of marijuana on private property (Metropolitan Police Department 2015). The sale of marijuana and driving under the influence of any drugs or alcohol remain illegal.

AP Images/Brennan Linsley

As states move to liberalize marijuana laws, a new problem has emerged: the public nuisance of secondhand marijuana smoke. Cynthia Hallett of the organization Americans for Nonsmokers' Rights argues that marijuana smoke produces the same harmful toxins as cigarette smoke, and more than half of Americans dislike the smell of marijuana (Hallett 2019). Frequent smoking has also led to increased conflicts between neighbors and to pushes for city and state governments to tighten restrictions on outdoor smoking (Rusch 2014).

Another model is legalization with regulation. In 2018, Canada approved a national policy of full legalization of medical and recreational marijuana. The production, sale, and distribution of the drug are tightly regulated by the government, much like the sale of alcohol (Bilefsky 2018). Colorado implemented a similar legalization policy in 2014 and, by 2019, had collected more than $1 billion in tax revenue from state-regulated marijuana sales (Paul 2019). State law requires that these tax revenues must be spent on health care, health education, substance abuse prevention and treatment programs, and law enforcement.

Proponents of drug legalization and decriminalization argue that these approaches reduce the social harms associated with drug abuse by shifting social resources away from prisons and toward public health and education measures, as well as by reducing the social stigma associated with drug use that often drives a cycle of addiction. For example, after 15 years of its experiment with decriminalization, Portugal experienced many benefits: drug use rates had not increased, incarceration rates had decreased, there were fewer deaths from drug overdoses, rates of HIV infections had declined, and rates of mental health and substance abuse treatment participation had increased (Drug Policy Alliance 2020). Similarly, a study of five U.S. states with marijuana decriminalization policies from 2007 to 2015 found a 75 percent reduction in drug-related arrests in those states, with no change in youth or adult marijuana use rates (Grucza et al. 2018).

Still, decriminalization does not solve all problems associated with drug use and may introduce new ones. In the United States, marijuana potency remains largely unregulated, and competition has resulted in lower marijuana prices in places where its recreational use is permitted (Hall et al. 2019). As a result of lower prices and greater availability, the proportion of users who use marijuana regularly has increased, even while the overall number of users has remained steady. Public health officials warn that the rapid pace of policy change has not allowed for sufficient research to understand the health effects of long-term, heavy use of high-potency marijuana, as well as related public health concerns such as driving under the influence and polydrug use interactions.

What do you think?

Marijuana legalization is often promoted as a strategy to combat racial inequalities in the criminal justice system. However, it has also introduced new forms of racial inequality in the burgeoning cannabis industry. Because marijuana remains illegal at the federal level, cannabis entrepreneurs often cannot secure business loans. Combined with the racial wealth gap (see Chapter 6), the result is a cannabis industry that is 80 percent White (Weiner 2019). Similarly, critics suggest there is a hypocrisy that allows White users "like the 'marijuana moms' to go viral for the very behavior that is stigmatized and even criminalized for their Black counterparts" (p. 1). What do you think should be done to address racial inequality in the burgeoning legal cannabis industry?

Medicalization Strategy: Addiction as Disease Management

Inpatient and Outpatient Treatment. Inpatient treatment refers to treatment of drug dependence in a hospital and, most importantly, includes medical supervision of detoxification. Most inpatient programs last between 30 and 90 days and target individuals whose withdrawal symptoms require close monitoring (e.g., alcoholics, cocaine addicts). Some drug-dependent patients, however, can be safely treated as outpatients. Outpatient treatment allows individuals to remain in their home and work environments and is often less expensive. In outpatient treatment, patients are under the care of a physician who evaluates their progress regularly, prescribes needed medication, and watches for signs of a relapse.

The extent to which treatment is successful depends on a number of variables. The National Institute on Drug Abuse (NIDA 2018b) has found that relapse rates for addiction are similar to those of other chronic diseases such as diabetes, hypertension, and asthma. Although drug addiction treatment is costly—approximately $600 billion per year—this is less expensive than alternatives, such as incarceration. For example, one year's worth of methadone treatment, a medication used to prevent opioid withdrawal symptoms, costs approximately $4,700 per patient. In contrast, one year of incarceration costs approximately $24,000 per person. NIDA provides a comprehensive review of evidence-based best practices for drug addiction treatment, including: (1) It is essential to match treatment strategies and services to an individual's particular addiction and social problems; (2) treatment needs to be readily available and accessible; (3) effective treatment needs to be tailored to psychological, social, medical, vocation, and legal issues as well as an individual's age, gender, ethnicity, and culture; (4) medications, when combined with behavioral therapy, enhance the effectiveness of treatment; (5) treatment does not need to be voluntary to be effective; and (6) addiction commonly occurs alongside mental and physical health problems and should be treated comprehensively.

Saloner and LeCook (2013) report that race is also a significant predictor of treatment completion. Using national data, they found that Black and Hispanic individuals are less likely to complete alcohol and drug treatment and that Native Americans are less likely to complete alcohol treatment. Only Asians were more likely to complete both treatments than White individuals. Because the failure of Black and Hispanic individuals to complete treatment was largely due to socioeconomic variables, the authors suggest that the Affordable Care Act may mediate differences in completion rates (see Chapter 2).

Twelve-Step Programs. Both Alcoholics Anonymous (AA) and Narcotics Anonymous (NA) are voluntary associations whose only membership requirement is the desire to stop drinking or taking drugs. AA and NA are self-help groups in that nonprofessionals operate them; they offer "sponsors" to each new member and proceed along a continuum of 12 steps to recovery. Members are immediately immersed in a fellowship of caring individuals with whom they meet daily or weekly to affirm their commitment. Some have argued that AA and NA members trade their addiction to drugs for feelings of interpersonal connectedness by bonding with other group members. AA boasts more than 60,000 groups with more than a million Americans regularly attending (Flanagin 2014).

Symbolic interactionists emphasize that AA and NA provide social contexts in which people develop new meanings. Others who offer positive labels, encouragement, and social support for sobriety surround abusers. Sponsors tell the new members that they can be successful in controlling alcohol and/or drugs "one day at a time" and provide regular interpersonal reinforcement for doing so. Some research indicates that mutual support programs work. For example, Kelly and Heppner (2013) found that AA has some impact on the number of days a respondent was abstinent and the number of drinks per drinking days. Others, however, are highly critical of 12-step programs in general and, in particular, Alcoholics Anonymous. In *The Sober Truth: Debunking the Bad Science behind Twelve-Step Programs and the Rehab Industry* (2014), Dodes and Dodes state that "peer-reviewed studies peg the success rate of AA somewhere between 5 and 10 percent. That is, about one of every 15 people who enter these programs is able to become and stay sober" (p. 1).

Community-Based Prevention and Public Health Strategies. Increasingly, public health experts promote evidence-based, health-focused strategies that seek to minimize the harm caused by drug abuse, rather than prioritizing ending drug use. **Harm reduction** strategies are focused on using evidence-based practices for reducing the number of people who die from drug overdoses and increasing the number of people who access treatment. Some strategies include Good Samaritan laws that provide legal immunity for calling 911 during a drug overdose, along with needle exchange or supervised consumption site programs (see this chapter's *Social Problems Research Up Close* feature).

In 2018, SAMHSA published an Opioid Overdose Prevention tool kit that recommended a national expansion of the use of Naloxone, an opioid overdose reversal drug. In particular, the tool kit suggested that Naloxone be made widely available among first responders, family members of people with substance use disorders, and business owners in communities with high opioid addiction rates. Despite being supported by the U.S. surgeon general, National Institutes of Health, and the Centers for Disease Control and Prevention, implementation of Naloxone training and distribution programs met with considerable resistance. Opponents argued that the availability of the drug only further encouraged drug abuse and promoted consequent free high-risk behavior (*Journal of Urgent Care Medicine* 2018).

> "Somebody suffering from substance use disorder or addiction is actually a fragile person. They are hopeless and the way we treat them can make their situation better or worse … . So we need to treat them with kindness and compassion and show them that they are a good person and that they do deserve to get better, because they can."
>
> **–JAN RADER, FIRE CHIEF OF HUNTINGTON, WEST VIRGINIA**

Prevention. Prevention of alcohol, tobacco, and prescription drug use requires reducing variables that put an individual at risk or, alternatively, increasing variables that protect the individual. Some variables which put a person at risk, for example, ethnicity or having an alcoholic parent, are ascribed and although they cannot be altered, prevention efforts can be directed toward specific at-risk populations. Other risk factors are micro-sociological (e.g., negative peer pressure) and/or macro-sociological (e.g., availability of alcohol to underage drinkers) and addressing such variables requires maximizing both micro-sociological protective variables (e.g., counteracting negative peer pressure) and macro-sociological protective variables (e.g., restricting the availability of alcohol to underage drinkers). In various ways, the prevention methods discussed next do some or all of these.

The World Health Organization's MPOWER initiative, one of the most successful initiatives in the global fight against tobacco use (WHO 2019), is a good example of decreasing risk factors (e.g., banning advertising) and increasing protective factors (e.g., drug education). According to the World Health Organization, the initiative has had a significant impact in reducing the number of deaths from tobacco use and the number of people exposed to secondhand smoke. Additional alcohol, tobacco, and prescription drug misuse prevention initiatives include public education campaigns, warning labels, and school- and community-based prevention programs.

harm reduction A recent public health position that advocates reducing the harmful consequences of drug use for the user as well as for society as a whole.

Public Education Campaign. Educating the public about the risks of alcohol, tobacco, and prescription drug use is an important component of substance abuse prevention. Tobacco and alcohol companies spend billions of dollars every year in advertising, and

social problems
RESEARCH
UP CLOSE

Moral Opposition to Needle Exchange Programs

Over the past decade, a plethora of medical and social science research has demonstrated that harm reduction strategies are more effective at improving the public health outcomes related to drug and alcohol addiction, compared with abstinence-only approaches. Needle exchange programs (NEPs) are one type of harm reduction strategy. These programs provide sites in the community where intravenous (IV) drug users can deposit used needles and pick up clean ones (Christie et al. 2019). IV drug users commonly reuse and share needles, leading to increased risk of spreading bloodborne illnesses such as HIV and Hepatitis C. Approximately half of all drug-related deaths in the world are a result of untreated Hepatitis C (World Drug Report [WDR] 2019). In 2016, the Surgeon General of the United States called for an expansion of harm reduction programs, specifically citing NEPs as a cost-effective way to reduce the transmission of disease that did not increase the likelihood of intravenous drug use (Surgeon General US 2016). However, the implementation of these programs is often met with resistance at the local level. One team of researchers wanted to understand the moral psychology that underlies community resistance to implementing NEPs. By understanding the moral foundations that lead to NEP opposition, public health agencies will be better equipped to partner with communities to implement effective harm reduction strategies in socially sustainable ways.

Sample and Methods

The researchers used an online platform to recruit a sample of 5,369 adult participants. The participants were asked to complete two questionnaires. One was the Moral Foundations Questionnaire (MFQ), a commonly used tool in psychological research to measure the extent to which a person's behavior is motivated by one of five main types of moral foundations: (1) Care, (2) Fairness, (3) Loyalty, (4) Authority, and (5) Purity. The researchers also developed another seven-item questionnaire that asked participants to rank, on a scale of strongly agree to strongly disagree, the extent to which they thought it was appropriate to provide drug users with access to specific services that are commonly provided at NEP locations, for example, access to clean needles, condoms, and the overdose reversal drug naloxone. The researchers also collected demographic details from the participants such as their sex, age, political affiliation, education, religion, and socioeconomic status. Two-thirds of the participants were from the United States, 7 percent from Canada, 6 percent from the U.K., and the remaining participants from other countries.

The researchers used regression analysis to determine the extent to which particular moral foundations predict attitudes about NEPs. In other words, how likely is it that any given individual with a specific moral foundation will be opposed to NEPs?

Findings and Conclusion

The results of the regression analysis showed that only two types of moral foundations predicted attitudes about NEPs: Care and Purity. The analysis predicted that those with a moral foundation rooted in Care would have positive attitudes about NEPs, while those with a moral foundation rooted in Purity would have negative attitudes about NEPs, even after controlling for demographic factors such as religious and political identification.

The researchers argue that while these results are "unsurprising," making the connection between moral foundations and attitudes about harm reduction strategies has direct policy implications. For example, concerns related to "moral contamination" led to a 2016 policy compromise in which federal funds could be used to support NEPs but not to purchase the needles themselves; such a "restriction may reduce the conceived contamination by putting psychological separation between the taxpayer and the purity violation of injections" (p. 240).

In order to persuade opponents of NEPs, it is not enough to argue about the evidence of the effectiveness of these programs or the need to care for the vulnerable in their communities. Rather, public health experts can address the Purity concern in the ways they promote these programs at the local level. For example, the researchers recommend deemphasizing imagery around injection and instead promoting imagery and language that is high in "positive purity content (e.g., clean, well-lit images with bright colors)" (p. 240). Additionally, public health advocates can emphasize the ways in which these programs reduce the spread of illness, therefore promoting cleanliness and purity in terms of broader community health. Ultimately, it is not enough to know that a harm reduction program is effective. Program implementation requires community support, and gaining support means understanding attitudes: "If official organizations like the National Institutes of Health (NIH), the Centers for Disease Control (CDC), and the World Health Organization (WHO) are correct in their assessment of needle exchange programs, then it is not an exaggeration to say that attitudes about the programs are a matter of life or death" (p. 240).

SOURCE: Christie et al. 2019.

mass media campaigns can help to counter that message. The social media campaign *Above the Influence* delivers anti-alcohol and anti-drug messages by encouraging youth to be aware of negative peer pressure and to develop strategies to critically engage with alcohol, tobacco, and drug messaging (Above the Influence 2020).

The FDA's *The Real Cost*, launched in 2014, focuses on educating 12- to 17-year-old at-risk youth about the negative consequences of tobacco use in the hope of reducing initiation rates and habitual use by those already experimenting with tobacco (FDA 2020). The campaign uses print media, television ads, radio public service announcements, as well as social media—Facebook, YouTube, and Twitter—to reach the targeted population. Since 2018, the campaign has focused its attention primarily on e-cigarette use prevention.

Lastly, the Partnership for Drug-Free Kids' Medicine Abuse Project (MAP) is a good example of a multifocused campaign that provides resources for caretakers, health care providers, educators, and community and law enforcement officials (MAP 2019). Its programs include National Prescription Drug Take Back Day to provide safe disposal sites for unused medications, as well as social media campaigns such as a video titled *Life of a Pill*, which uses Google Maps technology to help families understand the importance of safely disposing of unused medication.

Warning Labels. The use of warning labels on alcoholic beverages in the United States was mandated as part of the Anti-Drug Abuse Act of 1988. It is not, however, required in many other countries, including England and Canada (Taylor 2014). The impact of warning labels on drinking behavior is mixed. Although a review of the literature concludes that alcohol warning labels are "popular with the public, their effectiveness for changing drinking behavior is limited" (Thomas et al. 2014, p. 91).

Graphic warning labels depicting images of blackened lungs, underweight babies, and rotting teeth are common in many countries. One cross-national study compared reactions to pictorial warning labels on cigarette packaging among smokers and nonsmokers from 100 different countries (Hammond et al. 2019). The results found that graphic images, combined with text, were effective at reducing intentions to smoke among both smokers and nonsmokers. Additionally, they found that the effectiveness of this messaging was increased when combined with personal testimonies from people harmed by smoking.

In 2009, Congress passed a law requiring specific warning labels on cigarette packaging. The implementation of this policy was halted by more than a decade, however, as tobacco companies appealed the law, arguing it violated First Amendment free speech right (Public Health and Tobacco Policy Center 2020). Although appeals are still pending, the FDA plans to mandate graphic warning labels on cigarette packaging as soon as final federal authorization is approved (FDA 2020).

Family-, School-, and Community-Based Prevention Programs. A body of research suggests that school-based interventions may reduce alcohol and/or tobacco use. One analysis examined the effectiveness of 30 different school-based programs for adolescent marijuana use prevention in North America (Lize et al. 2016). The results indicated that these programs are most effective when they are implemented in middle school, are interactive in nature, and delivered by teachers rather than by youth peers or other adults from the community.

There is also strong evidence that school- and home-based prevention programs in combination reduce the likelihood of nonmedical prescription drug use (Crowley et al. 2014). A sample of students were asked in the 6th grade and at the end of each year through 12th grade, "Have you ever used Vicodin, codeine, Percocet or OxyContin not prescribed by a doctor?" Students and their families were assigned to either a treatment group or control group. Members of the treatment group received one of three school-based intervention programs with or without a family-based intervention program. In the absence of family intervention, only one of the three school-based intervention programs was associated with reduced opiate use. However, all of the school-based intervention programs when used in conjunction with the family-based interventions were successful in significantly reducing the nonmedical use of opiates.

Research indicates that labels warning about the health risks of cigarettes are most effective when they combine images, text, and personal testimonies from those affected.

Understanding Alcohol and Other Drug Use

In summary, substance abuse—that is, drugs and their use—is socially defined. As the structure of society changes, the acceptability of one drug or another changes as well. As theorists assert, the status of a drug as legal or illegal is intricately linked to those who have the power to define acceptable and unacceptable drug use. There is also little doubt that rapid social change, anomie, alienation, and inequality further drug use and abuse. Symbolic interaction also plays a role in the process: If people are labeled "drug users" and are expected to behave accordingly, then drug use is likely to continue. Thus, the theories of drug use complement rather than contradict one another.

Two issues need to be addressed in understanding drug use. The first is at the micro level: why does a given individual use alcohol or other drugs? Many individuals at high risk for drug use have been "failed by society"—they are living in poverty, unemployed, victims of abuse, dependents of addicted and neglectful parents, and the like. Despite the social origins of drug use, many treatment alternatives, emanating from a clinical model of drug use, assume that the origin of the problem lies within the individual rather than in the structure and culture of society. Although the problem may admittedly lie within the individual when treatment occurs, policies that address the social causes of drug abuse must be a priority in dealing with the drug problem in the United States.

The second question, related to the first, is why does drug use vary so dramatically across societies, often independent of a country's drug policies? The United States has historically some of the most severe penalties for drug violations in the world, but has one of the highest rates of marijuana and cocaine use. On the other hand, as mentioned earlier, Portugal decriminalized personal possession of all drugs, and youth drug use has decreased (Hollersen 2013). Most compellingly, a 2008 World Health Organization survey of 17 countries concluded that there is no link between the harshness of drug policies and the consumption rates of its citizenry (Degenhardt et al. 2008).

That said, what is needed is a more balanced approach—one that acknowledges that not all drugs have the same impact on society or on the individuals who use them. Increasingly, major U.S. and global health agencies are advocating for approaches to drug abuse that focus on harm reduction, community-based prevention efforts, and an understanding of the socioeconomic and physiological conditions that drive substance abuse.

Although punitive approaches to drug use and drug trafficking dominate American public policy, the election of Joe Biden may signal a shift in balance in this approach. The plan proposed by the Biden administration includes greater emphasis on substance use treatment and federal funding for mental health and addiction treatment programs, while also strengthening Department of Justice efforts to prosecute drug traffickers and hold pharmaceutical companies financially responsible for the aftermath of the opioid crisis (Biden-Harris 2020a). The new administration also proposes to federally decriminalize marijuana use and expunge the convictions of low-level marijuana offenders, while leaving decisions about recreational legalization up to individual states (Biden-Harris 2020b). Only time will tell if this new approach to drug control, one that reflects both a public health and criminal justice position while also creating a national patchwork of state-level policies, will be more successful than previous strategies.

Chapter Review

- **What is the difference between drug abuse and substance abuse disorder?**
  Drug abuse is a sociological term referring to the violation of acceptable social standards of drug use, resulting in adverse physiological, psychological, and/or social consequences. Substance abuse disorder refers to a medical diagnosis characterizing the relationship between recurrent use of alcohol and/or drugs and significant

health problems, disabilities, and inability to meet major life responsibilities.

- **How do the three sociological theories of society explain drug use?**
  Structural functionalists argue that drug abuse is a response to the weakening of norms in society, leading to a condition known as anomie or normlessness. From a conflict perspective, drug use occurs as a response to the inequality perpetuated by a capitalist system as societal members respond to alienation from their work, family, and friends. Symbolic interactionism focuses on the social meanings associated with drug use. An individual's likelihood of using or abusing drugs is influenced by interactions with friends, family, and the media.

- **What are the most frequently used legal and illegal drugs?**
  Alcohol is the most commonly used and abused legal drug in the United States, with more than half of the adult population reporting current alcohol use. Although tobacco use in the United States has been declining, the use of tobacco products is globally very high, with 80 percent of the world's over one billion smokers living in low- or middle-income countries.

- **What are the consequences of drug use?**
  The consequences of drug use are fivefold. First is the cost to the family and children, often manifesting itself in higher rates of divorce, spouse abuse, child abuse, and child neglect. Second is the relationship between drugs and crime. Those arrested have disproportionately higher rates of drug use. Although drug users commit more crimes, sociologists disagree as to whether drugs actually "cause" crime or whether, instead, criminal activity leads to drug involvement. Third are the economic costs (e.g., loss of productivity), which are in the billions. Then there are the health costs of abusing drugs, including shortened life expectancy; higher morbidity (e.g., cirrhosis of the liver and lung cancer); exposure to HIV infection, hepatitis, and other diseases through shared needles; a weakened immune system; birth defects such as fetal alcohol syndrome; drug addiction in children; and higher death rates. Finally, illegal drug production takes its toll on the environment, which impacts all Americans.

- **What treatment alternatives are available for drug users?**
  Although there are many ways to treat drug abuse, three methods stand out: The inpatient–outpatient model entails medical supervision of detoxification and may or may not include hospitalization. Twelve-step programs such as Alcoholics Anonymous (AA) and Narcotics Anonymous (NA) use a supportive community and peer leaders to help members reduce substance abuse, and drug courts divert offenders from the criminal just system through a multipronged treatment program.

- **What can be done about the drug problem?**
  First, there are government regulations limiting the use (e.g., the law establishing the 21-year-old drinking age) and distribution (e.g., prohibitions about importing drugs) of legal and illegal drugs. The government also imposes sanctions on those who violate drug regulations and provides treatment facilities for other offenders. Economic incentives (e.g., cost) and prevention programs have also been found to impact consumption rates. Finally, legal action holding companies responsible for the consequences for their product—for example, class-action suits against tobacco producers—have been fairly successful.

Test Yourself

1. Globally, approximately two-thirds of all deaths attributed to drug use disorders are associated with:
   a. opioids.
   b. methamphetamine.
   c. cocaine.
   d. marijuana.
2. More than 80 percent of people who smoke cigarettes live in high-income countries.
   a. True
   b. False
3. Symbolic interactionists would explain the increase in alcohol use and opioid overdoses that coincided with the COVID-19 global pandemic as a result of
   a. profiteering by drug manufacturers.
   b. public misperceptions that drugs and alcohol can kill the virus.
   c. the inequality that results from capitalism.
   d. feelings of social disconnection, isolation, and despair.
4. What theory would argue that the continued legality of alcohol is a consequence of corporate greed?
   a. Structural functionalism
   b. Symbolic interactionism
   c. Reinforcement theory
   d. Conflict theory
5. Men are more likely than women to engage in binge drinking behavior.
   a. True
   b. False
6. Cigarette smoking is
   a. the third leading cause of preventable death in the United States.
   b. not addictive.
   c. the most common use of tobacco products.
   d. increasing in the United States.
7. Which European country decriminalized personal possession of all drugs, including marijuana, cocaine, heroin, and methamphetamine?
   a. Spain
   b. France
   c. Portugal
   d. The Netherlands

8. Harm reduction strategies focus on
   a. reducing drug overdose deaths and increasing drug treatment access.
   b. removing drug users from the community.
   c. increasing the criminal penalties associated with drug use.
   d. eliminating the criminal penalties associated with drug use.

9. One long-term impact of the war on drugs in the United States was
   a. an increase in rates of illicit drug use.
   b. a shift in public perception about the harm of drug use.
   c. reduced criminal penalties for nonviolent drug users.
   d. an increase in rates of incarceration.

10. Decriminalization refers to the removal of penalties for certain drugs.
   a. True
   b. False

Answers: 1. A; 2. B; 3. D; 4. D; 5. A; 6. A; 7. C; 8. A; 9. D; 10. A.

Key Terms

binge drinking 84
chemical dependency 78
crack 78
crime displacement 98
demand reduction 103
drug 74
drug abuse 78
drug courts 104
e-cigarettes 86
fetal alcohol syndrome (FAS) 95
gateway drug 87
harm reduction 107
heavy drinking 84
psychotherapeutic drugs 90
polydrug abuse 92
substance use disorder 78
supply reduction 103
war on drugs 77

APU GOMES/AFP/Getty Images

> Unjust social arrangements are themselves a kind of extortion, even violence."
>
> **JOHN RAWLS**
> A Theory of Justice

4

Crime and Social Control

Chapter Outline

Learning Objectives

After studying this chapter, you will be able to ...

1 List patterns of crime and criminals that are consistent across global regions.

2 Argue which of the various methods of measuring crime is the most accurate.

3 Identify the crime prevention implications of each sociological theory of crime.

4 Differentiate among the types of crimes discussed.

5 Using each of the demographic variables discussed, explain their relationship to the likelihood of criminal behavior and victimization.

6 Assess the relative consequences to society of crimes in the streets and crimes in the suites.

7 Explain the origins of the Black Lives Matter protests against police brutality.

8 Analyze the various social forces that lead to and away from a punitive criminal justice policies.

AP Images/Martha Irvine

Chicago area high school students participate in a moment of silence at a Peace Rally to honor victims of gun violence. Gary Mitchell Tinder was one of numerous innocent victims of gun violence who die each year in the Chicago area.

GARY WAS ONE OF THOSE ANNOYING PEOPLE who was good at everything. He was also, by all accounts, one of the nicest guys you'd ever want to meet and was "unassumingly hilarious." He went to DePaul University after graduating from high school and, not surprisingly given his love of math, majored in accounting and maintained a 4.0 GPA. He worked at Starbucks as a barista to help pay for school. One night, walking home from work, a man exited an SUV and began following him—but Gary couldn't have known that—he had earbuds on and was glued to his phone. Just two blocks from his house, the man shot and killed Gary, who was only 20 years old (Lee and Oxnevad 2020). The alleged shooter was caught and charged with first-degree murder. No one knows why he was following Gary that night; there doesn't seem to be a link between the two men. What we do know is that in 2020, over Father's Day weekend in Chicago, 104 people were shot, 15 of them fatally (*Sun-Times Wire* 2020). One of them was Gary Mitchell Tinder.

The Chicago Police Department (CPD), CPD homicide detectives, CPD Office of Communications, the Cook County State Attorney's office, and the Circuit Court of Cook County, which set bail for the accused at $1 million, are all part of the larger bureaucracy we call the criminal justice system. In this chapter, we examine the criminal justice system as well as theories, types, and demographic patterns of criminal behavior. The economic, social, and psychological costs of crime are also examined. The chapter concludes with a discussion of social control, including policies and prevention programs designed to reduce crime in America.

The Global Context: International Crime and Social Control

Several facts about crime are true throughout the world. First, crime is ubiquitous—there is no country where crime does not exist. Second, most countries have the same components in their criminal justice systems: police, courts, and prisons. Third, adult males make up the largest category of crime suspects worldwide. Fourth, in all countries, theft is the most common crime committed, whereas violent crime is a relatively rare event.

Despite these similarities, dramatic differences exist in international crime rates, although comparisons are made difficult by variations in measurement and crime definitions. Nonetheless, looking at global statistics as a whole, we can make some statements about crime with confidence.

First, violent crime rates vary significantly by region and by country. Examining homicide rates is a case in point. Because homicide rates are correlated with other violent crimes—for example, as robbery rates increase, homicide rates increase—global homicide rates are very telling. In 2018, the most recent year for which data are available at the regional level, the Americas (i.e., North, Central, and South America) had the highest homicide rate in the world, 15.9 per 100,000 population, followed by Africa. The high murder rate in the Americas is driven by Central and South America, which have the highest violent crime rates in the world (United Nations Crime Trend Survey [UNCTS] 2019).

The remaining regions—Asia, Europe, and Oceania—have homicide rates between 2.1 and 2.8 per 100,000 population. Not surprisingly, robbery, which is also a violent crime, is also highest in the Americas with, for example, Brazil having a 2018 robbery rate of 696 per 100,000 population, compared to Australia in the Oceania region with a robbery rate of 41 per 100,000 population (UNCTS 2019).

Crime rates, usually expressed per 100,000 population, take into consideration differences in population size and allow more accurate comparisons than comparing raw numbers. Homicide rates are often linked to the socioeconomic development of a country: in general, the higher the level of socioeconomic development, the lower the homicide rate. Some countries, however, have higher homicide rates than would be expected given their level of development. In the Americas, for example, where firearms are readily available and gangs and organized crime are rampant, homicide rates are higher than one might predict given their high level of socioeconomic development (United Nations Office on Drugs and Crimes [UNODC] 2019). As in the United States, globally, homicides were most often committed by males: about 90 percent across all regions. Males were also more likely to be the victims of homicide, although women comprise the vast majority of intimate partner deaths (UNODC 2019).

Unlike homicide, which is categorized as a violent crime, property crimes usually entail some kind of theft such as burglary of a home or motor vehicle theft. As with violent crime, there are vast differences within and between regions. For example, car theft rates are below 10 per 100,000 population in Myanmar, Armenia, Syria, Indonesia, the Philippines, Kenya, and Honduras. Few people in these countries own cars and thus the opportunity to commit motor vehicle theft is low. In Honduras and Uruguay, motor vehicle theft is 3 and 560 per 100,000 population, respectively, despite both being in the Americas region, the first in Central America and the second in South America (UNCTS 2019).

Violent crimes and property crimes represent just two types of crime that take place worldwide. Although we are concerned about these types of crimes and the possibility of victimization, globalization and technological know-how have fueled the development of new categories of crimes (e.g., human trafficking, organ trade) and have expanded the prevalence of others (e.g., counterfeiting and pirated goods, cybercrime).

transnational crime Criminal activity that occurs across one or more national borders.

Transnational crime is organized criminal activity that crosses one or more national boundaries for the "purpose of obtaining power, influence, monetary and/or commercial gains ... through a pattern of corruption and/or violence" (U.S. Department of Justice [USDOJ] 2015, p. 1). The significance of transnational crime should not be minimized. As May (2017) notes, the estimated revenues of between $1.6 trillion and $2.2 trillion per year "not only line the pockets of the perpetrators but also finance violence, corruption, and other abuses" that undermine "local and national economies, destroy the environment, and jeopardize the health and well-being of the public" (p. xi).

As Table 4.1 indicates, transnational crime primarily functions to provide illicit goods and services. Drug trafficking is one of the most lucrative examples, representing more than one-third of the total revenue of transnational crime. The global market for drug traffickers lies primarily in marijuana, cocaine, opiates, and amphetamine-type stimulants (May 2017) (see Chapter 3). Drug trafficking organizations tend to flourish in developing countries where governments and security forces are too weak to put the cartels out of business.

AP Images/Pavel Rahman

A young boy, a victim of human trafficking, is forced to work in a balloon factory outside of Dhaka, Bangladesh. Human trafficking is estimated to be a multibillion-dollar enterprise, approaching the international value of drug and arms trafficking according to Interpol.

TABLE 4.1 The Retail Value of Transnational Crime, 2017

Transnational Crime	Estimated Annual Value (US$)
Drug trafficking	$426 billion to $652 billion
Small arms and light weapons trafficking	$1.7 billion to $3.5 billion
Human trafficking	$150.2 billion
Organ trafficking	$840 million to $1.7 billion
Trafficking in cultural property	$1.2 billion to $1.6 billion
Counterfeiting	$923 billion to $1.13 trillion
Illegal wildlife trade	$5 billion to $23 billion
IUU fishing*	$15.5 billion to $36.4 billion
Illegal logging	$52 billion to $157 billion
Illegal mining	$12 billion to $48 billion
Crude oil theft	$5.2 billion to $11.9 billion
Total	**$1.6 trillion to $2.2 trillion**

*Illegal, unreported, and unregulated
SOURCE: May 2017.

Human trafficking is another example of transnational crime, and it is on the rise. Despite common misconceptions, a victim need not be transported from one location to another to constitute human trafficking. The *Trafficking Victims Protection Act of 2000*, which has been amended nine times, defines "severe forms of trafficking in persons" as including (USDOJ 2020a, p. 1):

- "sex trafficking in which a commercial sex act is induced by force, fraud, or coercion, or in which the person induced to perform such an act has not attained 18 years of age; or
- the recruitment, harboring, transportation, provision, or obtaining of a person for labor or services, through the use of force, fraud, or coercion for the purpose of subjection to involuntary servitude, peonage, debt bondage, or slavery."

Globally, it is estimated that as many as 25 million men, women, and children are victims of forced labor, including coerced participation in the commercial sex industry, domestic housekeeping, construction, manufacturing, and agriculture and fishing. With the exception of the commercial sex industry and domestic work, men are most likely to be the victims in the other work sectors. Forced labor is thought to be the highest in Asia and the Pacific region, followed by Africa (International Labour Office [ILO] 2017). Based on victims' accounts and acknowledging regional differences, research indicates that trafficking for forced labor has decreased over the last decade, while human trafficking for sexual exploitation, having decreased in the early 2000s, is now increasing (Migration Portal 2020).

The United States, along with Mexico and the Philippines, was ranked as one of the worst places for human trafficking in the world with hundreds of thousands of victims (Pasley 2019). Based on calls to the National Human Trafficking Hotline, there are three types of human trafficking cases in the United States. The most common is sex trafficking, which includes escort services, residential-based services, and pornography, followed by labor trafficking (e.g., agriculture), and, third, a combination of sex and labor trafficking such as working at an illegal massage parlor, bar, or strip club (Polaris Project 2019).

Sources of Crime Statistics

The U.S. government spends millions of dollars annually to compile and analyze crime statistics. A **crime** is a violation of a federal, state, or local criminal law. For a violation to be a crime, however, the offender must have acted voluntarily and with intent and have no legally acceptable excuse (e.g., insanity) or justification (e.g., self-defense) for his or her behavior. The three major types of statistics used to measure crime are official statistics, victimization surveys, and self-report offender surveys.

crime An act, or the omission of an act, that is a violation of a federal, state, or local criminal law for which the state can apply sanctions.

Official Statistics

Local sheriffs' departments and police departments throughout the United States collect information on the number of reported crimes and arrests and voluntarily report them to the Federal Bureau of Investigation (FBI). The FBI then compiles these statistics annually and publishes them, in summary form, in the Uniform Crime Reports (UCR). The UCR lists **crime rates** (the number of crimes committed per 100,000 population), the actual number of crimes, the percentage of change over time, and clearance rates. **Clearance rates** measure the percentage of cases in which an arrest and official charge have been made and the case has been turned over to the courts.

These statistics have several shortcomings. First and most importantly, a great deal of crime goes unreported. For example, a 2019 survey by the U.S. Department of Justice indicates that fewer than half of all victims of violent crime reported the offense to the police (Morgan and Truman 2020). Even fewer property crimes were reported, 33 percent overall. Of all property and violent crimes measured by the FBI, the most likely to be reported is motor vehicle theft, no doubt because insurance companies require it, and the least likely to be reported is larceny, i.e., simple theft.

Even if a crime is reported, the police may not record it. Alternatively, some rates may be exaggerated. Motivation for such distortions may come from the public (e.g., demanding that something be done), from political officials (e.g., election of a sheriff), and/or from organizational pressures (e.g., budget requests). For example, a police department may "crack down" on drug-related crimes in an election year. The result is an increase in the recorded number of these offenses for that year. Such an increase reflects a change in the behavior of law enforcement personnel, not a change in the number of drug violations. Thus, official crime statistics may be a better indicator of what police are doing rather than of what criminals are doing.

What do you think?

Both property crime and violent crime in 25 of America's largest cities decreased between July 2019 and July 2020. Property crimes were down in 18 of the 25 cities sampled, and violent crime was down in 11 of the 25 cities. But in an unusual occurrence, despite violent crime rates decreasing in nearly half of the cities, the murder rate is up in 20 of the 25 cities. It is so unusual that it has only occurred in four years since 1960 (Asher and Horwitz 2020). What do you think is responsible for this unusual crime pattern?

The National Incident-Based Reporting System (NIBRS), "implemented to improve the overall quality of crime data collected by law enforcement[,] ... captures details on each single crime incident—as well as on separate offenses within the same incident" (FBI 2020, p. 1). Unlike the UCR, data on "victims, known offenders, relationships between victims and offenders, arrestees, and property involved in the crimes" are collected (p. 1). In fact, in 2021, the URC program was phased out and only NIBRS data were collected. In this chapter, both UCR and the NIBRS statistics will be reported where appropriate.

Victimization Surveys

Acknowledging the **dark figure of crime**, that is, the tendency for so many crimes to go unreported and thus undetected by the UCR—the U.S. Department of Justice conducts the National Crime Victimization Survey (NCVS). Begun in 1973, the NCVS annually collects data from a representative sample of approximately 150,000 people 12 and older, in about 95,000 households. Interviewers collect a variety of information, including the victim's background (e.g., age, race and ethnicity, sex, marital status, education, and area of residence), relationship to the offender (stranger or non-stranger), characteristics of the offender (e.g., age, race, gender), and the extent to which the victim was harmed (Morgan and Truman 2020).

In 2019, the latest year for which victimization data are available, there were 5.8 million violent crime incidents and 12.8 million property crimes—nearly five and two times more, respectively, than estimates using official statistics (Morgan and Truman 2020; FBI 2020a). Eight and a half million households, 6.8 percent of all households, were victimized by one or more property crimes in 2019. In the same year, 3.1 million people, or 1.1 percent of all persons aged 12 or older, experienced at least one violent victimization.

crime rate The number of crimes committed per 100,000 population.

clearance rate The percentage of crimes in which an arrest and official charge have been made and the case has been turned over to the courts.

dark figure of crime The volume of crime that goes unreported.

Self-report Offender Surveys

Self-report surveys ask offenders about their criminal behavior. The sample may consist of a population with known police records, such as a prison population, or it may include respondents from the general population, such as college students.

Self-report data compensate for many of the problems associated with official statistics but are still subject to exaggeration and concealment. In general, self-report surveys reveal that virtually every adult has engaged in some type of criminal activity. Why, then, is only a fraction of the population labeled criminal? Like a funnel, which is large at one end and small at the other, only a small proportion of the total population of law violators are ever convicted of a crime. For individuals to be officially labeled criminals, (1) the behavior must become known to have occurred; (2) the behavior must come to the attention of the police, who then file a report, conduct an investigation, and make an arrest; and (3) arrestees must go through a preliminary hearing, an arraignment, and a trial, at which they may or may not be convicted. At every stage of the process, offenders may be "funneled" out.

Sociological Theories of Crime

Some explanations of crime focus on the psychological aspects of the offenders, such as psychopathic personalities, unhealthy relationships with parents, and mental illness. Other crime theories focus on the role of biological variables, such as central nervous system malfunctioning, stress hormones, vitamin or mineral deficiencies, chromosomal abnormalities, and a genetic predisposition toward aggression. Sociological theories of crime and violence emphasize the role of social factors in criminal behavior and societal responses to it.

Structural-Functionalist Perspective

According to Durkheim and other structural functionalists, crime is functional for society. One of the functions of crime and other deviant behavior is that it strengthens group cohesion: "The deviant individual violates rules of conduct that the rest of the community holds in high respect; and when these people come together to express their outrage over the offense, they develop a tighter bond of solidarity than existed earlier" (Erikson 1966, p. 4).

Crime can also lead to social change. For example, an episode of local violence may "achieve broad improvements in city services, be a catalyst for making public agencies more effective and responsive, for strengthening families and social institutions, and for creating public–private partnerships" (National Research Council 1994, pp. 9–10).

Although structural functionalism as a theoretical perspective deals directly with some aspects of crime, it is not a theory of crime per se. However, three major theories of crime have developed from structural functionalism.

Anomie Theory. Anomie theory was developed by Robert Merton (1957) and uses Durkheim's concept of *anomie*, or normlessness. Merton argued that, when the structure of society limits legitimate means (e.g., a job) of acquiring culturally defined goals (e.g., money), the resulting strain may lead to crime. For example, rapid economic social change in Russia, specifically high rates of unemployment, led to increases in the homicide rates (Pridemore and Kim 2007).

Individuals, then, must adapt to the inconsistency between means and goals in a society that socializes everyone into wanting the same thing but that provides opportunities for only some (see Figure 4.1). *Conformity* occurs when individuals accept the culturally defined goals and the socially legitimate means of achieving them. Merton suggested that most individuals, even those who do not have easy access to the means and goals, remain conformists. *Innovation* occurs when an individual accepts the goals of society but rejects or lacks the socially legitimate means of achieving them. Innovation, the mode of

Rebel: rejects wealth and work; substitutes new goals and new means	**Culturally Defined Goals:**	
Structurally Defined Means:	Acceptance of (+)	Rejection of (−)
Accepts/has access to (+)	Conformist: desires wealth, works hard, saves money	Ritualist: rejects wealth, goes through motions of working hard
Rejects/lacks access to (−)	Innovator: desires wealth, teller at bank, embezzles money	Retreatist: rejects wealth and means to achieve it, turns to drugs

Figure 4.1 Merton's Modes of Adaptations Applied to Crime
SOURCE: Adapted from Merton, 1957.

adaptation most associated with criminal behavior, explains the high rate of crime committed by uneducated and poor individuals who do not have access to legitimate means of achieving the social goals of wealth and power.

Another adaptation is *ritualism*, in which, for example, individuals accept a lifestyle of hard work but reject the cultural goal of monetary rewards. Ritualists go through the motions of getting an education and working hard, yet they are not committed to the goal of accumulating wealth or power. *Retreatism* involves rejecting both the cultural goal of success and the socially legitimate means of achieving it. Retreatists withdraw or retreat from society and may become alcoholics, drug addicts, or vagrants. Finally, *rebellion* occurs when individuals reject both culturally defined goals and means and substitute new goals and means. For example, rebels may use social or political activism to replace the goal of personal wealth with the goal of social justice and equality.

General Strain Theory (GST). Robert Agnew developed a General Strain Theory (GST) based, in part, on Merton's anomie theory. Agnew argues that when a person experiences strain—that is, a negative condition or event—it increases the likelihood of criminal behavior (Agnew 2013). Strains vary in magnitude, duration, frequency, and so forth, as does the likelihood of a criminal response. For example, being fired (strain) may lead to anger and frustration (negative emotional responses), which in turn lead to stealing from the former employer (crime). Whether a criminal response is forthcoming, however, depends on the strength of the individual's support system, whether there are criminal incentives, the level of social control, ties to criminal or non-criminal significant others, and alternative coping mechanisms (Agnew 2013).

Whereas strain theory often explains criminal behavior as a result of blocked opportunities, subcultural theories argue that certain groups or subcultures in society have values and attitudes that are conducive to crime and violence. Members of these groups and subcultures, as well as other individuals who interact with them, may adopt the crime-promoting attitudes and values of the group. For example, Kubrin and Weitzer (2003) found that retaliatory homicide is a response to subcultural norms of violence that exist in some neighborhoods. However, other theorists argue that subcultural theorists incorrectly assume that subcultures are "coherent, consistent, and homogeneous ... with stable sets of shared beliefs, practices and identities" (Debies-Carl 2013, p. 118).

Control Theory. Control theory addresses one of the limitations of strain theory. If blocked opportunities and subcultural values are responsible for crime, why don't all members of the affected group become criminals? Consistent with Durkheim's emphasis on social solidarity, Hirschi (1969) suggests that a strong **social bond** between individuals and the social order constrains some individuals from violating social norms. Hirschi identified four elements of the social bond: *attachment* to significant others, *commitment* to conventional goals, *involvement* in conventional activities, and *belief* in the moral

social bond The bond between individuals and the social order that constrains some individuals from violating social norms (i.e., committing crime).

> When a man is denied the right to live the life he believes in, he has no choice but to become an outlaw."
>
> **–NELSON MANDELA, ANTI-APARTHEID ACTIVIST, FORMER PRESIDENT OF SOUTH AFRICA**

standards of society. Several empirical tests of Hirschi's theory support the notion that the higher the attachment, commitment, involvement, and belief, the higher the social bond and the lower the probability of criminal behavior. Bell (2009) reports that a weaker attachment to parents is associated with a greater likelihood of gang membership for both males and females. Similarly, Gault-Sherman (2012), after analyzing the responses of a national sample of more than 12,500 adolescents, reports a negative association between a youth's attachment to parents and self-reported delinquency: the higher the attachment, the lower the prevalence of delinquency.

Conflict Perspective

Conflict theories of crime suggest that deviance is inevitable whenever two groups have differing degrees of power; in addition, the more inequality there is in a society, the greater the crime rate in that society. Social inequality leads individuals to commit crimes such as larceny and burglary as a means of economic survival. Other individuals, who are angry and frustrated by their low position in the socioeconomic hierarchy, express their rage and frustration through crimes such as drug use, assault, and homicide. In Argentina, for example, the soaring violent crime rate of the late 1990s was hypothesized to be "a product of the enormous imbalance in income distribution between the rich and the poor" (Pertossi 2000). Furthermore, in an examination of global homicide rates, research indicates that the greater the income inequality, the higher the homicide rate (United Nations [UN] 2016).

What *do you* think?

Jamie Mosley runs a correctional facility in Kentucky and, like his fellow jailers, has benefited from the "e-cigarette boom, either by forming companies that sell vaping products to inmates in other jails or by handing lucrative businesses in their own facilities to friends and family" (Dunlop 2020, p. 1). In 2018, Mr. Mosley estimated that his company would make $3.5 million in e-cigarette sales. Do you think employees of correctional facilities should be able to profit by selling goods or services to inmates?

As conflict theorists are quick to note, privately owned prison corporations earn billions of dollars annually. Although the Obama administration moved away from using for-profit prisons, soon after President Trump's inauguration, the Department of Justice reopened bidding for private prisons to house federal inmates and migrants being held in detention. It is not surprising, from a conflict perspective, that President Trump's political action committee and his inaugural committee benefited from financial contributions from private prison corporations (Ahmed 2019).

According to the conflict perspective, those in power define what is criminal and what is not, and these definitions reflect the interests of the ruling class. Laws against vagrancy, for example, penalize individuals who do not contribute to the capitalist system of work and consumerism. Furthermore, D'Alessio and Stolzenberg (2002, p. 178) found that "in cities with high unemployment, unemployed defendants have a substantially higher probability of pretrial detention" than employed defendants. Rather than viewing law as a mechanism that protects all members of society, conflict theorists focus on how laws are created and enforced by those in power to protect the ruling class. To that end, oil and utility corporations and banks have donated millions of dollars to police foundations (Frazin 2020).

Women, too, are the victims of power differentials. Female prostitutes are more likely to be arrested than are the men who seek their services. Similarly, "rape myths" are perpetuated by the male-dominated culture to foster the belief that women are to blame for their own victimization, thereby, in the minds of many, exonerating the offenders. Such beliefs have very real consequences. Men who believe that "women secretly want to be raped," "rape is not harmful," and "some women deserve to be raped" are more likely to commit rape than men who do not endorse such beliefs (Mouilso and Calhoun 2013).

Symbolic Interactionist Perspective

Two important theories of crime emanate from the symbolic interactionist perspective. The first, labeling theory, focuses on two questions: How do crime and deviance come to be defined as such, and what are the effects of being labeled criminal or deviant? Second, differential association, examines the way in which meanings and definitions are transmitted between and within small groups.

Labeling Theory. According to Howard Becker (1963), often considered the architect of labeling theory:

> Social groups create deviance by making rules whose infractions constitute deviance, and by applying those rules to particular people and labeling them as outsiders. From this point of view, deviance is not a quality of the act a person commits, but rather a consequence of the application by others of rules and sanctions to an "offender." The deviant is one to whom the label has successfully been applied; deviant behavior is behavior that people so label. (p. 238)

Labeling theorists make a distinction between *primary deviance*, which is deviant behavior committed before a person is caught and labeled an offender, and *secondary deviance*, which is deviance that results from being caught and labeled. After a person violates the law and is apprehended, that person is stigmatized as a criminal. This deviant label often dominates the social identity of the person to whom it is applied and becomes the person's "master status"—that is, the primary basis on which the person is defined by others.

In research conducted by Pickett et al. (2012), White respondents were asked to estimate the percentage of criminals who are Black that commit various serious crimes including robbery with a firearm, residential burglary, commercial burglary, selling illegal drugs, juvenile crime, and violent crime in general. In each case, based on comparisons with official data, respondents overestimated the percentage of Black people involved in crime. For example, when asked, "If you think about crime and criminals, what percent of criminals who commit violent crime in the country are Black?" averaged a *perception* rating 46.4 percent compared to the actual percentage of 21.0.

Being labeled as deviant often leads to further deviant behavior because (1) the person who is labeled as deviant is often denied opportunities for engaging in non-deviant behavior and (2) the labeled person internalizes the deviant label, adopts a deviant self-concept, and acts accordingly. For example, a teenager who is caught selling drugs at school may be expelled and thus denied opportunities to participate in non-deviant school activities (e.g., sports and clubs) and to associate with non-deviant peer groups. A review of the literature on labeling and juvenile delinquency lends support for the theory. One researcher concludes, "[R]ather than discourage participation in conventional activities by labeling and isolating offenders, juvenile crime policy should be remedial and foster reintegration" (Ascani 2012, p. 83).

Differential Association. The assignment of meaning and definitions learned from others is also central to the second symbolic interactionist theory of crime, differential association. Edwin Sutherland (1939) proposed that, through interaction with others, individuals learn the values and attitudes associated with crime as well as the techniques and motivations for criminal behavior. Individuals who are exposed to more definitions favorable to law violation (e.g., "Crime pays") than to unfavorable ones (e.g., "Do the crime, you'll do the time") are more likely to engage in criminal behavior.

Unfavorable definitions come from a variety of sources. Of particular concern in recent years is the role of video games in promoting criminal or violent behavior. Chang and Bushman (2019) assigned pairs of children between the ages of 8 and 12 to play one of three versions of the video game *Minecraft*. In one version, children could use a gun to kill the monsters. In the second version, a child could use a sword to kill the monsters, and in the third there were no weapons and no monsters. Children who watched the two violent versions, compared to those who did not, were more likely to touch, hold, and pull the trigger of a disabled handgun that had been placed in a room where the children were instructed to play. Further, children who watched the gun violence version, on the average, had a higher total number of trigger pulls including trigger pulls while pointing the gun at themselves or their partners than those who had watched the sword or non-violent versions.

Types of Crime

The FBI identifies eight index offenses as the most serious crimes in the United States. The **index offenses**, or street crimes, as they are often called, can be against a person (called violent or personal crimes) or against property (see Table 4.2). Other types of crime include vice crime (such as drug use, gambling, and prostitution), organized crime, white-collar crime, computer crime, and juvenile delinquency. Hate crimes are discussed in Chapter 9.

index offenses Crimes identified by the FBI as the most serious, including personal or violent crimes (homicide, assault, rape, and robbery) and property crimes (larceny, motor vehicle theft, burglary, and arson).

Street Crime: Violent Offenses

The most recent data available from the FBI's Uniform Crime Report indicate that the violent crime rate decreased by 9.3 percent between 2010 and 2019 (FBI 2020a). Remember, however, that crime statistics represent only those crimes *reported* to the police: 1.2 million violent

TABLE 4.2 Index Crime Rates, Percentage Change, Clearance Rates, 2019

	Rate per 100,000, 2019	Percentage Change in Rate, 2010–2019	Percentage Cleared, 2019
Violent crime			
Murder	5.0	5.1	61.4
Forcible rape*	29.9	8.1	32.9
Robbery	81.6	−31.6	30.5
Aggravated assault	250.2	−1.0	52.3
Total	366.7	-9.3	45.5
Property crime			
Burglary	340.5	−51.4	14.1
Larceny/theft	1,549.5	−22.7	18.4
Motor vehicle theft	219.9	-8.0	13.8
Arson	NA*	NA	23.8
Total†	2,109.9	−28.4	17.2

*The legacy definition of rape was used for comparison purposes; arson rates per 100,000 are calculated independently because population coverage for arson is lower than for other index offenses.

†Property crime totals do not include arson.

SOURCE: FBI 2019.

crimes in 2019 (FBI 2020a). The reasons why people do not report violent victimization include feeling that (1) nothing will or can be done about it, (2) the offender(s) might retaliate, (3) the victimization isn't important enough to report, (4) the offender would get in trouble, and (5) there is a better way to handle it (Greenwood 2015; Morgan and Truman 2020).

Violent crime includes homicide, assault, rape, and robbery. The 2019 violent crime total increased above the 2015 level but was lower than the 2010 level (FBI 2020a). *Murder* is the willful or non-negligent killing of one human being by another individual or group of individuals. Although murder is the most serious of the violent crimes, it is also the least common, accounting for 1.4 percent of all violent crimes in 2019 (FBI 2020a). A typical homicide scenario includes a male killing a male with a handgun after a heated argument. The victim and offender are disproportionately young and of minority status. When a woman is murdered and the victim–offender relationship is known, she is most likely to have been killed by her husband or boyfriend (FBI 2019a).

Mass murders have more than one victim in a killing event. In 2013, Adam Lanza entered Sandy Hook Elementary School in Newtown, Connecticut, and killed 20 first-grade students and six staff members. The Sandy Hook killings led to a national debate about gun control and specifically the availability and use of high-capacity ammunition magazines and assault weapons (see "Gun Control" section). In 2019, 51 worshipers were murdered and 40 wounded at two mosques in New Zealand. Brenton Tarrant, a 29-year-old self-proclaimed white supremacist from Australia, pled guilty to all counts, including one count of terrorism (Allyn 2020).

Unlike mass murder, serial murder involves the killing of three or more victims by the same offender in different criminal events, usually over a long period of time. The most well-known serial killers, who are responsible for some of the most horrific episodes of homicide, are Ted Bundy, Jeffrey Dahmer, and Gary Ridgeway, also known as the Green River Killer. In 2020, James D'Angelo, known as the Golden State Killer, pled guilty to 13 counts of first-degree murder and 13 counts of kidnapping (Dickman 2020). The bulk of the former policeman's crimes, which in addition to murder included robbery, assault, and rape, took place between 1975 and 1986. This chapter's *Social Problems Research Up Close* compares media presentations of serial killers to real-life serial killing events.

Another form of violent crime, *aggravated assault*, involves attacking a person with the intent to cause serious bodily injury. Like homicide, aggravated assault occurs most often between members of the same race as a result of some kind of argument (FBI 2020a) and, as with violent crime in general, is more likely to occur in the warm-weather months. In 2019, the assault rate was 50 times the murder rate, with assaults making up an estimated 68.2 percent of all violent crimes (FBI 2020a).

It is estimated that nearly 80 percent of all rapes are **acquaintance rapes**—rapes committed by a family member, intimate partner, friend, or acquaintance (Rape, Abuse and Incest National Network [RAINN] 2017). Although acquaintance rapes are the most likely to occur, they are the least likely to be reported and the most difficult to prosecute. Unless the rape is what Williams (1984) calls a **classic rape**—that is, the rapist was a stranger who used a weapon, and the attack resulted in serious bodily injury—women hesitate to report the crime out of fear of not being believed. The increased use of "rape drugs," such as Rohypnol, may lower reporting levels even further.

Robbery, unlike simple theft, also involves force or the threat of force or putting a victim in fear and is thus considered a violent crime. Officially, in 2019, more than 267,988 robberies took place in the United States. Robberies are most often committed using "strong-arm" tactics, followed by the use of firearms or a knife or other sharp instrument (FBI 2020a). According to the FBI, in 2019, the average dollar value lost per robbery was $1,797, with banks reporting the highest average loss per incident—$4,213 (FBI 2020a). In 2019, eight men dressed as police officers stole $30 million worth of gold and other precious metals from an airport cargo facility in San Pablo, Brazil (Dixon 2019).

acquaintance rape Rape committed by someone known to the victim.

classic rape Rape committed by a stranger, with the use of a weapon, resulting in serious bodily injury to the victim.

Street Crime: Property Offenses

Property crimes are those in which someone's property is damaged, destroyed, or stolen; they include larceny, motor vehicle theft, burglary, and arson. The number of property crimes

social problems
RESEARCH
UP CLOSE

Media Presentations of Serial Killers

Serial killers, although a small proportion of total murderers, represent some of the most notorious criminals in the United States including Son of Sam, Jeffrey Dahmer, Ted Bundy, Jack the Ripper, and Richard Ramirez, the "Night Stalker." The present research by Call (2019), unlike previous analyses of crime in the mass media, examines the extent to which film depictions are accurate reflections of reality or contradict official records of serial killers, their victims, and the social context of their murders.

Sample and Methods

The present research analyzed the portrayal of serial murderers in films over a 35-year period. The films used were obtained from the Internet Movie Database (IMDb), using the search terms "serial murder," "serial killer," and "serial homicide." The sample criteria included feature-length, English-language movies that had been reviewed at least 100 times by IMDb users and that were released after January 1980. IMDb plot summaries were assessed to make sure the film took place in the United States, only portrayed fictional killers, contained no fantasy elements, and included only the initial appearance of a serial killer in a film. Of the 503-film sampling frame, 50 films were randomly selected.

The author measured sex, race, age, employment status and occupation, motivation, mobility, relationship status, "cooling off period," and fate of the pictured serial killer(s). Notable fictional serial killers included Michael Myers in *Halloween* (2007), Roy Burns in *Friday the 13th: In the Beginning* (1985), and Claywood Foster and Corey Holland in *The Town That Dreaded Sundown* (2014). Measured characteristics of the murder victims included sex, race, age, relationship to the killer, method and location of death, and occupation.

Findings and Conclusions

As portrayed in the films, the 53 serial killers were overwhelmingly adult, White males, who were most often employed as law enforcement officers or as first responders. In a clear majority of the films, the murderer's motivation for killing was revenge, and in three-quarters of the films there was no "cooling off period" between killings, and the murderer was most often killed rather than apprehended by authorities. Similarly, the majority of the 361 murder victims, like their killers, were White males. The victim's relationship to the murderer in two-thirds of the films was that of a stranger, and the most common method of death was by stabbing or cutting. The most frequently portrayed location of the killings was in a residence, either the victim's, another character's, or the serial killer's.

When comparing the results of the content analysis to the reality of serial killings, there were some striking discrepancies as well as accurate reflections. First, in the analysis of films, there were no minority serial killers. In fact, research indicates that serial killers are more likely to be Black than White despite the public's misconception of it being a "White" crime. Second, the FBI defines a serial killing as the murder of three or more people separated in time by a "cooling off period." Perhaps because of the time limitations of film, nearly half of the serial killings portrayed had no cooling off period. Third is the misrepresentation of the killing event. Serial killers are most often motivated by the desire to kill, not revenge; they kill women, not men; they shoot rather than stab their victims; and they are rarely killed prior to being captured.

Consistent with real-life serial killing events, serial killers in the films analyzed were disproportionately male. Second, killings were local events, i.e., serial killers did not travel long distances to commit their crimes. Again, contrary to public belief, serial killers do not usually operate in multiple states. Other accurate reflections of empirical research include the relative age of the victims, adults, and the fact that serial killers and their victims are most often strangers.

Finally, the author notes some limitations of the present research. Top grossing films were not identified and analyzed separately. International films, films where the killing event(s) took place outside of the United States, and documentaries were excluded. Measuring the *inaccuracies* in documentaries of real-life serial killers could be instrumental in identifying the origins of public myths surrounding serial killers since audiences are more likely to believe nonfiction rather than fictionalized accounts. It is yet to be determined whether the inaccuracies identified in the present films are the source of such myths, or whether other media such as crime-time television also contribute to the discrepancies between fact and fiction.

SOURCE: Call 2019.

has declined 24 percent since 2010, with a 4.1 percent decrease between 2018 and 2019 (FBI 2020a). **Larceny**, or simple theft (e.g., shoplifting, purse snatching), accounts for nearly three-fourths of all property arrests and is the most common index offense. Of the known categories, the most common theft is from a motor vehicle, followed by shoplifting from a building. In 2019, the average dollar value lost per larceny incident was $1,162 (FBI 2020a).

larceny Larceny is simple theft; it does not entail force or the use of force or breaking and entering.

Because of the cost involved, *motor vehicle theft* is considered a separate index offense. Numbering over 721,000 in 2019, the motor vehicle theft rate has decreased 8 percent since 2010 (FBI 2020a). The average value of a motor vehicle stolen in the

United States in 2019 was $8,866 (FBI 2020a). For model years 2016 to 2018, the most frequently stolen motor vehicles in 2019 were the Dodge HEMI Charger, the Dodge Challenger SRT Hellcat, the Infinity Q50, the Infinity QX80, and the GMC Sierra 1500 crew cab (Insurance Institute for Highway Safety 2019). For the third year in a row, Albuquerque, New Mexico, remains the car theft capital of the United States.

Burglary, which is the second most common index offense after larceny, entails entering a structure, usually a house, with the intent to commit a crime while inside. The number of burglaries decreased by 29.6 percent between 2015 and 2019 (FBI 2020a). Official statistics indicate that, in 2019, over 1.1 million burglaries occurred. Most burglaries are residential rather than commercial and take place during the day when houses are unoccupied.

What do you think?

Some crimes have increased during the COVID-19 pandemic, such as online fraud, extortion, and domestic violence. Other crimes, many of them violent, have decreased due to a lack of available victims. What other crimes do you think have been impacted by quarantines and closings, and do you think they've increased or decreased?

Arson involves the malicious burning of the property of another. Estimating the frequency and nature of arson is difficult given the legal requirement of "maliciousness." Of the reported cases of arson, almost half involved a structure, most of which were residential, and about a quarter involved movable property (e.g., boat or car), with the remainder being miscellaneous property (e.g., crops or timber). In 2019, the average dollar amount of damage as a result of arson was $16,371 (FBI 2020a).

Vice Crime

Vice crimes, often thought of as crimes against morality, are illegal activities that have no complaining participant(s) and are often called **victimless crimes**. Examples of vice crimes include using illegal drugs, engaging in or soliciting prostitution, illegal gambling, and pornography.

Most Americans view illicit drug use, with the exception of marijuana, as socially disruptive (see Chapter 3). There is less consensus, however, nationally or internationally, that gambling and prostitution are problematic. For example, Germany legalized prostitution in 2002, and it is estimated that there are now more than 400,000 sex workers, two-thirds of whom are from overseas. That said, one of the arguments commonly voiced against prostitution is that it makes it easier for women and girls to be forced into prostitution by traffickers. In a study of 116 countries, Cho et al. (2013) conclude that countries where prostitution is legal have higher rates of human trafficking than countries in which prostitution is illegal.

In Canada, it is illegal to receive benefits for the exchange of sexual services, to engage a person for the purposes of sexual services in exchange for material benefits, or to advertise the sale of sexual services (Haak 2018). In the United States, prostitution is illegal with the exception of several counties in Nevada, although currently several states, including Maine, Massachusetts, New York, and Vermont, are considering decriminalizing it (McKinley 2019; Rathke 2020).

Despite its illegal status, it is a multimillion-dollar industry, with 26,713 people arrested for prostitution and commercial vice in 2019 (FBI 2020a). Human trafficking for purposes of prostitution occurs both *between* the United States and other countries as well as *within* the United States. Children are particularly vulnerable.

> Offenders of this crime who are commonly referred to as traffickers, or pimps, target vulnerable children and gain control over them using a variety of manipulative methods. Victims frequently fall prey to traffickers who lure them in with an offer of food, clothes, attention, friendship, love, and a seemingly safe place to sleep. After cultivating a relationship with the child and engendering a false sense of trust, the trafficker will begin engaging the child in prostitution, and use physical, emotional, and psychological abuse to keep the child trapped in a life of prostitution. (USDOJ 2020b, p. 1)

victimless crimes Illegal activity that has no complaining participant(s), often thought to be a crime against morality.

In 2019, the FBI-led Operation Independence Day, a nationwide weekend sweep by members of over 400 law enforcement agencies, led to the arrest of 67 suspected traffickers, the rescue of 103 juveniles, and the opening of 60 new federal sex trafficking investigations (FBI 2019b).

Gambling is legal in many U.S. states, including casinos in Nevada, New Jersey, Connecticut, North Carolina, and other states, as well as state lotteries, bingo parlors, horse and dog racing, and jai alai. In 1961, Congress passed the *Federal Wire Act*, which prohibited interstate gambling, but, in 2011, the law was interpreted as permitting online gambling with the exception of sports betting. However, in 2019, the Department of Justice rejected the 2011 Obama-era interpretation, arguing that the 1961 Act applies to all forms of interstate gambling including sports betting. The debate over the scope of the *Federal Wire Act* is presently in the federal courts (Gatto 2019).

“Liquor prohibition led to the rise of organized crime in America, and drug prohibition has led to the rise of the gang problems we have now.”

–DREW CAREY, COMEDIAN

Opponents of legalized online gambling, primarily casino owners, argue that legalization would result in the furtherance of organized crime, corrupt minors, and increase gambling addiction rates. Proponents argue that casino owners, rather than truly being concerned about the safety and morality of the country, are simply trying to protect their financial interests. Some have argued that there is little difference between gambling and other risky ventures such as investing in the stock market, other than societal definitions of acceptable and unacceptable behavior. Conflict theorists are quick to note that the difference is who is making the wager.

Pornography, particularly Internet pornography, is a growing international problem. Regulation is made difficult by fears of government censorship and legal wrangling as to what constitutes “obscenity.” For many, the concern with pornography is not its consumption per se but the possible effects of viewing or reading pornography—increased aggression. Malamuth et al. (2012), using a representative sample of American men, concluded that some men actively seek out violent pornographic material, which in turn leads to sexually aggressive attitudes toward women. Further, in an investigation of tenth graders, Rostad et al. (2019) report that, taken as a whole, “our findings suggest that exposure to violent pornography may be a significant correlate of all types of TDV [teen dating violence] perpetration and victimization, particularly for male adolescents” (Rostad et al. 2019, p. 2144).

Organized Crime

Traditionally, **organized crime** refers to criminal activity conducted by members of a hierarchically arranged structure devoted primarily to making money through illegal means. For many people, organized crime is synonymous with the Mafia, a national band of interlocked Italian families, or the “Irish mob” made famous by such movies as *Road to Perdition*, *Gangs of New York*, and *The Irishman*. The Irish mob is one of the oldest organized crime groups in the United States. Fitting the stereotype of the “gangster” is James (Whitey) Bulger who, in 2013, was charged with 19 murders, drug trafficking, extortion, bribery, bookmaking, and loan sharking, just to name a few of the crimes listed in the 32-count indictment. He was convicted of 31 counts and sentenced to two consecutive life terms in prison (Murphy and Valencia 2013; Seelye 2013), where he died in 2018.

organized crime Criminal activity conducted by members of a hierarchically arranged structure devoted primarily to making money through illegal means.

white-collar crime Includes *occupational crime*, in which individuals commit crimes in the course of their employment; *corporate crime*, in which corporations violate the law in the interest of maximizing profit, and *political crime*, in which government actors, by their actions or inactions, commit a crime in their own self-interest to the detriment of the state.

Organized crime also occurs in other countries. In 2018, just one day after the “Godfather” of the Sicilian mafia was arrested, raids targeted suspected members of ’Ndrangheta in Italy, Belgium, the Netherlands, and Germany (Hada, Smith-Spark, and Schmidt 2018). Officials noted that the multinational operation was the largest ever to take place in Europe and that this mafia-styled organized crime group “is one of the most powerful criminal networks in the world, and controls much of Europe’s cocaine trade, combined with systematic money laundering, bribery and violent acts” (p. 1). It is estimated that ’Ndrangheta’s revenues were as high as $60 billion a year.

White-Collar Crime

According to the FBI, **white-collar crime** includes a:

> full range of fraud committed by business and government professionals. These crimes are characterized by deceit, concealment, or violation of trust and are

not dependent on the application or threat of physical force or violence; the motivation behind these crimes is financial—to obtain or avoid losing money, property, or services or to secure a personal or business advantage. (FBI 2019c, p. 1)

Thus, white-collar crime includes *occupational crime*, in which individuals commit crimes in the course of their employment; *corporate crime*, in which corporations violate the law in the interest of maximizing profit; and *political crime*, in which government actors, by their actions or inactions, commit a crime in their own self-interest to the detriment of the state (see Table 4.3).

TABLE 4.3 Types of White-Collar Crime

Crimes against Consumers	Crimes against Employees
Deceptive advertising	Health and safety violations
Antitrust violations	Wage and hour violations
Dangerous products	Discriminatory hiring practices
Manufacturer kickbacks	Illegal labor practices
Physician insurance fraud	Unlawful surveillance practices
Crimes against the Public	**Crimes against Employers**
Toxic waste disposal	Embezzlement
Pollution violations	Pilferage
Treason	Destruction of property
Security violations	Counterfeit production of goods
Police brutality	Business credit card fraud

Occupational Crime. Occupational crime is motivated by individual gain. Employee theft of merchandise, embezzlement, and insurance fraud are examples of occupational crime. In 2020, former president of the United Auto Workers (UAW), one of the largest labor unions in the United States, pled guilty to embezzling over $1 million in union money. Gary Jones acknowledged that he had used union funds for vacations, clothing, liquor, golfing, and expensive meals and had bought $60,000 worth of cigars and smoking paraphernalia (Boudette 2020).

In a related case, a top-ranking labor negotiator for Fiat Chrysler pled guilty to using union funds to purchase a Ferrari and renovate his nearly 7,000-square-foot house (Boudette 2020). His arrest, as Jones's, was part of a larger federal investigation that began years ago and is still active. Corruption in the UAW and related organizations are at such a grand scale that the U.S. Attorney's office is considering a federal takeover of the union. Over 30 years ago, the federal government took over the operations of the International Brotherhood of Teamsters. In February 2020, federal oversight of the Teamsters ended.

Corporate Crime. Price fixing, antitrust violations, and security fraud are all examples of corporate crime, that is, crime that benefits the organization. Officials of the French bank BNP Paribas pled guilty to criminal charges that they had violated U.S. money laundering laws. The bank was fined nearly $9 billion (Thompson and Perez 2015). Leaked files from the Switzerland subsidiary of the British bank HSBC, the second-largest bank in the world, revealed that bank officials had "helped wealthy customers dodge taxes and conceal millions of dollars of assets, doling out bundles of untraceable cash and advising clients on how to circumvent domestic tax authorities" (Leigh et al. 2015, p. 1). In 2019, the subsidiary admitted helping customers evade taxes and agreed to pay $192 million in fines (Michaels 2019).

Corporate Violence. **Corporate violence**, a form of corporate crime, refers to the production of unsafe products and the failure of corporations to provide a safe working environment for their employees. Corporate violence is the result of negligence, the pursuit of profit at any cost, and intentional violations of health, safety, and environmental regulations. The automobile industry provides several examples of corporate violence.

In the largest automobile recall in history, nearly 70 million cars across 14 different automakers were recalled by the National Highway and Traffic Safety Administration and/or car manufacturers for dangerous airbags (Krisher 2020). The injuries from the airbags, which exploded during minor accidents or, in some cases, for no apparent reason, were so traumatic that investigating police officers often thought that the victims had been shot or stabbed (Isidore 2014). To date, there have been 25 deaths and nearly 200 injuries as a result of the airbags deploying (Kageyama 2017; Krisher 2020). Workers at Takata Corporation, makers of the airbags, say officials knew of the problem as early as 2004 when after-hours and weekend tests were performed in secret. Upon

corporate violence The production of unsafe products and the failure of corporations to provide a safe working environment for their employees.

discovery of the problem, instead of alerting safety officials, "Takata executives discounted the results and ordered the lab technicians to delete the testing data from their computers and dispose of the airbag inflators in the trash" (Tabuchi 2014).

Lastly, dozens of people were killed, thousands of buildings demolished, and the town of Paradise in northern Carolina nearly destroyed as a result of the 2018 Camp Fire. A year-long investigation revealed that, despite repeated warnings, Pacific Gas and Electric (PG&E) executives failed to replace outdated power lines, resulting in the sparks that caused the horrific fire. In 2020, PG&E pled guilty to 84 counts of involuntary manslaughter and one felony count of unlawfully starting a fire. Bob Johnson, the CEO, after watching photographs and hearing the names of each victim replied, "Guilty, Your Honor"—84 times. The company was fined $3.5 million, and a $25.5 billion settlement was established to compensate victims, their families, and agencies in the county where the fire took place. Said the district attorney who prosecuted the case, "[T]his is the first time that PG&E, or any major utility, has been charged with homicide as a result of a record fire" (Romo 2020).

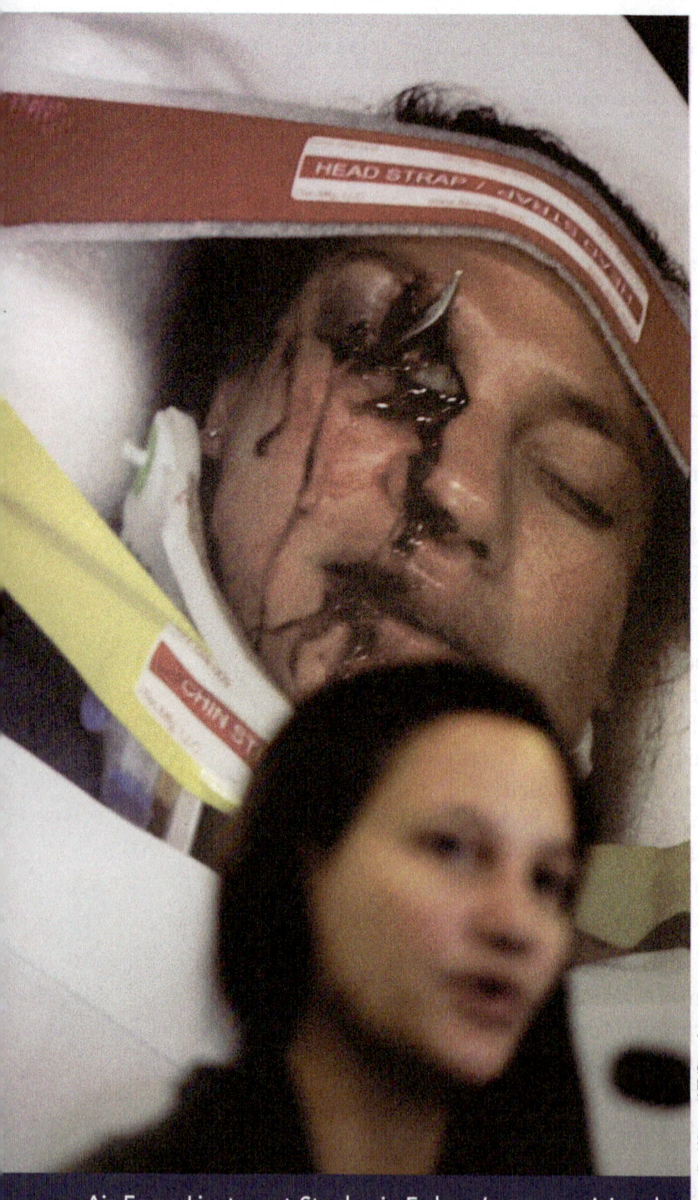

Jim Watson/AFP/Getty Images

Air Force Lieutenant Stephanie Erdman's eye was injured by shrapnel when the airbag in her 2002 Honda Civic exploded. Lieutenant Erdman testified before the U.S. Senate Committee on Commerce, Science, and Transportation on the defective Takata airbags and the need for vehicle recalls. In 2020, 10 million cars were recalled, bringing the total number of cars recalled to approximate 70 million (Krisher 2020).

Political Crime. Only a member of a municipal, county, state, or national government can commit a political crime. In 2020, for example, former Louisiana state senator Wesley Bishop pled guilty to making false statements to the U.S. Department of Housing and Urban Development in relation to rental property he owned (Zurik and Lillich 2020). Duncan Hunter was sentenced to 11 months in prison after pleading guilty to stealing campaign funds. The 60-count indictment documented his decade-old criminal activity including use of campaign funds for personal expenses, which were then written off as contributions to the non-profit veteran organization, Wounded Warriors (Watson 2020).

Senator Richard Burr, called the pandemic profiteer by his critics, was recorded warning his donors about the coronavirus, saying that it is "much more aggressive in its transmission than anything we have seen in recent history," while publicly denying its severity (quoted in Sexton 2020, p. 1). Compounding this breach of trust, Senator Burr sold nearly $2 million in stocks, in over 30 separate transactions, just one week before the stock market plunged as a result of COVID-19 fears. Despite being a member of the Senate Intelligence Committee and therefore having first-hand information about the pandemic, said Senator Burr in his defense, "I relied solely on public news reports to guide my decision regarding the sale of stocks" (p. 1). On January 19, 2021, the U.S. Department of Justice announced it would not pursue insider trading charges against Senator Burr (Fandos and Benner 2021).

Lastly, there is a long list of political crimes and alleged political crimes committed by former President Trump and/or his associates. After less than six months in office, Donald Trump became the focus of an investigation into whether there was Russian interference in the 2016 presidential election and, if so, whether there were links that would directly tie him or his associates to Russian officials. After a two-year investigation, the debate continues as to the culpability of the president and/or his associates. Although special prosecutor Mueller's Report (1) found "sweeping and systematic" interference by the Russian government, (2) identified links between the Russian government and the Trump campaign, (3) resulted in 37 indictments and seven guilty pleas, and (4) found evidence that the president had obstructed justice on at least five occasions, Mueller followed the guidelines of the Department of Justice that a sitting president cannot be indicted. The report, however, did not exonerate former President Trump (Presidential Investigation Education Project 2019).

As a result of the Mueller investigation, several Trump associates have been charged, convicted, and sentenced for a variety of political crimes, including witness tampering, obstruction of justice, and/or making false statements to the FBI or to Congress. These associates include but are not limited to Michael Cohen, Paul Manafort, Rick Gates, George Papadopoulos, Michael Flynn, and Roger Stone (Weiss, Cranley, and Panetta 2020). Days before his prison sentence was to begin and despite Mr. Stone's confession, in 2020 Donald Trump commuted his sentence. Tweeted Republican Senator Mitt Romney, "Unprecedented, historic corruption: an American president commutes the sentence of a person convicted by a jury of lying to shield that very president" (Shaw 2020, p. 1).

deferred prosecution agreement (DPA) An agreement that does not require a guilty plea and, if the requirements of the DPA are fulfilled, there usually is no further legal action.

What *do you* think?

In 2019 and 2021, former President Trump was impeached by the U.S. House of Representatives, charged with, respectively, abuse of power and obstruction of Congress, and incitement of insurrection. He was acquitted in both proceedings, although in the second, several prominent Republicans voted to convict. Many legal scholars argue that the former president violated federal law which prohibits two or more people [the former president and his supporters] using force [the attack on the U.S. Capitol] "to prevent, hinder, or delay" the lawful execution of any U.S. statute [the certification of the 2020 election] (18 USC 2384). Do you think Donald Trump should be held *criminally* responsible for violating federal law regardless of the Senate's failure to convict?

White-collar criminals go unpunished for a variety of reasons. First, many companies, organizations, or government entities, not wishing the bad publicity surrounding a scandal, simply dismiss the parties involved. Second, many white-collar crimes, as with traditional crimes, go undetected in part because they occur away from public scrutiny, often taking place in complex bureaucracies. Third, federal prosecutions of white-collar criminals have generally decreased as a result of the time- and resource-intensive nature of prosecution. **Deferred prosecution agreements (DPAs)**, which do not require a guilty plea from the defendant, are alternatives to adjudication. If the offender fulfills a prosecutor's requirement for a DPA, usually a large fine, the state takes no further action (Khan 2018).

SEEKING INFORMATION

VIOLENCE AT THE UNITED STATES CAPITOL

WASHINGTON, D.C.
JANUARY 06, 2021

Photograph #31

Photograph #32

Photograph #33

Photograph #34

Photograph #35

Photograph #36

Photograph #37

Photograph #38

Photograph #39

Photograph #40

DETAILS

The Federal Bureau of Investigation's (FBI) Washington Field Office is seeking the public's assistance in identifying individuals who made unlawful entry into the United States Capitol Building on January 6, 2021, in Washington, D.C.

Anyone with information regarding these individuals, or anyone who witnessed any unlawful violent actions at the Capitol or near the area, is asked to contact the FBI's Toll-Free Tipline at 1-800-CALL-FBI (1-800-225-5324) to verbally report tips. You may also submit any information, photos, or videos that could be relevant online at fbi.gov/USCapitol. You may also contact your local FBI office or the nearest American Embassy or Consulate.

Field Office: Washington D.C.

www.fbi.gov

Federal Bureau Of Investigation

Computer Crime

Computer crime refers to any violation of the law in which a computer is the target or means of criminal activity and thus may or may not include white-collar crimes. Sometimes called *cybercrime*, computer crime is one of the fastest-growing crimes in the United States (see Chapter 14). Hacking, or unauthorized computer intrusion, is one type of computer crime. In 2019, hackers accessed over 8 billion consumer records; some of the largest violations coming from online stores, banks, and credit card companies (Leonhardt 2019). Linked to Russian operatives, in 2020, thousands of networks were hacked globally, including those in the U.S. government, compromising not only individual privacy but national security (Sebastian 2021).

The Consumer Sentinel Network logged nearly 3.2 million consumer complaints in 2019, with identity theft being the second-largest named category of complaints (Federal Trade Commission 2020). **Identity theft**, the second type of computer crime, is the use of someone else's identification (e.g., Social Security number or birth date) to obtain credit or other economic rewards. After one of the largest data breaches in history, in 2020, four members of the Chinese military were charged with breaking into Equifax, a credit reporting agency, and stealing the personal information of 150 million Americans (Geller 2020).

A third type of computer crime is Internet fraud. According to the Internet Crime Complaint Center (ICCC 2019a), common types of Internet fraud include non-delivery of goods and services, advance fee schemes, and employment or business opportunity scams. One of the most disturbing Internet crimes of late is online businesses (individuals posing as businesses) citing "fake 'newly enacted' COVID-19 shipping laws, regulations, or insurance requirements" and charging additional delivery fees (ICCC 2020). Individuals who have low levels of self-control, risk avoidance, and self-awareness, along with high levels of trust, are more susceptible to Internet fraud (Williams, Beardmore, and Joinson 2017).

computer crime Any violation of the law in which a computer is the target or means of criminal activity.

identity theft The use of someone else's identification (e.g., Social Security number, birth date) to obtain credit or other economic rewards.

ransomware A form of malware intrusion in which a criminal holds an individual's or company's computer "hostage."

Ransomware, a fourth kind of computer crime, is a form of malware intrusion in which a criminal holds an individual's or company's computer "hostage" until the victim meets the "kidnappers'" demands. In 2019, over 200,000 organizations reported ransomware attacks, with some victims paying as much as $1 million to the hackers (Popper 2020). The FBI recommends that victims not pay the ransom, given that, in documented cases, a valid decryption key was not provided even after payment was made. Of late, "ransomware attacks are becoming more targeted, sophisticated, and costly" (ICCC 2019b, p. 1).

Finally, online child pornography and child sexual exploitation is the fifth kind of computer crime. In 2019, the CyberTipline of the National Center for Missing and Exploited Children (NCMEC) received reports that included "69.1 million images, videos and other files related to child sexual exploitation" (NCMEC 2020, p. 1). Online gaming is just one method that offenders use to meet potential victims. In 2019, an 11-year-old boy was coerced into producing child pornography after meeting a predator online who convinced him to communicate through a private chat room. The 29-year-old offender was sentenced to 14 years in prison (ICCC 2019c). Table 4.4 displays the results of a survey of corporate executives who were asked what information they would most like to protect, i.e., what information is most valuable to cybercriminals.

TABLE 4.4 Value of Information to Cybercriminals Ranked by Corporate Executives, 1 (most valuable) to 10 (least valuable)

Type of Information	Rank
Customer information	1
Financial information	2
Strategic plans	3
Board member information	4
Customer passwords	5
Research and Development information	6
Mergers and Acquisitions information	7
Intellectual property	8
Non-patented Intellectual Property	9
Supplier information	10

SOURCE: Global Information Security Survey 2020.

Juvenile Delinquency and Gangs

In general, children younger than age 18 are handled by the juvenile courts, either as status offenders or as delinquent offenders. A *status offense* is a violation that can be committed only by a juvenile, such as running away from home, truancy, and underage drinking. A *delinquent offense* is an offense that would be a crime if committed by an adult, such as the eight index offenses. The most common status offenses handled in juvenile court are underage drinking, truancy, and running away. In 2019, 7 percent of all arrests (excluding traffic violations) were of offenders younger than age 18 (FBI 2020a). As is the case with adults, juveniles commit more property crimes than violent crimes, and males are more likely to be arrested than females.

Originating in Los Angeles in the 1970s by Salvadoran refugees, one of the largest and most notorious gangs in the United States is Mara Salvatrucha, or MS13. Its estimated membership is between 50,000 and 70,000, and it has affiliates in half a dozen countries around the world, including El Salvador, Spain, and Guatemala. The gang has no national leaders, only councils that control local territories, and some areas are externally controlled. For example, the Los Angeles MS13 takes orders from the prison gang, Mexican Mafia (Dudley 2019). Research indicates that MS13 is first and foremost a social organization and, second, a criminal organization. A survey of MS13 members in El Salvador indicates the socioemotional reasons for joining the gang. Nearly half of the members joined because they wanted to "hang out" with gang members, and 17 percent reported that the gang gave them close friends and "brothers" (Cruz et al. 2017). Further, in conversations with one another, gang members often refer to *el barrio*, a word that means "neighborhood" and carries with it an intimate connotation (Dudley 2019). Thus MS13, as with so many other gangs, provides a "collective identity that is constructed and reinforced by shared, often criminal experiences, especially acts of violence and expressions of social control" (Dudley 2019, p. 3).

Demographic Patterns of Crime

Everyone violates a law at some time; however, individuals with certain demographic characteristics are disproportionately represented in the crime statistics (see Figure 4.2). The probability of being an offender varies by gender, age, race, social class, and region, as does the likelihood of being a victim.

Gender and Crime

It is a universal truth that women everywhere are less likely to commit crime than men. In the United States, both official statistics and self-report data indicate that females

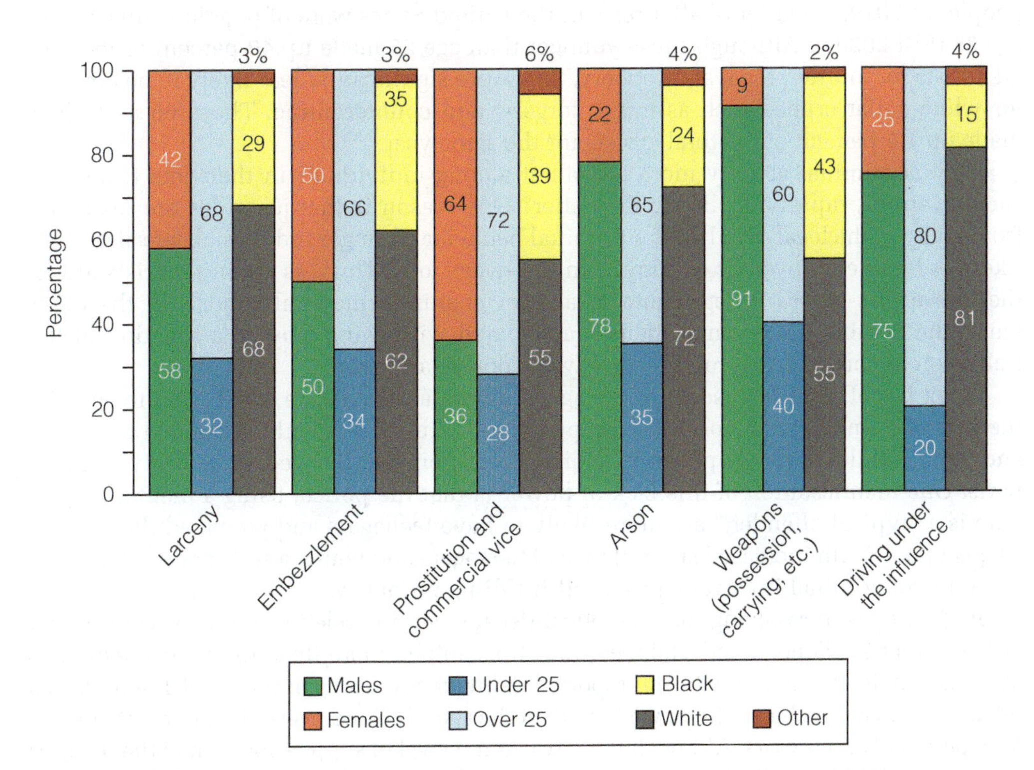

Figure 4.2 Percentage of Arrests for Selected Crimes by Sex, Age, and Race, 2019*

*Numbers have been rounded to equal 100.

SOURCE: FBI 2020a.

commit fewer violent crimes than males. In 2019, males accounted for three-quarters of all arrests (FBI 2020a). Not only are females less likely than males to commit serious crimes, the monetary value of female involvement in theft, property damage, and illegal drugs is typically far less than that for similar offenses committed by males.

The rates of female criminality, while remaining lower than those for males, have been increasing. Between 2015 and 2019, arrest rates for women increased for three of the four violent crimes. Not unrelated, during the same time period, weapons offenses such as carrying or possessing a firearm and drug abuse violations increased by 16 and 17 percent, respectively (FBI 2020a). Despite these increases, the gender gap in offending remains, and the more serious the crime is, the larger the gap. For example, males comprised 79 percent of all arrests for violent crimes but just 62 percent of all property crimes (FBI 2020a).

The increase in crimes committed by females has led to the growth of feminist criminology. **Feminist criminology** focuses on how the subordinate position of women in society affects their criminal behavior and victimization. For example, arrest rates for runaway juvenile females are higher than those for males not only because girls are more likely to run away as a consequence of sexual abuse in the home but also because police with paternalistic attitudes are more likely to arrest female runaways than male runaways (Sheldon 2013; Zahn et al. 2010; Pasko 2017). Feminist criminology, concentrating on gender inequality in society, thus adds insights into understanding crime and violence that are often neglected by traditional theories of crime (Malibon, Carson, and Yates 2018; Mallicoat 2020). Feminist criminology has also had an impact on public policy. Mandatory arrest for domestic violence offenders, the development of rape shield laws, public support for battered women's shelters, laws against sexual harassment, and the repeal of the spousal exception in rape cases are all, according to Winslow and Zhang (2008), outcomes of feminist criminology.

Age and Crime

In general, criminal activity is more prevalent among younger people than among older people. In 2019, a quarter of all arrests in the United States were of people younger than age 25 (FBI 2020a). Although those younger than age 25 made up 40 percent or more of all arrests for murder, rape, and robbery, they were significantly less likely to be arrested for white-collar crimes such as fraud, forgery, and counterfeiting. Those 65 and older made up 1.5 percent of the total arrests for the same year.

Why is criminal activity more prevalent among individuals in their late teens and early twenties, rapidly declining thereafter? One reason is that juveniles are insulated from many of the legal penalties for criminal behavior. Younger individuals are also more likely to be unemployed or employed in low-wage jobs. Thus, as strain theorists argue, they have less access to legitimate means for acquiring material goods. On the other hand, the decline in criminal offenses associated with aging may be a function of the transition to conventional roles—employee, spouse, and parent.

Other hypothesized reasons for the age–crime relationship are also linked to specific theories of criminal behavior. For example, conflict theorists would argue that teenagers and young adults have less power in society than their middle-aged and elderly counterparts. One manifestation of this lack of power is that the police, using a mental map of who is a "typical offender," are more likely to have teenagers and young adults in their suspect pool. With increased surveillance of teenagers and young adults comes increased detection of criminal involvement—a self-fulfilling prophecy.

In the hopes of resolving the theoretical debate over the relationship between age and crime, Sweeten, Piquero, and Steinberg (2013) examined more than 40 independent variables and their association with self-reported delinquency. Using a sample of 1,300 serious offenders, the researchers interviewed youth at the age of 16 and every six months thereafter for a period of seven years. Although there was some level of support for each of the theories tested, the strongest theoretical explanation of the relationship between age and delinquency was Sutherland's learning theory—the differential association theory—confirming the positive association between criminality and antisocial peers.

feminist criminology An approach that focuses on how the subordinate position of women in society affects their criminal behavior and victimization.

Race, Social Class, and Crime

Race is a factor in who gets arrested, convicted, and incarcerated. For example, Black people represent just 13.0 percent of the population but account for 36.4 percent of all arrests for violent index offenses and for 29.8 percent of all arrests for property index offenses (FBI 2020a). Black individuals have 3.6 times the arrest rate of non-minorities for possession of marijuana, and despite an increase in the number of states that have legalized marijuana, the racial gap has remained the same since 2010 (American Civil Liberties Union [ACLU] 2020). Nevertheless, it would be inaccurate to conclude that race and crime are causally related.

First, there is research evidence suggesting that the differences in arrest rates are a consequence of differences in police practices in Black and White neighborhoods (see this chapter's section on "Rethinking Law Enforcement"). In an analysis of 300 drug arrest reports from the St. Louis police between 2009 and 2013, Gaston (2019, p. 324) concludes:

> Unlike drug arrests in White neighborhoods or of White citizens that primarily stem from reactive policing, drug arrests in Black and racially mixed neighborhoods and of Black citizens result from officers' greater use of discretionary stops based on neighborhood conditions, suspicion of ambiguous demeanor, or minor infractions.

Second, the high rate of arrests, convictions, and incarceration of minorities may also be a consequence of structural variables. Velez, Lyons, and Santoro (2015), using neighborhood-level data from 87 U.S. cities, report that the relationship between race and violent crime is mediated by the political context of the city. In cities where there was, for example, a Black mayor or a civilian police review board, the percentage of Black residents in a neighborhood had "no detectable role ... in predicting violent crime" (Velez, Lyons, and Santoro 2015, p. 110).

Racial bias, sometimes called **racial profiling**, is the practice of targeting suspects on the basis of race. Proponents of the practice argue that because race, like gender, is a significant predictor of who commits crime, the practice should be allowed. Opponents hold that racial profiling is little more than discrimination, often based on stereotypes, and thus should be abolished. A survey of Seattle residents lends support to such a contention (Drakulich 2013). The results indicate that "crime stereotypes about racial and ethnic minorities are associated with reduced perceptions of neighborhood safety and increased anxieties about victimization" among White respondents (p. 322).

racial profiling The law enforcement practice of targeting suspects on the basis of race.

A third reason for the higher rates of Black individuals' involvement in the criminal justice system is that non-White individuals are overrepresented in the lower classes. Because lower-class members lack legitimate means to acquire material goods, they may turn to instrumental, or economically motivated, crimes. In addition, although the haves typically earn social respect through their socioeconomic status, educational achievement, and occupational role, the have-nots more often live in communities where respect is based on physical strength and violence,

Jeff J Mitchell/Getty Images News/Getty Images

In 2020, the killing of George Floyd by a Minneapolis police officer led to protests all over the world in support of Black Lives Matter (BLM). Here, protestors in Glasgow, Scotland, as throughout the United Kingdom, participate in a BLM demonstration.

as subcultural theorists argue. For example, Kubrin (2005) examined the "street code" of inner-city Black neighborhoods by analyzing rap music lyrics. Her results indicate that "lyrics instruct listeners that toughness and the willingness to use violence are central to establishing viable masculine identity, gaining respect, and building a reputation" (p. 375).

Region and Crime

In general, crime rates and, in particular, violent crime rates increase as population size increases. In 2019, the violent crime rate in metropolitan statistical areas (MSAs) was 395.2 per 100,000 inhabitants; in cities outside metropolitan areas, it was 390.8 per 100,000 inhabitants; and, finally, in non-metropolitan counties, the crime rate was 207.5 per 100,000 inhabitants (FBI 2020a). Furthermore, serious violent victimization, including simple and aggravated assault, robbery, and rape, is higher in urban areas than in rural areas for both men and women of all age groups. Despite these urban–rural differences, over the last two decades rural crime has been increasing (Office of Victims of Crime 2017; Mahtani 2018).

@Oregon GovBrown

Last night, the world was watching **Portland**. Here's what they saw: Federal troops left downtown. Local officials protected free speech. And Oregonians spoke out for Black Lives Matter, racial justice, and police accountability through peaceful, non-violent **protest**.

-Governor Kate Brown

Higher crime rates in urban areas result from several factors. First, social control is a function of small intimate groups that socialize their members to engage in law-abiding behavior, expressing approval for their doing so and disapproval for their non-compliance. In large urban areas, people are less likely to know one another and thus are not influenced by the approval or disapproval of strangers. Demographic factors also explain why crime rates are higher in urban areas: Large cities have large concentrations of poor, unemployed, and minority individuals. Some cities, including the ten most dangerous cities in the United States, have crime rates as much as three times the national average (Edwards 2020).

An examination of violent crime rates in April and May 2020, when compared to a previous three-year average for the same two months, indicates that homicide rates declined by 21.5 percent and 9.9 percent, respectively, after increasing in the first three months of the year. The results are consistent with the routine activities approach, which argues that crime statistics, to a large extent, depend on the everyday routines of people. The results of shelter-in-place orders and business closures due to COVID-19 are likely the reasons for the initial reduction in homicides. However, with the continued economic downturn and mounting stress, homicide rates increased dramatically, particularly in large cities, by the end of 2020 (Corley 2021; Abt and Lopez 2020).

Crime rates also vary by region of the country. In 2019, both violent and property crimes were highest in Southern states, with 48.7 percent of all murders, 37.3 percent of all rapes, and 42 percent of all aggravated assaults recorded in the South (FBI 2020a). The high rate of Southern lethal violence has been linked to high rates of poverty and minority populations in the South, a Southern "subculture of violence," higher rates of gun ownership, and a warmer climate that facilitates victimization by increasing the frequency of social interaction.

What *do you think?*

In August of 2020, Attorney General William Barr announced the launch of a new initiative, Operation Legend, named after 4-year-old LeGend Talifferro, who was shot and killed while sleeping in his home in Kansas City, Missouri (U.S. Department of Justice 2020c). The campaign entails sending federal law enforcement officers to large cities with escalating crime rates. At least initially, the program was highly controversial, following on the heels of federal troops being sent to Portland, Oregon, to quell Black Lives Matter protests. Do you think federal law enforcement officers should be used to fight local criminal activity?

Crime and Victimization

Property and violent crime victimization decreased between 2015 and 2019 according to the National Crime Victimization Survey (NCVS), mirroring the dramatic decline in both over the last 20 years (Morgan and Truman 2020). Of violent victimizations reported to the police, females have a higher rate than males, racial and ethnic minorities have a higher rate than White non-Hispanics, and people between the ages of 18 and 24 have the highest rate of violent victimization when compared to other age groups.

Because minorities are disproportionately offenders, and violent victimization is often directed toward family members, friends, and acquaintances, it is not surprising that, for example, Black people are disproportionately victimized. According to the NCVS, when the victim was Black, 70 percent of the time the offender was also Black (Morgan and Truman 2020). Similarly, the majority of violent offences against White people was committed by members of the same racial group.

The overall decrease in violent victimization between 2018 and 2019 was, in part, a function of the decline in rape or sexual assault victimizations (Morgan and Truman 2020). Rape and sexual assaults, along with simple assaults, are the violent crimes least likely to be reported to the authorities. According to the NCVS, in 2019, there were over 459,000 rapes and sexual assaults and more than 4 million simple assaults. Aggravated assault, unlike simple assaults, usually involves the use of a weapon or some other means to cause severe bodily harm.

Females are particularly vulnerable to certain kinds of crimes, most notably those that entail sexual violence. In one of the most horrific cases of its type to date, in 2013, Ariel Castro was charged "with 512 counts of kidnapping, 446 cases of rape, seven counts of gross sexual imposition, six counts of felonious assault, [and] three counts of child endangerment" (Welsh-Huggins 2013, p. 1). Castro, who had held three young women as captives for nearly ten years, was sentenced to life in prison but committed suicide just a month after sentencing (Boyle 2015). Said Michelle Knight 11 years after her captivity, "[I]t was difficult. I had to go blank anytime he was doing anything to me. I had to put myself in a different place" (Yang and Diaz 2020, p. 1).

The Societal Costs of Crime and Social Control

As this chapter has demonstrated, there are a variety of types of crimes and criminals. Nonetheless, often based on media stereotypes, many people think of crimes and criminals as predominantly "young black men living in poor urban neighborhoods committing violent and drug-related crimes" (Leverentz 2012, p. 348). The cost of white-collar crime, however, dwarfs the financial losses resulting from street crimes and is responsible for more death and destruction (Reiman and Leighton 2020). The costs of crime—and many of them are incalculable—go far beyond those perpetrated by what is thought to be the "typical offender." The following section discusses the costs associated with crime and criminals, including physical injury and loss of life, economic losses, social and psychological costs, and the costs to families and children.

Physical Injury and the Loss of Life

Crime often results in physical injury and the loss of life. Homicide is the third-most common cause of death among 20- to 24-year-olds, exceeded only by accidental deaths and suicides (Heron 2019). In 2019, 16,425 people were the victims of known homicides in the United States (FBI 2020a). White-collar crimes that result in death and injury include occupational hazards (e.g., black lung disease), environmental violations (e.g., pollution), medical malpractice (e.g., unnecessary surgery), and consumer safety violations (e.g., defective products) (Mokhiber 2007; Reiman and Leighton 2020).

Green criminologists study the impact of white-collar crime and corporate violence on the environment. In a review of the literature, Katz (2012) identifies several important findings from the existing research on corporate pollution and health, specifically cancer mortality:

> First, a variety of research illustrates multinational corporate culpability in the proliferation of environmental pollution both in the USA and globally. Second, this has resulted in increasing cancer mortality rates across the globe among minority communities as well as non-western developing nation states. Third, corporate environmental pollution has been facilitated through the proliferation of international free trade agreements and international financial loans from international financial institutions.

Using a conflict theory perspective, with its emphasis on the problems created by capitalism, Katz (2012) examined cancer mortality rates in relationship to the locations of the richest international corporations in the world. She reports that ten of the 152 countries studied have significantly higher cancer mortality rates than the remainder and, of that number, six are the locations of the headquarters of the largest transnational corporations. On closer examination, Katz (2012) argues that the data reveal that most of the industries in the ten nations where the cancer mortality rate is the highest are chemical, energy, water, and oil. Furthermore, Dow Chemical Company, a U.S.-based corporation, which Katz (2012) calls "the primary global polluter," has operations in nine of the ten nations.

Finally, the U.S. Public Health Service now defines violence as one of the top health concerns facing Americans. Health initiatives related to crime include reducing drug and alcohol use and the deaths and diseases associated with them, lowering rates of domestic violence, preventing child abuse and neglect, and reducing violence through public health interventions. It must also be noted that crime has mental as well as physical health consequences (see the section titled "Social and Psychological Costs").

The High Price of Crime

Conklin (2007, p. 50) suggests that the financial costs of crime can be classified into at least six categories. First are *direct losses* from crime, such as the destruction of buildings through arson, of private property through vandalism, and of the environment by polluters. In 2019, the average dollar loss as a result of destroyed or damaged property due to arson, per incident, was $16,371 (FBI 2020a). Computer crime also incurs direct costs for those who are victimized. According to the ICCC, in 2018, victims of Internet crime lost $2.7 billion (ICCC 2019d).

Second are costs associated with the *transferring of property*. Bank robbers, car thieves, and embezzlers have all taken property from its rightful owner at tremendous expense to the victims and society. The estimated cost of property crime is $15 billion, and the estimated cost of white-collar crime 30 times that—$500 billion (Khan 2018). Motor vehicle thefts, which include the theft of "sport utility vehicles, automobiles, trucks, buses, motorcycles, motor scooters, all-terrain vehicles, and snowmobiles," cost the American taxpayers over $6 billion in 2019. The average dollar loss per vehicle was $8,886 (FBI 2020a).

A third major cost of crime is that associated with *criminal violence*, including the medical cost of treating crime victims. Victims of sexual violence incur an economic burden of $122,461 over a lifetime in lost wages, health expenses, criminal justice costs, and property damage (Peterson et al. 2017). Just one violent event—although one of the most horrific, the 2017 Las Vegas massacre that killed 58 people and injured more than 800—had an economic toll on the community of more than $600 million (Giffords Law Center [GLC] 2018). Further, in 2020, a federal judge approved a $800 million settlement for the victims and relatives of the mass murder (Associated Press 2020).

Fourth are the costs associated with the production and sale of illegal goods and services—that is, *illegal expenditures*. The expenditure of money on drugs, gambling, and prostitution diverts funds away from the legitimate economy and enterprises, and lowers property values in high-crime neighborhoods. In 2016, an estimated $150 billion was spent on cocaine, heroin, marijuana, and methamphetamine purchases in the United States (Midgette et al. 2019).

Fifth is the cost of *prevention and protection*—the billions of dollars spent on locks and safes, surveillance cameras, self-defense products, security systems, guard dogs, insurance, counseling and rehabilitation programs, and the like. Sales of smart home surveillance cameras such as *Ring* and *Geeni*, now readily affordable, are expected to grow with a 2023 estimated market value of $9.7 billion (Narcotta 2018).

Finally, there is the cost of *controlling crime*. The United States spends more than $80 billion annually in correctional costs (i.e., jails, prisons) (Lewis and Lockwood 2019) and an additional 142.5 billion on law enforcement and 64.7 billion on the judicial and legal systems. Local governments spend more than state governments, and state governments spend more than the federal government (Hayes 2020). As indicated by Figure 4.3, however, criminal justice spending changes over time increasing in some functions (i.e., federal law enforcement) and decreasing in others (i.e., criminal justice assistance) (Congressional Research Services 2017).

Social and Psychological Costs

Crime entails social and psychological costs as well as economic costs. One such cost is fear. As Warner and Thrash (2020) state, fear of crime "has significant consequences for individuals experiencing that emotion, as fear can shape the social and protective actions that individual take, leading them to constrain their movements, behaviors, and social choices" (p. 302) (see this chapter's *Self and Society*). The Gallup Organization regularly asks a sample of Americans, "Is there any area near where you live—that is, within a mile—where you would be afraid to walk alone at night?" (Gallup 2020a). In 2019, 37 percent of respondents answered yes, up from 30 percent in 2017.

The Gallup Organization also regularly asks a sample of U.S. adults how serious the crime problem is in the United States (McCarthy 2019). Despite a general decline in violent crime since the 1990s, 64 percent of Americans believe crime is higher than it was a year ago, and over half of all Americans believe that crime is an extremely serious or very serious problem. Interestingly, Americans are much more likely to describe crime in the United States as a serious problem compared to crime in their local communities, with women and urban dwellers perceiving crime as a greater problem in their local areas than their male and suburban or rural counterparts.

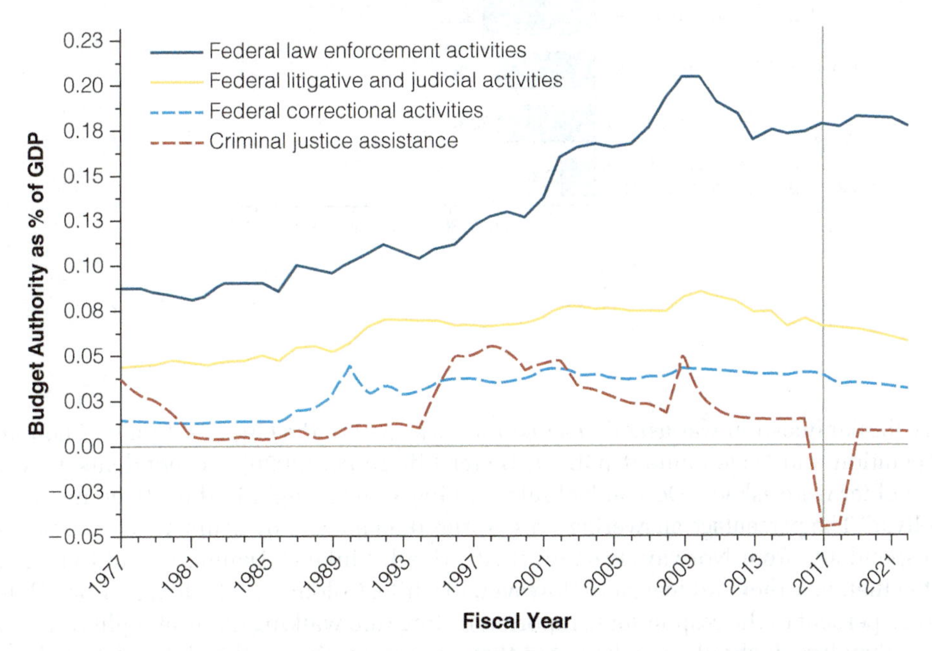

Figure 4.3 U.S. Administration of Justice by Subfunctions, Funding Year 1977 to 2022 (projected)
SOURCE: Congressional Research Services 2017.

Fear of Crime Assessment

Rank the following items from one (1), meaning the crime you worry the most about, to ten (10), meaning the crime you worry the least about. When you are finished, compare your rankings to that of a national sample of respondents that follows.

	RANK
1. Being the victim of a hate crime	____
2. Getting mugged	____
3. Having your car stolen or broken into	____
4. Being the victim of terrorism	____
5. Getting murdered	____
6. Being assaulted/killed by a coworker/employee where you work	____
7. Having your home burglarized when you are not there	____
8. Being attacked while driving your car	____
9. Having your personal, credit card, or financial information stolen by computer hackers	____
10. Being sexually assaulted	____

Here are the rankings of 1,520 U.S. respondents, 18 years and older, in 2019. Respondents were asked "How often do you, yourself, worry about the following things—frequently, occasionally, rarely, or never? How about ________." The percentage after the bar indicates the percentage of respondents who answered "Frequently" and "Occasionally" and are ranked from the most feared to the least feared.

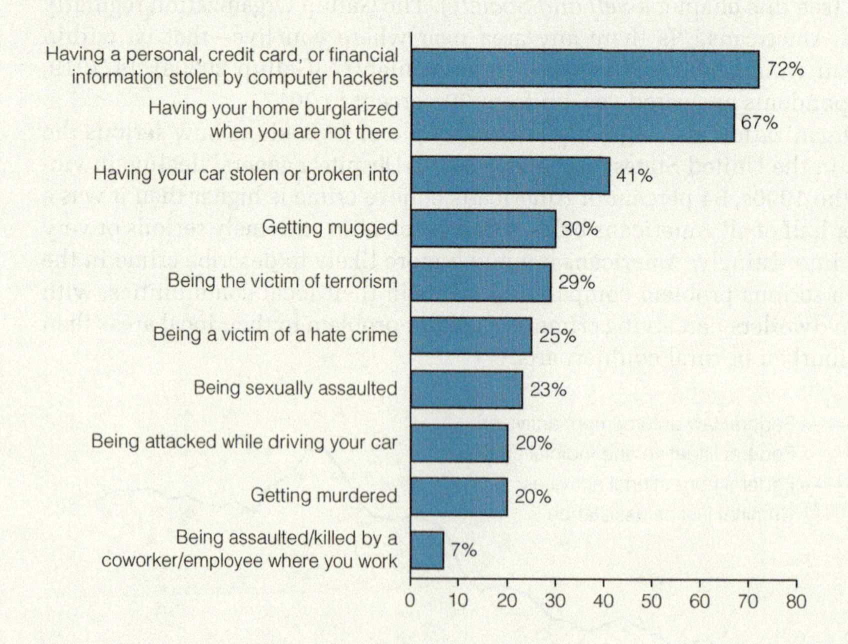

SOURCE: Statista 2019.

> "Rape is a more heinous crime than murder since the rape victim dies throughout the period she lives."
>
> **–AMIT ABRAHAM, PSYCHOLOGIST**

A similar question was asked in 41 countries as part of the Organization for Economic Cooperation and Development (OECD) Better Life Index (2020). Respondents 15 years old or older were asked, "Do you feel safe walking alone at night in the city or area where you live?" The percentage answering "yes" varies dramatically by country. Over 80 percent of respondents from Norway, Iceland, Switzerland, Finland, Denmark, and Canada reported that, yes, they did feel safe. However, in Chile, Colombia, Mexico, and Brazil, less than 50 percent of the respondents reported feeling safe walking alone at night in the city where they live. It should also be noted that an examination of the global data indicates that there are **safety gender gaps** in the countries surveyed; i.e., women expressed significantly lower rates of feeling safe than men.

Over the last several decades, crime rates have declined so dramatically that criminologists regularly call the period after the 1990s "the great crime decline" (LaFree 2018). Nonetheless, in 21 Gallup polls conducted since 1989, a majority of Americans responded that crime rates had risen since the year prior to the survey. Why do you think there is such a disconnect between the reality of crime and the perception of crime in the United States?

What do you think?

In 2014, the U.S. Department of Justice conducted one of the few national studies on the socioemotional problems (SEP) of violent crime using data from the NCVS (Langton and Truman 2014). Results of the investigation indicate that (1) 57.0 percent of violent crime victims experienced SEP; (2) emotional symptoms (e.g., angry, worried, depressed) of SEP were more common than physical symptoms (e.g., insomnia, upset stomach, headaches); (3) victims of intimate/partner violence compared to other victim–offender relationships had the highest rates of SEP; (4) regardless of the type of crime or victim–offender relationship, females were more likely to report SEP than males; and (5) experiencing SEP was associated with higher rates of reporting the crime to the police and receiving victim services. In 2019, only 8 percent of victims of violent crime received assistance from a victim service agency (Morgan and Truman 2019).

Although socioemotional responses vary by type of offense, it should not be concluded that white-collar crimes do not carry a social and psychological toll. In 2009, Bernard Madoff pled guilty to 11 felony counts related to a massive Ponzi scheme, the largest in history, run through his investment firm. The scheme to defraud investors of $65 billion took place over a 20-year period and involved thousands of victims. The 71-year-old Madoff was sentenced to 150 years in prison, leaving behind him a trail of misery. In 2020, Mr. Madoff, now in failing health, requested compassionate release from prison. Said the judge, "When I sentenced Mr. Madoff in 2009, it was fully my intent that he live out the rest of his life in prison. Nothing's happened in the 11 years since to change my thinking" (quoted in Reuters 2020, p. 1).

The following are excerpts from just six of the thousands of victims impacted by his crime. They attest to the social and psychological costs of white-collar crime (Victim Statements 2009):

- "He robbed us not only of our money, but of our faith in humanity, and in the systems in place that were supposed to protect us."
- "I can't tell you how scattered we feel—it goes beyond financially. It reaches the core and affects your general faith in humanity, our government and basic trust in our financial system."
- "I am constantly nervous and anxious about my future. I jump at the slightest noise. I can't sleep and all I do is worry about what will happen to us."
- "I don't know which emotion is more destructive, the fear and anxiety or the major depression that I experienced daily."
- "[W]hen you sentence Madoff, I trust that you touch on the loss of money, the loss of dignity, the loss of freedom from financial worries and possible financial ruin."
- "How do I live the rest of my life?"

The Cost to Children and Families

In addition to the obvious impact of death or injury of a loved one or financial loss, some costs take place slowly over long periods of time and are not the result of direct criminal victimization. For example, half of all adults in the United States have experienced having a family member in jail or prison. This is the result of the dramatic increase in incarceration rates in recent years that more often impacts women and children, people of color, and members of low-income households (Elder et al. 2018).

First, there is the financial cost to the family. Costs include money to post bail, court fees and fines, the cost to stay in touch (e.g., gas, lodging, long-distance calls), and the loss of income and child support if the incarcerated family member was the primary wage earner.

safety gender gap The difference between women's expressed fear and men's expressed fear of crime.

Once the family member leaves prison, there may be the cost of "victim restitution, monthly supervision fees, and other obligations that must be met to remain in compliance with parole requirements and avoid [having the family member] being sent back to prison" (p. 34).

Second, there are physical and mental health consequences for the family members of an incarcerated loved one—depression, high blood pressure, obesity, and diabetes—particularly for mothers who have incarcerated children. Although research is mixed, in general, children of incarcerated fathers have higher rates of depression, irritability, and anxiety than children who do not have incarcerated fathers. The strongest and most consistent relationship is between fathers' incarceration and behavioral problems, particularly among boys, including delinquency and aggression (Travis et al. 2014).

As previously noted, in recent years, there's been a dramatic increase in the number of incarcerated women. The result is an increase in the number of children with incarcerated mothers, three-quarters of whom lived with their children prior to confinement. The consequences of an incarcerated mother have only recently been investigated. What research exists suggests that maternal incarceration is associated with poor academic performance, school suspension, and higher rates of delinquency (Travis et al. 2014). In a review of the literature, Smyth (2012) concludes that "children with incarcerated mothers are at heightened risk for attachment disturbance, leading to depression, anxiety, and other traumatic-related stress ... [and] ... are often subject to frequent changing of caregivers within the foster care system, which exacerbates these problems" (p. 33).

Finally, incarceration impacts the stability of the families with someone in jail or prison. Men who are incarcerated are more likely not to get married or, if married, to get a divorce. Further, as the incarceration rate of women increases, the number of children in foster care also increases. One way that incarceration directly causes family separation is the *Adoption and Safe Families Act*, which requires that a state terminate the parent–child relationship if a child has been in foster care for 15 or more of the previous 22 months. The average number of months an inmate is incarcerated is 29 (Elder et al. 2018). Ironically, remaining in contact with family members and having a stable family support system increase the likelihood of non-reoffending.

Strategies for Action: Crime and Social Control

The extent to which crime is prevented or recidivism is reduced is difficult to assess. On the one hand, if a crime prevention program is initiated, and the crime rate goes down, it would be tempting to conclude that the program was a success. But what if the crime rate would have gone down anyway, with or without the prevention program? On the other hand, if a crime prevention program is initiated, and the crime rate goes up, many would conclude that the program was a failure. But what if the crime rate was lower than it would have been if the program had not been initiated?

Nonetheless, as a society, we want to be able to "do" something about the crime problem. In addition to suggested cultural changes (e.g., strengthen the family) and structural changes (e.g., eliminate poverty), numerous policies and programs have been initiated to alleviate the crime problem. These policies and programs include local crime-fighting initiatives, rethinking law enforcement, criminal justice policies, legislative action, and international efforts in the fight against crime.

Local Crime-Fighting Initiatives

Increasingly, municipalities are embracing technological innovations in their efforts to prevent crime. Additionally, local youth programs such as the Boys and Girls Clubs and community programs that involve families and schools are an effective "first line of defense" against crime and juvenile delinquency.

Technology and Crime. Police are turning to technology in the fight against crime. Byrne and Marx (2011) make a needed distinction between hard and soft technologies. *Hard* *technologies*, in the context of crime prevention, include such things as metal detectors in schools, home security systems, and police drones. *Soft technologies* "involve the strategic use of information to prevent crime ... and to improve the performance of the police" (p. 19). Crime analysis programs, Amber alerts, facial recognition software, and social media are examples of soft technologies.

In varying degrees, cities throughout the United States are using these innovations to fight crime. PredPol predicts crime based on an algorithm using three sociological variables—crime type, crime location, and crime date and time (PredPol 2020). Socio-economic information (e.g., household income) and demographic information (racial distribution of a neighborhood) are never used. Predictions are displayed on Google maps through a web interface, and officers are then dispatched to the identified locations.

The ShotSpotter system, through the placement of microphones throughout an area, can detect gunshots and calculate their location within 40 to 50 feet. It lowers the amount of time it takes the police to respond to a shooting incident and therefore not only increases the likelihood of apprehending the shooter but decreases response and transport time by EMS (Goldenberg et al. 2019). Further, the use of Facebook to post pictures of suspects, along with GPS devices and smartphone tracking techniques, makes finding criminals and documenting their locations easier.

One controversial type of "hard technology" is the use of video cameras in public places. Chicago, for example, has more than 32,000 cameras distributed throughout the city, and all of the cameras, both public and private, are linked into a single system called Operation Virtual Shield (OVS) (Glanton 2019). Initially, OVS was designed as just a camera surveillance system; however, recently biometric facial recognition abilities were added. **Biometric surveillance** is used to identify a specific person through the imaging of their distinct physical characteristics (Thakkar 2019).

Youth Programs. Early intervention programs acknowledge that preventing crime is better than "curing" it once it has occurred. Fight Crime: Invest in Kids is a non-partisan, non-profit anticrime organization made up of more than 5,000 law enforcement leaders, prosecutors, and violence survivors. According to the organization, afterschool programs reduce crime, improve academic outcomes, shape normative behavior, lead to healthier habits, and save money. For every dollar invested in afterschool programs, at least $3 are saved by reducing future crime and welfare costs and by increasing potential earnings through better academic outcomes (Fight Crime 2019).

The HighScope Perry Preschool Project (PPP) is another example of an early intervention program. After a sample of 123 Black children was randomly assigned to either a control group or an experimental group, the experimental group members received academically oriented interventions for one to two years, frequent home visits, and weekly parent–teacher conferences. The control group members received no interventions. The control group members and the experimental group members were then compared at various points in time between the ages of 3 and 40. As adults, experimental group members had higher incomes, were more likely to graduate from high school, had higher employment and home ownership rates, and significantly lower violent and property crime rates (PPP 2020). Fifty years from the initial experiment, the results of the PPP continue to impact preschool education and provide a benchmark for assessing the effectiveness of preschool programs.

Community Programs. Neighborhood watch programs involve local residents in crime prevention strategies. For example, MAD DADS (Men against Destruction—Defending against Drugs and Social Disorder) patrol the streets in high-crime areas of the city on weekend nights, providing positive adult role models and fun community activities for troubled children. Members also report crime and drug sales to police, paint over gang graffiti, organize gun buyback programs, and counsel incarcerated fathers. The organization has 75,000 men, women, and children in 60 chapters in 17 states, as well as international chapters in 18 countries (MAD DADS 2019).

The National Association of Town Watch (NATW) is a "nonprofit organization dedicated to enhancing the communities in which we live through an established network of law enforcement agencies, neighborhood watch groups, civic groups, state and regional crime

biometric surveillance
Used to identify a specific person through the imaging of their distinct physical characteristics.

prevention associations, and volunteers across the nation" (NATW 2020a, p. 1). NATW began the National Night Out event in 1984, and, 36 years later, over 38 million people in more than 16,000 communities in 50 states participate in the event in which citizens, businesses, neighborhood organizations, and local officials join together in outdoor activities to heighten awareness of neighborhood problems, promote anticrime messages, and strengthen community ties (NATW 2020b).

Office of Community Oriented Policing Services

Police officer Jackelyn Burgos receives hugs and well wishes after Gang Resistance Education and Training classes in the Cleveland Public Schools. This photo, one of the former winners in a COPS (Community Oriented Policing Services) photo contest, depicts the importance of the COPS programs—integration of police into the community. The COPS program dates back to 1994 with the passage of the Violent Crime Control and Law Enforcement Act.

Rethinking Law Enforcement

The United States had nearly 1 million full-time law enforcement officers and full-time civilian employees in 2018 (e.g., clerks, meter attendants, correctional guards). Seventy percent were sworn officers, yielding an average of 2.4 police officers per 1,000 inhabitants (FBI 2019a). There are more than 18,000 law enforcement agencies in the United States, including municipal (e.g., city police), county (e.g., sheriff's department), state (e.g., highway patrol), and federal agencies (e.g., FBI), often with overlapping jurisdictions.

In 2015, the latest year for which national data are available, 21 percent of the U.S. resident population 16 and older had contact with the police—76 million interactions (Davis, Whyde, and Langston 2018). According to a police–citizen survey, contacts were equally initiated by citizens (e.g., requests for police assistance) and by police (e.g., traffic stops). Males experienced higher rates of traffic stops than females, and drivers between the ages of 18 and 24 were more likely than any other age group to be stopped by the police.

Black citizens were significantly more likely to be detained by police in a traffic stop and to experience a "street stop," which includes being stopped by the police while in a public place or while sitting in a parked vehicle. Further, Black and Hispanic individuals were more than twice as likely to experience the threat or use of physical force when compared to their White counterparts. Lastly, independent of any other police–citizen contact, Black citizens were statistically more likely to be arrested than White or Hispanic citizens (Davis, Whyde, and Langston 2018).

Black Lives Matter Protests. For decades, accusations of racial profiling, police brutality, and discriminatory arrest practices have made police–citizen cooperation in the fight against crime difficult. Concerns have been fed by such highly publicized cases as the killings of Michael Brown, Eric Garner, Tamir Rice, Freddie Gray and, in 2020, Breonna Taylor, George Floyd, Rayshard Brooks, as well as others. People all over the world watched in horror as images of Minneapolis police officer Derek Chauvin kneeled on Mr. Floyd's neck as he pled for his life—"I can't breathe!" Eight minutes and 46 seconds later, George Floyd was dead. Chauvin was eventually charged with second-degree murder, and three other officers, two of whom helped hold Floyd down and one who failed to intervene, were charged with aiding and abetting second-degree murder and aiding and abetting second-degree manslaughter (Campbell, Sidner, and Levenson 2020).

In 2020, hundreds of thousands of protesters demonstrated against police use of lethal force against unarmed Black men and women. Portraits of George Floyd were hung on fences around U.S. embassies. People gathered outside the American consulate in Milan, Italy, kneeling with signs in front of them that read, "I can't breathe," each holding their hands around their necks in the universe choking sign (Rahim and Picheta 2020).

In Denmark, crowds of people chanted "No Justice, No Peace," and in Brazil, protesters held signs that read, "Vidas Negras Importam [Black Lives Matter]."

In the United States, in a single day, half a million people in over 500 locations turned out to demonstrate against the killing of George Floyd and in support of the Black Lives Matter (BLM) movement (Buchanan, Bui, and Patel 2020). Protests spread from large cities to smaller towns, suburbs, and rural areas. Demonstrators were younger, wealthier, and whiter than in the civil rights movement of the 1960s. Freeways were shut down by protestors in Los Angeles, Atlanta, Milwaukee, and Dallas. Curfews were imposed, but the protests continued.

In cities such as Denver, Boise, Louisville, Chicago, Los Angeles, New York City, and Portland, demonstrators began to address the larger issues of systemic racism, economic inequality, and police brutality (see Chapter 9). Portland, Oregon, became the new epicenter of the BLM movement, and, although initially peaceful, when confronted by federal officers, the demonstration devolved into chaos (Sanchez 2020). After federal officers left the city and as protests continued, dozens of Portland police officers were deputized as federal marshals (Grzeszczak 2020).

What do you think?

Qualified immunity for police officers protects them against lawsuits if, at the time of their alleged misconduct, they did not know their behavior was illegal. Do you think police officers should be accountable for their behavior even if they did not know their actions were illegal?

Response to BLM Protests. The 2020 BLM demonstrations against racial inequality in general, and police brutality in particular, were initially supported by a majority of the public (Hamel et al. 2020). However, as protests continued in some cities, and spread to others as a result of the announcement that no police officers were charged in Breonna Taylor's death, support for the demonstrations declined, particularly among White Americans and Republicans (Morrison and Stafford 2020). In fact, in Kenosha, Wisconsin, after the killing of Jacob Blake, a Black man, an armed teen who had come to Kenosha to help the police, shot and killed two protestors (Naylor 2020).

qualified immunity A legal principle that protects police officers from lawsuits if, at the time of their alleged misconduct, they did not know their behavior was unlawful.

More than 40 percent of the counties in the United States had BLM protests in 2020, and of that number, 95 percent were in counties with majority White populations (Buchanan, Bui, and Patel 2020). Further, according to a Kaiser Family Foundation survey, one in ten Americans attended a "rally, protest or demonstration" in support of BLM (Hamel et al. 2020). Had it not been for COVID-19, those numbers would likely have been much higher. Those who attended a protest were disproportionately Democrats or Independents, between the ages of 18 and 29, and college educated.

State and municipal responses to the protests were varied depending on the size and level of violence of the demonstrations. Municipal police were often outnumbered and overwhelmed. Twenty-one governors called in the National Guard to assist local police (Taylor 2020). In some cities, Seattle for example, police were accused of using excessive force on peaceful demonstrators, resulting in a court order prohibiting the use of pepper spray, tear gas, and blast balls (Carter 2020). When accused of violating the court order, the Seattle Police Department claimed, through its lawyers, that among the mostly peaceful crowd of 7,000, there "were armed and armored individuals with a plan and agenda to do harm in public, injure police and damage property along the way" (quoted in Carter 2020, p. 1).

Accusations of violence in need of heavy-handed police tactics also came from the former Trump administration. Just nine days after George Floyd was killed, President Trump criticized Minnesota officials for their failure to get Minneapolis under control and threatened to send in the military, tweeting, "[W]hen the looting starts, the shooting starts." Ironically, the former president did not know the origin of the phrase, which was used by a Miami police chief during the civil unrest of the 1960s (Sprunt 2020).

@taylorswift13

After stoking the fires of white supremacy and racism your entire presidency, you have the nerve to feign moral superiority before threatening violence? 'When the looting starts the shooting starts'??? We will vote you out in November. @realdonaldtrump

-Taylor Swift

Sent by the Trump administration to protect federal buildings in Portland, the arrival of three units of U.S. Department of Homeland Security (DHS) forces heightened tensions between protestors and local and federal police. Protesters decried the use of "camouflage-clad officers without clear identification badges using force and unmarked vehicles to transport arrested protesters" (Hosenball 2020, p. 1). The Oregon governor and Portland mayor complained of federal overreach and, several weeks after their arrival, following repeated requests by the governor and mayor, DHS personnel began withdrawing from the city. Both President Biden and Vice President Harris have been critical of the federal response to the protests (Bates 2020).

Police Use of Force. In 2018, the U.S. Commission on Civil Rights initiated a report to examine the use of police force and other police practices (U.S. Commission on Civil Rights 2018). Using available national and academic data as well as media sources, the Commission concluded that the "best available evidence reflects high rates of use of force nationally, and increased likelihood of police use of force against people of color" (p. 2).

Picheta and Pettersson (2020) compared police use of force in the United States to police use of force in other wealthy, developed democracies. First, the rate of suspects dying in police custody in the United States was twice that of Australia and six times higher than in the United Kingdom. Second, the police in the United States killed more people than any other developed country. In 2018, for example, police in the United States fatally shot 1,000 people. In Germany, Australia, and Sweden, the numbers of fatal police shootings were 11, eight, and six, respectively. Figure 4.4 displays the estimates of police use of excessive force by survey respondents' race and political party.

Edwards, Lee, and Esposito's (2019) analysis indicates that racial and ethnic minorities are more likely to be killed than White men or women in all age groups. However, Streeter (2019) concludes that, when taking into consideration the contextual variables of a police killing such as threat to the police and characteristics of the deceased, "racial disparity in the rate of lethal force is most likely driven by higher rates of police contact among African-Americans rather than ... officer bias in the application of lethal force" (Streeter 2019, p. 1124). Significantly, Nix et al. (2017) report that non-White people in lethal confrontations with the police were *less* likely to be acting aggressively and 50 percent *less* likely to be armed.

Finally, Mentch (2020) compares the racial distribution of police shooting victims to the racial distribution of the resident population where the shooting took place. Results indicate that the racial distribution of the shooting victims, who were disproportionately Black, was not consistent with the racial distribution of the population in the shooting location but rather was consistent with the racial distribution of the *arrestee* population in the shooting location.

Several complementary hypotheses explain these generally consistent research results. First, there is evidence of **overt differential law enforcement**. Overt differential law enforcement occurs when a police officer or any member of the criminal justice system treats one person differently than another because of that person's characteristics, whether it be race, socioeconomic status, age, gender, or sexual orientation. Overt differential law enforcement, then, is the result of a police officer's or any criminal justice actor's own biases.

overt differential law enforcement Criminal justice actors treating one person differently than another because of that person's characteristics, for example, race.

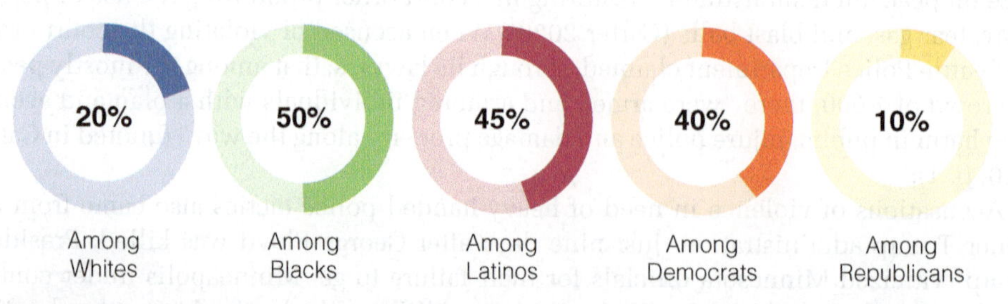

Figure 4.4 Estimates of the Percent of Police Officers Who Use Excessive Force, by Respondents' Race and Political Party, 2020
SOURCE: CATO 2020.

Alternatively, there is also evidence of **covert differential law enforcement**, i.e., law enforcement that results from structural and cultural variables rather than the actions of an individual. For example, assigning more patrol cars to neighborhoods with high crime rates makes sense—it's just what we would expect a police chief to do. But, in doing so, police surveillance increases, as does the likelihood of police–resident contact, leading to more arrests and thus the possibility of police force. As arrest rates remain high, law enforcement surveillance remains high, and the cycle continues.

Finally, the **differential involvement** hypothesis holds that non-White individuals are *differently* involved in crime compared to White individuals (cf. Loeber et al. 2015). If that is true, young Black males have higher frequencies of police interactions as a result of their own behavior, which increases the opportunity to be victimized by the police. Unfortunately, the use of official statistics to measure criminal behavior introduces an "other" (e.g., police officer, jailer) into the crime control process, making it often difficult to assess the relative weight of each of these three hypotheses.

Police Reform. Not surprisingly given the foregoing, in a 2019 survey, Black Americans were half as likely as White Americans to positively rate police treatment of racial and ethnic minorities and police use of force (Desilver, Lipka, and Fahmy 2020). Further, a majority of the survey respondents, although Black respondents more so than White respondents, responded that Black Americans are treated less fairly than White Americans by the police. However, holding race constant, there are large partisan differences. Eighty-eight percent of Democrats compared to 43 percent of Republicans responded that Black people are treated less favorably than White people by the police.

The call for police reforms is not new, but its urgency has been heightened by the BLM demonstrations of 2020. The militarization of the police, which began in the 1960s as a result of the civil rights movement, student protests, and the war on drugs, has social scientists and others advocating for demilitarization (Balko 2013; Lieblich and Shinar 2018; Nolan 2020). First, stun guns, tear gas, drones, and armored vehicles increase "the likelihood of violent attacks and non-negotiable-force over compromise, mediation, and peaceful conflict resolution" (Nolan 2020, p. 2).

Second and not unrelated, the militarization of the police has created a mind-set wherein anyone who is perceived as a threat to public order is seen as an enemy and thus encourages police to think like soldiers. Third, there is little evidence that the militarization of the police has enhanced safety or reduced crime. There is evidence, however, that special weapons and tactics (SWAT) units are disproportionately deployed to minority neighborhoods (Mummolo 2018) and that the militarization of police has resulted in higher death rates of in-custody suspects (Lawson 2018).

The majority of Americans believe that police officers who injure or kill civilians are treated too leniently, and thus several suggested reforms are directed toward making police more accountable for their behavior. To that end, as Figure 4.5 indicates, the majority of Americans support the use of body cams, requiring officers to report misconduct of their peers, and prosecuting and/or penalizing police officers who use excessive force and/or act in a racially biased way. Note, however, that just half of Americans support limiting the use of military equipment (Associated Press [AP]–National Opinion Research Center [NORC] 2020).

covert differential law enforcement Law enforcement that results in treating people differently based on structural and cultural variables rather than the characteristics of an individual.

differential involvement Certain groups of people, for example, males compared to females, are more likely to be involved in crime.

JASON REDMOND/AFP/Getty Images

The militarization of the police was apparent during the 2020 Black Lives Matter demonstrations. Here a Seattle SWAT officer rides in an armored vehicle during the Youth Day of Action and Solidarity with Portland demonstration in Seattle, Washington, on July 25, 2020.

Of late, however, public attitudes toward get-tough measures and the skyrocketing incarceration rates, which increased by 500 percent in the last 40 years (Sentencing Project 2020), have changed (Gotoff and Lake 2020). Seventy percent of Americans believe that the criminal justice system needs a complete overhaul or major changes (AP–NORC 2020). Since 2012, some 6,000 California inmates with life sentences imposed under the 1990s-era three strikes policy have been freed or had their sentence reduced (Cramer 2020). Similarly, in 2020, a Louisiana court ruled that inmates could challenge sentencing on the basis of it being "excessive or unjust." The case concerned Desert Storm veteran Derek Harris who had received a life sentence without the possibility of parole for selling $30 worth of marijuana. After serving 12 years, Derek Harris was released from prison in 2020 (Jones and Andrew 2020).

These shifts in opinion coincide with concerns of politicians, academicians, and criminal justice officials about the efficacy of mass incarceration. Their concerns can be classified into five categories. First, research indicates that incarceration may not deter crime. Using a sample of drug offenders, Mitchell et al. (2017) conclude that drug offenders in prison are no more likely to reoffend than those receiving non-prison sentences (e.g., probation) (Mitchell et al. 2017). Further, Oregonians passed Measure 11, a tough mandatory minimum sentencing policy, in the hopes of deterring crime. The law had no discernible impact on property or violent crime (Sundt and Boppre 2020).

Second is the accusation that get-tough measures are unevenly applied. Although the rate of imprisonment of Black individuals in 2019 was the lowest it has been since 1989, Black males were imprisoned at six times the rate of White males (Carson 2020). Representing only 13 percent of the U.S. population, Black Americans make up 40 percent of the incarcerated population (Sawyer and Wagner 2020). In general, Black Americans are more likely than White Americans to be arrested, convicted, and receive more severe sentences compared to White Americans.

breeding ground hypothesis A hypothesis that argues that incarceration serves to increase criminal behavior through the transmission of criminal skills, techniques, and motivations.

Third, there are concerns that putting people in prison may actually escalate their criminal behaviors once released. Inmates may learn new and better techniques of committing crime and, surrounded by other criminals, further internalize the attitudes toward and motivation for continued criminal behavior. Such a possibility is called the **breeding ground hypothesis**. Furthermore, as labeling theorists suggest, the stigma associated with being in prison makes reentry into society difficult. Unable to find employment, return to school, and meet new friends, former prisoners may feel they have no choice but to return to crime.

Andrew Burton/Getty Images

Recently, the percentage of prisoners over the age of 55 has been steadily increasing as a result of mandatory sentencing practices of the past. Although many inmates have not committed a violent crime, they have been sentenced to life in prison as a result of three strikes policies. Prisoners over the age of 55 are particularly vulnerable to COVID-19, not only because of their advanced age but because of the lack of health care facilities in many correctional facilities and the difficulty of maintaining social distance from other prisoners.

Fourth, in an environment of budget deficits and legislative cuts, states simply can no longer afford the policies of decades ago. State expenditures on corrections are in the billions, increasing from $7 billion in 1985 to $60 billion in 2017 (Sentencing Project 2019). The average annual cost of incarcerating an offender, more than $37,499 for federal inmates, ranges from a low of $14,000 per inmate in Indiana to over $60,000 in New York and California (Federal Register 2019; Kincade 2018). Additional costs, often not part of the calculations, include the cost of pensions, health care, and other benefits for correctional employees, construction and renovation costs, administrative and legal expenses, and the cost of rehabilitation programs, hospital care, and private prisons for inmates.

Finally, some are questioning the logic of increased criminal justice spending when the crime rate has significantly declined for nearly three decades. Despite these sharp decreases in both violent and property crime, incarceration rates have continued to increase since 2000 (Byrne, Pattavina, and Taxman 2015; Gramlich 2019). Increases in incarceration rates are a consequence of changing sentencing policies that result in more people being admitted to prison and longer sentences (Mauer 2018).

For some, the inverse relationship between incarceration rates and crime is an indication that get-tough policies work. Yet researchers and criminal justice experts, armed with 20 years of crime and incarceration data, conclude that since "about 1990, the effectiveness of increased incarceration on bringing down crime has been essentially zero" (Roeder et al. 2015, p. 23).

The Biden administration's plan for reforming the criminal justice system addresses many of these concerns (Biden 2020a). President Biden wants to "strengthen America's commitment to justice" by (1) reducing racial and ethnic bias, (2) addressing drug abuse and mental health needs of offenders, (3) lowering incarceration rates by, for example, emphasizing rehabilitation rather than incapacitation and eliminating mandatory minimum sentencing for non-violent offenders, and (4) expanding federal funding for crime prevention and diversion programs. Once these reforms are in place, theoretically, both the crime and incarceration rates should decline saving millions of taxpayer dollars in criminal justice costs.

Corrections. Over the last decade, the number of federal and state inmates decreased by 15 percent, the largest decreases among Black and Hispanic individuals (Carson 2020). The majority of state prisoners are serving time for violent offenses, compared to federal inmates who are predominantly drug and public order offenders (e.g., weapons, immigration) (Sawyer and Wagoner 2020). Males make up more than 95 percent of the prison population. At year-end 2018, 22 states, including Louisiana, Oklahoma, and Mississippi, had imprisonment rates higher than the national average while Minnesota, Maine, and Massachusetts had the lowest imprisonment rates in the United States (Carson 2020).

In 2018, the *First Step Act*, designed to institute criminal justice reform in the federal system by shortening prison sentences and improving prison conditions, was signed into law. While the first goal appears to have been implemented, the second remains unrealized. The system used to assess the likelihood of an inmate reoffending, which determines the possibility of, for example, a transfer to a halfway house, was changed to make it more difficult for a prisoner to achieve "low-risk" status. It is the same system used to determine whether an inmate can be transferred out of a federal prison during the COVID-19 pandemic (Grawert 2020). Figure 4.6 portrays the distribution of prisoners by type of correctional facility and type of crime committed.

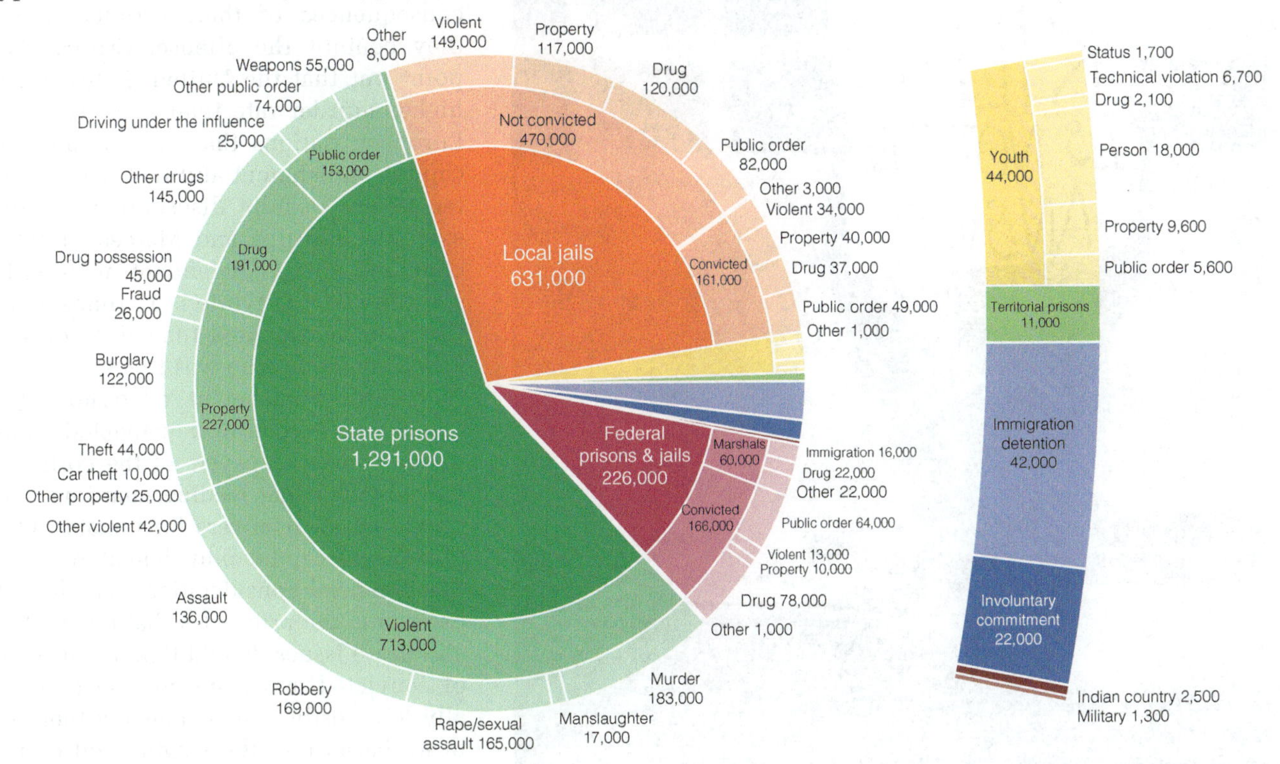

Figure 4.6 Prisoners by Correctional Facility Type and Crime, 2018
SOURCE: Sawyer and Wagner 2020.

probation The conditional release of an offender who, for a specific time period and subject to certain conditions, remains under court supervision in the community.

parole Release from prison, for a specific time period and subject to certain conditions, before the inmate's sentence is finished.

capital punishment The state (the federal government or a state) takes the life of a person as punishment for a crime.

"To me justice doesn't mean looking for the criminals. It means looking for the innocent."

–MWANANDEKE KINDEMBO, AUTHOR

Probation entails the conditional release of an offender who, for a specific time period and subject to certain conditions, remains under court supervision in the community. **Parole** entails release from prison, for a specific time period and subject to certain conditions, before the inmate's sentence is finished. Varying by race, age, and gender, nearly 4.3 million people were on probation or parole in the United States at the end of 2018 (Kaeble and Alper 2020). The reincarceration rate of parolees is nearly twice the reincarceration rate of probationers.

Capital Punishment. With **capital punishment**, the federal or state government takes the life of a person as punishment for a crime. More than 70 percent of the world's nations have abolished capital punishment either by law or in practice (Death Penalty Information Center [DPIC] 2020). In 2019, there were more than 657 executions in 20 countries, a decrease from the previous year. That number, however, does not include the thousands of estimated executions that take place in China annually. Of the nearly 200 countries in the world, the United States ranks sixth in the number of executions carried out in 2019.

Since 1977, over 1,525 inmates have been executed in the United States. Eighty percent of the executions occurred in Southern states, and Black people continue to be disproportionately represented in executions and on death row—34 percent and 42 percent, respectively. Today, half of the states have abolished or have a moratorium on executions (DPIC 2020). In 2020, after a 17-year hiatus, for the first time the federal government had more executions than all the states combined (DPIC 2021).

Proponents of capital punishment argue that executions of convicted murderers are necessary to convey public disapproval and intolerance for such heinous crimes. Those against capital punishment believe that no one, including the state, has the right to take another person's life and that putting convicted murderers behind bars for life is a "social death" that conveys the necessary societal disapproval.

Proponents of capital punishment also argue that it deters individuals from committing murder. Critics of capital punishment hold, however, that because most homicides are situational and are not planned, offenders do not consider the consequences of their actions before they commit the offense. Critics also point out that the United States has a higher murder rate than most Western European nations that do not practice capital punishment and that death sentences are racially discriminatory. For example, Phillips and Marceau (2020) conclude that a defendant who kills someone who is White is 17 times more likely to be executed than if he or she killed someone who was Black.

© Rick Wood/Milwaukee Journal Sentinel via Imagn Content Services, LLC

A protestor in Wauwatosa, Wisconsin, demonstrates against racial inequality in the criminal justice system. In 2020, 42 percent of death row inmates were Black, 42 percent White, and 13 percent Hispanic.

Capital punishment advocates also argue that executing a convicted murderer relieves taxpayers of the costs involved in housing, feeding, guarding, and providing medical care for inmates. Opponents of capital punishment respond that financial considerations should not determine the principles that decide life and death issues. In addition, taking care of convicted murderers for life is actually less costly than sentencing them to death because of the lengthy and costly appeals process for capital punishment cases (Johnson and Quigley 2019). In a

2020 survey, a record low percentage of Americans, 54 percent, found the death penalty morally acceptable down from 71 percent in 2006 (Brenan 2020).

Those in favor of capital punishment argue that it protects society by preventing convicted individuals from committing another crime, including the murder of another inmate or prison official. Opponents contend that capital punishment may result in innocent people being sentenced to death. According to the Innocence Project, there have been 375 exonerations using DNA evidence since 1989. The most common reasons for wrongful convictions are (1) eyewitness misidentification, (2) misused forensic science, (3) false confessions, (4) informants lying, (5) inadequate defenses, and (6) restricted access to DNA testing (Innocence Project 2020). President Biden is opposed to capital punishment and has stated that he will work to eliminate the death penalty (Biden 2020a).

Federal and State Legislative Action

Legislative action is one of the most powerful methods of fighting crime. Federal and state legislatures establish criminal justice policy by the laws they pass, the funds they allocate, and the programs they embrace.

Gun Control. According to a report by the FBI, in 2019, firearms were used in 73.7 percent of the nation's murders, 36.4 percent of robberies, and 27.6 percent of aggravated assaults (FBI 2020a). There are approximately 857 million civilian-owned guns in the world, and less than 12 percent are registered. More civilians own guns in the United States than in any other country in the world—120.5 firearms for every 100 civilians. The second highest civilian gun-owning country in the world is Yemen with 52.8 guns per 100 civilians. Put another way, Americans own four out of every ten handguns in the world (Karp 2018).

What do you think?

The Biden administration has pledged to reduce gun violence in the United States. Among other proposals, President Biden wants to ban the manufacture and sale of assault weapons and high-capacity magazines. As the president notes, "Federal law prevents hunters from hunting migratory game birds with more than three shells in their shotgun. That means our federal law does more to protect ducks than children" (Biden 2020b, p. 1). What do you think about banning assault weapons and high-capacity magazines?

Gun ownership is predictable using sociological variables. Gun ownership is higher in rural Southern areas, among older, Republican, less educated White people, who believe gun ownership is an essential right (Parker et al. 2017). For those who own a single gun, a handgun is the most common gun to own, and the most common reason for owning a gun is protection. Not surprisingly, Democrats, women, young people, and college graduates believe that gun laws in the United States should be stricter compared to their demographic counterparts (Gramlich and Schaeffer 2019).

Gun control is a contentious issue in the United States. Those who argue against gun control, including the National Rifle Association (NRA), argue that they have a constitutional right, based on the Second Amendment, to own firearms (NRA 2020). Those in favor of gun control argue that a reduction in the number of guns would lead to a variety of positive outcomes. In the single largest review of gun-related literature to date, the RAND Corporation (2020) reviewed thousands of scientific studies, and, although they concluded that much of the research was inconclusive, they were able to make the following statements:

- Child access prevention laws are strongly associated with a decrease in unintentional injuries and deaths and suicides.
- Background checks and waiting periods are moderately associated with a decrease in violent crime and suicides.

- Restrictions that impact the likelihood of domestic violence abusers or people with a mental illness obtaining a firearm are moderately and weakly, respectively, associated with a decrease in violent crime.
- Stand-your-ground laws are strongly associated with an increase in violent crime.
- Concealed carry laws are weakly associated with an increase in violent crime.

In a 2020 survey conducted by the Gallup Poll, a sample of U.S. adults was asked whether they had a gun in their homes. Thirty-seven percent reported "yes," a significant decline from the early 1990s when half of the respondents reported having a firearm at home. The majority of Americans, 64 percent, responded that the laws covering the sale of firearms should be stricter, 28 percent said the laws should be kept as they are, and 7 percent thought that laws governing firearms sales should be less strict (Gallup 2020b). Table 4.5 displays the states with the five highest and five lowest gun-related death rates. Note the relationship between the percentage of households with guns and the gun death rates per 100,000 population.

Other Crime and Social Control Legislation. There have been several landmark legislative initiatives, including the 1994 *Violent Crime Control and Law Enforcement Act*, which created community policing, and the 2006 *Adam Walsh Child Protection and Safety Act*, which, when enacted, created a national registry of substantiated cases of child abuse and neglect. A sample of significant crime-related legislation presently before Congress includes the following bills:

- *Concealed Carry Reciprocity Act of 2019.* This bill, among other things, would (1) allow persons with valid concealed weapon permits from their states of residence to lawfully carry a concealed firearm in jurisdictions that do not issue concealed gun permits and (2) override state laws prohibiting guns in certain locations.
- *End Racial Profiling Act of 2019.* If passed, this bill would prohibit racial profiling on the basis of actual or perceived race, ethnicity, national origin, religion, gender, gender identity, or sexual orientation by federal, state, county, and municipal law enforcement.
- *Dismantle Mass Incarceration for Public Health Act of 2020.* If passed, federal funds would be used to persuade state and local governments to release (e.g., to home confinement) inmates awaiting trial, those serving misdemeanor sentences, pregnant women, immigrants in detention, and inmates over 55 or those medically compromised, among others, for up to a year after the pandemic ends.

What do you think?

Executives of the National Rifle Association are being sued by the New York attorney general for (1) diverting millions of dollars in donations for their own personal use, (2) benefiting from organizations the NRA does business with, and (3) failing to report abuse and other questionable transactions to regulatory agencies (Mittendorf 2020). If the accusations are true, do their actions constitute occupational crime or corporate crime?

TABLE 4.5 Five States with Highest and Lowest Gun Death Rates by Household Gun Ownership, 2019

States with the Five *Highest* Gun Death Rates				States with the Five *Lowest* Gun Death Rates			
Rank	State	Household Gun Ownership	Gun Death Rate per 100,000	Rank	State	Household Gun Ownership	Gun Death Rate per 100,000
1	Alaska	56.4%	24.33	50	Hawaii	12.5%	2.73
2	Montana	67.5%	23.23	49	Massachusetts	14.3%	3.82
3	Alabama	49.5%	23.06	48	New York	22.2%	3.89
4	Louisiana	49.0%	21.52	47	Rhode Island	15.9%	4.06
5	Missouri	43.9%	21.38	46	Connecticut	22.2%	5.24

SOURCE: Violence Policy Center 2019.

International Efforts in the Fight against Crime

Europol is the European law enforcement organization that handles criminal intelligence. Unlike the FBI, Europol officers do not have the power to arrest; they predominantly provide support services for law enforcement agencies of countries that are members of the European Union. For example, Europol coordinates the dissemination of information, provides operational analysis and technical support, and generates strategic reports (Europol 2020a). Europol, in conjunction with law enforcement agencies in member states, fights against transnational crimes such as facilitating illegal immigration, illicit drug trafficking, cybercrime, child pornography, human trafficking, money laundering, and counterfeiting of the euro (Europol 2020b).

Interpol, the International Criminal Police Organization, was established in 1923 and is the world's largest international police organization, with 194 member countries, including the United States (Interpol 2020). Similar to Europol, Interpol provides support services for law enforcement agencies of member nations. It has three main crime programs. First, Interpol assists member countries in the prevention and disruption of terrorist activities. Second, Interpol fights international criminal networks. Finally, Interpol supports member countries in the fight against cyberattack through prevention efforts and investigative assistance.

The International Centre for the Prevention of Crime (ICPC 2020) is a consortium of policy makers, academicians, police, governmental officials, and non-governmental agencies from all over the world. The mission of the ICPC is to "support the development and implementation of practical and effective policies, programs and projects, designed to reduce crime and delinquency in communities, cities and other geographic units, and to reinforce a sense of safety" (p. 1). Since the beginning, ICPC has sought to raise awareness about crime prevention and community safety policies.

Understanding Crime and Social Control

What can we conclude from the information presented in this chapter? Research on crime and violence supports the contentions of both structural functionalists and conflict theorists. Inequality in society, along with the emphasis on material well-being and corporate profit, produces societal strains and individual frustrations. Poverty, unemployment, urban decay, and substandard schools—the symptoms of social inequality—in turn lead to the development of criminal subcultures and conditions favorable to law violation. Furthermore, criminal behavior is encouraged by the continued weakening of social bonds among members of society and between individuals and society as a whole, the labeling of some acts and actors as "deviant," and the differential treatment of minority groups by the criminal justice system.

Interpol The largest international police organization in the world.

Recently, there has been a general decline in crime, making it tempting to conclude that get-tough criminal justice policies are responsible for the reductions. Other valid explanations exist and are likely to have contributed to the falling rates: changing demographics, rising incomes, community policing, stricter gun control, the COVID-19 pandemic, and a reduction in the use of crack cocaine.

@CBS SportsNBA

LeBron James' voting rights group donating $100,000 to pay fines and fees so ex-felons can vote in Florida

cbssports.com/nba/news/lebro...

-CBS Sports NBA

Concerns over the cost of "nail 'em and jail 'em" policies, overcrowded prisons, and high recidivism rates have some policy makers looking elsewhere. At least one expert suggests that the uniquely punitive nature of the American penal system "may be converging toward humanitarian norms—dignity, proportionality, legitimacy, and rehabilitation—that are prevalent in European nations, Canada, and various other liberal democracies" (Mugambi 2019, p. 706). On the face of it, it appears that the Biden administration is moving in that direction.

Rather than getting tough on crime after the fact, some advocate getting serious about prevention. Prevention programs are not only preferable to dealing with the wreckage crime leaves behind, but they are also cost-effective. For example, a high-quality publicly funded preschool for all 3- and 4-year-olds would lead to financial benefits that would surpass the cost of the program within eight years

"I Don't Want My Body Anymore"

In 2015, Chanel Miller was raped by Stanford college freshmen Brock Turner. She was unconscious at the time. In 2016, Turner was found guilty and, despite recommendations by prosecutors of six years, he was sentenced to just six months in jail and only served three months. The following is an excerpt from the victim's impact statement. In 2019, Ms. Miller released her memoir of the ordeal entitled, *Know My Name*.

Your Honor,

If it is all right, for the majority of this statement I would like to address the defendant directly.

You don't know me, but you've been inside me, and that's why we're here today.

... On January 17th, 2015, it was a quiet Saturday night at home. My dad made some dinner and I sat at the table with my younger sister who was visiting for the weekend.

... Then, I decided it was my only night with her, I had nothing better to do, so why not, there's a dumb party ten minutes from my house, I would go, dance weird like a fool, and embarrass my younger sister.

... The next thing I remember I was in a gurney in a hallway. I had dried blood and bandages on the backs of my hands and elbow. ... A deputy explained I had been assaulted. I still remained calm, assured he was speaking to the wrong person. I knew no one at this party. When I was finally allowed to use the restroom, I pulled down the hospital pants they had given me, went to pull down my underwear, and felt nothing. I still remember the feeling of my hands touching my skin and grabbing nothing. I looked down and there was nothing. The thin piece of fabric, the only thing between my vagina and anything else, was missing and everything inside me was silenced.

... I was asked to sign papers that said "Rape Victim" and I thought something has really happened. My clothes were confiscated and I stood naked while the nurses held a ruler to various abrasions on my body and photographed them. The three of us worked to comb the pine needles out of my hair, six hands to fill one paper bag. ... I had multiple swabs inserted into my vagina and anus, needles for shots, pills, had a Nikon pointed right into my spread legs. I had long, pointed beaks inside me and had my vagina smeared with cold, blue paint to check for abrasions.

After a few hours of this, they let me shower. I stood there examining my body beneath the stream of water and decided, I don't want my body anymore. I was terrified of it, I didn't know what had been in it, if it had been contaminated, who had touched it. I wanted to take off my body like a jacket and leave it at the hospital with everything else.

... One day, I was ... scrolling through the news on my phone, and came across an article. In it, I read and learned for the first time about how I was found unconscious, with my hair disheveled, long necklace wrapped around my neck, bra pulled out of my dress, dress pulled off over my shoulders and pulled up above my waist, that I was butt naked all the way down to my boots, legs spread apart, and had been penetrated by a foreign object by someone I did not recognize. This was how I learned what happened to me .... I learned what happened to me the same time everyone else in the world learned what happened to me.

At the bottom of the article, after I learned about the graphic details of my own sexual assault, the article listed his swimming times. She was found breathing, unresponsive with her underwear six inches away from her bare stomach curled in fetal position. By the way, he's really good at swimming.

... I thought there's no way this is going to trial; there were witnesses, there was dirt in my body, he ran but was caught. He's going to settle, formally apologize, and we will both move on. Instead, I was told he hired a powerful attorney, expert witnesses, private investigators who were going to try and find details about my personal life to use against me. ...

... Instead of taking time to heal, I was taking time to recall the night in excruciating detail, in order to prepare for the attorney's questions that would

of its implementation. Governments would save about $27 billion and the public $115 billion by the year 2050. Additional savings to society would include "the value of material losses and the pain-and-suffering that would otherwise be experienced by the victims of juvenile crime, adult crime, and child abuse and neglect" (Lynch and Vaghul 2015, n.p.).

restorative justice A philosophy primarily concerned with reconciling conflict among the victim, the offender, and the community.

Lastly, the movement toward **restorative justice**, a philosophy primarily concerned with repairing the victim–offender–community relation, is a direct response to the concerns of an adversarial criminal justice system that encourages offenders to deny, justify, or otherwise avoid taking responsibility for their actions (see this chapter's *The Human Side*). Restorative justice holds that the justice system, rather than relying on "punishment, stigma, and disgrace" (Siegel 2006, p. 275), should "repair the harm" (Sherman 2003, p. 10). Key components of restorative justice include restitution to the victim, remedying the harm to the community, and mediation.

be invasive, aggressive, and designed to steer me off course, to contradict myself, my sister, phrased in ways to manipulate my answers … .

I was pummeled with narrowed, pointed questions that dissected my personal life, love life, past life, family life, inane questions, accumulating trivial details to try and find an excuse for this guy who didn't even take the time to ask me for my name, who had me naked a handful of minutes after seeing me.

And then it came time for him to testify. This is where I became revictimized. I want to remind you, the night after it happened, he said he never planned to take me back to his dorm. He said he didn't know why we were behind a dumpster. He got up to leave because he wasn't feeling well when he was suddenly chased and attacked. Then he learned I could not remember.

So one year later, as predicted, a new dialogue emerged. Brock had a strange new story, almost sounded like a poorly written young adult novel with kissing and dancing and hand holding and lovingly tumbling onto the ground, and most importantly in this new story, there was suddenly consent.

… Next in the story, two people approached you. You ran because you said you felt scared… The idea that you thought you were being attacked out of the blue was ludicrous. That it had nothing to do with you being on top my unconscious body. You were caught red handed, with no explanation. … When the policeman arrived and interviewed the evil Swede who tackled you, he was crying so hard he couldn't speak because of what he'd seen.

… To sit under oath and inform all of us, that yes, I wanted it, yes, I permitted it, and that you are the true victim attacked by guys for reasons unknown to you is sick, is demented, is selfish, is stupid. It shows that you were willing to go to any length, to discredit me, invalidate me, and explain why it was okay to hurt me. You tried unyieldingly to save yourself, your reputation, at my expense.

… You are guilty. Twelve jurors convicted you guilty of three felony counts beyond reasonable doubt, that's twelve votes per count, thirty-six yeses confirming guilt, that's one hundred percent, unanimous guilt. And I thought finally it is over. …

I want to say this. All the crying, the hurting you have imposed on me, I can take it. But when I see my younger sister hurting, … when she is crying so hard on the phone she is barely breathing, telling me over and over she is sorry for leaving me alone that night, sorry sorry sorry, when she feels more guilt than you, then I do not forgive you.

… It is deeply offensive that he would try and dilute rape with a suggestion of promiscuity. By definition rape is the absence of promiscuity, rape is the absence of consent, and it perturbs me deeply that he can't even see that distinction.

… He is a lifetime sex registrant. That doesn't expire. Just like what he did to me doesn't expire, doesn't just go away after a set number of years. It stays with me, it's part of my identity, it has forever changed the way I carry myself, the way I live the rest of my life.

… To conclude, I want to say thank you. To everyone … to my incredible parents who teach me how to turn pain into strength, to my friends who remind me how to be happy, to my boyfriend who is patient and loving, to my unconquerable sister who is the other half of my heart… Thank you to girls across the nation that wrote cards to my DA to give to me, so many strangers who cared for me.

And finally, to girls everywhere, I am with you. On nights when you feel alone, I am with you. When people doubt you or dismiss you, I am with you. I fought every day for you. So never stop fighting, I believe you. *Lighthouses don't go running all over an island looking for boats to save; they just stand there shining.* Although I can't save every boat, I hope that by speaking today, you absorbed a small amount of light, a small knowing that you can't be silenced, a small satisfaction that justice was served, a small assurance that we are getting somewhere, and a big, big knowing that you are important, unquestionably, you are untouchable, you are beautiful, you are to be valued, respected, undeniably, every minute of every day, you are powerful and nobody can take that away from you. To girls everywhere, I am with you.

SOURCE: Public Document 2016.

Chapter Review

- **Are there any similarities between crime in the United States and crime in other countries?**
  All societies have crime and have a process by which they deal with crime and criminals; that is, they have police, courts, and correctional facilities. Worldwide, most offenders are young males, and the most common offense is theft; the least common offense is murder.
- **How can we measure crime?**
  There are three primary sources of crime statistics. First are official statistics—for example, the FBI's Uniform Crime Reports, which are published annually. Second are victimization surveys designed to get at the "dark figure" of crime, crime that official statistics miss. Finally, self-report studies have all the problems of any survey research. Investigators must be cautious about whom they survey and how they ask the questions.
- **What sociological theory of criminal behavior blames the schism between the culture and structure of society for crime?**
  Strain theory was developed by Robert Merton (1957) and uses Durkheim's concept of *anomie*, or normlessness. Merton argued that, when the structure of society limits legitimate means (e.g., a job) of acquiring culturally defined goals (e.g., money), the resulting strain may lead to crime. Individuals,

then, must adapt to the inconsistency between means and goals in a society that socializes everyone into wanting the same thing but provides opportunities for only some.

- **What are index offenses?**
  Index offenses, as defined by the FBI, include two categories of crime: violent crime and property crime. Violent crimes include murder, robbery, assault, and rape; property crimes include larceny, car theft, burglary, and arson. Property crimes, although less serious than violent crimes, are the most numerous.

- **What is meant by white-collar crime?**
  White-collar crime includes three categories: occupational crime—i.e., crime committed in the course of one's occupation; corporate crime, in which corporations violate the law in the interest of maximizing profits; and political crime wherein government officials act in their own best interest to the detriment of the state.

- **How do social class and race affect the likelihood of criminal behavior?**
  Official statistics indicate that minorities are disproportionately represented in the offender population. Nevertheless, it is inaccurate to conclude that race and crime are causally related. First, official statistics reflect the behaviors and policies of criminal justice actors. Thus the high rate of arrests, conviction, and incarceration of minorities may be a consequence of individual and institutional bias against minorities. Second, race and social class are closely related in that non-White people are overrepresented in the lower classes. Because lower-class members lack legitimate means to acquire material goods, they may turn to instrumental, or economically motivated, crimes. Thus the apparent relationship between race and crime may, in part, be a consequence of the relationship between these variables and social class.

- **What are some of the economic costs of crime?**
  First are direct losses from crime, such as the destruction of buildings through arson or of the environment by polluters. Second are costs associated with the transferring of property (e.g., embezzlement). A third major cost of crime is that associated with criminal violence (e.g., the medical cost of treating crime victims). Fourth are the costs associated with the production and sale of illegal goods and services. Fifth is the cost of prevention and protection. Finally, there is the cost of the criminal justice system, law enforcement, litigation and judicial activities, corrections, and victims' assistance.

- **What led to the BLM protests of 2020, and what are some of the police reforms being considered?**
  The BLM protests began in Minneapolis as a result of the killing of an unarmed Black man, George Floyd, during an arrest. As a result of the protests, some of the police reforms being considered include a ban on neck restraints, the demilitarization of the police, increased racial and ethnic diversity of police departments, repeal of qualified immunity laws, and the addition of mental health care professionals and social workers when calls to the police deal with issues of mental health, substance abuse, and homelessness.

- **What is the present legal status of capital punishment in this country?**
  Half the states have abolished or have a moratorium on the death penalty. Concerns with the death penalty include racial bias, the lack of deterrence, the higher cost than life in prison, and the execution of an innocent person. In 2020, the federal government resumed the death penalty.

Test Yourself

1. The United States has the highest incarceration rate of developed countries.
   a. True
   b. False
2. The Uniform Crime Reports is a compilation of data from
   a. the U.S. Census Bureau.
   b. law enforcement agencies.
   c. victimization surveys.
   d. the Department of Justice.
3. According to _____, crime results from the absence of legitimate opportunities as limited by the social structure of society.
   a. Hirschi
   b. Marx
   c. Merton
   d. Becker
4. Which of the following is not an index offense?
   a. Drug possession
   b. Homicide
   c. Rape
   d. Burglary
5. The economic costs of white-collar crime outweigh the costs of traditional street crime.
   a. True
   b. False
6. Women everywhere commit less crime than men.
   a. True
   b. False
7. Probation entails
   a. early release from prison.
   b. a suspended sentence.
   c. court supervision in the community in lieu of incarceration.
   d. incapacitation of the offender.
8. Europol is an advisory and support law enforcement agency for European Union members.
   a. True
   b. False
9. The militarization of the police does not include which of the following?
   a. Creating a mind-set that regards the public as the enemy
   b. The addition of SWAT teams to police forces
   c. Surplus military equipment being allocated to police departments
   d. Use of the U.S. Army to control violent demonstrations
10. The crime rate in the United States has steadily increased over the last several decades.
    a. True
    b. False

Answers: 1. A; 2. B; 3. C; 4. A; 5. A; 6. A; 7. C; 8. A; 9. D; 10. B.

Key Terms

acquaintance rape 125
breeding ground hypothesis 150
biometric surveillance 143
capital punishment 152
classic rape 125
clearance rate 119
computer crime 132
corporate violence 129
covert differential law enforcement 147
crime 118
crime rate 119
dark figure of crime 119
deferred prosecution agreement (DPA) 131
deterrence 149
differential involvement 147
feminist criminology 134
identity theft 132
incapacitation 149
index offenses 124
Interpol 155
larceny 126
organized crime 128
overt differential law enforcement 146
parole 152
probation 152
qualified immunity 145
racial profiling 135
ransomware 132
recidivism 149
rehabilitation 149
restorative justice 156
safety gender gap 140
social bond 121
transnational crime 117
victimless crimes 127
white-collar crime 128

Lehn Dominis/The LIFE Picture Collection/Getty Images

The family is the basis of society. As the family is, so is the society, and it is human beings who make a family—not the quantity of them, but the quality of them."

ASHLEY MONTAGU
anthropologist

5

Family Problems

Chapter Outline

Learning Objectives

After studying this chapter, you will be able to ...

1. Identify examples of how family forms and norms vary around the world.
2. Compare and contrast arranged and forced marriages.
3. Describe at least five changes in family relationships and in the lives of children in the United States.
4. Explain how structural functionalism, conflict theory, and symbolic interactionism help us understand the family institution, access to family planning and abortion, divorce, and domestic violence and abuse.
5. Compare and contrast the marital decline perspective and the marital resiliency perspective.
6. Describe the factors that limit access to contraception and abortion, as well as the role of contraception and abortion in family planning.
7. Discuss the social causes of divorce and its consequences for children and adults.
8. Identify the different forms of abuse in relationships.
9. Describe the effects of violence and abuse on victims and the factors that contribute to domestic violence and abuse.
10. Discuss strategies to strengthen families, including expanding the definition of family, workplace and economic supports, sex education and access to contraception, relationship literacy education, divorce mediation, divorce education programs, and domestic violence and abuse prevention and policies.
11. Identify what many family scholars say is the common denominator in solving family problems.

"My parents took my whole life from me, but now I'm taking my life back. I'm a fighter. ... I fought to become the person I am."—Daughter of David and Louise Turpin

UPI/Alamy Stock Photo

ON JANUARY 14, 2018, a 17-year-old girl escaped her home in a California suburb and called 911 to seek help for her 12 siblings still in the house. "I live in a family of 15 people and my parents are abusive. They abuse us, and my two little sisters right now are chained up" [transcript of 911 call]. When police arrived at the home, they found the children in filthy conditions, having been regularly beaten, choked, and tortured. One of the children was chained to a bed and two others had just been released from chains. The children ranged in age from 2 to 29 but were so malnourished that they all appeared to be under 18. Neighbors rarely saw the children as they did not attend school and were not allowed to play outside. For a time, their mother regularly sent photographs to family members of the children smiling and dressed in matching clothing. Their parents, David Allen Turpin and Louise Anna Turpin, were arrested and eventually pled guilty to charges of torture, willful child cruelty, false imprisonment, and cruelty to an adult dependent. They were sentenced to life in prison with the possibility of parole in 25 years (Gutman and Shapiro 2019).

The highly publicized story of the Turpins brought national attention to the problem of domestic violence. In this chapter, we turn our attention to family problems, focusing on the ability to control when one has children, violence and abuse in intimate and family relationships, and problems associated with divorce and its aftermath. Many of the problems families face—concerns about health, poverty, job-related issues, drug and alcohol abuse, discrimination, and military deployment of a spouse—are dealt with in other chapters in this text. We begin by sampling the global diversity in family life and noting patterns, trends, and variations in contemporary U.S. families.

The Global Context: Family Forms and Norms around the World

The U.S. Census Bureau defines *family* as a group of two or more people related by blood, marriage, or adoption. Sociology offers a broader definition of family: A **family** is a kinship system of all relatives living together or recognized as a social unit, including adopted members. This broader definition recognizes foster families, unmarried same-sex and opposite-sex couples with or without children, and any relationships that function and feel like a family. Next, we describe several global variations in family forms and norms.

family A kinship system of all relatives living together or recognized as a social unit, including adopted people.

monogamy Marriage between two partners; the only legal form of marriage in the United States. Also refers to the restriction of sexual behavior to between two partners.

Monogamy and Polygamy

In many countries, including the United States, the only legal form of marriage is **monogamy**—a marriage between two partners. The term *monogamy* also refers to sexual relationships in which both partners have sex only with each other. In U.S. culture and throughout much of the world, monogamy is considered morally superior to other forms of marital and sexual relationships. A common variation of monogamy is **serial monogamy**—a succession of marriages in which a person has more than one spouse over a lifetime but is legally married to only one person at a time.

In some cultures, law or custom allows the practice of **polygamy**—a form of marriage in which a person has more than one spouse. **Polygyny**, the most common form of polygamy, involves one husband having more than one wife and is most common in traditional African cultures and in some Muslim populations. **Polyandry**—the concurrent marriage of one woman to two or more men—is less common than polygyny and is found in northern India, Nepal, and Tibet, as well as in 53 other societies. Polyandrous cultures tend to be small-scale egalitarian societies that obtain food by hunting and gathering, foraging, and horticulture and that have an imbalanced sex ratio favoring males (Starkweather and Hames 2012).

The U.S. Congress outlawed polygamy in the late 1800s. Under U.S. law, being married to more than one person at a time is a crime referred to as **bigamy**. Although the Mormon Church has officially banned polygamy, it is still practiced among some members of some fundamentalist Mormon splinter groups (Harrison 2017). Polygamy in the United States also occurs among some immigrants who come from countries where polygamy is accepted, such as West African countries. Immigrants who practice polygamy generally keep their lifestyle a secret because polygamy is grounds for deportation under U.S. immigration law (Lines 2016).

serial monogamy A succession of marriages in which a person has more than one spouse over a lifetime but is legally married to only one person at a time.

polygamy A form of marriage in which one person may have two or more spouses.

polygyny A form of marriage in which one husband has more than one wife.

polyandry The concurrent marriage of one woman to two or more men.

bigamy The criminal offense in the United States of marrying one person while still legally married to another.

Arranged Marriages versus Self-Choice Marriages

In some countries, young adults do not choose their spouses; rather, their parents or other third parties arrange their marriages. Each year about eight million brides marry men chosen by their parents and/or other family members. These are referred to as **arranged marriages**. In many arranged marriages, the bride and groom do not meet until a few weeks before the wedding, or on the wedding day. A 2013 survey found that 75 percent of people in India between the ages of 18 and 35 still prefer an arranged marriage. Parents arrange marriages based on caste membership, wealth, occupation, and other factors (Harris 2015). In recent decades, young adults living in societies that have traditionally practiced arranged marriage are increasingly selecting their own spouses. Factors associated with this transition from traditional to modern marriage norms include increased education, urbanization, and exposure to Western cultural norms and values (Allendorf 2013).

Reflecting a move in this direction, in about 25 percent of arranged marriages in India, marriage websites play an important role with profiles scrutinized by parents and prospective brides and grooms (Harris 2015). In most Western societies, individuals choose their own marriage or romantic partners, with 90 percent of adults in the United States citing love as the reason they got married or moved in with a partner (Horowitz, Graf, and Livingston 2019). Among Indian Americans and some other immigrant groups, arranged marriage still occurs in the United States although it is far less common than in their home countries.

It is important to distinguish between arranged marriages and **forced marriages**. Forced marriages are marriages that take place without the consent of one or both of the people involved. Forced marriages may occur when a family member or prospective spouse uses physical or emotional abuse, threats, or deception to force an individual to marry. This chapter's *The Human Side* is an example of a forced child marriage that occurred in the United States.

arranged marriage A type of marriage in which the bride and groom are selected by individuals other than the couple themselves, typically by family members such as parents. The bride and groom enter into the arrangement consensually.

forced marriage A marriage that takes place without the consent of one or both people in the marriage.

What do you think?

Divorce rates among people who have arranged marriages are lower than those in self-choice marriages (Dholakia 2015). Why do you think this is? Could there be a benefit to arranged marriages that makes them last longer?

Julia's Story

I have always kept my arranged marriage a secret, it's not often you find someone in the United States who would understand. Most people think that sort of thing doesn't happen in developed countries. It's ancient, archaic, something that happens somewhere else, not here. Not where we marry for *love*.

See, when you're a child you must rely on adults for just about everything. As we grow, we slowly gain responsibility and power of our own lives … . The natural progression is increased independence. But this is all disrupted when you are coerced or forced into an arranged marriage. Marriage is an adult decision. And when this decision is made for you, as a child, the damages are unparalleled.

But the worst part is the secrecy. Not wanting to tell anyone how me and my husband came to be. We are still married. We've been through unyielding challenges together. We've grown to be best friends, love, and respect each other. We have a family now. We escaped our abusive situation together. But we keep it to ourselves, 'cause we never thought anyone else would understand. Who could possibly understand, that we grew up in a religious cult and were told at the ages of 15 that we were to marry each other? We were told we were "made" for each other, meant to be together by God. We were just like everyone else in our commune, our partners were picked for us by the elders.

For many years I tried to say no, I said I didn't want to marry someone from the commune. I wanted to leave and have experiences. But the brutality was just too great. Friends and family wouldn't speak to me. I would get screamed at, people would come to my house late at night and yell at my mother about me. I was told if I wanted to stay in the commune, I had to marry my partner. I had no family outside the commune, and I loved my family very much. I didn't see a path for myself where I didn't marry him. So I surrendered.

I was one of the lucky ones, my partner was kind and gentle. We had always been friends, and we bonded over the obscurity of our lives. As our living conditions worsened in the commune, we decided to break free together. We got new jobs, moved, and cut ties. After years of therapy, we have begun to rebuild our lives outside the cult.

But the fact remains that we didn't have a choice in the biggest adult decision either of us would ever make. And that has lasting effects.

It wasn't until I found Unchained At Last that I knew anyone like me existed. I was sure I was the only one, besides the others I grew up with in the commune. … My strategy has been to hide, shield myself, blend in amongst the crowds, and keep my history to myself. But Unchained has inspired me to remove my veil, and stand amongst the other young women who were coerced into arranged marriage before adulthood. Because it turns out, it is more common than we think.

SOURCE: Julia's Story. (n.d.).

Division of Power in the Family

In many societies, male dominance in the larger society is reflected in the authority of husbands over their wives (see also Chapter 10, "Gender Inequality"). This form of societal structure is called **patriarchy**. For example, in some societies, men make decisions about their wife's health care, when their wife may visit relatives, and/or household purchases (Population Reference Bureau 2011). In Saudi Arabia, every woman legally must have a male guardian who controls her life from birth to death (Human Rights Watch 2019). She cannot travel, marry, seek health care, or be released from prison without her guardian's approval. Further, many women and men believe that it is acceptable for a man to beat his wife if she argues with him and/or refuses to have sex (Population Reference Bureau 2011). A recent survey of teenagers found that more than 35 percent of boys and girls in African countries believe wife beating is justified with, surprisingly, adolescent girls more supportive than boys (UNICEF 2020a).

In developed Western countries, although gender inequality persists (see Chapter 10), marriages between men and women have become more egalitarian, which means they view each other as equal partners who share decision making, housework, and child care. As couples become more egalitarian, the roles of mother and father often converge with mothers contributing to the income for the family and fathers doing more housework and child care (Parker and Wang 2013). However, women still spend more time on housework than men in all racial and ethnic groups with the gender gap being larger for Hispanic and Asian partnered women than for White and Black partnered women. The strongest predictor of how much housework men do is if their wives work more hours in a day than they do (Wight, Bianchi, and Hunt 2013).

patriarchy A male-dominated system in which men have primary power over major decision making.

Studies of same-sex couples find that they are more egalitarian in their division of housework and child-care responsibilities than **different-sex couples**, or what have historically been referred to as heterosexual couples (Goldberg, Smith, and Perry-Jenkins 2012; Brewster 2016). Gender and family scholars are moving away from referring to couples by sexual orientation (i.e., gay or lesbian and heterosexual), because sexual orientation is a characteristic of individuals not of couples (see Chapter 11).

Social Norms Related to Childbearing

In less developed societies, where social expectations for women to have children are strong, women on average have four to five children in their lifetime. In developed societies, where women may view having children as optional—as a personal choice—women have, on average, fewer than two children.

Norms regarding appropriate childbearing age also vary culturally. More than half of women in less developed countries are married before age 18—many before they turn 15 (UNICEF 2020b). In countries where child marriages are common, teenage births are considered normal even though they are associated with increased risk of health problems for the mother and baby and with poverty. Lack of access to contraception is a significant factor in teenage births. In the United States and other developed countries, teenage childbearing is discouraged.

What do you think?

In the United States, 23 states have no minimum age below which a child cannot marry. Most other states allow marriage under 18 under certain conditions. In 2017, a study in 38 states found that between 2000 and 2010, over 248,000 children under the age of 18, most girls and some as young as 12 years old, had been married in the United States (Reiss 2019). The majority, 77 percent, were married to adult men. What do you think is the appropriate age for a child, male or female, to be able to legally marry without their parent's permission?

Norms about childbirth out of wedlock vary worldwide (YaleGlobal Online 2017). The percent of children born to unmarried mothers ranges from 1 percent in India to 84 percent in Colombia. In many poor countries nonmarital childbirths are rare because girls often marry before they are 18 years old. Rates of nonmarital childbearing are highest in Central and South America where as many as 60 percent of births are to unwed mothers. In Northern and Western Europe between one-third and one-half of children are born to unmarried women. However, this may not mean that these children are living in single-parent homes. In many European countries, the norm is for couples to have children before marriage but while cohabiting. In the United States, four in ten births in 2018 were to unmarried women with over 50 percent belonging to women in cohabiting couples (Wildsmith, Manlove, and Cook 2018; Martin et al. 2017).

@katyperry

Really encouraged to be an American today... Love should live beyond labels & intolerance! #LoveWins #EqualityForAll

-Katy Perry

Same-Sex Couples

Norms, policies, and attitudes concerning same-sex intimate relationships also vary around the world. In some countries, gay and lesbian sexuality is punishable by imprisonment or even death. First legalized in the Netherlands in 2001, same-sex marriages are, as of this writing, now legal in at least 29 countries. Four countries—Costa Rica, Austria, Taiwan, and Ecuador—legalized same-sex marriage as recently as 2019 and 2020 (Human Rights Campaign 2020). In 2015, the U.S. Supreme Court ruled 5-to-4 that same-sex couples have the constitutional right to legally marry in all 50 states. See Chapter 11, "Sexual Orientation and the Struggle for Equality," for in-depth information about same-sex couples and families.

Looking at families from a global perspective underscores the fact that families are shaped by the social and cultural context in which they exist. As we discuss problems related to divorce, family violence, and abuse, we refer to social and cultural forces that shape these events and the attitudes surrounding them. Next, we look at patterns, trends, and variations in U.S. families.

different-sex couples Often referred to as heterosexual couples. The term different-sex better takes into account the reality that individuals' sexual orientation may not predict their relationship composition.

Contemporary U.S. Families: Changes in Relationships and in the Lives of Children

Family forms and norms vary not only across societies; they also vary over time. In the United States, more adults are delaying or opting out of marriage, a growing share of children are living with an unmarried parent, the number of stepkin is at an all-time high, and same-sex marriages are legal in all 50 states. Attitudes about these changes are shifting, too. A 2019 poll of U.S. adults found that most respondents thought these changes in family structure either made no difference or were a good thing. Just ten years ago a similar poll found that people were far more divided in their opinions (see Figure 5.1, Thomas 2020).

Changing Relationships

Americans marry, divorce, and cohabit more than in any other Western society. They also stop and start relationships more quickly (Cherlin 2009). While individuals still tend to gravitate toward partners who resemble themselves in terms of race, religion, and socioeconomic status, some of these norms are shifting with individuals crossing social boundaries in increasing numbers.

Delayed Marriage and Increased Diversity in Marriage Partners. By 2018, a 40-year decline took the marriage rate in the United States to its lowest number since 1867, the first year the government started tracking such figures. While in 1978, 59 percent of adults between the ages of 18 and 34 were married, by 2018 that number had decreased to under a third (Galvin 2020). U.S. women and men are staying single longer and marrying later in life, if at all. Between the mid-1950s and 2018, the median age for first marriages for U.S. women rose from 20 years old to 28 years old. For men, it rose from 23 years old to 30 years old (see Figure 5.2, U.S. Census Bureau 2018).

An increasing number of people also state that they do not intend to get married; 14 percent of never married adults say they do not want to get married, 27 percent say they are not sure, and just over half say they want to get married someday (Parker and Stepler 2017). However, after the legalization of same-sex marriage, by 2019, 61 percent of same-sex couples who were

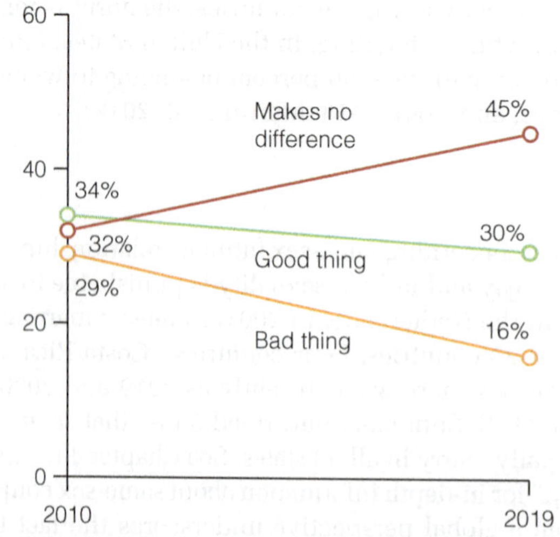

Figure 5.1 Shifting Views on a Growing Variety of Family Arrangements in the United States, 2019
SOURCE: Thomas DEJA 2020.

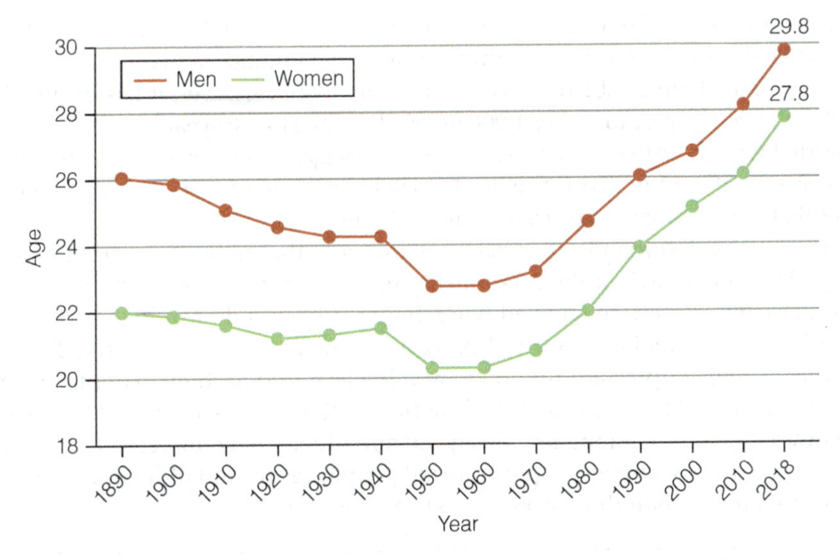

Figure 5.2 Median Age at First Marriage Has Been Increasing since the 1960s
SOURCE: U.S. Census 2018.

living together were married (Geiger and Livingston 2019), suggesting that marriage is still a social institution that is valued by many. This is also evidenced by the fact that one in four marriages in the United States involves at least one previously married partner (Wang and Parker 2014), i.e., even people who get divorced are often willing to remarry (Lewis and Kreider 2015).

While in the past married couples tended to be **homogamous**, there is more variation among married couples today than ever before. Homogamy refers to couples who are similar in their demographic characteristics (e.g., education, race, religion, etc.). Between 2000 and 2010, the percentage of U.S. different-sex marriages that involved husbands and wives of different races or Hispanic ethnicity increased from 7 percent to 10.2 percent (Rico, Kreider, and Anderson 2018) (see also Chapter 9). The percentage of inter-racial or inter-ethnic couples is even higher for unmarried different-sex couples and unmarried same-sex couples.

There has also been an increase in people marrying outside of their religion. One-quarter of Millennials (those born between 1981 and 1996) say they were raised in a religiously mixed family, as compared to 20 percent of Gen Xers (those born between 1965 and 1980), 19 percent of Baby Boomers (those born between 1946 and 1964), and 13 percent of the Silent Generation (those born before 1946) (Pew Research Center 2016). Thus, the longer ago the marriage, the lower the likelihood the marriage is between individuals of different faiths.

Increasing Cohabitation. Alongside the decrease in marriage rates has been an increase in cohabitation among different-sex couples. In fact, the percentage of adults between the ages of 18 and 44 who have lived with someone (59 percent) is now higher than the percentage who have ever been married (50 percent) (Horowitz, Graf, and Livingston 2019). Importantly, while in the past cohabitation usually ended in marriage, today the percentage of cohabiting unmarried unions lasting at least 5 years has grown to over 40 percent (Lamidi et al. 2019). Unfortunately, cohabitation is more likely to end with the breakup of the relationship than in previous years (Kuo and Raley 2016). Most U.S. adults believe it is acceptable for a couple to cohabit even if they don't plan to get married (69 percent), although just over half say that society would be better off if couples who want to stay together eventually got married (Horowitz, Graf, and Livingston 2019).

homogamous Relationships between two people with the same sociodemographic characteristics, such as race, religion, education, etc.

polyamory Multiple intimate sexual and/or loving relationships with the knowledge and consent of all partners involved.

What do you think?

In the United States, monogamy is considered the norm for couples. However, the United States is also one of the countries in which polyamory is the most common (Sheff 2014). **Polyamory** refers to having multiple intimate sexual and/or loving relationships with the consent of all partners involved. Many poly people are in the "closet," as openly polyamorous individuals are subject to stigmatization and discrimination. If everyone consents, do you think polyamorous relationships are acceptable? Why or why not?

Declining Divorce Rate. Contrary to popular belief, the divorce rate in the United States has declined since the 1980s. The **refined divorce rate**—the number of divorces per 1,000 married women—more than doubled between 1960 and 1980. However, it has declined from a high of 23 divorces per year for every 1,000 married women to a 40-year low of 16 divorces per 1,000 married women (Allred 2019). Research now suggests that between 25 and 39 percent of all marriages will end in divorce, with the risk being higher for second- and higher-order marriages than for first marriages (Dennison 2017; Luscombe 2018).

One reason the divorce rate has been declining is that more unmarried couples are living together, and so, when the relationship ends, there's no need for a divorce. In addition, Americans are getting married when they're older, and have higher levels of education and greater financial stability. Taking into account these "protective factors" and the fact that young couples are divorcing at a lower rate than their parents' generation, a recent study predicts a continued decline in the divorce rate (Cohen 2019). However, **gray divorces**—divorces among people age 50 and older—have increased. Although less likely to divorce than younger people, married people age 50 and older are twice as likely to get divorced today than they were in 1990 (Stepler 2017).

What do you think?

A national survey found that more than half (54 percent) of divorced women said they do not want to remarry compared to only 30 percent of divorced men (Wang and Parker 2014). Why do you think divorced women are less likely than divorced men to want to remarry?

Changing Lives of Children

As the life choices of parents have changed, so have the lives of children. A greater share of children today are growing up in households with parents in their thirties and forties, rather than twenties and thirties. Many children spend at least some of their childhood in a single-parent home. Children of White mothers in particular are more likely to grow up in dual-income households than in the past, while Black mothers have always been more likely to work outside the homes. Most children now have stepparents and/or stepsiblings, and a small but significant minority of children will experience at least one foster placement during their lifetime. While to parents and grandparents, it may seem that families have changed quite a bit, family historian Stephanie Coontz (2004) reminds us:

> [M]any things that seem new in family life are actually quite traditional. Two-provider families, for example, were the norm through most of history. Stepfamilies were more numerous in much of history than they are today. There have been several times and places when cohabitation, out-of-wedlock births, or nonmarital sex were more widespread than they are today. (p. 974)

Delayed Childbearing or Remaining Child Free. The average age for first-time mothers in the United States rose from 21.4 years old in 1970 to 26.4 years old in 2015 (Martin et al. 2017). Women across all racial and ethnic groups are now waiting until their mid- to late twenties to have children, although Asian and non-Hispanic White women have the oldest average age at first birth of a child.

Delayed childbearing enables women to pursue education and careers, thus establishing more financial security and independence. On the negative side, women having their first child after age 40 are at higher risk for health complications for themselves and their infant. Among U.S. adults ages 18 to 29, 74 percent say they want children, 19 percent aren't sure they want children, and 7 percent say they don't want to have children (Wang and Taylor 2011). In 2018, the U.S. fertility rate, the number of births per 1,000 women between the ages of 15 and 54, dropped to an all-time low for all racial and ethnic categories (National Center for Health Statistics 2019).

refined divorce rate The number of divorces per 1,000 married women.

gray divorces Divorces among people ages 50 and older.

Single-Parent Households. The majority of children in the United States live with two parents, 65 percent with parents who are married and 7 percent with cohabiting parents. However, as a result of births to unmarried or divorced women, about a quarter of

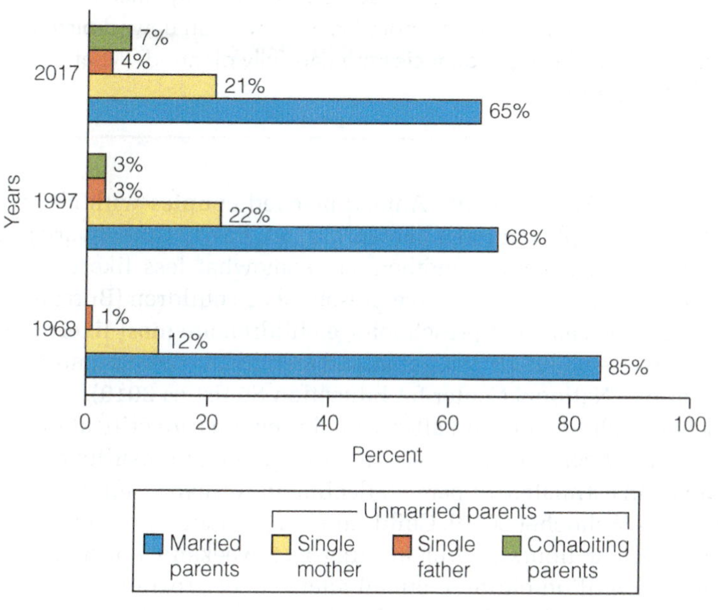

Figure 5.3 Children's Family Arrangements, 1968–2017
SOURCE: Livingston 2018.

all children live in single-parent households. Most of these live in a single-mother household (21 percent) while an increasing number live with their single fathers (4 percent of all children) (see Figure 5.3) (Livingston 2018). The United States has the highest percentage of children living in single-parent households of any country in the world (Kramer 2019).

The majority of children whose parents were married at the time of their birth grow up in intact families. However, by their tenth birthday, 27.6 percent of first-born children experience the divorce of their parents. More than half of children born to cohabiting parents will experience the breakup of their parents' relationship (Livingston 2018). After preschool, Black and Hispanic children are more likely to experience their parents' divorce than are White children (Stykes 2015). The recent trend toward lower divorce rates for younger adults may suggest that, in the coming years, fewer children will experience their parents' divorce (Amato and Patterson 2017).

Children born to unwed mothers are even more likely to grow up in a single-parent household. After a steady increase since 1960, nonmarital births hit their peak in 2008 and have decreased 14 percent since then. In 2018, about four in ten U.S. births were to unmarried women (Martin et al. 2017). The highest rates of nonmarital births are among Black, American Indian/Alaskan native, and Hispanic women. Children born to unwed mothers are more likely to experience poverty, poorer physical and mental health, and what Sawhill (2014) calls the "family-go-round": instability, less parenting, parents' multiple partners, or a series of cohabiting partners. Six in ten U.S. adults say having a baby outside marriage is "morally acceptable"—up from 45 percent in 2002 (Swift 2016).

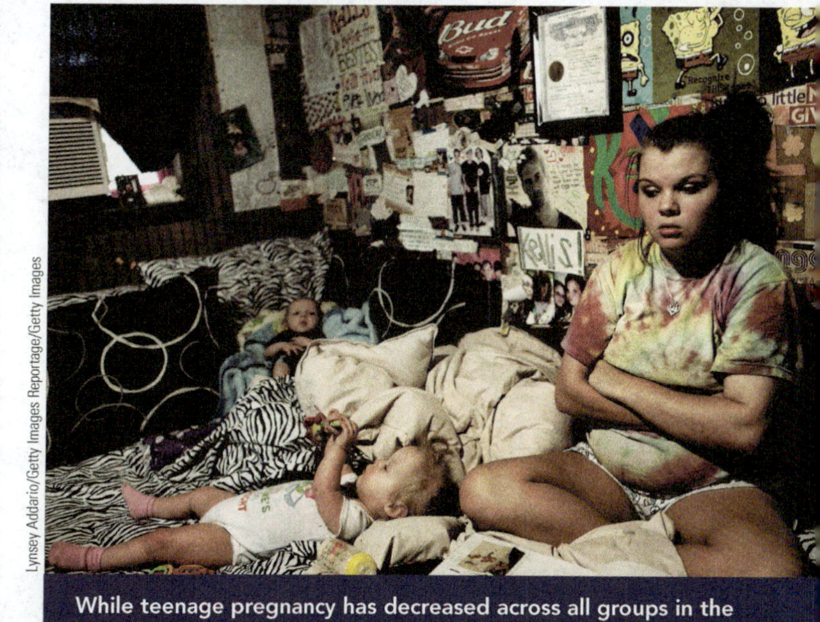

Lynsey Addario/Getty Images Reportage/Getty Images

While teenage pregnancy has decreased across all groups in the United States, teen moms face many hardships. They are less likely to complete high school, more likely to live in poverty, and more likely to have another child out of wedlock than women who have their first birth as adults.

What do you think? Births to teenagers and young adults have declined sharply since about 2005. However, women over the age of 30 are more likely to have an unwed birth now than ever before. Many of these pregnancies are carefully planned. What factors do you think explain this trend?

Increases in Mothers' Employment. Among married couples with children, over two-thirds of mothers are employed (69.9 percent), as are nearly three-quarters of unmarried mothers (73.2 percent). However, mothers are somewhat less likely to work full-time than fathers, particularly when they have preschool age children (Bureau of Labor Statistics 2019). It is still the case that preschool-age children are most likely to be in the care of their mothers, while about 29 percent attend a child care center and 19 percent are in the care of a relative (National Center for Education Statistics 2019).

Reliable schools, child care, and after-school programs are critical to women's ability to work for pay. *Child care deserts*, areas in which there are insufficient day care facilities, contribute to an estimated 12 percent decline in women's employment with little to no impact on men's (Schochet 2019). Child care is a particular problem for women whose work requires irregular hours and/or evening and weekend hours. At the end of the school day, over 10 million children attend after-school programs. Research shows that these programs can provide children with safety, support their academic achievement, and help to reduce health disparities by providing nutrition as well as opportunities for physical exercise (Centers for Disease Control and Prevention 2020a). Under the former Trump administration funding for after-school programs had repeatedly been threatened. The Biden administration has pledged to expand federal funding for after-school programs and community centers (Biden 2020a). Additional information on issues related to balancing work and family is presented in Chapter 7.

blended families Also known as *stepfamilies*, families involving children from a previous relationship.

Blended Families. Remarriages that involve children from a previous relationship are known as **blended families** or stepfamilies. Blended families have become the norm in the United States. The latest data available suggest that about 16 percent of U.S. children live in a blended family with a stepparent, stepsibling, and/or half-sibling (Pew Research Center 2015), and 40 percent of all married couples with children are stepfamilies (Lin, Brown, and Cupka 2017).

Henry Rose/Photodisc/Getty Images

About 1,300 new stepfamilies are formed every day in the United States (Stepfamily Foundation 2020).

Sixty-two percent of married and cohabiting couples under age 55 have at least one stepfamily member within three generations (child, sibling, and/or parent). Stepfamilies tend to be up to 40 percent larger than families without stepkin (Wiemers et al. 2019). As Figure 5.4 demonstrates, blended families are relatively equally distributed across racial groups; however, Asian children are less than half as likely to live in a blended family than are other children.

Stepparents and stepchildren do not have the same legal rights and responsibilities as biological/adopted children and biological/adoptive parents. Stepchildren do not automatically inherit from their stepparents, and courts have

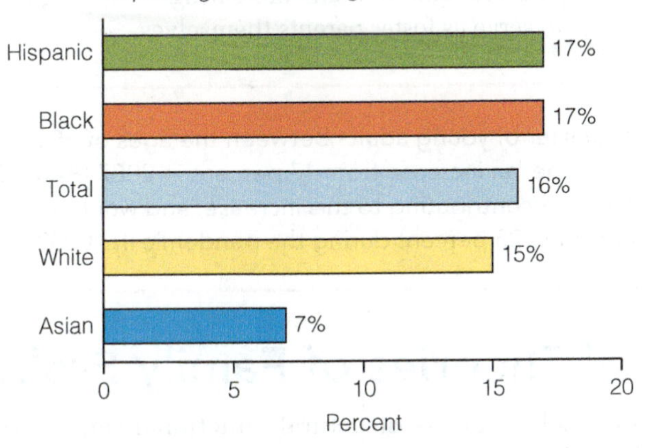

Figure 5.4 U.S. Children Living in Blended Families by Race
SOURCE: Pew Research Center 2015.

been reluctant to give stepparents legal access to stepchildren in the event of a divorce (Sweeney 2010). Wiemers et al. (2019) also found that married women give less time to assisting their stepparents and their adult stepchildren, and stepparents receive less assistance from their stepchildren in old age.

Foster Families. Foster care placement is a more common experience than one might expect. On any given day, about 428,000 children reside in foster care (Font et al. 2018). Over the course of their childhood (birth to age 18), 6 percent of all U.S. children will experience at least one foster care placement, with the rate being twice as high for Black children. Thirty-percent of foster care placements are with extended family members, while 70 percent of children are placed with individuals outside the family. Although many children spend only days in foster care, the average length of time in foster care is just over two years (FosterMore 2020).

Despite the fact that most children entering foster care have experienced neglect or abuse by a parent(s), about half of the children who exit the foster care system are reunified with their biological families. Approximately 30 percent of these children experience maltreatment again within three years, and about a quarter re-enter foster care within five years (Font et al. 2018). Over half of children adopted out of foster care are adopted by foster parents not previously related to them, and another third are adopted by relatives (FosterMore 2020). However, there is a shortage of people willing to adopt foster children because foster children tend to be older and are more likely to have emotional or physical problems (Koch 2009).

Young adults who age out of foster care face challenges as they do not have the support system in place to encourage them through college and/or into employment, although recent research finds that, in some cases, they fare better than those who have been reunified with their parent(s) (Font et al. 2018). In recognition of these struggles, some states have established support systems to assist teens aging out of the system, and some colleges and universities have programs to support foster youth.

Other Family Forms. About 3 percent of children live with relatives other than their mother or father, often grandparents. One in ten U.S. children live with a grandparent (with one parent, two parents, or no parents living with them) (U.S. Census Bureau 2016). Although grandparents may live with their adult children because the grandparent needs care and assistance with daily living (see also Chapter 12), more commonly grandparents are providing needed assistance with child care. **Grandfamilies**—families in which children reside with and are being raised by grandparents—include multigenerational families where grandparents care for grandchildren so the parent(s) can work or go to school. Grandfamilies also form in

grandfamilies Families in which children reside with and are being raised by grandparents; parents may or may not also live in the household.

response to the parent(s) experiencing job loss, out-of-state employment, military deployment, divorce, deportation, illness, death, substance abuse, incarceration, or mental illness (Generations United 2014). Some grandparents are caring for their grandchildren to keep them out of foster care or to serve as foster parents themselves.

What do you think? In 2018, almost a quarter of young adults between the ages of 25 and 29 were living in a parent's home, an increase from 17 percent in 2007 (Payne 2019). What factors do you think are contributing to this increase, and would you predict that it was higher or lower than 25 percent during the pandemic in 2020?

Sociological Theories of Family Problems

Three major sociological theories—structural functionalism, conflict and feminist theories, and symbolic interactionism—help to explain different aspects of the family institution and the problems in families today.

marital decline perspective A view of the current state of marriage that includes the beliefs that (1) personal happiness has become more important than marital commitment and family obligations and (2) the decline in lifelong marriage and the increase in single-parent families have contributed to a variety of social problems.

Structural-Functionalist Perspective

The structural-functionalist perspective views the family as a social institution that performs important functions for society, including producing and socializing new members, regulating sexual activity and procreation, and providing physical and emotional care for family members. This perspective views the high rate of divorce and the rising number of single-parent households as constituting a "breakdown" of the family as an institution. According to this perspective, the **marital decline perspective**, (1) personal happiness has become more important than marital commitment and family obligations, and (2) the decline in lifelong marriage and the increase in single-parent families have contributed to a variety of social problems, such as poverty, delinquency, substance abuse, violence, and the erosion of neighborhoods and communities (Amato 2004).

Kerem Yucel/AFP/Getty Images

Families with children are among the increasing number of people experiencing homelessness in the United States. This tent camp is in Minneapolis, Minnesota, the home to five Fortune 500 companies and the fifth largest hub of major corporate headquarters in the United States.

Further, structural functionalists argue that traditional gender roles contribute to family functioning. Women perform the *expressive role* of managing household tasks and providing emotional care and nurturing to family members, and men perform the *instrumental role* of earning income and making major family decisions. According to this view, families have been disrupted and weakened by the change in gender roles, particularly women's participation in the workforce—an assertion that has been criticized and refuted but is nonetheless held by "family values" scholars and others who advocate a return to traditional gender roles in the family.

Structural functionalism also looks at how changes in other social institutions affect families. For example, research has found that changes in the economy—specifically falling wages among unskilled and semiskilled men as well as cuts in employee benefits—have contributed to the increase in poverty and homelessness in the United States. While in the past an individual—typically a man—making minimum wage could afford to provide for a family, today many working families live in poverty, just one crisis away from being homeless. In 2018, about one in six children, 12 million in total, were living in poverty in the United States and children of single mothers were five times more likely to be poor than children from intact families (Boghani 2020) (see Chapter 6).

Conflict and Feminist Perspectives

Conflict theory focuses on how capitalism, social class, and power influence marriages and families. Feminist theory is concerned with how gender inequalities influence and are influenced by marriages and families. Feminists are critical of the traditional male domination of families—known as patriarchy—that is reflected in the tradition of wives taking their husband's last name and children taking their father's name. Patriarchy implies that wives and children are the property of husbands and fathers.

The overlap between conflict and feminist perspectives is evident in views on how industrialism and capitalism have contributed to gender inequality. With the onset of factory production during industrialization, workers—mainly men—left the home to earn incomes and women stayed home to do unpaid child-care and domestic work. This arrangement resulted in families founded on what Friedrich Engels calls "domestic slavery of the wife" (quoted by Carrington 2002, p. 32). Modern society, according to Engels, rests on gender-based slavery, with women doing household labor for which they receive neither income nor status, whereas men leave the home to earn an income. Times have changed since Engels made his observations, with most wives today leaving the home to earn incomes. However, wives employed full-time still do the bulk of unpaid domestic labor, and women are more likely than men to compromise their occupational achievement to take on child-care and other domestic responsibilities.

gendered distribution of labor The assumption in society that certain types of jobs and activities will be completed by men and others by women.

The impact of this **gendered distribution of labor** can be seen in the fact that female workers in the United States have been most negatively affected by the COVID-19 pandemic (see Chapter 10). While women were more likely to be employed in "essential jobs" (e.g., nurses, retail clerks, etc.), most of those did not allow for work from home (Kurtzleben 2020). With schools and child care centers closed, 13 percent of U.S. parents—mostly women—quit their jobs or reduced their working hours between March and June 2020 due to a lack of child care. Moreover, many parents who relied on grandparents for child care decided it was too risky given the impact of COVID-19 on the elderly. Thus, the responsibility for providing child care and supervising home schooling fell primarily on women. Even women who could work from home were more likely than men to be juggling child care while trying to work (Long 2020).

> As the mom of two boys under 6, Karin Brownawell has come to dread videoconferences. In late March, on her first video chat with her top bosses, she told her sons to play in the backyard and not come inside unless it was an emergency. Ten minutes into the call, her 3-year-old burst into the kitchen and yelled at the top of his lungs, "Mom, I have to poop!" (Long 2020)

Experts warn that women's lower work productivity during the pandemic is likely to have a long-lasting impact on the gender gap in pay (Kurtzleben 2020).

Brendon Thorne/Getty Images News/Getty Images

During the COVID-19 pandemic, women took on the greatest share of the child care and homeschooling required when schools closed across the globe. This woman considered herself fortunate to be able to work from home, while many women were forced to quit their jobs to stay home with their children.

Conflict theorists emphasize that powerful and wealthy segments of society largely shape social programs and policies that affect families. During the 1900s, hundreds of thousands of people in the United States were sterilized either against their will or without their knowledge as part of **eugenics** programs, many of them funded by the federal government. These "undesirable" populations included immigrants, people of color, poor people, unmarried others, delinquents, the disabled, and the mentally ill. California's eugenics program inspired none other than Adolf Hitler to write: "There is today one state in which at least weak beginnings toward a better conception [of citizenship] are noticeable. Of course, it is not our model German Republic, but the United States" (Hitler 1925).

While most of these programs ended by the late 1970s, a recent report found that the state of California coerced nearly 150 female inmates into sterilization as recently as 2006–2010 (Jindia 2020). Beginning in 2014, the Chinese government instituted a new policy focused on the Uigher population and other ethnic minorities living in the regions of Hotan, Kashgar, and Xinjiang (Associated Press 2020). Ethnic minority women in these regions are regularly subjected to pregnancy checks, forced sterilization, IUDs, and abortions. People who have too many children are sent to detention camps where they are subjected to political and religious re-education and forced labor, while their children are sent to orphanages. The goal of the program, begun under the new authoritarian leadership of President Xi Jinping, is to limit the size of the rural religious population who the Xinping government blame for Islamic extremism. As Figure 5.5 demonstrates, the sterilization rates in the Xinjiang—predominately Uighur—province are far greater than that of the national average and increasing over time. Scholar Joanne Smith Finley described it as "slow, painful, creeping genocide" (Associated Press 2020).

According to conflict theorists, the interests of corporations and businesses are often at odds with the needs of families. Private prison contractors have spent millions of dollars on lobbying and campaign contributions in order to influence legislators to support detention-focused policies aimed at undocumented immigrants and their families. In order to defeat Hillary Clinton, who was publicly opposed to private prisons, private prison companies donated hundreds of thousands of dollars to PACs supporting the Trump campaign. Their investment paid off; shortly after his inauguration, former President Trump signed two executive orders increasing the use of detention against undocumented immigrants. CoreCivic Inc. and GEO Group, Inc.—which managed more than half of the private prison contracts in the United States—earned more than $4 billion in 2017 (see Chapter 4). Immigration policies promoting detention have had catastrophic impacts on families on both sides of the southern U.S. border (Luan 2018). Read about one study of these impacts in this chapter's *Social Problems Research Up Close.*

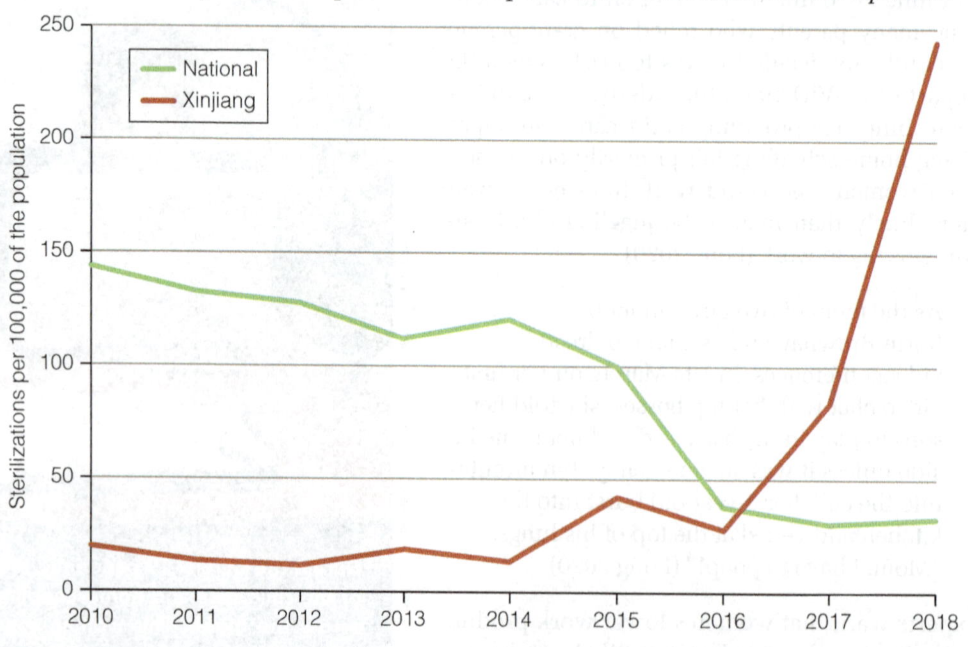

Figure 5.5 Sterilization Rates in Xinjiang Province Compared to the National Trends in China, 2010–2018
SOURCE: Zenz 2020.

eugenics The practice of selectively mating people with specific traits considered to be desirable with the belief that it will "improve" the human race.

social problems RESEARCH UP CLOSE Not Really Single

Deportations of working-class Latino men have reached historic levels in the United States. A quarter of those deported have U.S.-born children, and most have no criminal record. In this study, sociologist April Schueths focuses on a specific group of "mixed-status" families with deported husbands and citizen wives. The experiences of these families are a consequence of what has been called the Gendered Racial Removal Project (GRRP)—a series of post-911 immigration policies targeted at demonizing brown men previously viewed as simply "unthreatening reserve labor force" and turning them into "a dangerous quasi-criminal element embedded in American society" (Provine and Doty 2011, p. 266). These racialized, gender stereotypes serve to justify the removal of even employed men without criminal records supporting families. Schueths examines the transition of their mostly non-Hispanic citizen wives to single parents and sole providers, focusing on the gendered, classed, and racialized consequences of deportation.

Sample and Methods

Participants were recruited through e-mails and social media posts to immigration advocacy groups and social service organizations. Communications were in Spanish and English. Other participants were gained through snowball sampling. **Snowball sampling** is a technique in which one participant recommends others who might be interested in participating. It is a common sampling strategy when trying to study a population that is not easily identifiable. Semi-structured interviews were conducted with 17 women and two of their husbands. Most interviews were conducted in English. Interviews were conducted over the phone, in person, and through videoconferencing. They lasted about 90 minutes each.

Findings and Conclusions

Schueths reveals that deportations of husbands forced their wives into the role of single parent and breadwinner. Most women could not afford to join their husbands, and in some cases stepfamily situations kept them from being able to take their children out of the country. So they were forced to stay behind. While they were in stable financial situations before deportation, many had to go on public assistance for the first time in their lives, further increasing the impact of government agencies on their lives and stigmatizing them. The expense of pursuing legal status for their husbands was an added financial burden with which women struggled. The women interpreted these immigration actions as motivated by race. One woman summarized the pain this way:

> We have just the bare necessity that we need to live. When I say stressed, I'm always stressed. ... [I]t's very frustrating because, at the same time, this wasn't something that was caused by a relationship gone bad or—this is the government basically pulling our family apart (p. 1084).

SOURCE: Schueths, April. 2018. "Not Really Single: The Deportation to Welfare Path-way for U.S. Citizen Mothers in Mixed-Status Marriage." Critical Sociology. 45(7-8): 1075-1092

Symbolic Interactionist Perspective

The symbolic interactionist perspective seeks to understand how the meaning of marriage and family has changed. According to the **marital resiliency perspective**, rather than viewing divorce as a sign of the decline of marriage, divorce may be viewed as resulting from placing a high value on marriage, such that a less than satisfactory marriage is unacceptable. People who divorce may be viewed not as incapable of commitment but as those who would not settle for a bad marriage. Indeed, the expectations that young women and men have of marriage have changed. Whereas once the main purpose of marriage was to have and raise children, today women and men want marriage to provide adult intimacy and companionship (Geiger and Livingston 2019). In interviews with low-income single women with children, most women said they would like to be married but just have not found "Mr. Right" (Edin 2000).

> Interestingly, mothers say they reject entering into economically risky marital unions out of respect for the institution of marriage, rather than because of a rejection of the marriage norm. (Edin 2000, p. 130)

In a review of recent research, family scholar Andrew Cherlin concludes "as the practical importance of being married has declined, its symbolic value has remained high and may even have increased. It has become a marker of a successful personal life" (2020, p. 73).

The symbolic interactionist perspective emphasizes that interaction with family members, including parents, grandparents, siblings, and spouses, has a powerful effect on our

marital resiliency perspective A view of the current state of marriage that includes the beliefs that (1) marriage continues to be valued by many in society (as evidenced by same-sex marriages and remarriages) and (2) marriage is valued as a symbol of a successful personal life rather than for the tangible resources it provides.

snowball sampling A technique in which one participant in a study recommends others who might be interested in participating.

David Grossman/Alamy Stock Photo

U.S. deportations of undocumented immigrants have reached unprecedented levels in the last 25 years. These actions have disproportionately targeted Hispanic men, 25 percent of whom have U.S.-born children (Koball et al. 2015).

self-concepts. For example, negative self-concepts may result from verbal abuse in the family, whereas positive self-concepts may develop in families in which interactions are supportive and loving. Emotional abuse often involves using negative labels (e.g., *stupid, whore*) to define a partner or family member. Such labels negatively affect the self-concept of abuse victims, often convincing them that they deserve the abuse. Sibling violence is pervasive because it is commonly viewed as "a harmless and inconsequential form of family aggression" (Khan and Rogers 2015, p. 437); hence it is widely tolerated.

The symbolic interactionist insight that labels affect meaning and behavior can be applied to issues related to divorce. For example, when a noncustodial divorced parent (usually a father) is awarded "visitation" rights, he may view himself as a visitor in his children's lives. The meaning attached to the visitor status can be an obstacle to the father's involvement because the label *visitor* minimizes the importance of the noncustodial parent's role (Pasley and Minton 2001). Many states have dropped the term *custody* in favor of more family-friendly terms such as *parenting plan* to reduce conflict among parents who might otherwise fight over who "wins" custody.

Problems Associated with Restricted Access to Contraception and Abortion

Having control over when one has children has been shown to be a critical factor in the well-being of both adults and children. Experts refer to this as **family planning**. Reductions or delays in having children are associated with increases in education for women and men, increases in women's participation in the workforce, and a narrowing of the gender gap in pay. Unintended pregnancies can result in negative outcomes for mother and baby, including lack of prenatal care, low birthweight, complications during childbirth, and maternal depression. Pregnant teens are at higher risk of many birth complications, and their babies are at higher risk of low birth weight, preterm birth, and neonatal conditions (World Health Organization 2020).

Unintended pregnancies also drive up the number of single-parent families—families at the greatest risk of poverty (Livingston 2018). The rate of unintended pregnancy is also tied closely to the abortion rate since about 40 percent of unplanned pregnancies result in abortion (Sawhill and Guyot 2019). The rate of abortions in the United States declined by 25 percent between 2008 and 2014 when the unintended pregnancy rate declined. It reached an historic low as a result of access to contraception, more effective contraception, and a reduction in teen sexual intercourse (Guttmacher 2019a).

family planning Having control over when one has children, typically through the use of contraception but also through abstinence.

contraception Otherwise known as birth control, interferes with the process of ovulation, fertilization, and/or implantation of a fertilized egg with a goal of preventing pregnancy.

Factors Affecting Access to Contraception

In many parts of the world, lack of access to safe, effective birth control plays a significant role in whether pregnancies are intended or unintended. The Population Reference Bureau found that in the countries of Mozambique and Chad, 19 percent of married women have access to **contraception**. Lack of knowledge about methods of contraception or where to get them is also a significant barrier in countries like the Democratic Republic of Congo (DRC) and Nigeria (Population Reference Bureau 2019).

In 10 percent of countries a woman must be married to get maternity health care and more than 25 percent of countries require a woman to get the consent of her husband before receiving contraception (Wurfhorst 2020). Teens in developing countries face significant barriers to accessing contraception, including lack of knowledge, as well as restrictive laws, clinicians who are unwilling to provide teens with contraception, and a lack of transportation or money (World Health Organization 2020).

In the United States, lack of knowledge about contraception is also a factor, particularly among teenagers. In the late 1990s, as part of "welfare reform," the U.S. government began funneling money to abstinence-only programs within the United States and abroad (Lindberg, Maddow-Zimet, and Boonstra 2016). Even though research has documented that abstinence-only programs are ineffective in delaying sex, reducing sexual risk behaviors, or improving teen pregnancy rates, the federal government continues to spend tens of millions of dollars on these programs—$85 million in 2016 (Donovan 2017).

Parents in the United States also provide little sex education with the most common instruction from parents, if provided, being "how to say no" (Lindberg, Maddow-Zimet, and Boonstra 2016). A quarter of girls and a third of boys receive no information from their parents on how to communicate sexual boundaries or on sexually transmitted diseases, birth control, or sex itself. In fact, many teens don't receive information about birth control until after they have already had sexual intercourse (Donovan 2017). Thus, it is not surprising that both teenagers and adults in the United States tend to use contraception incorrectly or inconsistently, leading to unintended pregnancies (Sonfield, Hasstedt, and Gold 2014).

What do you think?

In the Netherlands, sex education begins in childhood, with children as young as four learning about relationships and appropriate touching (Rough 2018). By the age of 11, Dutch students are expected to be able to discuss reproduction, safer sex, and sexual abuse. Dutch teens do not have sex at an earlier age than teens from other European countries and they have the lowest rate of teen pregnancy in the world. Should such a program be instituted in schools in the United States?

In the United States, almost 5 percent of women between the ages of 15 and 44 have an unintended pregnancy each year, making up about 45 percent of the pregnancies in the United States. The good news is that this number is at an all-time low. However, unintended pregnancies are most common among groups that are already vulnerable: teenagers, low-income women, and women who have not completed high school (Finer and Zolna 2016). In fact, the United States has the highest rate of teen pregnancy among developed countries excluding the former Soviet Bloc countries.

Among these vulnerable groups the lack of access to effective contraception is a significant factor in unwanted pregnancies. For example, in the United States, low-income women are less likely to have insurance that covers contraception and often must rely on publicly subsidized programs through Title IX and Medicaid. These programs are under attack by conservative religious groups and the former Trump administration. In its quest to undermine the *Affordable Care Act* (ACA), the former administration expanded religious exemptions that allow organizations and employers to deny employees insurance coverage for birth control on religious grounds, a position upheld by the U.S. Supreme Court in 2020 (Nelson and Webber 2020). However, in contrast to the Trump administration, President Biden has pledged to expand the ACA including the addition of a public insurance option that will fully cover contraceptive needs (Biden 2020b).

> "I have a daughter and I have granddaughters and I will never vote to let a group of backward-looking ideologues cut women's access to birth control. We have lived in that world, and we are not going back, not ever."
>
> **–ELIZABETH WARREN, U.S. SENATOR**

The most effective contraceptive options—long-acting reversible contraceptives (LARCs) like the IUD and implants—tend to be the most expensive, ranging from $500 to $1,000, often putting them out of reach for many low-income women. Across developed countries, there is little difference in sexual activity rates among teenagers. However, there is a great deal of difference in access to contraception. Not surprisingly, countries such as the United States in which teenagers have the most difficulty accessing birth control have the highest rates of teen pregnancy and, subsequently, higher rates of single parenthood (Guttmacher Institute 2020).

Factors Affecting Access to Abortion

As previously noted, unintended pregnancy is closely tied to the abortion rate. **Abortion** is the removal of an embryo or fetus from a woman's uterus before it can survive on its own. In developed countries most legal abortions are performed through an outpatient surgical procedure or more recently through a medication protocol typically consisting of two pills, one commonly known as RU-486, the "abortion pill." Medication abortion is widely used in France, Great Britain, China, and the United States, where it currently accounts for over 40 percent of all abortions due to its high level of effectiveness, low rate of complications, low cost, and privacy (Kaiser Family Foundation 2020).

Globally, access to abortion has increased over the past 25 years, with almost 50 countries liberalizing their abortion laws. For example, in 2018, Ireland legalized abortion under any circumstances when previously it had only been legal in the case of threat to the woman's life. However, 90 million women in the world live in countries in which abortion is prohibited altogether (Center for Reproductive Rights 2020b).

In the United States, abortion has been legal since the U.S. Supreme Court ruled in *Roe v. Wade* in 1973 that a woman's right to an abortion was protected by the Fourteenth Amendment's right to privacy. However, several Supreme Court decisions since have limited the scope of the *Roe v. Wade* decision (e.g., *Planned Parenthood of Southeastern Pennsylvania v. Casey*). Historically, abortions have been banned when the fetus is considered viable. While this has long been considered 22 to 26 weeks from conception, in 2019, nine states imposed new gestational age bans ranging from when a fetal heartbeat can be detected (which could be as early as 6 weeks) to 18 weeks of pregnancy. These bans are being litigated in courts across the country (Guttmacher Institute 2019b).

Since *Roe v. Wade*, individual states have passed a patchwork of laws making abortion access more difficult. The pace of these types of legislation has increased in recent years with more than 350 pieces of legislation restricting abortions being introduced across the country in just the year 2019 (Guttmacher Institute 2019b). Many of these restrictions impose targeted regulations against abortion providers and thus are referred to as **TRAP laws** by abortion rights activists. For example, many states have enacted laws that require abortion providers to operate hospital-style surgical centers, regulations that would close many clinics. The U.S. Supreme Court struck down two such laws in Texas and Louisiana in 2016 and 2020, respectively. Other enacted state-level restrictions include (1) that an ultrasound be performed and shown to the patient, (2) the right of individual medical practitioners or institutions to refuse to perform an abortion, (3) state-mandated counseling, (4) waiting periods, (5) restrictions on state spending for abortions, (6) restrictions on private insurance coverage for abortions, (7) restrictions on the use of telemedicine for medication abortions, (8) restrictions on abortion if the fetus is diagnosed with Down syndrome, and (9) parental permission if involving a minor (Guttmacher Institute 2020).

Most recently, during the COVID-19 pandemic, governors in nine states across the South and Midwest issued executive orders to restrict access to abortion, interpreting them as "non-essential health procedures." Multiple bans in the state of Texas alone led hundreds of women to drive or fly to neighboring states to seek abortions, including women with fetal anomalies (National Public Radio 2020). As a result of lawsuits filed by women's health clinic and reproductive rights organizations, these bans were successfully blocked.

abortion The intentional termination of a pregnancy.

TRAP laws Laws designed to restrict access to abortion through targeted restrictions on abortion providers.

Abortion is a complex issue for societies, which must respond to the pressures of conflicting attitudes toward abortion and the reality of high rates of unintended and unwanted pregnancy. Despite media and political portrayals of individuals as either "pro-choice" or "pro-life," surveys in the United States indicate that individuals' attitudes on abortion are more complex. In 2020, 53 percent of U.S. adults surveyed agreed that abortion should be legal under some circumstances, 25 percent said they believe abortion should be legal under all circumstances, and 21 percent said it should be illegal under all circumstances (see Figure 5.6). However, on the same survey, 48 percent identified as "pro-life" and 46 percent as "pro-choice," suggesting that most people who identify as pro-life are not against abortion in all circumstances (Gallup 2020).

Erik McGregor/LightRocket/Getty Images

When women do not have access to contraception and safe, legal abortion, they are often left with no choice but to endure the risks associated with pregnancy. The United States has the highest maternal mortality rate of any developed country in the world, and the rate is rising.

Advocates of the pro-choice movement hold that freedom of choice is a central human value, that procreation choices must be free of government interference, and that women have a right to self-determination. Alternatively, the pro-life movement argues that an unborn fetus has a right to live and be protected, that abortion is immoral, and that alternative means of resolving an unwanted pregnancy should be found. In support of their goals, in 2017, former President Donald Trump reinstituted the global ban on U.S. funding assistance to family planning agencies that provide abortion services (CBSNews 2017). He also expanded the ban to include funding to all organizations receiving global health assistance from the United States, negatively impacting access to all types of health care for women in low- and middle-income countries. In contrast, President Biden, in order to ensure a woman's right to have an abortion, repealed the global ban on funding for organizations which perform abortions (the "Mexico City agreement") and advocates passage of a federal law that would protect abortion rights even if *Roe v. Wade* was overturned. The Biden administration also supports repealing

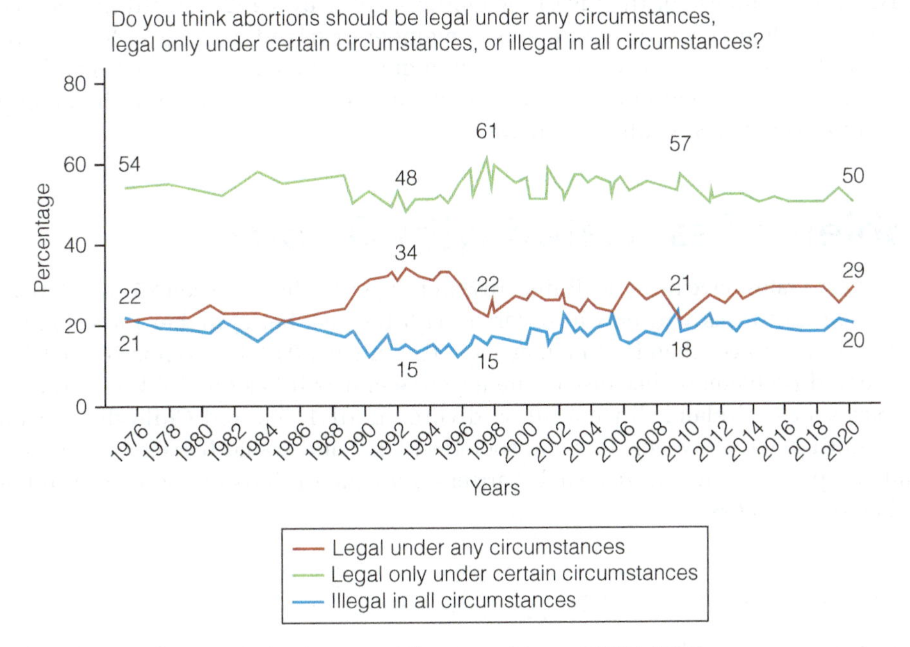

Figure 5.6 Attitudes on Abortion in the United States, 1976–2020
SOURCE: Gallup 2020.

Beliefs about Abortion

Decisions about abortion are deeply personal. Many of us never have to contemplate the range of difficult situations women sometimes find themselves in when pregnant. Sometimes these decisions are made when a pregnancy was unplanned and unwanted; sometimes these decisions have to be made when the pregnancy is very much wanted. For decades, the General Social Survey has been asking adults in the United States their opinions on whether abortion should be legal under certain circumstances. What do you think?

Please tell me whether you believe a pregnant woman should be able to obtain a legal abortion …

1. If she is not married and does not want to marry the man?	Yes	No
2. If she/the family has a very low income and cannot afford any or any more children?	Yes	No
3. If there is a strong chance of a serious defect in the baby?	Yes	No
4. If she is married and does not want any more children?	Yes	No
5. If the woman's own health is seriously endangered by the pregnancy?	Yes	No
6. If she became pregnant as the result of rape?	Yes	No
7. If she is in an abusive relationship?	Yes	No
8. If the woman wants it for any reason?	Yes	No

2018 General Social Survey Results:	%	%
1. If she is not married … ?	46	54
2. If she/the family cannot afford … ?	49	51
3. If there is a strong chance of a serious defect … ?	77	23
4. If she is married and does not want any more children?	51	49
5. If the woman's health is seriously endangered … ?	90	10
6. If she became pregnant as the result of rape?	80	20
7. If she is in an abusive relationship?	Not available in 2018	
8. If the woman wants it for any reason?	50	50

SOURCE: GSS, NORC at the University of Chicago

the Hyde amendment, which prohibits use of taxpayer funds for abortions except in the cases of rape, incest, or the life of the mother (BBC News 2020). Until they are faced with the issue themselves or with a family member or friend, most people don't think through all the scenarios in which a woman might consider an abortion. Consider your own opinions by completing the *Self & Society* quiz in this chapter, and then compare them with those of U.S. adults in general.

Problems Associated with Divorce

While divorce rates have been declining, research suggests that between 25 and 39 percent of all recent marriages will end in divorce, with the risk being higher for second- and higher-order marriages than for first marriages (Dennison 2017; Luscombe 2018). Divorce is considered problematic because of the negative effects it has on children, adults, and the larger society. Ireland did not allow divorce until 1995, and Chile did not allow divorce until 2004. However, in some societies, legal and social barriers to divorce are considered problematic because such barriers limit the options of spouses in unhappy and abusive marriages.

Social Causes of Divorce

When people think of divorce, we typically assume a number of individual and relationship factors may have contributed to the marital breakup: incompatibility in values or goals; poor communication; lack of conflict resolution skills; sexual incompatibility;

extramarital relationships; substance abuse; emotional or physical abuse or neglect; boredom; jealousy; and difficulty coping with change or stress related to parenting, employment, finances, in-laws, and illness. As discussed below, from a larger perspective, divorce is linked to a number of social and cultural factors.

> "When two people decide to get a divorce, it isn't a sign that they 'don't understand' one another, but a sign that they have, at last, begun to."
>
> –HELEN ROWLAND, AMERICAN JOURNALIST AND HUMORIST (1875–1950)

Changing Function of Marriage. Before the Industrial Revolution, marriage functioned as a unit of economic production and consumption that was largely organized around producing, socializing, and educating children. However, the institution of marriage has changed over the last few generations:

> Marriage changed from a formal institution that meets the needs of the larger society to a companionate relationship that meets the needs of the couple and their children and then to a private pact that meets the psychological needs of individual spouses. (Amato et al. 2007, p. 70)

U.S. society is characterized by **individualism**—the tendency to focus on one's individual self-interests and personal happiness rather than on the interests of one's family and community. As a result, "Marital commitment lasts only as long as people are happy and feel that their own needs are being met" (Amato 2004, p. 960). Belief in the right to be happy, even if it means getting divorced, is reflected in social attitudes toward divorce. Nearly three-quarters of U.S. adults (72 percent) say that divorce is morally acceptable (Swift 2016). When spouses do not feel that their social psychological needs for emotional support, intimacy, affection, love, or personal growth are being met in the marriage, they may consider divorce with the hope of finding a new partner to fulfill those needs. By contrast, in India, where an estimated 90 percent of marriages are arranged, only 1 percent of marriages end in divorce (Dholakia 2015).

Increased Economic Autonomy of Women. Before 1940, most wives were not employed outside the home and depended on their husband's income. This was particularly true of White women. Today, a majority of married women across all racial and ethnic groups are in the labor force and, in 2018, 29 percent of wives earned more than their husbands (U.S. Census Bureau 2019). A wife who is unhappy in her marriage is more likely to leave the marriage if she has the economic means to support herself (Jalovaara 2003). An unhappy husband may also be more likely to leave a marriage if his wife is self-sufficient and can contribute to the support of the children.

Increased Work Demands and Economic Stress. Another factor influencing divorce is increased work demands and the stresses of balancing work and family roles. Some workers are putting in longer hours, often working overtime or taking second jobs, while others face job loss and unemployment. As discussed in Chapter 2, Chapter 6, and Chapter 7, many families struggle to earn enough money to pay for rising housing, health care, and child-care costs. Financial stress can cause marital problems. Couples with low credit scores and couples with mismatched individual credit scores (one high and one low) have increased likelihoods of divorce (Dokko, Li, and Hayes 2015).

individualism The tendency to focus on one's individual self-interests and personal happiness rather than on the interests of one's family and community.

second shift The household work and child care that employed parents (usually women) do when they return home from their jobs.

no-fault divorce A divorce that is granted based on the claim that irreconcilable differences exist within a marriage (as opposed to one spouse being legally at fault for the marital breakup).

Inequality in Marital Division of Labor. Many employed parents, particularly mothers, come home to work a **second shift**—the work involved in caring for children and household chores (Hochschild 1989) (see Chapter 10). Women are more likely than men to spend time, on an average day, doing housework, food preparation, and child care (Bureau of Labor Statistics 2016). This unequal division of household chores negatively affects marital satisfaction among wives, particularly those who believe that marital roles should be more equal (Ogolsky et al. 2014).

No-Fault Divorce Laws. Before 1970, the law required a couple who wanted a divorce to prove that one of the spouses was at fault and had committed an act defined by the state as grounds for divorce—adultery, cruelty, or desertion. In 1969, California became the first state to initiate **no-fault divorce**, which permitted a divorce based on the claim that there were "irreconcilable differences" in the marriage. Today, all 50 states recognize some form of no-fault divorce. The availability of no-fault divorces has contributed to the U.S. divorce rate by making divorce easier to obtain.

Inter-Generational Patterns. Adults who have grown up with parents who are divorced are more likely to get a divorce. Cohabiting couples are also more likely to end their relationship if one or both of the partners' parents were divorced (Amato and Patterson 2016). There are a number of theories as to why this is the case. The first is that divorce may be a more acceptable alternative within a family that has experienced divorce compared to a family that has not. Others argue that children of divorced parents may not have witnessed positive conflict resolution techniques and therefore may not have learned how to work through problems in a relationship. However, more recent research finds that the link between parents' divorce and the divorce of an adult child has weakened (Pinsker 2019).

In the early 1970s, married people with divorced parents were about twice as likely as married people from intact families to get a divorce; today, that number is 1.2 times as likely. The stigma associated with getting a divorce in the 1970s necessitated a higher threshold of conflict before either party would file for a divorce, and it is that conflict that likely led to the higher rates of adult children getting a divorce (Pinsker 2019). This theory is consistent with recent evidence that children who grow up with both parents but whose parents are in a highly conflictual marriage are more likely to get divorced themselves. Alternatively, adult children brought up in a family where a parent left a highly conflictual marriage are no more likely to divorce than children who grew up in an intact, low-conflict family (Gager, Yabiku, and Linver 2016).

Another and perhaps more likely explanation for the intergenerational pattern is that children "inherit" the risk factors associated with divorce. Parents who married at an early age, did not complete high school, had low income levels, and/or had a baby before marriage are more likely to have adult children with the same characteristics, all of which are associated with higher risk of divorce or separation. Thus, it is the intergenerational transmission of risk factors, not the parental divorce itself, which explains the pattern. However, recent trends suggest that young adults are altering the divorce trends, with a ten-point reduction in the divorce rate between the ages of 20 and about 35 between 2008 and 2017 (see Figure 5.7). New research will be required to understand why divorce has declined so significantly in this group, as well as what their futures will hold.

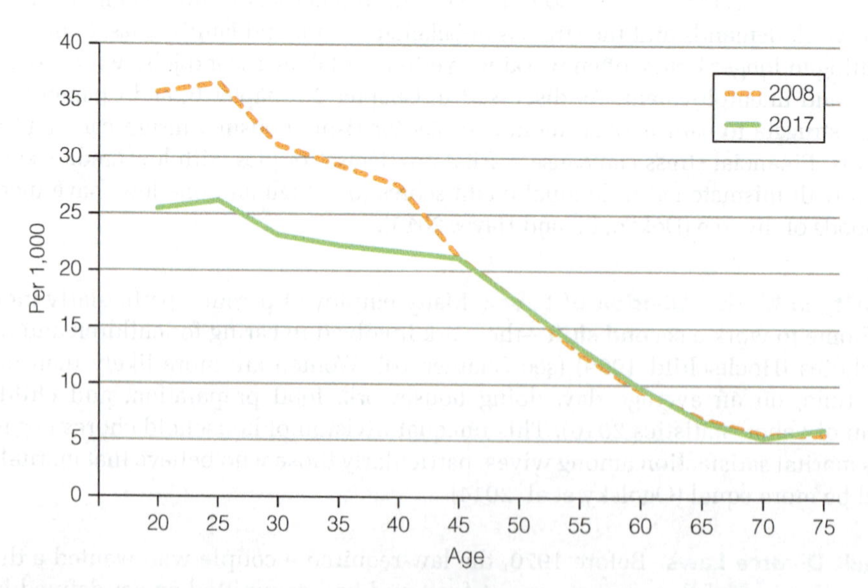

Figure 5.7 Age-Specific Divorce Rates, 2008 and 2017; Divorces per 1,000 Married Women
SOURCE: Cohen 2019.

Longer Life Expectancy. Finally, more marriages today end in divorce, in part, because people live longer than they did in previous generations. In the late 1800s when the United States started tracking life expectancy, the average person died at the age of 39, while today the average person lives into their late seventies. With an average age of first marriage in the low twenties, persons born in the late 1800s might be married for only 15–20 years before they died. Today it is not unforeseeable for a couple to be married for more than 40 years (National Center for Health Statistics 2017). "'Til death do us part" involves a longer commitment than it once did.

> A study of 20,000 Swedish adults who had been adopted as children found that their likelihood of divorce resembled that of their biological parents, even if they had been separated from those parents at a young age. The researchers conclude that there is a genetic predisposition to divorce, likely through personality traits inherited from one's biological parents (Salvatore et al. 2018). What are the implications for those working with couples hoping to avoid divorce?
>
> **What *do you* think?**

Consequences of Divorce

When parents have bitter and unresolved conflict or if one parent is abusing a child or the other parent, divorce may offer a solution to family problems. However, divorce often has negative effects for ex-spouses and their children and also contributes to problems that affect society as a whole.

Physical and Mental Health Consequences. Numerous studies show that divorced individuals have more health problems and a higher risk of mortality than married individuals. Divorced individuals also experience lower levels of psychological well-being, including more unhappiness, depression, anxiety, and poorer self-concepts (Amato 2003; Kamp Dush 2013). Unmarried people are more likely than married people to have low social attachment, low emotional support, and increased economic hardship (Walker 2001). Some research suggests that divorce leads to higher levels of depressive symptoms for women but not for men (Kalmijn and Monden 2006), especially when young children are in the family (Williams and Dunne-Bryant 2006). This finding is probably due to the increased financial and parenting strains experienced by divorced mothers who have custody of their young children.

On the other hand, some studies have found that divorced individuals report higher levels of autonomy and personal growth than married individuals do (Amato 2003). For example, many divorced mothers report improvements in career opportunities, social lives, and happiness after divorce; some divorced women report more self-confidence; and some men report more interpersonal skills and a greater willingness to self-disclose. For people in a poor-quality marriage, divorce has a less negative or even a positive effect on well-being (Amato 2003; Kalmijn and Monden 2006). However, leaving a bad marriage does not always result in increased well-being because "divorce is a trigger for even more problems after the divorce" (Kalmijn and Monden 2006, p. 1210). In sum, some men and women experience a decline in well-being after divorce; others experience an improvement.

Economic Consequences. Following divorce, there tends to be a dramatic drop in women's income and a slight drop in men's income (Gadalla 2009; de Vaus et al. 2017). Compared with married individuals, divorced individuals have a lower standard of living, have less wealth, and experience greater economic hardship, although this difference is considerably greater for women than for men (Mortelmans 2020). The economic costs of divorce are often greater for women and children because women tend to earn less than men (see Chapter 10), are more likely to have physical custody of children, and devote substantially more time to household and child-care tasks than fathers do. The time women invest in this unpaid labor restricts their educational and job opportunities as well as their income.

After divorce, both parents are typically responsible for providing economic resources for their children. In divorce cases in which one parent is granted primary custody of the children, the noncustodial parent is often assigned child support. In shared-custody situations, there may be no assignment of child support if the parents are dividing time and/or expenses relatively equally. In cases where one parent has more responsibility than the other, there might be proportional child support. In 2015, less than half of parents entitled to financial support received the full amount: 31 percent received nothing, and another 26 percent received something but less than the full amount (Grall 2020). In 2018, over $118 billion was owed in past-due child support payments (Meyer, Cancian, and Waring 2019).

Although not the only reason, most parents who fail to make child support payments do so as a result of insufficient income. Given declining male wages nationwide combined with required minimum child support amounts in many states, low-income fathers may be required to pay more than 50 percent of their earnings in child support (Brinig and Garrison 2018). Low-income fathers are 2.5 times as likely to not pay required child support than high-income fathers. Among fathers who do pay, high-income fathers pay more regularly (Astone, Karas, and Stolte 2016). For the average custodial parent who is entitled to receive a child-support payment, that payment reflects just under 10 percent of their personal income. For custodial parents under the poverty level, it represents more than 50 percent of their income (Grall 2020). Thus, nonpayment of child support has a significant impact on the economic well-being of children despite custodial parents attempting to make up the difference as they are able.

While most research has focused on women's economic well-being while still raising children, divorced women are often also negatively impacted when they reach retirement age, as they may not have accrued sufficient pension funds and/or social security to support themselves in old age. A marriage must last at least ten years for a spouse to be eligible for social security spousal benefits on their ex-spouse's earnings record as long as the divorced spouse is at least age 62 and has remained single. Under current rules, this is irrespective of whether the other spouse has remarried. This is a particular problem for women of the baby boom generation and above, who were less likely to work during their marriage (Butrica and Smith 2012).

In earlier decades, when most women worked in the home raising children, divorced women were commonly awarded permanent alimony—payments from the ex-husband that continued until death, the woman's remarriage, or a significant change in circumstances. Today, nearly half the workforce is women, and more than a third of employed married women earn more than their spouse. Although alimony laws vary by state, the general trend is for judges to award **rehabilitative alimony** for a specified length of time to allow the recipient time to find a job or to complete education or job training (Amundsen and Kelly 2014).

Effects on Children and Young Adults. Parental divorce is a stressful event for children that often involves continuing conflict between parents, a decline in the standard of living, the possibility of moving and changing schools, separation from the noncustodial parent (usually the father), and/or parental remarriage. These stressors place children of divorce at higher risk for a variety of emotional and behavioral problems, including lower academic performance, psychological adjustment, self-concept, social competence, and long-term health, as well as higher levels of aggressive behavior and depression (Amato 2003; Wallerstein 2003).

Recent research documents that shared-custody arrangements result in better outcomes for children; thus, the fact that shared custody is now the norm rather than the exception in many states should reduce negative impacts of divorce on children. Compared with children who live primarily with one divorced parent, children who live in shared parenting families tend to have higher levels of emotional, behavioral, and psychological well-being, better physical health, and better relationships with their mothers and fathers (Nielsen 2014). However, shared-custody arrangements are less common among low-income couples who may not be able to afford two separate residences large enough for children. As a result, low-income children are less likely to receive the benefits of shared custody (Meyer, Cancian, and Cook 2017). Some research suggests that the number of transitions associated

rehabilitative alimony
Alimony that is paid to an ex-spouse for a specified length of time to allow the recipient time to find a job or to complete education or job training.

with the divorce (e.g., moving, changing schools, etc.) are the most important predictor of negative impacts: the fewer the transitions, the better (Cavanagh and Fomby 2019).

Experiencing parental divorce during one's adult years can also affect an individual's well-being. While some adult children of divorce are not traumatized by their parents' divorce and some are even relieved, others struggle emotionally, especially when they are put in the middle of their parents' conflict or are forced to take sides between parents (Greenwood 2014).

Despite the adverse effects of divorce on children, research findings suggest that "most children from divorced families are resilient, that is, they do not suffer from serious psychological problems" (Emery et al. 2005, p. 24). Other researchers conclude that "most offspring with divorced parents develop into well-adjusted adults," despite the pain they feel associated with the divorce (Amato and Cheadle 2005, p. 191).

Divorce can also have positive consequences for children and young adults. In highly conflictual marriages, divorce may improve the emotional well-being of children relative to staying in a conflicted home environment (Jekielek 1998; Gager et al. 2016). In interviews with 173 grown children whose parents divorced years earlier, Ahrons (2004) found that most of the young adults reported positive outcomes for their parents and for themselves. Although many young adults who have divorced parents fear that they, too, will have an unhappy marriage (Dennison and Koerner 2008), such a fear can also lead young adults to think carefully about their choices regarding marriage.

Effects on Parent–Child Relationships. The research on the impact of divorce on the relationships between parents and children yields mixed results. Some research has found that young adults whose parents divorced are less likely to report having a close relationship with their nonresidential parent compared with children whose parents are together (DeCuzzi et al. 2004). Given trends in primary custody decisions, the nonresidential parent is most likely to be a father. Nonresidential fathers tend to spend less time with their children than married and shared-custody fathers, which likely negatively impacts their relationships with children (Nielsen 2014). Some nonresidential fathers pull away completely. As Figure 5.8 depicts, nonresidential fathers spent less time engaging in all activities with their preschool-age children as compared to residential fathers. Nonresidential fathers also tend to spend less time with school-age children. In highly

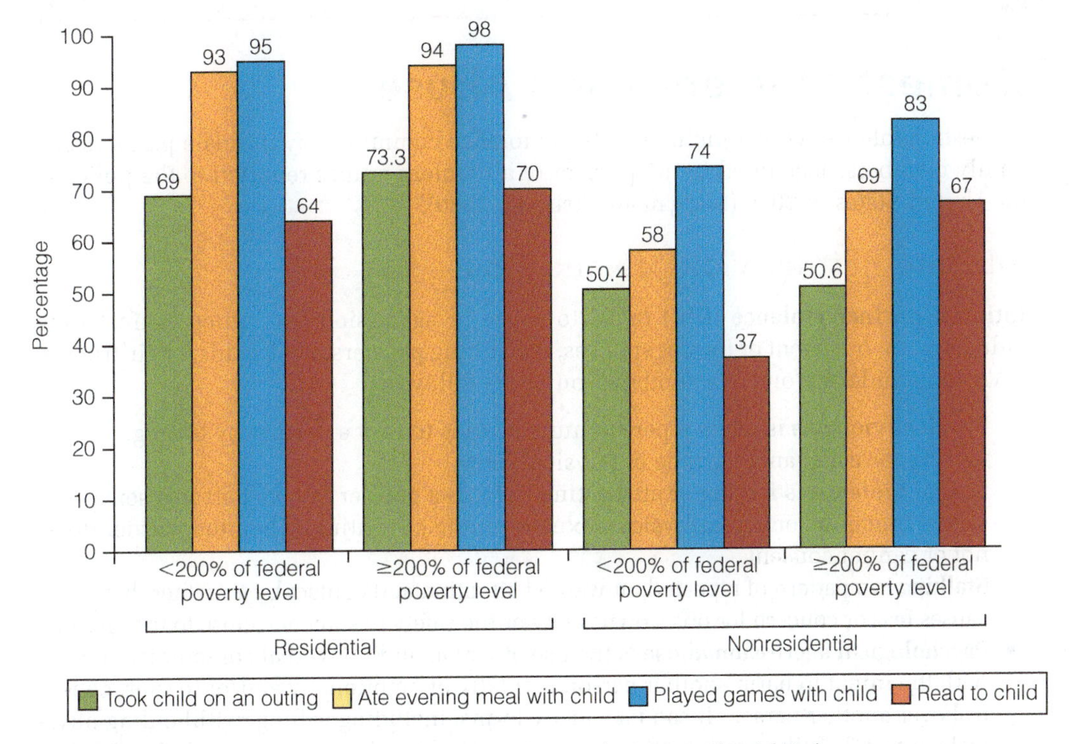

Figure 5.8 Weekly Activities with Children under Age 5 by Nonresidential Fathers
SOURCE: Aston, Karas, and Stolte 2016.

conflictual parent relationships, one parent may intentionally try to turn a child against the other parent (Bernet and Baker 2013).

Nonetheless, recent trends toward shared-custody arrangements have resulted in increased time and better relationships between fathers and children, even when the parents' relationship is combative (Nielsen 2015). A comparison of children in shared-custody arrangements with children with nonresidential fathers found that almost 90 percent of shared-custody fathers attended school events while only 60 percent of nonresidential fathers did. As a result of compelling evidence of the benefits of such arrangements, the consensus among psychologists is that shared custody has benefits for children and parents. However, even in the absence of shared custody, many adult children of divorce report their relationships with their fathers improved after the divorce (Ahrons 2004). Children may benefit from having more quality time with their fathers after parental divorce, and some fathers report that they became more active in the father role post-divorce. One study found that the older the child is at the time of parental separation, the greater the amount of time fathers spend with their children (Swiss and Le Bourdais 2009). Divorced dad Elliot Katz concluded:

> Becoming a single dad made me a better father as it forces me to step forward and take responsibility for dealing with situations that in the past I probably would have just left for my wife to handle or to tell me what to do. (Vinopal 2019)

As we have seen, the effects of divorce on adults and children are mixed and variable. In a review of research on the consequences of divorce for children and adults, Amato (2003) concluded that "divorce benefits some individuals, leads others to experience temporary decrements in well-being that improve over time, and forces others on a downward cycle from which they might never fully recover" (p. 206).

What do you think?

Although shared custody is on the rise, in some states, judges still default to primary physical custody (i.e., one parent or the other) arrangements unless the couple specifically asks for shared custody. Yet studies show that shared custody works even when one or more of the spouses are initially resistant. Why do judges so often default to giving custody to the mother?

Domestic Violence and Abuse

Domestic violence, which includes acts of violence committed by intimate partners and family members, accounted for 52 percent of all violent crimes reported to the police in the United States in 2019 (Morgan and Truman 2020).

intimate partner violence (IPV) Actual or threatened violent crimes committed against individuals by their current or former spouses, cohabiting partners, boyfriends, or girlfriends.

physical violence When a person hurts or tries to hurt a partner by hitting, kicking, or using another type of physical force.

sexual violence Forcing or attempting to force a partner to take part in a sex act, sexual touching, or a nonphysical sexual event (e.g., sexting) when the partner does not or cannot consent.

Intimate Partner Violence and Abuse

Intimate partner violence (IPV) refers to abuse or aggression committed against individuals by their current or former spouses, cohabiting partners, boyfriends, or girlfriends. IPV can include any one or a combination of the following:

- **Physical violence** is when a person hurts or tries to hurt a partner by hitting, kicking, or using another type of physical force.
- **Sexual violence** is forcing or attempting to force a partner to take part in a sex act, sexual touching, or a nonphysical sexual event (e.g., sexting) when the partner does not or cannot consent.
- **Stalking** is a pattern of repeated, unwanted attention and contact by a partner that causes fear or concern for one's own safety or the safety of someone close to the victim.
- **Psychological aggression/abuse** is the use of verbal and nonverbal communication with the intent to harm another person mentally or emotionally and/or to exert control over another person. It can take many forms, including yelling, withholding physical contact, belittling or insulting a person, and restricting a person's activities and/or access to friends and family (Centers for Disease Control and Prevention 2020b).

Figure 5.9 Prevalence of Stalking in the United States
SOURCE: Centers for Disease Control and Prevention 2020.

Intimate partner violence is widespread. About one in four women and one in ten men in the United States have experienced sexual violence, physical violence, and/or stalking during their lifetime (see Figure 5.9 for rates of stalking alone, Centers for Disease Control 2020b). Globally, one woman in every three has experienced physical or sexual violence by a partner (World Health Organization 2014b). Violence against women is condoned by many governments; one in four women do not have the legal right to say no to sex (Wurfhorst 2020).

These statistics don't take into account the psychological aggression/abuse which also factors into IPV. Although most research on IPV has focused on different-sex couples, studies indicate that gay and bisexual men and women experience IPV at about the same rate, although there is some evidence that severe violence may be even more common among same-sex couples (Rollè et al. 2018). IPV starts early in the form of **teen dating violence**, especially for women. A study of young U.S. adults ages 17 to 24 found that four in ten experienced physical violence and five in ten experienced verbal abuse in their current or most recent relationship (Halpern-Meekin et al. 2013).

Four Types of Intimate Partner Violence. Over the years, many typologies of IPV have been developed and subsequently revised. Building off these models, the following four types of partner violence have been identified (Kelly and Johnson 2008):

1. *Situational couple violence* refers to occasional acts of violence that are not connected to a general pattern of control in the relationship. Situational couple violence usually does not escalate and is less likely to involve severe violence. However, when situational couple violence is frequent, it is more likely to lead to serious injuries. This is the most common type of IPV reported in surveys and is the type of IPV that both men and women commit.
2. *Coercive controlling violence* is motivated by a wish to control one's partner and involves not only violence but also nonviolent behaviors with a motive of control. Coercive controlling violence is most commonly perpetrated by men and is more likely to escalate over time and to involve serious injury. This type of IPV has also been referred to as battering and intimate terrorism. It is less likely to be reported by survey respondents but is the type of violence most likely to lead victims to go to emergency rooms and domestic violence shelters. For female victims, coercive controlling violence can be fatal. However, this is less likely to be the case when a male is the victim.

stalking A pattern of repeated, unwanted attention and contact by a partner that causes fear or concern for one's own safety or the safety of someone close to the victim.

psychological aggression/abuse The use of verbal and nonverbal communication (e.g., yelling, belittling, insulting, etc.) with the intent to harm another person mentally or emotionally and/or to exert control over another person.

teen dating violence Intimate partner violence that occurs before the age of 18.

@JuVee Prods

Not everyone is safer at home. There are people living in situations around this country and abroad that we couldn't possibly imagine. If you are in a crisis, call the **National Domestic Violence Hotline** at 1-800-799-7233 or text LOVEIS to 22522.

-Viola Davis, Academy Award winning actress, philanthropist, and CEO/Cofounder of @JuVeeProds

3. *Violent resistance* refers to acts of violence that are committed in self-defense or in defense of others—typically children—against a violent partner. Acts of violent resistance that receive the most attention are when a woman kills her violent partner. These typically result from a pattern of escalating coercive controlling violence including forced abusive sexual acts, failed attempts by the victim to escape the relationship, and thoughts or attempts of suicide. Most such killings happen as defensive actions during a brutal attack (Ferraro 2006).
4. *Separation-instigated violence* is violent behavior triggered by the departure of an intimate partner, when no history of violence exists in the relationship. The violence is triggered by a separation that is perceived to be sudden or humiliating. The violence is atypical behavior during which the perpetrator appears to have "lost control" and is typically limited to one or two episodes during the separation period. This type of behavior is engaged in by both men and women.

The differences in gendered patterns across the four types of IPV may explain the debate over whether IPV is perpetrated more by women or by men. U.S. government data find that one in four women and one in ten men report having experienced physical violence, sexual violence, or stalking by an intimate partner in their lifetime (Centers for Disease Control 2020b). However, survey research indicates that females report engaging in acts of IPV at rates equivalent to or exceeding the rates reported by male perpetrators (Cui et al. 2013; Fincham et al. 2013). It may be that women are simply more likely than men to report their acts of IPV because less stigma is associated with women's violence against men. Men who abuse women are depicted in American culture as dangerous brutes whereas women's violence against men is considered funny or justified (Anderson 2013). Some research has found that when women assault their male partners, these assaults tend to be acts of retaliation, in self-defense, or are intended to protect others—particularly children (Simiao et al. 2015; Swan et al. 2008; Johnson 2001).

Effects of Intimate Partner Violence and Abuse. IPV results in bruises, broken bones, traumatic brain injury, cuts, burns, other physical injuries, and even death. About 41 percent of female IPV survivors and 14 percent of male IPV survivors experience some form of physical injury and about one in six homicide victims are killed by an intimate partner (Centers for Disease Control 2020b). When women are abused during pregnancy, the result is often miscarriage or birth defects in the child. Victims of intimate partner violence are at higher risk for depression, anxiety, suicidal thoughts and attempts, lowered self-esteem, inability to trust men, fear of intimacy, substance abuse, and harsh parenting practices (Gustafsson and Cox 2012).

Bilgin Sasmaz/Anadolu Agency/Getty Images

Activists raise awareness of domestic violence by holding events such as this annual ceremony in Chicago. Each candle represents one of the 2,500 victims of domestic violence in the city.

Battering also interferes with women's employment. Some abusers prohibit their partners from working. Other abusers "deliberately undermine women's employment by depriving them of transportation, harassing them at work, turning off alarm clocks, beating them before job interviews, and disappearing when they promise to provide child-care" (Johnson and Ferraro 2003, p. 508). Battering also undermines employment by causing repeated absences, impairing the woman's ability to concentrate, and lowering their self-esteem and aspirations.

Abuse is also a factor in many divorces, often resulting in a loss of economic resources. As one means of controlling an intimate partner is through overseeing their finances, women who flee an abusive home often have limited or no economic resources. That combined with the fact that abusers often isolate their victims from friends and family contributes to the fact that many abused women find themselves homeless. Some landlords even refuse to rent to domestic violence victims (American Civil Liberties Union 2020).

Children who witness domestic violence are at risk for emotional, behavioral, and academic problems as well as for future violence in their own adult relationships (Kitzmann et al. 2003;

THE WORLD in quarantine — When Home Isn't Safe

On March 24, 2020, the day after lockdown began in England, a woman we will refer to only as M.V. was tiptoeing back into her house after work when her husband became enraged. Her husband had forbidden her from leaving the house, claiming that she would bring the virus back with her. "He was screaming that I had disobeyed," she said. She locked herself in the bathroom, but her husband crashed through the door, then punched her and spat on her face, until she managed to break free and flee to a bedroom and lock the door. "I was so scared," she said, "but I didn't know what to do."

During the first month after the lockdown began in late March, 16 women and girls were killed in suspected domestic homicides in Great Britain—more than triple the number from the same period in 2019.

With schools and workplaces closing during the COVID-19 pandemic, many victims of domestic violence were forced to shelter in place with their abuser. The result was a surge in domestic violence across the globe. In April 2020, United Nations Secretary General António Guterres issued a call for "peace at home," stating "violence is not confined to the battlefield. For many women and girls, the threat looms largest where they should be safest—in their own homes" (United Nations 2020).

The pandemic created conditions that were ripe for increased domestic violence. Missed work, lowered incomes, increased financial stress, looming threats of eviction, coupled with the increased demands of child care and homeschooling children put families on edge. On top of that, shelter-in-place orders restricted victims to the home with their abuser.

Further exacerbating the situation, many shelters closed their doors as a result of COVID-19 restrictions, fearing the spread of the virus. According to a *New York Times* report, in Great Britain, where hotel rooms were being opened up to get the homeless off the streets, no provisions were being made for domestic violence victims who needed a safe place to escape their abuser. With courtrooms closed, victims were unable to pursue protective orders. In some cases, judges refused to enforce protective orders keeping the abuser away from the home because of concerns that the abuser would have nowhere to shelter during the pandemic. One man was sent home to his wife because he had coronavirus, despite the fact that she had a restraining order in place. Court dates on criminal charges had to be postponed, sometimes leaving the abuser in the home.

Domestic violence hotlines, underfunded for years, scrambled to meet increased volumes, while also having to find the resources to provide cell phones to staff and volunteers so that they could receive calls at home while sheltering in place. Guterres reported that hotlines across the globe had seen volume double or triple (United Nations 2020). After months of requests for support, the British government allocated funds targeted at addressing domestic violence. By July 2020, less than 3 percent of the funds promised had been dispersed.

In late April, a woman we will call Hajrah attempted suicide by taking an overdose of painkillers after enduring severe emotional and physical abuse from her father during lockdown. When her father and brother found her unconscious, they did nothing. She got help only because a friend alerted the police. When she was released from the hospital, landlords were reluctant to take new residents due to the pandemic. It looked like she would be forced to go home. Fortunately, a local, privately funded domestic violence charity found her a room.

SOURCE: Taub. Amanda and Jane Bradley. 2020. "As Domestic Abuse Rises, U.K. Failings Leave Victims in Peril." New York Times. June 2, 2020. Available at www.nytimes.com

Parker et al. 2000). Children may also commit violent acts against a parent's abusing partner. Children who witness domestic violence are also at increased risk of being violent toward animals (Phillips 2014).

Child Abuse

Child abuse refers to the physical or mental injury, sexual abuse, negligent treatment, or maltreatment of a child under the age of 18 by a person who is responsible for the child's welfare. Worldwide, nearly one in four adults have been physically abused as a child (World Health Organization 2014a). In low- and middle-income countries, three-quarters of children between 2 and 14 experience violent discipline in the home (Levtov et al. 2015). The most common form of child maltreatment is **neglect**—the caregiver's failure to provide adequate attention and supervision, food and nutrition, hygiene, medical care, and a safe and clean living environment (see Figure 5.10).

The highest rates of victimization are for children under age 1 and for children who are American Indian/Alaska Native, Black, or multiracial. Very low rates of child abuse are found among Asian children and among children living in female same-sex couple

child abuse The physical or mental injury, sexual abuse, negligent treatment, or maltreatment of a child younger than age 18 by a person who is responsible for the child's welfare.

neglect A form of abuse involving the failure to provide adequate attention, supervision, nutrition, hygiene, health care, and a safe and clean living environment for a minor child or a dependent elderly individual.

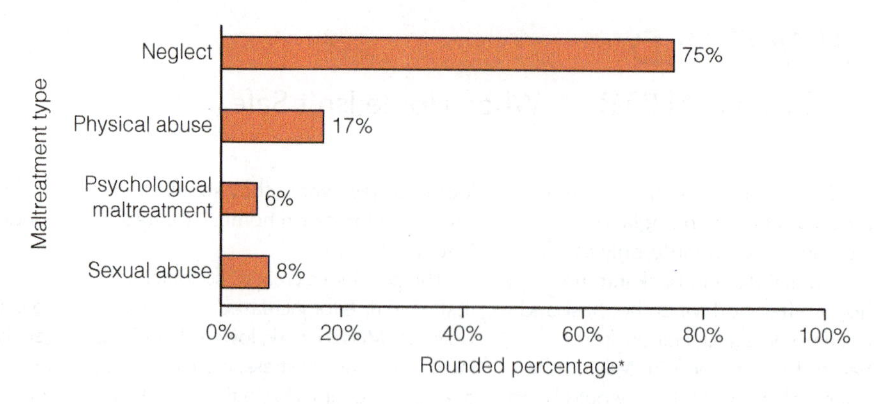

*Percentages sum to more than 100% because children may experience more than one type of maltreatment.

Figure 5.10 U.S. Distribution of Child Maltreatment, by Type (2015)
SOURCE: U.S. Department of Health and Human Services 2017.

families (Gartrell et al. 2010; National Center for Injury Prevention and Control 2014a). Research suggests that racial/ethnic differences in child abuse in the U.S. are primarily explained by poverty rates (Kim and Drake 2018).

Perpetrators of child abuse are most often the parents of the victim. Abuse is more common in families where there is a lot of stress that can result from a family history of violence, drug or alcohol abuse, poverty, chronic health problems, and social isolation.

Replying to @ava @Oprah Young activist Elijah Lee is leading his 3rd Annual Child Abuse Awareness March. He would love your support. For his efforts, Elijah was featured on the Marvel Hero Project on Disney Plus and the Kelly Clarkson Show. Elijah is working to make a change.

-Elijah Lee

Effects of Child Abuse. Physical injuries sustained from child abuse routinely cause pain, disfigurement, scarring, physical disability, and death. In 2018, an estimated 1,770 U.S. children died of abuse and/or neglect (Child Welfare Information Gateway 2020). One of the deadliest forms of child abuse is **abusive head trauma**, commonly known as **shaken baby syndrome (SBS)**. SBS is a form of traumatic brain injury resulting from dropping, throwing, or violently shaking a child, or from blunt force trauma. Shaking a baby causes the brain to move back and forth against the child's skull causing bruising, swelling, and bleeding which can lead to death. Abusive head trauma is the primary cause of traumatic death in children under two years old.

A majority of abusive head trauma victims are under 1 year old, typically between the ages of 3 and 8 months. These injuries have been observed to occur in children up to 5 years old (Joyce and Huecker 2020; Reese et al. 2014). Those who survive may suffer from severe brain damage and/or permanent disabilities, including cerebral palsy, seizure disorders, and blindness. Babies younger than 4 months old are at the greatest risk as inconsolable crying is the most common trigger for shaking a baby (Joyce and Huecker 2020; James 2014).

Adults who were abused as children have an increased risk of a number of problems, including PTSD, depression, smoking, alcohol and drug abuse, eating disorders, obesity, high-risk sexual behavior, and suicide (Centers for Disease Control and Prevention 2014; Hailes et al. 2019). Sexual abuse of young girls is associated with decreased self-esteem, increased levels of depression, running away from home, alcohol and drug use, and multiple sexual partners (Jasinski et al. 2000; Whiffen et al. 2000). Sexual abuse of boys produces many of the same reactions that sexually abused girls experience, including depression, sexual dysfunction, anger, self-blame, suicidal feelings, guilt, and flashbacks (Daniel 2005).

Married adults who were physically and sexually abused as children report lower marital satisfaction, higher stress, and lower family cohesion than married adults with no abuse history (Nelson and Wampler 2000). However, children do not have to be targets of violence themselves to be affected by IPV. A recent study found that children exposed to violence against their mother had a 22 percent higher likelihood of substance use than those who were not exposed (James et al. 2018).

abusive head trauma
A form of inflicted brain injury resulting from violent shaking or blunt impact; a leading cause of death in children under 1 year (*see also* shaken baby syndrome).

shaken baby syndrome
A form of potentially fatal brain damage resulting from violently shaking a baby (*see also* abusive head trauma).

Elder, Parent, and Sibling Abuse

Domestic violence and abuse may involve adults abusing their elderly parents or grandparents, children abusing their parents, and siblings abusing one another.

Elder Abuse. **Elder abuse** includes physical abuse, sexual abuse, psychological abuse, financial abuse (such as improper use of the elder person's financial resources), and neglect. The most common form of elder abuse is neglect—failure to provide basic health and hygiene needs, such as clean clothes, doctor visits, medication, and adequate nutrition. Neglect also involves unreasonable confinement, isolation of elderly family members, lack of supervision, and abandonment.

Approximately 1 in 10 Americans over the age of 60 have experienced some form of elder abuse (National Council on Aging 2021). Older women are far more likely than older men to suffer from abuse or neglect. Two out of every three cases of elder abuse reported to state adult protective services involve women. Although elder abuse also occurs in nursing homes, most cases of elder abuse occur in domestic settings. The most likely perpetrators of elder abuse are adult children, followed by other family members and spouses or intimate partners.

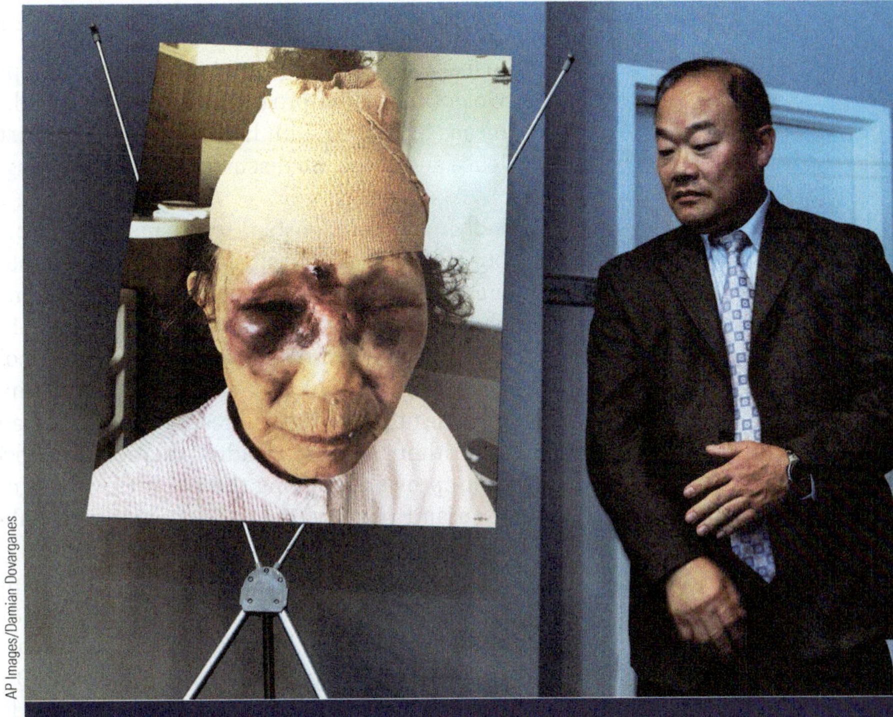

AP Images/Damian Dovarganes

Elderly women are at most risk of elder abuse, most often by a member of their own family. If you know a vulnerable senior citizen being affected by domestic violence, contact your Area Agency on Aging or other support agency.

Parent Abuse. Parent abuse, committed by children and teenagers under the age of 18, includes physical violence, threats, intimidation, and damage to property. Parent abuse is most commonly committed by teenage sons against single mothers (Santich 2014). In some cases, when a child is acting violently toward a parent, the parent has a history of being violent toward the child. However, many abused parents have done nothing to provoke the abuse. Rosemary Pate reportedly did everything she could to help her 19-year-old son who, after years of abusing his mother, fatally stabbed her in her own bed (Davis 2014). In response to Rosemary's tragic death, Florida lawmakers considered a bill that would have (1) established a separate legal category for parental abuse by a minor child, (2) required mandatory reporting of suspected parental abuse to the authorities, and (3) allocated additional resources and police protection for abused parents (Santich 2014). The bill died in the Florida Senate.

Sibling Abuse. Also referred to as "sibling bullying," sibling abuse is "a widespread and serious problem ... and, arguably, the most frequent type of aggression in American society" (Skinner and Kowalski 2013, p. 1727). It is estimated that 80 percent of youth experience some sort of sibling mistreatment, a number higher than spousal and child abuse combined (Lancer 2020). Because of the tendency to not report sibling abuse, often because of the belief that it is just normal "sibling rivalry," it is likely that true amount of sibling abuse is unknown. Sexual abuse also occurs between siblings. Over one-third of sex offenses against children are committed by other minors, and 93 percent of those offenses involve brothers abusing younger sisters.

elder abuse The physical, sexual or psychological abuse, financial exploitation, or medical abuse or neglect of the elderly.

What do you think?

In recent years, the problem of bullying among peers in schools has received increased attention and those concerned about bullying have worked to spread the message that bullying is not an acceptable behavior. In contrast, sibling bullying is commonly viewed as acceptable and expected (Skinner and Kowalski 2013), leading some to call it a "hidden epidemic" (Lancer 2020). Why do you think peer bullying is viewed as a serious social problem, whereas sibling bullying is generally not viewed as a serious social concern?

Factors Contributing to Domestic Violence and Abuse

In many ways, U.S. culture tolerates and even promotes violence through sports (e.g., football), violent television programs and movies, violent video games, and the use of conflict as a means of dealing with international disputes. Other factors that contribute to domestic violence and abuse are discussed in the following sections.

Individual, Relationship, and Family Factors. Although we all have the potential to "lose our cool" and act abusively toward an intimate partner or family member, some individuals are at higher risk for being abusive. Risk factors include having witnessed or been a victim of abuse as a child, past violent or aggressive behavior, lack of employment and other stressful life events or circumstances, and drug and alcohol use (National Center for Injury Prevention and Control 2014b). Alcohol and other drugs increase aggression in some individuals and enable the offender to avoid responsibility by blaming the violent behavior on drugs or alcohol.

Although abuse in adult relationships occurs among all socioeconomic groups, it is more prevalent among the poor. However, most poor people do not maltreat their children and poverty, per se, does not cause abuse and neglect. The correlates of poverty, including stress, inadequate access to child care, and insufficient resources for food and medical care, increase the likelihood of maltreatment. Child neglect increases in states which cut their welfare benefits, putting additional strain on parents and forcing women to choose between leaving children alone or in unsafe care versus not having the financial resources to pay their bills (Francis 2000). A study found that the stress of increases in the unemployment rate during the Great Recession of 2007–2009 was associated with increases in mother's experiences of abusive behavior even when the perpetrator himself had not lost his job (Schneider, Harknett, and McLanahan 2016).

Gender Inequality and Gender Socialization. In the United States before the late 19th century, a married woman was considered the property of her husband. A husband had a legal right and marital obligation to discipline and control his wife through the use of physical force. This traditional view of women as property may contribute to men doing with their "property" as they wish. Donald Trump's comment that with women "you can do anything … Grab 'em by the p***y" (Bullock 2016) reflects the continuing objectification of women.

The view of women and children as property also explains marital rape and father–daughter incest. Historically, the penalties for rape were based on property rights laws designed to protect a man's property—his wife or daughter—from rape by other men (Russell 1990). Although a husband or father "taking" his own property in the past was not considered rape, today, marital rape and incest is considered a crime in all 50 states. Acquaintance and date rape can be explained in part by the fact that U.S. culture "teaches people not to be raped instead of teaching people not to rape" (Tanner 2019).

Traditional male gender roles have taught men to be aggressive and to be dominant in male–female relationships. Traditional male gender socialization also discourages men from verbally expressing their feelings, which increases the potential for violence and abusive behavior (Umberson et al. 2003). Traditional female gender roles have also taught some women to be submissive to their male partner's control.

Acceptance of Corporal Punishment. **Corporal punishment**—the intentional infliction of pain intended to change or control behavior—is widely accepted as a parenting practice. In the United States, most parents do not use corporal punishment, but it is not uncommon. In a study of spanking over the last 12 months, 49 percent of parents reported they had spanked a child between the ages of 0 and 9 years of age, while 23 percent of parents reported they had spanked a child between 10 and 17 years (Finkelhor et al. 2019).

Although not everyone agrees that all instances of corporal punishment constitute abuse, some episodes of parental "discipline" are undoubtedly abusive. Many mental health and child development specialists advise against physical punishment arguing that it is ineffective and potentially harmful to the child (Straus 2000). A study of 3,870 families found that children who were spanked at age 1 were more likely to be aggressive at age 3 and depressed, anxious, and/or withdrawn by age 5 (Gromoske and Maguire-Jack 2012).

corporal punishment The intentional infliction of pain intended to change or control behavior.

All of these factors contribute to why abused adults stay in abusive relationships. Many worry that they can't make it on their own (financially and/or emotionally); they may feel committed to the relationship and emotionally attached to their abuser; and they have often been led to feel responsible for their abuser's behavior and guilty for the failings in their relationship. Some leave to protect their children, while others stay for fear of losing custody of their children. Leaving an abusive relationship is a process which usually takes several attempts (Murray, Crowe and Flasch 2015).

Strategies for Action: Strengthening Families

Each chapter in this text discusses problems that affect families. Thus, addressing social problems such as inadequate health care (Chapter 2), substance abuse (Chapter 3), crime (Chapter 4), gender inequality (Chapter 10), racial and ethnic discrimination (Chapter 9), and so forth, also strengthens families. Our discussion here focuses on expanding the definition of family, access to sex education and contraception, workplace and economic supports (see also Chapter 6 and Chapter 7), reducing unplanned nonmarital pregnancies, divorce education programs, divorce mediation, and domestic violence prevention and policies.

Expand the Definition of Family

Increasingly, government and workplace definitions of "family" are expanding to include unmarried couples so that they and their children can access benefits and protections. Legal recognition of varied family forms is important for the well-being of children, as more than a third of unmarried cohabiting couples have children under age 18 living in the home, and many cohabiting same-sex couples are raising children. When children are denied a legal relationship to both parents because of the parents' unmarried status, they may be denied Social Security survivor benefits, health care insurance, or the ability to have either parent authorize medical treatment in an emergency, among other protections.

Some states, cities, counties, and employers allow unmarried partners (same-sex and/or different-sex partners) to apply for a **domestic partnership** designation, which grants them some legal entitlements, such as health insurance benefits and inheritance rights that have traditionally been reserved for married couples. Some states have revised definitions of family members, partners, and intimate relationships to provide domestic violence protection to victims regardless of marital status, sexual orientation, or gender identity.

Reduce Unplanned Nonmarital Childbearing

One way to reduce the negative effects of nonmarital childbearing and single parenting (discussed earlier) is to reduce the number of unplanned and nonmarital births. Sawhill (2014) argues that instead of planning when to have children, young women tend to "drift" into parenthood. The "drifters" could be turned into "planners" by providing them with long-acting reversible contraception, allowing women to delay childbearing until they are better able to provide a more stable home environment for the child.

Sex Education. A recent study by the World Health Organization found that two-thirds of women in 36 countries who wanted to delay childbearing stopped using contraception, in part because they underestimated the likelihood of getting pregnant. Eighty-five percent of women who stop using contraception get pregnant within a year (World Health Organization 2019). Knowledge is a critical component in individuals' ability to plan when they have children. Research shows that globally and within the United States, the majority of young people do not have access to comprehensive sex education to prepare them to make informed decisions around sex.

Sex education has been on the decline in the United States over the last 15 years (Shapiro and Brown 2018) despite the fact that research indicates that **comprehensive sex education**

domestic partnership A status that some states, counties, cities, and workplaces grant to unmarried couples, including same-sex couples, that conveys various rights and responsibilities.

comprehensive sex education Educational programs that include information about sexuality, sexual consent, reproduction, contraception, and STD prevention. These can occur within or outside the school system.

The Packard Foundation/Getty Images Publicity/Getty Images

Students who learn how to correctly use contraception through comprehensive sex education programs are less likely to experience an unintended pregnancy than students in abstinence-only programs.

programs have the greatest influence in reducing teen pregnancy. In fact, only 24 states and the District of Columbia require sex education in public schools. In 2018, the U.S. Department of Health and Human Services announced that a program called the Teen Pregnancy Prevention (TPP) Program, established by the Obama administration in 2010, would hereafter only provide federal funds to organizations promoting abstinence-only programs (Shapiro and Brown 2018).

Prior to the Trump administration's redirection of funding, the TPP program contributed to a substantial decline in teen pregnancy between 2007 and 2017 and hopefully will do so again. In contrast to the former administration, it is likely that the Biden administration will redirect funds to educational programs that have demonstrated to be effective in reducing teen pregnancy and STDs.

Increased Access to Contraception for Low-Income Women. The vast majority of unintended pregnancies happen to women who do not use contraception or use it inconsistently or incorrectly. Removing financial and/or other barriers to access to long-acting and reversible contraceptives (LARC) is an effective way to reduce unintended pregnancies. As discussed earlier, the Biden administration has pledged to ensure all women have no-cost contraception insurance coverage independent of their employer's faith with exemptions for nonprofits and religious organizations (Long, Ramaswamy, and Salganicoff 2020).

State programs include, for example, a Delaware program that led to a 15 percent decrease in unplanned pregnancies simply by increasing the number of trained practitioners in the area, improving billing processes, and fixing state reimbursement policies that disincentivized doctors from inserting IUDs and implants (Weese 2018). The program was funded by a nonprofit organization, but every public dollar spent helping women avoid unintended pregnancies saved about $5 in Medicaid expenditures for pregnancy-related care, which was a net government saving of $10.5 billion per year (Frost et al. 2014).

@Mayors forAGI

We will never rectify racial injustice until we address **poverty** and invest in the economic security that keeps families in their homes, **children** fed, and promotes mobility. A guaranteed income would do that. I stand with @MayorsforAGI.

-Eric Garcetti, 42nd Mayor of the City of Los Angeles

Workplace and Economic Supports

The most important strategies for forming and strengthening families may be those that maximize employment and earnings. A third of never married U.S. adults said that financial insecurity was the main reason for not being married (Wang and Parker 2014).

Supports such as job training, employment assistance, flexible workplace policies that decrease work–family conflict, and other economic supports are discussed in Chapter 6 and Chapter 7.

The need for workplace and economic support is greater than ever as COVID-19 leads to record unemployment, reduced wages, and the loss of child-care alternatives at a time when many schools are closed. Some parents have had to quit their jobs or put their careers or education on hold to take care of their children. However, the Biden administration has promised to "make substantial investments in the infrastructure of care in our country [in order] to make childcare more affordable and accessible for working families" (Biden 2020c). Additionally, President Biden supports increasing the federal minimum wage to $15 an hour, raising the wages of over 17 million workers who otherwise would be earning less than $15 an hour (Buchwald 2020).

Research shows that a single parent with two children needs to earn a minimum of $24.19 an hour full-time—or about $50,315 a year—to provide for the family's basic needs (Glasmeier 2021). Yet the highest state minimum wage is $15.00 per hour (New York), and the federal minimum wage is only $7.25. If the minimum wage in 1968 had kept up with productivity, as it once did, it would have reached $24.00 in 2020 (Baker 2020). Should the federal government raise the minimum wage to a level sufficient to provide for a family's basic needs?

What do you think?

Strategies to Strengthen Families during and after Divorce

Children are more traumatized by post-divorce conflict between parents than by the divorce itself (Ahrons 2004). Children benefit when divorced (or divorcing) parents minimize their conflict and develop a cooperative coparenting relationship, which may be achieved through divorce education programs and divorce mediation.

Divorce Education Programs. Divorce education programs are designed to help parents who are divorced or planning to divorce reduce parental conflict, foster cooperative parenting, and understand and respond to their children's reactions to the divorce. Many counties and states require divorcing spouses to attend a divorce education program, whereas it is optional in other jurisdictions. Despite the expectation that cooperative postdivorce parenting benefits children, one study found that parents' perceptions of their children's psychological, behavioral, and social well-being were not associated with the parents' perceptions of their coparenting relationship (Beckmeyer et al. 2014). The researchers concluded that "divorcing parents can effectively rear children even when coparenting is limited or conflictual" (p. 533) and that what is more important for children's functioning is the parents' individual parenting skills and quality or the quality of the parent–child relationship. This is reassuring news for divorced parents who are unable to establish peaceful, positive, and cooperative coparenting with their ex-spouses.

Divorce Mediation. In **divorce mediation**, divorcing couples meet with a neutral third party, a mediator, who helps them resolve issues of property division, child and spousal support, and assist in developing a parenting plan in a way that minimizes conflict and encourages cooperation. It is generally best for parents and their children if the parents can agree on a parenting plan. Otherwise, "judges routinely decide where the children of divorced parents will attend school, worship and receive medical care; judges may even decide whether they play soccer or take piano lessons" (Emery 2014, n.p.).

Mediation can not only speed settlement, save money on attorney and court fees, and increase compliance, it can also result in improved relationships between nonresidential parents and children, as well as between divorced parents long after the dispute settlement (Emery et al. 2005). Some research indicates that mediation, when compared to other alternative dispute resolution techniques, for example, negotiation, results in a more satisfactory outcome (Bogacz, Pun, and Klimecki 2020).

Programs to Increase Child Support Payments. Experts agree that many of the systems in place to ensure child support is paid are often so complex that parents don't understand their obligation to pay. Further, the complexity of the process may actually serve as a disincentive to payment. To encourage the development of interventions to address these problems, the Office of Child Support Enforcement of the U.S. Department of Health and Human Services provided $13.4 million in grants to eight states to develop behavioral intervention programs to improve compliance with child support orders (Lovering 2019).

The goal was to identify and reduce "behavioral bottlenecks" that decrease the likelihood that individuals will comply. The focus of many of the programs was on (1) engaging with parents early on rather than only after noncompliance, and (2) simplifying communications with individuals under child support orders. In one Texas county, staff met with individuals one-on-one to review payment options and set up individualized plans. In California, they discontinued their use of a 41-page support order that was particularly

divorce mediation A process in which divorcing couples meet with a neutral third party (mediator) who assists the individuals in resolving issues such as property division, child custody, child support, and spousal support in a way that minimizes conflict and encourages cooperation.

confusing for English as a second language speakers. Instead, they adopted simplified and personalized mailings and phone calls. Both states saw an increase in child support payments and/or payment amounts and plan to expand their programs (Lovering 2019).

Domestic Violence and Abuse Prevention Strategies

Abuse prevention strategies include reducing violence-provoking stress by reducing poverty and unemployment and providing adequate housing, child-care programs and facilities, nutrition, medical care, and educational opportunities. Strengthening the supports for poor families with children reduces violence-provoking stress and minimizes neglect that results from inaccessible or unaffordable community services. Public education and media campaigns may help to reduce domestic violence by conveying the criminal nature of domestic assault and offering ways to prevent abuse. And parent education teaches parents realistic expectations about infant and child behavior and methods of child discipline that do not involve corporal punishment.

At the global level, an international campaign called MenCare promotes men and boys' involvement in equitable, nonviolent caregiving. With initiatives in 45 countries, MenCare promotes positive parenting, violence prevention, and sharing domestic and caregiving work with women (MenCare 2020).

Banning Corporal Punishment. Corporal punishment can be physically and emotionally abusive and reinforces the cultural acceptance of violence in family relationships. In 1979, Sweden became the first country in the world to ban corporal punishment in all settings, including the home. By 2020, 60 countries banned corporal punishment, and 28 more committed to legally prohibiting it (Global Initiative to End All Corporal Punishment of Children 2020). It is legal in all 50 U.S. states for a parent to spank, hit, belt, paddle, whip, or otherwise inflict punitive pain on a child, so long as the corporal punishment does not meet the individual state's definition of child abuse. Corporal punishment in schools is currently legal in 19 states, and over 160,000 children who go to school in those states are subject to corporal punishment each year (Gershoff and Font 2018).

Help for Abuse Victims

Domestic violence and abuse victims may seek help from local domestic violence programs. The National Domestic Violence Hotline **(1-800-799-SAFE)** is a 24-hour, toll-free service that provides information, support, and referrals to local domestic violence programs, which offer emergency shelter; transitional housing; legal services; counseling; and/or assistance with employment, transportation, food, medical care, and child care. Due to inadequate funding and staff, local domestic violence programs receive thousands of requests for housing and other services that cannot be provided because the programs do not have the resources (National Network to End Domestic Violence 2016). When services are not available, abuse victims often return to the abuser, become homeless, or live in their cars. In 2019, the Trump administration quietly changed the federal definitions of domestic violence and sexual assault so that only physical violence constitutes a felony or misdemeanor. This change may restrict access to services and reduce prosecutions for domestic and sexual violence. However, President Biden, author and advocate of the 1994 *Violence against Women Act* that was reauthorized in 2019, has vowed to expand support for domestic and sexual violence survivors, "helping victims secure housing, gain economic stability, and recover from the trauma of abuse" (Biden 2020c).

In 48 states, the law requires members of certain professions (e.g., teachers, doctors, counselors) to report suspected child abuse and/or neglect, and in 18 states, any person who suspects child abuse or neglect must report it to a law enforcement authority via a child welfare hotline or to Child Protective Services. Child Protective Services, sometimes referred to as the Department of Children and Family Services, is the state agency that investigates reports of child abuse and neglect. Unfortunately, the child protective services system is understaffed and suffers from an overload of cases. An investigation by the Associated Press revealed that over a six-year period, at least 768 U.S. children died from child abuse or neglect while their families were being investigated or while the children were under some form of protection services because of abuse or neglect in the home (Mohr and Burke 2014).

Legal Action against Abusers

Domestic violence and abuse are crimes for which individuals can be arrested, jailed, and/or ordered to leave the home or enter a treatment program. Many states and Washington, D.C., have laws that require police to arrest abusers, and prosecutors to prosecute those arrested for abuse, even if the victim does not want to press charges. This removes responsibility for arrest and prosecution from the victim, who may fear that taking legal action against their abuser may result in retaliation. Abuse victims may also obtain a restraining order that prohibits the perpetrator from going near the abused partner.

Abusers may also be required by courts to receive treatment, although some abusers may enter treatment voluntarily. Treatment for abusers typically involves group and/or individual counseling, substance abuse counseling, and/or training in communication, conflict resolution, and anger management. The success rate of these intervention programs is relatively low, and some experts say that treatment should not replace punishment (Cohn 2014).

Understanding Family Problems

Families are influenced by the larger society and culture of which they are a part. Just as societies vary around the world and change across time, so do family norms and structures vary from region to region and from generation to generation.

What we consider to be family problems also varies over time, across cultures, and even within cultures. For example, many Americans view physical punishment of children as necessary for effective child discipline; others view physical punishment as harmful to child development. For some, the solution to family problems implies encouraging marriage and discouraging other family forms, such as single parenting, cohabitation, and same-sex unions. However, many family scholars argue that the fundamental issue is making sure that children are well cared for, regardless of their parents' marital status, sexual orientation, or lifestyle. "Refocusing laws, regulations, and policies on children, rather than the adults to whom they are related, would far better serve the diverse families that actually live in the United States today" (Sheff 2014, p. 285).

Strengthening marriage is a worthy goal if the goal is strong relationships that benefit individuals and their children. However, as research on conflictual marriages demonstrates, marriage for marriage sake is not necessarily a positive step. The reality is that contemporary families come in many forms, each with its strengths, needs, and challenges. Given the diversity of families today, social historian Stephanie Coontz (2016) suggested that "the growing diversity of living arrangements and intimate relationships should cause us to rethink social policies that still assume the universality of lifelong marriage as the main source of caregiving" (p. xxxvi).

The family problems emphasized in this chapter—lack of access to contraception and abortion, problems of divorce, and domestic violence and abuse—have one common denominator—poverty which can be a contributing factor and a consequence of each of these problems. These have led to what sociologists Sara McLanahan and Wade Jacobsen (2014) refer to as the "diverging destinies" of children in the United States, shaped by an increasing gap between the resources parents have at their disposal with which to raise their children. In the next chapter, we turn our attention to poverty and economic inequality, problems that are at the heart of many other social ills.

Chapter Review

- **What are some examples of diversity in families around the world?**
Some societies recognize monogamy as the only legal form of marriage, whereas other societies permit polygamy. Societies also vary in their norms and policies regarding arranged marriage versus self-chosen marriage, same-sex couples, childbearing, and the division of power in the family.

- **What are some patterns, trends, and variations in contemporary U.S. families?**
Patterns, trends, and variations in U.S. families include increased singlehood and older age at first marriage; more interracial and interethnic relationships; delayed childbearing or remaining child free; increased different-sex and same-sex cohabitation; increased births to unmarried women; stabilized divorce rate; increased "gray" divorces; more blended families; more foster families; greater education, employment, and income of wives and mothers; increase in grandfamilies; fewer children living in "traditional" families.

- **How do the three main sociological theories (structural functionalism, conflict theory, and symbolic interactionism) view family problems?**
Structural functionalism views the family as a social institution that performs important functions for society and views traditional gender roles as contributing to family functioning. It also looks at how changes in other social institutions affect families. Conflict theory focuses on how capitalism, social class, and power influence families and emphasizes that powerful and wealthy groups shape social programs and policies that affect families. Both feminist and conflict theories are concerned with how gender inequality influences and results from marriage and family patterns. Symbolic interaction looks at how family interactions affect self-concept and draws attention to the effects of labels and meanings on self-concept and behavior related to family forms, domestic violence, and issues related to divorce.

- **What is the marital decline perspective? What is the marital resiliency perspective?**
According to the marital decline perspective, the recent transformations in American families signify a collapse of marriage and family in the United States. According to the marital resiliency perspective, divorce may be an indicator that people value marriage for its symbolic value and are not willing to stay in bad marriages just for the sake of being married. Same-sex marriages and remarriage also suggest that people still value the institution of marriage.

- **What factors affect access to contraception and abortion in the United States?**
Lack of knowledge as a result of inadequate sex education is a major factor affecting access to contraception and teen pregnancy rates in the United States. The United States has the highest teen pregnancy rate among developed countries. Cost is also a significant factor in access to effective contraception. State-imposed TRAP laws are designed to restrict women's access to abortion in many states, although the Supreme Court has ruled that many such laws are unconstitutional. Most Americans believe abortion should be legal in at least some situations, even though Americans are split relatively equally between "pro-life" and "pro-choice" groups.

- **What are the social causes of divorce discussed in your text?**
Social causes of divorce include (1) the changing function of marriage, (2) increased economic autonomy of women, (3) greater work demands and economic stress, (4) inequality in marital division of labor, (5) no-fault divorce laws, (6) increased individualism, (7) intergenerational patterns, and (8) longer life expectancy.

- **What are some of the effects of divorce on children?**
Reviews of research on the consequences of divorce for children find that children with divorced parents score lower on measures of academic success, psychological adjustment, self-concept, social competence, and long-term health and that they have higher levels of aggressive behavior and depression. Such effects are related to the economic hardship associated with divorce, the reduced parental supervision resulting from divorce, and parental conflict during and after divorce. In highly conflictual marriages, divorce may actually improve children's emotional well-being relative to staying in a conflicted home environment.

- **What are the four types of partner violence that Kelly and Johnson (2008) identified?**
The four patterns of partner violence are (1) situational couple violence (occasional acts of violence arising from arguments that get "out of hand"), (2) coercive controlling violence (violence that is motivated by a wish to control one's partner), (3) violent resistance (acts of violence that are committed in self-defense), and (4) separation-instigated violence (atypical violence which occurs during the separation process).

- **Why do some abused adults stay in abusive relationships?**
Adult victims of abuse are commonly blamed for choosing to stay in their abusive relationships. From the point of view of the victim, reasons to stay in the relationship include economic dependency, emotional attachment, commitment to the relationship, guilt, fear, hope that things will get better, and the view that violence is legitimate because they "deserve" it. Some victims with children leave to protect the children, but others stay because they need help raising the children or fear losing custody of them if they leave. Leaving is a process which usually takes several attempts.

- **What is the most common form of child abuse? What is one of the most deadly forms of child abuse and the leading cause of death in children under age 1 year?**
The most common form of child abuse is neglect. One of the most deadly forms of child abuse, particularly in children under age 1, is abusive head trauma, also known as shaken baby syndrome.

- **What factors contribute to domestic violence and abuse?**
  Factors that contribute to domestic violence and abuse include (1) individual, relationship, and family factors (drug/alcohol use, stressful life events, having witnessed or experienced child abuse as a child); (2) gender inequality and gender socialization; and (3) acceptance of corporal punishment.

- **What strategies to strengthen families are discussed in your text?**
  Strategies to strengthen families include (1) expanding the definition of family; (2) workplace and economic supports; (3) improving sex education and access to contraception; (4) divorce mediation, and divorce education programs; and (5) domestic violence and abuse prevention and policies.

Test Yourself

1. The United States has the highest teenage pregnancy of any developed country in the world.
   a. True
   b. False
2. Most children live with
   a. a mother only.
   b. two parents.
   c. a father only.
   d. at least one grandparent.
3. Two perspectives on the state of marriage in the United States are the marital decline perspective and the marital ________ perspective.
   a. health
   b. resiliency
   c. incline
   d. stability
4. Self-choice marriages have lower rates of divorce than arranged marriages.
   a. True
   b. False
5. The majority of U.S. adults view spanking as
   a. appropriate only for children under the age of 10.
   b. harmful.
   c. sometimes necessary.
   d. the responsibility of the father, not the mother.
6. Most U.S. adults believe that divorce is morally acceptable.
   a. True
   b. False
7. In the majority of married-couple families with children under 18, both parents are employed.
   a. True
   b. False
8. The purpose of divorce mediation is to help couples who are considering divorce repair their relationship and stay together.
   a. True
   b. False
9. TRAP laws are laws specifically designed to
   a. reduce teenage pregnancy.
   b. catch domestic violence offenders.
   c. limit women's access to abortion.
   d. enforce child support arrangements.
10. The divorce rate in the United States has decreased and stabilized. This trend is the result of which of the following factors?
    a. More couples are cohabiting.
    b. When couples cohabit before getting married, they simply break up instead of divorcing.
    c. People are waiting longer to get married.
    d. All of the above

Answers: 1. A; 2. B; 3. B; 4. B; 5. C; 6. A; 7. A; 8. B; 9. C; 10. D.

Key Terms

abortion 178
abusive head trauma 190
arranged marriage 163
bigamy 163
blended families 170
child abuse 189
comprehensive sex education 193
contraception 176
corporal punishment 192
different-sex couples 165
divorce mediation 195
domestic partnership 193
elder abuse 191
eugenics 174
family 162
family planning 176
forced marriage 163
gendered distribution of labor 173
grandfamilies 171
gray divorces 168
homogamous 167
individualism 181
intimate partner violence (IPV) 186
marital decline perspective 172
marital resiliency perspective 175
monogamy 162
neglect 189
no-fault divorce 181
patriarchy 164
physical violence 186
polyamory 167
polyandry 163
polygamy 163
polygyny 163
psychological aggression/abuse 186
refined divorce rate 168
rehabilitative alimony 184
second shift 181
serial monogamy 163
sexual violence 186
shaken baby syndrome 190
snowball sampling 175
stalking 186
teen dating violence 187
TRAP laws 178

Lucas Jackson/Reuters

In a country well governed, poverty is something to be ashamed of. In a country badly governed, wealth is something to be ashamed of."

CONFUCIUS

6

Economic Inequality, Wealth, and Poverty

Chapter Outline

Learning Objectives

After studying this chapter, you will be able to...

1. Describe patterns of economic inequality, poverty, and wealth distribution around the world.
2. Summarize different ways of defining and measuring poverty.
3. Explain criticisms of the definition of the U.S. poverty line.
4. Explain how each of the major sociological theories would view the nature, causes, and consequences of economic inequality.
5. Describe patterns of economic inequality, poverty, and wealth in the United States.
6. Explain how economic inequality contributes to a range of social problems in the areas of health, housing, legal and political outcomes, war and social conflict, vulnerability to natural disasters, education, and family life.
7. Identify international and U.S. strategies aimed at reducing poverty and economic inequality.
8. Identify the major U.S. public assistance and welfare programs.
9. Analyze evidence that counters common myths about welfare in the United States.
10. Explain the implications of alleviating poverty and reducing economic inequality for societal stability and well-being.

Patrick T. Fallon/Reuters

A long line for food distribution is just one of the many signs of the economic disruption caused by the COVID-19 pandemic. Millions of American families, just like Sergine and Dave's, faced a financial crisis during 2020.

"I FEEL LIKE A FAILURE," Sergine Lucien told the police officer who knocked on the window of the car where Sergine was living with her husband Dave and their two children, Jayden and Phoenix. "I wish I could get you out this situation," the officer told her. "My heart hurts for you" (Jaffe 2020, p. 1). Like many of the more than half a million Americans who are experiencing homelessness on any given night, Sergine's sense of personal failure is embedded within a broader system of institutional and cultural failures. Sergine and Dave's family moved to Florida from New York seven years earlier, attracted by the availability of service jobs in the tourist industry surrounding Orlando. They planned to stay in motels for just a few months while they saved up for a better place to live. Although they both worked every chance they could get—usually multiple jobs—they could never seem to dig themselves out of a hole. Being poor is expensive. Motels and restaurant food cost more than rent and a home cooked meal, so they struggled to save enough for a deposit. Motel-hopping also required paying to keep a storage unit, so they didn't need to start from scratch every time they found a place where they could settle down for a while. Sergine's dream was a clean RV in a safe trailer park. After six years of living in motels, they finally achieved this dream. Only a few months later, however, the COVID-19 pandemic shut down the tourist industry in central Florida, and they lost their jobs, forcing them to live in their car. After a month, Dave got another job making $14 an hour in construction, but he wouldn't see a paycheck for two weeks. In the meantime, the hole they were in seemed to only get deeper. Dave qualified for unemployment, but the system was backlogged and the phone lines constantly busy. They also qualified for $3,400 through the government's COVID-19 financial relief legislation, but without a home address, they were unable to collect it. Homeless shelters were full, and charitable organizations were overwhelmed. Still, Sergine and Dave blame themselves for their inability to get ahead. "We get a little money and we rush out. That's what we did with the RV. We rush into things without planning because we want out so bad" (p. 1). Finally, Dave's first paycheck came through. They were able to get out of the car and back into a motel room. They were back to where they started seven years earlier.

wealth The total assets of an individual or household minus liabilities.

absolute poverty The lack of resources necessary for material well-being—most important, food and water, but also housing, sanitation, education, and health care.

relative poverty The lack of material and economic resources compared with some other population.

The Global Context: Economic Inequality, Wealth, and Poverty around the World

Approximately 10 percent of the world's population—734 million people—are extremely poor, living on less than $1.90 a day (World Bank 2020a). The World Bank estimates that the global COVID-19 pandemic will push an additional 40 to 60 million into extreme poverty in the next year. The majority of the world's extremely poor live in sub-Saharan African or South Asia. The very poor suffer from myriad problems including malnutrition, lack of access to clean water and sanitation, inadequate housing, lack of education, and poor health—problems we discuss later in this chapter.

In contrast to the extreme poverty that plagues 734 million people living on this planet, there were 2,153 billionaires on earth in 2019 (Oxfam 2020). These billionaires—who account for 0.0000003 percent of the world's population—hold the same amount of wealth as 4.6 billion people, or 60 percent of the world's population. The majority of the world's billionaires live in the United States, China, and Western Europe (Forbes 2020). Living opulent, lavish lifestyles that include luxuries most of us can only imagine, billionaires have access to anything money can buy: yachts and private jets, multiple luxurious homes around the world, the best education and health care, and, as we discuss later, influence in political affairs.

Economic inequality—the wide gap that divides the rich and the poor—includes inequality in both income and wealth. **Wealth** refers to the total assets of an individual or household minus liabilities (mortgages, loans, and debts). Wealth includes the value of a home, investments, real estate, cars, unincorporated business, life insurance (cash value), stocks, bonds, mutual funds, trusts, checking and savings accounts, individual retirement accounts (IRAs), and valuable collectibles.

Although rising incomes have dramatically reduced the number of people living in extreme poverty over the past 30 years—dropping from 1.9 billion in 1990 to 734 million in 2020—the level of wealth inequality in the world has risen sharply (World Bank 2020a). In other words, although fewer people are suffering from the most extreme hardships of poverty like starvation and inadequate shelter, an ever-growing share of the world's wealth is being concentrated in the hands of a very small number of individuals. The stark nature of this inequality is highlighted in statistics from Oxfam's (2020) *Time to Care* report:

- The combined wealth of the 22 richest men in the world is more than the combined wealth of all the women in the continent of Africa.
- The richest 1 percent of the world's population have twice as much wealth as 6.9 billion people.
- If you saved $10,000 a day since the building of the pyramids in Egypt, you would have only one-fifth the average fortune of the world's five richest billionaires.
- Taxing an additional 0.5 percent of the wealth of the richest 1 percent over the next 10 years is equivalent to the investments needed to create 117 million jobs in education, health, elder care, and other care work sectors.

@RBReich

In the middle of a pandemic and the worst economic crisis since the Great Depression, America's billionaires have made a fortune:

Bezos: +$76B
Musk: +$43B
Zuckerberg: +$42B
Balmer: +$18B
Ellison: +$12B

If this doesn't convince you our system is broken, I don't know what will.

-Robert Reich

extreme poverty Living on less than $1.90 a day.

Defining and Measuring Poverty

Absolute poverty refers to the lack of resources necessary for well-being—most importantly food and water, but also housing, sanitation, education, and health care. In contrast, **relative poverty** refers to the lack of material and economic resources compared with some other population. If you are a struggling college student living on a limited budget, you may feel as though you are "poor" compared with the middle- or upper-middle-class lifestyle to which you may aspire. However, if you have a roof over your head; access to clean water, toilets, and medical care; and enough to eat, you are not absolutely poor; indeed, you have a level of well-being that millions of people living in absolute poverty may never achieve.

Image Press Agency/Alamy Stock Photo

The world's richest man, Amazon founder and CEO Jeff Bezos, has an estimated net worth of $206 billion (Forbes 2020). This is slightly more than the 2019 GDP of the entire country of New Zealand (World Bank 2020). In contrast, the typical Amazon warehouse worker earns about $15 per hour.

Global Measures of Poverty

The most widely used standard to measure **extreme poverty** at the global level is a daily income level of $1.90 (PPP) or less in income per day. This level is measured in terms of **purchasing power parity (PPP)**, which is a metric used to standardize differences in currency values and cost of living in order to make international comparisons in levels of wealth and poverty. A dollar amount in PPP indicates its value in the United States.

Olivier Asselin/Alamy Stock Photo

People who experience absolute poverty, like the people living in this slum in Monrovia, Liberia, lack access to sufficient shelter and sanitation.

In other words, the 734 million people in the world who are classified as extremely poor live each day on less than what $1.90 would buy you in the United States. This is certainly an indicator of absolute poverty, as $1.90 in the United States would not be enough to provide for sufficient food, let alone shelter or sanitation.

A measure of relative poverty is based on comparing the income of a household to the median household income in a specific country. According to this relative poverty measure, members of a household are considered poor if their household income is less than 50 percent of the median household income in that country.

Low income is only one indicator of impoverishment. The **Multidimensional Poverty Index** is a measure of serious deprivation in the dimensions of health, education, and living standards that combines the number of the deprived and the intensity of their deprivation (see Figure 6.1). About 1.3 billion people in the world are multidimensionally poor (United Nations Development Programme [UNDP] 2020).

What do you think?

The next section describes how poverty is measured in the United States by comparing the annual pretax income of a household to the official U.S. poverty line—a dollar amount that determines who is considered poor. Before reading further, answer these questions: How much annual income do you think a household with one adult needs to earn to avoid living in poverty? What about a household with two adults? One adult and one child? Two adults and one child? Compare your answers with the official poverty thresholds in Table 6.1.

purchasing power parity (PPP) An economic metric used to standardize differences in currency values and standards of living in order to make international comparisons in levels of wealth and poverty; $1 PPP is equivalent to the purchasing value of $1 in the United States.

Multidimensional Poverty Index A measure of serious deprivation in the dimensions of health, education, and living standards that combines the number of deprived and the intensity of their deprivation.

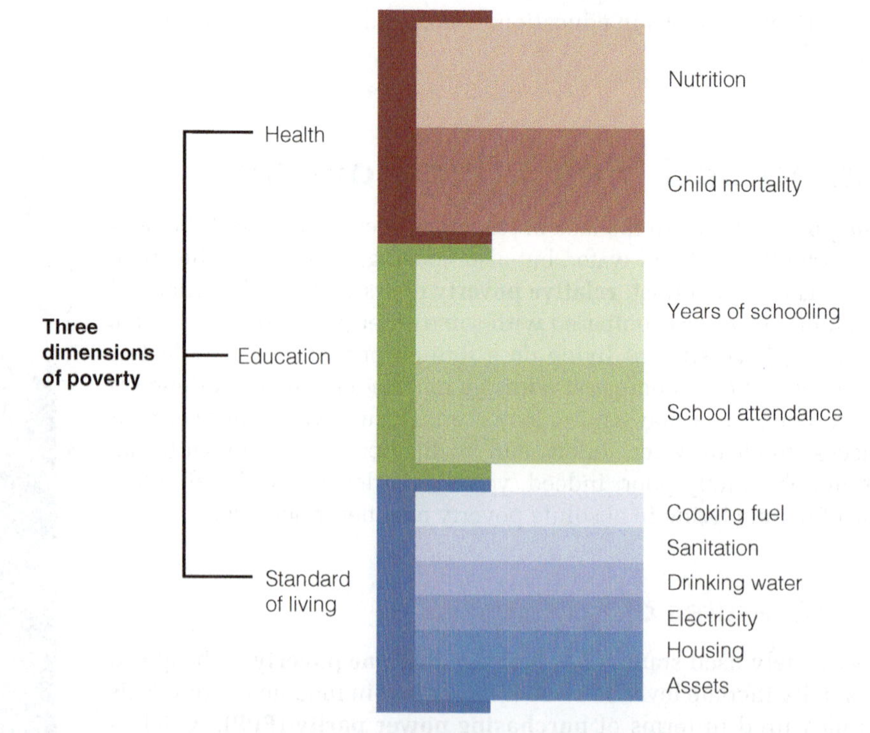

Figure 6.1 Structure of the Global Multidimensional Poverty Index
People are counted as multidimensionally poor if they are deprived in one-third or more of these ten indicators.
SOURCE: (UNDP) 2020.

U.S. Measures of Poverty

TABLE 6.1 Poverty Thresholds, 2020 (householder younger than 65 years)

Household Makeup	Poverty Threshold
One adult	$13,465
Two adults	$17,331
One adult, one child	$17,839
Two adults, one child	$20,832
Two adults, two children	$26,246

SOURCE: U.S. Census Bureau 2021.

In 1964, the Social Security Administration devised a poverty index based on data indicating that families spent about one-third of their income on food. The official poverty level was set by multiplying food costs by three. Since then, the poverty level has been updated annually for inflation; it differs by the number of adults and children in a household and by the age of the head of household, but it is the same across the continental United States (see Table 6.1). Anyone living in a household with pretax income below the official poverty line is considered "poor." Individuals living in households with incomes that are above the poverty line but not very much above it are classified as "near-poor," and those living in households with income below 50 percent of the poverty line live in "deep poverty," also referred to as "severe poverty." A common working definition of low-income households is households with incomes that are between 100 percent and 200 percent of the federal poverty line or up to twice the poverty level.

The U.S. poverty line has been criticized on several grounds. First, the official poverty line is based on pretax income, so tax burdens, as well as tax credits, are disregarded. Family wealth, including savings and property, is also excluded in official poverty calculations, and noncash government benefits that assist low-income families—food assistance, Medicaid, and housing and child-care assistance—are not taken into account. In addition, the federal poverty line is a national standard that does not reflect the significant variation in the cost of living from state to state and between urban and rural areas. Finally, the poverty line underestimates the extent of material hardship in the United States because it is based on the assumption that low-income families spend one-third of their household income on food. That was true in the 1950s, but because housing, medical care, child care, and transportation costs have risen more rapidly than food costs, low-income families today spend far less than one-third of their income on food.

The Economic Policy Institute's Family Budget Calculator measures the income that families of different sizes need to obtain "a secure yet modest living standard" based on estimated costs of housing, food, child care, transportation, health care, other necessities, and taxes in specific geographic locations (Economic Policy Institute [EPI] 2018).

To secure a modest yet adequate standard of living for a two-parent, two-child family, the EPI estimates that a family would need at least $82,555 per year in Birmingham, Alabama, $115,583 per year in Honolulu, Hawaii, and $123,975 per year in Washington, DC. The Family Budget Calculator finds that families need, at a *minimum*, twice the poverty-level income to meet basic needs.

Sociological Theories of Economic Inequality, Wealth, and Poverty

Americans are taught that we live in a **meritocracy**—a social system in which individuals get ahead and earn rewards based on their individual efforts and abilities (McNamee and Miller 2009). In a meritocracy, everyone has an equal chance to succeed; those who are "$ucce$$ful" are smart and talented and have worked hard and deserve their success, while those who fail to "make it" have only themselves to blame. This individualistic perspective views economic inequality as the result of some people developing their potential and working hard and earning their success, while others don't measure up, make bad choices, don't work hard enough, and have only themselves to blame for their predicament. In contrast to the individualistic perspective, structural functionalism, conflict theory, and symbolic interactionism offer sociological insights into the nature, causes, and consequences of poverty and economic inequality.

meritocracy A social system in which individuals get ahead and earn rewards based on their individual efforts and abilities.

Structural-Functionalist Perspective

According to the structural-functionalist perspective, poverty results from institutional breakdown: economic institutions that fail to provide sufficient jobs and pay, educational institutions that fail to equip members of society with the skills they need for employment, family institutions that do not provide two parents, and government institutions that do not provide sufficient public support. Sociologist William Julius Wilson explains:

> Where jobs are scarce... and where there is a disruptive or degraded school life purporting to prepare youngsters for eventual participation in the workforce, many people eventually lose their feeling of connectedness to work in the formal economy; they no longer expect work to be a regular, and regulating, force in their lives. (Wilson 1996, pp. 52–53)

More than 60 years ago, Davis and Moore (1945) presented a structural-functionalist explanation for economic inequality, arguing that a system of unequal pay motivates people to achieve higher levels of training and education and to take on jobs that are more important and difficult by offering higher rewards for higher achievements. However, this argument is criticized on the grounds that many important occupational roles, such as child-care workers and nurse assistants, have low salaries, whereas many individuals in nonessential roles (e.g., professional sports stars and entertainers) earn outrageous sums of money. The structural-functionalist argument that CEO pay is high in order to reward strong performance is shattered by the fact that CEOs are paid huge salaries and bonuses even when they contribute to the economic failure of their corporation and/or to the problem of unemployment.

In his classic article "The Positive Functions of Poverty," sociologist Herbert Gans (1972) draws on the structural-functionalist perspective to identify ways in which poverty can be viewed as functional for the nonpoor segments of society. For example, having a poor population ensures that society has a pool of low-cost laborers willing to do unpleasant jobs, providing a labor pool for jobs ranging from the military to prostitution. Poor populations also provide labor for the affluent in the form of domestic work, such as maids and gardeners. The poor help keep others employed in jobs such as policing, prison work, and social work. And the poor provide a pool of consumers for used goods and second-rate service providers. However, the structural-functionalist view of poverty also highlights the ways in which poverty and economic inequality are *dysfunctional* for society. For example, as discussed later in this chapter, poverty and economic inequality are linked to crime and violence, family instability, and social conflict and war (see also Chapter 4, Chapter 5, and Chapter 16).

Conflict Perspective

Karl Marx (1818–1883) proposed that economic inequality results from the domination of the *bourgeoisie* (owners of the factories, or "means of production") over the *proletariat* (workers). In a capitalistic economy, the bourgeoisie accumulate wealth as they profit from the labor of the proletariat, who earn wages far below the earnings of the bourgeoisie. Marx predicted that inequality resulting from capitalism would lead to the collapse of society. Based on an ambitious analysis of data on income and wealth in 20 countries over a period of three centuries, French economist Thomas Piketty (2014) concluded that inequality is intrinsic to capitalism and is likely to increase to levels that threaten democracy.

Modern conflict theorists recognize that the power to influence economic outcomes arises not only from ownership of the means of production but also from management positions, interlocking board memberships, control of media, financial contributions to politicians, and lobbying. The conflict perspective views money as a tool that can be used to achieve political interests. Wealthy corporations and individuals use financial political contributions to influence political elections and policies in ways that benefit the wealthy. The interests of the wealthy include things such as keeping taxes low on capital gains, and the wealthy are more likely than the general public to oppose increasing the minimum wage and other policies that would create upward mobility among low-income Americans (Callahan and Cha 2013).

The power of the wealthy to influence political outcomes was reinforced by the 2010 Supreme Court ruling (five to four) in *Citizens United v. Federal Election Commission* that corporations have a First Amendment right to spend unlimited amounts of money to support or oppose candidates for elected office. Senator Bernie Sanders, who wants to overturn the Supreme Court's *Citizens* ruling, explained:

> What the Supreme Court did in Citizens United is to tell billionaires like the Koch brothers and Sheldon Adelson, "You own and control Wall Street. You own and control coal companies. You own and control oil companies. Now, for a very small percentage of your wealth, we're going to give you the opportunity to own and control the United States government." That is the essence of what Citizens United is all about. (quoted in Easley 2015, n. p.)

Citizens United allowed the creation of Super PACs, which are political action committees that are allowed to raise and spend unlimited amounts of money for the purpose of supporting or defeating a political candidate, as long as the monies are not given to any political candidate's campaign. Although Super PACs are prohibited from coordinating directly with a campaign, they often operate on behalf of the candidate in ways that narrowly avoid those restrictions and raise concerns about conflicts of interest. For example, America First Action—a Super PAC that supported the Trump campaign—gave a combined $18.8 million to a law firm representing former President Trump on a range of different lawsuits, including a sexual harassment suit brought by a former campaign staffer, a set of personal injury lawsuits brought by demonstrators who claimed they were beaten at a Trump rally, and a civil suit that claims the Trump campaign illegally distributed information obtained by Russian hackers during the 2016 election (Lipton 2020). In these cases, the *Citizens United* ruling contributed to the blurring of lines between public and private funding, as well as public and private interests of the candidate holding office.

> "[I]f one person controls the activities and voting patterns of a member of Congress, that is in direct contravention to everything we stand for and believe in … . [T]here will be scandals … . [T]here's too much money washing around the political arena today. And it's just, it's inevitable … as the sun will come up tomorrow. … I feel a great sense of disappointment and sorrow."
>
> **–SENATOR JOHN McCAIN (R-AZ), SPEAKING ABOUT THE *CITIZENS UNITED* DECISION**

Laws and policies that favor the rich, such as tax breaks that benefit the wealthy, are sometimes referred to as **wealthfare**. For example, during the Trump presidency, the former administration touted the 2017 tax policy overhaul, referred to as the *Tax Cuts and Jobs Act*, as promoting wage growth for the middle class. The legislation included a reduction in effective corporate tax rates from 17.2 percent to 8.8 percent, which would allow corporations to shift about $664 billion in profits held in overseas accounts back to the United States, leading to a reinvestment in American businesses, job growth, and wage increases. However, an independent analysis from the nonpartisan Congressional Research Service found that the majority of these funds were directed toward the corporate repurchasing of shares, and "relatively little was directed to paying worker bonuses, which had been announced by some firms" (Gravelle and Marples 2019, p. i). The analysis found that while GDP had grown by 2.9 percent in the first year following the implementation of the tax cuts, the wages of production and nonsupervisory workers grew by only 1.2 percent; in other words, the majority of the economic benefits from the tax cuts went to corporate profits rather than into workers' paychecks. The Biden administration has pledged to reverse many of the features of the so-called "Trump tax cuts," namely by increasing the tax rate on the wealthiest 1 percent of Americans, imposing a tax penalty on corporations that outsource their labor to other countries, and raising the corporate tax rate (Biden-Harris 2020).

Corporate welfare refers to government subsidies to corporations, including direct payments and tax breaks. The profitable oil and gas industries receive large federal subsidies, and many states give corporations tax breaks as part of their economic development efforts to entice businesses to locate operations in their state. Legal tax loopholes

wealthfare Laws and policies that benefit the rich.

corporate welfare Laws and policies that benefit corporations.

allow many corporations to pay little or no federal corporate tax at all in a given year. One example of a corporate tax loophole is the practice known as **corporate tax inversion**, in which a company lowers its taxes by merging with a foreign company and changing to an offshore address. Inversions largely occur on paper and typically do not involve moving operations overseas.

Conflict theorists also note that "free-market" trade and investment economic policies, which some claim to be a solution to poverty, primarily benefit wealthy corporations. Trade and investment agreements enable corporations to (1) expand production and increase economic development in poor countries, and (2) sell their products and services to consumers around the world, thus increasing poor populations' access to goods and services. Yet such policies also enable corporations to relocate production to countries with abundant supplies of cheap labor, which leads to a lowering of wages and a resultant decrease in consumer spending, which leads to more industries closing plants, going bankrupt, and/or laying off workers.

When corporations claim that their products or services are essential in the fight against poverty, a conflict perspective might reveal a different story. For example, powerful food and biotech corporations such as Monsanto, Cargill, and Archer Daniels Midland have used their economic and political power to impose a system of agriculture based on intensive chemical use and on patented and genetically modified seeds (McDonagh 2013). These corporations assert that their model of agriculture, which requires farmers to purchase their chemicals and seeds, yields more and better food and thus is important in the global fight against hunger and poverty. Yet this corporate control of agriculture has resulted in farmers' dependence and debt (and an epidemic of suicides among poor farmers), environmental degradation (through the increased use of chemicals), and health risks associated with chemicals and genetically modified foods.

Symbolic Interactionist Perspective

Symbolic interactionism focuses on how meanings, labels, and definitions affect and are affected by social life. This view calls attention to ways in which wealth and poverty are defined and the consequences of being labeled "poor." Individuals who are poor are often viewed as undeserving of help or sympathy; their poverty is viewed as due to laziness, immorality, irresponsibility, lack of motivation, or personal deficiency (Katz 2013). Wealthy individuals, on the other hand, tend to be viewed as capable, motivated, hardworking, and deserving of their wealth. The term *welfare* itself may be a "dirty word," according to one study which found that support for government spending on the poor was higher when it was described as "spending on the poor" instead of "welfare" (Harell et al. 2008).

Culture shapes how people interpret what it means to "deserve" public assistance, thereby shaping attitudes about welfare policy. In one study comparing perceptions of Danish and American citizens, researchers found that when welfare recipients were characterized as lazy, respondents from both countries were more likely to express opposition to welfare spending and that when welfare recipients were characterized as unlucky, respondents were more likely to express support for welfare spending (Aarøe and Petersen 2014). While this association is not surprising on its own, the researchers found that overall support for social welfare spending was significantly higher in Denmark compared to the United States. The authors conclude that while there is a general tendency among all people to associate deservingness with luck, cultural values that are unique to each society shape public perceptions. In other words, features of Danish culture contribute to the perception that a person's status is largely determined by luck, whereas features of American culture contribute to the perception that a person's status is largely determined by effort.

corporate tax inversion The practice in which a company lowers its taxes by merging with a foreign company and changing to an offshore address.

The symbolic interactionist perspective is concerned with how social conditions come to be viewed as social problems. The fallout from the economic crisis of 2008 and subsequent taxpayer-funded bailouts of the banking industry prompted widespread public protests and increased attention to issues of economic inequality in the media and among politicians. One of these movements, known as Occupy Wall Street, symbolized these concerns with a focus of the "99 percent"—that is, the concerns of regular, hardworking folks versus the "1 percent"—the wealthy. The publication of French economist Thomas Piketty's (2014) book *Capital in the Twenty-first Century* has further increased public awareness of economic inequality as a social problem. Piketty's book essentially argues that economic inequality will continue to increase to levels that threaten social stability unless governments take such actions as enforcing a global wealth tax.

Robert K. Chin/Alamy Stock Photo

Occupy Wall Street protestors frequently use symbols associated with greed and luxury to focus attention on economic inequality.

In 2020, economic inequality emerged as one of the top ten issues identified by voters as "very important" to their vote in the 2020 election (Pew Research Center 2020). In the 2016 election, economic inequality did not break the 40 percent threshold to make it into the list of top issues for voters. Polling also indicates a decline in the strength of Americans' belief in meritocracy; although a majority in 2018 expressed satisfaction with the opportunity for a person to get ahead by working hard, this level of satisfaction was 13 percentage points lower than in 2001 and showed a sharp dip during the Great Recession of 2008–2010 (see Figure 6.2).

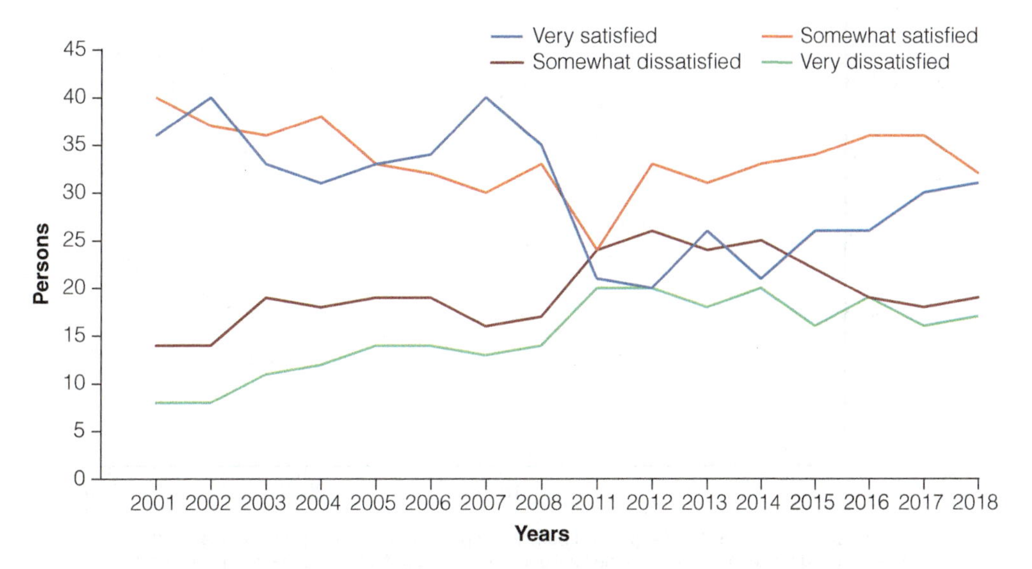

Figure 6.2 Americans' Views of Meritocracy, 2001–2018

Respondents Reporting Level of Satisfaction with the Opportunity for a Person in This Nation to Get Ahead by Working Hard.

NOTE: Numbers do not add up to 100 due to "no opinion" responses.

SOURCE: Gallup News Service 2018.

Economic Inequality, Wealth, and Poverty in the United States

The United States has the greatest degree of income inequality and the highest rate of poverty of any industrialized nation. After looking at U.S. income and wealth inequality and the "1 percent," we look at patterns of U.S. poverty.

U.S. Income Inequality

In 2017, the top 1 percent of U.S. taxpayers earned 21 percent of all U.S. income (York 2020). Historically, workers' income has been tied to overall productivity. Since the 1970s, however, a startling pattern in income inequality has emerged: although worker productivity—the amount of goods and services produced per hour-worker—has continued to climb steadily, workers' wages have mostly flattened (see Figure 6.3). From 1948 to 1979, worker productivity increased 108.1 percent and wages increased nearly as much (93 percent). But from 1979 to 2018, although worker productivity increased by 69.6 percent, compensation (including both wages and benefits) rose by only 11.6 percent (Gould 2020). In other words, productivity has risen six times more than workers' pay. The wage stagnation of middle- and low-income earners is in stark contrast to the huge increase in income of the top earners. From 1979 to 2018, income of the top 1 percent grew 157.8 percent, and the income of the top 0.1 percent grew by 340.7 percent. In contrast, income for the bottom 99 percent rose by only 23.9 percent.

One reason why workers' wages have not increased in sync with their productivity is that CEOs are taking a bigger piece of the pie. The most extreme wage inequality is found between the compensation (salaries, bonuses, stock options, and so on) of chief executive officers (CEOs) and the average employee. In 2019, CEOs at the top 350 U.S. corporations received, in salaries and other compensation (such as bonuses and stocks), 320 times the average compensation of U.S. workers (Mishel and Kandra 2020). That means that a typical worker would have to work 320 years to earn what a CEO makes in 1 year! Table 6.2 shows the dramatic increase in the ratio of CEO pay to average worker pay since 1965. Since 1978, CEO compensation has grown by 940 percent (Mishel and Wolfe 2019). Most of this growth is due to the shift toward CEO compensation packages that are mostly tied to stock options rather than to wages.

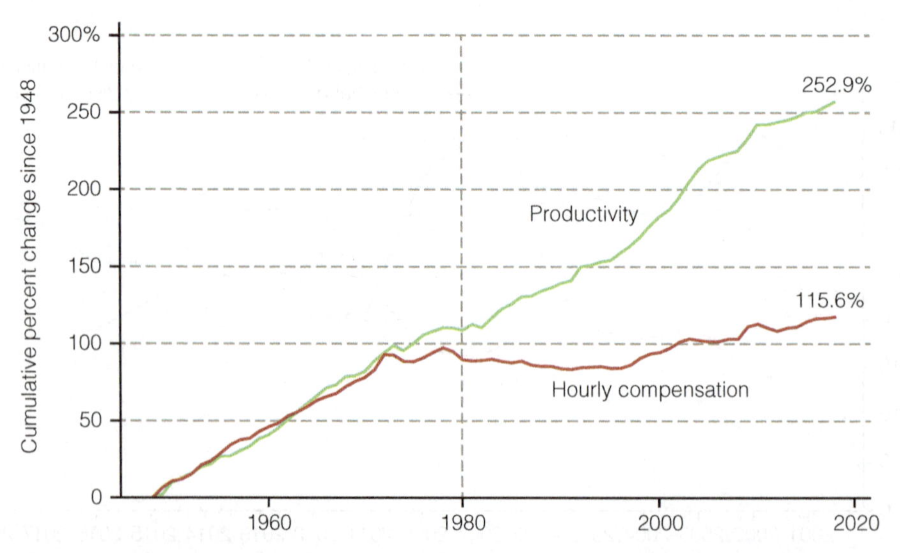

Notes: Data are for compensation (wages and benefits) of production/nonsupervisory workers in the private sector and for net productivity of the total economy. "Net productivity" is the growth of output of goods and services less depreciation per hour worked.

Figure 6.3 Growth in Productivity and Hourly Compensation, 1948–2018
SOURCE: Gould 2020.

Most Americans agree that CEOs should earn more than typical workers, but how much more? Nearly two-thirds of Americans believe that there should be a maximum cap on how much CEOs can earn, and the majority of Americans think that CEOs should not be permitted to be paid more than between 6 and 17 times more than the average worker (Larcker et al. 2016). What do you think the ideal ratio of CEO pay to worker pay should be, and why?

What do you think?

U.S. Wealth Inequality

Wealth in the United States, as in the rest of the world, is unevenly distributed and concentrated at the top (see Figure 6.4). The wealthiest 1 percent has received much attention as a result of the Occupy Wall Street's "99 percent versus 1 percent" slogan. But inequality exists even among the rich. More than 40 percent of U.S. wealth in 2016 was owned by the top 1 percent, and more than half of that wealth was owned by the top 0.1 percent (Zucman 2019).

There is a saying: "The best way to make a million dollars is to start out with $900,000!" Wealth tends to snowball, and the bigger the snowball you start off with, the bigger it grows. Consider that between 1963 and 2016 (Urban Institute 2017),

- Families at the 99th percentile saw their wealth increase sevenfold.
- Families at the 90th percentile (those wealthier than 90 percent of families) saw their wealth increase fivefold.
- Families in the *middle* of the wealth distribution more than doubled their wealth.
- Families at the *bottom* 10 percent of the wealth distribution went from having no wealth, on average, to being about $1,000 in debt.

TABLE 6.2 Ratio of CEO Compensation* to Average Worker Pay, U.S., 1965–2019

Year	Ratio
1965	20:1
1978	30:1
1995	123:1
2000	376:1
2015	276:1
2019	320:1

*Rounded; includes pay and stock options; based on the top 350 U.S. firms, excluding Facebook due to its high-outlier CEO compensation.
SOURCE: Mishel and Kanda 2020.

Disparities in income and wealth show an economic advantage of White people over racial/ethnic minorities. In 2019, median household income for non-Hispanic White households was $76,057 compared with $56,113 for Hispanic households and $45,438 for Black households (Semega et al. 2020). The median wealth of White households is nearly 8 times the median wealth of Black households and more than 5 times the wealth of Hispanic households (see Figure 6.5). White families are more likely than Black or Hispanic families to own homes, which for many Americans, is their most valuable asset. And White families are five times more likely than Hispanic or Black families to receive large gifts or inheritances, which can be used to pay for college, a down payment on a home, and other wealth-building assets (Urban Institute 2017).

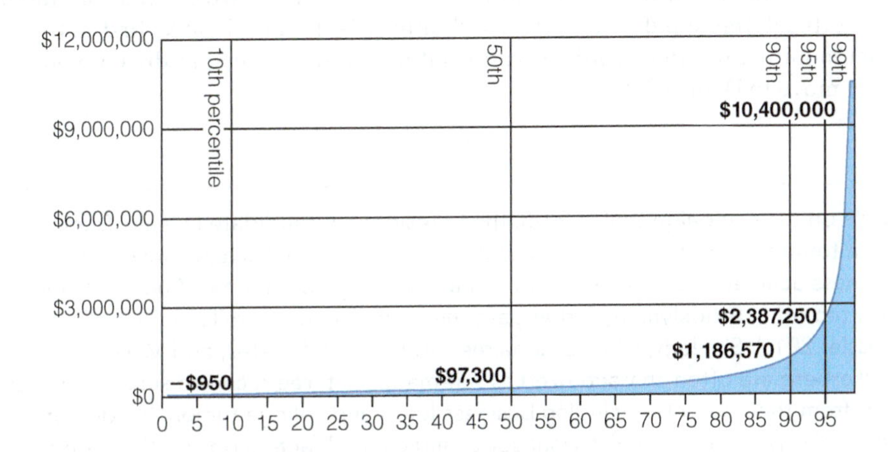

Figure 6.4 Average Wealth of U.S. Families, 2016
SOURCE: Based on Urban Institute 2017.

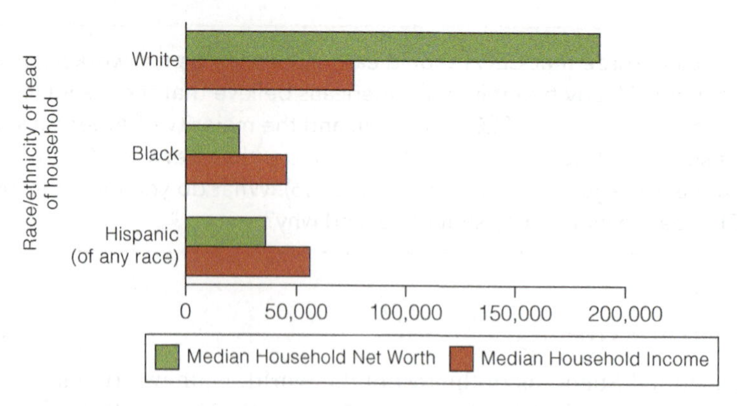

Figure 6.5 Median Household Income and Wealth by Race and Hispanic Origin, 2019
SOURCE: Board of Governors of the Federal Reserve System 2020, Semega et al. 2020.

The Wealthiest Americans

The United States is home to eight of the world's ten wealthiest people, including both the world's wealthiest man (Jeff Bezos, $206 billion) and wealthiest woman (Alice Walton, $69 billion) (Forbes 2020). The famous Forbes 400 list of the wealthiest Americans describes the youngest billionaire in the world—21-year-old Kylie Jenner—as "self-made." Although the term *self-made* suggests achieving financial success on one's own, without assistance from family or society, the Forbes definition of *self-made* is set on a scale, ranging from 1 to 10, which ranks individuals on the basis of how much of their wealth was inherited and how many hardships they had to overcome (Dolan and Kroll 2019). A person who inherited everything scores a 1, whereas a person like Oprah Winfrey, who grew up in poverty and encountered substantial obstacles, scores a 10. Kylie Jenner, who came from a wealthy family and who was able to leverage the fame of her mother and older sister to build her successful cosmetics brand, scores a 7.

Still, the Forbes "self-made" ranking provides only an individual-level glimpse into the pathway to wealth and does not provide insight into the role that the societal-level dynamics—such as educational inequalities and tax policies—play in creating wealthy individuals. There are, indeed, true "rags to riches" success stories in the United States that exemplify the idea that anyone can achieve the American dream. However, such stories are the exception rather than the rule. **Social mobility** data, which measure the likelihood and the extent to which persons will change their socioeconomic status over the course of their lives and across generations, indicates that social mobility in the United States is relatively low compared to other wealthy, industrialized nations. The World Economic Forum (2020) ranks the United States 27th in the world in social mobility, behind Portugal, the Republic of Korea, and Lithuania. Denmark ranks first in the world in social mobility. In other words, if you want to live the American dream, you should probably move to Denmark.

social mobility The likelihood and extent to which persons will change their socioeconomic status over the course of the life and across generations.

What *do you* think?

While on the campaign trail in 2015, then presidential candidate Donald Trump often touted his route to billionaire status as a function of his hard work and business acumen. He famously said, "It has not been easy for me. And you know I started off in Brooklyn, my father gave me a small loan of a million dollars" (Kessler 2016). Similarly, while billionaires Jeff Bezos, Bill Gates, and Mark Zuckerberg are often characterized as self-made, all three received substantial support from their parents, including business loans and college tuition. In addition to financial support from family, what social and cultural factors do you think are most influential in determining social mobility to billionaire status?

Patterns of Poverty in the United States

Although poverty is not as widespread or severe in the United States as it is in many other parts of the world, the United States has the highest rate of poverty among wealthy countries belonging to the Organisation for Economic Cooperation and Development (OECD). In 2019, 34 million Americans—10.5 percent of the U.S. population—lived below the poverty line (Semega et al. 2020). The U.S. poverty rate showed a steady decline from 2014 to 2019 as the economic recovery from the Great Recession continued; however, experts warn that the economic aftermath of the global COVID-19 pandemic will contribute to rising poverty rates over the next few years (Giannarelli et al. 2020; World Bank 2020b). Poverty is not isolated to one segment of the population; between the ages of 25 and 60, nearly 62 percent of the population will spend at least one year in relative poverty, and 42 percent will experience a year of absolute poverty (Rank and Hirschl 2015).

> "Some people are born on third base and go through life thinking they hit a triple."
>
> **–AUTHOR UNKNOWN**

Age and Poverty. If the poverty statistics for adults are troubling, the statistics for children are even worse (see Figure 6.6). Approximately one in five children in the United States lives in poverty, which is one of the highest rates of child poverty among wealthy nations. In fact, among OECD countries, only Romania, Costa Rica, and South Africa have higher rates of child poverty than the United States (OECD 2020).

Childhood poverty is particularly problematic because "[g]rowing up in poverty can cast a shadow over the rest of a person's life" (Golden 2013, n.p.).

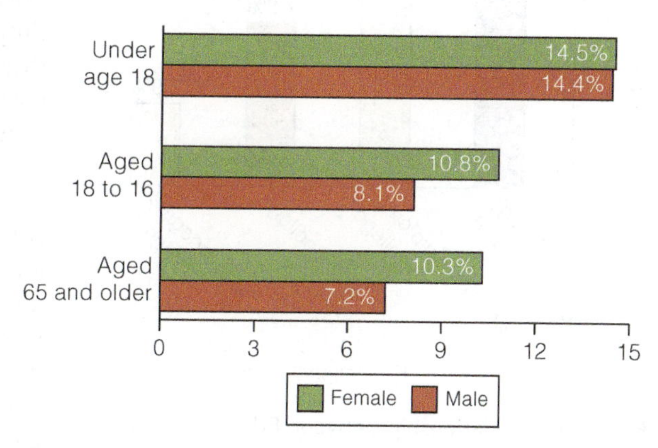

Figure 6.6 U.S. Poverty Rates by Age and Sex, 2019
SOURCE: Semega et al. 2020.

Sex and Poverty. Women are slightly more likely than men to live below the poverty line—a phenomenon referred to as the **feminization of poverty**. In 2019, 10.8 percent of women and 8.1 percent of men were living below the poverty line (Semega et al. 2020). As discussed in Chapter 10, women are less likely than men to pursue advanced educational degrees and tend to have low-paying jobs, such as service and clerical jobs. However, even with the same level of education and the same occupational roles, women still earn significantly less than men. Women who are of racial or ethnic minorities and/or who are single mothers are at increased risk of being poor.

feminization of poverty The disproportionate distribution of poverty among women.

Boaz Rotem/Alamy Stock Photo

The United States has one of the highest child poverty rates among industrialized nations.

Education and Poverty. Education is one of the best insurance policies for protecting an individual against living in poverty. In general, the higher a person's level of educational attainment is, the less likely that person is to be poor (see Figure 6.7). The relationship between educational attainment and poverty points to the importance of fixing our educational system so that students from all socioeconomic backgrounds have access to quality education

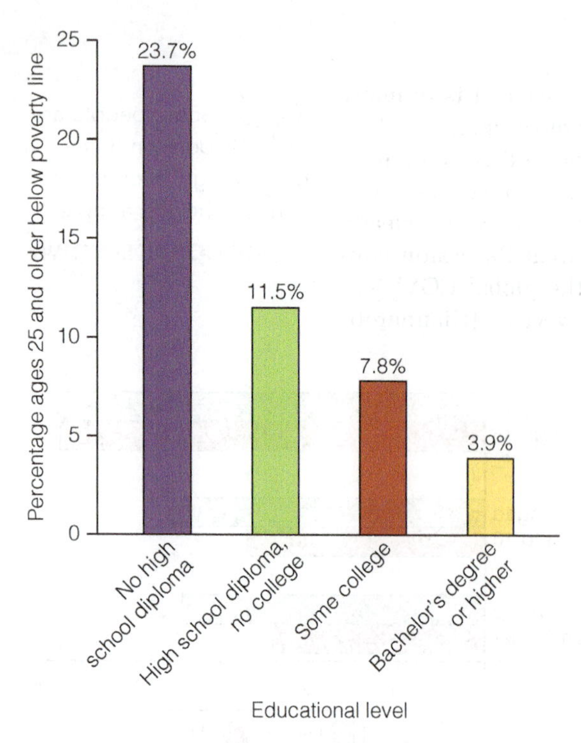

Figure 6.7 Relationship between Education and Poverty, 2019
SOURCE: Semega et al. 2020.

(see also Chapter 8). But we also need to consider the ways the economy is structured to provide opportunities for accessing both high-quality education and good-paying jobs.

Throughout most of the 20th century, the manufacturing sector provided good-paying jobs that secured a middle-class lifestyle for much of the workforce that did not have a college degree. However, the decline in manufacturing jobs, combined with the increasing costs of college in recent decades, has made it increasingly difficult for those who come from less economically privileged backgrounds to access the educational opportunities that are needed to provide a pathway out of poverty. Employment projections through 2020 predict that nearly two-thirds of new jobs will require a post–high school education (Carnevale et al. 2013). However, the costs of college are rising much faster than wages. For example, data from the National Center for Education Statistics (NCES 2020) show that, in 1985, the average annual tuition at a four-year college or university (in 2018 dollars) was $5,505. In 2018, that cost was $27,357—an increase of 397 percent! In response to these rising costs, more and more families take out student loans with the hope of later securing a good-paying job. For many, however, student debt presents just one more obstacle to getting out of poverty (see this chapter's *Human Side: College, Debt, and Economic Opportunity*). Wright and Rogers (2015) suggest that "poverty in a rich society does not simply reflect a failure of equal opportunity to acquire a good education; it reflects a social failure in the creation of sufficient jobs to provide an adequate standard of living for all people regardless of their education or levels of skills" (p. 286).

Labor Force Participation and Poverty. The lack of jobs that provide an adequate standard of living contributes to the numbers of Americans who are classified as **working poor**—individuals who spend at least 27 weeks per year in the labor force (working or looking for work) but whose income falls below the official poverty level. A common image of the poor is that they are jobless and unable or unwilling to work. Although the poor in the United States are primarily children and adults who are not in the labor force, the Bureau of Labor Statistics estimates that in 2018, 4.5 percent of people working in the labor force—approximately 7 million workers—had incomes that fell below the official poverty threshold (Bureau of Labor Statistics [BLS] 2020).

What do you think?

President Biden has vowed to increase the federal minimum wage from $7.25 an hour to $15 an hour. This would mean that the annual yearly income for a full-time, year-round minimum wage worker would increase from approximately $14,500 to about $30,000 per year. This would be slightly above the poverty threshold for a family of four. Do you think the federal minimum wage should be high enough to ensure that a household with at least one full-time worker doesn't fall below the poverty line? What do you think the minimum wage should be?

working poor Individuals who spend at least 27 weeks per year in the labor force (working or looking for work) but whose income falls below the official poverty level.

Family Structure and Poverty. Poverty is much more prevalent among female-headed single-parent households than among other types of family structures (see Figure 6.8). In other industrialized countries, poverty rates of female-headed families are lower than those in the United States. Unlike the United States, other developed countries offer a variety of supports for single mothers, such as income supplements, tax breaks,

College, Debt, and Economic Opportunity

The American Dream is premised on the idea of meritocracy: work hard and get a good education, and you can secure a comfortable middle-class life. However, the rising costs of college has meant that this pathway to economic security is increasingly out of reach to those coming from low-income backgrounds. The following stories highlight the obstacles to achieving economic security from those who relied on student loans to get a college education.

> The first in my family to attain a graduate degree, I did so only by borrowing heavily (my cumulative student loan debt exceeds $200,000). Average pay in my profession is approximately $40,000. … My monthly payment amount, supposedly based upon income, exceeds my monthly rent on my apartment—over $800.00 per month. There simply is no way to afford this. … College costs keep increasing every year—even at public universities and colleges, and the student loan companies are the ones profiting, rather than college/university students like myself, who mistakenly believed that a higher education was the key to success.
>
> —Brian S., Texas

I went into my master's program thinking I would come out with a great job—but the recession happened, and it has taken me nearly 10 years to get where I should have been 5 years ago. … I worked hard, but I didn't make much, and during my phase of unemployment I had to defer loan payments. … Education is only getting more expensive, meaning there are fewer opportunities for regular American kids who want to excel, who want to be a part of society, work hard and contribute to the greater good. We cannot shackle another generation to crippling student debt and assume that all will be well.

—Elizabeth P., Texas

I am currently in school for my graduate degree, and have come to the conclusion that student debt never goes away unless you are rich or you hit the lottery. I am already $20K in the hole and will have more debt owed once I complete my graduate degree. I still live at home with my parent, I am not able to make an adequate living based off of what I get paid now, and I can barely afford healthcare. … The American Dream: get an education, start a career, have a family and live life. How are we supposed to do that when one student loan payment for some of us is over $300 a month? On a salary of less than $40K? I really hope that these politicians who are supposed to work for us in office think about what it is like for the average American student to pay for college.

—Ashley C., California

Student debt has hurt both our sons and ourselves. Because we were working middle class people, we got little student aid help, just loans. This debt, added to our home mortgage, was an essential element in losing our home at the 2008 crash. Our children are harnessed also with massive college debt. It's a travesty and something must be done about it.

—Ellen P., Ohio

I have over $20,000 in debt for a degree that will only qualify me for a career that pays no more than just over minimum wage. I ask how that is to improve my future and that of my children when I couldn't find a full-time job for over a year. The payments are unaffordable on a loan that size and would take over 10 years to pay off.

—Naomi S., Illinois

SOURCE: Higher Ed Not Debt 2018.

universal child care, national health care, and higher wages for female-dominated occupations.

In general, same-sex couples are more likely than heterosexual couples to be poor, even when controlling for well-known poverty predictors like education and race (Schneebaum and Badgett 2019). Poverty rates among households headed by same-sex couples are more closely tied to gender than to sexual orientation; same-sex couples are more likely to be poor than comparable different-sex married couples, and lesbian couples are more likely to be poor than unmarried different-sex couples.

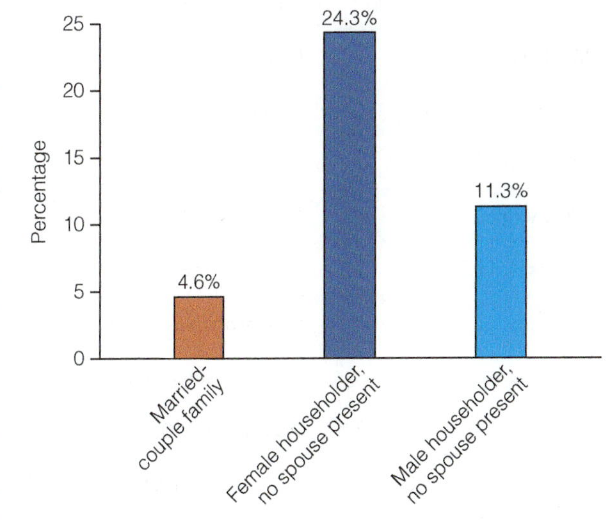

Figure 6.8 U.S. Poverty Rates by Family Structure, 2019
SOURCE: Semega et al. 2020.

Race or Ethnicity and Poverty. As displayed in Figure 6.9, poverty rates are higher among racial and ethnic minority groups than among non-Hispanic White people. As discussed in Chapter 9, past and present discrimination has contributed to the persistence of poverty among minorities. Other contributing factors include the loss of manufacturing

@Tay Zonday

Being poor now just leads to being more poor later. Can't pay to clean your teeth? Next year, pay for a root canal. Can't pay for a new mattress? Next year, pay for back surgery. Can't pay to get that lump checked out? Next year, pay for stage 3 cancer. Poverty charges interest.

-Tay Zonday

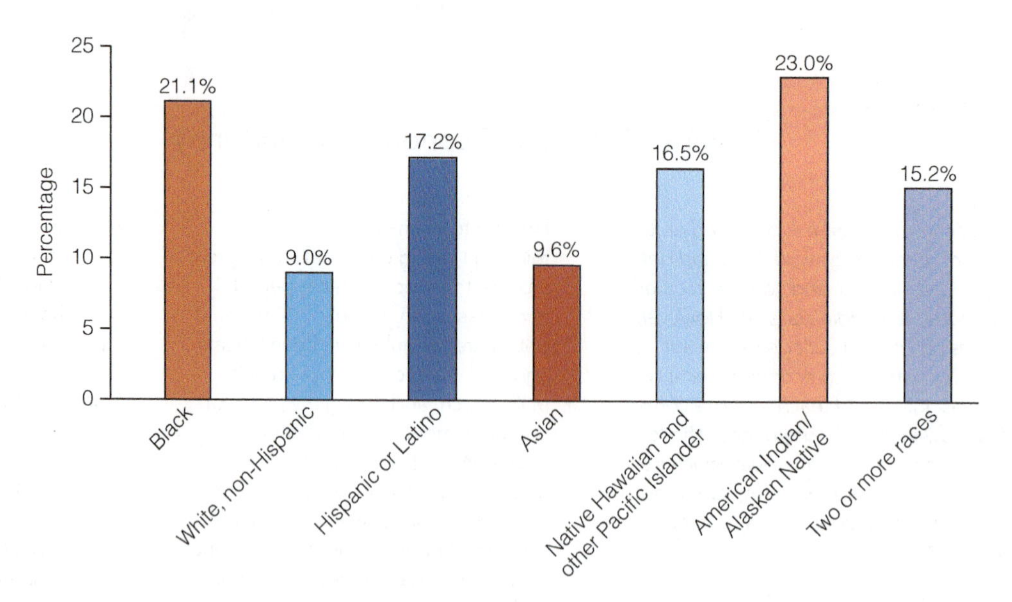

Figure 6.9 U.S. Poverty Rates by Race and Hispanic Origin, 2019
SOURCE: U.S. Census Bureau 2019b.

jobs from the inner city, the movement of White and middle-class Black families out of the inner city, and the resulting concentration of poverty in predominantly minority inner-city neighborhoods (Massey 1991; Wilson 1987, 1996). Finally, Black and Hispanic families are more likely to live in female-headed households with no spouse present—a family structure that is associated with high rates of poverty.

Region and Poverty. Poverty rates vary considerably by region of the United States, with the highest rates being in the South and West, and the lowest rates being in the Northeast. In 2015, the rates of poverty ranged from a low of 8.2 percent in New Hampshire to a high of 24.0 percent in Mississippi (see Table 6.3).

Poverty is increasingly found in the suburbs. The number of suburban poor surpassed the number of urban poor in the 2000s; there are now 3 million more poor people living in suburbs than living in cities (Maher 2018). Compared with poor urban dwellers, the suburban poor have less access to public transit and social safety-net programs.

TABLE 6.3 States with the Highest and Lowest Poverty Rates, 2018

Highest Poverty Rates (%)	Lowest Poverty Rates (%)
Mississippi: 19.7	New Hampshire: 7.6
New Mexico: 19.5	Maryland: 9.0
Louisiana: 18.6	Utah: 9.0
West Virginia: 17.8	New Jersey: 9.5
Arkansas: 17.2	Colorado: 9.6
Kentucky: 16.9	Minnesota: 9.6
Alabama: 16.8	Massachusetts: 10.0
District of Colombia: 16.2	Washington: 10.3
South Carolina: 15.3	Connecticut: 10.4
Tennessee: 15.3	North Dakota: 10.7

SOURCE: U.S. Census Bureau 2019.

Consequences of Economic Inequality and Poverty

From one point of view, economic inequality and poverty are problematic because they contradict the values of fairness, justice, and equality of opportunity, and they constitute a moral violation of basic human rights. Economic inequality and poverty are also viewed as problems because they have economic and social consequences that affect the whole society. For example, when income is concentrated toward the top, less money circulates in the local economy because, while money earned by low- and middle-income households is likely to be spent on goods and services that benefit the local economy, money among the rich is often invested in other regions and spent on luxuries (Talberth et al. 2013). The larger the segment of the population that is in the lowest income brackets, the more our society is affected by problems that plague the poor, but that also affect us all—problems discussed in the following sections.

Health Problems, Hunger, and Poverty

In developing countries, absolute poverty is associated with high rates of maternal and infant deaths, indoor air pollution from heating and cooking fumes, and unsafe water and sanitation (see also Chapter 2). In wealthy countries, such as the United States, we take for granted the availability of bathrooms and toilets, as well as safe drinking water that is piped into our homes. Globally, about one in three people do not have access to safe drinking water, and six in ten people do not have access to improved sanitation facilities—those that ensure hygienic separation of human excreta from human contact (WHO/UNICEF Joint Monitoring Programme for Water Supply and Sanitation 2019). Lack of access to clean water and sanitation facilities is a major cause of disease and death. Inadequate sanitation and hygiene contribute to the spread of diseases such as Ebola, and poor sanitation causes diarrheal diseases, which along with malnutrition, are among the leading cause of death among young children in developing countries (World Health Organization [WHO] 2020).

Living in poverty is also linked to hunger and malnourishment. In 2019, approximately 9 percent of the global population was chronically undernourished, with the highest rate of hunger in sub-Saharan Africa, where 19 percent are undernourished (Food and Agriculture Organization 2020). Inadequate nutrition hampers the ability to work and generate income, and it can produce irreversible health problems such as blindness (from vitamin A deficiency) and physical stunting (from protein deficiency).

Hunger in the United States is measured by the percentage of households that are "food insecure," which means that the household has difficulty providing enough food for all its members due to a lack of resources. In 2019, nearly 10.5 percent of U.S. households were food insecure at some time during the year (Coleman-Jensen et al. 2020). The COVID-19 global pandemic has exacerbated food insecurity in the United States and around the world. Feeding America estimates that in 2020, one in six adults and one in four children experienced food insecurity. Assess your own degree of food security in this chapter's *Self and Society: Food Security Scale*.

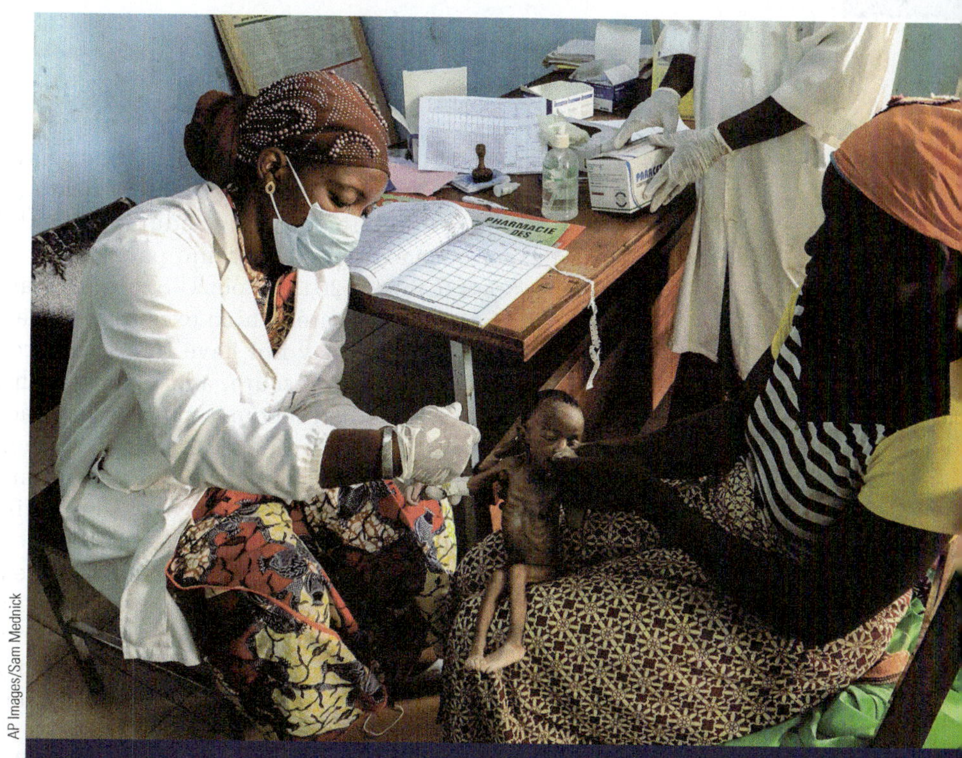

AP Images/Sam Mednick

High rates of child malnourishment in sub-Saharan Africa are expected to worsen as the COVID-19 global pandemic isolates small villages from food markets and medical care.

SELF and society: Food Security Scale

The U.S. Department of Agriculture conducts national surveys to assess the degree to which U.S. households experience food security, food insecurity, and food insecurity with hunger. To assess your own level of food security, respond to the following items and use the scoring key to interpret your results:

1. In the last 12 months, the food that (I/we) bought just didn't last, and (I/we) didn't have money to get more.
   - **a. Often true**
   - **b. Sometimes true**
   - c. Never true
2. In the last 12 months, (I/we) couldn't afford to eat balanced meals.
   - **a. Often true**
   - **b. Sometimes true**
   - c. Never true
3. In the last 12 months, did you ever cut the size of your meals or skip meals because there wasn't enough money for food?
   - **a. Yes**
   - b. No (skip Question 4)
4. If you answered yes to Question 3, how often did this happen in the last 12 months?
   - **a. Almost every month**
   - **b. Some months but not every month**
   - c. Only 1 or 2 months
5. In the last 12 months, did you ever eat less than you felt you should because there wasn't enough money to buy food?
   - **a. Yes**
   - b. No
6. In the last 12 months, were you ever hungry but didn't eat because you couldn't afford enough food?
   - **a. Yes**
   - b. No

Scoring and Interpretation

The answer responses in **boldface type** indicate affirmative responses. Count the number of affirmative responses you gave to the items, and use the following scoring key to interpret your results.

Number of Affirmative Responses and Interpretation

0 or 1 item: *Food secure* (In the last year, you have had access to enough food for an active, healthy life.)

2, 3, or 4 items: *Food insecure* (In the last year, you have had limited or uncertain availability of food and have been worried or unsure you would get enough to eat.)

5 or 6 items: *Food insecure with hunger evident* (In the last year, you have experienced more than isolated occasions of involuntary hunger as a result of not being able to afford enough food.)

If you scored as food insecure (with or without hunger), you might consider exploring whether you are eligible for public food assistance (e.g., food stamps) or whether there is a local food assistance program (e.g., food pantry or soup kitchen) that you could use.

SOURCE: Based on the short form of the 12-month Food Security Scale found in Bickel et al. 2000.

In the United States, low-wage earners have higher rates of obesity, hypertension, diabetes, arthritis, and premature death (Buszkiewicz et al. 2020). Compared with middle- and upper-income adults, U.S. adults living in poverty are more likely to experience extreme or chronic pain, worry and mental distress, sadness, anger, and stress (Graham 2015). Poor U.S. children and adults tend to receive inadequate and inferior health care, which exacerbates their health problems. Minimal income means that people may not have funds to purchase medicine to control their cholesterol, high blood pressure, and other health problems. As discussed in Chapter 2, people with limited incomes may not have access to or be able to afford healthier foods such as fresh produce, which tend to be more expensive than processed convenience foods that are higher in calories, sugar, salt, and fats. Finally, many people partially assess their self-worth based on their income, and long-term feelings of low self-worth also have negative consequences for health (Leigh and Bruen 2013).

@ProfRG Wilkinson

Where people are ranked by social status there will always be discrimination and prejudice against those lower down as we judge each other by position. The solution is to get rid of the material inequality which creates the framework for hierarchy, ranking & downward prejudice.

-Richard G Wilkinson

Economic inequality is also linked to a wide range of health and social problems. Epidemiologists Richard Wilkinson and Kate Pickett have studied the relationship between inequality and health for decades and found that even in very wealthy countries, there is a clear association between greater inequality and a wide range of problems including overall death rates, maternal mortality, heart disease, suicide rates, domestic violence, literacy, mental illness, drug and alcohol addiction, incarceration rates, and social trust (Wilkinson and Pickett 2009, 2020). These findings hold true among even middle- and high-income individuals who live in highly unequal societies. In other words, inequality harms *everyone* in a highly unequal society, not just those at the bottom of the ladder. Wilkinson and Pickett explain this association as an outcome of *social evaluative threat*, in other words, inequality increases stress and anxiety, which contribute to health problems (through increased levels of the stress hormone cortisol) and social problems (through feelings of inadequacy and depression).

Carmen K. Sisson/Cloudybright/Alamy Stock Photo

Approximately 6 million American households are considered to be living in substandard housing, meaning they lack sufficient kitchen, plumbing, and structural integrity to be considered safe for residence (National Center for Healthy Housing [NCHH] 2020).

Substandard Housing and Homelessness

Having a roof over one's head is considered a basic necessity. However, for the poor, that roof may be literally caving in. In addition to having leaky roofs, housing units of the poor often have holes in the floor and open cracks in the walls or ceiling. Low-income housing units often lack central heating and air conditioning, sewer or septic systems, and electric outlets in one or more rooms. Housing for the poor is also often located in areas with high crime rates and high levels of pollution.

Concentrated areas of poverty and poor housing in urban areas are called **slums**. One-third of urban populations in developing regions are living in slums. The United Nations estimates that 1 billion people in the world live in slums (UN-Habitat 2018).

Over the course of a lifetime, an estimated 6 percent of the U.S. population will experience homelessness, however, these rates vary dramatically by race (Fusaro et al. 2018). Nearly 5 percent of White people, 17 percent of Black people, and 8 percent of Hispanic people experience homelessness at some point in their lives. In January 2019, there were 567,715 homeless people in the United States: of those, 30 percent were people in families, 6.5 percent were veterans, and almost 6 percent were unaccompanied youth (National Alliance to End Homelessness 2020). Although the majority of the homeless population stays in shelters or transitional housing, about a third lives on the street, in a car, in an abandoned building, or in places not meant for human habitation, such as storage units or makeshift dwellings made of a variety of discarded materials such as pieces of wood and boards, cardboard, mattresses, fabric, and plastic tarps. This chapter's *Social Problems Research Up Close: Housing First* examines the effectiveness of Housing First policies in alleviating homelessness.

slums Concentrated areas of poverty and poor housing in urban areas.

Many cities have installed "hostile architecture" in an effort to deter homeless people from sleeping or resting in public spaces. Have you noticed hostile architecture in your community? Do you think these features improve public safety, or do they unfairly target people experiencing homelessness?

What *do you* think?

social problems RESEARCH UP CLOSE

Housing First

What is the most effective way to end homelessness? Traditional approaches to addressing homelessness, particularly among those who have experienced long-term or repeated homelessness, have focused on providing housing on a conditional basis. These conditions may include requiring an individual to attend drug and alcohol addiction treatment, attend job training, or take part in mental health support programs, in order to receive or maintain housing assistance. In contrast, a Housing First (HF) approach "views housing as the foundation for life improvement and enables access to permanent housing without prerequisites beyond those of a typical renter" (National Alliance to End Homelessness 2016, p. 1). While support services for drug treatment, mental health, and job placement are available to participants in HF programs, participation in these services is not mandated as research indicates that these services are more effective when individuals engage in them voluntarily.

A wealth of research has established that the Housing First approach is more cost-effective and leads to more sustainable pathways out of poverty than traditional shelter and service-conditional housing programs (see Ly and Latimer 2015 for a review). As HF gains traction in cities around the United States, service providers and funders are increasingly concerned with the complexities of large-scale implementation of these programs across a wide range of geographies, housing markets, and service agencies.

The veterans' housing program through the Department of Housing and Urban Development and the Department of Veterans' Affairs Supportive Housing services (referred to as HUD-VASH) serves 85,000 veterans experiencing housing instability. Beginning in 2012, the Department of Veterans' affairs issued a policy change mandating that HUD-VASH shift to a Housing First model by eliminating all preconditions for housing services and prioritizing housing for the most vulnerable and chronically homeless veterans.

The national scale of this policy offered a team of researchers affiliated with the Department of Veterans' Affairs a "naturalistic window" to study a large-scale and organizationally complex implementation of Housing First (Kertesz et al. 2017). The results of this study provide insight for other agencies, state, and municipal governments into the most effective strategies in HF implementation.

Sample and Methods

The research team selected a sample of eight VA medical centers (VAMCs). At each center, they conducted interviews of leadership and staff working in housing support services. The sample was selected purposively in order to ensure appropriate variation in geographic region, VAMC size, local rental market conditions, and size of the local homeless and veteran populations. Interviews took place twice: first, at the beginning of the policy change in late 2011 and early 2012. The interviewers conducted a follow-up interview 12 to 16 months later, in late 2013 and early 2014. A total of 175 interviews were conducted over the two cycles.

The interviews focused on assessing the extent to which each VAMC housing support services adhered to HF criteria, referred to as program fidelity. These criteria, drawn from previous validated research, identified 20 key criteria under five domains: (1) no sobriety or treatment preconditions; (2) rapid placement into

Examples of hostile architecture include metal spikes next to sidewalks, armrests in the middle of bench seats, and doorway awnings with gaps to allow rain to drip through. These features are designed to make it difficult for people who lack shelter to rest in public spaces.

Jansos/Alamy Stock Photo

Causes of homelessness include poverty, unemployment, eviction, domestic violence, mental illness and substance abuse, and lack of affordable housing. Housing is considered affordable when a household pays no more than 30 percent of its income on housing expenses. In 2020, 8 million renters paid more than half of their income on housing (National Alliance to End Homelessness 2020).

For people living on the street, every day can be a struggle for survival. In recent years, there has been a surge in unprovoked violent and sometimes fatal attacks against homeless individuals (National Coalition for the Homeless 2016). In most cases, the attacks are by teenage and young adult males. Many acts of violence toward the homeless are not reported to the police, so documented cases may be just the tip of the iceberg. During the years he lived homeless on the street, David Pirtle was attacked five times, and he did not report the attacks to police. "I was struck on the back, kicked, urinated on, spray-painted. ... A lot of people who are homeless go through it, and it's just the way it is" (quoted in Dvorak 2009, p. DZ01).

permanent housing; (3) prioritization of the most vulnerable homeless clients; (4) sufficient supportive services available in a community context; and (5) a modern recovery philosophy guiding all services.

The researchers used a directed content analysis approach to study the interview field notes. They scored the interviews using the five domains and 20 criteria of Housing First. Researchers assigned program fidelity scores to each site ranging for each of the criteria from 4 (element of HF is present and consistently used as intended) to 1 (element of HF is not present or only partly in place and not being used as intended). Based on the follow-up interviews, the researchers then determined the extent to which the VAMC demonstrated a change in fidelity, ranging from high performance (baseline fidelity showed substantial improvement) to low performance (baseline fidelity showed substantial decline).

Results

The results of this analysis revealed that a Housing Approach has many advantages in effectively addressing problems of chronic homelessness, as well as indicating some challenges in implementation.

The domain with strongest fidelity across the research sites was in removing sobriety and treatment preconditions. However, the ease of transitioning to this mandate was primarily driven by the lack of resources for clinical support services. Most of the VAMCs fell short in meeting the criteria for case manager ratios and staff training. Fidelity to a modern recovery philosophy was also low, largely because staff did not have a sufficient understanding of concepts like harm reduction and motivational interviewing techniques.

Although most VAMCs demonstrated high fidelity to prioritizing the most chronically homeless clients, they struggled to identify those with the highest levels of medical or psychological vulnerability. As these populations require the most consistent and intensive support services, the lack of sufficient clinical support services and staff training, combined with an insufficient prioritization of the most vulnerable, contributes to a key vulnerability in widespread implementation of Housing First.

The researchers present three key ingredients for HF success learned from this study that communities seeking to adopt a Housing First approach need to incorporate: resources, guidance, and leadership.

First, the biggest challenge to long-term program effectiveness is upholding clinical recovery. This requires that sufficient funding and managerial leadership be in place before program implementation.

Second, clear, consistent, and centralized bureaucratic guidance is essential for success. A key advantage of the VA system is its ability to provide formal manuals, training, and requirement checklists on a national scale.

Finally, a transformation in policy requires credible leadership mandating a change from the top down. The hierarchical structure of the VA gives it an advantage in enacting swift and consistent policy change. The researchers warn that the coordinating bodies seeking to implement changes to housing policy in most communities "are not likely to have sufficient political power to deliver such major change unless their efforts are linked to and buttressed by the leverage that can come from political leaders, business collaborative, philanthropic agencies (e.g. United Way), and others, working together" (Kertesz et al. 2017, p. 126).

Despite the documented benefits of Housing First approaches to ending homelessness, its success depends upon the determination of leaders to allocate resources to housing and to overturn traditional ideas about what it means to deserve, or be "ready," for housing.

What *do you* think?

Under hate crime laws, violators are subject to harsher legal penalties if their crime is motivated by the victim's race, religion, national origin, or sexual orientation. The number of fatal attacks on the homeless is nearly triple the number of hate crime deaths based on race, religion, ethnicity, and sexual orientation combined (National Coalition for the Homeless 2020). A handful of states have added homeless status to their hate crime law, and several cities and counties have also taken measures to recognize homeless status in their laws or procedures. Proposed legislation to add homelessness to the federal hate crime law has not, as of this date, passed. Do you think that violent acts toward homeless individuals should be categorized as hate crimes and be subject to harsher penalties? Why or why not?

Legal Inequality

The American ideal of "justice for all" may be more accurately described as "justice for those who can afford to pay for it." Many poor defendants are held in pretrial detention because they cannot afford to post bail (Human Rights Watch 2020). Although the Supreme Court ruled in 1963 (*Gideon v. Wainwright*) that criminal defendants who cannot afford to hire an attorney have the constitutional right to a public defense, public defender offices are overworked and underfunded, and public defenders often spend only minutes per case due to their unrealistic caseloads (Giovanni and Patel 2013). Without the resources for effective legal representation, poor defendants often accept unfair plea bargains, and "the systemic result is harsher outcomes for defendants and more people tangled in our costly criminal justice system" (Giovanni and Patel 2013, p. 1). The economic inequality embedded in the U.S. legal system is problematic not only for the poor but for the entire society, as it contributes to the social and economic costs of mass incarceration in the United States (see also Chapter 4).

VStock/Alamy Stock Photo

The invisible wounds of war, including PTSD and traumatic brain injury, is a key factor contributing to housing instability among veterans.

Political Inequality and Alienation

Economic inequality also contributes to political inequality, as expressed in a version of the Golden Rule: "He who has the gold makes the rules." Although the United States represents itself as a democracy whose government represents all citizens, research shows that a small group of economic elite has more influence over political outcomes than do ordinary citizens (Gilens and Page 2014). Thus, instead of being a true democracy, the United States can be described as a **plutocracy**: a country governed by the wealthy. The poor and even middle classes feel that their interests are not represented by their elected politicians. In countries around the world, people at the bottom of the inequality spectrum often feel as though they do not have a say in the policies that govern them. Hence, those in the lower socioeconomic classes are vulnerable to experiencing **political alienation**—a rejection of or estrangement from the political system accompanied by a sense of powerlessness in influencing government. The poor face obstacles in running for political office, as money and connections are needed to run for office. The poor are less likely than the affluent to vote: In 2012, fewer than half of eligible voters with family incomes under $20,000 voted, compared with about 80 percent of those with annual incomes of $100,000 or more (Weeks 2014). The poor have a lower voting turnout than the wealthier segments of the population, in part due to political alienation, but also because of obstacles such as the difficulty taking time off from work, transportation problems getting to the polls, and lack of a required form of identification.

Crime, Social Conflict, and War

Poverty and economic inequality are linked to crime and violence (see also Chapter 4 and Chapter 16). For example, inequality predicts murder rates better than any other variable—including poverty and drug addiction (Daly 2017). Both within and between countries, approximately half of the variance in murder rates can be accounted for by level of inequality. The denial of a political voice or influence to masses of people at the bottom of the wealth distribution can cause social tensions, political instability, and violent conflict (United Nations 2020). Economic inequality and poverty are often root causes of conflict and war within and between nations. Poorer countries are more likely than wealthier countries to be involved in civil war, and countries that experience civil war tend to become and/or remain poor.

Armed conflict and civil war are generally more likely to occur in countries with extreme and growing inequalities among ethnic groups, and countries with higher levels of equality are more likely to be peaceful (Institute for Economics & Peace 2020). Not only does poverty breed conflict and war, but war also contributes to poverty. War devastates infrastructures, homes, businesses, and transportation systems. In the wake of war, populations often experience hunger and homelessness (see Chapter 15).

plutocracy A country governed by the wealthy.

political alienation A rejection of or estrangement from the political system accompanied by a sense of powerlessness in influencing government.

Natural Disasters, Economic Inequality, and Poverty

Although natural disasters such as hurricanes, tsunamis, floods, and earthquakes strike indiscriminately—rich and poor alike—poverty increases vulnerability to devastation from such disasters. In 2010, both Chile and Haiti experienced major earthquakes, but the damage in Haiti was much more severe, with the death toll magnitudes higher than those in Chile. The reason Haiti suffered more was, in part, due to the fact that Haiti is much poorer than Chile. Chileans had the advantage of having homes and offices with steel skeletons designed to withstand earthquakes—even low-income housing was built to be earthquake resistant. In contrast, there is no building code in Haiti, and homes crumbled and collapsed in the earthquake (Bajak 2010). Wealthy countries also have more resources than poor countries for natural disaster relief efforts, such as rebuilding infrastructure, providing medical care for the injured, and offering food and shelter for people who have lost their homes.

But even in wealthy countries, the poor are more vulnerable to natural disasters, while the more affluent have resources that enable them to cope. The effects of climate change present ever increasing risks of natural disaster—record hurricanes, wildfires, coastal flooding, droughts, and heat waves—which will ultimately push millions of Americans to move. Journalist Abrahm Lustgarten (2020) warns:

> Once you accept that climate change is fast making large parts of the United States uninhabitable, the future looks like this: ... Something like a tenth of the people who live in the South and the Southwest ... decide to move north in search of a better economy and a more temperate environment. Those who stay behind are disproportionately poor and elderly. In these places, heat alone will cause as many as 80 additional deaths per 100,000 people—the nation's opioid crisis, by comparison, produces 15 additional deaths per 100,000. The most affected people, meanwhile, will pay 20 percent more for energy, and their crops will yield half as much food or in some cases virtually none at all. That collective burden will drag down regional incomes by roughly 10 percent, amounting to one of the largest transfers of wealth in American history, as people who live farther north will benefit from that change and see their fortunes rise. (n.p.)

Educational Problems and Poverty

In many countries, children from the poorest households have little or no schooling, and enter their adult lives without basic literacy skills (see also Chapter 8). In the United States, children from low-incomes homes have lower levels of academic achievement than those from middle- and high-income homes (Olszewski-Kubilius and Corwith 2018). Children who grow up in poverty tend to receive lower grades, receive lower scores on standardized tests, are less likely to finish high school, and are less likely to attend or graduate from college than their nonpoor peers.

The poor often attend schools that are characterized by lower-quality facilities, overcrowded classrooms, and a higher teacher turnover rate (see also Chapter 8). Although other rich countries invest more money in education for disadvantaged children, the United States spends more on schools in wealthy districts because public schools are funded largely by local property tax money (Baker 2020).

AP Images/Ravell Call

Tens of thousands of homes have been destroyed by fire in California and other Western states in recent years, as climate changes have contributed to the increasing frequency and intensity of wildfires. People with low incomes bear a disproportionate burden of the harms caused by natural disasters.

Children who grow up in poverty suffer more health problems that contribute to their lower academic achievement. Because poor parents have less schooling on average than do nonpoor parents, they may be less able to encourage and help their children succeed in school. Children from poor households have limited access to high-quality preschools, books and computers, and enriching after-school and summer activities including tutoring, travel, lessons (music, dance, sports, etc.), and camps (Olszewski-Kubilius and Corwith 2018). With the skyrocketing costs of tuition and other fees, many poor parents cannot afford to send their children to college. Although some students have wealthy parents who write out tuition checks, other students are graduating from college with substantial college debt (see this chapter's *Human Side: College, Debt, and Economic Opportunity*).

Marriage Opportunity Gap and Family Problems Associated with Poverty and Economic Inequality

The erosion of legal prohibitions against same-sex marriage in recent years reflects, in part, the cultural value and belief that everyone should have the opportunity to marry and to have children within a stable, socially recognized family household. But poverty and economic inequality create a marriage opportunity gap. The top one-third of U.S. households are more likely to enjoy the benefits of being a "two-two-two-one household"—having two parents, two college degrees, two incomes, and one stable marriage (Blankenhorn et al. 2015). But among lower- and even middle-income households, marriage rates are low, divorce rates are high, and nonmarital childbearing rates are high (see also Chapter 5).

The stresses associated with low income contribute to substance abuse, domestic violence, and child abuse and neglect (see also Chapter 3 and Chapter 5). Children in poor households are more likely to experience harsh or neglectful parenting (Lanier et al. 2014). Without access to affordable child care and medical care, poor parents may leave their children at home without adult supervision or fail to provide needed medical care.

Poverty is also linked to teenage pregnancy and unintended childbearing. Poor women are more than five times as likely as affluent women to have an unintended birth because they are less likely to use contraception and are less likely to have an abortion once pregnant (Reeves and Venator 2015). Poor adolescent teenagers are at higher risk of having babies than their nonpoor peers. Early childbearing is associated with increased risk of premature babies or babies with low birth weight, dropping out of school, and lower future earning potential as a result of lack of academic achievement. Luker (1996) noted that "the high rate of early childbearing is a measure of how bleak life is for young people who are living in poor communities and who have no obvious arenas for success" (p. 189).

> Having a baby is a lottery ticket for many teenagers: It brings with it at least the dream of something better, and if the dream fails, not much is lost. ... In a few cases it leads to marriage or a stable relationship; in many others it motivates a woman to push herself for her baby's sake; and in still other cases it enhances the woman's self-esteem, since it enables her to do something productive, something nurturing and socially responsible. (Luker 1996, p. 182)

Intergenerational Poverty

Problems associated with poverty, such as health and educational problems, create a cycle of poverty from one generation to the next. Nearly half of U.S. children born to low-income parents will become low-income adults (McEwan and McEwan 2017). Poverty that is transmitted from one generation to the next is called **intergenerational poverty**.

intergenerational poverty Poverty that is transmitted from one generation to the next.

Intergenerational poverty creates a persistently poor and socially disadvantaged population, referred to as the underclass. Although the underclass is stereotyped as being composed of minorities living in inner-city areas, the underclass is a heterogeneous population that includes poor White people living in urban and nonurban communities (Alex-Assensoh 1995). Intergenerational poverty and the underclass are linked to a variety of social factors, including the decrease in good-paying jobs and their movement out of urban areas, the resultant decline in the availability of marriageable males able to support a family, dropping marriage rates and an increase in out-of-wedlock births, the migration of the middle class to the suburbs, and the effect of deteriorating neighborhoods on children and youth (Wilson 1987, 1996).

Strategies for Action: Reducing Poverty and Economic Inequality

Because poverty and economic inequality are primary social problems that cause many other social problems, strategies to reduce or alleviate problems related to poverty and economic inequality include those discussed in other chapters of this text that deal with such issues as health (Chapter 2), work (Chapter 7), education (Chapter 8), racial discrimination (Chapter 9), and gender discrimination (Chapter 10). Here we briefly outline a number of strategies aimed at reducing poverty and economic inequality in the United States and internationally.

International Responses to Poverty and Economic Inequality

In 2000, leaders from 191 United Nations member countries pledged to achieve eight Millennium Development Goals (MDGs)—an international agenda for reducing poverty and improving lives. One of the MDGs was to halve, between 1990 and 2015, the proportion of people who live in severe poverty and who suffer from hunger. The MDG poverty reduction goal was met in 2010—five years ahead of schedule. The MDGs expired at the end of 2015 and were replaced with a new, expanded set of 17 goals, collectively called the **Sustainable Development Goals (SDGs)** (see Table 6.4).

In 2014, Oxfam International launched a global campaign called Even It Up, calling for governments, corporations, and institutions to reduce economic inequality. Next we discuss some approaches for reducing poverty and economic inequality throughout the world.

Taxes on the Wealthy. Oxfam (2020) calculated that adding a 1.5 percent tax on the world's billionaires could raise $74 billion in tax revenue—enough to fill the annual gaps in funding needed to provide education to every child and deliver health care services in the poorest 49 countries. Another way to gain tax revenue that could both alleviate poverty and reduce economic inequality is to close tax loopholes that enable corporations and the wealthy to avoid paying taxes on much of their income. Each year developing countries lose an estimated $1 billion to corporations that use tax havens, tax breaks, and exemptions (Oxfam 2017).

Economic Development. Increasing the economic output or the gross domestic product of a country can bring needed economic resources into poor countries. However, economic development does not always reduce poverty; in some cases, it increases it. Policies that involve cutting government spending, privatizing basic services, liberalizing trade, and producing goods primarily for export may increase economic growth at the national level, but the wealth ends up in the hands of the political and corporate elite at the expense of the poor. Economic growth does not help poverty reduction when public spending is diverted away from meeting the needs of the poor and instead is used to pay international debt, to finance military operations, and to support corporations that do not pay workers fair wages.

Another problem with economic development is that the environment and natural resources are often destroyed and depleted in the process of economic growth. The environmental problems caused by economic growth can be minimized if governments and corporations embrace **green growth**, which is economic growth that is environmentally

Sustainable Development Goals A set of 17 goals that comprise an international agenda for reducing poverty and economic inequality and improving lives.

green growth Economic growth that is environmentally sustainable.

TABLE 6.4 Sustainable Development Goals

Goal
Goal 1. End poverty in all its forms everywhere.
Goal 2. End hunger, achieve food security and improved nutrition and promote sustainable agriculture.
Goal 3. Ensure healthy lives and promote well-being for all at all ages.
Goal 4. Ensure inclusive and equitable quality education and promote lifelong learning opportunities for all.
Goal 5. Achieve gender equality and empower all women and girls.
Goal 6. Ensure availability and sustainable management of water and sanitation for all.
Goal 7. Ensure access to affordable, reliable, sustainable and modern energy for all.
Goal 8. Promote sustained, inclusive and sustainable economic growth, full and productive employment and decent work for all.
Goal 9. Build resilient infrastructure, promote inclusive and sustainable industrialization and foster innovation.
Goal 10. Reduce inequality within and among countries.
Goal 11. Make cities and human settlements inclusive, safe, resilient and sustainable.
Goal 12. Ensure sustainable consumption and production patterns.
Goal 13. Take urgent action to combat climate change and its impacts.
Goal 14. Conserve and sustainably use the oceans, seas and marine resources for sustainable development.
Goal 15. Protect, restore and promote sustainable use of terrestrial ecosystems, sustainably manage forests, combat desertification, and halt and reverse land degradation and halt biodiversity loss.
Goal 16. Promote peaceful and inclusive societies for sustainable development, provide access to justice for all and build effective, accountable and inclusive institutions at all levels.
Goal 17. Strengthen the means of implementation and revitalize the global partnership for sustainable development.

SOURCE: United Nations 2014.

sustainable (World Bank Group and International Monetary Fund 2015) (see also Chapter 13). Recent analyses, however, have questioned the potential for green growth arguing that it is a "misguided objective" that relies on overly optimistic assumptions about our technological abilities to promote increases in GDP while simultaneously reducing carbon emissions (Hickel and Kallis 2020). Other critics have argued that wealthy countries put an unfair burden on poorer countries to implement green growth strategies, when wealthy countries achieved their own economic rise without using sustainability strategies and continue to consume most of the world's natural resources and are responsible for the majority of the world's carbon footprint (Iqbal and Pierson 2017).

Human Development. Unlike the economic development approach to poverty alleviation, the human development approach views people—not money—as the real wealth of a nation:

> The central contention of the human development approach... is that well-being is about much more than money. ... Income is critical but so are having access to education and being able to lead a long and healthy life, to influence the decisions of society, and to live in a society that respects and values everyone. (UNDP 2010, p. 114)

In many poor countries, large segments of the population are illiterate and without job skills and/or are malnourished and in poor health. Investments in human development involve programs and policies that provide adequate nutrition, sanitation, housing, health care (including reproductive health care and family planning), and educational and job training.

Microcredit Programs. The old saying "It takes money to make money" explains why many poor people are stuck in poverty: They have no access to financial resources and services. **Microcredit programs** refer to the provision of loans to people who are generally excluded from traditional credit services because of their low socioeconomic status. Microcredit programs give poor people the financial resources they need to become self-sufficient and to contribute to their local economies.

The Grameen Bank in Bangladesh, started in 1976, has become a model for the more than 3,000 microcredit programs that have served millions of poor clients (Roseland and Soots 2007). One international comparison of microcredit programs in Asia, Africa, Latin America, and Europe found that microcredit significantly reduced poverty and also contributed to improvements in nutrition, health, education, and employment (Saeed and Jan 2016).

Reducing U.S. Poverty and Economic Inequality

Strategies to reduce poverty and economic inequality include those discussed in other chapters, such as improving the quality and *equality* of health care and education to ensure that these services and resources are not unfairly skewed toward children and young adults from more affluent families (see Chapter 2 and Chapter 8). Chapter 7 discusses strategies that also impact poverty and economic inequality, such as job creation and training, and strengthening labor unions. Here, we address issues concerning minimum wage and living wages, tax reform efforts, political reforms, and efforts to reduce wage theft.

Tax Reforms. The wealthy investor Warren Buffet pointed to the unfairness of the U.S. tax system when he famously remarked that he paid a lower tax rate than his secretary because his capital gains earnings are taxed at a lower rate than ordinary income. While most politicians agree that the tax system needs fixing—that it is currently too complicated and/or unfair—there is ongoing partisan disagreement about how to reform the tax system.

One way to reduce the gap between the top and the bottom of the economic system is to make the tax system more progressive. **Progressive taxes** are those in which the tax rate increases as income increases, so that those who have higher incomes are taxed at higher rates. A more progressive tax system would increase taxes on the wealthy. Other tax reforms that could help reduce economic inequality include increasing estate taxes (labeled by opponents as "death taxes") and gift taxes, as well as capital gains taxes. Other tax reform proposals include limiting itemized deductions for the wealthy (such as the mortgage interest deduction), increasing the cap on Social Security taxes (in 2020, Social Security taxes applied only to the first $137,700 of earned income), and closing corporate tax loopholes that enable many corporations to pay less than their "fair share" of taxes. Increasing taxes on corporations and the rich does not necessarily result in simple redistribution of income or wealth from the rich to the poor. Rather, revenue from increasing taxes on the rich could be directed to public projects that would provide more equal access to education, health care, public transportation, and other services that would give low-income Americans the resources they need to improve their economic situation (McNamee and Miller 2009).

microcredit programs The provision of loans to people who are generally excluded from traditional credit services because of their low socioeconomic status.

progressive taxes Taxes in which the tax rate increases as income increases, so that those who have higher incomes are taxed at higher rates.

What do you think?

Former Presidential candidate and United States Senator Bernie Sanders has repeatedly called for a tax on what he calls "extreme wealth": an increase of the tax rate between 2 to 8 percent on all net worth above $32 million for a married couple. In other words, a household with less than $32 million in net worth would not see any tax increase under this plan. Do you consider a net worth of $32 million to be "extreme wealth"? Where would you draw the line of "extreme wealth" at which point a household would be required to pay higher taxes?

Political Reform. Given the unfair advantage the wealthy have in influencing the political process, another key strategy in reducing economic inequality is to reduce the influence of money in politics. Some lawmakers are calling for an amendment to the Constitution to overturn the 2010 *Citizens United* ruling that gave corporations unlimited political spending power. Reducing the political inequality that perpetuates economic inequality also necessitates enacting limits on the amount of money that wealthy individuals can spend on politics.

Minimum Wage Increase and "Living Wage" Laws. As of this writing, the federal minimum wage is $7.25. At this hourly rate, a person working full-time with two children earns $14,500 (before taxes)—an income well below the poverty line. Although the federal minimum wage has not been increased since 2009, costs of living continue to rise. In other words, each year that the minimum wage does not increase, low-wage workers are essentially taking a pay cut.

Many cities and counties throughout the United States have **living wage laws** that require state or municipal contractors, recipients of public subsidies or tax breaks, or, in some cases, all businesses to pay employee wages that are significantly above the federal minimum, enabling families to live above the poverty line. Research findings show that businesses that pay their employees a living wage have lower worker turnover and absenteeism, reduced training costs, higher morale and productivity, and a stronger consumer market (Kraut et al. 2000). In 2012, the advocacy group Fight for 15 began organizing fast food workers to go on strike and demand an increase to state minimum wages to $15 per hour—what they considered to be a living wage. Although the federal minimum wage has not increased, as of January 1, 2021, 29 states and the District of Columbia have mandated a minimum wage that is higher than the federal $7.25, and 18 states have minimum wages of at least $10 per hour.

Reduce Wage Theft. **Wage theft** is the failure to pay what workers are legally entitled to. Wage theft occurs in a variety of ways, including when employers require workers to work off the clock or refuse to pay them for overtime when their weekly work hours exceed 40. Wage theft also occurs when employers pay less than the state-mandated minimum wage or steal tips from service workers. Wage theft is widespread: In one survey, approximately 26 percent of workers reported being paid less than the minimum wage in the previous week, and nearly two-thirds reported at least one wage theft violation such as failure to be paid overtime, not being paid for all hours worked, and stolen tips (Huizar 2019). The Economic Policy Institute estimates that workers lose more than $15 billion per year to wage theft (Cooper and Kroeger 2017).

Because wage theft affects primarily low-wage workers, reducing this illegal practice would help lift the incomes of those at the bottom. Combating wage theft calls for increasing penalties to deter companies and employers from engaging in this practice; denying federal contracts to companies found guilty of wage and hour violations; and increasing the number of investigators working for the Department of Labor's Wage and Hour Division (Cooper and Kroeger 2017).

living wage laws Laws that require state or municipal contractors, recipients of public subsidies or tax breaks, or, in some cases, all businesses to pay employees wages that are significantly above the federal minimum, enabling families to live above the poverty line.

wage theft Occurs when employers "steal" workers' wages by requiring them to work off the clock or refusing to pay them for overtime.

means-tested programs Assistance programs that have eligibility requirements based on income.

earned income tax credit (EITC) A refundable tax credit based on a working family's income and number of children.

The Safety Net: Public Assistance and Welfare Programs in the United States

Public assistance, or "welfare," programs in the United States are aimed at providing a safety net for adults and children who are economically disadvantaged. Many assistance programs are **means-tested programs** that have eligibility requirements based on income and/or assets. Public assistance programs designed to help the poor include the earned income tax credit, Supplemental Security Income, Temporary Assistance for Needy Families, food programs, housing assistance, medical care, educational assistance, and child-care assistance.

Earned Income Tax Credit. The federal **earned income tax credit (EITC)**, created in 1975, is a refundable tax credit based on a person's income, marital status, and number of children. The EITC is designed to offset Social Security and Medicare payroll taxes on working poor families and to encourage and reward work; it lifts millions of U.S. children and adults out of poverty each year. About half the states have their own EITCs to supplement the federal EITC.

Supplemental Security Income. Supplemental Security Income (SSI), administered by the Social Security Administration, provides a minimum income to poor people who are age 65 or older, blind, or disabled. SSI is not the same as Social Security: A millionaire can collect Social Security, but a person must be either elderly or disabled *and* must have limited income and assets to collect SSI.

Temporary Assistance for Needy Families. In 1996, the U.S. social welfare system was dramatically changed with the passage of the *Personal Responsibility and Work Opportunity Reconciliation Act* (PRWORA), which replaced the cash assistance program Aid to Families with Dependent Children (AFDC) with a new program, **Temporary Assistance for Needy Families (TANF)**. Although the previous AFCD program provided a more reliable safety net to the poorest of Americans, the current TANF program is a cash assistance program for the poor that offers more limited assistance, with time limits and work requirements. Within two years of receiving benefits, adult TANF recipients must be either employed or involved in work-related activities, such as on-the-job training, job search, and vocational education. A federal lifetime limit of five years is set for families receiving benefits, and able-bodied recipients age 18 to 50 without dependents have a two-year lifetime limit. Some exceptions to these rules are made for individuals with disabilities, victims of domestic violence, residents in high unemployment areas, and those caring for young children. The success of the TANF program is measured not by how many low-income families move into careers that provide a living wage but rather by the number of people leaving the TANF program, regardless of their reason for doing so or their well-being thereafter (Green 2013).

Food Assistance. The largest food assistance program in the United States is the **Supplemental Nutrition Assistance Program (SNAP)** (formerly known as the Food Stamp Program), followed by school meals and the Special Supplemental Nutrition Program for Women, Infants, and Children (WIC). SNAP issues monthly benefits through coupons or an electronic debit card. The U.S. Census Bureau (2020) reported that nearly one in five households with children receives SNAP benefits.

In 2019, the average benefit for an individual receiving SNAP was equal to $129.83 per month, or about $4.32 per day (USDA Food and Nutrition Service 2020). Unemployed adults aged 18 to 49 who are not physically or mentally disabled or caring for a minor child are allowed to receive only three months of SNAP benefits in a three-year period, although states can allow a temporary waiver of the three-month limit in areas with persistent high unemployment (Bolen 2015). To supplement SNAP, school meals, and WIC, many communities have food pantries (which distribute food to poor households), "soup kitchens" (which provide cooked meals on site), and food assistance programs for the elderly population (such as Meals on Wheels).

Temporary Assistance for Needy Families (TANF) A federal cash welfare program that involves work requirements and a five-year lifetime limit.

Supplemental Nutrition Assistance Program (SNAP) The largest U.S. food assistance program.

What do you think?

SNAP benefits cannot be used to purchase any hot foods (such as a rotisserie chicken from the grocery store deli) or any nonfood items such as diapers, toilet paper, soap, toothpaste, or feminine sanitary products. Other prohibited items include alcohol, tobacco, vitamins, or medicine. Do you think any of these items should be eligible for SNAP benefits? If so, which ones?

Housing Assistance. The biggest expense for most families is housing. Lack of affordable housing is not just a problem for the poor living in urban areas. A comparison of household earnings to housing costs indicates that there is a national shortage of 7.2 million affordable homes (National Low Income Housing Coalition 2020).

Federal housing assistance programs include public housing, Section 8 housing, and other private project–based housing. The **public housing** program, initiated in 1937, provides federally subsidized housing that is owned and operated by local public housing authorities (PHAs). To save costs and avoid public opposition, high-rise public housing units were built in inner-city projects. These have been plagued by poor construction, managerial neglect, inadequate maintenance, and rampant vandalism. Poor-quality public housing has serious costs for its residents and for society:

> Distressed public housing subjects families and children to dangerous and damaging living environments that raise the risks of ill health, school failure, teen parenting, delinquency, and crime—all of which generate long-term costs that taxpayers ultimately bear. ... These severely distressed developments are not just old, outmoded, or run down. Rather, many have become virtually uninhabitable for all but the most vulnerable and desperate families. (Turner et al. 2005, pp. 1–2)

Section 8 housing involves federal rent subsidies provided either to tenants (in the form of certificates and vouchers) or to private landlords. Unlike public housing that confines low-income families to high-poverty neighborhoods, the aim with Section 8 housing is to disperse low-income families throughout the community. However, because of opposition by residents in middle-class neighborhoods, most Section 8 housing units remain in low-income areas.

A major barrier to building affordable housing is zoning regulations that set minimum lot size requirements, density restrictions, and other controls. Such zoning regulations serve the interests of upper-middle-class suburbanites who want to maintain their property values and keep out the "riffraff"—the lower-income segment of society who would presumably hurt the character of the community. Thus one answer to the housing problem is to change zoning regulations that exclude affordable housing and to require developers to reserve a percentage of units for affordable housing (Grunwald 2006).

Alleviating and Preventing Homelessness. Programs to temporarily alleviate homelessness include "homeless shelters" that provide emergency shelter beds and transitional housing programs that provide time-limited housing and services designed to help individuals gain employment, increase their income, and resolve substance abuse and other health problems. However, the number of homeless individuals exceeds the number of beds available (National Alliance to End Homelessness 2020). Resolving homelessness also requires strategies to *prevent* homelessness from occurring in the first place, such as increasing employment and living wages, providing tax benefits to renters (not just to homeowners), providing more affordable housing, protecting homeowners and renters against foreclosures, providing treatment for mental illness and substance abuse, and expanding programs to house victims of domestic violence.

Earlier we mentioned that some cities have laws that prohibit the homeless from sleeping, asking for money ("panhandling"), sitting, or "loitering" in public places and that install "hostile architecture" to prevent people experiencing homelessness from resting in public areas. In addition, at least 70 cities have rules that restrict or prohibit individuals or groups (e.g., faith-based and other nonprofit organizations) from providing food to homeless individuals in public places (Low Income Housing Authority 2019).

public housing Federally subsidized housing that is owned and operated by local public housing authorities (PHAs).

Section 8 housing A housing assistance program in which federal rent subsidies are provided either to tenants (in the form of certificates and vouchers) or to private landlords.

Medicaid. Medicaid is a government program that provides medical services and hospital care for the poor through reimbursements to physicians and hospitals (see also Chapter 2). States vary in rules about who is eligible for Medicaid; many low-income individuals and families do not qualify for Medicaid.

Educational Assistance. Educational assistance includes Head Start and Early Head Start programs and college assistance programs (see also Chapter 8). Head Start and Early Head Start programs provide educational services for disadvantaged infants, toddlers, and preschool-age children and their parents and are designed to improve children's cognitive, language, and social-emotional development and strengthen parenting skills (Administration for Children and Families 2002).

To help low-income individuals wanting to attend college, the federal government offers grants, loans, and work opportunities. The Pell Grant program aids students from low-income families. The federal college work–study program provides jobs for students with "demonstrated need." The guaranteed student loan program enables college students and their families to obtain low-interest loans with deferred interest payments. However, mounting student debt has reached disturbing levels. A total of 44.7 million Americans hold a total of $1.56 trillion in student loan debt, with an average debt of $32,731 (Friedman 2020).

Child-Care Assistance. In the United States, lack of affordable, high quality child care is a major obstacle to employment for single parents and a tremendous burden on dual-income families and employed single parents. Child Care Aware of America (2019) reported the following:

- The cost of full-time center-based care for two children is the highest single household expense for families living in the Northeast, Midwest, and South. In the West, child-care costs for two children are second only to average housing costs.
- The annual cost of infant child care ranges from $5,760 in Mississippi to $16,452 in California.
- According to the U.S. Department of Health and Human Services, any child care that costs more than 7 percent of a family's income is unaffordable. Based on this standard, infant care in a day care center is unaffordable in 49 states and the District of Columbia.

Some public policies provide limited assistance with child care, such as tax relief related to child-care expenses and public funding for child-care services for the poor (in conjunction with mandatory work requirements). However, child-care assistance is inadequate; many states have waiting lists for child-care assistance. Only about one in every seven eligible children receive federally funded child-care assistance (Child Care Aware of America 2019). Finally, many families earn more than the eligibility limit but not enough to afford child-care expenses.

@Rev DrBarber

MLK once made the case that the only group that could form the massive restructure of American society would be Blacks, Poor Whites, Progressive Whites, Working-Class folks & even recipients of welfare.

-Rev. Dr. William J. Barber II

Welfare in the United States: Myths and Realities

As we have seen in the previous section, the welfare system in the United States is a complex web of agencies, programs, and policies. Historian Michael Katz wrote:

> The American welfare state resembles a massive watch that fails to keep very accurate time. Some of its components are rusty and outmoded; others were poorly designed; some work very well. They were fabricated by different craftsmen who usually did not consult with one another; they interact imperfectly; and at times they work at cross-purposes. (2001, pp. 9–10)

Further complicating the effectiveness of the welfare system is the widespread negative public opinion about welfare programs and welfare recipients (Epstein 2004). But these negative images are grounded in myths and misconceptions about welfare. In this section, we examine some of these myths.

Myth 1. People who receive welfare are lazy, have no work ethic, and prefer to have a "free ride" on welfare rather than work.

Reality. More than 70 percent of recipients of TANF are children; 41 percent of TANF cases are child-only cases where no adult is involved in the benefit calculation and only children are aided (Congressional Research Service [CRS] 2020). Because we do not expect children to work, we can hardly think of children in need of assistance as "lazy."

Are adults receiving public assistance lazy? To be eligible to receive TANF, adults must meet strict work requirements, including job training, education, and documenting their job searches. Unemployed adult welfare recipients experience a number of barriers that prevent them from working, including disability and poor health, job scarcity, lack of transportation, and lack of education. Parents with infants or young children may be unable to work because they cannot afford child care. Finally, most adult welfare recipients would rather be able to support themselves and their families than rely on public assistance.

Many TANF families also have employment earnings, but these earnings are insufficient to meet the household's basic needs. TANF funds are based on household finances and size, and families receive the maximum amount only if they have no other source of income. The maximum amount a family can receive in a month is set by the state, which ranges (for a family of three) from $1,039 in New Hampshire to $170 in Mississippi. The image of a welfare "freeloader" lounging around enjoying life is far from the reality of the day-to-day struggles and challenges of supporting a household on a monthly TANF check of $170.

Myth 2. Fraud and abuse are rampant in the welfare system.

Reality. Fraud and abuse in the welfare system is difficult to precisely measure because the government only reports data on "improper payments" without distinguishing the reason for these payments. Still, the evidence available indicates that the vast majority of improper payments occur as a result of bureaucratic error, not due to misrepresentation of need on the part of the recipient. One analysis found that of all the improper payments from welfare programs, only 2 percent were due to fraud on the part of welfare recipients; 98 percent were due to bureaucratic error on the part of the government (Schnurer 2013). Those who receive improper overpayments are required to pay the government back.

Welfare fraud certainly does occur, but most of the cost associated with this fraud is due to fraudulent action on the part of administrators of the programs, not on the part of recipients. In February 2020, the former welfare director for the state of Mississippi, John Davis, along with five others (including a Christian evangelist minister and former pro wrestler "Iron" Mike DiBiase) were arrested and charged for an embezzlement scheme in which they fraudulently funneled millions of dollars out of the state welfare fund and into their own pockets (Chappell 2020). Furthermore, the majority of fraudulent payments within the Medicaid system is attributed to overcharging on the part of health care practitioners, not misrepresentation on the part of Medicaid recipients (Gilman 2020).

Myth 3. Welfare benefits are granted to many people who are not really poor or eligible to receive them.

Reality. Although some people obtain welfare benefits through fraudulent means, it is much more common for people who are eligible to receive welfare not to receive benefits. The TANF-to-poverty ratio measures the number of eligible families (households with children living below the poverty line) compared to the number of families actually receiving those benefits. In 2018, this ratio hit its lowest point in history, with only 22 out of every 100 eligible families receiving benefits, down from 68 out of every 100 families when TANF was enacted in 1996 (Center on Budget and Policy Priorities [CBPP] 2020; see Figure 6.10). In 16 states, this ratio is 10 or fewer families per every 100 eligible; in Louisiana, Texas, Arkansas, and Georgia, the ratio was 5 or fewer per 100 eligible.

One reason for not receiving benefits is lack of information; some people do not know about various public assistance programs, or even if they know about a program, they do not know they are eligible. Another reason that many people who are eligible for public assistance do not apply for it is because they desire personal independence and do not want to be stigmatized as lazy people who just want a "free ride" at the taxpayers' expense. Others have difficulty navigating the complex administrative processes involved

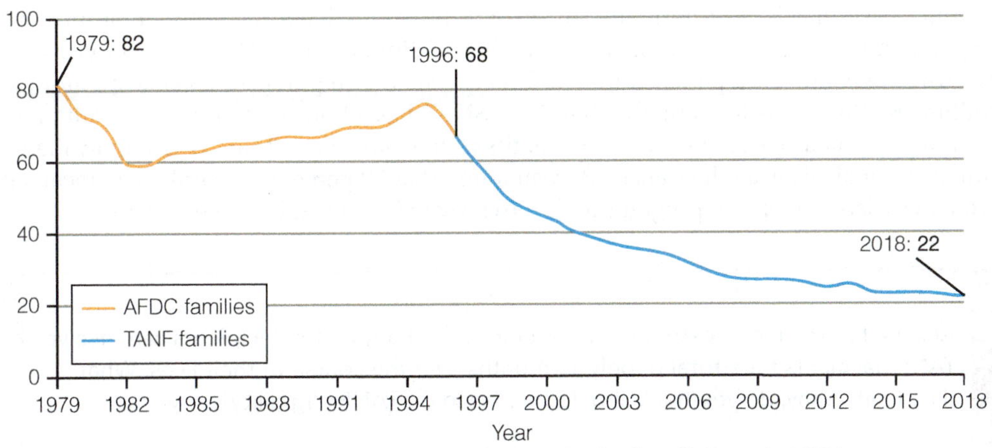

Figure 6.10 TANF-to-Poverty Ratio
SOURCE: Center on Budget and Policy Priorities (CBPP) 2020.

in applying for assistance. Assistance programs are administered through separate offices at different locations, have various application procedures and renewal deadlines, and require different sets of documentation. Green (2013) explains:

> [a]s low-income mothers struggle to meet the intense demands of balancing work and family, they also have to continue the time-intensive task of piecing together in-kind and cash benefits to pad their low wages. Doing so involves traveling from one office to another; repeatedly disclosing intimate and personal information; and documenting, in a detailed paper trail, the legitimacy of one's story. … Although there is little time to spare in this world, each office treats clients as if they have endless time to waste. Furthermore, poor families are often at the mercy of buses that are late, babysitters who do not show up, overworked caseworkers who misplace documents, and other similar barriers to the successful performance of the role of "good client." This situation can lead to extreme levels of personal frustration, which add to the hardship and defeatism experienced while engaging this system. (p. 55)

Navigating through the "system"—getting time off work to meet with caseworkers and finding child care and transportation—can produce so much frustration that some people who are eligible for assistance just give up on the system.

Finally, some individuals who are eligible for public assistance do not receive it because it is not available. In cities across the United States, thousands of eligible low-income households are on waiting lists for public housing assistance because not enough public housing units are available, and some cities have even stopped accepting housing applications. Even when people receive benefits, using them may be difficult. For example, individuals with Section 8 vouchers for housing may have a hard time finding a landlord who will accept them, even though it is against the law to refuse a Section 8 renter. Individuals who have Medicaid may have difficulty finding a doctor who will take Medicaid patients. Low-income parents who receive child-care assistance vouchers are often unable to find an available "voucher slot" and may be on a waiting list at a child-care center for more than a year.

Myth 4. Welfare makes people dependent on the government and prevents them from becoming contributing citizens.

Reality. The anti-dependency ideology embedded in the design of the welfare system is evident in the naming of the programs themselves. For example, titles like "Temporary"

Assistance to Needy Families and "Supplemental" Nutrition Assistance Program imply that recipients are not intended to rely on the programs for the long term or even for meeting basic needs. Furthermore, all welfare programs have restrictive qualification requirements and limits on their use. The federal limitation on TANF is five years total over the course of a person's life, and recipients are required to meet work and job training requirements to maintain eligibility. SNAP has a household resource limitation, meaning that families cannot receive benefits if they have more than $2,250 in available funds. An analysis from the Census Bureau found that 79 percent of people who received welfare assistance left the program within two years (Irving and Loveless 2015).

What do you think?

Arizona has the most restrictive TANF time limit of any state: Recipients can receive TANF payments for a total of only 12 months over the course of their lives. What time limit, if any, do you think should be placed on TANF eligibility?

Myth 5. Immigrants place a huge burden on our welfare system.

Reality. Low-income noncitizen immigrants, including adults and children, are less likely to receive public benefits than those who are native born. Moreover, when noncitizen immigrants receive benefits, the value of benefits they receive is lower than the value of benefits received by those born in the United States (Ku and Bruen 2013). Federal rules restrict immigrants' eligibility for public benefit programs, and undocumented immigrants are generally ineligible to receive benefits from Medicaid, SNAP, and TANF, although some benefit programs, such as the National School Lunch Program, the Women, Infants, and Children Nutrition Program (WIC), and Head Start, do not include immigration status as an eligibility factor. Although children born in the United States are considered citizens and are therefore eligible for public assistance, undocumented parents often do not apply for assistance for their children because they either do not know their children can receive benefits, or they fear that applying for benefits for their children will result in their deportation (see also Chapter 9).

Understanding Economic Inequality, Wealth, and Poverty

As we have seen in this chapter, the quality of our lives is intricately related to the economic resources we have—resources that buy access to goods and services such as housing, food, education, health care, resources that influence virtually every aspect of our lives. On a positive note, significant gains have been made in improving the standard of living for populations living in absolute poverty. But at the same time, economic inequality has reached unprecedented levels in the world and in the United States, as the "rich get richer."

A common belief among U.S. adults is that the rich are deserving and the poor are failures. Blaming poverty on the individual rather than on structural and cultural factors implies not only that poor individuals are responsible for their plight but also that they are responsible for improving their condition. If we hold individuals accountable for their poverty, we fail to make society accountable for making investments in human development that are necessary to alleviate poverty, such as providing health care, adequate food and housing, education, child care, job training and job opportunities, and living wages. Lastly, blaming the poor for their condition diverts attention away from the recognition that the wealthy—individuals and corporations—receive far more benefits in the form of wealthfare or corporate welfare, without the stigma of welfare.

Efforts to alleviate poverty and reduce economic inequality are often motivated by a sense of moral responsibility. But alleviating poverty and reducing economic inequality also make sense from an economic standpoint. The Poor People's Campaign, a social justice movement that extended from the Poor People's March on Washington that Martin Luther King Jr. was organizing just before he was assassinated, summarizes some of the economic benefits of anti-poverty programs (2020):

© Stephen Pavey, Hope In Focus

The Poor People's campaign is a social justice activist movement that advocates for antipoverty programs using a framework of morality.

- Raising the federal minimum wage from $7.25 to $15 per hour would put an additional $328 billion into the hands of low-income working households who will spend most of the money back into their local economy.
- Every $1 invested in early childhood education leads to $7.30 in savings from increased earnings, better health, and lower incarceration rates.
- For every $1 per hour that wages rise among workers in the bottom 60 percent of earners, spending on government assistance programs falls by roughly $5.2 billion.
- Universal health care would save our economy $278 billion per year.
- Raising the minimum wage by $2 could have prevented more than 57,000 suicides from 1990 to 2015.

Ending or reducing poverty begins with the recognition that doing so is a worthy ideal and an attainable goal. Imagine a world where everyone had comfortable shelter, plentiful food, clean water and sanitation, adequate medical care, and education. If this imaginary world were achieved and if absolute poverty were effectively eliminated, what would be the effects on social problems such as crime, drug abuse, family problems (e.g., domestic violence, child abuse, and divorce), health problems, prejudice and racism, and international conflict? In the current global climate of conflict and terrorism, we might consider that "reducing poverty and the hopelessness that comes with human deprivation is perhaps the most effective way of promoting long-term peace and security" (World Bank 2005). Instead of asking if we can afford to eradicate poverty, we might consider: Can we afford not to?

> "I am, somehow, less interested in the weight and convolutions of Einstein's brain than in the near certainty that people of equal talent have lived and died in cotton fields and sweatshops."
>
> **–STEPHEN JAY GOULD**

Chapter Review

- **What is the extent of poverty, wealth, and economic inequality in the world?**
  More than one in ten people are extremely poor, living on less than $1.90 a day. In stark contrast to this extreme poverty, in 2019, there were 2,153 billionaires in the world, and the richest 1 percent of adults (ages 20+) owned more than half of global wealth.

- **How is poverty measured?**
  The most widely used standard to measure extreme poverty in the developing world is $1.90 per day. According to measures of relative poverty, members of a household are considered poor if their household income is less than 50 percent of the median household income in that country. The Multidimensional Poverty Index considers health,

education, and living standards in measuring who is poor. Each year, the U.S. federal government establishes "poverty thresholds" that differ by the number of adults and children in a family and by the age of the family head of household. Anyone living in a household with pretax income below the official poverty line is considered "poor." The Economic Policy Institute's Family Budget Calculator finds that U.S. families need at minimum twice the poverty-level income to meet a family's basic economic needs.

- **How do the structural-functionalist, conflict, and symbolic interactionist perspectives view wealth, poverty, and economic inequality?**
According to the structural-functionalist perspective, poverty results from institutional breakdown: economic institutions that fail to provide sufficient jobs and pay, educational institutions that fail to equip members of society with the skills they need for employment, family institutions that do not provide two parents, and government institutions that do not provide sufficient public support. The structural-functionalist perspective also focuses on both the functions and dysfunctions of poverty.
The conflict perspective is concerned with how capitalism has contributed to poverty and economic inequality and how wealthy corporations and individuals use their wealth to influence elections and policies in ways that benefit the wealthy. The conflict perspective is also critical of how trade and investment policies benefit wealthy corporations and negatively affect the poor.
The symbolic interactionist perspective calls attention to ways in which wealth and poverty are defined and the effects of being labeled "poor." This perspective is also concerned with how economic inequality has come to be more widely recognized as a social problem as a result of media attention stemming from the Occupy Wall Street movement and the recent publication of Thomas Piketty's book *Capital in the Twenty-first Century* (2014).

- **What is the extent of economic inequality and poverty in the United States?**
The United States has the highest level of income inequality and the highest rate of poverty of any industrialized nation. In 2017, the top 1 percent of U.S. taxpayers earned 21 percent of all U.S. income, and the wealthiest 1 percent owned 40 percent of U.S. wealth; more than half of that wealth was owned by just the top 0.1 percent.
In the United States, groups that are at increased risk of poverty include children, women (especially those who are single parents), those with low levels of educational attainment, and racial and ethnic minorities.

- **What are some of the consequences of poverty and economic inequality for individuals, families, and societies?**
Poverty is associated with health problems and hunger; substandard housing and homelessness; unequal treatment in the legal system; political inequality and alienation; crime, social conflict, and war; increased vulnerability to natural disasters; problems in education; and unequal marriage opportunity and other family problems including higher rates of divorce, child abuse/neglect, and teenage and unintended pregnancy. These various problems are interrelated and contribute to the perpetuation of poverty across generations, feeding a cycle of intergenerational poverty.

- **What are some global strategies for reducing poverty and economic inequality?**
Global approaches for reducing poverty and economic inequality include increasing taxes on the wealthy, promoting economic development, investing in human development, and providing microcredit programs that provide loans to poor people.

- **What are some strategies to reduce U.S. poverty and economic inequality?**
Strategies to reduce U.S. poverty and economic inequality include the earned income tax credit, tax reform efforts that make taxes more progressive and that close corporate tax loopholes, political reforms to reduce the influence of money in politics, increasing the minimum wage and establishing living wage laws, and reducing wage theft so that workers receive the income they are legally entitled to.

- **What are some public assistance and welfare programs in the United States?**
U.S. public assistance and welfare programs include Supplemental Security Income, Temporary Assistance for Needy Families (TANF), food programs (such as school meal programs and SNAP), housing assistance, Medicaid, educational assistance (such as Pell Grants), child-care assistance, and the earned income tax credit (EITC).

- **What are five common myths about welfare and welfare recipients?**
Common myths about welfare and welfare recipients are (1) that welfare recipients are lazy, have no work ethic, and prefer to take a "free ride" on welfare rather than work; (2) that fraud and abuse are rampant within the welfare system; (3) that welfare benefits are granted to many people who are not really poor or eligible to receive them; (4) that welfare makes people dependent on the government and prevents them from becoming contributing citizens; and (5) that immigrants place an enormous burden on our welfare system.

Test Yourself

1. The ______ Poverty Index is a measure of serious deprivation in the dimensions of health, education, and living standards that combines the number of deprived and the intensity of their deprivation.
   a. Relative
   b. Human
   c. International
   d. Multidimensional
2. According to the 2019 official U.S. poverty threshold guidelines, a single adult earning $13,000 a year is considered "poor."
   a. True
   b. False
3. In a meritocracy, individuals get ahead and earn rewards based on their
   a. political power.
   b. inherited wealth.
   c. luck.
   d. efforts and abilities.
4. In 2019, a typical U.S. worker would have to work ______ years to make what a typical CEO of a large U.S. corporation earns in one year.
   a. 15
   b. 99
   c. 276
   d. 320
5. Corporate welfare refers to which of the following?
   a. The taxes that corporations pay providing most of the funding for federal welfare programs for the poor
   b. Tax-deductible contributions that corporations make to charitable organizations
   c. Laws and policies that benefit corporations
   d. Employee assistance programs offered by corporations to help employees who are struggling with debt
6. What age group in the United States has the highest rate of poverty?
   a. Younger than 18
   b. 30–44
   c. 45–64
   d. Older than 65
7. According to the text, the wealthy are hardest hit by natural disasters because they have more to lose than do the poor.
   a. True
   b. False
8. A plutocracy is a country ruled by
   a. the wealthy.
   b. the people.
   c. women.
   d. religion.
9. SNAP benefits can be used to purchase
   a. alcohol.
   b. tobacco.
   c. diapers.
   d. none of the above.
10. The majority of improper welfare payments are the result of recipients fraudulently misrepresenting their needs.
    a. True
    b. False

Answers: 1. D; 2. B; 3. D; 4. D; 5. C; 6. A; 7. B; 8. A; 9. D; 10. B.

Key Terms

Sarah Silbiger/Getty Images News/Getty Images

> “When you go to work, if your name is on the building, you’re rich. If your name is on your desk, you’re middle class. If your name is on your shirt, you’re poor.”
>
> **RICH HALL**
> Writer and performer

7

Work and Unemployment

Chapter Outline

The Global Context: The New Global Economy

- **Self and Society:** How Do You View Capitalism and Socialism?

Sociological Theories of Work and the Economy

Problems of Work and Unemployment

- **The Human Side:** Life as a Child Worker in a Garment Factory
- **The World in Quarantine:** When the Motherhood Penalty Collides with a Pandemic
- **Social Problems Research Up Close:** What Impact Do Right-to-Work Laws Have on Workers and the Economy?

Strategies for Action: Responses to Problems of Work and Unemployment

Understanding Work and Unemployment

Chapter Review

Learning Objectives

After studying this chapter, you will be able to…

1. Describe the lasting impact of the Great Recession on the global economy.
2. Explain the differences between capitalism and socialism.
3. Identify the pros and cons of free trade agreements and transnational corporations.
4. Explain how structural functionalism, conflict theory, and symbolic interactionism view work and the economic institution.
5. Identify problems associated with work in the United States and around the world, including unemployment, slavery, sweatshop labor, child labor, health and safety in the workplace, work/life conflict, alienation, job stress, and issues concerning labor unions and workers' rights.
6. Describe strategies for reducing unemployment; creating alternatives to capitalism through worker cooperatives; ending slavery, child labor, and sweatshop labor; making the workplace safer; helping workers achieve work/life balance; and strengthening labor unions.
7. Describe the reasons for workplace-based human rights violations.

"WE WERE TREATED WORSE THAN ANIMALS," said Bellaliz Gonzalez (Rose and Peñaloza 2020, p. 1). Gonzalez, an asylum-seeker from Venezuela, came to the United States in 2018 and found work for a disaster recovery company that cleans up after floods and fires. In Michigan in the spring of 2020, two disasters coincided: the failure of two dams, which caused widespread flooding in the city of Midland, and the COVID-19 pandemic, which led to a statewide shutdown order. A disaster recovery contractor brought Gonzalez, along with dozens of other workers from Texas and Florida, into Michigan to clean up the flooded basement and morgue of the MidMichigan Medical Center. Despite repeated requests for a safety plan and personal protective gear, Gonzalez says their work crew were never provided with even basic safety needs like temperature checks, face masks, or gloves. They were required to stay four to five in cramped hotel rooms and were shuttled to the work site each day in a crowded 16-passenger van. Complicating matters further, many of the workers spoke no English and were undocumented or, like Gonzalez, on temporary visas in the United States with an uncertain status; they were afraid to speak out against the unsafe conditions. Their requests for improved safety conditions were ignored, even after numerous workers on the crew fell ill with COVID-19. Saket Soni, the director of a nonprofit that advocates for recovery workers, says that the structure of the disaster recovery industry is the root of the problem. With its layers of contractors and subcontractors, employers can easily avoid accountability for upholding workplace safety laws. Workers in meatpacking plants and on farms face similar problems; she says: "These workers are essential, but no one behaves like it" (quoted in Rose and Peñaloza 2020, p. 1).

Bruno Iyda Saggese

This cartoon, titled "The Romanticization of the Quarantine Is a Class Privilege," illustrates the inequality in health and safety issues faced by workers with different locations in the economy.

Health and safety hazards in the workplace, often due to employers' willful violations of health and safety regulations, are among the work-related problems discussed in this chapter. Other problems we examine include unemployment, forced labor, child labor, sweatshop labor, alienation, work/life conflict, and declining labor union strength and representation. We set the stage with a brief look at the global economy.

The Global Context: The New Global Economy

global economy An interconnected network of economic activity that transcends national borders.

In recent decades, innovations in communication and information technology have spawned the emergence of a **global economy**—an interconnected network of economic activity that transcends national borders and spans the world. The globalization of economic activity means that our jobs, the products and services we buy, and our nation's economic policies and agendas influence and are influenced by economic activities occurring around the world.

The **Great Recession**, the period of global financial crisis that lasted from 2007 to 2009, demonstrated the interconnectedness of the global economy. The crisis originated in the United States, triggered by years of risky investments across the banking sector, particularly in the housing and mortgage markets. As the housing boom turned to bust and adjustable rate mortgages were reset to higher rates, millions of homeowners were unable to keep up with mortgage payments, and foreclosures skyrocketed. Homeowners lost their homes, renters lost their leases, and banks suffered because the foreclosed homes they now owned were often worth less than the mortgages owed on them. As banks lost revenue they had less money to lend, so credit froze, consumer spending plummeted, businesses went bust, and stockholders watched their investments and retirement accounts take a nosedive. The whole banking system was faltering—some got bailed out; others (e.g., Bear Stearns) went bust. All this happened in the United States, but in this new global economy, what happens in Vegas does not stay in Vegas—the crisis spread around the world. This is because those risky subprime and adjustable rate mortgages were packaged and resold as "mortgage-backed securities" to financial institutions around the world.

Although the period of recession officially lasted from 2007 to 2009, its effects were felt for a decade afterward. By 2020, global trade was booming again, and unemployment rates were at historic lows. The emergence of the global COVID-19 pandemic, first identified in China in 2019 and then spreading throughout the world over the course of the next year, triggered another global financial crisis. To help stem the spread of the virus, governments ordered shutdowns and quarantines. Production and trade came grinding to a halt in many parts of the world, and unemployment skyrocketed as workplaces shuttered.

These financial crises illustrate the globalization of the **economic institution**. Drops in consumer spending in one country trigger the loss of jobs and revenue in others. Stalled production and revenue losses in one place lead to financial crisis on the other side of the world.

Global economic crises like the Great Recession and the impact of the COVID-19 pandemic have reignited debate between those who view capitalism as the cause of economic problems in the world and those who hail capitalism as "the greatest engine of economic progress and prosperity known to mankind" (Ebeling 2009). After summarizing capitalism and socialism—the two main economic systems in the world—we look at the emergence of free trade agreements and transnational corporations.

> "Globalization is a fact of life. But I believe we have underestimated its fragility. The problem is this. The spread of markets outpaces the ability of societies and their political systems to adjust to them, let alone guide the course they take."
>
> **–KOFI ANNAN**

Great Recession The period of global economic crisis that lasted from 2007 to 2009 that was characterized by high levels of unemployment, widespread home foreclosures, the collapse of the banking industry, and plummeting global trade.

economic institution The structure and means by which a society produces, distributes, and consumes goods and services.

Capitalism and Socialism

Under **capitalism**, private individuals or groups invest capital (money, technology, machines) to produce goods and services to sell for a profit in a competitive market. Capitalism is characterized by economic motivation through profit, the determination of prices and wages primarily through supply and demand, and the absence of government intervention in the economy. **Socialism** is an economic system in which the means of production (factories, machinery, land, stores, offices, etc.) are socially owned by the public, worker collectives, or the state. In a socialist economy, theoretically, goods and services are distributed according to the needs of the citizens, with some goods and services—such as water, education, health care, and/or child care—provided to all citizens through a socially funded system of taxation. Whereas capitalism emphasizes individualistic pursuit of profit and individual freedom, socialism emphasizes collective well-being and social equality.

Socialism is often confused with communism, and some people use the two terms interchangeably. But socialism is more accurately an economic system that most often exists with some variation of a democratic political system. Bernie Sanders describes himself as a "democratic socialist." **Communism** is a political system that usually takes on the form of totalitarianism—a one-party system where the state controls all aspects of public and private life, permitting no individual freedom. In communism, there is no

capitalism An economic system characterized by private ownership of the means of production and distribution of goods and services for profit in a competitive market.

socialism An economic system characterized by state ownership of the means of production and distribution of goods and services.

communism A political system that usually takes on the form of totalitarianism—a one-party system where the state controls all aspects of public and private life, permitting no individual freedom.

private property (all property is commonly owned), there are no class distinctions among people (goods and services are distributed equally so that no one has more or less than anyone else), and goods and services are distributed directly to people without the use of money.

There are no pure socialist or capitalistic economies. Rather, most countries have mixed economies, incorporating elements of both capitalism and socialism. Most developed countries, for example, have both privately owned and publicly owned enterprises, as well as a social welfare system. The U.S. economy, dominated by capitalism, also includes elements of socialism, such as the provision of fire and police services, public roads, education, a postal service, public assistance programs for the poor, government subsidies and low-interest loans to industry, fiscal stimulus money, and bailout money to the auto industry and banks.

@Bernie Sanders

When we talk about "democratic socialism," we're talking about making sure everybody can live with dignity in the richest country on Earth. That means making sure everyone has the basic necessities of life: decent housing, good health care, good education and a good retirement.

-Bernie Sanders

Critics of socialism argue that it creates excessive government control and lowers the standard of living, while proponents of socialism argue that it enhances overall well-being by reducing inequality and all of its associated social problems (see Chapter 6). A national survey of U.S. adults found that only 39 percent say they have a positive image of socialism (Newport 2020). Democrats are much more likely than Republicans to view socialism positively (65 percent vs. 9 percent).

Recent polls also show a sharp generational divide in views of capitalism and socialism (see Figure 7.1). Over the past decade, the percentage of Millennials and Gen Zers who viewed socialism positively has stayed mostly steady, but positive views of capitalism have declined from 66 percent in 2010 to 51 percent in 2019 (Saad 2019). Over the same period, older generations have demonstrated an increase in viewing capitalism positively. What is behind our attitudes about capitalism and socialism? See this chapter's *Self and Society* feature.

Critics of capitalism point to a number of social ills linked to this economic system, including high levels of inequality; economic instability; job insecurity; pollution and depletion of natural resources; and corporate dominance of media, culture, and politics. Capitalism is also criticized as violating the principles of democracy by allowing (1) private wealth to affect access to political power (see also Chapter 6); (2) private owners of property to make decisions that affect the public (such as when the owner of a factory decides to move the factory to another country); and (3) workplace dictatorships where workers have little say in their working conditions, thus violating the democratic principle that people should participate in collective decisions that significantly affect their lives (Wright 2015). Wolff (2013a, 2013b) explains that major shareholders

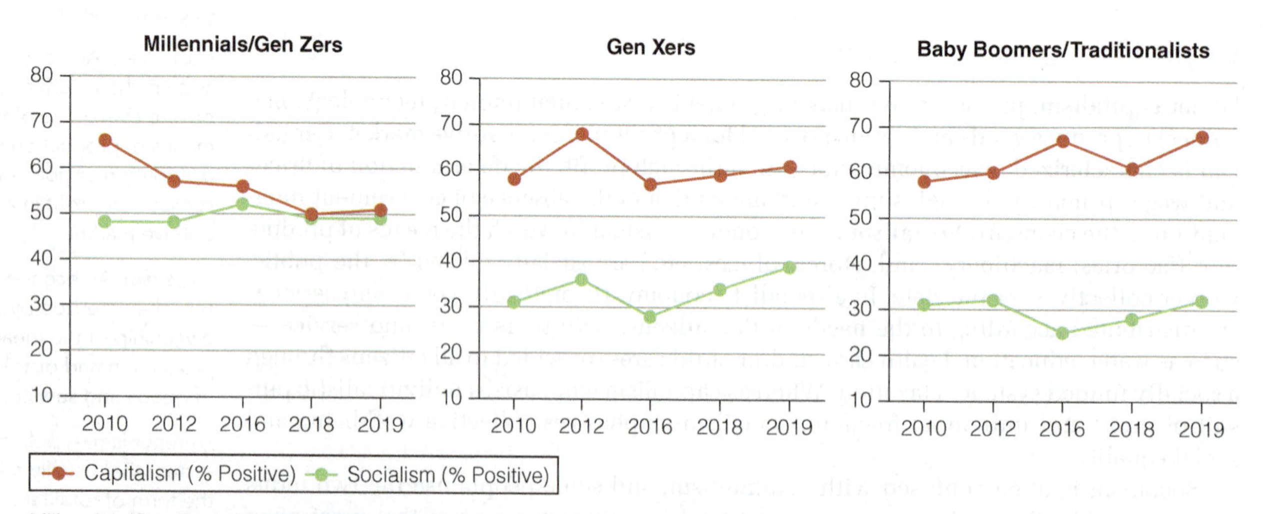

Millennials/Gen Zers: Ages 18 to 39 in 2019
Gen Xers: Ages 40 to 54 in 2019
Baby boomers/Traditionalists: Ages 55 and older in 2019

Figure 7.1 Trends in Positive Views toward Capitalism and Socialism, by Generation
SOURCE: Saad Lydia. 2019.

and boards of directors within corporations make key decisions about what products the corporation will produce, what technologies will be used, where production will occur, and how the revenues will be distributed—decisions that profoundly affect workers who have no say in these decisions:

> The most important activity of an adult's life in this country is work. It's what we do five days out of every seven. If democracy belongs anywhere, it belongs in the workplace. Yet we accept, as if it were a given, that once we cross the threshold of our store, factory, or office, we give up all democratic rights. (Wolff, quoted in Barsamian 2012, p. 12)

Although a slight majority (51 percent) of Millennials and Gen Zers view capitalism positively, this is a steep drop over the past decade—down from 66 percent in 2010. Among older generations, positive views of capitalism have slightly increased over the past decade. What do you think contributes to this generational divide in views of capitalism and socialism?

What *do you* think?

The Globalization of Trade and Free Trade Agreements

The globalization of trade refers to raw materials, manufactured goods, and agricultural products being bought and sold across national and hemispheric borders. The first set of global trade rules were adopted through the General Agreement on Tariffs and Trade (GATT) in 1947. In 1995, the World Trade Organization (WTO) replaced GATT as the organization overseeing the multilateral trading system.

In the 1980s and early 1990s, U.S. officials began negotiating regional free trade agreements that would open doors to U.S. goods in neighboring countries and reduce the growing U.S. trade deficit. A **free trade agreement (FTA)** is a pact between two countries or among a group of countries that makes it easier to trade goods across national boundaries. Free trade agreements reduce or eliminate foreign restrictions on exports, reduce or eliminate tariffs (or taxes) on imported goods, and prevent technology from being copied and used by competitors through protection of "intellectual property rights." Both Bernie Sanders and Donald Trump have argued that trade deals cost America jobs, especially in manufacturing. The United States currently has 14 free trade agreements with 20 countries. During his first week in office, former President Trump withdrew from the Trans-Pacific Partnership trade agreement which was negotiated under the previous Obama administration.

In January 2020, the Trump administration announced that the North American Free Trade Agreement (NAFTA)—a signature trade deal of the Clinton administration—would be replaced with a new deal: the United States–Mexico–Canada Agreement (USMCA). President Biden has publicly stated that although he does not see the USMCA agreement as ideal, he does see it as better than NAFTA and promises to work to continue to improve the deal in a way that would promote "Made in America" manufacturing (*Washington Post* 2020).

The controversy over NAFTA illustrates the economic tensions inherent to FTAs. Although free trade agreements have expanded trading opportunities, benefiting large export manufacturing and service industries in the global north, they have also undermined the ability of national, state, and local governments to implement laws designed to protect workers, consumers, and the environment. Trade agreements include provisions that supersede national and local laws, including the U.S. Constitution, and give foreign corporations the right to sue governments for lost profits due to environmental, worker safety, food or product safety, and other laws that hurt corporate profits (see also Chapter 6 and Chapter 13; Korten 2015; Scott and Ratner 2005).

Free trade agreements have also hurt both U.S. and foreign workers. The NAFTA allowed U.S. corn growers to sell their corn in Mexico, but Mexican corn farmers could not compete with the cheap price of U.S. corn, which put many Mexican corn growers

@realDonaldTrump

... Remember, NAFTA was one of the WORST Trade Deals ever made. The U.S. lost thousands of businesses and millions of jobs. We were far better off before NAFTA - should never have been signed. Even the Vat Tax was not accounted for. We make new deal or go back to pre-NAFTA!

-Donald J. Trump

free trade agreement (FTA) A pact between two or more countries that makes it easier to trade goods across national boundaries by reducing or eliminating restrictions on exports and tariffs (or taxes) on imported goods and protecting intellectual property rights.

How Do You View Capitalism and Socialism?

While slightly more than half of Americans have a negative view of socialism, attitudes about capitalism and socialism are increasingly showing a generational divide (see Figure 7.1). What is behind our opinions about capitalism and socialism? Respond to the following questions, and then compare your results to a national survey of more than 10,000 adults in the United States (Pew Research Center 2019).

1. Would you say you have a positive or negative impression of capitalism?
   [ ] Very positive
   [ ] Somewhat positive
   [ ] Somewhat negative
   [ ] Very negative
2. Would you say you have a positive or negative impression of capitalism?
   [ ] Very positive
   [ ] Somewhat positive
   [ ] Somewhat negative
   [ ] Very negative
3. If you had a very or somewhat positive view of socialism, which of the following reasons best describes why you hold that view?
   [ ] It creates a fairer and more generous system.
   [ ] It builds upon and improves capitalism.
   [ ] In historical and international comparison, countries with socialist policies (e.g., Finland and Denmark) take care of their populations better than in the United States.
   [ ] In general, it seems good, or it is better than capitalism.
   [ ] Another reason
4. If you had a very or somewhat negative view of socialism, which of the following reasons best describes why you hold that view?
   [ ] It undermines work ethic and increases reliance on the government.
   [ ] In historical and international comparison, countries with socialist policies (e.g., Venezuela and Russia) are worse off than in the United States.
   [ ] It undermines democracy and is not right for the United States.
   [ ] In general, it is not good, or it is worse than capitalism.
   [ ] Another reason
5. If you had a very or somewhat positive view of capitalism, which of the following reasons best describes why you hold that view?
   [ ] It promotes individual opportunity, work ethic, and self-reliance.
   [ ] It is not a perfect system but has proven to work better than other systems.

out of business (Scott and Ratner 2005). Free trade agreements have also made it easier for U.S. companies to move jobs "offshore," usually to countries where wages are low and there are few environmental, health, or safety regulations with which to comply. Although offshoring jobs increases profits to corporations, it also takes jobs away from U.S. workers.

The new USMCA trade agreement that replaced NAFTA won bipartisan support after nearly three years of negotiations. The new provisions address some of the concerns raised by critics of NAFTA, most notably, restrictions of country-of-origin rules and labor provisions stipulating that for cars to be exempt from tariffs, at least 75 percent of their components must be manufactured in the three partner countries and at least 40 percent of the automobile parts must be made by workers earning at least $16 per hour (Office of the United States Trade Representative 2020). These provisions are intended to prevent the widespread off-shoring of manufacturing jobs that occurred after NAFTA. Critics argue, however, that these provisions do not provide enough labor protection for workers and that the deal does not address the environmental regulation concerns long raised by many opponents of NAFTA.

[ ] It is essential to the American ideal of freedom and its economic strength.
[ ] Another reason

6. If you had a very or somewhat negative view of capitalism, which of the following reasons best describes why you hold that view?
   [ ] It primarily benefits only a few and results in an unequal distribution of wealth.
   [ ] It is exploitative or corrupt in nature.
   [ ] It undermines the democratic process.
   [ ] It could work but needs better regulation.
   [ ] Another reason

Results:

Among those who had a very or somewhat positive view of socialism:

31% It creates a fairer and more generous system.
20% It builds upon and improves capitalism.
6% In historical and international comparison, countries with socialist policies (e.g., Finland and Denmark) take care of their populations better than in the United States.
8% In general, it seems good, or it is better than capitalism.
35% Another reason

Among those who had a very or somewhat negative view of socialism:

19% It undermines work ethic and increases reliance on the government.
18% In historical and international comparison, countries with socialist policies (e.g., Venezuela and Russia) are worse off than in the United States.
17% It undermines democracy and is not right for the United States.
21% In general, it is not good or is worse than capitalism.
25% Another reason (or refused to answer)

Among those who had a very or somewhat positive view of capitalism:

24% It promotes individual opportunity, work ethic, and self-reliance.
18% It is not a perfect system but has proven to work better than other systems.
20% It is essential to the American ideal of freedom and its economic strength.
38% Another reason (or refused to answer)

Among those who had a very or somewhat negative view of capitalism:

23% It primarily benefits only a few and results in an unequal distribution of wealth.
20% It is exploitative or corrupt in nature.
8% It undermines the democratic process.
4% It could work but needs better regulation.
45% Another reason (or refused to answer)

Transnational Corporations

Although free trade agreements have increased business competition around the world, resulting in lower prices for consumers for some goods, they have also opened markets to monopolies (and higher prices) because they have facilitated the development of large-scale transnational corporations. **Transnational corporations**, also known as *multinational corporations*, are corporations that have their home base in one country and branches, or affiliates, in other countries.

Transnational corporations provide jobs for U.S. managers, secure profits for U.S. investors, and help the United States compete in the global economy. Transnational corporations benefit from increased access to raw materials, cheap foreign labor, and the avoidance of government regulations. They can also avoid or reduce tax liabilities by moving their headquarters to a "tax haven." But the savings that transnational companies reap from cheap labor and reduced taxes are not passed on to consumers. "Corporations do not outsource to far-off regions so that U.S. consumers can save money. They outsource in order to increase their margin of profit" (Parenti 2007). For example, shoes made by

transnational corporations Also known as *multinational corporations*, corporations that have their home base in one country and branches, or affiliates, in other countries.

Indonesian children working 12-hour days for 13 cents an hour cost only $2.60 but are still sold for $100 or more in the United States.

Transnational corporations contribute to the trade deficit in that more goods are produced and exported from outside the United States than from within. Transnational corporations also contribute to the budget deficit, because the United States does not get tax income from U.S. corporations abroad, yet transnational corporations pressure the government to protect their foreign interests; as a result, military spending increases. Transnational corporations contribute to U.S. unemployment by letting workers in other countries perform labor that U.S. employees could perform. Finally, transnational corporations are implicated in an array of other social problems, such as poverty resulting from fewer jobs, urban decline resulting from factories moving away, and racial and ethnic tensions resulting from competition for jobs.

Sociological Theories of Work and the Economy

In sociology, structural functionalism, conflict theory, and symbolic interactionism serve as theoretical lenses through which we may better understand work and economic issues and activities.

Structural-Functionalist Perspective

According to the structural-functionalist perspective, the economic institution is one of the most important of all social institutions, as it functions to provide the basic necessities common to all human societies, including food, clothing, and shelter and thus contributes to social stability. After the basic survival needs of a society are met, surplus materials and wealth may be allocated to other social uses, such as maintaining military protection from enemies, supporting political and religious leaders, providing formal education, supporting an expanding population, and providing entertainment and recreational activities. Societal development is dependent on an economic surplus in a society (Lenski and Lenski 1987).

As noted in Chapter 6, structural-functionalist theorists Davis and Moore (1945) argued that jobs differ in their pay so that workers will be motivated to achieve higher levels of education and training and that jobs that offer higher rewards are those that are more important and difficult. This argument falls apart when one considers that many very important jobs have low pay. Consider, for example, that most of the jobs deemed "essential" during the COVID-19 pandemic were primarily low-wage jobs in places like grocery stores and meatpacking plants. As discussed in Chapter 6, the inequality associated with salaries and wages is often considered dysfunctional rather than functional for society. The economic institution can also be dysfunctional when it fails to provide members with jobs and the goods and services they need; when the production, distribution, and consumption of goods and services deplete and pollute the environment; when participation in the labor force leads to alienation and work/life conflict (discussed later in this chapter); and when it includes practices that violate basic human rights, such as slavery and unsafe working conditions (also discussed later in this chapter).

The structural-functionalist perspective also focuses on how the economy affects and is affected by changes in society. For example, as societies become more economically developed, infant and child death rates drop, women have fewer children, and life expectancy increases, which lead to the aging of the world's population. Population aging affects the economy because as the proportion of older people increases, there are fewer working-age adults to fill jobs and to support the elderly population (see also Chapter 12).

Which jobs are the most important to society? Structural functionalists argue that pay inequalities are necessary in order to motivate people to go into the most difficult and important occupations. Brainstorm a list of occupations that meet these criteria: (1) perform an extremely important role in society and (2) require a high level of education and training. Then, go to the Bureau of Labor Statistics (BLS) website for Occupational Employment and Wage Estimates. Are the jobs you listed among the most highly paid? Do you think the occupations that have the highest average salaries (as listed by BLS) meet both of these criteria?

What do you think?

Conflict Perspective

According to the conflict perspective, the ruling class controls the economic system for its own benefit and exploits and oppresses the working masses. Conflict theorists use the term **disaster capitalism** to describe the pattern that occurs when elites use disruptive events—such as natural disasters or high unemployment—to push for policies that the public would be unlikely to accept in normal circumstances, such as lower wages, privatization, and deregulation. For example, many rural states facing high unemployment after local manufacturing plants send operations overseas are likely to offer tax exemptions and loosen labor protections in order to attract new manufacturers to the state.

The conflict perspective is also critical of ways that the government caters to the interests of big business at the expense of workers, consumers, and the public interest. This system of government that serves the interests of corporations—known as **corporatocracy**—involves ties between government and business. For example, Donald Trump, a wealthy businessman himself, chose a vice president and cabinet members who have strong ties to corporate interests. Former vice president Mike Pence has ties to billionaire Republican David Koch, co-owner of Koch Industries. One of the biggest companies in the United States, Koch Industries operates a number of businesses including those involved with petroleum, chemicals, pulp and paper, plastics, ranching, and finance. During his political career, Pence has also received generous monetary support from the finance sector and from the construction, pharmaceutical, and chemical industries. Other high-level officials with corporate ties under the Trump administration included former Secretary of State Rex Tillerson who was a prior CEO of Exxon Mobile and former head of the Department of the Treasury Steven Mnuchin who had been a senior executive at Goldman Sachs, a multinational finance company. Critics have likewise raised alarms over the corporate connections of many of Biden's cabinet appointees. Secretary of the Treasury Janet Yellen, Secretary of State Antony Blinken, and Director of National Intelligence Avril Haines collectively made millions of dollars in speaking and consulting fees from Wall Street banks, executives, and large corporations shortly before being confirmed to Biden's cabinet (Thompson and Meyer 2021).

The pervasive influence of corporate power in government exists worldwide. The policies of the International Monetary Fund (IMF) and the World Bank pressure developing countries to open their economies to foreign corporations, promoting export production at the expense of local consumption, encouraging the exploitation of labor as a means of attracting foreign investment, and hastening the degradation of natural resources as countries sell their forests and minerals to earn money to pay back loans. In his book *Confessions of an Economic Hit Man*, John Perkins (2004) described his prior job as an "economic hit man"—a highly paid professional who would convince leaders of poor countries to accept huge loans (primarily from the World Bank) that were much bigger than the country could possibly repay. The loans would be used to help develop the country by paying for needed infrastructure, such as roads, electrical plants, airports, shipping ports, and industrial plants. One of the conditions of the loan was that the borrowing country had to give 90 percent of the loan back to U.S. companies (such as Halliburton or Bechtel) to build the infrastructure. The result: The wealthiest families in the country benefit from additional infrastructure, and the poor masses are stuck with a debt they cannot repay. The United States uses the debt as leverage

disaster capitalism An economic pattern that occurs when elites use disruptive events—such as natural disasters or high unemployment—to push for policies that the public would be unlikely to accept in normal circumstances, such as lower wages, privatization, and deregulation.

corporatocracy A system of government that serves the interests of corporations and that involves ties between government and business.

Spencer Platt/Getty Images News/Getty Images

Protestors demonstrating outside of Trump Tower in New York City advocate for closing tax loopholes that allow transnational corporations to avoid paying taxes by holding their profits in offshore tax havens. An analysis from the Institute on Taxation and Economic Policy (Gardner, Roque, and Wamhoff 2019) found that in 2018, 91 of the most profitable corporations from the Fortune 500 paid no federal income taxes, including Amazon, Chevron, Halliburton, and IBM.

to ask for "favors," such as land for a military base or access to natural resources such as oil. According to Perkins, large corporations want "control over the entire world and its resources, along with a military that enforces that control" (quoted by MacEnulty 2005, p. 10).

Symbolic Interactionist Perspective

According to symbolic interactionism, the work role is a central part of a person's self-concept and social identity. When making a new social acquaintance, one of the first questions we usually ask is, "What do you do?" The answer largely defines for us who that person is. An individual's occupation is one of the person's most important statuses; for many, it represents a "master status," that is, the most significant status in a person's social identity.

Symbolic interactionism emphasizes the fact that attitudes and behavior are influenced by interaction with others. The applications of symbolic interactionism in the workplace are numerous: Employers and managers use interpersonal interaction techniques to elicit the attitudes and behaviors they want from their employees; union organizers use interpersonal interaction techniques to persuade workers to unionize. Parents teach their young adult children important lessons about work and unemployment through interaction with them.

Symbolic interactionism also focuses on the importance of labels. In contrast to the structural-functionalist perspective, symbolic interactionists emphasize that social prestige is an important component of occupations. Thus some jobs that are difficult, socially important, and yet relatively underpaid—for example, teachers, nurses, police, and firefighters—are still highly sought after because they are accompanied by a high degree of social prestige. Conversely, social labels may also make it more difficult for some workers to successfully negotiate for higher pay or better working conditions. For example, stereotypes about women's natural caretaking instincts and assumptions that caring for others is intrinsically satisfying is often used to justify lower wages in female-dominated occupations such as teaching and nursing (England 2005; Hebson, Rubery, and Grimshaw 2015; Dill, Erickson, and Diefendorff 2016). In other words, these stereotypes perpetuate the myth that the satisfaction of contributing to the betterment of society should be considered a reward in and of itself and that workers in these fields do not deserve higher pay.

Problems of Work and Unemployment

In this section, we examine unemployment and other problems associated with work. Poverty, minimum wage and living wage issues, workplace discrimination, and retirement and employment concerns of older adults are discussed in other chapters. We discuss problems concerning unemployment, child labor, forced labor, sweatshop labor, health and safety hazards in the workplace, alienation, job stress, work/life conflict, and labor unions and the struggle for workers' rights.

Unemployment and Vulnerable Employment

Prior to the outbreak of the COVID-19 pandemic in late 2019 and early 2020, 5.4 percent of the global population—approximately 188 million people worldwide—were unemployed (International Labour Office [ILO] 2020a). A large share of workers worldwide—45 percent—were in "vulnerable" forms of employment in 2019, with limited access to forms of social protection. Nearly one in five workers in the world live in "working poverty," earning less than US$3.20 per day, with most of these workers concentrated in Southern Asia and sub-Saharan Africa. Unemployment rates are higher among women than among men, but women experience a higher rate of extreme working poverty than men.

The COVID-19 pandemic created shockwaves throughout global job markets in 2020. The ILO (2020b) estimates that 94 percent of the world's workers in 2020 were living in countries with some form of workplace closure measures in place. The widespread job losses and reduced work hours resulting from these workplace closures have disproportionately impacted younger workers, a pattern that ILO warns could lead to a "lockdown generation" of workers who feel the impact of the social and economic consequences of the pandemic throughout their working lives. The ILO calls on world governments to address the employment crisis sparked by the global pandemic by focusing on five key challenges (ILO 2020c):

1. *Finding the right balance of economic and social policy interventions.* Premature reopening of economic activities can lead to surging cases of the virus, which will then prolong and worsen the long-term impact of the pandemic on the labor market.
2. *Implementing large-scale policy interventions.* The immense magnitude of the labor market disruptions will require widespread and financially intensive policy responses to counter the dangers of growing poverty, inequality, joblessness, and exclusion from economic activity.
3. *Policy measures targeted and prioritized toward the groups hit hardest by the labor market disruptions.* These groups are women, young people, migrants, and informal workers.
4. *Global cooperation to fill the "stimulus gap."* In this gap, wealthy countries are able to fund income support during mandated business closures, whereas governments in low- and middle-income countries cannot, thus furthering the global economic development divide.
5. *Robust social dialogue.* This is needed to meet the increasingly complex challenges of the pandemic.

In the United States, the official **unemployment** rate includes individuals who are currently unemployed, are available for employment, and have looked for work in the past four weeks. The U.S. unemployment rate dipped to a historic low of 3.5 percent in February 2020, before spiking up to 14.7 percent in April as a result of the COVID-19 pandemic (see Figure 7.2). In contrast, previous historically high levels of unemployment

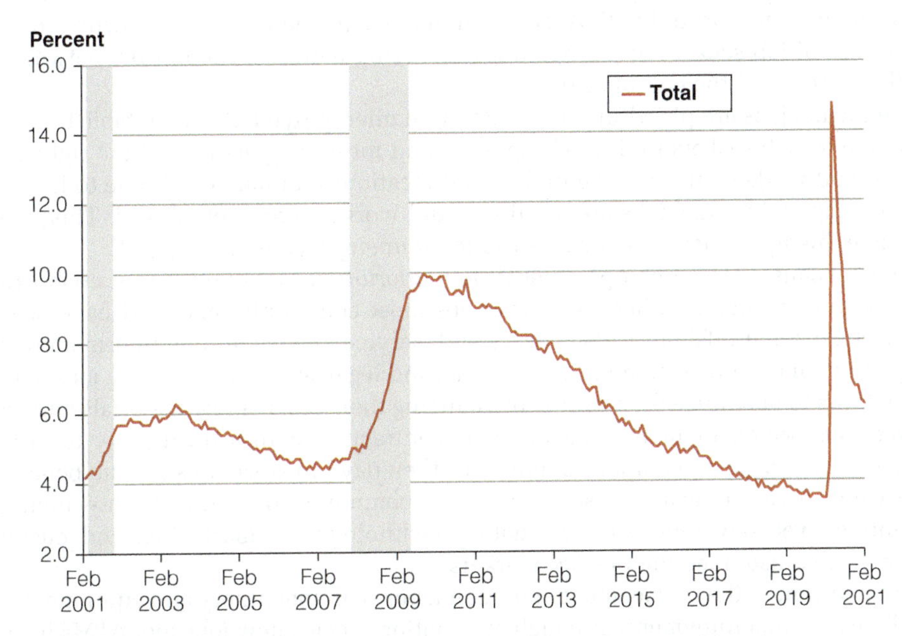

Figure 7.2 Unemployment Rate among U.S. Civilian Workers Age 20 and Older, 2000–2020
SOURCE: U.S. Bureau of Labor Statistics 2021.

unemployment To be currently without employment, actively seeking employment, and available for employment, according to U.S. measures of unemployment.

occurred during the Great Recession (10 percent in 2009) and the Great Depression (24.9 percent in 1933).

A **recession** refers to a significant decline in economic activity spread across the economy and lasting for at least six months. During times of economic recession, rates of unemployment are disproportionately high among certain groups, including racial and ethnic minorities (see also Chapter 9) and among those with lower levels of education. In April 2020, the unemployment rate among White workers in the United States was 13.3 percent, compared to 16.8 percent among Black workers and 17.6 percent among Hispanic workers (U.S. Bureau of Labor Statistics [BLS] 2020a).

The official unemployment rate does not include people who are "marginally attached" to the labor force—individuals who want a job, have searched for work in the past 12 months, but have not looked for a job in the past four weeks. These "marginally attached" persons may have not looked for a job in the past four weeks because they are discouraged over job prospects or because they have family responsibilities, school or job training commitments, ill health, or disability. The official unemployment rate also does not include involuntary part-time workers—those who want to work full-time but who settle for a part-time job, nor does it include highly skilled and highly educated individuals who are working in low-skill, low-paying jobs. Because the official unemployment rate does not include these marginally attached and involuntary part-time workers, it only partly represents the percentage of people whose employment needs are not being met.

Causes of Unemployment. A primary cause of unemployment is lack of available jobs. In February 2020, jobs were plentiful: the ratio of job seekers to job openings was 0.8 to 1, meaning there were not enough people seeking employment to fill all the jobs that were available. In April of 2020, that ratio spiked to 4.6 job seekers for every job that was available (BLS 2020a). At the peak of the Great Recession in July 2009, there were 6.4 job seekers for every available job. At that time, however, the unemployment rate was hovering close to 10 percent, lower than the 2020 peak of 14.7 unemployment that was accompanied by a lower ratio of 4.6 job seekers per job. This is because the COVID-19 pandemic dissuaded many unemployed people from seeking new jobs, which illustrates the fact that the official unemployment rate doesn't necessarily reflect the realities of labor force participation among the whole population. This chapter's *World in Quarantine* feature focuses on the plight of working mothers who felt forced out of the labor market due to school and day care closures. Still, the years between the end of the Great Recession and the start of the COVID-19 pandemic showed an overall downward trend in the ratio of job seekers to job openings, hovering between 1.3 and 0.8 between the middle of 2015 and the end of 2019.

Even when jobs are plentiful, some amount of unemployment is inevitable because it takes time for job seekers to find the right job and for employers to find the right workers. Jobs and workers may also be in different locations and may need time to find each other, during which time jobs are not filled, and workers are unemployed. This type of inevitable unemployment is known as *frictional unemployment.*

Another cause of U.S. unemployment is *job exportation*, also referred to as **offshoring**—the relocation of jobs to other countries. Jobs most commonly offshored have been in manufacturing, but offshoring also occurs with service jobs, including information technology, human resources, finance, purchasing, and legal services. Exporting jobs enables corporations to maximize their profits by reducing their costs of raw materials and labor. **Outsourcing**, which involves a business subcontracting with a third party to provide business services, saves companies money as they pay lower salaries and no benefits to those who provide outsourced services. Many commonly outsourced jobs—including accounting, web development, information technology, telemarketing, and customer support—are outsourced to non-U.S. workers.

Automation, or the replacement of human labor with machinery and equipment, also contributes to unemployment, although automation creates new jobs too. ATMs have reduced the need for bank tellers, but at the same time, they have created jobs for workers who produce and service ATMs. Emerging 3-D printing technologies—which involve

recession A significant decline in economic activity spread across the economy and lasting for at least six months.

offshoring The relocation of jobs to other countries where products can be produced more cheaply.

outsourcing A practice in which a business subcontracts with a third party to provide business services.

automation The replacement of human labor with machinery and equipment.

downloading a digital design for a product and then producing the product through specialized plastic or other raw materials—illustrate this trend. When 3-D printing technology first emerged, economic analysts were concerned that it would lead to massive disruptions in the global assembly line workforce. Recent evidence indicates, however, that large manufacturers have not adopted the technology for large-scale production. Instead, 3-D printing is having the biggest impact on those industries that manufacture small quantities of high-value and highly customizable products, such as medical devices and cosmetics (Dreyfus 2019). Instead of displacing large sectors of the manufacturing workforce as originally anticipated, 3-D printing technology has led many companies to invest in high-skill technology training for their workforces.

(Left) PA Images/Alamy Stock Photo; (Right) Traimak Ivan/Alamy Stock Photo

An automotive manufacturing plant in 1950 (left) and 2019 (right) illustrates the impact that automation has had on the labor force over the past century.

Another cause of unemployment is increased global and domestic competition. Mass layoffs in the U.S. automobile industry have occurred, in part, due to competition from makers of foreign cars who, unlike U.S. automakers, do not have the burden of providing health insurance for their employees. Unemployment also results from mass layoffs that occur when plants close and companies downsize or go out of business.

Effects of Unemployment on Individuals, Families, and Societies. One study found that poor-quality jobs—those with high demands, low control over decision making, high job insecurity, and low pay—had more negative effects on mental health than having no job at all (Butterworth et al. 2011). Nevertheless, unemployment has been linked to depression, low self-esteem, and increased mortality rates, and these negative health effects can be offset by increasing unemployment benefits (Shahidi et al. 2019). One study found that, among workers with no preexisting health conditions, losing a job because of a business closure increased the risk for a new health problem by 85 percent, with the most common health problems including hypertension, heart disease, and arthritis (Strully 2009). Unemployment is also a risk factor for homelessness, substance abuse, and crime, as some unemployed individuals turn to illegitimate, criminal sources of income, such as theft, drug dealing, and prostitution.

Long-term unemployment—which is defined as being unemployed for 27 weeks or longer—can have lasting effects, such as increased debt, diminished retirement and savings accounts (which are depleted to meet living expenses), home foreclosure, and/or relocation from secure housing and communities to unfamiliar places to find a job. But, even when individuals who are laid off find another job in one or two weeks, they still suffer damage to their self-esteem from having been told they are no longer wanted or needed at their workplace. And being fired affects worker trust and loyalty in future jobs. Employees who are not fired during a mass layoff are also affected, as they worry that "their job could be next" (Uchitelle 2006).

In families, unemployment is also a risk factor for child and spousal abuse and marital instability. When an adult is unemployed, other family members are often compelled to work more hours to keep the family afloat. And unemployed noncustodial parents, usually fathers, fall behind on their child support payments. Emerging research on the employment disruptions caused by the COVID-19 pandemic has been linked to increases in domestic violence, child abuse, drug and alcohol abuse, anxiety, depression, and suicide (Khan et al. 2020).

Plant closings and large-scale layoffs affect communities by lowering property values and depressing community living standards. High numbers of unemployed adults create a drain on societies that provide support to those without jobs. Job displacement is linked to disruptions in a person's social network connections, contributing to greater isolation and declines in physical and psychological well-being (Brand 2015; Rözer et al. 2020). Globally, the high numbers of young adults

long-term unemployment Unemployment that lasts for 27 weeks or more.

without jobs create a risk for crime, violence, and political conflict (World Economic Forum 2020). Unemployment creates a vicious cycle: The unemployed (as well as those who fear job loss) cut back on spending, which hurts businesses that then must cut jobs to stay afloat.

Barry Chin/Boston Globe/Getty Images

The class of 2020 is considered an "unlucky cohort" (Schwandt and von Wachter 2019); by entering the labor market during an economic recession, they are likely to experience higher unemployment and lower wages over the next decade compared to those who graduated in 2019.

Employment Concerns of Recent College Grads

In general, adults (ages 25 and older) with a two- or four-year college degree are less likely to be unemployed than adults who have not attended college. But recent college grads today face a challenging situation: Decades of stagnating wages, a labor market increasingly polarized into high-skill and low-skill jobs, and higher competition among college graduates have slowed career advancement opportunities. The economic recession prompted by the COVID-19 pandemic has added an additional layer of obstacles to recent college grads: Graduates entering the labor market for the first time during an economic recession show higher rates of unemployment and lower earnings for at least 10 to 15 years following the end of the recession, compared with graduates entering the labor force just one or two years earlier (Schwandt and von Wachter 2019).

Recent college grads (ages 21 to 24) also face stagnant wages and declining pension benefits. The average hourly wage of a recent college grad was $20.74 in 2019 ($43,100 for a full-time, full-year worker)—just pennies more than in 2000 (Gould, Mokhiber, and Wolfe 2019). This average masks the gender inequality in wages of recent college grads. Between 2000 and 2019, the average gender wage gap between recent college graduates increased from 10.7 percent to 12.9 percent. This gap translates into approximately $5,900 per year in lower earnings for women compared to men. The researchers conducting this analysis argue that even though women earn bachelor's degrees at higher rates than men:

> this advantage is doing little to insulate them from wage gaps within educational categories. And different amounts of experience cannot account for these wage gaps: Given the tight age restriction in our estimates, both groups should have relatively similar work experience on average. The particularly disquieting fact is that these gaps exist from the very beginning of women's careers. (Gould, Mokhiber, and Wolfe 2019, p. 1)

For economic reasons, including unemployment, low wages, and student loans and other college debts, many young adults end up living with their parents or other family, instead of on their own. High rates of unemployment in the summer of 2020 contributed to the dramatic increase in the number of young adults (age 18 to 29 years) living with their parents. While rates of young adults living with their parents had been steadily increasing since the 1960s—growing from 31 percent in 1970 to 46 percent in 2019—it jumped to 52 percent in July 2020, even higher than the previous historical high of 48 percent during the Great Depression in the 1930s (Fry, Passel, and Cohn 2020).

Decades of stagnating wages and perceptions of instability in labor due to outsourcing, offshoring, and automation have combined with the new high-tech global economy to create a new pattern in the labor force: the **gig economy**. While freelance and short-term forms of labor are nothing new, the Internet has prompted new forms of gig labor such as ride sharing, selling crafts through online retailers, and renting out property. The Federal Reserve (2020) estimates that nearly one-third of adults earned money through gig work in 2019, and for approximately one in ten adult workers, their gig occupied at least 20 hours of their workweek. While gig work can provide many advantages to workers, such as being able to control their own work hours, the evidence indicates that most of these workers are doing so out of need rather than preference. Among those surveyed by the Federal Reserve who performed gig work in the previous month, 47 percent had full-time jobs and 33 percent had part-time jobs.

Slavery

Slavery, also known as **forced labor**, refers to any work that is performed under the threat of punishment and is undertaken involuntarily. An estimated 5.4 out of every 1,000 people throughout the world live in slavery, with North Korea, Eritrea, Burundi, the Central African Republic, and Afghanistan having the highest prevalence of forced labor (*Global Slavery Index* 2018). Slaves are found working in agriculture, domestic work, food service, construction, garment manufacturing, the sex industry, factories, mining, and other sectors. In South Asia, where sex slavery is most common, either girls are forced into prostitution by their own husbands, fathers, and brothers to earn money to pay family debts, or they are lured by offers of good jobs and then are forced to work in brothels under the threat of violence.

The form of slavery most people are familiar with is **chattel slavery**, in which slaves are considered property that can be bought and sold. Although chattel slavery still exists in some areas, most forced laborers today are not "owned" but are rather controlled by violence, the threat of violence, and/or debt. The most common form of forced labor today is called *bonded labor*. Bonded laborers are poor individuals who take out a loan simply to survive or to pay for a wedding, funeral, medicines, fertilizer, or other necessities. Debtors must work for the creditor to pay back the loan, but they are often unable to repay it. Creditors can keep debtors in bondage indefinitely by charging the debtors illegal fines (for workplace "violations" or for poorly performed work) or charge laborers for food, tools, and transportation to the work site while keeping wages too low for the debt to ever be repaid.

There are an estimated 403,000 slaves in the United States, most of whom were trafficked into the country and forced into slavery, most commonly in domestic work, farm labor, and the sex industry (*Global Slavery Index* 2018; Skinner 2008). Migrant workers are tricked into working for little or no pay as a means of repaying debts from their transport across the U.S. border. Migrant workers are particularly vulnerable because if they try to escape and report their abuse, they risk deportation. Traffickers posing as employment agents lure women into the United States with the promise of good jobs and education but then place them in "jobs" where they are forced to do domestic or sex work.

In the United States, the Thirteenth Amendment provides that slavery or involuntary servitude may be used as a punishment for crime. As a result, many inmates in U.S. state and federal prisons work for little or no pay, often under unsafe conditions. In the federal prison system, inmates earn between 12 and 40 cents per hour (Federal Bureau of Prisons 2020). Prison labor is deeply woven into the U.S. economy. The nonprofit group Worth Rises estimates that more than 4,100 corporations in the United States profit from the use of prison labor (Worth Rises 2020). The precise impact of prison labor on the U.S. economy is unknown because it is not tracked by government employment numbers and private prisons are not required to report their labor statistics. The most recent census of prisons, conducted in 2005, found that manufacturing work performed by inmates accounted for 4 percent of all manufacturing jobs in the United States (National Public Radio [NPR] 2020).

gig economy A labor force pattern characterized by people performing free-lance or temporary work for other individuals, typically connected through an on-line platform.

slavery Any work that is performed under the threat of punishment and is undertaken involuntarily; also known as *forced labor*.

forced labor See **slavery**.

chattel slavery A form of slavery in which slaves are considered property that can be bought and sold.

What do you think?

Although inmates technically volunteer to work, most say prison rules compel them to work, even against their will. For example, refusing to work is considered a rule violation, which can result in being ineligible for parole, being placed in solitary confinement, or being placed in housing units with rampant violence (NPR 2020). Furthermore, inmates are pressured to work in order to pay for things like phone calls or hygiene items from the prison canteen. Prison administrators justify the extraordinarily low wages of inmates by arguing that they defray the costs of imprisonment: In most cases, inmates are actually paid fair market value for the labor, but their paychecks are cut by 80 percent to pay for the cost of their room and board. Do you think prisoners should be compelled to work to pay for the cost of their incarceration?

Sweatshop Labor

sweatshops Work environments that are characterized by less than minimum wage pay, excessively long hours of work (often without overtime pay), unsafe or inhumane working conditions, abusive treatment of workers by employers, and/or the lack of worker organizations aimed to negotiate better working conditions.

Millions of people worldwide work in **sweatshops**—work environments that are characterized by less than minimum wage pay, excessively long hours of work (often without overtime pay), unsafe or inhumane working conditions, abusive and discriminatory treatment of workers by employers, and/or the lack of worker organizations aimed at negotiating better working conditions. Sweatshop labor conditions occur in a wide variety of industries, including garment production, manufacturing, mining, and agriculture.

Sweatshop Labor in the United States. Many Americans believe they can avoid buying unethically produced goods by only buying products with a "Made in America" label. However, investigations by the U.S. Department of Labor found that many high-profile clothing retailers, such as Forever 21, TJ Maxx, and Fashion Nova, were able to avoid complying with labor laws by acquiring their garments through a complex labyrinth of factory middlemen (Kitroeff 2019). One worker cited in the investigation stated that she earned $270 for a seven-day workweek, in dirty conditions with rats and cockroaches. Most garment workers in the United States are immigrant women who typically work 60 to 80 hours a week, often earning less than minimum wage with no overtime, and many face verbal and physical abuse.

Go Nakamura/Bloomberg/Getty Images

These firefighters are inmates in the California Department of Corrections, earning $1.90 per day. For decades, the state of California has used inmate firefighters during its annual wildfire season but did not allow those with violent criminal records to work as civilian firefighters after their term was completed. In response to mounting public pressure during the unprecedented 2020 wildfire season, California governor Gavin Newsom signed a law to provide a pathway toward firefighter careers for those inmates who completed the prison firefighting program.

Immigrant farmworkers, who process most of the fruits and vegetables grown in the United States, also work under sweatshop conditions. Many live in substandard and crowded housing provided by their employers and lack access to safe drinking water as well as bathing and sanitary toilet facilities (see Chapter 9). Farmworkers commonly suffer from heat exhaustion, back and muscle strains, injuries resulting from the use of sharp and heavy farm equipment, and illness resulting from pesticide exposure (Austin 2002). Working long hours under hazardous conditions, farmworkers have the lowest annual family incomes of any U.S. wage and salary workers, and more than one-third of them live in poverty (U.S. Department of Labor 2018).

Child Labor

Child labor involves a child performing work that is hazardous, that interferes with a child's education, or that harms a child's health or physical, mental, social, or moral development. Even though virtually every country in the world has laws that prohibit or limit the extent to which children can be employed, child labor persists throughout the world. The most recent report from the International Labour Organization (ILO 2017) estimates that 152 million children are engaged in child labor, of whom 78 million worked in hazardous conditions. Nearly half of child laborers were under the age of 12, and 71 percent worked in agriculture. Some of the worst forms of child labor are forced labor, drug trafficking, sexual exploitation, and armed conflict.

AP Images/Brian Inganga

When her mother lost her job as a school custodian due to the COVID-19 pandemic, 10-year-old Irene Wanzila, along with her mother and two siblings, had to take jobs breaking rocks at a Kenyan quarry. In Africa, nearly 20 percent of children are engaged in child labor, a rate that is two times the global average (ILO 2017). The United Nations warns that the COVID-19 pandemic will significantly reduce the gains that have been made in recent years in the fight against child labor.

Child labor is involved in many of the products we buy, wear, use, and eat. Child laborers work in factories, workshops, construction sites, mines, quarries, fields, and on fishing boats. Child laborers make bricks, shoes, soccer balls, fireworks and matches, furniture, toys, rugs, and clothing. They work in the manufacturing of brass, leather goods, and glass. They tend livestock and pick crops. Child laborers work long hours with few (or no) breaks or days off, often in unsafe conditions where they are exposed to toxic chemicals and/or excessive heat, and they endure beatings and other forms of mistreatment from their employers, all for as little as a dollar a day. This chapter's *Human Side* features stories of child laborers working in garment factories.

Child labor also exists in the United States in restaurants, grocery stores, meatpacking plants, garment factories, and agriculture. In most jobs, U.S. law requires children to be at least 16 years old to work, with a few exceptions, such as that 14- and 15-year-olds can work as cashiers, grocery baggers, and car washers. In farmwork, children are legally allowed to work as young as age 10 (with parental permission and outside of school hours, children over the age of 14 can work without parental permission). While it is difficult to accurately estimate the number of child farmworkers in the United States, researchers estimate that there are tens of thousands legally employed child farmworkers and likely tens of thousands more working illegally (Arnold et al. 2020). Child farmworkers are typically children of immigrants; many are U.S. citizens. Farmworkers—both adults and children—are exposed to chemicals, sun, and temperature extremes; they work with sharp tools and heavy machinery, climb tall ladders, and carry heavy buckets and sacks. They frequently work 12-hour days without adequate access to drinking water, toilets, or hand-washing facilities. Although children are not legally permitted to purchase tobacco products, they are allowed to work on tobacco farms where they are vulnerable to "green tobacco sickness," a poisoning that occurs when workers absorb nicotine through the skin while handling tobacco plants. Child farmworkers are also especially vulnerable to heat-related illnesses, such as heat exhaustion and heat stroke (Arnold et al. 2020).

child labor Involves a child performing work that is hazardous, that interferes with a child's education, or that harms a child's health or physical, mental, social, or moral development.

Life as a Child Worker in a Garment Factory

Globally, nearly 10 percent of children are engaged in child labor (International Labour Organization [ILO] 2017). Although child labor is formally illegal in most countries, high poverty and lack of access to free public education drive many children to work to help support their families. The demand for low-cost clothing in North America and Europe is a primary driver of child labor around the world. The following are the stories of three children working in garment factories in Southeast Asia.

Feroza was sent by her parents to work in a garment factory when she was 12. Now a mother of six, she has since sent her two oldest daughters—Bithi and Dola—to work in the garment industry when they turned 12. "As a mother I feel sad," Feroza says, "but I still have to be realistic." Feroza's husband was injured and couldn't work, and the family often went hungry. "There was no food, not even rice. I cry when I remember those days. I thought it's better for us to die than not to have food." Bithi and Dola work sewing pockets on blue jeans. Bithi says she can sew 60 pockets an hour, helping to create 480 pairs of pants per day. These jeans will be sold in designer shops in the United States and Canada. For this work, they earn $1 per day (Nonkes 2015, p. 1).

When Mai's father was injured in a warehouse accident, she was forced to drop out of the eighth grade and start working to support her family. Due to her age, Mai was not legally allowed to work. However, because her parents and two younger sisters depended on her income, Mai lived in fear of losing her job:

> I wanted to become a teacher, but ended up as a worker in a garment factory due to my family's situation. ... I felt insecure whenever I heard the news that there would be an inspection in the factory. The factory simply warned us that if we were found, we would be laid off immediately. I often had to run and seek a hiding place such as a toilet, a store or a firewood pile at the backyard of the factory in order to not lose my job and income.

Mai also describes the grueling work conditions:

> In times of high demand, the workers have to work "all night" with a one-hour break for a nap on the concrete floor at dawn and then continue to work the next whole day. The line leader often came and checked my work speed and if she was not satisfied or found an error, she twisted my ear. It hurt me a lot. (Fair to Wear Foundation [FWF] 2018, p. 27)

Pria was forced to leave school at age 12 when her parents could no longer afford her school fees. First she began working alongside her parents harvesting green beans but then decided to apply to a garment factory in the hopes of earning more money to help support her family. Her job was to feed shirt buttons into tiny holes in the fabric, which required her to stand without rest from 8:00 a.m. until 6:00 p.m. every day. "My fingertips became swollen and painful," she says, "but I had to work standing constantly without even having a few minutes to sit and rest. My legs grew stiff and painful. When the electricity blackouts happened and there was no supervisor nearby, the whole room was suddenly dark and I got a chance to sit, but I had to immediately stand up when the electricity came on again" (FWF 2018, p. 35).

Health and Safety in the U.S. Workplace

Although many workplaces are safer today than in generations past, fatal and disabling occupational injuries and illnesses still occur in troubling numbers. In 2018, 5,250 U.S. workers—most of whom were men—died of fatal work-related injuries (BLS 2020b). The most common type of job-related fatality involves transportation accidents (see Figure 7.3). Nonfatal occupational injuries and illnesses in U.S. workplaces include sprains, strains, and cuts. Workers who do repeated motions such as typing or assembly line work are prone to repetitive strain injuries.

The incidence of occupational injury and illnesses is probably much higher than the reported statistics show. Government data on workplace illness and injury exclude many categories of workers including self-employed workers, federal government agency employees, farms with fewer than 11 workers, and private household workers (AFL-CIO 2020). Long-term illnesses caused by, for example, exposure to carcinogens often are difficult to relate to the workplace and are not adequately recognized and reported. Employees don't always report workplace injuries or illnesses for fear of being disciplined, losing their jobs, or, in the case of undocumented immigrants, fear of being deported. Employers discourage reporting of injuries by offering financial rewards to workers and departments when no injuries are reported for a specified period of time. Even when workers report their injuries, employers don't always keep accurate records because they

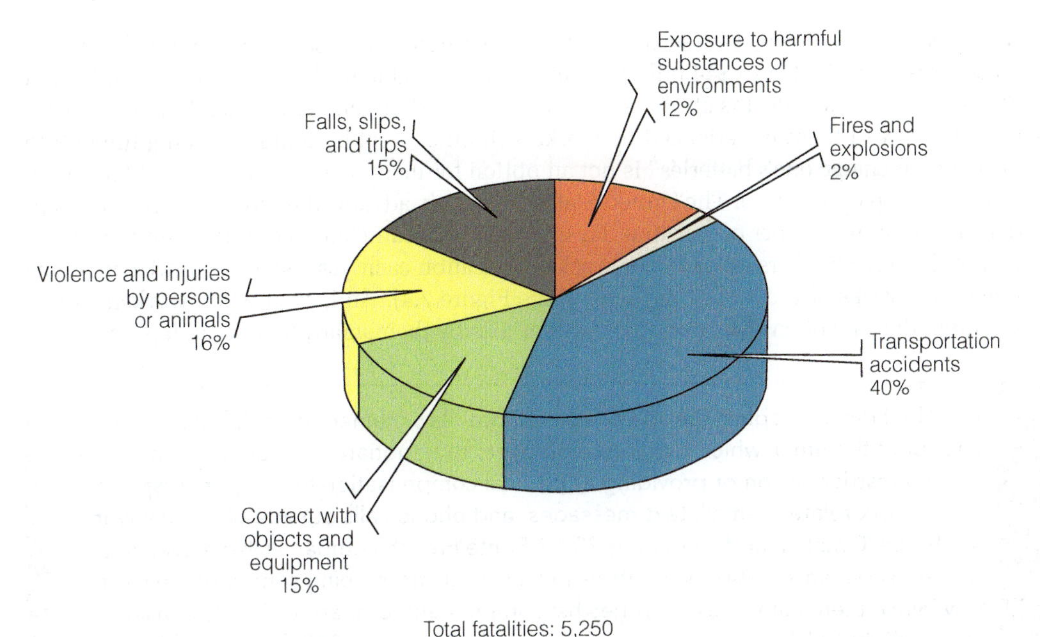

Figure 7.3 Causes of Workplace Fatalities, 2018
Note: Total fatalities: 5,250
SOURCE: Bureau of Labor Statistics 2020.

fear increased workers' compensation costs, increased scrutiny from government workplace inspectors, and fines for possible violations of health and safety regulations.

Workplace inspections reveal that many workplace injuries, illnesses, and deaths are due to employers' willful violations of health and safety regulations. The Occupational Safety and Health Administration (OSHA), created in 1970, develops, monitors, and enforces health and safety regulations in the workplace. But inadequate funding, hiring freezes, and failure to fill open positions have left OSHA unable to provide enough workplace inspectors to do the job effectively. The International Labour Organization recommends one inspector per 10,000 workers; in 2019, there was one federal or state OSHA inspector for every 88,149 workers. At current staffing levels, it would take federal OSHA inspectors, on average, 162 years to inspect each workplace just once (AFL-CIO 2020).

What *do you* think?

Suppose that a corporation is guilty of a serious violation of health and safety laws, in which "serious violation" is defined as one that poses a substantial probability of death or serious physical harm to workers. What penalty do you think that corporations should pay for such a violation? Serious violations of workplace health and safety laws in 2019 carried an average federal penalty of $3,717 or a state penalty of $2,032 (AFL-CIO 2020). Even when violations result in worker fatalities, the median penalty in 2019 was only $17,830 for federal and state penalties combined. Under criminal law, a willful violation that results in a worker's death is considered a misdemeanor. Since 1970, only 99 cases of willful workplace health and safety violations have been prosecuted, with defendants serving a total of 112 months in jail (AFL-CIO 2020). Why do you think penalties for violating workplace health and safety laws are so weak?

Job Stress

A Gallup Poll found that 26 percent of U.S. workers are completely or somewhat dissatisfied with the amount of stress in their jobs, making job stress workers' most common complaint (Gallup News 2020). Prolonged job stress, also known as **job burnout**, can cause or contribute to physical and mental health problems, such as high blood pressure, ulcers, headaches, anxiety, and depression. A U.S. study of ten common job stressors found that more than 120,000 people die each year from physical and mental health problems associated with

job burnout Prolonged job stress that can cause or contribute to high blood pressure, ulcers, headaches, anxiety, depression, and other health problems.

workplace-related stress (Goh et al. 2019). Furthermore, the researchers found that workplace stress contributed to $44 billion in annual preventable health care costs in the United States. These rates of deaths attributed to work-related stress and associated financial health costs far exceed those experienced by workers in other wealthy nations. Taking time off to heal and "recharge one's batteries" is not an option for many workers who do not have paid sick leave or paid vacation. The United States is the only advanced nation that does not mandate a minimum number of vacation days; in the European Union countries, employers are required to give workers at least four weeks of vacation each year (some countries mandate five or six weeks of vacation) (Maye 2019; see Figure 7.4). And, even when workers are on vacation, they are often still connected to their jobs by their smartphones and laptops.

What do you think?

In 2016, France adopted the so-called El Khomir law (named after the Minister of Labour at the time), which required employers to negotiate with labor unions on setting restrictions on or providing additional compensation for workers responding to work-related e-mail, text messages, and phone calls outside of normal working hours (Ornstein and Glassberg 2019). Some French companies have even set up e-mail systems that prevent after-hours e-mails from being sent until the next day. Since then, numerous countries have implemented or are currently considering similar Right to Disconnect laws, including Italy, Spain, the Philippines, Belgium, the Netherlands, Luxemburg, India, and Canada. Do you think the United States should implement a Right to Disconnect Law? If so, what types of rules and restrictions on after-hours work-related technology use would you want to see implemented?

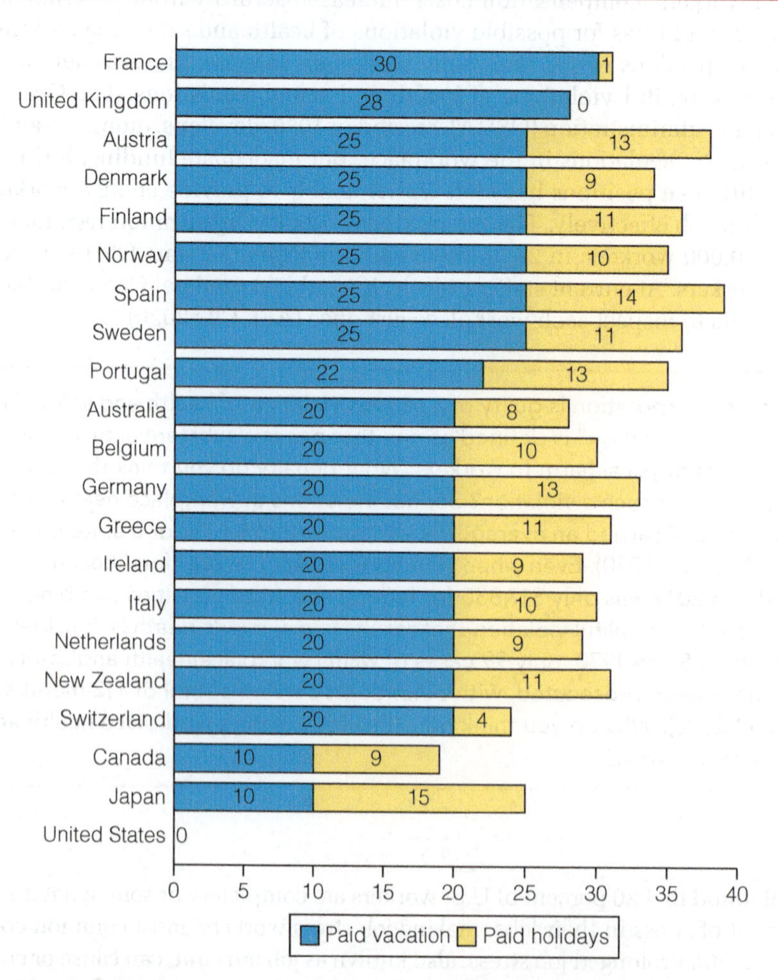

Figure 7.4 Minimum Mandatory Paid Working Leave Days, OECD Nations, 2019

SOURCE: Adewale Maye. (May, 2019). No-Vacation Nation, Revised. Retreived from https://cepr.net/images/stories/reports/no-vacation-nation-2019-05.pdf

Work/Life Conflict

A major source of stress for U.S. workers is **work/life conflict**—the day-to-day struggle to simultaneously meet the demands of work and other life responsibilities and goals, including family, education, exercise, and recreation. Employed spouses strategize ways to coordinate their work schedules to have vacation time together or simply to have meals together. Employed parents with young children must find and manage arrangements for child care and negotiate with their employers about taking time off to care for a sick child or to attend a child's school event or extracurricular activity. Some employers post their workers' weekly schedules only a few days in advance, making it difficult to arrange child care. In some two-parent households, spouses or partners work different shifts so that one adult can be home with the children. However, working different shifts strains marriage relationships because the partners rarely have time off together. Some employed parents who cannot find or afford child care leave their children with no adult supervision.

work/life conflict The day-to-day struggle to simultaneously meet the demands of work and other life responsibilities and goals, including family, education, exercise, and recreation.

The stress of managing work and child care falls disproportionately on mothers. When comparing fathers and mothers who work full-time, a Pew Research survey found that mothers were more likely than fathers to say that being a parent makes it harder for them to advance at work and that they have needed to reduce their work hours, felt like they couldn't give 100 percent at work, have had to turn down a promotion, have been passed over for a promotion or important assignment, or have been treated as if they weren't committed to their work (Horowitz 2019; see Table 7.1). Some of the stress of work/life conflict is likely due to the incongruence between personal preferences and cultural pressure. The same survey found that 84 percent of mothers who work full-time said working is personally best for them; however, only 33 percent of the general public agreed that it is ideal for women with young children to work full-time. The COVID-19 pandemic further exacerbated these work/life conflict issues, especially for mothers. This chapter's *World in Quarantine* discusses the pressures facing working mothers as they try to balance their own jobs with managing child care and virtual schooling in the face of widespread school and day care closures.

TABLE 7.1 Perceptions of Work/Life Challenges Faced by Mothers and Fathers, 2019

Percentage of employed parents agreeing with each statement:	Mothers	Fathers
Being a working parent has made it harder for me to …		
… advance in my job or career.	50	39
… be a good parent.	53	51
I have personally experienced ___ because I was balancing work and parenting responsibilities.		
… Needing to reduce work hours.	54	44
… Feeling like I couldn't give 100% at work.	51	43
… Turning down a promotion.	23	15
… Being treated as if I wasn't committed to my work.	27	20
… Being passed over for an important assignment.	19	14
… Being passed over for a promotion.	19	13

SOURCE: Horowitz 2019.

THE WORLD in quarantine

When the Motherhood Penalty Collides with a Pandemic

The COVID-19 pandemic upended many aspects of working life in the United States. While millions found themselves suddenly unemployed, many millions of others found themselves classified as "essential" employees with an increased workload and more job stress as they were confronted with longer hours and mounting health and safety concerns. Still others found themselves in the ultimate work–life balance conundrum as they began to work from home. For families with infants and school-aged children, widespread shutdowns required a rapid and dramatic reorganization of family life.

In 2020, more than 75 percent of employers reported having their employees work from home because of the pandemic (SHRM 2020). At the same time, an estimated 90 percent of children were doing some or all of their schoolwork from home (Psacharopoulos et al. 2020). Parents with children under the age of 18 make up one-third of all workers in the United States (Santhanam 2020). With schools and day cares closed, millions of working parents found themselves in an impossible situation. Attorney Keisha Hudson, with a toddler and an 8-year-old, said "We found it impossible to work our jobs and then be parenting and supervising and teaching. We're not giving 100 percent to our jobs, and we're not giving anything close to 100 percent to parenting at this moment" (Santhanam 2020, p. 1). The burden of this impossible task has fallen disproportionately on women.

One survey found that mothers reported high rates of feeling lonely or isolated (49 percent), feeling as though they did not have time for themselves (60 percent), feeling depressed (51 percent), and having recently cried due to feeling stressed, frustrated, or overwhelmed (42 percent). Fathers, on the other hand, reported these negative feelings at rates very close to nonparents (ranging between 19 and 41 percent on these indicators) (Cox and Abrams 2020).

The result has been that mothers are dropping out of the labor force in staggering numbers. Between August and September 2020, more than 80 percent of the people who left the labor force were women (Vesoulis 2020). Some of this pattern is due to layoffs in sectors like retail, food service, and child care—sectors that are dominated by female workers. But a large part of this pattern is due to women feeling pushed out of the labor force due to the demands of parenting and virtual learning.

The rapid drop in women's labor force participation is alarming many economists, who say school and child-care shutdowns will clearly exacerbate the existing **motherhood penalty**, the disadvantages in wages and hiring experienced by mothers compared to women without children. Dartmouth College economist Claudia Olivetti says that, "based on decades of research we know that there was one institution that was effective at limiting gender inequality and encouraging women's participation in the workplace, and it was early childhood education" (Taub 2020, p. 1). Not only will the pandemic lead to widening gender inequality, economists warn that women's reduced labor force participation will result in wider economic and educational inequality. In low-income or single-parent households, dropping out of the labor force means financial disaster. Instead, many working parents remain in the labor force and find they are unable to meet the emotional and educational needs of their children. While the long-term impact is not yet known, economists anticipate the results will be dire. Once out of the labor force, women struggle to get back in, and their earnings are adversely affected for the rest of their lives. It also has adverse effects on long-term financial recovery by decreasing economic output and increasing the relative cost of labor (Vesoulis 2020).

Stefan Wermuth/Bloomberg/Getty Images

The COVID-19 pandemic led to widespread school and day care shutdowns, leaving many working parents without child-care options. Mothers have been disproportionately impacted by this increasing intensification of work/life conflict.

The COVID-19 has laid bare a central weakness in the U.S. economy: There is no backup plan for parents. Without federally mandated paid leave or universal child-care options, working parents find themselves in an impossible dilemma when crisis hits.

motherhood penalty The tendency for women with children, particularly young children, to be disadvantaged in hiring, wages, and the like, compared to women without children.

Employees with elderly and/or ill parents worry about how they will provide care for their parents or arrange for and monitor their care, while putting in a 40-hour (or more) workweek. Although there have been some increases in the number of workplaces that offer benefits to new parents, such as paid parental leave or onsite lactation rooms, only 10 percent of workplaces provide some form of leave benefits for those who provide elder care to older family members (Society for Human Resource Management [SHRM] 2019).

Balancing the responsibilities of a job with the demands of school is also a challenge for many adults. Among the 20 million college students in the United States, approximately 70 percent of them work, including 3.3 million who attend school, go to work, and have children (Carnevale and Smith 2018).

> "Shonda, how do you do it all? The answer is this: I don't. Whenever you see me somewhere succeeding in one area of my life, that almost certainly means I am failing in another area of my life. If I am killing it on a *Scandal* script for work, I am probably missing bath and story time at home. ... If I am succeeding at one, I am inevitably failing at the other. That is the tradeoff. That is the Faustian bargain one makes with the devil that comes with being a powerful working woman who is also a powerful mother. You never feel a hundred percent OK. ... Something is always lost. Something is always missing."
>
> **–SHONDA RIMES**

Alienation

Work in industrialized societies is characterized by a high degree of division of labor and specialization of work roles. As a result, workers' tasks are repetitive and monotonous and often involve little or no creativity. Limited to specific tasks by their work roles, workers are unable to express and utilize their full potential—intellectual, emotional, and physical. According to Marx, when workers are merely cogs in a machine, they become estranged from their work, from the product they create, from other human beings, and from themselves. Marx called this estrangement "alienation."

Alienation in the workplace has four components: (1) *Powerlessness* results from working in an environment in which one has little or no control over the decisions that affect one's work; (2) *meaninglessness* results when workers do not find fulfillment in their work; (3) workers may experience *normlessness* if workplace norms are unclear or conflicting, such as when companies have nondiscrimination policies yet they practice discrimination; and (4) *self-estrangement* may stem from workers' inability to realize their full human potential in their work roles.

The fast-food industry epitomizes working conditions that lead to alienation. But the principles that characterize the fast-food industry also characterize many other workplaces, a phenomenon known as **McDonaldization** (Ritzer 1995). McDonaldization involves four principles:

1. *Efficiency.* Tasks are completed in the most efficient way possible by following prescribed steps in a process overseen by managers.
2. *Calculability.* Quantitative aspects of products and services (e.g., portion size, cost, and the time it takes to serve the product) are emphasized over quality.
3. *Predictability.* Products and services are uniform and standardized. A Big Mac in Albany is the same as a Big Mac in Tucson. Workers behave in predictable ways. For example, servers at McDonald's learn to follow a script when interacting with customers.
4. *Control through technology.* Automation and mechanization are used in the workplace to replace human labor.

What are the effects of McDonaldization on workers? In a McDonaldized workplace, employees are not permitted to use their full capabilities, be creative, or engage in genuine human interaction. Workers are not paid to think, just to follow a predetermined set of procedures. Because human interactions are unpredictable and inefficient (they waste time), "we're left with either no interaction at all, such as at ATMs, or a 'false fraternization.' Rule number 17 for Burger King workers is to smile at all times" (Ritzer, quoted by Jensen 2002, p. 41). Workers may also feel that they are merely extensions of the machines they operate. The alienation that workers feel—the powerlessness and meaninglessness that characterize a "McJob"—may lead to dissatisfaction with one's job and, more generally, with one's life.

McDonaldization The process by which principles of the fast-food industry (efficiency, calculability, predictability, and control through technology) are being applied to more sectors of society, particularly the workplace.

Labor Unions and the Struggle for Workers' Rights

Having a job is no guarantee of having favorable working conditions and receiving decent pay and benefits. **Labor unions** are worker advocacy organizations that have developed to protect workers and represent them at negotiations between management and labor.

labor unions Worker advocacy organizations that developed to protect workers and represent them at negotiations between management and labor.

Benefits and Disadvantages of Labor Unions to Workers. Labor unions have played an important role in fighting for fair wages and benefits, healthy and safe work environments, and other forms of worker advocacy. Compared with nonunion workers, unionized

© Joe Rondone/The Commercial Appeal via Imagn Content Services, LLC

Highly routinized and repetitive work, such as that performed by these Amazon fulfillment center employees, can lead to feelings of alienation—a general sense of disconnection and dissatisfaction.

workers tend to have higher earnings, better insurance and pension benefits, and more paid time off. Union members also have higher life satisfaction than nonunion members, which is due not to higher income but rather to the fact that union members (1) experience greater satisfaction with their work experiences, (2) feel greater job security, (3) have more opportunities for social interaction and integration, and (4) benefit from democratic participation and citizenship within their union (Flavin and Shufeldt 2014).

One of the disadvantages of unions is that members must pay dues and other fees and that these dues have been rising in recent years. Union members resent the high salaries that many union leaders make. Another disadvantage for unionized workers is the loss of individuality. Unionized workers are members of an overall bargaining unit in which the majority rules. Decisions made by the majority may conflict with individual employees' specific employment needs.

What *do you* think?

One strategy labor unions have historically used to advance efforts for higher wages and safer working conditions is the collective bargaining agreement, which bars employers in some labor sectors, such as mining, from employing workers who don't belong to a union. As of 2020, 28 states had enacted right-to-work (RTW) laws, which prohibit these types of agreements. Advocates of RTW laws say that these types of "compulsory unionization" rules inhibit worker freedom and employment opportunities (National Right to Work Legal Defense Foundation 2020). Opponents argue that RTW laws are not designed to protect workers' rights but rather are typically the result of lobbying efforts by large corporations to weaken the effectiveness of union bargaining power. Because unions are legally required to advocate for all workers represented by a union contract, whether or not they are members of the union, RTW laws mean that nonunion members benefit from the union's bargaining, whether or not they have paid union dues. RTW laws therefore reduce the funds available to unions to advocate for worker pay, benefits, and workplace protections (Jones and Shierholz 2018). Do you think RTW laws are harmful or beneficial to workers?

Declining Union Density. The strength and membership of unions in the United States have declined over the last several decades. **Union density**—the percentage of workers who belong to unions—grew in the 1930s and peaked in the 1940s and 1950s, when 35 percent of U.S. workers were unionized. In 2019, the percentage of U.S. workers belonging to unions had fallen to 10.3 percent, down from 20.1 percent in 1983 (BLS 2019). Union membership among public-sector workers is more than five times the rate of union membership among private-sector employees.

One reason for the decline in union representation is the loss of manufacturing jobs, which tend to have higher rates of unionization than other industries. In addition, globalization has led to layoffs and plant closings at many unionized work sites as a result of companies moving to other countries to find cheaper labor. Corporations also take active measures to discourage workers from unionizing.

union density The percentage of workers who belong to unions.

Weak U.S. Labor Laws and Anti-Union Legislation. The 1935 *National Labor Relations Act* (NLRA) guarantees the right to unionize, to bargain collectively, and to strike against private-sector employees. However, in addition to excluding public-sector workers, the law excludes agricultural and domestic workers, supervisors, railroad and airline employees, and independent contractors. As a result, millions of workers do not have the right under U.S. law to negotiate their wages, hours, or employment terms.

Anti-union legislation refers to any law that undermines or weakens unions. Republicans are more likely than Democrats to support anti-union legislation. Right-to-work laws, which prohibit collective bargaining agreements that would restrict employers from hiring nonunion workers, are one example of anti-union legislation. This chapter's *Social Problems Research Up Close* examines the impact of right-to-work laws on local economies and labor union effectiveness. Other examples include state policies that restrict unions for public workers (such as teachers) from bargaining for anything beyond wages, a policy passed by Republican governor Scott Walker in Wisconsin in 2011.

Sometimes employers engage in "union-busting" tactics, such as intimidation, spying on union meetings, or threatening retaliation against employees who talk to union representatives. Although these tactics are illegal under the *National Labor Relations Act*, a recent study from the Economic Policy Institute found that employers in the United States violated federal law in 41.5 percent of all union election campaigns and that one in five union election campaigns involved illegal firing of employees for engaging in union activity (McNicholas et al. 2019). Employers spend approximately $340 million every year on hiring "union avoidance" consultants. Because of weak laws and anti-union legislation, the United States has one of the lowest union density rates among wealthy, democratic nations (see Figure 7.5).

> "Trade unions have been an essential force for social change, without which a semblance of a decent and humane society is impossible under capitalism."
>
> **–POPE FRANCIS**

Labor Union Struggles around the World. In 1949, the International Labour Office established the Convention on the Right to Organise and Collective Bargaining. About half of the world's workforce lives in countries that have not ratified this convention, including China, India, Mexico, Canada, and the United States. According to the International Trade Union Confederation (2020), in 2019:

- 85 percent of countries violated the right to strike;
- 80 percent of countries violated the right to collective bargaining;
- 74 percent of countries excluded workers from establishing or joining a trade union;

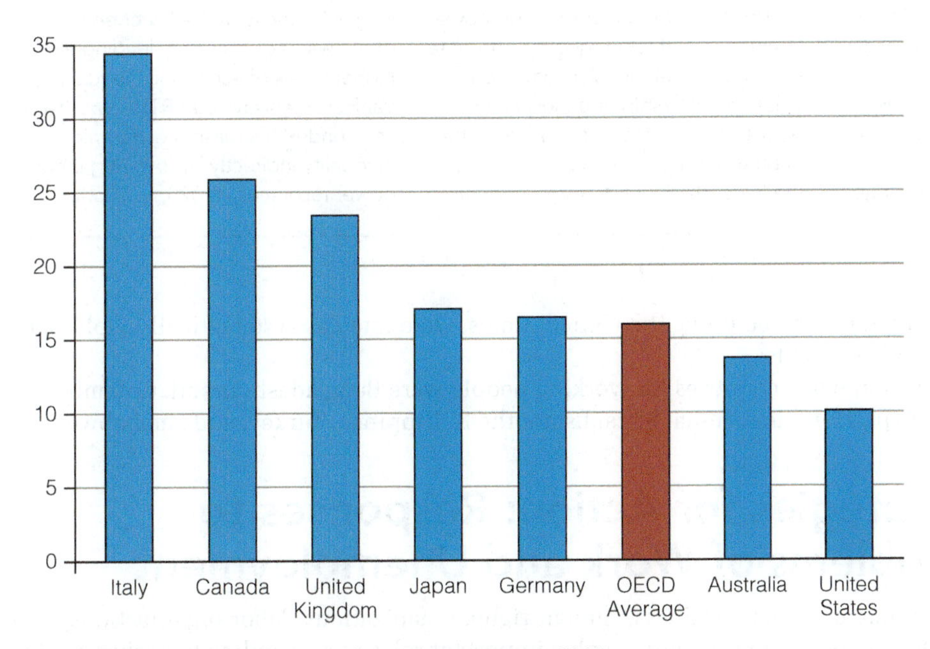

Figure 7.5 Union Density in Select OECD Nations, 2018
SOURCE: Organisation for Economic Co-operation and Development (OECD) 2019.

social problems RESEARCH UP CLOSE

What Impact Do Right-to-Work Laws Have on Workers and the Economy?

Proponents of right-to-work (RTW) laws often argue that these policies will lead to lower unemployment rates and stronger local economies by encouraging more hiring and business investment (Sherk 2011). Opponents of RTW argue that these laws depress wages and increase inequality by reducing the bargaining power of unions. RTW laws have existed since the 1940s and, as of 2020, are on the books in 28 states. This has provided a wealth of data to answer the question of what the impact of RTW laws are on workers and the local economy. University of Minnesota Sociologist Tom VanHeuvelen (2020) examined this data in order to understand how the passage of RTW laws affects the distribution of resources among workers within state economies.

Data and Variables

VanHeuvelen (2020) gathered national-level economic data from a variety of sources: the U.S. Census, the American Community Survey, the IRS, the U.S. Union Sourcebook, and the National Labor Relations Board. Using these sources, VanHeuvelen compiled economic data for each state over a 77-year period. The two main independent variables were (1) whether or not the state had passed an RTW law in a given year and (2) the percentage of the state's workforce that was in a union. The dependent variable was income inequality. In order to account for variations in local labor markets, the researcher also used a wide range of control variables including age, race, gender, population density, unemployment rate, political party in control of the state, and cost of living. VanHeuvelen analyzed the data using a two-way fixed effect regression model.

Findings

The results of the statistical analysis led the author to four main conclusions.

First, RTW laws operate differently in different political contexts. In states where union membership was high, the passage of RTW laws led to an increase in anti-union practices and filings of unfair labor practices compared to those states that passed RTW laws but had lower rates of union membership. In other words, "[F]irms tend to engage in these anti-union behaviors when RTW laws signal they may be effective, and stop when RTW laws signal that they may be unnecessary" (VanHeuvelen 2020, p. 1279).

Second, RTW laws have inconsistent direct associations with inequality. Although these laws do not universally contribute to higher levels of inequality, the author argues that this is because changes in bargaining power between unions and firms at the state level are not equally consequential.

Third, RTW laws remove the power of unions to affect inequality. The data show a clear relationship between union membership and inequality prior to the passage of RTW laws: the greater the union density in a state, the lower the level of wage inequality. However, once an RTW law is passed in a state, this association disappears. Thus "unions are consequential for reducing inequality outside RTW contexts, but not inside them" (VanHeuvelen 2020, p. 1281).

Fourth, RTW laws have different effects on the top and bottom ends of the wage distribution. The passage of RTW laws is associated with a decrease in inequality among those at the top of the wage distribution but an increase in inequality at the bottom of the wage distribution. Thus RTW laws do lead to higher wages for some workers but lower wages for others. This finding further emphasizes the decreased collective bargaining power available to workers at the lower end of the wage distribution after RTW laws are passed.

Conclusion

Based on these findings, VanHeuvelen concludes that RTW laws do have an effect on the distribution of economic resources but only in contexts where labor has something to lose. In local contexts where union participation is low and inequality is high, the passage of RTW laws has no discernible effect on the local economy. However, in places where union participation is high, RTW laws result in reduced wages for those at the lower end of the wage distribution, leading to higher levels of economic inequality. VanHeuvelen says that RTW laws "work as intended, increasing economic inequality indirectly by lowering labor power resources" (2020, p. 1300).

- 41 countries, including the United States, were rated as systematically violating workers' rights;
- the ten worst countries for working people were Bangladesh, Brazil, Colombia, Egypt, Honduras, India, Kazakhstan, the Philippines, Turkey, and Zimbabwe.

Strategies for Action: Responses to Problems of Work and Unemployment

Government, private business, human rights organizations, labor organizations, college student activists, and consumers play important roles in responding to problems of work and unemployment.

Strategies for Reducing Unemployment

Some economists argue that full employment should be a top priority of U.S. economic policy. **Full employment** is achieved when unemployment rates are below 5 percent, taking into account that some amount of unemployment is unavoidable (Paul et al. 2018). Full employment not only ensures that jobs are available for all workers but also creates labor scarcity, which gives workers more bargaining power for decent wages.

One policy option for achieving full employment is a federal job guarantee. During the Great Depression, the federal government implemented direct hiring programs such as the Works Progress Administration (WPA) and the Civilian Conservation Corps (CCC), which provided national economic relief, albeit only on a temporary basis. Advocates for a permanent federal job guarantee program argue that it would help to alleviate poverty and strain on social welfare systems, while also providing broad economic stimulus that would result in higher employment and wages in the private sector (Paul et al. 2018).

Another strategy for reducing unemployment is to ensure that current and future workers have the skills needed for employment. Technical and vocational high schools focus on occupational training in specific fields, such as health care, culinary arts, computing, construction, and business. In one survey, 73 percent of employers said that finding qualified candidates was somewhat or very difficult, and more than one-third said that colleges are not adequately preparing students for jobs (Bauer-Wolf 2019). In the same survey, 77 percent of college students expressed concern about whether they had the skills needed for a job. The skills employers said they most highly valued were listening skills, attention to detail, and effective communication.

Although educational attainment is often touted as the path to employment and economic security, an estimated 36 percent of U.S. jobs in 2020 will require only a high school diploma or less (Carnevale, Smith, and Strohl 2014); thus, it is important to ensure that all jobs pay a minimum "living wage" (see also Chapter 6). As long as our economy allows people who work full-time to earn poverty-level wages and fails to ensure affordable health care, housing, child care, and other necessities, having a job is not necessarily the answer to economic self-sufficiency.

> A job alone is not enough. Medical insurance alone is not enough. Good housing alone is not enough. Reliable transportation, careful family budgeting, effective parenting, effective schooling are not enough when each is achieved in isolation from the rest. There is no single variable that can be altered to help working people move away from the edge of poverty. Only where the full array of factors is attacked can America fulfill its promise. (Shipler 2005, p. 11)

Workforce Development. The federally funded *Workforce Investment Act* (WIA) and *Workforce Innovation and Opportunity Act* (WIOA) award stipends to some unemployed persons to help pay for further education and job training. However, the effectiveness of these programs is questionable. One evaluation study found that 30 months after enrolling, program participants did not exhibit higher rates of employment or earnings than nonparticipants (Fortson et al. 2017). Some workforce development programs focus on strategies to improve the employability of hard-to-employ individuals through providing targeted interventions such as substance abuse treatment, domestic violence services, prison release reintegration assistance, mental health services, and homelessness services, in combination with employment services (Institute for Research on Poverty 2019).

Job Creation and Preservation. Policy approaches to job creation and preservation vary depending on political orientation. Former President Trump campaigned on promises of revitalizing the coal industry and rolling back environmental regulations that he argued put unfair burdens on corporations and limited job growth. In contrast, the Biden administration has pledged to invest in research, technology, and low-carbon manufacturing in an effort to boost job growth in the clean energy sector, such as wind, solar, and nuclear energy. Part of this plan includes offering tax credits and subsidies for businesses to upgrade

full employment Exists when there are jobs available for all workers; usually when unemployment rates are below 5 percent.

global supply chains for garment and footwear production. This Transparency Pledge asks companies to annually disclose the locations and numbers of workers of every site involved in its supply chain, from the growers of the cotton, to the weaving of cloth, to the sewing and printing, and finally to the storefront. Since 2016, the number of companies committing to the Transparency Pledge has more than tripled, largely due to public pressure (Human Rights Watch 2019). Some prominent companies committed to the pledge are Nike, Lululemon, and Athletica.

What do you think? Human Rights Watch attributes the growing success of the Transparency Pledge to widespread public pressure on major corporations like Nike. Have you ever refused to buy a product due to reports of the company using sweatshop labor? How important is it to you to know that a company uses ethical labor practices in producing the goods that you buy?

@USAS

While @JeffBezos made over $55 billion in 2020, @amazon failed in key COVID-19 protection for workers and has fired workers for speaking out against unsafe and unfair working conditions! This #PrimeDay2020, USAS demands @JeffBezos do more to #ProtectAmazonWorkers!

-United Students Against Sweatshops

Student Activism. United Students Against Sweatshops (USAS), formed in 1997, is a grassroots organization of youth and students who fight against labor abuses and for the rights of workers around the world, including campus workers, fast-food workers, and garment workers who make collegiate licensed apparel. USAS student activists have influenced more than 180 colleges and universities to affiliate with the Worker Rights Consortium (WRC), which investigates factories that produce clothing and other goods with school logos to make sure that the factory meets the code of conduct developed by each school. A typical code of conduct includes fair wages, a safe working environment, a ban on child labor, and the right to be represented by a union or other form of employee representation. If the WRC investigation finds that a factory fails to meet the code of conduct, the companies—often well-known international brands—who purchase items from that factory are warned that their contract with the school will be terminated if working conditions at the factory do not improve. Most recently, local USAS chapters were involved in the successful 2020 ballot initiative to raise the Florida minimum wage to $15 per hour, successfully pressured the University of Illinois Chicago to provide PPE and hazard pay to student workers during the COVID-19 pandemic, and worked to ensure that Gonzaga University would publicly post their COVID-19 testing data (USAS 2020). Thinking beyond the 2020 election, USAS activists stated their agenda for the future: "As the student arm of the labor movement, we must think beyond our two-party system and towards a future where workers are in charge of their own destinies" (USAS 2020, n.p.).

Legislation. Perhaps the most effective strategy against sweatshop work conditions is legislation. Some U.S. states, cities, counties, and school districts have passed "sweat-free" procurement laws that prohibit public entities (e.g., schools, police, and fire departments) from purchasing uniforms and apparel made under sweatshop conditions (SweatFree Communities n.d.).

Establishing and enforcing labor laws to protect workers from sweatshop labor conditions is difficult in a political climate in which the leaders of transnational corporations hold more power and influence than their workers. This power imbalance leads to **policy drift**, or the failure to update labor laws to reflect changes in the broader society and economy. The *National Labor Relations Act* (NLRA) was enacted in 1935 and was last revised in 1959. As a result, "a system of regulation designed with New Deal-era industrial relations in mind governs a twenty-first century global economy" (Galvin and Hacker 2020, p. 224). Corporations with global reach and international supply chains are often able to skirt around anti-sweatshop policies through layers of subcontractors. The ever growing pace of globalization also makes it difficult for legislation to keep pace with changing conditions. While national governments are responsible for the enactment and enforcement of labor laws, transnational corporations have the flexibility to shift production sites when local labor policies prove unfavorable to their profits.

policy drift The failure to update labor laws to reflect changes in the broader society and economy.

behavior-based safety programs A strategy used by business management that attributes health and safety problems in the workplace to workers' behavior, rather than to work processes and conditions.

Responses to Workplace Health and Safety Concerns

In developing countries, governments fear that strict enforcement of workplace regulations will discourage foreign investment. Investment in workplace safety in developing countries, whether by domestic firms or foreign multinationals, is far below that in the rich countries. Unless global standards of worker safety are implemented and enforced in all countries, millions of workers throughout the world will continue to suffer under hazardous work conditions. Low unionization rates and workers' fears of losing their jobs—or their lives—if they demand health and safety protections leave most workers powerless to improve their working conditions.

In the United States, industry groups and politicians often argue that initiatives to strengthen workplace health and safety result in excessive regulation that is burdensome to business, hampers investment, and hurts job creation. Indeed, in his calls to reinvigorate the coal industry, former President Trump often touted the rolling back of "job-killing" mining regulations. One of these regulations was a policy that required that mine safety examinations be conducted before miners began working in a new area and required the reporting of any conditions that may adversely affect the health and safety of the miners. In 2018, the Trump administration enacted a change to this rule, saying that the safety examination could occur at the same time that miners began their work and that safety inspectors were not required to make their reports available to the miners. The rule change also involved removing the phrase "competent person" from the language of the regulation that specified who should be involved in conducting the safety inspection (Brookings Institute 2020). The Biden administration has publicly stated a plan to reinstate many of the workplace safety regulations that had been weakened or eliminated during the Trump administration, including increasing the number of OSHA and Mine Safety and Health Administration (MSHA) inspectors (Biden-Harris 2020b).

@Joe Biden

The 40-hour workweek
Minimum wage
Overtime pay
Health care
Workplace safety protections

They're all because of unions—and it's time we recognize that.

-Joe Biden

Behavior-Based Safety Programs. Instead of examining how work processes and conditions compromise health and safety on the job, **behavior-based safety programs** attribute workplace illnesses and injuries to workers' own carelessness and unsafe acts (Frederick and Lessin 2000). These programs focus on teaching employees and managers to identify, "discipline," and change unsafe worker behaviors that cause accidents and encourage a work culture that recognizes and rewards safe behaviors.

Critics contend that behavior-based safety programs divert attention away from the employers' failures to provide safe working conditions. They also say that the real goal of behavior-based safety programs is to discourage workers from reporting illness and injuries. Workers whose employers have implemented behavior-based safety programs describe an atmosphere of fear in the workplace, such that workers are reluctant to report injuries and illnesses for fear of being labeled "unsafe workers."

Ryan McGinnis/Alamy Stock Photo

Critics argue that behavior-based safety programs do not result in safer workplaces but serve primarily to discourage workers from reporting injuries.

Work/Life Policies and Programs

Policies that help women and men balance their work and family responsibilities are referred to by a number of terms, including *work/family*, *work/life*, and *family-friendly* policies. As shown in Figure 7.6, the United States lags far behind many other countries in national work/family provisions.

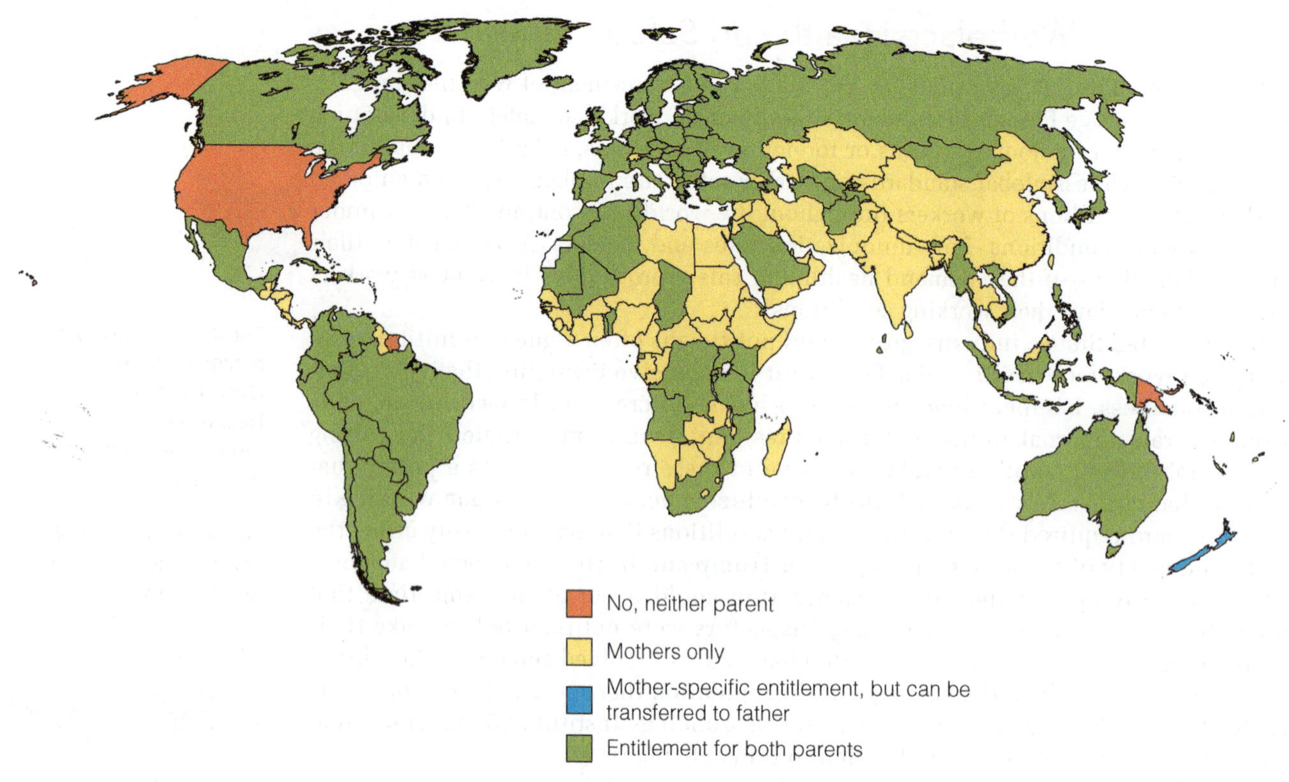

Figure 7.6 Global Availability of Paid Parental Leave, 2020
SOURCE: World Policy Center 2020.

Federal and State Family and Medical Leave Initiatives. The **Family and Medical Leave Act (FMLA)**, signed into law in 1993, requires employers (with 50 or more employees who each worked at least 1,250 hours in the preceding year) to provide up to 12 weeks of job-protected, *unpaid* leave so that they can care for a seriously ill child, spouse, or parent; stay home to care for their newborn, newly adopted, or newly placed foster child; or take time off when they are seriously ill. A 2008 amendment to the FMLA requires employers to provide up to 26 weeks of unpaid leave to employees to care for a seriously ill or injured family member who is in the armed forces, including the National Guard or Reserves. However, only about half of workers are eligible for FMLA benefits (Miller 2015). Some eligible workers do not use their FMLA benefit because they cannot afford to take leave without pay and/or they fear they will lose their job if they take time off. Some employers do not comply with the FMLA either because they are unaware of their responsibilities under FMLA or because they are deliberately violating the law. A national survey of employers found that one in four employers who are supposed to comply with FMLA fails to do so (Matos, Galinsky, and Bond 2017).

Family and Medical Leave Act (FMLA) A federal law that requires public agencies and companies with 50 or more employees to provide eligible workers with up to 12 weeks of job-protected, unpaid leave so that they can care for an ill child, spouse, or parent; stay home to care for their newborn, newly adopted, or newly placed foster child; or take time off when they are seriously ill, and up to 26 weeks of unpaid leave to care for a seriously ill or injured family member who is in the armed forces, including the National Guard or Reserves.

Out of 185 countries, only two lack guaranteed paid maternity leave: the United States and Papua New Guinea (*Economist* 2019). Within the United States, six states and the District of Colombia offer state-paid family leave insurance programs for some eligible workers, but for the most part, workers rely on employer-provided benefits for any form of paid leave (Donovan 2019). Only one-third of employers in the United States offer paid leave for new parents, and the typical length of paid parental leave is six weeks (SHRM 2019). The Congressional Research Service (Donovan 2019) estimates that only 16 percent of American workers have access to some form of paid family leave.

A bill introduced in Congress—the *Family and Medical Insurance Leave Act*—would create a national-paid family and medical leave program enabling workers to receive partial pay for up to 12 weeks to care for new children, ill family members, or their own health problems. Joint employee and employer payroll contributions of 2 cents per $10 in wages, or about $2.00 per week for a typical worker, would fund the program (National Partnership for Women and Families 2020a).

The United States has no federal policy that guarantees workers paid sick leave. The *Healthy Families Act* proposes a sick day policy that would allow workers to earn up to seven paid sick days a year if they or a family member is ill, has a medical appointment, or is seeking assistance with domestic violence, stalking, or sexual assault. Thirteen states, the District of Columbia, and 21 local jurisdictions have adopted laws mandating some paid sick days (National Partnership for Women and Families 2020b). Paid sick day policies benefit businesses by improving productivity, slowing the spread of infectious disease, and reducing absenteeism (by reducing the duration of illness) (Corley, Frothingham, and Bahn 2017).

What do you think?

In response to the COVID-19 pandemic, Congress passed the *Families First Coronavirus Response Act* (FFCRA), which required some employers to provide up to two weeks of paid leave to employees who were unable to work because they were ordered to quarantine or because they needed to care for a family member who was ordered to quarantine or to care for a child whose school or day care was closed due to the pandemic. However, many families found these provisions to be insufficient because school closures and medical issues related to COVID often lasted longer than two weeks. Additionally, FFCRA expired on December 31, 2020, and as of this writing, there is no plan in place to extend these provisions into 2021. What type of family and medical leave provisions do you think should be in place during a long-term public health emergency like COVID-19?

Employer-Based Work/Life Policies. Aside from government-mandated work/family policies, some corporations and employers have "family-friendly" work policies and programs, including unpaid or paid family and medical leave, child-care assistance, assistance with elderly parent care, and telecommuting options (see Table 7.2).

Efforts to Strengthen Labor

Nearly two-thirds of Americans say they approve of labor unions, but there is a stark partisan divide: 83 percent of Democrats and 45 percent of Republicans approve of labor

> "Do we know how long puppies are allowed to stay with their mothers after they have given birth? Eight weeks. ... So the market has decided that women and people who give birth deserve less time with their children than a dog. I think that, at its core, has shown that the market has failed to treat people with dignity and basic respect. So when that happens, I think it is our job as the public to redefine the rules of society and to treat people who give birth with the dignity they deserve."
>
> **–ALEXANDRIA OCASIO CORTEZ, U.S. HOUSE REPRESENTATIVE (D-NY)**

TABLE 7.2 Employer-Based Work/Life Benefits and Policies, U.S., 2019

Benefit or Policy	Percentage of Employers* That Provide Benefit/Policy
Paid holidays	96%
Paid time off including both vacation and sick time	62%
Onsite lactation/mother's room	51%
Paid time off to vote	43%
Telecommuting on a part-time basis	42%
Paid parental leave	27%
Bring child to work in emergency	25%
Paid family leave and elder care leave	24%
Parental leave above federal FMLA leave	20%
Elder care leave above federal FMLA leave	13%
Subsidized child-care center or program	4%

*Employers with 50 or more employees.
SOURCE: Society for Human Resource Management (SHRM) 2019.

unions (Brenan 2020). The lack of bipartisan support for labor unions is a challenge for those who want to strengthen labor unions in order to remedy many of the problems facing workers.

In an effort to strengthen their power, some labor unions have merged with one another. Labor union mergers result in higher membership numbers, thereby increasing the unions' financial resources, which are needed to recruit new members and to withstand long strikes. Because workers must fight for labor protections within a globalized economic system, their unions must cross national boundaries to build international cooperation and solidarity. Otherwise, employers can play working and poor people in different countries against each other.

Strengthening labor unions requires combating the threats and violence against workers who attempt to organize or who join unions. One way to do this is to pressure governments to apprehend and punish the perpetrators of such violence. Another tactic is to stop doing business with countries where government-sponsored violations of free trade union rights occur.

Understanding Work and Unemployment

On December 10, 1948, the General Assembly of the United Nations adopted and proclaimed the Universal Declaration of Human Rights. Among the articles of that declaration are the following:

> *Article 23*. Everyone has the right to work, to free choice of employment, to just and favourable conditions of work and to protection against unemployment.
>
> Everyone, without any discrimination, has the right to equal pay for equal work.
>
> Everyone who works has the right to just and favourable remuneration ensuring for himself and his family an existence worthy of human dignity, and supplemented, if necessary, by other means of social protection.
>
> Everyone has the right to form and to join trade unions for the protection of his interests.
>
> *Article 24*. Everyone has the right to rest and leisure, including reasonable limitation of working hours and periodic holidays with pay.

More than a half century later, workers around the world are still fighting for these basic rights as proclaimed in the Universal Declaration of Human Rights.

To understand the social problems associated with work and unemployment, we must first recognize that corporatocracy—the ties between government and corporations—serves the interests of corporations over the needs of workers. Although some people argue that the growth of multinational corporations brings economic growth, jobs, lower prices, and quality products to consumers throughout the world, others view global corporations as exploiting workers, harming the environment, dominating public policy, and degrading cultural values.

Decisions made by U.S. corporations about investment—what and where to invest—influence the quantity and quality of jobs available in the United States. As conflict theorists argue, such investment decisions are motivated by profit, which is part of a capitalist system. Profit is also a driving factor in deciding how and when technological devices will be used to replace workers and increase productivity. If goods and services are produced too efficiently, however, workers are laid off and high unemployment results. When people have no money to buy products, sales slump, recession ensues, and social welfare programs are needed to support the unemployed. When the government increases spending to pay for its social programs, it expands the deficit and increases the national debt. Deficit spending and a large national debt make it difficult to recover from the recession, and the cycle continues.

What can be done to break the cycle? Those adhering to the classic view of capitalism argue for limited government intervention on the premise that business will regulate itself by means of an "invisible hand" or "market forces." But Americans are growing increasingly skeptical of the notion that big business, which has caused many of the economic problems in our country and throughout the world, can solve the very problems it creates. New models of business, such as the workplace cooperative or workers' self-directed enterprises, offer visions of what the workplace could look like. The Occupy Wall Street social movement (see also Chapter 6) was instrumental in bringing attention to issues of job insecurity, worker exploitation, and extreme levels of wage inequality—aspects of U.S. jobs that are not consistent with the American dream. At the time of this writing, the transition between the Trump and Biden presidential administrations is just beginning. Joe Biden and Kamala Harris have outlined a plan to strengthen labor unions, invest in green energy to modernize manufacturing, and build a caregiving and education workforce to alleviate the pressures on working parents. However, facing a divided Congress and a politically polarized American public, it is unclear whether this administration will live up to its promise to "build back better."

Chapter Review

- **The economy has become globalized. What does that mean?**
  In recent decades, innovations in communication and information technology have led to the globalization of the economy. The global economy refers to an interconnected network of economic activity that transcends national borders and spans the world. The globalized economy means that our jobs, the products and services we buy, and our nation's economic policies and agendas influence and are influenced by economic activities occurring around the world.
- **What are some of the criticisms of capitalism?**
  Capitalism is criticized for creating high levels of inequality; economic instability; job insecurity; pollution and depletion of natural resources; and corporate dominance of media, culture, and politics. Capitalism is also criticized as violating the principles of democracy by allowing (1) private wealth to affect access to political power; (2) private owners of property to make decisions that affect the public (such as when the owner of a factory decides to move the factory to another country); and (3) workplace dictatorships where workers have little say in their working conditions, thus violating the democratic principle that people should participate in collective decisions that significantly affect their lives.
- **What are transnational corporations?**
  Transnational corporations are corporations that have their home base in one country and branches, or affiliates, in other countries.
- **How do structural functionalism, conflict theory, and symbolic interactionism view work and the economy?**
  The structural-functionalist perspective focuses on the functions and dysfunctions of the economic institution and ways in which the economy affects and is affected by changes in society. According to the conflict perspective, the ruling class controls the economic system for its own benefit and exploits and oppresses the working masses. The conflict perspective is critical of ways that the government caters to the interests of big business at the expense of workers, consumers, and the public interest. According to symbolic interactionism, the work role is an important part of a person's self-concept and social identity. Symbolic interactionism emphasizes that job-related attitudes and behaviors are influenced by social interaction.
- **What are some of the causes of unemployment?**
  Causes of unemployment include not enough jobs, job exportation or "offshoring" (the relocation of jobs to other countries), automation (the replacement of human labor with machinery and equipment), increased global competition, and mass layoffs as plants close and companies downsize or go out of business.
- **Does slavery still exist today? If so, where?**
  Slavery exists today all over the world, including the United States. Most forced laborers work in agriculture, mining, prostitution, and factories.
- **Why are reported statistics on injuries and illnesses in U.S. workplaces inaccurately low?**
  Government data on workplace illness and injury exclude many categories of workers. Long-term illnesses caused by, for example, exposure to carcinogens often are difficult to relate to the workplace and are not adequately recognized and reported. Employees don't always report workplace injuries or illnesses for fear of being disciplined, losing their jobs, or fear of being deported. Employers discourage reporting of injuries by offering financial rewards to workers and departments when no injuries are reported for a specified period of time. And employers don't always keep accurate records because they fear increased workers' compensation costs, increased scrutiny from government workplace inspectors, and fines for possible violations of health and safety regulations.
- **What is work/life conflict?**
  Work/life conflict is the day-to-day struggle to simultaneously meet the demands of work and other life responsibilities, including family, education, exercise, and recreation.

- **From the point of view of workers, what are the advantages and disadvantages of labor unions?**
Labor unions have benefitted workers by fighting for fair wages and benefits, healthy and safe work environments, and other forms of worker advocacy. Compared with non-union workers, unionized workers tend to have higher earnings, better insurance and pension benefits, more paid time off, and higher life satisfaction.

  One disadvantage of unions is that members must pay dues and other fees. Another disadvantage for unionized workers is the loss of individuality. Unionized workers are members of an overall bargaining unit in which the majority rules. Decisions made by the majority may conflict with individual employees' specific employment needs.

- **What are worker cooperatives?**
Worker cooperatives are businesses that are owned and democratically governed by their employees.

- **How does policy drift make it difficult to combat the use of sweatshop labor?**
Policy drift occurs when labor laws are not updated to reflect the changes in the broader society and economy. The primary labor law—the *National Labor Relations Act*—was designed in the 1930s–1950s. Since then, globalization has fundamentally altered the nature of work and economic production. Transnational corporations are often able to skirt labor regulations and sweatshop laws by using layers of subcontractors in their supply chains.

- **What is the federal Family and Medical Leave Act?**
The 1993 *Family and Medical Leave Act* (FMLA) requires all companies with 50 or more employees to provide eligible workers with up to 12 weeks of job-protected, unpaid leave so that they can care for a seriously ill child, spouse, or parent; stay home to care for their newborn, newly adopted, or newly placed foster child; or take time off when they are seriously ill. The United States is the only advanced nation that does not mandate paid leave for family and medical reasons.

Test Yourself

1. According to this text, corporations outsource labor to other countries so that U.S. consumers can save money on cheaper products.
   a. True
   b. False
2. The United States is the only advanced nation that does not mandate a minimum number of vacation days.
   a. True
   b. False
3. Which of the following is *not* one of the common outcomes of free trade agreements?
   a. Offshoring
   b. Reduced trade deficits
   c. Stronger copyright protections for intellectual property
   d. Increased tariffs
4. Which of the following is *not* one of the four principles of McDonaldization?
   a. Affordability
   b. Predictability
   c. Calculability
   d. Efficiency
5. The modern working environment is characterized by highly repetitive, monotonous labor that involves little or no creativity. Karl Marx says that these roles lead workers to feel like cogs in a machine, a pattern he referred to as ________________.
   a. alienation
   b. boredom
   c. burnout
   d. exhaustion
6. In states with right-to-work laws, workers are allowed to work at unionized workplaces
   a. only if they pay union dues.
   b. only if they agree not to go on strike.
   c. without having to join the union and pay union dues.
   d. both a and b
7. Worker cooperatives are businesses that are owned by
   a. investors.
   b. managers.
   c. workers.
   d. banks and other financial institutions.
8. Policy drift occurs when
   a. labor unions have more power than corporate leaders.
   b. legislation regarding working conditions changes so rapidly that businesses are unable to adapt.
   c. legislation for better working conditions does not keep up with the rapidly changing global economy.
   d. business and labor union leaders agree on regulations regarding working conditions and pay.
9. Sweatshop labor and slavery do not occur in the United States.
   a. True
   b. False
10. As of 2020, the United States did not guarantee workers paid maternity leave.
   a. True
   b. False

Answers: 1. A; 2. A; 3. D; 4. A; 5. A; 6. C; 7. C; 8. C; 9. B; 10. A.

Key Terms

automation 250
behavior-based safety programs 269
capitalism 241
chattel slavery 253
child labor 255
communism 241
corporatocracy 247
disaster capitalism 247
economic institution 241
Family and Medical Leave Act (FMLA) 270
forced labor 253
free trade agreement (FTA) 243
full employment 265
gig economy 253
global economy 240
Great Recession 241
job burnout 257
labor unions 261
long-term unemployment 251
McDonaldization 261
motherhood penalty 260
offshoring 250
outsourcing 250
policy drift 268
recession 250
slavery 253
socialism 241
sweatshops 254
transnational corporations 245
unemployment 249
union density 262
worker cooperatives 266
workers' self-directed enterprises 266
work/life conflict 258

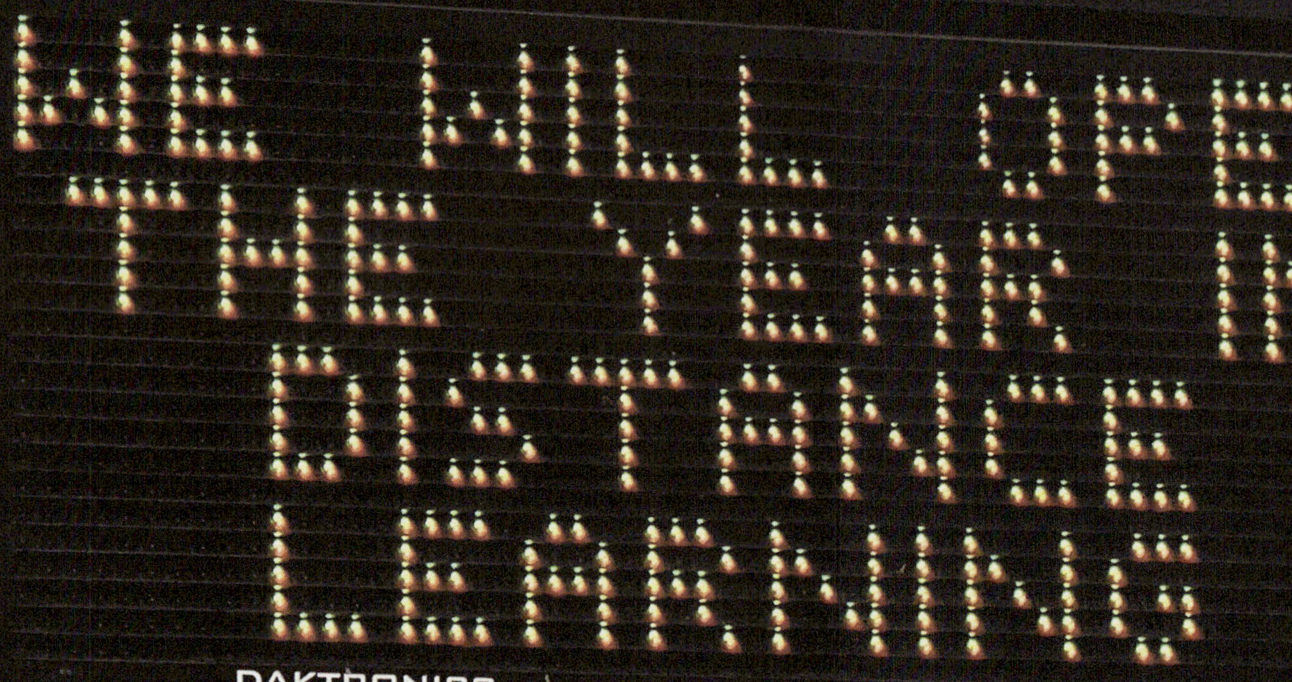

> Educating the mind without educating the heart is no education at all."
>
> **ARISTOTLE**
> Greek philosopher

8

Problems in Education

Chapter Outline

Learning Objectives

After studying this chapter, you will be able to …

1. Identify educational trends in countries that participate in the Organisation for Economic Cooperation and Development (OECD).
2. Analyze education in America using structural functionalism, conflict theory and symbolic interactionism.
3. Discuss each of the independent variables associated with school achievement.
4. Identify factors that contribute to the strains COVID-19 placed on students and teachers around the world.
5. Describe the social problems (e.g., the dropout rate) associated with schools, teachers, and students.
6. Summarize educational policies and practices in the United States.

AP Images/Scott Takushi/Pioneer Press

Teachers across the United States have taken to the streets to express their frustration with low pay; increases in non-teaching duties; and lack of resources like pencils, books, and paper.

AMANDA COFFMAN WAS a middle school teacher at Shawnee Mission School District in Kansas. Teacher contracts at the school expired in the 2018–2019 academic year, and negotiations between the district and teachers' union seemed to be at a standstill by late January. Rather than continue the negotiations, the school board unilaterally approved a three-year contract and told the teachers they could accept the new contract, continue under their old contract, or resign their positions. Amanda Coffman didn't just turn in her resignation. As the cameras rolled, she shared with the world her reasons for quitting at the next school board meeting. Her resignation went viral, touching a chord with others as she described her frustrations shared by so many other teachers, parents, and students.

> Several years ago, a good friend decided to leave education, and she said to me, "Amanda, teaching is like a bad marriage. You never get your needs met, but you stay in it for the kids." I didn't fully understand what she meant until this past month. Just like a bad relationship, our communication has broken down. You aren't listening. ... When you failed to show up for the conversation and sent your lawyer instead, I finally understood that this just isn't going to work. The kids and I deserve better (quoted in Guinness 2020).

In the United States, teacher salaries are about 55 percent of similarly educated workers (Strauss 2017), and, while starting salaries are typically competitive compared to other countries, salary increases for U.S. teachers are small leading to poorly paid experienced teachers. U.S. teachers also work longer hours than their counterparts in other countries, with middle-school teachers working 1,366 in-school hours per year compared with an OECD average of 1,135. Those hours don't include the additional time teachers in the United States routinely work before and after school, nor does it include the 21 hours per week they typically work throughout the summer months even though they are not under contract (Startz 2019a).

Class sizes are average compared to other OECD countries, but teachers in the United States are also required to satisfy state and federal mandates for documentation and instruction, leaving them less time to get to know individual students (OECD 2020). These are some of the issues that teachers in the Shawnee Mission School District and many other teachers around the country wanted to see resolved in their next contract.

In the United States and across the globe, education plays a primary role in determining people's life outcomes—whether they escape poverty, live healthy lives, participate in civic life, and find fulfillment in their work. Yet educational systems in many parts of the world, including in our own country, are under-resourced and struggling. Teachers feel undervalued and are undercompensated. Schools struggle to provide the resources children need to thrive.

In this chapter, we focus on the important role of education and on what is being called the educational crisis. Sociologists recognize education as a process that goes on within schools, **formal education**, and outside of schools throughout our everyday lives. Throughout most of this chapter, we use the term *education* as synonymous with formal education. We begin with a look at education around the world.

formal education Structured education that occurs within schools and is guided by a curriculum typically established by teachers.

The Global Context: Cross-Cultural Variations in Education

Looking only at the American educational system might lead one to conclude that most societies have developed some method of formal instruction for their members. After all, the U.S. educational system has more than 130,930 primary and secondary schools, 6,502 college and universities, 5.1 million primary and secondary school teachers and college faculty, and 77.4 million students including 1.3 million children in pre-kindergarten (National Center for Education Statistics [NCES] 2019). Many societies have no formal mechanism for educating their population. And, while literacy rates have been improving from one generation to the next, 770 million adults and young people lack basic literacy skills (United Nations 2020). Further, over half of children in low- and middle-income countries cannot read "proficiently" by the age of 10 (World Bank 2019). However, although Figure 8.1 indicates that over 59 million children who should be in elementary school are not, it also shows us that the number of out-of-school children and adolescents has decreased over time.

There is more reason for optimism. *Education at a Glance*, a publication of the Organisation for Economic Cooperation and Development (OECD), reports education statistics on more than 35 countries (OECD 2020). Young adults, those between the ages of 25 and 34, are better educated than in previous generations. In ten years, the percentage of young adults without upper secondary education, i.e., high school, decreased from 20 percent in 2009 to 15 percent in 2019. Although varying significantly by country, 45 percent of young adults in OECD countries have completed college. For example, 24 percent of young adults in Mexico have graduated from college compared to 70 percent in Ireland (OECD 2020).

> Sesame Street is broadcast in over 130 countries with millions of children watching daily. In a meta-analysis of over 20 studies encompassing 10,000 children in 15 countries, Mares and Pan (2013) report that children with higher exposure to Sesame Street had higher learning outcomes (e.g., number recognition). On more than one occasion, in order to reduce federal spending, politicians have suggested cutting funds to public television which, among other programming, airs Sesame Street (Tepper 2017; Stanton 2019). Do you think funding for public broadcasting should be cut from the federal budget?
>
> **What do you think?**

One of the primary reasons individuals and communities invest in education is the strong relationship between education and employment, and between education

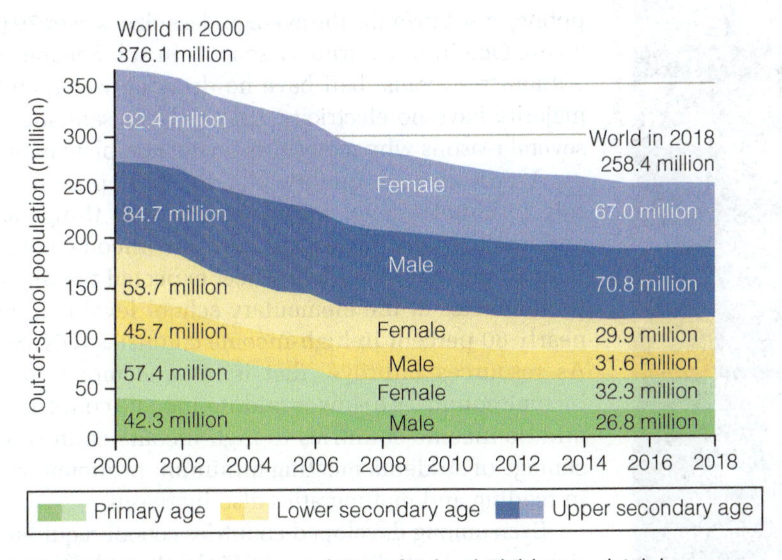

Figure 8.1 Global Number of Out-of-School Children and Adolescents, 2000–2018

SOURCE: UNESCO Institute for Statistics database.

"The more you read, the more things you will know, the more that you learn, the more places you'll go."

–DR. SEUSS, CHILDREN'S AUTHOR, POET

and income. The more education people have, the greater the likelihood they will be employed and the higher their incomes. Consequently, in most countries, the unemployment rate of college-educated people is less than half of those with only a high school degree (OECD 2020). In countries struggling with recessions, unemployment for those without a high school education is as high as 35 percent.

Globally, the income advantages of investing in education are significant. Full-time workers with college degrees earn 54 percent more than those who have only a high school degree. Further, the income gap increases over time as incomes of the college educated increase at a faster rate than the incomes of those without a college degree (OECD 2020). This is referred to as the "**earnings premium**," the tendency for the benefits of having a college degree to far outweigh the cost of getting one. The earning premium, however, is greater for men than for women as a result of the gender wage gap (see Chapter 10).

In some countries, Denmark, Finland, and Norway for example, college education is fully paid by the government and thus the earnings premium for graduates is realized quite quickly (OECD 2020). In other countries such as the United States, where students pay either all (private) or part (public) of their college education, it takes longer to recoup the initial investment. However, while the cost of a college education has increased in recent years, college graduates in the United States earn, on average, $30,000 more annually than individuals with only a high school degree (Abel and Deitz 2019).

Education is not only a good investment for individuals, but also for societies as a whole. Higher incomes are associated with additional tax revenue, and lower unemployment is associated with reduced social welfare costs. Education is also positively associated with health (see Chapter 2) and negatively associated with both substance abuse (see Chapter 3) and criminal behavior (see Chapter 4), further reducing the costs to society. People with higher levels of education are also less likely to divorce and women with higher levels of education have fewer children (see Chapter 5). Thus, while OECD countries spend an average of $11,200 per year per pupil on education, the economic and social benefits outweigh the direct public costs (OECD 2020). And, while the public cost of a tertiary, or college, education varies widely between countries, the average financial return to society for a college education is 8 percent for men and 6 percent for women.

earnings premium The benefits of having a college degree far outweigh the cost of getting one.

The experiences of formal education vary across societies. The average student–teacher ratio in primary schools in OECD-participating countries is less than 15:1. However, the average primary school class sizes range from a high of 37 students in China to a low of 15 in Latvia and Estonia (OECD 2020). In the Philippines some schools are divided into two shifts to address problems of overcrowding. Nonetheless, teachers often have as many as 65 students in a class (Sison 2020). In much of sub-Saharan Africa, the average primary class size is 50 students; in Malawi, the Central African Republic, and Tanzania, the average class size is over 70 (UNESCO 2016). One in three primary schools in sub-Saharan Africa are without restrooms, half have no drinking water, and the vast majority have no electricity. Lack of basic sanitation is one of several reasons why many girls do not complete high school.

AP Images/Héctor Alfaro/Agencia Press South/NurPhoto

Each year thousands of students in Mexico sit for the National Autonomous University of Mexico (UNAM) entrance exam administered in UNAM's soccer stadium. Due to the COVID-19 pandemic, students were assigned seats and appropriately spaced for the duration of the exam while wearing face masks and face shields. Of the nearly 84,000 prospective students who took the exam, only about 6,000 will be accepted.

Access to and quality of education play a significant role in country-level indicators on such things as reading and math scores. For example, in low-income countries less than 20 percent of students meet expected reading and math proficiencies at the elementary school level compared with nearly 80 percent in high-income countries (UNESCO 2017). As resources improve, that is, as one moves from lower-income countries to lower-middle-income countries to upper-middle-income countries to high-income countries, the percentage of students meeting minimum proficiency standards in reading and mathematics also increases.

Even among developed countries school requirements, curriculum, and activities vary. In Finland, students aren't graded and only take one standardized test when they are seniors in high school (Colagrossi 2018). Primary school students have more time devoted to recess and spend less time in school during the day when compared to the average primary school student in the United States. Further, teachers who have their Master's degree and are part of the national teacher pool in Finland have a great deal of flexibility in the classroom. By contrast, in South Korea, the educational environment is highly competitive (Mani and Trines 2018). Students typically spend 8 hours in school, go home for dinner, and then attend a private school for tutoring followed by time to do homework. In the United States, most children attend school for 6.75 hours a day, standardized testing is the norm, and just two-thirds of U.S. school districts require recess for elementary school students (London 2019).

Sociological Theories of Education

The three major sociological perspectives—structural functionalism, conflict theory, and symbolic interactionism—are important in explaining different aspects of American education.

Structural-Functionalist Perspective

According to structural functionalism, education is a fundamental institution in society, performing important tasks such as instruction, socialization, status conferral, and providing custodial care (Ballantine, Hammack, and Stuber 2017). Many social problems, such as unemployment, crime and delinquency, and poverty, can be linked to the failure of the educational institution to fulfill these basic functions (see Chapter 4, Chapter 6, and Chapter 7). Structural functionalists also examine the reciprocal influences of the educational institution on other social institutions, including the family, political institutions, and economic institutions.

Instruction. A major function of education is to teach students the knowledge and skills that are necessary for future occupational roles, self-development, and social functioning. Although some parents teach their children basic knowledge and skills at home, most parents rely on schools to teach their children to read, spell, write, tell time, count money, and use computers. As discussed later, many U.S. students display low levels of academic achievement. The failure of many schools to instruct students in basic knowledge and skills both causes and results from many other social problems.

Socialization. The socialization function of education involves teaching students to respect authority—behavior that is essential for social organization (Merton 1968). Students learn to respond to authority by asking permission to leave the classroom, sitting quietly at their desks, and raising their hands before asking a question. They are recognized for perfect attendance in elementary school award ceremonies, underscoring their adherence to rules. The idea being that students who do not learn to respect and obey teachers may later disrespect and disobey employers, police officers, and judges.

The educational institution also socializes youth into the dominant culture. Schools attempt to instill and maintain the norms, values, traditions, and symbols of the culture in a variety of ways, such as celebrating holidays (e.g., Martin Luther King Jr. Day and Thanksgiving); requiring students to speak and write in standard English; displaying the American flag; and discouraging violence, drug use, and cheating. The conferring of grades, calculation of GPAs, and recognition of school valedictorian and salutatorian are all mechanisms by which schools promote competition and individual achievement as cultural norms. In contrast to some societies in which individual accomplishments are not publicly recognized in school, in the United States these achievements are often celebrated in award ceremonies and the like, even as early as elementary school.

Status Conferral. Sociologist Randall Collins (1979) argued that modern societies increasingly rely on credentials such as degrees, diplomas, and certificates as evidence of ability and therefore readiness for employment. Schools are essential to this credentialing process. The achievements of various levels of education serve as credentials which sort people into different statuses—for example, "high school graduate," "Harvard alumna,"

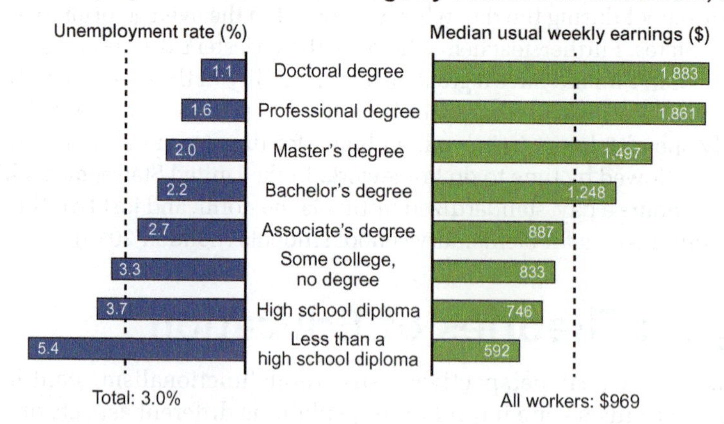

Figure 8.2 Earnings and Unemployment Rates by Educational Attainment, 2019
SOURCE: Bureau of Labor Statistics 2019.

and "English major." These credentials have become increasingly important screening tools as workplaces increase in size. In addition, schools sort individuals into professional statuses by awarding degrees in fields such as medicine, engineering, and law. The significance of such statuses lies in their association with occupational prestige and income—the higher one's education, the higher one's income. Furthermore, unemployment rates and earnings are tied to educational status, as seen in Figure 8.2.

Custodial Care. The educational system also serves the function of providing custodial care by providing supervision and care for children and adolescents until they are 18 years old (Merton 1968). Recently, several countries have extended their instructional time by extending the length of the school day (OECD 2020). In the United States, there are two lines of thought concerning the length and time of the school day. The first is to start school later in the morning given evidence that students perform better academically with more sleep—especially teenagers (Dunster et al. 2018). Parents, however, are concerned about such changes impacting their work schedules (Dunietz et al. 2017).

The second line of thinking calls for lengthening the school day arguing that it is out of sync with the lives of working parents (Brown, Boser, and Baffour 2016). Excluding summer school, the time students are out of class during the year often exceeds the vacation time provided by many employers. To address this issue in low-income school districts, former senator Kamala Harris introduced the 2019 *Family Friendly Schools* that would lengthen school days by 3 hours not only better aligning children's and parents' schedules but reducing the cost of child care (Senate Bill S.2784 2019). It is estimated that aligning work and school schedules would save the United States $55 billion in lost productivity (Brown, Boser, and Baffour 2016).

Conflict Perspective

Conflict theorists argue that the institution of education engages in the process of **social reproduction**—the replication of the social class structure from one generation to the next. Rather than education providing opportunities for upward mobility, conflict theorists view education as a means by which the elite control the masses. Therefore, they argue, educational opportunities and the quality of education are not equally distributed.

Conflict theorists point out that the socialization function of education is really indoctrination into a capitalist ideology (Ballantine, Hammack, and Stuber 2017). Students in upper-class schools are encouraged to be creative, to question assumptions, and to take on leadership positions; students in lower-class schools are taught to be obedient, not question authority, and do what they are told. Thereby replicating, conflict theorists would argue, the roles they are expected to play in later life. Even so-called "no excuses" charter schools that focus on raising the academic achievement of low-income, often minority children in urban

social reproduction A process through which the social class structure is repeated from one generation to the next.

areas rely on authoritarianism to achieve their goals. In her study of one such award-winning school, Golann (2015) found that students learned that they were not trusted to make decisions, their opinions were not valued, and that they must defer to authority. Golann (2015) concludes, "No-excuses schools thus promote academic achievement while reinforcing inequality in cultural skills" (p. 17).

In addition, education serves as a mechanism for **cultural imperialism**, or the indoctrination into the dominant culture of a society. When cultural imperialism exists, the norms, values, traditions, histories, and languages of minority groups have been historically ignored. The Black Lives Matter movement has invigorated conversation on the gaps in the public-school curriculum around the lives and history of racial, ethnic, and religious minorities in the United States. A 2017 study found that while middle- and high-school teachers reported integrating Black history into their curriculum, in fact, approximately only 9 percent of class time was devoted to it (King 2017).

John Moore/Getty Images News/Getty Images

A Stamford High School student in Stamford, Connecticut, walks past a wall mural that represents the history of Black Americans since the early beginnings of slavery in this country.

While some states mandate Black history be included in the curriculum, critics argue these mandates have little teeth and tend to focus on a limited number of topics such as slavery and the civil rights movement. Similarly, only 16 states require that the Holocaust be taught in public schools (U.S. Holocaust Memorial Museum 2020). Not surprisingly, a 2020 survey revealed that 10 percent of adults under 40 years old did not recall ever hearing the word *Holocaust* (Claims Conference 2020). **Multicultural education**—that is, education that includes all racial and ethnic groups in the school curriculum—promotes awareness and appreciation for cultural diversity and, ideally, leads to **cultural competence** among teachers and students alike (also see Chapter 9).

The conflict perspective also focuses on what Kozol (1991) called the "savage inequalities" in education that perpetuate racial disparities. Kozol documented gross inequities in the quality of education in poorer districts, largely composed of minorities, compared with districts that serve predominantly White middle-class and upper-middle-class families. This pattern continues across the United States today, with school districts having high concentrations of students living in poverty more than twice as likely to have a gap in basic funding than high-income districts (The Century Foundation 2020).

Because local taxes make up 45 percent of U.S. funding, school districts vary dramatically based on socioeconomic variables (NPR 2016). Local expenditures on schools come from taxes, usually property taxes, and as housing prices decline, property taxes decline. Although states provide additional funding to supplement local taxes, this funding is not always enough to lift schools in poorer districts to a level that even approaches the funding available to schools in wealthier districts.

Overall, the United States underfunds public schools by an estimated $150 billion annually with underfunded schools serving more than 30 million children (The Century Foundation 2020). As a result of these inequalities, poor and/or minority children are more likely to attend schools that are understaffed, have limited textbooks and computers, don't offer advanced placement courses, and lack intervention services (Knoff 2019). Figure 8.3 illustrates the impact of this inequality in the Arkansas public schools.

Symbolic Interactionist Perspective

Whereas structural functionalism and conflict theory focus on macro-level issues, such as institutional influences and power relations, symbolic interactionism examines education from a micro-level perspective. This perspective is concerned with individual and small-group issues, such as teacher–student interactions and the self-fulfilling prophecy.

cultural imperialism The indoctrination into the dominant culture of a society.

multicultural education Education that includes all racial and ethnic groups in the school curriculum, thereby promoting awareness and appreciation for cultural diversity.

cultural competence In reference to education, refers to having an awareness of and appreciation for different cultural groups in order to inform student–teacher and student–student interaction.

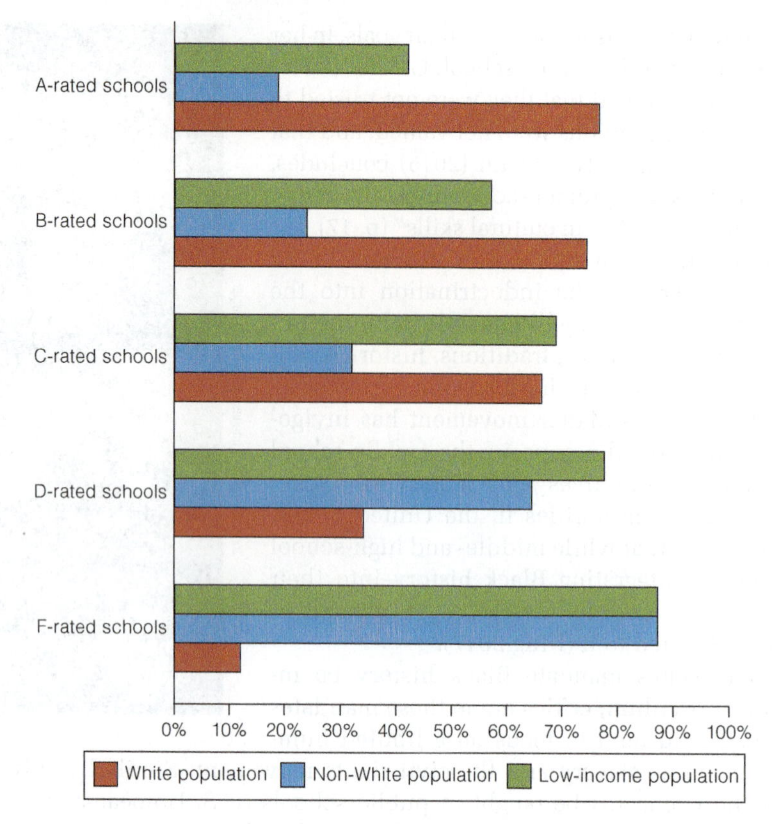

Figure 8.3 Race and Socioeconomic Status by State School Report Card Rating, Arkansas 2017–2018 Academic Year
SOURCE: Knoff 2019.

> "There is nothing in a caterpillar that tells you it's going to be a butterfly."
>
> **–BUCKMINSTER FULLER, AMERICAN ARCHITECT, AUTHOR**

Teacher–Student Interactions. Symbolic interactionists examine the consequences of the social meaning conveyed in teacher–student interactions. From the teachers' point of view, middle-class students may be defined as easy and fun to teach. They grasp the material quickly, do their homework, and are more likely to "value" the educational process. Students from economically disadvantaged homes often bring fewer social and verbal skills to those same middle-class teachers, who may, inadvertently, hold up social mirrors of disapproval. The differences in interactions may be subtle and are often unconscious. Consider the impact of two only slightly different statements: "You only have a B?" (message: you can do better) versus "You got a B!" (message: you can't do much better). Teacher disapproval contributes to lower self-esteem among disadvantaged and troubled youth. This chapter's *The Human Side* is a call to teachers to acknowledge the traumatic impact of the pandemic on children and to set aside curricular expectations in favor of compassion.

At the college level, first-generation students often struggle to understand the language and unspoken rules of the college culture. Do faculty really want students to attend office hours? What is a syllabus? Many faculty assume that students have family members who have attended college and that they understand "how things work" and will speak up when they have questions, but first-generation students are less likely to do so for fear of insulting the professor or exposing their status (Longwell-Grice et al. 2016).

Self-fulfilling prophecy A concept referring to the tendency for people to act in a manner consistent with the expectations of others.

Self-Fulfilling Prophecy. The **self-fulfilling prophecy** refers to a tendency for people to act in a manner consistent with the expectations of others. A classic study by Rosenthal and Jacobson (1968) provides empirical evidence of the self-fulfilling prophecy in the public school system. Five elementary school students in a San Francisco school were selected at random and identified for their teachers as "spurters." Such a label implied that they had superior intelligence and academic ability.

Educating the Children of the Pandemic

This letter to educators was written by Teresa Thayer Snyder, a retired superintendent of schools in upstate New York. Dr. Snyder passionately appeals to teachers to meet the emotional rather than academic needs of students—a cohort of students who have undergone significant disruptions in their lives as a result of the pandemic. Children's routines have been disrupted, the continuity of their learning broken, significant life events missed, and a sense of security and safety lost (Centers for Disease Prevention and Control [CDC] 2020).

As a result, as Dr. Snyder advocates, returning students need to be met "where they are, not where we think they should be," embraced, listened to, empowered.

Dear Friends and Colleagues:

I am writing today about the children of this pandemic. After a lifetime of working among the young, I feel compelled to address the concerns that are being expressed by so many of my peers about the deficits the children will demonstrate when they finally return to school. My goodness, what a disconcerting thing to be concerned about in the face of a pandemic which is affecting millions of people around the country and the world. It speaks to one of my biggest fears for the children when they return. In our determination to "catch them up," I fear that we will lose who they are and what they have learned during this unprecedented era. What on earth are we trying to catch them up on? The models no longer apply, the benchmarks are no longer valid, the trend analyses have been interrupted. We must not forget that those arbitrary measures were established by people, not ordained by God. We can make those invalid measures as obsolete as a crank up telephone! They simply do not apply. When the children return to school, they will have returned with a new history that we will need to help them identify and make sense of. When the children return to school, we will need to listen to them. Let their stories be told. They have endured a year that has no parallel in modern times. There is no assessment that applies to who they are or what they have learned. Remember, their brains did not go into hibernation during this year. Their brains may not have been focused on traditional school material, but they did not stop either. Their brains may have been focused on where their next meal is coming from, or how to care for a younger sibling, or how to deal with missing grandma, or how it feels to have to surrender a beloved pet, or how to deal with death. Our job is to welcome them back and help them write that history.

I sincerely plead with my colleagues, to surrender the artificial constructs that measure achievement and greet the children where they are, not where we think they "should be." Greet them with art supplies and writing materials, and music and dance and so many other avenues to help them express what has happened to them in their lives during this horrific year. Greet them with stories and books that will help them make sense of an upside-down world. They missed you. They did not miss the test prep. They did not miss the worksheets. They did not miss the reading groups. They did not miss the homework. They missed you.

Resist the pressure from whatever "powers that be" who are in a hurry to "fix" kids and make up for the "lost" time. The time was not lost, it was invested in surviving an historic period of time in their lives—in our lives. The children do not need to be fixed. They are not broken. They need to be heard. They need be given as many tools as we can provide to nurture resilience and help them adjust to a post pandemic world.

Being a teacher is an essential connection between what is and what can be. Please, let what can be demonstrate that our children have so much to share about the world they live in and in helping them make sense of what, for all of us has been unimaginable. This will help them—and us—achieve a lot more than can be measured by any assessment tool ever devised. Peace to all who work with the children!

Retired Superintendent Teresa Thayer Snyder

SOURCE: Diane Ravitch. (December 12, 2020).

In reality, they were no different from the other students in their classes. At the end of the school year, however, these five students scored higher on their intelligence quotient (IQ) tests and made higher grades than their classmates who were not labeled as spurters. In addition, the teachers rated the spurters as more curious, interested, and happy, and more likely to succeed than the nonspurters. Because the teachers expected the spurters to do well, they treated the students in a way that encouraged better school performance.

A recent study tested the self-fulfilling prophecy using expectations formed by teachers at the beginning of the school year (Gentrup et al. 2020). The study revealed that teachers who defined a student as strong were more likely to provide that student with

David Paul Morris/Bloomberg/Getty Images

Students in this elementary classroom are engaged in a hybrid approach to instruction. While the teacher guides the students through the lesson in the classroom, the remaining students follow the lesson remotely, as shown on the monitor.

positive feedback on their work and less likely to make negative comments about it even when warranted. By contrast, teachers were more likely to provide negative feedback on academic performance and behavior to students they had not labelled as strong. Further, teachers' negative expectations, even if inaccurate, predicted poor learning outcomes (Gentrup et al. 2020).

For years, ability grouping—which takes place within classes, in contrast to tracking, which takes place between classes—has been criticized for its potential negative effect on students and, consequently, went out of favor several decades ago (Loveless 2013). The fear, still held by some today, was that ability grouping not only reflects differences in race and class but also contributes to those differences; that is, it perpetuates inequality. Nonetheless, the use of ability grouping is on the rise. A review of the literature concludes that it is not a one-size-fits-all approach and that specific "types of ability grouping may be more beneficial or harmful than others both academically and psychologically, depending on students' particular backgrounds and levels" (Bolick and Rogowsky 2016).

Who Succeeds? The Inequality of Educational Attainment

Figure 8.4 shows the variation in highest level of education attained by individuals 25 years of age and older in the United States. However, these educational achievements are not equally distributed across different social groups in society. Educational inequality is based on social class and family background, race and ethnicity, and gender. Each of these factors influences who succeeds in school.

Social Class and Family Background

@College Board

New research shows that AP Computer Science Principles students - especially Black, Hispanic, female, & first-generation - are more likely to choose a computer science major in college, a critical step toward a STEM career.

Read more: spr.ly/6016H3CBA. #APCSP #CSEdWeek

-The College Board

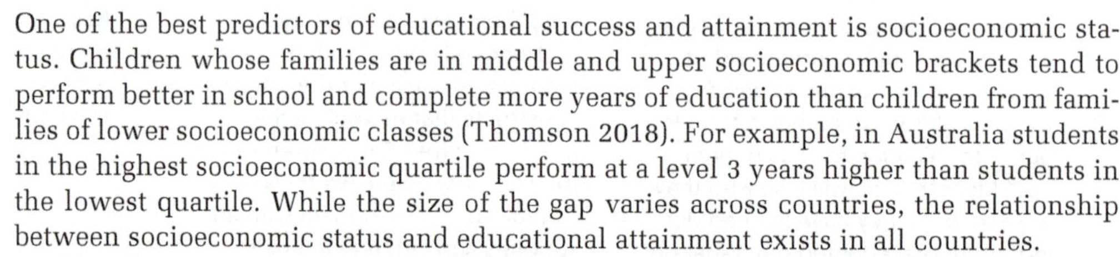

One of the best predictors of educational success and attainment is socioeconomic status. Children whose families are in middle and upper socioeconomic brackets tend to perform better in school and complete more years of education than children from families of lower socioeconomic classes (Thomson 2018). For example, in Australia students in the highest socioeconomic quartile perform at a level 3 years higher than students in the lowest quartile. While the size of the gap varies across countries, the relationship between socioeconomic status and educational attainment exists in all countries.

Why this relationship exists is an important question. Some argue that students from low-income households are at a disadvantage in school because they lack an enriching academic home life. For example, globally, teenagers who live in homes with an extensive book collection tend to perform better on literacy and numeracy tests than do those with few or no books (Flood 2018). Parents with high levels of education are also more likely to expose their children to educational material at home and through family activities, such as attending museums, visiting historical sites, and attending lectures. In China, for example, parents' education has a larger impact on a student's school performance in urban areas compared to rural areas suggesting that where there are opportunities for enrichment, by virtue of location, educated parents take advantage of the opportunities (Li and Qiu 2018).

In the United States, parents who have less education and lower income levels are less likely to be involved in school-related activities such as attending a school event or volunteering at a school (NCES 2016). Furthermore, low-income students are less likely to be engaged in the kind of activities that build the "soft skills" (e.g., teamwork) necessary to be successful as an adult (Venator and Reeves 2015).

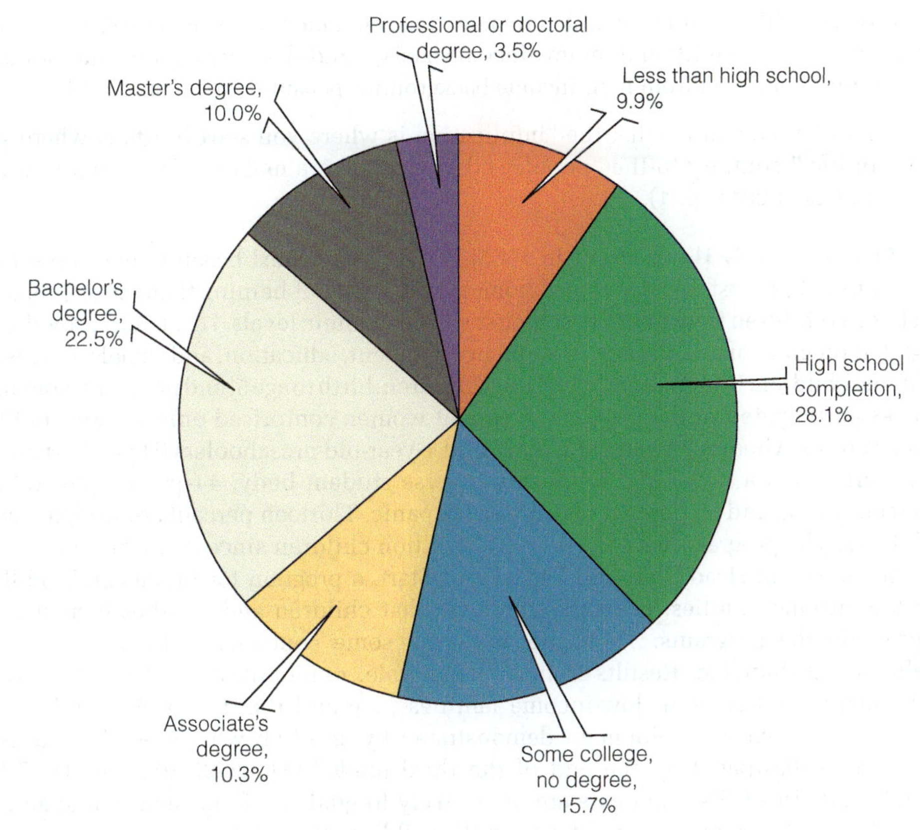

Note: Percentages do not sum to 100 percent because of rounding.

Figure 8.4 Highest Level of Education Attained by Individuals Age 25 Years Old and Older, 2019
SOURCE: U.S. Census Bureau 2019.

School environment is also influential. Students who live in low-income school districts are more likely to attend schools that are underfunded and have fewer educational resources. Nationally, students in low-income school districts are more likely to have teachers who are less qualified than students in high-income districts (Flannery 2018). Teachers may also have biased perceptions of students from low-income backgrounds. A study found teachers viewed homeless youth as less competent and engaged than other youth, independent of objective learning outcomes (LaFavor et al. 2020).

> Reardon (2013) argues that the achievement gap is a consequence of "rich students ... increasingly entering kindergarten much better prepared to succeed in school" (p. 1), thus creating a "rug rat race" (Ramey and Ramey 2010). President Biden has vowed to invest in universal preschool for all three- and four-year-olds (Biden 2020). Do you think the federal government should pay for universal preschool, regardless of income?
>
> **What *do you* think?**

In one of the most significant studies on the relationship between family background and educational success, sociologist Carl Alexander and his colleagues followed nearly 800 Baltimore students from 1982 when they entered first grade until young adulthood (Alexander, Entwistle, and Olsen 2014). Over the course of 25 years, the researchers obtained school records and interviewed and surveyed the children, their parents, and teachers. Research findings include the following:

- Only 4 percent of children from low-income backgrounds completed college compared to 45 percent of children from high-income backgrounds.
- Of those who did not attend college, White men from low-income backgrounds secured higher-paying blue-collar jobs than Black men from low-income backgrounds.
- White women from low-income backgrounds had higher incomes than Black women from low-income backgrounds.

- Nearly half of the children sampled remained in the same socioeconomic status as their parents—only 33 children from low-income backgrounds became high-income adults and only 19 children from high-income backgrounds became low-income adults.

The lead author concludes that the "implication is where you start in life is where you end up in life," contrary to the "popular ethos that we are makers of our own fortune" (quoted in Rosen 2014, p. 1).

> "Public schools power our community, but their strength depends on us addressing the systemic inequities that our students and communities face."
>
> –ROCIO INCLÁN, NEA CENTER FOR SOCIAL JUSTICE

Head Start and Early Head Start. In 1965, Project **Head Start** began to help preschool children from the most disadvantaged homes, with a goal of helping them be as prepared for school as children from families at higher socioeconomic levels. Head Start provides an integrated program of health care, parental involvement, education, and social services for qualifying children. In 2019, nearly 1 million children, birth to age 5, and pregnant women in 50 states participated in the program. Pregnant women comprised only 1 percent of the serviced group, whereas 72 percent were 3- and 4-year-old preschoolers (Head Start 2019). Head Start serves a racial and ethnically diverse student body; 44 percent are White, 30 percent Black, and 37 percent identify as Hispanic. Thirteen percent are students with disabilities. The program has served over 37 million children since its inception.

Assessments of Head Start and Early Head Start, a program for infants and toddlers from low-income families, generally conclude that children and families benefit from inclusion in the programs. There are, however, some concerns as to whether those benefits are sustainable. Results from a "large-scale, randomized, controlled study of nearly 5,000 children from low-income families ... found that the positive effects on literacy and language development demonstrated by children who entered Head Start at age 4 had dissipated by the end of the third grade" (Maxwell 2013, p. 1). Other research finds Head Start students are more likely to graduate from high school and attend college and display higher self-esteem than siblings who did not attend the program (Schanzenbach, Whitmore, and Bauer 2016). Children of Head Start participants are also more likely to finish high school and enroll in college and are less likely to become teen parents or to be involved in the criminal justice system (Barr and Gibbs 2017).

Race and Ethnicity

The U.S. public school population is racially and ethnically diverse, with racial and ethnic minorities comprising 54 percent of the pre-kindergarten through 12th grade public school student population (NCES 2020a). In 2017, the non-Hispanic White student population made up 48 percent of all students, Black students 15 percent, Hispanic students 27 percent, Asian students 6 percent, Biracial students 4 percent, and American Indian/Alaskan Native students 1 percent. The representation of Hispanic and biracial students is expected to continue to rise through 2029. This chapter's *Social Problems Research Up Close* examines the relationship between income segregation and racial disparities in student achievement.

Compared to White students, Black, Hispanic, and Native American students are less likely to succeed at every level of education. As early as the start of kindergarten, Black children have lower reading, math, and working memory scores when compared to their White counterparts, much of the difference explained by a student's socioeconomic status and school quality (Quinn 2015). By fourth grade, approximately 80 percent of Black, Hispanic, and Native American students are reading below grade level compared to 56 percent of White students and 45 percent of Asian students (Kids Count 2020). Similarly, by eighth grade, 87 percent of Black, 85 percent of Native American, and 81 percent of Hispanic students compared to 57 percent of White and 39 percent of Asian students are below grade level in mathematics. As Table 8.1 indicates, although educational attainment has increased over time for all groups, racial and ethnic disparities remain.

It is important to note that socioeconomic status interacts with race and ethnicity. Because race and ethnicity are so closely tied to socioeconomic status—that is, a disproportionate number of racial and ethnic minorities are poor—it *appears* that race or ethnicity determines school success. For example, cities have the highest proportion of minority students as well as the highest proportion of students receiving free or reduced-price lunch, a proxy for socioeconomic status (NCES 2018). Although race and ethnicity

Head Start Begun in 1965 to help preschool children from the most disadvantaged homes, Head Start provides an integrated program of health care, parental involvement, education, and social services for qualifying children.

social problems **RESEARCH** UP CLOSE

Income Segregation in School Districts and Inequality in Student Achievement

Inequalities in learning outcomes between high- and low-income children have increased over the last 50 years (Reardon 2013), yet research indicates low-income parents invest more in educational resources and time than ever before (Bassok et al. 2016). During this same period, racial differences in educational achievement between Black and White students have remained large and stable. The author of this study examines whether income segregation among school districts explains these gaps, suggesting that inequality in economic and social resources within school districts impacts the educational experience of students over and above their individual and family characteristics.

Sample and Methods

This research uses data from the Panel Study on Income Dynamics, a national longitudinal study of families, which includes information on family income, children's race, children's test scores (math and reading) at two time periods, and metropolitan area of residence. The study was limited to the 1,202 children who were attending public schools and were 8 or older. The researcher connected the child's data with data from the School District Demographic System (SDDS), data produced by the National Center for Education Statistics. The SDDS includes counts by income category of families with children attending schools in the district. Using the SDDS data, the researcher calculated a measure of income segregation for the school district in which the child attended school. She also took into account several characteristics of the metropolitan area including income inequality, racial and ethnic composition, multiracial segregation of students between school districts, median income, and private school enrollment.

The dependent variables being predicted were math and reading test scores at time 2 of the data collection process, controlling for math and reading test scores at time 1. As a result, the researcher essentially predicted whether a student improved, remained the same, or lost ground in math and reading performance.

Findings and Conclusions

Results of the study indicate that income segregation among school districts has a positive association with math and reading test scores at time 2, but *only* for students from affluent parents, those with the top 20 percent of family incomes. This positive association is over and above the positive influence of the students' own socioeconomic background and that of the metropolitan area in which they live. Surprisingly, income segregation has no negative effect on students from poor families, with the lowest 20 percent of incomes, over and above the impact of their own socioeconomic background and that of the area in which they live.

Examining the race gap in educational achievement, the researcher found that White students perform better on math and reading tests the more income segregated their schools. Black students, on the other hand, earn lower scores in reading, although highly not math, when they attend a highly income segregated school.

This study found that students from the wealthiest families earn even higher test scores when they attend a school in a highly income-segregated school district. White students also gained ground in math and reading when attending an income-segregated school. By contrast, the lowest-income students do not lose ground in math or science when attending an income-segregated school, while Black students lost ground in reading the more income segregated their school. Her results indicate that income segregation in schools contributes to the achievement gap largely because already advantaged students pull ahead even more when enrolled in these contexts.

This research has important implications for educational policy aimed at decreasing the educational achievement gap (Owens 2018, p. 18).

> To reduce the income achievement gap, policy makers and researchers must understand what characteristics of affluent contexts are beneficial for children and how those benefits can be reproduced. If instructional resources, high-quality curriculum, higher teacher salaries, or state-of-the-art facilities produce high math scores, school finance policies providing compensatory (beyond simply adequate) funding to low-income districts could equalize outcomes.

On the other hand, as the author points out, if it is the social resources—like parents' social networks, information, or cultural differences—in these income-segregated districts that predict achievement, then socioeconomic integration across school districts would be necessary to narrow the achievement gap.

SOURCE: Owens 2018.

may have an independent effect on educational achievement, their relationship is largely a result of the association between race and ethnicity, as well as socioeconomic status.

In addition to the socioeconomic variables, there are several reasons that minority students have academic difficulty. First, minority children may be English language learners (ELLs). The National Center for Education Statistics estimates that the number of ELLs in the United States has grown to over 5 million. In ten states, including Alaska, California, Colorado, Florida, Illinois, Kansas, Nevada, New Mexico, and Texas, and the District of Columbia, 10 percent or more of the public school population are ELLs (NCES 2020b). ELLs come from over 400 different language

TABLE 8.1 Educational Attainment among Persons Age 25 and over, by Race/Ethnicity and Sex, 1970 and 2019*

	1970		2019	
	Males	Females	Males	Females
High school completion or higher				
White	54.0	55.0	96.2	96.4
Black	30.1	32.5	88.9	93.0
Hispanic	37.9	34.2	84.6	88.4
Asian and Pacific Islander	61.3	63.1	97.0	97.1
Total	51.9	52.8	92.7	94.3
Bachelor's or higher degree				
White	14.4	8.4	40.8	49.2
Black	4.2	4.6	28.4	29.5
Hispanic	7.8	4.3	18.2	23.1
Asian and Pacific Islander	23.5	17.3	69.5	70.6
Total	13.5	8.1	35.7	41.8

*2019 figures for Whites are non-Hispanic Whites only.
SOURCE: NCES 2020.

backgrounds, the overwhelming majority being native Spanish speakers. Most are elementary age students.

Now imagine, as Goldenberg (2008) suggests,

> you are in second grade and don't speak English very well and are expected to learn in one year ... irregular spelling patterns, diphthongs, syllabication rules, regular and irregular plurals, common prefixes and suffixes, antonyms and synonyms; how to follow written instructions, interpret words with multiple meanings, locate information in expository texts ... read fluently and correctly at least 80 words per minute, add approximately 3,000 words to your vocabulary ... and write narratives and friendly letters using appropriate forms, organization, critical elements, capitalization, and punctuation, revising as needed. (p. 8)

Not surprisingly, ELLs score significantly below non-ELLs on standardized tests in both reading and mathematics (NCES 2020b).

To help ELLs, some educators advocate **bilingual education**, teaching children in both English and their non-English native language. Advocates argue that bilingual education results in better academic performance of minority students, enriches all students by exposing them to different languages and cultures, and enhances the self-esteem of minority students. Critics counter that bilingual education limits minority students and places them at a disadvantage when they compete outside the classroom, reduces the English skills of minority group members, costs money, and leads to hostility with other minorities who are also competing for scarce resources.

bilingual education In the United States, teaching children in both English and their non-English native language.

The second reason that racial and ethnic minorities don't perform well in school, and compounding the difficulty ELLs have, is that many of the tests used to assess academic achievement and ability are biased against minorities (Rosales 2018). Questions on standardized tests often require students to have knowledge that is specific to White middle-class culture—knowledge that racial and ethnic minorities may not have.

The disadvantages that minority students face on standardized tests have not gone unnoticed. In 2019, a series of lawsuits were filed against the University of California (UC) system alleging that the SAT requirement put minorities at a disadvantage in the admissions process (Watanabe 2019). The UC Board of Regents had already been reviewing the use of the SAT and ACT in admissions when in May 2020 they announced that these tests would no longer be required. In a press conference, the UC–Berkeley chancellor explained that "[R]esearch had convinced ... [the Board of Regents] that performance on the SAT and ACT was so strongly influenced by family income, parents' education, and race that using them for high-stakes admission decisions was simply wrong" (quoted in Soares 2020).

An analysis of UC applicants found that while Black and Hispanic students were 5 percent of the top decile for test scores, they were 23 percent of the top decile in high school GPAs (Soares 2020), suggesting that college and university student bodies would be more racially diverse if admissions decisions did not factor in test scores. More than 1,000 universities had made standardized test scores optional before the COVID-19 pandemic (Watanabe 2019), with the majority dropping them for 2021 admissions decisions as a result of the pandemic.

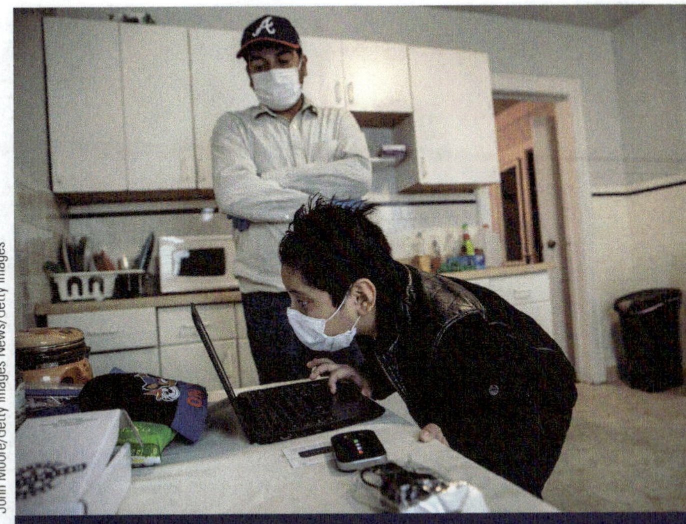
John Moore/Getty Images News/Getty Images

As COVID-19 impacted the nation, for this Guatemalan refugee family seeking asylum, the pandemic hit especially hard. Both Junior and his father ended up testing positive, forcing Junior to quarantine at home. Fortunately, Junior's bilingual/ELL teacher was able to help him continue with his schoolwork remotely.

A third factor that hinders minority students' academic achievement is overt racism and discrimination. Discrimination against minority students may take the form of unequal funding, as discussed earlier, as well as racial profiling, school segregation, and teacher and peer bias. In a qualitative study of racial discrimination in the classroom, Black students recounted incidents of being disciplined more severely than White students for the same behavior, the expectation of failure by their teachers, and a lack of institutional support for racial inclusiveness (Hope et al. 2015). Experiences of racial discrimination in school are associated with reduced academic curiosity and lower persistence in school (Leath et al. 2019).

Racial Integration. In 1954, the U.S. Supreme Court ruled in *Brown v. Board of Education* that segregated education was unconstitutional because it was inherently unequal. In 1966, a landmark study titled *Equality of Educational Opportunity* (Coleman et al. 1966) revealed that almost 80 percent of all U.S. schools attended by White students contained 10 percent or fewer Black students and that, with the exception of Asian American students, White students outperformed minorities on standardized tests. Coleman and colleagues emphasized that the only way to achieve quality education for all racial groups was to desegregate the schools. This recommendation, known as the **integration hypothesis**, advocated busing to achieve racial balance.

Despite the Coleman report, court-ordered busing, and a societal emphasis on the equality of education, U.S. public schools remain largely segregated. The majority of Black and Hispanic students attend schools that are predominantly minority in enrollment (Garcia 2020a). Furthermore, 54 percent of low-income White students compared to 81 percent of low-income Black students attend a "high-poverty" school in which 50 percent or more of the students are eligible for the Free or Reduced Lunch program (Garcia 2017). As a result, White students in New York, for example, had a 230 percent higher likelihood of having the opportunity to earn college credit while in high school than Black or Hispanic students (New York Equality Coalition 2018). Sixty-five percent of Black Americans support integrated schools "even if it means some students don't go to school in their local community" compared to 35 percent of White Americans (Horowitz 2019).

integration hypothesis A theory that the only way to achieve quality education for all racial and ethnic groups is to desegregate the schools.

Socioeconomic Integration. In 2007, the U.S. Supreme Court held that public school systems "cannot seek to achieve or maintain integration through measures that take explicit account of a student's race" (Greenhouse 2007, p. 1). At the time, the Court's decision reflected a general trend toward using socioeconomic or income-based integration

Katherine Frey/The Washington Post/Getty Images

Increasingly, school classrooms will be characterized by racial and ethnic diversity. Fifty-four percent of pre-K through 12th grade children are members of a racial or ethnic minority group (NCES 2020a).

rather than race-based integration variables. Unlike race-based integration that is subject to "strict scrutiny" by the government, school assignments based on socioeconomic status are perfectly legal.

This approach leads to a number of positive outcomes. First, socioeconomic integration leads to higher academic achievement. Low-income students attending more-affluent schools scored almost two years ahead of low-income students in high poverty schools (Kahlenberg, Potter, and Quick 2019). Further, low-income students in socioeconomically integrated schools are 68 percent more likely to attend college. Finally, evidence suggests socioeconomic integration is a more cost-effective means of raising student achievement than spending additional dollars in high-poverty schools (Kahlenberg, Potter, and Quick 2019).

Second, because of the relationship between race and income, socioeconomic integration achieves racial integration (Carlson et al. 2019), and racial integration, in turn, fosters acceptance of racial diversity and social cohesion. In 2020, the U.S. House of Representatives passed *The Strength in Diversity Act*, which, if signed into law, would provide grants to local municipalities to develop or implement plans to reduce racial and/or socioeconomic segregation in public schools (H.R. 2639 2019).

@Real FactsNC

Great news: The Wake County school system is expanding anti-discrimination protection to its transgender students and employees.

-Real Facts NC

Gender

Not only do women comprise two-thirds of the world's illiterate adults, but 132 million girls do not attend school (UNESCO 2019b). Although progress in reducing the education gender gap has been made, gender parity in primary and secondary schools has not been achieved. Sixty-six percent of countries for which there is data have achieved gender parity at the elementary school level, but only 45 percent have reached gender parity at the lower secondary level and 25 percent at the upper secondary level (UNICEF 2020).

Historically, U.S. schools have discriminated against women. Before the 1830s, U.S. colleges accepted only male students. In 1833, Oberlin College in Ohio became the first college to admit women. Even so, in 1833, female students at Oberlin were required to wash male students' clothes, clean their rooms, and serve their meals and were forbidden to speak at public assemblies (Fletcher 1943; Flexner 1972).

In the 1960s, the women's movement sought to end sexism in education. Title IX of the Education Amendments of 1972 states that no person shall be discriminated against on the basis of sex in any educational program receiving federal funds (see Chapter 10, "Gender Inequality"). These guidelines were designed to end sexism in the hiring and promoting of teachers and administrators. Title IX also sought to end sex discrimination in granting admission to college and awarding financial aid. Finally, the guidelines called for an increase in opportunities for female athletes by making more funds available for their programs. In the United States, the push toward equality has had considerable effect. For example, in 1970, nearly twice as many men as women had four or more years of college; however, in 2019, more women earned four-year college degrees or higher than men (see Table 8.1).

While access to education continues to be a problem globally, when girls are allowed to attend school, they outperform boys in reading, mathematics, and science in 70 percent of countries (Stoet and Geary 2015). In the United States, scores on the National Assessment of Educational Progress (NAEP) exam indicate that the gender gap in both mathematics and reading scores has decreased over the last few decades (National Assessment of Educational Progress 2019). However, differences remain and, in general, following educational stereotypes—boys outscore girls in mathematics, and girls outscore boys in reading. One explanation is that teachers' expectations of male

The arguments for single-sex classrooms (SSCs) are fairly simple: Girls would participate more in the learning process, particularly in **STEM** courses (i.e., those in science, technology, engineering, and math), without being intimidated by confident, overbearing boys. Boys would be less distracted and less likely to feel that they can't compete with their higher-performing female classmates, particularly in reading skills. Despite a growing trend toward single-sex education, research on the relative benefits of it are mixed (Chen 2020). What do you think about single-sex education?

What *do you* think?

and female students and the sorting of teachers by gender into different disciplines (e.g., male teachers in sciences and females in language arts) influence student performance (Meinck and Brese 2019).

Much of the research on gender inequality in the schools focuses on how female students have been disadvantaged in the educational system. But what about male students? Boys are more likely to lag behind girls in the classroom, be diagnosed with attention-deficit/hyperactivity disorder (ADHD), have learning disabilities, feel alienated from the learning process, and drop out or be expelled from school (Owens and McLanahan 2020; NCES 2019). Furthermore, Black and Hispanic males compared to White males score lower on the NAEP, are less likely to be in gifted and talented programs or to be in advanced placement (AP) classes, and are less likely to graduate from high school or college (Schott Report 2015).

The gap in academic achievement between boys and girls is largest among children from low-income backgrounds, suggesting that boys benefit most from the resources higher-income schools can provide (Autor et al. 2016). One important factor may be the absence of recess in many low-income schools (Ramstetter and Murray 2017). Recess is an important component of the school day that has been shown to help boys in particular (Reilly 2017). Thus the problems boys have in school may indeed require schools to devote more resources and attention to them. In the past decade, a number of school districts have opened all-boy public schools to address these concerns. In fact, more all-boy public schools exist in the United States than all-girls schools (Education Week 2017).

Problems in the American Educational System

In 2020, the COVID-19 pandemic created unprecedented problems for educational systems around the world. Teachers, administrators, students, and parents grappled with the transition from in-person classroom instruction to online education. Local and state governments faced difficult decisions about when and how to safely resume in-person instruction. However, even before the pandemic, the U.S. educational system was plagued by a number of problems. The 52nd Annual PDK Poll of public attitudes toward schools, taken just before the U.S. COVID-19 outbreak, found that more than half of U.S. adults disapproved of the Trump administration's performance on education (PDK 2020).

Thus we now turn our attention to some of the problems associated with U.S. schools, including inadequate funding, facilities, and school mental health services; low levels of educational achievement; problems in recruiting and retaining quality teachers; and crime, violence, and school discipline.

Inadequate Funding, School Facilities, and Mental Health Services

Results of the 52nd Annual PDK Poll indicate that for the 19th consecutive year, lack of financial support was identified as the top problem facing public schools (PDK 2020). "Money matters for schools ... in determining school quality and student outcomes" (Baker 2020, p. 1). During the 2020 COVID-19 pandemic, school closures highlighted the need for increased funding for the provision of computers and high-speed Internet service to students across the country (Quirk 2020).

STEM An acronym for science, technology, engineering, and mathematics.

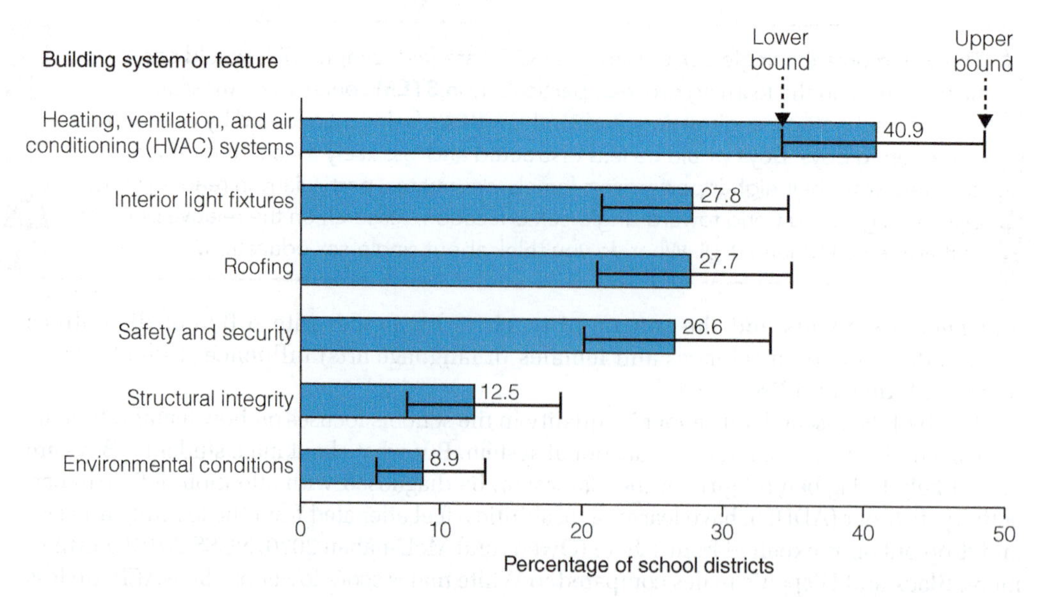

Figure 8.5 Percentage of Public School Districts Where at Least Half the Schools Need Updates or Replacements of School Building Systems and Features, 2019
SOURCE: Government Accountability Office 2020.

Per pupil spending on public school students varies dramatically by state. For example, in 2020 New York spent more than $23,000 per pupil per year, while Utah spent just over $7,000 per student (World Population Review 2020a). Per pupil spending includes money spent on teacher salaries and benefits, materials, and equipment for classrooms and school programs, as well as administrative expenses and support staff.

The Century Foundation (2020) reports that K–12 public schools are underfunded by $150 billion annually—the estimated amount needed for U.S. students to achieve average proficiency on reading and math assessments. Funding gaps are more likely and are larger in school districts that are (1) low income, (2) located in the U.S. Southwest and Southeast, and (3) predominantly Black or Hispanic. As disparities in school funding disproportionately hurt low-income and minority youth, increasing school funding and reducing funding disparities would particularly benefit these student populations.

Inadequate School Facilities. Inadequate school funding is a barrier to fixing and upgrading the physical building facilities of schools, especially those in low-income and predominantly minority districts. Many schools in the United States were built in the 1940s and 1950s and need costly repairs and renovations to provide safe, healthy, and modern learning environments. According to the American Society of Civil Engineers (ASCE), a quarter of U.S. schools are in "fair or poor" condition, and more than half of schools need improvements to attain the designation of "good" (ASCE 2017). Figure 8.5 graphically portrays the estimated percent range of districts where at least half the schools need improvements.

Pennsylvania Department of Education awarded Schuylkill Valley Community Library $134,712 in a Keystone Grant for building improvements, including replacing the roof, improving ADA accessibility and adding restrooms, @SenJudySchwank announced. ow.ly/YUEE50ClHpy

-bctv.org

Examples of needed upgrades include replacing HVAC systems (which, if not repaired, can lead to mold and unhealthy indoor air quality), lighting, roofing, safety and security systems, plumbing, windows, cable, and Wi-Fi. One school in Florida had multiple buckets throughout the school to collect water that leaked through the roof, and the principal commented that it frequently "rained" in her office (Government Accountability Office 2020). Because many school buildings also serve as emergency shelters during natural disasters, school building improvements are also needed to provide safe shelter in high winds and earthquakes.

The physical conditions of school facilities include outdoor air quality. In a study of Baltimore City schools, those that had high levels of outdoor pollutants (due to high traffic emissions and/or near-school industrial pollutants) had higher rates of absenteeism, which may be due to higher levels of respiratory illness among students at those schools (Berman et al. 2018).

Research documents the relationship between the school physical environment and academic achievement. In a review of the literature, researchers concluded that a "building's structural facilities profoundly influence learning. Inadequate lighting, noise, low air quality, and deficient heating in the classroom are significantly related to worse student achievement" (Cheryan et al. 2014, p. 4).

© Joshua A. Bickel/The Columbus Dispatch via USA TODAY Network

This first grader was dropped off on her first day of school amidst numerous COVID-19 restrictions including mandatory facemasks and social distancing. The fear and isolation these young children have experienced, both in school and at home, are cause for concern among mental health professionals.

Inadequate Mental Health Services. Another problem related to inadequate school funding is that many schools are not able to hire school counselors to provide mental health services to students. School principals cite lack of funding as the most significant barrier to school-based mental health services (Quirk 2020). In 2019, 14 million U.S. students attended school with a police officer on duty but no school counselor, social worker, or nurse psychologist. Providing mental health services to youth is critical, as stress, trauma, anxiety, and depression are not uncommon among youth (see also Chapter 2). Many children and teens struggle with problems such as parental abuse/neglect, parental divorce, experienced and witnessed racism, teen pregnancy, gender and sexual identity issues, community violence, and substance abuse.

During the COVID-19 pandemic, children have experienced social isolation and family separation, COVID-19–related illness and deaths of loved ones, family job loss, and housing insecurity. At the same time children have also been exposed to unprecedented levels of social hostility based on differing political views, including conflicting views about whether COVID-19 was a pandemic or a "hoax." Psychological trauma related to COVID-19 is likely to persist long after the pandemic, which further underscores the need for school-based mental health services.

The failure to provide adequate school-based mental health services for children places youth at risk for other problems, such as substance abuse and violent and criminal behavior. And, untreated mental health problems in youth increase the risk for low academic achievement (Agnafors, Barmark, and Sydsjo 2020).

Low Levels of Academic Achievement

The Educational Research Center uses several indicators to measure K–12 achievement in public schools, including current levels of performance, improvement over time, and the achievement gap between poor and non-poor learners called the *poverty gap* (Hightower 2013). Based on a 100-point scale, the achievement average for the nation in 2019 was 75.6 (C) and ranged from a high of 88.4 (B) for Massachusetts to a low of 66.4 (D) for New Mexico (Lloyd and Harwin 2019).

National trends indicate that reading and mathematics scores, as a whole, improved between 1992 and 2019, but not for all students (National Assessment of Educational Progress [NAEP] 2020). In general, Black and Hispanic students made larger gains during the same time period than White students, narrowing the reading and math gap between White and minority populations. Nonetheless, Black and Hispanic students remain behind their White counterparts in both reading and math. Schools with higher percentages of Black and Hispanic students have lower educational achievement levels as a result of, in part, higher levels of poverty (Spector 2019). The gender gap has also narrowed over time although, in 2019, females still outperformed males in reading, and males outperformed females in mathematics (NAEP 2020).

School Dropouts. Graduating from high school is considered an important milestone, but not all students complete high school. The *status dropout rate* is the percentage of 16- to 24-year-olds who are not in school and have not earned a high school degree or its equivalent. The overall status dropout rate declined from nearly 10 percent in 2006 to just over 5 percent in 2018 (U.S. Department of Education Center for Education Statistics 2020). Nonetheless, in 2018, there were 2.1 million status dropouts. Asian Americans have the lowest status dropout rate at just under 2 percent compared to, in order from low to high, Whites, individuals reporting two or more races, Blacks, Hispanics, Pacific Islanders, and American Indian/Alaska Natives with a rate that approaches 10 percent. Students with disabilities and English language learners also have higher dropout rates (Bustamante 2019). Reasons for dropping out of school include chronic absenteeism, getting poor grades, becoming pregnant, having to care for a family member or support family, feeling unsafe at school, and thinking it would be easier to get a GED.

Compared with high school graduates, high school dropouts have limited employment opportunities; poorer health and lower life expectancy; and higher rates of poverty, dependence on public assistance, single parenthood, and incarceration (Bustamante 2019; Segal 2013). These problems affect not only those individuals who have not competed high school but their communities as well.

Second-chance initiatives such as GED certification allow students to complete their high school requirements. Other dropout interventions include early-college programs that allow dropouts to enroll in community colleges or, in some cases, four-year degree programs. There they receive a secondary school education, earn a high school diploma, and often accrue college credits. There are also efforts, at the state and federal levels, to increase the age of compulsory school attendance in the hopes that it will deter students from dropping out of school.

What do you think?

In some states, students are allowed to drop out of school at age 16, but in other states, the minimum age for dropping out of school is 17 and in some states, 18. What do you think the minimum legal age for dropping out of school should be? Additionally, there are maximum age limits on eligibility for a free education. The limits range from a low of 17 in Alabama to high of 22 in West Virginia, South Carolina, and Massachusetts (NCES 2019). What do you think the maximum legal age for receiving a free public education should be?

U.S. Ranking in Educational Achievement. U.S. students are outperformed by many of their foreign counterparts—something particularly troubling in a knowledge-based global economy. The most recent results from *Education at a Glance* indicate that the United States ranks above the Organisation for Economic Cooperation and Development (OECD) average of 80 percent. However, with a high school graduation rate of 86 percent, an all-time high, the United States still ranks below several countries including Chile, Finland, Ireland, Israel, Italy, Korea, Slovenia, Spain, and New Zealand (OECD 2020). Nonetheless, it is encouraging that the U.S. graduation rate has, in general, increased consecutively for the last 14 years. These additional graduates "produce benefits to the nation's economy, health, and civic society and position themselves to pursue the American dream" (Atwell et al. 2020, p. 1).

Crime, Violence, and School Discipline

As noted earlier, one reason why some students drop out of school is that they don't feel safe. In 2017, 4 percent of students between the ages of 12 and 18 reported "being afraid of attack or harm at school during the school year," and 3 percent reported being "afraid of attack or harm away from school during the school year" (Wang et al. 2020, p. 87).

Crime and Violence against Students. In the 2018–2019 academic year, students reported over 800,000 types of victimizations at school and over 400,000 victimizations away from school. These incidences include both property (e.g., vandalism) and violent (e.g., assault) offenses. Students most likely to report criminal victimization at school were male, American Indian/Alaska Natives, ninth graders, and students attending urban public schools.

Despite the horrors of school shootings such as those at Columbine, Sandy Hook Elementary School, and Marjory Stoneman Douglas High School, the chance of a student dying at school is quite low. Less than 3 percent of the total youth homicide rate, and proportionately even fewer suicides, take place on school property (Wang et al. 2020). Even so, in 2018–2019, there was a record total of 66 school shootings resulting in a record 29 deaths.

In 2018, 71 percent of public schools recorded at least one or more violent incident; 21 percent recorded one or more serious violent incident, 33.4 percent recorded one or more theft, and 59.8 percent recorded one or more other criminal offense. Elementary schools report lower percentages of violent crime, serious violent crime, and other crimes when compared to middle schools and high schools (Wang et al. 2020).

Sexual Assault on Campus. In a survey of 33 colleges and universities conducted by the Association of American Universities (AAU), 13 percent of students reported experiencing nonconsensual sexual contact during their enrollment at their school (Cantor et al. 2020). The rates for women and transgender, genderqueer, and non-binary students were significantly higher than for men. Rates of sexual assault victimization among undergraduate women were nearly three times higher than for women graduate and professional students.

About half of incidents of nonconsensual penetration occurred by physical force; the other half involved the inability to consent (the victim was asleep, passed out, or incapacitated due to alcohol or drugs). Two filmmakers who interviewed sexual assault victims and assailants for their film *The Hunting Ground* concluded that the assaults are usually premeditated; the "survivor was picked out, plied with alcohol and set up to be assaulted" (quoted in Dockterman 2015).

The AAU survey of sexual assault on campuses found that most students who had been victimized did not report the assault to local police or to campus resources (Cantor et al. 2020). The top reason for not reporting is that victims didn't believe the incident was serious enough to warrant further action. But some victims of sexual assault don't report because they fear retaliation from the perpetrator and/or they want to avoid the shame, blame, and stress they may experience from the reporting process. A controversial rule, passed under the leadership of then Secretary of Education Betsy DeVos, may exacerbate the fear and discomfort that victims experience when reporting a sexual assault. The new rule, finalized in 2020, changes Title IX regulations for sexual misconduct, giving the accused the right to an in-person hearing and to cross-examine the accuser, previously not allowed (Strauss 2020a). Sexual assault advocates are against this rule, as it is likely to discourage sexual assault victims to report the incident. President Biden has indicated that he plans to revise these policies to protect the rights of accusers (Kingkade 2020).

School Discipline. In 2018, the most common school offense prompting disciplinary action by school officials was physical fights or attacks, followed by illegal drug offenses, alcohol offenses, and use or possession of a weapon (Wang et al. 2020). Out-of-school suspensions mean lost opportunities for learning. In the 2015–2016 school year, U.S. students lost more than 11 million days of instruction due to out-of-school suspensions (Losen and Martinez 2020). Black students—particularly boys—were much more likely to be suspended than students of other racial/ethnic groups; however, suspensions are also more common among Hispanic, Native American, and Hawaiian/Pacific Islander students than among their White peers.

Do these disparities reflect differences in students' behavior? Do Black students, for example, misbehave more than White students? A study of the racial gap in school suspensions and expulsions among students ages 5 to 9 found that much of the racial disparities in school discipline were due to teachers' differential treatment of White and Black students (Owens and McLanahan 2020). Losen and Martinez (2020) suggest that "we cannot expect to close the racial achievement gap if we do not close the discipline gap" (p. vi).

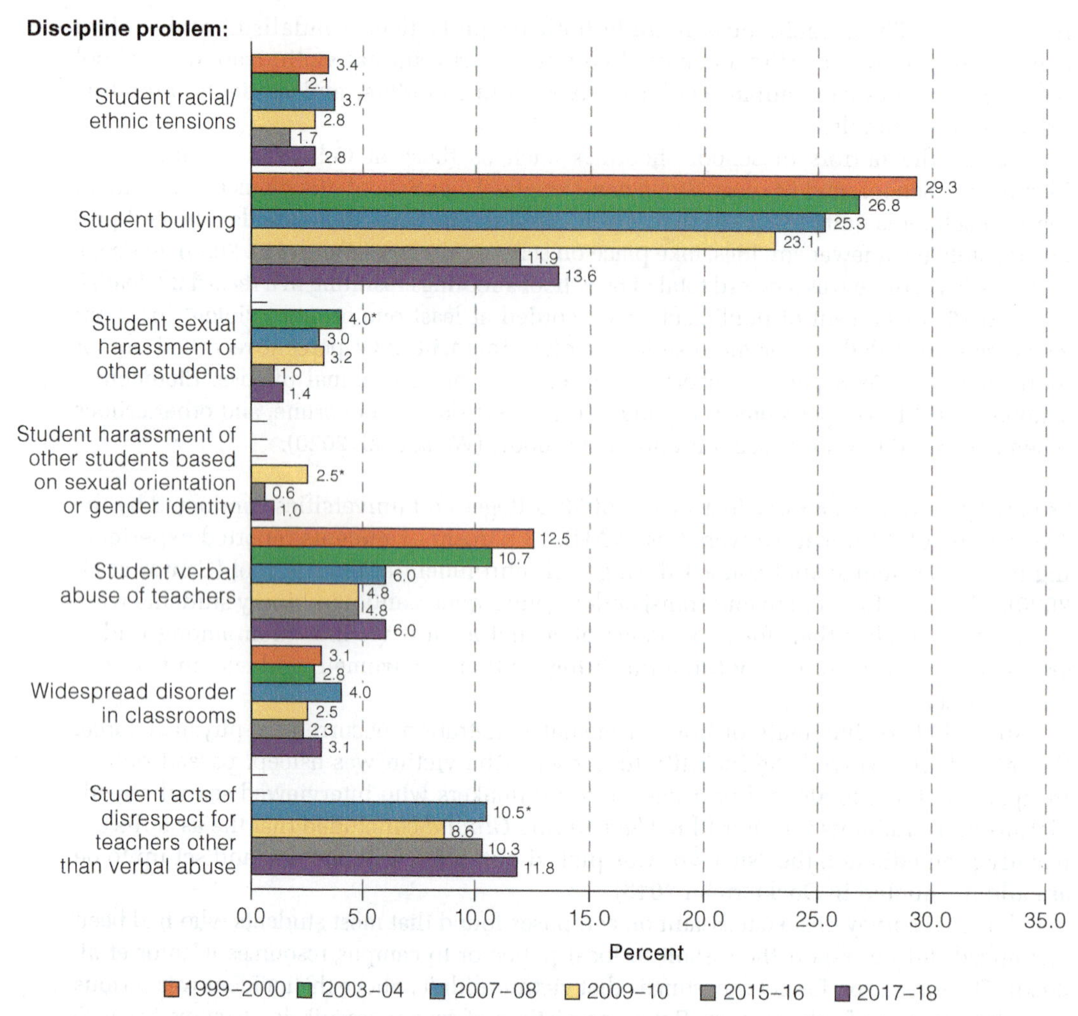

Figure 8.6 Percentage of Public Schools Reporting Discipline Problems That Occur at Least Once a Week, by School Years

*The first year of data collection for this item
SOURCE: Wang et al. 2020.

> "He who opens a school door, closes a prison."
>
> **–VICTOR HUGO, FRENCH NOVELIST, POET**

school-to-prison pipeline The established relationship between severe disciplinary practices, increased rates of dropping out of school, lowered academic achievement, and court or juvenile detention involvement.

bullying Bullying occurs when "a physically stronger or socially more prominent person (ab)uses her/his power to threaten, demean, or belittle another" (Juvonen and Graham 2014, p. 161).

Disciplinary practices differ between schools and school districts with some expelling or suspending 90 percent of the student body at least once. Such rates are the result of *zero-tolerance disciplinary policies.* Although the origin of the phrase is from the 1980s and fears over drug use and students bringing guns to school, today's zero-tolerance policies may include suspensions or expulsions for "bringing a cell phone to school, public displays of affection, truancy, or repeated tardiness" (National Public Radio 2013, p. 1).

Increasingly, educators, counselors, teachers, and psychologists "denounce such practices [suspensions and expulsions] as harmful to students academically and socially, useless as prevention tools, and unevenly applied" (Shah 2013, p. 1). Further motivation for school reform comes from the **school-to-prison pipeline**—the established relationship among severe disciplinary practices, increased rates of dropping out of school, lowered academic achievement, and court or juvenile detention involvement. The school-to-prison pipeline disproportionately hurts minority students who are more likely to be suspended or expelled and when "forced out of school become stigmatized and fall behind in their studies, ... drop out of school altogether, and ... may commit crimes in the community" (Amurao 2013, p. 1).

Bullying. Bullying "involves targeted intimidation or humiliation ... [A] physically stronger or socially more prominent person (ab)uses her/his power to threaten, demean, or belittle another" (Juvonen and Graham 2014, p. 161). In 2018, one in five students ages

12–18 reported being bullied at school, including on a school bus and going to and from school (Wang et al. 2020) (see Figure 8.6). The most common experiences of being bullied involved being the subject of rumors and being made fun of, called names, or insulted. Other forms of being bullied include being excluded, having one's property destroyed, being threatened with harm, and being pushed, tripped, or spit on. Cyberbullying—another common form of bullying—is discussed in Chapter 14.

What do you think?

During the COVID-19 pandemic, before schools shut down and after they re-opened, Asian American students were taunted and bullied by classmates who blamed them for the COVID-19 pandemic (Kam 2020; Quirk, 2020). Do you think the bullying behavior of Asian American students resulted from the terminology repeatedly used by the former administration including "China virus," the "Wuhan virus," and the "kung-flu"?

Recruitment and Retention of Quality Teachers

Schools across the country—especially those in high poverty areas—are having difficulty hiring and retaining teachers. In the 2015–2016 school year, 9 percent of U.S. schools tried to fill a teaching vacancy but could not, an increase of 3 percent from the 2011–2012 academic year (Garcia and Weiss 2019a). In the same time period, the percentage of schools that found it "very difficult" to fill a vacancy nearly doubled, from 20 percent to 36 percent. Schools are also having difficulty retaining teachers, as more than one in ten teachers are leaving their jobs or the teaching profession altogether.

High teacher turnover is costly. To replace a teacher who leaves, schools must spend money to recruit and train new teachers, not to speak of the time and effort involved in recruiting, hiring, and training new teachers. More importantly, teacher turnover has lasting, negative effects on the quality of instruction and on student achievement, which then exacerbates turnover in future years (Sorensen and Ladd 2020). New teachers who replace those who leave are more likely to have low levels of experience and to lack full licensure and certification in the subject they are hired to teach, thus lowering the quality of education and achievement levels for students.

Low Teacher Pay. Teacher walkouts and protests in recent years have brought attention to the problem of low teacher pay. Median pay for kindergarten and elementary school teachers in 2019 was about $59,000 a year; for high school teachers, it was $61,660 a year (Bureau of Labor Statistics 2020). Salaries for beginning teachers trend lower and also vary by state.

Low salaries for K–12 teachers are a major reason for high teacher turnover. In the 1960s and 1970s, female teachers were paid more than comparably educated women in other jobs. But this wage premium has been replaced by a growing wage penalty. In 2019, female teachers earned 13 percent *less* than women with similar levels of education working in other occupations (Allegretto and Mishel 2020). Male teachers, in 2019, earned 30 percent less than similarly educated college graduates who worked in other professions, which helps to explain why only one in four teachers is male. Low teacher pay is one of the reasons for the declining numbers of college students enrolled in education degree programs or teacher preparation programs (Garcia and Weiss 2019a).

@technologylaura

Teacher Humor brought to you by me 💜 @WeAreTeachers

2 degrees
8+ years of teaching
4 Instructional Preps

Biggest Challenge------>
copy machine.

-Laura

Difficult Working Conditions for Teachers. Another reason why teacher turnover is so high—particularly in high-poverty schools—is the difficult working conditions. Roughly one-quarter of teachers say that students are coming to school unprepared to learn and with inadequate parental involvement. More than one in five teachers report they have been threatened at school, and one in eight say they have been physically attacked by a student at school (Garcia and Weiss 2019b) (see Figure 8.6).

Cheney Orr/Bloomberg/Getty Images

As teachers continued to contend with difficult working conditions, the impact of the COVID-19 pandemic created even more obstacles. U.S. public school enrollment declined for the 2020–2021 academic year as a result of the pandemic.

Many teachers feel the relationships between teachers, administrators, and parents are not supportive. Teachers feel that they are not recognized for their hard work and that they lack influence over what and how they teach (Garcia 2020b). Although most teachers do not teach in the summer, unless it is as a second job, teachers put in long hours during "off-times," including nights and weekends, preparing lessons and grading papers. A 2018 poll of U.S. adults reveals that most adults do not consider teaching to be a desirable profession with more than half of respondents saying they would not want their child to become a teacher (Garcia 2020).

Teacher Effectiveness. A controversial strategy for measuring teacher effectiveness is the use of **value-added measurement (VAM)**, which involves the use of student achievement data to assess a teacher's effectiveness. Value-added evaluations of teachers are based on whether their students' test performances improved or worsened over the academic year, which suggests whether the teacher "added value" to the students' academic year (Chen 2019). Against the urging of the American Federation of Teachers, teachers' value-added evaluations are made public, enabling parents to determine the ability of their children's teachers.

Assessing teachers based on student performance assumes all else is constant and ignores the reality of student differences in such nonschool factors as family life, poverty, emotional and physical obstacles, and access to computers and wi-fi (see Figure 8.7 on p. 304). In addition, there are concerns that teachers, fearing for their jobs and concerned about merit-based pay, may begin "teaching to the test" and worse. In 2015, ten former Atlanta public school educators were convicted for providing students with answers to standardized test questions and changing incorrect answers to correct ones.

To place quality teachers in the classroom, many states have implemented mandatory competency testing (e.g., the Praxis Series). The need for teachers who are officially classified as "highly qualified" is tied to federal mandates that emphasize the importance of having licensed teachers in the classroom. Additionally, teachers who have a bachelor's degree and have been in the classroom for three or more years are also eligible for national board certification. Some studies indicate that students of "highly qualified" teachers and/or board-certified teachers perform better on standardized tests and have shown greater testing gains than students of teachers who are not "highly qualified" and/or board certified. For example, a 2017 study of schools in Mississippi found that kindergarten and third-grade students with reading teachers who had national board certification performed, on average, significantly higher on literacy assessments than students with reading teachers who were not board certified (National Strategic Planning and Analysis Research Center 2017).

The Challenges of Higher Education

Although there are many types of postsecondary education, higher education usually refers to two- or four-year, public or private, degree-granting institutions. In 2018, there were nearly over 4,200 degree-granting colleges and universities in the United States with an enrollment of 19.6 million undergraduate and graduate students (Moody 2019; NCES 2018).

Over the last two decades, women and racial and ethnic minorities disproportionately contributed to college and university enrollment growth (NCES 2018). However, enrollment began to decline in 2015 and is expected to continue to decline through 2024 because of demographic changes—there are fewer students currently in elementary, middle, and high schools compared to recent generations (NCES 2019).

value-added measurement (VAM) The use of student achievement data to assess teacher effectiveness.

Over the last decade, there has also been a significant decrease in full-time tenured or tenure-track faculty, once the "core" of academia, and significant increases in noninstructional staff (e.g., administrators) and non-tenure-track, part- or full-time instructors. In general, as academic rank increases—that is, from instructor to assistant professor, to associate professor, to full professor—salaries increase and the proportion of women and minorities decreases. In the academic year 2018–2019, the average salary for female faculty was $79,995; for male faculty, $96,369 (NCES 2019).

Cost of Higher Education. In the academic year 2018–2019, the total annual cost for a full-time, four-year college student who lived on campus, paid in-state tuition, bought books and supplies, and incurred other expenses such as transportation costs was $20,598 for a public college and $44,662 for a private college (NCES 2019). Out-of-state students' costs are even higher. For many students, without financial aid, the expense of a four-year degree would make a college degree unobtainable. On average, students today are paying more than twice what their parents would have paid for a college education.

Nearly two-thirds of college students graduated with school-related debt in 2019, averaging $31,172 (Issa 2019). In the first quarter of 2019, U.S. student debt approached the $2 trillion mark, a total larger than the sum of all auto and credit card debt (Issa 2019). School debt is so inescapable that many retirees are still paying on their loans, and COVID-19 has only exacerbated the problem (see this chapter's "The Impact of COVID-19 on Education" section).

During the COVID-19 pandemic, former Department of Education Secretary Betsy DeVos paused student loan payments through January 2021, which the Biden administration then extended to September 30, 2021 (Deliso 2021). However, during her time in office, DeVos proposed to dismantle the Public Service Loan Forgiveness program, which allows individuals to have a portion of their debt cancelled if they work at least ten years in government or public service. Further, her office approved only 6 percent of applicants' requests for loan forgiveness (Douglas-Gabriel 2020) leading a federal judge to order that the Department clear any unresolved claims and pay 30 percent of each applicant's loans, per month, if the process of resolving the claims took more than 18 months.

DeVos also failed to enforce the *borrower defense to repayment rule* put in place by President Obama. The rule provides students with loan repayment protections if a predatory institution failed to provide them with an education. This program was put in place after several for-profit institutions closed, leaving students with high student loans and no degree. Lastly, Secretary DeVos loosened restrictions on student loan vendors, reopening the door to so-called predatory lenders who charge high-interest rates. President Biden has proposed forgiving up to $50,000 in student loans over a five-year period if an individual has performed public service (Biden 2020).

@ewarren

Our next economic relief package can boost our economy from the grassroots up, put money directly in the hands of working families, and fight the racial wealth gap.

How? By cancelling student loan debt.

-Elizabeth Warren

"Degrees of Inequality." In the now classic book, *Degrees of Inequality: How the Politics of Higher Education Sabotaged the American Dream* (Mettler 2014), the author argues that higher education perpetuates inequality rather than expanding opportunities. Although the book outlines many ways in which this is the case, the central arguments include (1) the growth and expense of for-profit colleges and universities (e.g., the University of Phoenix, Kaplan Higher Education) that predominantly serve students from low-income families; (2) the dismal graduation rates—25 percent on average—at for-profit colleges and universities; (3) the increased cost of going to college at a time when student aid programs cover less and less of the expense; and (4) the underfunding of community colleges at a time when federal loans to students at for-profit institutions increases student debt (i.e., they are more expensive than nonprofit institutions), often without a degree, and costs taxpayers millions of dollars in defaulted loans.

From a college or university's perspective, spreading what aid there is to more people in smaller amounts both increases enrollment rates and the institution's reputation as grade point averages and the proportion of incoming freshmen who graduate increases. By definition, those who receive merit-based aid are high achievers and more likely to have graduated from better-funded high schools where enrollment for students from low-income backgrounds is small. Thus not only is need-based aid for low-income high school graduates less available, but these graduates are less likely to be eligible for merit-based aid.

@safe returnRI

"The problem is that high-stakes standardized **testing** has not only failed at achieving racial **equality**, its proliferation has only exacerbated racial inequality and worsened the **education** for students of color."

-Safe Return to School RI

Because socioeconomic status and race/ethnicity are related, the same social forces behind "degrees of inequality" that impact low-income students disproportionately affect racial and ethnic minorities. Of the total fall enrollment at four-year degree-granting institutions in 2018, only 13.0 percent of students were Black and 19 percent Hispanic (NCES 2020d). It is important to note that Black and Hispanic students are three times more likely to attend for-profit institutions than White students (Body 2019).

On standardized tests such as the SAT and the ACT, "children from the lowest-income families have the lowest average test scores, with an incremental rise in family income associated with a rise in test scores" (Corbett et al. 2008, p. 3). Muller and Schiller (2000) report that students from higher socioeconomic backgrounds are more likely to enroll in advanced courses for mathematics credit and to graduate from high school—two indicators of future educational and occupational success. After a wave of negative publicity on the impact of private SAT tutoring among high-income students, the College Board collaborated with Khan Academy to create free SAT and preparatory courses (Koenig 2019).

Additionally, although college enrollments of racial and ethnic minorities have increased over the last several decades, the increases have been accompanied by "separate and unequal" access to "selective and well-funded four-year colleges" (Carnevale and Strohl 2013, p. 6). Between 1995 and 2009, 82 percent of incoming White college students were enrolled at the top 468 colleges and universities in the United States, compared to 9 percent of Black college students and 13 percent of Hispanic college students. Lastly, African American and Hispanic students with A averages in high school are more likely to be enrolled in community colleges compared to their similarly qualified White counterparts. The 468 most selective four-year colleges where White students disproportionately attend have (1) more financial resources, (2) higher graduation rates, (3) higher enrollment in and completion of graduate degrees, and (4) graduates with greater future earnings (Carnevale and Strohl 2013). (See Chapter 9 for a discussion of race, ethnicity, and affirmative action.)

Community Colleges. An increasing number of people in the United States see community colleges as contributing to a strong workforce and being worth the cost (Nguyen 2018). Yet, enrollment in two-year colleges decreased between 2010 and 2018 after dramatic increases between 2000 and 2010. Over 1 million associate degrees were awarded in the 2017–2018 academic year, more than 60 percent to women (NCES 2019).

Community colleges play a vital role in U.S. educational policy. They are starting points for many Americans who hope to eventually transfer to baccalaureate institutions. Minority and low-income students and women are more likely to enroll in community colleges than their White, male, middle-class counterparts for a variety of reasons. Community colleges are often closer to where students live, and they offer a more flexible schedule, allowing students to work full- or part-time while attending school and avoiding the cost of room and board by living at home. An increasing number of community colleges offer dual-enrollment options for high school students, expanding the variety of courses available to them and helping them to move ahead with college or vocational careers.

Community college students often attend school to earn a degree in a vocational trade, thus preparing them to make a valuable contribution to the workforce upon graduation.

During the 2020 Democratic primary, several candidates proposed nationwide tuition-free community colleges (Startz 2019b). In 2017, San Francisco became the first city to offer tuition-free community college for all city residents, regardless of income (Berndtson 2017). In the same year, New York State created the Excelsior Scholarship, which covers tuition for any student pursuing a two- or four-year college degree at any of the city or state colleges as long as the individual or family earns less than $125,000 annually. Since then, 11 states

Career Readiness

The National Association of Colleges and Employers (NACE) asked employers to identify skills and qualities they want to see in college graduates they are considering for jobs. Using the following scale, answer each of the questions by placing the number that corresponds to your level in the blank provided.

1 = not a strength, 2 = improving in this area, 3 = definitely a strength

To what extent is each of the following a strength for you?	Your Response (1–3)
1. Communication skills (written)	____
2. Problem-solving skills	____
3. Ability to work in a team	____
4. Initiative	____
5. Analytical/quantitative skills	____
6. Strong work ethic	____
7. Communication skills (verbal)	____
8. Leadership	____
9. Detail-oriented	____
10. Technical skills	____
11. Flexibility/adaptability	____
12. Computer skills	____
13. Interpersonal skills (relates well to others)	____

All of these qualities were selected by at least half of the employers who responded, with more than three-quarters identifying written communication skills, problem-solving skills, and ability to work in a team as qualities they were seeking. Other skills identified by employers include organizational ability, strategic planning skills, tactfulness, creativity, friendly/outgoing personality, entrepreneurial skills/risk taker, and fluency in a foreign language.

Fortunately, the vast majority of these skills can be developed while in college. Now that you know what employers are looking for, take this list to your advisor and ask what opportunities exist on campus and/or in your academic program to develop them.

Good News! If you identify as an introvert, less than 25 percent of employers listed friendly/outgoing personalities as an attribute they are looking for in an employee.

SOURCE: National Association of Colleges and Employers 2019.

have put "free college" plans in place. Most are "last dollar" scholarships, meaning they pick up the remaining costs after financial aid and grants have paid all they will. In 2020, New York announced that new Excelsior Scholarships might not be available for the coming year given the impact of the COVID-19 pandemic on the state's budget.

In 2019, a California appeals court upheld a previous jury's decision that San Francisco State University retaliated against a professor by denying her tenure after she complained about the climate on campus for minority women (Flaherty 2019). At the college level, tenure assures that professors cannot be fired for, as examples, teaching unpopular ideas or holding political views deemed by others as "radical." Tenure provides university faculty academic freedom. Do you think the system of tenure in public schools, including colleges and universities, should be eliminated or retained?

What *do you* think?

The Impact of COVID-19 on Education

In early 2020, the emerging COVID-19 pandemic challenged most schools, colleges, and universities across the globe to abruptly pivot instruction to protect students, faculty, and staff, as well as their respective families and communities, from infection. The crisis exposed many of the inadequacies and inequalities in educational systems across the world and across the United States.

K–12 Schools

By March 2020, all 46 OECD countries had closed some or all of their primary and secondary schools (OECD 2020b). Countries used a variety of strategies to maintain some instruction during this period. Low-income countries relied heavily on radio and television, as well as assignments distributed on paper. High-income and middle-income countries were more likely to transition to online instruction with many European and Scandinavian countries relying on previous agreements with private companies for distance learning (OECD 2020b).

In the United States, school strategies varied dramatically. Private schools moved to remote instruction primarily through synchronous Internet-based video instruction as well as materials provided through Khan Academy (Harris et al. 2020). Private school teachers tended to have little experience with distance education, but their students were more likely to have home computers and Internet access.

@andre aducas

"More than 2,600 students in Jersey City were still without a device and/or internet connection to access their virtual classrooms as of Nov. 4, NJ Spotlight reported, citing state Department of Education data."

-Andrea Ducas

For public schools maintaining instruction was more complicated. Harris et al. (2020) reports that two-thirds of schools used a platform to offer online instruction. However, few schools monitored attendance or required students to participate. Forty percent of schools made printed packets available for instruction, 50 percent made PCs or tablets available to students, 31 percent provided referrals to free Wi-Fi providers, and 7 percent provided home hotspots to students in need. Further, more than half of schools continued to provide meals to students at pickup locations while others provided meals through a district-level distribution center. Interestingly, Harris et al. (2020) conclude that low-income school districts provided the greatest variety of options for student participation, and were more likely to provide assistance with computers and Internet access than higher-income schools.

Early experiences with the pandemic revealed the difficulties that students, parents, and teachers faced moving forward. As Figure 8.7 reveals, the long-standing digital divide meant low-income students were often left behind despite school efforts to provide support (see Chapter 14). Seventy percent of parents with young children—23.5 million working parents—had no caregiver in the home (Bateman 2020). Parents with "essential" jobs were faced with the dilemma of how to supervise their children and keep their jobs, while parents who could work from home juggled child care and paid work from makeshift home offices.

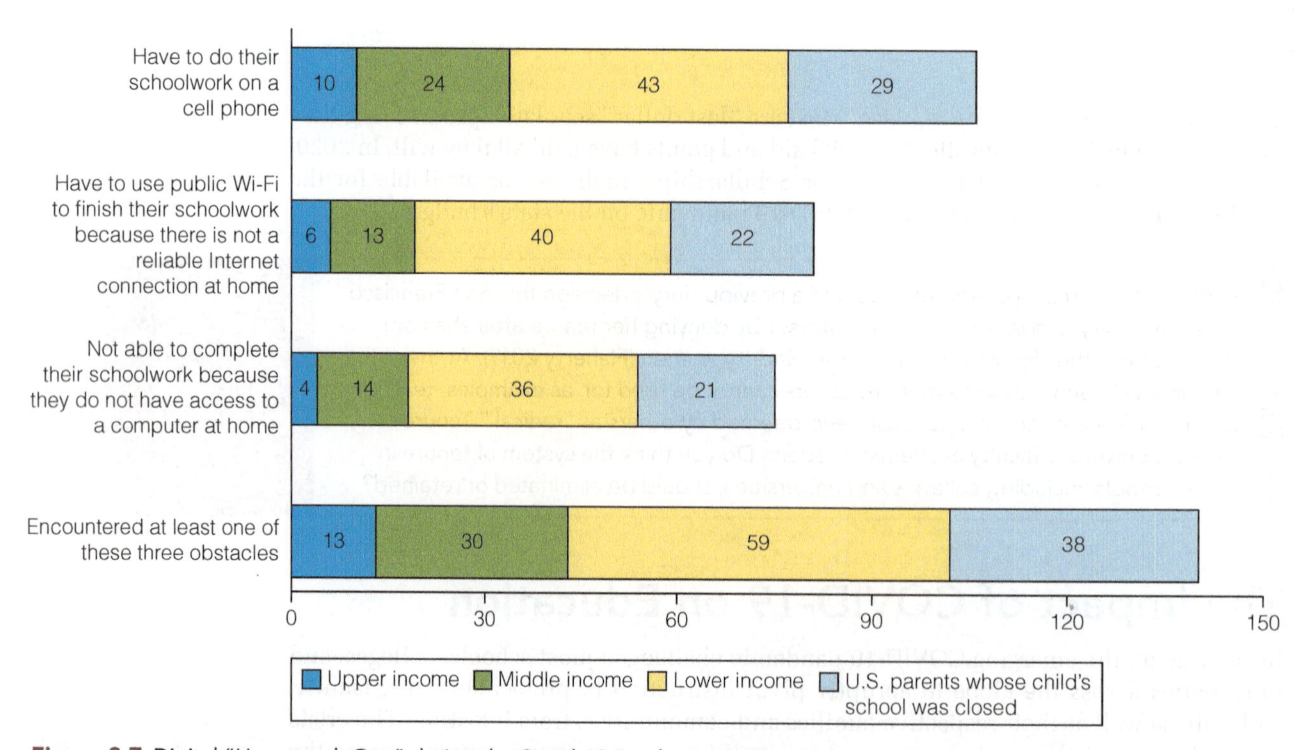

Figure 8.7 Digital "Homework Gap" during the Covid-19 Pandemic, 2020
SOURCE: Vogels 2020.

Reopening amid Protests. The Centers for Disease Control (CDC) released guidance on when and under what conditions schools could safely reopen. Former President Trump was quick to criticize the guidelines and pointed to other countries that had successfully reopened their schools. However, the countries used as examples had adhered to strict protocols that had met or exceeded CDC recommendations (OECD 2020b). When Japanese schools reopened, for example, all students and teachers wore masks and sat behind plexiglass shields, and classroom windows were kept open (Craft 2020). In the United States, school districts were left to make their fall 2020 decisions with no clear guidance at the federal level or, in many cases, at the state level.

AP Images/Rick Bowmer

Students and parents alike have been seen throughout the nation protesting the closure of schools during the COVID-19 pandemic. Deciding when and how to reopen schools safely has been a difficult decision for school administrations and government officials.

Complicating these decisions was a growing divide among parents, with one group adamant that children should be allowed to return to school and the other convinced that to do so was dangerous to students, teachers, families, and communities. For some in the "reopen schools" camp, opinions were fueled by concerns that (1) their children were not learning in the virtual format, (2) they could not supervise their children and keep their jobs, and (3) absence from school and interactions with teachers and other students was impairing their children's social and emotional well-being (Holcombe 2020). Furthermore, a vocal minority sided with the former administration that the seriousness of the pandemic was exaggerated, if not a hoax, and that students were not at risk.

Across the country, protesters from both contingents carried signs, one side appealing to the American value of freedom (e.g., "Freedom over Fear") and the other to health concerns (e.g., "We will not be your experiment") (Hay and O'Brien 2020; Strauss 2020b). A third constituency, teachers, protested face-to-face instruction with slogans like, "Don't Make Me Choose Between My Students and My Health!"

Schools across the country scrambled to make reopening plans ranging from traditional face-to-face instruction five days a week to being fully closed with virtual instruction only and everything in between. A school in New Jersey even opened up evening classes when they realized that many young children were unable to complete homework on days their parents were working (Pomrenze and Gokodryga 2020). Schools with "hotspots" retreated to Zoom classes, and others tried to open their doors. The sheer complexity of what teachers were asked to do, often with little notice, was staggering, putting many teachers, some with underlying health conditions, at risk. This chapter's *The World in Quarantine* feature discusses the impact of COVID-19 on the teaching profession.

Higher Education

Globally, colleges and universities were also faced with moving to online instruction quickly. In the United States, where many students lived in residence halls on campus, the sudden move to online education occurred at the same time students were informed they must leave their dormitories. Harvard University, for example, announced in March 2020 that all classes were moving online and gave students five days to vacate their dorm rooms (Hess 2020).

Some colleges and universities made exceptions for international students, homeless students, and students with compromised immune systems, but others did not. Many countries, including the United States, provided financial assistance to students to help with unexpected expenses (OECD 2020b), and some gave students prorated refunds on dormitory and meal plan costs (Kerr 2020).

While living on or near campus typically provides students with Internet and computer access, a significant minority of college students went home to complete the spring 2020 semester online without a computer, tablet, and/or Internet access. Sixteen percent

> "Everyone in education knew that the lack of technology was a problem. We've never had what we needed. But now we're supposed to just make it work in the middle of a pandemic?"
>
> –JESSYCA MATHEWS, ENGLISH TEACHER

THE WORLD in quarantine

Those Who Can, Teach

In the United States, recruiting and retaining teachers has always been challenging. Every year, 8 percent of teachers leave the profession while another 8 percent choose to leave their assigned post for another while remaining in the profession (Wang 2019). Thus, given teacher attrition *and* the decrease in the number of college students pursuing education degrees, the United States is experiencing a shortage of teachers (Garcia and Weiss 2019a). Presently, the gap between supply and demand is over 110,000 teachers, and the shortage is projected to increase to 300,000 by 2025 (Garcia and Weis 2020c). Further, as a result of the pandemic, there are fears of a mass exodus of teachers, although others argue that those fears may be unwarranted (Will, Gewertz, and Schwartz 2020).

Prior to the pandemic, teachers were leaving the profession in response to excessive testing and accountability requirements, a lack of administrative support and resources, challenging working conditions, and a general dissatisfaction with the profession (Clays 2020; Wang 2019). Financial considerations, disrespect, and violence in the classroom were also major factors in teachers' decisions to leave the profession, along with expectations to perform tasks outside their core duties such as administering medications (Alsup 2020). Suddenly, however, with little to no warning, teachers were expected to maintain instruction from their homes as a result of the COVID-19 pandemic.

While teachers and professors alike suffered from stress as a result of the pandemic, some educators had more to deal with than others. Midcareer teachers—those most likely to have children at home—struggled to balance their work responsibilities with their home lives (Kraft and Simon 2020). Early career teachers struggled with self-confidence having little in-class experience and now being forced to "perform" on Zoom.

Parents and their children weren't the only ones uncomfortable with the new technologies. Veteran teachers were three times more likely than early career teachers to report struggling with technology while teaching from home. And, for those teaching face-to-face, if only two or three days a week, there was the stress from fear of contracting the virus. One teacher reported that 55 of her colleagues in Leesburg, Virginia, had tested positive for COVID-19 by June 2020. By the end of the school year, several surveys estimated that between 10 percent and 20 percent of teachers were considering leaving the profession prior to the beginning of the next school year (Reilly 2020b).

As the 2020–2021 school year approached, teachers across the country faced two equally difficult options: maintain instruction from home or assume the risk of teaching in person. Although the American Academy of Pediatrics recommended that schools reopen and the former administration pressured schools to reopen, school districts' responses to the pandemic were varied (Reilly 2020b). As the pandemic became politicized and a talking point in the 2020 election, state officials pushed school administrators to open schools or risk cuts in much-needed funding. Given the partisan influence on health-related mandates for schools, even the idea of requiring students to wear masks was considered controversial in some schools.

In Connecticut, a union survey found that 43 percent of teachers felt they were at a high risk for contracting the virus. While some districts were able to accommodate teachers' requests to work from home, others told teachers they had no choice but to come back to the classroom (Kamenetz 2020). A middle school teacher in Atlanta, Georgia, returned to the classroom for the 2020–2021 school year but decided to keep her own children at home out of fear that the school wasn't safe. Each day she wore a mask and a face shield, taught from behind a plexiglass divider, and then showered and changed her clothes before going home.

Teacher morale also declined. Teaching virtually or in socially distanced classrooms is exhausting and left teachers overwhelmed (Will, Gewertz, and Schwartz 2020). Many felt attacked by parents and friends who posted demoralizing comments on social media, comments that questioned whether anything had been accomplished since the pandemic sent everyone home. These new criticisms were particularly frustrating given how hard teachers had worked to move online to support their students and their schools.

By the end of the summer 2020, 751 teachers in Arizona had left their jobs, an increase of 75 percent from the same time in the previous year. About 43 percent cited COVID-19 as the reason for leaving (Will, Gewertz, and Schwartz 2020). Some late career teachers, many with high-risk conditions or caretakers of people with underlying conditions, chose to retire earlier than expected. In the state of New York, retirements were 7 percent higher than the year before (Esposito 2020). However, in other parts of the country, retirements were down as teachers faced new or worsening economic pressures at home that left them with no choice but to continue to work (Will, Gewertz, and Schwartz 2020).

In contrast, teachers who felt supported by their school districts reported feeling good about what they had accomplished during the spring transition (Kraft and Simon 2020). Although the public feared and experts predicted record high teacher turnover before the start of the 2020–2021 school year, evidence supporting the many fears and predictions is surprisingly low in many districts. In New York City, for example, there was a 20 percent drop in teacher retirements from the same period a year earlier despite the 7 percent increase across the state.

It's impossible to know what lasting impact the pandemic will have on an already struggling teaching profession in the United States. For one social studies and special education teacher in Staten Island, New York, the pandemic renewed her commitment to teaching. Said Alessia Quintana when asked about resigning, "It never crossed my mind. If anything, [the pandemic] did the opposite—it made me realize how much more students need us right now. They're dealing with trauma and they need our support" (quoted in Will, Gewertz, and Schwartz 2020).

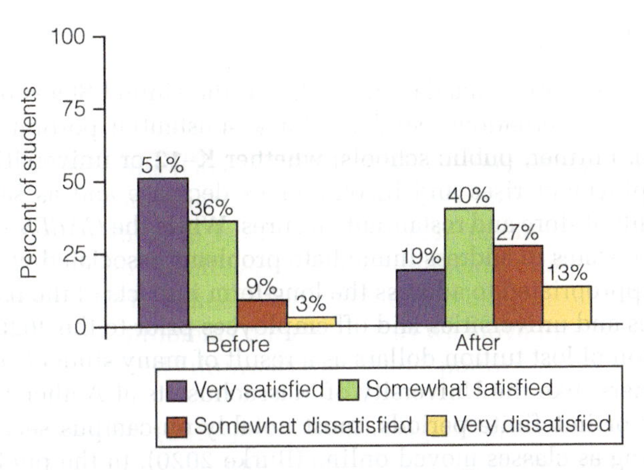

Figure 8.8 Students' Satisfaction with College Courses Before and After Move to Remote Instruction as a Result of COVID-19, 2020
SOURCE: Digital Promise 2020.

of college students reported that they had problems with accessing their online courses "often" or "very often" (Means and Neisler 2020). Internet or other technology problems were more common among low-income and Hispanic students. Hispanic students were also more likely to report having difficulty locating a quiet place to concentrate and fitting coursework around work and family responsibilities.

Overall, student satisfaction with courses after they transitioned online was low with a majority of students reporting they had more trouble staying interested in the course, felt opportunities to collaborate with other students were limited, and felt that help with classes was not always available (see Figure 8.8) (Means and Neisler 2020). While in general college and university professors were better prepared to move online than primary and secondary school teachers, many had never taught online before, and most courses had not been planned for online delivery (Means and Neisler 2020).

As with the K–12 school systems, colleges and universities instituted a variety of plans for the fall 2020 semester with some opening in person with full or limited residence hall occupancy. Others opened virtually. Some universities modified their schedules to start classes early and end before Thanksgiving. Others chose to open face-to-face for a limited group of students, for example, providing first-year students with the opportunity to have an on-campus experience or keeping students who needed access to labs on campus. A number of universities that had opened with face-to-face instruction initially transitioned quickly to online as the number of COVID-19 cases on campus rose (Higgins-Dunn 2020).

International Students. The pandemic had a unique impact on international students, with many having to decide whether to leave their host country before travel bans made it impossible to leave or visa statuses changed (Schleicher 2020). Some countries like Canada and the United Kingdom offered leniency on visa rules while others did not. Many U.S. students studying abroad had trouble finding flights home as countries closed airports to international travel. Most eventually came home through circuitous routes or flights chartered by their universities or the Department of State (Redden 2020).

For international students who remained in their host countries, on-campus employment opportunities, typically the only work they qualified for, disappeared, as did support from home as parents lost their jobs (Quinton 2020). International students didn't qualify for public assistance, even though in many cases they could not get home due to travel bans. In Australia the situation became so severe that it was called a "humanitarian crisis" with many international students losing their homes and having no money for food (Kinsella 2020).

protesting the role of business leaders (e.g., the Gates Foundation) in education policy, the continued increase in classroom size, the Federal Charter Program's misuse of taxpayer's dollars, high-stakes testing and their use in teacher evaluations, and school privatization.

Character Education

Character education entails teaching students to act morally and ethically, including the ability to "develop just and caring relationships, contribute to community, and assume the responsibilities of democratic citizenship" (Lickona and Davidson 2005). Despite most schools' emphasis on academic achievement, knowledge without character is potentially devastating. Sanford McDonnell (2009), the former CEO of McDonnell Aircraft Corporation and chairman emeritus of the Character Education Partnership, recounts a letter written by a principal and former concentration camp survivor to his teachers at the start of a new school year:

> My eyes saw what no person should witness: gas chambers built by learned engineers, children poisoned by educated physicians, infants killed by trained nurses, women and babies shot and burned by high school and college graduates. Your efforts must never produce learned monsters and skilled psychopaths. Reading, writing, and arithmetic are important only if they serve to make our children more humane. (p. 1)

@mayalau

Postmaster General Louis DeJoy's son, who was not a top tennis prospect, made it onto Duke's competitive tennis team while DeJoy & his wife donated to Duke's athletic dept. In all they gave at least $2.2M latimes.com/world-nation/s...

-Maya Lau

It is difficult, however, to create "a culture of integrity; [it] is not quick or easy, particularly when a long-standing and well-entrenched culture of cheating is already in place" (Stephens and Wangaard 2013).

In 2010, the National Collegiate Athletic Association (NCAA) began an investigation into the University of North Carolina (UNC) football program focusing on "impermissible benefits from sports agents," and in 2012, they sanctioned the football program. In the same year, an investigation began into one academic department's role in fraudulently assigning grades. The results indicated that over the course of two decades, there were over 200 "make believe" courses and more than 500 unexplained grade changes involving over 1,500 student-athletes, and several administrators, faculty members, academic advisors, and coaches (Beard 2014; Friedlander 2014). In 2016, two of the teams in the men's basketball final four had been sanctioned or investigated by the NCAA, Syracuse University and UNC (New 2016). Although the decision was criticized, the NCAA took no further action, arguing that it was not in a position to judge the integrity of academic programs (Tracy 2017).

Unfortunately, the scandal at UNC, although highly publicized, is just one of many examples of academic integrity violations that occur at all education levels. In 2020, 59 cadets at West Point, out of a suspected 73, admitted to cheating on a calculus exam (Romo and Bowman 2020). Globally, between 50 and 80 percent of students admit to frequently committing some kind of academic misconduct, plagiarism or cheating on an exam, at some time during their university career (Lacquaio and Ives 2020). Furthermore, technological advances and the increase in distance learning have made academic dishonesty easier and more common.

character education Education that emphasizes the moral and ethical aspects of an individual.

What do you think?

Cheating detection companies have made millions of dollars during the pandemic acting as online proctors, watching students via computer webcams when they are taking online tests. At North Carolina AT&T State University, a professor sent an e-mail to his/her class that read, in part, "[A] lot of head and eye movements for a short time period. A student in six minutes had 776 head and eye movements. ... That is an indication of eyes moving away from the screen. ... I would hate to have to write you up for online cheating which gets filed in the Dean's office" (quoted in Harwell 2020, p. 1). Do you think cheating detection programs deter academic misconduct?

Finally, in 2019, federal prosecutors brought charges against 50 people who were involved in a nationwide college admission's scam called Operation Varsity Blues. A college counselor, William Singer, conspired with wealthy clients (e.g., Felicity Huffman and Lori Loughlin) to bribe coaches at major universities (e.g., Yale, Stanford) to admit students to athletic programs who had no experience in the sport and to falsify SAT and ACT test scores (Nadworney and Treviño 2020). Charges against Mr. Singer also included money laundering, document fabrication, wire fraud, and obstruction of justice (Shamsian and McLaughlin 2020). Bribery payments were as high as $25 million.

The Debate over School Choice

Historically, children have gone to public schools in the districts where they live. However, in 2020, more than half of U.S. families had alternative school options available to them, with various levels of satisfaction (see Figure 8.10) (DiPerna, Catt, and Shaw 2020). Further, several recent surveys indicate bipartisan support for school choice, with some results indicating that as many as 80 percent of the voting public are in favor of alternatives to traditional public schools (Ruszkowski 2020). School vouchers, charter schools, and private schools provide parents with alternative school choices for their children.

School Vouchers. School vouchers and state tax credits allow families to send their children to private religious or nonreligious schools. **School vouchers** are state-funded "scholarships" that allow public school students to attend private schools. Eligibility for vouchers is usually confined to special populations, for example, low-income students, special education students, and students in chronically low-performing schools. Similarly, "tax credit programs ... allow businesses or individuals to contribute to organizations that distribute private-school scholarships to low-income families" (Henderson et al. 2015, p. 15). In the 2019–2020 academic year, nearly 190,000 U.S. students in 15 states and Washington, D.C. used vouchers to attend K–12 schools (EdChoice 2020).

Proponents of the voucher system argue that it increases the quality of schools by creating competition for students. Those who oppose the voucher system argue that it drains needed funds and the best students away from public schools (Johnson 2019). Opponents also argue that vouchers increase segregation because White parents use the vouchers to send their children to private schools with few minority students and that public funding should not be used for religious private schools.

school vouchers State-funded "scholarships" paid directly to parents that can be used to send qualifying public school students (e. g., low-income students) to private schools.

The Milwaukee Parental Choice Program is the "oldest and largest publicly funded voucher program in the United States," beginning in 1991 and continuing to the present (Cowen et al. 2013; Wisconsin Department of Public Instruction 2020). Wisconsin state law requires an evaluation of the voucher program and its impact on academic achievement. To assess attainment, students attending voucher schools were matched

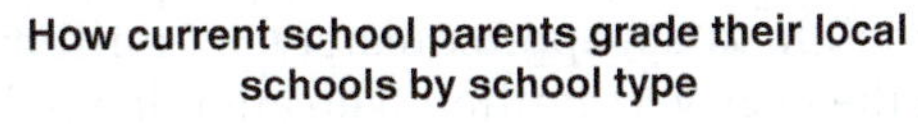

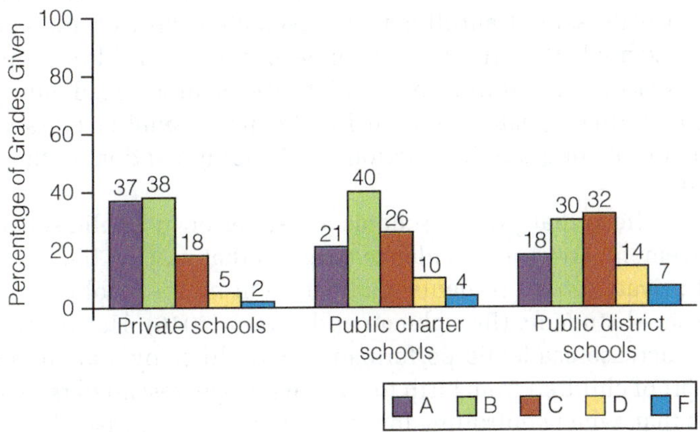

Figure 8.10 How Current School Parents Grade their Local Schools, 2019
SOURCE: DiPerna, Catt, and Shaw 2020.

(e.g., percentage Black, percentage female, etc.) with those attending public schools so that an accurate comparison could be made. The results indicate that students who were exposed to a voucher environment in the eighth and ninth grades were more likely to graduate from high school, enroll at a four-year institution, and stay in school beyond their first year of college (Cowen et al. 2013).

Given that disadvantaged students are the most likely to leave voucher programs (Carlson and Cowan 2015) and that disadvantaged students are more likely to perform poorly academically, metrics such as graduation from high school or enrolling in a four-year institution may be artificially inflated. Further, evaluations of other school choice programs such as the D.C. Opportunity Scholarship Program found *no* significant difference between program participants and public school students and, in the case of Louisiana's voucher program, there was a statistically significant *negative* effect on participants' test scores (Catt et al. 2020).

Charter Schools. In some states, vouchers can be used for charter schools. **Charter schools** originate in contracts, or charters, that articulate a plan of instruction that local or state authorities must approve. Although foundations, universities, private benefactors, and entrepreneurs can fund charter schools, many are supported by tax dollars. While the former are private charter schools in that they are privately managed, the latter are considered public charter schools (Study International 2019). Unlike private schools, charter schools are tuition free. They educate just a small fraction of the nation's over 50 million public school students, approximately 6 percent, although as high as 30 percent in some cities (Prothero 2018).

Charter schools, like school vouchers, were designed to expand schooling options and to increase the quality of education through competition. Like vouchers, charter schools have come under heavy criticism for increasing school segregation, reducing public school resources, and "stealing away" top students. Proponents such as former Secretary of Education Betsy DeVos argue that charter schools encourage innovation and reform and increase student learning outcomes.

In 2020, Secretary DeVos approved $131 million in grant funding to expand the number of charter schools. Said the former secretary, "All too many students, particularly the most vulnerable, have fallen further behind because the one-size-fits-all system couldn't transition and adapt to their needs" as a result of the COVID-19 pandemic (U.S. Department of Education 2020). President Biden supports school choice although he is against for-profit charter schools and charter schools that are performing poorly. He has stated that charter schools, like all schools, must be held accountable and is not in favor of public funds going to private schools (Greenberg 2020).

Privatization of Schools. Another school choice parents can make is to send their children to a private school. In 2020, approximately 5.7 million students attended private schools compared to the nearly 51 million who attended public schools (NCES 2020d). The decline in private school enrollment is associated with an increase in students attending charter schools (Ewert 2013). Parents send their children to private schools for a variety of reasons, including the availability of more academic programs and extracurricular activities, greater parental involvement, smaller class size and lower student–teacher ratio, religious instruction, and dissatisfaction with public schools (Barrington 2019).

Many people believe that private schools are superior to public schools in terms of academic achievement, and some evidence suggests that may be the case. For example, 97.2 percent of private school students graduate from high school compared to 82 percent of public school students (Broughman and Swaim 2016). However, whether or not private schools increase academic performance may differ by student subgroup. After following a cohort of children from birth to 15 years of age, researchers Pianta and Ansari (2018) conclude that, when controlling for social demographic variables such as income, there was no academic advantage of private school attendance.

In addition to traditional private schools, there's a growing movement for local or state governments to "contract out" schools and other educational services to private

charter schools Schools that originate in contracts, or charters, which articulate a plan of instruction that local or state authorities must approve.

for-profit corporations. Research comparing Michigan public schools, nonprofit-managed charter schools, and for-profit charter schools operated by education management organizations (EMOs) indicates that (1) charter schools, in general, have a higher proportion of Black students than traditional public schools; (2) charter schools have a lower proportion of Hispanic students than traditional public schools; and (3) for-profit charter schools are less likely to enroll poor students than nonprofit charter schools. The authors suggest that the different enrollment patterns of poor students between nonprofit and for-profit charter schools may be a result of the latter's need to maximize revenues and minimize costs (Ertas and Roch 2014).

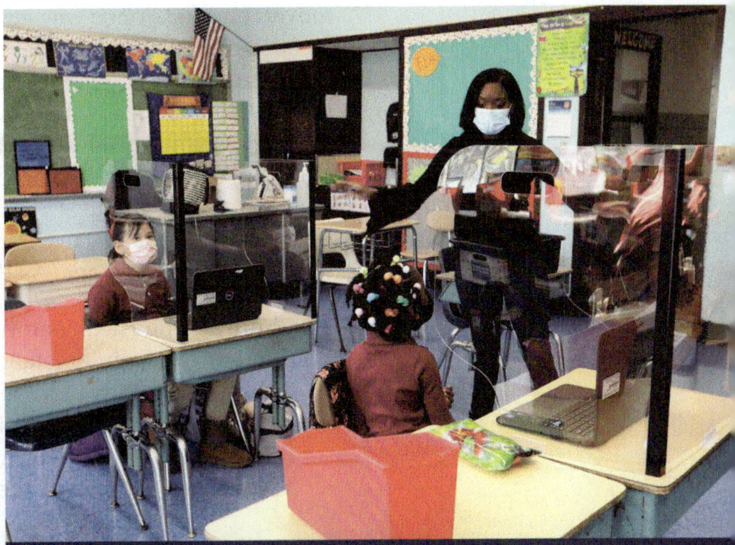

There has been a decline in private school enrollment, of late, in part due to costs. There has also been a decline in the number of private colleges as a result of financial pressure, competition from state schools, changing demographics, and so-called free tuition programs (West 2018). Some predict, however, that private K–12 enrollment may increase as a result of Secretary DeVos's $180 million voucher program set aside for private and religious schools. The federal funds used for the program were diverted from funds earmarked for COVID-19 relief. In a radio interview, when the former secretary was asked whether she was taking advantage of the crisis to help private schools, her answer was, "Absolutely" (quoted in Flannery 2020).

As schools reopened under the Biden administration, emphasis was placed on health, safety, adequate funding, high-quality education, and closing the "zip code" achievement gap.

Understanding Problems in Education

Educational reform continues to be the focus of legislators and governments across the country and the world. Although there is disagreement as to what needs to be done and how, all can agree that significant reform is needed to meet the needs of a global economy in the 21st century and, perhaps more importantly, to fulfill Horace Mann's dream of education as the "balanced wheel of social machinery," equalizing social differences among members of an immigrant nation.

First, we must invest in teacher education and in teaching practices that have been empirically documented to work in improving student outcomes. Teachers' salaries, as President Biden has noted, need to better reflect the priority Americans place on children and education, and should not be tied to student performance (Biden 2020). As a society, we need to address the obstacles that have led to a growing teacher dropout crisis.

Second, the "savage inequalities" in education, primarily based on race, ethnicity, and socioeconomic status, must be addressed. Segregation, rather than decreasing, is increasing—a reflection of housing patterns, local school districts' heavy reliance on property taxes, school choice, and immigration patterns. Public schools should provide all U.S. children with the academic and social foundations necessary to participate in society in a productive and meaningful way; however, for many children, schools perpetuate an endless downward cycle of failure, alienation, and hopelessness.

Third, the general public needs to become involved, not just in their children's education but also in the *institution* of education. An uneducated and unthinking populace hurts all of society, particularly in terms of global competitiveness. As Kohn (2011) observes, children from low-income families continue to be taught "the pedagogy of poverty" (Haberman 1991) or what has been called "McEducation" (Hopkinson 2011). Like Big Macs, children are being packaged in one-size-fits-all wrappers—with learning how to think, explore, question, and debate being replaced by worksheets and standardized tests. Sadly, as education historian Diane Ravitch notes, the present reform movement that "once was an effort to improve the quality of education [has] turned into an accounting strategy" (Ravitch 2010, p. 16).

Fourth, as conflict theorists would note, we must be wary of market principles in schools and of those who advocate them. Educational policy in the United States is increasingly influenced by K–12 corporate providers, many of which have political ties and vested interests. For-profit education is a lucrative business and "one of the largest U.S. investment markets" (Rawls 2015).

Finally, as a society, we must attend to and be cognizant of the importance of early childhood development. As Poliakoff (2006) notes:

> Children's physical, emotional, and cognitive development are profoundly shaped by the circumstances of their preschool years. Before some children are even born, birth weight, lead poisoning, and nutrition have taken a toll on their capacity for academic achievement. Other factors—excessive television watching, little exposure to conversation or books, parents who are absent or distracted, inadequate nutrition—further compromise their early development. (p. 10)

We must provide support to families so that children grow up in healthy, safe, and nurturing environments. Children are the future of our nation and of the world. Whatever resources we provide to improve the lives and education of children are sure to be wise investments in our collective future.

Chapter Review

- **Do all countries educate their citizens?**
  No. Many societies have no formal mechanism for educating the masses. As a result, millions of adults around the world are illiterate. The problem of illiteracy is greater in developing countries than in developed nations and, worldwide, disproportionately affects women more than men.

- **According to the structural-functionalist perspective, what are the functions of education?**
  Education has four major functions. The first is instruction—that is, teaching students knowledge and skills. The second is socialization that, for example, teaches students to respect authority. The third is sorting individuals into statuses by providing them with credentials, i.e., status conferral. The fourth function is custodial care—a babysitting agency, of sorts.

- **What is a self-fulfilling prophecy?**
  A self-fulfilling prophecy occurs when people act in a manner consistent with the expectations of others.

- **What variables predict school success?**
  Three variables tend to predict school success. Socioeconomic status predicts school success; the higher the socioeconomic status, the higher the likelihood of school success. Race and ethnicity predict school success, with non-White and Hispanic students having more academic difficulty than White and non-Hispanic students. Gender also predicts success, although it varies by grade level and subject.

- **What are the four reasons why Black and Hispanic Americans, in general, do not perform as well in school as their White and Asian counterparts?**
  First, because race and ethnicity are so closely tied to socioeconomic status, it appears that race or ethnicity alone can determine school success when, in fact, it may be socioeconomic status. Second, many minorities are not native English speakers, making academic achievement significantly more difficult. Third, standardized tests have been demonstrated to be culturally biased favoring those in the upper and middle classes. Finally, racial and ethnic minorities may be the victims of racism and discrimination.

- **What are some of the ways COVID-19 impacted education?**
  After initial shutdowns, many K–12 schools, as well as colleges and universities, either resumed classes online or offered hybrid classes, which were face-to-face a few days a week and virtual the remaining time. Many students, particularly low-income and minority students, did not have access to technology to participate in school. The pandemic was particularly difficult for international students who had to make a decision as to whether to remain in or return to their countries of origin. Colleges and universities were also impacted, as enrollment and thus revenues decreased.

- **What are some of the problems associated with the American school system?**
  One of the main problems is the lack of funding and the resulting reduction in programs and personnel (e.g., mental health professionals). Low levels of academic achievement in our schools are also of some concern—particularly when U.S. data are compared with data from other industrialized countries. Minority dropout rates are high, and school violence, crime, and discipline problems continue to be a threat. School facilities are in need of repair and renovations, and personnel,

including teachers, have been found to be deficient. Problems in higher education also include high costs and student debt, the perpetuation of inequality, and sexual assault on campus.

- **What is meant by value-added measurement (VAM), and why are there concerns about its use?**
VAM is the use of student achievement data to assess teachers' effectiveness. Critics of VAM argue that assessing teachers based on student performance assumes all else is constant and ignores the reality of student differences in such nonschool factors as family life, poverty, emotional and physical obstacles, and access to computers and wi-fi.

- **What are the arguments for and against school choice?**
Proponents of school choice programs argue that they reduce segregation and that schools that have to compete with one another will be of a higher quality. Opponents argue that school choice programs increase segregation and treat disadvantaged students unfairly. Low-income students cannot afford to go to private schools, even with vouchers. Furthermore, those opposed to school choice are quick to note that using government vouchers to help pay for religious schools is unconstitutional.

Test Yourself

1. All societies have some formal mechanism to educate their citizenry.
   a. True
   b. False
2. According to structural-functionalists, which of the following is not a major function of education?
   a. Teach students knowledge and skills
   b. Socialize students into the dominant culture
   c. Indoctrinate students into the capitalist ideology
   d. Provide custodial care for children
3. Common Core State Standards have been embraced by and implemented in all states.
   a. True
   b. False
4. Which of the following statements is true about dropouts in the United States?
   a. Dropout rates in the United States are increasing.
   b. Students who drop out of school, on the average, make more money because they've been working longer than their high school graduate counterparts.
   c. Dropout rates in the United States are similar between racial and ethnic subgroups.
   d. Dropouts have poorer health and shorter life expectancies.
5. Which of the following statements about bullying is not true?
   a. Approximately 20 percent of students between the ages of 12 and 18 report being bullied at school.
   b. Bullying can include intimidation and/or humiliation.
   c. The most common category of bullying is physical violence.
   d. Bullying can occur face-to-face or online.
6. Academic misconduct by most college and university students is less than 10 percent.
   a. True
   b. False
7. Which of the following statements is correct?
   a. The teaching population is more diverse than the student population in terms of race and ethnicity.
   b. Low-income students are more likely to engage in "soft skills."
   c. The structure of schools has very little impact on student academic performance.
   d. Title IX prohibits discrimination on the basis of sex by schools receiving federal funds.
8. As a result of the COVID-19 pandemic, many colleges and universities
   a. hired more faculty to meet the needs of increased enrollment.
   b. provided Internet access for almost all students living on or near campus.
   c. were encouraged by former President Trump to remain closed.
   d. had increased revenues as, for example, more students chose to live in residence halls.
9. The number of students attending private schools is increasing.
   a. True
   b. False
10. School vouchers are state-funded "scholarships" that are transferred to the private schools that parents select for their children.
   a. True
   b. False

Answers: 1. B; 2. C; 3. B; 4. D; 5. C; 6. B; 7. D; 8. B; 9. B; 10. A.

Key Terms

bilingual education 290
bullying 298
character education 312
charter schools 314
cultural competence 283
cultural imperialism 283
earnings premium 280
formal education 278
Head Start 288
integration hypothesis 291
multicultural education 283
parent trigger laws 311
school-to-prison pipeline 298
school vouchers 313
self-fulfilling prophecy 284
social reproduction 282
STEM 292
value-added measurement (VAM) 300

Albert Cesare/The Enquirer via Imagn Content Services, LLC

> “No one is born hating another person because of the color of his skin, or his background, or his religion. People must learn to hate. And if they can learn to hate, they can learn to love.”
>
> **NELSON MANDELA**

9

Race, Ethnicity, and Immigration

Chapter Outline

The Global Context: Diversity Worldwide

- **Self and Society:** The Value of Diversity

Racial and Ethnic Diversity in the United States

Immigrants in the United States

Sociological Theories of Race and Ethnic Relations

- **The World in Quarantine:** A Tale of Two Pandemics
- **The Human Side:** Experiencing Racism at School

Prejudice and Racism

- **Social Research Problems Up Close:** Using Implicit Bias Training to Reduce Racial Disparities

Discrimination against Racial and Ethnic Minorities

Strategies for Action: Responding to Prejudice, Racism, and Discrimination

Understanding Race, Ethnicity, and Immigration

Chapter Review

Learning Objectives

After studying this chapter, you will be able to …

1. Explain the social construction of race and ethnicity.
2. Describe patterns of racial and ethnic diversity in the United States.
3. Explain the relationship between historical patterns of immigration and current immigration policies.
4. Compare and contrast views on issues related to race, ethnicity, and/or immigration from the perspectives of structural functionalism, conflict theory, and symbolic interactionism.
5. Identify examples of different forms of racism and prejudice.
6. Explain the differences between individual and institutional forms of discrimination.
7. Evaluate political and cultural strategies for promoting racial and ethnic equity.

WHO IS RESPONSIBLE for atoning for America's "original sin"? In 1860, 283 White residents of the city of Asheville, North Carolina, owned more than 1,900 slaves (Dwyer 2020). In July 2020, the city council of Asheville voted unanimously to formally apologize for the city's role in slavery, to remove a Confederate monument erected on the site of an old slave auction, and to pay financial restitution—in the form of community investments—to the city's African American residents. Reparations have a long and complicated history in the United States. At the end of the Civil War, freed slaves were promised "40 acres and a mule" to start their new lives; however, the order was quickly overturned and only about 1 percent of freedmen ever received the promised reparations (Gates 2013). Jerry Green, chairman of the local Republican Party organization, opposed the city council's vote and asked, "[H]ow do you make restitution for people that are not here? People that were hurt are gone. In my opinion, we ought to look toward the future and work from here and quit dwelling on the past" (quoted in Dwyer 2020, p. 1). Those who supported the resolution argue that the goal of working "from here" is impossible, however, as the long history of slavery, residential segregation, and employment discrimination in the city has contributed to a massive wealth gap. At the same time, the financial success of the city is inextricably linked to the unpaid labor provided by slaves. Rob Thomas, a community liaison for the Racial Justice Coalition that brought the issue before the city council, argues that the resolution is "asking you to look at the facts, and saying ... this happened. ... This much money was taken out of the Black community and it would equal this much today. We're just asking people to do what's right" (quoted in Romo 2020, p. 1).

Manny Ceneta/Getty Images News/Getty Images

For decades, the unfulfilled promise of "40 Acres and a Mule" has served as a rallying cry for protestors advocating for financial reparations for the harms of slavery and for the accumulation of financial losses that occurred over centuries of employment, housing, and educational discrimination.

Throughout history, humans have encountered other humans who are different from them—different in language, beliefs, norms, and physical appearance. And we don't have to travel to another continent or country to find these differences: we find religious, racial, and ethnic diversity in our own communities, and even in our own families. This chapter is concerned with racial and ethnic differences globally and within the United States. We focus on the nature and consequences of racism, prejudice, and discrimination toward minority groups, and strategies to reduce these problems. A **minority group** is a category of people who have unequal access to positions of power, prestige, and wealth and are targets of prejudice and discrimination. Minority status is not based on numerical representation in society but rather on social status. For example, although Hispanic individuals outnumber non-Hispanic White individuals in California, Texas, and New Mexico, they are considered a "minority" because they are underrepresented in positions of power, prestige, and wealth, and because they are targets of racism, prejudice, and discrimination. In this chapter, we also examine issues related to U.S. immigration, because immigrants often bear the double burden of being minorities *and* foreigners who are not welcomed by many native-born Americans. Note that some very important issues related to race are not covered in this chapter because they are discussed in other chapters. Chapter 4, for example, discusses race and crime, treatment by police, and the criminal justice system, and Chapter 6 covers race and poverty.

minority group A category of people who have unequal access to positions of power, prestige, and wealth in a society and who tend to be targets of prejudice and discrimination.

The Global Context: Diversity Worldwide

All human beings belong to the same species. Yet, because of differences in physical appearance and culture, humans are often classified into categories based on race and ethnicity. After examining the social construction of race and ethnicity, we review patterns of interaction among racial and ethnic groups and examine racial and ethnic diversity in the United States.

The Social Construction of Race and Ethnicity

The concept of "race" has been described as "one of the most misunderstood, misused, and often dangerous concepts of the modern world" (Marger 2012, p. 12). The term *race* has been used to describe people of a particular nationality (the Mexican "race"), religion (the Jewish "race"), skin color (the White "race"), and even the entire human species (the human "race"). Confusion around the term *race* stems from the fact that it has both biological and social meanings.

scientific racism The belief, which dominated racial thinking in the biological and social sciences in the 19th and early 20th centuries, that human groups could be ranked into hierarchies on the basis of observable characteristics, such as nose width and skin color.

Race as a Biological Concept. As a biological concept, *race* refers to a classification of people based on hereditary physical characteristics such as skin color, hair texture, and the size and shape of the eyes, lips, and nose. But there are no clear guidelines for distinguishing racial categories on the basis of visible traits. Skin color is not black or white but rather ranges from dark to light with many gradations of shades. Noses are not either broad or narrow but come in a range of shapes. Physical traits come in an infinite number of combinations. For example, a person with dark skin can have a broad nose (a common combination in West Africa), a narrow nose (a common combination in East Africa), or even blond hair (a combination found in Australia and New Guinea).

Another problem with race as a biological concept is that the physical traits used to mark a person's race are arbitrary. What if we classified people into racial categories based on eye color instead of skin color? Or hair color? Or blood type? What if all dark-haired individuals were considered to belong to one race, and all light-haired people to another race? Is there any scientific reason for selecting certain traits over others in determining racial categories? The answer is no. As a biological concept, "[R]aces are not scientifically valid because there are no objective, reliable, meaningful criteria scientists can use to construct or identify racial groupings" (Mukhopadhyay et al. 2007, p. 5).

Historically, however, scientists did try to assign meaningful criteria to perceived biological differences between racial groups. **Scientific racism**, also called biological racism, refers to the body of theories that dominated racial thinking in the social and biological sciences in the 19th and early 20th centuries. This view held that observable biological characteristics—such as the width of a person's nose or the color of their skin—were associated with traits such as intelligence. By recording and categorizing these characteristics, scientists could (presumably) rank human groups into a racial hierarchy.

Ernst Krause

Theories of scientific racism have historically been used to justify the oppression and exploitation of racial and ethnic minorities. This photo depicts Nazi Anthropologist Bruno Berger measuring the facial features of a Tibetan man in 1938 in order to determine the extent to which Tibetan people had so-called "Aryan" characteristics.

By the mid-20th century, however, support for scientific racism largely declined within the scientific community. This shift in thinking occurred for two reasons. First, the horror of witnessing the application of scientific racist theories within the Nazi regime and the Holocaust caused many to question the foundations of biological theories of race. Second, emerging evidence from the science of genetics in the second half of the 20th century challenged biological notions of race. Geneticists have discovered that the genes of any two unrelated people, chosen at random from around the globe, are 99.9 percent alike (Ossorio and Duster 2005). Furthermore, "most human genetic variation—approximately 85 percent—can be found between any two individuals from the same group (racial, ethnic, religious, etc.). Thus, the vast majority of variation is within-group variation" (Ossorio and Duster 2005, p. 117). Finally, classifying people into different races fails to recognize that, over the course of human history, migration and intermarriage have resulted in the blending of genetically transmitted traits:

> To summarize, races are unstable, unreliable, arbitrary, culturally created divisions of humanity. This is why scientists ... have concluded that race, as scientifically valid biological divisions of the human species, is fiction not fact. (Mukhopadhyay et al. 2007, p. 14)

> "There is no such thing as race. Racism is a construct; a social construct. And it has benefits. Money can be made off of it. People who don't like themselves can feel better because of it. It can describe certain kinds of behavior that are wrong or misleading. So [racism] has a social function. But race can only be defined as a human being."
>
> —TONI MORRISON, AUTHOR

Race as a Social Concept. The idea that race is socially created is one of the most important lessons in understanding race from a sociological perspective. The social construction of race means that "the actual meaning of race lies not in people's physical characteristics, but in the historical treatment of different groups and the significance that society gives to what is believed to differentiate so-called racial groups" (Higginbotham and Andersen 2012, p. 3). The concept of race grew out of social institutions and practices in which groups defined as "races" have been enslaved or otherwise exploited.

People learn to perceive others according to whatever racial classification system exists in their culture. Systems of racial classification vary across societies and change over time. For example, as late as the 1920s, Italians, Greeks, Jews, Irish, and other "White" ethnic groups in the United States were not considered to be White. Over time, the category of "White" changed so that it included these groups. As an example of cross-cultural variation in racial categories, Brazilians use dozens of terms to racially categorize people based on various combinations of physical characteristics, although officially, the major racial categories in Brazil are *brancos* (white), *pardos* (brown or mulatto), *pretos* (black), and *amarelos* (yellow).

Incorporating both biological and social meanings of race, we define **race** as a category of people who are perceived to share distinct physical characteristics that are deemed socially significant. The significance of race is not biological but social and political because race is used to separate "us" from "them" and becomes a basis for unequal treatment of one group by another. Despite increasing acceptance that "there is no biological justification for the concept of 'race'" (Brace 2005, p. 4), its social significance continues throughout the world.

What do you think?

In the tiny town of East Jackson, Ohio, race is about more than skin color. The majority of the town's residents identify as Black, even though they would be visibly perceived of as White in most places in the United States (Shah 2019). This occurred through a complicated history of residential segregation during the Jim Crow era. Shortly after the Civil War, the nearby town of Waverly forced most newcomers—including laborers, housekeepers, and those with dark skin—to be designated as "Black" on legal documents and required them to live in the newly created town of East Jackson. Over time, interracial marriage led most of the residents of East Jackson to appear White. But the history of imposed identity, family legacy, and connection to place has led many residents to hold fiercely to their Black identity. Which social and cultural characteristics do you think are most important in determining racial identity in American society?

race A category of people who are perceived to share distinct physical characteristics that are deemed socially significant.

Ethnicity as a Social Construction. **Ethnicity**, which refers to a shared cultural heritage, nationality, or lineage, is also partly socially constructed. Ethnicity can be distinguished on the basis of language, forms of family structures and roles of family members, religious beliefs and practices, dietary customs, forms of artistic expression such as music and dance, and national origin or origin of one's parents.

Although the Census Bureau defines Hispanic or Latino as "a person of Cuban, Mexican, Puerto Rican, South or Central American, or other Spanish culture or origin regardless of race" (U.S. Census Bureau 2020, p. 1), when it comes down to collecting census data on the U.S. population, a person is Hispanic if they say they are Hispanic (see Figure 9.1). Hence, both ethnicity and race are socially constructed.

Maddie McGarvey/The Guardian

East Jackson, Ohio, resident Roberta Oiler and her two daughters identify as Black. Oiler says, "You don't have to look black to be black. ... Maybe the black has run out of my bloodstream, I don't know. But I still consider myself as what my mom put me as [on my birth certificate], and that's exactly what I say I am" (Shah 2019, p. 1).

Patterns of Racial and Ethnic Group Interaction

When two or more racial or ethnic groups come into contact, one of several patterns of interaction occurs; these include genocide, expulsion, segregation, acculturation, pluralism, and assimilation.

Genocide refers to the deliberate, systematic annihilation of an entire nation or people. The European invasion of the Americas, beginning in the 16th century, resulted in the decimation of most of the original inhabitants of North and South America. Some native groups were intentionally killed; others fell victim to diseases brought by the Europeans. In the 20th century, Hitler led the Nazi extermination of 12 million people, including 6 million Jews, in what is known as the Holocaust. In the early 1990s, ethnic Serbs attempted to eliminate Muslims from parts of Bosnia—a process they called "ethnic cleansing." In 1994, genocide took place in Rwanda when Hutus slaughtered hundreds of thousands of Tutsis. Genocide is continuing in many places in the world, typically orchestrated by governments who argue that ethnic and religious minority groups present a terrorist threat within their countries. This is the case with the Buddhist dominated government against the minority Muslim Rohingya group in Myanmar (Burma), the Alawite dominant government against Sunni opposition groups in Syria, and the Chinese government against the Muslim Uyghurs. A recent increase in ethnic cleansing events and civil war have led to largest refugee crisis since the end of the Second World War (see also Chapter 15).

Expulsion occurs when a dominant group forces a subordinate group to leave the country or to live only in designated areas of the country. The *1830 Indian Removal Act* called for the relocation of eastern tribes to land west of the Mississippi River. The movement, lasting more than a decade, has been called the Trail of Tears because tribes were forced to leave their ancestral lands and endure harsh conditions of inadequate supplies and epidemics that caused illness and death. After Japan's attack on Pearl Harbor in 1941, 120,000 Japanese Americans, who became viewed as threats to national security, were forced from their homes and into concentration camps surrounded by barbed wire.

Segregation refers to the physical separation of two groups in residence, workplace, and social functions. Segregation can be *de jure* (Latin meaning "by law") or *de facto* ("in fact"). Between 1890 and 1910, a series of U.S. laws, which came to be known as *Jim Crow laws,* were enacted to separate Black people from White people by prohibiting Black individuals from using "White" buses, hotels, restaurants, and drinking fountains. In 1896, the U.S. Supreme Court (in *Plessy v. Ferguson*) supported de jure segregation of Black and White people by declaring that "separate but equal" facilities were constitutional. Black families were forced to live in separate neighborhoods and attend separate schools. Beginning in the 1950s, various rulings overturned these Jim Crow laws, making it illegal to enforce racial segregation. Although de jure segregation is illegal in the United States, de facto segregation still exists in the tendency for racial and ethnic groups to live and go to school in segregated neighborhoods.

ethnicity A shared cultural heritage or nationality.

genocide The deliberate, systematic annihilation of an entire nation or people.

expulsion Occurs when a dominant group forces a subordinate group to leave the country or to live only in designated areas of the country.

segregation The physical separation of two groups in residence, workplace, and social functions.

➜ **NOTE: Please answer BOTH Question 8 about Hispanic origin and Question 9 about race. For this census, Hispanic origins are not races.**

8. Is person 1 of Hispanic, Latino, or Spanish origin?

☐ **No,** not of Hispanic, Latino, or Spanish origin
☐ Yes, Mexican, Mexican Am., Chicano
☐ Yes, Puerto Rican
☐ Yes, Cuban
☐ Yes, another Hispanic, Latino, or Spanish origin—*Print, for example, Salvadoran, Dominican, Colombian, Guatemalan, Spaniard, Ecuadorian, etc.*

9. What is person 1's race?
Mark ☒ *one or more boxes* ***AND*** *print origins.*

☐ White—*Print, for example, German, Irish, English, Italian, Lebanese, Egyptian, etc.*

☐ Black or African Am.—*Print, for example, African American, Jamaican, Haitian Nigerian, Ethiopian, Somali, etc.*

☐ American Indian or Alaska Native—*Print name of enrolled or principal tribe(s), for example, Navajo Nation, Blackfeet Tribe, Mayan, Aztec, Native Village of Barrow Inupiat Traditional Government, Nome Eskimo Community, etc.*

☐ Chinese ☐ Vietnamese ☐ Native Hawaiian
☐ Filipino ☐ Korean ☐ Samoan
☐ Asian Indian ☐ Japanese ☐ Chamorro
☐ Other Asian—*Print, for example, Pakistani, Cambodian, Hmong, etc.*
☐ Other Pacific Islander—*Print, for example, Tongan, Fijian, Marshallses, etc.*

☐ Some other race—*Print race or origin.*

Figure 9.1 U.S. Census Bureau Approach to Defining Race and Ethnicity
SOURCE: Jones 2015.

Acculturation refers to adopting the culture of a group different from the one in which a person was originally raised. Acculturation may involve learning the dominant language, adopting new values and behaviors, and changing the spelling of the family name. In some instances, acculturation may be forced. In the 1800s, the U.S. federal government began allowing religious institutions and other private organizations to forcibly remove Native American children from their homes and transport them to boarding schools hundreds or even thousands of miles away. By 1887, over 14,000 Indian children were living in the more than 200 boarding schools where the children were forced to abandon their Indian language, culture, and religion and adopt White ways. "The premise of these schools was that Indian children needed a 'proper' education and that Indian religion and culture were inferior to white religion and culture" (American Civil Liberties Union 2014, p. 56).

acculturation The process of adopting the culture of a group different from the one in which a person was originally raised.

Pluralism refers to a state in which racial and ethnic groups maintain their distinctness but respect each other and have equal access to social resources. In Switzerland, for example, four ethnic groups—French, Italians, Swiss Germans, and Romansch—maintain their distinct cultural heritage and group identity in an atmosphere of mutual respect and social equality.

Assimilation is the process by which formerly distinct and separate groups merge and become integrated as one. *Primary assimilation* occurs when members of different racial or ethnic groups are integrated in personal, intimate associations, as with friends, family, and spouses. *Secondary assimilation* occurs when different groups become integrated in public areas and in social institutions, such as neighborhoods, schools, the workplace, and in government.

akg-images/The Image Works

These Native American children were forced to adopt White U.S. culture as part of their "education" in White-run schools.

Assimilation is sometimes referred to as the "melting pot," whereby different groups come together and contribute equally to a new, common culture. Although the United States has been referred to as a melting pot, in reality, many minorities have been excluded or limited in their cultural contributions to the predominant White Anglo-Saxon Protestant culture. In other words, some groups have been actively prevented from "melting."

Immigration scholar Lawrence H. Fuchs referred to this pattern as **coercive pluralism**, in which racial and ethnic minority groups are excluded from the dominant culture and thereby from social and political power, through three main mechanisms (Fuchs 1990). One pattern is known as *predatory pluralism*, in which indigenous inhabitants are pushed aside through colonial conquest and come to be defined as foreigners in their own land. Another pattern is known as *caste pluralism*, primarily experienced by African American people in the United States, in which both de jure and de facto segregation prevent a minority group from accessing and contributing to the dominant culture. The third pattern is known as *sojourner pluralism*, which occurs when immigration policy limits entry among some ethnic and racial minority groups to a temporary basis, typically in order to fulfill a domestic labor shortage. In the United States, Latino and Asian groups have been most affected by this form of coercive pluralism, resulting in perceptions within the dominant culture that members of these groups are fundamentally different and unable to assimilate. Thus, different minority groups within a society do not experience opportunities for or restrictions preventing assimilation in the same way (see Table 9.1).

TABLE 9.1 Fuchs's Model of Coercive Pluralism

	Strategy of Coercion	Negative Stereotypes within Dominant Culture	Examples of Groups Affected in the United States
Predatory pluralism	Expulsion	Foreigners in their own land	Native Americans
Caste pluralism	Segregation	The outsider within	African Americans
Sojourner pluralism	Exclusion	Unable or unwilling to assimilate	Latinos and Asians

SOURCE: Fuchs 1990.

pluralism A state in which racial and ethnic groups maintain their distinctness but respect each other and have equal access to social resources.

assimilation The process by which formerly distinct and separate groups merge and become integrated as one.

coercive pluralism A pattern in which racial and ethnic minority groups are excluded from the dominant culture and thereby from social and political power; includes predatory, caste, and sojourner pluralism.

The Value of Diversity

Directions: The U.S. Census Bureau projects that by 2060, more than half the American population will be non-White (Vespa et al. 2020). How do Americans feel about this growing racial and ethnic diversity? Respond to these questions, then compare your results to the sample of 6,637 adults in the United States surveyed by the Pew Research Center in 2019.

"The fact that the U.S. population is made up of people of many different races and ethnicities has a ___ impact on the country's culture."

____ Positive ____ Negative ____ Doesn't make much difference

"The fact that the U.S. population is made up of people of many different races and ethnicities makes it ___ to solve the country's problems."

____ Easier for policy makers ____ Harder for policy makers ____ Doesn't make much difference

"It is __ for companies and organizations to promote racial and ethnic diversity in their workplace."

____ Not important at all ____ Not too important ____ Somewhat important ____ Very Important

"When it comes to decisions about hiring and promotions, companies and organizations should ..."

____ Only take qualifications into account, even if it results in less diversity.

____ Also take race and ethnicity into account in order to increase diversity.

"Students should ..."

____ Go to schools in their local communities, even if it means most schools are not racially and ethnically mixed.

____ Go to schools that are racially and ethnically mixed, even if it means some students don't go to school in their local communities.

"It would bother me __ to hear people speak a language other than English in a public place."

____ A lot ____ Some ____ Not much ____ Not at all

Racial and Ethnic Group Diversity in the United States

> **@amy klobuchar**
>
> @KamalaHarris, the first African American, Asian American, and woman to become Vice President, is sworn in by Justice Sonia Sotomayor, the first Latina on the Supreme Court. This is what breaking the glass ceiling looks like.
>
> -Amy Klobuchar

The first census in 1790 divided the U.S. population into four groups: free White males, free White females, slaves, and other people (including free Blacks and Indians). To increase the size of the slave population, the *one-drop rule* specified that even one drop of "negroid" blood defined a person as Black and therefore eligible for slavery. The "one-drop rule" is still operative today: Biracial individuals are typically seen as a member of whichever group has the lowest status (Wise 2009). For example, former President Barack Obama and current Vice President Kamala Harris are commonly referred to as the first Black president and vice president, respectively, despite the fact that both are biracial; Obama's mother was a White American and his father Nigerian, while Harris's parents (both of whom immigrated to the United States) are Jamaican and Indian.

In 1960, the census recognized only two categories: White and non-White. In 1970, the census categories consisted of White, Black, and "other" (Hodgkinson 1995). In 1990, the U.S. Census Bureau recognized four racial classifications: (1) White; (2) Black; (3) American Indian, Aleut, or Eskimo; and (4) Asian or Pacific Islander. The 1990 census also included the category of "other." Beginning with the 2000 census, racial categories expanded to include Native Hawaiian or other Pacific Islander, and also allowed

Results

Question	All Adults	White	Black	Hispanic
"The fact that the U.S. population is made up of people of many different races and ethnicities has a ___ impact on the country's culture."				
Positive	64%	64%	58%	70%
Negative	12%	14%	9%	8%
Doesn't make much difference	23%	22%	31%	22%
"The fact that the U.S. population is made up of people of many different races and ethnicities makes it ___ to solve the country's problems."				
Easier for policymakers	7%	4%	11%	14%
Harder for policymakers	47%	52%	30%	42%
Doesn't make much difference	45%	44%	59%	43%
"It is __ for companies and organizations to promote racial and ethnic diversity in their workplace."				
Not at all/not too important	24%	—	—	—
Somewhat/very important	75%	73%	81%	75%
"When it comes to decisions about hiring and promotions, companies and organizations should ..."				
Only take qualifications into account, even if it results in less diversity.	74%	78%	54%	69%
Also take race and ethnicity into account in order to increase diversity.	24%	21%	37%	27%
"Students should ..."				
Go to schools in their local communities, even if it means most schools are not racially and ethnically mixed.	54%	62%	28%	47%
Go to schools that are racially and ethnically mixed, even if it means some students don't go to school in their local communities.	42%	35%	68%	49%
"It would bother me __ to hear people speak a language other than English in a public place."				
A lot	11%	14%	9%	3%
Some	18%	20%	16%	10%
Not much	24%	25%	26%	17%
Not at all	47%	41%	48%	68%

Note: Share of respondents who didn't offer an answer not shown. Whites and Blacks include those who report being only one race and non-Hispanic. Hispanics are of any race.
SOURCE: Horowitz 2019.

individuals the option of identifying themselves as being more than one race rather than checking only one racial category (see Figure 9.1).

Although U.S. citizens come from a variety of ethnic backgrounds, the largest ethnic population in the United States is of Hispanic origin. The Census Bureau began collecting data on the U.S. Hispanic population in 1970.

As you read the following section on U.S. Census data on race and Hispanic origin, keep in mind that the use of racial and ethnic labels is often misleading and imprecise. The ethnic classification of "Hispanic/Latino," for example, lumps together disparate groups such as Puerto Ricans, Mexicans, Cubans, Venezuelans, Colombians, and others from Latin American countries. The racial term *American Indian* includes more than 300 separate tribal groups that differ enormously in language, tradition, and social structure. The racial label *Asian* includes individuals from China, Japan, Korea, India, the Philippines, or one of the countries of Southeast Asia. The term *Asian American* is used to describe people with Asian racial features who are born in the United States, as well as those who immigrate to the United States.

U.S. Census Data on Race and Hispanic Origin

Census data show that the U.S. population is becoming increasingly diverse. In 2016, non-Hispanic Whites comprised over 61 percent of the population. By 2030, that number is expected to shrink to less than 56 percent and by 2060 is projected to be just over

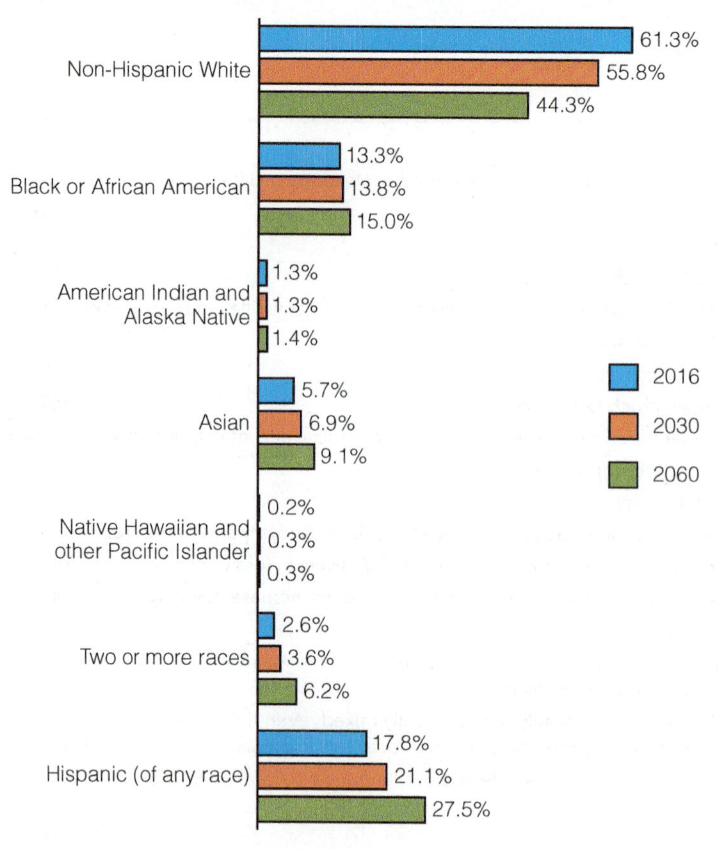

Figure 9.2 U.S. Population by Race and Hispanic/Latino Origin, 2016, 2030, and 2060 (projected)
SOURCE: Vespa et al. 2020.

> "Surrounded as I am now by wonderful children and grandchildren, not a day goes by that I don't think of Richard and our love, our right to marry, and how much it meant to me to have that freedom to marry the person precious to me. ... I am proud that Richard's and my name is on a court case that can help reinforce the love, the commitment, the fairness, and the family that so many people, black or white, young or old, gay or straight seek in life. I support the freedom to marry for all. That's what Loving, and loving, are all about."
>
> **–MILDRED LOVING, 2007**

44 percent (Vespa et al. 2020). During this same period, the proportion of the population that identifies as Hispanic is expected to increase from less than 18 percent to over 27 percent of the total population, and the population that identifies as two or more races is expected to increase from less than 3 percent to slightly more than 7 percent. Figure 9.2 shows the U.S. population by race and Hispanic origin in 2016 and projected in 2060.

These changing demographic trends indicate that by the time today's college graduates are entering their retirement years, the United States will be a majority-minority nation (see this chapter's *Self and Society* feature). As discussed at the beginning of this chapter, minority status does not refer to numerical representation but rather to representation among social positions of status, power, and prestige. For example, the state of Texas is a minority-majority state; only 41 percent of the state's population is non-Hispanic White, nearly 40 percent are Hispanic or Latino, and nearly 14 percent are Black (U.S. Census Bureau 2019). Despite this, 80 percent of elected officials are non-Hispanic White, and only 20 percent belong to racial or ethnic minority groups (Reflective Democracy Campaign 2020).

More than half of the U.S. Hispanic population is Mexican, with other Hispanics having ties to Puerto Rico, Central America, South America, Cuba, and other countries. More than half of the U.S. Hispanic population lives in just three states: California, Texas, and Florida.

One of the most common confusions about Hispanic origin is the question of whether "Hispanic" is a race or an ethnicity. According to the federal government, Hispanic origin is considered an ethnicity, not a race. But many people—Hispanic and non-Hispanic—think otherwise. Although the 2020 census included instructions that stated, "for this census, Hispanic origins are not races," many Hispanics identified their race as "Latino," "Mexican," "Puerto Rican," "Salvadoran," or other ethnicity or national origin, which the census classified in the category of "Some Other Race" (see Table 9.2). The current Census Bureau classification system does not allow people of mixed Hispanic or Latino ethnicity to identify themselves as such. Although they may choose to identify as "two or more races" or "some other race," individuals with one Hispanic and one non-Hispanic parent still must say that they are either Hispanic or not Hispanic.

TABLE 9.2 Racial Identification of Hispanics/Latinos in the United States, 2015

White	65.8%
Black or African American	2.1%
American Indian/Alaska Native	0.9%
Asian	0.3%
Native Hawaiian/Pacific Islander	0.1%
Some other race	26.2%
Two or more races	4.6%

SOURCE: U.S. Census Bureau 2020.

Mixed-Race Identity

Until 2000, individuals did not have the option to identify as more than one race on the U.S. Census. Although there was an option to identify as "Other" on Census forms from 1970 through 1990, an individual still had to choose only one racial identity. This restrictive understanding of racial identity is, in part, tied to the history of **antimiscegenation laws**, which banned interracial marriage of White and non-White individuals—primarily Blacks, but in some cases, also Native Americans and Asians. Over time, many states overturned these laws, but 16 states had active antimiscegenation in 1967, when the Supreme Court declared these laws unconstitutional in the landmark case *Loving v. Virginia*.

antimiscegenation laws Laws banning interracial marriage until 1967, when the Supreme Court (in *Loving v. Virginia*) declared these laws unconstitutional.

The multiracial population has grown as mixed-race marriages have increased over recent years. In 1970, the first Census after the *Loving* decision, fewer than 1 percent of all U.S. married couples consisted of spouses with different racial identities (Wang 2015). By 2016, that number had increased to 10.2 percent (Rico et al. 2018, see also Figure 9.3).

Attitudes toward interracial marriages have changed dramatically over the last few generations. The percentage of U.S. adults who disapproved of marriage between Blacks and Whites decreased from 94 percent in 1958, to 29 percent in 2002, to 11 percent in 2013 (Gallup Organization 2020a). Individuals most likely to have positive attitudes about intermarriage are minorities, younger, more educated, liberal, and living in urban areas (Livingston and Brown 2017).

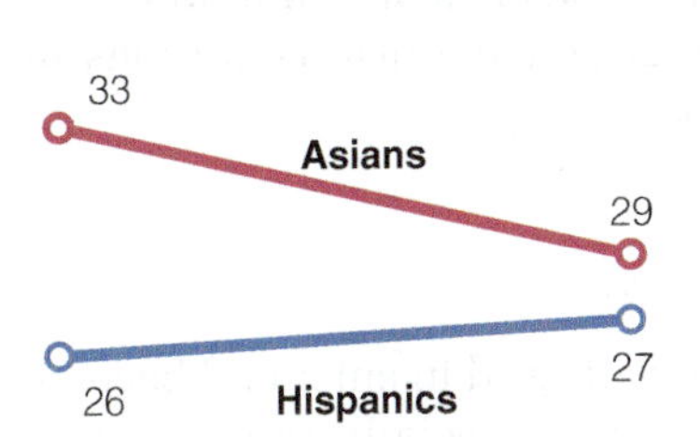

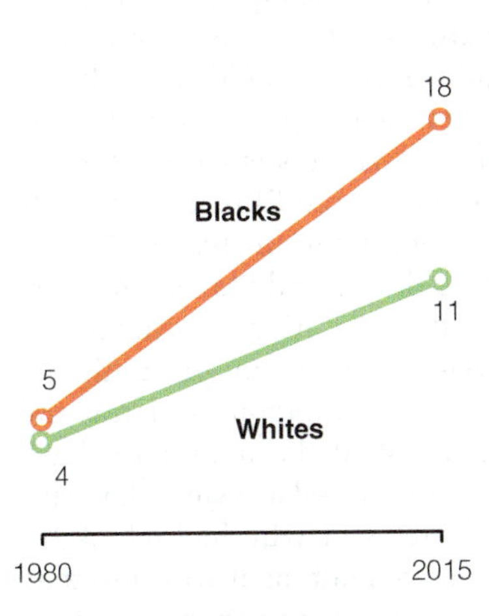

Figure 9.3 Percentage of U.S. Newlyweds in Interracial Marriages by Race, 1980–2015
SOURCE: Livingston and Brown 2017.

Race and Ethnic Group Relations in the United States

In a 2019 Gallup Poll, 40 percent of Americans said they personally worry "a great deal" about race relations in the United States—up from 13 percent in 2009 (Gallup Organization 2020a). The surge in worry over race relations is likely linked to (1) incidents of police shooting unarmed Black men and the tense protests that follow, and (2) race-related comments made by Donald Trump (Swift 2017). The Gallup Organization (2020a) also asked a national sample of U.S. adults to rate relations between various groups in the United States. Table 9.3 presents the results of this survey. Relations between various racial and ethnic groups are influenced by prejudice and discrimination (discussed later in this chapter). Race and ethnic relations are also complicated by issues concerning immigration—the topic we turn to next.

Free Lance–Star, via Associated Press

Richard and Mildred Loving with their three children. The Loving's legal challenge in Virginia led the Supreme Court to declare antimiscegenation laws unconstitutional in 1967.

TABLE 9.3 Perceptions of Race and Ethnic Relations in the United States, 2020

A national sample of U.S. adults was asked to rate relations between various groups in the United States. The results are depicted in this table.

	Very or Somewhat Good	Very or Somewhat Bad
Whites and Blacks	44%	55%
Whites and Hispanics*	62%	33%
Blacks and Hispanics	66%	30%
Whites and Asians	78%	11%

*Question was last asked in 2018.
Note: Percentages do not add up to 100 because of "no opinion" responses.
SOURCE: Gallup Organization 2020.

What do you think?

A 2009 poll asked U.S. adults, "If blacks and whites honestly expressed their true feelings about race relations, do you think this would do more to bring races together or cause greater racial division?" More than half (56 percent) replied "bring races together"; about a third (37 percent) said "cause greater division" (7 percent had no opinion) (Gallup Organization 2020a). If this poll were taken today, how do you think the results would be different?

Immigrants in the United States

The growing racial and ethnic diversity of the United States is largely the result of immigration as well as the higher average birthrates among many minority groups. Immigration generally results from a combination of "push" and "pull" factors. Adverse social, economic, and/or political conditions in a given country "push" some individuals to leave that country, whereas favorable social, economic, and/or political conditions in other countries "pull" some individuals to those countries.

U.S. Immigration: A Historical Perspective

For the first 100 years of U.S. history, all immigrants were allowed to enter and become permanent residents. The continuing influx of immigrants, especially those coming from non-White, non-European countries, created fear and resentment among native-born Americans, who competed with immigrants for jobs and who held racist views toward some racial and ethnic immigrant populations. America's open-door policy on immigration ended in 1882 with the *Chinese Exclusion Act*, which suspended the entrance of the Chinese to the United States for ten years and declared Chinese ineligible for U.S. citizenship (this act was repealed in 1943). The *Immigration Act of 1917* required all immigrants to pass a literacy test before entering the United States. Immigration legislation passed in the 1920s established a quota system, limiting the numbers of immigrants from specific countries. Under the quota system, 70 percent of immigrant slots were allotted to people from just three countries: United Kingdom, Ireland, and Germany. In 1965, the passage of the *Hart-Celler Act* abolished the national origins quota system that had been in place since the 1920s, and it instituted a system that gave preference to immigrants who had family in the United States or who had job skills. The *Hart-Celler Act* was an extension of the civil rights movement, as it was designed to end discrimination based on race and ethnicity, and it continues to shape immigration law today.

The percentage of the U.S. population that is foreign-born increased from 6.2 percent in 1980 to 14.2 percent in 2019 (U.S. Census Bureau 2020). The proportion of the

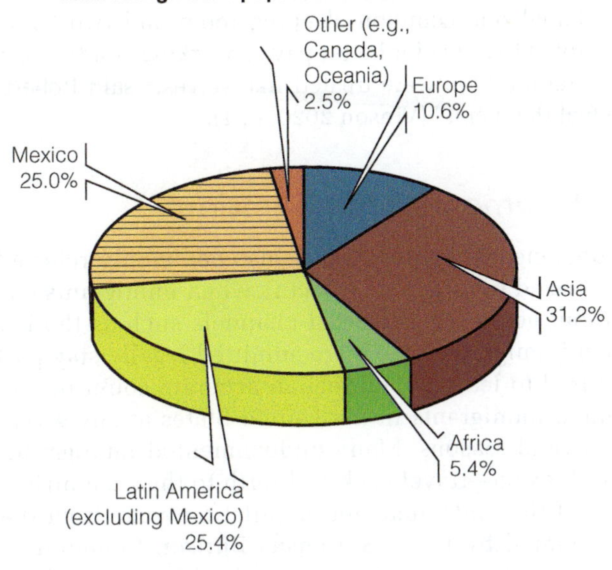

Figure 9.4 U.S. Foreign-Born Residents by Region of Birth, 2018
SOURCE: U.S. Census Bureau 2020c.

population that is foreign born is projected to increase by at least 58 percent by 2060 (Vespa et al. 2020). More than half of the U.S. foreign-born population came from either Mexico or South or East Asia (see Figure 9.4).

Guest Worker Program

The United States has two guest worker programs that allow employers to import unskilled labor for temporary or seasonal work: the H-2A program for seasonal agricultural work and the H-2B program for nonagricultural work. H-2 visas generally do not permit guest workers to bring their families to the United States.

Immigrant "guest workers" are hardly treated like "guests"; these workers are systematically exploited and abused. Guest workers are bound to the employers who "import" them so they are not able to change jobs if they are mistreated. Guest workers are often cheated out of pay and forced to live in squalid conditions; and although they perform some of the most difficult and dangerous jobs in the United States, many who are injured on the job are unable to obtain medical treatment and workers' compensation benefits. Immigrant women working at low-wage jobs are often targets of sexual violence. If guest workers complain about mistreatment, they are threatened with deportation or being "blacklisted" and unable to find another job. Rampant with labor and human rights abuses, the guest worker program was described as a modern-day system of indentured servitude (Southern Poverty Law Center [SPLC] 2013).

In August 2020, the SPLC filed a class-action lawsuit on behalf of hundreds of guest workers employed in Louisiana by Lowry Farms, Inc. (Vinson 2020). The lawsuit alleges that workers were paid "piece rate"—meaning their pay was based on the number of acres harvested—rather than the $10.73 minimum wage as required by the H-2A visa contract. Roberto, one of the plaintiffs named in the lawsuit, says that his pay averaged $4.60 per hour for a typical 60-hour work week. Workers are also routinely expected to pay for their own transportation to the United States and to pay fees associated with acquiring the visa and finding a work placement. Because most of these workers must take out loans to pay these fees, and then are routinely underpaid by their employers, many end up accumulating debt that they cannot repay.

The SPLC lawsuit also alleges unsafe working and living conditions. Workers were routinely required to work at least 10 hours per day, 7 days a week, in all weather

Andrew Lichtenstein/Corbis Historical/Getty Images

Immigrant guest workers are often forced to live in crowded and substandard housing.

conditions. The housing they were provided were trailers, in which 35 workers shared one common sleeping room and two bathrooms. Workers who attempted to argue for better pay or working conditions were threatened with deportation. "We were treated like slaves," said Roberto, "that's what made me feel the worst" (Vinson 2020, p. 1).

Undocumented Immigration

Undocumented immigration—also commonly referred to as illegal or unauthorized immigration—occurs when immigrants enter the United States without going through legal channels such as the H-2 visa program, and when immigrants who were admitted legally stay past the date they were required to leave. Obtaining an accurate count of the number of undocumented immigrants in the United States at any given time is challenging for several reasons. Many undocumented immigrants—particularly those from Mexico—travel back and forth to their countries of origin seasonally. Most estimates of the undocumented population come from the American Community Survey, operated by the U.S. Census Bureau. Community isolation and fears that contact with government agencies may lead to deportation can suppress survey responses, leading to an undercount. Still, the most recent estimates from the Department of Homeland Security and U.S. Census Bureau indicate that there are currently between 10 and 12 million undocumented immigrants living in the United States, approximately 7.6 million of whom work in the U.S. labor force (Kamarck and Stenglein 2019; Radford 2019).

What do you think?

In 2019, the Trump administration pushed the U.S. Census Bureau to include a question in 2020 asking whether the respondent was a legal citizen of the United States. Civil rights groups mounted a legal challenge to adding the question, arguing that it was intended to intimidate undocumented residents from responding to the Census, leading to an undercount and thereby reducing the amount of federal funding that would be available to the communities in which they reside. The administration dropped the request after mounting legal challenges (Wines 2019). Do you think the next Census should include a citizenship question?

@rep johnlewis

There is no such thing as an illegal human being. We may have all come here on different ships, but we're in the same boat now.

-John Lewis

The majority of undocumented immigrants (65 percent) come from Mexico and Central America. More than half of the unauthorized immigrant population lives in five states: California, Texas, Florida, New York, and New Jersey (Passel and Cohn 2019). Although these states have historically had the highest numbers of undocumented immigrant population, the geographic distribution of this population is showing rapid change (see Figure 9.5). North and South Dakota, Louisiana, Maryland, and Massachusetts have shown the greatest relative increases in unauthorized immigrant populations in the past decade.

Border Crossing. U.S. Customs and Border Protection, an agency within the Department of Homeland Security, has nearly 20,000 border patrol agents who patrol 6,000 miles of Canadian and Mexican borders and more than 2,000 miles of coastal waters around Florida and Puerto Rico (U.S. Customs and Border Protection 2020). Despite the border patrol, along with a "fence" at the U.S.–Mexico border, people continue to find ways to illegally cross the U.S. border.

Some people cross (or attempt to cross) the U.S.–Mexican border with the help of a *coyote*—a hired guide who typically charges $3,000 to $5,000 to lead people across the border (Maril 2011). Crossing the border illegally involves a number of risks, including death from drowning (e.g., while trying to cross the Rio Grande) or dehydration.

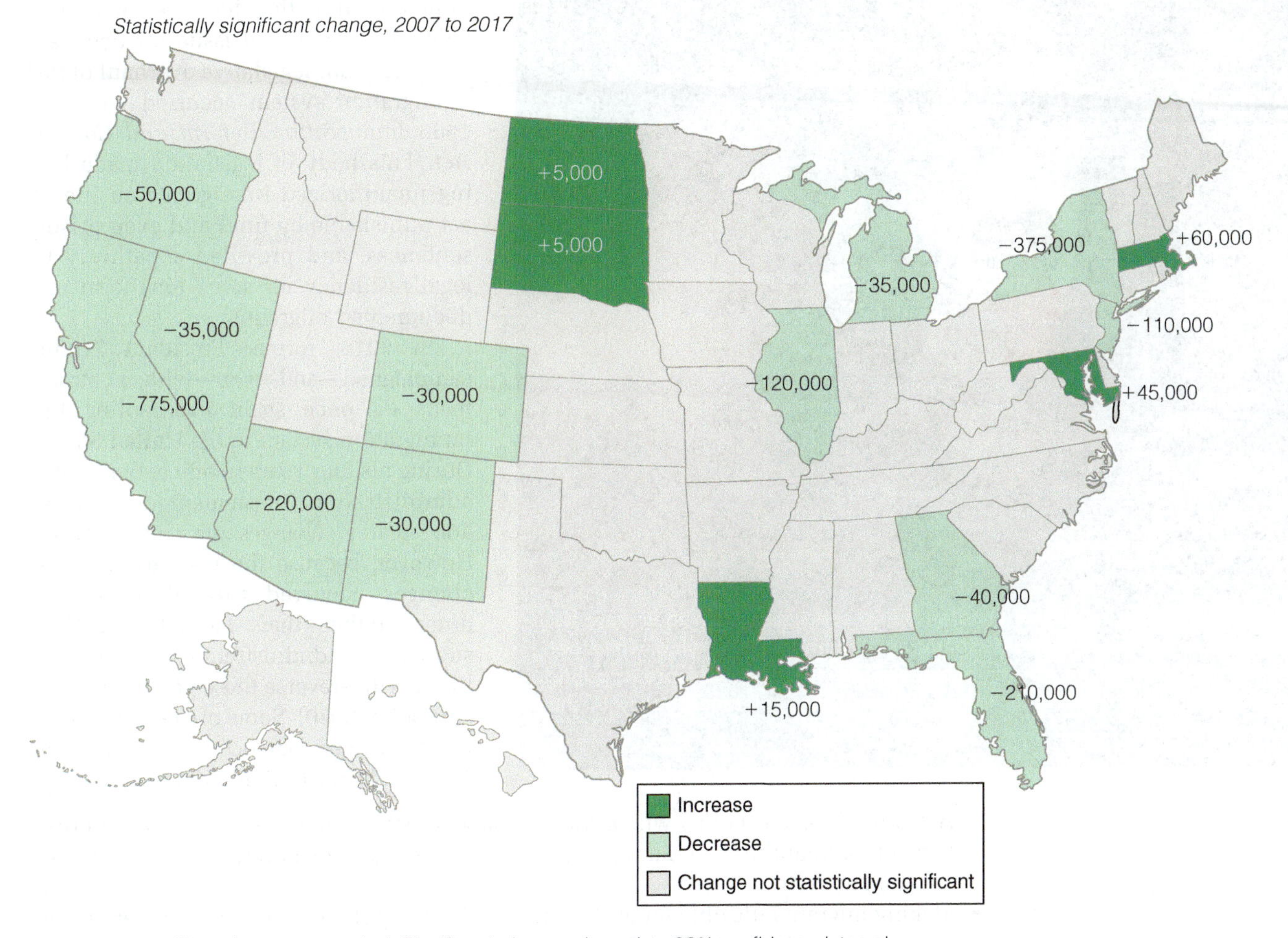

Figure 9.5 Unauthorized Immigrant Population Changes, 2007–2017
SOURCE: Passel and Cohn 2019.

Border crossers also risk encounters with members of **nativist extremist groups**—organizations that not only advocate restrictive immigration policy but also encourage their members to use vigilante tactics to confront or harass suspected unauthorized immigrants.

Unauthorized Immigrants in the Workforce. In 2016, there were 7.8 million undocumented immigrants in the U.S. labor force, comprising about 5 percent of the U.S. workforce (Passel and Cohn 2018). Sociologist Robert Maril (2004) noted that "[t]he vast majority of illegal immigrants leave their home countries to work hard, save their money, then return to their homeland. ... These individuals do not travel their difficult and dangerous journeys searching for a welfare handout; they immigrate to work" (pp. 11–12).

Undocumented workers often work in sweatshop conditions (see Chapter 7), where they are exploited and mistreated by employers. Americans who feel that undocumented workers do not deserve protection from labor rights and human rights abuses might consider the argument made by economist David Cooper (2015):

> Failure to protect any group of workers—even those without lawful immigration status—is damaging to all workers. When businesses can exploit immigrant workers who cannot speak out for fear of deportation, it lowers the wages of other workers in the same or similar fields. (n.p.)

nativist extremist groups Organizations that not only advocate restrictive immigration policy, but also encourage their members to use vigilante tactics to confront or harass suspected undocumented immigrants.

John Moore/Getty Images News/Getty Images

Heart-wrenching images of children being separated from their parents, like this one of a 2-year-old Honduran asylum seeker crying while her mother is arrested by border patrol agents, sparked widespread public protests against the Trump administration's zero-tolerance policy in the summer of 2018.

Policies Regarding Illegal Immigration. Immigration policy has been a highly contentious political issue for decades. The last major legislative overhaul of the immigration system occurred with the 1986 *Immigration Reform and Control Act*. This body of legislation made hiring unauthorized immigrants an illegal act punishable by fines and even prison sentences, and provided a pathway to legal residency for some long-term undocumented migrants.

In 2016, former President Trump campaigned—and won—with a strong focus on once again overhauling the immigration system in the United States. During his four years in office, the Trump administration implemented at least 400 policy changes on immigration. However, because the majority of these changes occurred through executive orders rather than legislative action, subsequent administrations could—and likely will—reverse these policies (Pierce and Bolter 2020). Some of the most prominent changes that occurred during the Trump administration included:

- A "zero-tolerance" policy which directed law enforcement agencies to make apprehension of unauthorized immigrants a priority for arrest, even if they have no other criminal activity.
- Making migrants illegible for asylum if they fail to apply for asylum elsewhere prior to their arrival in the United States.
- Requiring asylum seekers from Latin America to stay in Mexico while awaiting their asylum hearings.
- Lowering the annual maximum number of refugees permitted to settle in the United States to 18,000, down from nearly 85,000 in 2016.
- Increasing funding, resources, and personnel for Immigration and Border Patrol and Immigration and Customs Enforcement agencies.
- A "public charge" rule that allowed applications for permanent residency or green cards to be denied based on past or potential future use of government benefits.

The Biden administration has vowed to reverse all of these policies through executive action within the first 100 days of taking office (Biden-Harris 2020a). Additionally, President Biden's campaign included a pledge to overhaul the immigration system through legislation and modernize the 1986 *Immigration Reform and Control Act*. However, President Biden is likely to face the same legislative obstacles as his predecessor. At the time of this writing, the future of immigration policy in the United States remains uncertain.

In addition to these policy changes, the Trump administration also attempted to end the Deferred Action for Childhood Arrivals (DACA) program, an Obama-era policy that allowed undocumented residents who had been brought to the United States as children to access work permits, drivers' licenses, and federal benefits such as financial aid for college. In June 2020, the Supreme Court ruled that the administration's attempt to end the program was unconstitutional. The Trump administration responded by altering the policy by denying all new first-time applications and by reducing renewals from two-year to one-year periods (Pierce and Bolter 2020). The Biden administration has pledged to "reinstate the DACA program, and ... explore all legal options to protect their families from inhumane separation" (Biden-Harris 2020a n.p.). However, this vague promise does not provide a clear path forward for these so-called "dreamers" to establish permanent and legal residency in the United States.

Perhaps the most controversial change implemented by the Trump administration was the family separation and child detention policy. Because the zero-tolerance policy pushed for the swift detention and prosecution of anyone who entered the country illegally and the Justice Department does not permit children to be detained in adult facilities, the Department of Homeland Security began the process of separating adults and children at border detention centers in May 2018. After widespread public outcry, the Trump administration issued an executive order to end the practice of child separation in June 2020. However, the practice of family separation continues as the executive order included a provision that family separations could continue "for-cause" and gave Border Patrol agents wide discretion to determine whether cause for separation was warranted. From the start of the family separation policy in June 2018 until January 2020, at least 5,500 children have been forcibly separated from their parents at the U.S.–Mexican border; at least 1,100 of these separations occurred after Trump's June 2020 executive order to end family separations. The Biden administration has called this policy "inhumane" and vowed to prioritize family reunification (Biden-Harris 2020a). Biden's nominee to head the Department of Homeland Security, Alejandro Mayorkas, would be the first Latino and first immigrant to lead the department.

What *do you* think?

President Trump's 2016 campaign was based on a promise to build a wall that would stem the flow of illegal immigrants across the U.S.–Mexico border, at an estimated cost of at least $18 billion. However, Border Patrol agents and Congressional representatives in border districts have argued that additional border security is needed, not a wall. Instead, they have suggested that investments in more sophisticated surveillance technology and personnel increases would be more effective at reducing illegal border crossings than a physical barrier (Nixon 2018). If you were in charge of the border control budget, what steps would you prioritize to improve border security?

Some local governments have implemented their own policies to counter the surge of increasingly restrictive immigration policies. **Sanctuary cities** refer to jurisdictions that have implemented rules to limit local law enforcement from cooperating with federal immigration authorities, such as ICE, in order to prevent the detention or deportation of undocumented residents (Cooke and Hesson 2020). Similarly, some schools and college campuses have established sanctuary policies such as banning the collection of student data on immigration status and limiting access by immigration authorities' to student records or physical access to student spaces in order to prevent the apprehension of undocumented students (Tsu 2020).

Public opinion polls indicate that immigration policy will continue to be a politically contentious issue for years to come. According to the most recent polls conducted by Gallup (2020b), 36 percent of Americans say they personally worry about illegal immigration a "great deal," and 47 percent say that large numbers of undocumented immigrants entering the United States poses a critical threat to the country's vital interests. Forty percent favor the construction of a border wall, while 60 percent oppose it. Although Americans are sharply divided on many issues surrounding immigration policy, there is strong support for the DACA program; 83 percent responded in favor of allowing immigrants who were brought to the United States illegally as children the chance to become citizens if they meet certain requirements.

@The EllenShow

For me, America is great because of all the people who came here. Not in spite of them. #NoBan

-Ellen DeGeneres

sanctuary city A jurisdiction that has implemented rules to limit local law enforcement from cooperating with federal immigration authorities in order to prevent the detention or deportation of undocumented residents.

naturalized citizens Immigrants who apply for and meet the requirements for U.S. citizenship.

Becoming a U.S. Citizen

More than half of the 44.7 million foreign-born U.S. residents in 2018 were **naturalized citizens** (immigrants who applied and met the requirements for U.S. citizenship) (Batalova and Zong 2020). Requirements to become a U.S. citizen include

(1) having resided continuously as a lawful permanent U.S. resident for at least five years (three years for a spouse of a U.S. resident); (2) being able to read, write, speak, and understand basic English (certain exemptions apply); (3) being "a person of good moral character" (cannot have a record of criminal offenses such as prostitution, illegal gambling, failure to pay child support, drug violations, and violent crime); (4) demonstrating willingness to support and defend the U.S. Constitution by taking the Oath of Allegiance; and (5) passing an examination on English (speaking, reading, and writing) and U.S. government and history (U.S. Citizenship and Immigration Services 2020).

Myths about Immigration and Immigrants

Many foreign-born U.S. residents work hard to succeed educationally and occupationally. The percentage of foreign-born adults (age 25 and older) with at least a bachelor's degree (32 percent) nearly matches that of native-born U.S. adults (33 percent) (Batalova Zong 2020). Despite the achievements and contributions of immigrants, many myths about immigration and immigrants persist, largely perpetuated by anti-immigrant groups and campaigns:

Myth 1. Immigrants increase unemployment and lower wages among native workers.

Reality: Most academic economists agree that immigration overall has a small but positive impact on the wages of native-born workers. Although immigration results in slightly lower wages among workers within the same jobs, they also consume goods and services which creates more jobs and leads to higher overall wages in other employment sectors (Edo 2019). Additionally, the small decrease in wages is experienced almost exclusively by prior immigrant entrants working in the same industry, not U.S. citizens (National Academies of Science, Engineering and Medicine [NASEM] 2017). Immigrants also start their own businesses at a higher rate than native U.S. residents, which increases demand for business-related supplies (such as computers and office furniture) and service providers (such as accountants and lawyers) (Pollin 2011). Overall, the presence of unauthorized immigrants in the U.S. economy is associated with significant reductions in the price of consumer goods and services, especially in child care, food preparation, and construction (NASEM 2017).

Myth 2. Immigrants drain the public welfare system and our public schools.

Reality: Unauthorized and temporary immigrants are ineligible for major federal benefit programs, and even legal immigrants may face eligibility restrictions. Two benefit programs that do not have restrictions against unauthorized immigrants are the Special Supplemental Nutrition Program for Women, Infants, and Children (WIC) and the National School Lunch Program. Still, the U.S. citizen children of undocumented immigrants are less likely to receive welfare benefits than other children because of the "chilling effect" of deportation threats to their family members (Bernstein et al. 2020).

Regarding public education, a 1982 Supreme Court case (*Plyer v. Doe*) held that states cannot deny students access to public education, even if they are not legal U.S. residents. The Court ruled that denying public education could impose a lifetime of hardship "on a discrete class of children not accountable for their disabling status" (Armario 2011). There are approximately 1 million undocumented children enrolled in school in the United States, which is less than 2 percent of the total enrollment of public elementary and secondary schools (Migration Policy Institute 2020).

Although the states bear the cost of education, as well as some social services and medical services for the immigrant population, research suggests that the economic benefits that immigrants provide for the states outweigh the costs associated with supporting them. The Institute on Taxation and Economic Policy (2017) reported that undocumented immigrants collectively pay an estimated $11.7 billion a year in state and local taxes. For example, they pay sales and excise taxes on goods and services they purchase. They also pay property taxes—a large source of funding for public schools—directly on their homes or indirectly as renters. Approximately 10 percent of total Social Security contributions come from undocumented immigrants, who are themselves ineligible from receiving those benefits. Furthermore, the long-term economic contributions of immigrant families outweigh the short-term costs. An economic analysis from NASEM (2017) found that a first-generation immigrant family costs state budgets an average of $1,600 per year, but that second- and third-generation families created net gains to state budgets of approximately $1,700 per year.

@Joe Biden

This pandemic has put a bright light on the disparities that have plagued our nation for too long. We've got to make sure these communities get the help they need now—and root out the systemic racism that created these inequalities in the first place.

-Joe Biden

Myth 3. Immigrants do not want to learn English.

Reality: In 2018, 47 percent of the U.S. foreign-born population (age 5 and older) spoke English less than "very well" (Batalova et al. 2020). While children benefit from public school programs that teach English as a Second Language (see also Chapter 8), adults who want to learn English have limited educational opportunities, and there are not enough adult English education programs to meet the demand. The majority of programs designed to teach English to adults with limited English proficiency have waiting lists, with wait times ranging from a few weeks to more than three years (Wilson 2014). Although English proficiency is one of the requirements of becoming a naturalized citizen, the United States does not have an official language.

Myth 4. Undocumented immigrants have children in the United States as a means of gaining legal status.

Reality: Under the Fourteenth Amendment of the U.S. Constitution, any child born in the United States is automatically granted U.S. citizenship. But having children who are U.S. citizens does not provide immigrants with a means of gaining legal status in the United States. Children under 21 are not allowed to petition for their parents' U.S. citizenship. Nevertheless, some legislators have called for ending birthright citizenship by amending the Constitution or enacting state law to limit citizenship to children who have at least one authorized parent (Dwyer 2011).

Myth 5. Immigrants have high rates of criminal behavior.

Reality: Former President Trump infamously referred to undocumented immigrants as "bad hombres" and has called some Mexican immigrants "rapists" and "criminals." Trump claimed that immigrants pose a threat to public safety and national security. But studies find that immigrants, including unauthorized immigrants, are less likely than natives to commit crimes (Nowrasteh 2015; Orrenius and Zavodny 2019). Furthermore, increasing pathways for unauthorized immigrants to gain legal status is associated with a decrease in crime (Orrenius and Zavodny 2019). Because they risk deportation, undocumented immigrants have a strong motivation to avoid involvement with the law. Criminologist Jack Levin said, "If you want to find a safe city, first determine the size of the immigrant population. If the immigrant community represents a large proportion of the population, you're likely in one of the country's safer cities" (quoted in Balko 2009).

Sociological Theories of Race and Ethnic Relations

Some theories of race and ethnic relations suggest that individuals with certain personality types are more likely to be prejudiced or to direct hostility toward minority group members. Sociologists, however, focus on how the structure and culture of society affect race and ethnic relations.

Structural-Functionalist Perspective

The structural-functionalist perspective focuses on how parts of the whole are interconnected. This perspective reminds us that we cannot fully understand the history of U.S. civil rights in a vacuum; we need to consider how forces outside the United States affected U.S. policies and culture regarding race relations. In *Cold War Civil Rights: Race*

THE WORLD in quarantine

A Tale of Two Pandemics

Jason Hargrove was frustrated. The 50-year-old Detroit bus driver, husband, and father of six was considered an "essential worker" and continued to drive his route every day, despite the surge in COVID-19 cases in his city. He had implored riders who stepped on his bus to wear a mask, cover their coughs, and stay six feet apart, but he continually felt ignored and disrespected. After one woman refused to cover her cough despite his repeated pleas, Hargrove vented his frustrations in a video posted to Facebook, which quickly went viral:

> We're out here as public workers, doing our job, trying to make an honest living to take care of our families. But for you to get on the bus, and stand on the bus, and cough several times without covering up your mouth, and you know that we're in the middle of a pandemic, that lets me know that some folks don't care. (quoted in Levenson 2020, p. 1)

Eleven days later, Jason Hargrove died from COVID-19.

Hargrove's story is tragically similar to so many victims of COVID-19. As a Black American and service worker, however, Hargrove was at greater risk than White Americans of contracting and dying from COVID-19—3.57 times more likely, to be exact (Wood 2020). Similar disparities are also evident in Native American and Hispanic communities (Yong 2020).

But how can a virus be racist? While all individuals can potentially be infected with the virus, the likelihood of exposure and risk of serious illness and death are not evenly distributed. The structural conditions of society—including employment conditions, local health resources, and residential patterns—mean that some groups will experience greater risk. If racial disparities exist within any part of the structure of society, those disparities will manifest in health outcomes. "We know that these racial ethnic disparities in COVID-19 are the result of pre-pandemic realities," says Dr. Marcella Nunez-Smith, director of the Equity Research and Innovation Center at Yale School of Medicine. "It's a legacy of structural discrimination that has limited access to health and wealth for people of color" (quoted in Godoy and Wood 2020, p. 1).

Public health experts emphasize three particular pre-pandemic realities that have driven the current crisis. First, racial and ethnic minorities are more likely to be exposed to the virus through occupational and residential segregation (Godoy and Wood 2020). Front-line or "essential" jobs—work that can't be done remotely, primarily in the service, transportation, and food production industries—are disproportionately performed by racial and ethnic minorities. For example, while Hispanics make up less than 17 percent of the total U.S. workforce, they comprise 21 percent of the front-line workforce, including 28 percent of food and agricultural production (McNicholas and Poydock 2020). Hispanic and Black Americans are also much more likely than White Americans to live in crowded housing conditions, in multigenerational households, and in population-dense urban areas (Godoy and Wood 2020). In the Navajo Nation, which has experienced some of the highest COVID-19 infection rates in the country, one-third of residents can't easily wash their hands because of lack of access to clean water (Yong 2020). Thus, the ability to follow recommended social distancing and hygiene guidelines is itself a racial privilege.

Second, a robust body of research has long documented that racial and ethnic minority groups in the United States are more likely than White people to suffer from long-term chronic health conditions such as high blood pressure, diabetes, and obesity, even after controlling for individual behavioral factors. Health experts refer to these patterns as "weathering processes" that culminate from lifelong stressors caused by experiencing discrimination. In other words, racism literally gets "under the skin" (Das 2013). These underlying conditions increase the chance that an individual will suffer serious illness or death after contracting COVID-19.

Finally, minorities have less access, compared to White Americans, to high-quality health care. Racial stereotypes contribute to minorities receiving poorer-quality care than White patients—for example, having doctors refuse to offer pain treatments because of the erroneous stereotype that Black individuals have a higher threshold for pain (Tello 2017). As a result, many Black Americans avoid going to the doctor and are less likely to seek preventative care. Unequal access also occurs at the community level. After the Civil War, Jim Crow laws led most hospitals in the South to be built far away from Black communities. Today, former slave states have among the lowest investments in public health and the highest proportions of Black population (Yong 2020).

The combination of surging cases of COVID-19 and racial justice protests across the country in the summer of 2020 prompted dozens of state and local governments, along with the American Medical Association (AMA) and the American Academy of Pediatrics (AAP), to declare racism a public health crisis (Vestal 2020). In a statement to the U.S. House of Representatives in June 2020, the American Medical Association pleaded for more resources to study and analyze the racial and socioeconomic disparities in COVID-19, saying the "pandemic has revealed starkly disproportionate impact of the virus on communities of color. The causes … are rooted in this country's historical and structural racism and the social, economic, and health inequities that have resulted, and continue to result, in adverse health outcomes" (American Medical Association [AMA] 2020).

and the Image of American Democracy, Mary Dudziak (2000) links the 1965 passage of U.S. civil rights legislation to the United States' efforts to win the Cold War. Following World War II, international public opinion was critical of the extreme racial inequality in the United States. Racial discrimination was damaging to the United States' credibility as a democracy and to U.S. foreign relations. Although some legislators who supported the passage of civil rights legislation wanted to end the injustice of discrimination, they were also responding to international pressure and seeking to bolster an image of the United States as a democracy and world leader.

The structural-functionalist perspective considers how aspects of social life are functional or dysfunctional—that is, how they contribute to or interfere with social stability. Racial and ethnic inequality are functional in that keeping minority groups in a disadvantaged position ensures that there are workers who will do menial jobs for low pay. Most sociologists emphasize the ways in which racial and ethnic inequality are dysfunctional—a society that practices discrimination fails to develop and utilize the resources of minority members (Williams and Morris 1993). Prejudice and discrimination aggravate social problems, such as crime and violence, war, unemployment and poverty, health problems, family problems, urban decay, and drug use—problems that cause human suffering as well as impose financial burdens on individuals and society. Picca and Feagin (2007) explain:

> [T]he system of racial oppression in the United States affects not only Americans of color but white Americans and society as a whole. ... Whites lose when they have to pay huge taxes to keep people of color in prisons because they are not willing to remedy patterns of unjust enrichment and ... to pay to expand education, jobs, or drug-treatment programs that would be less costly. They lose by driving long commutes so they do not have to live next to people of color in cities. ... They lose when white politicians use racist ideas and arguments to keep from passing legislation that would improve the social welfare of all Americans. Most of all, whites lose ... by not having in practice the democracy that they often celebrate to the world in their personal and public rhetoric. (p. 271)

@Official MLK3

It has been 57 years since my father gave his "I Have a Dream" speech, but we still have so much justice to demand. We must demilitarize the police, dismantle mass incarceration, and declare as determinedly as we can that Black Lives Matter. #BLM

-Martin Luther King III

The structural-functionalist analysis of manifest and latent functions also sheds light on issues of race and ethnic relations. For example, the manifest function of the civil rights legislation in the 1960s was, in part, to improve conditions for racial minorities. However, civil rights legislation produced an unexpected negative consequence, or *latent dysfunction*. Because civil rights legislation supposedly ended racial discrimination, White Americans were more likely to blame Black Americans for their social disadvantages and thus perpetuate negative stereotypes such as "blacks lack motivation" and "blacks have less ability" (Schuman and Krysan 1999).

Conflict Perspective

The conflict perspective examines how competition over wealth, power, and prestige contributes to racial and ethnic group tensions. Consistent with this perspective, the "racial threat" hypothesis views White racism as a response to perceived or actual threats by minorities to White people's economic well-being or cultural dominance.

For example, between 1840 and 1870, large numbers of Chinese immigrants came to the United States to work in mining (the California Gold Rush of 1848), railroads (the transcontinental railroad, completed in 1869), and construction. As Chinese workers displaced White workers, anti-Chinese sentiment rose, resulting in increased prejudice and discrimination and the eventual passage of the *Chinese Exclusion Act of 1882*, which restricted Chinese immigration until 1924. More recently, White support for Proposition 209—a 1996 resolution passed in California that ended state affirmative action programs—was higher in areas with larger Latino, African American, or Asian American populations, even after controlling for other factors (Tolbert and Grummel 2003). In other words, opposition to affirmative action programs that help minorities was higher in areas with greater racial and ethnic diversity, suggesting that Whites living in diverse areas felt more threatened by the minorities.

Racial resentment has been offered as one explanation of Donald Trump's election to the presidency in 2016. Researchers using data from American National Election Studies surveys showed who scored high on measures of racial resentment—for example, agreeing with statements that racial and ethnic minorities get more than they deserve—were more strongly supportive of Trump as a candidate (Abramowitz and McCoy 2019). Those who voted for Trump were also more likely than those who didn't to hold anti-immigration views and to express that their family finances and economic opportunities were worse than in the past. The authors conclude that the "deep pessimism evinced by so many of Trump's supporters appears to be based largely on unhappiness with nation's changing demographics and values" (p. 155).

Conflict theorists also suggest that racial antagonisms exist because it serves the interests of the wealthy and powerful. By fostering negative attitudes toward minorities and maintaining racial and ethnic tensions among workers, there is less chance that workers will join forces to advance their own interests at the expense of the capitalists. The "haves" perpetuate racial and ethnic tensions among the "have-nots" to deflect attention away from their own greed and exploitation of workers.

One way this exploitation occurs is through the existence of a surplus labor force—that is, by having more workers than are needed. A surplus labor force ensures that wages will remain low because someone is always available to take a disgruntled worker's place. Minorities who are disproportionately unemployed serve the interests of the business owners by providing surplus labor, keeping wages low, and consequently enabling them to maximize profits.

Symbolic Interactionist Perspective

The symbolic interactionist perspective focuses on the social construction of race and ethnicity—how we learn conceptions and meanings of racial and ethnic distinctions through interaction with others—and how meanings, labels, and definitions affect racial and ethnic groups and intergroup interaction. We have already explained that contemporary race scholars agree that there is no scientific, biological basis for racial categorizations. However, people have learned to think of racial categories as real, and, as the *Thomas theorem* suggests, if things are defined as real, they are real in their consequences. Ossorio and Duster (2005) explain:

> People often interact with each other on the basis of their beliefs that race reflects physical, intellectual, moral, or spiritual superiority or inferiority. ... By acting on their beliefs about race, people create a society in which individuals of one group have greater access to the goods of society—such as high-status jobs, good schooling, good housing, and good medical care—than do individuals of another group. (p. 119)

The labeling perspective directs us to consider how negative stereotypes affect minorities. **Stereotypes** are exaggerations or generalizations about the characteristics and behavior of a particular group. Negative stereotyping of minorities can lead to a self-fulfilling prophecy—a process in which a false definition of a situation leads to behavior that, in turn, makes the originally falsely defined situation come true (see also Chapter 8). Marger (2012) provides the following example of how the negative stereotype of Blacks as less intelligent than Whites can lead to a self-fulfilling prophecy:

> If blacks are considered inherently less intelligent, fewer community resources will be used to support schools attended primarily by blacks on the assumption that such support would only be wasted. Poorer-quality schools, then, will inevitably turn out less capable students, who will score lower on intelligence tests. The poorer performance on these tests will "confirm" the original belief about black inferiority. Hence, the self-fulfilling prophecy. (p. 15)

stereotypes Exaggerations or generalizations about the characteristics and behavior of a particular group.

Even stereotypes that appear to be positive can have negative effects. The view of Asian Americans as a "model minority" involves the stereotypes of Asian Americans as excelling in academics and occupational success. These stereotypes mask the struggles and discrimination that many Asian Americans experience and also put enormous pressure on Asian American youth to live up to the social expectation of being a high academic achiever (Tanneeru 2007).

The symbolic interactionist perspective also draws attention to the importance of symbols in race relations. One such symbol is the Confederate flag, which many Americans find offensive because it symbolizes slavery, racism, and oppression, while others defend the Confederate flag as a symbol of Southern pride. Following the 2015 fatal shootings of nine Black churchgoers in Charleston, South Carolina, lawmakers in that state passed a bill to remove the Confederate flag from the state capitol. Ongoing debates about monuments to Confederate soldiers and military bases named after Confederate leaders further highlight the power of symbols in understanding racial identity. (See this chapter's *The Human Side* feature.)

Zach D Roberts/NurPhoto/Getty Images

Symbolic interactionists focus on the ongoing debates over the removal of Confederate monuments as an issue of the divergent meanings that people of different backgrounds attribute to their symbolism. Protestors in Richmond, Virginia, projected the image of George Floyd, a Black man who died after being restrained by police, onto a monument of Confederate General Robert E. Lee.

The symbolic interactionist perspective is concerned with how individuals learn negative stereotypes and prejudicial attitudes through language. Different connotations of the colors white and black, for example, may contribute to negative attitudes toward people of color. The white knight is good, and the black knight is evil; angel food cake is white, and devil's food cake is black. Other negative terms associated with black include *black sheep*, *black plague*, *black magic*, *black mass*, *blackballed*, and *blacklisted*. The continued use of derogatory terms for racial and ethnic groups confirms the power of language in perpetuating negative attitudes toward minority group members. For example, advocates for immigrant rights suggest that the terms *illegal aliens* and *illegal immigrants* are derogatory and stigmatize and criminalize people rather than their actions. In 2013, the Associated Press (AP) news agency announced it would not use the term *illegal immigrant*. AP journalists are now instructed to "use *illegal* only to refer to an action, not a person: *illegal immigration*, but not *illegal immigrant* (unless the term is used in direct quotations)" (Colford 2013).

What's the difference between *black* and *Black*? In the wake of widespread racial justice protests in the summer of 2020, multiple news outlets, including *The New York Times* and the Associated Press, announced a change to their style guidelines: the word *Black*, when used in reference to the racial identity, should be treated as a proper noun and therefore capitalized (Coleman 2020). The rationale for the change was that *Black* referred to a distinct cultural identity that arose because the history of slavery in the United States cut Black Americans off from connections to their African heritage and from full inclusion into a cultural identity as Americans. One senior staff editor who consulted *The New York Times* on the change said she thought "the capital B makes sense as it describes a race, a cultural group, and that is very different from a color in a box of crayons" (p. 1). Do you think treating Black racial identity as a proper noun will change the way people think about the symbolism of blackness?

What *do you think?*

Experiencing Racism at School

In the following testimonials, people of color share stories of their personal experiences with racism at school, and the impact those experiences had on them.

The story that I wanted to share was I went to Clemson University. My senior year, 2007… . [a] group of white kids threw a Martin Luther King party. … A lot of white kids showed up with their skin darkened in some way or with, you know, other stereotypical things like foil on their teeth or 40s in their hand or pads in their butt to make their butt bigger. And it got out on Facebook. … Like, you're—often, you might be the only black person in classes there. And the school—they barely wanted to do anything at all. Let's keep the peace was the bigger response. … I'm trying to be a good member of society. And at every turn, it's like you're reminded that you're nothing. And you're always going to be thought of as a joke. I guess every time something like this happens, it takes a chunk out of you.

~

I'm Hmong American. My siblings —my younger sister, my older sister and my little brother—we made up basically the Asian student body. You know, at first, I didn't really think about it. I just felt like it was school. Then the other children started singing songs. And then they would make these gestures. They would slant their eyes up. And then they would slant it down. … And they would do that every single day. When we're walking, they would slant their eyes and say, can you see? Can you see? And, you know, you're supposed to go to school. And you're supposed to feel safe. And I didn't feel safe. I felt tormented. And now looking back, I just feel—I feel angry. I've never, ever spoken about this for the past two decades—two and a half decades.

~

I'm a member of the Pawnee tribe. And I also am Ponca and Kiowa. You know, growing up in Oklahoma, what we would do is we would celebrate land run days. Basically, they take you out on schoolyard. And you got this little stick with a flag on it. And you just go stake your claim. You know, the girls have little bonnets on. There's these little wagons. They look like covered wagons. I mean, it's just nonsense. We would celebrate Thanksgiving, you know? And you either put the little, black belt buckle on or the hat. Or you do a headdress. … I can remember as a child picking … . I'd be the pilgrim, you know? To demean your own culture like that—a headdress has a meaning. In Ponca culture, there's only a few people that can wear headdresses. … It's a symbol of authority and respect. But it's just been so stereotyped into a joke. You know, you've got kids cutting out construction paper and putting it on their head and just kind of going around making fools of Indians. … [T]he harmful aspect of that is no one ever takes it seriously because of that.

SOURCE: Balaban 2019.

Bonilla-Silva (2012) gives examples of what he calls the "racial grammar" that shapes how we see or don't see race and how we frame matters as racial or not race related. Racial grammar involves conveying and perpetuating racial meanings not only through what we say but also by what we *don't* say. For example, "[I]n the USA one can talk about HBCUs (historically black colleges and universities), but not about HWCUs (historically white colleges and universities) or one can refer to black movies and black TV shows but not label movies and TV shows white when in fact most are" (Bonilla-Silva 2012, p. 173).

In the next section, we explore the concepts of racism and prejudice in more depth and discuss ways in which socialization and the media perpetuate negative stereotypes.

Prejudice and Racism

prejudice Preconceived opinion or bias that could be positive or negative.

racism The belief that race accounts for differences in human character and ability and that a particular race is superior to others.

Prejudice refers to a preconceived opinion or bias that could be negative or positive. Prejudice can be directed toward individuals of a particular religion, sexual orientation, political affiliation, age, physical appearance, social class, sex, race, or ethnicity. Most references to prejudice refer to negative opinions and biases, as these tend to be most problematic.

Racism is generally defined as the belief that race accounts for differences in human character and ability and that a particular race is superior to others. According to this general definition, anyone of any race can be racist. However, many scholars argue that an essential element of racism is power, and so only the dominant social group—in the

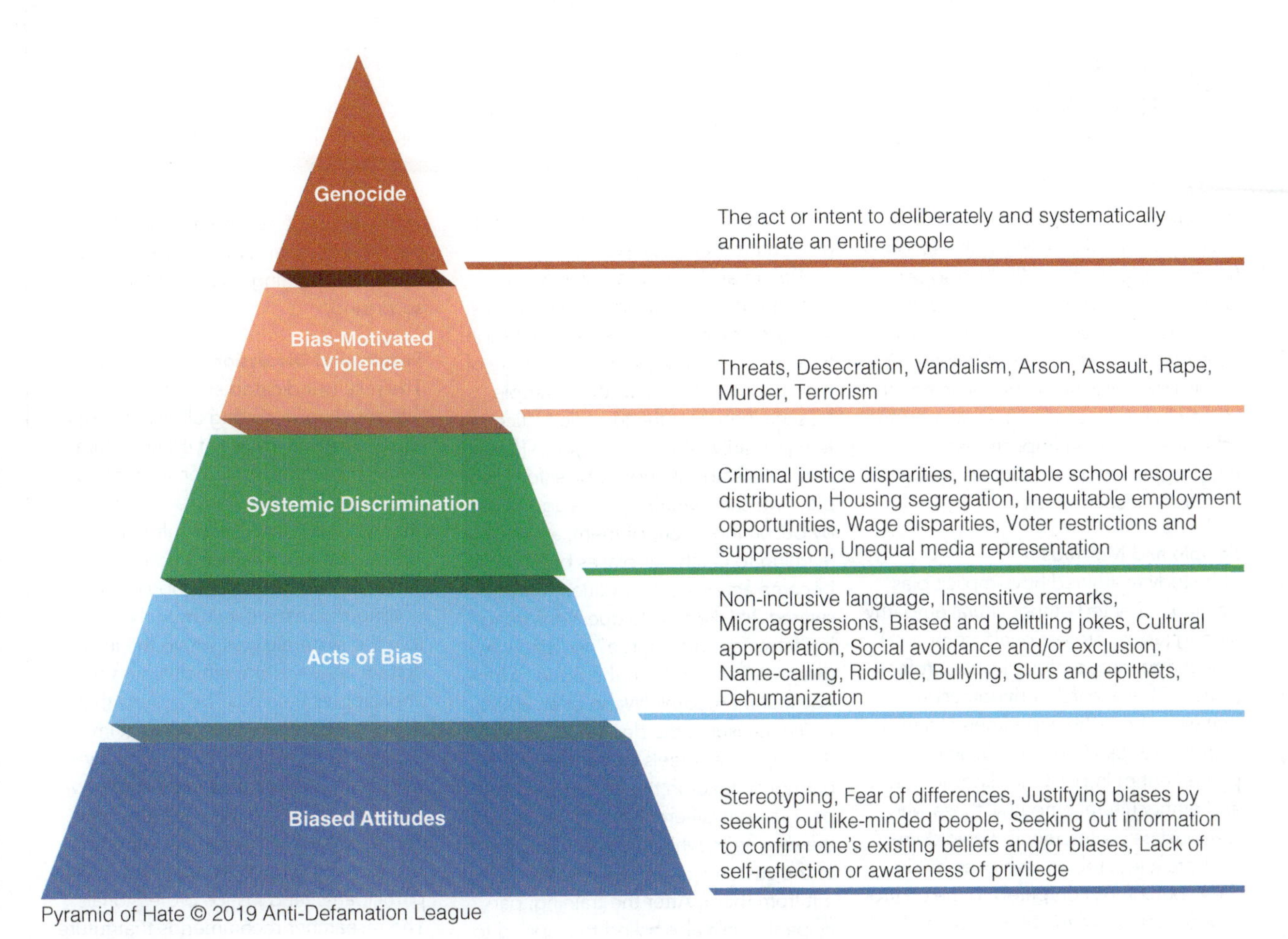

Figure 9.6 Pyramid of Hate
SOURCE: Allport 1954 and Anti-Defamation League (ADL) 2019.

United States, that would be White people—can be racist. According to the "racism = prejudice + power" definition, Black individuals who view White people as inferior are prejudiced but not racist because Black people as a group do not control the institutions that affect the opportunities and resources of White people. The term **institutional racism** refers to the systematic distribution of power, resources, and opportunity in ways that benefit the White community and disadvantage minorities (see also "Individual versus Institutional Discrimination" discussed later in this chapter).

Forms of Racism and Prejudice

Compared with traditional, "old-fashioned" prejudice and racism, which are blatant, direct, and conscious, contemporary forms are often subtle, indirect, and unconscious.

institutional racism The systematic distribution of power, resources, and opportunity in ways that benefit White people and disadvantage minorities.

implicit prejudice A prevalent form of racial bias over which a person has little or no conscious awareness or control.

Implicit Prejudice. **Implicit prejudice** is a prevalent form of racial bias over which a person has little or no conscious awareness or control (Sinclair et al. 2014). You can go online to www.implicit.harvard.edu and take the Implicit Association Test (IAT), which measures implicit racial prejudices. After journalist Chris Mooney (2014) took the IAT, he wrote:

> Taking the IAT made me realize that we can't just draw some arbitrary line between prejudiced people and unprejudiced people, and declare ourselves to be on the side of the angels. Biases have slipped into all of our brains. And that means we all have a responsibility to recognize those biases—and work to change them.

social problems RESEARCH UP CLOSE

Using Implicit Bias Training to Reduce Racial Disparities

Implicit bias is a key component in explaining racial disparities in the school-to-prison pipeline (Fix 2020; also see Chapter 8 of this text, "Problems in Education"). In order to find effective strategies to address these disparities, Johns Hopkins University public health researcher Rebecca L. Fix evaluated the effectiveness of an implicit bias training program for professionals working within the justice system.

Sample and Methods

This study examined how implicit bias training impacted ethnocultural empathy among justice professionals, compared to a control group of nonjustice professionals. The researcher hypothesized that while both the experimental and control groups would exhibit an improvement of implicit bias knowledge and ethnocultural empathy, those who were justice professionals would demonstrate a less robust improvement in these outcomes compared to the control group because of the high degree of systemic bias within the justice system.

The researcher recruited participants from a group of professionals who were prescribed implicit bias training by their workplace. Most were White (92 percent) and female (74 percent). The experimental group consisted of 243 justice professionals, primarily working in pretrial court, police department, and specialty court settings. The control group consisted of 274 professionals in nonjustice settings, the majority of whom worked in health care, business, retail, and education. The participants were informed about the voluntary nature of the study and their right to withdraw at any time. They were not offered any compensation for their participation, as most were state government employees.

Before and after the training, the participants were asked to complete a survey that asked 18 true-false questions to assess the participant's knowledge of facts about implicit bias (for example, "discrimination in the juvenile justice system primarily affects white youth [false]") and beliefs about implicit bias (for example, "we can manage microaggressions by becoming aware of them, and slowly learning to catch our biases before they become actions"). In addition, they answered 24 Likert-scale questions (ranging from "strongly agree" to "strongly disagree") drawn from the Scale of Ethnocultural Empathy (Wang et al. 2003), which measures the degree to which the respondent feels and expresses empathy, can take on the perspectives of people from different backgrounds, and the degree to which they are accepting of the customs of groups who are different from them. After the training, participants were also asked to respond to questions about what they appreciated about the training and what they found most surprising.

The training was led by a Black man and consisted of lectures, discussion, case study exercises, activities, and completion of the Implicit Association Test.

Pre- and post-training surveys were compared using one-way ANOVA statistical testing. This measure indicates whether the training had a significant effect on participant's ethnocultural empathy and knowledge of implicit bias based on the participant's status as a justice or nonjustice professional, controlling for workplace setting, age, gender, and race. Chi-square tests were also used to test whether working within a justice setting and gender impacted what participants appreciated about the training and what they found surprising.

Results and Discussion

The results indicated statistically significant change among all participants for all three measures of ethnocultural empathy overall; however, justice professionals did not exhibit a significant change in acceptance of cultural differences. The results among nonjustice professionals were more robust, supporting the researcher's hypothesis that the systemic bias within the justice system weakens the effectiveness of implicit bias training. The groups that experienced the most significant improvement in ethnocultural empathy were men, Whites, and nonjustice professionals. Women, Black participants, and nonjustice professionals reported liking the training more than men, White participants, and justice professionals. The researcher recommends that future implicit bias training should focus more extensively on acceptance of cultural differences among justice professionals.

Overall, this study provides evidence that implicit bias training can help increase ethnocultural empathy among all professionals and provide an important first step in addressing issues of racial bias within the juvenile justice system. By becoming aware of one's own unconscious biases and learning strategies to identify and address those biases in everyday interactions, individual professionals have the ability to reduce racial disparities in a wide variety of social institutions.

SOURCE: Fix 2020.

This chapter's *Social Problems Research Up Close* feature examines the effectiveness of using implicit association training to reduce racial bias among people working in criminal justice professions.

aversive racism A subtle form of prejudice that involves feelings of discomfort, uneasiness, disgust, fear, and pro-White attitudes.

Aversive Racism. **Aversive racism** represents a subtle, often unintentional, form of prejudice exhibited by many well-intentioned White Americans who possess strong egalitarian values and who view themselves as unprejudiced. The negative feelings that aversive racists have toward Black individuals and other minority groups are not feelings of hostility or hate but rather feelings of discomfort, uneasiness, disgust, and sometimes fear (Gaertner and Dovidio 2000). Aversive racists may not be fully aware that they harbor these negative racial feelings; indeed, they disapprove of individuals who are prejudiced and would feel falsely accused if they were labeled prejudiced. "Aversive racists find blacks 'aversive,' while at the same time find any suggestion that they might be prejudiced 'aversive' as well" (Gaertner and Dovidio 2000, p. 14).

Another aspect of aversive racism is the presence of pro-White attitudes, as opposed to anti-Black attitudes. In several studies, respondents did not indicate that Blacks were worse than Whites, only that Whites were better than Blacks (Gaertner and Dovidio 2000). For example, Blacks were not rated as being lazier than Whites, but Whites were rated as being more ambitious than Blacks. Gaertner and Dovidio (2000) explain that "aversive racists would not characterize Blacks more negatively than Whites because that response could readily be interpreted by others or oneself to reflect racial prejudice" (p. 27). Compared with anti-Black attitudes, pro-White attitudes reflect a more subtle prejudice that, although less overtly negative, is still racial bias.

Color-blind Racism. Many White people claim that they don't see color, just people, and that minorities (especially Blacks) are responsible for whatever racial tension exists in the United States (Bonilla-Silva 2013). **Color-blind racism** is based on the belief that paying attention to race is, itself, racism and that therefore people should ignore race. Color-blindness assumes we are living in a postracial world where race no longer matters, when in reality, race continues to be a significant issue. Color-blindness is a form of racism because it prevents acknowledgment of privilege and disadvantage associated with race, and therefore it allows the continuation of cultural and structural forms of racial bias. A college student at Washington University explains how she learned the color-blind mindset (Frieden 2013):

> Growing up as a white girl in New Hampshire, I was raised with this kind of a "color-blind" mindset: a mindset that says that race should be ignored, and that racism exists because some people simply refuse to ignore race like they should. (p. 1)

Interestingly, research has found that "people who claim to be 'color-blind' and go to great pains to avoid talking about race during social interactions, are in fact perceived as more prejudiced by black observers than people who openly acknowledge race" (Apfelbaum 2011).

Learning to Be Prejudiced: The Role of Socialization and the Media

Psychological theories of prejudice focus on forces within individuals that give rise to prejudice. For example, the frustration-aggression theory of prejudice (also known as the scapegoating theory) suggests that prejudice is a form of hostility that results from frustration. According to this theory, minority groups serve as convenient targets of displaced aggression. The authoritarian-personality theory of prejudice suggests that prejudice arises in people with a certain personality type. According to this theory, people with an authoritarian personality—who are highly conformist, intolerant, cynical, and preoccupied with power—are prone to being prejudiced.

Rather than focus on individuals, sociologists focus on social forces that contribute to prejudice. Earlier, we explained how intergroup conflict over wealth, power, and prestige gives rise to negative feelings and attitudes that serve to protect and enhance dominant group interests. Prejudice is also learned through socialization and the media.

Learning Prejudice through Socialization. In the socialization process, individuals adopt the values, beliefs, and perceptions of their family, peers, culture, and social groups. Prejudice is taught and learned through socialization, although it need not be taught directly and intentionally. White parents who teach their children to not be prejudiced yet live in an all-White neighborhood, attend an all-White church, and have only White friends may be indirectly

color-blind racism A form of racism that is based on the idea that overcoming racism means ignoring race, but color-blindness is, in itself, a form of racism because it prevents acknowledgment of privilege and disadvantage associated with race, and therefore allows the continuation of institutional forms of racial bias.

AF archive/Alamy Stock Photo

Because the comic book character Storm is described as Nigerian and portrayed with dark skin, some media critics argued that the choice to cast Halle Berry in the role for the *X-Men* films was an example of colorism.

elections experts. Furthermore, in an August 2020 interview, Trump stated his intent to block Congressional requests for increased funding for the U.S. Postal Service as a strategy to prevent mail-in voting (Associated Press 2020). The disproportionate impact of COVID-19 on minority communities (see this chapter's *World in Quarantine* feature), combined with more restrictive access to in-person polling options in minority-majority communities, means that limitations on mail-in ballots will disproportionately impact voting access to racial and ethnic minorities (Shapiro and Knight 2020).

Colorism

Colorism refers to prejudice or discrimination based on skin tone. Discussions of colorism tend to focus on the preference for lighter skin among Blacks, framing colorism as intraracial "Black-on-Black" prejudice and discrimination. But Whites also engage in colorism: One study found that Whites are more likely to view African Americans and Hispanics with lighter skin as intelligent (Hannon 2015). It is important to recognize White colorism as a variation of White racism. Hannon (2015) explains:

> Consider, for example, a hypothetical case where a white employer discriminates against darker skinned African Americans for customer relations positions. Claiming racism would be insufficient; such a claim could be countered with evidence of past (lighter skinned) African American hires. (p. 19)

In one recent poll, 64 percent of Hispanics with darker skin reported experiencing discrimination on the basis of their race or ethnicity, compared with 50 percent of Hispanics with lighter skin (Gonzalez-Barrera 2019).

What *do you* think?

Although representation of non-White actors in film and television has increased in recent decades, critics point out that performers with lighter skin tones are typically chosen for the most prominent roles, even when the characters they are based on had darker skin (Onyejiaka 2017). Recently, controversies have emerged over the casting of Afro-Latinx actress Zoe Saldana as singer Nina Simone and the castings of Halle Berry and Alexandra Shipp as the comic book character Storm (who is described as Nigerian in the original comic books) in the *X-Men* films. For some cultural critics, these casting choices indicate "that Hollywood still overwhelmingly believes that a black woman must possess non-black ancestry or features to be considered beautiful or valuable" (p. 1). Do you think skin tone should be taken into consideration when casting for a film or television role?

Employment Discrimination

In a months-long search for a job, José Zamora sent out between 50 and 100 résumés a day, applying online to every job he found that he felt qualified for. Day after day he got no responses. So he decided to drop one letter in his name: *José* became *Joe*. A week after he began applying for jobs as "Joe," the responses started pouring in with requests for interviews. In a Buzzfeed video, Zamora explained, "Sometimes I don't even think people know or are conscious or aware that they're judging—even if it's by a name—but I think we do it all the time" (quoted in Matthews 2014).

colorism Prejudice or discrimination based on skin tone.

Despite laws against it, discrimination against minorities occurs today in all phases of the employment process, from recruitment to interview, job offer, salary, promotion, and firing decisions. One way that sociologists study employment discrimination is through audit studies, in which researchers measure differences in callback rates between job applications that are identical for all information except for one factor, such as race. One audit study found that otherwise identical applications with White-sounding names (such as *Emily* and *Greg*) were 50 percent more likely to receive a callback than those with Black-sounding names (such as *Lakisha* and *Jamal*) (Bertrand and Mullainatham 2004). Another study examining the relationship between race and criminal record found that White applicants with no criminal record were the most likely to be called back for an interview (34 percent) and that Black applicants with a criminal record were the least likely to be called back (5 percent). But surprisingly, White applicants with a criminal record (17 percent) were more likely to be called back for a job interview than were Black applicants *without* a criminal record (14 percent). The researcher concluded that "the powerful effects of race thus continue to direct employment decisions in ways that contribute to persisting racial inequality" (Pager 2003, p. 960).

An analysis of 24 different employment discrimination studies representing over 55,000 job applications conducted from 1989 to 2014 found that racial discrimination in hiring practices has exhibited no change over the 25-year period of the study (Quillian et al. 2017). These studies consistently found that White job applicants received 36 percent more callbacks than Black applicants and 24 percent more callbacks than Latino applicants, even when controlling for applicant education, gender, study method, occupational groups, and local labor market conditions.

Discrimination in hiring may be unintended. For example, many businesses rely on their existing employees to refer new recruits when a position opens up. Word-of-mouth recruitment is inexpensive and efficient; some companies offer bonuses to employees who bring in new recruits. But this traditional recruitment practice tends to exclude minority workers because they often do not have a network of friends and family members in higher positions of employment who can recruit them (Schiller 2004).

Employment discrimination contributes to the higher rates of unemployment and lower incomes of Black and Hispanic Americans compared with those of Whites (see Chapter 6 and Chapter 7). Lower levels of educational attainment among minority groups account for some but not all of the disadvantages they experience in employment and income. As shown in Figure 9.7, mean earnings of Whites are higher than those for Blacks, Hispanics, and Asians

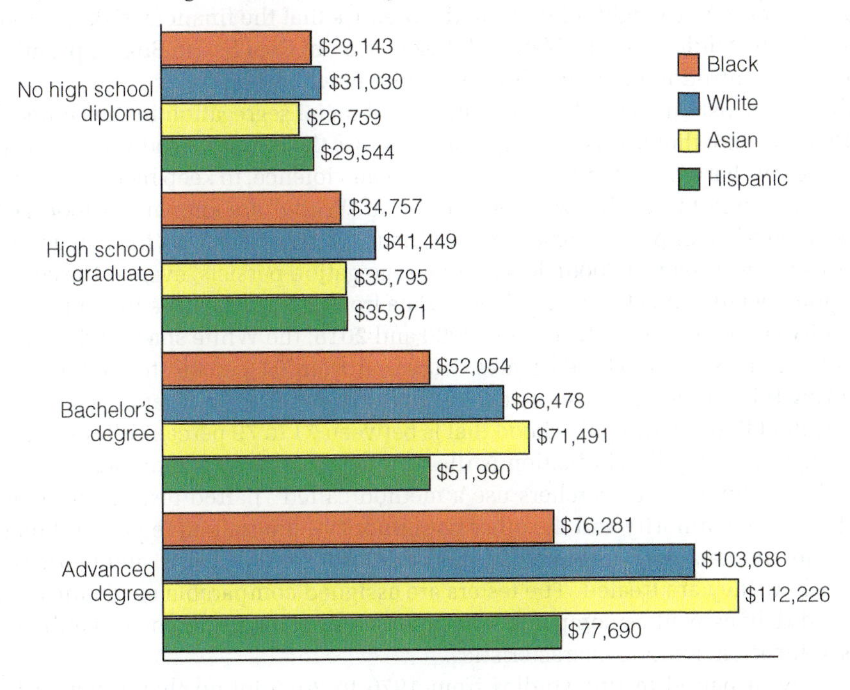

Figure 9.7 Median Annual Earnings of Workers 25 Years and over by Educational Attainment, Race, and Hispanic Origin, 2018

Note: Black, White, and Asian race categories are all non-Hispanic. Hispanic includes all races. Only includes those who worked full-time, year-round.

SOURCE: Current Population Survey (CPS) (2020).

at most levels of educational attainment. Furthermore, the racial gap in earnings between White and Asian workers compared to Black and Hispanic workers widens as educational attainment increases. Rather than education serving as the great equalizer, these data indicate that other factors are at play, such as "minorities not having equal access to the informal professional networks that often lead to job opportunities, and/or discrimination against racial and ethnic minorities" (Shierholz et al. 2013, p. 11).

Workplace discrimination also includes unfair treatment and harassment. The Equal Employment Opportunity Commission (EEOC) defines race discrimination as "treating someone (an applicant or employee) unfavorably because he/she is of a certain race or because of characteristics associated with race (such as hair texture, skin color, or certain facial features)" (2020, p. 1). In 2019, the EEOC received nearly 24,000 filings of workplace racial discrimination.

What do you think?

Chastity Jones, a Black woman from Alabama, filed an EEOC racial discrimination complaint when her job offer was rescinded after she refused the hiring manager's demands to get rid of her dreadlocks. The EEOC's lawsuit argued that the manager's request violated anti-discrimination law because dreadlocks are a hairstyle that is "psychologically and culturally" associated with Black Americans. In 2018, the 11th U.S. Circuit Court of Appeals ruled against Ms. Jones's lawsuit, finding that the company's grooming policy was "race neutral" and that hair is not a "racial characteristic." The ruling prompted several states to initiate legislation that would ban discrimination on the basis of hair texture or style (Kaur 2020). Do you think schools and employers should be allowed to have grooming policies that ban specific types of hairstyles, such as dreadlocks?

Housing Discrimination and Segregation

Before the *1968 Fair Housing Act* and the *1974 Equal Credit Opportunity Act*, discrimination against minorities in housing and mortgage lending was common. Banks and mortgage companies commonly engaged in "redlining"—the practice of denying mortgage loans in minority neighborhoods on the premise that the financial risk was too great, and the ethical standards of the National Association of Real Estate Boards prohibited its members from introducing minorities into White neighborhoods.

Redlining is just one part of the long history of racial segregation in the United States. From 1890 to 1968, thousands of communities across the country used various means, including the law, harassment, intimidation, and even violence, to keep racial and ethnic minorities out. Communities that were purposely "all-White" are known as **sundown towns** because some of them posted signs warning Black visitors "Don't let the sun go down on you in this town" (Loewen 2006). Residential segregation persists, even as the country becomes more racially and ethnically diverse. Data from the U.S. Census Bureau's American Community survey found that, between 2000 and 2018, the White share of the 200 largest metropolitan areas in the United States decreased from 64 percent to 55 percent. However, during this period there was no change in the residential segregation gap; the typical White resident lives in a neighborhood that is between 71 to 79 percent White (Frey 2020).

Although housing discrimination is illegal today, it is not uncommon. To assess discrimination in housing, researchers use a method called "paired testing" in which two individuals—one minority and the other nonminority—are trained to pose as home seekers, and they interact with real estate agents, landlords, rental agents, and mortgage lenders to see how they are treated. The testers are assigned comparable or identical income, assets, and debt as well as comparable or identical housing preferences, family circumstances, education, and job characteristics.

sundown towns Communities that are purposely "all-White" and that have used various means to deliberately keep racial and ethnic minorities out.

A review of paired testing studies from 1976 to 2016 found that although housing discrimination has declined overall during the 30-year period of the study, racial discrimination remains common. For example, when compared to equally qualified White applicants, the probability of receiving a response to an initial inquiry is 8 percentage points lower among Blacks, 4 percentage points lower among Hispanics, and 3 percentage points lower among Asians (Quillian et al. 2020). Black and Hispanic borrowers were also more likely than equally qualified White borrowers to be rejected for a home loan and were more likely to receive a high-cost mortgage.

Racial disparities in homeownership and property values are also a form of housing discrimination. For example, one couple in Jacksonville, Florida, claims they experienced racial discrimination when they applied for refinancing for their home in a predominantly White neighborhood. After receiving an appraisal that was more than $100,000 less than what similar homes in the neighborhood sold for, Abena Horton (who is Black) and her husband Alex (who is White) suspected racism was playing a role (Kamin 2020). The couple requested a second appraisal, but this time, they removed all signs of Blackness: they took down family photos and holiday cards except for those depicting White family and friends, removed artwork featuring Black subjects and books by Black authors. Abena Horton even removed herself and their biracial son—leaving Mr. Horton alone to greet the appraiser. The second appraisal came in $135,000 higher than the original.

Educational Discrimination and Segregation

Both institutional discrimination and individual discrimination in education negatively affect racial and ethnic minorities and help to explain why minorities (with the exception of Asian Americans) tend to achieve lower levels of academic attainment and success (see also Chapter 8). Institutional discrimination is evidenced by inequalities in school funding—a practice that disproportionately hurts minority students (Kozol 1991). Because minorities are more likely than Whites to live in economically disadvantaged areas, they are more likely to go to schools that receive inadequate funding. Inner-city schools, which serve primarily minority students, receive less funding per student than do schools in more affluent, primarily White areas.

Another institutional education policy that is advantageous to Whites is the policy that gives preference to college applicants whose parents or grandparents are alumni. The overwhelming majority of alumni at the highest-ranked universities and colleges are White. Thus, White college applicants are the primary beneficiaries of these so-called legacy admissions policies. About 10–15 percent of students in most Ivy League colleges and universities are children of alumni, and legacy students have a much higher rate of admission than nonlegacy applicants ("Ivy League Legacy Admissions" 2020). In 2004, Texas A&M University became the first public college to abandon its legacy admittance policy.

Minorities also experience individual discrimination in the schools as a result of continuing prejudice among teachers. For example, the Department of Education Civil Rights Data Collection (CRDC 2020) database shows that although Black students comprise 15.4 percent of all public school students, they make up 36.1 percent of school-related arrests, 40.6 percent of suspensions, 34.8 percent of expulsions, and 31 percent of referrals to law enforcement. White teachers' lower educational expectations of Black students and more negative evaluations of Black students' behavior may help explain the disproportionately high use of discipline against Black students (Sablich 2016). In one study of implicit bias among teachers, researchers found that not only are teachers disproportionately White (80 percent) compared to the general population, 77 percent of the teachers studied were found to hold anti-Black implicit bias (Stark et al. 2020). Furthermore, county-level analysis of teacher implicit bias indicates a strong association with the average level of anti-Black implicit bias among teachers and rates of school discipline among Black students (Chin et al. 2020).

Racial and ethnic minorities are also treated unfairly in educational materials, such as textbooks, which often distort the history and heritages of people of color (King 2000). For example, Columbus and his successors are portrayed in educational materials as navigators and discoverers rather than as participants in Native American genocide. Recently, a McGraw-Hill geography textbook adopted by the Texas Board of Education came under fire when a parent posted to social media a map of "patterns of immigration" from the text with a caption that read "the Atlantic slave trade brought millions of workers from Africa to the southern United States to work on agricultural plantations" (quoted in Isensee 2015). This chapter's *Human Side* feature provides accounts of how individuals' experiences with racism at school impacted them.

Finally, racial and ethnic minorities are largely isolated from Whites in a largely segregated school system. Black children are five times more likely than White children to attend a highly segregated school (defined as school that is majority non-White) and more than seven times more likely to attend a high-poverty school (Garcia 2020). School segregation is largely due to the persistence of housing segregation and the termination of court-ordered desegregation plans. Court-mandated busing became a means to achieve equality of education and school integration in the early 1970s, after the Supreme Court (in *Swann v. Charlotte Mecklenburg*) endorsed busing to desegregate schools. But in the 1990s, lower courts lifted desegregation orders in dozens of school districts (Winter 2003). And in 2007, the U.S. Supreme Court issued a landmark ruling, that race cannot be a factor in the assignment of children to public schools. Schools today are more segregated than they were in the 1970s, with New York and California having the highest percentages of students attending segregated schools (Frankenberg et al. 2019).

Racial Microaggressions

In the summer of 2020, publicist Autumn Lewis, fed up with the daily experience of hearing charged comments, started an Instagram account called @OverheardWhileBlack, where people could anonymously share stories of the "casual cruelty and pointed remarks black people endure everyday" (Women's Health Editors 2020). In academic literature, these everyday experiences of casual cruelty are referred to as **racial microaggressions**, defined as "brief and commonplace daily verbal, behavioral, or environmental indignities, whether intentional or unintentional, that communicate hostile, derogatory, or negative racial slights and insults toward the target person or group" (Sue et al. 2007, p. 271). Some of the microaggressions collected on @OverheardWhileBlack include:

- "I was told they understand the struggles of being black because they wrote a research paper about it in college."
- "You don't look anything like what you sound like on the phone. You're so well spoken."
- "Good thing your parents didn't give you one of those super black names."
- "You'd be prettier if you relaxed your hair."
- "But your hair and the way you dress ... you're basically white."
- "No but where you *from from*?"
- "Wow this is your house? I always thought your family lived in the projects."
- "You probably only got in here because of affirmative action, though."
- "You're tall and brown, why aren't you playing ball?"

racial microaggressions Brief and commonplace daily verbal, behavioral, or environmental indignities, whether intentional or unintentional, that communicate hostile, derogatory, or negative racial slights and insults toward the target person or group.

Although Whites may experience racial microaggressions, they do so much less frequently than racial and ethnic minorities (see Table 9.4). Research has found a relationship between experiencing racial microaggressions and low self-esteem (Nadal et al. 2014).

TABLE 9.4 Experiences with Microaggressions, by Racial Group
Percent reporting experiencing each event "often" or "very often" in their day-to-day life in the past 12 months.

	Black Adults	Hispanic Adults	Asian Adults	White Adults
People acted as if they were better than you.	32%	21%	17%	10%
People acted as if they thought you were not smart.	25%	12%	9%	5%
You were treated with less courtesy than other people.	22%	8%	7%	4%
You were treated with less respect than other people.	20%	7%	5%	4%
People acted as if you were dishonest.	19%	4%	4%	1%
People acted as if they were afraid of you.	18%	3%	4%	2%
You received worse service than other people at restaurants or stores.	14%	4%	4%	2%

SOURCE: Gallup Organization 2020.

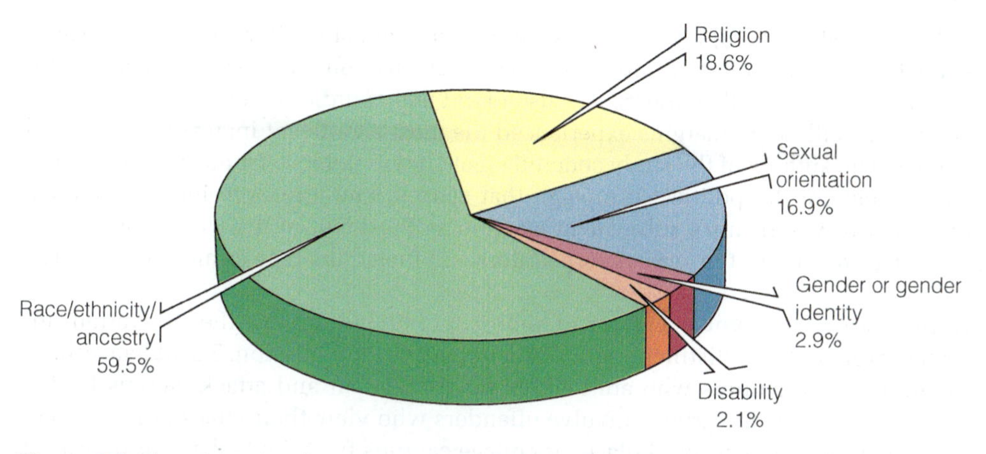

Figure 9.8 Hate Crime Incidence by Category of Bias, 2018 (rounded to nearest percent)
SOURCE: FBI 2020.

Hate Crimes

On August 3, 2019, 21-year-old Patrick Wood Crusius walked into a Walmart in El Paso, Texas, and opened fire with an assault rifle. The shooting left 22 people dead and 23 others injured. A federal grand jury indicted Crusius on charges of committing a **hate crime**—an unlawful act of violence motivated by prejudice or bias. Examples of hate crimes, also known as "bias-motivated crimes," include intimidation (e.g., threats), destruction of or damage to property, physical assault, and murder. The indictment alleges that shortly before the shooting, Crusius posted a document online in which he stated that his planned attack "is a response to the Hispanic invasion of Texas. They are the instigators, not me. I am simply defending my country from cultural and ethnic replacement brought on by the invasion" (quoted in Department of Justice 2020, p. 1).

hate crime An unlawful act of violence motivated by prejudice or bias.

Islamophobia Anti-Muslim and anti-Islam bias.

According to the Federal Bureau of Investigation (FBI), between 6,000 and 10,000 hate crimes occur in the United States each year. However, FBI hate crime data undercount the actual number of hate crimes because (1) not all U.S. jurisdictions report hate crimes to the FBI (reporting is voluntary), (2) it is difficult to prove that crimes are motivated by hate or prejudice, (3) law enforcement agencies shy away from classifying crimes as hate crimes because it makes their community "look bad," and (4) victims are often reluctant to report hate crimes to the authorities. The National Crime Victimization Survey reveals that more than half (54 percent in 2011–2015) of hate crimes are not reported to the police (Langton 2017).

Shay Horse/NurPhoto/Getty Images

In August 2017, hundreds of neo-Nazis, Alt-Right, and white supremacists gathered in Charlottesville, Virginia, for a rally dubbed Unite the Right. Richard Spencer, one of the neo-Nazi organizers of the rally, was a staunch supporter of Trump's 2016 campaign. He said that the Charlottesville rally "wouldn't have occurred without Trump. ... [T]he alt-right found something in Trump. He changed the paradigm and made this kind of public presence of the alt-right possible" (quoted in Daugherty 2019, p. 1).

FBI hate crime data reveal that most hate crimes are based on racial bias, primarily against Blacks, followed by Whites (see Figure 9.8). Most hate crimes motivated by religious bias target the Jewish religion, followed by Islam. After the terrorist attacks of September 11, 2001, hate crimes against individuals perceived to be Muslim or Middle Eastern increased significantly. Anti-Muslim and anti-Islam bias—commonly known as **Islamophobia**—was rekindled after Trump's election to the presidency. Using time series and geographic statistical analyses, researchers

Edwards and Rushin (2018) found a statistically significant surge in reported hate crimes across the United States after Trump's November 2016 election, even when controlling for alternative explanations. The researchers also found that counties that voted for President Trump by especially high margins experienced the most significant increases in reported hate crimes. The authors of the study conclude that "it was not just Trump's inflammatory rhetoric throughout the political campaign that caused hate crimes to increase. Rather, we argue that it was Trump's subsequent election as President of the United States that validated that rhetoric in the eyes of perpetrators and fueled the hate crime surge" (p. 1).

Motivations for Hate Crimes. Levin and McDevitt (1995) found that the motivations for hate crimes were of three distinct types: thrill, defensive, and mission. Thrill hate crimes are committed by offenders who are looking for excitement and attack victims for the "fun of it." Defensive hate crimes involve offenders who view their attacks as necessary to protect their community, workplace, or college campus from "outsiders" or to protect their racial and cultural purity from being "contaminated" by interracial marriage and childbearing. Mission hate crimes are perpetrated by white supremacist group members or other offenders who have dedicated their lives to bigotry. Hate groups known to engage in violent crimes include the Ku Klux Klan, the Identity Church Movement, the neo-Nazis, and the skinheads. The Southern Poverty Law Center (SPLC) identified 940 active hate group chapters in 2019—down from a peak of 1,018 in 2011 (SPLC 2020).

@rep johnlewis

Do not get lost in a sea of despair. Be hopeful, be optimistic. Our struggle is not the struggle of a day, a week, a month, or a year, it is the struggle of a lifetime. Never, ever be afraid to make some noise and get in good trouble, necessary trouble. #goodtrouble

-John Lewis

The FBI has recently warned that violence from white nationalist extremist groups is one of the most pressing national security threats facing the United States (McGarrity 2019). These groups are often associated with the term "alt-right" short for "alternative right" and include a range of groups and individuals on the extreme right who embrace racist ideology, white nationalism, and/or white supremacy. One of the most recent groups emerging within the broader alt-right movement is the "boogaloos"—a play on the term "big luau" in reference to the Hawaiian-patterned shirts group members often wear. Some individuals identifying with the boogaloo movement claim the group does not have racist motivations but rather identify with a staunchly libertarian and anti-government ideology. However, analyses of the online presence of boogaloo groups have shown strong anti-Black and anti-immigrant racial rhetoric, admiration for prominent neo-Nazi figures, and a desire to instigate racial conflict between anti-racist protestors and police in order to spark a racialized civil war that adherents believe will usher in a new age of white supremacist government (Miller 2020).

AP Images/Emilio Morenati

The Black Lives Matter movement has brought global attention to the issue of racial bias in policing. Waves of protest, in the United States and throughout the world, exploded in the summer of 2020 after the high-profile deaths of George Floyd and Breonna Taylor at the hands of police.

Hate on Campus. Amid the COVID-19 pandemic in the spring of 2020, Oklahoma City University held its graduation ceremony online, on the digital meeting platform Zoom. During the opening benediction, viewers saw their screens switch over to an image of a swastika, with the word *n****r* displayed (Campus Racial Incidents 2020). Racist Zoom-bombing is just one new example of the ways in which college campuses are the targets of racist propaganda. From 2017 to 2019, the number of incidents of white supremacist propaganda being distributed on college campuses more than doubled, accounting for more than one-fourth of all incidents of white supremacist propaganda in the United States in 2019 (Center on Extremism 2020). According to the FBI (2020), nearly one in ten hate crimes occur at schools or colleges.

What do you think?

Schools and college campuses have reported a sharp increase in incidents of racially motivated bullying, hate crimes, and white supremacist propaganda since 2016. Many analysts argue that former President Trump's racially inflammatory rhetoric—for example, calling for vigilante violence against anti-racist protestors and referring to COVID-19 as the Wuhan flu and the China virus—has contributed to this increase. Do you think the language a president uses to talk about race affects race relations in the United States?

Strategies for Action: Responding to Prejudice, Racism, and Discrimination

In the summer of 2020, the deaths of George Floyd in Minneapolis, Michigan, Breonna Taylor in Louisville, Kentucky, and the shooting of Jacob Blake in Kenosha, Wisconsin—all incidents involving alleged inappropriate use of deadly force by police against Black people—sparked waves of protests across the country. The racial justice movement Black Lives Matter (BLM) had been organizing to address institutional discrimination against people of color since 2013, when George Zimmerman was acquitted of the murder of 17-year-old Trayvon Martin. The events of the summer of 2020 served as a flashpoint that mobilized a diverse coalition of activists to join with BLM to protest racial injustice and seek to address problems of institutional racism in the United States. Counterprotests and backlash movements also emerged at this time, and amidst campaigns for the 2020 presidential election, a tense and tumultuous conversation among the American public focused on the question, "What should we do about racial and ethnic inequality in the United States?"

Efforts to achieve racial equality in the criminal justice system are discussed in Chapter 4. In the following sections, we discuss the Equal Employment Opportunity Commission's role in responding to employment discrimination and examine the issue of affirmative action in the United States. We also discuss educational strategies to promote diversity and multicultural awareness and appreciation in schools. Finally, we look at apologies and reparations as a means of achieving racial reconciliation.

The Equal Employment Opportunity Commission

The **Equal Employment Opportunity Commission (EEOC)**, a U.S. federal agency charged with ending employment discrimination in the United States, is responsible for enforcing laws against discrimination, including Title VII of the *1964 Civil Rights Act* that prohibits employment discrimination on the basis of race, color, religion, sex, or national origin. The EEOC investigates, mediates, and may file lawsuits against private employers on behalf of alleged victims of discrimination. The most frequently filed claims with the EEOC are allegations of race discrimination, racial harassment, or retaliation from opposition to racial discrimination. In 2019, the EEOC filed 87 lawsuits under Title VII and resolved more than 26,000 claims of racial discrimination (EEOC 2020).

In 2007, the EEOC launched a national initiative to combat racial discrimination in the workplace. The goals of this initiative, called E-RACE (Eradicating Racism and Colorism from Employment), are to (1) identify factors that contribute to race and color discrimination, (2) explore strategies to improve the administrative processing and litigation of race and color discrimination cases, and (3) increase public awareness of race and color discrimination in employment.

Equal Employment Opportunity Commission (EEOC) A U.S. federal agency charged with ending employment discrimination in the United States and responsible for enforcing laws against discrimination, including Title VII of the 1964 Civil Rights Act that prohibits employment discrimination on the basis of race, color, religion, sex, or national origin.

Affirmative Action

Affirmative action refers to a broad range of policies and practices in the workplace and educational institutions to promote equal opportunity and diversity. Affirmative action

affirmative action A broad range of policies and practices in the workplace and educational institutions to promote equal opportunity as well as diversity.

is an attempt to compensate for the effects of past discrimination and prevent current discrimination against women and racial and ethnic minorities. Veterans and people with disabilities may also qualify under affirmative action policies. Federal affirmative action policies developed in the 1960s required any employer (universities as well as businesses) who received contracts from the federal government to make "good faith efforts" to increase the pool of qualified minorities and women by expanding recruitment and training programs (U.S. Department of Labor 2002).

In higher education, affirmative action policies seek to improve access to education for those groups that have been historically excluded or underrepresented, such as women and racial/ethnic minorities. Supporters of affirmative action in higher education view the underrepresentation of minorities in higher education as a problem. Consider that, in 1965, only 5 percent of undergraduate students, 1 percent of law students, and 2 percent of medical students in the country were African American (National Conference of State Legislatures 2013). Over the last several decades, colleges and universities have implemented affirmative action policies designed to recruit and admit more minority students, and these policies have been successful in increasing the enrollment of minority students, although racial/ethnic gaps in college enrollment remain (see also Chapter 8).

In 1978, the Supreme Court's ruling in *Regents of the University of California v. Bakke* marked the beginning of a series of several court challenges to affirmative action. Although the Court ruled that affirmative action programs could not use fixed quotas in admission, hiring, or promotion policies, it affirmed the right for universities and employers to consider race as a factor in admission, hiring, and promotion to achieve diversity. Since the *Bakke* case, numerous legal battles have challenged affirmative action. Beginning in the 1990s, states began to pass legislation that banned the use of affirmative action in public universities. One comprehensive study of the long-term effects of this legislation found that affirmative action bans at 19 public universities resulted in a statistically significant decline in the number of Black and Hispanic college enrollees (Long and Bateman 2020). States with affirmative action bans exhibited a gap of 14 percentage points between Black and Hispanic college student enrollment and the representation of these groups in the general population. Furthermore, the analysis showed that this enrollment decline impacted the long-term earnings of Black and Hispanic youth, contributing to the increase in economic inequality over the past 25 years. Although this gap has started to slowly close in recent years, the authors project that at the current pace, it will take Black and Hispanic youth at least 50 years to reach economic parity with White youth.

In 2013, the Supreme Court ruled in *Fisher v. University of Texas* that universities may seek racial diversity, but they must demonstrate that race-neutral alternatives to achieving diversity are not sufficient (Kahlenberg 2013). A number of colleges and universities have already developed creative alternatives to race-based admissions policies to increase their minority student populations. Some universities are using a "holistic admissions" approach that considers the unique circumstances of each student, prioritizing academics and treating all other factors, including race, equally. Other universities use the so-called 10 percent plan as a way to maintain minority enrollment. Graduates in the top 10 percent of their high school classes are admitted automatically to the public college or university of their choice; standardized test scores and other factors are not considered. Yet another approach is to increase financial aid packages to low-income students, which benefits all students from economically disadvantaged backgrounds, regardless of race/ethnicity.

Educational Strategies

Schools and universities play an important role in whether minorities succeed in school and in the job market. One way to improve minorities' chances of academic success is to reduce or eliminate disparities in school funding. As noted earlier, schools in poor districts—which predominantly serve minority students—have traditionally received less funding per pupil than do schools in middle- and upper-class districts (which predominantly serve White students). Other educational strategies focus on reducing prejudice, racism, and discrimination and on fostering awareness and appreciation of racial and ethnic diversity. These strategies include multicultural education, "Whiteness studies," and efforts to increase diversity among student populations.

Multicultural Education in Schools and Communities. In schools across the nation, multicultural education, which encompasses a broad range of programs and strategies, works to dispel myths, stereotypes, and ignorance about minorities; to promote tolerance and appreciation of diversity; and to include minority groups in the school curriculum (see also Chapter 8). With multicultural education, the school curriculum reflects the diversity of U.S. society and fosters an awareness and appreciation of the contributions of different racial and ethnic groups to U.S. culture. The Southern Poverty Law Center's program Teaching Tolerance publishes and distributes materials and videos designed to promote better human relations among diverse groups. These materials are sent to schools, colleges, religious organizations, and a variety of community groups across the nation.

Many colleges and universities promote awareness and appreciation of diversity by offering courses and degree programs in racial and ethnic studies and by sponsoring multicultural events and student organizations. Some colleges mandate that students take a certain number of required diversity courses to graduate.

What do you think?

As of August 2020, the California state legislature is considering two bills that would require students in public K–12 and at public colleges and universities to take at least one course in an area of ethnic studies, such as African American, Chicano, or Jewish studies. In the meantime, California State's Board of Trustees voted to require all students to take a course in ethnic or social justice studies. One professor commented "At a time when racism and anti-Semitism are rising in our country, an ethnic-studies and social-justice requirement should provide students the tools to understand and confront this danger" (quoted in Parry 2020). Do you think required ethnic studies courses will reduce racism in the future?

Whiteness Studies. Traditionally, studies of and courses on race have focused on the social disadvantages that racial minorities experience, while ignoring or minimizing the other side of the racial inequality equation—the social advantages conferred upon Whites. Courses in Whiteness studies, which are being offered in many colleges and universities, focus on understanding "Whiteness" as a social construction and fostering critical awareness of White privilege—an awareness that is limited among White students (Yeung et al. 2013). "White privilege is so fundamental as to be largely invisible, expected, and normalized" (Picca and Feagin 2007, p. 243).

In an often cited paper called "White Privilege: Unpacking the Invisible Knapsack," Peggy McIntosh (1990) likened White privilege to an "invisible weightless knapsack" that Whites carry with them without awareness of the many benefits inside the knapsack. Here are just a few of the many benefits McIntosh associates with being White:

- Go shopping without being followed or harassed.
- Be assured that my skin color will not convey that I am financially unreliable when I use checks or credit cards.
- Swear, or dress in secondhand clothes, or not answer letters without having people attribute these choices to the bad morals, poverty, or the illiteracy of my race.
- Avoid ever being asked to speak for all the people of my racial group.
- Be sure that, if a traffic cop pulls me over or if the IRS audits my tax return, it is not because I have been singled out because of my race.
- Take a job with an affirmative action employer without having coworkers on the job suspect that I got it because of race.

Diversification of College Student Populations. Recruiting and admitting racial and ethnic minorities in institutions of higher education can foster positive relationships among diverse groups and enrich the educational experience of all students—minority and nonminority alike (American Council on Education and American Association of University Professors 2000). Psychologist Gordon Allport's (1954) "contact hypothesis" suggested that contact between groups is necessary for the reduction of prejudice between group members. Having a diverse student population can provide students with opportunities for contact with different groups and can thereby reduce prejudice. In a study of 2,000 UCLA students, researchers found that students who were randomly assigned to roommates of a different race or ethnicity developed more favorable attitudes toward students of different backgrounds (Sidanius et al. 2010).

Retrospective Justice Initiatives: Apologies and Reparations

Various governments around the world have issued official apologies for racial and ethnic oppression. After World War II, West Germany signed a reparations agreement with Israel in which West Germany agreed to pay Israel for the enslavement and persecution of Jews during the Holocaust and to compensate for Jewish property that the Nazis stole. In 2008, the Australian government issued a formal apology for the treatment of the country's aboriginal people—specifically, for the decades during which the government removed aboriginal children from their families in a forced acculturation program.

In the United States, President Gerald Ford and Congress apologized to Japanese Americans in 1976 for their internment during World War II, and reparations of $20,000 were granted to each surviving internee who was a U.S. citizen or legal resident alien at time of internment. In 1993, President Bill Clinton apologized to native Hawaiians for overthrowing the government of their nation. In 1994, the state of Florida offered monetary compensation to the survivors and descendants of the 1923 Rosewood massacre, in which a White mob attacked and murdered Black residents in Rosewood, Florida, and set fire to the town. And in 1997, the U.S. government offered monetary reparations to surviving victims of the Tuskegee syphilis study, in which Blacks suffering from syphilis were denied medical treatment.

Although various forms of reparations were offered to Native American tribes to compensate for land that had been taken by force or deception, it wasn't until 2008 that the Senate Committee on Indian Affairs passed a resolution apologizing to all Indian tribes for the mistreatment and violence committed against them. In 2008, the U.S. House of Representatives issued a formal apology to African Americans for slavery; and in 2009, the U.S. Senate passed a resolution apologizing for slavery, adding the stipulation that the official apology cannot be used to support claims for restitution. The resolution "acknowledges the fundamental injustice, cruelty, brutality, and inhumanity of slavery and Jim Crow laws" and "apologizes to African Americans ... for the wrongs committed against them and their ancestors who suffered under slavery and Jim Crow laws" (CNN 2009).

Opponents of reparations typically argue that "preoccupation with past injustice is a distraction from the challenge of present injustice" (Brown University Steering Committee on Slavery and Justice 2007, p. 39). Supporters of the reparative justice movement, however, believe that the granting of apologies and reparations to groups that have been mistreated promotes dialogue and healing, increases awareness of present inequalities, and stimulates political action to remedy current injustices. In making the case for monetary reparations to Black Americans for slavery, Ta-Nahisi Coates argues:

> We cannot escape our history. All of our solutions to the great problems of health care, education, housing, and economic inequality are troubled by what must go unspoken. ... An America that asks what it owes its most vulnerable citizens is improved and humane. An America that looks away is ignoring not just the sins of the past but the sins of the present and the certain sins of the future. More important than any single check cut to any African American, the payment of

reparations would represent America's maturation out of the childhood myth of its innocence into a wisdom worthy of its founders. (2014, p. 71)

Understanding Race, Ethnicity, and Immigration

After considering the material presented in this chapter, what understanding about race and ethnic relations are we left with? First, we have seen that racial and ethnic categories are socially constructed; they are largely arbitrary, imprecise, and misleading. Although some scholars suggest that we abandon racial and ethnic labels, others advocate adding new categories—multiethnic and multiracial—to reflect the identities of a growing segment of the U.S. and world population.

Conflict theorists and structural functionalists agree that prejudice, discrimination, and racism have benefited certain groups in society. But racial and ethnic disharmony has created tensions that disrupt social equilibrium. Symbolic interactionists note that negative labeling of minority group members, which is learned through interaction with others, contributes to the subordinate position of minorities.

Prejudice, racism, and discrimination are debilitating forces in the lives of minorities and immigrants. Despite these negative forces, many minority group members succeed in living productive, meaningful, and prosperous lives. But many others cannot overcome the social disadvantages associated with their minority status and become victims of a cycle of poverty (see Chapter 6). Minorities are disproportionately poor, receive inferior education and health care, and, with continued discrimination in the workplace, have difficulty improving their standard of living.

Intentional and hateful racism still exist, but much racism is more subtle and unintentional. Attempts to be "color-blind" may be well intentioned, but if we don't "see" race, then we also do not see racial injustices or the need to remedy such injustices.

Achieving racial and ethnic equality requires first *seeing* how discrimination is institutionalized in the structure of society and then making changes in society to eliminate institutionalized discrimination and provide opportunities for minorities—in education, employment and income, and political participation. In addition, policy makers concerned with racial and ethnic equality must find ways to reduce the racial and ethnic wealth gap and to foster wealth accumulation among minorities (Conley 1999). Social class is a central issue in race and ethnic relations. Professor and activist bell hooks (2000) (who spells her name in all lowercase) warned that focusing on issues of race and gender can deflect attention away from the larger issue of class division that increasingly separates the haves from the have-nots. Addressing class inequality must, suggests hooks, be part of any meaningful strategy to reduce inequalities that minority groups suffer.

Making change requires members of society to recognize that change is necessary, that there is a problem that needs rectifying:

> One has to perceive the problem to embrace the solutions. If you think racism isn't harmful unless it wears sheets or burns crosses or bars blacks from motels and restaurants, you will support only the crudest antidiscrimination laws and not the more refined methods of affirmative action and diversity training. (Shipler 1998, p. 2)

The highly contentious election between Donald Trump and Joe Biden brought the American public into direct confrontation with contrasting visions of the future of race relations in the United States. The spike in hate crimes toward and harassment of minorities and immigrants that occurred in the years following his election suggests that Trump's rhetoric emboldened white supremacists and their sympathizers. But it also emboldened citizens all across the country who are raising their voices, calling for racial and ethnic justice in all aspects of U.S. culture. Whether the election of a new president and the first biracial and female vice president signals a permanent shift in the dynamics of race relations in the United States is yet to be seen.

Chapter Review

- **What is a minority group?**
A minority group is a category of people who have unequal access to positions of power, prestige, and wealth in a society and who tend to be targets of prejudice and discrimination. Minority status is not based on numerical representation in society but rather on social status.

- **What is meant by the idea that race is socially constructed?**
The concept of race refers to a category of people who are perceived to share distinct physical characteristics that are deemed socially significant. The significance of race is not biological but social and political because race is used to separate "us" from "them" and becomes a basis for unequal treatment of one group by another. Races are cultural and social inventions; they are not scientifically valid because there are no objective, reliable, meaningful criteria scientists can use to identify racial groupings. Different societies construct different systems of racial classification, and these systems change over time.

- **What are the various patterns of interaction that may occur when two or more racial or ethnic groups come into contact?**
When two or more racial or ethnic groups come into contact, one of several patterns of interaction occurs, including genocide, expulsion, segregation, acculturation, pluralism, and assimilation.

- **What is an ethnic group?**
An ethnic group is a population that has a shared cultural heritage, nationality, or lineage. Ethnic groups can be distinguished on the basis of language, forms of family structures and roles of family members, religious beliefs and practices, dietary customs, forms of artistic expression such as music and dance, and national origin. The largest ethnic population in the United States is Hispanics, or Latinos.

- **How does the U.S. Census Bureau define race and ethnicity?**
The Census asks a person to identify their ethnicity as either Hispanic or non-Hispanic as well as a racial category of White, Black, American Indian or Alaskan Native, Asian, Hawaiian or Pacific Islander, or some other race. People can choose as many races as they identify with, and a person's racial and ethnic category is what they say are.

- **How is the racial and ethnic makeup of the United States expected to change over the next 40 years?**
The White population in the United States is expected to drop below 50 percent by 2060. Hispanics (of any race) are currently the largest minority group in the United States (17.8), and their share is expected to grow to 27.5 percent by 2060.
From 2020 to 2060, the proportion of the population that is foreign born is expected to increase by at least 58 percent.

- **How has immigration policy changed over time?**
For the first 100 years of its history, the United States had an open immigration system. The first restrictions on immigration occurred with the 1882 *Chinese Exclusion Act*. In the early 20th century, a quota system favored White European immigrations, contributing to the racial makeup of the United States today. The 1965 *Hart-Celler Act* eliminated the quota system and led to an increasing diversification of the U.S. population. The last major reform to the immigration system occurred in 1986, which made it much more difficult for undocumented migrants to work in the United States. While the Trump administration issued more than 400 policy changes on immigration, including the highly controversial zero-tolerance policy, the Biden administration is likely to reverse these.

- **What were the manifest function and latent dysfunction of the civil rights movement?**
The manifest function of the civil rights legislation in the 1960s was to improve conditions for racial minorities. However, civil rights legislation produced an unexpected consequence, or latent dysfunction. Because civil rights legislation supposedly ended racial discrimination, Whites were more likely to blame Blacks for their social disadvantages and thus perpetuate negative stereotypes such as "Blacks lack motivation" and "Blacks have less ability."

- **Is it possible for an individual to discriminate without being prejudiced?**
Yes. In overt discrimination, individuals discriminate because of their own prejudicial attitudes. But sometimes individuals who are not prejudiced discriminate because of someone else's prejudice. For example, a store clerk may watch Black customers more closely because the store manager is prejudiced against Black people and has instructed the employee to follow Black customers in the store closely. Discrimination based on someone else's prejudice is called adaptive discrimination.

- **How is color-blindness a form of racism?**
Color-blindness is a form of racism because it prevents acknowledgment of privilege and disadvantage associated with race and therefore allows the continuation of institutional racism.

- **Are U.S. schools segregated?**
Most White public school students attend schools that are largely White, and most minority public students attend schools where the majority of students are minorities. Racial and ethnic minorities are largely isolated from Whites in an increasingly segregated school system. The upward trend in school segregation is due to large increases in minority student enrollment, continuing White flight from urban areas, the persistence of housing segregation, and the termination of court-ordered desegregation plans.

- **What are the recent trends in hate crimes?**
Racial bias is the most common form of hate crime reported by the FBI. Incidents of hate crimes surged after the election of Donald Trump in 2016, and many analysts have attributed this increase to the president's use of racially inflammatory rhetoric.

- **What is the role of the Equal Employment Opportunity Commission (EEOC) in combating employment discrimination?**
  The Equal Employment Opportunity Commission (EEOC) is responsible for enforcing laws against discrimination, including Title VII of the 1964 *Civil Rights Act* that prohibits employment discrimination on the basis of race, color, religion, sex, or national origin. The EEOC investigates, mediates, and may file lawsuits against private employers on behalf of alleged victims of discrimination.

- **What are Whiteness studies?**
  Courses in Whiteness studies, which are being offered in many colleges and universities, focus on increasing awareness of White privilege—an awareness that is limited among White students.

Test Yourself

1. When it comes down to collecting census data on the U.S. population, a person can select any of the racial and ethnic identities that they feel describes them.
   a. True
   b. False
2. Which of the following occurs when a person adopts the culture of a group different from the one in which that person was originally raised?
   a. Secondary assimilation
   b. Acculturation
   c. Genocide
   d. Pluralism
3. According to census data, the largest minority group in the United States is
   a. Hispanic.
   b. Black.
   c. Asian.
   d. two or more races.
4. Compared with the U.S.-born population, immigrants have high rates of criminal behavior.
   a. True
   b. False
5. Which minority group in the United States is considered a "model minority"?
   a. Women
   b. Black women
   c. Scandinavian immigrants
   d. Asian Americans
6. Which of the following has/have moved "backstage"?
   a. Racist behavior
   b. Undocumented immigrants
   c. Multicultural education
   d. Affirmative action
7. According to the FBI, most reported hate crimes are based on
   a. political bias.
   b. anti-immigrant bias.
   c. religious bias.
   d. racial bias.
8. Which of the following is responsible for enforcing laws against discrimination?
   a. Local police departments
   b. U.S. Department of Labor
   c. Equal Employment Opportunity Commission
   d. The Supreme Court
9. Colleges and universities are prohibited from requiring students to take diversity courses as a graduation requirement.
   a. True
   b. False
10. The U.S. Congress has issued an official apology to African Americans for slavery and Jim Crow laws.
   a. True
   b. False

Answers: 1. A; 2. B; 3. D; 4. B; 5. D; 6. A; 7. D; 8. C; 9. B; 10. A.

Key Terms

acculturation 324
adaptive discrimination 347
affirmative action 355
antimiscegenation laws 329
assimilation 325
aversive racism 344
coercive pluralism 325
color-blind racism 345
colorism 348
cyber-racism 346
discrimination 346
Equal Employment Opportunity Commission (EEOC) 355
ethnicity 322
expulsion 323
genocide 323
hate crime 353
implicit prejudice 343
individual discrimination 347
institutional discrimination 347
institutional racism 343
Islamophobia 353
minority group 320
nativist extremist groups 333
naturalized citizens 335
overt discrimination 347
pluralism 325
prejudice 342
race 322
racial microaggressions 352
racism 342
sanctuary city 335
scientific racism 321
segregation 323
stereotypes 340
sundown towns 350

AP Images/Alex Milan Tracy/Sipa USA

> “Both men and women should feel free to be sensitive. Both men and women should feel free to be strong. … [I]t is time that we all perceive gender on a spectrum not as two opposing sets of ideas.”
>
> **EMMA WATSON**
> actress, model, activist

10

Gender Inequality

Chapter Outline

Learning Objectives

After studying this chapter, you will be able to …

1. Describe the ways gender inequality exists around the world.
2. Identify what leads to gender inequality according to each of the three sociological and feminist theories.
3. Analyze how the structure of society contributes to gender inequality.
4. Analyze how the culture of society contributes to gender inequality.
5. Describe how traditional gender roles lead to the social problems discussed in this chapter.
6. Evaluate the effectiveness of U.S. laws and policies directed toward reducing gender inequality.

People of all ages, races, and genders marched in the Trans Pride March in Portland, OR in support of trans rights and protections for adults and children like Max.

CLASSIFIED AS MALE AT BIRTH, Max has lived most of his childhood as a boy despite feeling deeply "different than … a boy for all of [his] … life" (Brooks 2018, p. 1). In elementary school, Max explored his gender identity by "hanging out with girls and dressing in pink boas." At age 11, a classmate called Max a "girl-boy" which led to thoughts of suicide and ultimately a call to a suicide prevention counselor who talked Max off a third-floor balcony and back to safety. By age 13, Max had discovered that his gender identity could not be neatly categorized as male or female but rather neither male nor female, i.e., Max is **nonbinary** (also referred to as *gender neutral* and *agender*). The term *nonbinary* is a catch-all phrase that includes anyone who, as Max, identifies neither as male nor female or people whose gender identity is both male and female or, alternatively, sometimes male and sometimes female. As a nonbinary youth, Max has embraced the pronouns they/them/their, although they admit it is often difficult for friends and family to use the right pronoun. While presently on medication that prevents Max from going through puberty, soon Max will have to confront something few of us have to face: deciding whether to go through male or female puberty (Brooks 2018).

About 60 percent of U.S. adults report hearing about the use of gender-neutral pronouns, and nearly 20 percent of Americans report that they personally know someone who prefers a gender pronoun other than the traditional "he" or "she" (Pew Research Institute 2019). In fact, the White House contact form allows users to select their preferred pronoun including she/her, he/him, they/them, other, and "prefer not to say" (Maxouris 2021).

Sex refers to one's biological classification, whereas **gender** refers to the social definitions and expectations associated with being female or male. Today, researchers, educators, and parents alike are challenging the binary concepts of sex and gender, and, just as our evolving notion of sexual orientation, it has led to feelings of uneasiness for some (see Chapter 11). For example, words like *transgender* and *transsexual*, *gender variant*, *androgyny*, *intersexed*, *boi* and *birl*, *metrosexual*, *two-spirited*, *nonbinary*, and *third sex* were not part of the American lexicon just decades ago. In most Western cultures, we take for granted that there are two categories of gender. However, in many other societies, three and four genders have been recognized:

> On nearly every continent, and for all of recorded history, thriving cultures have recognized, revered, and integrated more than two genders. Terms such as transgender and gay are strictly new constructs that assume three things: that there are only two sexes (male/female), as many as two sexualities (gay/straight), and only two genders (man/woman). (National Public Radio 2011, p. 1)

Gender identity refers to a person's "deeply felt internal and individual experience of gender, which may or may not correspond with the sex assigned at birth, including the personal sense of the body … and other expressions of gender, including dress, speech, and mannerisms" (Castro-Peraza et al. 2019, p. 1). A **transgender individual** (sometimes called trans or gender nonconforming) is a person whose sense of gender identity is inconsistent with her or his birth sex. A **cisgender** individual is a person whose sense of gender identity is consistent with her or his birth sex. Transgender is not a sexual orientation, and transgender individuals just as cisgender individuals may have any sexual orientation—heterosexual, gay, or bisexual. Lastly, **gender expression** includes the sum of an individual's external presentation of self (e.g., dress, grooming, behaviors) as masculine, feminine, or somewhere on the continuum between the two.

sex A person's biological classification as male or female.

gender The societal definitions and expectations associated with being female or male.

transgender individual A person whose sense of gender identity is inconsistent with his or her birth (sometimes called chromosomal) sex (male or female).

cisgender An individual whose sense of gender identity is consistent with his or her birth sex.

gender expression The way in which a person presents her- or himself as a gendered individual (i.e., masculine, feminine, or androgynous) in society. A person could, for example, have a gender identity as male but present their gender as female.

There is documented movement away from a binary emphasis on sex and gender. For example, in a national survey about one-third of respondents reported that their sex classification at birth (i.e., male or female) was not completely consistent with their gender identification (i.e., masculine or feminine) (Magliozzi, Saperstein, and Westbrook 2016). In fact, 7 percent of the sample defined themselves as equally masculine and feminine, and nearly four percent responded in ways that did not "match" their sex at birth (i.e., females who saw themselves as more masculine than feminine, or males who saw themselves as more feminine than masculine). Further, using an experimental research design, Medeiros, Forest, and Ohberg (2020) conclude that use of a binary question in a survey was viewed more negatively than use of a nonbinary question.

Nonetheless, most Americans still think in terms of females and males, and of women and men, in much the same way they continue to use such false dichotomies as Black and White, gay and straight, and young and old. Although empirically inaccurate, this *social shorthand* makes conversation easier and fulfills our need to "know" and respond "appropriately." Thus, upon meeting someone, we quickly attach the label of gay, White male, or elderly Black female, though knowing little about their social biographies. Similarly, this chapter emphasizes **sexism** and gender inequality as traditionally defined—that is, as the inequality between women and men—but, wherever possible, information on transgender individuals will be included in the discussion. Note that civil rights issues for transgender individuals are discussed in Chapter 11.

sexism The belief that innate psychological, behavioral, and/or intellectual differences exist between women and men and that these differences connote the superiority of one group and the inferiority of the other.

What *do you* think?

The Hungarian government is in the process of preparing legislation that would end the legal recognition of their transgender citizens by defining gender as "biological sex based on primary sex characteristics and chromosomes" (Walker 2020, p. 1). The new law will stand in violation of European human rights law, which may lead to legal challenges. What do you think? Should transgender women and men have the same legal rights as cisgender women and men?

The Global Context: The Status of Women and Men

There is no country in the world in which women and men have equal status. Although much progress has been made in closing the gender gap in areas such as education, health care, employment, and government, gender inequality is still prevalent throughout the world. It is so prevalent that a survey in Great Britain of girls and young women between the ages of 11 and 21 indicates that 87 percent believe that "schools should be assessed to ensure girls and boys get equal treatment in subject choices and career advice" (Girls' Attitudes Survey 2019). Further, a global survey of women and men indicates that nearly 90 percent of the respondents were biased against women in some way (United Nations Development Program 2020).

The World Economic Forum (WEF 2020) assessed the gender gap in 153 countries by measuring the extent to which women have achieved equality with men in four areas: economic participation and opportunity, educational attainment, health and survival, and political empowerment. Table 10.1 presents the overall index score and subindex scores for regions around the world. Note that the scores approximate the proportion of the gender gap closed by region. For example, Western Europe with a score of 0.767 has closed 76.7 percent of the gender gap overall. Across regions, health and survival (95.8) and educational attainment (95.7) have the greatest gender parity and political empowerment the lowest (24.1).

For the 11th year in a row, Iceland is the most gender-equal nation in the world, followed by Norway, Finland, Sweden, Nicaragua, New Zealand, Ireland, and Spain. Chad, Iran, Democratic Republic of Congo, Syria, Pakistan, Iraq, and Yemen have the lowest gender equality ranking of those countries included in the index. Globally, there has been progress toward gender equality with 31.4 percent of the average gender gap yet to be closed. Regionally, at the current rate, the gender gap could be closed in 54 years in Western Europe, at one end of the continuum, and in 163 years at the other end of the continuum, i.e., in East Asia and the Pacific (World Economic Forum [WEF] 2020, p. 6).

Gender inequality varies across cultures, not only in its extent or degree but also in its forms. In the United States, gender inequality in family roles commonly takes the form of an unequal division of household labor and child care, with women bearing the heavier responsibility for these tasks. In other countries, forms of gender inequality in the family include the expectation that wives ask their husbands for permission to use birth control (see Chapter 2), unequal penalties for spouses who commit adultery, with wives receiving harsher punishment, and the practice of aborting female fetuses in cultures that value male children over female children.

In a book entitled *Unnatural Selection*, author Mara Hvistendahl (2011) documented how medical technology, and specifically the increased availability of ultrasounds, has made sex-selection abortions commonplace around the world. According to the United Nations (UN), there are between 113 and 200 million "missing" girls, more than the entire female population of the United States. One researcher, upon visiting a peasant family in rural China, sits with the mother-in law as her son's wife is about to give birth.

> There was a low sob, and then a man's gruff voice said accusingly: "useless thing."
>
> I thought I heard a slight movement in the slop pail behind me, and automatically glanced toward it. ... To my absolute horror, I saw a tiny foot poking out of the pail. ... Then the tiny foot twitched! The midwife must have dropped that tiny baby alive into the slop pail!
>
> We sat in silence while I stared, sickened, at the pail. ... The little foot was still now. ...
>
> "Doing a baby girl is not a big thing around here," ... the older woman said.
>
> "That's a living child!" I said in a shaking voice. ...
>
> "It's not a child," she corrected me.
>
> "What do you mean, it's not a child? I saw it."
>
> "It's not a child. ... If it was, we would be looking after it, wouldn't we? ... It's a girl baby." (pp. 27–28)

Benoit Tessier/Reuters

When former Olympic star Bruce Jenner transitioned to Caitlin Jenner in 2015, it was described by one academic as a "trans tipping point" (Brown 2019). Pictured here is 28-year-old model Andreja Pejic who started violating conventional gender norms over ten years ago as an androgynous male model. Today, Pejic has transitioned and is a top fashion model. She recently appeared in the movie *The Girl in the Spider's Web* opposite English actress Claire Foy.

A global perspective on gender inequality must also consider the different ways in which such inequality is viewed. For example, many non-Muslims view the practice of Muslim women wearing a headscarf in public as a symbol of female subordination and oppression. To Muslims who embrace this practice, and not all Muslims do, wearing a headscarf reflects the high status of women and represents the view that women should be respected and not treated as sexual objects.

Similarly, cultures differ in how they view the practice of female genital mutilation, also known as female genital cutting or female circumcision (FGM/C). There are several forms of FGM/C, ranging from a symbolic nicking of the clitoris to removal of the clitoris and labia and partial closure of the vaginal opening by stitching the two sides of the vulva together, leaving only a small opening for the passage of urine and menstrual blood. After marriage, the sealed opening is reopened to permit intercourse and childbearing.

> "Treating women as second-class citizens is a bad tradition. It holds you back. There's no excuse for sexual assault or domestic violence. There's no reason that young girls should suffer genital mutilation. There's no place in civilized society for the early or forced marriage of children. These traditions may date back centuries; they have no place in the 21st century."
>
> **–PRESIDENT BARACK OBAMA**

Nonmedical personnel perform most FGM/C procedures using unsterilized blades or string. Health risks associated with FGM/C include pain, hemorrhaging, shock, problems urinating, infections, complications during childbirth, and even death. Today, over 200 million girls and women, predominantly in Africa, the Middle East, and Asia, have undergone FGM/C (World Health Organization [WHO] 2020).

People from countries in which FGM/C is not the norm generally view this practice as a barbaric form of violence against women. In countries where it commonly occurs, FGM/C is viewed as an important and useful practice. Although having no health benefit, in some countries it is considered a rite of passage that enhances a woman's status. In other countries, it is aesthetically pleasing. For others, FGM/C is believed to be a moral imperative based on religious beliefs. However, in all cases, it "reflects deep-rooted inequality between the sexes, and constitutes an extreme form of discrimination against women" (WHO 2020, p. 2).

Gender Inequality in the United States

Although attitudes toward gender equality are becoming increasingly liberal, the United States has a long history of gender inequality. Women have had to fight for equality: the

TABLE 10.1 Gender Gap Regional Performances, Overall and by Subindex, 2020

		Subindexes			
	Overall Index	Economic Participation and Opportunity	Educational Attainment	Health and Survival	Political Empowerment
Western Europe	0.767	0.693	0.993	0.972	0.409
North America	0.729	0.756	1.000	0.975	0.184
Latin American and the Caribbean	0.721	0.642	0.996	0.979	0.269
Eastern Europe and Central Asia	0.715	0.732	0.998	0.979	0.150
East Asia and the Pacific	0.685	0.663	0.976	0.943	0.159
Sub-Saharan Africa	0.680	0.666	0.872	0.972	0.211
South Asia	0.661	0.365	0.943	0.947	0.387
Middle East and North Africa	0.611	0.425	0.950	0.969	0.102
Global average	0.685	0.582	0.957	0.958	0.241

SOURCE: WEF Global Gender Gap Report 2020.

right to vote, equal pay for comparable work, quality education, entrance into male-dominated occupations, and legal equality.

Based on the World Economic Forum's assessment of inequality between women and men in economic participation and opportunity, political empowerment, educational attainment, and health and survival, the United States ranks 53rd among 153 countries (World Economic Forum [WEF] 2020). Having a larger gender gap on the overall measure of gender inequality than Jamaica, Zambia, Mexico, Cuba, and Bangladesh, the United States ranks 26th in economic participation and opportunity, 34th in educational attainment, 70th in health and survival, and 86th in political empowerment. Consistent with these results, women have lower incomes, hold fewer prestigious jobs, earn fewer graduate degrees, and are more likely than men to live in poverty.

Men are also victims of gender inequality. In 1963, sociologist Erving Goffman wrote that in the United States, there is only

> one complete unblushing male ... a young, married, white, urban, northern heterosexual, Protestant father of college education, fully employed, of good complexion, weight and height, and a recent record in sports. ... Any male who fails to qualify in one of these ways is likely to view himself ... as unworthy, incomplete, and inferior. (p. 128)

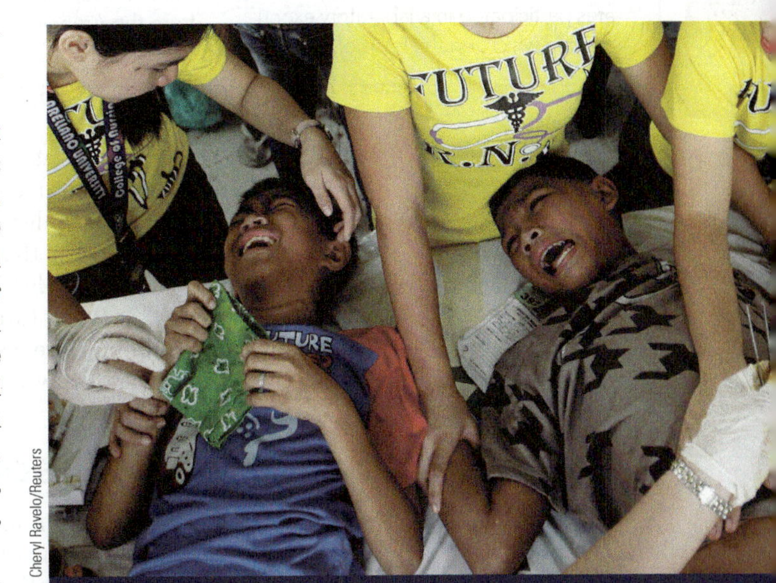

Cheryl Ravelo/Reuters

Every year in the Philippines, thousands of boys are circumcised as a rite of passage. There is even a term for boys who do not undergo the procedure, *supot*, which means coward, or not being a *real* man for being unable to withstand the pain of the procedure (Pugh 2019). Of late, the medical procedure has come under fire as questions arise about its safety, its necessity, and, similarly to FGM/C, its abuse of a child's anatomical integrity.

Although standards of masculinity have relaxed, Williams (2000) argues that masculinity is still based on "success"—at work, on the athletic field, on the streets, and at home. Similarly, Vandello et al. (2008) concludes "[t]he view that manhood is tenuous, and therefore requires public proof, is consistent with research across multiple areas" (p. 1326). (See this chapter's *Social Problems Research Up Close*.)

social problems RESEARCH UP CLOSE

Overdoing Gender

In a now classic study, Willer et al. (2013) test what is called the **masculine overcompensation thesis**—the assertion "that men react to masculinity threats with extreme demonstrations of masculinity" (p. 980). The researchers evaluate this thesis using four research designs, three of which are discussed here. Two are laboratory experiments, and the third looks at the relationship between masculinity threat and attitudes associated with dominance using a national sample of American adults.

Sample and Methods

Experiments are unique methods of research in that the researchers manipulate the independent variable and measure its effect on the dependent variable. In studies 1 and 2, female and male undergraduates participated in a laboratory experiment in which, on the basis of a "gender identity survey" administered, they were randomly assigned to one of two conditions of the independent variable. The independent variable is *gender identity threat.* In the first condition, males and females were given feedback that they were in the average female range (males' masculinity threatened). In the second condition, males and females were told they were in the average male range (females' femininity threatened). In the first study, dependent variables were assessed on a "political views survey." In the second study, which was similarly designed to the first laboratory experiment, social dominance variables were measured.

In the third study, to assess the reliability of the results of the first two studies with a larger and more diverse sample, Willer et al. (2013) used data from the American Values Survey. Masculinity threat was measured by respondents' feelings that social change is threatening the status of men. The dependent variables were a variety of attitudinal questions related to masculinity and dominance.

Findings and Conclusions

Study 1 examined the relationship between masculinity threat and femininity threat and support for the war in Iraq, attitudes toward homosexuality, desirability of owning an SUV, and how much a subject was willing to pay for an SUV. Males who had received feedback that they had scored in the average femininity range (i.e., had their masculinity threatened) were significantly more likely when compared to male subjects who did not have their masculinity threatened to voice support for the war

When U.S. college students were asked to list the best and worst things about being the opposite sex, the same qualities, although in opposite categories, emerged (Cohen 2001). For example, what males listed as the best thing about being female (e.g., free to be emotional), females list as the worst thing about being male (e.g., not free to be emotional). Similarly, what females listed as the best thing about being male (e.g., higher pay), males listed as the worst thing about being female (e.g., lower pay). As Cohen (2001) noted, although "some differences are exaggerated or oversimplified,... we identif[ied] a host of ways in which we 'win' or 'lose' simply because we are male or female" (p. 3).

Sociological Theories of Gender Inequality

Both structural functionalism and conflict theory concentrate on how the structure of society and, specifically, its institutions contribute to gender inequality. However, these two theoretical perspectives offer opposing views of the development and maintenance of gender inequality. Symbolic interactionism, on the other hand, focuses on the culture of society and how gender roles are learned through the socialization process.

Structural-Functionalist Perspective

Structural functionalists argue that preindustrial society required a division of labor based on gender. Women, out of biological necessity, remained in the home performing functions such as bearing, nursing, and caring for children. Men, who were physically stronger and could be away from home for long periods of time, were responsible for providing food, clothing, and shelter for their families. This division of labor was functional for society and became defined as both normal and natural over time.

masculine overcompensation thesis The thesis that men have a tendency to act out in an exaggerated male role when believing their masculinity is threatened.

Industrialization rendered the traditional division of labor less functional, although remnants of the supporting belief system still persist. With increased control over reproduction (e.g., contraception), declining birthrates, and fewer jobs dependent on physical size and strength, women's opportunities for education and workforce participation in Iraq and express negative attitudes toward homosexuality.

Men whose masculinity was threatened also reported that an SUV was more desirable than men who did not have their masculinity threatened, and they were willing to pay, on the average, $7,320 more for the SUV than nonthreatened men. Femininity threat was unrelated to any of the dependent variable indicators—support for the war in Iraq, negative views of homosexuality, SUV desirability, or SUV purchase price.

The second study looked at the relationship between gender threat and dominance attitudes, including such items as "Superior groups should dominate inferior groups" and "In getting what your group wants, it is sometimes necessary to use force against other groups." Participants were also administered scales designed to measure political conservatism (e.g., support for the U.S. military, affirmative action, etc.), system justification (e.g., "Everyone has a fair shot at wealth and happiness"), and traditionalism (e.g., "It's better to stick with what you have than to keep trying new uncertain things").

The results of study 2 indicate that men—although not women—whose gender identity was threatened scored higher on the dominance scale than men whose gender identity was not threatened, but they did not score significantly higher on political conservatism, system justification, or traditionalism. Study 3 yielded similar results.

Using a national sample of 2,210 American adults, the findings indicate that "[t]he more men felt that the status of their gender was threatened by social changes, the more they tended to support the Iraq war, hold negative views of homosexuality, believe in male superiority, and hold strong dominance attitudes" (p. 1001).

The results of these three studies lend consistent support for the masculine overcompensation thesis. Alternatively, they do not support a feminine overcompensation thesis. The authors note, however, that this does not mean that feminine overcompensation does not exist. It may simply mean that the dependent variables used in these studies—for example, support for the war on Iraq—were not sufficient to detect feminine overcompensation. Willer et al. (2013) conclude

> that extreme masculine behaviors may in fact serve as telltale signs of threats and insecurity. Perhaps those men who appear most assuredly masculine, who in their actions communicate strength, power, and dominance at great levels, may actually be acting to conceal underlying concerns that they lack exactly those qualities they strive to project. (p. 1016)

SOURCE: Willer et al. 2013

increased (Wood and Eagly 2002). Thus, modern conceptions of the family have, to some extent, replaced traditional ones—families have evolved from extended to nuclear, authority is more egalitarian, more women work outside the home, and greater role variation exists in the division of labor. Structural functionalists argue, therefore, that as the needs of society change, the associated institutional arrangements also change.

Conflict Perspective

Many conflict theorists hold that male dominance and female subordination are shaped by the relationships men and women have to the production process. During the hunting-and-gathering stage of development, males and females were economic equals, both controlling their own labor and producing needed subsistence. As society evolved to agricultural and industrial modes of production, private property developed and men gained control of the modes of production, whereas women remained in the home to bear and care for children. Inheritance laws that ensured that ownership would remain in their hands furthered male domination. Laws that regarded women as property ensured that women would remain confined to the home.

As industrialization continued and the production of goods and services moved away from the home, the gaps between females and males continued to grow—women had less education, lower incomes, fewer occupational skills, and were rarely owners. World War II necessitated the entry of a large number of women into the labor force, but in contrast with previous periods, many of them did not return to the home at the end of the war. They had established their own place in the workforce and, facilitated by the changing nature of work and technological advances, now competed directly with men for jobs and wages.

Conflict theorists also argue that continued domination by males requires a belief system that supports gender inequality. Two such beliefs are (1) that women are inferior outside the home (e.g., they are less intelligent, less reliable, and less rational) and (2) that women are more valuable in the home (e.g., they have maternal instincts and are naturally nurturing). Even when initiatives appear to be subverting these beliefs, patriarchy remains intact and often extended. Jitha (2013), in an investigation of micro-credit loans to women in India, concludes that women who step outside traditional gender roles are excluded from the program, the loans emphasize financial reward thereby demeaning unpaid household labor, and "it is in fact men who decide if the women can join the group, if the loan is required at all, and for what purpose it should be used" (p. 268). Thus, unlike structural functionalists, conflict theorists hold that the subordinate position of women in society is a consequence of social inducement rather than biological differences that led to the traditional division of labor.

Alexi Rosenfeld/Getty Images Entertainment/Getty Images

A person with a fist painted with the transgender flag attends the Black Lives Matter Protest in New York City in June 2020. According to symbolic interactionists, symbols, including language, gestures, and objects, convey social meaning. Designed by Monica Helms, the transgender flag is pink and blue—the colors associated traditionally with males and females, with white in the middle signifying those who are somewhere in-between—intersexed, transitioning, or nonbinary.

Symbolic Interactionist Perspective

Although some scientists argue that gender differences are innate, symbolic interactionists emphasize that, through the socialization process, both females and males are taught the meanings associated with being feminine and masculine. For example, in a study of fathers' interactions with their toddler sons and daughters, it was revealed that fathers of sons are more likely to engage in "rough and tumble play," and use more analytical and achievement-related language than fathers of daughters. However, when interacting with their little girls, fathers of daughters were significantly more likely to talk about emotions such as sadness, and to whistle or sing (Mascaro et al. 2017). Gender assignment begins at birth as a child is classified as either female or male. However, the learning of gender roles is a lifelong process whereby individuals acquire society's definitions of appropriate and inappropriate gender behavior.

Gender roles are taught by the family, in the school, in peer groups, and by media presentations of girls and boys and women and men (see the discussion on the social construction of gender roles later in this chapter). Most important, however, gender roles are learned through symbolic interaction as the messages that others send us reaffirm or challenge our gender performances. In an examination of parent–child interactions, Tenenbaum (2009) found that discussions regarding course selections for high school followed gender-stereotyped patterns. Here, a father talks to his fifth grade daughter (how the conversation was coded by the researchers is in brackets):

> But you know, spelling is English, right? That's what English is. For the most part generally speaking, girls do better with those kinds of skills, they have a harder time with math, generally, you know? [Code: Lack of ability.] Now, of course, you know it's nice to do the things you're really good at too, and you like to do. But, sometimes when you're trying to be, when you grow up to be someone in life, you also gotta take classes that are, not really, how do I say? You're not really good at, you have to put more practice in, right? [Code: Lack of ability.] So, I picked, I selected algebra, 'cause that's kinda, high school, mathematics. (pp. 458–459)

Although the father encouraged his daughter to take mathematics, he twice conveyed to his daughter that she is (and girls in general are) not very good in math.

Feminist theory, although also consistent with a conflict perspective, incorporates many aspects of symbolic interactionism. Feminists argue that conceptions of gender are socially constructed as societal expectations dictate what it means to be female or what it means to be male. Thus, women are generally socialized into **expressive roles** (i.e., nurturing and emotionally supportive roles), and males are more often socialized into **instrumental roles** (i.e., task-oriented roles). These roles are then acted out in countless daily interactions as boss and secretary, doctor and nurse, football player and cheerleader "do gender."

expressive roles Roles into which women are traditionally socialized (i.e., nurturing and emotionally supportive roles).

instrumental roles Roles into which men are traditionally socialized (i.e., task-oriented roles).

Ridgeway (2011), in her book *Framed by Gender*, begins by asking the research question, "How, in the modern world, does gender manage to persist as a basis or principal for inequality?" (p. 3). Her answer lies in the socially constructed meanings associated with gender. Despite the tremendous advances that have been made, whenever people encounter gender in a social relationship, circumstance, or situation that they are unsure of, they rely on traditional definitions as an organizing principle whereby "[g]ender inequality is rewritten into new economic and social arrangements as they emerge, preserving that inequality in modified form" (Ridgeway 2011, p. 7).

Noting that the impact of the structure and culture of society is not the same for all women and all men, feminists encourage research on gender that takes into consideration the intersection of class, race, ethnicity, and sexual orientation. In other words, to fully understand the experiences of others, we must acknowledge that it cannot be accomplished by concentrating on just one subordinate status.

In Afghanistan, little girls in families with no male children often assume the persona of little boys—short hair, traditional male clothing, and an appropriate male name (McKenzie 2019). In addition to the social pressure to have a boy child, the decision to raise a female child as a "bacha posh" is pragmatic—such a child can help his father on the farm, travel independently, and work outside the home. Bacha posh children return to being females sometime around puberty. Do you think gender, not sex, is a consequence of nature or a consequence of nurture?

What do you think?

Gender Stratification: Structural Sexism

As structural functionalists and conflict theorists agree, the social structure underlies and perpetuates much of the sexism in society. **Structural sexism**, also known as *institutional sexism*, refers to the ways the organization of society and specifically its institutions subordinate individuals and groups based on their sex classification. Structural sexism has resulted in significant differences in the education and income levels, occupational and political involvement, and civil rights of women and men (see Figure 10.1).

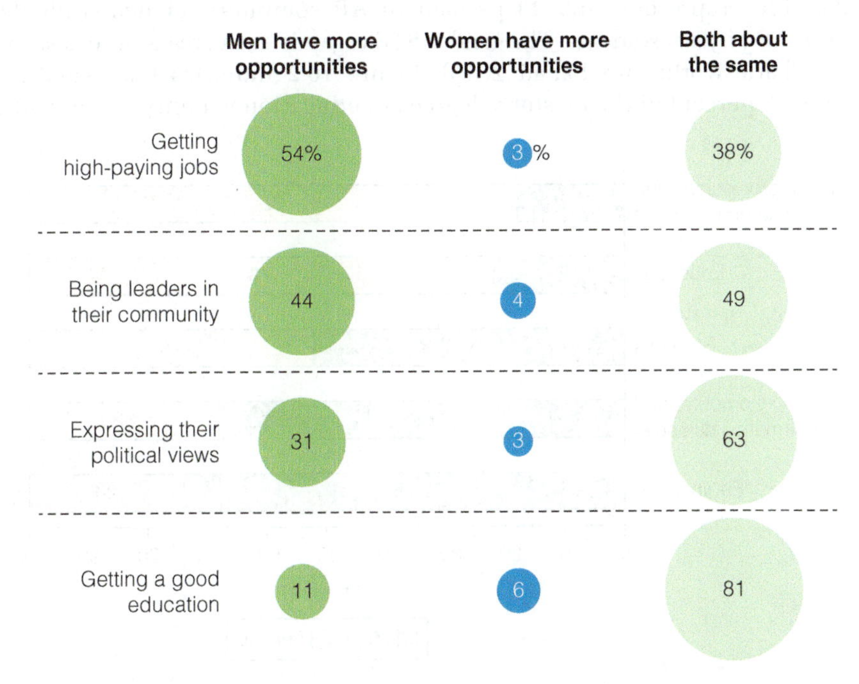

Figure 10.1 Percent Responding Men or Women Have More Opportunities in Selected Areas, Median Responses for 34 Countries, 2019
SOURCE: Horowitz and Fetterolf 2020.

structural sexism The ways in which the organization of society, and specifically its institutions, subordinate individuals and groups based on their sex classification.

Education and Structural Sexism

As discussed in Chapter 8, women comprise nearly two-thirds of the world's illiterate population (United Nations Girls Education Initiative [UNGEI] 2019). One hundred and thirty million girls worldwide are not in school, and, in poor countries, less than two-thirds of girls complete primary school and only one-third complete secondary school (Global Partnership 2019). Further, girls "are more likely to remain excluded from education while out-of-school boys stand a greater chance of eventually entering school" (United Nations Educational, Scientific and Cultural Organisation [UNESCO] 2016, p. 3).

Because children born to educated mothers are less likely to die at a young age, there is an **education dividend** associated with educating women (see Chapter 2). Globally, between 1990 and 2009, improved education for women of reproductive years saved over 2 million children under the age of 5 (UNESCO 2014). Educated women are more likely to have their children immunized and are better informed about dietary requirements. If all mothers completed primary education the maternal death rate would decline by two-thirds and with each additional year of education, the probability of contracting HIV would decline by nearly 7.0 percent. Additionally, in those countries with the highest rates of child marriages, girls who complete secondary school are five times less likely to be child brides than girls who have little to no education (Global Partnership 2019).

In 2018, few differences existed between men and women in their completion rates of high school and college degrees (National Center for Educational Statistics [NCES] 2020a, 2020b). In fact, in recent years, most U.S. colleges and universities have had a higher percentage of women than men enrolling directly from high school (see Chapter 8). Not surprisingly, this trend has led to an increase in wives having higher levels of education than their husbands. This reversal in the spousal gender gap in education, once linked to higher rates of divorce, is no longer associated with marital instability (Bavel, Schwartz, and Esteve 2018). There remains, however, concerns that many American men may not have the education they need to compete in today's global economy.

Although women have made strides in earning postsecondary and graduate degrees, concerns over the continued lack of women in STEM (science, technology, engineering, and mathematics) persist. As early as middle school, more than twice as many boys than girls express a desire to work in science- or engineering-related fields, and although U.S. females comprise a higher proportion of advanced placement (AP) high school students than males, they represent only 23 percent of AP computer science students and 29 percent of AP physics students. The gender STEM gap is even greater in postsecondary education (Charlesworth and Banaji 2019). Figure 10.2 indicates that females earn, for example, 78 percent of the master's degrees in education yet only 32 percent of the

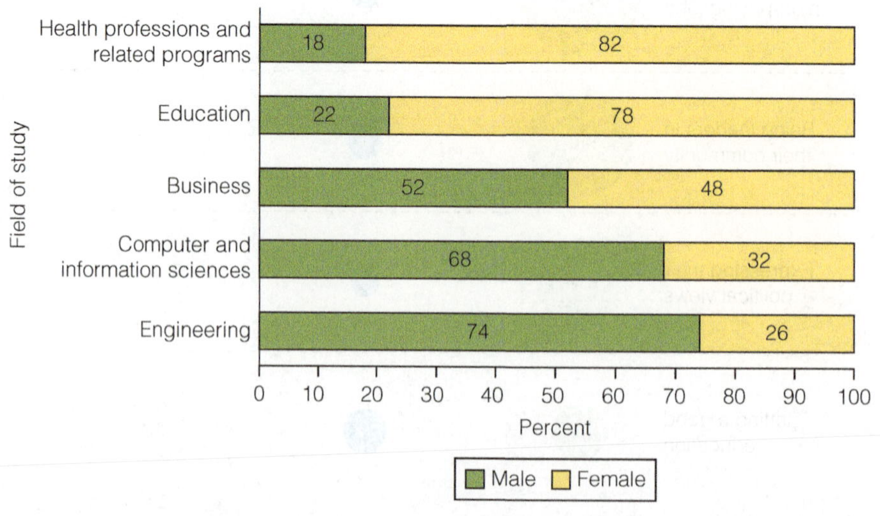

Figure 10.2 Percentage of Master's Degrees Conferred by Postsecondary Institutions in Selected Fields of Study by Sex, Academic Year 2017–2018
SOURCE: National Center for Educational Statistics (NCES) 2020b.

education dividend The associated benefits of educating women including, for example, reducing the mortality rate of children under 5 years old.

master's degrees in computer and information sciences (NCES 2020a). Reasons for the STEM gender disparity, although decreasing, include reliance on gender stereotyping ("Boys are better in math and science than girls!"), a lack of female STEM role models, little encouragement to follow STEM pursuits, and a lack of awareness about women in STEM fields.

Women earn 53.5 percent of doctorate degrees (e.g., JDs, MDs, PhDs) (NCES 2020a). Nonetheless, unlike many of their male counterparts, many women "opt out" of labor force participation. Hersch (2013) reports that female graduates of elite institutions, those most likely to find satisfying employment in their field of study, are more likely to "opt out" of the workforce than peers from less selective institutions. Furthermore, female graduates from elite institutions who have children under the age of 18 have significantly lower levels of labor force involvement than their academic equals without children under the age of 18.

Work and Structural Sexism

According to an International Labour Organization (ILO) report, over the last two decades, the employment gender gap has declined by less than 2 percent. Women are less likely to be employed than men—26 percent less likely, with 700 million women unemployed globally. Women also have higher unemployment rates because of the differences in educational attainment between men and women, occupational segregation, unpaid caregiving responsibilities, and higher rates of exiting and reentering the labor force as a result of family obligations (ILO 2019).

Women are disproportionately employed in what is called **vulnerable employment**. Such is the case in developing countries where women are often "contributing family workers," and in developed countries where women are disproportionately working in low-wage service occupations and/or self-employed (ILO 2019). Vulnerable employment is characterized by informal working arrangements, little job security, few benefits, and little recourse in the face of an unreasonable demand.

Finally, women are more likely to hold positions of little or no authority within the work environment and to receive lower wages than men. Worldwide, only 27.1 percent of managerial positions are held by women ranging from a high of 39.0 percent in the Americas to a low of 11.1 percent in the Middle East (ILO 2019). No matter what the job, if a woman does it, it is likely to be valued less than if a man does it. For example, in the early 1800s, 90 percent of all clerks were men and being a clerk was a prestigious profession. As the job became more routine, in part because of the advent of the typewriter, the pay and prestige of the job declined, and the number of female clerks increased. Presently, 83.1 percent of clerks are female (Bureau of Labor Statistics [BLS] 2020), and the position is one of relatively low pay and prestige.

The concentration of women in certain occupations and men in other occupations is referred to as **occupational sex segregation**. Although occupational sex segregation remains high, as indicated by Table 10.2, it has decreased in recent years for some occupations. Between 1983 and 2019, the percentage of female physicians and surgeons more than doubled from 16 percent to 41 percent, female dentists increased from 7 percent to 34 percent, and female clergy increased from 6 percent to 21 percent (BLS 2020; U.S. Census Bureau 2009).

Although the pace is slower, men are increasingly applying for jobs that women have traditionally held. Spurred by the loss of jobs in the manufacturing sector and the economic crisis of 2007, and the pandemic of 2020, many jobs traditionally defined as male (e.g., auto worker, construction worker) have been lost. Thus, over the last 20 years, there has been a significant increase in the number of men transitioning to traditionally held female jobs, particularly if unemployed (Yavorsky and Dill 2020).

Some evidence suggests that men in traditionally female-held jobs have an advantage in hiring, promotion, and salaries called the **glass escalator effect**. For example, teacher assistants are overwhelmingly female—89 percent. Yet the average median weekly salary for male teacher assistants in the United States is $707 compared to $579 for female teaching assistants (Institute for Women's Policy Research [IWPR] 2020). Similarly, male nurses in Great Britain, working in a predominantly female occupation, were found to have higher salaries and to be more likely to receive a promotion (Punshon et al. 2019).

@Empower_Women

Flawed and sexist economic systems have disregarded the unpaid and underpaid #carework done primarily by women and girls around the 🌍

-Empower Women

vulnerable employment Employment that is characterized by informal working arrangements, little job security, few benefits, and little recourse in the face of an unreasonable demand.

occupational sex segregation The concentration of women in certain occupations and men in other occupations.

glass escalator effect The tendency for men seeking or working in traditionally female occupations to benefit from their minority status.

TABLE 10.2 Top 10 Lowest- and Highest-Paying Jobs, Percentage Female Workers and Earnings, 2019

Job Title	% Female Workers	Median Weekly Earnings
Lowest-Paying Jobs		
Laundry and dry-cleaning workers	70.4%	$432.00
Combined food preparation and serving workers, including fast food	56.5%	$436.00
Dishwashers	19.8%	$441.00
Dining room and cafeteria attendants and bartender helpers	49.2%	$447.00
Food preparation workers	56.4%	$463.00
Cashiers	72.9%	$465.00
Maids and housekeeping cleaners	87.6%	$467.00
Hosts and hostesses, restaurant, lounge, and coffee shop	76.5%	$471.00
Food servers, nonrestaurant	69.8%	$484.00
Cooks	39.3%	$486.00
Highest-Paying Jobs		
Chief Executive	28.0%	$2,291.00
Architectural and engineering managers	12.8%	$2,226.00
Pharmacists	62.7%	$2,071.00
Physicians and surgeons	42.6%	$2,001.00
Lawyers	40.3%	$1,947.00
Nurse practitioners	84.8%	$1,894.00
Dentists	50.0%	$1,884.00
Computer and information systems managers	25.5%	$1,877.00
Software developers, applications and systems software	20.3%	$1,864.00
Chemical engineers	12.3%	$1,858.00

SOURCE: American Association of University Women (AAUW) 2020a.

Persistence of the Occupational Sex Segregation. Sex segregation in occupations continues for several reasons. First, cultural beliefs about what is an "appropriate" job for a man or a woman still exist. Snyder and Green's (2008) analysis of nurses in the United States is a case in point. Using survey data and in-depth interviews, the researchers identified patterns of sex segregation. Over 88 percent of all patient-care nurses were in sex-specific specialties (e.g., intensive care and psychiatry for male nurses, and labor or delivery and outpatient services for female nurses). Interestingly, although women rarely mentioned gender as a reason for their choice of specialty, male nurses frequently did so, acknowledging the "process of gender affirmation that led them to seek out 'masculine' positions within what was otherwise construed to be a women's profession" (p. 291).

Second, opportunity structures for men and women differ often as a result of stereotyping that impacts perceived competence. Women and men, in school and upon career entry, are often channeled by teachers and employers into gender-specific coursework or jobs that carry different wages and promotion opportunities. Cimpian et al. (2016), using longitudinal data from elementary school children, conclude that teachers consistently rate girls' mathematical proficiency lower than boys even when boys and girls have equal levels of mathematical achievement. Similarly, female medical students report that male attending physicians discourage women from entering into a career as a surgeon (Bruce et al. 2015). Even women in higher-paying jobs may be victimized by a **glass ceiling**—an invisible barrier that prevents women and other minorities from moving into top corporate positions.

Working mothers are particularly vulnerable. Female lawyers returning from maternity leave found their career mobility stalled after being reassigned to less prestigious cases (Williams 2000). Using an experimental design, Correll et al. (2007) report that, even when qualifications, background, and work experience were held constant, "evaluators rated mothers as less competent and committed to paid work than non-mothers" (p. 1332). Other examples of the "**motherhood penalty**" include women who feel pressured to choose professions that permit flexible hours and career paths, sometimes known as *mommy tracks* (Moen and Yu 2000; Elsesser 2019). Thus, women dominate the field of elementary education, which permits them to be home when their children are not in school. Nursing, also dominated by women, often offers flexible hours. Although the type of career pursued may be the woman's choice, it is a **structured choice**—a choice among limited options as a result of the structure of society.

Finally, Blau and Kahn (2013) argue that the comparatively low rates of female labor force participation and female labor force participation growth are the result of a lack of U.S. worker-friendly and perhaps, more important, female-friendly, employment policies. An analysis of policy data indicates that "most other countries have enacted parental leave, part-time work, and child care policies that are more extensive than in the United States, and the gap has grown over time" (p. 4). The authors conclude that such policies make part-time work more attractive and make it easier for women to "have it all"—that is, combine work and family life.

glass ceiling An invisible barrier that prevents women and other minorities from moving into top corporate positions.

motherhood penalty The tendency for women with children, particularly young children, to be disadvantaged in hiring, wages, and the like, compared to women without children.

structured choice Choices that are limited by the structure of society.

What do you think?

Research indicates that paternal leave exclusively reserved for fathers and non-transferable to mothers is associated with a decrease in mothers' unemployment, and with an increase in full-compared to part-time participation in the labor force as well as an increase in maternal wages (Dunatchik and Ozcan 2019). Do you think men should be required to take paternity leave after the birth of a child?

Income and Structural Sexism

In 2019, full-time working women in the United States earned, on the average, 82 percent of the median annual earnings of full-time working men (National Women's Law Center [NWLC] 2019a). Of the 120 occupations for which there is sufficient labor force data, there are only five occupations in which women's median weekly earnings are higher than men's (Hegewisch and Barsi 2020). Women "are affected by the wage gap as soon as they enter the labor force and the gap continues to expand over the course of a woman's career" (p. 1).

The gender pay gap varies over time. By decade, in 1980, 1990, 2000, 2010, and 2018, annual earnings of women over the age of 16 as a percentage of men's earnings increased from 64 percent, to 76 percent, to 77 percent, to 84 percent, and to 85 percent, respectively (Graf, Brown, and Patten 2019). As indicated, closing the gender gap has slowed down since the early 1990s. At the present rate of progress, it is predicted that, compared to White non-Hispanic men, White women will not see equal pay until the year 2055, Black women 2130, and Hispanic women 2224 (Lacorte and Hayes 2019).

Racial differences also exist. Although women, in general, earned 82 percent as much as men in 2019, Black American women earned just 62 percent of White men's salaries, and Hispanic American women earn just 54 percent of White men's salaries (NWLC 2019). Even among celebrities, significant income gaps exist. For example, with the exception of tennis, the gender pay gap in sports is staggering (Kaufman 2019). The highest base salary in the Women's National Basketball Association is $117,500; the highest base salary in the NBA is $40 million.

The gender gap not only impacts women but their families as well. Over 7 million families are headed by single working mothers with children, and more than 25 percent of them are poor (see Chapter 6). Furthermore, at some point in 2013, nearly 1.5 million married couples with children and 4.2 million married couples with no children were dependent on the woman's earnings alone. For the average woman in 2019, closing the gender wage gap would mean an additional $10,194 per year which could be used for

- three months of groceries, $1,935;
- three months' child-care payments, $2,450;
- three months' rent, $2,547;
- three months' health insurance premiums, $1,810;
- four months' student loan payments, $1,088;
- seven tanks of gas, $345 (NWLC 2019).

The motherhood penalty impacts not only occupational sex segregation but and in association with it, the pay gap between mothers and nonmothers. Globally, the motherhood pay gap is larger in developing countries than in developed ones, and it increases with the number of children. There is also some evidence that the motherhood pay gap is smaller when there is a female child who may take on some of the household labor and child care (ILO 2015). Although the number of theoretical explanations of the motherhood pay gap are considerable, many social scientists hold that the gap is a consequence of child-care needs that necessitate child-friendly jobs such as part-time employment, leaving and reentering the labor force after pregnancy, stereotypical decision making by employers, and occupational ghettos where women, in general, are paid less.

Why Does the Gender Pay Gap Exist? There are several arguments as to why the gender pay gap exists. One, the **human capital hypothesis**, holds that pay differences between females and males are a function of differences in women's and men's levels of education, skills, training, and work experience. Bertrand et al. (2009) report that the "presence of children is associated with less accumulation of job experience, more career interruptions, and shorter work hours for female MBAs but not for male MBAs" (p. 24). Based on their analysis, the authors conclude that a decade after graduation, female MBAs earn an average annual salary of $243,481, and male MBAs earn an average annual salary of $442,353.

Lower incomes over time create a significant deficit later in life. This is particularly true given a woman's higher life expectancy and the exhaustion of household savings when her husband becomes ill. Over a lifetime, a woman who works full-time for 40 years will see a deficit of $407,760 compared to her male counterpart (NWLC 2019).

Human capital theorists also argue that women make educational choices (e.g., school attended, major, etc.) that limit their occupational opportunities and future earnings. Women, for example, are more likely to major in the humanities, education, or the social sciences rather than science and engineering, which results in reduced incomes (NCES 2020a). Research also indicates, however, that after controlling for "college major, occupation, economic sector, hours worked, months unemployed since graduation, GPA, type of undergraduate institution, institution selectivity, age, geographical region, and marital status... a 7 percent difference in the earnings of male and female college graduates one year after graduation was still unexplained" (Corbett and Hill 2012, p. 8).

human capital hypothesis The hypothesis that pay differences between females and males are a function of differences in women's and men's levels of education, skills, training, and work experience.

devaluation hypothesis The hypothesis that women are paid less because the work they perform is socially defined as less valuable than the work men perform.

Further support for the human capital thesis comes from a survey of unemployed women and men between the ages of 25 and 54. Results indicate that 43 percent of "homemakers" who were able to work wanted a job outside the home. However, the majority indicated that they would prefer part- rather than full-time employment, and 61 percent compared to 37 percentage of men responded that they were not presently employed due to family responsibilities (Hamel et al. 2014). Although the percentage of working women in high-income countries has dramatically increased between 2000 and 2019 (World Bank 2020), women's labor force participation in the United States reached its height in 1999 at 60 percent, declined to 57 percent in 2019, and is projected to decline to 52 percent by the year 2060 (Catalyst Research 2019).

The second explanation for the gender gap is called the **devaluation hypothesis**. It argues that women are paid less because the work they perform is socially defined as less valuable than the work men perform. Guy and Newman (2004) argue that these jobs are undervalued in part because they include a significant amount of **emotion work**—that is, work that involves caring, negotiating, and empathizing with people, which is rarely specified in job descriptions or performance evaluations.

Finally, there is evidence that, even when women and men have equal education and experience (and, therefore, not a matter of human capital differences) and are in the same occupations (and, therefore, not a matter of women's work being devalued), pay differences remain, suggesting discriminatory occupational practices. In 2020, a California state judge allowed a class action suit against Oracle, the technology giant, to go forward stating that "education, years of prior job experience, tenure at Oracle, and performance review scores do not explain the gender gap faced by women in the same job code as men" (Rosenblatt and Burnson 2020, p. 1).

What *do you* think?

At the end of 2020, countries around the world expressed fears of a second wave of COVID-19 (see Chapter 2). Interestingly, research documents that even after controlling for size of the urban population and the percentage of the elderly, for example, countries with female leaders had fewer positive COVID-19 cases and fewer COVID-19-related deaths than countries led by men (Janjuha-Jivraj 2020). One of the reasons for the gender gap in COVID-19 cases and deaths was that female leaders (e.g., Angela Merkel of Germany) were less likely to take into consideration the impact of shutting down the country on the economy and therefore acted more decisively than their male counterparts. Why do you think that was the case?

Politics and Structural Sexism

Women received the right to vote in the United States in 1920, with the passage of the Nineteenth Amendment. Even though this amendment went into effect over a hundred years ago, women still play a rather minor role in the political arena. In general, the more important the political office is, the lower the probability that a woman will hold it. Although women constitute half of the population, no woman has ever been president of the United States, and it was not until 2021 that a woman assumed the office of the vice president of the United States. Vice President Kamala Harris is also the first Black and Asian American to be elected to the executive office. Other high-ranking women who have served in the U.S. government include former Secretaries of State Madeleine Albright, Condoleezza Rice, and Hillary Clinton, the first female presidential candidate from a major political party. In 2020, there were only nine female governors, and women held just 23.7 percent of all U.S. congressional seats leading to a political inequality rank of 86 out of 155 countries (Center for Women in Politics [CAWP] 2020; WEF 2020).

@Carolyn BMaloney

Now more than ever, our country needs to elect incredible women who can work together to get things done.

-Carolyn B. Maloney

In state legislatures, women held 29.2 percent of seats in 2020, with Nevada (53.4 percent) having the highest female representation and West Virginia (13.4 percent) the lowest (CAWP 2020). Globally, women represent just 24.3 percent of representatives to legislative bodies, an increase from 11.3 percent in 1995 (United Nations [UN] 2019). Government ministers (who are similar to secretaries of departments in the United States such as the secretary of education) are 20.7 percent female, and most often head departments responsible for social affairs, the family, children, the elderly, and the disabled. In response to the underrepresentation of women in the political arena, today half of the world's countries have some type of legislative electoral quota in the hopes of equalizing gender representation.

The Underrepresentation of Women in Politics. The relative absence of women in politics, as in higher education and in high-paying, high-prestige jobs in general, is a consequence of structural limitations and cultural definitions of traditional gender roles (see Figure 10.3). Running for office requires large sums of money, the political backing of powerful individuals and interest groups, and a willingness of the voting public to elect women. In fact, in a survey of present and former female politicians, money was identified as one of the greatest barriers

emotion work Work that involves caring for, negotiating, and empathizing with people.

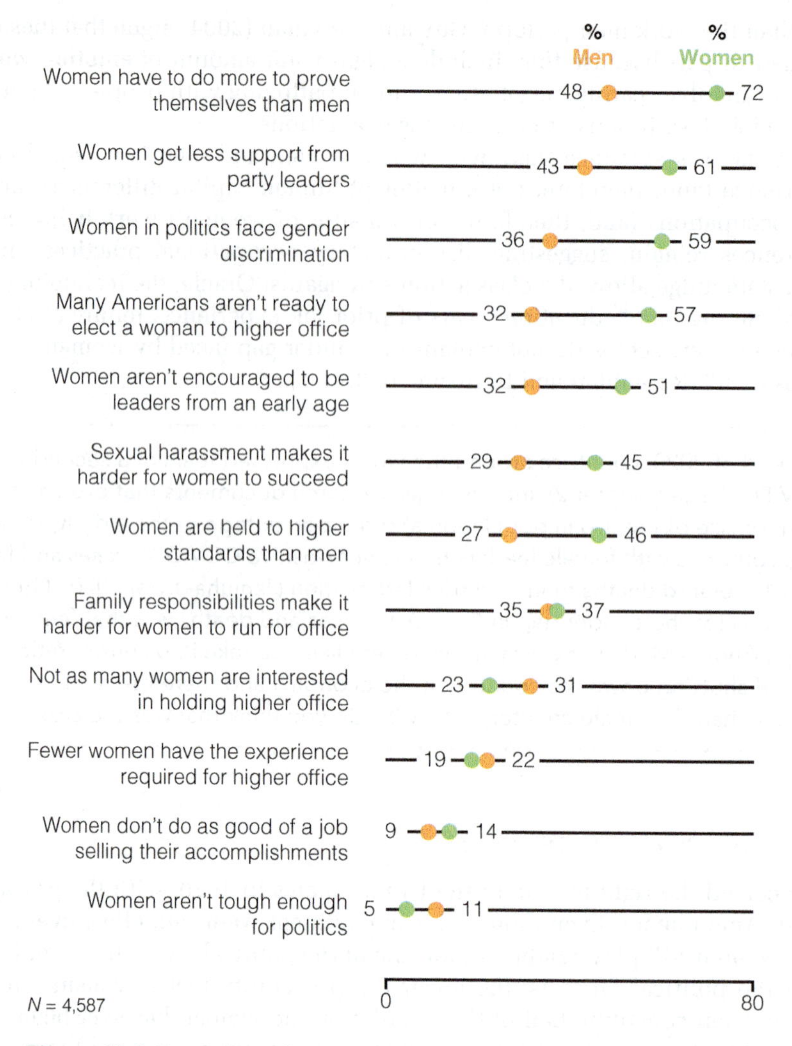

Figure 10.3 Percentage of Women and Men Saying Each Is a Major Reason Why There Are Fewer Women Than Men in High Political Offices, 2019
SOURCE: Thomas 2019.

to running for office (Green 2018). Disproportionately lacking these resources, minority women have even greater structural barriers to election and, not surprisingly, represent an even smaller percentage of elected officials. Of the 535 congressional representatives in 2020, women of color represent just 9.0 percent of the total (CAWP 2020).

There is also evidence of gender discrimination against female candidates. In an experiment where two congressional candidates' credentials were presented to a sample of respondents—in one case as Ann Clark and in the other as Andrew Clark—Republican respondents were significantly more likely to say they would vote for a father with young children rather than a mother with young children. They were also more likely to vote for women without small children than women with small children—additional evidence of the "motherhood penalty." The opposite pattern was detected for Democratic respondents (Morin and Taylor 2008).

Not only do voters discriminate based on gender, but political parties do so as well. Although once overt, today's discrimination takes place covertly as political parties overrely on established networks to recruit candidates that often exclude women. Political gatekeepers who are more likely to be men have a tendency to recruit candidates who are similar to themselves (Political Parity 2018). Thus, men recruit men to run for political office and, once candidates are elected, particularly in higher political positions, it becomes difficult to unseat them. Incumbents have a 90 percent probability of being reelected leading one female political leader to remark, "I do not believe there is a glass ceiling for women in politics, only a thick layer of men" (Political Parity 2018, p. 40).

Civil Rights, the Law, and Structural Sexism

In many countries, victims of gender discrimination cannot bring their cases to court. This is not true in the United States. The 1963 *Equal Pay Act* and Title VII of the 1964 *Civil Rights Act* make it illegal for employers to discriminate in wages or employment (e.g., hiring, firing, promotions, layoffs, etc.) based on sex.

Despite the Trump administration's "broad-based effort to eliminate transgender rights across the government, in education, housing, the military, ... and healthcare," in 2020 the U.S. Supreme Court held that the civil rights protection against sex discrimination in the workplace afforded cisgender males and females also includes transgender employees (Sanger-Katz and Green 2020, p. 1). Thus, rather than defining sex as something determined at birth, the court adopted a more inclusive definition of sex, that determined by gender identity, a position the Biden administration appears to embrace (Levin 2020).

AFP/Getty Images

Some child brides are as young as 5 years old and rarely know the ramifications of what's happening to them. Here, a 16-year-old boy waits to marry a much younger girl during a mass child wedding ceremony in Indore, India.

Despite being illegal, sex discrimination continues to occur as evidenced by the thousands of grievances filed each year with the Equal Employment Opportunity Commission (EEOC). In 2019, 23,532 grievances were filed and resolved (EEOC 2020a). One technique employers use to justify differences in pay is the use of different job titles for the same type of work—for example, janitor and housekeeper. The courts have repeatedly ruled, however, that jobs that are "substantially equal," regardless of title, must result in equal pay. In addition to individual discrimination, discrimination may take place at the institutional level (see Chapter 9).

Institutional discrimination includes screening devices designed for men, hiring preferences for veterans, the practice of promoting from within an organization based on seniority, male-dominated recruiting networks, and, in many countries, laws that prohibit women from owning or inheriting assets (Head et al. 2014; Reskin and McBrier 2000). In the United States, women's lower incomes, shorter work histories, and less collateral make it difficult for some women to obtain home mortgages, rental property, or loans.

What do you think?

There are more than 650 million child brides in the world—12 million added every year—and hundreds of thousands of them are in the United States (Taylor 2019; Brown, Riviera, and Crawford 2019). Ironically, education is one of the best deterrents of early marriage, yet early marriage most often signals the end of educational opportunities as the "skills and knowledge for a healthy transition to adulthood give way to the responsibilities of being a wife, a mother, a daughter-in-law" (Warner et al. 2014, p. 3). What do you think would be the most effective way to end this cycle?

The Social Construction of Gender Roles: Cultural Sexism

As social constructionists note, structural sexism is supported by a system of cultural sexism that perpetuates beliefs about the differences between women and men. **Cultural sexism** refers to the ways the culture of society—its norms, values, beliefs, and symbols—perpetuate the subordination of an individual or group because of the sex classification of that individual or group. Cultural sexism takes place in a variety of settings, including the family, the school, and the media, as well as in everyday interactions.

cultural sexism The ways in which the culture of society perpetuates the subordination of an individual or group based on the sex classification of that individual or group.

Ulrich Baumgarten/Getty Images

Societal definitions of the appropriateness of gender roles have traditionally restricted women and men in terms of educational, occupational, and leisure-time pursuits. Fifty years ago, little boys playing with dolls and displaying nurturing behaviors would have been unheard of.

Family Relations and Cultural Sexism

From birth, males and females are treated differently. **Gender roles** are patterns of socially defined behaviors and expectations, associated with being female or male.

Toys are one way that what is considered appropriate gender role behavior is often conveyed to children. A study by Professor Becky Francis of Roehampton University in England examined the impact of educational toys on 3- to 5-year-olds' learning. When parents were asked their child's favorite toys, boys' toys "involved action, construction, and machinery," whereas girls' toys were more often "dolls and perceived 'feminine' interests, such as hairdressing" (Lepkowska 2008, p. 1). After purchasing and analyzing the toys parents selected, Francis concluded that girls' toys had limited "learning potential," whereas boys' toys "were far more diverse" and propelled boys "into a world of action as well as technology" designed to "be exciting and stimulating" (quoted in Lepkowska 2008, p. 1).

Household Division of Labor Globally, women and girls continue to be responsible for household maintenance including cooking, gathering firewood and fetching water, and taking care of younger siblings (European Social Survey [ESS] 2013). In the United States the division of household labor is also largely stereotypical. Survey results from a sample of heterosexual married or cohabitating couples indicate that women are more often responsible for laundry, cleaning the house, and cooking, while men are more often responsible for yardwork and car maintenance (Brenan 2020).

The traditional division of labor within the home is a consequence not only of individual choice (e.g., "opting out") but of structural constraints (see this chapter's *Self and Society*). Using a survey experimental design, Pedulla and Thebaud (2015) asked a sample of 329 unmarried 18- to 32-year-olds what their preferred relationship structure was—male breadwinner, female breadwinner, or egalitarian—under various levels of work constraints. Regardless of educational level, when told to imagine that their work environment was family-friendly (e.g., had subsidized child care, flexible work options, working from home), both men and women preferred a relationship structure wherein both partners were equally responsible financially and for household duties and child care. Similarly, Rehel (2014) concludes that in the absence of work constraints, fathers equally experience the responsibility of child care and become co-parents rather than simply "helpers." Discussing paternity leave, one father notes:

> Those five weeks went by so fast—we were constantly taking care of [our son]. Really made me realize to what extent taking care of a child is more than a full-time job. You don't get your 15-minute breaks, your half-hour break when you want. You don't get time off. You don't have a switch off like you do at work. Really, your attention is always—especially with a newborn—100 percent on him. (p. 122)

A sample of U.S. 12th-grade students was used to examine preferences of husband–wife household division of labor arrangements from 1976 to 2014 (Domberger and Popin 2020). Respondents were asked, "[I]magine you are married and have one or more preschool children. How would you feel about each of the following working arrangements," followed by six household scenarios. Although there was a general movement toward more egalitarian work arrangements over time, the *husband works full-time/wife doesn't work* arrangement remained the most desirable and *husband doesn't work/wife works full-time* was the least desirable.

gender roles Patterns of socially defined behaviors and expectations associated with being female or male.

The Household Division of Labor

Each of the following questions should be answered based on your present household status or, if more appropriate, the household in which you grew up. When you are done, compare your answers to that of a random sample of U.S. adults from all 50 states.

DIRECTIONS: For each of the areas listed here, who is more likely to do each of the following in your household?

	Woman more likely	Man more likely	Both equally	Neither/Other/No opinion
1. Make decisions about furniture and decorations	______	______	______	______
2. Wash dishes	______	______	______	______
3. Keep car in good condition	______	______	______	______
4. Laundry	______	______	______	______
5. Pay bills	______	______	______	______
6. Plan family activities	______	______	______	______
7. Clean the house	______	______	______	______
8. Yardwork	______	______	______	______
9. Care for children on daily basis	______	______	______	______
10. Grocery shopping	______	______	______	______
11. Prepare meals	______	______	______	______
12. Make decisions about savings or investment	______	______	______	______

Compare your answers to the figure below. Note that the order of the household areas in the figure has been rearranged to reflect respondents' answers. The first row had the highest responses that women were more likely to engage in that behavior ("Make Decisions about Furniture and Decorations"), and the last row had the highest responses that men were more likely to engage in that behavior ("Yardwork").

Roles of Men and Women in U.S. Households*

Who is more likely to do each of the following in your household?

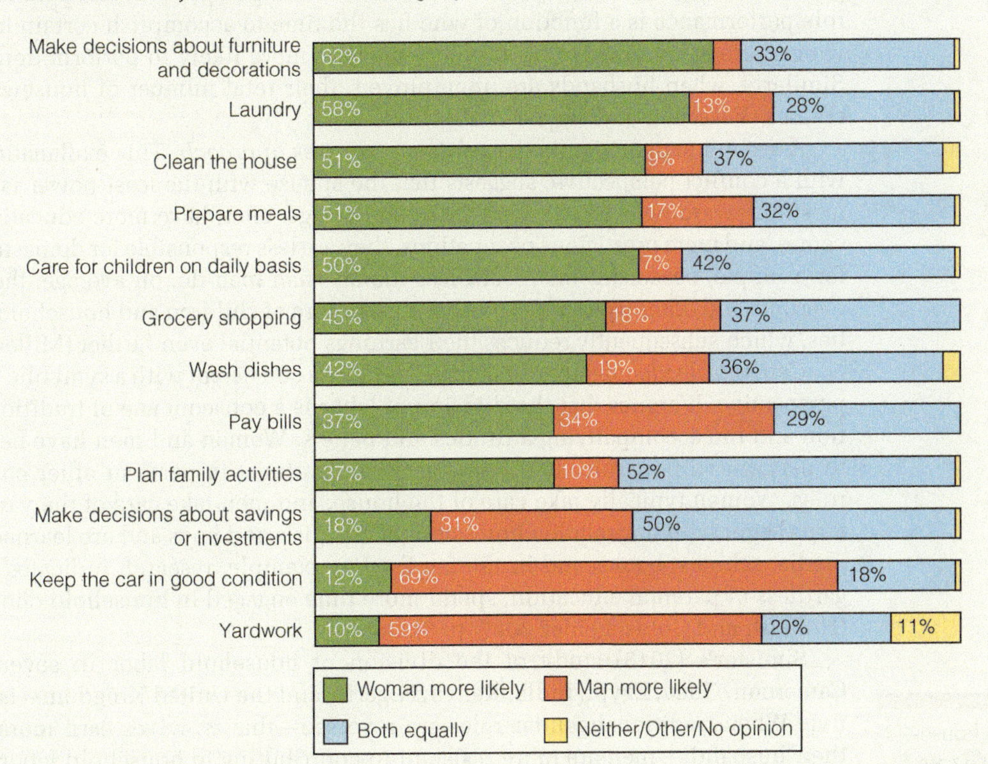

*Among heterosexual couples who are married or living together
N = 3,062

SOURCE: Gallup 2020.

@Empower_Women

Many #COVID19 frontliners are women. Not only that, women do 3x as much unpaid care work as men.

Let's call for collective action amid the #Coronavirus crisis to not exacerbate gender inequality in the workplace, marketplace and community.

-Empower Women

At the other end of the age continuum, Leopold and Skopek (2014) investigated childcare duties of more than 5,000 grandparent couples from 10 countries. The results indicate that even when both grandparents are working, grandmothers contribute more hours to caring for their grandchildren than grandfathers. Although men's share of the household labor has increased over the years and women's has decreased (Cohen 2004), the fact that women, even when working full-time, contribute significantly more hours to home care than men is known as the **second shift** (Hochschild 1989).

There has been some speculation that the global pandemic of 2020 (see Chapter 2) may, if only temporarily, change the dynamics of the "second shift" as women, disproportionately "essential" workers (e.g., nurses), continue to go to work while their furloughed, laid-off, or "working from home" husbands take on a greater share of child care and household duties (Schulte and Swenson 2020). Research by Heilman, Bernardini, and Pfeifer (2020) does suggest that, as a result of COVID-19, the number of hours men engage in child care and household labor has increased compared to pre-pandemic hours. However, it has also increased for women as parents respond to stay-at-home orders and social distancing requirements. Importantly, Heilman, Bernardini, and Pfeifer (2020) report that "respondents of both sexes are united in their view that though the unpaid care and domestic workload has increased ... the *distribution* [emphasis added] of that work had not changed dramatically as a result of the pandemic" (p. 15). Further, there is evidence that COVID-19 is disproportionately impacting professional women. For example, editors of academic journals report that, although submissions for publication are down in general, submissions from women, particularly female single authors, have dramatically declined (Flaherty 2020).

Explanations for the Traditional Division of Household Labor. Although some changes have occurred, the traditional division of labor remains to a large extent the one preferred after marriage (Parker, Horowitz and Stepler 2017; Demberger and Pepin 2020). Three explanations emerge from the literature. The first explanation is the *time-availability approach.* Consistent with the structural-functionalist perspective, this position claims that role performance is a function of who has the time to accomplish certain tasks. Because women are more likely to be at home, they are more likely to perform domestic chores. Similarly, when husbands are unemployed, their total number of housework hours increases (Vossemer and Heyne 2019).

A second explanation is the *relative resources approach.* This explanation, consistent with a conflict perspective, suggests that the spouse with the least power is relegated the most unrewarding tasks. Because men, on average, often have more education, higher incomes, and more prestigious occupations, they are less responsible for domestic labor. Thus, for example, because women earn less money than men do, on average, they turn down overtime and other work opportunities to take care of children and household responsibilities, which subsequently reduces their earnings potential even further (Miller 2020).

Gender role ideology, the final explanation, is consistent with a symbolic interactionist perspective. It argues that the division of labor is a consequence of traditional socialization and the accompanying attitudes and beliefs. Women and men have been socialized to perform various roles and to expect their partners to perform other complementary roles. Women typically take care of the house, and men take care of the yard. These patterns begin with household chores assigned to girls and boys and are learned through the media, schools, books, and in the family. For example, research indicates that girls, regardless of parental education, spend more time engaged in household chores than boys (Hofferth and Goldscheider 2017, p. 8).

Simister's (2013) study of the division of household labor in seven countries—Cameroon, Chad, Egypt, India, Kenya, Nigeria, and the United Kingdom—is also illustrative. When traditional gender roles are reversed—that is, wives earn more money than their husbands—men are more resistant to contributing to household labor than when a husband's income is greater. Similarly, Schneider (2012) reports that men in traditional female occupations spend more time on "men's work" at home; women who are in traditionally male occupations spend more time on "women's work" at home.

second shift The household work and child care that employed parents (usually women) do when they return home from their jobs.

The findings of both these studies are consistent with the **gender deviance hypothesis**. When there is "gender deviance" (income or occupation inconsistent with traditional gender roles), techniques are employed to neutralize the deviance (engage in traditional household division of labor), to bring it back into alignment, thereby reclaiming what is perceived to be what it means to be male and what it means to be female in American society. Note that the findings of both of these studies contradict the time availability approach and the relative resources approach.

The School Experience and Cultural Sexism

Around the world, cultural norms that further gender inequality persist. Norms favoring child marriage, quick conception, and large families, coupled with a lack of birth control, makes schooling difficult for many girls and young women, particularly in middle- and low-income countries. Even when education is equally available for females and males, continued gender bias in instructional materials and student–teacher interactions, and in school programs and policies may contribute to a self-fulfilling prophecy.

Instructional Materials and Students–Teacher Interactions. The bulk of research on gender images in books and other instructional materials documents the stereotypical way males and females are portrayed. In a study of 200 "top-selling" children's picture books, women and girls were significantly underrepresented, with twice as many male title and main characters. Males were also more likely to be in the illustrations, to be pictured in the outdoors and, if an adult, to be visibly portrayed as employed outside the home. Both men and women were more than nine times more likely to be pictured in traditional rather than in nontraditional occupations, and "female main characters… were more than three times more likely than were male main characters… to perform nurturing or caring behaviors" (Hamilton et al. 2006, p. 761).

Research indicates that teachers who communicate gender stereotypes impact student outcomes. Carlana (2019) measured gender stereotypes held by teachers (the independent variable) and their relationship to standardized test scores of eighth-grade Italian boys and girls (the dependent variable). Girls assigned to teachers with stronger boy-math biases had lower math scores and lower confidence in their math ability, and they were more likely to choose a vocational high school track rather than an academic one.

There is also convincing evidence that elementary and secondary school teachers are more responsive to boys than to girls—talking to them more, asking them more questions, listening to them more, counseling them more, giving them more extended directions, and criticizing and rewarding them more frequently. In *Still Failing at Fairness,* Sadker and Zittleman (2009) recount a fifth-grade teacher's instructions to her students. "There are too many of us here to all shout out at once. I want you to raise your hands, and then I'll call on you. If you shout out, I'll pick somebody else." The discussion on presidents continues, with Stephen calling out:

Stephen: I think Lincoln was the best president. He held the country together during a war.

Teacher: A lot of historians would agree with you.

Kelvin [seeing that nothing happened to Stephen, calls out]: I don't. Lincoln was OK but my Dad liked Reagan. He always said Reagan was a great president.

David [calling out]: Reagan? Are you kidding?

Teacher: Who do you think our best president was, David?

David: FDR. He saved us from the Depression.

Max [calling out]: I don't think it's right to pick one best president. There were a lot of good ones.

Teacher: That's interesting.

gender deviance hypothesis
The tendency to overconform to gender norms after an act(s) of gender deviance; a method of neutralization.

poverty rates for women, particularly women of color, were predicted to increase worldwide as the result of the COVID-19 pandemic (Cox 2020).

The Social-Psychological Costs of Gender Socialization

How we feel about ourselves begins in early childhood. Significant others, through the socialization process, expect certain behaviors and prohibit others based on our birth sex. Both girls and boys, often because of these expectations and the ability to perform them, feel varying degrees of self-esteem, autonomy, depression, and life dissatisfaction. For example, women generally have higher rates of depression and/or anxiety (see Chapter 2). Although it may be tempting to attribute such differences to biology, research by Platt et al. (2016) suggests that anxiety and depression in women are more likely a function of the gender pay gap in the workplace. When female incomes were *greater* than those of their matched male counterparts, the likelihood of depression and anxiety was significantly reduced.

Similarly, Freeman and Freeman (2013), in their book *Stressed Sex: Uncovering the Truth about Men, Women, and Mental Health*, conclude that the higher rates of mental health problems in women is a result of "life events."

> Being judged on one's appearance and the degree to which one conforms to a largely unattainable physical "ideal," shouldering the burden of responsibility for family, home and career, growing up in a society that routinely valorizes masculinity while belittling femininity, and having to run the gauntlet of everyday sexism—all of these factors are likely to help lower women's self-esteem, increase their level of stress and leave them vulnerable to mental health problems. (p. 1)

> "A gender-equal society would be one where the word "gender" does not exist: where everyone can be themselves."
>
> –GLORIA STEINEM, FEMINIST

Transgender individuals also suffer from self-esteem issues and depression. A study by the Centers for Disease Control and Prevention found that 2 percent of high school students in the United States identify as transgender and that one-third of them have attempted suicide (Johns et al. 2019) (see Chapter 11). In a review of the literature, Drydakis (2019) concludes that changing one's appearance to match one's gender identity, i.e., transitioning, is linked to higher rates of reported life satisfaction, job satisfaction, optimism about the future, and self-esteem. However, even a positive transition experience can be mediated by family rejection, negative health care experiences, employment and housing discrimination, and internalized stigma. Consistent with these results, a 2020 letter signed by over 200 medical professionals, counselors, and social workers sent to state legislatures considering limiting transgender minors' rights, stated that "[T]o put it plainly, gender affirming care saves lives and allows trans young people to thrive" (Andrew 2020, p. 1).

Pressure to conform to traditional gender roles exists not only at the individual level but at the societal level as well. After administering a questionnaire to a sample of more than 6,500 undergraduates in 13 countries, Arrindell et al. (2013) conclude that in societies where there is a strong emphasis on masculinity (i.e., in "tough countries"), masculine gender role stress is significantly higher than in countries where a masculine identity is less emphasized (i.e., in "soft countries").

The traditional male gender role places enormous cultural pressure on men to be successful in their work and to earn high incomes. Sanchez and Crocker (2005) found that, among college-aged women *and* men, the more participants were invested in traditional ideals of gender, the *lower* their self-concept and psychological well-being. Traditional male socialization also discourages males from expressing emotion and asking for help—part of what William Pollack (2000) calls the **boy code**.

boy code A set of societal expectations that discourages males from expressing emotion, weakness, or vulnerability, or asking for help.

It must not go unsaid, however, that although women have higher rates of some types of stress-related mental illnesses, men are more likely to abuse alcohol and drugs (see Chapter 3) and to commit suicide (see Chapter 2) (see "Gender-Based Violence" in this chapter). In the United States, in 2018, the suicide rate for men was nearly 4 times the rate for females, although suicide rates for women are increasing at a faster pace than suicide rates for men (Centers for Disease Control and Prevention [CDC] 2020).

Although girls and women report facing more pressure to be physically attractive than men, men report significant gender pressures as well. Men between the ages of 18 and 36 compared to older men (i.e., those 37 years old and older) are more likely to report pressure to "throw a punch if provoked," "have many sexual partners," and "join in when other men are talking about women in a sexual way" (Parker, Horowitz, and Stepler 2017, p. 2). Why do you think, at a time when the concept of gender is becoming more fluid, younger men feel more pressure to conform to the traditional notion of masculinity than older men?

What *do you* think?

Gender Role Socialization and Health Outcomes

There is a gender gap in health outcomes throughout the world; women live longer than men, by about five years, and the difference is increasing (White and Witty 2009; Harvard Men's Health Watch [HMHW] 2019) (see Chapter 2). Men are less likely to go to a doctor or to have an annual physical examination than women for a variety of structural and cultural reasons. Men, for example, are more likely to be working full-time compared to women, making it difficult to see a doctor or attend preventive medicine programs that are often only available during the day.

Traditionally defined gender roles for men are linked to high rates of cirrhosis of the liver (e.g., alcohol consumption), many cancers (e.g., tobacco use), and cardiovascular diseases (e.g., stress). Men are also more likely to engage in self-destructive behaviors—poor diets, lack of exercise, higher drug use, refusal to ask for help or wear a seat belt, and stress-related activities—more often than women (HMHW 2019). Men are also more likely to contract COVID-19 than women; they take the pandemic less seriously and are less likely to maintain social distance, wear protective gear, and wash their hands (Kahn 2020).

The World Health Organization (WHO 2009, 2011, 2018) has identified several ways traditional definitions of gender impact the health and well-being of women and girls. Primarily responsible for household duties, women are exposed to hundreds of pollutants as they cook that contribute to their disproportionately high rates of death from chronic obstructive pulmonary disease (COPD). Deaths from lung cancer and other tobacco-related illnesses are expected to rise as the tobacco industry targets women in developing countries. Finally, many women and girls throughout the world have a higher probability of suffering or dying from a variety of diseases because of their gender: They are more likely to be poor, are less likely to be seen as worthy of care when resources are short, and, in many countries, are forbidden to travel unaccompanied by a male, making access to a hospital difficult.

gender-based violence Any harm that is perpetrated against a person because of power inequalities based on gender roles.

Almost all maternal deaths take place in developing countries (WHO 2019). Although the maternal mortality rate has declined, approximately 810 women die every day while giving birth (WHO 2019), and many others "suffer serious complications from pregnancy, labor, and delivery, which can result in long-term disabilities" (U.S. Department of State 2012, p. 1). Maternal morbidity is linked to child marriage and the resulting adolescent pregnancy. Each year, 12 million girls under the age of 18 marry, the highest proportion in sub-Saharan Africa (UNICEF 2020).

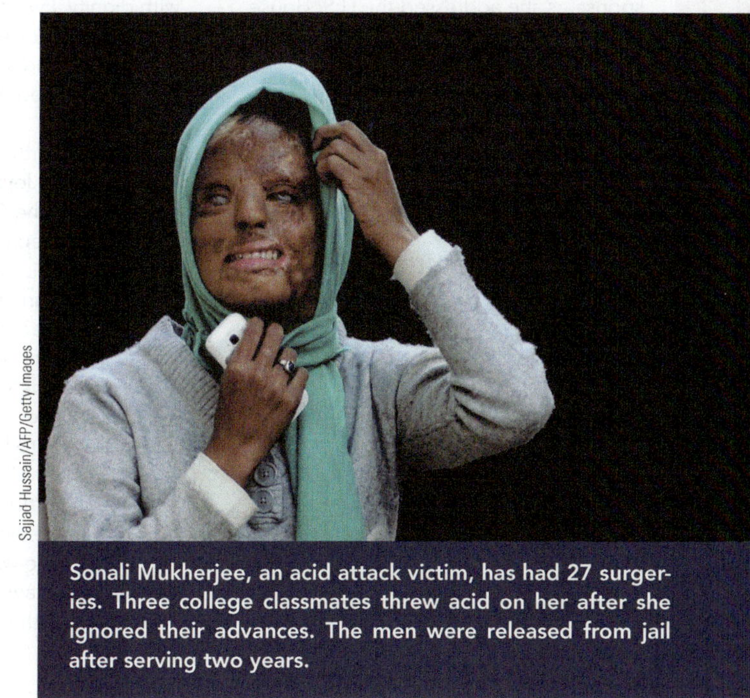

Sajjad Hussain/AFP/Getty Images

Sonali Mukherjee, an acid attack victim, has had 27 surgeries. Three college classmates threw acid on her after she ignored their advances. The men were released from jail after serving two years.

Gender-Based Violence

Gender-based violence is a general term for "any harm that is perpetrated against a person's will and that results from power inequalities based on gender roles"

(Wirtz et al. 2020, p. 1). Gender-based violence includes domestic violence and marital rape (see Chapter 5), sexual violence and sexual exploitation (see "Online Activism" in this chapter, and Chapter 4), honor killings, and dowry killings, as well as child marriage and FGM/C, which have already been discussed. Young or old, male or female, cisgender or transgender, straight or gay, rich or poor, people of all races, religions, and ethnicities are victims of gender-based violence (see this chapter's *The Human Side*).

Men are more likely than women to be involved in violence—to kill and be killed; to wage war and die both as combatants and noncombatants; to take their own lives, usually with the use of a firearm; to engage in violent crimes of all types; to bully, harass, and abuse. As sociologist Michael Kimmel (2012) notes in his discussion of the mass killings at Sandy Hook Elementary School, unlike girls,

> boys learn that violence is not only an acceptable form of conflict resolution, but the one that is admired. ...They learn it from their fathers, ...from a media that glorifies it, from sports heroes who commit felonies and get big contracts, from a culture saturated in images of heroic and redemptive violence. They learn it from each other.

"I Don't Want My Body Anymore"

On July 20, 2020, Florida Representative Todd Yoho approached Congresswoman Alexandria Ocasio-Cortez of New York and, in a "brief but heated exchange," insulted her several times and, minutes later, used a sexist slur (Lillis 2020). A day later, Mr. Yoho apologized for the "abrupt manner of the conversation" but denied using the slur despite confirmation from others (p. 1). Following is the transcript of Ms. Ocasio-Cortez's response as she addressed the U.S. House of Representatives.

> Thank you, Madam Speaker. And I would also like to thank many of my colleagues for the opportunity to not only speak today but for the many members from both sides of the aisle who have reached out to me for support following the incident this week.
>
> About two days ago, I was walking up the steps of the Capitol when Representative Yoho suddenly turned a corner ... and accosted me on the steps right here in front of our nation's capital. I was minding my own business, walking up the steps, and Rep. Yoho put his finger in my face: he called me 'disgusting', he called me 'crazy', he called me 'out of my mind', and he called me 'dangerous'. And then he took a few more steps, and after I had recognized his comments as rude, he walked away and said 'I'm rude? You're calling me rude?'
>
> I took a few steps ahead and I walked inside and cast my vote, because my constituents send me here each and every day to fight for them and to make sure they are able to keep a roof over their head, that they are able to feed their families, and that they are able to carry their lives with dignity. I walked back out and there were reporters in the front of the Capitol, and in front of reporters, Rep. Yoho called me, and I quote, 'a f*****g b****.'
>
> These were the words that Representative Yoho levied against a Congresswoman, the Congresswoman that not only represents New York's 14th Congressional District, but every congresswoman and every woman in this country, because all of us have had to deal with this in some form, some way, some shape, at some point in our lives.
>
> And I want to be clear that Representative Yoho's comments were not deeply hurtful or piercing to me, because I have worked a working-class job. I have waited tables in restaurants, I have ridden the subway, I have walked the streets in New York City. And this kind of language is not new. I have encountered words uttered by Mr. Yoho and men uttering the same words as Mr. Yoho while I was being harassed in restaurants. I have tossed men out of bars that have used language like Mr. Yoho's, and I have encountered this type of harassment riding the subway in NYC. This is not new. And that is the problem.
>
> Mr. Yoho was not alone. He was walking shoulder to shoulder with Representative Roger Williams. And that's when we start to see that this issue is not about one incident. It is cultural. It is a culture of lack of impunity, of accepting of violence and violent language against women, an entire structure of power that supports men, because not only have I been spoken to disrespectfully, particularly by members of the Republican party and elected officials in the Republican Party, not just here, but the [former] President of the United States last year told me to 'go home to another country,' with the implication that I don't even belong in America. The Governor of Florida, Governor DeSantis, before I even was sworn in, called me a 'whatever-that-is.'

Kimmel (2011) also argues that violence against transgender and gay youth is rooted in notions of masculinity and the threat that gender nonconformity poses to "real men." Consistent with this assertion, in a study of male aggression toward feminine gay, feminine heterosexual, masculine gay, and masculine heterosexual men, only feminine heterosexual men elicited a significantly higher level of aggression, suggesting a greater violation of gender conformity than that attributed to gay men (Sloan et al. 2015). Not surprisingly, transgender youth report high rates of being harassed, physically assaulted, and sexually assaulted at school (Johns et al. 2019).

Attacks on women's and girls' bodies are fairly routine, often taking place in the name of religion, war, or honor. Globally, one in three women and girls will be the victim of gender-based violence (United Nations Population Fund 2020). For example, every 90 minutes an **honor killing**—a murder, often public, because of a female dishonoring or being perceived to have dishonored her family or community—takes place. Although illegal throughout the world, even in Muslim countries where they occur the most frequently, laws prohibiting such behaviors are often not enforced or contain loopholes

honor killings Murders, often public, as a result of a female dishonoring, or being perceived to have dishonored, her family or community.

Dehumanizing language is not new. And what we are seeing is that incidents like these are happening in a pattern. This is a pattern of an attitude towards women and de-humanization of others. So while I was not deeply hurt or offended by little comments that were made, when I was reflecting on this, I honestly thought that I was just gonna pack it up and go home. It's just another day, right?

But then yesterday, Representative Yoho decided to come to the floor of the House of Representatives and make excuses for his behavior. And that I could not let go. I could not allow my nieces, I could not allow the little girls that I go home to, I could not allow victims of verbal abuse and worse, to see that—to see that excuse—and to see our Congress accept it as legitimate, and accept it as an apology, and to accept silence as a form of acceptance. I could not allow that to stand, which is why I am rising today to raise this point of personal privilege.

And I do not need Representative Yoho to apologize to me. Clearly, he does not want to. Clearly, when given the opportunity, he will not, and I will not stay up late at night waiting for an apology from a man who has no remorse over calling women and using abusive language towards women. But what I do have issue with is using women, wives, and daughters, as shields and excuses for poor behavior.

Mr. Yoho mentioned that he has a wife and two daughters. I am two years younger than Mr. Yoho's youngest daughter. I am someone's daughter too. My father, thankfully, is not alive to see how Mr. Yoho treated his daughter. My mother got to see Mr. Yoho's disrespect on the floor of this House towards me on television. And I am here because I have to show my parents that I am their daughter, and that they did not raise me to accept abuse from men.

Now, what I am here to say is that this harm that Mr. Yoho levied—tried to levy—against me was not just an incident directed at me, but when you do that to any woman, what Mr. Yoho did was give permission to other men to do that to his daughters. In using that language in front of the press, he gave permission to use that language against his wife, his daughters, women in his community, and I am here to stand up to say 'that is not acceptable.' I do not care what your views are. It does not matter how much I disagree, or how much it incenses me, or how much I feel people are dehumanizing others. I will not do that myself. I will not allow people to change and create hatred in our hearts.

And so, what I believe is that having a daughter does not make a man decent. Having a wife does not make a decent man. Treating people with dignity and respect makes a decent man. And when a decent man messes up—as we all are bound to do—he tries his best and does apologize. Not to save face, not to win a vote—he apologizes genuinely to repair and acknowledge the harm done so that we can all move on.

Lastly, what I want to express to Mr. Yoho is gratitude. I want to thank him for showing the world that you can be a powerful man and accost women. You can have daughters and accost women without remorse. You can be married and accost women. You can take photos and project an image to the world of being a family man and accost women without remorse and with a sense of impunity. It happens every day in this country. It happened here on the steps of our nation's Capital. It happens when individuals who hold the highest office in this land admit—admit—to hurting women and using this language against all of us.

But once again, I thank my colleagues for joining us today. I will reserve the hour of my time and I will yield to my colleague, Rep. Jayapal of Washington. Thank you.

SOURCE: Public Document.

that will permit the killer to go unpunished (Kristof 2016). Honor killings also take place in Western countries. In 2003, 17-year-old Shafilea Ahme, who was born in England, was murdered by her parents for refusing to agree to an arranged marriage to a man in Pakistan. It took nine years to convict her parents of the murder (Gill 2019).

To assess contemporary attitudes toward honor killing, Eisner and Ghuneim (2013) administered a survey to 856 ninth graders in Amman, a city of 2.5 million and the capital of Jordan. The results indicate that nearly half of the boys and 20 percent of the girls surveyed condone honor killings—that is, believe that killing a woman who has dishonored her family is justifiable. The authors conclude that "three... factors are important for understanding attitudes toward honor killings, namely traditionalism, the belief in female chastity as a precious good, and a general tendency to morally neutralize aggressive behavior" (p. 413).

Dowry killings involve a woman being killed by her husband or in-laws when she or her family are unable to give money or valuables to her husband's parents—a gift to the new family (United Nations Development Fund for Women 2007; Selby 2018). Although the act of giving or receiving a dowry has been illegal in India since 1961, nearly 8,000 dowry deaths occur each year, 20 every day, and the number is increasing (Selby 2018; Bundhun 2017). Despite changes in marriage practices in India, marrying later and fewer arranged marriages, the custom continues based on the cultural premise that women are to be valued less than men and that, therefore, the husband's parents should be compensated for taking the wife into their family.

Misogyny and Men's Rights Groups. Globally, one of the most "severe forms of violence against women and girls" is human trafficking (ILO 2014, p. 1)—predominantly for domestic servitude and sexual slavery (see Chapter 4). Human trafficking of women and girls, as well as other forms of violence against females, is rooted in gender inequality and the lingering notion that women and girls are *property.* **Misogyny**, defined as hatred of women (Aron 2019), is not only part and parcel of hate groups in general, particularly neo-Nazi and white supremacy groups, male supremacy groups have recently become so common that the Southern Poverty Law Center [SPLC] classifies them as a "hate movement" in the United States (Hendrix 2019).

Forty-year-old Scott Paul Beierle was a confirmed woman hater. His stories, poems, and song lyrics, beginning in eighth grade, described horrific fantasies of rape and murder. He was banned from his college campus, booted out of the Army, and fired from several jobs for talking to and/or touching women inappropriately. In 2018, he drove 250 miles to Tallahassee Florida where he entered a yoga studio and shot six women, two of them fatally, before taking his own life.

What do you think?

A 2019 indictment reads that Jeffrey Epstein, "willfully and knowingly did combine, conspire, confederate, and agree ... to commit an offense against the United States, to wit, sex trafficking of minors" (Indictment 2019, p. 1). However, prior to his trial, Epstein was found hanging in his prison cell in an apparent suicide. A compensation fund for his victims has been set up out of the multimillionaire's estate. On what basis do you think awards should be made?

Beierle was a member of a group known as *incels*—a subculture of online men who are involuntarily celibate and blame women for their celibacy. It would be comforting to think that Beierle's behavior was the act of a single deranged individual, an anomaly or outlier, but, following the attack, Beierle's actions were hailed by other incels who described his targets as "spandex wearing yoga whores" (quoted in Hendrix 2019).

Similar rampages, fueled by the same motivation, have taken place in recent years. A second attack occurred in 2018 when the driver of a van jumped the curb of a Toronto sidewalk, killing ten people. When the driver was arrested, he admitted that he was a "violent misogynist" inspired by an online group of men calling themselves "incels" (Cecco 2019). In 2020, a 17-year-old entered a Canadian establishment and fatally stabbed a woman and injured another in what police have called a hate-motivated crime; evidence

dowry killing The killing of a woman by her husband or in-laws when she or her family are unable to give money or valuables to her husband's parents.

Misogyny Hatred of women.

was uncovered that linked the teenager to the incel movement (Godin 2020). Such behaviors are thought to fall under the rubric of **toxic masculinity**, extreme forms of aggression, violence, and misogyny, socially induced and culturally specific (Salter 2019).

Misogyny is a "dangerous and underestimated component of extremism" (Anti-Defamation League [ADL] 2018) fueled by the belief that men are superior to women and, therefore, are owed a superior status in society. However, not all "men's rights" groups are, of course, violently misogynistic. Groups such as The Red Pill, MGTOW (i.e., men going their own way), a Voice for Men, and Justice for Men and Boys, in varying degrees, use their platforms to support important gender-based concerns of men: unfair custody agreements, gender-based court alimony policies, domestic violence against men, deconstruction of the male gender role, and so forth. However, most men's rights groups today, in contrast to men's equality groups, call for the return of patriarchy one way or another.

Stephanie Keith/Getty Images News/Getty Images

In August 2019, Jeffrey Epstein was found hanging in his New York City jail cell after guards repeatedly failed to check on him every two hours as instructed. The multimillionaire was charged with sex trafficking of underage girls and was being held without bail. Here a group of women protest in front of the federal courthouse just a month before his death.

Strategies for Action: Toward Gender Equality

In recent decades, awareness of the need to increase gender equality has grown throughout the world (see Figure 10.4). Strategies to achieve this end have focused on empowering women in social, educational, economic, and political spheres and improving women's access to education, nutrition, health care, and basic human rights. But as we will see in the following section on social movements, online activism, government policies, and international efforts, concern with gender inequities also requires addressing issues facing men.

Social Movements

Efforts to achieve gender equality in the United States have been largely fueled by the feminist movement. Despite the conservative backlash of men's supremacy groups and other anti-feminist organizations, feminists have made some gains in reducing structural and cultural sexism.

> "Human rights are women's rights, and women's rights are human rights."
>
> **–HILLARY CLINTON, PRESIDENTIAL CANDIDATE**

Feminism and the Women's Movement. **Feminism** is the belief that women and men should have equal rights and responsibilities. The U.S. feminist movement began in Seneca Falls, New York, in 1848, when a group of women wrote and adopted a women's rights manifesto modeled after the Declaration of Independence. Although many of the early feminists were primarily concerned with suffrage, feminism has its "political origins ...in the abolitionist movement of the 1830s," when women learned to question the assumption of "natural superiority" (Andersen 1997, p. 305). Early feminists were also involved in the temperance movement, which advocated restricting the sale and consumption of alcohol, although their greatest success was the passing of the Nineteenth Amendment in 1920, which recognized women's right to vote.

The rebirth of feminism almost 50 years later was facilitated by a number of interacting forces: an increase in the number of women in the labor force, the publication of Betty Friedan's book *The Feminine Mystique* (1963), an escalating divorce rate, the socially and politically liberal climate of the 1960s, student activism, and the establishment of the Commission on the Status of Women by John F. Kennedy. The National Organization for Women (NOW) was established in 1966, and it remains one of the

toxic masculinity An extreme forms of aggression, violence, and misogyny, socially induced and culturally specific.

feminism The belief that men and women should have equal rights and responsibilities.

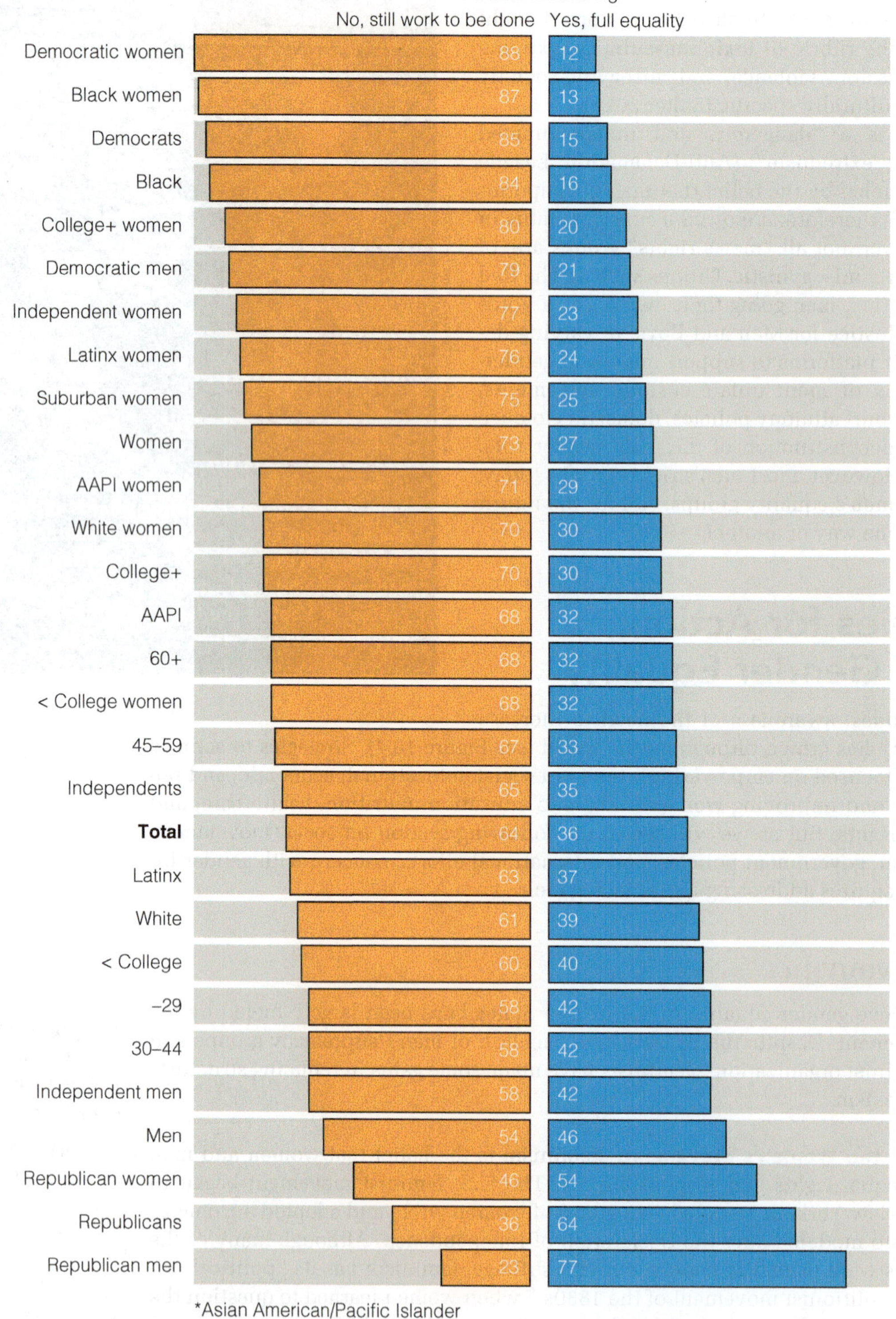

Figure 10.4 Percentage Answering "Do you think there is full equality for women in work, life, and politics or is there still work to be done?" by Demographic Variables, 2019*

*N=1,912; 2020 likely voters

SOURCE: Supermajority 2019.

largest feminist organizations in the United States, with hundreds of thousands of members in over 500 chapters across the country. The purpose of NOW is "to promote feminist ideals, lead societal change, eliminate discrimination, and achieve and protect the equal rights of all women and girls in all aspects of social, political, and economic life" (NOW 2020, p. 1).

One of NOW's hardest fought battles is the struggle to win ratification of the **equal rights amendment (ERA)**, which states that "equality of rights under the law shall not be denied or abridged by the United States, or by any state, on account of sex." The proposed twenty-eighth amendment to the Constitution passed both the House of Representatives and the Senate in 1972 but failed to be ratified by the *required 38* states by the 1979 deadline, which was later extended to 1982.

In 2019, resolutions were introduced in both houses of Congress to eliminate the 1982 deadline for the ratification of the ERA, and, in 2020, the U.S. House of Representatives voted to remove the deadline (Kurtzleben 2020). The debate over the removal of the deadline has recently grown in significance as Virginia became the 38th and last state needed to adopt the amendment (Schneider and Vozzella 2020). Despite the former administration's position that the time for the ERA had expired, President Biden has stated that he will advocate for the ERA now that the required number of states has ratified the amendment (Biden 2020).

Proponents of the amendment, such as Speaker of the U.S. House of Representatives Nancy Pelosi, argue that "the ERA will strengthen America, unleashing the full power of women in our economy and upholding the value of equality in our democracy" (quoted in Kurtzleben 2020, p. 1). Proponents also argue that "without the explicit wording and intention of women's rights documented in the principles of our government, women remain second-class citizens" (Cook 2009, p. 1). Forty-six percent of women between the ages of 18 and 35 consider themselves feminists, and, when asked what feminism means, a majority of respondents simply replied "equality of the sexes" (Tin 2018).

Black Feminism. Sociology must not only examine the feminist movement from the perspective of the White, middle-class women who predominantly populate it but also analyze the unique experiences of women of color at the intersection of race, class, and gender. It is only then that researchers can look at the "interlocking systems of oppression which function within Black women's experiences as Black, as women, and as having a particular class location" within the social structure (Walkington 2019, p. 51). The gender pay gap, occupational segregation, the underrepresentation of women in politics, and violence against transgender women disproportionately impact poor, working-class women of color.

Ironically, the women's rights movement, as well as the Black Lives Matter movement, heavily influenced by women of color, has disenfranchised Black women in favor of White women and Black men, respectively (Runyan 2018). The very origin of the women's movement and the feminism of the 1960s and 1970s focused on married, White women's desires to shed the housewife role supported by what Friedan (1963) called a **feminine mystique**, i.e., the false belief that women can only be fulfilled by domestic life (Biana 2020). In a critique of Friedan, Black feminist theorist Hooks (1984) states that Friedan failed to "speak of the needs of women without men, without children, without homes. She ignored the existence of all non-white women and poor white women" (pp. 1–2).

Four central themes distinguish Black feminist thought from the whitewashed versions of early feminists: oppression as a function of power differentials (i.e., racism, sexism, and classism), commitment to social justice and resistance to oppression, ownership of Black narratives, and the intersection of marginal identities (Collins 2015). While the narratives of Black women's lives have too often focused on stereotypical notions of the single welfare mother, there is a dearth of research on, for example, the racialized gender violence against Black women. Despite the women's movement and feminism and despite the #MeToo campaign and #Black Lives Matter, "we are still living at a moment when violence against Black women often fails to register as a pressing social justice issue" (Williamson 2020, p. 525).

Marching for Equality. In 2020, women across Mexico joined together in a national strike called a *Day without Women* to protest the government's inaction against increasingly high rates of **femicide**, that is, the killing of women and girls targeted because of their gender. In 2015, 426 women were the victims of femicide; in 2019, more than twice that number. Over 80,000 people joined the march, Mexico's largest, with similarly themed marches previously held in Chile, Argentina, Poland, and Spain (Averbuch 2020).

equal rights amendment (ERA) The proposed Twenty-eighth Amendment to the Constitution, which states that "equality of rights under the law shall not be denied or abridged by the United States, or by any state, on account of sex."

feminine mystique The false belief that women can only be fulfilled by domestic life.

femicide The killing of women and girls targeted because of their gender.

In response to the 2016 election of Donald Trump, the largest protest in U.S. history took place just days after his inauguration. More than 2 million people demonstrated across the country representing a diverse cross section of Americans. According to the Women's March organizers (Women's March 2020, p. 1):

> We must create a society in which women—including Black women, Indigenous women, poor women, immigrant women, disabled women, Jewish women, Muslim women, Latinx women, Asian and Pacific Islander women, lesbian, bi, queer, and trans women—are free and able to care for and nurture their families, however they are formed, in safe and healthy environments free from structural impediments.

However, when a sample of 516 participants in the 2017 Washington, D.C., march, the largest in the country, were asked to self-identify their motivation for attending the event, only 53 percent responded, "women's rights." Other top answers included "equality," "reproductive rights," "the environment," and "racial justice." Closer examination of the data revealed that motivation for attending the march was significantly associated with the respondents' social identities based on gender, race, ethnicity, and sexual orientation (Fisher, Dow, and Ray 2017). Thus, women compared to men were more likely to identify women's rights as their motivation for attendance, Black individuals were more likely to identify racial justice, and so forth.

Similar national and international women's marches have taken place in subsequent years including 2018, 2019, and 2020. In-between marches, supporters stay in contact via social media. Using a sample of 134 Instagram posts containing the hashtag #womensmarch, Einwohner and Rochford's (2019) content analysis reveals that the "majority of the Instagram posts ... contained content intended to teach the viewer about some topic and/or to persuade the viewer to take some kind of action" (p. 1100). The researchers reject the label of *slacktivist* to describe people who protest online rather than in person, arguing rather that "virtual space is an extension of physical space" and that online activism is useful and can act as a conduit between physical events.

Finally, in 2020, in the wake of the Black Lives Matter protests, tens of thousands of Americans rallied across the United States in support of Black transgender women and men. The New York City rally alone, sometimes called Brooklyn Liberation, had an estimated 15,000 people in attendance. With placards reading "Black Trans Lives Matter" and "Say Their Names," the rallies were, in part, a reaction to the suspected murder of two transgender women, as well as the Trump administration's repeal of health care protections for transgender individuals that were put in place during the Obama administration (Miller 2020) (also see Chapter 11).

The Men's Equality Movement. As a consequence of the women's rights movement, men began to reevaluate their own gender status. As with any social movement, the men's liberation movement has a variety of factions. One of the early branches of the men's movement is known as the mythopoetic men's movement, which began after the publication of Robert Bly's (1990) *Iron John*—a fairy tale about men's wounded masculinity that was on the *New York Times* best-seller list for more than 60 weeks (Zakrzewski 2005). Participants in the men's mythopoetic movement met in men-only workshops and retreats to explore their internal masculine nature, male identity, and emotional experiences through the use of stories, drumming, dance, music, and discussion.

Today, there is a sharp distinction to be made between the *men's rights movement* that view gender equality as a zero-sum game—i.e., if women win, men lose—and the *men's liberation groups* that advocate for gender equality and work to make men more accountable for sexism, violence, and homophobia. Men's liberation groups today continue to critique the meaning of the male gender role, the "straitjacketed definition of masculinity ... [and] the myths that construct[ed] men as naturally dominant, aggressive and emotionally detached" (Priscott 2018, p. 1).

One such group is the National Organization for Men against Sexism (NOMAS), founded in 1975. NOMAS "advocates a perspective that is pro-feminist, gay affirmative, anti-racist, dedicated to enhancing men's lives, and committed to justice on a broad range of social issues including class, age, religion, and physical abilities" (NOMAS 2020, p. 1). Similarly, the main objective of the National Coalition for Men (NCFM) is to "promote awareness of how gender-based expectations limit men legally, socially, and psychologically" (NCFM 2020, p. 1).

Men often do not feel free to express their emotions, reject pressures to be successful, play a hands-on role in their children's lives, and choose a nontraditional career without being stigmatized for participating in "feminine" work (Priscott 2018). A study of men in nontraditional work roles found that these men commonly experience embarrassment, discomfort, shame, and disapproval from friends and peers (Sayman 2007; Simpson 2005). Similarly, Dunn et al. (2013), after interviewing working women with stay-at-home husbands, report that the wives' coworkers and relatives often made negative comments about their husbands.

@Alyssa_Milano

If you've been sexually harassed or assaulted write 'me too' as a reply to this tweet.

Me too.

Suggested by a friend: "If all the women who have been sexually harassed or assaulted wrote 'Me too.' as a status, we might give people a sense of the magnitude of the problem."

-Alyssa Milano

Online Activism

Online activism "takes many forms, from symbolic signaling of one's stance on a politicized issue (e.g., changing one's social media profile picture) to more complex engagement (e.g., writing detailed posts about a social issue)" (Greijdanus et al 2020, p. 49) (also see Chapter 14). In October of 2017, #MeToo began trending after actress Alyssa Milano posted sexual assault allegations against Hollywood producer Harvey Weinstein (Mendes, Ringrose, and Keller 2018) and encouraged other women to do the same. Original coined by Tarana Burke in 2006 to call attention to sexual violence against women and girls, particularly women and girls of color, within 24 hours the hashtag had been used 1 million times (Williams, Singh, and Mezey 2019). One year later, the hashtag had been used 19 million times, an average of 55,319 times per day (Anderson and Toor 2018). Figure 10.5 graphically portrays the number of #MeToo twitter posts between Alyssa Milano's first post and Christine Blasey Ford's testimony before the Senate, alleging sexual misconduct by Brett Kavanaugh, then Supreme Court nominee.

online activism Activism that entails the use of social media to embrace a particular social or political issue.

The first well publicized conviction attributed to the "cultural shift" as a result of #MeToo, was Bill Cosby, a noted comedian and actor, who was found guilty of aggravated indecent assault and sentenced to three to ten years in prison. Fifty women testified against Cosby. After the verdict was read, one of Cosby's victims said, "Today this jury has shown that what the MeToo movement has [been] saying is that women are worthy of being believed. And I thank the jury. I thank the prosecution" (quoted in Levenson 2018, p. 1). In 2020, the Pennsylvania Supreme Court granted Cosby an appeal based on procedural issues.

Although men from the entertainment industry are the most well-known casualties of #MeToo activism, at "least 200 prominent men have lost their jobs after public allegations of sexual harassment" and are facing criminal charges (Carlsen et al. 2018, p. 1). In 2020, Harvey Weinstein was convicted of several sex-related crimes including rape and was sentenced to 23 years in prison (Dwyer 2020). In the same year, U.S.A. Gymnastics offered $215 million to the victims of U.S. Olympics' doctor Larry Nasar, who was convicted of sexually abusing more than 200 women and girls under the guise of medical treatment. Nasar was sentenced to 40 to 125 years in prison (Levenson 2020).

Results of a 2018 survey by Atwater et al. (2019) indicate that, although there have been some positive effects of the #MeToo campaign (e.g., nearly three-quarters of women said they would be more likely to report harassment), there is some evidence of a negative

AP Images/Ahn Young-joon

International Women's Day is celebrated annually on March 8 in South Korea. Participants at the rally hold placards echoing the U.S. #MeToo campaign protesting sexual violence against women. As of 2020, despite rising awareness of sexual violence against women, no significant laws have been enacted or policies changed (Maresca 2020).

Common examples of sexual harassment include unwanted touching, the invasion of personal space, making sexual comments about a person's body or attire, and telling sexual jokes (Uggen and Blackstone 2004). Sexual harassment occurs in a variety of settings, including the workplace, schools, military academies, and college campuses. Keplinger et al. (2019), using the EEOC definition, operationalized sexual harassment as involving three categories of behavior: sexual coercion, unwanted sexual attention, and gender harassment. The researchers surveyed a cross section of U.S. women in September 2016 ($N = 513$) (as the authors note, before Donald Trump was elected or the #MeToo campaign had begun), and again in September 2018 ($N = 263$).

Although upward of 80 percent of women in both samples reported at least one type of sexual harassment in the workplace, very few women reported sexual coercion. Reports of sexual coercion, the most serious of the sexual harassment behaviors, decreased between 2016 and 2018 although levels of gender harassment increased between the two years. Interestingly, in 2018, following the #MeToo campaign, the EEOC reported that charges of workplace discrimination and sexual harassment increased for the first time in the decade (Williams, Singh, and Mezey 2019).

Affirmative Action. As discussed in Chapter 8 and Chapter 9, **affirmative action** refers to a broad range of policies and practices to promote equal opportunity as well as diversity in the workplace and on campuses. Affirmative action policies, developed from federal legislation in the 1960s, require that any employer, universities as well as businesses, that receives contracts from the federal government must make "good faith efforts" to increase the number of female and other minority applicants. Such efforts can be made through expanding recruitment and training programs and by making hiring decisions on a nondiscriminatory basis.

However, a 1996 California ballot initiative, the first of its kind, abolished race and sex preferences in government programs, which included state colleges and universities. Over the years, several other states followed suit. In 2003, the U.S. Supreme Court held that universities have a "compelling interest" in a diverse student population and therefore may take minority status into consideration when making admissions decisions. In 2016, the "compelling interest" argument was reaffirmed in *Fisher v. The University of Texas* (Barnes 2016). Not surprisingly, in a study of opinions of affirmative action, female medical students rated the importance of affirmative action programs higher than their male medical student counterparts (Siller et al. 2019).

What do you think?

Following the success of #MeToo, over 300 women, primarily in the entertainment industry, established #TimesUp, which focuses on gender employment disparities. #TimesUp advocates argue that you must first eradicate workplace inequities before you can address sexual harassment and other issues related to power differences. Do you think the organizers of #TimesUp are correct that gender power differentials must be addressed before sexual harassment?

International Efforts

International efforts to address problems of gender inequality date back to the 1979 Convention to Eliminate All Forms of Discrimination against Women (CEDAW), often referred to as the International Women's Bill of Rights, adopted by the United Nations in 1979. To date, 188 out of 194 countries have ratified this bill of rights, including every country in Europe and South and Central America. Only six countries have not ratified CEDAW: Iran, South Sudan, Somalia, Palau, Tonga, and the United States.

Another significant international effort occurred in 1995, when representatives from 189 countries adopted the Beijing Declaration and Platform for Action at the Fourth World Conference on Women sponsored by the United Nations. The platform reflects an

affirmative action A broad range of policies and practices in the workplace and educational institutions to promote equal opportunity as well as diversity.

international commitment to the goals of equality, development, and peace for women everywhere. The platform identifies strategies to address critical areas of concern related to women and girls, including poverty, education, health, violence, armed conflict, and human rights.

In addition to the CEDAW and the Beijing Platform, in 2000, all of the members of the United Nations adopted the Millennium Declaration. One of the eight Millennium Development Goals, as stated in the Millennium Declaration, was the promotion of gender equality and women's empowerment by 2015, which has not been met. The secretary-general of the United Nations, on the occasion of the International Women's day, stated that "to be truly transformative, the post-2015 development agenda must prioritize gender equality and women's empowerment. The world will never realize 100 per cent of its goals if 50 per cent of its people cannot realize their full potential" (UN 2015, p. 1).

Directed toward preventing the millions of acts of violence against women and girls annually is the *International Violence against Women Act* (I- VAWA 2019). The act, which is largely based on the *U.S. Strategy to Prevent and Respond to Gender-Based Violence Globally* that was released in 2012, was written by Amnesty International and a consortium of other concerns. The act, which also protects men and boys, would ensure that "the U.S. government has a strategy to efficiently and effectively coordinate existing cross-governmental efforts to prevent and respond to ... [gender-based violence] globally" (Coalition to End Violence against Women and Girls Globally 2017, p. 1).

Finally, online activism is not unique to the United States. The equivalent of #MeToo has appeared in 46 different languages around the world which and has been searched for in 196 countries, and new hashtags associated with sexual assault and abuse have emerged in several languages. Through social media, discussions of sexual violence against women are now an "unprecedented global conversation, where previously silenced voices have been amplified, supporters around the world have been united, and resistance has gained steam" (Williams, Singh, and Mezey 2019, p. 371).

Understanding Gender Inequality

The traditional gender roles of men and women, and the binary concepts of sex and gender, are weakening. Women who have traditionally been expected to give domestic life first priority are now finding it more acceptable to seek a career outside the home; the gender pay gap is narrowing, albeit slowly; and significant improvements in educational disparities between men and women, boys and girls, have been made. Further, several states have passed legislation designed to (Beitsch 2018; North 2019; Johnson, Menefee, and Sekaran 2019):

- limit the use of nondisclosure agreements (NDA);
- improve the processing of rape kits;
- reexamine workplace sexual harassment policies;
- extend the statutes of limitation for sex crimes;
- expand requirements for harassment training and prevention; and
- remove caps for monetary damages in lawsuits.

These legislative initiatives, which can be directly tied to #MeToo activism, have been made more palatable by a cultural readiness in the United States to deal with the issues of sexual harassment and assault.

However, despite the global nature of #MeToo, in many countries its success has been limited. Chinese feminists are "battling headwinds in a political environment where the ruling Communist Party's control over the Internet, media and independent activism is tighter than it has been in 30 years" (Wang 2020). In Japan, where just over half of the population favor more gender equality (Poushter and Fetterole 2019), women who share their stories of sexual assault and harassment online are publicly

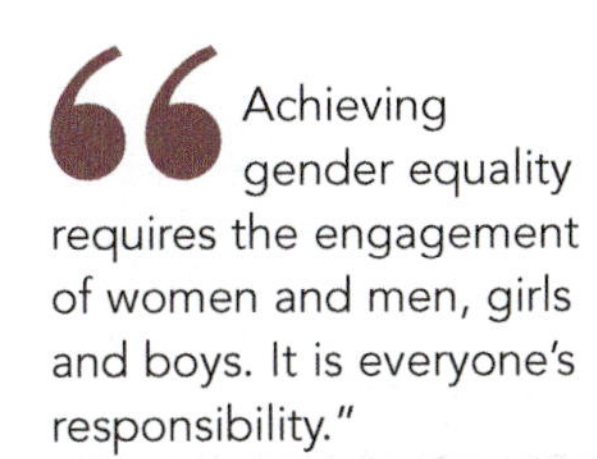

Achieving gender equality requires the engagement of women and men, girls and boys. It is everyone's responsibility."

–BAN KI-MOON, FORMER UN SECRETARY GENERAL

humiliated (Ishikawa 2020). #MeToo accounts of sexual misconduct published in Australian media outlets have been successfully met with million-dollar lawsuits by the accused. In Australia, neither free speech nor freedom of the press is legally protected (Funnell 2020).

Federally, the *BE HEARD (Bringing an End to Harassment by Enhancing Accountability and Rejecting Discrimination) in the Workplace Act* was introduced during the 2019–2020 legislative session, and it too is in direct response to #MeToo activism. This bill would give "much-needed guidance to the federal courts, which too often have excused objectively abusive conduct in the workplace, depriving victims of harassment of relief, and dissuading other from seeking legal redress" (Leveille and Lapidus 2019).

The *BE HEARD in the Workplace Act* limits the behaviors it governs to those in the workplace, however, and fails to protect victims of sexual assault and harassment in schools (see Chapter 8), families (see Chapter 5), and the military (see Chapter 15). Nonetheless, it is a beginning and a testament to the overwhelming support for gender justice policies in the United States. Over 85 percent of registered voters in a recent national survey considered "preventing sexual-harassment and assault" an important issue for Congress to work on (NWLC 2019b). If passed by Congress, President Biden has pledged to sign the *BE HEARD Act* into law (Biden 2020).

Gender equality is not just good for individuals; it is good for families and society as a whole. Increasing women's level of education, labor force participation, and incomes contributes to the well-being of a society, reducing infant mortality rates, poverty, the transmission of HIV, and many other social problems. In a poll of unemployed U.S. adults between the ages of 25 and 54, 61 percent of women compared to 37 percent of men reported not working because of family responsibilities. However, many reported that they would go back to work if they could work from home or had flexible hours (Hamel et al. 2014). Higher rates of female employment in developed countries—for example, Germany, Canada, France, and Great Britain—are linked to family-friendly policies such as subsidized child care, paid parental leave, and taxation of individuals rather than families (Blau and Kahn 2013). Adopting such policies in the United States would go a long way toward gender equality.

Men are also victimized by discrimination and gender stereotypes that define what they "should" do rather than what they are capable, interested, and willing to do. The National Coalition for Men (NCFM 2020) has incorporated this view into its philosophy:

> We have heard in some detail from the women's movement how such sex-stereotyping has limited the potential of women. More recently, men have become increasingly aware that they too are assigned limiting roles which they are expected to fulfill regardless of their individual abilities, interests, physical/emotional constitutions or needs. Men have few or no effective choices in many critical areas of life. They face injustices under the law. And typically, they have been handicapped by socially defined "shoulds" in expressing themselves in other than stereotypical ways. (p. 1)

Increasingly, people are embracing **androgyny**—the blending of traditionally defined masculine and feminine characteristics. The concept of androgyny implies that both masculine and feminine characteristics and roles are *equally valued*. However, "achieving gender equality… is a grindingly slow process, since it challenges one of the most deeply entrenched of all human attitudes" (Lopez-Claros and Zahidi 2005, p. 1).

Regardless of whether traditional gender roles emerged out of biological necessity, as the structural functionalists argue, or out of economic oppression, as the conflict theorists hold, or both, it is clear that today, gender inequality carries a high price: poverty, loss of human capital, feelings of worthlessness, violence, physical and mental illness, and death. Perhaps we have reached a time when the costs of traditional gender roles are simply too high.

androgyny Having both traditionally defined feminine and masculine characteristics.

Chapter Review

- **Does gender inequality exist worldwide?**
There is no country in the world in which men and women are treated equally. Although women suffer in terms of income, education, and occupational prestige, men are more likely to suffer in terms of mental and physical health, mortality, and the quality of their relationships.

- **How do the three major sociological theories view gender inequality?**
Structural functionalists argue that the traditional division of labor was functional for preindustrial society and has become defined as both normal and natural over time. Today, however, modern conceptions of the family have replaced traditional ones to some extent. Conflict theorists hold that male dominance and female subordination evolved in relation to the means of production—from hunting-and-gathering societies in which females and males were economic equals, to industrial societies in which females were subordinate to males. Symbolic interactionists emphasize that, through the socialization process, both females and males are taught the meanings associated with being feminine and masculine.

- **What is meant by the terms *structural sexism* and *cultural sexism*?**
Structural sexism refers to the ways in which the organization of society, and specifically its institutions, subordinate individuals and groups based on their sex classification. Structural sexism has resulted in significant differences between education and income levels, occupational and political involvement, and civil rights of women and men. Structural sexism is supported by a system of cultural sexism that perpetuates beliefs about the differences between women and men. *Cultural sexism* refers to the ways the culture of society—its norms, values, beliefs, and symbols—perpetuate the subordination of an individual or group because of the sex classification of that individual or group.

- **What is the difference between the glass ceiling and the glass escalator?**
The glass ceiling is an invisible barrier that prevents women and other minorities from moving into top corporate positions. The glass escalator refers to the tendency for men seeking traditionally female jobs to have an edge in hiring and promotion practices.

- **What are some of the problems caused by traditional gender roles?**
First is the feminization of poverty. Women are socialized to put family ahead of education and careers, a belief that is reflected in their less prestigious occupations and lower incomes. Second are social-psychological costs. Women and transgender individuals tend to have lower self-esteem and higher rates of depression than men. Men and boys are often subject to the emotional restrictions of the "boy code." Third, traditional gender roles carry health costs in terms of death and illness. Finally, gender-based violence is responsible for the deaths of men, women, and transgender individuals.

- **What strategies can be used to end gender inequality?**
Social movements, such as feminism, the women's rights movement, and the men's equality movement, have made significant inroads in the fight against gender inequality. Online activism has also led to policy changes. Other laws addressing gender inequality include the *Equal Pay Act of 1963*, Title VII of the *Civil Rights Act of 1964*, Title IX of the *Education Amendments of 1972*, and the 2009 *Ledbetter Equal Pay Act*. Besides these national efforts, international efforts continue as well. One of the most important is the Convention to Eliminate All Forms of Discrimination against Women (CEDAW), also known as the International Women's Bill of Rights, which the United Nations adopted in 1979.

Test Yourself

1. The United States is the most gender-equal nation in the world.
   a. True
   b. False
2. Symbolic interactionists argue that
   a. male domination is a consequence of men's relationship to the production process.
   b. gender inequality is functional for society.
   c. gender roles are learned in the family, in the school, in peer groups, and in the media.
   d. women are more valuable in the home than in the workplace.
3. Recent trends indicate that more males enter college from high school than females.
   a. True
   b. False
4. The devaluation hypothesis argues that female and male pay differences are a function of women's and men's different levels of education, skills, training, and work experience.
   a. True
   b. False
5. The glass ceiling is an invisible barrier that prevents women and other minorities from moving into top corporate positions.
   a. True
   b. False
6. Which of the following statements is true about women and U.S. politics?
   a. There are more female than male governors.
   b. Kamala Harris is vice president of the United States and the first woman to be elected to the executive branch.

   c. The highest-ranking woman in the U.S. government is the former Secretary of State Hillary Clinton.
   d. Women received the right to vote with the passage of the Twenty-first Amendment.
7. Which of the following statements is true?
   a. Online activism is limited to the United States.
   b. Male supremacy groups often embrace misogyny.
   c. Black feminism is supportive of Friedan's concept of the *feminine mystique.*
   d. An analysis of movie posters reveals that women are more often portrayed in leadership roles than men.
8. The feminization of poverty refers to
   a. the tendency for women to be caretakers of the poor.
   b. feminists' criticism of public policy on poverty.
   c. the disproportionate number of women who are poor.
   d. gender role socialization of poor women.
9. *Quid pro quo* sexual harassment refers to the existence of a hostile working environment.
   a. True
   b. False
10. Feminists would argue which of the following?
   a. The ERA should be part of the Constitution.
   b. The passage of the Nineteenth Amendment was a mistake.
   c. Occupational sex segregation benefits everyone.
   d. NOMAS is a radical, misogynistic organization.

Answers: 1. B; 2. C; 3. B; 4. B; 5. A; 6. B; 7. B; 8. C; 9. B; 10. A.

Key Terms

affirmative action 400
androgyny 402
attributional gender bias 384
boy code 388
cisgender 364
cultural sexism 379
devaluation hypothesis 376
dowry killings 392
education dividend 372
emotion work 376
equal rights amendment (ERA) 395
expressive roles 370
femicide 395
feminine mystique 395
feminism 393
gender roles 380
gender 364
gender deviance hypothesis 383
gender expression 364
gender-based violence 389
glass ceiling 374
glass escalator effect 373
honor killings 391
human capital hypothesis 376
instrumental roles 370
masculine overcompensation thesis 368
misogyny 392
motherhood penalty 374
occupational sex segregation 373
online activism 397
second shift 382
sex 364
sexism 365
sexual harassment 399
structural sexism 371
structured choice 374
toxic masculinity 393
transgender individual 364
vulnerable employment 373

> "It takes no compromise to give people their rights. ... [I]t takes no money to respect the individual. It takes no political deal to give people freedom. It takes no survey to remove repression."
>
> **HARVEY MILK,**
> Politician, LGBT activist

11

Sexual Orientation and the Struggle for Equality

Chapter Outline

Learning Objectives

After studying this chapter, you will be able to . . .

1. Hypothesize reasons for the differences in acceptance of LGBTQ people by regions of the world.
2. Using each of the sociological theories, explain societal reactions to non-heterosexual, non-cisgender women and men.
3. Explain the origins of anti-LGBTQ bias.
4. Hypothesize reasons for the relationship between LGBTQ prejudice and the social variables sex, age, income, and education.
5. List the consequences of anti-LGBTQ bias.
6. Evaluate the various programs and policies set in place to achieve LGBTQ equality.
7. Discuss the possible future impact of the passage of proposed LGBTQ-related legislation (e.g., *Do No Harm Act*, the *Equality Act*).

AP Images/Ted S. Warren

As Helen Grace James before her, Major Margaret Witt had been dishonorably discharged from the U.S. Air Force and, like Dr. James, Major Witt challenged her dismissal which was based solely on her sexuality, i.e., a gay woman. Four years later a federal judge overturned the decision, holding that the dismissal violated Witt's constitutional rights.

IT WAS 1955 WHEN Airman 2nd Class Ellen Grace James left the base to meet a friend, but shortly after stopping to eat, a floodlight washed over the inside of her car. She had been followed. Within a few days, she was arrested, interrogated, and, under threat of having her sexual orientation revealed to her family, confessed to being a lesbian. Ms. James, who had served her country with distinction for three years, was discharged from the Air Force as an "undesirable." She received no pension, no veteran's health care, and no help from the GI Bill (Simon 2018). Sixty years later, Ellen Grace James, at the age of 90, and after decades of petitioning the U.S. Air Force for a change in her discharge status, finally received an honorable discharge. Said her attorney, this "has crippled her throughout her life." Said Ms. James, despite a successful career as a professor, "I [just] ... wasn't whole" (quoted in Swenson 2018).

> "Remember, bisexuality doesn't mean halfway between gay or straight. It is its own identity."
>
> —EVAN RACHEL WOOD, ACTRESS

The *lavender scare,* parallel to McCarthyism and the Red Scare of the 1940s and 1950s, resulted in thousands of gay men and women being dismissed from the federal government for fear they were security risks.

Although today may seem light years away from the arrests and interrogations of the lavender scare, the fight for equality based on sexual orientation and gender identity continues to be met with opposition. The term *sexual orientation* refers to a person's emotional and sexual attractions, relationships, self-identity, and behavior. **Heterosexuality** refers to the predominance of emotional, cognitive, and sexual attraction to individuals of the opposite sex. In the past, *homosexuality* was the preferred term used to refer to the predominance of emotional, cognitive, and sexual attraction to individuals of the same sex; however, this term also carries negative clinical associations that are outdated (see section on the history of this term in Non-Heterosexuality as Pathology). As such, the current preferred terms are **lesbian**, which refers to women who are attracted to same-sex partners and **gay**, which refers to either women or men who are attracted to same-sex partners. **Bisexuality** is the emotional, cognitive, and sexual attraction to members of both sexes. The terms *lesbian*, *gay*, and *bisexual*, collectively, refer to sexual minorities. **Pansexual** refers to anyone whose sexual orientation is not limited by their partner's birth sex, gender, or gender identity, the prefix *pan* meaning "all." Table 11.1 shows the distribution of undergraduate students from a national sample by self-described sexual orientation and gender identity.

TABLE 11.1 Self-described Sexual Orientation and Gender Identity of U.S. College Undergraduate Students, 2019*

Straight/heterosexual	81.0%
Gay/lesbian/queer	4.2%
Bisexual/pansexual	11.1%
Questioning	2.1%
Transgender/nonbinary	3.0%

*0.7 percent described themselves as asexual and 0.5 percent as "identity not listed"; based on sample of 30,084 undergraduate students at 58 institutions.
SOURCE: American College Health Association 2019.

Much of the current literature on the treatment of individuals who are gay, lesbian, and bisexual includes transgender individuals (LGBT) while others do not (LGB). As discussed in Chapter 10, *transgender* (sometimes called *trans*), is the appropriate term used to refer to individuals whose sense of gender identity as male or female is inconsistent with their assigned birth sex. Alternatively, **cisgender**—*cis* the Latin for "on the side of" as opposed to *trans*, Latin for "on the other side of"—refers to a person whose gender identity is consistent with their birth sex. **Gender nonconforming** refers to displays of gender that are inconsistent with society's expectations of what is appropriate given an individual's gender.

Research indicates that gender is a fluid rather than immutable concept. Increasingly, "[C]ultures and countries from all over the globe recognize in law and in cultural traditions genders other than male and female" (UN 2018, p. 2). Presently, several countries have begun the legislative process to recognize gender based on self-identification, including Norway (2016), Belgium (2017), and Austria (2018). Do you think the United States would benefit from recognizing gender as self-identified rather than assigned at birth?

What *do you* think?

The acronyms **LGBT, LGBTQ, and LGBTQI**, as well as an assortment of other variations, are used to refer collectively to individuals who are lesbian, gay, bisexual, transgender, questioning or "queer," and *intersexed*, i.e., people born with both male and female anatomy. The word **queer**, although originally a pejorative term for gay males and lesbians, has been reclaimed by the LGBTQ community to mean anyone who is not heterosexual or cisgender and thus, by its very definition, recognizes the fluidity of sexual orientation and gender identity. Sometimes one will encounter the acronym accompanied by the letter *A*, which often stands for *allies.* Which set of acronyms are used in this chapter, LGB, LGBT, or LGBTQ, depends on the particular research being discussed but, in general, includes gay, bisexual, and transgender men and women, and those who identify as queer.

It is important to acknowledge that both sexual orientation and gender identity exist on a continuum, and that, over a lifetime, individuals may embrace various combinations of orientations and gender identities. However, our discussion of sexual orientation is limited by the available research that has traditionally used a three category classification scheme—gay (same-sex or non-heterosexual), heterosexual (straight, different-sex, or non-gay), and bisexual—rather than a nuanced identity system.

It is also impossible to discuss the struggle for same-sex equality without acknowledging transgender individuals who have been part of the larger LGBTQ civil rights struggle. Thus, some of the research reported in this chapter is on lesbians, gays, and bisexuals, whereas other studies include transgender people as participants in addition to lesbians, gays, and bisexuals. Furthermore, many laws and policies discussed in this chapter do not specifically refer to bisexuals but would, by default, apply if they were in a same-sex relationship.

In this chapter, we focus primarily on Western conceptions of diversity in sexual orientation. It is beyond the scope of this chapter to explore in depth how sexual orientation and its cultural meanings vary throughout the world. The global legal status of lesbians and gay men, however, will be summarized. We also discuss the prevalence of LGBTQ men and women in the United States, the beliefs about the origins of sexual orientation, and then apply sociological theories to better understand societal reactions to non-heterosexuals. After discussing the cultural origins of anti-LGBTQ bias and the ways in which gay and transgender people are victimized by prejudice and discrimination, we end the chapter with a discussion of strategies to reduce anti-gay and anti-trans prejudice and discrimination.

heterosexuality The predominance of emotional, cognitive, and sexual attraction to individuals of the opposite sex.

lesbian A term referring to women who are emotionally, cognitively, and sexually attracted to women.

gay A term referring to women or men who are emotionally, cognitively, and sexually attracted to individuals of the same sex.

bisexuality The emotional, cognitive, and sexual attraction to members of both sexes.

pansexual Anyone whose sexual orientation is not limited by their partner's birth sex, gender, or gender identity, the prefix "pan" meaning *all*.

cisgender Refers to a person whose gender identity is consistent with his or her birth sex.

AP Images/Jun Yasukawa

Here, same-sex couples celebrate their weddings in Taipei, Taiwan, on May 24, 2019. The Taiwanese court ruled that defining marriage as being only between a man and a woman was unconstitutional. In 2017, Taiwan became the first Asian country to make same-sex marriages legal.

The Global Context: A Worldview of Sexual and Gender Minorities

Same-sex sexual behavior has existed throughout human history and in most, perhaps all, human societies. A global perspective on laws and social attitudes regarding gay and lesbian sexuality reveals that countries vary tremendously

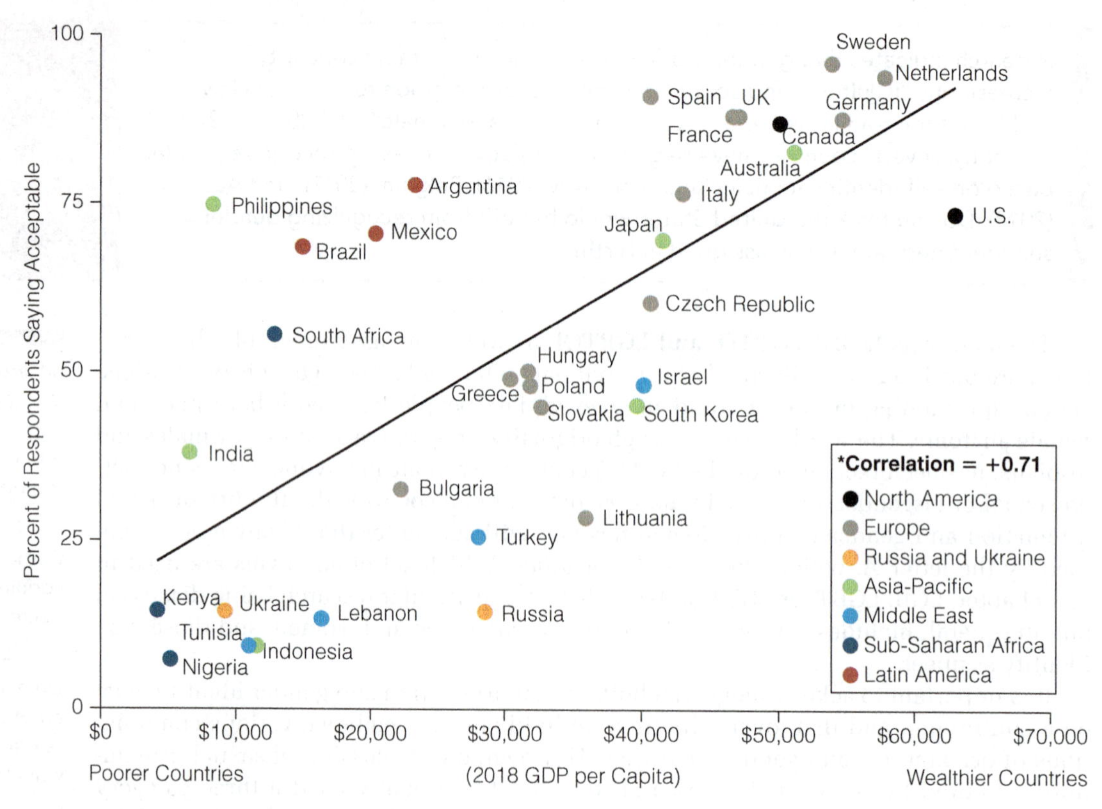

Figure 11.1 Acceptability of Same-Sex Sexual Behavior in Countries by Country's Wealth, 2019*

*The correlation of +0.71 indicates that there is a strong association between wealth of a country (independent variable) and acceptance of same-sex sexual behavior (dependent variable), i.e., as wealth increases, acceptance increases (see Appendix A).

SOURCE: Poushter and Kent 2020.

in their treatment of same-sex sexual behavior—from intolerance, criminalization, and even death, to acceptance and legal protection. Sixty-four percent of the United Nations member countries have decriminalized non-heterosexual sexual activity, although it varies dramatically by region. In general, the wealthier the country, the greater the acceptance of gay and bisexual men and women (see Figure 11.1). For example, the United States and Canada have legalized same-sex activity. However, in Africa only 21 of 54 (39 percent) countries have decriminalized same-sex relations, and only 20 of 42 (48 percent) countries in Asia (Mendos 2019).

Legal penalties vary for violating laws that prohibit same-sex sexual behavior. In some countries, same-sex sexual behavior is punishable by prison sentences (e.g., up to 14 years in Gambia, 10 years in Barbados) and/or corporal punishment, such as whipping or lashing (e.g., Brunei and Malawi) (Mendos 2019). In 12 nations, all predominantly Muslim, someone found guilty of engaging in same-sex sexual behavior may receive the death penalty. It should also be noted, however, that in some countries, regardless of official penalties, gay men and women are routinely executed, as in the case of Qatar. There are also several countries in which male-male sexual behavior is prohibited although female-female sexual behavior is not (e.g., India, Lebanon) and, where both are illegal, men receive more severe punishment than women (e.g., Oman) (Carroll 2016).

Globally, same-sex marriage is legal in 29 of 195 countries including Costa Rica, New Zealand, Taiwan, Columbia, and Norway. According to the nonprofit advocacy group *Freedom to Marry*, only 15 percent of the world's population live in countries where gay women and men are free to marry their same-sex partners (Freedom to Marry 2020). Several other countries have legal protections for same-sex and different-sex relationships. Often called domestic partnerships or civil unions, these often federally recognized relationships convey most but not all the rights of marriage.

gender nonconforming Often used synonymously with transgender (and sometimes called gender variant), displays of gender that are inconsistent with society's expectations.

LGBT, LGBTQ, and LGBTQI Collectively, lesbian, gay, bisexual, transgender, questioning or "queer," and/or intersexed individuals.

queer Although originally a pejorative term for gay men and women, the term has been reclaimed by the LGBTQ community to mean anyone who is not heterosexual or cisgender.

In the southern Russian republic of Chechnya, as a result of an increase in political and religious conservatism, LGBTQ relatives bring shame to the entire family. The shame is so great that the gay, bisexual, or transgender relative is often a victim of an honor killing to "wash away the shame" and allow the family to resume living normally. Before honor killing takes place, the family tries to change their LGBTQ relative through "various forms of violence: beatings, placement under house arrest for months or years, and imposed isolation" (ILGA 2020, p. 157). Do you think these behaviors constitute human rights violations, and, if so, are interventions by other countries warranted?

What *do you* think?

In general, countries throughout the world are moving toward increased legal protection of non-heterosexuals, as discrimination based on sexual orientation and gender identity has become part of a broader international human rights agenda. In 1996, South Africa became the first country in the world to include in its constitution a clause banning discrimination based on sexual orientation. Following South Africa's lead, today four additional countries—New Zealand, Portugal, Mexico, and Sweden—have national constitutions banning discrimination based on sexual orientation. Bolivia, Ecuador, Fiji, Malta, and the United Kingdom have national constitutions that ban discrimination based on sexual orientation *and* gender identity (Hutt 2018). Further, in 2019, the World Health Organization removed "gender identity disorder" as a mental illness in its *Global Diagnostic Manual* and redefined it as "gender incongruence," which now appears as part of the chapter on sexual health. The new guidelines will go into effect in 2022 (Lewis 2019).

Despite these general international trends toward LGBTQ human rights, there have been recent setbacks in many regions of the world. In 2013, Russia, which decriminalized same-sex sexual behavior in 1993, established laws prohibiting "propaganda of nontraditional sexual relations to minors"—a prohibition so broadly defined as to include public displays of affection between same-sex partners (Guillory 2013). Further, the results of a national survey indicate that one-third of Russian citizens want to "isolate" gay men and women from Russian society, compared to only 9 percent who want to "assist" them (Kuhr 2020). Finally, in 2020, a constitutional amendment banning gay marriages was approved by Russian voters (Venkatraman 2020), and the newly elected president of Poland came to office after running a campaign based on an anti-LGBTQ agenda (Gera and Thiesing 2020).

Of late, several countries of the world in regions that have traditionally been hostile to LGBTQ people—Africa, Asia, and the Caribbean—have renewed their antigay efforts. In 2019, Uganda's 2014 *Anti-Homosexuality Act*, which defined same-sex relations as a capital offense, was scheduled to be reintroduced into Parliament and was only tabled after the United States and the European Union, among others, threatened to withdraw financial aid (Bhalla 2019; *African News* 2019). The Botswana government is appealing a high court decision to decriminalize same-sex relations, and in Kenya the court ruled against repealing the prohibition of same-sex relations. As one observer noted, these are "significant setbacks ... in Africa, where homosexuality and gender nonconformity tend to be regarded as 'unnatural' Western behaviors that have been imported from abroad" (Mithika 2019 p. 1).

Further, traditionally open societies in Europe have regressed and now report high incidences of violence, anti-LGBTQ legislation, and/or repeal of equality safeguards. Rainbow Index–Europe, which measures LGBTI equality on six dimensions, reports that in 2019, for the first time in ten years, "countries are moving backwards as existing laws and policies disappear" (Rainbow Europe 2019). The result, even in countries that have traditionally embraced LGBTQ equality, is an increase in hate speech and physical violence. For example, in part as a result of the populist movement surrounding Brexit, the number of hate crimes and related anti-LGBTQ incidences in England and Wales increased from 5,807 in 2015 to 13,530 in 2019. Similarly, in some European countries, there has been a move to ban LGBTQ public events, an increase in anti-LGBTQ activity and graffiti including neo-Nazi protests, and an increase in anti-gay and anti-transgender speech by government officials and religious leaders (ILGA Europe 2020).

The LGBTQ Population in the United States

Reliable estimates of the size and demographic characteristics of the LGBT population in the United States are difficult to come by. Historically, LGBT individuals specifically, and the larger queer community in general, are undercounted in part because respondents are often hesitant to self-identify as gay, bisexual, or transgender. For the first time, the decennial census, which was conducted in April 2020, replaced the response of "husband or wife" with "opposite-sex husband/wife/spouse" or "same-sex husband/wife/spouse." Additionally, the response "unmarried partner" was replaced with "opposite-sex unmarried partner" or "same-sex unmarried partner" (Redmond-Palmer 2020). The use of census data only in the aggregate protects respondents' anonymity.

There were nearly 1 million same-sex couple households in 2017, the most recent year for which data are available (U.S. Census 2019). There are more female same-sex households than male same-sex households, although it varies considerably based on geographical location. For example, 3 percent of all households in San Francisco are same-sex households, over three times the national average, and of that number, over two-thirds are male same-sex households. Of the total number of same-sex households in the United States, nearly 60 percent of household same-sex couples are married.

According to a Gallup poll, the percentage of adults in the United States who identify as LGBT increased from 4.5 percent in 2017 to 5.6 percent in 2020 (Jones 2021). Those most likely to self-identify as LGBT are in Generation Z (those born between 1997 and 2002) at 16 percent compared to Millennials (those born between 1981 and 1996) at just 9 percent. Not surprisingly, as age increases the percent of respondents who self-identify as LGBT decreases, with 2 percent or less of those born before 1947 claiming an LGBT identity. Across all age groups, over half of the LGBT respondents self-identify as bisexual.

The Gallup survey found no significant differences in LGBT identification by educational levels. However, identifying as LGBT was higher among women than men, and among political liberals compared to moderates, and among moderates compared to conservatives. Yet, as discussed in Chapter 1, it should be remembered that the reliability of survey data is wholly dependent upon the veracity of respondents' answers. It may be that those who identify with more conservative political ideologies are simply less likely to *report* being gay or bisexual.

The Origins of Sexual Orientation

One of the most common questions regarding sexual orientation centers on its origin or "cause." Questions about the causes of sexual orientation are typically concerned with the origins of same-sex attraction because different-sex attraction is considered normative and "natural." A wealth of biomedical and social science research has investigated the possible genetic, hormonal, developmental, social, and cultural influences on sexual orientation, yet no conclusive findings have suggested that sexual orientation can be completely explained by any particular factor or interplay of factors (Ganna et al. 2019).

In the absence of compelling findings, many practitioners and professionals believe that sexual orientation might be determined by the interplay of environmental and biological variables. Aside from what "causes" same-sex attraction, sociologists are interested in what people *believe* about the "causes" of non-heterosexuality. People rely heavily on ideology, religion, and life experiences to form their beliefs about the causes of same-sex attraction (Haider-Markel and Joslyn 2008). A 2019 Gallup poll addressed this issue. Respondents were asked, "In your view, is being gay or lesbian something a person is born with, or due to factors such as upbringing and environment?" (Gallup 2020, p. 19). A decade ago, about one-third of U.S. adults believed that being gay was determined at birth; today, about half (49 percent) of Americans believe that being gay is something someone is born as, 32 percent believe that it is due to environmental factors, and 11 percent believe that it is a combination of both (Gallup 2020). Independent variables predictive of beliefs about the origins of sexual orientation include race, sex, political party affiliation, and income with white people, women, liberal Democrats, and high earners being more likely to embrace a nature rather than nurture argument (McCarthy 2014).

Research by Spence, Helwig, and Cosentino (2018) documents the relationship between beliefs about the origin of sexuality and their impact on evaluations of same-sex relations, even among children. Canadian respondents between the ages of 5 and 11 (children), and between 13 and 14 (adolescence), were interviewed about the acceptability of same-sex sexual behavior, laws regulating same-sex romantic relationships, and beliefs about the origin of non-heterosexual attraction. Results indicate that the belief that same-sex attraction is biologically determined increased the likelihood of evaluating same-sex romantic relationships and same-sex marriages positively for 13- and 14-year-olds, although the relationship was not significant for other age groups. Interestingly, the authors report a "relative absence in our sample of references to same-sex attraction and sexual behavior as 'unnatural'" (p. 1001).

Can Sexual Orientation Be Changed?

Given the contentious nature of the debate over LGBTQ rights, research often focuses on trying to identify variables that explain an individual's location on the sexuality and gender continuums. Often the motivation for such research is, if we can know the cause then we can institute a change. Note, however, that the desire to change an individual's sexual orientation is confined to gay and bisexual men and women. The desire to change a heterosexual's orientation has yet to be suggested.

Imagine being heterosexual and being asked, in therapy, what do you think is responsible for your heterosexuality; when did you first realize you were a heterosexual; to whom have you revealed your heterosexual tendencies; and what are you doing to change your sexual orientation? For a majority of Americans, such questions might be met with, "What do you mean, what *caused* my heterosexuality?" followed by "I was born heterosexual!" Such a sense of indignation reflects **heteronormativity**—the cultural presumption that heterosexuality is the norm whereby other orientations, by default, become abnormal and something to change.

> "What is straight? A line can be straight, or a street, but the human heart, oh, no, it's curved like a road through mountains."
>
> **–TENNESSEE WILLIAMS, AUTHOR**

> One method of "conversion," **corrective rape**, is most often used on gay and bisexual women and transgender men (Doan-Minh 2019). Corrective rape is "committed because the perpetrator intends to punish the victim for failing to conform to gender and sexuality norms" (p. 177). Despite the prevalence of the behavior, it is rarely treated as a hate crime, even in the United States, and "[M]ost victims of corrective rape will never see their case resolved in the legal system" (p. 181). Do you think that corrective rape should be defined as a hate crime?
>
> **What do you think?**

Various **sexual orientation change efforts (SOCE)**—whether *reparative therapy*, *conversion therapy*, or *reorientation therapy*—are dedicated to changing the sexual orientation of individuals who are non-heterosexual. Globally, "treatments" generally fall into three categories: *psychological* (e.g., aversion therapy through the use of electric shocks), *medical* (e.g., hormone or steroid therapy), or *faith based* (e.g., religious counseling, beatings, exorcism) (UN 2020). In other cases, adolescents are "kidnapped" by their parents and sent to "conversion camps" or group homes.

Eighteen states ban conversion therapy for minors, and legislation is pending in 20 states (Allen 2020). Critics of SOCE include the American Medical Association, American Academy of Pediatrics, American Academy of Child Adolescent Psychiatry, American College of Physicians, American Counseling Association, American Psychological Association, American Association of Marriage and Family Therapy, National Association of Social Workers, the American School Counselor Association, and the American Psychiatric Association. All agree that gay and lesbian sexuality is not a mental disorder, that sexual orientation cannot be changed, and that efforts to change sexual orientation may be harmful (Human Rights Campaign [HRC] 2020a).

Leelah Alcom, a 17-year-old transgender youth, walked in front of a semitruck on December 27, 2014, after posting a suicide note on Tumblr. Forced by her parents to attend

heteronormativity The cultural presumption that heterosexuality is the norm whereby other orientations, by default, become abnormal and something to change.

corrective rape A method of "conversion" most often used on gay and bisexual women, and transgender men. The perpetrators are usually straight cisgender men.

sexual orientation change efforts (SOCE) Collectively refers to reparative, conversion, and reorientation therapies.

religious conversion therapy sessions, Leelah, defined as male at birth, also left a simple handwritten note: "I've had enough" (quoted in Kutner 2015, p. 1). In response to her death, a petition signed by more than 120,000 signatories, requesting a federal ban on SOCE, was sent to President Obama. In support of the ban, former President Obama stated:

> Tonight, somewhere in America, a young person, let's say a young man, will struggle to fall to sleep, wrestling alone with a secret he's held as long as he can remember. Soon, perhaps, he will decide it's time to let that secret out. What happens next depends on him, his family, as well as his friends and his teachers and his community. But it also depends on us—on the kind of society we engender, the kind of future we build. (Obama 2015, p. 1)

In contrast to former President Obama's support for banning conversion therapy for minors, the 2016 Republican platform "tacitly endorse[d] the right of parents to send their teens to conversion programs" (Ross and Epstein 2017, p. 1) and, 2020, the Republican National Committee (RNC) voted to re-adopt the 2016 platform for the presidential election (Crump 2020). Further, in the same year, Republicans in Florida, where conversion therapy is banned in several municipalities, introduced a bill that, if passed, would not only legalize conversion therapy in the state but would impose prison sentences of up to 15 years for physicians who provide medical care to transitioning transgender youth (Allen 2020).

Sociological Theories of Sexual Orientation Inequality

Sociological theories do not explain the origin or "cause" of sexual orientation diversity; rather, they help to explain societal reactions to non-heterosexuals and ways in which sexual orientations are socially constructed.

Structural-Functionalist Perspective

Structural functionalists, consistent with their emphasis on institutions and the functions they fulfill, emphasize the importance of monogamous heterosexual relationships for the reproduction, nurturance, and socialization of children. From a structural-functionalist perspective, same-sex relations, as well as heterosexual nonmarital relations, are "deviant" because they do not fulfill the main function of the family institution—producing and rearing children. Clearly, however, this argument is less salient in a society in which (1) other institutions, most notably schools, have supplemented the traditional functions of the family, (2) reducing (rather than increasing) population is a societal goal, and (3) same-sex couples can and do raise children.

Some structural functionalists argue that antagonisms between individuals based on sexual orientation or gender identity disrupt the natural state, or equilibrium, of society. Durkheim (1993 [1938]), however, recognized that deviation from society's norms can also be functional. Specifically, the LGBTQ rights movement has motivated many people to reexamine their treatment of sexual orientation and gender identity minorities and has produced a sense of cohesion and solidarity in the LGBTQ population. Gay activism has also been instrumental in advocating HIV/AIDS prevention strategies and health services that benefit society, and the recognition that birth sex and gender incongruence has expanded our thinking on the nature of gender and gender roles.

The structural-functionalist perspective is concerned with how changes in one part of society affect other aspects. For example, the worldwide increase in legal and social support for gay, lesbian, bisexual persons is likely the result of other cultural and structural changes in society. First, concern with the equality of the sexes has led to a relaxing of traditional gender roles, thereby supporting the varied expressions of female and male sexuality (see Chapter 10). Second, as gender roles have changed, so has the institution of marriage (see Chapter 5). Once an inherently unequal contractual relationship, heterosexual marriages today are most often a partnership whereby two people share, in varying degrees, household chores, financial responsibilities, and child rearing.

Third, LGBTQ persons are more visible today than in any other time in history, leading to larger numbers of people who can "put a face" on the cause. Nearly 80 percent of Americans know someone who is gay, and 20 percent someone who is transgender (Rosenberg 2018). The rejection of stereotypes (e.g., gay people don't believe in monogamy) becomes easier in the face of contradictory personal experience and is predictive of support for LGB equality (Reyna et al. 2014).

Lastly, globalization permits the international community to influence individual nations. Those advocating LGBTQ equality, whether in the United States or England or South Africa, are seen and heard through traditional and social media—strategies shared, petitions signed, and rage felt. Governments, corporations, or organizations that support LGBTQ rights may also bring pressure to bear on those that don't. In 2019, activist George Clooney wrote that "the nation of Brunei will begin stoning and whipping to death any of its citizens that are proved to be gay. Let that sink in. In the onslaught of news where we see the world backsliding into authoritarianism this stands alone" (Clooney 2019, p. 1). Coupled with months of international pressure, the government of Brunei withdrew the death penalty proposal.

The structural-functionalist perspective is also concerned with latent functions, or unintended consequences. For example, the movement toward LGBTQ equality and, more specifically, gay marriage led to a backlash whereby over 30 states amended their state constitutions to define marriage as exclusively between a man and a woman. A second unintended consequence, although a positive one, is the economic boom enjoyed by states since the Supreme Court's decision. It is estimated that in the five years following the legalization of gay marriage, wedding spending by same-sex couples and their guests has "boosted state and local economies by an estimated $3.8 billion and generated an estimated $244.1 million in state and local sales tax revenue" (Mallory and Sears 2020, p. 1).

Conflict Perspective

The conflict perspective frames the gay rights movement and the opposition to it as a struggle over power, prestige, and economic resources. Sexual and gender minorities want to be recognized as full citizens deserving of all the legal rights and protections entitled to individuals who are heterosexual and gender conforming. Gay and lesbian individuals have been waging a political battle to win civil rights protections against discrimination and, successfully in the United States, have fought to be allowed to marry a same-sex partner.

Conflict theory helps to explain why many business owners and corporate leaders support nondiscrimination policies. First, gay-friendly work policies help employers maintain a competitive edge in recruiting and retaining a talented and productive workforce. Second, many LGBTQ as well as heterosexual consumers prefer to purchase products and services from businesses that provide workplace protections for LGBTQ employees. Recent trends toward increased social acceptance of same-sex relationships may, in part, reflect the corporate world's competition for the gay and lesbian consumer dollar.

Business interests in the United States are galvanized in response to states that pass **religious freedom laws**. Opponents to religious freedom laws argue that, passed by conservative lawmakers, such laws legalize discrimination against LGBTQ consumers based on religious grounds. Indiana's passage of such a measure was met with vocal opposition, including threats of boycotts from organizations as diverse as the National Collegiate Athletic Association (NCAA), Angie's List, and Marriot (Socarides 2015). Over 75 executives in the technology industry representing such companies as Yelp, Apple, Twitter, Netflix, and PayPal signed an open letter that, in part, states (Kaufman 2015):

> To ensure no one faces discrimination and ensure everyone preserves their right to live out their faith, we call on all legislatures to add sexual orientation and gender identity as protected classes to their civil rights laws and to explicitly forbid discrimination or denial of services to anyone. (p. 1)

religious freedom laws Laws that protect business owners who discriminate against customers (e.g., gay men and women) based on religious grounds.

AP Images/Michael Conroy/File

In 2015, in reaction to the legalization of same-sex marriage, some states passed "religious freedom" bills ostensibly to protect the religious beliefs of Americans who were opposed to same-sex marriage. Similarly, the *First Amendment Defense Act* was introduced into Congress in 2018 and, if passed, would allow individuals or groups to refuse services or products to same-sex couples based on their religious belief that marriage should be between a man and a woman.

After intense pressure from business, civic and sports leaders, Indiana's *Religious Freedom Restoration Act* was amended to prohibit discrimination against LGBT consumers (Cook et al. 2015). To date, 21 states have religious freedom laws (London and Siddiqi 2019).

Furthermore, the ***First Amendment Defense Act*** **(FADA)**, introduced in the U.S. Senate in 2018 and presently in committee, prohibits the government from penalizing individuals or organizations that discriminate against same-sex couples if the individual or group has a "sincerely held religious belief or moral conviction that marriage is or should be recognized as a union of one man one woman" (S. 2525). Critics argue that the bill legalizes discrimination against LGBTQ individuals by permitting lawsuits against the federal government for interfering when discrimination takes place, theoretically, on religious grounds. It also holds that in such a suit, the attorney general of the United States would represent the individual or group that discriminated. Although the former administration was in favor of the bill, the Biden administration opposes it (see the "Law and Public Policy" section in this chapter).

What *do you* think? The federal government grants tax-exempt status to nonprofit organizations and colleges. In 1970, the IRS denied tax-exempt status for private colleges that continued to discriminate on the basis of race, for example, prohibiting interracial dating. Do you think that tax-exempt status should be revoked from private colleges that have anti-gay programs and policies (e.g., expelling students in a same-sex marriage)?

Symbolic Interactionist Perspective

Symbolic interactionism focuses on the meanings of heterosexuality, non-heterosexuality, and bisexuality; how these meanings are socially constructed; and how they influence the social statuses, self-concepts, and well-being of societal members. Examining the meaning of sexual orientations is part of a larger research area, the sociolinguistics of gay and lesbian lexicon. Just as "queer language" is learned, evolves, and is shared with others through social interaction, the meanings we associate with same-sex relations are learned—through society and its institutions.

First Amendment Defense Act An act, if passed, that would prohibit the government from imposing penalties for discriminating against married same-sex couples if it is demonstrated that the individual or organization held a "religious belief or moral conviction" against such marriages.

A study of children raised by gay or lesbian parents is a case in point. Sasnett (2015) interviewed 20 adults raised in same-sex households. Although the construction of identities as children of lesbian and gay parents was sometimes difficult, the problem was not the sexual orientation of their parents. It was *when* they learned of their biological parent's sexual orientation that proved to be the "critical component for how they were able to create meanings in their lives" (p. 196).

Children born into same-sex households who had never known any other type of family structure experienced less stress than those who, after a divorce, learned of their parent's sexual orientation. However, the stress was often externally imposed as others offered their own definitions of the situation. Ann, 43 years old, remarks:

> My mom and dad were married for 20 years, and I was fourteen when my mom left my dad for another woman. ... My dad supported my mom's decision to leave, but my school counselor freaked out. When she found out about the divorce she said, "You cannot live with your mom." Her negative response sent a clear message that I could not discuss this with people. So it wasn't my parents' response, but instead the response of the counselor at my school that scared me and made me think that my family wasn't normal. (p. 213)

Similarly, Haines et al. (2018) note that the question of the *legitimacy* of the family is raised when one or more of its members are non-heterosexual and/or non-cisgender (see this chapter's *Social Problems Research Up Close*).

The negative meaning associated with being gay, lesbian, or bisexual is also reflected in the current use of the phrases "that's so gay" and "you're so gay," and the tag line, "no homo" at the end of a sentence. In an investigation of such microaggressions (see Chapter 9), gay and bisexual male and female undergraduate students who reported higher frequencies of hearing such phrases on campus were more likely to report feeling "left out" (Woodford et al. 2012) and to engage in risk-taking behaviors including binge drinking and illicit drug use (Winberg et al. 2019).

Sociological research has also shown that the use of such words as *fag* and *queer* by heterosexuals to insult one another facilitates social acceptance by peers, particularly among heterosexual males. The results of a study by Slaatten and Gabrys (2014) on gay-related name-calling among ninth-grade students suggest that "gay-related name-calling among boys is more frequently used as a way of regulating unwanted expressions of masculinity than as a way of actively hurting or teasing someone" (p. 31). Nonetheless, gay-related name-calling comes at the expense of non-heterosexuals who are witness to or targets of such interactions.

The symbolic interactionist perspective also points to the effects of labeling on individuals. Language is power. Once individuals become identified or labeled as lesbian, gay, bisexual, or transgender, that label tends to become their **master status**. In other words, the dominant heterosexual cisgender community tends to view "gay," "lesbian," "bisexual," or "transgender" as the most socially significant statuses of individuals who are identified as such. These labels, in turn, often imbued with negative social meanings, impact an individual's self-definition and may lead to increased risk for a variety of, for example, mental and physical health problems, including substance abuse (Goldbach et al. 2014) and domestic violence (Hines 2014) (see this chapter's "The Consequences of Anti-LGBTQ Bias").

Cultural Origins of Anti-LGBTQ Bias

Anti-LGBTQ bias, although declining, has existed for hundreds of years, as countries around the world defined LGBTQ behavior as a symptom of mental illness, a sin, or both. In the United States, anti-LGBTQ bias has its roots in various aspects of American culture, including religion, psychology, rigid gender roles, and the myths and negative stereotypes about gay, bisexual, and transgender women and men.

Religious Beliefs

Organized religion has been both a source of comfort and distress for many gay, bisexual, and transgender Americans. Countless LGBTQ individuals have been forced to leave their faith communities due to the condemnation embedded in doctrine and practice. Not surprisingly, LGBTQ Americans are more likely to leave their childhood religion than their heterosexual or cisgender counterparts (Jones et al. 2014; Scheitle and Wolf 2017). Research has also found that higher levels of religiosity and religious attendance (Brown and Henriquez 2008; Lipka 2014; Pew 2013b; Pew 2019a; Shackelford and Besser 2007; Hoffarth, Hodson, and Molnar 2018), religious fundamentalism

master status The status that is considered the most significant in a person's social identity.

social problems
RESEARCH
UP CLOSE

Microaggressions toward Sexual and Gender Minority Families

Research documents the way that sexual orientation and gender identity (SOGI) minorities are subjected to microaggressions, i.e., everyday verbal or behavioral slights against members of an oppressed group (e.g., name calling, eye rolling). The following research is unique in that it looks at microaggressions directed toward a group, in this case the family, rather than at individuals.

Sample and Methods

Participants in the present study self-identified as being a member of a SOGI family. Respondents were between the ages of 18 and 58 with an average age of 27 and represented all regions of the United States. Sample respondents were recruited online and completed an open-ended survey. Seventy-two percent were female and 67.4 percent White, and they came from a variety of socioeconomic and educational backgrounds. Over two-thirds of the sample were gay or bisexual, and half were part of a SOGI couple. Other family configurations included but were not limited to SOGI couples with children living at home, SOGI couples with children but who were not living at home, SOGI singles, SOGI singles with children living at home, and SOGI singles with children who were not living at home.

Participants responded to questions about their microaggression experiences within the family context. Three types of microaggression experiences were identified. First, *micro-assaults*, defined as any "explicit derogation with the intention of harm through name-calling." Second, *micro-insults*, or "communications that convey rudeness and insensitivity and that demean a person's identity." And, finally, *micro-invalidations*, or communications that discount the "thoughts, feelings, or experiential reality of the marginalized group" (p. 1139). Researchers met to discuss classification of the responses and to come to a consensus on their coding. Participants were given the opportunity to clarify demographic information and their role in the family.

Findings and Discussion

Data analysis revealed three prominent themes related to microaggressions directed toward SOGI families. The first general theme was *questioning the legitimacy* of the SOGI family. This was done in a variety of ways including a lack of recognition of family relationships. For example, a 25-year-old gay man in a same-sex relationship, stated, "Everybody in our housing complex [were] straight couples or singles and they all received a Christmas present from our landlord except us" (p. 1143). Similarly, a transgender parent who was not called when her daughter was sick at school stated, "Because the county had forms that were not inclusive of all family types my [daughter] was made to feel bad" (p. 1144).

The second way that the legitimacy of families was questioned was through the refusal to recognize lawful marriages by referring to them as "not real." One young man commented that his family did not recognize his same-sex marriage as legitimate for the first year. Further, a queer 26-year-old female recounted that "my aunt told me that my marriage wasn't a 'real' marriage" (p. 1144). When the respondent asked why, her aunt explained that real marriages were between a man and a woman and that the respondent should "face reality."

The final way that the legitimacy of families was questioned was by attacking the notion that for a family to exist, members must be related by law or blood. A White woman in a same-sex relationship reported that her status as "aunt" to two African American boys adopted by another same-sex couple was challenged by others first on the basis of race and then on the lack of law or blood ties. A 43-year-old gay man stated:

> A number of friends/family refuse to call us "family," even though we always refer to ourselves that way. Just because we choose to be family in a non-traditional way doesn't negate the fact that we are committed to each other in the pursuit of life-long, living-together, sharing-everything relationships ...
>
> "[C]hosen family" is tried and true. (p. 1145)

The second general theme of microaggressions that emerged from the responses was that SOGI families *conflicted with real family values*. One gay woman stated, "My partner's family has disowned and verbally attacked our family due to our relationship and the addition of our children ... calling our family a travesty, calling us molesters, unfit to raise a family" (p. 1146). A heterosexual woman remembered:

> A friend in elementary school once said she didn't feel comfortable staying over because my mom likes girls. At first, I was hurt and embarrassed, but then I became angry. My mom called the girl's mom to make sure the family understood that the difference between homosexuality and pedophilia is a vast one. (p. 1146)

The final major theme that evolved from the analysis was the *presumed lack of traditional gender roles* in the family. Heteronormativity dictates that there is a man's role and a woman's role. One gay female was asked, "[W]ho is the man in the relationship?" and stated that she "was annoyed and tried to explain that there are no traditional gender roles for many lesbian couples" (p. 1146). As the authors note, microaggression as well as other forms of discrimination against sexual orientation and gender minorities often focus on gender nonconformity, particularly in the case of trans women and men who may be misgendered as society continues to classify people on the basis of birth sex.

The authors conclude that the research has limitations, most notably recruitment of respondents online that led to a relatively small sample size ($N = 46$) and disproportionately educated, middle-class, White respondents. If the experiences of more SOGI families of color had been included, the incidences of microaggression might have been more numerous and more negative, given the nexus between race and sexual orientation and gender identity. The researchers also call for future research on microaggression that explores the differences between those directed toward gay and bisexual women and men compared to microaggressions directed toward transgender women and men. The results of this study indicate that microaggressions based on gender identity were substantively different from those based on sexual orientation.

SOURCE: Haines et al. 2018.

(Pew 2013c; Summers 2010), and more conservative political beliefs (Jones et al. 2014; McCarthy 2014; Pew 2019b) are consistently associated with more negative attitudes toward sexual and gender minorities.

> Pete Buttigieg, formerly the mayor of South Bend, Indiana, was the first openly gay candidate from a major party to run for president of the United States. His husband, Chasten, regularly campaigned for him and at one event, explained, that when the "Mayor" asked him if he should run for president, he said yes because "Pete got me to believe in myself again ... and ... I knew there were other kids sitting out there in this country who needed to believe in themselves too" (LeBlanc and Merica 2020, p. 1). Do you think having a gay man run a successful primary campaign for president of the United States impacted attitudes toward gay men and women?
>
> **What do you think?**

As recently as 2003, every major religious group in the United States was opposed to marriage equality. Today, there is variability within Judeo-Christian traditions regarding attitudes toward LGBTQ individuals. For example, in a survey of Anglican youth in England, nearly half believe that same-sex marriage is acceptable compared to 38 percent in 2013 (Gallagher 2020). However, despite the legalization of same-sex marriage in the United States, America's largest mainline Protestant group, the United Methodist Church, is facing a divide over gay marriages and the ordination of LGBTQ ministers (Associated Press 2020a). Figure 11.2 displays trends in attitudes toward gay marriage in the United States by religion from 2000 to 2019.

The official position of a religion on same-sex marriage or on LGBTQ people, in general, does not necessarily represent the beliefs of the followers, in part because of the multiple statuses (e.g., White married father, college graduate) that people hold. Despite the Vatican's prohibition against same-sex marriage, the majority of Catholics—61 percent—support gay marriage (Pew 2019a). The extent of acceptance, however, varies by not only religion—Muslims are more likely to disapprove of same-sex relationships than Jews—but also within religious category based on demographic variables (Adamczyk 2017). For example, younger people tend to be more accepting of same-sex relationships than older people, and liberals tend to be more accepting of same-sex relationships than conservatives.

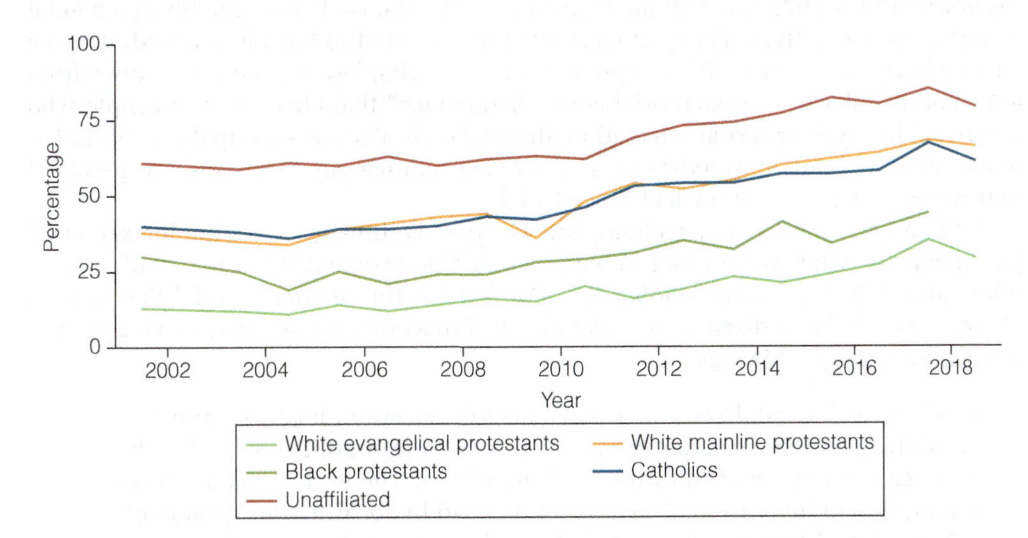

Figure 11.2 Percent of U.S. Adults Who Favor Same-Sex Marriage, by Religion, 2001–2019
SOURCE: Pew 2019a.

This billboard first appeared along Interstate 15 outside of Orem, Utah, in June 2015. Today, similar billboards appear in Salt Lake City, Utah, and Dearborn Heights, Michigan. In 2020–2021, the billboard is scheduled to appear in Raleigh, Greensboro, and Fayetteville, North Carolina.

In response to anti-LGBT and the resulting rejection from existing religious institutions, in 1968 a Pentecostal minister established the first of today's nearly 300 congregations in 22 countries (Metropolitan Community Church [MCC] 2020a). Metropolitan Community Churches are LGBT affirming, congregationally led denominations that offer "a safe and open community for people to worship, learn, and grow in their faith" (MCC 2020b, p. 1). In her research, Kane (2013) argues that the formation of MCCs was an act of protest challenging the assumption that non-heterosexuality and Christianity are inherently incompatible. Confirming this belief, when the United Federation of Metropolitan Churches petitioned to join the National Council of Churches (NCC), "essentially requesting full inclusion in U.S. religious life," they were denied by the leadership of the NCC (Kane 2013, p. 137). The MCC remains excluded from the National Council of Churches.

Non-Heterosexuality as Pathology

Medical and mental health professionals (e.g., psychologists, psychiatrists), to a large extent, are responsible for anti-LGBTQ bias. For decades, "homosexuality" and "transvestism", as they were once called, were labeled as "abnormal, pathological, and deviant" (Baughey-Gill 2011, p. 1). The origin of **homophobia**, broadly defined as the hate or fear of or prejudice toward gay and bisexual men and women, can be traced back to early Christianity which, out of historical necessity, defined sex as only appropriate for the purposes of procreation. "Homosexuality" was thus sinful but, following a "long tradition in medicine and psychiatry, which in the 19th century appropriated homosexuality from the Church … [It was] promoted … from sin to mental disorder" (Burton 2015, p. 1).

In 1968, homosexuality first appeared in the *Diagnostic and Statistical Manual* of the *American Psychiatric Association* (APA), which officially classified it as a mental disorder. At their 1973 convention, the APA voted to remove homosexuality as a mental disorder. Although the majority of members present voted to have it removed, it was a slim majority, and so the APA compromised by changing lesbian and gay statuses from a mental disorder to a "sexual orientation disturbance" that characterized people who were troubled by their sexual orientation (Burton 2015). Thus, the group that contributed to the feelings of guilt and anxiety often suffered by non-heterosexuals was now prepared to treat the very problem they had exacerbated.

Ironically, it was a psychologist, George Weinberg—a Columbia-schooled psychoanalyst who was trained to believe that "homosexuality" was pathological—who coined the term *homophobia* in 1972. Unlike the *medical model* that treated "homosexuality" as a disease and, like physical diseases, had a *cause* and a *cure*, the concept of homophobia was

> a milestone. It crystallized the experiences of rejection, hostility, and invisibility that homosexual men and women in mid-20th century North America had experienced throughout their lives. The term stood a central assumption of heterosexual society on its head by locating the "problem" of homosexuality not in homosexual people, but in heterosexuals who were intolerant of gay men and lesbians. (Herek 2004, p. 8)

homophobia Broadly defined as the hate or fear of or prejudice toward homosexuals.

Commonly used words such as *transphobia*, *biphobia*, and *heterophobia* are derivatives of homophobia and can be attributed to Weinberg's groundbreaking work.

Despite the evolution of mental health professionals in the United States, in many parts of the world today the behavior and identities of gay, bisexual, and transgender men and women are still defined as mental illness. In a global survey of over 1,600 victims of conversion therapy, the most common perpetrator of the "treatment" was medical/mental health providers (UN 2020). In 2019, the American Psychoanalytic Association apologized for the practice of defining "homosexuality" and "transvestism" as pathological. In a news release on the 50th anniversary of the **Stonewall Uprising** in New York City, where LGBTQ patrons of a now historic tavern fought back against police brutality, the president of the Association stated, "[R]egrettably, some of that era's understanding of homosexuality and gender identity can be attributed to the American Psychoanalytic establishment ... [I]t is long past time to recognize and apologize for our role in the discrimination and trauma caused by our profession" (American Psychoanalytic Association 2019, p. 1).

@George Takei

The LGBTQ, civil rights movement began 51 years ago with the Stonewall Riots, led by trans POC heroes. Today we mark another milestone in our struggle for equality with a victory in the Supreme Court, extending Title VII nondiscrimination protections to LGBTQs.

O happy day!
-George Takei

Rigid Gender Roles

Disapproval of gay, bisexual, and transgender men and women also stems from rigid gender roles. Kimmel (2011), a sociologist who specializes in research on masculinity, writes about the relationship between perceived masculinity and bullying:

> Why are some students targeted? Because they're gay or even "seem" gay—which may be just as disastrous for a teenage boy. After all, the most common put-down in American high schools today is "that's so gay," or calling someone a "fag." It refers to anything and everything: what kind of sneakers you have on, what you're eating for lunch, some comment you made in class, who your friends are or what sports team you like. ... Calling someone gay or a fag has become so universal that it's become synonymous with dumb, stupid, or wrong. (p. 10)

From a conflict perspective, heterosexual men's subordination and devaluation of gay men reinforces gender inequality. "By devaluing gay men ... heterosexual men devalue the feminine and anything associated with it" (Price and Dalecki 1998, pp. 155–156). Negative views toward lesbians also reinforce the patriarchal system of male dominance. Social disapproval of lesbians is a form of punishment for women who relinquish traditional female sexual and economic dependence on men.

Not surprisingly, research findings suggest that individuals with traditional gender role attitudes tend to hold more negative views toward non-heterosexuality than those with more modern perspectives (DeCarlo 2014; Falomir-Pichastor et al. 2010). Further, results of an online survey of 6,063 respondents from Belgium indicate that binary gender role beliefs are significant predictors of negative attitudes toward gay, bisexual, *and* transgender individuals (Dierckx, Meier, and Motmans 2017). Younger and more educated respondents were less likely to hold traditional gender role beliefs and thus were less likely to be homophobic, biphobic, and transphobic than their older and less educated counterparts.

Myths and Negative Stereotypes

The stigma associated with gay, bisexual, or transgender men and women also comes from myths and negative stereotypes. One negative myth about non-heterosexuals is that they are sexually promiscuous and lack "family values," such as monogamy and commitment to a relationship. Although some gay men and lesbians do engage in casual sex, as do some heterosexuals, many same-sex couples develop and maintain long-term committed relationships. Sixty-one percent of same-sex couples who live together are married, and gay, bisexual, and transgender men and women, as their straight and cisgender counterparts, are most likely to cite love as the most important reason to get married (Masci, Brown, and Kylie 2019).

Stonewall Uprising In 1969, patrons of this now historic Greenwich Village gay bar fought back against police brutality; often thought to be the beginning of the gay rights movement.

oppression The use of power to create inequality and limit access to resources, which impedes the physical and/or emotional well-being of individuals or groups of people.

privileged When a group has a special advantage or benefits as a result of cultural, economic, societal, legal, and political factors.

heterosexism A form of oppression that refers to a belief system that gives power and privilege to heterosexuals, while depriving, oppressing, stigmatizing, and devaluing people who are not heterosexual.

Another myth is that non-heterosexuals, as a group, are a threat to children—most notably as child molesters. In other words, people confuse same-sex attraction with pedophilia, or sexual activity with prepubescent children. Having a non-heterosexual orientation is unrelated to pedophilia. Research has *not* demonstrated a connection between an adult's same-sex attraction and an increased likelihood of molesting children and teenagers. In fact, a pedophile does not have an adult sexual orientation because they are fixated on children and teenagers (Schlatter and Steinback 2014).

Finally, gay men and women are often stereotyped in terms of their appearance and mannerisms—gay men as feminine and gay women as masculine. There are, of course, gay men and women who conform to the stereotype but, statistically, very few. The persistence of the opposite-gender characterization of gay men and women is a result of several sociological processes. First, gay men and women who fit the stereotype are noticed (e.g., gay feminine male hairdresser), while those who don't are not (e.g., gay masculine male construction worker), reinforcing the stereotype that gay men are feminine and work in certain professions. Interestingly, the feminine-looking man may not be gay but the stereotype will be reinforced nonetheless. Second, wanting to avoid that feeling of discomfort when confronted with dissonant information (e.g., gay men are masculine), we reject the possibility with reassuring confidence—"No, he can't be gay—he's too masculine!"

What do you think?

Anti-LGBTQ forces did not operate independently of one another but rather acted synergistically. For example, of the 71 countries in the world that still outlaw non-heterosexual behavior, half of them were British colonies or protectorates (Westcott 2018). Colonial administrators, who held the Victorian attitude that any sexual behavior other than for procreation was taboo, imposed their beliefs on locals. Interestingly, many of the colonies, particularly in Africa, had few normative definitions about appropriate sexuality and gender roles prior to colonization (Fisher 2013). Why do you think so many African countries remain homophobic, while countries in the United Kingdom are some of the most progressive in the world in regard to LGBTQ equality?

Prejudice against Lesbians, Gay Men, and Bisexuals

Oppression refers to the use of power to create inequality and limit access to resources, which impedes the physical and/or emotional well-being of individuals or groups of people. A person or group is **privileged** when they have a special advantage or benefits as a result of cultural, economic, societal, legal, and political factors (Guadalupe and Lum 2005). **Heterosexism** is a form of oppression and refers to a belief system that gives power and privilege to heterosexuals, while depriving, oppressing, stigmatizing, and devaluing people who are not heterosexual (Herek 2004; Szymanski et al. 2008). **Cisgenderism** is a form of oppression that refers to a belief system that gives power and privilege to those whose gender identities align with their assigned birth sex, while subordinating gender variant people.

cisgenderism A form of oppression that refers to a belief system that gives power and privilege to those whose gender identities align with their assigned birth sex, while subordinating gender variant people.

prejudice Negative attitudes and feelings toward or about an entire category of people.

discrimination Actions or practices that result in differential treatment of categories of individuals.

The belief that being straight and cisgender are superior to being gay or transgender results in prejudice and discrimination against gay, bisexual, and transgender men and women. **Prejudice** refers to negative attitudes, whereas **discrimination** (discussed in the next section) refers to behavior that denies individuals or groups of individuals equality of treatment (see Chapter 9). In turn, this often leads non-heterosexual and transgender individuals to question the legitimacy of their same-sex attractions and gender identity. Before reading further, you may wish to complete this chapter's *Self and Society* feature, which assesses your attitudes toward gay and transgender people and allows you to compare your answers to respondents from all over the world.

Attitudes toward Sexual and Gender Minorities around the World

For each of the following statements, indicate whether you *agree* with the statement, *disagree* with the statement, or *neither* agree nor disagree. For questions 9 and 10, indicate whether you are *comfortable* socializing, *uncomfortable* socializing, or *neither* comfortable nor uncomfortable. When finished, compare your responses to those from a survey of 116,000 respondents from 77 countries representing all regions of the world.

	Agree	Disagree	Neither
1. Equal rights and protections should be applied to everyone including gay people.	___	___	___
2. Equal rights and protections should be applied to everyone including transgender people.	___	___	___
3. It is possible to respect my religion and be accepting of gay people.	___	___	___
4. It is possible to respect my religion and be accepting of transgender people.	___	___	___
5. All workers, including those who are gay, should be protected from workplace discrimination.	___	___	___
6. All workers, including those who are transgender, should be protected from workplace discrimination.	___	___	___
7. Transgender adults should be granted full legal recognition of the identity they declare.	___	___	___
8. People who engage in same-sex relationships should be charged as criminals.	___	___	___

	Comfortable	Uncomfortable	Neither
9. How comfortable are you socializing with gay people?	___	___	___
10. How comfortable are you socializing with transgender people?	___	___	___

	Percent who:		
	Agreed	Disagreed	Neither
1. Equal rights and protections should be applied to everyone including gay people.	55	25	20
2. Equal rights and protections should be applied to everyone including transgender people.	59	21	20
3. It is possible to respect my religion and be accepting of gay people.	49	28	23
4. It is possible to respect my religion and be accepting of transgender people.	54	23	24
5. All workers, including those who are gay, should be protected from workplace discrimination.	57	23	20
6. All workers, including those who are transgender, should be protected from workplace discrimination.	59	21	43
7. Transgender adults should be granted full legal recognition of the identity they declare.	50	15	25
8. People who engage in same-sex relationships should be charged as criminals.	28	49	23

	Percent who were:		
	Comfortable	Uncomfortable	Neither
9. How comfortable are you socializing with gay people?	54	33	29
10. How comfortable are you socializing with transgender people?	42	28	30

*May not sum to 100 percent because of rounding; not all respondents answered all questions resulting in unequal *N*s; questions have been shortened for the purposes of this exercise. Respondents came from 15 countries in the Americas, 22 European countries, 15 African countries, 23 countries in Asia, and two in Oceania.

SOURCE: IGLA-RIWI 2018.

Prejudice toward non-heterosexual individuals varies significantly by social variables globally and in the United States. A 2019 *Global Attitudes Survey* indicates that, in general, younger people are more accepting of gay and bisexual men and women than older respondents (Poushter and Kent 2020). Women tend to be more accepting of gay and lesbian sexuality than men, although, just as acceptance in general, it varies by country. For example, the acceptance percentage difference between men and women in South Korea is 14 percent, while in the Netherlands where acceptance is almost universal, men are only 5 percent less likely to accept gay and bisexual men and women than women. Acceptance also increases with the wealth (see Figure 11.1) and education of a country and with the percentage of the population that is left-leaning politically (Poushter and Kent 2020).

The reduction of LGB prejudice is key in the larger fight for LGB and transgender rights. As more LGBT Americans "come out" to their family, friends, and coworkers, heterosexuals and gender-conforming individuals have more personal contact with LGBT individuals. Psychologist Gordon Allport (1954) asserted that contact between groups is necessary for the reduction of prejudice—an idea known as the **contact hypothesis**. Research has shown that straight people have more favorable attitudes toward gay men and women if they have had prior contact with gay men and women.

There are also several indicators of the growing acceptance of non-heterosexuals (see Figure 11.3). A national opinion poll indicates that the majority of Americans—72 percent—now believe that gay men and women should be accepted by society (Poushter and Kent 2020). Similarly, research that examined prejudice toward minority groups found that, between 2007 and 2016, anti-gay bias decreased by 33 percent. The researchers conclude that the change "is not only fast, but is also steady. The model predicts consistent decreases over time, such that anti-gay bias could reach complete neutrality (zero bias) between 2025 and 2045" (Charlesworth and Banaji 2019).

One reason for the reduction in gay prejudice is the legalization of same-sex marriage in the United States and elsewhere. An examination of over 1 million residents in various states over a 12-year period indicates that anti-gay bias, although decreasing over time in general, declined significantly once a respondent's state passed same-sex marriage legislation (Ofosu et al. 2019). Similarly, Figure 11.3 indicates that between 2010 and 2015, the percentage of Americans saying that same-sex sexual behavior should be accepted increased by just 2 percent. However, after the legalization of same-sex marriages nationwide in 2015, the acceptability of same-sex sexual behavior in American society increased from 60 percent to 72 percent (Poushter and Kent 2020).

contact hypothesis The idea that contact between groups is necessary for the reduction of prejudice.

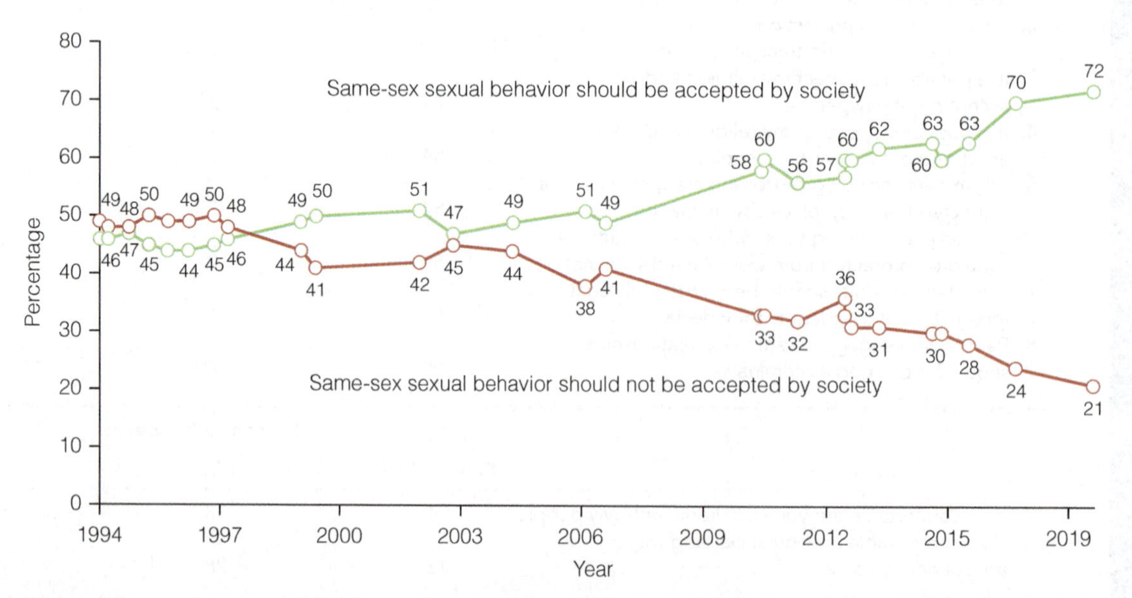

Figure 11.3 Percent of Americans Saying That Same-Sex Sexual Behavior Should Be Accepted by Society, 1994–2020*

*The term "homosexuality" was used in the original surveys. Numbers may not sum to 100 due to missing data.
SOURCE: Pew Research Center 2020.

Another indicator of the greater acceptance of LGBTQ people is the election of non-heterosexual, non-cisgender officials to political offices. In total, there are 854 LGBTQ elected officials in the United States, including two governors, nine U.S. Congress members, 46 mayors, and 160 state legislators (Victory Institute 2020). The importance of the increase in the number of elected LGBTQ officials in the United States lies not only in the greater acceptance it signifies, but, as former U.S. Congressman Barney Frank once said, "If you are not at the table, you are likely on the menu."

Discrimination against Lesbians, Gay Men, and Bisexuals

In June 2003, a Supreme Court decision in *Lawrence v. Texas* invalidated state laws that criminalize sodomy—oral and anal sexual acts. This historic decision overruled a 1986 Supreme Court case (*Bowers v. Hardwick*), which upheld a Georgia sodomy law as constitutional. The 2003 ruling, which found that sodomy laws were discriminatory and unconstitutional, removed the legal stigma and criminal branding that sodomy laws have long placed on LGB individuals. Nonetheless, as of 2019, sodomy or "unnatural acts" are still illegal in 15 states and are primarily used against gay and bisexual men and women (Chibbaro 2020).

Figure 11.4 summarizes LGBT equality on several dimensions including, but not limited to, nondiscrimination in the workplace, marriage equality, parental rights, religious freedom exemptions (as already discussed), health care, and criminal justice protections. Policies were assigned positive or negative numbers based on their relative impact on the non-heterosexual and gender variant communities and then summed. The higher the number, the greater that state's current equality score. The scores were then divided into five categories (i.e., no, low, fair, medium, and high) representing the LGBT policy climate in each state (Movement Advancement Project [MAP] 2020a). Before discussing some of the dimensions used to measure equality, or the lack thereof, it is important to note that in 2010 just over 5 percent of gay, bisexual, and transgender women and men lived in states with high policy equality; today, nearly half of all sexual and gender minorities live in states with high policy equality scores.

Workplace Discrimination and Harassment

In 2015, an executive order signed by President Obama prohibiting discrimination on the basis of sexual orientation or gender identity went into effect. The order protected federal workers and millions of employees of government contractors. However, with the election of Donald Trump, LGBTQ rights, including the prohibition against workplace discrimination, slowly eroded.

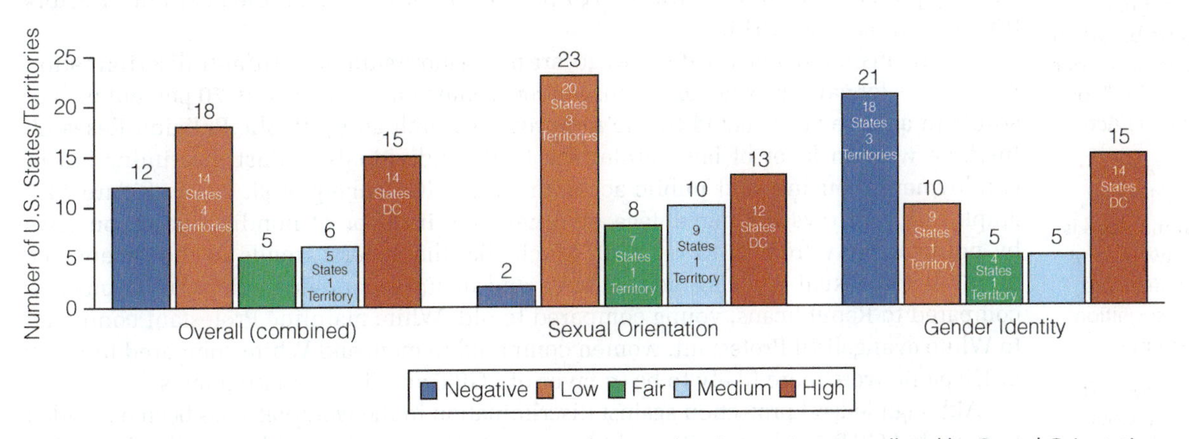

Figure 11.4 Number of U.S. States and Territories in Policy Equality Categories, Overall and by Sexual Orientation, and Gender Identity, 2020
SOURCE: MAP 2020a.

In 2017, an edited memo from the then-secretary of the interior to his department staff regarding ethical behavior of the agency was obtained through the *Freedom of Information Act*. The memo appeared as here: "13. I shall adhere to all laws and regulations that provide equal opportunity for all Americans regardless of race, color, religion, sex, national origin, age, ~~sexual orientation~~ or handicap" (reproduced in D'Angelo 2019, p. 1).

Although the deletion of "sexual orientation" was explained by an agency spokesperson as a simple clarification in language (i.e., the use of the word sex implies gender, gender identity, transgender status, and sexual orientation), several months later, the U.S. Department of Justice filed a brief with the U.S. Supreme Court asking it to find that the termination of transgender employees was not a violation of Title VII of the U.S. *Civil Rights Act* of 1964. The administration's argument was that an individual's right against employment termination is protected on the basis of sex but that "sex," as used in Title VII, refers to birth sex only. The same year, the Trump administration announced a transgender military ban. However, President Biden overturned the former administration's transgender military ban in 2021 (Singman 2021).

What do you think?

In 2016, North Carolina became infamous for its "bathroom bill," which outlawed trans men and women from using the bathroom associated with their gender identity. In 2019, a federal judge prohibited North Carolina "from banning transgender people from using bathrooms in state buildings that match their gender identity, ending a years-long legal battle that prompted a divisive cultural debate" (Levin 2019). Do you think the court made the right decision?

In 2020, the U.S. Supreme Court ruled in a 6–3 decision that the 1964 civil rights law protects gay and transgender workers against discrimination in the workplace. Until this decision, it was legal in more than half the states to fire an employee for no other reason than their sexual orientation or transgender status (Liptak 2020). The logic of the argument, as succinctly argued by Justice Neal Gorsuch writing for the majority, was that "an individual's homosexuality or transgender status is not relevant to employment decisions. That's because it is impossible to discriminate against a person for being homosexual or transgender without discriminating against that individual based on sex" (*Bostock v. Clayton County, Georgia* 2020).

Prior to the ***Bostock v. Clayton County, Georgia*** decision, states, counties, and municipalities had a patchwork of laws and regulations governing LGBTQ discrimination in the workplace. The likelihood of having gay-friendly policies increased if (1) the company was in a state with progressive LGBT laws, (2) the company's competitors had gay-friendly policies, or (3) there was a high proportion of women on the board of directors (Everly and Schwarz 2015).

Bostock v. Clayton County, Georgia A U.S. Supreme Court decision that held that discrimination based on sexual orientation or gender identity violates Title VII of the 1964 *Civil Rights Act.*

Defense of Marriage Act (DOMA) Federal legislation stating that marriage is a legal union between one man and one woman and denies federal recognition of same-sex marriage.

Obergefell v. Hodges The 2015 U.S. Supreme Court decision that legalized same-sex marriage in the United States.

Americans, compared to a decade ago, are much more supportive of anti-discrimination protections for gay, bisexual, and transgender women and men. Nearly 70 percent of U.S. adults in a survey conducted by the nonprofit research group Public Religion Research Institute were in favor of laws protecting LGBT individuals against discrimination in employment, housing, and public accommodation (Greenberg et al. 2019). Figure 11.5 graphically portrays the percentage of Americans in favor of nondiscrimination laws by political party from 2015 to 2019. Much like the general trends in the acceptance of non-heterosexual and cisgender women and men, respondents who were Democrats compared to Republicans, young compared to old, White mainline Protestant compared to White evangelical Protestant, women compared to men, and White compared to Black or Hispanic were more likely to be in favor of LGBT nondiscrimination laws.

Although federal protection against discrimination in the workplace has been expanded to include LGBT employees, it would be premature to say the problem of discrimination against LGBT Americans has been solved. Two million students 15 years old and older live in states where there is no protection against discrimination in schools; 5.4 million LGBT

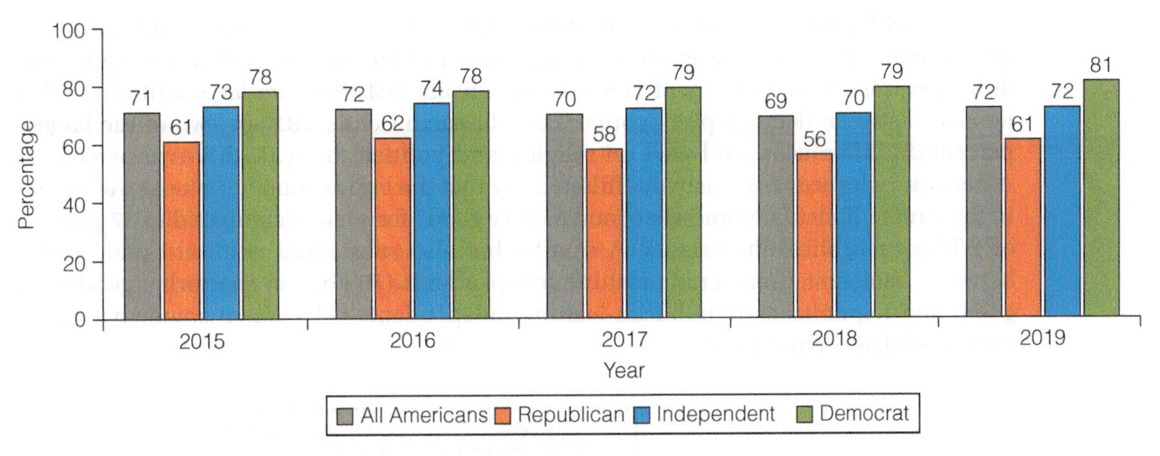

Figure 11.5 Percent of U.S. Adult Population in Favor of Laws to Protect LGBT People from Discrimination in Jobs, Public Accommodations, and Housing, by Political Party, 2015–2019
SOURCE: Public Religion Research Institute (PRRI) 2020.

adults live in states where there is no protection against discrimination in housing; and approximately 6.5 million Americans 13 years old and older live in states where there is no protection against discrimination in public accommodations on the basis of sexual orientation or gender identity (Conron and Goldberg 2020). Over half of Americans report that new laws are needed to reduce discrimination against LGBT people (Gallup Poll 2020).

> "Marriage is a magic word. And it is magic throughout the world. It has to do with our dignity as human beings, to be who we are openly."
>
> **–EDIE WINDSOR, DEFENDANT, *U.S. V. WINDSOR***

Marriage Inequality

As of 2020, same-sex marriages are legal in just 29 countries (Human Rights Campaign [HRC] 2020a), disproportionately in Western democracies (Felter and Renwick 2020). The legalization of same-sex marriage is a relatively new phenomena, taking place just over the last 17 years between 2001 (the Netherlands) and 2020 (Costa Rica). In 1996, the U.S. Congress passed and President Clinton signed the ***Defense of Marriage Act* (DOMA)**, which (1) stated that marriage is a "legal union between one man and one woman"; (2) denies federal recognition of same-sex marriage; and (3) allows states to either recognize or not recognize same-sex marriages performed in other states. In 2003, Massachusetts became the first state to recognize same-sex marriages; and, over the course of the next 12 years, as a result of legislation, court decisions, or ballot initiatives, 31 additional states had followed suit.

At the federal level, the U.S. Supreme Court issued two key rulings affecting same-sex marriage in the United States. In the first ruling, in 2013, the parts of DOMA that denied the same equal benefits and recognition to legally married same-sex spouses that married heterosexual spouses have were deemed unconstitutional (*U.S. v. Windsor* 2013). The second ruling, ***Obergefell v. Hodges*** (2015), resulted in marriage equality throughout the United States which, theoretically, assures that all same-sex married couples can now receive the more than 1,000 federal rights, benefits, and responsibilities of marriage heretofore reserved for different-sex married couples (see this chapter's section "Law and Public Policy").

Bill Aron/PhotoEdit

Two gay Jewish men wearing yarmulkes stand under a chuppah during their outdoor wedding ceremony. The reformed Jewish community was an early supporter of LGBT equality and allows the ordination of gay and transgender male and female rabbis.

both parents *should* be considered the legal parents of a child from birth on as is the case with heterosexual married couples. In the case of adoption, some states allow same-sex married couples to jointly adopt and/or allow the non-biological parent to adopt the child or children of his or her spouse. In other states, the partner of a biological or adoptive parent is not legally a parent of the child, and has no or few rights to the child or children if the relationship ends (Family Equality Council 2014). There is some speculation, however, that with the legalization of gay marriage, legal restrictions such as these may be abolished (Giambrone 2015).

Several respected national organizations—including the American Academy of Family Physicians, American Academy of Pediatrics, American Bar Association, American Medical Association, American Psychological Association, Child Welfare League of America, National Adoption Center, the National Association of Social Workers, the North American Council on Adoptable Children, and Voice for Adoption—have taken the position that a parent's sexual orientation has nothing to do with his or her ability to be a good parent (American Medical Association 2011; U.S. Court of Appeals 2010; *Position Paper* 2015). Furthermore, research indicates that the majority of Americans, regardless of age or political affiliation, favor adoption rights for same-sex couples (Gallup 2020).

That said, a sizeable minority, 23 percent, is opposed. Those opposed traditionally cite concerns that same-sex parents will lead to dysfunctional families and gender role confusion in children. For example, since Mormons hold that gender roles are "specific, complementary, and essential," and determined by God, two women or two men raising children is "unnatural" although they recently changed their policy to allow children of same-sex couples to be baptized (Sumerau and Cragun 2014; Dias 2019).

The results of an investigation into parenting of 41 heterosexual couples (mothers and fathers) and 48 male couples (half biological fathers, half adoptive fathers) may call into question the assumption of "natural" gender roles (Abraham et al. 2014). All couples were in committed relationships and parents for the first time. Each of the gay couples had used surrogacy to become parents and the biological mothers were not involved in the care of the child.

After videotaping each parent interacting with their child in the home, researchers examined MRI images of the brain when subjects later watched the interactions. When mothers, all primary caregivers, watched their babies, there was increased activity in the *emotion* processing portion of the brain. When heterosexual fathers, none of whom were primary caregivers, watched their babies the *cognitive* portion of their brain, the part that interprets and understands social behavior, increased in activity. When gay primary-care fathers watched their babies, *both* the emotional *and* cognitive portions of their brains were triggered, and the more time spent with his child, the better the connectivity between the two spheres. The authors conclude that the "connectivity between the two networks in primary-caregiving fathers suggests that, although only mothers experience pregnancy, birth, and lactation, ... [there are] other pathways for adaptation to the parental role in human fathers, and these alternative pathways come with practice, attunement, and day-by-day caregiving" (p. 9795).

In 2015, the U.S. Supreme Court heard arguments both for and against the legalization of gay marriages as part of an appeal by petitioners in Ohio, Tennessee, Michigan, and Kentucky. The petitioners, including James Obergefell, successfully challenged the right of each state to deny legal recognition of their marriage from a different state and/or refusal of the state to issue a marriage license. Within this context, the American Sociological Association (ASA 2015) filed an *amicus* brief outlining social science research on children raised in same-sex families compared to children raised in different-sex families. Child outcome categories included academic performance and cognitive development (e.g., GPA), social development (e.g., number of friends), mental health (e.g., depression), early sexual activity (e.g., age of first intercourse), and substance abuse and behavioral problems (e.g., alcohol abuse). The ASA brief concluded that, consistently and with a consensus, these "studies reveal that children raised by same-sex parents fare just as well as children raised by different-sex parents across a wide spectrum of factors used by social scientists to measure child wellbeing" (ASA 2015, p. 5).

Same-sex couples are seven times more likely to be raising adopted or foster children than different-sex couples. Of all couples, same-sex and different sex, male and female, married and not married, married male same-sex couples are the most likely to be raising adopted or foster children (Goldberg and Conron 2018). Sixteen percent of all same-sex couples are raising children. As with different-sex couples, the majority of children being raised in same-sex households are biological children. Of all same-sex couples raising children, it is estimated that about half, as of 2016, are married. The percentage of married male same-sex couples, married female same-sex couples, and married different-sex couples raising children are 13 percent, 30 percent, and 39 percent, respectively.

What do you think?

Eight states are presently considering legislation to "deny gender-affirming medical care to transgender youth" (Conron and O'Neill 2020, p. 1). Most of the bills make it a crime for physicians to engage in gender-affirming surgeries and, in two states, parents could be reported to child welfare agencies for facilitating their son's or daughter's gender transitions. What do you think? Should the state be able to tell a minor and/or a parent and/or their physician when and if transgender child can transition?

The Health Care Industry

In 2019, the Trump administration proposed changes to the *Affordable Care Act*, popularly known as Obamacare, which would narrow the definition of sex discrimination to exclude transgender patients. Long advocated by political and religious conservatives, the new rules "erase protections for transgender patients against discrimination by doctors, hospitals, and health care insurance companies" (Sanger-Katz and Weiland 2020, p. 1). This is particularly alarming given the ongoing COVID-19 crisis and research that indicates that LGBTQ men and women are not only more vulnerable to its economic impact but have greater risks of health complications (HRC 2020). Said the director of the National Center for Transgender Equality, it is "horrendous to gut non-discrimination protections, but to gut non-discrimination protections in the middle of a pandemic" is really horrendous (quoted in Sanger-Katz and Weiland 2020).

@LGBTfdn

LGBT abuse has more than doubled during lockdown. 22 extra people call our helpline every day at a point of crisis.

-LGBT Foundation

Health care for transgender men and women is often stigmatized and misunderstood ([NCTE] 2020). Critics argue that the medical care required to transitioning transgender men and women is unnecessary since, so they argue, the need to transition from biological sex to gender identity is unnecessary. Experts, however, argue that for many trans people, transitioning is a matter of mental health. Major medical groups such as the American Medical Association, the American Psychological Association, and the American Academy of Pediatrics have "detailed guidelines for providers working with youth and adults who are transgender, and those same groups opposed any discrimination against trans patients, including restrictions on the care or coverage a transgender person can receive" (NCTE 2020, p. 1). Although some religious organizations support the proposal to remove nondiscrimination protections in health care for trans men and women, the president and chief executive officer of the Catholic Health Association of the United States, the largest nonprofit provider of health care in America, stated:

> While we welcome the efforts to reaffirm the unique mission of faith-based health care providers, refusing to provide medical assistance or health care services merely because of discomfort with or animus against an individual on the basis of how that person understands or expresses gender or sexuality is unacceptable. (Quoted in Kaplan 2020, p. 1)

Whether or not the U.S. Supreme Court's decision to define the prohibition against sex discrimination as inclusive of sexual orientation and gender identity is yet to be legally determined.

AP Images/Ben Curtis

Protestors in Kenya demonstrate against Uganda's increasingly punitive policies against gay men and women. In 2014, the Ugandan High Commission passed a bill that made "aggravated" same-sex sexual behavior punishable by life in prison. The act was repealed as a result of international pressure. Today, as a result of the draconian laws of the early British colonists, being gay in Uganda remains against the law.

Violence, Hate, and Criminal Victimization

According to the Federal Bureau of Investigation (FBI), 16.7 percent of reported hate crimes in the United States in 2018 were motivated by sexual orientation bias and 2.3 percent because of gender identity (FBI 2019). It must be noted, however, that crime statistics consistently underestimate the incidence of hate crimes and of crimes in general (see Chapter 4). For example, only 23 states and Washington, D.C., define hate-motivated crimes on the basis of sexual orientation and gender identity. An additional 11 include sexual orientation but not gender identity in their hate crime statutes (MAP 2020c) (see Figure 11.8 later in this chapter).

The National Coalition of Antiviolence Programs (NCAVP), in their 2018 annual report, the latest year for which data are available, states that in 2017, there was a 21-year high in the number of LGBTQ hate-motivated homicides (NCAVP 2018). People of color comprised 71 percent of all LGBTQ victims, and transgender or gender-nonconforming people comprised 52 percent of all LGBTQ victims. As in the previous five years, the most common victim of an LGBTQ hate-motivated crime was a transgender woman of color. According to the NCVAP report, the most common perpetrators of LGBTQ hate crime survivors were, in rank order: (1) employers and coworkers, (2) relatives, and (3) landlords, tenants, and neighbors. In 2020, as part of the *Black Lives Matter* movement, protests were organized in response to the murder of two Black trans women in New York City (see Chapter 10). One of the most horrific LGBTQ hate crimes to date was the 2016 mass murder of 49 and injury of 53 men and women in a gay bar in Orlando, Florida.

Anti-LGBTQ violence is not unique to the United States. Hate crimes based on sexual orientation and gender identity remain a serious problem in many regions of the world. In Jamaica, often called the most anti-gay country in the world, newspapers routinely post incidents of anti-LGBTQ-motivated hate crimes including assaults ("gay bashing") and murder, which are all relatively accepted (West and Cowell 2015; Equality and Justice Alliance [EJA] 2020). Between 2011 and 2019, LGBT hate crimes in the United Kingdom increased 300 percent. Disaggregating the data reveals an even more alarming trend—transgender violence in the United Kingdom increased 700 percent during the same time period (EJA 2020).

Valencia, Williams, and Pettis (2019) investigated the relationship between the legalization of same-sex marriages and LGBT hate-motivated crimes in the United States. The **legitimacy hypothesis** predicts that the legalization of same-sex marriages signifies approval of LGBT behaviors and therefore should reduce LGBT hate-motivated crimes. Alternatively, the **polarization hypothesis** argues that the legalization of same-sex marriages solidifies respective group members' beliefs—i.e., those who approve of same-sex marriage and those who disapprove of same-sex marriage—and creates a greater gulf between the two. Analyzing county-level state data, the researchers conclude that the pro-LGBT message conveyed by the announcement of the legalization of same-sex marriage is associated with a decrease in violent LGBT hate crimes and, to a lesser extent, property crimes.

legitimacy hypothesis A model that predicts that the legalization of same-sex marriages signifies approval of LGBT behaviors and therefore should reduce LGBT hate-motivated crimes.

polarization hypothesis A model that argues that the legalization of same-sex marriages solidifies respective group member's beliefs—i.e., those who approve of same-sex marriage and those who disapprove of same-sex marriage—and creates a greater gulf between the two.

Anti-LGBQ Hate and Harassment in Schools and on Campuses. Steven Caruso, a gay 16-year-old high school student, left a suicide note. The note read, in part (Caruso n.d.):

> I am sorry to the people that I love but I can't ... take it anymore.
>
> So I am gay. Why does everyone hate me because of that. ... I have been punched and spit on and called faggot, queer, loser, pussy, fag boy. Some asshole painted faggot on my locker. Some people do not talk to me. ... I am so ... tired of the shit. I have received hate letters telling me to leave school ... that faggots aren't welcome. (p. 1)

Steven survived his suicide attempt, but many other LGBTQ youth and young adults do not. In fact, LGBTQ youth have higher rates of suicide attempts resulting in the need for medical care than their heterosexual and gender-conforming peers (Center for Disease Control and Prevention [CDC] 2018).

Hostile school environments, characterized by anti-LGBTQ attitudes, remarks, and actions by students and teachers, have been documented to exist as early as elementary school. A survey was conducted of more than 23,001 students between the ages of 13 and 21, in grades 6 through 12, from all 50 states, the District of Columbia, and the five U.S. territories. Results from the *National School Climate Survey* (Kosciw et al. 2018) indicate that in the 2016–2017 school year,

- 70.1 percent of LGBTQ students reported being verbally harassed at school because of their sexual orientation, 59.1 percent based on gender expression, and 53.2 percent based on gender identity.
- 95.3 percent of LGBTQ students heard homophobic remarks, such as "faggot" or "dyke," frequently or often at school; 94.0 percent heard negative comments about gender expression (e.g., not acting masculine enough or feminine enough) frequently or often.
- Over half of LGBTQ students, 59.5 percent, reported that they felt unsafe in school because of their sexual orientation; 44.6 percent because of gender expression.
- 34.8 percent of LGBTQ students missed at least one day of school in the previous month because of safety concerns or feeling uncomfortable; 10.5 percent missed four or more days in the past month.

LGBTQ students are also victimized on college campuses. A survey by the Association of American Universities (AAU) on sexual assault and misconduct on college campuses, the largest ever conducted, found that 23.1 percent of transgender and gender-nonconforming undergraduates were the victims of unwanted sexual contact at some point since enrolling in school (AAU 2019). Further, since enrolling in school, 65.1 percent of undergraduate transgender and gender-nonconforming students reported experiencing sexual harassment.

Police Mistreatment. There is "a significant history of mistreatment of LGBT people by law enforcement in the United States, which included profiling, entrapment, discrimination and harassment" (Mallory et al. 2015, p. 1). Consequently, due to fear of further victimization, many cases of LGBT violence are not reported to the police (Ciarlante and Fountain 2010; NCAVP 2018). In fact, in a 2018 report by the NCAVP, only 43 percent of surviving victims of LGBTQ violent hate crimes report interacting with the police as a result of their victimization.

As with heterosexuals, intimate partner violence and sexual assault occur in the lives of LGBTQ individuals. Many cases of LGBTQ violence are not reported to the police out of fear of further victimization and trepidation in formally acknowledging the nature of the relationship (Ciarlante and Fountain 2010; NCAVP 2018). In 2017, over 2,000 cases of intimate partner violence were reported to the NCAVP, but, of that number, only 60 percent were reported to law enforcement. Of those cases, a significant number of surviving victims reported police misconduct including excessive force and unjustifiable arrest.

The Consequences of Anti-LGBTQ Bias

Sexual and gender minorities, as many other minorities, suffer the negative consequences of harassment, prejudice and discrimination, violence, and isolation. However, unlike, for example, racial and ethnic minorities, gay, bisexual, and transgender men and women are often rejected by their schools and religious institutions, communities, families, and friends. In a qualitative study of LGBTQ youth, when asked about negative factors in their lives, they were most often related to "families, schools, religious institutions, and community and neighborhood" (Higa et al. 2014, p. 663).

Physical and Mental Health

Lesbian, gay, and bisexual Americans have higher rates of depression, particularly young adults between the ages of 18 and 25, compared to the general population, and the disparities are getting greater over time (Substance Abuse and Mental Health Services Administration [SAMHSA] 2020). Thus, it is not terribly surprising that LGB women and men have higher rates of suicidal thoughts, attempted suicide, and completed suicides.

Transgender women and men also have higher rates of suicide, in part as a result of prejudice and discrimination that create stressful life events. For example, trans men and women are more likely to be the victims of violent crime, and being the victim of a violent crime increases the likelihood of suicidal thoughts and attempts (Herman, Brown, and Haas 2019). Transgender men and women who experience family rejection, as is more often the case than with cisgender men and women, have higher rates of reported suicide attempts than their gender-conforming counterparts. Lastly, transgender respondents who live in states where gender identity is protected by law and/or who receive hormone therapy and/or surgical medical care have lower self-reported rates of previous-year suicidal thoughts and attempts.

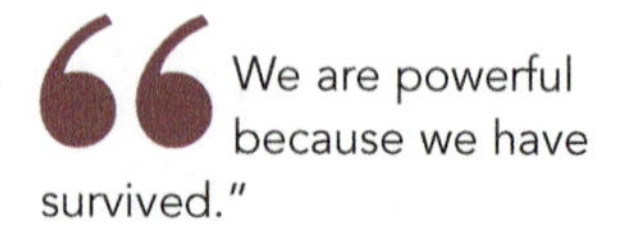

"We are powerful because we have survived."

–AUDRE LORDE, FEMINIST POET

Results of a 2019 national survey of high school students indicate that gay and bisexual students are more likely to report feeling sad or hopeless compared to their straight classmates (SAMHSA 2020). Nearly half of gay and bisexual students reported seriously thinking about committing suicide compared to 13 percent of non-LGB students. LGB students were also more likely to make a plan to commit suicide and to attempt suicide (SAMHSA 2020). In all three scenarios, seriously thinking about suicide, making a plan to commit suicide, and attempting suicide, gay and bisexual females had higher rates than gay and bisexual males, perhaps because of their double minority status. Moreover, gay adolescents whose parents tried to change their sexual orientation by forcing them to undergo conversion therapy were twice as likely to attempt suicide as teenagers not forced to go conversion therapy (Fadel 2019).

LGB high school students are also less likely to engage in healthy behaviors such as regular exercise and getting eight hours of sleep (CDC 2018). Gay, bisexual, and transgender men and women are more likely to rate their health as fair or poor compared to straight or cisgender men and women (Gates 2017). In a review of the literature, Hafeez et al. (2017) conclude that gay, bisexual, and transgender men and women are at a higher risk of a variety of physical illnesses. For example, gay men are more likely to be diagnosed with various types of cancers including prostate and testicular cancer and are at a higher risk of sexually transmitted diseases and hepatitis. Alternatively, gay and bisexual women have higher than average risks of breast, ovarian, and endometrial cancer, in part as a result of fewer pregnancies, higher rates of obesity, and fewer mammograms.

Although difficult to discern how much it contributes to these increased risks, research documents that gay, bisexual, and transgender patients are hesitant to reveal their sexual orientation and gender identity to their health care providers and are more likely to report having negative health care experiences (Elliot et al. 2015; Hafeez et al. 2017; Goldsen 2018; Burton et al. 2019). It is also true that LGBT men and women are less likely to have health insurance as a result of higher unemployment and lower incomes (see this chapter's section "Economic Inequality, Poverty, and Homelessness"). Twenty-two percent of self-identified LGBT adults compared to 15 percent of non-LGBT adults report that they did not have enough money for health care in the previous year (Gates 2017).

Finally, based on a national survey of LGBT youth between the ages of 13 and 24 in the United States, the greatest barrier to receiving mental health care was the inability to pay for it (Trevor Project 2020).

Substance Abuse

The relationship between LGBT status and increased risk of substance abuse is rooted in sociological variables. Heck et al. (2014) examined the association between school environment and LGBT high school students' use of illicit drugs and misuse of prescription drugs. An online survey was completed by 475 LGBT high school students. School environment was measured by the absence or presence of a high school GSA, called Gay-Straight Alliance at the time, a school-based club for all students, and an indicator of how supportive a school is of LGBT students.

The results support previous studies that indicate that the absence of a school GSA is associated with increased risk for drug use. Controlling for such demographic variables as ethnicity, gender, and school type (private versus public), students in schools without a GSA reported higher lifetime use of illegal drugs in general and had higher likelihoods of using cocaine, hallucinogens, and marijuana, and misusing prescription pain killers and ADHD medication. The authors conclude that one possible reason for the relationship between the presence of a school GSA and LGBT students' reduced use of drugs may be the moderating variable victimization. LGBT students in schools without GSAs report higher rates of school-based victimization, and higher rates of victimization are associated with increased drug use. In addition to victimization, a lack of social support, stress, bad coming-out experiences, and a negative housing status have been found to be predictive of LGBT substance abuse (Goldbach et al. 2014).

Similarly, a study by Canadian researchers indicates that community-level support, including organizations (e.g., Gender and Sexualities Alliance), events (e.g., Pride celebrations), resources (e.g., health clinics), and programs (e.g., anti-bullying) for LGBTQ youths, is associated with lower lifetime odds of illegal drug use by both gay and bisexual boys and girls (Watson at al. 2020). These results notwithstanding, in the Netherlands, a country with a documented LGB-friendly environment, Dutch LGB youth had higher rates of tobacco and marijuana consumption than non-LGB youth (Kiekens et al. 2020).

Using survey data, Jabson and colleagues (2014) examined substance abuse in a national sample of LGB and heterosexual respondents. The researchers were interested in comparing not only the prevalence of substance abuse between non-heterosexual and heterosexual subsamples but also the role of stress as a possible mediating variable. The results indicate that gay men and women, particularly bisexuals, have higher rates of current cigarette smoking, lifetime use of marijuana, and lifetime use of substances other than marijuana (including cocaine, crack cocaine, heroin, or methamphetamine) when compared to heterosexuals. Interestingly, stress did contribute to the relationship between sexual orientation and substance abuse, but only for bisexuals.

Finally, two national research projects, one surveying students in grades 9 through 12 (CDC 2018) and the other respondents 12 years of age or older (SAMSHA 20) provide summary data about LGB populations in the United States. The first, the *Youth Behavioral Risk Surveillance* survey, indicates that LGB youth are more likely to use both legal and illegal substances compared to their non-LGB classmates. LGB youth were not only more likely to smoke cigarettes and marijuana and to consume alcohol but to have begun doing so at an earlier age and to do so more frequently. They also have higher rates of ever using cocaine, heroin, or methamphetamine than nonsexual minorities.

The second report indicates that self-reported illicit drug use among the adult U.S. LGB population is highest for marijuana, followed by psychotherapeutic drugs, hallucinogens, and cocaine (see Chapter 3). The rate of use by the LGB population is more than twice that of the general population, 37.6 percent and 16.2 percent, respectively. Although opiate misuse among LGB 18- to 25-year-olds has significantly decreased between 2015 and 2018, self-reported misuse among respondents 26 years old and older has increased. Finally, substance use is more frequent among LGB adults who have been diagnosed with mental health issues.

Economic Inequality, Poverty, and Homelessness

There are many stereotypes about sexual minorities, most of them negative. However, there is at least one stereotype that is positive, but untrue—that gay men and women are disproportionately wealthy (McDermott 2014). In fact, as noted earlier, when comparing LGBT and non-LGBT men and women, LGBT men and women have lower standards of living, are less likely to be able to afford basic necessities, have higher unemployment rates, and are more likely to worry about finances (Gates 2014; Williams Institute 2016; Yochim 2020).

Because of prejudice and discrimination, LGBT men and women are more likely to be living in poverty. Despite the fact that in 2020, the Supreme Court held that workplace discrimination on the basis of LGBT status is a violation of Title VII of the 1964 *Civil Rights Act*, generations of LGBT employees have been subject to employment discrimination. This is particularly true for gay and bisexual women who bear the burden of two minority group memberships. In fact, gay women, on the average, earn 12 percent less than their heterosexual counterparts. Similarly, regardless of race or ethnicity, male same-sex couples have higher annual incomes than female same-sex couples (Yochim 2020).

A report by the Williams Institute at UCLA's School of Law summarizes the impact of sexual orientation and gender identity on economic well-being (Badgett, Choi, and Wilson 2019). The survey data from 35 states indicates that LGBT respondents have a higher poverty rate than cisgender straight respondents, 21 percent and 15.7 percent, respectively. However, transgender respondents had the highest rates of poverty—nearly 30 percent (see Figure 11.7). Interestingly, as Figure 11.7 indicates, gay and straight men have lower poverty rates than gay and straight women, and gay and straight women have lower poverty rates than bisexual men and women.

LGBT poverty rates also vary by race and ethnicity and by geographical location. As with the straight population, White LGBT individuals have lower poverty rates than racial or ethnic LGBT women and men. LGBT people in rural areas have higher poverty than LGBT individuals in urban areas, but both have higher rates of poverty than cisgender straight people in rural or urban areas. Like gay women, Black, Asian, or

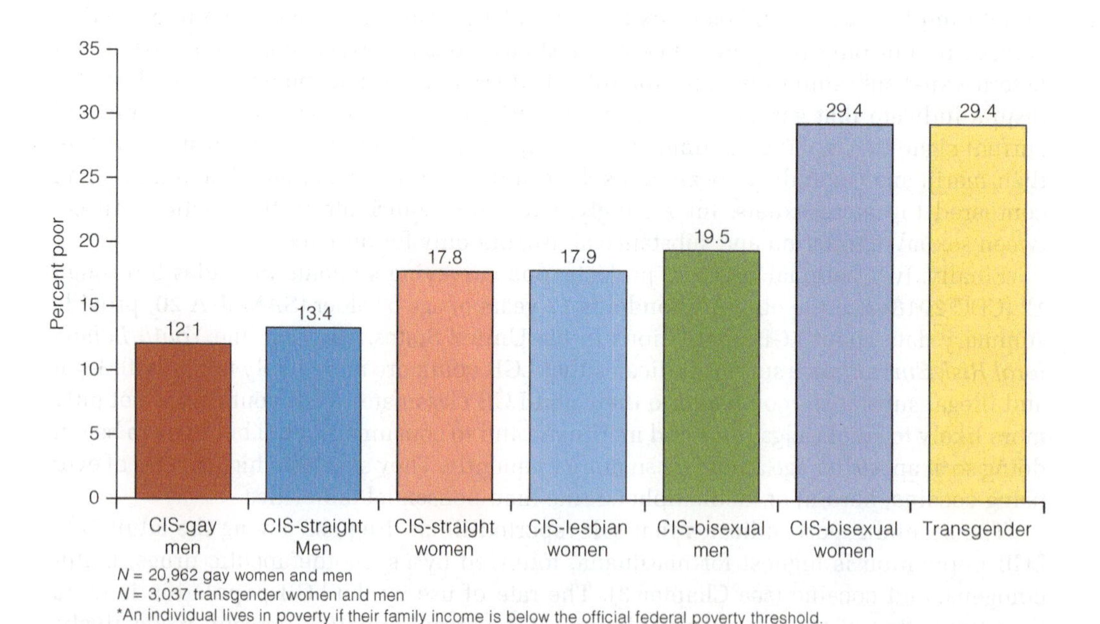

Figure 11.7 Percent of U.S. Adults Living in Poverty by Birth Sex, Sexual Orientation, and Gender Identity

SOURCE: Badgett, Choi, and Wilson 2019.

multiple-race LGBT respondents may be victims of double discrimination. Additionally, as Badgett, Choi, and Williams (2019) explain, LGBT respondents were also more likely to have characteristics related to poverty, i.e., being younger and having a disability. Even after controlling for variables that might explain the relationship between LGBT status and poverty, such as employment and educational differences, lesbian, gay, bisexual, and transgender men and women were still more likely to live in poverty than their cisgender heterosexual counterparts.

GLADD is an advocacy organization that examines the representation of LGBT characters in media and culture. As part of the group's charge, in 2019 they conducted a national survey of nearly 1,800 adults asking how comfortable they would be in a variety of scenarios such as learning that your child's teacher, a family member, or your doctor is gay, bisexual, or transgender (GLAAD Media Institute 2020a). Although the results have remained fairly stable over time, the 2019 results indicate a general decline in the comfort level and acceptance of LGBT people by respondents between the ages of 18 and 34. Why do you think that's true?

What *do you* think?

Lastly, LGBT youth are disproportionally homeless when compared to their non-LGBT counterparts (Trevor Project 2020). A survey of homeless youth providers indicates that 20 percent of youth seeking services are gay and that 3 percent are transgender (Choi et al. 2015). LGBT adults are also disproportionately homeless. Using three nationally representative surveys of U.S. adults collected between 2016 and 2019, Wilson et al. (2020) examined rates of homelessness of sexual orientation and gender identity minorities and the general population. The percentage of LGBT respondents who had experienced homelessness at least once in their lifetime was 17 percent compared to 6.2 percent of the general population (Wilson et al. 2020). Given the relationship between poverty and homelessness, these results, although perhaps more dramatic, are not unexpected.

Aging and Retirement

Despite the fact that by 2060 the number of older LGBTQ adults will exceed 5 million, an estimated 20 million if those who do not publicly identify themselves as LGBTQ are included, LGBTQ seniors are an underserved population (Goldsen 2018). Aging LGBT individuals are at a higher risk for a range of negative life outcomes. First, LGBT elders are more likely to suffer from a variety of health concerns when compared to non-LGBT older adults (Frediksen-Goldsen and Espinoza 2014). The stress associated with being LGBT, often the result of years of discrimination and victimization, increases the risk of cardiovascular disease and shortens life expectancy. LGBT elders are also more likely to suffer from anxiety, diabetes, obesity, asthma, high blood pressure, and cognitive impairment when compared to their non-LGBT counterparts (Services and Advocacy for GLBT Elders [SAGE] 2015; Choi and Meyer 2016; Goldsen 2018).

Second, because gay, bisexual, and transgender men and women have higher rates of poverty than their heterosexual and gender-conforming peers, over the course of their working history, they have less disposable income to save for the future (UBS 2018). Compounding the lack of retirement income, studies indicate that

> LGBT older people are twice as likely as their heterosexual peers to live alone. They are half as likely to have life partners or significant others; half as likely to have close relatives to call for help; and four times less likely to have children to provide care. Heartbreakingly, almost a quarter of LGBT older people have no one to call in case of an emergency. (SAGE 2019, p. 1)

Eighty percent of long-term health care in the United States is provided by family members without which access to long-term care facilities becomes increasingly important and yet problematic. Due to higher rates of poverty, LGBT seniors may not be able to afford the skyrocketing costs of retirement communities and nursing homes. Additionally, as Michael Adams CEO of SAGE observes, 85 percent of retirement facilities are faith based, leading to fears that religious freedom laws will legally permit long-term care facilities to discriminate on the basis of sexual orientation and gender identity (Brady 2020).

Lastly, prior to the *Obergefell* decision, unless married *and* living in a state that recognizes the marriage, the surviving LGBT partner was ineligible for Social Security survivor benefits or for their partners' retirement fund or pension plan (Frediksen-Goldsen and Espinoza 2015; SAGE 2015). However, following the *Windsor* decision in 2013 and *Obergefell* in 2015, gay married couples now have the same benefits as straight married couples in terms of Social Security survivor benefits, the right to a partner's retirement fund or pension plan, and eligibility for Medicare or Medicaid. However, in 2017 the Trump administration withdrew a proposal that would have required long-term care facilities receiving federal aid to treat same-sex spouses the same as different-sex spouses (Kates et al. 2018).

Strategies for Action: Toward Equality for All

@closeted_thots

I came out to my parents today. It didn't go perfectly, and my mom probably won't talk to me for a while, but I did It.

I have a right to live my authentic life. And I don't have to feel bad if someone doesn't understand.

I'm feeling ok, but any love would be appreciated.

-More LDS Gay Thoughts

As discussed in this chapter, attitudes toward gay, bisexual, and transgender men and women in the United States have become more accepting over the years, and support for protecting civil rights of gay men and lesbians is increasing. Many of the efforts to change policies and attitudes regarding non-heterosexuals and gender-nonconforming individuals have been spearheaded by organizations that specifically advocate for LGBT rights including the Human Rights Campaign (HRC), the (NGLTF), the Gay and Lesbian Alliance Against Defamation (GLAAD), the Gay, Lesbian and Straight Education Network (GLSEN), and Amnesty International. In addition to organizational efforts, the media, local, state, and federal governments, and educational policies offer "strategies for action."

LGBTQ Status and the Media

For the third year in a row, the representation of gay and bisexual male and female characters in major motion pictures increased between 2018 and 2019, although screen time was limited, there was an absence of transgender characters, and racial diversity of LGBTQ characters dropped dramatically. Gay men were nearly twice as likely to appear in films as gay women (GLAAD Media Institute 2020b).

The media has been instrumental in the lives of LGBTQ individuals for several reasons. First, it has provided LGBTQ individuals with role models. Gomillion and Giuliano (2011) report that media role models are not only a source of pride and comfort but have influenced respondents' self-realization of a gay identity and the coming-out process. In an episode of the Emmy-winning television series *Schitt's Creek*, one of the characters comes out to his parents in such a touching and memorable way that thousands of fans sent tweets, posted on Facebook, and sent letters to the show's actors and writers telling them that the episode gave them the strength to come out to their parents. **Coming out** is a term used to describe an ongoing process whereby an LGBTQ individual becomes aware of his or her sexual and/or nonconforming gender identity, accepts and incorporates it into his or her overall sense of self, and shares that information with others such as family, friends, and coworkers. The coming-out process only exists because heterosexuality and gender conformity are considered normative in society.

coming out The ongoing process whereby a lesbian, gay, or bisexual individual becomes aware of his or her sexuality, accepts and incorporates it into his or her overall sense of self, and shares that information with others such as family, friends, and coworkers.

the HUMAN side

Billy Porter—Better than Ever!

© Guardian News & Media Limited

In addition to being a Broadway and television star and recording artist, he and his husband are LGBTQ rights activists. Said Mr. Porter, "As a Black gay man [in the 80s], there were no examples to validate my existence. So, I vowed at that age to try to be that for the generations behind me if I had a chance." He recently starred in Netflix's miniseries *Pose.*

Billy Porter won a Tony Award for Best Actor in a Musical in 2013 and a 2014 Grammy Award for Best Musical album for the Broadway musical Kinky Boots. *He has also appeared in* Grease, Jesus Christ Superstar, *and* Dreamgirls; *made a cameo appearance in Taylor Swift's video "You Need to Calm Down"; and was named by* Time Magazine *as one of the 100 Most Influential People of 2020. In the following interview, Mr. Porter discusses being a Black gay man in America.*

I became a household name around the world because I wore a dress. I refer to it as BOAO: Before Oscar, After Oscar. My life before the Oscars and my life now are completely different. The last couple of years felt like a rocket. It's like that image of Indiana Jones, running from the boulder behind him. My life has been like that, just trying to make sure the boulder doesn't run me over.

I have survivor's guilt, but not in a debilitating way. I've realized what it is. I can recognize it. I can allow it and I can move through it. I'm a little black queen who grew up in the ghettos of Pittsburgh. Why did I get out? That's why the kind of work that I do and the kind of imagery I put out in the world is so important to me, because I'm here for a reason.

When I started out in this business, I came to New York City and played my trump card, which was my singing voice. I call it extreme singing. This was the late 1980s, early 1990s, and nobody knew where to put that kind of energy in a mainstream way. So I became a clown; I became the flamboyant queen.

In the late 90s, I made the decision that that was no longer acceptable for me, and the work dried up. When I asked for what I wanted, I was dismissed. Nobody wanted to see me do anything but be the fairy clown. That was what motivated me to do things differently. What you're seeing in the past two years is the result of me making that choice, all those years ago.

I don't care what you think about me because I'm wearing a dress to the Oscars. I give zero fucks about what anybody thinks that I'm doing. That doesn't happen when you're 20 – I had to live long enough. You can say whatever you want, you can tweet whatever you want, you can write in my comments. All that stuff is of no consequence to me, because I'm going to continue to do me.

Sesame Street asked me to come on in the Christian Siriano dress, and they wrote a special song about friendship with me and a penguin. When they put up a picture of me standing on the steps, there was a backlash from the south. The governor of Arkansas threatened to de-fund PBS in Arkansas if they ran the episode, because it's perverted, it's the gay agenda, and I'm gonna come into their home and molest their children. This is the fearmongering that still exists.

I have lived as a black gay man for 50 years in America. Nothing shocks me. I'm not surprised by anything that's going on right now. I wish I was surprised, but I'm not. "Eternal vigilance is the price of liberty." Justice is 200-plus years of work, pissed away in three-and-a-half years by these fucking assholes.

I grew up in the pentecostal church and that wasn't very supportive for me. So my chosen family became my support system. Without them I would not be here. I am spiritual, but religion is man-made, and I'm not having that any more.

I never thought that marriage would be possible. The day it became legal in the U.S., I wept for an hour. I didn't know that lawful validation of my love was something I wanted or that I actually needed. Then I got married, stood in front of my friends and my witnesses, and said my vows to my husband, and I was like, oh, right, *this* is what we didn't really get. It's about community. That's what was taken away from us.

SOURCE: Copyright Guardian News & Media Ltd 2020.

Second, in addition to providing role models, the media increases LGBTQ visibility and counteracts stereotypes (GLAAD 2015; Wilcox and Wolpert 2000). Since Ellen DeGeneres's 1998 coming-out episode in her sitcom *Ellen*, many television viewers have watched shows depicting LGBTQ characters in more realistic ways and that take a supportive stance on LGBTQ rights issues. Anti-LGBTQ-bias shows include *The Fosters, Buffy the Vampire Slayer, Glee, The Big Gay Sketch Show, The Handmaid's Tale, True*

Blood, Will & Grace, Unbreakable Kimmy Schmidt, Modern Family, Orange Is the New Black, Brooklyn Nine-Nine, Please Like Me, How to Get Away with Murder, Schitt's Creek, Atypical, and *Billions* (Ahlgrim 2020). Importantly, the results of several studies indicate college students reported lower levels of anti-gay prejudice after watching television or movies showing gay characters in a positive light (e.g., Schiappa et al. 2005; Madžarević and Soto-Sanfiel 2018). Ironically, however, some research suggests that, although exposure to gay television characters reduces negative attitudes, there is also evidence that it crystallizes stereotypes (McLaughlin and Rodriguez 2017).

The visibility of famous gay, lesbian, and bisexual individuals also has a societal impact. Many television and movie celebrities (e.g., Don Lemon, Anderson Cooper, Wanda Sykes, Neil Patrick Harris, Dan Levy, Billy Porter), singers (e.g., Elton John, Lil Nas X, Lady Gaga, Melissa Etheridge), political leaders (e.g., Tammy Baldwin, Pete Buttigieg), and sports figures (e.g., Jason Collins, Megan Rapinoe, Michael Sam) have come out over the years. Similarly, many public figures have actively supported gay rights. For example, former New York Giants defensive end and current Fox NFL analyst Michael Strahan filmed a public service announcement (PSA) supporting the legalization of same-sex marriage.

Finally, social media have been important in addressing the needs of LGBTQ youth. Following several gay teenage suicides that appeared in the news during September 2010, Dan Savage and his partner created a project called It Gets Better (IGB). The IGB project turned into a global viral cyber-movement, inspiring thousands of user-created videos that instill messages of hope and support for LGBTQ youth who have been bullied, feel that they must hide in shame, or who experience rejection from family members. To date, hundreds of thousands of people have taken the pledge:

> I believe in a world where hope outshines fear. I commit to stand up and speak out against hate and intolerance. My support for LGBTQ youth will be steadfast. I am part of a global community that is proud and resolute in its efforts to create a brighter, more inclusive world for all people. I know it will get better. (IGB 2020)

Law and Public Policy

As is often the case with other social problems, federal, state, and local governmental agencies have tackled the issues surrounding LGBTQ inequality. Various policies and laws have been passed or proposed that address LGBTQ discrimination in the workplace, marriage inequality, restrictive adoption laws, and hate crimes.

Ending Workplace Discrimination. Although both conservatives (Amy Coney Barrett was not yet on the Court), Trump's appointments to the U.S. Supreme Court, Neal Gorsuch in 2017 and Brett Kavanaugh in 2018 split on *Bostock v. Clayton County, Georgia* (2020), which held that workplace discrimination on the basis of LGBT status is illegal. The decision, as previously noted, applies to both public and private employers, nationally, all federal agencies, and labor unions.

Despite being a landmark decision and a win for the LGBT community, the decision does have limitations. First, it only applies to workplaces that employ 15 people or more. It does not apply to employers of small businesses, i.e., those with fewer than 15 employees, who would remain free to discriminate based on sexual orientation and gender identity.

Second, the *Bostock* decision applies only to employers who are subject to Title VII of the *Civil Rights Act*, which mandates that an employer may not discriminate in the "term, condition, or privilege of employment." This includes such things as recruiting, hiring, firing, assigning and evaluating work, and so forth (Justia 2020). It does not include, for example, who can or cannot use a particular bathroom, an important and contentious issue for transgender men and women. Given the limitations of *Bostock*, an employer could insist that a transgender employee use the bathroom of their birth sex. That being the case, transgender employees may decide that they are unable to remain working in that kind of environment and may decide to quit. Although the employer did not fire the transgender employee, the result is the same.

***Religious Freedom Restoration Act of 1993* (RFRA)** An act stating that the "government shall not substantially burden a person's exercise of religion" (RFRA 1993).

Do No Harm Act An amendment to the RFRA that would ensure that it was not used to sidestep federal nondiscrimination laws and that, if so, such behaviors would be unlawful.

Lastly and memorialized in the decision, Justice Gorsuch noted that the ***Religious Freedom Restoration Act of 1993*** **(RFRA)** could limit the *Bostock* decision. The RFRA states that the "government shall not substantially burden a person's exercise of religion" (RFRA 1993). The question remains then, can an employer, based on religious beliefs, not hire someone who is gay, bisexual, or transgender? As one legal expert said, "[T]here is a possibility that while the court with one hand extends statutory protections to LGBT people, it might with the other hand gut those same protections by expanding religious freedom defenses" (Moreau 2020).

@Pharrell

Equality = Freedom.
Freedom = Love
#LoveWins

-Pharrell Williams

Marriage Equality. Gay marriages are legal in 29 countries, the most recent being Taiwan in 2019, the first country in Asia to legislate marriage equality, and Costa Rica in 2020, the first Latin American nation to sanction same-sex marriages (Perper 2020).

What do you think?

Catering for the rehearsal dinner, a wedding cake and flowers for the reception, and photos of the couples' arguably happiest life event—getting married. Each of these things, in different states and by different businesses, has been denied same-sex couples based on religious beliefs. Further, each of the states where the couples were denied a product or service had nondiscrimination laws. Do you think business owners should be able to violate a state's nondiscrimination clause based on a religious exemption? What if the refusal was based on race or nationality rather than sexual orientation?

The U.S. Supreme Court decision in *Obergefell v. Hodges* (2015) was announced on June 26, 2015. James Obergefell and his partner John Arthur had been in a committed relationship for over 20 years when they received the devastating news that John had ALS, a fatal nerve disorder. With time running out, the couple married on an airport tarmac in Maryland, John too ill to be moved. When they learned that their home state of Ohio would not recognize their marriage and therefore would not list James as spouse on John's death certificate, the couple decided to file a federal lawsuit (Rosenwald 2015). After his partner's death, Obergefell continued to fight for equal recognition of his marriage, resulting in the 5–4 U.S. Supreme court decision legalizing same-sex marriage throughout the United States (*Obergefell v. Hodges* 2015).

Although the Supreme Court is considered the "court of last resort," the decision has fueled a backlash from conservatives and faith-based groups. Those who oppose the decision include conservative religious leaders as well as many members of the former administration. Some have argued that the *Religious Freedom Restoration Act of 1993 (RFRA)*, like similar legislation in 21 states, is being used by individuals and organizations (see Chapter 5) to sidestep federal nondiscrimination laws that protect classes of individuals on the basis of, for example, race, sex, age, and, of particular interest here, sexual orientation and gender identity.

In response to such accusations, the ***Do No Harm Act***, which would amend the RFRA, was introduced into Congress in 2019. Supporters argue that the new legislation is needed to "protect civil rights and otherwise prevent meaningful harm to third parties" (H.R. 1450)

Alex Wong/Getty Images News/Getty Images

Jim Obergefell speaks to the media after the U.S. Supreme Court decision in *Obergefell v. Hodges*. Toobin (2015) notes that the two great causes of the gay rights movement have been the right to serve in the military and marriage equality. Furthermore, he observed that when the *Obergefell* decision was announced outside the Supreme Court, supporters joined hands and sang "The Star Spangled Banner"—a song "for people who believe in the United States, who want to participate in American society more than they want to transform it" (p. 1).

as a result of the encroachment of the RFRA and the Trump administration's broad interpretation of religious freedom (London and Siddiqi 2019). Critics of the proposed law argue that it is unnecessary because the RFRA provides a neutral three-part test as to whether an individual or organization's claim of immunity from federal nondiscrimination laws is valid (Sharp 2019).

Religious exemption laws are not the only legal quagmire for same-sex couples. There are questions concerning the implications of the *Obergefell* decision. In 2017, the Texas Supreme Court *dismissed* a lower court's ruling that spouses of public employees who are in same-sex marriages are entitled to the same marriage benefits a spouse in a different-sex marriage would be entitled to (Ura 2017). Anthony Gonzalez and Mark Johnson lived together as a committed couple for 15 years and married in 2013 once gay marriage was legal in New Mexico where they lived. Six months after they married, Johnson died of cancer. His widower was denied Social Security benefits because they had been married for less than nine months, a requirement of the *Social Security Act.* Those who agree with the decision argue that the regulation is being applied equally, i.e., surviving spouses in same-sex marriages or different-sex marriages cannot receive benefits unless they have been married nine months.

Critics of the decision argue that because Gonzales and Johnson were legally prohibited from getting married, the nine-month rule should not apply. Says Gonzales, "[W]e established a joint checking account, named each other as our beneficiaries, and cared for each other when sick—basically all the things that committed couples do. ... We got married as soon as humanly possible, but I'm still barred from receiving the same benefits as other widowers even though my husband worked hard and paid into the Social Security system with every paycheck" (Moreau 2018, p. 1). Lawsuits against the Social Security Administration have been filed on behalf of Gonzales and other similarly situated plaintiffs.

Religious leaders argue that a Court cannot alter "God's law" (i.e., marriage is between one man and one woman) and replace it with government-made law (i.e., the Fourteenth Amendment guarantees equal protection). Some political leaders, despite the separation of church and state, make the same argument but also question the right of the federal government to impose its will in the face of opposing state law. Finally, despite a long history of what some have called "judicial activism," one of the primary concerns of the dissenting judges in the marriage equality case was the majority justices' use of the Constitution to "invent a new right and impose that right on the rest of the country" (*Obergefell v. Hodges*, p. 102).

Parental Rights. The *Obergefell* decision was based on cases from different jurisdictions, each challenging the state-level prohibition against same-sex marriage. Two of the cases involved same-sex couples in which one of the partners "sought legal recognition as parent which was not directly addressed by the Court" (Knauer 2019, p. 8). However, in 2020, the Trump administration's Justice Department, although not a party to the action, directly spoke to the issue of child placement with same-sex couples. The Department of Justice voluntarily submitted a brief to the U.S. Supreme Court in support of Catholic Social Services (CSS). CSS appealed the City of Philadelphia's termination of their contract after they violated the nondiscrimination clause by refusing to consider same-sex couples as foster parents (Child Welfare League of America 2020). The U.S. Supreme Court will not hear this case until a later term.

***Every Child Deserves a Family Act* (ECDFA)**
Federal legislation that, if passed, would remove obstacles to non-heterosexual individuals or couples providing homes for adoption or foster care by prohibiting public child welfare agencies from discriminating on the basis of sexual orientation, gender identity, or marital status.

In the United States, ***Every Child Deserves a Family Act* (ECDFA)** has been introduced into Congress for several years, most recently in 2019. But, to date, ECDFA has yet to receive legislative approval. If passed, the act would remove obstacles to non-heterosexual individuals or couples providing homes for adoption or foster care by prohibiting public child welfare agencies from discriminating on the basis of sexual orientation, gender identity, or marital status. Agencies that do discriminate would be in danger of losing federal financial assistance (S. 1791). Presently 11 states permit licensed child welfare agencies to refuse to place children or provide services to children and families on the basis of sexual orientation or gender identity if it conflicts with their religious beliefs (MAP 2020b).

Laws regulating adoption by LGB couples vary not only by state but by nation. Twenty-six countries have provisions that allow joint adoption by same-sex couples, and 23 countries allow for second parent adoption, i.e., if a gay man or woman has a child, the partner may adopt the child as a co-parent (ILGA 2016). In countries where second parent adoption is illegal, child care can be difficult. As one man from Cuba states, "I've held him in my arms since birth ... [but] if the baby has to go to the hospital ... I have no legal authority to decide anything about his illness" (Gonzalez 2013, p. 1).

Hate Crimes Legislation. In 2009, the ***Matthew Shepard and James Byrd, Jr. Hate Crimes Prevention Act* (HCPA)** was signed into law. This law expands the 1969 federal hate crimes law to include hate crimes based on actual or perceived sexual orientation, gender, gender identity, and disability. The law was named after Matthew Shepard who was brutally murdered and James Byrd Jr., a Black man who was attacked, chained to a vehicle, and dragged to his death in Texas. In 2020, the *Justice for Victims of Hate Crimes*

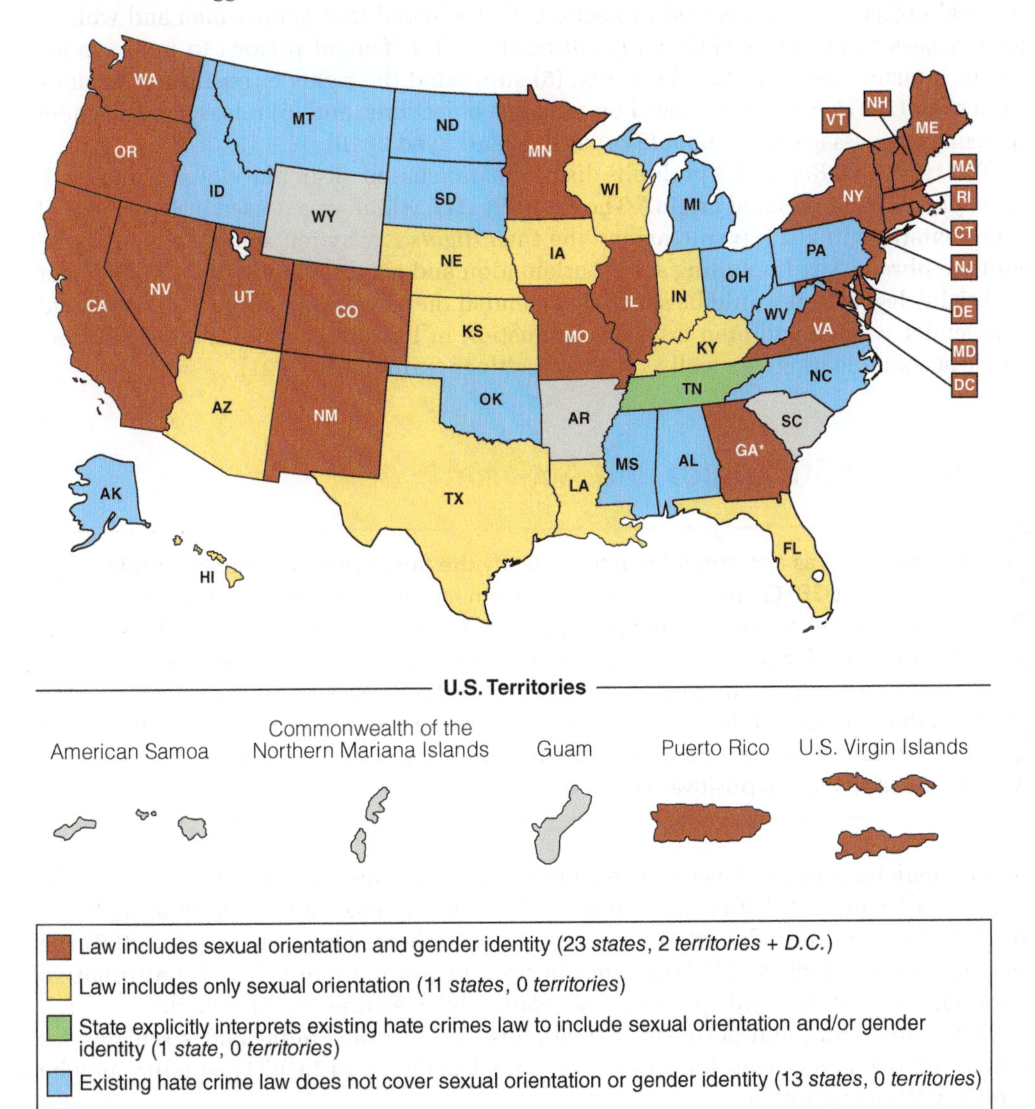

*Georgia's state law explicitly enumerates sexual orientation, sex, and gender. "Sex" was added to the bill after the June 2020 U.S. Supreme Court decision in Bostock, which affirmed that sex applies to both sexual orientation and gender identity.
Note: Tennessee state law explicitly enumerates sexual orientation, but not gender identity. However, the law does enumerate "gender," and the state attorney general affirms that this means transgender people are also protected.

Figure 11.8 Hate Crime Laws with and without Sexual Orientation and Gender Identity Protections by State, 2020
SOURCE: MAP 2020c.

***Matthew Shepard and James Byrd, Jr. Hate Crimes Prevention Act* (HCPA)** This law expands the original 1969 federal hate crimes law to cover hate crimes based on actual or perceived sexual orientation, gender, gender identity, and disability.

Act, a bipartisan bill, was introduced into Congress. The bill, if passed, would strengthen the enforcement of the HCPA and bring greater uniformity to state laws (Gupta 2020). Further, President Biden supports passage of the *NO HATE Act*, a bill intended to promote more reliable collection of hate crime data (Biden 2020).

The Equality Act. Despite courting the LGBTQ vote during his campaign and releasing a statement in the first month of his presidency that he was "determined to protect the rights of all Americans, including the LGBTQ community" (quoted in Berg and Syed 2019), President Trump and his former administration consistently took the position that regulations that prohibit discrimination on the basis of sex do not protect gay, bisexual, and transgender Americans.

Between 2017 and 2020, the Trump administration, among numerous other actions, (1) revoked the Obama-era regulation that allowed transgender students to use the bathroom consistent with their gender identity, (2) expanded protections for federally funded medical personnel who refuse to provide health care to LGBT patients based on religious or moral objections, (3) repealed protections that allowed transgender men and women equal access to homeless shelters, (4) directed staff at federal prisons to house transgender inmates based on their birth sex, (5) supported the rights of businesses to deny services to LGBTQ customers based on religious objections, and (6) released a statement opposing the passage of the *Equality Act* (Berg and Syed 2019).

The 1964 *Civil Rights Act* prohibits discrimination based on an individual's "race, color, religion, sex, or national origin." The ***Equality Act***, which was passed by the House of Representatives in 2021, would amend the *Civil Rights Act* by replacing the word "sex" with the phrase "sex (including sexual orientation and gender identity)" (H.R. 5). Other civil rights legislation would be similarly amended thereby protecting gay, bisexual, and transgender women and men from discrimination in housing, employment, education, public accommodations, as well as in other settings (Kurtzleben 2021).

Educational Strategies and Activism

What do you think?

While not listed as a strategy for action, clearly the family plays an important role in addressing LGBTQ family members. Research has identified over 100 accepting and rejecting behaviors engaged in by parents, family members, and caretakers that impact LGBTQ youth outcomes. While rejecting behaviors increase the likelihood of negative outcomes (e.g., depression, substance use) accepting behaviors shield the LGBTQ child from these risks and promote well-being (Family Acceptance Project 2020). How do we change the behaviors of families in order to maximize the likelihood of positive outcomes?

Educational institutions bear the responsibility of promoting the health and well-being of all students. Thus, they must address the needs and promote acceptance of LGBTQ youth through certain policies and programs. The strategies for attaining these goals are to include LGBTQ issues in sex education, include LGBTQ-affirmative classroom curricula, and promote tolerance in learning environments through policies, education, and activism. Nonetheless, as in other important strategies for action, the perceived need for greater religious freedom and LGBTQ equality may be at odds with one another.

Sex Education and LGBTQ-Affirmative Classroom Curricula. The censorship of LGBTQ current issues, historical figures and events, and sexual health both in the classroom and in school libraries is a heated debate across the nation. Whether LGBTQ themes can be brought into public school classrooms varies considerably between and within states. In 2020, the Illinois legislature passed a bill that requires LGBTQ history be taught in public schools and that inclusive and nondiscriminatory textbooks be used in the classroom.

Equality Act An act that would amend the 1964 *Civil Rights Act* by replacing the word "sex" with the phrase "sex (including sexual orientation and gender identity)."

Illinois is the fifth state to pass such a law. Other states include California, Colorado, New Jersey, and Oregon (Leone 2019).

At the other end of the continuum, some states prohibit teachers from talking about same-sex sexual relations. For example, in South Carolina, it is against the law for "public school sex education classes to mention anything other than heterosexual relationships, unless the talks involve sexually transmitted diseases" (Associated Press 2020b, p. 1). The law also says that any teacher who allows "a discussion of alternative sexual lifestyles" can be fired (p. 1). Two national banks threatened to withhold financial support for Florida's school voucher program as a result of taxpayer dollars going to 156 private Christian schools with anti-gay policies (Strauss 2020). In Alabama, sex education teachers are required to state that being gay is "not a lifestyle acceptable to the general public and that homosexual conduct is a criminal offense under the laws of the state" (quoted in Dunigan 2019). In 2019, in Huntsville, Alabama, 15-year-old Nigel Shelby committed suicide after being bullied at school for being gay. After Nigel's death, a sheriff's deputy posted on Facebook, "Liberty Guns Bible Trump BBQ, That's my kind of LGBTQ" (Aviles and Griffith 2019).

School climate (e.g., supportive educators, GSAs, inclusive curricula), in general, has consistently been found to be related to the well-being of LGBTQ youth (Johns et al. 2019). For example, results of a national survey indicate that LGBTQ youth who reported having at least one in-person LGBTQ-affirming space had lower rates of attempted suicide in the past year when compared to those who did not (Trevor Project 2020). Inclusive curriculum is important as well. After interviewing 9th-grade students and observing ninth-grade sex education classes and interviewing teachers, Jarpe-Ratner (2020) concludes that students want a *more* inclusive LGBTQ+ curriculum—one that pays more attention to identity topics such as gender identity formation and development and an integrated discussion of sexuality.

Camika Shelby

Fifteen-year-old Nigel Shelby from Huntsville, Alabama, committed suicide after being bullied and harassed in his school. Suicides by LGBTQ youth are higher than for non-LGBTQ youth and even higher for young LGBTQ people of color. Wanting to go to school in a nonhostile environment is not part of some kind of "gay agenda."

One of the earliest and most successful examples of curriculum integration took place in 2012, when then California governor Jerry Brown signed legislation called the *Fair, Accurate, Inclusive, and Respectful Act*, known as the *FAIR Act*. Initially, the act called for a public school curriculum that included LGBTQ persons and people with disabilities in the state's civic and social science classes. Since that time, activists and allies have successfully advocated for the inclusion of gay, bisexual, and transgender women and men in the history and social science curricula. The report, *Making the Framework FAIR*, calls not only for "greater LGBT content in curriculum but a transformation in how history and social science are taught by using gender and sexuality as lenses for understanding the past and present" (Donahue 2019, p. 1). Most importantly, California straight *and* gay high school and middle school students who attended schools with an inclusive curriculum reported feeling safer and had lower rates of bullying than straight *and* gay students attending schools without an inclusive environment (Snapp et al. 2015).

Promotion of Tolerance in Learning Environments. Concerns over harassment and/or bullying of students have become the focus of many states, often the result of highly publicized student suicides. The *Student Non-Discrimination Act* was introduced into

> Never be bullied into silence. Never allow yourself to be made a victim. Accept no one's definition of your life; define yourself."
>
> –HARVEY FIERSTEIN, ACTOR PLAYWRIGHT, TONY WINNER

Congress in 2018 (H.R. 5374). While still tabled, if passed, the act would prohibit discrimination in schools based on real or perceived sexual orientation or gender identity. It also defines harassment as a form of discrimination and authorizes the federal government to cut off funds to schools that fail to enforce the new law. Finally, it asserts the rights of victims of discrimination on the basis of sexual orientation or gender identity to seek redress in the courts.

The Gay, Lesbian, and Straight Education Network (GLSEN) is a national organization that collaborates with educators, policy makers, community leaders, and students. Their research indicates that, as in California, curriculums that are inclusive of LGBTQ people, history, and experiences lead to better academic outcomes, more positive environments, and a greater sense of well-being by LGBTQ students (GLSEN 2020). Among other things, GLSEN has identified four important ways for schools to create a supportive environment for all students: (1) develop supportive educators who will advocate for LGBTQ students, (2) pass and implement comprehensive policies around key LGBTQ student issues (e.g., anti-bullying and nondiscrimination policies), (3) advocate for LGBTQ inclusive and affirming curriculum, and (4) support Gender and Sexuality Alliance (GSA) student groups (GLSEN 2020).

Campus Programs and Policies. In addition to university-wide nondiscrimination policies, other measures to support the LGBTQ college student population include gay and lesbian studies programs, social centers, and support groups, as well as campus events and activities that celebrate diversity. Some campuses have a "Lavender Graduation" ceremony in which LGBTQ graduates are honored and receive rainbow tassels for their mortarboards. Many campuses also have Safe Zone programs designed to visibly identify students, staff, and faculty who support the LGBTQ population.

Transgender and gender nonconforming (TGNC) students often face a unique set of obstacles on college campuses. The results of a study of 507 TGNC students from across the United States indicate that (1) community colleges and religious-affiliated schools are less supportive of TGNC students and (2) supportive policies of TGNC students included gender-inclusive bathrooms, gender identity nondiscrimination policies, and the ability to change your name on campus records without having to go through the legal process. As with gay and bisexual students, TGNC students reported a greater sense of belonging and a more positive environment at institutions with trans-supportive policies (Goldberg, Beemyn, and Smith 2019).

According to Rankin, Garvey, and Dunn (2020), in their retrospective analysis of LGBTQ issues on college and university campuses over the last decade, as a result of a greater acceptance of sexual and gender minorities over the last 30 years, LGBTQ students, faculty, and staff are more visible on college campuses than ever before. Nonetheless, as the authors also note, a climate of fear still exists on college campuses today and visibility of gay, bisexual, and transgender students, faculty and staff should not be taken as an indication that LGBTQ equality has been achieved.

Understanding Sexual Orientation and the Struggle for Equality

As both structural functionalists and conflict theorists note, non-heterosexuality challenges traditional definitions of family, child rearing, and gender roles. Thus, every victory in achieving legal protection and social recognition for gay men and women fuels the backlash against them by groups who are determined to maintain traditional notions of family and gender, often informed by religious ideology. And, as symbolic interactionists note, the meanings associated with same-sex relations are learned, often embedded with stereotypes resulting from myths about sexual orientation. Powerful individuals and groups opposed to gay rights focus their efforts on maintaining these negative meanings of non-heterosexuality to keep the gay, lesbian, and bisexual population marginalized.

In what can only be called landmark decisions, as noted previously, the U.S. Supreme Court held that same-sex marriage (*Obergefell*) and freedom from discrimination in the workplace based on sex (including sexual orientation and gender identity) (*Bostock*) are fundamental rights. There is no doubt that these decisions represent a turning point in gay, bisexual, and transgender equality. However, with conservatives dominating state governments (NCLS 2020), it is not surprising that, of the 396 pro-LGBT equality bills introduced in state legislatures across the nation in 2019, only 84 became law (Byrnes, Harrington, and Suneson 2020).

What do you think?

Gallup conducted a national survey of Americans for one month prior to the 2016 presidential election asking respondents to rate their well-being, from 1 to 10 (Gates 2016). They then conducted a second survey, with different respondents, for one month following the election and asked the same question. Both samples were then divided into LGBT respondents and non-LGBT respondents. Ratings of LGBT respondents after President Trump's election fell 10 percentage points when compared to LGBT ratings before the election. Non-LGBT respondents' rankings remained the same. Why do you think LGBT respondents' sense of well-being fell so dramatically?

If the *Equality Act* was signed into law and if sex was legally defined to include differences based on sexual orientation and gender identity, it would signify legitimacy and a greater acceptance for millions of gay, bisexual, and transgender men and women (e.g., Valencia, Williams, and Pettis 2019). Understanding "acceptance and rejection of LGBT people lies at the heart of understanding violence, discrimination, and a multitude of negative consequences arising from exclusion and unfair treatment" (Flores 2019, p. 2). Reduced stigma is associated with a reduction in hate crimes, not just for LGBTQ individuals but for straight cisgender individuals who may be *perceived* as gay or transgender. Passage of the *Equality Act* would also mean reduced rates of stress, depression, fear, substance abuse, homelessness, and suicide for LGBTQ men and women and consequently less anxiety for friends and family of LGBTQ youth.

@iam wandasykes

The #EqualityAct has passed the House and its now on to the Senate. Call your reps! We can't be equal until we are ALL equal! Text COSPONSOR to 472472 or click below to get involved!! @HRC

-Wanda Sykes

If the *Equality Act* was passed, same-sex marriages would not be, as Justice Ginsburg noted, "skim milk marriages." Same-sex couples would be afforded all the rights and responsibilities of different sex-couples. They would be less likely to live in poverty, have better mental and physical health, and thus be better positioned to provide a safe and healthy environment for their children. A study of LGBT elders who remained single, often out of residual fear and anxiety from past decades, indicates that they are at "a disadvantage across nearly every social-economic, social, and health indicator" (Goldsen 2017, p. 1). If same-sex marriages were "whole milk marriages," children would be legally protected from being removed from their biological, foster, or adoptive families, and hundreds of thousands of children presently in foster care or available for adoption would have homes.

If the *Equality Act* was passed, it would strengthen state laws that ban conversion therapy and encourage other states to legislate similar bans (Gonzales and Gavulic 2020). Transgender men and women would benefit from gender-affirming medical care. A comprehensive review of research on the impact of trans-friendly health care links it to increased mental health, lower suicide rates, and, among trans youth, reduced fear and fewer incidents of school harassment and bullying (Cornell 2017). Stigmatization early in life often has long-lasting effects, and public health research consistently shows that "peer victimization of LGBT students in K–12 schools is associated with elevated levels of self-harm, suicidal ideations, depression, and anxiety disorders" (Conron and O'Neill 2020, p. 801).

Of course, even when laws are passed to protect LGBTQ individuals, it doesn't ensure behavioral changes. LGBTQ individuals employed at workplaces with anti-discrimination policies still experience harassment and rejection from their coworkers; students in schools with policies prohibiting bullying are still subjected to anti-gay taunts; hate

© Kellogg Company, Battle Creek Enquirer via Imagn Content Services, LLC

Big business supports the passage of the *Equality Act* as exemplified by Kellogg's limited release of *All Together* cereal, made up of six different cereal brands as indicated by the mascots on the front of the box. The cereal was sold in honor of #SpiritDay, a day devoted to awareness of, and efforts to stop, LGBTQ bullying sponsored by GLAAD.

crime statutes don't prevent LGBTQ victimization; and countries around the world continue to beat, imprison, and execute gay and transgender men and women. But passage of the *Equality Act* would serve two additional important functions:

First, it would provide legal recourse for violations of the *Equality Act* where, in many states, none exist. To date, in 27 states, men and women, simply because of their sexual orientation or gender identity, can be denied housing and in 28 states denied access to public accommodations such as restaurants, stores, and movie theaters. Additionally, only 22 states prohibit harassment and/or bullying of gay, bisexual, or transgender boys and girls in public schools, leaving parents with few legal remedies. Finally, just 22 states protect youth from conversion therapy "through licensing restrictions which prevent licensed mental health service professionals from conducting conversion therapy on youth under age 18" (HRC 2020b, p. 1).

Second, and perhaps more importantly, passage of the *Equality Act* is what most Americans want. Upward of 70 percent of Americans support prohibiting discrimination based on sexual orientation or gender identity (Kaiser Family Foundation [KFF] 2020). Further, corporate America supports the act. A coalition of over 200 businesses support the passage of the *Equality Act* representing all 50 states with combined revenues of $4.5 trillion. This coalition includes such companies as Abercrombie and Fitch, Amazon, Apple, AT&T, Bank of America, Best Buy, Coca-Cola, CBS, Domino's Pizza, Facebook, Food Lion, General Motors, Hershey's, Hyatt Hotels, IBM, Johnson & Johnson, Kellogg's, Levi-Strauss, Morgan Stanley, Netflix, Nike, Pepsi-Cola, Pet Smart, Sony Electronics, Spotify, Starbucks, Target, Tesla, Twitter, Uber, Under Armour, United Airlines, UPS, Verizon, Whirlpool, Xerox, Yelp, and Zillow (HRC 2019).

Embracing the outdated concepts of the binary nature of gender and sexuality, many continue to fight for the subjugation of gay and transgender men and women. But the consensus of scientific research is that LGBTQ statuses are *ascribed statuses* and can be changed neither by choice nor by the efforts of others. Thus, sexual orientation and gender identity are akin to other ascribed statuses—gender, race, and ethnicity—and, as we are finally beginning to acknowledge, diversity, although difficult, is good. Diversity

> enhances creativity ... encourages the search for novel information and perspectives, leading to better decision making and problem solving. ... This is not just wishful thinking: it is the conclusion ... from decades of research from organizational scientists, psychologists, sociologists, economists and demographers. (Phillips 2014, p. 1)

Imagine the lost contributions of women, racial and ethnic minorities, and gay and transgender women and men who for hundreds of years were unable to reach their full potential as a consequence of stereotyping, prejudice, and discrimination. The *Equality Act* "could provide—fully, uncompromisingly and on a national scale—a corrective soft lighting to identities that shouldn't have been disparaged to begin with" (Tensley 2019, p. 1). Millions of Americans are hoping that President Biden and his administration will successfully guide the *Equality Act* through Congress.

Chapter Review

- **Are there any countries in which same-sex sexual behavior is illegal?**
  Yes, globally, numerous countries criminalize same-sex relations. Legal penalties vary for violating laws that prohibit same-sex sexual behavior. In some countries, same-sex sexual behavior is punishable by prison sentences and/or corporal punishment, such as whipping or lashing, and in others, predominately Muslim, people found guilty of engaging in same-sex sexual behavior may receive the death penalty. There has been, however, a general trend toward the legalization and acceptance of same-sex relations.
- **What is the relationship between beliefs about what "causes" same-sex attraction and attitudes toward same-sex attraction?**
  Individuals who believe that non-heterosexuality is biologically based or inborn tend to be more accepting of LGB individuals. In contrast, individuals who believe that lesbian, gay, and bisexual people choose their sexual orientation are less tolerant of LGB individuals. The consensus of scientific opinion is that sexual orientation is the result of a combination of biological, psychological, and sociological variables factors and cannot be changed by the individual or outside intervention.
- **How does each of the three sociological theories addresses the issue of LGBTQ inequality?**
  Structural functionalism argues that the increased acceptance of LGBTQ men and women is a consequence of greater gender equality, changing gender roles, and the increased visibility of LGBTQ individuals. Conflict theory holds that the struggle for LGBTQ equality is a struggle over power, prestige, and economic resources. Alternatively, symbolic interactionism emphasizes the social meanings of sexual orientations and gender identities and how they have evolved over time.
- **What are some of the cultural origins of anti-LGBTQ bias?**
  The origins of anti-LGBTQ bias include religious beliefs that define non-heterosexual behavior as a sin; psychology's early view that "homosexuality" and "transvestism" were symptoms of mental illness, rigid gender roles, and myths and negative stereotypes about gay, bisexual, and transgender women and men.
- **What are some of the consequences of anti-LGBTQ bias for individuals?**
  Sexual and gender minorities, as many other minorities, suffer the negative consequences of harassment, prejudice, and discrimination, violence, and isolation. Among others, the consequences include higher rates of suicide, physical illness, substance abuse, poverty, and homelessness, when compared to their non-LGBT counterparts, as well as an uncertain future as they age.
- **Can LGBT individuals be legally discriminated against in the workplace?**
  In a major victory for LGBT Americans, in 2020, the U.S. Supreme Court held that Title VII of the 1964 *Civil Rights Act*, which prohibits employment discrimination based on sex, includes protections for sexual orientation and gender identity minorities. The decision, however, is limited in its scope, among other things, applying only to businesses with 15 or more employees and acknowledging the role of religious exemptions from adherence.
- **What was the decision in *Obergefell v. Hodges*?**
  In a 5–4 decision, the Supreme Court of the United States held that it is unconstitutional for a state to ban same-sex marriages or to deny legal recognition of a same-sex marriage from another state. Although same-sex marriages are now legal in the United States, the decision does not ensure equal treatment of same-sex spouses compared to different-sex partners.
- **What are some of the arguments for and against the legalization of same-sex marriage?**
  Arguments for same-sex marriage include, but are not limited to, the American value of equality, increased stability for children with LGB parent(s), and the extension of federal benefits, including health care, to gay, bisexual, and transgender women and men and their children. Arguments against same-sex marriage include the belief that marriage should only be between a man and a woman and that gay and lesbian sexuality is immoral.
- **Why is the media important in the advancement of LGBTQ civil rights?**
  LGBTQ visibility in the media counteracts stereotypes of LGBTQ individuals and allows non-LGBTQ individuals to see them not as an abstraction, but as real people. This is consistent with Gordon Allport's contact hypothesis that suggests more contact with or exposure to a group results in the reduction of prejudice. The media has also provided LGBTQ individuals, most importantly youth, with role models for "coming out."
- **What are some of the strategies for action in dealing with LGBTQ inequality?**
  Strategies for action include creating more positive environments for LGBTQ youth in schools, policy changes that permit LGBTQ couples to foster and adopt children, 50-state adoption of hate crime laws that include sexual orientation and gender identity protections, and passage of the *Equality Act*, which would make it illegal to discriminate on the basis of sexual orientation and gender identity.

Test Yourself

1. Research has indicated that many individuals are not exclusively heterosexual or gay and that sexual orientation can be represented on a continuum.
   a. True
   b. False
2. A national study of U.S. college students found that what percentage identified as gay, lesbian, or queer?
   a. 1.2
   b. 4.2
   c. 7.5
   d. 10.1
3. In 2017, there were nearly 1 million same-sex couple households in the United States.
   a. True
   b. False
4. Individuals who believe that same-sex attraction is a consequence of birth
   a. tend to be in the minority.
   b. are more supportive of gay rights than those who believe it's a choice.
   c. are more likely to believe that same-sex attraction is not acceptable.
   d. are correct according to scientific research.
5. Sexual orientation change efforts (SOCE)
   a. have been found to be successful 60 percent of the time.
   b. are rooted in the belief that non-heterosexuality is immoral or pathological.
   c. are supported by many professional associations such as the American Medical Association.
   d. are also suitable for transgender individuals.
6. Heterosexism leads to
   a. prejudice.
   b. the oppression of heterosexuals.
   c. discrimination.
   d. both a and c.
7. The contact hypothesis suggests that the more gay men and women reveal their sexual orientation, the lower the prejudice against them.
   a. True
   b. False
8. The It Gets Better project was started because
   a. same-sex couples continue to encounter obstacles in second-parent and joint adoptions.
   b. LGBTQ youth continue to be bullied and harassed by their peers, resulting in shame, isolation, and even suicide.
   c. LGB servicemen and -women have been waiting a long time to serve openly in the military.
   d. American society seems to be growing in its acceptance of the legal recognition of same-sex couples.
9. The *Obergefell v. Hodges* Supreme Court decision held that states do not have the right to ban same-sex marriages.
   a. True
   b. False
10. The first country to legalize same-sex marriage was
   a. Norway.
   b. England.
   c. the Netherlands.
   d. Argentina.

Answers: 1. A; 2. C; 3. A; 4. B; 5. B; 6. D; 7. A; 8. B; 9. A; 10. C.

Key Terms

bisexuality 408
Bostock v. Clayton County, Georgia 426
cisgender 408
cisgenderism 422
coming out 438
contact hypothesis 424
corrective rape 413
Defense of Marriage Act (DOMA) 427
discrimination 422
Do No Harm Act 441
Equality Act 444
Every Child Deserves a Family Act (ECDFA) 442
First Amendment Defense Act 416
gay 408
gender nonconforming 408
heteronormativity 413
heterosexism 422
heterosexuality 408
homophobia 420
legitimacy hypothesis 432
lesbian 408
LGBT, LGBTQ, and LGBTQI 409
master status 417
Matthew Shepard and James Byrd, Jr. Hate Crimes Prevention Act (HCPA) 443
Obergefell v. Hodges 427
oppression 422
pansexual 408
polarization hypothesis 432
prejudice 422
privileged 422
queer 409
religious freedom laws 415
Religious Freedom Restoration Act (RFRA) 440
sexual orientation 408
sexual orientation change efforts (SOCE) 413
Stonewall Uprising 421

Vincent Isore/IP3/Getty Images News/Getty Images

> "There's a strong correlation between the risk of pandemic and human population density. We've done the math and we've proved it."
>
> **DR. PETER DASZAK,**
> A disease ecologist

12

Population Growth and Aging

Chapter Outline

Learning Objectives

After studying this chapter, you will be able to …

1. Describe the history, current trends, and future projections of population growth.
2. Explain the causes of the aging of the world's population.
3. Discuss ways in which structural functionalism, conflict theory, and symbolic interactionism can be applied to the study of population and aging.
4. Explain how population growth is related to poverty, unemployment, food insecurity, and global insecurity.
5. Describe problems related to the aging of the population, including ageism, employment and retirement concerns, and family caregiving.
6. Describe strategies to curb global population growth, increase population growth in some countries, combat ageism and age discrimination in the workplace, and reform Social Security.

Nelson Almeida/AFP/Getty Images

After months with no contact during the COVID-19 pandemic, some nursing home and assisted living facilities set up "hug tunnels" to allow family members contact with their parents and grandparents.

FOR THE FIRST TIME IN MONTHS, a woman finally gets a chance to put her arms around her 85-year-old mother. But because of COVID-19, the hug is through a "hug tunnel," a transparent plastic curtain between mother and daughter used to prevent transmission of the disease. After months of separation, a wife takes a dishwashing job in her husband's nursing home just so she can see him. A 5-year-old girl who misses her 93-year-old neighbor sends him a letter: "Hello—my name is Kirah. ... I just wanted to check to see if you are OK. I have drawn you a rainbow to remind you that you are not alone. Please write back if you can" (quoted in Brent 2020).

Thousands of elderly patients faced separation from families and friends as the virus spread throughout communal living facilities. Within 12 months of its arrival in the United States, over 350,000 elderly patients had died from COVID-19, often without family and loved ones nearby (Centers for Disease Control 2021; Kamp and Mathews 2020)

The COVID-19 pandemic highlighted the treatment of the elderly in the United States—isolated from family and friends, warehoused in long-term care facilities, and often abused and neglected. The negative treatment of the elderly in the United States reflects, in part, the American values of independence, self-reliance, and productivity, which the elderly may no longer represent. Yet living a long life is something most people aspire to and is thus the one minority group of which we want to be a member. As we will read in this chapter, people over the age of 65 are the fastest growing demographic group in the United States, and, as their numbers grow and the composition and size of the global population changes, the question is whether we will be able to adapt.

The Global Context: A Worldview of Population Growth and Aging

In the 1920s, demographer Raymond Pearl estimated that the Earth could support 2 billion people. Yet today, over 7.8 billion people populate the Earth. Reflecting on the Earth's current population, Jesse Ausubel, director of the Program for the Human Environment at Rockefeller University, notes, "Through the invention and diffusion of technology, humans alter and expand their niche, redefine resources, and violate population forecasts" (Kolitz 2019). How many people can the Earth sustain, and what are the implications of the population growth that has occurred and the distribution of the Earth's population today? In the following sections, we describe how the size and life span of human populations have increased over time.

World Population: History, Current Trends, and Future Projections

Humans have lived on planet Earth for at least 200,000 years. For 99 percent of human history, population growth was restricted by disease and lack of food. Around 8000 BCE, the development of agriculture and the domestication of animals led to increased food

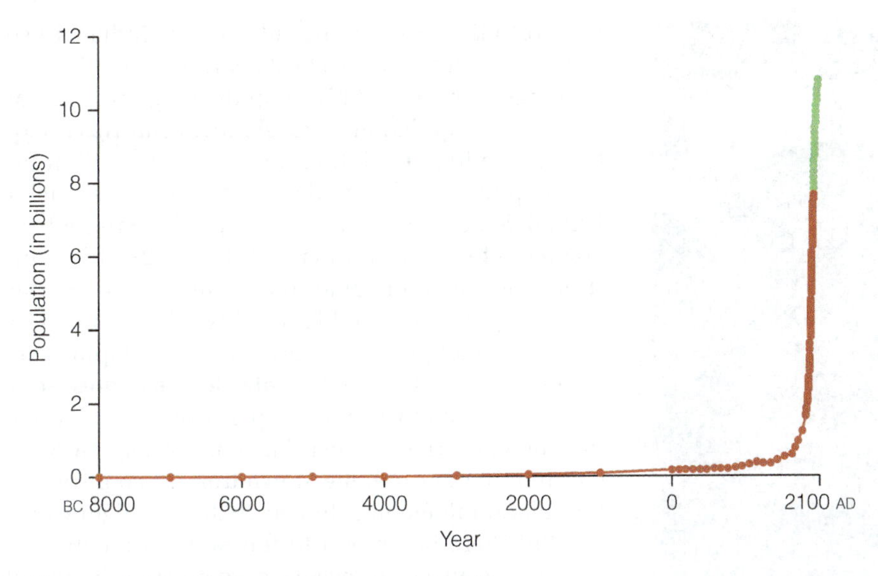

Figure 12.1 World Population throughout History
SOURCE: Roser, Ritchie, and Ortiz-Ospina, 2020.

supplies, and, despite continued harsh living conditions, some population growth occurred. This pattern continued until the mid-18th century when the Industrial Revolution improved the standard of living for much of the world's population through better and safer food production, cleaner drinking water, improved housing and sanitation, and advances in medical technology (e.g., antibiotics and vaccinations against infectious diseases)—all of which contributed to rapid population growth.

Although it took thousands of years for the world's population to reach 1 billion, population growth exploded around 1800 and the Earth's population grew from 1 billion to 6 billion over the next 200 years (see Figure 12.1). At the beginning of the 20th century the world's population was 1.6 billion; just 100 years later, it was six times greater. Increasing to 7.7 billion in 2019, the world population is projected to reach 9.7 billion by 2050 and 10.9 billion by 2100 (Roser, Ritchie, and Ortiz-Ospina 2019).

Most of the world's population lives in less developed countries, primarily in Asia and Africa. As Figure 12.2 illustrates, Asia is projected to continue to be the most populous region through 2100. However, by then, Africa's population will nearly be as large. The most populated country in the world is China where about one in five people in

> **@BioAnth FunFacts**
>
> About 7% percent of all humans who have ever lived are alive today There are over 7.5 billion people on the planet alive today. The **Population Reference Bureau** estimates that about 107 billion people have ever lived.
>
> -Hominid Evolution Fun Facts by Alexandra Kralick

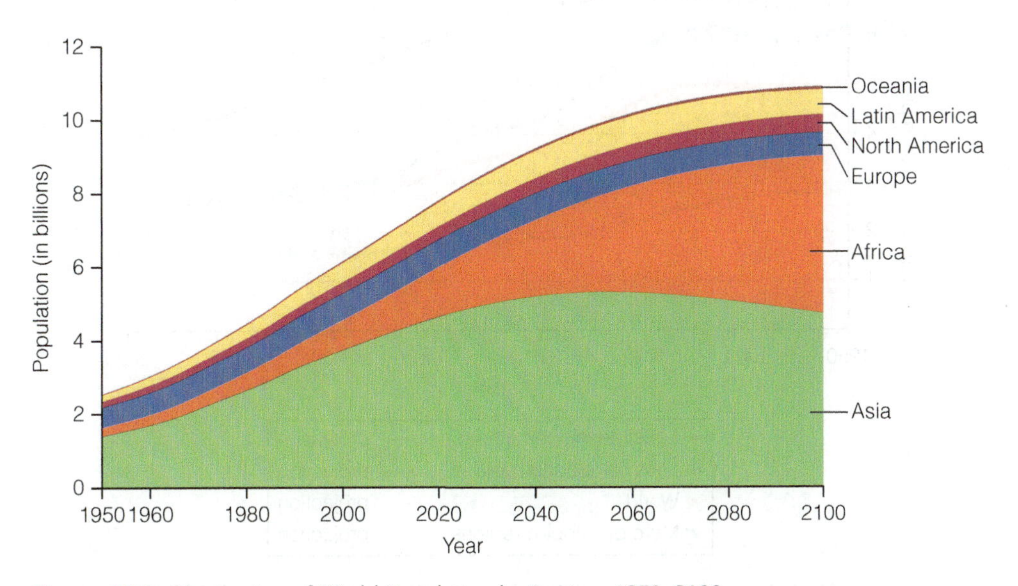

Figure 12.2 Distribution of World Population by Region, 1950–2100.
SOURCE: Roser, Ritchie, and Ortiz-Ospina 2019.

MOHAMED EL-SHAHED/AFP/Getty Images

Cairo, the capital of Egypt, is the third-largest city on the African continent, with 100 million residents. The rising poverty rate currently stands at one-third of the population, who are living on less than $1.50 per day.

the world live. Projections indicate that India will surpass China in population size by 2024 (Roser 2019).

Largely because of high population growth in Asia and Africa, the population of developing countries is expected to increase from 6.1 billion in 2015 to 9.9 billion in 2100. The population of more developed countries, such as the United States and most of Europe, is expected to grow gradually from 1.25 billion in 2015 to 1.28 billion in 2100. If not for migration from developing countries, these regions' populations would actually decline (United Nations 2019a). Higher population growth in developing countries is largely a result of higher **total fertility rates**—the average lifetime number of births per woman in a population. In nine countries, women have an average of 5 or more children in their lifetime, down from 21 countries in 2016 (Population Reference Bureau 2020) (see Figure 12.3).

Will there be an end to the rapid population growth that has occurred in recent decades? Will the population of the world stabilize? Researchers are increasingly confident that the global population will stabilize around the middle of the 21st century and may decline thereafter. **Population growth rate**—the number of people added to the population in a calendar year (e.g., births − deaths)—reached its peak in 1968 with a growth rate of 2.1 percent; today, the world's growth rate is about 1 percent per year (Roser 2019).

To reach population stabilization, fertility rates throughout the world would need to achieve what is called "replacement level" whereby births would replace but not outnumber deaths. **Replacement-level fertility** is 2.1 births per woman—slightly more than 2 because not all females born survive to reproductive age. In 2020, 91 countries—mostly in Europe and East Asia—had below-replacement fertility, an increase of eight countries over the previous five years (Population Reference Bureau 2016, 2020). Weeks (2015) explains that in these countries, "the motivation to have large families has disappeared, at least for the time being, and has been replaced by a propensity to try to improve the family's standard of living by limiting the number of children" (p. 46).

total fertility rates The average lifetime number of births per woman in a population.

population growth rate The number of people added to the population in a year (e.g., births − deaths).

replacement-level fertility The level of fertility at which a population exactly replaces itself from one generation to the next; currently, the number is 2.1 births per woman.

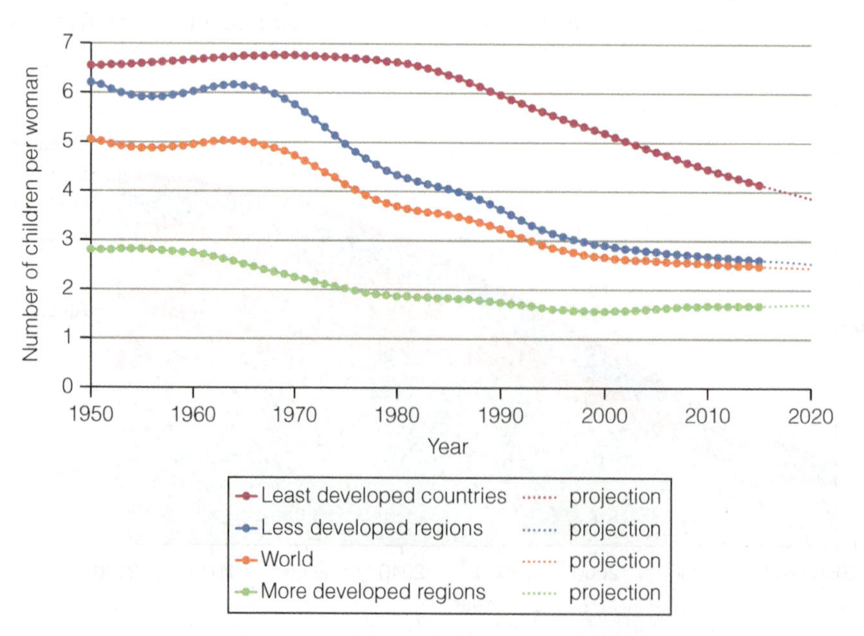

Figure 12.3 Average Number of Children per Woman, by Region, 1950–2017
SOURCE: Roser 2017.

In some countries with below-replacement fertility, populations will continue to grow for several decades because of **population momentum**—population growth due to previous high fertility rates that resulted in a large number of women who are currently entering childbearing years. For example, India's projected growth is occurring despite the fact its fertility rate dropped between 1990 and 2019, from an average of 4 children to 2.3 children per woman (Our World in Data 2020). It will take some time for its projected growth to peak in the late 2050s at around 1.7 billion people, at which point it is expected to begin a decline (Roser 2019). The United Nations (2019a) projects that 55 countries have populations that will decrease over the next three decades, with half seeing reductions of 10 percent of their population.

> "We all worry about the population explosion, but we don't worry about it at the right time."
>
> —ART HOPPE, FORMER SYNDICATED COLUMNIST

Worldwide, **migration**—the movement of people from one country or region to another—is also a factor in a country's population size. Between 2010 and 2020, 14 countries or regions, mostly in Europe, North America, North and West Africa, and Australia and New Zealand, were projected to see a net influx of more than 1 million migrants (i.e., **immigration**), while ten countries would see a net outflow of the same number (i.e., **emigration**) (United Nations 2019a).

Migration occurs predominantly as a result of the demand for migrant workers, people fleeing armed conflict and violence, and economic insecurity of their home countries (Ramos 2017). Historically, the United States has experienced population growth as a result of immigration even as fertility rates have dropped below replacement levels. Immigrants made up 13.6 percent of the U.S. population in 2017, up from a low of 4.7 percent in 1970 (Radford 2019). However, the number of new immigrants is falling, largely due to a decrease in illegal immigration with more, for example, Mexican immigrants leaving the United States than entering it.

A slow economic recovery from the Great Recession of 2008 and stricter border control and deportation measures under both Presidents Obama and Trump have contributed to this decline (Gonzalez-Barrera 2015; Radford 2019). Today, the largest group of U.S. immigrants are from Asia (see Figure 12.4). During the first six months of the COVID-19 pandemic, the former administration stepped up its efforts to restrict immigration passing more than 48 policy changes related to immigration (Zak 2020). While it remains to be seen how these policies will affect U.S. population growth long term, the Biden administration reversed many of the Trump immigration policies and encourages assimilation of immigrants into their communities (Biden 2020) (see Chapter 9).

In sum, two population trends are occurring simultaneously that appear to be contradictory: (1) The total number of people on this planet is rising and is expected to continue to increase over the coming decades, and (2) fertility rates are so low in some countries that the countries' populations are expected to decline over the coming years. As a result, the distribution of the world's population is shifting and the age distribution of many nations is changing. As discussed later in this chapter, each of these trends presents a set of problems and challenges.

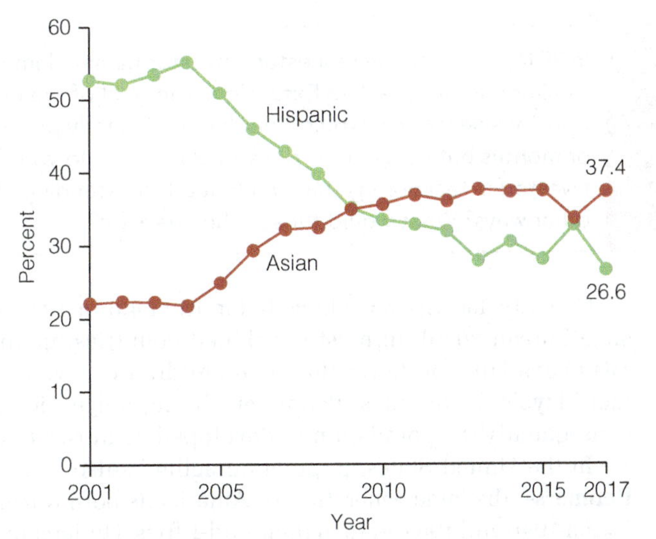

Figure 12.4 Percentage of Hispanic and Asian Immigrants in the United States, 2001–2017
SOURCE: Radford 2019.

The Aging of the World's Population

One demographic trend that presents its own set of opportunities and challenges is the increasing number and proportion of older individuals in the total population. Globally, the population aged 65 and older is the fastest-growing age group, making

population momentum Continued population growth as a result of past high fertility rates that have resulted in a large number of young women who are currently entering their childbearing years.

migration The movement of people from one country or region to another; includes immigration and emigration.

immigration The movement of people into a country or region from another.

emigration The movement of people out of one country or region to another.

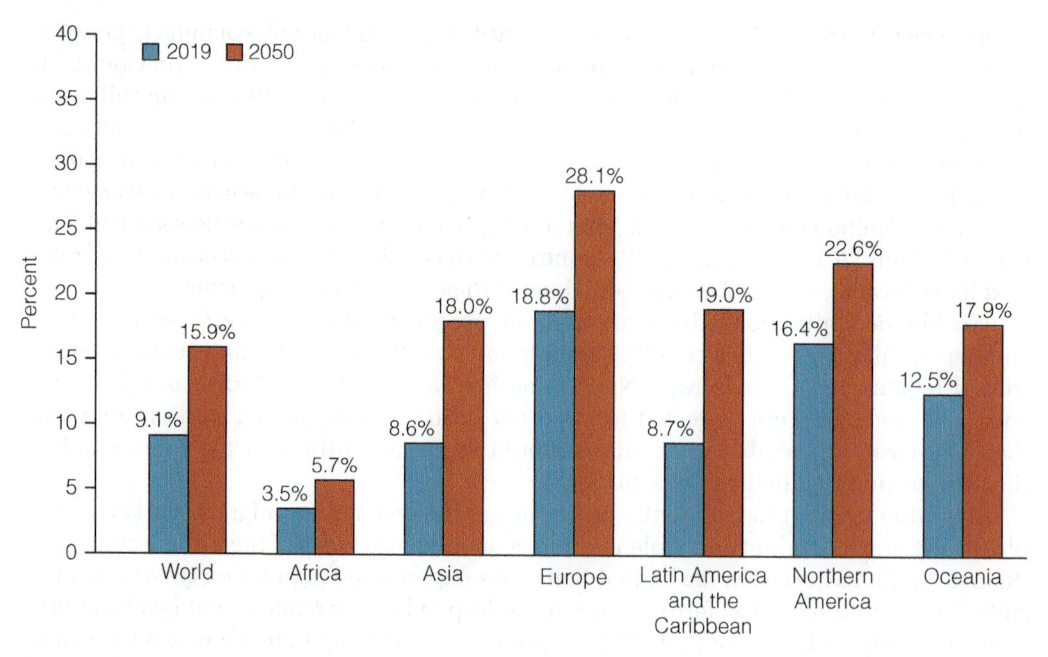

Figure 12.5 Percent of Population Age 65+, by Region, 2019–2050 (projected)
SOURCE: United Nations 2019.

up 9 percent of the world's population (United Nations 2019a). Between 2019 and 2050, the number of older individuals in the world is expected to double to 1.5 billion people. In 2018, the global population of people aged 65 years or over outnumbered the population of children under age 5 for the first time in history, and in approximately 60 years, people aged 65 and over will surpass children under 15 (Ritchie and Roser 2019). The 80 and older population is also growing and is expected to triple in number between 2019 and 2050 (United Nations 2020a). As illustrated in Figure 12.5, the growth of the elderly population is in all regions of the world, although some regions have a significantly greater proportion of the world's elderly than others.

The aging of the population is the result of both increased longevity and declining fertility rates. Global life expectancy increased from 46 years in 1950 to 72.6 years in 2020, and it is expected to increase to 77 years by 2050, and to 83 years by 2100 (United Nations 2019a; Roser, Ortiz-Ospina, and Ritchie 2019). In the United States, the current life expectancy at birth is about 79 years, although it varies significantly by gender and race (see Chapter 2).

What do you think?

In 2016, an American real estate investor named James Strole established the Coalition for Radical Life Extension. The goal of the organization is to garner support for science that would one day significantly prolong human life, not by weeks or months but by decades or even centuries. Do you think resources should be invested in technology that would lead to extending life as long as possible, or in other ways? What would those other ways be?

Globally, fertility rates have fallen from 4.6 in 1970 to 2.3 in 2019 (Population Reference Bureau 2020). In most developed countries, including the United States, fertility rates have fallen to fewer than two children per woman and have been below replacement levels for decades. People are living longer, fewer children are being born, and consequently the population in developed countries is getting older.

In the United States, population aging is also occurring as even the youngest **Baby Boomers**—the large generation of Americans born during a period of high birthrates between 1946 and 1964—are in their mid-fifties. Understanding the age and sex composition of a county or region is aided with a **population pyramid**—a graph that represents the number of males and females in each age group with a bar, younger people on the bottom of the graph and older people at the top.

Baby Boomers The generation of Americans born between 1946 and 1964, a period of high birthrates.

population pyramid A graph of a country's or region's population that represents the number of males and females in each age group with a bar.

Figure 12.6 is the U.S. population pyramid for the year 2020. Social scientists have known for decades that the aging of the Baby Boomer generation would create challenges. However, as illustrated in Figure 12.6, the children of the Baby Boomers is an even larger generation which, in 20 to 30 years, will bring additional challenges. This group is sometimes called the Baby Boomlet or the Echo Boom, and is 60 percent larger than the Baby Boom generation. The first members of this generation will become senior citizens in 25 years increasing the U.S. elderly population from 16.6 percent in 2020 to 22.4 percent in 2050 (United Nations 2020b).

Population aging increases pressure on a society's ability to support its elderly members because as the proportion of older people increases, there are fewer working-age adults to support the elderly population. A commonly used indicator of this pressure is the **elderly support ratio**, calculated as the number of working-age people divided by the number of people 65 or older. Globally, "working age" is considered to be between the ages of 15 and 64 although in most developed countries it is between 20 and 64.

Andrew Caballero-Reynolds/AFP/Getty Images

While the rate of marriage among younger people is decreasing, the number of new marriages among people over 65 has been growing in the United States as people are living longer. This couple, both over the age of 70, were married in a Zoomed ceremony during the Coronavirus pandemic.

The global elderly support ratio has been declining. In 1950, there were 12 working-age people for every one older person; in 2019, there were seven working-age people for every one older person, and in 2050, there will be only four working-age adults for every person 65 years and older (United Nations 2019a). In developed countries, the elderly support ratio is much lower than in less developed regions. In the United States, for example, there are 3.5 working-age adults for every one elderly adult, and that number is expected to decrease to 2.5 by 2050. Japan has the lowest elderly support ratio in the world at 1.7 (United Nations 2019a).

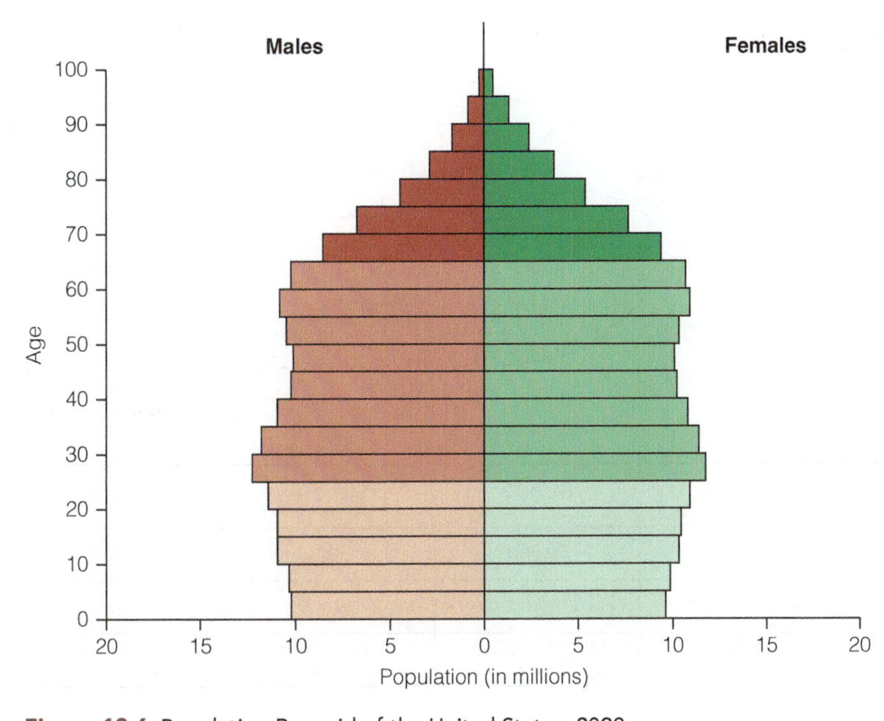

Figure 12.6 Population Pyramid of the United States, 2020
SOURCE: United Nations 2019a.

elderly support ratio The ratio of working-age adults (15 to 64) to adults aged 65 and older in a population.

> **What do you think?**
>
> A law in China requires that *all* adult children—not just those who volunteer—provide financial, emotional, and physical assistance to their elderly parents. This includes a requirement that adult children visit their parents. Violators face fines and jail time. In the Ukraine and other Soviet bloc countries, the elderly can sue their children for lack of support. Do you think this is fair?

The decrease in the elderly support ratio raises concerns about whether there will be enough workers to take care of the older population, not just financially but emotionally and physically. Population aging raises other concerns, as well: How will societies provide housing, medical care, transportation, and other needs for the increasing elderly population? Later in this chapter, we look at the problems, challenges, and opportunities associated with a growing elderly population.

demographic transition theory A theory that attributes population growth patterns to changes in birthrates and death rates associated with the process of industrialization.

birthrate The number of live births per 1,000 people in the population per calender year.

mortality rate The number of deaths per 1,000 people in the population per calender year; sometimes referred to as the death rate.

Sociological Theories of Population Growth and Aging

The three main sociological perspectives—structural functionalism, conflict theory, and symbolic interactionism—can be applied to the study of population and aging.

Structural-Functionalist Perspective

Structural functionalism focuses on how changes in one aspect of the social system affect other aspects of society. For example, the **demographic transition theory** of population describes how industrialization and economic development affect population growth by influencing **birthrates** and **mortality rates** (see Figure 12.7) (Roser, Ritchie, and Ortiz-Ospina 2019). The birthrate is the number of live births per 1,000 people in a population per calendar year. The mortality or death rate is the number of deaths

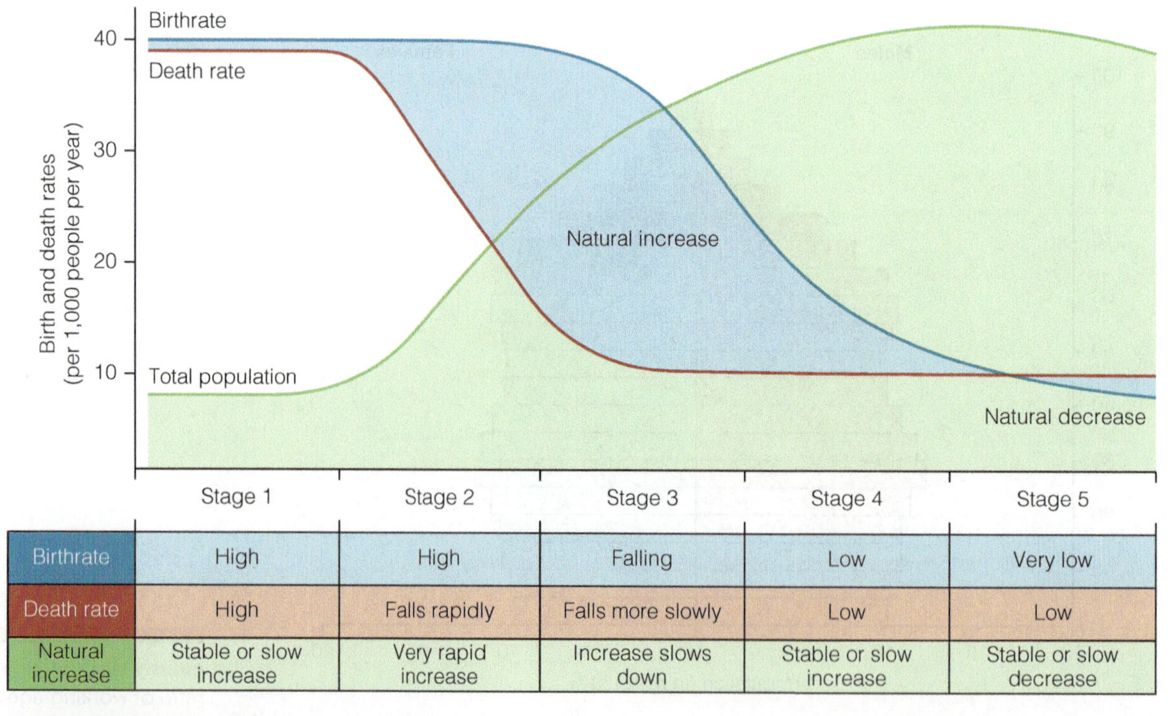

	Stage 1	Stage 2	Stage 3	Stage 4	Stage 5
Birthrate	High	High	Falling	Low	Very low
Death rate	High	Falls rapidly	Falls more slowly	Low	Low
Natural increase	Stable or slow increase	Very rapid increase	Increase slows down	Stable or slow increase	Stable or slow decrease

Figure 12.7 Demographic Transition Theory
SOURCE: Roser, Ritchie, and Ortiz-Ospina 2019.

social problems
RESEARCH UP CLOSE
Explaining Why Increasing Mother's Education Decreases Child Mortality

Over five million children under 5 die every year. In 2017, 5.4 million children under the age of 5 died; over 4 percent of all children in the world (Roser, Ritchie, and Dadonaite 2019). Research indicates that the likelihood of a child dying before the age of 5 is associated with a mother's education, although there are a number of interpretations as to why that is the case. Some theorize that maternal schooling benefits subsequent children by increasing the economic resources mothers have to care for their children, improving their access to food, shelter, and clean water. Others hypothesize that women who have been educated are more likely to take advantage of modern medicine and understand how to avoid exposure to disease. Still others argue that women who are educated are empowered to take a more active role in decision making and to resist relationships that expose children to domestic violence.

In this study, researchers Andriano and Monden take advantage of the Universal Primary Education (UPE) reforms that were implemented in Malawi and Uganda in the 1990s to examine what happened to the children born to the girls who benefitted from the expansion of primary education to all children in these countries. Malawi and Uganda are located in sub-Saharan Africa, a region in which in the 1990s more than 15 percent of children died before their fifth birthday.

Sample and Methods

The researchers used survey data collected from 19,835 women who were born between 1969 and 1980 and, therefore, were too old to have had been affected by the education reforms, and 36,866 women who were born between 1981 and 1991 and did have access to universal primary school through these reforms. The survey included information about the women's completed level of education, number of children, child deaths, wealth, ability to pay for medical care, contraceptive use, use of preventive medical care, whether they lived close to a health facility, questions on medical knowledge, decision-making power in the household, and beliefs on whether a man has the right to beat his wife.

Findings and Conclusion

The researchers found that the educational reforms did result in higher levels of education among the younger women, although the results depended on how rigorously their district implemented the program. Importantly, each additional year of completed education reduced the likelihood that a child would die by almost 10 percent in Malawi and by 16.6 percent in Uganda. Increases in women's education led to increased wealth in Uganda only but decreased the perception that money was a barrier to medical care in both countries. Education increased the use of contraception by women in both countries but increased the use of preventive health care in Uganda only. In Malawi increased education improved medical knowledge. In both countries, increased education had no impact on participation in household decision making, but it did impact beliefs about whether a man should beat his wife.

This research demonstrates that expanding girls' access to education, even modestly through the primary school level, has important health consequences for the next generation, although the mechanisms by which this effect occurs varies from country to country.

SOURCE: Andriano, Liliana, and Christian W. S. Monden. 2020. "The Causal Effect of Maternal Education on Child Mortality: Evidence from a Quasi-Experiment in Malawi and Uganda." Demography. 56: 1765–1790.

per 1,000 people in a population per calendar year. According to this theory, traditional agricultural societies have both high birthrates and high mortality rates. Families need many children to help support the family farm and many children don't survive to adulthood. As a result, fertility rates are high, but population growth is slow because births are balanced by deaths (Stage 1).

As a society becomes industrialized and urbanized, improved sanitation and food supply along with education lead to a decline in mortality especially for infants and children. The population growth rate increases because births are no longer balanced by deaths (Stage 2). This stage describes most European countries in the late-18th to mid-19th centuries, and fits some developing countries, such as Afghanistan today. As individuals move from working on farms to working in factories, children are no longer needed to help support the family. Education levels increase, especially for women. Both factors contribute to a decline in fertility, moving countries into Stage 3 of the demographic transition. Mexico is an example of a country at this stage of transition because its fertility rate is still in the process of declining. Stage 4 is characterized by both low birth and low death rates, as is the situation in the United States today. In addition to the decline in the economic value of children, the spread of the social acceptability and advantages of controlling fertility contribute to countries moving to this stage (Zaida and Morgan 2017).

Unlike the global trend of population growth, an increasing number of countries like Russia, Japan, and Sweden are experiencing decreases in population due to below replacement rate fertility, entering what some researchers speculate may be a fifth stage of development. Others have called this a second demographic transition linked to a rise in individualism and materialism, increased availability of effective contraception, and a loosening of the link between marriage and fertility (Zaidi and Morgan 2017).

The demographic transition theory is a generalized model that has not predicted the trajectories of all countries (Caldwell et al. 2006). It has been criticized for imposing a pattern of development characteristic of mostly Western countries onto the rest of the world. It also fails to adequately address the role of culture and the impact of natural disasters and health crises (e.g., AIDS, COVID-19) in the evolution of countries. It remains to be seen how countries will evolve in the future and what factors will predict their patterns.

Structural functionalists view support for the elderly as a critical function of the family. As children have moved further away from their parents as a result of improvements in transportation and education, the burden has been placed on other social institutions to fill in the gaps for what the family is no longer able to provide the elderly. One means of evaluating to what extent the well-being of the elderly in a country is a national priority is to examine how much of the country's **gross domestic product (GDP)** is dedicated to these types of issues.

Gross domestic product refers to the total value of goods produced and services provided in a country during one year. For example, in Sweden, 3.6 percent of the country's GDP is allocated to elder care; all citizens receive a pension from the government, health care is free, and home health care is publicly funded. Ninety percent of all Swedes "age in place," meaning they are able to remain in their homes as they grow older (Intriago 2020). Structural functionalists would point out that this is an example of the institutions of government and health care taking on what is in other countries and at other times in history the responsibility of the family.

gross domestic product A measure of the economic productivity of a country; the total value of goods produced and services provided in a country during one year.

World System Theory A theory arguing that wealth and power are divided unequally among the core countries, periphery countries, and semi-periphery countries in the world.

Conflict Perspective

The conflict perspective focuses on how wealth and power, or the lack thereof, affect population problems. **World System Theory** addresses the question of unequal distribution of wealth and power at the global level. Wallerstein (1974) argued that countries could be divided into three groups: the core countries, the periphery countries, and the semi-periphery countries. Core countries like the United States, Western European countries, and Japan exploit the labor and resources of periphery countries in order to develop economically and maintain power. These high technology countries accumulate capital over time.

The periphery countries in Africa, much of Asia, and Latin America have economies that are reliant on the labor of their residents and the natural resources they possess. They are dependent on selling resources to the core to sustain their economies. Semi-periphery countries like Brazil, Mexico, South Korea, and India are those who share some of the characteristics of each, being exploited by some countries but exploiting the resources of others. The system creates inequality as surplus resources are transferred from periphery countries to the core, in a worldwide capitalist economy. Periphery countries are slower to develop because they fail to accumulate surplus capital to allow their citizens to benefit from education, improved sanitation, lower fertility rates, and the like.

Eric Lafforgue/Art in all of Us/Corbis News/Getty Images

Angola has a booming oil industry; yet many of its urban residents live in slums like this one, earning less than $2 per day. Seventeen percent of children in Angola die before age 5.

While countries' economies may improve over time, the gap between the core and the periphery countries persists (Chase-Dunn and Hall 2018). World System Theory has been credited with expanding our attention from the level of inequality between individuals or groups of individuals to inequality between and among nations. It also calls into question why core countries invest in family-planning programs in periphery countries rather than investing in programs to improve economies in these countries. The first of these strategies would limit population numbers, whereas the second would increase independence and ultimately power in periphery countries. In Chapter 13, we discuss the impact of global systems of inequality on the environment.

What do you think?

The United States and other developed countries provide the lion's share of the funding for the World Health Organization (WHO), which provides programs to eradicate polio, malaria, TB, HIV/AIDS, as well as providing basic health care in developing countries. However, in 2020, while in the midst of the COVID-19 pandemic, the Trump administration withdrew the United States from the WHO and declined to provide continued global support (Rogers and Mandavilli 2020). Do you think high-income countries like the United States should assist countries that are struggling to meet the health needs of their populations? Why or why not?

Some conflict theorists view the elderly population as a special interest group that competes with younger populations for scarce resources. Debates about funding programs for the elderly (e.g., Social Security and Medicare) versus funding for youth programs (e.g., public schools and child health programs) largely represent conflicting interests of the young versus the old. A growing elderly population means that senior citizens have increased power on political issues. In the United States, adults ages 60 and older have the highest rate of voting of any age group (U.S. Census Bureau 2017), and their growing size makes them a significant political force when they have common interests.

Symbolic Interactionist Perspective

The symbolic interactionist perspective focuses on how meanings, labels, and definitions learned through interaction affect population problems and issues concerning aging. For example, many societies are characterized by **pronatalism**—a cultural value that promotes having children. In ancient times, many religions worshipped fertility and viewed it as essential to the survival of the human race. We see this in the recognition of gods and saints to both male and female fertility throughout history (Neto et al. 2019; Behjati-Ardakani et al. 2016).

pronatalism A cultural value that promotes having children.

Today many religions continue to teach that procreation is intrinsically tied to the functions of marriage, especially women's roles in marriage (Carroll 2018; Ataullahjan, Mumtaz, and Valliantos 2019; Shain 2019). Researchers continue to find that women who are more religious tend to have more children than those who report lower levels of religiosity, suggesting that pronatalist messages continue to permeate religious teachings (Peri-Rotem 2016; Shain 2019). Women who use contraception in communities in which family planning is not socially accepted face ostracism by their community, disdain from relatives and friends, and even violence, divorce, and abandonment by their husbands (Kabagenyi et al. 2016). By contrast, women (and men) who fulfill the fertility expectations of their religious communities and social groups are reinforced for their behavior with praise, gifts, and social inclusion (Shain 2019). These beliefs, when coupled with few economic opportunities, have led to significant regions of poverty across the globe. However, symbolic interactionists and others would point out that understanding the cultural meaning of fertility in these regions is critical to any intervention efforts.

Stefan Heunis/AFP/Getty Images

Each year thousands of people attend the yearly Osun-Osugbo Festival in Nigeria to celebrate and make sacrifices to the river goddess Osun in return for fertility and good fortune.

The symbolic interactionist perspective also emphasizes the importance of examining social meanings and definitions associated with aging. "Old age" is largely a social

© Vivium

De Hogewyk is a self-contained care village in the Netherlands designed to allow people with dementia maximum freedom. It includes apartments, an outpatient care center, community center, restaurants, shops, and entertainment facilities.

construct; there is no biological marker that indicates when a person is "old." Indeed, physical quality of life varies tremendously in old age with some battling physical impairments compared to others running marathons. In the United States, and much of the world, people are considered to be "old" (i.e., senior citizens) when they reach age 65, as this is the age that company pension plans, Medicare, and Social Security have used to define when a person typically retires and collects benefits. Despite the fact that more and more people over 65 work for pay and the retirement age for receiving full Social Security benefits has been increased to 67 for those born after 1960, we continue to use 65 to define those who are considered *elderly*.

However, as longevity has increased, some are starting to make a distinction between the "young-old" (65 to 80 years of age) or "third agers" and the "old-old" (80+ years of age) or "fourth agers" (Lev, Wurm, and Ayalon 2018), the first group being those who are active and enjoying life and the second group being those who are assumed to be inactive and in decline (Loos and Ivan 2018). These new labels reflect the symbolic interactionist perspective that the cultural interpretations of symbols are dynamic across time and place.

We also learn both positive and negative meanings associated with old age, such as "wise" and "experienced," as well as "frail" and "impaired" (Kornadt and Rothermund 2010). In U.S. culture, negative labels of older people, such as "crone," "old geezer," and "old biddy," are predominant and reflect ageism—a topic we discuss later in this chapter. Negative stereotypes of aging have implications for health and older peoples' own perceptions of themselves as burdens to others.

What *do you* think?

Bette Davis, a famous Hollywood actress of the 1930s and 1940s, once said, "Getting old is not for sissies." Today, however, some are saying, "60 is the new 50." In fact, one demographer argues, "The old-age thresholds of 60 or 65 are inconsistent with the new reality in which people are living longer and healthier" (quoted in Jefferson 2017, p. 1). As life expectancy increases and people redefine 60 as today's 40, do you think that the designation of "senior citizen," "old," or "elderly" will change from 60 or 65 to a higher number? What age do you think of as the beginning of old age?

Social Problems Related to Population Growth and Aging

Social problems related to population growth include (1) poverty, unemployment, and global insecurity, (2) food insecurity and environmental problems, (3) ageism, (4) family caregiving, and (5) financial insecurity for the growing elderly population. Maternal deaths related to pregnancy and childbirth, the leading causes of mortality for reproductive-age women in developing countries, increase as fertility rates increase and thus are also associated with population growth (see Chapter 2 and Chapter 10).

Poverty, Unemployment, and Global Insecurity

In 2015, researchers at Harvard University estimated that the world economy would have to create 734 million new jobs by 2030 in order to decrease current unemployment rates, meet increasing demands for female employment, and address the anticipated

21 percent increase in the working age population (Bloom and McKenna 2015). Ninety-one percent of those jobs would need to be created in low- and lower-middle-income countries where most of the population growth is occurring.

Youth unemployment is a particular problem in many countries with rates exceeding 30 percent in, for example, Greece, Italy, Spain, Iran, and Saudi Arabia, and many countries in sub-Saharan Africa (United Nations 2019b). By contrast Japan, which is already experiencing population decline due to low fertility, has had to relax its immigration laws because the number of new workers is not keeping pace with the number of older workers who are retiring.

Poverty is a significant problem in countries with high population growth. The relationship is reciprocal: Poverty affects population size and population size affects poverty. High fertility exacerbates poverty because families have more children to support and the natural resources available are strained to meet the needs of the growing population. On the other hand, in an impoverished country people cannot invest in education, economic growth, and health care, which are key requirements for increasing rights for women and increasing contraceptive use. Economic opportunities are limited and wages are low. In many of these countries even young people who are working are earning poverty-level wages (United Nations 2019b).

Rapid population growth is a contributing factor to global insecurity including civil unrest, war, and terrorism (Weiland 2005). Although world population is, overall, aging, some countries in Africa and the Middle East are experiencing a "youth bulge"—a high proportion of 15- to 29-year-olds relative to the adult population. Youth bulges result from high fertility rates and declining **infant mortality rates** or deaths among children in the first year of life, a common pattern in developing countries today. The combination of a youth bulge with other characteristics of rapidly growing populations, such as resource scarcity, high unemployment rates, poverty, and rapid urbanization, set the stage for political unrest (Pritchard and Pakes 2014).

> Large groups of unemployed young people, combined with overcrowded cities and lack of access to farmland and water, create a population that is angry and frustrated with the status quo and thus is more likely to resort to violence to bring about change (Weiland 2005, p. 3).

Population growth is straining the Earth's resources to the breaking point, and educating girls is the single most important factor in stabilizing that. That, plus helping women gain political and economic power and safeguarding their reproductive rights."

–AL GORE, FORMER VICE PRESIDENT

Food Insecurity and Environmental Problems

About 690 million people in the world, 9 percent of the population, suffer from **undernourishment** while one in four people are moderately to severely **food insecure**; i.e., they don't have reliable sources of food (United Nations 2020b). In 1798, Thomas Malthus predicted that the population would grow at a faster rate than the food supply and that masses of people were destined to be poor and hungry. According to Malthusian theory, food shortages would lead to war, disease, and starvation which would eventually slow population growth. Will population growth continue to contribute to food insecurity and world hunger before it hits its peak?

On the one hand, data demonstrate that countries with the most population growth have the greatest problems with hunger as measured by the **Global Hunger Index**. The Index takes into account four indicators related to hunger: undernourishment, **child wasting**, **child stunting**, and child mortality. Child wasting is a condition in which children weigh less than they should for their age while child stunting is when a child is shorter than typical for their age. Both are products of undernourishment and are predictors of poor health and child mortality. This seems to support Malthus's perspective—as Figure 12.8 indicates, the greater the population growth the higher the hunger index.

However, incidences of deaths due to famine have not increased as population size has increased. In fact, the largest number of deaths from famines took place in the late 1800s. Moreover, famine deaths have become less common in the last 50 years when population growth has been high. Currently about 300,000 people per year die of undernourishment globally, compared to 3.5 million in the 1970s. While hunger is highest in countries with the most population growth, it has also improved the most in these

infant mortality rates The number of deaths among children in the first year of life per 1,000 live births in a calender year.

undernourishment Having a caloric intake that is insufficient to meet minimum energy requirements.

food insecure A situation in which people lack secure access to sufficient safe and nutritious food for normal growth and development and an active life.

Global Hunger Index A measure of hunger in a country that takes into account four indicators related to hunger: undernourishment, child wasting, child stunting, and child mortality.

child wasting A condition in which children weigh less than they should for their age.

child stunting A condition in which a child is shorter than is typical for their age.

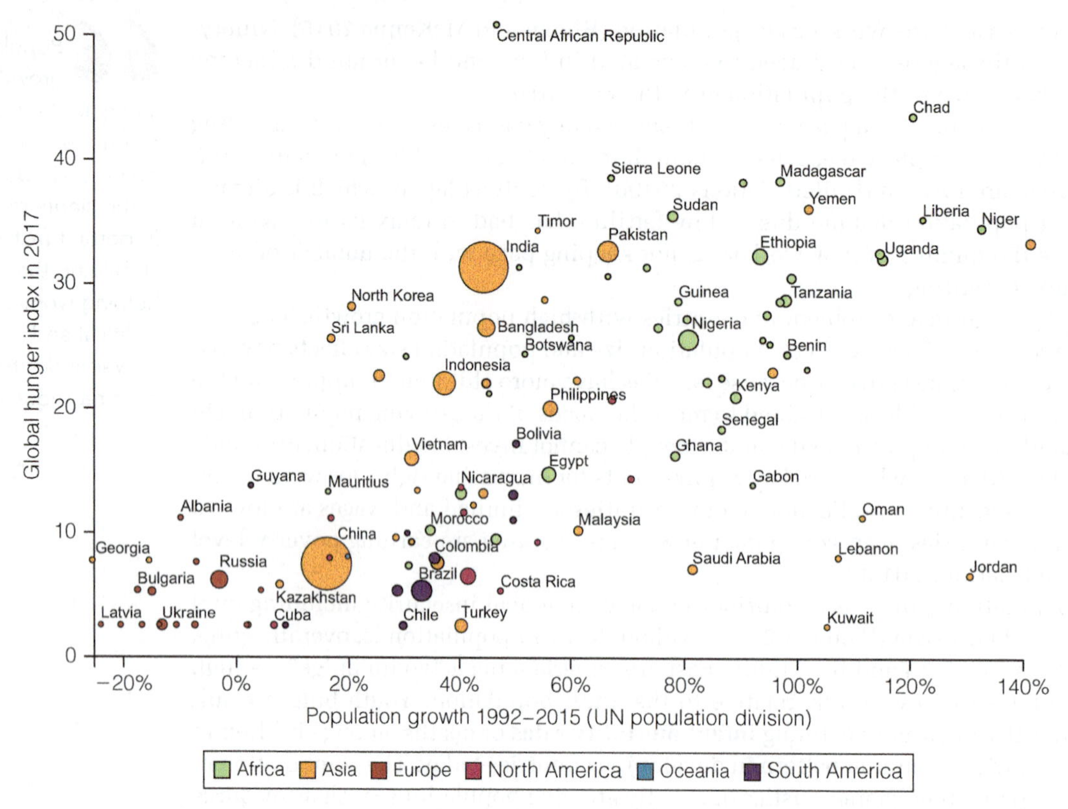

Figure 12.8 Global Hunger Index in 2017 versus Population Growth, 1992–2015
Note: The size of the circle represents the relative population size of the country.
SOURCE: Hassell 2018.

countries (Hassell 2018). Most contemporary famines have been the result of political oppression, corruption, or conflict, not food insecurity *per se*. Furthermore, most scholars agree that the world has enough resources to support its population; however, some regions like North America and Europe have surpluses, while others live with scarcity (e.g., sub-Saharan Africa and Southern Asia).

What do you think?

Before the COVID-19 pandemic, 40 million households in the United States were already struggling to put food on the table. About a third of college students also reported that they cut their portions or skipped meals due to lack of money for food (Lee 2020). In the meantime, grocery stores and restaurants were throwing out surplus food at the end of each night. What do you think could be done to end food insecurity in your community?

Environmental problems, including climate change, pose a threat to food security, and population growth contributes to many of today's environmental problems. For example, according to the United Nations Environment Programme (UNEP), in 2025, two-thirds of the world's population will be living in countries with water scarcity. Without intervention the world's fisheries will be depleted by the middle of this century (Engelman 2011).

Further, the demand for farmland has never been higher with approximately 40 percent of the Earth's land, an area the size of the continent of South America, already given over to crop production (Harris 2017). The countries that suffer most from shortages of water, farmland, and food are developing countries with the highest population growth rates. However, countries with the largest populations do not necessarily have the largest negative impact on the environment. This is because the demands that humanity makes on the Earth's natural resources—each person's **environmental footprint**—are determined by the patterns of production and consumption in that person's culture. The environmental footprint of an average person in a high-income country is much larger than that of someone in a low-income country. Hence, although population growth is a contributing factor in

environmental footprint
The demands that humanity makes on the earth's natural resources.

environmental problems, patterns of production and consumption are at least as important in influencing the effects of population on the environment (see Chapter 13).

There is a body of research suggesting that, in the long run, it isn't the large growth countries that need to be concerned about food insecurity but rather those with declining populations. With fewer children being born as the population ages there will be fewer workers to produce food (Naso, Lanz, and Swanson 2020). These countries will have to find a way to meet labor demands, either through innovation, immigration, or increasing fertility rates, or their populations will starve.

Justin Sullivan/Getty Images News/Getty Images

Food insecurity is not limited to developing countries. In the United States, 5.3 million seniors do not have consistent access to food, like these seniors lined up at a food bank in San Francisco.

Ageism: Prejudice and Discrimination toward the Elderly

Ageism refers to negative stereotyping, prejudice, and discrimination based on a person's or group's perceived chronological age. Ageism is a global phenomenon. For example, in a survey of adults from 57 countries, 60 percent reported they did not believe older adults were well respected in their countries, with "respect" being the lowest in high-income countries such as the United States (Officer et al. 2016).

A number of factors contribute to ageism. Although historically elders held a revered status as the repositories of culture (e.g., stories, knowledge); with the invention of the printing press and other method of communication that status was lost (Nelson 2011). Ageism also stems from fear and anxiety surrounding the aging process and, ultimately, death (Lev, Wurm, and Ayalon 2018). Americans are uncomfortable discussing death and don't like to acknowledge that death is a natural part of the life cycle. We use euphemisms to avoid talking about death: We say someone "passed away," "has gone to a better place," "is resting in peace," or "has departed," and so on. Old people are reminders of our own mortality and, as such, take on negative social meanings. Before reading further, you may want to complete the "Ageism Survey" in this chapter's *Self and Society* feature.

Ageism is reflected in negative stereotypes of the elderly. They are often labeled as slow, set in their ways, grumpy, poor drivers, unwilling or unable to learn new things, and physically and/or cognitively impaired (Nelson 2011). Old age is stereotypically viewed as a negative time in one's life wherein physical and cognitive ability declines and dependence on others increases. People are viewed as aging "successfully" when they are perceived to have few or none of these attributes (Lev, Wurm, and Ayalon 2018). As a result of ageism, older individuals are spoken to differently—with raised voices and in simple terms as the speaker assumes deafness and cognitive decline. Younger people rush to "help" the elderly with all good intentions but often spurred by cultural stereotypes of weakness and frailty. Older adults may also be denied employment because of their age just as others are denied work because of less acceptable "isms" such as racism, sexism, and heterosexism (Kita 2019).

ageism Negative stereotyping, prejudice, and discrimination based on a person's or group's perceived chronological age.

Imagine a world in which a rapidly spreading virus is mostly infecting people under the age of 50. Now, imagine that the death toll is highest among children and that, as of today, the United States had reported more than 28 million confirmed cases and over 500,000 deaths—mostly among elementary-age children. Imagine that scientists suspect elders are at lower risk based on past exposure to similar viruses. How would you react to a disease that, worldwide, was disproportionately killing young people? Is it different from how you've felt about the COVID-19 pandemic? Why?

What *do you* think?

SELF and society Facts on Aging Quiz

Circle "T" (True) or "F" (False) next to each statement without looking at the answers. Check your answers at the end to assess your knowledge of the elderly population in the United States.

1. Research has shown that old age truly begins at 65. T F
2. As people grow older, their intelligence declines significantly. T F
3. It is very difficult for older adults to learn new things. T F
4. As adults grow older, reaction time increases. T F
5. All medical schools require students to take courses in geriatrics and gerontology. T F
6. Physical strength declines in old age. T F
7. Retirement is often detrimental to health—i.e., people frequently seem to become ill or die soon after retirement. T F
8. Most older people are living in nursing homes. T F
9. The modern family no longer takes care of its elderly. T F
10. Most older drivers are quite capable of safely operating a motor vehicle. T F
11. Grandparents today take less responsibility for rearing grandchildren than ever before. T F
12. Most older adults consider their health to be good or excellent. T F
13. Poverty is no longer a significant problem for most older Americans. T F
14. Depression is more common among older than younger people. T F
15. Most old people lose interest in and capacity for sexual relations. T F

How many did you get correct?
10–15: Excellent!
9 or less: Learning more about the elderly will help you to work in any social setting, given the aging of our population.

Answers:

1. F; 2. F; 3. F; 4. T; 5. F; 6. T; 7. F; 8. F; 9. F; 10. T; 11. F; 12. T; 13. F; 14. F; 15. F

SOURCE: Breytspraak and Badura 2015

Visual Ageism. One specific type of ageism—**visual ageism**—occurs when older adults are underrepresented in the media, or represented in prejudiced or stereotypical ways. In the last two decades the frequency of media portrayals of the "young-old" have increased and have been more positive (Loos and Ivan 2018). For example, advertising has diversified its models to include older women. In 2020, *Sports Illustrated* made the news for having a 56-year-old swimsuit model in its iconic swimsuit edition. Other models, such as Catherine Deneuve, Daphne Selfe, Jenni Rhodes, and Jan de Villeneuve, continue to work and are over 70 years old. This may reflect a cultural shift, but critics point out that these older models don't look their age (Karpf 2014). Furthermore, these models are all thin, are discouraged from looking "too sexy," and are often used specifically to dispel an age stereotype rather than in an "age neutral" manner (Karpf 2014).

> "My face carries all my memories. Why would I erase them?"
>
> —DIANE VON FURSTENBERG, FASHION DESIGNER

Jacky O'Shaugnessy, who modeled for American Apparel at the age of 62, was posed fully clothed with her legs wide open. She describes the response she received from her pose: "A morning talk show host said it was distasteful to see a woman my age—love that!—in that position. Here, I thought, was rampant ageism" (Quoted in Karpf 2014).

Older-old adults continue to be underrepresented in the media relative to their representation in the population (Loos and Ivan 2018). This is an example of **ageism by invisibility**. Although Hollywood makes more movies about the elderly than ever before, when they are represented, they are typically depicted as burdens and rarely in major television or movie roles (*The Economist* 2017).

visual ageism Underrepresentation of older adults in the media or representation in prejudiced or stereotypical ways.

ageism by invisibility The underrepresentation of older adults in advertising and educational materials.

Visual ageism reinforces our negative view of the outward manifestation of aging—wrinkles, gray hair, and a stooped appearance. In the United States, 7 million wrinkle reduction treatments (e.g., Botox) occurred in 2016 for an average cost of $385 totaling more than $2.5 billion (Rossman 2017). Millions of Americans purchase products or treatments to make themselves look younger, spending substantial sums of money and undergoing unnecessary and often risky medical procedures all in the search for the fountain of youth.

Ageism in Medicine. Ageism is not limited to the general public. Despite widespread assumptions that the elderly struggle with health problems, medical schools in the United States provide little training on the specific health needs of people over 65. Aspiring physicians receive weeks of training on pediatric care, years on adults, in general, but there is no required training in **geriatrics**; the branch of medicine focused on health care for the elderly (Aronson 2020). The American Geriatrics Society estimates that the country needs 14,000 more geriatricians to adequately address the health needs of the elderly community in the United States today and 24,000 more by 2030 (American Geriatrics Society 2020). There is also a significant need for geriatric nurses, nurse practitioners, social workers, and pharmacists.

Ageist assumptions have had a significant impact on how the elderly and their caregivers have been treated during the COVID-19 pandemic. In late April 2020, the governor of California advised hospitals to prioritize younger people for care during the coronavirus pandemic based on the incorrect assumption that elderly people are less healthy than younger people. The governor quickly retracted his statement after he was roundly criticized on social media (Luna 2020). But his statement was not without precedent. Arguing for lessening restrictions in March 2020, Texas Lieutenant Governor Dan Patrick suggested grandparents should be willing to risk their lives to save the American economy for future generations (Morris and Garrett 2020). On social media, #BoomerRemover in reference to COVID-19 gained popularity in 2020 (Lichtenstein 2020). Some have even argued that the elderly should not be counted in COVID-19 death statistics because "they were going to die anyway" (Woodward 2020).

The low priority given to the elderly population has been evident in Washington, D.C. as well. Although tens of thousands of assisted living and nursing home residents had tested positive for COVID-19, just three percent of a $175 billion federal package, $5 billion, was directed toward helping health care providers in these facilities (Rau 2020). As a result, by mid-June 2020, many nursing home workers had only cloth or used masks and were wearing plastic ponchos or trash bags to work, which contributed to the deaths of over 600 staff members (Whorisky et al. 2020). As of July, 2020, approximately 45 percent of all COVID-19 deaths in the United States were residents in nursing homes and assisted living facilities. By contrast, in Hong Kong, which had quickly implemented strict containment measures, there had been no COVID-19 deaths in nursing homes by that time (Khazan 2020).

> "The pandemic has led to a lot of language that homogenizes older people and also treats them as if they are "the other"—the group that should be walled off."
>
> **–DR. BECCY LEVY, YALE SCHOOL OF PUBLIC HEALTH**

In spite of the negative views of aging, research across 160 countries documents that in high-income, English-speaking countries such as the United States, people get happier, less stressed, and more satisfied with their life as they age. It is people between the ages of 45 and 54, those who are considered middle-aged, who are weak on these dimensions of well-being (Steptoe, Deaton, and Stone 2015). Although 77 percent of the elderly have two or more chronic health conditions (American Psychological Association 2020), it is not until the age of 85 that some of the elderly, only 20 percent, require assistance with everyday life activities. Even after the age of 85, only 40 percent of men and 53 percent of women require assistance.

Many people in their later years are productive and fulfilled, engaging in paid work, volunteering, caring for grandchildren, participating in physical exercise, hobbies, and the like (see Table 12.1). In the 2020 U.S. presidential election, both candidates were over the age of 70. How the elderly fare in their later years is a product of genes, physical and cognitive engagement throughout their life, and the long-term consequences of inequality earlier in life.

Employment Age Discrimination Nearly half of U.S. "retirees" are currently working, have worked, or plan to work during retirement, contributing to an increase in the number of adults over the age of 55 still in the workforce (Gerontological Society of America [GSA] 2018) (see Figure 12.9). More than one-third of non-retired U.S. adults expect to retire *after* age 65, either out of financial necessity or by choice (Riffkin 2015). In 1967, Congress passed the *Age Discrimination in Employment Act* (ADEA) but, more than 50 years later, older workers continue to face age discrimination in the workplace.

geriatrics The branch of medical care focused on health care for the elderly.

TABLE 12.1 Accomplishments of Famous Older Individuals

At 64, Diana Nyad was the first person to swim the 110 miles from Cuba to Florida without a shark cage. It took her 53 hours.
At 72, feminist author Betty Friedan published *The Fountain of Age*, where she debunks misconceptions about aging.
At 75, Nelson Mandela was elected as the first president of a democratic South Africa.
At 81, Benjamin Franklin facilitated the compromise that led to the adoption of the U.S. Constitution.
At 81, actor Denzel Washington received the Screen Actors Guild Lifetime Achievement Award. In the following three years he appeared in six more movies.
At 85, actress Mae West starred in the film *Sextette*.
At 85, Coco Chanel was the head of a fashion design firm.
At 87, Mary Baker Eddy created the newspaper *Christian Science Monitor.*
At 88, actress Betty White became the oldest person to ever host *Saturday Night Live*.
At 89, Doris Haddock, also known as "Granny D," began a 3,200-mile walk from Los Angeles to Washington, D.C., to raise awareness for the issue of campaign finance reform. She walked 10 miles a day for 14 months and completed her cross-country walk at age 90.
At 90, Pablo Picasso was producing drawings and engravings.
At 90, Harry Belafonte acted in *Spike Lee's BlacKkKlansman*, playing an elderly civil rights activist—a role mirroring his own history of working with Martin Luther King, Jr.
At 94, philosopher Bertrand Russell was active in promoting peace in the Middle East.
At 100, Grandma Moses, noted for her rural American landscapes, was painting. She only started painting at age 78, but by the time she died at 101, she had created over 1,500 works of art.

A survey by the American Association of Retired Persons (AARP) found that one in four people over the age of 45 had been subjected to negative comments about their age from supervisors or coworkers (Kita 2019). In an interview with an AARP reporter, an attorney specializing in employee rights commented, "Age discrimination is so pervasive that people don't even recognize it's illegal" (quoted in Kita 2019). A study that sent 40,000 fictitious applications for over 13,000 positions in 12 American cities found that older adult applications were the least likely to receive a callback across a variety of different job types when compared to middle-aged and young adults (Neumark, Burn, and Button 2017). The pattern was especially true for female applicants, i.e., age mattered more for women than for men. Similarly, Carlsson and Eriksson (2019) found more age discrimination against female applicants than male, and across an even wider range of occupations with callbacks being rare among all applicants close to retirement age.

Older workers may be more vulnerable to being "let go" because, although they have seniority on the job, they may also have higher salaries that include costly benefits packages. Businesses that need to cut payroll expenses can save more money by letting higher-salaried personnel go (Gosselin 2018). IBM took this strategy between 2013 and 2018 when they dismissed over 20,000 employees age 40 and older and replaced them with younger, less experienced, and lower-paid employees (Gosselin and Tobin 2018). The company mostly avoided scrutiny under anti-age-discrimination laws by forcing employees to sign away their rights in order to receive severance pay or forcing them to take voluntary retirement rather than risk being laid off without benefits.

Older adults who lose their jobs stay unemployed longer than younger job seekers and typically end up in new jobs earning less than they earned before leading to significant financial hardship (Gosselin 2018). Prospective employers may view older job applicants

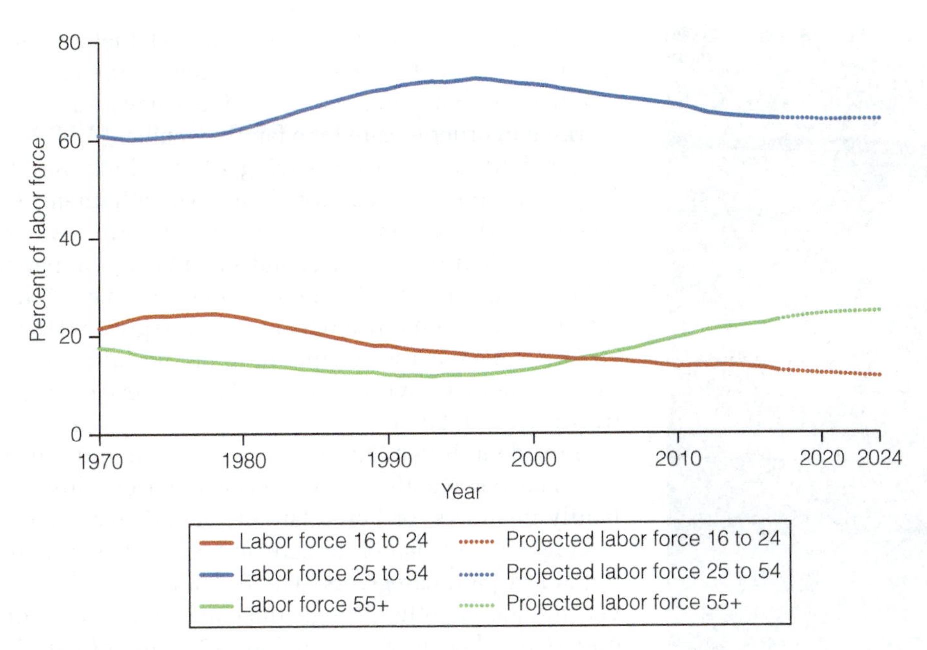

Figure 12.9 Trends in U.S. Labor Force by Age, 1970–2024
SOURCE: Toossi and Torpey 2017.

as "overqualified" for entry-level positions, less productive than younger workers, and/or more likely to have health problems that could affect not only their productivity but also the cost of employer-based group insurance premiums. Employers also perceive older workers to be less able to learn new tasks, less flexible, and less ambitious (Carlsson and Eriksson 2019).

Family Caregiving for Our Elders

Many adults have or will provide care and/or financial support for aging spouses, parents, grandparents, and in-laws. Although caring for older adults has been an important function of the family for thousands of years, due to social, economic, and cultural changes, it has become more difficult for families throughout the world to care for aging relatives. Across the globe, caregiving most often falls on women and, as women's employment has expanded, many of these caregivers are combining paid work outside the home with the unpaid work of caregiving. Increased migration of family members for employment also means caregiving is not as likely to be shared across multiple family members.

sandwich generation A generation of people who care for their aging parents while also taking care of their own children.

In the United States, about a quarter of caregivers are elderly adults taking care of a spouse (see Figure 12.10). Another half are children taking care of aging parents (GSA 2018). Due to delayed childbearing, an increasing number of adults who care for aging relatives are also taking care of their own children; referred to as members of the **sandwich generation** where they are "sandwiched" in between providing care for both parents and children. An increasing number of caregivers, 40 percent, live with the individual(s) for whom they are responsible (AARP 2020; National Alliance for Caregiving 2020). Less than 50 percent of caregivers have another unpaid person who helps them with their caregiving responsibilities. For those with help, the unpaid assistance comes from a child under age 18. Only three in ten caregivers have outside paid help. Family caregiving in the United States is estimated to represent $500 billion in unpaid labor each year (GSA 2018).

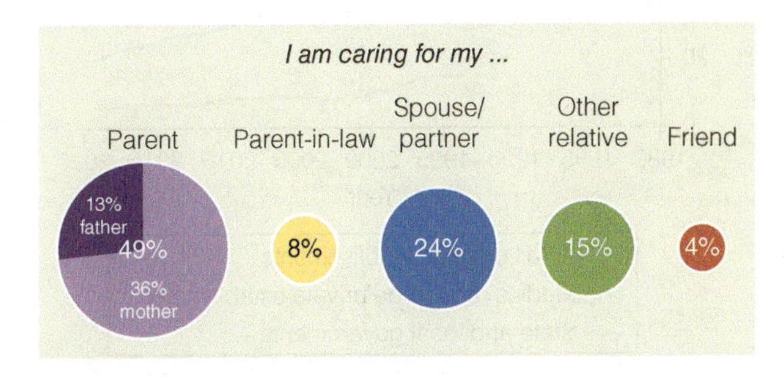

Figure 12.10 Unpaid Caregiving of Adults, United States, 2017
SOURCE: Gerontological Society of America 2018.

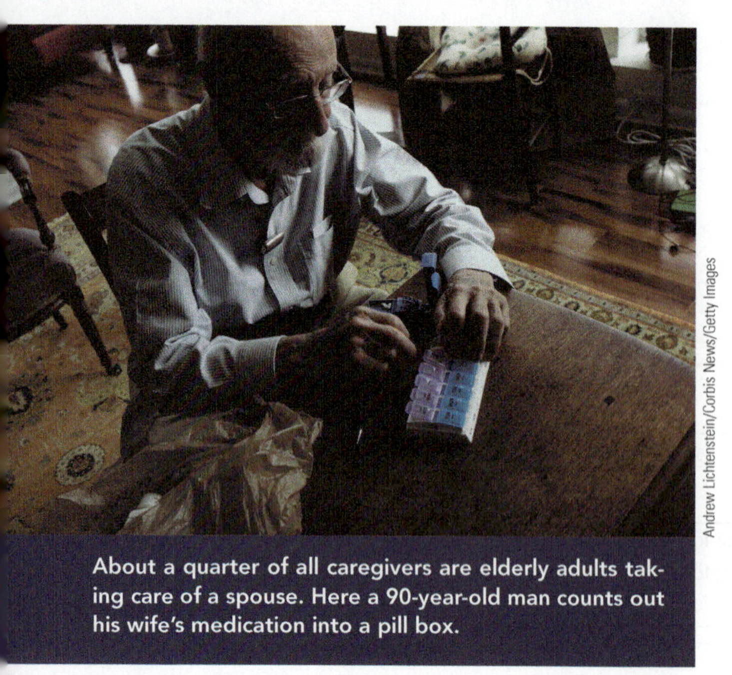

Andrew Lichtenstein/Corbis News/Getty Images

About a quarter of all caregivers are elderly adults taking care of a spouse. Here a 90-year-old man counts out his wife's medication into a pill box.

Of increasing concern is the impact that caregiving and its associated stressors are having on the caregiving population. Ten percent of caregivers have had to stop working in order to care for a family member (AARP 2020; National Alliance for Caregiving 2020) which may have long-term consequences for their own retirement. Caregivers report financial strain with less opportunity to contribute to their own savings and some taking on personal debt. Caregivers are also more prone to depression and insomnia than are those who are not caregivers (Hopps et al. 2017), even across people in high-, middle-, and low-income countries with a variety of social services systems (Koyangi et al. 2018).

President Biden has proposed several initiatives to help families whether they are caring for children, older family members, or both. The president's plan does not distinguish between child care and elder care or between unpaid or paid caregivers. The Biden administration advocates, among other things, the expansion of the *Family and Medical Leave Act* to include 12 weeks of paid leave and a proposed tax cut "for as much as $5000 to reimburse families for expenses associated with unpaid caregiving" (Span 2020). The Biden plan would also give family members Social Security credits for the time they spend out of work caring for family members.

@elizabeth rosner

Breaking news: full-time caregiving for an increasingly frail parent is definitely not for sissies.

-Elizabeth Rosner

Financial Security of Older Americans

Financial security in retirement has traditionally been thought of as a "three-legged stool," with retirees relying on a combination of social security, pension benefits, and/or individual savings/assets to subsidize their needs in retirement. However, each leg of this stool is becoming less stable. In 2018, 9.7 percent of elderly adults in the United States were living in poverty (see Chapter 6). Poverty is particularly high among seniors who are members of minority groups with 18.8 percent of Black and 19.5 percent of Hispanic seniors living below the poverty threshold (Romig 2020a). More than 40 percent of U.S. working adults expect they will not have enough money to live comfortably in retirement (Brenan 2019).

Decreasing Availability of Pension Plans Traditional pensions, which are **defined benefit plans** in which retirees receive a specified annual amount until their death, are becoming less common among businesses and government agencies. Between 1983 and 2016, the percentage of workers covered by traditional pension plans decreased from 62 percent to 17 percent (Rutledge and Sanzenbacher 2019) (see Figure 12.11). Further, many pension plans have been under-funded or mismanaged leaving millions of retirees vulnerable to cuts in benefits.

The federal Pension Benefit Guaranty Corporation (PBGC) insures pensions when a private pension plan fails. However, it guarantees individual payments only up to a certain maximum amount which may be as low as 20 percent of an original plan. Thus, a retired worker could still find that their retirement benefits have been significantly reduced if the PBGC needs to take over (Sword 2020). Some 125 pension plans insured by the PBGC are projected to run out of money by 2040. These plans cover 1.3 million

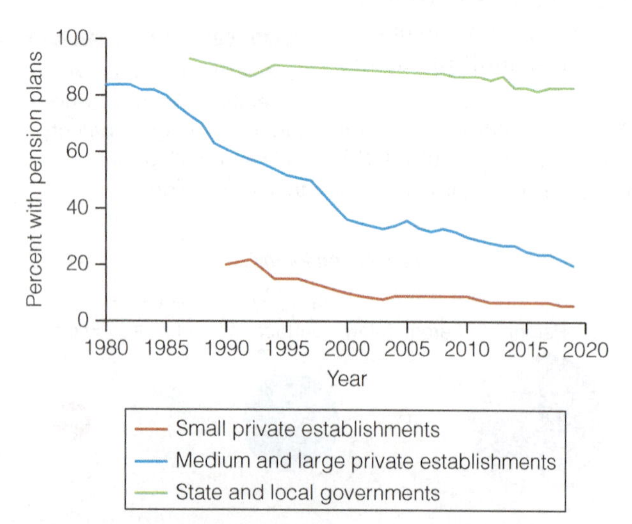

Figure 12.11 Percentage with Defined Pension Plans, United States, 1980–2020
SOURCE: Van De Water and Romig 2020.

members. Sadly, it is estimated that the PBGC will run out of funds by 2025 unless some action is taken soon.

State pension plans are generally more secure than private employer plans, although a handful of states have underfunded their pension plans significantly over the last two decades. For example, in 2017, the state of Kentucky's pension fund was only 33.9 percent funded. Legislators sought to address the issue by cutting the pension benefits of current state employees but were stopped by the state Supreme Court on a procedural technicality (Walker 2018). Puerto Rico's state pension is less than 2 percent funded. Whether and to what extent individual workers' state pension payments are protected varies from state to state depending on whether pensions are guaranteed by the state's constitution, by contract law, by property law, or by other means. In at least 21 states, workers' future pension benefits can be altered (Ergungor 2017), meaning an employee could work for decades believing they are working toward a financially secure retirement only to find the state has cut the retirement income they were promised when they were hired.

Drew Angerer/Bloomberg/Getty Images

Millions of Americans are in danger of having their pension benefits cut or reduced. The House of Representatives passed a bill in July 2020 to shore up pension funds, but Republicans in the Senate refused to bring it to the floor for discussion.

Increasing Reliance on Defined Contribution Plans In place of pension plans that provide a specific amount of money after retirement, many public and private employers have changed to what are called **defined contribution plans**. Seventy-three percent of workers are now covered by defined contribution plans, up from 12 percent in 1983 (Rutledge and Sanzenbacher 2019). In these plans, workers contribute money to 401(k) plans or individual retirement accounts (IRAs) without any guarantee of what their future benefits will be. In many cases, employers will match worker's contributions up to a certain percentage of income, thus doubling the contribution at no cost to the employee. Individuals can also open up their own investment accounts for retirement independent of their employer.

Defined contribution retirement plans (i.e., 401[k]s and IRAs) are risky because they involve investments in the stock market, and thus their values fluctuate. During the Great Recession of 2007–2009, the stock market was volatile, producing losses that decimated the IRAs and 401(k)s of millions of older Americans. Many workers who were planning to retire could no longer afford to stop working, and many others who were recent retirees had to reenter the labor force (Population Reference Bureau 2015). Economic downturns can also lead companies to reduce or eliminate their employer contributions to these plans.

By 2020, more than 64 private companies, including BestBuy, Dell, Choice Hotels, La-Z-Boy, and Stein Mart, had suspended their 401(k) matching contributions as a way of coping with reduced revenues during the COVID-19 pandemic (Center for Retirement Research 2020). Many colleges and universities across the country also announced temporary suspensions on retirement contributions (Flaherty 2020). These changes reflect how much easier it is for employers to cut costs of defined contribution plans as opposed to defined benefit plans. While preferable to layoffs, these cuts will have long-term consequences for workers' assets going into retirement. Further, the brunt of the impact of the economic downturn on retirement savings has to be absorbed by households under defined contribution plans. In contrast, employees of private companies and local, state, or federal governments under defined benefit plans are not impacted as severely.

A Teetering Social Security System **Social Security**, actually titled *Old Age, Survivors, Disability, and Health Insurance,* is a federal insurance program established in 1935 that protects against loss of income due to retirement, disability, or death. The amount

defined benefit plan A retirement plan in which retirees receive a specified annual amount until their death, typically calculated on the basis of their highest salary and the number of years they have worked for the employer.

defined contribution plan A retirement plan in which workers contribute money to 401(k) plans or individual retirement accounts (IRAs), without any guarantee of what their future benefits will be. Employers may match worker's contributions up to a certain percentage of income.

Social Security Also called "Old Age, Survivors, Disability, and Health Insurance," a federal program that protects against loss of income due to retirement, disability, or death.

@deasy_ diane

Women could be damaging their state pension entitlement by taking a break to look after children ... oh so its women's fault 😡 #CEDAWPeoplesTribunal #50sWomen

-Diane D 1959 #50sWomen

people receive from Social Security is based on how much they earned during their working years—higher lifetime earnings result in higher benefits. Benefit payments also depend on the age at which a person retires. The minimum age for receiving full benefits was 65 for many years but, in 1983, Congress phased in a gradual increase in the full retirement age from age 65 to age 67. People born in 1960 and later are subject to the new retirement age of 67 (Social Security Administration 2020a). Retirees can claim reduced benefits as early as age 62; however, they receive a larger benefit if they wait until age 70. In the United States, spouses may choose to receive their deceased partners' benefits if those benefits are higher than their own regardless of their own work histories and Social Security contributions. A divorced spouse also has some rights to social security benefits earned by their former partner if the marriage lasted at least ten years.

As of mid-year 2020, the average monthly Social Security benefit to retired workers was $1,513, or $18,156 annually (Brandon 2020). When the program was developed, Social Security was not intended to be a person's sole economic support in old age but rather supplemental along with other savings and assets. Today, however, for many older Americans, it is their major source of support. In fact, for over half of Americans age 65 and older, Social Security provides more than half of their income.

Without Social Security, more than a third of all seniors would be living in poverty (see Table 12.2). Instead, as noted in Chapter 6, poverty rates for U.S. adults ages 65 and older are lower than for any other age group. However, because Social Security payments are based on the number of years of paid work and pre-retirement earnings, women and minorities, who often earn less during their employed years, average less in retirement benefits. Even so, without Social Security more than half of Black seniors and 47.5 percent of Hispanic seniors would be living in poverty. This chapter's *The Human Side* provides some perspective on what it is like to be elderly and homeless in the United States today.

Forty-two percent of U.S. working adults believe that Social Security will not provide them with any retirement benefits, while another 42 percent believe benefits will be reduced by the time they retire (Parker, Morin, and Horowitz 2019). A number of factors threaten the long-term ability of the Social Security system to meet its financial obligations to future retirees. These include (1) the retirement of the Baby Boomers, (2) increasing life expectancy, (3) the declining elderly support ratio, that is, fewer

TABLE 12.2 Effect of Social Security on Elderly Poverty by Sex and Race, 2018

	Percent in Poverty		
Demographic Group	Excluding Social Security	Including Social Security	Number Lifted Out of Poverty by Social Security
Sex			
Men	33.6%	8.1%	6,104,000
Women	41.3%	11.1%	8,707,000
Race/Ethnicity			
White	35.4%	7.3%	11,287,000
Black	50.5%	18.8%	1,525,000
Latino	47.2%	19.5%	1,260,000
Other	35.6%	12.7%	738,000
Total, Age 65+	**37.8%**	**9.7%**	**14,810,000**

SOURCE: Romig 2020.

Elderly and Homeless in the United States

The homeless population in the United States is aging. People over age 50 now make up about one-third of the homeless population (Culhane et al. 2019). Many have been on the streets for a generation, while others have more recently been forced to learn to live on the streets after losing jobs in the last recession. Some simply don't have enough income on Social Security to afford food, medication, and housing. It remains to be seen how many more senior citizens will join their ranks as a result of job losses during the COVID-19 pandemic. Programs designed to assist the homeless struggle to address the specific needs of an aging population, especially as homeless people tend to have health issues frequently seen in people 20 years older (Culhane et al. 2019). Here are the voices of elderly homeless people in America collected by Invisible People, a nonprofit group that educates about and advocates for homeless people around the world.

> I've been homeless for approximately five years now due to health conditions. I'm drawing my Social Security, but it's not enough to pay for my rent and my medications right now. So, I'm on the street until I can find the resources necessary for help. Problem is there are very limited resources. ... I was just released from the hospital yesterday. I had an aortic abdominal aneurism that they had to fix. ... I'm back out here because I had no other choice. Normally I try to sleep during the day because I can't find a place to sleep at night that's safe.
>
> —David (66-year-old homeless man in Denver)

> It's been about two years. I came up here to see my daughter and she didn't want to see me. ... My wife died of brain cancer. ... Charlene, the love of my life. It hurts me so bad. ... I come over here to Union Station during the day 'cause I got friends over here. And at night, Second and D, I got a place to sleep. But the bed bugs. ... You know I wish I was dead.
>
> —Bruce (homeless man in D.C. who was a lawyer for 33 years and sleeps in a homeless shelter)

> I found a luxury little apartment right here in downtown Harrisburg, perfect little area, right in downtown, 700 dollars a month all utilities included. ... I could afford that, pay it, but can I get it because I'm homeless? Look at me. ... "[A]ffordable housing" is not affordable, the housing that is affordable, they don't want you. What are you supposed to do? I don't know. ... So I'm living in a tent ... trying to stay safe and trying to stay warm. ... You know I get tired, I get tired of walking and carrying shit around. I really do, I'm tired of it. ... I don't have a give up spirit ... even when I feel like I just can't do this anymore, which I have felt in the last couple of days, quite frankly.
>
> —Eileen (61-year-old homeless woman in Harrisburg, Pennsylvania, who worked in theater)

SOURCE: Invisible People https://invisiblepeople.tv/.

workers per Social Security beneficiary, (4) high rates of unemployment, and (5) wage stagnation.

Social Security is funded by workers through a payroll tax called the *Federal Insurance Contributions Act* (FICA) that comprises 12.4 percent of a worker's wages (6.2 percent is deducted from the worker's paycheck and 6.2 percent is paid by the employer). Self-employed workers pay the entire FICA tax. Another source of funding for Social Security is a tax on higher-income beneficiaries. Under tax laws passed in 2019, Social Security benefits are taxed for all but the lowest-income recipients, those living on less than $25,000 a year (Social Security Administration 2020b).

Social Security funds are held in a trust fund and invested in securities guaranteed as to both principal and interest by the federal government. Social Security benefits are paid out of this trust fund. Since 1984, surpluses have been accumulating in the Social Security trust fund creating significant reserves for the Baby Boomers' retirement. It is anticipated that this surplus will be depleted by 2034, leaving the system reliant on taxes paid by current workers from that point forward unless long-term changes are made (Social Security Board of Trustees 2020).

Where Do Today's Workers Stand in Terms of Retirement? Only a quarter of working adults and adults with working spouses in the United States believe they are saving enough for old age. Eighteen percent report they have saved nothing for retirement (see Figure 12.10; Brenan 2019). Experts agree that Americans are not saving enough to prepare for retirement, with one study indicating that the collective gap between what has

Evan Hurd Photography/Corbis Historical/Getty Images

As a result of gender inequality in pay and interruptions in their work histories for caregiving, experts believe the average woman between the ages of 60 and 64 needs to save an additional $310,635 to be prepared for retirement.

been saved and what needs to be saved totals $3.83 trillion (VanDerhei 2019). Women are particularly at risk, with the average *annual* retirement shortfall for 60- to 64-year-old single women, not including widows, averaging $62,127 compared to $24,905 for single men (see Chapter 10). In October 2018, Kara M. Stein, then a member of the U.S. Securities and Exchange Commission, commented on the "retirement crisis":

> The retirement crisis is a tsunami that is rapidly approaching. We can already see it and, indeed, we are starting to feel its effects. Americans are having to work past traditional retirement age. And the number of bankruptcies for those over the age of 65 has increased dramatically. The size and speed of the tsunami is likely to increase as it gets closer and closer to us. Our population is aging and the cost of medical care—an important factor for retirees—is increasing. We must address this problem before we are collectively underwater ... If these trends continue, the three-legged stool may look more like the Leaning Tower of Pisa.

Strategies for Action: Responding to Problems of Population Growth and Aging

Next we look at efforts to curb population growth as well as efforts to increase population in countries experiencing population decline. We discuss the important role of economic growth on fertility in both high- and low-growth countries. We end the chapter with a discussion of strategies that address the social problems associated with aging.

Efforts to Curb Population Growth: Reducing Fertility

Although worldwide fertility rates have fallen significantly since the 1970s, they are still high in many less developed regions. Approaches to reducing fertility in high-fertility regions include family planning, improving the status of women, providing access to safe abortion, coercive government strategies, and voluntary childlessness.

Family Planning and Contraception. Since the 1950s, governments and nongovernmental organizations such as the International Planned Parenthood Federation have sought to lower fertility through family planning programs that provide reproductive health services and access to contraceptive information and methods. Yet there are still 214 million women in developing countries who want to delay or stop childbearing who are not using effective contraception (World Health Organization 2019a).

Globally, 40 percent of pregnancies are unintended; half of these pregnancies result in abortion, 13 percent end by miscarriage, and 38 percent result in an unplanned birth (Sedgh et al. 2014). The United States has historically been one of the largest contributors to international family planning efforts, although under the Trump administration, funding for several of these initiatives was withheld (Kaiser Family Foundation 2019). Although former President Trump discontinued federal support for the World Health Organization (British Broadcasting Corporation 2020a), which promotes family planning and maternal and child health care, President Biden rejoined the global health organization in his first month in office (Deliso 2021) (see Chapter 2).

Lack of access to family planning is only one reason for not using contraception. Other reasons for not using modern contraception include: (1) lack of knowledge, (2) underestimation of the risk of getting pregnant, (3) cost, (4) side effects or health concerns, (5) disapproval of male partner, and (6) the belief that God should decide how many children a woman has (Ryerson 2011; Population Reference Bureau 2019). In some countries it is illegal for a woman to receive contraception without the consent of a third-party, typically her husband (Kaiser Family Foundation 2019).

In the United States, less than 10 percent of women at risk of getting pregnant don't use contraception, although 5 percent rely on the highly ineffective method of withdrawal (Guttmacher Institute 2020). However, the rate is double that among sexually active U.S. teenagers with 18 percent reporting they don't use any birth control method (see Chapter 5). Unintended pregnancies also happen because contraception is used inconsistently or incorrectly. Encouraging couples to use modern contraception not only requires providing access to affordable contraceptive methods, it requires education to dispel the myths about the dangers of using contraception and an understanding on how to use contraception effectively. Moreover, the consistent and correct use of contraception also requires that couples understand the health and economic benefits of delayed, spaced, and limited childbearing.

Victor Biro/Alamy Stock Photo

The Bill & Melinda Gates Foundation has a Maternal, Newborn & Child Health Division that funds projects that increase access to family planning and health services for women and children around the world (Bill & Melinda Gates Foundation 2020).

The Status of Women: The Importance of Education and Employment. Throughout the developing world, the primary status of women is that of wife and mother. Women in developing countries traditionally have not been encouraged to seek education or employment. Instead, they are encouraged to marry early and have children. In fact, child marriage remains common in many sub-Saharan African countries today (United Nations 2018) (see Chapter 10).

Improving the status of women by providing educational and occupational opportunities is vital to curbing population growth (Ní Bhrolcháin and Beaujouan 2012). However, the United Nations has concluded that developing countries would need to commit about a quarter of their national budgets on education in order to meet the goal of universal education (Sohngen 2017). Such an investment is simply unachievable in many poor countries.

One means by which the United Nations is trying to make up the difference is through their United Nations Global Compact program which engages business in working toward United Nation's goals including universal education (United Nations Global Compact 2020). Increasing access to education and school completion in developing countries and the United States has also been the focus of efforts by some American celebrities, including Oprah Winfrey and Chaka Khan (Sayei 2017).

Primary school enrollment of both young girls and boys is associated with declines in fertility rates. However, in many parts of the world, children live too far from a school, and getting qualified teachers is difficult (Rueckert 2019). Even where schools are available in developing countries, many families cannot afford school fees or need children to remain at home. If a family has limited funds,

Gallo Images/Sunday Times/Kevin Sutherland/Getty Images

A number of celebrities and other individuals with means have initiated programs to improve the lives and health of girls and women in developing countries. In 2007, Oprah Winfrey contributed $40 million to open a school for girls in South Africa. She is actively involved with the school, as seen here at a graduation ceremony.

@Zosia Bielski

What if instead of asking women why they don't have kids, we asked mothers why they do? In non-pandemic news, I look at "pronatalism," the lingering assumption that all normal roads to adulthood lead through motherhood, and what that means for women.

-Zosia Bielski

they are more likely to send their sons to school than their daughters who are presumed to marry early and leave home. In some areas of the world, girls are not sent to school due to safety concerns as in Nigeria where hundreds of schoolgirls were kidnapped by the extremist group Boko Haram in 2014 (Sohngen 2017). Even when girls have access to school, they often miss many days for the simple fact that they have no supplies to deal with their menstrual cycles (Rueckert 2019).

Small interventions can make a difference. In western Kenya, the government subsidized required primary school uniforms for boys and girls in the mid-2000s. This led to a decrease in both school dropout rates and marital fertility across the region (Duflo, Dupas, and Kremer 2015). In countries where primary school enrollment is widespread or nearly universal, fertility declines more rapidly because (1) schools educate students on the benefits of family planning, and (2) couples have fewer children as they are unable to pay the required school fees for a large family (Roser 2017).

The education of women, in particular, leads to declines in fertility. Educated women tend to marry later, are more likely to work outside the home, want smaller families, and use contraception. Data at both the country and the individual levels demonstrate that increases in a woman's education help to reduce the fertility rate. Across multiple countries that increased women's education levels from 0 to 6 years between about 1950 and 2010, the fertility rate decreased by an average of 40 percent (Roser 2017). Even in countries like Mali where the norm was to have many children, six or more, women who completed secondary school had an average of four to five children while women who went to college had three children on average (Roser 2017). See this chapter's *Social Problems Research Up Close* on page 461 for a discussion of why this relationship exists.

Educated women also tend to have healthier children. As a result, the number of births is decreased long term because women don't need to have as many children in order for some to survive. Further, as opposed to men, women who work outside the home invest more of their money into improving the well-being of their offspring and their communities (Organization for Economic Co-operation and Development 2020). While educational opportunities for women and girls have improved in many countries (Roser and Ortiz-Ospina 2020), see Chapter 8 for a discussion of the continuing restrictions women and girls face around the world.

Another important component of family planning and reproductive health programs involves changing male attitudes toward women. According to traditional male gender attitudes: (1) a woman's most important role is as wife and mother, (2) a husband has the right to demand sex, (3) it is a husband's right to refuse to use condoms, and (4) it is a husband's right to forbid his wife to use any other form of contraception. Further, in 18 countries, a husband can legally prevent his wife from working (World Bank 2018) and in many additional countries it is just normative to do so.

There is also evidence that programs focused on HIV prevention have had an impact on women's beliefs in their rights within intimate relationships (Fedor, Kohler, and McMahon 2016). Although many initiatives focus on empowering girls and women themselves, a number of programs work with groups of boys and young men to change traditional male gender attitudes. For example, the United Nations has trained men in Cambodia and Zimbabwe to challenge beliefs about violence against women. The international Walk a Mile in Her Shoes campaign strives to raise awareness and break down stereotypes and assumptions about women's roles in society.

What do you think?

The idea of increased contraceptive options for men has been around for more than 40 years. Many studies indicate that men, especially young men, are receptive to the idea of using a contraceptive pill or other birth control method. Despite the fact that several options have been developed, the pharmaceutical industry has committed few resources to clinical trials to test them (Mullin 2016). Why do you think this is?

Access to Safe Abortion. Abortion is a sensitive and controversial issue that has religious, moral, cultural, legal, political, and health implications (see also Chapter 5).

Although abortions are on the decline in the United States, almost one in four women have had an abortion by age 45 (Jones and Jerman 2017).

One-quarter of women live in countries where abortion is prohibited or allowed only to save the life of the mother (Center for Reproductive Rights 2020). Yet, abortions are most common in countries where they are the most restricted (Bearak et al. 2020), likely because countries that restrict abortion also often restrict access to contraception. Where abortions are illegal, unsafe abortions are very common. Globally, about 25 million unsafe abortions occur annually including those performed in unhygienic conditions by unskilled providers and those that are self-induced by a woman inserting a foreign object into her uterus (World Health Organization 2019b). Of the estimated 73 million abortions around the world each year, 23,000 girls and women die and tens of thousands more experience significant health complications (Center for Reproductive Rights 2020). In recent years, deaths due to abortions have declined in some regions where they are illegal or inaccessible because women, via the Internet, have gained access to abortion-inducing medications (Singh et al. 2018; Khazan 2018). However, this is not an option in all areas of the world. Almost all unsafe abortions take place in developing countries, especially African countries (World Health Organization 2019b).

Qilai Shen/In Pictures/Getty Images

As a result of China's one-child policy, most couples of child-rearing age today are themselves only children. As a result, far fewer children in China have biological aunts or uncles or cousins (Swanson 2015) than in other countries.

Coercive Family Planning Policies. **Coercive family planning policies** are government or institutional policies that dictate how many children people should have. These can include policies that restrict whether or how many children a woman can give birth to, as well as policies that restrict women's access to contraception or abortion in an effort to increase population size—potentially against the woman's will.

In the last four decades, China has had two campaigns to restrict how many children were born. One is the current coercive and violent campaign using forced sterilization and forced abortion, and imprisonment against the Uyghurs and other minorities. This policy and its impacts are discussed in greater detail in Chapter 5. The other campaign was the well-known one-child policy which began in 1979 as a nationwide effort to curb population growth (Sotamayor 2020). The one-child policy required women to have an IUD inserted after the birth of their first child and to be sterilized if they had a second child.

Families that had more than one child without authorization faced fines. Further, women who refused to comply could lose their government employment as well as access to education and health services for their children. There were exceptions allowed under the rule (e.g., if the first child was disabled), but estimates indicate that about one-third of the country was subjected to the one-child limit. The one-child policy was highly effective in reducing population growth in China (see Figure 12.12) which has the highest rate of modern contraceptive use in the world, tied with the United Kingdom (Population Reference Bureau 2019). However, in 2013, China modified its policies to allow families to have two children due to growing concerns about the rapid aging of the population, the financial costs of pensions and health care for the aging population, and the shortage of younger workers to fill jobs and help support the older population.

As China's population pyramid illustrates, the number of girls and women born during the period in which the one-child policy was enforced is unusually lower than the number of boys and men in the same age groups, particularly for ages 10 to 25. This suggests, as many have theorized, that selective abortion and/or female infanticide was occurring when the firstborn child was a female (YaleHRJ 2018). Chinese tradition dictates that boys support their parents while girls go to the homes of their husband's family.

coercive family planning policies Government or institutional policies that dictate how many children people should have by restricting access to resources.

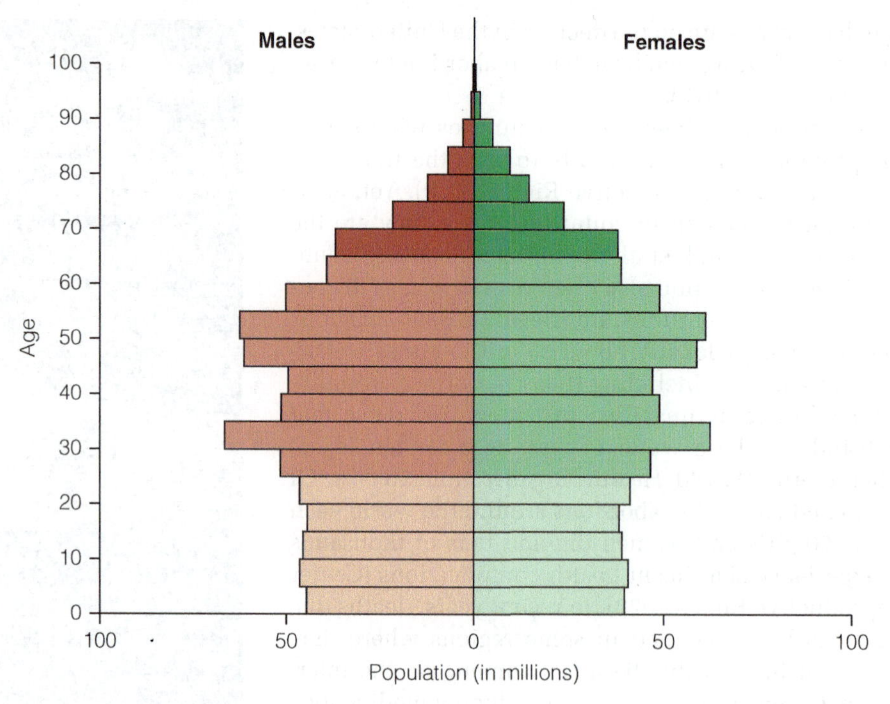

Figure 12.12 Population Pyramid of China, 2020.
SOURCE: United Nations 2019a.

> "I figured it was selfish for us to pour our resources into making our "own" babies when those very resources and energy could not only help children already here, but through advocacy and service transform the world into a place where no child ever needs to be born into poverty and abuse again."
>
> –ASHLEY JUDD, ACTOR

It is difficult to estimate the full impact on female births because, in some cases, families did not report female children being born. This is evidenced by the fact that 4 million more children appeared on later census records, after the one-child policy was relaxed, compared to earlier census records (Steger 2016). Nonetheless, it is believed that over 10 million and as many as 30 million female children were aborted or left to die shortly after birth. Thousands more were abandoned in orphanages or adopted by families in Western countries, including the United States (Sotamayar 2020).

Voluntary Childlessness. In the United States, as in other countries of the world, the cultural norm is for women and couples to want to have children. However, a small but growing segment of U.S. women and men does not want children and chooses to be childless. In the United States, about 15 percent of women ages 40–49 are childless, down from 20 percent in 2005 (Duffin 2019). Research suggests half are childless by choice (Notkin 2013), or **childfree**. Other women who are childless either are unable to have children or are waiting for the right relationship or life situation before motherhood. In general, childfree women are more educated, have higher incomes, live in urban areas, are less religious, and do not adhere to traditional gender ideology (Frejka 2017).

In a study of childless-by-choice men and women, respondents reported a number of reasons why they made the decision not to have children (Blackstone 2019). These include that they wanted to keep being able to do the activities they enjoyed, valued their freedom and independence, and did not want to take on the responsibility of a child. Another reason some individuals choose not to have children is concern for overpopulation and a deep caring for the health of the planet. Yet, ironically, voluntarily childless individuals are often criticized as being selfish and individualistic, as well as less well adjusted and nurturing (McQuillan et al. 2012).

childfree Term used to describe individuals or couples who choose not to have children, to distinguish those from those who are involuntarily childless.

Efforts to Increase Population

Whereas some countries are struggling to slow population growth, others are challenged with maintaining or even increasing their populations. One way to increase population

THE WORLD in quarantine

Will the COVID-19 Pandemic Decrease the Size of the U.S. Population Long Term?

History has shown us that large-scale societal disruptions—including economic recessions, natural disasters, and pandemics—can have impacts on population size. For example, during the 1918 Flu Pandemic, the U.S. population shrank for the only time in history. It is unknown at this point how large and long-lasting the impact of the COVID-19 pandemic will be, but there surely will be impacts through not just *mortality* but also *fertility* and *migration*.

As of the end of February 2021, over 2.5 million people had died of COVID-19 worldwide (Johns Hopkins University 2021). The United States was leading the world in both confirmed cases of the virus and number of deaths, over 500,000 by March 1, 2021. These death figures are likely underrepresentative of the total number of COVID-19 deaths, especially early in the pandemic before widespread testing was available (Weinberger, Chen, and Cohen 2020). It remains to be seen how many more people will lose their lives to COVID-19 either during the illness period itself or to the significant long-term physical effects that have been documented in many cases.

Because the death rate is significantly lower among people of childbearing age, deaths associated with COVID-19 should not significantly impact future births. However, there is growing evidence that the pandemic is having impacts on fertility plans among women in the United States. In a survey of over 2,000 women, the Guttmacher Institute found that a third of women reported the pandemic had led them to decide to delay having children and/or to have fewer children (Lindberg et al. 2020). Almost half of Black and Hispanic women reported intentions to delay or reduce their fertility plans. A third of women agreed with the statement "because of the COVID-19 (coronavirus) pandemic I am being more careful than I used to be about using contraception every time I have sex." In contrast, 17 percent of women reported they wanted to have a child sooner or wanted more children as a result of the pandemic. Experts estimate there will be 300,000–500,000 fewer babies born in the coming year as a result of the pandemic (Kearney and Levine 2020). If the economic impacts of the pandemic continue into 2021, as most expect they will, fertility rates may continue to decline until would-be parents have more confidence in their economic situations.

The second contributor to population growth is migration. As discussed earlier, the population growth in the United States is largely fueled by immigration. Between 2018 and 2019, immigration added 595,000 people to the U.S. population (Knapp 2019). In May 2019, the State Department issued nearly 40,000 visas for permanent immigration into the United States. However, a year later, fewer than 700 visas were issued (Pinsker 2020), and the former administration banned immigration through the end of 2020.

In violation of the international laws that govern immigration policies during a pandemic, the Trump administration blocked most asylum seekers from entering the United States, ironically many coming from countries with lower rates of COVID-19 than the U.S. (Somin 2020). These restrictions may have a significant impact on population growth, but will almost certainly limit the available pool of laborers and, as such, will contribute to the further decline of the U.S. economy. And, as already noted, the health of a nation's economy is intricately tied to family fertility plans (see this chapter's *The Key Role of the Economy on Family Planning)*.

is to increase immigration. As noted previously, Japan has eased immigration restrictions in order to bring in younger workers. Several European countries have also eased restrictions on immigration to increase their young working population, but large waves of refugees and economic migrants from African countries, beginning in 2015, led to a tightening of immigration policies (Chamie and Mirkin 2020) (see Chapter 9).

While in the 1980s only 13 countries had efforts in place to promote fertility, today more than a third of the world's governments have adopted pronatalist policies to increase their domestic populations. These range from coercive measures such as restricting access to birth control and abortion to family-friendly policies and cash incentives. For example, after two decades of family planning campaigns which reduced the fertility rate from 6

ITAR-TASS News Agency/Alamy Stock Photo

In an effort to increase fertility in Russia, Putin recognizes families with seven or more children with the Order of Parental Glory in an annual ceremony.

to almost 2.1, the government of Iran reversed course outlawing voluntary sterilization and restricting women's access to effective methods of contraception (Dérer 2019). Turkey has also adopted coercive policies to further its slogan of "minimum three children" by restricting access to contraception and abortion in public hospitals and clinics (Kiliç 2017).

Fertility rates tend to be the lowest in high-income countries that embrace the male breadwinner model and, therefore, provide limited public support for the mother's employment (Cooke and Baxter 2010). In these countries, working women struggle the most to balance work and family obligations. To promote fertility, a range of countries including South Korea, Singapore, France, Australia, Canada, Russia, and Poland offer "baby bonuses" for each child born (Howe 2019).

Many European countries have generous family leave policies and universal child care. France has the highest fertility rate of any European country, 1.92 births per woman. Its policies include a payment of more than $1,000 at the birth of each child, monthly stipends, paid parental leave, and subsidized childcare (British Broadcasting Corporation 2020b). Promotional campaigns are also common in many countries. In an attempt to address his country's low fertility rate of 1.48 births per woman, Russian President Vladimir Putin recognizes families with seven or more children with the Order of Parental Glory. At a recognition ceremony in 2019, President Putin restated his commitment to the policies his government had instituted to encourage high fertility rates, including reduced taxes, discounted mortgage programs, and increased childcare capacity: "Such families, people like you are an asset and a blessing to this country. And we will strive for them to grow in numbers" (Putin 2019).

Japan has offered funding for local governments to sponsor matchmaking events (Howe 2019). In Denmark, concern over the implications of a below replacement birthrate of just 1.7 children is widely recognized. A Danish travel agency created a tongue-in-cheek advertisement campaign, "Do It for Denmark," offering travel as the solution to this national problem by playing up the positive impacts of travel on sexual activity and thus fertility.

Analysts agree, however, that one-time payments have done little to increase fertility rates long term, while family-friendly policies including child care and parental leave have a longer-term impact (Sobotka, Matysiak, and Brzozowska 2019; Howe 2019; British Broadcasting Corporation 2020b). Nonetheless, in times of economic uncertainty would-be parents are still cautious when it comes to making the financial commitment of having children.

What do you think?

One strategy for encouraging childbearing in European countries with low fertility rates is to provide work–family supports to make it easier for women to combine childbearing with employment. If the United States offered more generous work–family benefits, such as paid parenting leave and government-supported child care, would the U.S. birthrate increase? Would such policies affect the number of children you would want to have?

"One thing you can say is that in places where women are in charge of their bodies, where they have the vote, where they are allowed to dictate what they do and what they want, whether it's proper medical facilities for birth control, the birth rate falls."

–SIR DAVID ATTENBOROUGH, NATURALIST AND BROADCASTER

The Key Role of the Economy in Family Planning

Whether they are studying countries with high fertility rates or low fertility rates, analysts agree that one of the most important factors in fertility decisions is the economy. There is a reciprocal relationship between fertility rates and economic well-being. As fertility rates decrease in high-fertility countries, often coupled with women's education, the economy improves (Roser 2017). As the economy improves, sanitation and health care improve. This means children are more likely to survive in the long term making fewer births necessary.

As economies grow, the likelihood of a country having universal education increases along with the costs to families who must now shoulder the added burden of paying school fees and the loss of free child labor at home. As a result, over time, fertility rates decrease (Roser 2019). Further, with universal education, women are more likely to pursue a career thereby delaying childbirth and, again, lowering fertility. Thus, improving the economy in high-fertility countries should lead to a decrease in fertility.

Alternatively, in countries with low fertility rates, rising unemployment and deteriorating economic conditions have been identified as factors contributing to fertility decline (Sobotka, Matysiak, and Brzozowska 2019). Economic uncertainty negatively impacts first birth intentions, especially for men, leading to delayed fertility and ultimately fewer children. Rising costs of housing also make it increasingly difficult for couples to move forward with plans to start a family. In a study of women in 17 low- to moderately low-fertility countries, the average desired number of children was between 2.10 and 2.24—at or above replacement levels. However, the actual average number of children born was less, suggesting factors other than intention played a role in the number of children a couple had (Sobotka, Matysiak, and Brzozowska 2019). While infertility is a factor, analysts point out that the gap between desired and realized fertility is the highest during economic recessions.

J.D. Pooley/Getty Images

Nearly half of seniors in the United States work full- or part-time, like this 72-year-old man in Ohio.

Thus, a healthy economy is an important variable in determining the number of children a couple has in both high- and low-fertility countries. This suggests that governments would benefit from focusing on efforts to stabilize their economies, no matter what their goals with respect to population growth.

Supporting Employment among the Elderly

Many elderly want or need to continue to work for pay. Nearly half of U.S. retirees are currently working, have worked, or plan to work in retirement (GSA 2018). These include individuals who retired from previous positions and have begun working full-time elsewhere, those who have started a business or consulting firm, and those who have already retired but have taken a "bridge job" from which they eventually plan to retire full-time.

The 1967 *Age Discrimination in Employment Act* (ADEA) was designed to ensure continued employment for people between the ages of 40 and 65. In 1986, the upper age limit was removed thus making mandatory retirement illegal. But there are exceptions to this rule. In the private sector, certain executives, such as partners in accounting firms, can be subject to mandatory retirement. Firefighters, law enforcement officers, military personnel, pilots, air traffic controllers, and, in some states, judges are also subject to mandatory retirement (Grandjean and Grell 2019).

Mandatory retirement is justified by the argument that certain occupations require high levels of physical and/or mental ability and that age-related decline could jeopardize the safety or well-being of the public and/or the worker. Because mandatory retirement policies are often based on a fixed age and not on an evaluation of the worker's abilities, the American Association of Retired Persons (AARP), the country's largest organization representing people over 50, argues that mandatory retirement is a form of age discrimination (Kita 2019).

Under ADEA, it is illegal to discriminate against people because of their age with respect to hiring, firing, promotion, layoff, compensation, benefits, job assignments, and training. Age discrimination is difficult to prove. Nonetheless, thousands of age discrimination cases are filed annually with the Equal Employment Opportunity Commission (EEOC). Many individual states also have laws protecting individuals from age-based discrimination. Nuemark et al. (2019) find that callback rates for older job applicants were greater in states that had their own nondiscrimination policies.

Continued employment among the elderly is beneficial in some societies today and will be necessary in even more countries in the future. Many countries with aging populations need elderly workers to remain in the workforce, as they don't have enough younger workers to fill vacancies. Some have developed incentive structures to discourage the elderly from retiring (GSA 2018). Singapore has instituted a program that offers tax breaks to employers who hire anyone aged 65 and older, by offering a 3 percent offset on the employees' monthly wages. Germany's Initiative 50 Plus program encourages older workers to stay in the workforce by offering training programs and incentives to take positions with low salaries. Japan has a public–private partnership called Silver Center Workshops, which assist the elderly in finding part-time jobs.

Options for Reforming Social Security

Many are unaware that Social Security supports not only retirees but also disabled working-age adults and the children and surviving spouses of deceased workers. Thus, it is a critical source of income to millions of Americans. Social Security remains the most solvent part of the U.S. government. It is funded through its own separate tax and interest from a trust fund. No other government program or agency is fully funded. Social Security is also a very efficient program. Although it collects taxes from more than 90 percent of the workforce and sends benefits to more than 50 million Americans, Social Security spends less than one cent of every dollar on administrative costs (Center on Budget and Policy Initiatives 2020).

Nevertheless, unless changes are made in years to come, Social Security won't have enough funds to provide full payouts to future recipients. Estimates indicate that the trust fund reserves will be depleted sometime between 2031 and 2065 (Romig 2020b), after which benefits would be paid using the tax revenue collected from current workers paying into the system. Those funds are expected to cover 79 percent of benefits for another 75 years and then decrease to 73 percent. These are, of course, estimates based on assumptions of how many people will be paying into the system and how many people will be receiving benefits.

The COVID-19 pandemic and its economic impacts are illustrative of the kinds of phenomena that make projections challenging. Increased unemployment as a result of COVID-19 has decreased taxes paid into the Social Security system. In addition, individuals who contract COVID-19 may experience long-term disability which would add them to the number of people receiving benefits. At present, it is unknown what the impacts of the pandemic will be or how long they will last (Romig 2020b).

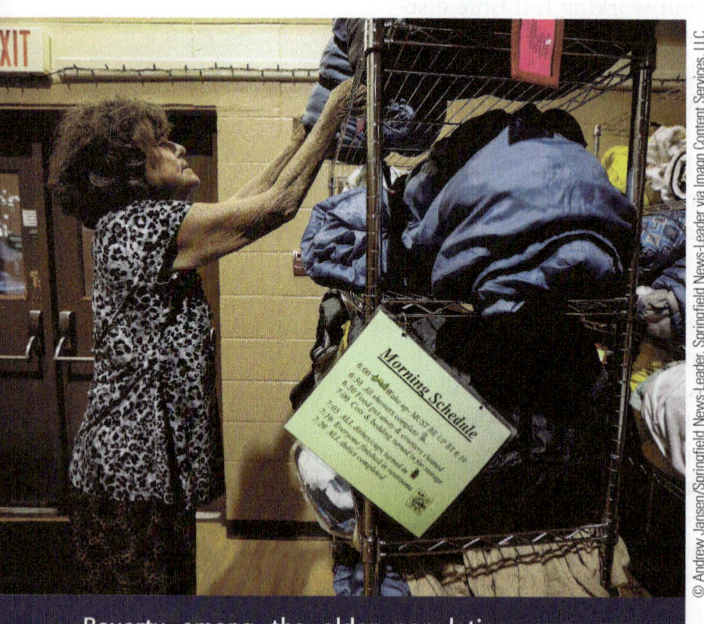

© Andrew Jansen/Springfield News-Leader via Imagn Content Services, LLC

Poverty among the older population, even many receiving Social Security, contributes to the fact that people over 50 make up over 30 percent of the U.S. homeless population. Lack of affordable housing has contributed to seniors, like this woman, seeking beds in homeless shelters across the country.

Options for reforming Social Security include cutting Social Security benefits, increasing Social Security revenue, and expanding Social Security benefits. The following outlines these options (Edwards et al. 2012; Romig 2020b).

Cut Social Security Benefits. Some reform proposals call for cuts in Social Security benefits, which would create a significant financial burden on households that depend on Social Security. As previously discussed, Social Security payments are quite modest, so cutting benefits is an unpopular choice, although some have proposed cutting benefits only for those over a certain income threshold from supplemental sources (e.g., pension or 401[k] income).

Aside from simply reducing the amount of benefits paid to recipients, another way to cut benefits is to increase the retirement age. This approach was used in 1983, when the retirement age was raised from 65 to 67 to be phased in over the next 23 years (Social Security Administration 2020a). Some policy makers suggest raising the retirement age even further, perhaps to 70. However, due to the link between higher socioeconomic status and longer life expectancy (Sanzenbacher et al. 2017), raising retirement age imposes the greatest burden on low earners who have lower life expectancies and, therefore, fewer years to collect Social Security. In addition, older workers already face employment discrimination making finding or keeping a job difficult for seniors. Finally, the older workers become, the higher the likelihood that they will experience health problems and hence file disability claims.

> "A population that does not take care of the elderly and of children and the young has no future, because it abuses both its memory and its promise."
>
> –POPE FRANCIS

Increase Social Security Revenue. Instead of cutting benefits, many advocates for the elderly suggest raising revenues. One option for increasing Social Security revenue is simply to raise the tax that funds Social Security (i.e., the payroll or FICA tax) from its current rate of 12.4 percent (6.2 percent from employer; 6.2 percent from employee) to a higher rate. Indeed, this was proposed in a bill considered by Congress in 2019 which would have raised the tax rate to 14.8 percent (O'Brien 2019). The bill was ultimately revised to only increase the cap on income which is taxed by FICA, as discussed below.

To offset gains in life expectancy, Social Security taxes increased 19 times, from 1 percent between 1937 and 1949, to 6.2 percent in 1990. To date, Social Security taxes have not increased since that time, the longest period without an increase to the payroll tax in history. The downside to raising the payroll tax is that it could result in fewer jobs as employers have a higher financial burden. Paying a higher payroll tax would also disproportionately burden lower-wage earners. Thus, President Biden has proposed raising the payroll tax but only for those whose incomes exceed $400,000 annually (Gleckman 2020).

Another option is to raise the tax cap so that more earnings are taxed, providing more funds to the Social Security program. In 2019, Congress used this approach increasing the annual Social Security tax cap from the 2017 level of $127,200 to $137,700 in 2020 (AARP 2020). Thus, in 2020, the maximum income subject to Social Security taxes was $137,700. As a result, the maximum Social Security tax paid per worker in 2020 was $17,074.80, half paid by the employee and half paid by the employer. Self-employed individuals must pay the entire $17,074.80 themselves.

Finally, any policies that increase employment and wages of all workers would lead to more revenue for Social Security. Because health insurance costs are deducted from taxable income, policies to reduce health care costs would also indirectly increase Social Security funding. Substantial increases in health insurance premiums have been cited as one reason the tax base for Social Security has decreased in recent years (Romig 2020b).

Expand Social Security Benefits. Options for increasing Social Security benefits include raising the minimum benefit amount, offering wage credits for unemployed parents who are taking care of their children, and increasing the benefits for the very old (85 years and older). Another proposal involves restoring the student benefit which existed between 1965 and 1985 so that children of the retired, deceased, or disabled could continue to receive benefits until age 22 if they are attending college or vocational school (National Committee to Preserve Social Security & Medicare 2018). At present, these children receive benefits only up to age 18.

Not surprisingly, proposals to cut Social Security benefits or raise retirement age for receiving benefits are vehemently opposed by seniors and advocacy groups for older adults, like the American Association of Retired Persons (AARP). Raising the tax cap would go far to ensure the future financial viability of Social Security, but this option is generally not supported by the wealthy segment of the population who would bear the burden of higher taxes. Until legislators enact changes to prevent the long-term Social Security deficit, there is little chance of seeing Social Security benefits increase.

Understanding Problems of Population Growth and Aging

What can we conclude from our analysis of population growth and aging? First, although many countries are experiencing a decline in their fertility rates, world population will continue to grow for several decades. This growth will occur largely

> "The problem is that the population is growing the fastest where people are less able to deal with it. So it's in the very poorest places that you're going to have a tripling in population by 2050. And we've got to make sure that we help out with the tools now so that they don't have an impossible situation later."
>
> **–BILL GATES, BUSINESS LEADER AND PHILANTHROPIST**

in developing regions, while other parts of the world will see their population shift, with an increasing proportion over the age of 65. Both trends have their challenges for the world's population as a whole and for individual nations.

Many governments recognize the value of controlling population size and supporting family planning programs. However, there are still significant unmet needs in the area of contraception. Efforts to control population must go beyond providing safe, effective, and affordable methods of birth control. Slowing population growth necessitates interventions that improve the status of women and increase education and economic opportunities for youth.

As fertility rates decline and life expectancy increases, the United States and many other countries are experiencing population aging. Although most elderly people in developed countries live independently, an aging population necessitates societal adaptations to support caregiving needs and financial security in old age. With fewer young people in society, it is increasingly clear that the family cannot be the only support elders have in their last years. Some societies have made significant advancements in this area, while others lag behind. Prejudice toward and discrimination against older people impacts quality of life and hinders progress in this area.

Fears of global overpopulation have been around since Thomas Malthus back in the 1700s. His concern was our ability to provide the food needed to support the world's population. Of more recent concern is the impact of population density on the potential for future pandemics. Finally, of growing and now more immediate concern is the impact of climate change on our ability to sustain life on this planet. While climate change is related to population growth, it is not population growth that is having the greatest impact on climate change. As Betsy Hartmann points out, "There's a big difference between a poor peasant farming the land and a fossil fuel corporate executive. The overpopulation framework tends to lump all humans together into one broad category, not differentiating between their differential impacts on the planet" (quoted in Kolitz 2019).

Chapter Review

- **How long did it take for the world's population to reach 1 billion? How long did it take for it to reach 6 billion?**
  It took thousands of years for the world's population to reach 1 billion, and just another 200 years for the population to grow from 1 billion to 6 billion.
- **Is the world population still increasing? Is it decreasing? Or is it remaining stable?**
  World population is still growing. It is projected to grow from 7.7 billion in 2019, to 9.7 billion in 2050, and 10.9 billion by 2100.
- **Where is most of the world's population growth occurring?**
  Most world population growth is in developing countries, primarily in Africa and Asia.
- **By 2050, how is the number of older individuals (ages 65 and over) expected to change globally?**
  Globally, the population aged 65 and older is the fastest growing age group. Between 2019 and 2050, the number of older individuals (aged 65 and over) in the world population is expected to double to 1.5 billion people.
- **What factors are contributing to population aging?**
  Population aging is a function of lowered fertility as well as increased longevity. In the United States, population aging is also occurring because the Baby Boomers are reaching their senior years.
- **What is the demographic transition theory?**
  The demographic transition theory of population describes how industrialization and economic development affect population growth by influencing birth and death rates. According to this theory, traditional agricultural societies have both high birthrates and high death rates. As a society becomes industrialized and urbanized, improved sanitation, health, and education lead to a decline in mortality. The increased survival rate of infants and children along with the declining economic value of children leads to a decline in birthrates. The demographic transition theory is a generalized model that does not apply to all countries and does not account for population change due to HIV/AIDS, war, migration, and changes in gender roles and equality.

- **What is the argument of World System Theory?**
World System Theory argues that countries can be divided into three groups: the core countries, the periphery countries, and the semi-periphery countries. Core countries exploit the labor and resources of periphery countries in order to develop economically and maintain power. The periphery countries are dependent on selling resources to the core to sustain their economies. The theory helps us to understand inequality at the global level.

- **Many countries are experiencing below-replacement fertility (fewer than 2.1 children born to each woman). Why are some countries concerned about their low fertility?**
In countries with below-replacement fertility, there are or will be fewer workers to support a growing number of elderly retirees and to maintain a productive economy.

- **What kinds of problems are associated with population growth?**
Population growth is associated with high levels of unemployment and poverty. Societies with high fertility rates tend to experience lower levels of maternal and child health. High population growth is also associated with food insecurity.

- **Why is population growth considered a threat to global security?**
In developing countries, rapid population growth results in a "youth bulge"—a high proportion of 15- to 29-year-olds relative to the adult population. The combination of a youth bulge with other characteristics of rapidly growing populations—such as resource scarcity, high unemployment rates, poverty, and rapid urbanization—sets the stage for civil unrest, war, and terrorism, because large groups of unemployed young people resort to violence in an attempt to improve their living conditions.

- **How is ageism different from other "isms" (racism, sexism, heterosexism)?**
Ageism is more widely accepted than other isms, and, unlike other isms, everyone is vulnerable to experiencing ageism if they live long enough.

- **How have sources of retirement income changed for American adults over the last few decades?**
Both private and public employers have shifted away from traditional pension plans to defined contribution plans. This means that workers do not have a guaranteed income when they retire. Defined contribution plans can be impacted by downturns in the market, leaving both current workers and current retirees with less in their accounts.

- **How important is Social Security income for senior citizens in the United States?**
For more than half of Americans over age 65, Social Security provides more than half of their income; and without Social Security income, nearly half of all seniors would be living in poverty.

- **Is Social Security in crisis?**
In the short term, Social Security is not "broke" and there is no immediate crisis. But long-term changes to the Social Security system will be needed to ensure that the program is able to meet its financial obligations to future retirees.

- **What is the "sandwich generation"?**
The "sandwich generation" refers to adults who care for their aging parents while also taking care of their own children—they are "sandwiched" between taking care of both parents and children.

- **Efforts to curb population growth include what strategies?**
Efforts to curb population growth include strategies to reduce fertility by providing access to family planning services including safe and legal abortion, involving men in family planning, and improving the status of women by providing educational and employment opportunities. Increased economic opportunity is also associated with reductions in fertility.

- **What are three general options for reforming Social Security?**
Options for reforming Social Security include strategies for increasing Social Security revenue, cutting benefits, and expanding Social Security benefits.

Test Yourself

1. In 2019, world population was 7.7 billion. In 2050, world population is projected to be ___ billion.
   a. 8.8
   b. 9.7
   c. 10.9
   d. 13

2. Increases in population size have led to an increase in deaths due to famine.
   a. True
   b. False

3. The value of a defined contribution plan can change as a result of changes in the economy.
   a. True
   b. False

4. World System Theory helps us to understand
   a. how countries work together for the good of all humankind.
   b. why the aging population is not equal in countries across the globe.
   c. how some countries benefit from the resources and labor of other countries, creating country-level inequality.
   d. how the fertility rate declined in some countries faster than in other countries.
5. Research shows that women's education contributes to declines in childhood mortality.
   a. True
   b. False
6. What would happen if every country in the world achieved below-replacement fertility rates?
   a. Population growth would stop and world population would remain stable.
   b. World population would immediately begin to decline.
   c. World population would continue to grow for several decades.
   d. World population would decline, but then go up again.
7. Pronatalism is a cultural value that promotes which of the following?
   a. Looking young
   b. Abstaining from sex until one is married
   c. Having children
   d. Urban living
8. According to the Facts on Aging Quiz, most old people lose interest in and capacity for sexual relations.
   a. True
   b. False
9. A study of age discrimination in employment found that women were ________ to be discriminated against in callbacks for job applications as men.
   a. equally as likely
   b. more likely
   c. less likely
10. Most family caregivers of elderly adults have someone else to help them.
   a. True
   b. False

Answers: 1. B; 2. B; 3. A; 4. C; 5. A; 6. C; 7. C; 8. B; 9. B; 10. B.

Key Terms

ageism 467
ageism by invisibility 468
Baby Boomers 458
birthrate 460
childfree 480
child wasting 465
child stunting 465
coercive family planning policies 479
defined benefit plan 472
defined contribution plan 473
demographic transition theory 460
elderly support ratio 459
emigration 457
environmental footprint 466
food insecure 465
geriatrics 469
Global Hunger Index 465
gross domestic product 462
immigration 457
infant mortality rate 465
migration 457
mortality rate 460
population momentum 457
population growth rate 456
population pyramid 458
pronatalism 463
replacement-level fertility 456
sandwich generation 471
Social Security 473
total fertility rates 456
undernourishment 465
visual ageism 468
World System Theory 462

I want you to panic. I want you to feel the fear I feel every day. And then I want you to act."

GRETA THUNBERG
Environmental activist

13

Environmental Problems

Chapter Outline

Learning Objectives

After studying this chapter, you will be able to ...

1. Identify the evidence of global warming and climate change.
2. Discuss the impacts of global warming and climate change.
3. Identify the factors contributing to global warming and climate change.
4. Give examples of how structural functionalism, conflict theory, and symbolic interactionism can be applied to our understanding of environmental problems.
5. Identify the pros and cons of the various sources of the world's energy.
6. Briefly describe the following environmental problems: dwindling water supplies; threats to biodiversity; depletion of natural resources; air, land, water, and light pollution; and environmental illness.
7. Explain how population growth, industrialization and economic development, human supremacy, individualism, consumerism, and militarism contribute to environmental problems.
8. Describe various strategies and efforts to restore and protect the environment.
9. Identify the challenges that must be overcome in order to restore and protect the environment.

The smoke from the wildfires in California during fall 2020 was so dense and spread so far that San Francisco's Golden Gate Bridge was bathed in red smoke during the day.

Harold Postic/AFP/Getty Images

OVER THE 2020 LABOR DAY WEEKEND, California experienced record-breaking temperatures of 105 to 120 degrees Fahrenheit while 14,800 fire fighters battled 23 major fires. More than 200 Labor Day vacationers had to be airlifted to safety by National Guard helicopters after the road to their site was blocked by fire. The pilots, who had been advised not to attempt the rescue, had to don night vision goggles in order to see through the smoke. Pilot Chief Warrant Officer Joseph Rosamond said: "Conditions were pretty extreme. ... There were points along the route where we were just about ready to say 'enough'" (Sisk 2020). Even so, they returned twice to get everyone out. Between August 15 and September 6, 2020, 900 wildfires burned 1.5 million acres, destroyed 3,300 structures, and caused eight deaths (Sanchez and Weber 2020). Just two years earlier, the Camp Fire had made history as the single most destructive and deadliest fire in California history, killing at least 85 people (Brekke 2019). By late October the 2020 wildfire season surpassed the ten-year average for number of fires and for acres burned, with 46,681 wildfires that burned over 8,608,646 acres (Center for Disaster Philanthropy 2020).

Extreme heat has been the top weather-related killer in the United States over the past 30 years. The increasing number and severity of wildfires are among the many impacts of global warming and climate change (Union of Concerned Scientists 2020a)—the most challenging environmental problem of our time. In this chapter, we discuss the causes and consequences of global warming, climate change, and pollution that threaten the lives and well-being of people, plants, and animals all over the world—today and in future generations. After examining how globalization affects environmental problems, we view environmental issues through the lens of structural functionalism, conflict theory, and symbolic interactionism. We then present an overview of major environmental problems, examining their social causes and exploring strategies to reduce or alleviate them.

The Global Context: Global Warming and Climate Change

In looking at environmental problems from a global perspective, we see that many environmental problems have causes and consequences that cross international borders. Scientists agree that the single greatest environmental threat to the globe is global warming (National Aeronautics and Space Administration [NASA] 2020). Global warming has far-reaching effects leading to changes in climate, increasing and worsening natural disasters, and rising sea levels (which will be discussed later in this chapter).

global warming The increasing average temperature of the Earth's atmosphere, water, and land, caused mainly by the accumulation of various gases (greenhouse gases) that collect in the atmosphere.

Global warming refers to the increasing average global temperature of the Earth's atmosphere, water, and land. The world's average global temperature in 2020 was 0.98 degrees Celsius (1.76 Fahrenheit), above the 20th-century average (NOAA 2021). As Figure 13.1 illustrates, average global temperature has increased by more than 1.2 degrees Celsius since 1901, with most of that increase occurring after 1980. Nineteen of the 20 warmest years have occurred since 2001.

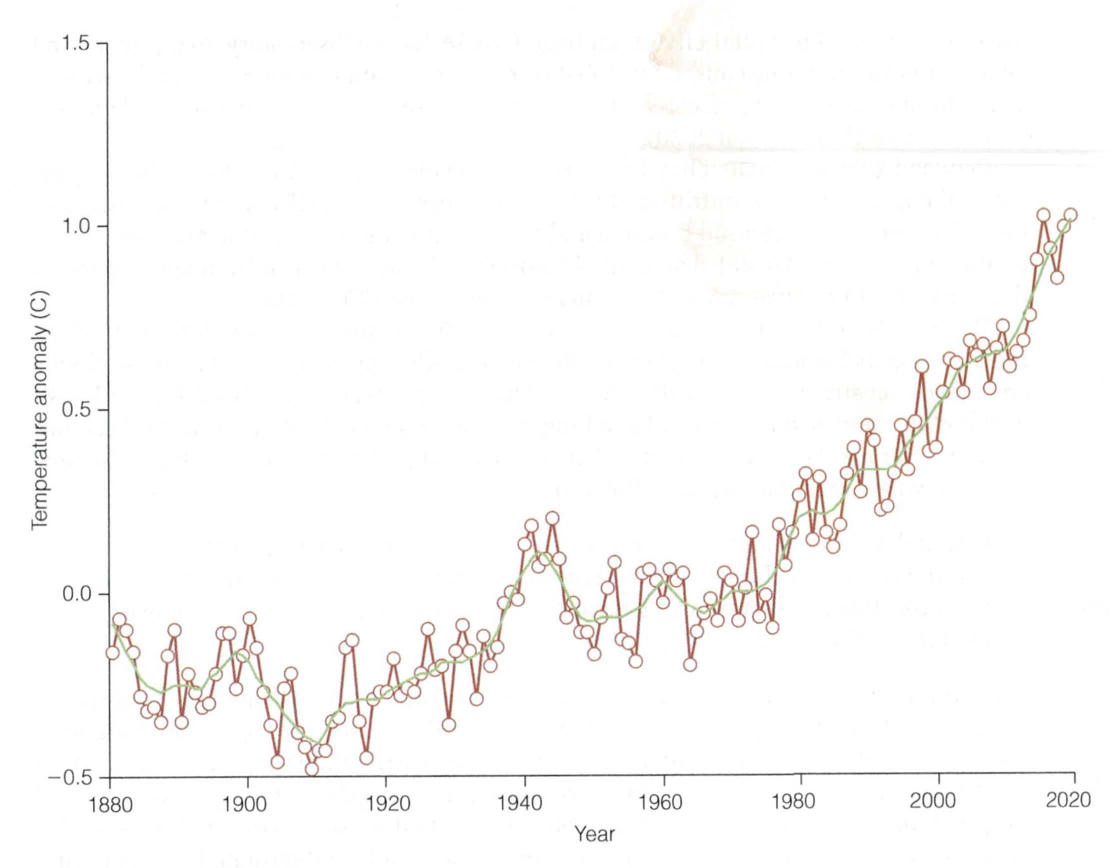

Figure 13.1 Average Global Temperature by Year, 1880–2019
SOURCE: NASA Goddard Institute for Space Studies. 2020. "Graphic: Global Land-Ocean Temperature Index." Available at www.climate.nasa.gov.

The link between global warming and climate change has been recognized for decades. In 1988, the United Nations acknowledged climate change as an issue of global importance with the creation of the Intergovernmental Panel on Climate Change (IPCC) (IPCC 2020). The IPCC is charged with continuously reviewing the state of scientific knowledge about climate change, social and economic impacts of climate change, and strategies for addressing the issue through international policy (IPCC 2020). Four years later, the United Nations Framework Convention on Climate Change (UNFCCC) was adopted with a specific goal of addressing the problem of climate change through reducing greenhouse gas concentrations, which contribute to global warming (UNFCCC 2020a).

Impacts of Global Warming and Climate Change

Significant impacts of global warming and the changes in climate associated with it are occurring around the world. As shown in Figure 13.1, the world today is more than 1.2 degrees warmer than it was in the late 19th century (UNFCCC 2020b), and its temperature is increasing at nearly twice the rate it was 50 years ago (National Geographic 2019). Due to global warming, the average global sea level has risen 8 to 9 inches from 1880 to 2020 and is expected to rise at least 12 inches by 2100 even if greenhouse gas emissions are greatly reduced, and up to 8.2 feet if emissions remain high (Lindsey 2020). By 2100, an estimated 48 islands will be lost due to rising sea levels (Podesta 2019). Sea level rise is occurring as a result of (1) thermal expansion of seawater (water expands as it warms) and (2) the melting of glaciers and of the Greenland and polar ice sheets.

Eastern Siberia, the Russian region well known for its frigid temperatures, recorded temperatures over 100 degrees Fahrenheit in summer 2020, sparking wildfires that spread smoke as far as Alaska (NASA Earth Observatory 2020). A drought in Chili has lasted more than a decade, reducing by more than one half the water level in the main reservoir serving the capital city of Santiago (NASA Earth Observatory 2020). Increased heavy rains and flooding caused by global warming contribute to increases in drownings and to human exposure to insect- and water-related diseases, such as malaria and cholera (World Health Organization 2020).

Between 2030 and 2050, climate change is expected to cause 250,000 deaths per year across the globe, from malnutrition (due to crop failure), malaria (from increased mosquitos), diarrhea (from increased flooding and water-borne diseases), and heat stress (World Health Organization 2018a). The United Nations estimates that 1 million species are on the brink of extinction as a result of climate change (UNFCCC 2020b).

The tiny island nation of Tuvala, home to 11,000 people, encapsulates many of the impacts of global warming. Tuvala is on the brink of disappearing; two of its nine islands are already mostly underwater (Roy 2019). The fish that islanders rely on for food have declined in number due to **coral bleaching**, which happens when ocean water becomes too warm. Tapua Pasuma returned to Tuvala after completing her college degree in New Zealand; when she returned, she observed:

> I immediately noticed the difference. The heat is sometimes unbearable now, and the erosion is also dramatic. Some of my favourite spots have disappeared. I feel like this is a part of who I am and I shouldn't just run away from it, even though it's disappearing. (Roy 2019)

"The climate crisis is our third world war. It needs a bold response."

–JOSEPH STIGLER, NOBEL PRIZE–WINNING ECONOMIST

Plans under consideration to address the crisis include the construction of a seawall, dredging and reclaiming land, or the potential construction of a floating island. Neighboring countries like Fiji and Australia have offered refuge to the entire population; offers that the prime minister points out fail to address the climate change issues that have created the problem in the first place (Roy 2019). More than 2,000 people have already migrated to New Zealand to avoid the flooding, extreme heat, and contaminated underground water supplies. They are among the growing number of climate migrants or "climate refugees" across the globe, escaping climate change impacts such as flooding, drought, and fire (Podesta 2019). Each year, millions of people across the globe are forced to move due to weather-related conditions that have been caused or exacerbated by global warming.

coral bleaching When ocean water becomes too warm, coral will expel algae, causing the coral to turn white.

In the United States, the impacts of climate change can be seen in the increasing frequency and intensity of natural disasters like hurricanes, fires, flooding, and drought. In 2017 Hurricane Harvey led to widespread flooding and knocked out power to over 300,000 customers in Texas, causing ripple effects on hospitals, water and wastewater treatment plants, and oil and gas refineries (U.S. Global Change Research Program 2018). Harvey contributed to over 100 deaths with damages totaling an estimated $125 billion, ranking as the second most costly hurricane to hit the U.S. mainland since 1900 (Huber 2018).

In 2018, the Department of Defense reported that 10 percent of its installations in the United States were being affected by extreme temperatures and another 6 percent by flooding due to storm surge and by wildfires (U.S. Global Change Research Program 2018). In spring 2020, sections of the James River in South Dakota were flooded over its banks for more than a year (NASA 2020), and in Alaska thawing of the permafrost has led to costly damage to roads, buildings, and pipelines (U.S. Global Change Research Program 2018).

AP Images/Kyodo

The island nation of Tuvala, home to 11,000 people, is on the front lines of the impacts of climate change. Two of its nine islands are already almost completely underwater.

Causes of Global Warming and Climate Change

Natural variability such as changes in solar output and volcanic emissions have contributed to climate change over short periods of time (e.g., a year or a decade) (U.S. Global

Change Research Program 2018). However, climate scientists agree that the long-term pattern of global warming is primarily due to human activity.

> Multiple studies published in peer-reviewed scientific journals show that 97 percent or more of actively publishing climate scientists agree: Climate-warming trends over the past century are extremely likely due to human activities. In addition, most of the leading scientific organizations worldwide have issued public statements endorsing this position. (National Aeronautics and Space Association 2020)

US Air Force Photo/Alamy Stock Photo

Flooding as a result of tropical storms and hurricanes has been increasing across the globe. In 2017, Hurricane Harvey damaged or destroyed nearly 135,000 homes and about a million cars. Several dozen schools remained closed for more than a month due to flooding (Huber 2018).

In 2018, the U.S. Global Change Research Program (2018) concluded that "Earth's climate is now changing faster than at any point in the history of modern civilization, primarily as a result of human activities." In its 2019 Annual Report, the United Nations Framework Convention on Climate Change concluded that "we must alter course before it is too late. ... [W]e need to accept that while we might not yet understand the full extent of future changes to the climate, there is enough scientific certainty to warrant sweeping action, particularly when faced with the heat of larger-scale and irreversible damage" (UNFCCC 2020b).

The primary human contributor to global warming has come from the increase in global atmospheric concentrations of greenhouse gases since industrialization began. In the 1850s, scientists demonstrated that carbon dioxide (CO_2) and other gases in the atmosphere prevent heat from escaping the Earth's atmosphere. This became known as the **greenhouse effect** (U.S. Global Change Research Program 2018). **Greenhouse gases**—primarily CO_2, methane, and nitrous oxide—accumulate in the atmosphere and act like the glass in a greenhouse, holding heat from the sun close to the Earth.

Global increases in greenhouse gases are caused primarily by humans, particularly through the use of fossil fuels, but also by population growth, economic activity, lifestyles, technology, and land use patterns. For example, deforestation—the mass cutting of forests—is a major contributor to increasing levels of carbon dioxide in the atmosphere because trees and other plant life use carbon dioxide and release oxygen into the air. As forests are cut down or burned, they release the carbon dioxide they have absorbed back into the atmosphere (Pappas 2020). Deforestation releases nearly a billion tons of carbon into the atmosphere every year. Deforestation also reduces the number of trees available to absorb carbon dioxide in the future. In 2020, the Trump Administration's Department of Agriculture lifted restrictions on logging in the Tongass National Forest (Wamsley and Neuman 2020). The Tongass stores an estimated 400 million metric tons of CO_2 and takes in another 3 million metric tons per year, the equivalent of taking about 650,000 cars off the road each year.

China and the United States account for 33 percent of the world's carbon emissions as a result of fuel combustion (Union of Concerned Scientists 2020b). Although China produces almost twice as many total carbon emissions than the United States, the United States' per capita emissions are more than twice that of China's. The United States is fourth in the world in carbon emissions per person, surpassed only by Saudi Arabia (1), Kazakhstan (2), and Australia (3) which is nearly tied with the United States. China has shown a steady and significant increase in its CO_2 emissions over the last 20 years (International Energy Agency 2020a), while the European Union countries have been steadily declining in emissions since about 2007 (see Figure 13.2).

Even if greenhouse gases are stabilized, global air temperature and sea level are expected to continue to rise for hundreds of years. That is because global warming that has already occurred contributes to further warming of the planet—a process known as a *positive feedback loop*. For example, the melting of ice and snow due to global warming exposes more land and ocean area, which absorbs more heat than ice and snow, further warming the planet.

greenhouse effect The accumulation of gases in the Earth's atmosphere, which then acts like the glass in a greenhouse, holding heat from the sun close to the Earth.

greenhouse gases The gases (primarily carbon dioxide, methane, and nitrous oxide) that accumulate in the atmosphere and act like the glass in a greenhouse, holding heat from the sun close to the Earth.

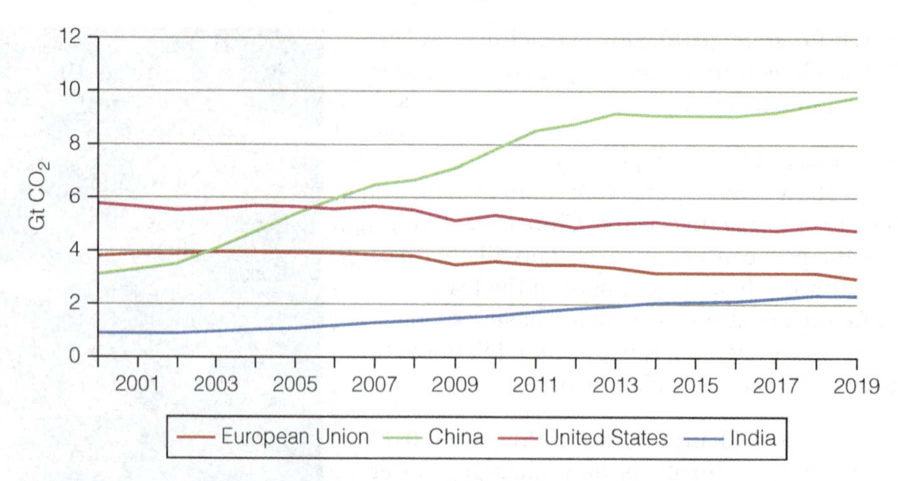

Figure 13.2 CO_2 Emissions for Selected Economies by Year, 2000–2019
SOURCE: International Energy Agency. 2020a. *Global CO_2 Emissions from Fuel Combustion: Overview 2020.* Available at www.iea.org.

Experts argue that annual emissions need to fall by 50 percent between 2020 and 2030 and reach net zero by 2050, a goal shared by President Biden, in order to limit global warming to 1.5 degrees Celsius (Intergovernmental Panel on Climate Change [IPCC] 2018). Even at that level, impacts will be felt but the United Nations IPCC report concludes, "limiting global warming to 1.5C, compared with 2C, could reduce the number of people both exposed to climate-related risks and susceptible to poverty by up to several hundred million by 2050."

climate deniers People who do not accept the scientific consensus that human-caused global warming and climate change are scientific facts.

Beliefs and Attitudes about Global Warming and Climate Change

Sixty-three percent of Americans believe that climate change is affecting their local community a great deal or some; this is especially true of people who live near the coast (Kennedy 2020). Beliefs about the impact of human activity on climate vary across partisan lines with Democrats more likely to believe human activity has at least "some" impact on climate change. More than a third (35 percent) of Republicans and 6 percent of Democrats are "**climate deniers**"—people who do not accept the scientific consensus that human-caused global warming and climate change are scientific facts (see Figure 13.3).

Wolfgang Kaehler/LightRocket/Getty Images

The Tongass National Forest in Alaska is the largest national forest in the United States. Known as America's Climate Forest (Bowe 2020), it stores an estimated 400 million metric tons of CO_2 and takes in another 3 million metric tons per year, the equivalent of taking 650,000 cars off the road each year. Even so, the Trump administration lifted restrictions on logging in the forest in September 2020.

Former President Trump, while in office, referred to the climate crisis as a "hoax," whereas President Biden has called climate change an "existential threat—not just to our environment, but to our health, our communities, our national security, and our economic well-being" (Biden 2020). More than half of Democrats *and* half of Republicans, including conservative Republicans, believe that the government needs to take more action to address global climate change and support initiatives such as planting trees, providing tax credits for carbon capture, instituting tougher regulations on power plant emissions, and developing alternative energy sources. And more than 45 percent of Republicans and Democrats support taxing corporations based on their carbon emissions and instituting tougher fuel efficiency standards for cars. In 2019, 65 percent of U.S. adults polled agreed that "protection of the environment should be given priority, even at the risk of curbing economic growth" (Saad 2020).

However, the fossil fuel industry has engaged in aggressive misinformation campaigns and lobbying efforts to influence policy makers to not take action to reduce carbon emissions, even going so far as to start their own climate change denial think tanks such as the Heartland Institute. The Heartland Institute has spent millions spreading disinformation through conferences, sponsored panels, op-eds, and mailing climate denial books to teachers, along with providing financial support to political candidates who support policies driven by climate denial (Banerjee 2017).

One of Heartland Institute's "policy experts," Kathleen Hartnett White, wrote an op-ed in 2016 stating that carbon dioxide—the most toxic greenhouse gas—is "the gas of life." Ms. White, whose education is in East Asian studies and comparative religion, was appointed by then President Trump to be the chairperson of the White House Council on Environmental Quality. The well-funded and aggressive misinformation campaign backed by the fossil fuel industry and exemplified by the work of the Heartland Institute has developed into a **climate denial machine**. Bolstered by this misinformation campaign, 150 members of the 116th Congress—all Republicans—have indicated they do not believe that human activity is leading to climate change (Hardin and Moser 2019), in direct opposition to the public they serve. In 2017, Scott Pruitt, then head of the Environmental Protection Agency (EPA) under the Trump administration, said he was not persuaded that human activity is a primary factor in global warming (Dennis and Mooney 2017) and defended the continued reliance on coal and oil by stating that the bible says we should "harvest the natural resources that we've been blessed with to truly bless our fellow mankind" (Brody 2018).

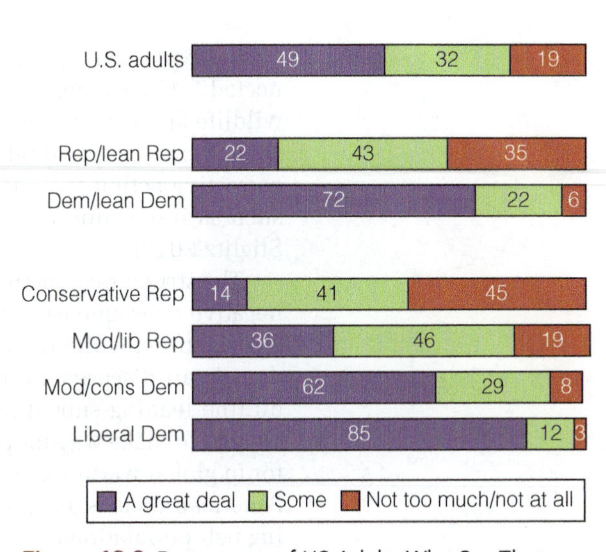

Figure 13.3 Percentage of US Adults Who Say That Human Activity Contributes to Climate Change
SOURCE: Tyson and Kennedy 2020.

@Fire DrillFriday

The **climate crisis** is here. How long will it take for world leaders to take meaningful action to stop it?

-Fire Drill Fridays

Sociological Theories of Environmental Problems

Climate change as a result of global warming is the most serious environmental problem facing humankind, but it is not the only environmental problem that must be tackled. Air, water, and land pollution; depletion of natural resources; loss of biodiversity; and other problems are discussed later in this chapter. Next we explore how the three main sociological theories—structural functionalism, conflict theory, and symbolic interactionism—can be applied to our understanding of environmental problems.

Structural-Functionalist Perspective

Structural functionalism views social systems (e.g., families, workplaces, societies) as composed of different parts that work together to keep the whole system functioning. Likewise, humans are part of a larger **ecosystem**, which consists of all the organisms living in a particular area, as well as all the nonliving, physical components of the environment—such as air, water, soil, and sunlight—that interact to keep the whole ecosystem functioning. Each living and nonliving part of the ecosystem plays a vital role in maintaining the whole; disrupt or eliminate one element of the ecosystem, and every other part could be affected. Thus, the human activity that has led to global warming is having negative effects not only on human beings but also on all of Earth's life forms (UNFCCC 2020b).

The structural functionalist view brings attention to how environmental issues affect and are affected by various social institutions, such as the economy. President Biden

climate denial machine A well-funded and aggressive misinformation campaign run by the fossil fuel industry and its allies that involves attacking and discrediting climate science, scientists, and scientific institutions.

ecosystem A biological environment consisting of all the organisms living in a particular area, as well as all the nonliving, physical components of the environment such as air, water, soil, and sunlight, that interact to keep the whole ecosystem functioning.

(2020) explained, "[O]ur environment and our economy are completely and totally connected." For example, as croplands become scarce or degraded, as forests shrink, as wildlife faces extinction, and as marine life dwindles, millions of people who make their living from these natural resources must find alternative livelihoods. On the other hand, corrective action to address climate change can lead to new job opportunities in areas such as innovation and production of clean energy and electric vehicles (Biden 2020; Stiglitz 2019).

The structural-functionalist perspective also raises awareness of latent dysfunctions—negative consequences of social actions that are unintended and not always recognized. For example, while the U.S. ban on trans fats in foods (see Chapter 2) is expected to reduce heart disease and improve health, the ban also means increased demand for palm oil (the leading substitute for trans fat), which means more deforestation as forests are cleared to make way for palm oil plantations (Worland 2015). Deforestation is a contributor to global warming. Similarly, medical advice to reduce red meat and increase fish in American diets is a contributor to the current problem of overfishing, which is threatening fish populations.

Conflict Perspective

The conflict perspective focuses on the roles that wealth, power, and the pursuit of profit play in environmental problems and solutions. The fossil fuel industry has used its wealth to support political candidates and policy makers who favor supporting the fossil fuel industry through such means as subsidies and deregulation. The fossil fuel industry gave millions of dollars to the reelection efforts of Donald Trump (Stone 2020). Before taking the position at the EPA in the previous administration, Pruitt received over $1.25 million in campaign donations from the fossil fuel industry (Drugmand 2020). In 2019, oil, gas, and coal companies spent $190 million lobbying members of Congress (Merkley 2020). These investments have paid off for the industry. In 2019, the fossil fuel industry received over $15 billion in federal subsidies (Coleman and Dietz 2019). And the Trump administration took 159 actions aimed at eliminating or reducing regulation of the fossil fuel industry. These included rolling back auto emission standards, opening the Arctic National Wildlife Refuge to drilling, rolling back regulations on methane leaks, and exiting the United States from the **Paris Climate Accord**, a 2015 international agreement to combat global warming (Loeb, Lavelle, and Feldman 2020).

Paris Climate Accord An international agreement which calls for the countries of the world to work together to reduce greenhouse gas emissions to ward off what could become extinction-level global warming and climate change.

In contrast, President Biden is one of more than 2,800 U.S. politicians who signed the "No Fossil Fuel Money Pledge," committing to not take more than $200 from the fossil fuel industry "and instead prioritize the health of our families, climate, and democracy over fossil fuel industry profits" (No Fossil Fuel Money 2020). Climate activists have questioned whether President Biden has strictly adhered to this pledge and have expressed alarm over his choice of Cedric Richmond as a senior advisor and liaison between climate activists and business interests. Representative Richmond received $341,000 from the fossil fuel industry during his ten years in Congress (Rosane 2020).

What do you think?

In 2019, one in four members of Congress and their spouses held stock in the fossil fuel industry (Kotch 2020). Republican members of Congress owned about $60.4 million in fossil fuel stock, and Democrats owned about $32.3 million. In 2019, Senators Brown and Merkeley introduced the *Ban Conflicted Trading Act*, which would prohibit members of Congress from buying or selling individual stocks. Brown argued, "Members of Congress serve the American people, not their stock portfolios" (Kotch 2020). Do you believe that members of Congress should be allowed to buy and trade stock while in office? Why or why not?

The capitalistic pursuit of profit encourages making money from industry, often disregarding the damage done to the environment. To maximize sales, manufacturers design products intended to become obsolete. As a result of this **planned obsolescence**, consumers continually get rid of used products and purchase replacements. For example, smartphone companies have been accused of a practice of "pushing" upgrades that slow down phones and making it virtually impossible for even tech-savvy consumers to fix phones themselves, leading customers to seek updated models (Elder 2019). The average life span of computer technologies in general has been rapidly decreasing, with a current overall average of four to five years. Some countries have adopted laws making it illegal for a company to knowingly shorten the life span of a product (Sustainability for All 2020).

Perceived obsolescence—the *perception* that a product is obsolete—is a marketing tool used to convince consumers to replace certain items even though the items are still functional. Fashion is a prime example as consumers are encouraged to buy the latest trends in clothing style every season, even though their current clothing may still be in good condition. As a result of this so-called fast fashion, the fashion industry is the second largest polluter in the world (Philip et al. 2020). Similarly, every year, car and smartphone companies release new models often with very slight upgrades but large marketing campaigns. The global market for electronic waste management and recycling alone is expected to reach $55.3 billion by 2022 (Elder 2019). Both planned and perceived obsolescence benefit industry profits, but at the expense of the environment, which must sustain the constant production and absorb ever-increasing amounts of waste.

The conflict perspective is also concerned with **environmental injustice** (also known as *environmental racism*)—the tendency for marginalized populations and communities to disproportionately experience adversity because of environmental problems. For example, 90 percent of public lands in northern New Mexico have been leased by the federal government to private companies for oil and gas drilling, despite the threat fracking poses to sacred Indian artifacts and the health of the local Native American communities (Nelson 2020). These deals increased in number under the Trump administration (see Figure 13.4).

In the United States, polluting industries and industrial and hazardous waste facilities are more often located in minority communities and communities with high rates of poverty (Kramar et al. 2018). As a result, non-White communities are disproportionately

> "Environmental injustice is about [the state] creating sacrifice zones where we place everything which no one else wants."
>
> **–MUSTAFA ALFI, FORMER HEAD OF THE EPA ENVIRONMENTAL JUSTICE PROGRAM**

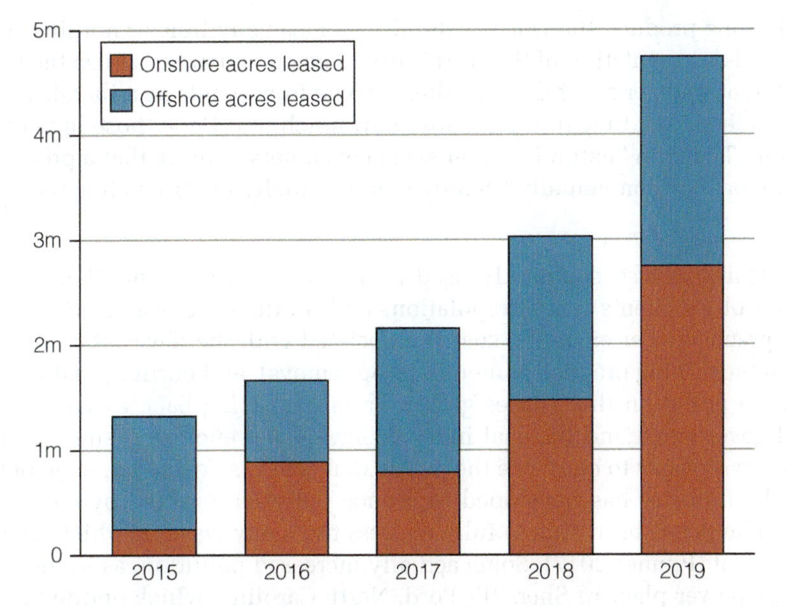

Figure 13.4 Public Lands and Water Leased by the Government to Private Companies for Oil and Gas Drilling
SOURCE: Holden 2020.

planned obsolescence The manufacturing of products that are intended to become inoperative or outdated in a fairly short period of time.

perceived obsolescence The perception that a product is obsolete; used as a marketing tool to convince consumers to replace certain items even though the items are still repairable.

environmental injustice Also known as *environmental racism*, the tendency for marginalized populations and communities to disproportionately experience adversity due to environmental problems.

exposed to air pollution and the health burdens associated with pollution (Mikata et al. 2018). President Biden's climate plan includes a focus on environmental justice and the establishment of an Environmental and Climate Justice Division of the U.S. Department of Justice (Berardelli 2020).

One factor that contributes to environmental injustice is known as "**Not in My Backyard**," or **NIMBY**, which refers to opposition by local residents to a proposed new development in their community. Residents who have wealth and political clout may be more successful at stopping developments, such as oil or gas drilling sites or hazardous waste storage facilities, "in their backyard," which are then located in more impoverished areas where residents have little political influence (Doshkin 2016). This chapter's *Social Problems Research Up Close* explores what happens when people decide they do want oil development in their backyards.

Symbolic Interactionist Perspective

The symbolic interactionist perspective focuses on how meanings, labels, and definitions learned through interaction and through the media affect environmental problems. The words we use reflect and influence how we relate to the environment. Native American Robin Wall Kimmerer (2013) explains that in her native language (Potawatomi), "[W]e speak of the land as *emingoyak*: that which has been given to us. In English, we speak of the land as 'natural resources' or 'ecosystem services,' as if the lives of other beings were our property" (p. 383). The indigenous language that conveys that land is a gift teaches people to be grateful for the Earth and its bounty, to respect and care for the Earth, and to use the Earth's gifts (such as plants that provide food and water that sustains life) for the benefit of all rather than for individual profit or gain.

Large corporations and industries that are environmentally damaging often use marketing and public relations strategies to portray their corporation, industry, or products as environmentally friendly—a practice known as **greenwashing**. For example, some brands of household items such as toilet paper and dish soap are advertised as "green," "all natural," or "earth-friendly." Marketing companies know that "green" sells, as consumers are becoming more eco-minded. But how valid are the environmental claims made on the labels of the products we buy? The Environmental Working Group's *Guide to Healthy Cleaning* warns:

> On a cleaning product, the word "natural" can mean anything or nothing at all—there is no regulation of the word's use. Some manufacturers use the term to mean that some or all of the ingredients come from plants or minerals rather than petroleum, but they rarely disclose how much or little of those ingredients is present. The term "natural" can mislead consumers to think that a product is safer or more environmentally friendly than it actually is. ("Decoding the Labels" 2015)

Greenwashing is also commonly used by public relations firms that specialize in damage control for clients whose reputations and profits have been hurt by poor environmental practices. For example, coal is associated with the devastation of communities through the mining practice of mountaintop removal, and burning coal is the biggest contributor to pollution that causes global warming. Dating back as far as the 1980s, the federal government and the coal industry have spent enormous sums of money on "clean coal" campaigns to convince the public that coal is safe. In fact, none of the many processes the industry has developed to reduce pollution emitted by coal mines and coal power plants has been able to fully address the many ways in which coal pollutes the environment (Plumer 2017). Some actually increased pollution, as in the case of the Duke Energy power plant in Sherrill's Ford, North Carolina, which emitted more nitrogen oxides *after* switching in 2012 to "refined" or "clean" coal (McLaughlin 2018). And one of the carcinogenic chemicals used to refine the coal leached into nearby waterways, contaminating the water supply for more than a million people. Thus, environmentalists argue, coal can hardly be called "clean."

"Not in My Backyard" Opposition by local residents to a proposed new development in their community, also known as NIMBY.

greenwashing The way in which environmentally and socially damaging companies portray their corporate image and products as being "environmentally friendly" or socially responsible.

social problems RESEARCH UP CLOSE
Please in My Backyard

Many research studies on development projects find that opposition among local residents is high, with many taking on a NIMBY perspective—Not in My Back Yard. In this study, researchers investigated why residents of a rural Pennsylvania community invited fracking into their backyards and supported it through what they call *quiet mobilization* (Jerolmack and Walker 2018). During the study, residents also described the impacts fracking was having on their land, water, and quality of life.

Sample and Methods

Much of rural Pennsylvania sits on top of the Marcellus shale formation, with an estimated 300–500 trillion cubic feet of natural gas. This has made it a prime site for development by oil companies and an ideal location for a study on how communities respond to new oil development projects. The researchers engaged in extensive participant observation between 2013 and 2018, attending community events and meetings, attending open houses at local colleges, volunteering at the local state forest, and observing occasional protest events. They also interviewed 36 White landowners, 26 of whom had leased land to gas companies for the purpose of fracking. About half of the participants were working-class and "land-poor," having inherited land but having fewer other resources by which to support themselves.

Findings

The researchers found that many in the community viewed fracking as beneficial for its monetary rewards to landowners as well as for the creation of jobs in the community. The community—once a lumber capital—had seen hard economic times. Residents were not unaware of some of the risk of fracking but prioritized short-term monetary gains; some did brush off what they had seen on TV about the risks as "liberal propaganda." Through *quiet mobilization*—everyday conversations with neighbors, local civic meetings, discussions at churches, and other forms of routine civic engagement—community members decided to stick together, placing what they viewed as individual sovereignty and private property rights above potential impacts on neighbors. Even many who chose not to lease their own land supported neighbors who did.

Landowners who leased did experience some financial benefits through lease bonuses and royalties, although few made enough money to do more than stabilize their financial position. However, landowners quickly learned that the companies had not been fully transparent with them. Damage to property was much more extensive than companies had led them to believe with their glossy brochures containing photographs of pristine landscapes. Landowners were surprised to find that fine print in the contracts allowed companies to extend their five-year leases in perpetuity once drilling began. Thus, they lost the right to reclaim their land. Many landowners found, once they signed the contract, that the company behaved as if the land belonged to them, driving across private sections of the property, installing security cameras without notice, and even threatening to have landowners arrested if they stepped onto the well pad on their own land. Five residents who had not leased found their well water contaminated by methane from wells on a neighbor's property. Unable to use their own water, they fought the company for years to get clean water. Community members also complained of loud truck traffic on their rural roadways seven days a week—even on Christmas. Finally, landowners ultimately filed a class-action lawsuit against one of the companies for $5 million in "post-production cost" fraudulent deductions in their royalty checks. Despite the community's distrust of government, the state did step in to investigate and passed legislation to protect lessors from these types of fraudulent deductions in the future.

SOURCE: Jerolmack and Walker 2018.

Energy Use and Environmental Problems

Environmental problems include depletion of natural resources; air, land, and water pollution; problems associated with fracking; global warming and climate change; threats to biodiversity; light pollution; and environmental illness. Because many of these environmental problems are related to the ways that humans produce and consume energy, we begin with an overview of global energy use and its impacts. The difficult reality is that all forms of energy have environmental impacts, and some have negative impacts on health. The question is how to minimize environmental and health impacts of energy production and use.

More than two-thirds of the world's energy comes from fossil fuels, which include oil, natural gas, and coal (see Figure 13.5). Many of the major environmental problems facing the world today—air, land, and water pollution; destruction of habitats; biodiversity loss; global warming and climate change; and environmental illness—are linked to the production and use of these fossil fuels. Fossil fuels are referred to as nonrenewable sources of

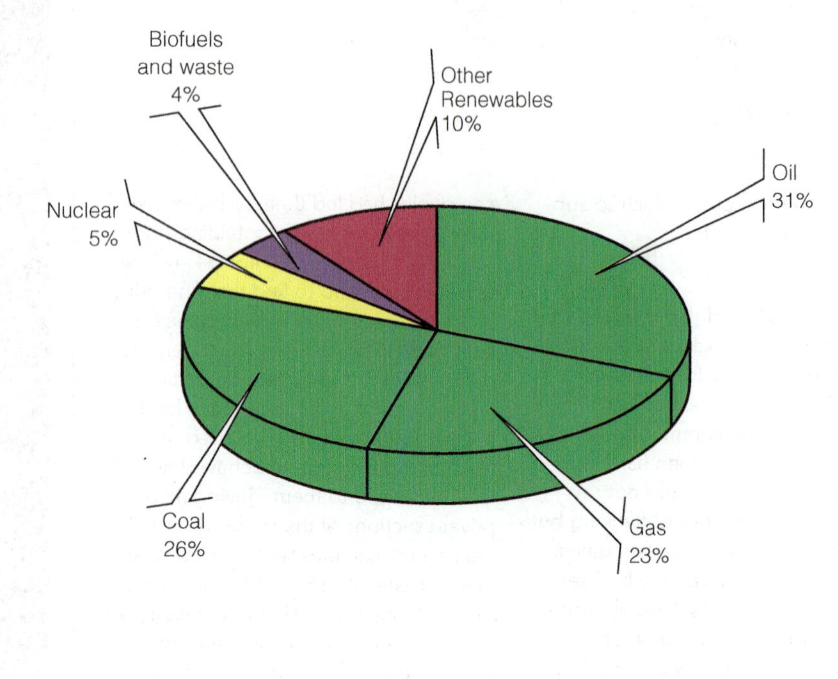

Figure 13.5 World Energy Supply by Source, 2019
SOURCE: International Energy Agency 2020b.

energy because they take millions of years to create and therefore, once used, cannot be replenished. Nuclear power makes up only 5 percent of the world's energy supplies, but its use is expanding as some countries are turning to nuclear to reduce their CO_2 emissions. While nuclear does have advantages in this regard, it has its own environmental impacts and safety risks. Biofuels and other renewable sources of energy, which have the least environmental impacts, currently make up less than 15 percent of the world's energy supply, but the Paris Climate Accord is leading to increased research and investment into these alternatives, as we shall discuss.

Crude Oil/Petroleum

In 2018, the United States led global oil-production producing 15 percent of the world's oil supply in 2019 (United States Energy Information Administration 2020a). Texas is the largest oil-producing state, generating 41 percent of all U.S. crude oil, although 16 percent takes place in the Gulf of Mexico where thousands of oil and natural gas platforms operate in waters up to 6,000 feet deep.

Extracting petroleum is one of the most environmentally hazardous processes in the world (Turgeon and Morse 2018). Conventional oil extraction involves using traditional vertical drilling and pumping techniques to extract oil from underground land or offshore reservoirs. While oil spills can happen on land or sea, spills are especially hazardous to the environment when they happen in the water as millions of barrels of oil can spill into the ocean before the well is capped. For example, a 2010 explosion on the Deepwater Horizon oil platform led to 4 million gallons of oil flooding into the Gulf of Mexico. The explosion left 11 people dead and 17 injured, killed over 6,000 animals, and resulted in the largest environmental damage settlement in history (NOAA 2017; Turgeon and Morse 2018). The leak began on April 20 and was not capped until July 15 while every day an average 40,000 barrels of oil flowed into the ocean.

Unconventional or newer techniques include extracting oil from **tar sands**, which are large, naturally occurring deposits of sand, clay, water, and a dense form of petroleum that looks like tar. Most of the world's tar sands are in the eastern part of Alberta, Canada. **Tar sands oil** has been referred to as the world's dirtiest oil (Natural Resources Defense Council 2017). Converting tar sands into liquid fuel requires energy and generates high levels of greenhouse gases, and the process also leaves behind large amounts of toxic waste. Estimates indicate that the extraction of oil from tar sands releases 17 percent more greenhouse gases than other extraction techniques (Natural Resources Defense Council 2017).

tar sands Large, naturally occurring deposits of sand, clay, water, and a dense form of petroleum that looks like tar.

tar sands oil Oil that results from converting tar sands into liquid fuel. It is known as the world's dirtiest oil because producing it requires energy, generates high levels of greenhouse gases, and leaves behind large amounts of toxic waste.

The mining of tar sands also requires large amounts of water and involves destruction of forests and wetlands, which disrupts wildlife habitats. Further, transporting tar sands oil through a pipeline involves the risk of leakage. As this type of oil is denser than others, it must be mixed with other hazardous chemicals in order to be transported through pipes making any leaks particularly dangerous to safe drinking water (National Academies of Sciences, Engineering, and Medicine 2016).

Because tar sand oils are denser than traditional oils, they are also harder to clean up when leaks do occur (Natural Resources Defense Council 2017). The controversial Keystone XL pipeline is slated to carry 830,000 barrels a day of tar sands oil from Canada to the Gulf of Mexico (Borunda 2020), crossing over land and water sources on the sovereign land of the Standing Rock Sioux Indians and that of other tribes. The Trump administration granted approval of the Keystone XL pipeline in 2017; however, several legal suits between 2017 and 2020 led to repeated stops and starts of construction. In November 2019, an estimated 383,000 gallons of oil leaked from the pipeline onto wetlands in North Dakota (Public Broadcasting Service 2019). Subsequently, in 2020 a judge ordered that the company must seek permits for every location at which the pipeline is slated to cross bodies of water or wetlands (Borunda 2020). A judge also ordered that use of the Dakota Access Pipeline be halted while its environmental impacts are reviewed. On his first day in office, President Biden issued an order blocking the Keystone XL pipeline (Deliso 2021).

AP Images/David Goldman

Military veterans and indigenous people joined in one of many protests against the Keystone XL Pipeline project. In an effort to stop protestors, South Dakota has twice passed laws attempting to restrict demonstrations. The American Civil Liberties Union has filed suit arguing the laws infringe on First Amendment rights (American Civil Liberties Union [ACLU] 2020).

Coal

Coal, which is nothing more than combustible rock, contributes to 26 percent of the world's energy. About two-thirds of U.S. coal production is through surface mining, sometimes called **strip mining**, by which large machines are used to remove topsoil and layers of rock to expose coal less than 200 feet underground (United States Energy Information Administration 2019a). This includes a process called **mountaintop removal**, whereby the tops of mountains are dynamited and removed to access the coal below. The other third of coal production happens through underground mining in which miners ride elevators down to tunnels which run thousands of feet under the surface of the Earth.

The mining of coal releases methane gas which must be vented out of mines in order to avoid explosions. In 2017, methane emissions from coal mining and abandoned coal mines accounted for about 9 percent of methane gas emissions in the country (United States Energy Information Administration 2019a). Mining also leads to the drainage of contaminated water into surrounding areas. Coal processing generates a liquid called *slurry* which is typically held in containment ponds on site and contains arsenic, barium, lead, and manganese (Greene and McGinley 2019). This liquid contaminates nearby wells, streams, and water supplies. A meta-analysis of 28 studies conducted around the globe concluded that living in the vicinity of coal mining is associated with increased morbidity and mortality from 78 recognized medical conditions, including cancers, prenatal conditions, chromosomal conditions, lung and heart disease (Cortes-Ramirez et al. 2018).

It has been well documented that coal miners often develop a condition known as **black lung disease** which damages their lungs and makes it difficult to breathe, often leading to death (Cohen et al. 2016). The condition is the result of exposure to coal dust. Since as early as 1969, federal regulations have been in place that require dust monitoring and protective equipment within coal mines (Greene and McGinley 2019).

In the 1990s, however, evidence emerged that incidences of black lung had increased, even among young miners with relatively short exposure. After over 20 years of evidence, the Mine Safety and Health Administration finally introduced tighter regulations in 2010. In the same year, autopsies on miners killed in a 2010 mine explosion revealed that almost three-quarters—even young miners with less than five years' exposure—had black lung disease, a rate much higher than previously documented.

A multiyear investigation by National Public Radio and the PBS program *Frontline* concluded that coal company managers had engaged in "intentional cheating to avoid compliance with mine dust regulatory standards" (Berkes 2012). They also found that government regulators failed to take action when they did have evidence of excessive

> "They're going to kill us with their greed."
>
> **–SAM SAGE, ADMINISTRATOR OF THE COUNSELOR CHAPTER HOUSE OF THE NAVAJO NATION**

strip mining A process of coal mining by which large machines are used to remove topsoil and layers of rock to expose coal less than 200 feet underground.

mountaintop removal A process of coal mining in which the tops of mountains are dynamited and removed to access coal seams below.

black lung disease A condition resulting from exposure to coal dust, which leaves the lungs of coal miners scarred, shriveled, and black, making it difficult for them to breathe and causing debilitating coughing.

Mandel Ngan/AFP/Getty Image

The most common method of coal mining in the United States today is mountaintop removal. The top of a mountain is blasted with explosives, and the debris is bulldozed into the surrounding valleys. This mountaintop in West Virginia once looked like the surrounding mountains.

and toxic mine dust exposure. In 2019, the Department of Justice charged multiple coal mine managers with fabricating dust monitoring results (Greene and McGinley 2019).

Mountain top removal mining has been particularly active in the last 30 years in the Appalachian Mountains of West Virginia and Ohio (U.S. Energy Information Administration 2019). This relatively new practice avoids many of the regulations for underground coal mines and requires fewer employees thereby cutting costs. The practice first requires deforestation of the area, releasing CO_2 into the atmosphere. The explosions used for mountaintop removal spread polluted rock and dirt into local streams, harming aquatic wildlife.

Repeated explosions cause cracks in the walls of nearby homes and produce air and noise pollution that affect local residents (Strobo 2012). Huge machines called draglines then push the remaining rock and dirt into surrounding valleys and streams to uncover the coal beneath (Earth Justice 2020). An estimated 2,000 miles of streams and headwaters that provide drinking water for millions of residents have been buried in the past few decades. While companies are required by law to engage in reclamation of the land, most plant only non-native grasses rather than mature trees, meaning that the Appalachian region may soon change from a net carbon sink area to a net carbon source (Appalachian Voices 2020).

Studies have linked the practice of mountaintop removal mining to increased cancer and heart disease deaths (Krometis et al. 2017), increased birth defects (Holzman 2011), and increased depression (Canu et al. 2017) in the Appalachian region. A comprehensive study by the National Academy of Sciences into the health impacts of mountaintop removal coal mining was authorized by the Obama administration in 2016, but the Trump administration cancelled the study without explanation in mid-2017 (Greene and McGinley 2019).

Natural Gas

Like oil, natural gas is the result of millions of years of heat and pressure that turned plant and animal remains deep under the surface of the land and ocean floor into oil and natural gas. Natural gas can be recovered through conventional drilling and it is also recovered during the process of drilling for oil (United States Energy Information Administration 2019b). However, companies have developed a new technique to remove natural gas found in the small pores within formations of shale, sandstone, and other sedimentary rocks. This is called hydraulic fracturing, or **fracking**—a process that involves injecting at high pressure a mixture of water, sand, and chemicals into deep underground wells to break apart the shale rock and release gas. Gas production in the United States has nearly doubled since 2005, largely due to hydraulic fracturing techniques (United States Energy Information Administration 2019b). Fracking is most common in the United States, although it also occurs in China and Canada (Steiner and Schwartz 2019). Within the United States, Texas and Pennsylvania are the two biggest producers of natural gas through fracking (Gross 2020). Natural gas now makes up 23 percent of the world's energy sources, up from 21 percent in 2014 (International Energy Agency 2020b).

fracking Hydraulic fracturing, commonly referred to as fracking, involves injecting a mixture of water, sand, and chemicals into drilled wells to crack shale rock and release natural gas into the well.

Fracking leads to significant environmental problems (Inglis and Rumpler 2015; Ridlington, Norman, and Richardson 2016). It involves the use of chemicals that contaminate land, water, and air. Oil and gas companies do not have to disclose to the federal government the chemicals they use in fracking, yet scientists have analyzed fracking water and found hundreds of chemicals, including hydrochloric acid, methanol, benzene, formaldehyde, ethylene glycol, and sodium hydroxide. Many of these chemicals cause cancer and others are associated with brain, lung, heart, kidney, and reproductive organ damage, as well as birth defects and eye and skin irritations (Ridlington, Norman, and Richardson 2016).

Fracking also extracts materials from underground that can be equally or more toxic than the hydraulic fracturing fluid, including radioactive material, arsenic, and lead. Fracking operations have triggered earthquakes in at least five states. Fracking also contributes to global warming through the release of methane from fracking wells and waste storage facilities; in fact, the single greatest leak of methane gas in U.S. history came from a failing storage well in Southern California, which released more than 100,000 tons of methane (Ridlington, Norman, and Richardson 2016).

Erik McGregor/LightRocket/Getty Images

Fracking generates a large amount of hazardous and highly volatile liquids that must be disposed of. Residents and city officials of Exton, Pennsylvania, have expressed strong opposition to the SUNOCO Mariner II East Pipeline, which is designed to transport used fracking liquid through their community on its way to be shipped overseas for plastic manufacturing.

Fracking requires massive amounts of water, which is problematic in drought-stricken areas such as California, Oregon, and Nevada (Ridlington, Norman, and Richardson 2016). This can cause water shortages for farmers who must compete with oil and gas companies for irrigation sources. Billions of gallons of toxic fracking wastewater produced each year must be stored, transported, and disposed of. But leaks, overflows, and illegal dumping of wastewater allow chemicals to flow into local land and water supplies (Johnston and Cushing 2020). An Environmental Protection Agency report concluded that fracking can impact drinking water, although it argued that there are insufficient data to conclude how much of a problem this is (EPA 2016).

In fact, the accumulated evidence documents significant health impacts to animals and humans as a result of chemical exposure, noise and light pollution, radioactive material, and seismic activity. Pets, farm animals, and wildlife raised near fracking sites have become sick and died, apparently as a result of contact with and/or consumption of contaminated water from fracking sites (Bamberger and Oswald 2014; Malin and DeMaster 2016). Beef cattle and other animals raised for human consumption are not tested for contamination from fracking chemicals, raising concerns about whether humans are eating contaminated animal products. People who live or work near fracking sites are at increased risk for migraines, chronic rhinosinusitis, severe fatigue, and asthma (Gorski and Schwartz 2019).

Babies born to mothers living near fracking sites are significantly more likely to have birth defects and to be born underweight. Evidence also suggests an association with certain cancers and neurodegenerative diseases. Land and home values close to fracking sites also decline (Muehlenbachs, Spiller, and Timmins 2015). When individuals seek compensation from oil and gas companies for harm done to their health and/or property due to fracking, the companies typically settle out of court for a release of indemnity for damages and a nondisclosure agreement (Malin and DeMaster 2016) (see Figure 13.6).

There is virtually no federal regulation of fracking in the United States. A loophole inserted into the 2005 *Energy Policy Act* exempts fracking from sections of the *Safe Drinking Water Act* (1974), the *Clean Air Act* (1970), the *Clean Water Act* (1972), and the *Resource Conservation and Recovery Act* (*RCRA*) (Gorski and Schwartz 2019). This loophole is known as the Halliburton Loophole because then Vice President Dick Cheney who presided over the deliberations had been CEO of Halliburton, which was one of several companies that benefited from the legislation. While the Obama administration put into place Interior Department regulations that would close several of these loopholes, these were overturned in court in 2016 on the argument that states should control these regulations. Yet states vary in their regulations (Schipani 2017). For example, some states limit or ban methane venting, while other states have no regulation on methane venting, despite the fact that methane has a direct impact on global warming. Texas, the largest producer of natural gas in the United States, has no regulations on methane emissions (Gross 2020).

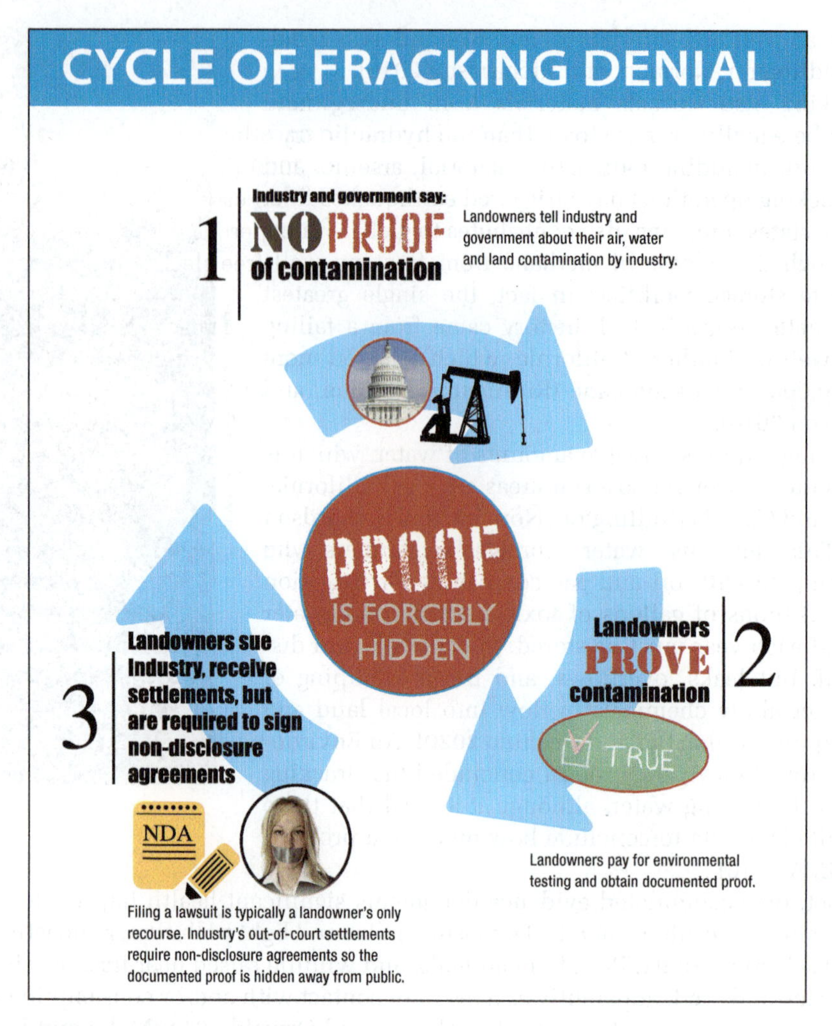

Figure 13.6 Cycle of Fracking Denial
SOURCE: Earth Works 2013.

Only 28 states require disclosure of the chemicals used in fracking, allowing companies to withhold chemical names under the argument of "trade secrets." Also, no legislation governs interstate waste disposal issues, allowing a company to carry wastewater from a state with stricter regulations to a state with looser regulations for disposal (Gorski and Schwartz 2019).

Because of health and environmental concerns, some countries—including France, Germany, Wales, and Scotland—have banned fracking. It has also been banned in New York, Massachusetts, Maryland, and Vermont. Hundreds of towns, cities, and counties across the United States have also passed bans or moratoriums on fracking (Food & Water Watch 2020). In 2018, the Permanent Peoples' Tribunal (PPT), an internationally recognized coalition of human rights lawyers and academics, conducted hearings to determine whether fracking violates human rights. In its Final Advisory Opinion released on April 12, 2019, the PPT Council recommended a worldwide ban on the practice based on its assessment that it violates the right to life, to water, and to full information and participation (Permanent People's Tribunal 2019). President Biden has said he will not ban fracking, but he supports no new fracking on federal land (Berardelli 2020).

Nuclear Power

Nuclear power generates approximately 10 percent of the world's electricity and is the second largest global source of low carbon electricity after hydropower (Krikorian 2020). As of 2019, there were 450 operable nuclear power reactors in the world, with another

53 under construction. Four countries—Bangladesh, Belarus, Turkey, and United Arab Emirates—were building their first nuclear power plants in 2019. In contrast, 15 countries have pledged not to rely on nuclear power, and eight have pledged to phase it out.

The United States has more nuclear reactors than any other country, with 95 operable U.S. nuclear reactors and two under construction in 2020 (United States Energy Information Administration 2020b). Most are located in plants east of the Mississippi River (see Figure 13.7). Nuclear power plants produce about 20 percent of the electricity in the United States and over 60 percent of the country's carbon-free electricity (World Nuclear Association 2020a).

Nuclear power is highly controversial. Some have argued that nuclear power is one solution to the problem of global warming as it does not contribute to carbon emissions and it produces predictable levels of energy at a relatively low cost once the reactors have been built (Rhodes 2018). However, nuclear power plants have powerful negative impacts on the environment (Asaff 2020). Nuclear power requires uranium, a highly unstable element that must be mined, transported, and stored under strictly monitored conditions (World Nuclear Association 2020b). Uranium mining releases arsenic and radon into the environment, causing health problems among miners and local populations.

Nuclear power plants use an extraordinary amount of water to cool fuel rods (Asaff 2020), and the water released from power plants is superheated and damages the chemistry of the ocean or lake into which it is discharged, killing plants and animals. Nuclear plants also produce radioactive waste which remains dangerous to all life on Earth for hundreds of years. Few countries have developed long-term solutions to the problem of nuclear power radioactive waste (World Nuclear Association 2020c). In the United States, radioactive waste associated with military operations is disposed of in a deep geological repository in New Mexico, but no long-term disposal sites have been developed for waste from nuclear power plants. Some reactors currently under construction in other countries are designed to utilize accumulated nuclear waste, thus eliminating the risks associated with storing it indefinitely (Wallach 2020).

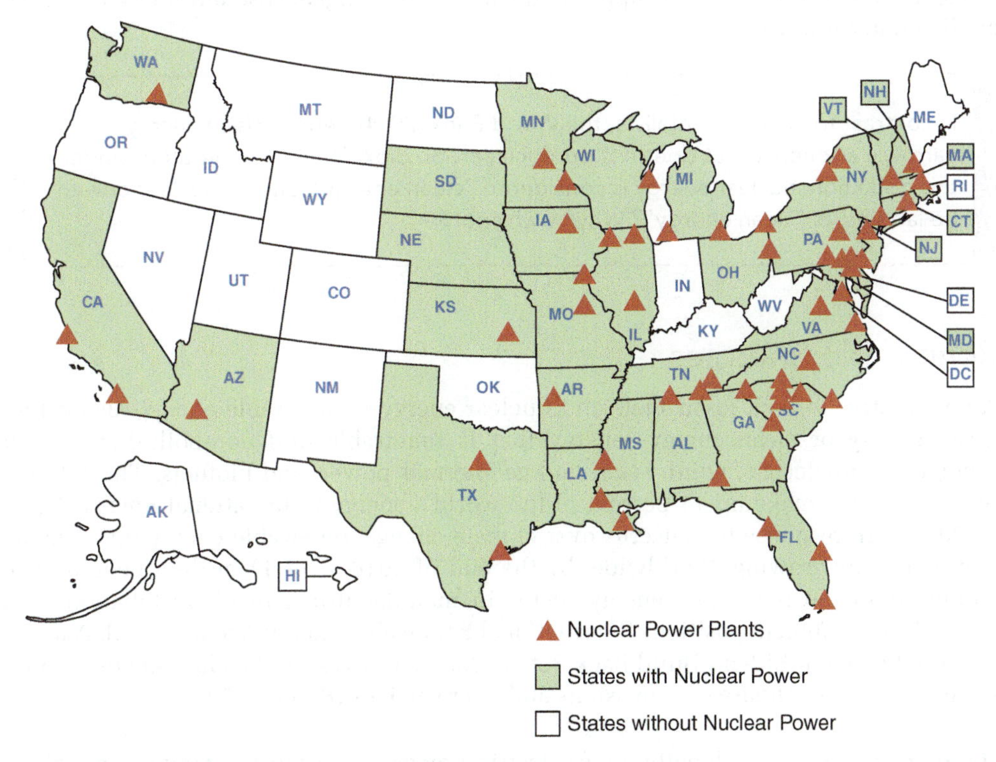

Figure 13.7 Location of Nuclear Power Plants in the United States
SOURCE: United States Energy Information Administration 2020a.

Public perception of nuclear power is mixed, with almost half of Americans believing it is unsafe (Reinhart 2019). The primary fears are that equipment failures, natural disasters, or terrorist attacks could result in leaks of harmful radiation.

The deadliest nuclear power plant accident in history occurred in 1986 when an explosion at the Chernobyl Nuclear Power Plant in the Ukraine killed 31 workers due to direct exposure to the explosion and/or radiation in the plant. Estimates of deaths due to long-term radiation exposure from the plant range from 4,000 to 60,000 (Ritchie 2017). The accident resulted from poor construction and lack of compliance with safety regulations, but the number of deaths would undoubtedly have been reduced if the Ukrainian government had not waited days to evacuate residents.

A more recent nuclear accident occurred in 2011, when a 9.0 magnitude earthquake and tsunami caused a series of equipment failures and nuclear core meltdowns at the Fukushima Daiichi nuclear power plant in Japan, forcing the evacuation of 200,000 people living in communities around the plant (World Nuclear Association 2020d). Unlike the Chernobyl disaster, the Japanese government acted swiftly and while approximately 574 deaths have been attributed to the Fukushima disaster, only one was the result of exposure to radiation (Ritchie 2017). Two months after the Fukushima Daiichi accident, the Japanese government raised the nationally recognized level of acceptable radiation exposure to 20 times the international standard, allowing them to progressively lift evacuation orders (Little 2019). Citizens and experts alike question whether these levels are safe for long-term living, especially for children. Odaka, the closest village to the plant, stayed closed to residents until 2019; only a quarter had returned by summer 2020.

A full cleanup of the Fukushima accident is expected to take decades, cost billions of dollars, and require a specially designed generation of robots because high gamma radiation destroyed the first robots that were used (Martin 2019). And, the Japanese government is faced with a dilemma of what to do with tens of thousands of gallons of contaminated water used to cool the reactor. Running out of sufficient storage space, they have proposed releasing the water containing radioactive carbon into the ocean where it could negatively impact sea and even human life (Woodyatt and Wakatsuki 2020). Not surprisingly, trust in the Japanese government's decision-making around the risks associated with radiation is low, making the future of nuclear power in Japan uncertain (Little 2019). President Biden (2020) supports research to investigate cost and safety issues regarding nuclear power.

What do you think?

In contrast to environmentalists, the World Bank (2014) defines *clean energy* narrowly as energy that does not produce carbon dioxide when generated. Under this definition, nuclear energy is considered "clean energy." Should nuclear power be labeled as "clean energy"? Why or why not?

Renewable Energies

An alternative to both fossil fuels and nuclear energy is renewable energy, also called **green energy** or clean energy—energy that is renewable *and* non-polluting—which includes hydroelectric, wind, solar, and geothermal power, and biofuels. These forms of green energy make up 14 percent of the world's energy (International Energy Agency 2020b). In response to concerns over climate change, renewable energy investments worldwide are growing. Worldwide, by the end of 2018, over 11 million people were employed in the renewable energy sector in manufacturing, retail, installation, and research/development positions (International Renewable Energy Agency 2019). A major pillar of President Biden's Build Back Better Plan is to invest in the clean energy economy to reduce greenhouse gas emissions and to create jobs (Biden 2020).

green energy Also known as clean energy—energy that is nonpolluting and/or renewable, such as solar power, wind power, biofuel, and hydrogen.

Hydroelectric Power. Globally, hydroelectric power is the most common form of renewable energy (see Figure 13.8). Using the power of water for energy dates back before the Industrial Revolution. Modern hydroelectric power involves generating electricity from water moving through a turbine. The turbine may be located in a dam, a diversion canal above a waterfall, or a penstock connecting two reservoirs (one higher than the other). In the latter construction, when demand for electricity is high (e.g., during hot summer afternoons), water can be pumped back up from the lower reservoir to fuel the turbines again.

China generates more hydropower than any other country in the world, three times that of Brazil and the United States, which are in second and third place (International Hydropower Association 2020). Hydropower has gained momentum in the last two decades as a means for developing countries to meet their energy needs, as well as the climate goals outlined in the Paris Climate Accord (Law and Troja 2019).

Trevor Mogg/Alamy Stock Photo

In 2011, the Fukushima Daiichi nuclear power plant in Japan was hit by a tsunami that resulted in multiple core meltdowns and other equipment failures. Although the government acted quickly to evacuate 200,000 people from the surrounding communities, residents question whether the new level of "acceptable radiation exposure" set months after the accident is safe for children long term.

The Hoover Dam and the Grand Coulee Dam were both constructed in the United States during the Great Depression as a part of Roosevelt's New Deal (Bureau of Reclamation 2020). By the 1940s, hydropower accounted for 40 percent of the nation's electricity. Today, the Grand Coulee Dam generates enough electricity to power 4.2 million households.

Although historically thought to be a nonpolluting and cost-efficient way to generate power, hydropower can have a number of negative environmental impacts. Water in reservoirs and artificial lakes tends to evaporate faster than water in flowing rivers; in fact, evaporation from the world's reservoirs accounts for 7 percent of all freshwater consumed by human activities (Earth Law Center 2017). Reservoirs also contribute to the emission of greenhouse gases as the vegetation growing on the bottom of the reservoir breaks down and sediment-rich water no longer makes it to the ocean to feed carbon dioxide–consuming algae.

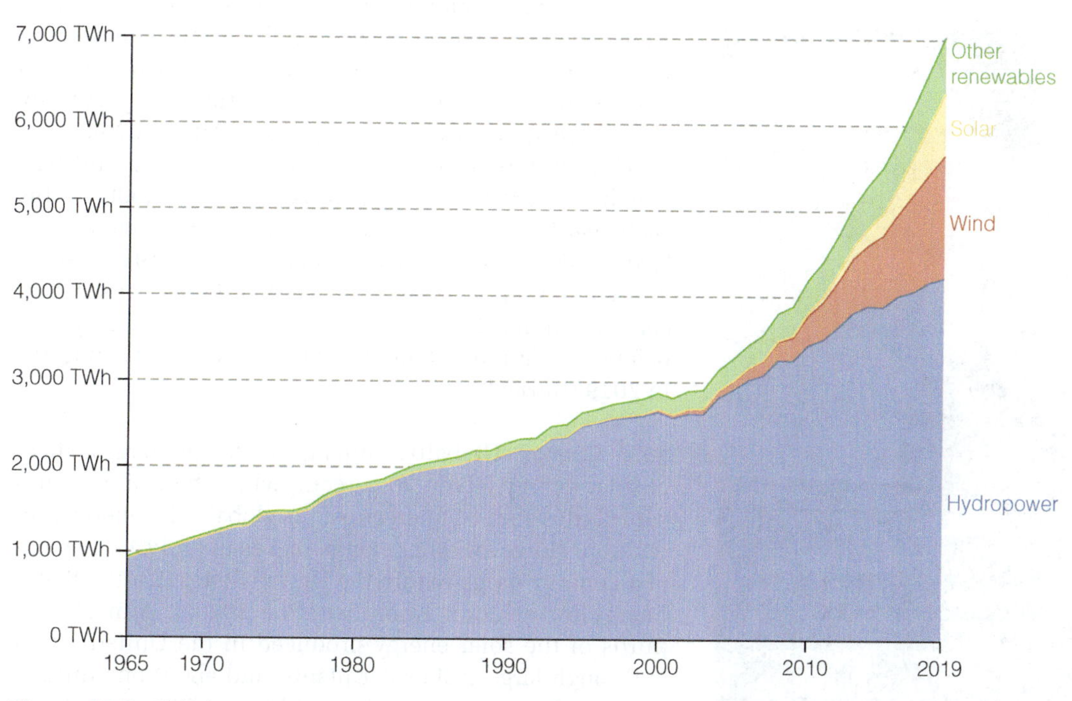

Figure 13.8 Global Renewable Energy Generation, 1965–2019
SOURCE: Ritchie and Roser 2019.

Most importantly, hydropower is criticized for disrupting local ecosystems and displacing people and wildlife (Bagher et al. 2015). China's Three Gorges Dam, the world's largest hydropower plant, led to the displacement of 1.4 million people and the destruction and flooding of more than 1,000 towns and villages (Tran and Provost 2012). It also negatively impacted native fish populations (Cheng et al. 2018) and water supplies for those living downstream from the dam (Su et al. 2019). The Itapiu Dam between Brazil and Paraguay displaced over 40,000 families and destroyed 700 square kilometers of rainforest (Tran and Provost 2012). Around the world, aging dams pose a serious threat to human life and property. In 2017 almost 200,000 people living in communities near the Oroville Dam in northern California were evacuated due to fears that the dam's emergency spillway would fail, causing massive flooding downstream (Earth Law Center 2017). Also, in 2017 a dam in Nebraska failed flooding homes, farmland, and railway lines downstream. It is estimated that $64 billion is required to repair aging dams in the United States (Newburger 2020) and that $300 billion is needed to secure aging dams around the world (Earth Law Center 2017).

@Mike Hudema

The trains in the **Netherlands** are 100% #windpowered. Over 5,500 trips a day 100% #renewable.

We have the solutions, let's implement them.

#ActOnClimate #climateaction #climate #energy #go 100re #GreenNewDeal

-Mike Hudema

In the early 21st century, largely in response to widespread outcries about the impacts of hydropower plant construction, the World Commission on Dams acknowledged that greater attention needed to be paid to the social and environmental impacts of dam construction (Law and Troja 2019). What developed since is a Hydropower Sustainability Assessment Council representing multiple global stakeholders as well as the development of a Hydropower Sustainability Assessment Protocol which is used to assess projects against 26 social, environmental, and governance performance criteria (IHA Sustainability Ltd. 2020).

Wind Farming. Utilizing wind farming as a renewable source of energy was first identified as a goal in 1978 by then President Jimmy Carter. Although it has taken decades to achieve, the United States is the home to five of the largest wind farms in the world and generates enough electricity through wind farming to power over 28 million homes (Office of Energy Efficiency & Renewable Energy 2020). The U.S. Department of Energy (2015) projected that the country has the capacity to reach 35 percent wind power by 2050 should it choose to do so (U.S. Department of Energy 2015). Wind farms produce no carbon emissions; once they are in operation, most repay their own carbon footprints (from the construction process) within six months of operation. They also use no water and the average land-based wind farm leaves 98 percent of the land on which it is based undisturbed, leaving it free for farming or grazing by animals (American Wind Energy Association 2020).

However, wind turbines are a risk to birds and bats that live nearby, causing approximately 150,000 deaths each year in the United States, or about 1.5 percent of the bird deaths caused by cats annually (Fox 2019). There is no evidence that noise from wind turbines causes cancer, as claimed by former President Trump in a speech before the National Republican Congressional Committee in 2019 (Burke 2019).

Bettmann/Getty Images

In 1979, then President Jimmy Carter promoted the new solar industry by installing solar panels on the roof of the White House. The panels provided hot water to the staff wing of the White House. When Ronald Reagan took over, he had the panels removed as his campaign was allied with the fossil fuel industry.

Solar Energy. Globally, China generates the largest share of solar energy (about 36 percent) while the United States contributes about 11 percent of the global solar energy production. However, solar energy makes up only 1.8 percent of all energy usage within the United States (United States Energy Information Administration 2020c). Almost two-thirds of the solar energy produced in the United States is through large-scale solar plants, and about one-third is from small-scale systems designed to provide energy to individual buildings.

Photovoltaic power stations, sometimes called **solar farms** or solar parks, are typically large arrays owned by utility companies to contribute to their electricity supply, which is then sold to consumers (Marsh 2019). In 2019, the largest solar farm in the world was the Pavagada Solar Park in India, covering 2,500 acres of land and supplying energy to hundreds of thousands of households. However, Egypt and China are not far behind with their own giant solar farm projects, and Japan—having neither the space nor the cloudless days for large solar farms—is working on a plan to put solar panels in space (NBC News 2019).

On a smaller scale, in 2017 former President Jimmy Carter brokered a deal to have ten acres of his farmland in Plains, Georgia, converted to a solar farm which now provides half of the energy for the community (*Atlanta Journal Constitution* 2017). The president was an early supporter of solar power when he had 32 solar panels installed on the White House during his presidency. His successor, Ronald Reagan, had them removed declaring his support for the fossil fuel industry. In recent years, the idea of **community solar gardens** has gained popularity in the United States. These are smaller-scale solar farms of just an acre or more designed to support the power needs of a group of households. Customers may own a portion of the solar garden or lease energy from the system.

Solar farms produce no carbon emissions, no air pollution, and no water pollution (United States Energy Information Administration 2019b). However, large solar farms require the clearing of land, which may involve deforestation and the displacement of wildlife. Toxic materials and chemicals are used to construct the photovoltaic cells that convert sunlight into electricity, and some solar thermal systems use hazardous liquids to transfer heat. Leaks and disposal of these liquids could be hazardous to the environment. New concentrating solar power (CSP) technologies have made solar power technology far more efficient and cost-effective. One estimate by the U.S. Department of Energy found that the cost of utility-scale solar farms decreased by 70 percent from 2010 to 2018 (Hayes 2020). For homeowners, the cost of solar panels has also gone down, and most states provide tax credits for solar installation, reducing the cost even further.

Geothermal Energy. **Geothermal energy** is energy produced from the heat deep within the Earth. This heat is produced from the slow decay of radioactive particles within the inner core of the Earth (United States Energy Information Administration 2019c). The temperature within the inner core of the Earth is as hot as the surface of the sun; surrounding that core is a layer of hot molten rock called magma, which can be as hot as 7,000 degrees Fahrenheit. Because the Earth's crust is broken into pieces, there are geographic areas of the world where magma comes close to the surface and heats the rocks and water above. These include volcanoes, hot springs, and geysers. Geothermal power plants are located in these regions of the world; in the United States, this includes mostly western states and the state of Hawaii.

In the United States, geothermal energy is used to heat buildings, as well as for food dehydration, gold mining, and milk pasteurizing (United States Energy Information Administration 2019c). Geothermal energy provides 90 percent of the energy required for Iceland (International Renewable Energy Agency 2020) and a significant share of the electricity in New Zealand, Kenya, and the Philippines.

Historically, geothermal power plants have emitted only low levels of greenhouse gases in part due to the sites on which they have been located (Energy Sector Management Assistance Program 2016); however, the World Bank recently warned that planned future expansion of sites could include sites with naturally higher levels of greenhouse gases. Protocols need to be developed to inform selection of sites and control these emissions. Compared to their fossil fuel–powered alternatives, geothermal power plants generate very little air pollution (United States Energy Information Administration 2019d). Most geothermal plants inject the steam and water they use back into the earth, which helps to renew the resource.

Biomass. **Biomass**, or biofuel, is a renewable energy source utilizing organic materials that come from plants and animals, including wood and wood processing wastes (e.g., chips, pellets, sawdust), agricultural crops and waste materials, municipal solid waste

solar farms Large solar arrays owned by utility companies to contribute to their electricity supply, which is then sold to consumers. Also known as solar parks.

community solar gardens Small-scale solar farms designed to support the power needs of a group of households.

geothermal energy Energy produced from the heat deep within the Earth; produced from the slow decay of radioactive particles deep within the inner core of the Earth.

biomass Material derived from plants and animals, such as dung, wood, crop residues, and charcoal; used as fuel.

(e.g., paper, cotton, food and yard waste), and animal manure and human sewage (United States Energy Information Administration 2020d).

Biomass can be converted to energy through burning or conversion into solid, liquid, or gaseous fuels. For example, a chemical process can convert vegetable oils, animal fats, and greases into biodiesel. Biological conversion processes, including fermentation, convert biomass into ethanol for use in vehicles. In the United States, corn is frequently used to make ethanol (Alternative Fuels Data Center 2020a). While all gasoline has some ethanol mixed in (typically about 10 percent), ethanol is also available as an E85 grade fuel, which includes up to 85 percent ethanol for use in flexible fuel vehicles, and some have suggested that standard gasoline be increased to an E30 blend (Runge 2016). The U.S. government and a number of U.S. states have grant, loan, and tax incentive programs to encourage the development and use of ethanol fuel (Alternative Fuels Data Center 2020b).

The environmental impacts of increasing ethanol in fuel are mixed. While ethanol produces fewer carbon emissions than fossil fuel, it reduces fuel efficiency (MPG) and damages engines, requiring more frequent replacement. Increased use of corn for fuel also increases the cost of corn, which impacts consumers needing feed corn for animals and for human consumption. And ethanol-based particles in air are associated with asthma, lung disease, and heart disease. A study in Brazil—a country which has transitioned nearly all new passenger cars to high biofuel gasoline—found that high ethanol-based gasoline increases air pollution, and heart and lung disease, especially in urban areas (Scovronick et al. 2016). Intensive corn farming has also been associated with declines in bee populations across the Midwest (Runge 2016), which has its own negative environmental impacts.

Biomass was the largest source of energy used around the world until the mid-1800s and is still used as a major source of heating and cooking fuel in many countries (e.g., dung, wood, crop residues, and charcoal). In fact, its use is increasing in many developed countries as a means of reducing greenhouse gas emissions. It also reduces the amount of waste in landfills. However, the World Health Organization estimates that 4.3 million deaths occur each year as a result of indoor air pollution, much of this due to the use of biomass for heating and cooking in poorly ventilated homes (Stolark 2018). Women and children are the most affected. Biomass, or biofuels, currently make up about 4 percent of the world's energy supply (International Energy Agency 2020b).

Our Growing Environmental Footprint

The world's economy is built around the use of natural resources, raw materials like lumber, metals, plants, animals, and water. The demands that humanity makes on the Earth's natural resources are known as our **environmental footprint**. Beginning in about 1970, humanity's environmental footprint has exceeded the Earth's biocapacity—its capacity to produce useful resources such as water and crops and to absorb waste, such as carbon dioxide (CO_2) emissions. Globally, we exceed the Earth's biocapacity by more than 50 percent, meaning that we need 1.6 planet Earths to support our consumption (Earth Overshoot Day 2020). The environmental footprint of high-income countries that consume more goods and produce more carbon dioxide is six times more than that of low-income countries (Ritchie 2018). The countries of North America have the highest emissions per person, contributing 18 percent of the carbon emissions of the world, while having only 5 percent of the world's population. The largest component of humanity's environmental footprint is carbon, caused primarily from burning fossil fuels (World Wildlife Fund [WWF] 2020).

environmental footprint The demands that humanity makes on the earth's natural resources.

Earth Overshoot Day The approximate date on which humanity's annual demand on the planet's resources exceeds what our planet can renew in a year.

Every year the Global Footprint Network identifies **Earth Overshoot Day**—the approximate date in a given year on which humanity's annual demand on the planet's resources exceeds what our planet can renew in a year. In 2020, Earth Overshoot Day was August 22 meaning that in just under eight months we used as many natural resources as our planet can renew in a year. Importantly, this was a marked improvement over 2019 when the date fell on July 29. This improvement was the result of a 14.5 percent reduction in carbon emissions and an 8.4 percent decline in deforestation as a result of the COVID-19 pandemic (Earth Overshoot Day 2020). As Figure 13.9 illustrates, since 1970, we have moved from living "within our means" to borrowing against our natural resources for the equivalent of four months out of the year. The depletion of the Earth's resources, such as forests, water, minerals, and fossil fuels, is due to consumption patterns and population growth (see also Chapter 12).

NPeter/Shutterstock.com

The world's population uses the resources of 1.6 planet Earths each year. If everyone lived as we do in the United States, we would use up five planet Earths' worth of natural resources.

Dwindling Water Supplies

Sixty-eight countries experienced water stress or scarcity in 2018 (Dormido 2019). Almost a quarter of the world's population, about 1.8 billion people in 17 countries, are on the verge of a severe water crisis (Dormido 2019). Water supplies around the world are dwindling, while the demand for water continues to increase because of population growth, industrialization, rising living standards, and changing diets that include more food products that require larger amounts of water to produce, such as milk, eggs, chicken, and beef. With most water use going to agriculture, water shortages threaten food production and supply (WWF 2020) (see also Chapter 2 and Chapter 12).

Deforestation and Desertification

The world's forests are also being depleted due to the expansion of agricultural land, human settlements, wildfires, wood harvesting, and road building. The result is **deforestation**—the conversion of forestland to nonforestland. Between 1990 and 2015,

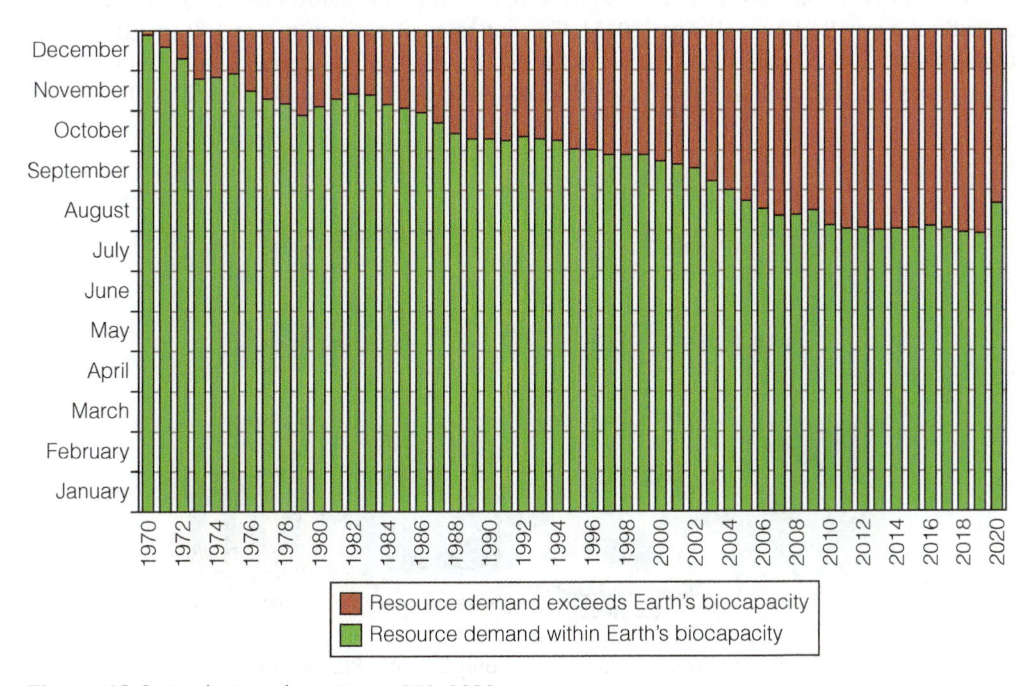

Figure 13.9 Earth Overshoot Day, 1970–2020
SOURCE: Earth Overshoot Day 2020.

deforestation The conversion of forestland to nonforestland.

the world's forests shrank by an area larger than South Africa (Derouin 2019). Every *second* an area of forest estimated to be the size of a soccer field is lost (Carrington et al. 2018), with the focus of deforestation currently in the tropics. Only about 15 percent of the forests that likely existed before human civilization remain intact today.

Deforestation displaces people and wild species from their habitats, leads to soil erosion that can cause severe flooding, and contributes to global warming by releasing carbon dioxide into the atmosphere. Deforestation affects water supply and quality because about three-quarters of the Earth's freshwater comes from forested watersheds (Food and Agriculture Organization 2018). It can also reduce rainfall; the Amazon rainforest, for example, generates about half of its own rainfall (McSweeney 2019).

Because trees prevent soil erosion and their roots hold water, deforestation contributes to **desertification**—the transformation of once fertile land into degraded land that is unusable for agriculture (McSweeney 2019). Climate change, overcultivation, and overgrazing of livestock are also factors in desertification. Desertification has led to an increase in sand and dust storms. As more land is degraded, populations can no longer sustain a livelihood on the land, and so they migrate to urban areas or other countries, contributing to social and political instability.

Threats to Biodiversity

@Prof Strachan

"A staggering fifth of countries globally are at risk of their ecosystems collapsing due to a decline in **biodiversity**"

"Natural services such as food, clean water & air, & flood protection have already been damaged by human activity"

@Team4Nature

-Professor Peter Strachan

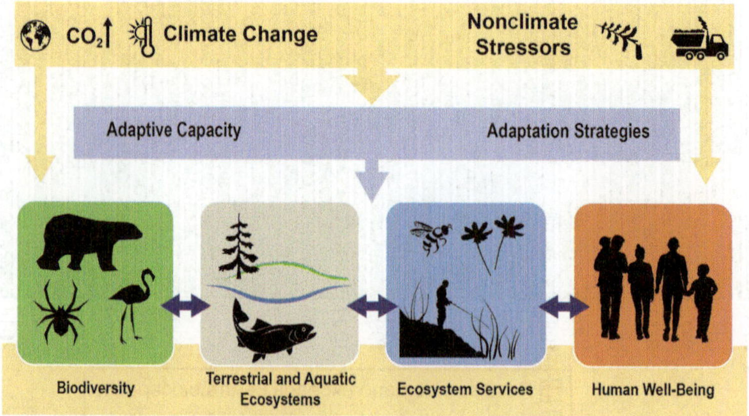

An estimated 8.7 million species of life live on Earth, 1.6 million of which have been named and cataloged (Bale 2019). Some scientists believe the number is much higher, with millions of microorganisms unaccounted for (Loceya and Lennona 2016). Humans are just one of these species, and we are dependent on other species for our well-being as they are dependent on us (see Figure 13.10). The enormous diversity of life, known as **biodiversity**, provides food, medicines, fibers, and fuel; purifies air and freshwater; pollinates crops and vegetation; and makes soils fertile.

Between 1970 and 2016, the global population of 21,000 mammals, birds, fish, reptiles, and amphibians monitored by the Living Planet Index declined by an average of 68 percent (World Wildlife Fund 2020). The International Union for Conservation of Nature (IUCN) (2020), which maintains the Red List, the globally accepted list of threatened species, has identified 32,441 species as critically endangered, endangered, or vulnerable to extinction (see Table 13.1).

Scientists estimate that 150 to 200 species of life on Earth (plant, insect, bird, or mammal) go extinct every 24 hours (Pearce 2015). Current rates of extinction are 100 to 1,000 times higher than what scientists consider to be the standard rate of extinction before human activity became a significant factor (World Wildlife Fund 2018). As shown in Table 13.1, more

Figure 13.10 The Interrelationships among Climate, Biodiversity, Ecosystems, and Human Well-Being
SOURCE: U.S. Global Change Research Program 2018.

desertification The degradation of semiarid land, which results in the expansion of desert land that is unusable for agriculture.

biodiversity The diversity of living organisms on Earth.

than 32,000 species worldwide are threatened with extinction. The loss associated with species extinction is profound. Sociologist Eileen Crist explains,

> When we drive a species to extinction, we're prematurely taking out of existence a unique, amazing manifestation of life that has never existed before and will never arise again, and we're extinguishing all possibilities of its evolution into new forms. (Quoted in Tonino 2020, p. 6)

TABLE 13.1 Threatened Species Worldwide, 2020

Category	Number of Threatened* Species
Mammals	1,299
Birds	1,486
Amphibians	2,276
Reptiles	1,406
Fishes (bony)	2,849
Insects	1,819
Crustaceans	734
Mollusks	2,275
Plants	17,507
Fungi and protists	199
Total	32,441

*Threatened species include those classified as critically endangered, endangered, and vulnerable.
SOURCE: International Union for Conservation of Nature (IUCN) 2020.

Scientists attribute these declines in biodiversity to the following five factors (in order of importance): (1) changes in land and sea use, (2) direct exploitation of organisms, (3) climate change, (4) pollution, and (5) invasive alien species (United Nations 2019). For example, over the course of human history, approximately 75 percent of the terrestrial environment and 66 percent of the marine environment on Earth have been severely altered by human actions, including an 85 percent reduction in wetlands. A third of marine fish stocks are being "overfished" at levels that will lead to extinction if not reversed. More than 245,000 square kilometers of coastline are "dead zones" due to the use of fertilizer (United Nations 2019), and an estimated 18 percent of tuna and swordfish in the Mediterranean have plastic debris in their stomachs (World Wildlife Fund 2018).

Almost half of terrestrial mammals and a quarter of birds have already been affected by climate change (United Nations 2019). Since 1970, the number of invasive alien species has increased by 70 percent across 21 countries that maintain records. Scientists believe that the Earth has experienced five mass extinction events during the last 450 million years, each destroying 70–95 percent of plants, animals, and microorganisms in existence (Ceballos, Ehrlich, and Raven 2020). Some believe the sixth mass extinction event has already begun and will continue to accelerate unless humanity takes decisive action.

Air Pollution

Perhaps the most concerning type of air pollution is greenhouse gas emissions—primarily CO_2, methane, and nitrous oxide—which contribute to global warming and climate change as discussed earlier in the chapter. Other forms of air pollution—including carbon monoxide, sulfur dioxide, arsenic, nitrogen dioxide, mercury, dioxins, and lead—largely result from transportation vehicles, fuel combustion, industrial processes (such as burning coal and processing minerals from mining), and solid waste disposal. Leaded aviation gasoline is one of the few fuels in the United States to still contain lead, and it is the single-largest source of lead emissions in the country (EPA 2020a).

Indoor air pollution is a particular problem for the approximately 3 billion people, mostly in low- and middle-income countries, who use biomass fuels—wood, charcoal, crop residues, and dung—for cooking and heating their homes (World Health Organization 2018b). Biomass is typically burned on open fires or stoves without chimneys, creating smoke and indoor air pollution. Exposure is particularly high among women and children, who spend the most time near the domestic hearth or stove. Each year, 3.8 million people die prematurely due to exposure to smoke from cooking fires (World Health Organization 2018b).

Liz Weber/Shutterstock.com

In 1986, a photographer took this photo of what was thought to be an Eastern Cougar. In fact, the last sighting of an Eastern Cougar—or "Ghost Cat"—was in the 1930s. The species was declared extinct 80 years later in 2018. Experts are reluctant to declare a species extinct until they are *really* sure. The Eastern Cougar succumbed to overhunting.

In affluent countries, much air pollution is invisible to the eye and exists where we least expect it—in our homes, schools, workplaces, and public buildings. An estimated one in every 15 homes in the United States has an elevated level of radon, a gas that forms when uranium, thorium, or radium break down in water, rocks, and soil, leaking the gas into the dirt beneath the home (American Lung Association 2020a). Radon is the second-leading cause of lung cancer in the United States, causing an estimated 21,000 deaths per year. Many building materials, pressed wood products, paints, carpeting, and cleaning products give off volatile organic compounds (VOCs) such as formaldehyde and benzene, which are harmful to human health (American Lung Association 2020b).

Other sources of indoor air pollution include pesticides used for pest management; secondhand tobacco smoke; and by-products of combustion (e.g., carbon monoxide) from furnaces, fireplaces, heaters, and dryers. Some air fresheners, deodorizers, and disinfectants emit the pesticide paradichlorobenzene. Older homes may contain lead dust (from old lead-based paint) or asbestos, both of which were banned from use in construction in the 1970s (American Lung Association 2020c).

Both outdoor and indoor air pollution are linked to health problems—including heart disease, lung cancer, stroke, chronic obstructive pulmonary emphysema, pneumonia, and other respiratory infections—and cause about 7 million premature deaths each year, or one in eight of total global deaths (World Health Organization 2018c). The World Health Organization estimates that nine out of ten people breathe air that exceeds its limits for pollutants; most of these are individuals in low- and middle-income countries. This figure confirms that air pollution is "the world's largest single environmental health risk" (World Health Assembly 2015). In the United States, five in ten people live in counties with unhealthy levels of either ozone (smog) or particulate pollution (soot), and this number is rising (American Lung Association 2020d). Low-income people are more likely to live in counties with high level of air pollution, and 57 percent of people of color in the United States live in such counties. Long-term exposure to air pollution was associated with higher rates of COVID-19 hospitalizations and deaths in 2020 (Wu et al. 2020).

Destruction of the Ozone Layer. The ozone layer of the Earth's atmosphere protects life from the sun's harmful ultraviolet rays. Yet the ozone layer has been weakened by the use of certain chemicals, particularly chlorofluorocarbons (CFCs), used in refrigerators, air conditioners, and spray cans. The impact of these chemicals created a large hole in the ozone layer over the Antarctic first discovered in the mid-1980s (Reiny 2018). While the

AP Images/Niranjan Shrestha

Narayan Maharjan/NurPhoto/Getty Images

One benefit of the COVID-19 pandemic was that air pollution across the globe decreased as a result of reductions in the use of automobiles, plane travel, and plant production. Normally the mountains surrounding Kathmandu, Nepal, are not visible from the city, which is ranked as one of the most polluted in the world.

hole can vary in size depending on weather conditions, it has been as large as the size of North America (NASA 2016). The 1987 Montreal Protocol—an international agreement to phase out the use of ozone-depleting CFCs—has resulted in the steady shrinking or "healing" of the ozone hole (Strahan and Douglass 2018; United Nations 2018). However, CFCs can linger in the atmosphere for 50–100 years, meaning that full repair of the ozone layer is not expected until 2060–2080 (Reiny 2018).

Reducing the use of CFCs not only helped the ozone layer to recover, but also slowed global warming (Harvey 2020). However, the chemicals that replaced CFCs—hydrofluorocarbons (HFCs)—were discovered to be potent greenhouse gases that contribute to global warming (Zaelke, Borgford-Parnell, and Andersen 2016). HFCs are widely used in the United States to cool refrigerators and cars; they are also used in fire suppressants on aircraft, bear repellant for hikers, and inhalers for asthma (Eilperin and Mufson 2020).

wcjohnston/iStock/Getty Images

Acid rain kills trees by damaging their leaves, dissolving the nutrients in the soil before they can be absorbed, and releasing toxic substances like aluminum that impact the ability of trees to take in water through their roots (Air Quality 2020).

The 2016 Kigali Amendment to the Montreal Protocol aims to phase out the use of HFCs (Harvey 2020). More than 100 countries and the EU have ratified the Kigali Amendment. Although the Trump administration refused to ratify it, President Biden supports the Kigali Amendment and its goal of reducing the use of HFCs. As this book goes to press, bipartisan efforts are underway in Congress to reach agreements on an energy bill that would phase out production and importation of HFCs.

Acid Rain. Air pollutants, such as sulfur dioxide and nitrogen oxide, mix with precipitation to form **acid rain**. Polluted rain, snow, and fog contaminate crops, forests, lakes, and rivers. Acid rain also deteriorates the surfaces of buildings, statues, and automotive coatings. As a result of the effects of acid rain, no fish remain in 177 lakes in New York's Adirondack Mountains due to early death and chemically induced infertility (American Fisheries Society 2020). Because winds carry pollutants in the air, industrial pollution in the Midwest falls back to the Earth as acid rain on southeast Canada and the northeast New England states.

China has some of the most acid rain in the world as a result of its reliance on burning coal for electricity; scientists in China have recorded pH levels of rain as low as 2.8, which is corrosive enough to damage stone (Morton 2018). In fact, one recent study concluded that a landslide that killed 74 people in China may have been the result of acid rain. Legislation like the *Clean Air Act* in the United States and the LRTAP Convention in Europe, passed in the 1970s and 1980s, have reduced acid rain (Menz and Seip 2004); however, pollutants from countries like China travel across country boundaries impacting nations thousands of miles away.

Land Pollution

About 30 percent of the world's surface is land, which provides soil to grow the food we eat. Increasingly, humans are polluting the land with nuclear waste, solid waste, and pesticides. In 2020, 1,874 hazardous waste sites in the United States (also called Superfund sites) were on the National Priorities List (Environmental Protection Agency 2020b). Superfund sites are areas with hazardous waste that have been identified by the Environmental Protection Agency as needing cleanup.

acid rain The mixture of precipitation with air pollutants, such as sulfur dioxide and nitrogen oxide.

Nuclear Waste. Nuclear waste, resulting from both nuclear weapons production and nuclear reactors or power plants, contains radioactive plutonium, a substance linked to cancer and genetic defects. Radioactive wastes and contaminated materials from nuclear power will remain potentially harmful to human and other life for 250,000 years (Nader 2013). The question of how to safely dispose of nuclear waste has not been resolved.

Nuclear plants produce the majority of the over 90,000 metric tons of nuclear waste in the United States (Jennewin 2018). It is stored in pools to cool for several years and then moved to above-ground concrete casks. These casks must be regularly monitored by personnel to ensure that no leaks have occurred, check temperatures are stable, and maintain a security presence to avoid theft. Even after ten years of decay, this type of waste could emit 100 times a fatal dose of radiation over the course of one hour (Jennewein 2018). Six states have prohibited the construction of nuclear power plants until a long-term solution is found for nuclear waste storage (National Conference of State Legislators 2017).

Power plants are not the only producers of nuclear waste. Millions of liters of radioactive liquid waste from weapons production are also sitting in storage containers, some of it since World War II (Jacoby 2020). Many of these storage containers are actively leaking radioactive waste into the soil and have been for years. While they wait for a more permanent solution, scientists are moving forward with a plan to stabilize the waste by converting it into a solid form, namely glass.

Since the late 1980s the federal government has been looking for an appropriate site to create an underground storage facility for radioactive waste (Jennewein 2018). In 2002 the Department of Energy selected the Yucca Mountain in Nevada as an appropriate site, but Nevadans protested, and the Obama Administration and Congress withdrew their support. In 2018, the Trump administration requested $150 million to restart the Yucca Mountain process, but was denied. The Biden administration opposes storing nuclear waste at Yucca Mountain and is exploring safe alternatives (World Nuclear News 2021).

Solid Waste. The United Nations predicts by the year 2025 that the world's cities will produce 2.2 billion tons of waste every year, more than three times the amount produced in 2009 (United Nations Environment Programme 2020). This figure does not include mining, agricultural, and industrial waste; demolition and construction wastes; junked autos; or obsolete equipment wastes. In the United States, the amount of household garbage each person produced each day increased from 2.7 pounds in 1960 to 4.51 pounds in 2017 (EPA 2020c). About 35 percent of household garbage is composted or recycled; the rest is dumped in landfills. Almost a quarter of what ends up in landfills is food waste; another 19 percent consists of plastics.

What do you think?

Most plastic shopping bags, commonly used in grocery and retail stores, are made from petroleum and require a lot of fossil fuel energy to produce. They also contain toxic chemicals and end up in landfills, where it takes 1,000 years for them to degrade, or in oceans, where marine life can choke or starve after swallowing them. Dozens of countries, including the European Union, South Africa, China, and India, have either bans or taxes on plastic shopping bags. In 2014, California became the first state to ban plastic shopping bags, followed by Hawaii in 2015 (exceptions include for medical and sanitary purposes). Would you support a ban or a tax on plastic bags in your community or state?

Solid waste includes discarded electrical appliances and electronic equipment, known as **e-waste**. It is estimated that in 2021 the world's population will discard 57 million tons of e-waste (Cho 2018). Until 2018, 70 percent of the world's electronic waste ended up in landfills in China; however, they stopped accepting e-waste due to environmental concerns. The estimated value of recoverable materials (e.g., gold and copper) in global e-waste is over $64 billion (Cho 2018) and recycling these materials costs less than mining them (Zeng, Mathews, and Li 2018). However, only 20 percent of e-waste is properly recycled to enable recovery of the valuable materials (Cho 2018). The main concern about dumping e-waste in landfills is that hazardous substances, such as lead, cadmium, barium, mercury, PCBs, and polyvinyl chloride, can leach out of e-waste and contaminate the soil and groundwater.

e-waste Discarded electrical appliances and electronic equipment.

Pesticides. Pesticides are used worldwide for crops and gardens; outdoor mosquito control; the care of lawns, parks, and golf courses; and indoor pest control. The Centers for Disease Control also recognize many common household cleaners as antimicrobial pesticides (Centers for Disease Control 2020). Pesticides contaminate food, water, and air and can be absorbed through the skin, swallowed, or inhaled. Many common pesticides are considered potential carcinogens and neurotoxins and may have other health impacts. A study of pesticide spray workers in Egypt found significant decreases, as compared to unexposed community members, in blood platelet volume, as well as decreased levels of testosterone and thyroxine, a hormone produced by the thyroid (Nassar, Salim, and Malhat 2016).

Even when a pesticide is found to be hazardous and is banned in the United States, other countries from which we import food may continue to use it. In a United States Department of Agriculture (2016) analysis of more than 10,000 food samples, only 15 percent of the samples tested were free of detectable pesticide residues. An estimated 90 percent of Americans have pesticides or their by-products in their bodies from consumption of foods and exposure to pesticides in homes, outside, and at work (Gross 2019). Tests of *washed* produce items by the U.S. Department of Agriculture found a dozen items routinely tested positive for two or more pesticides, and some for as many as 18 (Environmental Working Group [EWG] 2020; see Table 13.2). The "dirtiest" item tested was not a fresh fruit or vegetable but a dried one—raisins. To avoid pesticide exposure, these items are best purchased from organic farmers.

> "Every year US farmers use about a billion pounds of chemicals on crops, including the fruits, nuts, and vegetables many parents beg their kids to eat."
>
> **–LIZA GROSS, JOURNALIST**

Water Pollution

Water pollution is largely caused by fertilizers and other chemicals used in intensive agriculture, industrial production, mining, and untreated urban runoff and wastewater (World Water Assessment Program 2015). Other water pollutants include plastics, pesticides, vehicle exhaust, acid rain, oil spills, sewage, and military waste. Water pollution is the most severe in developing countries, where untreated sewage is commonly dumped directly into rivers, lakes, and seas that are also used for drinking and bathing.

In the United States, one indicator of water pollution is the thousands of fish advisories issued by the EPA that warn against the consumption of certain fish caught in local waters because of contamination with pollutants such as mercury and dioxin. The EPA and the U.S. Food and Drug Administration advise women who may become pregnant, pregnant women, nursing mothers, and young children to avoid eating certain fish altogether (swordfish, shark, king mackerel, and tilefish) because of their high levels of mercury (U.S. Food and Drug Administration 2019).

Pollutants in drinking water can cause serious health problems and even death. In 2013, in Flint, Michigan, a predominately Black community that used to be the home of General Motors, an emergency manager appointed by the governor switched the city's water system from water piped in from Detroit to water from the Flint River (Denchak 2018). This decision was made to save the city money while a new pipeline was being built. The city failed to treat the water prior to pumping it into people's homes so the corrosive water leached lead from the aging system of pipes into people's drinking, cooking, and bathing water. City officials ignored complaints from residents that something seemed to be wrong with the water, even when residents brought jars of brown tap water to city council meetings (Michigan Civil Rights Commission 2017). At one point, the state of Michigan office building in Flint brought in bottled water for state employees, but nothing was done for residents.

In 2015, researchers from Virginia Tech and from the Environmental Protection Agency documented that the water exceeded accepted federal levels for lead and a local

> "The Flint scandal showed the American people and the world that access to clean water in the US is not always a given. ... I want everyone to know that as of today the EPA has not kept its promises to fix the laws, and still allows states to cheat on water testing."
>
> **–LEEANNE WALTERS, FLINT RESIDENT AND ACTIVIST**

TABLE 13.2 "Dirty Dozen"—Fresh Produce Found to Have the Highest Levels of Pesticides

1. Strawberries	7. Peaches
2. Spinach	8. Cherries
3. Kale	9. Pears
4. Nectarines	10. Tomatoes
5. Apples	11. Celery
6. Grapes	12. Potatoes

SOURCE: Environmental Working Group 2020.

Jim Watson/AFP/Getty Images

Beginning in 2013 in Flint, Michigan, water coming into schools, homes, and businesses was contaminated with lead and bacteria, due to shortcuts by the city manager to save money. City officials ignored complaints from residents. Hundreds of children experienced lead poisoning with potentially long-term effects, and 12 people died of Legionnaire's disease.

physician reported that the incidence of elevated blood lead levels in children had more than doubled in one year. Lead consumption can affect the heart, kidneys, and nerves; in children, it can lead to impaired cognition, behavioral disorders, hearing problems, and delayed puberty (National Institute for Occupational Safety and Health 2018). Bacteria were found in the water as well, as the city had failed to adequately treat it with chlorine (Shulz 2018), contributing to an outbreak of Legionnaire's Disease, which led to at least 12 deaths. Local and state leaders continued to deny any problems with the water, until residents filed a federal class-action lawsuit. Fifteen officials were ultimately charged, including the state health director, who was charged with involuntary manslaughter; seven of the accused pled guilty, and eight cases are still outstanding (CNN Editorial Board 2020a). In a 129-page report, the Michigan Civil Rights Commission concluded "the people of Flint did not enjoy the equal protection of environmental or public health laws" (Michigan Civil Rights Commission 2017, p. 4) as a result of "deeply embedded institutional, systemic and historical racism" (p. 9). This is what researchers and activists refer to as environmental racism.

Another growing concern surrounds the increasing amount of plastic pollution found in the world's oceans. Every year about 8 million tons of plastic waste end up in the oceans, the equivalent of a garbage truck's worth of waste every day (Parker 2019). Most comes from middle- and low-income countries with poor waste management systems, although high-income countries like the United States generate more plastic pollution per person (Ritchie and Roser 2018). Much of this plastic is difficult to see because of its small size. Microplastics, which are fragments of plastic that measure less than 5 millimeters, come from the degradation of plastic products. These microplastics have been found all across the globe, including in community drinking water systems, and are virtually impossible to recover (Parker 2019).

Plastics tend to absorb toxins that are harmful to marine organisms and potentially harmful to humans who eat seafood (Seltenrich 2015). Millions of animals are killed by plastics every year, usually through entanglement or starvation, but sometimes from liver or other organ damage (Parker 2019). During the COVID-9 pandemic, increased use of plastics for personal protective equipment and food service items as well as the suspension of many recycling centers suggests that plastic pollution may have increased as much as 30 percent (Prata et al. 2020), some of which will end up in the world's water.

Sebnem Coskun/Anadolu Agency/Getty Images

Experts estimate that plastic water pollution has likely increased by as much as 30 percent during the COVID-19 pandemic (Prata et al. 2020). Millions of animals are killed each year by plastic pollution.

Chemicals, Carcinogens, and Health Problems

About 250 billion tonnes of toxic chemicals are released into the environment each year (SciNews 2017). About 15,000 accidental or illegal releases of toxic substances happen in the United States every year (Centers for Disease Control 2020). Exposure to toxic chemicals has been linked by the United Nations to 16 million deaths that occurred globally in 2016 (United Nations Environment Programme 2019a). Chemicals in the environment enter our bodies via the food and water we consume, the air we breathe, and the substances with which we come in contact. For example, in a recent study, individuals who ate at fast food restaurants were found to have elevated levels of PFAS (per- and polyfluoroalkyl substances) in their blood systems (Gibbens 2019). PFAS have been linked to cancer, thyroid disorders, hormonal changes, and weight gain. They are used in food wrappers and containers including in the fast food industry.

Many chemicals remain in the environment long after their use has been discontinued, like polychlorinated biphenyls (PCBs), which appear to be killing off the world's Orca population despite the fact that the chemical has not been in use for more than 40 years (Welch 2018). And most chemicals present in pregnant women's bodies pass through the placenta to the fetus, often at much greater levels (Kurtzman 2016). Pesticides are one factor, along with climate change and habitat loss, for the sharp decline in the bee population in the United States (Jacobo 2019). As honey bees are critical for the pollination of flowers, fruits, and vegetables, up to a 90% decline in bee colonies in some areas of the country is of great concern to those in agriculture. This chapter's *Human Side* features the perspectives of an activist in Kenya who worked to protect her community from chemicals being dumped into its water supply.

In its most recent report, the National Toxicology Program (2016) lists 248 chemical substances that are "known to be human carcinogens" or "reasonably anticipated to be human carcinogens," meaning that they are linked to cancer. These may constitute only a fraction of actual human carcinogens. When the *Toxic Substances Control Act* of 1976 was enacted, it "grandfathered" in the 62,000 chemicals then on the market. Since then, the EPA has restricted the use of only six of the 80,000 chemicals used in the United States; 95 percent of the chemicals in use have not been tested for safety (Milman 2019). Under a 1976 law, the EPA was given only 90 days to determine whether a new product poses a hazard before it can be marketed. A 2016 amendment known as the *Lautenberg Act* requires the EPA to evaluate all potentially risky chemicals but it has a significant backlog. Indeed, in 2019 it had only just added asbestos to its list of chemicals to review (Milman 2019).

In the European Union, legislation known as *Registration, Evaluation, and Authorization and Restriction of Chemicals* (*REACH*) requires chemical companies to conduct safety and environmental tests to prove that the chemicals they are producing are safe. If they cannot prove that a chemical is safe, it is banned from the market (Milman 2019). The European Union has become a world leader in environmental stewardship by placing the "precautionary principle" at the center of EU regulatory policy. The precautionary principle requires industry to prove that their products are safe. In contrast, in the United States, chemicals are assumed to be safe unless proven otherwise and the burden is put on the consumer, the public, or the government to prove that a chemical is harmful. As a result, chemicals that are banned in Europe, such as formaldehyde, asbestos, and parabens, are allowed in U.S. cosmetics.

What do you think?

In order for a medication to be approved for use by the public, the Food and Drug Administration requires evidence of its safety. Should the same standard be held to the use of chemicals in manufacturing? Do we have the right to know that the products we are exposed to are made with materials that won't harm us?

The environment in which we live and work is likely a significant contributor to the development of cancer—one of the leading causes of death globally (World Health Organization 2018d). Research studies have found a link between multiple types of cancer and air pollution (Hwang et al. 2020; Wong et al. 2016). Many of the chemicals we are exposed to in our daily lives cause not only cancer but also other health problems such as infertility, birth defects, and a number of childhood developmental and learning problems (Koman et al. 2019). The costs associated with the health impacts of environmental chemical exposures are estimated to be in the billions of dollars. While it is well known

Phyllis Omido Takes on the Kenyan Government to Protect Her Community from Contaminated Water

The experiences of the residents of Phyllis Omido's village in Kenya are eerily similar to those of the residents in Newark, New Jersey (2017) and Flint, Michigan (2014), where high lead levels in city water were ignored by government officials (Lindwall 2019; Denchak 2018). In all three communities, lead exposure is expected to result in long-term health problems and significant intellectual delays in children that may never be overcome. In every case, the activism of community residents eventually forced local, state, and national leaders to take action. In Omido's case the polluter was her employer and one of the children poisoned was her son.

I am a Community Organiser in Mombasa and Founder of the Centre for Justice, Governance and Environmental Action (CJGEA), an organisation focused on promoting environmental justice in Kenya's coastal region. Originally with a background in business, I founded the CJGEA in 2009 in order to address environmental and human rights challenges facing the urban poor in Kenya.

Early in 2009 the urban poor community woke up to thick, choking pungent plumes of smoke. In the days to follow the children became sick with coughing, fevers, fatigue and headaches. The trees shriveled and died, the chicken and other livestock and pets succumbed to the smoke and thick, filthy and smelly flu gasses and effluent directed into the community from the smelter. One day while Kelvin (a boy in the community) played in the football field his ball strayed into the drain. The act of fetching his ball left him with a scar that would cause him and his grandmother sleepless nights.

I ... educated the people about Lead (Pb) poisoning, human rights, the right to life, dignity and a clean environment. The Lead (Pb) smelter spewing toxic fumes and effluent [liquid waste], exposed the community to sickness and loss of life.

Because of the urgency of the issue, CJGEA engaged in ... lobbying, but it seemed to fall on deaf ears. In 2009, 5% of the children tested positive for severe lead poisoning. By 2011, Kelvin's blood lead level was at 32 ppm, much higher than the World Health Organisation upper limit of 5 ppm.

Both employees and community members succumbed to the poisoning ... CJGEA came under attack from the government. Police raids were conducted on our offices and my son and I were accosted and abused by gunmen while entering my house in late 2011. I was accused of funding illegal groups and being a terrorist. In 2012 while planning a public demo to lobby and protest the injustice, I was arrested alongside 17 other CJGEA employees and community members.

... Kelvin was by now suffering seizures, forgetfulness, fainting and by 2012 toxic BLL levels of 38 ppm.

When the factory and the Kenyan environment ministry refused to do anything, Omido organized demonstrations of thousands of residents. Their protests gained national and international attention leading to support from nongovernmental organizations (NGOs) like Human Rights Watch and Frontline Defenders. When the Kenyan government refused to act, the protestors turned to the East Africa Community, a regional body that eventually banned lead from Kenya, forcing the closure of this and other factories in the country. Ten years after her discovery, Phyllis Omido and other activists celebrated a $12 million court judgment to help fund a cleanup of the area and compensate for the deaths and ongoing health problems experienced by the community (Watts 2020). But the money won't bring back the estimated 100 children who died as a result of lead poisoning (Schlanger 2018); nor will it make up for the death threats, beatings, and arrests that forced Phyllis Omido to live in hiding during the court battle. For her work, Omido was awarded the Goldman Environmental Prize in 2015.

Source of Quote: Phyllis Omido Testimony. (n.d.).

that children and pregnant women are at particular risk from many toxic chemicals, the EPA has not identified them as subpopulations under consideration in moving through its backlog of chemical safety reviews (Koman et al. 2019).

multiple chemical sensitivity (MCS) Also known as "environmental illness," a condition whereby individuals experience adverse reactions when exposed to low levels of chemicals found in everyday substances.

Multiple chemical sensitivity (MCS), also known as environmental illness, is a condition whereby individuals experience adverse reactions when exposed to low levels of chemicals found in everyday substances (vehicle exhaust, fresh paint, housecleaning products, perfume and other fragrances, synthetic building materials, and numerous other petrochemical-based products). These individuals can be particularly affected by chemicals found in common household, personal, and commercial products, as well as fragrances found in a host of consumer products ranging from colognes to shampoos to tampons. Individuals with MCS often avoid public places and/or wear a protective breathing filter to avoid inhaling the many chemical substances in the environment. Researchers are not certain what triggers the condition, with some speculating it may be a strong and acute exposure, which then triggers higher sensitivity to future exposure and others theorizing MCS may be the result of long-term, low-dose exposure that increases sensitivity (Rossi and Pitidis 2018). Some in the medical field question the syndrome's existence altogether.

What do you think?

Some businesses, universities, hospitals, and local governments are voluntarily limiting fragrances to accommodate employees and consumers who experience ill effects from them. What do you think about banning fragrances in the workplace or other public places? If your college or university were considering instituting a ban on fragrances on campus, would you support the ban? Why or why not?

Light Pollution

The United States, like much of the rest of the world, has become increasingly "lit up" with artificial light. **Light pollution** refers to artificial lighting that is annoying, unnecessary, and/or harmful to life forms on Earth. Artificial light has negative impacts on the well-being of both humans and wildlife (Cox and Takahashi 2019), at least in part because it disrupts the circadian rhythm, which regulates the sleep/wake cycle and other natural processes. People who live in areas with a lot of outdoor artificial light are more likely to use insomnia medications (Min and Min 2018), and sleep disturbances are associated with hormonal changes, obesity, diabetes, and depression (Riemann et al. 2019; Foster 2020). Disruptions in humans' circadian rhythms have even been shown to increase risk of cancer, especially breast cancer (Hansen 2017). About 30 percent of vertebrates and 60 percent of invertebrates are nocturnal and vulnerable to light pollution, disrupting their patterns of mating, migration, feeding, and pollination (Bogard 2013). For example, artificial lights attract migrating birds who, disoriented, fly into buildings and are killed by the millions each year (Drake 2019). Baby sea turtles use moonlight reflecting off waves tops to guide them into the sea after they hatch, but artificial light from nearby buildings and streets can disorient them, leading to death if they haven't made it into the water by morning (Drake 2019).

> "We lose something essential; we lose a part of ourselves when we lose access to the night sky. We lose that sense of stillness and awe that should be right over our heads every night."
>
> **–AMANDA GORMLEY, HEAD OF THE INTERNATIONAL DARK-SKY ASSOCIATION**

Social Causes of Environmental Problems

Various structural and cultural factors have contributed to environmental problems. These include population growth, industrialization and economic development, as well as cultural values and attitudes such as human supremacy, individualism, consumerism, and militarism.

Population Growth

The world's population is growing, exceeding 7.8 billion in 2020 and projected to grow to 9.7 billion in 2050 (see Chapter 12). Population growth places increased demands on natural resources and results in increased waste. The relationship between population growth and environmental problems was recognized as early as 180 CE by writer and priest Tertullian, and by individuals as varied as the 14th Dalai Lama, actor Morgan Freeman, and physicist Stephen Hawking (Population Matters 2020). Digby McLaren, a famous Canadian geologist, famously summed up the problem:

> If an unseen intelligent being from somewhere else in our galaxy were to visit the Earth, perhaps the most incomprehensible phenomenon it could observe would be that the planet's apparently wise and competent dominant beings are totally ignorant of the life-support system they are destined to live within. They are, furthermore, unaware that their uncontrolled reproductive capacity has grown to the extent that it is rapidly destroying this system, while they fight among themselves to preserve their freedom to do so.

light pollution Artificial lighting that is annoying, unnecessary, and/or harmful to life forms on Earth.

It is true that the sheer size of the world's population is putting strain on farmland, fish and wildlife, waste disposal, air pollution, and forests. For example, the more food we need to feed our growing population, the more forests we cut down, which then affects air pollution. Slowing population growth could result in as much as a 40 percent reduction in greenhouse gas emissions (Bongaarts and O'Neill 2018).

However, population growth itself may not be as critical as the ways in which populations produce, distribute, and consume goods and services. For example, while population growth has increased greenhouse gas emissions, it is not as impactful as the rise in greenhouse gas emissions per person, particularly in wealthy countries (Global Footprint Network 2020a). If the entire world's population lived as people do in the United States, we would need the resources of five planet Earths to meet the demands, whereas if we all lived as they do in India, we would have excess resources (Global Footprint Network 2020a).

The reality is that both matter. As Dr. Muhtari Aminu-Kano, Director-General of the Nigerian Conservation Foundation explains:

> We should be talking about population and we should be talking about consumption that goes with population. It is true, the average Nigerian, as a single person, does less damage than the average American, British, European or Russian, or any of the others, but then a lot of us do a lot of damage as well. I think it is not "either or," it's "and with." It's not a binary issue, really.

Industrialization and Economic Development

Many of the environmental problems confronting the world are associated with industrialization, in large part because industrialized countries consume more energy and natural resources and contribute more pollution to the environment than poor countries. Although more developed countries, such as France, may have fewer and cleaner manufacturing industries compared with developing countries such as China, wealthier countries continue to support some pollution-causing manufacturing industries, to import goods from countries who heavily pollute the environment, and to operate such industries in multinational business operations.

As discussed, fossil fuel use has an enormous impact on global warming and on air, land, and water pollution. While some industrialized countries like Sweden and Norway have transitioned much of their energy production to renewable energies like wind and solar, other countries like Nigeria, Mozambique, and Venezuela export fossil fuels and rely on them as a major portion of their economy (Singh 2019). Similarly, South Africa and Mongolia rely heavily on coal consumption. Figure 13.11 illustrates how close each country is to transitioning to secure, sustainable, affordable, and reliable energy in accordance with the Paris Climate Accord. This is based on the Energy Transition Index, which takes into account not only the type of energy relied upon in the country but also the degree to which regulations and investments exist to support the transition. It also takes into account political instability, which calls into question the degree to which any changes can be counted on long term. Dark blue countries have made the most progress; dark orange have made the least. The United States is not in the darkest blue category because, at the time the Index was calculated, the United States still relied on nonrenewable energies and the Trump administration had rolled back many policies supporting renewable energies, making the political support for energy transition unstable.

Cultural Values and Attitudes

Cultural values and attitudes that contribute to environmental problems include human supremacy, individualism, consumerism, and militarism.

Human Supremacy. If, in the words of President Biden, the climate crisis poses an existential threat to our planet and to humanity, why has the global community been so slow

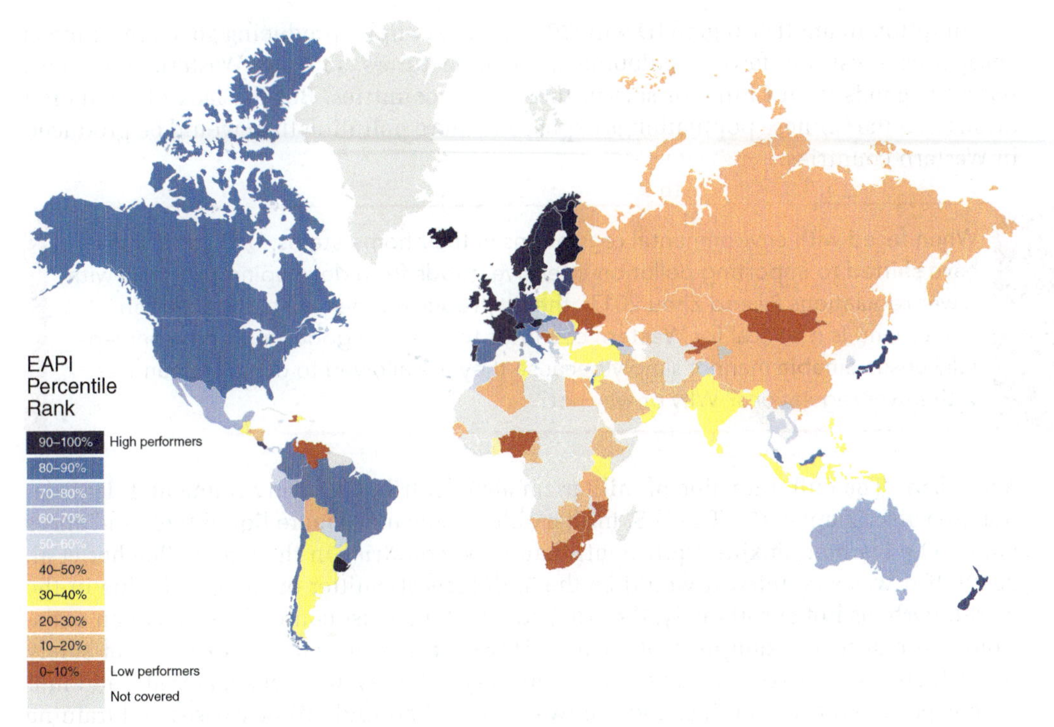

Figure 13.11 Energy Transition Index, 2019
SOURCE: Singh 2019.

to respond? According to sociologist Eileen Crist, the cause of our inaction is "human supremacy"—a largely unconscious belief that humans are the masters of creation rather than just one species among millions. Crist explains that human supremacy, also referred to as *anthropocentrism*, "is a widely shared, unconscious worldview that tells us we are superior to the rest of nature and thus entitled to treat nonhumans and their habitats however we please" (quoted in Tonino 2020, p. 7).

Human supremacy involves the belief that humans can and should dominate and control the natural environment and all its inhabitants. However, Crist warns that the forces of the Earth are bigger than we are and that "when climate change hits home, and when wild fish and coral reefs give way to an ocean of plastic, humanity will find out that our domination was a hollow illusion" (quoted in Tonino 2020 p. 11).

Individualism. Individualism, which is a characteristic of U.S. culture, puts individual interests over collective welfare (Rosenbaum 2018). Even though recycling is good for our collective environment, many individuals do not recycle because of the personal inconvenience involved in washing and sorting recyclable items (Kirakozian 2015). Similarly, individuals often indulge in countless behaviors that provide enjoyment and convenience for themselves at the expense of the environment: long showers, frequent meat eating, heavy use of air conditioning, and driving large, gas-guzzling SUVs, to name just a few. Across the globe, countries with cultures that value individualism also have the largest environmental footprints (Komatsu, Rappleye, and Silova 2019).

Consumerism. Consumerism—the belief that personal happiness depends on the purchasing of material possessions—also encourages individuals to continually purchase new items and throw away old ones. The media bombard us daily with advertisements that tell us life will be better if we purchase a particular product. Consumerism contributes to pollution and environmental degradation by supporting polluting and resource-depleting industries and by contributing to waste (Devlin 2017). Western consumers increase pollution not only in their own countries but more heavily in the countries from which they are purchasing cheap goods. About 22 percent of deaths related to air pollution are associated with the manufacture of goods produced in one region of the globe for

human supremacy A largely unconscious belief that humans are the masters of creation rather than just one species among millions. Also known as anthropocentrism.

consumption in another region (Devlin 2017). For example, producing goods in China is cheaper as a result of fewer regulations on manufacturers, so many Western companies purchase goods from China for sale in their home countries. But the lack of regulation means that the Chinese population is exposed to more pollution than would be produced in Western countries.

What do you think?

When faced with environmental regulations in their home states, U.S. companies have shifted to importing pollution-intensive goods from developing countries with fewer regulations (Li and Zhou 2017), thereby reducing their costs for pollution abatement. Is it ethical for Western countries to produce goods in an environmentally unsustainable manner simply because they are allowed to do so in countries with fewer regulations? Why or why not?

Militarism. The cultural value of militarism also contributes to environmental degradation (see also Chapter 15). The U.S. military alone consumes more liquid fuels and emits more CO_2e (carbon dioxide equivalent) than most countries in the world (Belcher et al. 2019). If it was a country, it would be the 55th biggest emitter of carbon dioxide in the world. Perhaps not surprisingly, the Air Force is the largest contributor to carbon emissions through its consumption of jet fuel. However, the military's impact is not solely through air travel. Toxic substances from military vehicles, weapons materials, and munitions pollute the air, land, and groundwater in and around military bases and training areas.

The use of toxic munitions and burn pits during the wars in Afghanistan and Iraq has led sections of the countries to be so toxic that increased cancers, birth defects, and infant mortality have been documented among the local population (Hussain 2019), and veterans of these wars face chronic illnesses as a result of exposure (McCarthy 2019). Ironically, the U.S. military is also one of the leading federal agencies investigating renewable energy, including solar and biofuels, and preparing for the effects of rising sea levels (Bigger and Neimark 2017; Belcher et al. 2019). Under the Trump administration, several of these initiatives were abandoned, however, including the Navy's Climate Change Task Force, which had been working on sustainable solutions since 2009 (Simkins 2019).

Strategies for Action: Responding to Environmental Problems

Actor Robert Redford (2015) said, "Our planet's resources are limited, but there is no limit to the human imagination and our capacity to solve our biggest problems." Efforts to solve environmental problems include environmental activism, environmental education, reduction of carbon emissions, the use of green energy and energy efficiency, modifications in consumer products and behavior, slowing population growth, and sustainable economic development.

Environmental Activism

During the 2019–2020 school year, Swedish teenager Greta Thunberg traveled the world to spread the word about climate change. Thunberg's "school strike" invigorated activism in general and particularly youth activism around the globe, with 6 million people taking to the streets in September 2019 to demand action on climate change (Tran, Watts, and Bartlett 2019). In some parts of the world, environmental activists have faced strong resistance. In Moscow, for the September 2019 strike, environmental activists were forced to take turns holding one sign raising awareness about climate change in the main Pushkin Square because officials would not grant a permit for a demonstration (Taylor, Watts, and Bartlett 2019). In Colombia an estimated 106 human rights activists were killed in 2019; about half of these were environmental activists (Tomassoni 2020). Environmental and other human rights activists routinely receive death threats, as well as threats of rape and other forms of violence.

Yet tens of millions of people have joined thousands of international, national, and local environmental organizations, making the environmental movement one of the largest social movements in the world. A 2018 Global Attitudes Survey found that, in most countries surveyed, a majority of people viewed climate change as a major threat (Fagan and Huang 2019). Fifty-nine percent of Americans agreed, and another 23 percent described it as a minor threat. A Gallup survey found that 17 percent of U.S. adults report being active participants in the environmental movement (Norman 2017) and that 42 percent consider themselves environmentalists (Gallup Organization 2020) (see Table 13.3).

TABLE 13.3 Involvement in the Environmental Movement*

Involvement	Percentage of U.S. Adults
Active participant	17
Sympathetic but not active	45
Neutral	28
Unsympathetic	8
No opinion	2

*In a 2017 Gallup survey, a national sample of U.S. adults was asked, "Do you think of yourself as an active participant in the environmental movement, sympathetic toward the movement, but not active, neutral, or unsympathetic toward the movement?"
SOURCE: Norman 2017.

Environmentalists and environmental organizations vary considerably in their goals and the methods that they use to achieve them. The Environmental Defense Fund works across virtually all environmental issues, has employees and offices around the world, and engages in science to develop solutions and influence policies around the world (Environmental Defense Fund 2020a). In contrast, Fridays for Future is a movement of individual activists focused on direct action aimed at educating the public and putting pressure on policy makers to address climate change (Fridays for Future 2020).

Environmental organizations exert pressure on government and private industry to initiate or intensify actions related to environmental protection. Environmentalist groups also design and implement their own projects and stage public protests and demonstrations. Environmental organizations disseminate information to the public about environmental issues and use the Internet and e-mail to send e-mail action alerts to members, informing them when Congress and other decision makers threaten the health of the environment. These members can then send e-mails and faxes to Congress, the president, and business leaders, urging them to support policies that protect the environment.

charismatic megafauna
Particular species that have popular appeal, such as the panda, polar bear, monarch butterfly, and bald eagle, and that are used to draw public attention to larger environmental issues.

Wildlife activism efforts are sometimes focused on **charismatic megafauna**—particular species that have popular appeal, such as the panda, polar bear, monarch butterfly, and bald eagle (Pariona 2017). These species are often used to draw public attention to a larger environmental issue. For example, the image of a stranded polar bear is often used to depict the melting polar region, drawing attention to the larger issue of global warming and climate change.

Religious Environmentalism. Since the 1990s, religious leaders and scholars have become increasingly concerned with environmental problems and climate change, and every major religion has issued statements about the importance of protecting the environment (United Nations Environment Programme 2019b).

In 2017, the United Nations launched the Faith for Earth Initiative, which engages with faith-based organizations to work toward the UN Sustainable Development Goals. In response to the growing problem of global warming and climate change, in 2015, the Archbishop of Canterbury, along with religious leaders from the Muslim, Sikh, Jewish, Catholic, and other faiths, signed a declaration urging the global community to transition to a low-carbon economy (Archbishop's Council 2015).

Swedish environmental activist Greta Thunberg began a global movement when she started her School Strike for Climate in August 2018. In the 16 months afterward, she met with heads of state and the Pope, addressed the World Economic Forum, the UN General Assembly, and the UN Climate Summit and inspired millions around the world to protest against global warming. She was named *Time's* 2019 Person of the Year.

In the same year, Pope Francis delivered a 184-page encyclical—a letter addressed to the more than 1 billion Catholics around the world—in which he lamented, "Never have we so hurt and mistreated our common home as we have in the last 200 years," expressing the urgent need to change our lifestyle, production, and consumption (Pope Francis 2015).

In 2015, the Islamic Declaration on Climate Change was completed, which stated, "We human beings are created to serve the Lord of all beings, to work the greatest good we can for all the species, individuals, and generations of God's creatures" (Islamic Foundation for Ecology and Environmental Sciences [IFEES] 2020). The Declaration also states that climate change has been caused by human activity. Jeane McKay, an English conservation biologist who has worked on conservation projects across the globe, points out that "(u)sing faith-based approaches can prove to be a positive way forward, and indeed has the potential to gain far-reaching benefits rather than staying confined to a conventionally science-based approach" (Rust 2017).

Radical Environmentalism. The **radical environmental movement** is a grassroots movement of individuals and groups that employs unconventional and often illegal means of protecting wildlife or the environment. Radical environmentalists believe in what is known as **deep ecology**—the view that maintaining the Earth's natural systems should take precedence over human needs, that nature has a value independent of human existence, and that humans have no right to dominate the Earth and its living inhabitants (Fellows 2019). Contemporary radical environmental groups include Earth First!, Extinction Rebellion, and Sea Shepherd–groups that typically engage in direct action. Sea Shepherd boats interfere with the work of whaling boats in order to protect whales from extinction. For six years, Earth First! and other activists lived in the trees of the Hambacher Forest in western Germany to prevent further deforestation by a coal-mining company, fighting off efforts to be removed by police (Mansel 2019). In 2016, thousands of opponents of the Dakota Access pipeline gathered in camps to block expansion of the pipeline; some activists turned valves on the pipeline to shut off oil flow (Brown 2019). A year later, activists began burning holes into the pipeline with welding tools.

radical environmental movement A grassroots movement of individuals and groups that employs unconventional and often illegal means of protecting wildlife or the environment.

deep ecology The view that maintaining the Earth's natural systems should take precedence over human needs, that nature has a value independent of human existence, and that humans have no right to dominate the Earth and its living inhabitants.

ecoterrorism The use or threatened use of illegal force by groups or individuals in order to protect environmental and/or animal rights.

Radical environmentalists are sometimes accused of **ecoterrorism**, defined as "the use or threatened use of illegal force by groups or individuals, in order to protect environmental and/or animal rights" (Yang et al. 2014, p. 11). The term *ecoterrorism* was coined after 9/11 when the fur and biomedical industries lobbied the FBI to target groups like the Earth Liberation Front (ELF) and the Animal Liberation Front (ALF) for damaging businesses that profited from the destruction of the environment or the abuse of animals. At that time, the FBI denounced radical environmentalists as the "number one domestic terror threat," above white supremacist groups, militias, and violent anti-abortion groups like Operation Rescue, which had killed several people (Brown 2019).

What do you think? Should motives be considered in imposing penalties on individuals who are convicted of acts of ecoterrorism? For example, should a person who sets fire to a business to protest that business's environmentally destructive activities receive the same penalty as a person who sets fire to a business in order to collect insurance money?

Hunters/Anglers and Environmentalism. Some people believe that hunting and fishing are incompatible with environmentalism. "For many environmentalists, the word hunter suggests a mindless brute, an enemy of nature who loves guns, kills for fun, and cares nothing for biodiversity or ecological integrity" (Cerulli 2014). Indeed, some commercial hunting and fishing practices contribute to the problem of species depletion. Over 100,000 African elephants were killed between 2014 and 2017 for their ivory tusks, putting them in danger of extinction (Actman 2019). And overfishing threatens fish populations, including the Atlantic halibut, the monkfish, all sharks, and blue fin tuna (Mok 2020). On the African continent, between 2009 and 2016, poachers killed almost 600 rangers who were attempting to protect animals.

Yet many hunters and anglers are avid conservationists who are committed to preserving natural habitats to maintain stable populations of animal and marine life. Hunting has also been used as a wildlife management tool to help mitigate the overpopulation of species such as deer, which can otherwise populate to unsustainable levels due to the lack of natural predators (Maryland Department of Natural Resources 2020). Additionally, the excise tax revenue from the purchase of guns, ammunition, angling gear, duck stamps, and hunting and fishing licenses is used to fund various environmental management policies (U.S. Fish and Wildlife Service 2020). Although some environmentalists, especially those who live a vegetarian or vegan lifestyle, are philosophically opposed to killing animals, the environmental interests of hunters/anglers and their opponents are often the same.

Environmental Education

One goal of environmental organizations and activists is to educate the public about environmental issues and the seriousness of environmental problems. Being informed about environmental issues is important because people who have higher levels of environmental knowledge tend to engage in higher levels of pro-environment behavior (Geiger, Geiger, and Wilhelm 2019). For example, environmentally knowledgeable people are more likely to save energy in the home, recycle, conserve water, purchase environmentally safe products, avoid using chemicals in yard care, support policies that protect the environment, and donate funds to conservation. In this chapter's *Self and Society* explore your own opinions on some of the environmental issues we've discussed in the chapter and compare them to the U.S. population as a whole.

A primary source of information about environmental issues is the media. Because corporations and wealthy individuals with corporate ties own the mainstream media, unbiased information about environmental impacts of corporate activities often cannot be readily found through these sources. For example, when the United Nations issued its fifth assessment report on climate change, half of the major print media outlets (e.g., *Washington Post*, *LA Times*, *Wall Street Journal*) gave the climate denier perspective significant coverage although the report indicated scientific consensus on climate change (Theel, Greenberg, and Robbins 2013).

During the 2016 presidential election debates, no questions were asked about climate change despite the fact that it was the hottest year ever recorded in human history (Holden 2019). However, climate change was a last-minute addition to questions in the first 2020 presidential debate (Cooper 2020). The growth of social media platforms in recent years has expanded access to information about the environment. Long-standing organizations like the World Wildlife Fund, Greenpeace, and Sea Shepherd Global have effectively used social media to increase their reach (d'Estries 2018).

New groups on social media, like A Focus on Nature and Next Generation Birders, and youth activists/social influencers, like Sweden's Greta Turnberg and China's Wang Junkai, are educating teenagers and young adults about the environment. However, not surprisingly, climate deniers also use social media. A study of the first six months of 2020 found that a definable group of lay people, not scientists, posted four times more climate denier social media messages than climate science posters, making climate denial messages "louder" (Khoo and Ryan 2020) (see Chapter 14). The same group went on to post COVID-19 conspiracy messages beginning in March 2020, and Black Lives Matter conspiracy posts beginning in June 2020. The group also shares QAnon posts. As the documentary *The Social Dilemma* (Netflix 2020) notes, the algorithms of social media thrive on fake news as they create "rabbit holes" that entice users to remain on the site.

Attitudes toward Environmental Issues and Climate Change

Answer each of the following questions:

1. Do you think air pollution is a big problem, a moderate problem, a small problem, or not a problem?
2. Do you think pollution of rivers, lakes, and oceans is a big problem, a moderate problem, a small problem, or not a problem?
3. Do you think extinction of plant and animal species is a big problem, a moderate problem, a small problem, or not a problem?
4. Do you think the amount of garbage, waste, and landfills is a big problem, a moderate problem, a small problem, or not a problem?
5. Do you think the loss of forests is a big problem, a moderate problem, a small problem, or not a problem?
6. Which of the following statements comes closer to reflecting your view, even if neither is exactly right?
   a. Protecting the environment should be given priority, even if it causes slower economic growth and some loss of jobs.
   b. Creating jobs should be the priority, even if the environment suffers to some extent.
7. Do you favor or oppose using more renewable energy in this country?
8. In your view, is global climate change a very serious problem, somewhat serious, not too serious, or not a problem?
9. Do you think the national government is doing too much, too little, or about the right amount to reduce the effects of climate change?
10. Do you think it is generally safe or unsafe to eat fruits and vegetables grown with pesticides?
11. Do you think it is generally safe or unsafe to eat food and drinks with artificial preservatives?

Results from the U.S. Population

Telephone interviews with a sample of 1,502 U.S. adults from across the nation found the following:

1. Air pollution is a big problem (63%), a moderate problem (23%), a small problem (9%), or not a problem (4%).
2. Pollution of rivers, lakes, and oceans is a big problem (75%), a moderate problem (15%), a small problem (6%), or not a problem (2%).
3. Extinction of plant and animal species is a big problem (60%), a moderate problem (21%), a small problem (10%), or not a problem (7%).
4. The amount of garbage, waste, and landfills is a big problem (72%), a moderate problem (19%), a small problem (6%), or not a problem (3%).
5. The loss of forests is a big problem (66%), a moderate problem (18%), a small problem (6%), or not a problem (8%).
6. Protecting the environment should be given priority, even if it causes slower economic growth and some loss of jobs (64%). Creating jobs should be the priority, even if the environment suffers to some extent (31%).
7. Using more renewable energy: favor (74%), oppose (24%).
8. Global climate change is a very serious problem (53%), somewhat serious (21%), not too serious (10%), or not a problem (15%).
9. The national government is doing too much (11%), too little (63%), or about the right amount (21%) to reduce the effects of climate change.
10. Do you think it's generally safe (26%) or unsafe (48%) to eat fruits and vegetables grown with pesticides? Don't know (25%).
11. Do you think it's generally safe (23%) or unsafe (45%) to eat food and drinks with artificial preservatives? Don't know (32%).

SOURCE: Adapted from the Pew Research Center's International Science Survey (Funk et al. 2020).

What do you think? First during and then after the 2020 presidential election, social media companies like Facebook and Twitter flagged posts that contained misinformation about the election. Some repeat offenders had their accounts blocked. Should social media companies block or flag posts that share "news" that is not factual, like climate denial posts? Why or why not?

Corporate propaganda is sometimes packaged as "environmental education." The American Petroleum Institute (API) distributes curriculum materials and lesson plans that question climate science and promote the value of fossil fuels to K–12 teachers (Mulvey and Shulman 2015). The Heartland Institute, a think tank started by the fossil fuel industry, sent 300,000 teachers around the country copies of a book they published called *Why Scientists Disagree about Global Warming*, which fails to inform teachers that more than 97 percent of scientists *agree* on the causes and impacts of global warming (Banerjee 2017).

One survey found that of the three-fourths of science teachers who cover the topic of climate change in the classroom, 30 percent teach students that global warming and climate change is "likely due to natural causes" and 31 percent teach that the science is unsettled (Boyle 2017). Several states have proposed bills or resolutions that allow or require teachers to teach global warming and climate change science as a theory rather than as a proven fact (Boyle 2017; Chen 2016). In contrast, Portland, Oregon public schools voted to ban textbooks and other materials that cast doubt on climate change and humans' role in global warming (Dicker 2016).

The Next Generation Science Standards (NGSS) finalized in 2013 are the first set of national guidelines to require that K–12 students be taught that climate change is a scientific fact and is mainly caused by burning fossil fuels (National Science Teaching Association 2020). The standards were developed as a collaboration between the National Research Council, the National Science Teachers Association, the American Association for the Advancement of Science, and many other state and national agencies, business leaders, and communities in an effort to support and improve science education. As of September 2020, these voluntary standards have been accepted in 20 states, and another 22 have revised their standards using the framework of the NGSS. There has been resistance in several states, especially those whose economy is dominated by the fossil fuel industry (Boyle 2017).

Reduction of Carbon Emissions

The UN's Intergovernmental Panel on Climate Change (2018) says that to reduce risks associated with climate change, developed countries must reduce greenhouse gas emissions to nearly zero by 2100 to keep global warming under 2 degrees Celsius—the IPCC's threshold for dangerous climate change.

In 1992, countries joined an international treaty, the United Nations Framework Convention on Climate Change, to cooperate in limiting global warming and in coping with climate change and its impacts. In 1997, delegates from 160 nations met in Kyoto, Japan, and forged the **Kyoto Protocol**—the first international agreement to place legally binding limits on greenhouse gas emissions from developed countries. The United States, the world's largest producer of greenhouse gas emissions, rejected the Kyoto Protocol in 2001 under the leadership of President George W. Bush (CNN 2020b).

In 2009, the U.S. House passed the *American Clean Energy and Security Act*, which sought to establish a federal **cap and trade system** modeled after the European Union Emission Trading Scheme. In cap and trade, power plants and other polluting industries buy credits, allowing them to emit a limited amount of carbon dioxide. They can sell leftover credits to other polluters, creating an economic incentive to reduce emissions. The bill was defeated in the Senate. In 2010, California adopted a statewide cap and trade program that led to a 13 percent decline in carbon emissions in its first ten years (Environmental Defense Fund 2020b).

In 2013, nine northeastern states organized in the Regional Greenhouse Gas Initiative (RGGI)—the first mandatory, cap and trade carbon emissions reduction program in the United States. In 2020, RGGI states had cut carbon emissions by 50 percent and had reduced the emission of other pollutants like mercury, sulfur dioxide, and nitrogen oxides (Ho 2020). Over time, the allowable emissions standards are being reduced further, providing incentives for companies to innovate and leading to thousands of new jobs. New Jersey and Virginia are joining the RGGI Market.

Following Kyoto, a series of international agreements to slow global warming and adapt to climate change have been forged, but none has gone far enough to stop global warming. The 2015 global summit on climate change in Paris resulted in the Paris Agreement—a global effort to combat climate change and adapt to its effects. The Paris Agreement set a target to limit global warming to well below 2 degrees Celsius, and ideally to 1.5 degrees Celsius. A 2017 report found that to meet the Paris goal, fossil fuel consumption needs to peak within the next decade and decline to below a quarter of the world's energy supply by 2100 (Smith 2017).

Kyoto Protocol The first international agreement to place legally binding limits on greenhouse gas emissions from developed countries.

cap and trade system A free-market approach that provides economic incentives to power plants and other industries for reducing carbon emissions.

In 2015, under President Obama, the EPA proposed the first-ever federal limits on carbon emissions from power plants—the nation's largest source of carbon pollution. Under Obama's **Clean Power Plan**, the EPA proposed state-by-state targets for carbon emissions reductions, allowing states flexibility in how to meet these targets. In 2017, President Trump signed the Energy Independence Executive Order, which rescinded several of the Obama administration's orders and policies concerning climate change and ordered a review or revision of "regulations that may place unnecessary, costly burdens on coal-fired electric utilities, coal miners, and oil and gas producers" (Environmental Protection Agency 2017). Under former President Trump's orders, the United States exited the Paris agreement in 2020, becoming the only country out of nearly 200 to do so since the agreement was adopted in 2015. In the hours after he was sworn into office, President Biden signed an executive order to rejoin the Paris agreement as part of his overall strategy to prioritize the reduction of carbon emissions (Daley 2020).

What do you think?

At the height of the 2020 COVID-19 pandemic, Pope Francis (2020) noted, "We put on our face masks to protect ourselves and others from a virus we cannot see. But what about all those other viruses we need to protect ourselves from? How will we deal with the hidden pandemics of this world, the pandemics of hunger and violence and climate change?" In what ways is climate change and our societal reactions to it similar to the COVID-19 pandemic?

Green Energy and Energy Efficiency

The fossil fuel industry has long pitted the environment against economic growth, but this is a false dichotomy. The U.S. Green Economy, the sector of the economy associated with low carbon and environmentally-friendly goods and services, is estimated to represent $1.3 trillion in annual sales revenue and to employ nearly 9.5 million workers (Georgeson and Maslin 2019). An analysis of the costs and benefits of the *Clean Air Act Amendments* of 1990 found that the associated reductions in health care costs and premature deaths have contributed $3.8 trillion to the U.S. economy (Mui and Levin 2020).

A study by the EPA found improvements in environmental quality were associated with positive impacts on the overall economy, in part due to improvements in human health (Ferris et al. 2017). Green energy jobs are among the fastest growing jobs in the country offering salaries above the median for the country (Kiersz and Akhtar 2019). The claim that environmental regulation hurts jobs in the long run is not supported by evidence, and the costs of regulations are far outweighed by the benefits to health, safety, and well-being.

A major part of President Biden's Build Back Better Plan involves investing in clean energy and energy efficiency, with the goal of carbon-free electricity production by 2035 (Biden 2020). Prior to President Biden's election, in 2019, a group of more than 60 representatives in the U.S. House of Representatives introduced House Resolution 109 advocating for what has come to be known as the Green New Deal (U.S. House of Representatives 2020).

Modeled after the New Deal introduced by President Roosevelt in 1932, the Resolution seeks, among other things, to eliminate greenhouse gases by developing green technologies and improving existing infrastructure across the country to improve energy efficiency. In so doing, it also proposes to create millions of jobs that would address problems of wage stagnation and income inequality in the United States. It also asserts that access to clean water, clean air, and healthy food is a basic human right. The Green New Deal is a broad resolution, not a specific plan. President Biden's climate plan is not as aggressive or as far-reaching as the Green New Deal, as he has tried to strike a balance "between trying to win the confidence of climate crusaders on the left while not alienating more moderate voters in the middle" (Berardelli 2020).

Clean Power Plan Establishes the first-ever federal limits on carbon emissions from U.S. power plants, establishing state-by-state targets for carbon emissions reductions and allowing states flexibility in how to meet these targets.

In March 2020, former President Trump signed the Safer Affordable Fuel-Efficient (SAFE) Vehicles Rule, which requires automakers to increase fuel efficiency by 1.5 percent per year through 2026, a reduction in goals from 5 percent per year under the Obama administration (Rott and Ludden 2020). Automotive industry leaders had pushed for the change arguing they could not meet the Obama administration standards in a cost-effective manner.

After the passage of the SAFE Vehicles Rule, California governor Gavin Newson signed an executive order requiring that by 2035 all passenger cars and trucks sold in California be electric or otherwise generate zero-emissions. Claiming that federal fuel efficiency standards supersede those of any state, the Trump administration, with the support of several automakers, initiated legal action to try to stop California from setting its own emissions rules. After President Biden's election in 2020, General Motors (GM) withdrew support for Trump's litigation against California saying that its plan to bring 30 new electric vehicles to market by 2025 aligns with Biden's endorsement of electric vehicles (LaReau 2020).

> "Our planet's resources are limited, but there is no limit to the human imagination and our capacity to solve our biggest problems."
>
> **–ROBERT REDFORD, ACTOR**

The federal government and most states offer incentive programs including tax credits and rebates to encourage energy efficiency and the use of renewable energy sources. For example, the federal government is offering tax credits for the residential installation of geothermal heat pumps, small wind turbines, and solar energy systems and for the purchase of electric and plug-in hybrid vehicles. Some states have offered tax credits for purchases of hybrid or electric cars, housing insulation, or solar panels (U.S. Department of Energy 2020).

In 2015, the state of Hawaii passed a law requiring the state to work toward transitioning to 100 percent renewable energy by 2045 (Fields 2019). Since then three other states—Maine, Minnesota, and Rhode Island—have set similar goals. Ten states have set goals to transition solely to clean energy sources, including Arizona, California, Connecticut, Nevada, New Jersey, New Mexico, New York, Virginia, Washington, and Wisconsin. Target dates range from 2030 to 2050. Clean energy commitments allow for the use of nuclear energy, while renewable energy commitments do not.

Hydrogen Power. Hydrogen can be used in a battery cell or as clean-burning fuel for electricity production, heating, cooling, and transportation. Hydrogen fuel from plant waste may one day replace gas for transportation vehicles (Connor 2015). California has 43 publicly accessible hydrogen-fueling stations, which allow for the first large-scale introduction of hydrogen fuel cell–powered cars in the United States (National Association of Convenience Stores [NACS] 2020). However, there are only five fueling stations outside of California. The infrastructure for hydrogen power has developed more quickly outside of the United States (Research and Markets 2019). Denmark has sufficient fueling stations to cover the entire country; Iceland is at over 80 percent. Germany has 45 stations, and Japan has over 100; both are home to major automotive manufacturers.

Modifications in Consumer Products and Behavior

In the United States and other industrialized countries, many consumers are making "green" choices in their behavior and purchases that reflect concern for the environment (Liobikiene, Madravickaite, and Bernatoniene 2016). In some cases, these choices carry a price tag, such as paying more for organically grown food or for clothing made from organic cotton. Consumers are also motivated to make green purchases that save money. Consumers often consider their utility bill when they choose energy-efficient appliances and electrical equipment.

Although some eco-minded individuals choose green products and services, others choose to reduce their overall consumption and "buy nothing" rather than "buy green" (Helm et al. 2019). For example, many consumers are repurposing clothing, furniture, and other goods rather than buy new. Many choose to carry water bottles and drink tap water instead of buying single-use plastic bottles. The switch from bottled to tap is partly fueled by the need to cut down on unnecessary spending in hard economic times, but environmental concerns are also a factor. The production and transportation of bottled water uses fossil fuels, and the disposal of plastic water bottles adds to our already overburdened landfills.

Although the average size of new housing in the United States has increased considerably, some homeowners are choosing to downsize their housing (Willis 2019). For some, the driving force behind housing downsizing is economic, but others are moving into smaller dwellings out of concern for the environment.

Table 13.4 presents tips for how consumers can reduce the amount of carbon dioxide each of us produces and reduce our impact on the environment through reducing pollution and saving water.

Slow Population Growth

Slowing population growth is an important component of efforts to protect the environment (Bailey 2019), although on its own, no model of declining population growth can save us from the negative effects of our carbon footprint in the near future (Schienman 2019). The impact of population growth (or decline) on the environment depends on region, with declines in Western countries having larger projected impacts on the environment than those in other parts of the world (Weber and Sciubba 2019). Most of these countries, apart from the United States, already have fertility rates under replacement level. Thus, it is unlikely that Western governments will promote additional decreases in fertility as a matter of policy, since they are faced with the challenges of aging populations and increasing dependency ratios, as discussed in Chapter 12. However, decreases in population growth in the developing world will still benefit the environment and have other positive impacts like greater economic well-being and gender equality (Bailey 2019).

TABLE 13.4 Ten Things You Can Do to Fight Global Warming and Pollution

Adopting the following "green" practices will help save the environment and reduce global warming. Plus, many of them will make you healthier in the long run!
1. Replace incandescent lightbulbs in your home/dorm room with light-emitting diode (LED) bulbs or with compact fluorescent bulbs (CFLs): Swapping 75-watt incandescent bulbs with 19-watt CFLs saves 275 pounds of CO_2. LED bulbs are even more energy efficient than CFLs (and LEDs produce less heat so they are less of a fire hazard and save on cooling costs).
2. Reduce, reuse, recycle. Only 3 percent to 5 percent of what is in the average student's trash can is actually garbage. Most things that are thrown out can be recycled or composted.
3. Sure, it may be hot, but get a fan, set your thermostat to 75 degrees Fahrenheit, and blow away 363 pounds of CO_2.
4. Enable the stand-by/sleep mode on your computer. The average computer uses much less power when it is in sleep mode.
5. Shortening your shower by 1 minute can save 1,000 gallons of water. Turning off the faucet while brushing your teeth will save about 2.5 gallons of water per minute.
6. Reduce plastic pollution by investing in a reusable water bottle and bringing it everywhere you go. You'll also stay better hydrated with one along.
7. Use notebooks made of 100 percent recycled paper, and conserve printer paper by printing on both sides when printing lecture notes or drafts of papers.
8. Cut down on red meat consumption. Producing 2.2 pounds of beef generates the same CO_2 emissions as an average car emits every 60 miles.
9. Leave the car at home, and take public transportation, carpool, or ride a bike to work or school. Cars in the United States emit 314 million metric tons of CO_2 per year, which is enough to fill a coal train that could circle the Earth twice.
10. Bring reusable bags with you everywhere you go to reduce the need for plastic bags when you're shopping.

SOURCES: UCLA Sustainability 2014.

And as the standard of living improves in the developing world, the global environmental footprint will likely increase, which would be catastrophic if population growth continued at its current rate.

There is also evidence that young adults are considering climate change as they make decisions about having children. In the United States, a 2018 poll found that one-third of men and women between 20 and 45 cited climate change as a factor in their decision to have fewer children (Schienman 2019). In the U.K., a group called BirthStrike includes people who have vowed to be child free until governments address the problem of global warming. Researchers estimate that in the developed world, a child has a carbon footprint of about 58.6 metric tonnes per year (Schienman 2019). However, the concerns of many young adults are not just what impact will their child(ren) have on the environment but what impact will the growing environment crisis have on their child(ren) (Scheinman 2019).

Stephen Goodwin/Alamy Stock Photo

In 2019, there was a 35 percent increase in electric vehicle charging stations on college and university campuses across the country (ChargePoint 2020).

The Role of Institutions of Higher Education

Colleges and universities can help protect the environment by encouraging use of bicycles on campus, using hybrid and electric vehicles, establishing recycling programs, involving students in organic gardening to provide food for the campus, using clean energy, and incorporating environmental education into the curricula. Universities are making progress. For example, in 2018 Arizona State University–Tempe made a fair trade pledge to sell and use products manufactured in an environmentally sustainable manner (Kowarski 2019). The College of the Atlantic has an initiative to Break Free from Plastic, limiting the use of nonrecyclable plastics, and the University of California–Irvine converted its central plant to a system that utilizes more treated wastewater than regular water to cool its buildings.

A growing number of colleges and universities are establishing **green revolving funds (GRFs)**, which are funds dedicated to financing energy efficiency upgrades and other projects that decrease resource use and minimize environmental impact (Harvard University 2020). Harvard University has a $12 million GRF to fund projects that will pay back to the university in energy savings in no more than five to ten years. A portion of those savings are then re-invested into the fund. The 200 projects funded have saved the university $4 million thus far.

Many colleges and universities have also joined the fossil fuel divestment movement that began in 2012—a movement that involves getting rid of stocks, bonds, and other investment funds from fossil fuel companies (Melia 2020). Students are putting pressure on more universities to follow through on their stated commitments to address climate change by divesting. In addition, hundreds of institutions, governments, faith-based organizations, individuals, philanthropic foundations, and schools have divested from fossil fuels.

The number of universities offering degrees in environmental science, environmental policy, environmental studies, and sustainability studies has also increased. For students who are interested in where colleges and universities stand on environmental issues; the *Princeton Review* now publishes a ranking of the Top 50 Green Colleges.

green revolving funds (GRFs) College and university funds that are dedicated to financing cost-saving energy-efficiency upgrades and other projects that decrease resource use and minimize environmental impacts.

Understanding Environmental Problems

Environmental problems are linked to reliance on fossil fuels for energy, the prioritization of corporate profits, the influence of industry over policy makers, rapid population growth, expanding world industrialization, and patterns of excessive consumption. The Global Footprint Network (2020) offers the following analysis of our current status:

> Our economies are currently running a fraudulent Ponzi (or pyramid) scheme with our planet. We are consuming natural resources faster than we can regenerate; we are using Earth's future resources to operate in the present; we are digging ourselves deeper and deeper into ecological debt. Prosperity can only last if we embrace the limits of our planet. Accepting limits allows us to build an economy that works forever. Ignoring limits leads to a finite, time-limited economy which ... erodes the planet it depends on.

Whether we understand environmental problems as resulting from a complex set of causes or from one, simple underlying cause such as consuming more than the Earth can provide, we cannot afford to ignore the growing evidence of the irreversible effects of global warming and loss of biodiversity and the adverse health effects of toxic waste and other forms of pollution. Never before has human civilization on a global scale been threatened, as it is now, by destruction of our ecosystems. To avoid the collapse of the global civilization, it is essential to reduce greenhouse gas emissions to nearly zero by the end of the century. This would require a major shift away from the use of fossil fuels—a shift the fossil fuel industry cannot support without jeopardizing its own profits. It is no wonder, therefore, that the industry has spent millions—likely billions—on political donations, lobbying, and misinformation campaigns to confuse the American public.

Many Americans believe in a "technological fix" for the environment—that science and technology will solve environmental problems. Paradoxically, the same environmental problems that have been caused by technological progress may be solved by technological innovations designed to clean up pollution, preserve natural resources and habitats, and provide clean forms of energy. For example, environmentalists and scientists are using drones to study wildlife in remote regions, to measure polar ice melting, to collect water samples for water quality monitoring, to plant trees to reduce deforestation, and to detect and apprehend illegal poachers (Carroll 2015). Advances in solar and wind power have made both more available and affordable to everyday consumers. Technology can be used to help the environment.

Leaders of government and industry must have the will to finance, develop, and use green technologies on a large scale. But the direction of technical innovation is largely in the hands of big corporations that place profits over environmental protection. Unless the global community challenges the power of transnational corporations to pursue profits at the expense of environmental and human health, corporate behavior will continue to take a heavy toll on the health of the planet and its inhabitants. As the United Nations (2020) has articulated, we must as a global community find the means by which we can achieve **sustainable development**, that balancing point at which all human beings can live healthy, equitable, and peaceful lives without degrading our natural environment.

Global cooperation is vital to resolving environmental concerns, but rich and poor countries have different economic development agendas. Developing poor countries struggle to survive and provide for the basic needs of their citizens; developed wealthy countries struggle to maintain their wealth and relatively high standard of living. Can both agendas be achieved without further pollution and destruction of the environment? With mounting concern about climate change; the health impacts of air, water, and land pollution; and the need to ensure energy access to all, the Paris Climate Accord made tremendous strides in bringing countries to the table to develop a plan for lasting change (Stern 2018).

If Donald Trump had won his bid for reelection in 2020, his withdrawal from the Paris agreement and his disregard for the science of global warming and climate change would have, in all likelihood, severely compromised the state of our national and global environment. Time will tell whether President Biden will live up to the promise to enact what has been called "the most ambitious climate plan of any President we've ever had in the United States" (Horowitz 2020). But the best efforts and policies of any one nation cannot, alone, stand against the tide of global warming and climate change, which is why President Biden created a new position in his administration—"special presidential envoy for climate"—to which he appointed John Kerry—former secretary of state, senator, and presidential candidate (McKenna 2020). Indeed, China, India, and all the nations of

sustainable development The balancing point at which all human beings can live healthy, equitable, and peaceful lives without degrading our natural environment.

the world must do their part to reduce carbon emissions. Whether we act out of gratitude, fear, or a sense of responsibility to future generations, we must act quickly if we are to protect the Earth and its living inhabitants. As one popular slogan in the environmental movement says: "There is no Planet B."

Chapter Review

- **How is global warming impacting the world, and what do we need to do to reverse it?**
The world today is 1 degree Celsius warmer than it was in the middle of the 9th century. Climate change is contributing to increases in sea levels, natural disasters, droughts, and species extinction. In order to stop and reverse global warming, we must reduce the carbon emissions that are contributing to the Greenhouse Effect. The primary greenhouse gas is carbon dioxide, which is released into the atmosphere by burning fossil fuels. Reducing global warming requires a global effort to change how we produce and use energy, as well as a halt on deforestation.

- **How does the conflict perspective view environmental problems and solutions?**
The conflict perspective focuses on the role that wealth, power, and the pursuit of profit play in environmental problems and solutions. For example, the fossil fuel industry has spent a great deal of money to influence lawmakers to support policies that benefit the fossil fuel industry and to discredit climate science.

- **What does the term *environmental injustice* refer to?**
Environmental injustice, also called *environmental racism,* refers to the tendency for marginalized populations and communities to disproportionately experience adversity due to environmental problems. For example, in the United States, polluting industries, industrial and waste facilities, and transportation arteries (which generate vehicle emissions pollution) are often located in minority communities.

- **What is greenwashing?**
Greenwashing refers to the ways in which environmentally and socially damaging companies portray their corporate image and products as being "environmentally friendly" or socially responsible.

- **Where does most of the world's energy come from?**
Most of the world's energy comes from fossil fuels, which include oil, coal, and natural gas. This is significant because many of the serious environmental problems in the world today, including global warming and climate change, biodiversity loss, and pollution, stem from the use of fossil fuels.

- **What are some problems associated with nuclear energy?**
Nuclear waste contains radioactive plutonium, a substance linked to cancer and genetic defects. Nuclear waste in the environment remains potentially harmful to human and other life for thousands of years, and disposing of nuclear waste is problematic. Accidents at nuclear power plants, such as the 2011 Fukushima disaster, and the potential for nuclear reactors to be targeted by terrorists add to the actual and potential dangers of nuclear power plants.

- **What is mountaintop removal mining?**
Mountaintop removal mining is a form of coal mining in which mountains are deforested and then their tops are removed using explosives. The debris is then pushed into the surrounding valleys, filling and contaminating waterways.

- **What is our environmental footprint?**
The demands that humanity makes on the Earth's natural resources are known as the environmental footprint. Since 1970, humanity's environmental footprint has exceeded the Earth's capacity to produce useful resources such as water and crops, and to absorb waste, such as CO_2 emissions.

- **What are the major causes and effects of deforestation?**
The major causes of deforestation are the expansion of agricultural land, human settlements, wood harvesting, and road building. Deforestation displaces people and wild species from their habitats, contributes to global warming, and leads to desertification, which results in the expansion of desert land that is unusable for agriculture. Soil erosion caused by deforestation can cause severe flooding.

- **What are the effects of air pollution on human health?**
Indoor and outdoor air pollution, which is linked to heart disease, lung cancer, and respiratory ailments, kills about 7 million people a year, or one in eight global deaths.

- **Why do U.S. politicians promote the "climate denier" view, despite overwhelming evidence to the contrary?**
In order to protect their economic interests, the fossil fuel industry has invested millions in promoting the view that global warming is not the result of greenhouse gas emissions. Much of this money is used to lobby members of Congress and members of the U.S. Cabinet. In return, the industry has received financial incentives and reductions in environmental regulations. The majority of the American public believes that climate change is primarily the result of human activity and that the federal government should do more to address it.

- **What are the major sources of water pollution?**
Water pollution is largely caused by fertilizers and other chemicals used in intensive agriculture, industrial production, mining, and untreated urban runoff and wastewater. Oil spills and plastics also contribute to water pollution.

- **Why is there increasing public concern over fracking?**
Fracking involves the use of chemicals that contaminate land, water, and air. Oil and gas companies do not have to disclose to the federal government the chemicals they use in fracking, but scientists find carcinogenic and toxic chemicals in fracking fluid and wastewater. Leaks, overflows, and illegal dumping of wastewater allow chemicals to flow into local land and water supplies. Fracking requires massive amounts of water, which is problematic in drought-stricken areas, and fracking operations have triggered earthquakes. People—as well as pets, farm animals, and wildlife near fracking sites—have become sick as a result of contact with and/or consumption of contaminated water from fracking sites.

- **What is the precautionary principle?**
The precautionary principle, adopted by the European Union, requires industries to prove that their products are safe. In contrast, in the United States, chemicals are assumed to be safe unless proven otherwise, and the burden is put on the consumer, the public, or the government to prove that a chemical causes harm.

- **What social and cultural factors contribute to environmental problems?**
Social and cultural factors that contribute to environmental problems include population growth, industrialization and economic development, and cultural values and attitudes such as individualism, consumerism, and militarism.

- **What are some of the strategies for alleviating environmental problems?**
Strategies for alleviating environmental problems include environmental activism, environmental education, reducing carbon emissions, the use of "green" energy and energy efficiency, modifications in consumer products and behavior, and slowing population growth.

- **What impact has environmental activist Greta Thunberg had on climate awareness around the world?**
During the 2019–2020 school year, Swedish teenager Greta Thunberg traveled the world to spread the word about climate change. Thunberg's "school strike" invigorated activism in general and particularly youth activism around the globe, with 6 million people taking to the streets in September 2019 to demand action on climate change.

Test Yourself

1. The Intergovernmental Panel on Climate Change says that the world must keep global warming under ___ degrees Celsius (from 1990 levels) in order to avoid dangerous climate change.
   a. 2
   b. 5
   c. 8
   d. 11
2. When car manufacturers like Volkswagen promote their vehicles as "environmentally responsible," they are engaging in ____________?
   a. global warming.
   b. greenwashing.
   c. climate denial.
   d. an alternative fuel.
3. In the United States, polluting industries and industrial waste facilities are often located in minority communities.
   a. True
   b. False
4. If greenhouse gases were to be stabilized today, global air temperature and sea level would be expected to
   a. remain at their current levels.
   b. decrease immediately.
   c. begin to decrease within 20 years.
   d. continue to rise for hundreds of years.
5. The United States has more operating nuclear reactors than any other country.
   a. True
   b. False
6. Fracking has been banned in which of the following states?
   a. Pennsylvania
   b. Texas
   c. New York
   d. Florida
7. In the United States, the EPA has required testing on all of the more than 80,000 chemicals that have been on the market since 1976.
   a. True
   b. False
8. The Paris Agreement is a global effort to combat which of the following?
   a. Loss of biodiversity
   b. Governmental environmental regulations
   c. Climate change
   d. Ecoterrorists
9. __________ is a gas that forms when uranium, thorium, or radium breaks down in water, rocks, and soil, leaking the gas into the dirt beneath the home.
   a. Carbon dioxide
   b. Nitrous oxide
   c. Radon
   d. Benzene
10. Under the Trump administration, states were allowed to set their own fuel efficiency standards.
    a. True
    b. False

Answers: 1. A; 2. B; 3. A; 4. D; 5. A; 6. C; 7. B; 8. C; 9. C; 10. B.

Key Terms

acid rain 517
biodiversity 514
biomass 511
black lung disease 503
cap and trade programs 531
charismatic megafauna 527
Clean Power Plan 532
climate denial machine 497
climate deniers 496
community solar gardens 511
coral bleaching 494
deep ecology 528
deforestation 513
desertification 514
e-waste 518
Earth Overshoot Day 512
ecosystem 497
ecoterrorism 528
environmental footprint 512
environmental injustice 499
fracking 504
geothermal energy 511
global warming 492
green energy 508
greenhouse effect 495
greenhouse gases 495
green revolving funds (GRFs) 535
greenwashing 500
human supremacy 523
Kyoto Protocol 531
light pollution 523
mountaintop removal 503
multiple chemical sensitivity (MCS) 522
"Not in My Backyard" (NIMBY) 500
Paris Climate Accord 498
perceived obsolescence 499
planned obsolescence 499
radical environmental movement 528
solar farms 511
strip mining 503
sustainable development 536
tar sands 502
tar sands oil 502

We live in a society exquisitely dependent on science and technology, in which hardly anyone knows anything about science and technology."

CARL SAGAN
Astronomer, Astrophysicist

14

Science and Technology

Chapter Outline

Learning Objectives

After studying this chapter, you will be able to ...

1. List the three types of technology.
2. Explain possible reasons why the United States has fallen behind the rest of the world in science and technology.
3. Argue that each of the sociological theories discussed contributes to our understanding of science and technology in society.
4. Identify the ways in which science and technology transform social relationships.
5. Assess the positive and negative effects of science and technology on society.
6. Give examples of how the Internet is used by individuals, groups, and countries to further their own ends.
7. Evaluate the various strategies used to control the negative effects of science and technology.

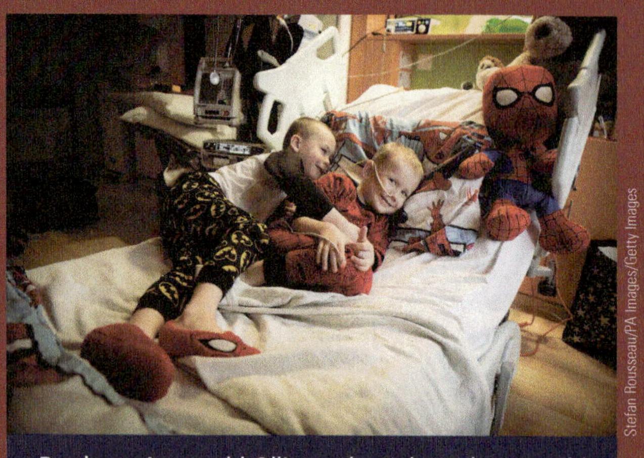

Stefan Rousseau/PA Images/Getty Images

Brothers, 4-year-old Ollie on the right and 6-year-old Finley on the left, play in the Royal Marsden Hospital in Surrey, England. Finley donated stem cells to his younger brother who suffers from acute myeloid leukemia, a type of cancer. Finley even shaved his head in empathy with his brother, and to raise money for the hospital.

THEIR 4-YEAR-OLD son had a fatal disease—Duchenne Muscular Dystrophy (DMD)—a progressive muscle-wasting disorder that only affects males. The doctor's advice? "Take your son home. Love him. ... Give him a good life and enjoy him because there's really not many options right now" (quoted in Hamilton 2020a, p. 1). But, like any parents, Connor's were committed to helping their son. They heard scientists had developed a delivery system that might transfer the missing gene that causes DMD into the patient via "a molecular FedEx truck" (Hamilton 2020a). At age 9, Connor received what he calls "muscle juice," and, within three weeks, he was standing, walking, and running. Although how long the effects will last is unknown, eight boys have now benefited from this treatment (Hamilton 2020b; Ridgefield 2020). Thanks to science, this awe-inspiring medical technology may help young boys, once destined to live their lives in wheelchairs and facing almost certain death before age 20, to run, jump, and play.

Technological dualism is the tendency for technology to have both positive and negative consequences. Clearly, advances in medical technology have helped people like Connor and around the world survive illnesses and injuries that would have been deadly only a few decades ago. On the other hand, medical interventions can themselves cause harm. Research suggests that the global overuse of medicines, screening and diagnostic tests, and therapeutic procedures are likely to "harm patients physically, psychologically, and financially" (Brownlee et al. 2017, p. 162).

technological dualism The tendency for technology to have both positive and negative consequences.

science The process of discovering, explaining, and predicting natural or social phenomena.

technology Activities that apply the principles of science and mechanics to the solutions of a specific problem.

mechanization Dominant in an agricultural society, the use of tools to accomplish tasks previously done by hand.

automation The replacement of human labor with machinery and equipment.

cybernation Dominant in postindustrial societies, the use of machines to control other machines.

Like medical technology, the 3-D printer is another example of technological dualism. Although efficient and cost-effective for manufacturers, 3-D printers may lead to higher rates of unemployment, particularly among assembly line laborers. Similarly, 3-D printers can be used to make a robotic hand for a child with amniotic band syndrome or to make a semiautomatic gun used by a mass murderer.

Science and technology go hand in hand. **Science** is the process of discovering, explaining, and predicting natural or social phenomena. A scientific approach to understanding COVID-19, for example, might include investigating the molecular structure of the virus, the means by which it is transmitted, and public attitudes about COVID-19. **Technology**, as a form of human cultural activity that applies the principles of science and mechanics to the solution of problems, is intended to accomplish a specific task—in this case, the development of a COVID-19 vaccine.

Societies differ in their level of technological sophistication and development. In agricultural societies, which emphasize the production of raw materials, the use of tools to accomplish tasks previously done by hand, or **mechanization**, dominates. As societies move toward industrialization and become more concerned with the mass production of goods, automation prevails. **Automation** involves the use of self-operating machines, as in an automated factory where autonomous robots assemble automobiles. Finally, as a society moves toward post-industrialization, it emphasizes service and information professions (Bell 1973). At this stage, technology shifts toward **cybernation**, whereby machines control machines—making production decisions, programming robots, and monitoring assembly performance.

What are the effects of science and technology on humans and their social world? How do science and technology help to remedy social problems, and how do they contribute to social problems? Is technology, as author Neil Postman (1992) argued in his classic book *Technopoly,* both a friend and a foe to humankind? We address each of these questions in this chapter.

The Global Context: The Technological Revolution

Less than 50 years ago, traveling across state lines was an arduous task, a long-distance phone call was a memorable event, and mail carriers brought belated news of friends and relatives from far away. Today, travelers journey between continents in a matter of hours, and for many, e-mail, faxes, instant messaging, texting, and cell phones have replaced previously conventional means of communication.

The world is a much smaller place than it used to be, and it will become even smaller as the technological revolution continues. In 2020, the Internet had over 4.8 billion users with 62 percent of the worlds' population online (Internet World Statistics 2020). Of all Internet users, the highest proportion come from Asia (50.3 percent), followed by Europe (15.9 percent), Africa (11.5 percent), Latin America and the Caribbean (10.1 percent), North America (7.6 percent), the Middle East (3.9 percent), and Oceania/Australia (0.6 percent) (Internet World Statistics 2020).

The **penetration rate** is the percentage of people who have access to and use the Internet in a particular area (see Figure 14.1). Although it is higher in industrialized countries, 94.6 percent in North America, there is movement toward the Internet becoming a truly global medium, as Africans and Middle Easterners increasingly get online. Although Internet use in North America grew 208 percent between 2000 and 2020, the number of Internet users increased by 2,109 percent in Africa and by 5,527 percent in the Middle East during the same time period. Despite such tremendous growth, nearly 3 billion people are offline, predominantly in developing and less developed countries (Iyengar 2018; Internet World Statistics 2020).

On the average, global Internet users spend 2.5 hours per day on social media sites (Buchholz 2020). Facebook, the most popular social media platform, has 2.4 billion users, one-third of the world's population and over half of all Internet users. Social media sites began in the early 2000s and have grown exponentially. For example, TikTok added

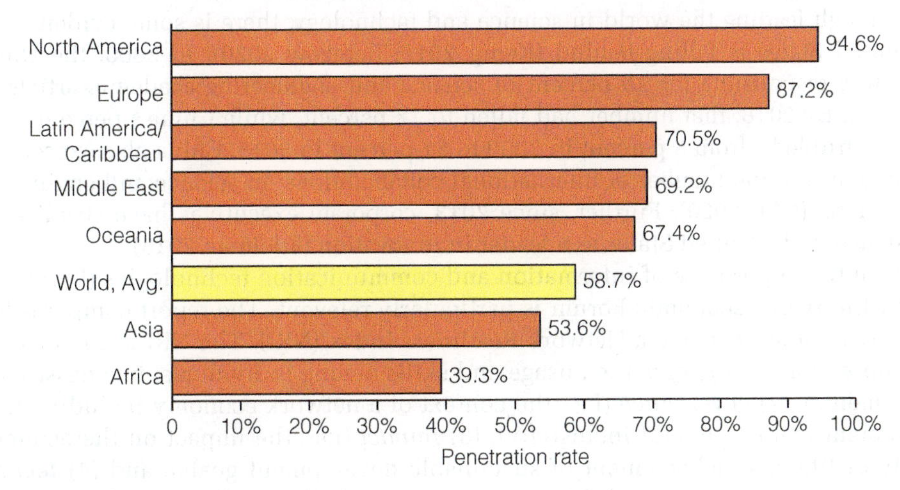

Figure 14.1 Internet World Penetration Rates by Geographical Region, January 1–March 31, 2020
SOURCE: Internet World Statistics 2020.

penetration rate The percentage of people who have access to and use the Internet in a particular area.

approximately 20 million *new users* a month in the period between September 2016 and mid-year 2018 (Cortez-Ospina 2019) and, in 2020, Triller had about 64 million active users every month, up 2 million users from three years ago (Lee 2020).

Social media use is likely to continue its rapid expansion around the world as technology corporations work to bring affordable Internet and wireless access to previously unconnected parts of the developing world. Google set up free Wi-Fi service in over 400 train stations in India, and Samsung built one of the world's largest mobile cell phone factories outside of New Delhi (Iyengar 2018).

The movement toward the globalization of science and technology is, of course, not limited to the use and expansion of the Internet. The world robot market and the U.S. share of it continue to expand, Microsoft's Internet platform and support products are sold all over the world, scientists collect skin and blood samples from remote islanders for genetic research, a global treaty regulating trade of genetically altered products was signed by more than 100 nations, and South Korea's Samsung and China's Huawei control over 40 percent of the cell phone market (Eadicicco 2019).

What do you think?

Despite an anti-science stance by the former administration, recently an interest in science, as it pertains to the global challenges we face, has grown. In fact, 1 in 10 U.S. adults took part in a citizen science activity in 2020 (Thigpen and Funk 2020). The procedures citizen scientists use are the same as those used by professional scientists, and often collaboration is possible. What do you think? Would you be interested in being a citizen scientist?

Science and Technology in the United States

To achieve scientific and technological innovations, sometimes called *research and development* (R&D), countries need material and economic resources. *Research* entails the pursuit of knowledge; *development* refers to the production of materials, systems, processes, or devices directed to the solution of practical problems. According to the National Science Foundation (NSF), the United States spends over $549 billion a year in research and development accounting for about 30 percent of the global total, with China a strong second at 21 percent. As in most other countries, U.S. funding sources are primarily from private industry, 73 percent of the total, followed by higher education, and the federal government (NSF 2020). However, over half of Americans support increased federal funding for scientific research and innovation (Funk 2019).

Although leading the world in science and technology, there is some evidence that the United States is falling behind (Kemp 2017; Berezow 2020). In 2000, the United States was responsible for 28 percent of science and engineering academic articles in the world. By 2018, that number had fallen to 17 percent, while China's percent of the total quadrupled—from 5 percent in 2000 to 21 percent in 2018. China also exceeds the United States in the number of international collaborations on science and engineering publications (NSF 2020). Further, since 2013, corporate executives have steadily lost confidence in the United States as a leader in innovation (Edelman 2018).

Given the importance of information and communication technologies (ICTs), a report by the World Economic Forum is particularly relevant. The report compares ICTs across 121 countries using a Network Readiness Index (NRI). The NRI is composed of four subsections: (1) *people* (i.e., usage and skills among individuals, businesses, and governments), (2) *governance* (i.e., the context of a network economy including trust, government regulation, and inclusivity), (3) *impact* (i.e., the impact on the economy, quality of life, and achievement of sustainable development goals), and (4) *technology* (i.e., access, content, and future). In the 2019 NRI, Nordic countries (Sweden, Netherlands, Norway, Switzerland, Denmark, and Finland) held six of the top ten index positions, with Singapore ranked second and the United States ranked eighth (Dutta and Lanvin 2020).

As the NSF notes (2020), the United States, although still a global leader in science and technology, is no longer viewed as an uncontested leader. The declining supremacy of the United States in science and technology is likely to be the result of several interacting forces. First, the federal government has been scaling back its investment in research and development, a decline that is "notable as federally funded R&D is an important source of support, particularly for higher education ... and for the nation's basic research enterprise" (NSF 2020, p. 1).

Second, corporations, the largest contributors to research and development, often focus on maximizing profits for shareholders, which can lead corporate executives to focus "myopically on short-term earnings ... at the expense of long-term performance ... and innovations" (Stout 2012). It should be noted, however, that a business roundtable of top CEOs from such companies as Apple and Pepsi recently issued a statement that the purpose of a corporation should no longer exclusively be to maximize profits for stockholders (Gelles and Yaffe-Bellany 2019).

Third, developing countries, most notably China, are expanding their scientific and technological capabilities at a faster rate than the United States. Chinese scientists publish more papers on artificial intelligence (AI) than U.S. researchers, China has nine times the science and technology college graduates that the United States has, investment in R&D is growing at a faster pace, and capital investment in Chinese start-up companies exceeds that in the United States (Dace 2020). Further, in 2020, China released a global plan entitled "China Standards 2035" with the "aim of influencing how the next generation of technologies, [from] telecommunications to artificial intelligence, will work" (Kharpal 2020, p. 1).

Fourth, there has been concern over science and math education in the United States, in terms of both quality and quantity. When comparing eighth-grade students on mathematics and science to other developed countries, the United States scores somewhere in the middle with very little growth in achievement over the last ten years. While the number of science and engineering degrees has doubled in China over the last decade, there has been very little growth in other science and engineering degree-granting nations (NSF 2020). Further, although the United States awards the highest number of science and engineering PhDs in the world, many of these foreign nationals do not stay in the United States after receiving their degrees (Trapani and Hale 2019; NSF 2020).

Finally, Mooney and Kirshenbaum (2009) document "unscientific America"—the tremendous disconnect between the citizenry, media, politicians, religious leaders, education, and the entertainment industry (e.g., *The Big Bang Theory, Silicon Valley, Avatar*), on the one hand, and science and scientists, on the other. Post–World War II America, in part because of the Cold War, invested in R&D, leading to such scientific and technological advances as the space program, the development of the Internet, and the decoding of the genome. Yet, despite these significant contributions, most Americans know very little about science or the scientific process (see this chapter's *Self and Society*). In a survey of U.S. adults, just over half were able to identify a hypothesis (Kennedy and Hefferon 2019).

> "The scientist is not a person who gives the right answers, he [or she] is the one who asks the right questions."
>
> **–CLAUDE LEVI-STRAUSS, ANTHROPOLOGIST**

Science and U.S. Policy. Many Americans believe that social problems can be resolved through a **technological fix** rather than through social engineering. Social engineering efforts to address the problem of date rape on college campuses, for example, might include sexual assault awareness campaigns and public service announcements about binge drinking and the definition of "consent." A technological fix for the same problem was developed by four students at North Carolina State University who invented a type of nail polish that, when a woman dips her finger into a drink, changes color if common date rape drugs are present (Sullivan 2014).

Both approaches, although through different methods, use *science* to address the problem. Despite the public's relative lack of scientific knowledge, Americans largely see science as having a positive impact on society and overwhelmingly believe that over the next 20 years, scientific developments will make people's lives better (Thigpen and Funk 2019). Americans believe that scientists are intelligent and focus on solving real problems (Funk and Hefferon 2019). Furthermore, the majority of Americans believe scientists should take an active role in policy decision making (Funk 2020).

technological fix The use of scientific principles and technology to solve social problems.

What Is Your Science and Technology IQ?

DIRECTIONS: Answer the following questions. When finished, compare the percentage of questions you answered correctly to the following distribution. For example, if you answered eight questions correctly, you scored better than 52 percent of the public (the sum of everything below eight correct answers: 5 + 5 + 6 + 6 + 6 + 7 + 8 + 9 = 52 percent), the same as 8 percent of the public, and lower than 39 percent of the public (the sum of everything above eight correct answers: 10 + 13 + 16 = 39 percent).

1. Oil, natural gas, and coal are examples of
   a. biofuels.
   b. fossil fuels.
   c. geothermal resources.
   d. renewable resources.

2. A scientist is conducting a study to determine how well a new medication treats ear infections. The scientist tells the participants to put ten drops in the infected ear each day. After two weeks, all participants' ear infections had healed. Which of the following changes to the design of this study would most improve the ability to test whether the new medication effectively treats ear infections?
   a. Have participants use eardrops for only one week
   b. Have participants put eardrops in both their infected ear and healthy ear
   c. Create a second group of participants with ear infections who do not use any eardrops
   d. Create a second group of participants with ear infections who use 15 drops today

3. Which of the following is an example of genetic engineering?
   a. Growing a whole plant from a single cell
   b. Finding the sequences of bases in plant DNA
   c. Inserting a gene into plants that makes them resistant to insects
   d. Attaching the root of one type of plant to the stem of another type of plant

4. What is the main cause of the seasons on the Earth?
   a. The speed that the Earth rotates around the Sun
   b. Changes in the amount of energy coming from the Sun
   c. The tilt of the Earth axis in relation to the Sun
   d. The distance between the Earth and the Sun

5. The time it takes a computer to start has increased dramatically. One possible explanation for this is that the computer is running out of memory. This explanation is a scientific
   a. hypothesis.
   b. observation.
   c. conclusion.
   d. experiments.

6. Many diseases have an incubation. Which of the following best describes what an incubation period is?
   a. The period during which someone builds up immunity to disease
   b. The effect of the disease on babies
   c. The period during which someone has an infection but is not showing symptoms
   d. The recovery period after being sick

7. When large areas of forest are removed so land can be converted for other uses, such as farming, which of the following occurs?
   a. Colder temperature
   b. Increased erosion
   c. Greater oxygen production
   d. Decreased carbon dioxide

According to the Office of Science and Technology Policy, America "leads global scientific progress by example, promoting core principles of freedom of inquiry, scientific integrity, collaboration, and openness" (Droegemeier 2019, p. 1). There is, however, evidence that rather than promoting science, the former administration attacked scientific inquiry in several areas including COVID-19 (see Chapter 2) and the environment (see Chapter 13). Figure 14.2 provides a categorical distribution of the 80 times, according to the Union of Concerned Scientists, former President Trump attacked science in the first two years of his administration (Carter et al. 2019).

8. An antacid relieves an overly acidic stomach because the main components of antacids are
   a. isotopes.
   b. bases.
   c. acids.
   d. neutral.

9. Which of these is a major concern about the overuse of antibiotics?
   a. It can lead to antibiotic-resistant bacteria.
   b. There will be an antibiotic shortage.
   c. Antibiotics can cause secondary infections.
   d. Antibiotics will get into the water system.

10. A car travels at a constant speed of 40 mph. How far does the car travel in 45 minutes?
    a. 25 miles
    b. 30 miles
    c. 35 miles
    d. 40 miles

11. Of the following four cities in the United States, which of the following has the greatest annual range of temperature?
    a. Chicago, Illinois, with a low temperature in the winter of 27 degrees Fahrenheit and a high temperature of 75 degrees Fahrenheit in the summer
    b. Los Angeles, California, with a low temperature in the winter of 57 degrees Fahrenheit and a high temperature of 70 degrees Fahrenheit in the summer
    c. New York, New York, with a low temperature of 35 degrees Fahrenheit in the winter and a high temperature of 77 degrees Fahrenheit in the summer
    d. Cleveland, Ohio, with a low temperature in the winter 10 degrees Fahrenheit and a high temperature of 92 degrees Fahrenheit in the summer

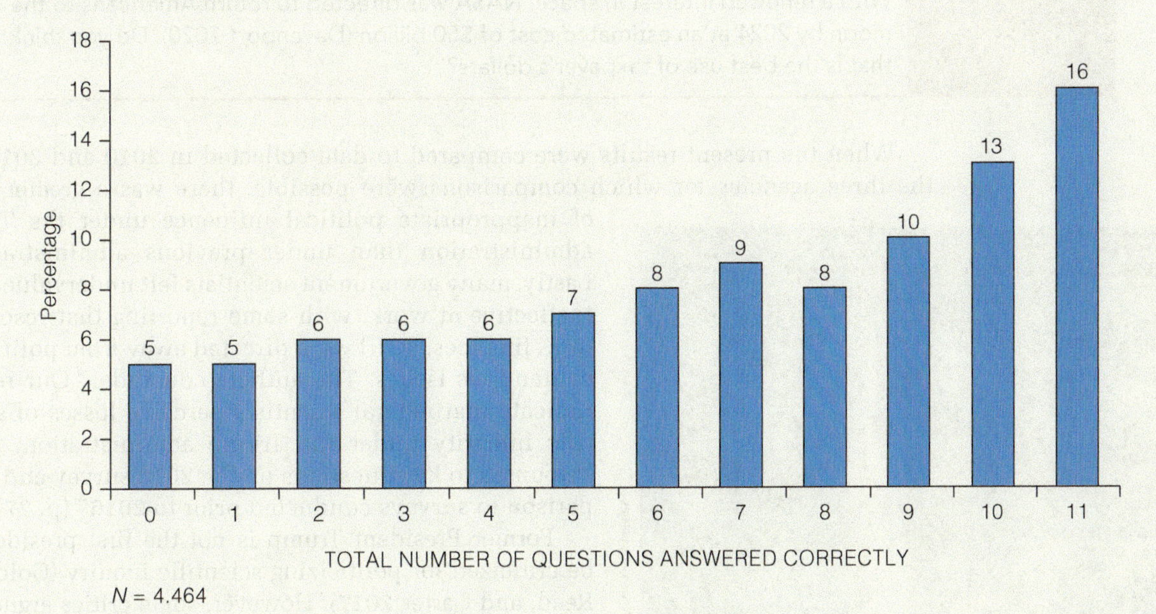

N = 4.464

SOURCE: "Science Knowledge Quiz." Pew Research Center, Washington, D.C. 2019.

Answers: 1. B; 2. C; 3. C; 4. C; 5. A; 6. C; 7. B; 8. B; 9. A; 10. B; 11. D

The Anti-Science Administration. In 2018, a survey was distributed to scientists in the former administration. Questions concerned scientific integrity in the federal agencies where the respondents worked (Goldman et al. 2020). In general, although there were variations between agencies, scientists reported the leadership in their respective agencies did not have the technical expertise or trustworthiness to direct scientific inquiry. Specifically, two-thirds of the respondents across the five agencies studied reported that the "influence of political appointees" and "absence of leadership with needed scientific expertise" were barriers to evidence-based research (p. 9).

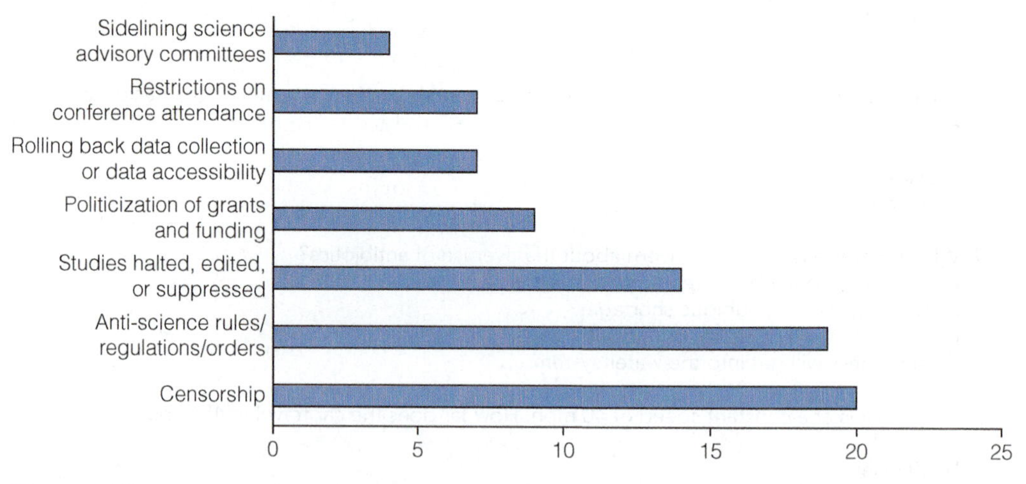

Figure 14.2 Frequency of Attacks on Science in First Two Years of Trump Presidency, January 2016–January 2018*

*As of August 2020, the number of attacks on science by the Trump administration, according to the Union of Concerned Scientists, has increased to 150 (Desikan 2020).

SOURCE: Carter et al. 2019.

What do you think?

The Space Force, the sixth branch of the U.S. Armed Services, will be used to "protect strategic American space infrastructure, including communications, navigation and spy satellites, from adversaries such as Russia and China" (Sprunt 2020, p. 1). With a renewed interest in space, NASA was directed to return Americans to the moon by 2024 at an estimated cost of $50 billion (Davenport 2020). Do you think that is the best use of taxpayer's dollars?

When the present results were compared to data collected in 2010 and 2015, for the three agencies for which comparisons were possible, there was a greater level of inappropriate political influence under the Trump administration than under previous administrations. Lastly, many government scientists felt undervalued and ineffective at work, with some reporting that resources (e.g., finances, staff) were directed away from politically contentious issues. The authors conclude: "Our results indicate that federal scientists perceive losses of scientific integrity under the Trump administration, given responses to key questions on the 2018 survey and comparison to surveys conducted prior to 2016" (p. 27).

Joe Raedle/Getty Images News/Getty Images

On April 11, 2019, at the NASA Kennedy Space Center, SpaceX Falcon lifted off carrying a Lockheed Martin communications satellite to be put into orbit. The rocket is the most powerful in existence and is the second launch by SpaceX founder and chief executive Elon Musk.

Former President Trump is not the first president to be criticized for politicizing scientific inquiry (Goldman, Reed, and Carter 2017). However, some critics argue that the former president's disregard for science was pervasive and exceeded that held by previous administrations (Plumer and Davenport 2019). As early as the summer of 2017, just six months after the inauguration, Nobel Laureates met in Germany to discuss what they called the "anti-science movement" (Levine 2017). The president of the Council's opening remarks sounded the alarm: "Scientists cannot ignore what is happening in the world, some rulers, and people, seem to feel threatened by progress and the fact-oriented power of science" (quoted in Levine 2017, p. 1).

Concerns resulted in open letters from organizations of scientists. For example, in 2016, 2018, and 2020, members of the U.S. National Academy of Science (NAS) authored and signed an open letter entitled, "Statement to Restore Science-Based Policy in Government." Among other things, the letter called on the federal government to "maintain scientific content on publicly accessible websites, to appoint qualified personnel to positions requiring scientific expertise, [and] to cease censorship and intimidation of government scientists" (NAS 2020, p. 1).

At least some critics have argued that the former Trump administration has "diminished the role of science in federal policy making while halting or disrupting research projects nationwide, marking a transformation of the federal government whose effects ... could reverberate for years" (Plumer and Davenport 2019, p. 1). The damage to the environment, and any deaths attributable to the administration's anti-science stance on COVID-19, cannot be reversed. Scientists, "forced out, sidelined, or muted," have left work in droves (Gowen et al. 2020, p. 1) resulting in the loss of expertise needed to make evidence-based decisions (Rosenberg and Rest 2018). The hope by many is that the Trump administration's 2021 budget, which contains proposed funding cuts to nearly all science-related agencies, is rejected and significantly revised by Congress during budget negotiations (Malakoff and Mervis 2020). The 2022 budget will be prepared by the incoming Biden administration.

The Science Administration. The election of Joe Biden and Kamala Harris was, in part, a response to the former administration's anti-science stance, particularly in reference to COVID-19. Present Biden's acceptance speech signified a return to science and a significant policy change from the former administration's, stating, "Americans have called on us to ... marshal the forces of science and the forces of hope in the great battles of our time" (quoted in Baron 2020).

Not surprisingly, an article published in *Science Magazine*, the journal of the American Association for the Advancement of Science, called on the new administration to address a number of science-based problems, perhaps none more important than the pandemic. President Biden has promised to "follow the science" and to "accelerate the development of treatments and vaccines" (Biden 2020a). His administration has already established a COVID-19 advisory board made up of leading researchers and health professionals, has expanded test and tracing programs, and has reached out to scientists and physicians around the world (Tollefson 2020) (see Chapter 2).

President Biden has also stated that he will ensure that the United States rejoins the 2015 Paris climate agreement and continues the fight against global warming, pledging to achieve net-zero admissions in the United States by 2050 (Tollefson 2020; Biden 2020b) (see Chapter 13). The Biden administration has also promised a "clean energy revolution," stating that after assuming the office, he will ask the federal government to invest $1.7 trillion over a ten-year period in new clean air technologies as well as increasing oversight of the fossil fuel industry (Malakoff 2020; Biden 2020b).

Perhaps most importantly, President Biden has vowed to remove political influence from research agencies such as the National Institute of Health, the Centers for Disease Control and Prevention, and the Food and Drug Administration, promising to "let science lead" (Malakoff 2020). The present administration's rejection of science and using it as a partisan tool reflects President Biden's stated emphasis on using science for the common good and in the public's best interest (Bardon 2020).

Sociological Theories of Science and Technology

Each of the three major sociological frameworks helps us to better understand the nature of science and technology in society.

Structural-Functionalist Perspective

Structural functionalists view science and technology as emerging in response to societal needs—that "science was born indicates that society needed it" (Durkheim 1973/1925). As societies become more complex and heterogeneous, finding a common and agreed-on knowledge base becomes more difficult. Science fulfills the need for an assumed objective measure of "truth" and provides a basis for making intelligent and rational decisions. In this regard, science and the resulting technologies are functional for society.

Scientific knowledge has grown at a more rapid rate over time; during each of three historical periods—from about 1650–1750, 1750–1950, to 1950–2012—scientific knowledge grew at triple the rate of the previous phase (Bornmann and Mutz 2015). If society changes too rapidly as a result of science and technology, however, problems may emerge. When the *material part* of culture (i.e., its physical elements) changes at a faster rate than the *nonmaterial part* (i.e., its beliefs and values), a **cultural lag** may develop (Ogburn 1957). For example, the typewriter, the conveyor belt, and the computer expanded opportunities for women to work outside the home. With the potential for economic independence, women were able to remain single or to leave unsatisfactory relationships and/or establish careers. But although new technologies have created new opportunities for women, beliefs about women's roles, expectations of female behavior, and values concerning equality, marriage, and divorce have lagged behind.

Robert Merton (1973), a structural functionalist and founder of the subdiscipline sociology of science, also argued that scientific discoveries or technological innovations may be dysfunctional for society and may create instability in the social system. The development of time-saving machines increases production, but it also displaces workers and contributes to higher rates of employee alienation. Defective technology can have disastrous effects on society. For example, faulty airbags from Japanese automobile parts provider Takata resulted in deaths and serious injuries and in the recall of more than 70 million cars (Krishner 2019).

Conflict Perspective

Conflict theorists, in general, argue that science and technology benefit a select few. For some conflict theorists, technological advances occur primarily as a response to capitalist needs for increased efficiency and productivity and thus are motivated by profit. As McDermott (1993) predicted, most decisions to increase technology are made by "the immediate practitioners of technology, their managerial cronies, and for the profits accruing to their corporations" (p. 93). In the United States, private industry spends more money on research and development than the federal government does. Shareholders accuse Boeing of not properly testing the flight-control system or adequately training airline pilots, which resulted in two fatal crashes of their 737 Max 8 and the death of over 300 people. The lawsuit alleges that top officials "effectively put profitability and growth ahead of airplane safety and honesty" (Picchi 2019, p. 1). Lawsuits have been filed by shareholders as well as by families of the victims.

Science and technology also further the interests of dominant groups to the detriment of others. The need for scientific research on AIDS was evident in the early 1980s, but the required large-scale funding was not made available so long as the virus was thought to be specific to gay men and intravenous drug users. Only when the virus became a threat to mainstream Americans were millions of dollars allocated to AIDS research. Hence, conflict theorists argue that agencies, by the projects they fund or don't fund, act as gatekeepers to scientific discoveries and technological innovations. These agencies are influenced by powerful interest groups and the marketability of the product rather than by the needs of society.

When the dominant group feels threatened, it may use technology as a means of social control. Chang and Lin (2020), in an investigation of 153 countries between 1995 and

cultural lag A condition in which the material part of culture changes at a faster rate than the nonmaterial part.

2018, conclude that the Internet, rather than contributing to the rise of civil society, is the means by which autocratic governments suppress civil society. In fact, Internet censorship is growing for the eighth consecutive year. Censorship by some governments is anything but subtle. In Saudi Arabia, for example, all Internet traffic is directed through a router that is monitored by the Ministry of the Interior, which is responsible for maintaining a record of blocked websites (Open Access 2019).

Finally, conflict theorists as well as feminists argue that technology is an extension of the patriarchal nature of society that promotes the interests of men and ignores the needs and interests of women. As in other aspects of life, women, particularly women of color, play a subordinate role in reference to technology in terms of both its creation and its use (see Figure 14.3). For example, washing machines, although time-saving devices, disrupted the communal telling of stories and the resulting friendships among women who gathered together to do their chores. Bush (1993) observed that, in a "society characterized by a sex-role division of labor, any tool or technique ... will have dramatically different effects on men than on women" (p. 204).

Symbolic Interactionist Perspective

Knowledge is relative. It changes over time, over circumstances, and between societies. We no longer believe that the world is flat or that the Earth is the center of the universe, but such beliefs once determined behavior because individuals responded to what they thought to be true. The scientific process is a social process in that "truths"—socially constructed truths—result from the interactions among scientists, researchers, and the lay public.

In a now classic statement, Kuhn (1973) argued that the process of scientific discovery begins with assumptions about a particular phenomenon (e.g., the world is flat). Because unanswered questions always remain about a topic (e.g., why don't the oceans drain?), science works to fill these gaps. When new information suggests that the initial assumptions were incorrect (e.g., the world is not flat), a new set of assumptions or framework emerges to replace the old one (e.g., the world is round). It then becomes the dominant belief or paradigm.

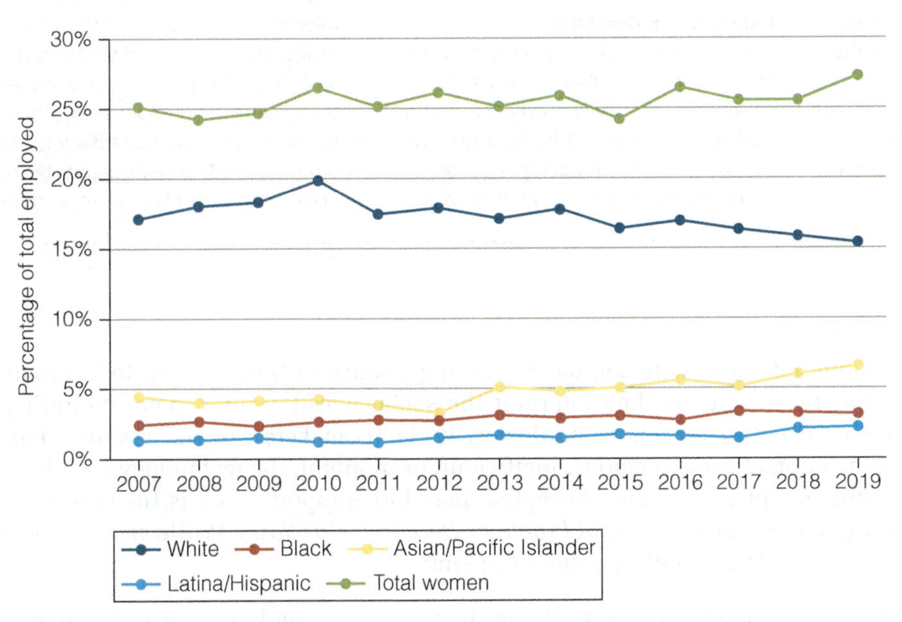

Figure 14.3 Women Employed in Computing-Related Occupations in the United States, by Ethnicity, 2007–2019
SOURCE: Statista 2020.

social problems RESEARCH UP CLOSE

Scientific Literacy, Conspiracy Mentality, and Belief in Disinformation

Protesting the lack of science-based policies and an administration that is viewed as anti-science, the first March for Science took place in April 2017 in Washington, D.C. and simultaneously in 600 cities around the world. Hundreds of thousands of scientists continue to march every year (and virtually in 2020) to raise awareness of the systematic efforts by some politicians to discredit scientific reasoning, logic, and methods. As Landrum and Olshansky (2019) observe, there are two ways to disagree with science—by rejecting well-established scientific theories or by embracing fabricated ones (Landrum and Olshansky 2019). Here the researchers investigate the role of conspiracy mentalities and scientific literacy in the acceptance of viral deceptive claims about science.

Sample and Methods

Landrum and Olshansky (2019) argue that the public's disagreement with science takes place in one of two ways, as previously noted. First, an individual may doubt the credibility of science believing rather in a conspiracy theory that justifies rejecting scientific evidence. Second, an individual may have a "belief system or political worldview" in which all authority is questioned, including scientists, finding authority figures and institutions as "inherently deceitful" (p. 194). The aim of the research, then, is to "examine who rejects well-supported scientific theories" and "who accepts viral deceptions about science" (p. 195).

Landrum and Olshansky (2019) used two samples. The first sample, which was part of a larger online project, contained 513 participants living in the United States ranging in age from 18 to 80. The average age of the participants was 49. Females comprised 56 percent of the sample, and 21.5 percent were Black American, Asian American, or Hispanic American. The median level of education was an associate's degree, meaning that half of the respondents had a two-year college degree or more, and half of the respondents had less than a two-year college degree.

The second sample consisted of 21 individuals who attended a Flat Earth International Conference (FEIC) in Raleigh, North Carolina and who, of the approximately 60 attendees, completed and returned an online survey. Many of the respondents did not complete all of the demographic information. However, with that caveat, sample statistics include an average age of 39, the most commonly reported educational level was a bachelor's degree, the majority reported being White, and there were nine males and seven females, with five participants failing to answer the gender question.

Independent Variables

The researchers state two hypotheses. Hypothesis 1 was that having a conspiracy mentality will predict rejecting well-supported scientific theories. Hypothesis 2 was that having a conspiracy mentality will predict evaluating viral deceptions about science, which are actually false, as true.

Conspiracy mentality was measured by a scale comprised of seven items. Participants were asked to respond to seven viral conspiracy theories as (1) definitely false, (2) likely false, (3) likely true, or (4) definitely true. Conspiracy theories included such statements as the "Apollo program never landed on the moon," and "Barack Obama was not born in the United States." Initial results indicate that FEIC participants, on the average, rated viral conspiracies as "likely true" while the national sample participants ranked the viral conspiracies as "likely untrue."

Other independent variables included political party affiliation (strong Democrat, Democrat, Independent, Republican, strong Republican, and unaffiliated/other), religiosity (i.e., how much do faith and religion guide your everyday lives; not at all to a great deal), and sample (i.e., whether a participant was from the national sample or from the FEIC sample).

Dependent Variables

Belief in well-supported scientific theories comprised the first dependent variable and was measured by two indicators. The first indicator asked participants to respond to the statement that "climate change is real and human caused." Response options included true, false because it is not human caused, false because it is not happening, or prefer not to answer.

Symbolic interactionists emphasize the importance of this process. Individuals socially construct reality as they interpret the social world around them, including the meaning assigned to various technologies. If claims makers can successfully define a product as impractical, dangerous, inefficient, or immoral, the technology is unlikely to gain public acceptance or, once accepted, may lose support. Such is the case with the vaccines (see this chapter's *Social Problems Research Up Close*). While vaccines are lifesaving biotechnologies, anti-vaccine sentiment:

> has been building for decades, a byproduct of an internet humming with rumor and misinformation; the backlash against Big Pharma; an infatuation with celebrities that give special credence to anti-immunization statements … and now, the [former] … administration's anti-science rhetoric. (p. 1)

The second indicator asked participants to respond to the statement that "humans evolved from earlier species of animal." The response set included true, false, or prefer not to answer (p. 197).

How participants evaluated "viral deceptions about science," including statements about "GMOs [genetically modified organisms], a cure for cancer, the Zika virus, and childhood vaccinations" (p. 197), was the second dependent variable. For example, one indicator read "a cure for most types of cancer has already been found, but medical circles prefer to keep getting research funding from governments and keep their findings a secret." The vaccination item read "childhood vaccinations are unsafe and cause disorders like autism." Respondents were asked, on a four-point scale, whether they thought the statement was (1) definitely true, (2) likely true, (3) likely false, or (4) definitely false (p. 197).

Results and Conclusions

Initial results indicate that the FEIC respondents were more likely to reject the notion that climate change is caused by humans and to reject the assertion that human evolution exists when compared to the national sample. In fact, all of the FEIC participants rejected *both* scientific-based conclusions. The FEIC participants were also more likely to endorse all of the viral conspiracy claims than the national sample participants, including that (1) the cause for cancer exists but is being hidden, (2) GMOs cause cancer, (3) the Zika virus was caused by genetically modified mosquitos, and (4) vaccinations are unsafe and cause health problems such as autism. Each of the six bivariate relationships, i.e., relationships between sample (FEIC or national) and rejection of the *two* scientific claims and acceptance of *four* viral conspiracy claims, was statistically significant.

The multivariate analysis indicates that, when science literacy is high for both Republicans and Democrats, the odds of rejecting the scientifically based claim that climate change is real and caused by humans is 4,105 percent greater for Republicans than for Democrats, evidence of the polarization between the two parties on such hot-button issues as the environment. Similarly, although less dramatic, among respondents who had low conspiracy mentalities, the odds of Republicans rejecting the theory of evolution was 33 percent greater for Republicans than for Democrats. Thus Republicans with high science literacy and low conspiracy mentality were more likely to reject evidence-based scientific theories than their Democratic counterparts.

The second aim of the research was to examine susceptibility to believing viral conspiracies. As hypothesized, higher conspiracy mentalities, in general, were associated with evaluating the deceptive claims (i.e., a cure for cancer is being suppressed, GMOs cause cancer, the Zika virus is caused by genetically modified mosquitoes, and childhood vaccinations are unsafe and cause health disorders such as autism) as true. Participants with lower scientific literacy were also more likely to embrace the deceptive claims. Further, people who reported higher levels of religiosity were more likely to accept the false claim that vaccinations are unsafe and are related to disorders like autism.

The authors also note several interactions. An *interaction* takes place when two variables act differently in the presence of each other than each would act separately. So, for example, as science literacy *increases* for Republicans, the likelihood of believing that climate change is real and caused by humans *decreases*. Alternatively, for Democrats, as science literacy *increases*, the likelihood of believing that climate change is real and caused by humans increases. Thus, there is an interaction between the variable science literacy and political affiliation on the climate science indicator.

Finally, researchers assess the hypotheses that conspiracy mentality predicts rejection of well-supported scientific theories (Hypothesis 1) and acceptance of the scientifically deceptive claims (Hypothesis 2). Although the evidence is mixed, the best predictor of rejecting well-supported scientific theories and accepting scientifically deceptive claims was, as predicted, conspiracy mentality. Scientific literacy, in general, was the second most important variable in predicting rejection of scientific theories and acceptance of scientifically deceptive claims. Thus, as the researchers conclude, increasing science literacy may contribute to reducing misinformation about science-related viral deceptions.

SOURCE: Landrum and Olshansky 2019.

The vaccine resistance movement can be traced to a 1982 NBC documentary that erroneously reported a link between the DPT vaccine, used to fight diphtheria, whooping cough, and tetanus, and seizures in children (Hoffman 2019). Although doctors quickly corrected the show's inaccuracies, fear continued to spread. Today, despite record numbers of confirmed cases and hundreds of thousands of deaths worldwide, a recent Gallup poll reveals that 35 percent of Americans would not take the COVID-19 vaccine even if it were available, Food and Drug Administration approved, and free (O'Keefe 2020).

Not only are technological innovations subject to social meaning, but who becomes involved in what aspects of science and technology is also socially defined. Women, for example, are less likely to major in computer science and more likely to be part of the "incredible shrinking pipeline" whereby they disproportionately "leak" out of computer science classes and careers. Even after male and female students were equally exposed to programming via a one-hour tutorial, males were more likely to appreciate the importance of learning how to code and reported a greater willingness to take future programming courses than females (Du and Wimmer 2019).

Technology and the Transformation of Society

A number of modern technologies are considerably more sophisticated than technological innovations of the past. Nevertheless, older technologies have influenced the nature of work as profoundly as the most mind-boggling modern inventions. Postman (1992) described how the clock—a relatively simple innovation that is taken for granted in today's world—profoundly influenced the workplace and with it the larger economic institution.

> What the monks [could] not foresee was that the clock is a means not merely of keeping track of the [canonical] hours but also of synchronizing and controlling the actions of men. And thus, by the middle of the fourteenth century, the clock had moved outside the walls of the monastery, and brought a new and precise regularity to the life of the workman and the merchant. ... In short, without the clock, capitalism would have been quite impossible. The paradox ... is that the clock was invented by men who wanted to devote themselves more rigorously to God; it ended as the technology of greatest use to men who wished to devote themselves to the accumulation of money. (pp. 14–15)

Today, technology continues to have far-reaching effects not only on the economy but also on every aspect of social life. In the following section, we discuss societal transformations resulting from various modern technologies, including workplace technology, computers, the Internet, and science and biotechnology.

Adam Lubroth/The Image Bank/Getty Images

Robots in automated factories can perform the labor originally provided by human workers as in this automobile company. When the work requires a "softer touch," humans work alongside collaborative robots, for example, in the case of picking fruit, which can be quite a delicate job given how easily fruit can be damaged.

Technology and the Workplace

All workplaces—from government offices to factories, supermarkets, and real estate agencies—have felt the impact of technology. The transformation has been a remarkable one from conveyer belts and typewriters to "automation and advances in artificial intelligence ... that perform complex cognitive tasks" (Council on Foreign Relations [CFR] 2018, p. 11). A survey of nearly 7,000 employees in seven major industries (e.g., education, health care, technology, and R&D) representing five global areas revealed that two-thirds of workers agreed or strongly agreed with the statement, "[T]echnology can help create good work–life balance which makes me motivated and engaged" (Research Now 2019, p. 21).

Robotics. Robotic technology has revolutionized manufacturing. Globally, *industrial robot* installations are predicted to increase 22 percent by 2021. The industries most likely to use robotic manufacturing are automotive, followed by electronics (e.g., computers, televisions), metals and machinery (e.g., welding, assembly), plastics and chemical products (e.g., textiles, paper), and food and beverages. The industrial robot market in China remains the largest in the world, exceeding the number of installations in Europe and America combined (International Federation of Robotics [IFR] 2019).

The second category of robots used in the workplace includes *professional service robots*. Service robots are defined as semi- or fully autonomous machines that can complete commercial tasks, excluding those in manufacturing operations. Examples of professional service robots include customer service robots that interact with customers and come in humanoid or nonhumanoid form; professional medical robots that are used in surgery and also to improve patient care delivery; defense robots used in military combat to gain a tactical advantage; and demolition robots used to safely and remotely destroy buildings, bridges, and the like (Robotic Industries Association 2020).

Of late, robots are becoming smarter and more autonomous as programmers code complex decision-making capabilities into machines. In 2017, John Deere, manufacturer of farm machinery for decades, purchased Blue River Technology, developer of See and Spray. The machine uses cameras, machine learning, and sophisticated robotics to differentiate between crops and weeds. Said the John Deere representative, "[T]he machine processes images at a rate of one image every 50 milliseconds … [and] compares those real images to a library of over 300,000 images, making sure that only the weeds are targeted" (quoted in Anandan 2019, p. 1).

What do you think?

Artificial intelligence (AI) is the science of making a machine smart, smart enough to perform tasks it hasn't been programmed to perform. Apple's personal assistant, Siri, is a form of AI. Even though she's never been programmed to tell you the phone number of your favorite pizza place, once she recognizes the trigger, "Hey Siri," she records what you're saying and then disaggregates the speech into words that she can then match to a database of words and phrases. What do you think are some other examples of AI devices that you own?

Software Robotics. Robot Process Automation (RPA) is a technology whereby computer software is programmed to imitate and integrate human actions into a digital format to accomplish tasks. They can, for example, sign into computer applications, manipulate files and folders, complete calculations, copy and paste information, and open e-mails and attachments. They're most useful in completing repetitive and mundane tasks where an employee's time is better spent elsewhere. They also reduce errors and increase output (Casey 2020).

Since RPA relies on people to program the software, it is called *attended automation*. When RPA is combined with artificial intelligence (AI), the result is Intelligent Process Automation (IPA), or what is called *unattended automation*. While RPA requires structured data (e.g., spreadsheets), involves programmed logic, and works with employees, IPA works independently, does not require structured data, and develops its own logic. IPA can think for itself; RPA has to be told what to think.

Worker Error and Technological Failure. Although technology in the workplace is used to increase productivity, it can also contribute to worker error. The use of computerized medical records, intended to improve efficiency and reduce medical error, has met with mixed results in the health care industry. In one case, a teenage patient was given a massive overdose of an antibiotic leading to a grand mal seizure that nearly led to his death. The error happened when the doctor ordering the medication through a computer system failed to notice that the drop-down menu on the computer screen was set to "milligrams per kilogram" rather than "milligrams." One doctor who worked on the case suggested that such medical errors are a result of a lag between the advancements of technology and the understanding of that technology among the humans who use it (Wachter 2015).

In 2016, Uber tested an autonomous car in San Francisco without getting approval from the California Department of Transportation. The car ran six red lights, but no one was injured. However, in 2018, a 49-year-old pedestrian was killed in Tempe, Arizona, while crossing the street with her bicycle. After a two-year investigation,

artificial intelligence (AI) The science of making a machine smart enough to perform tasks it hasn't been programmed to perform.

> "Men have become tools of their tools."
>
> –HENRY DAVID THOREAU, AUTHOR

the National Transportation Safety Board concluded that, although the "human safety driver" had been looking at her phone a third of the time the car was on the road, the onboard computer did not correctly identify the woman crossing the street as a person. Thus, the crash was a result of both employee error and technological failure (Lee 2019).

Finally, it should be noted that technology can be used to intentionally *increase* product failure (Hutson 2018). Although computers in self-driving cars use artificial intelligence, sometimes called machine intelligence, to recognize 3-D objects, competitors can create *adversarial images* so that the sign at the intersection looks like a stop sign rather than a garage sale sign, leading the car to incorrectly come to a complete stop. Adversarial images are pictures "engineered to trick machine vision software, incorporating special patterns that make AI systems flip out. ... [T]hink of them as optical illusions for computers" (Vincent 2017, p. 1).

Telecommute and Telepresence. In 2019, approximately 7 percent of the U.S. labor force had the flexibility at their workplace to telecommute (DeSilver 2020) (see Chapter 7). At that time, they were predominantly highly paid, white-collar executives, and professional workers such as lawyers, professors, and engineers. However, since the COVID-19 pandemic, as many as half of Americans are working from home, and there are predictions that the number will remain high even after the pandemic is over (Guyot and Sawhill 2020) (see Chapter 2 and Chapter 7). Twitter, for example, has told their employees that they can keep working from home permanently.

Although there is some indication that telecommuting increases worker productivity and job satisfaction, the incorporation of advanced technology such as AI can blur the distinction between work and personal life. Unlike structural functionalists who would argue that telecommuting provides a flexible work option for working parents, Glass and Noonan (2016), consistent with a conflict perspective, suggest that "telecommuting appears ... to have become instrumental in the general expansion of work hours, facilitating workers' needs for additional work time beyond the standard workweek and/or the ability of employers to increase or intensify work demand among their salaried employees" (p. 38).

telepresence A general term for a group of technologies that allow a person to feel that they are present or look like they are present in a remote location.

Remote workers need to be technologically independent, reporting they most often use their computers for writing and sending e-mails and Internet browsing, job-related productivity, and work meetings (Research Now 2019). **Telepresence** is a general term for a group of technologies that allow a person to feel that they are present or look like they are present in a remote location. Telepresence allows employees to "attend" a work meeting or participate in a conference call from remote and multiple locations. Of late, one of the most popular telepresence platforms is Zoom. As a result of the pandemic, the videoconferencing platform's stock has increased in value 330 percent in just the first half of 2020.

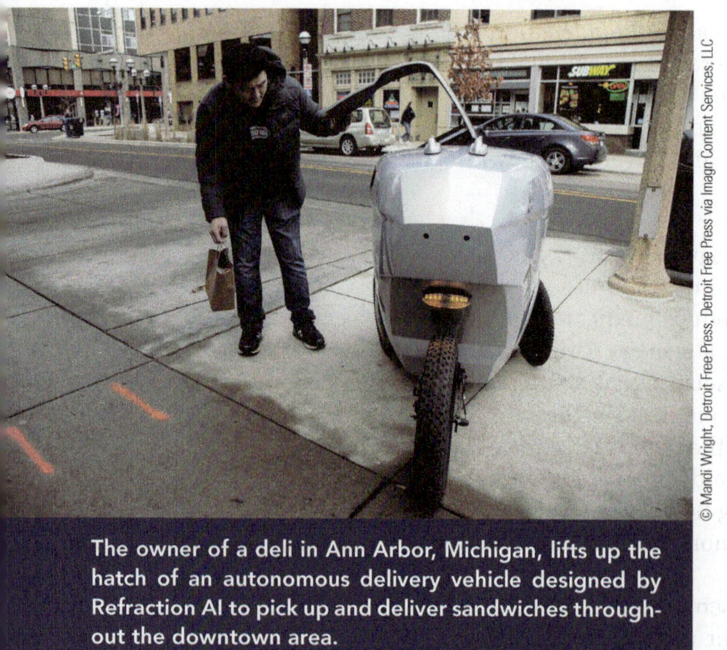

The owner of a deli in Ann Arbor, Michigan, lifts up the hatch of an autonomous delivery vehicle designed by Refraction AI to pick up and deliver sandwiches throughout the downtown area.

Technology and Social Control. Some technology lessens the need for supervisors and makes control by employers easier. Employees in the Department of Design and Construction in New York City must scan their hands each time they enter or leave the workplace. The use of identifying characteristics such as hands, faces, fingerprints, iris patterns, or voices is part of a technology called *biometrics* (Kloppenburg and van der Ploeg 2018). Union leaders "called the use of biometrics degrading, intrusive and unnecessary and said that experimenting with the technology could set the stage for a wider use of biometrics to keep tabs on all elements of the workday" (Chan 2007, p. 1).

Technology can also make workers more accountable by gathering information about their performance. In 2020, as a result of the surge of employees working at home, there was an increased demand for software that can monitor employees. *Hubstaff*, an employee-monitoring software program, takes a screenshot of websites accessed, documents written, and media sites visited every few minutes. It also maps your physical activity and location on your smartphone (Satariano 2020). Wearable technology can be used to improve workers' safety and overall health by monitoring the number of steps taken per day or the number of hours slept at night. It can also, however, be used to assess productivity, act as a surveillance device, and "extend managerial supervision to matters normally left to the employees' discretion" (Miele and Tirabeni 2020, p. 12).

The Computer Revolution

Early computers were much larger than the small machines we have today and were thought to have only esoteric uses among members of the scientific and military communities. In 1951, only about a half dozen computers existed (Ceruzzi 1993). The development of the silicon chip and sophisticated microelectronic technology allowed tens of thousands of components to be imprinted on a single chip smaller than a dime. This advancement made computers affordable and led to the development of laptops, cell phones, digital cameras, tablets and notebooks, e-readers, and portable DVDs. Although the first PC was developed only a little over 30 years ago, as of 2018, 92 percent of U.S. households had access to a computer and 84 percent had broadband Internet access (Statista 2020a). Despite almost universal access to a computer in the United States, worldwide, only about half of the households have a computer at home (Statista 2020b).

Technology use and ownership vary by important demographic variables (see Table 14.1). Most important, when examining the table, note the following points:

- Eighteen- to 29-year-olds are the most likely to use the Internet and social media and to have a smartphone and be dependent on it for Internet access.
- White non-Hispanics are the most likely to live in a house with broadband access.
- As educational attainment and income increase, Internet use, home broadband access, and smartphone ownership also increase.
- Minorities and those living in rural areas are more likely to be dependent on their smartphones for Internet access than city or suburban dwellers.

Although computer and Internet usage has grown rapidly in the past decade, the world is sharply divided into thirds in terms of computer and Internet access. In 2019, in *developed countries* (e.g., the United States, Germany), over 82 percent of households had both Internet and computer access at home. In *developing countries* (e.g., India, China), 46.7 percent of households had Internet access and 38.5 percent had computer access at home. Finally, in *less developed countries* (e.g., Chad, Cambodia), just 11.8 percent of households had Internet access at home and 9.5 percent computer access at home (International Telecommunication Union 2019) (see this chapter's section, "The Digital Divide").

Algorithms. An **algorithm** is a program that instructs a computer to follow a set of calculations in order to accomplish a task, i.e., it is a mathematical instruction. Today, algorithms are used in every facet of life. Facebook algorithms recommend news articles, Audible has algorithms that recommend books to read, Amazon's algorithms recommend products to buy, Netflix algorithms suggest movies to watch, and Spotify algorithms recommend music to listen to. There is some evidence that consumers, rather than being algorithm-averse as once thought, are overly dependent on algorithm recommendations, often adopting an inferior product simply because it was recommended (Banker and Khetani 2019).

algorithm A program that instructs a computer to follow a set of calculations in order to accomplish a task; i.e., it is a mathematical instruction.

TABLE 14.1 Technology Use and Ownership in the United States by Demographic Characteristics (%), 2019

Characteristics	Uses Internet	Home Broadband	Uses Social Media	Owns Smart-phone	Smartphone Dependent*
Total	**90**	**73**	**72**	**81**	**17**
Age					
18–29 years	100	77	90	96	22
30–49 years	97	77	82	92	18
50–64 years	88	79	69	79	14
65 years and older	73	59	40	53	12
Race and Hispanic origin					
White, non-Hispanic	92	79	73	82	12
Black, non-Hispanic	85	66	69	80	23
Hispanic (of any race)	86	61	70	79	25
Sex					
Male	90	73	65	84	17
Female	91	73	78	79	16
Community					
Urban	91	75	76	83	17
Suburban	94	79	72	83	13
Rural	85	63	66	71	20
Educational Attainment					
Less than high school graduate	71	46	N/A	66	32
High school graduate (includes equivalency)	84	59	64**	72	24
Some college or associate's degree	95	77	74	85	16
Bachelor's degree or higher	98	93	79	91	4
Income					
Less than $30,000	71	56	68	71	26
$30,000–$49,999	84	72	70	78	20
$50,000–$74,999	95	87	83	90	10
$75,000 +	98	92	78	95	6

*Uses smartphone as primary means of accessing the Internet, does not have broadband.
**Includes high school graduates or less.
SOURCES: Pew Research Center 2019a, 2019b, 2019c.

Although in each of these cases the use of algorithm-generated recommendations is likely not a matter of life or death, there are circumstances in which algorithms are used in life-altering ways. Automated decision making has been used in business (e.g., hiring), health care (e.g., diagnoses), banking (e.g., loan approval), risk assessment (e.g., breast cancer), government policies (e.g., immigration), social media (e.g., targeting politically susceptible users), industry (e.g., driverless cars), augmented and virtual reality (e.g., simulated environments), and the criminal justice system (e.g., predicting hotspots).

The use of algorithms in the criminal justice system is quite common. Algorithms are used in facial recognition technology in order to establish an individual's identity and/or location. Researchers at Dartmouth College are degrading high-quality images of license plates. These images are then catalogued as mathematical representations. When the police have a license plate they cannot read, its image can be compared to the stored mathematical representations of the degraded license plates until a match is found. Once there is a match, the low-quality image can be transformed to the high-quality image, and the plate can be read (Rigano 2019).

The criminal justice system also uses prediction algorithms to calculate flight risks, recidivism, and the likelihood of criminal victimization (Rigano 2019) (see Chapter 4). Algorithmic decision making may use statistics such as regression analysis to make predictions or AI, a more sophisticated method, in the hopes of avoiding human bias in data collection or programing. Although perhaps unintentional, **automated discrimination** may be a function of theoretical and mathematical models used by an AI algorithm. In 2019, the Department of Housing and Urban Development (HUD) charged Facebook with housing discrimination alleging that the social media giant used an advertising AI algorithm that systematically violated the *Fair Housing Act* by "encouraging, enabling, and causing" unlawful discrimination by restricting minorities from viewing housing advertisements (Jan and Dwoskin 2019, p. 1).

What do you think?

Del Grant was sentenced to life in prison after genotyping, based on probabilities, led a jury to find him guilty of first-degree murder (Ortiz 2020). After being incarcerated for nine years, a software program determined that Grant's DNA did not match the unknown male profile. A statistical algorithm had freed him. Do you think laws should require the use of algorithms in murder cases?

A study of the reliability of criminal risk assessment found that risk scores were "remarkably unreliable in forecasting violent crime: only 20 percent of the people predicted to commit violent crime actually went on to do so" (Angwin et al. 2016, p. 1). Further, Black defendants were *wrongly* identified as likely to reoffend at twice the rate of White defendants, and White defendants were *wrongly* identified as unlikely to reoffend compared to Black defendants. An analysis of the factors used in the algorithm indicated that racial bias was built into the equation by including such factors as "unemployed" and has "less than a high school degree," two variables we know are more likely to be characteristic of Black compared to White Americans as a result of systemic racism (see Chapter 9).

Not surprisingly, many Americans are skeptical of computer algorithms. In a recent survey, a majority of Americans agreed with the statement that computer programs will always reflect human bias. Furthermore, a majority of the respondents view the use of algorithms in criminal risk assessment, automated resume screening, analysis of job interviews, and calculation of financial scores, as "unacceptable" (Smith 2019).

"The saddest aspect of life right now is that science gathers knowledge faster than society gathers wisdom."

–ISAAC ASIMOV, BIOCHEMIST, AUTHOR

automated discrimination The bias that may result, intentionally or unintentionally, as a function of theoretical and mathematical models used by AI algorithms.

Computers as Big Business. Computers are big business, and the United States is one of the most successful manufacturers of computers in the world, boasting three of the six top producers—Hewlett-Packard (American), Acer (Taiwanese), Dell (American), Lenovo (Chinese), Toshiba (Japanese), and IBM (American) (Das 2020). The sale of desktop and laptop computers has steadily declined over the decades while the sale of mobile devices and tablets has increased. For example, a revenue breakdown of Apple Inc., ranked ninth in computer production worldwide, indicates that over half of its income comes from the iPhone and less than 10 percent from sales of its Mac computers (Wallach 2020).

Computer software is also big business and, in some cases, too big. Beginning in 1998, Microsoft battled the U.S. Department of Justice over antitrust violations. These laws prohibit unreasonable restraint of trade. At issue were Microsoft's Windows operating system and the vast array of Windows-based applications (e.g., spreadsheets, word processors, tax software)—applications that *only* work with Windows. Ultimately, the government won its case, giving start-ups such as Amazon, Google, and Facebook the opportunity to innovate and grow. Ironically, they too are now under government scrutiny (see this chapter's section, "Technology and Corporate America").

Information and Communication Technology and the Internet

Information and communication technology (ICT) refers to any technology that carries information. Most information technologies were developed within a 100-year span: taking pictures and telegraphy (1830s), rotary power printing (1840s), the typewriter (1860s), transatlantic cable (1866), the telephone (1876), motion pictures (1894), wireless telegraphy (1895), magnetic tape recording (1899), radio (1906), and television (1923) (Beniger 1993). The concept of an "information society" dates back to the 1950s, when an economist identified a work sector he called "the production and distribution of knowledge." In 1958, 31 percent of the labor force were employed in this sector; today, more than 50 percent are. When this figure is combined with those in service occupations, more than 75 percent of the labor force is involved in the information society. Figure 14.4 displays the adoption of various information technologies, as well as other innovations, in the United States between 1950 and 2019.

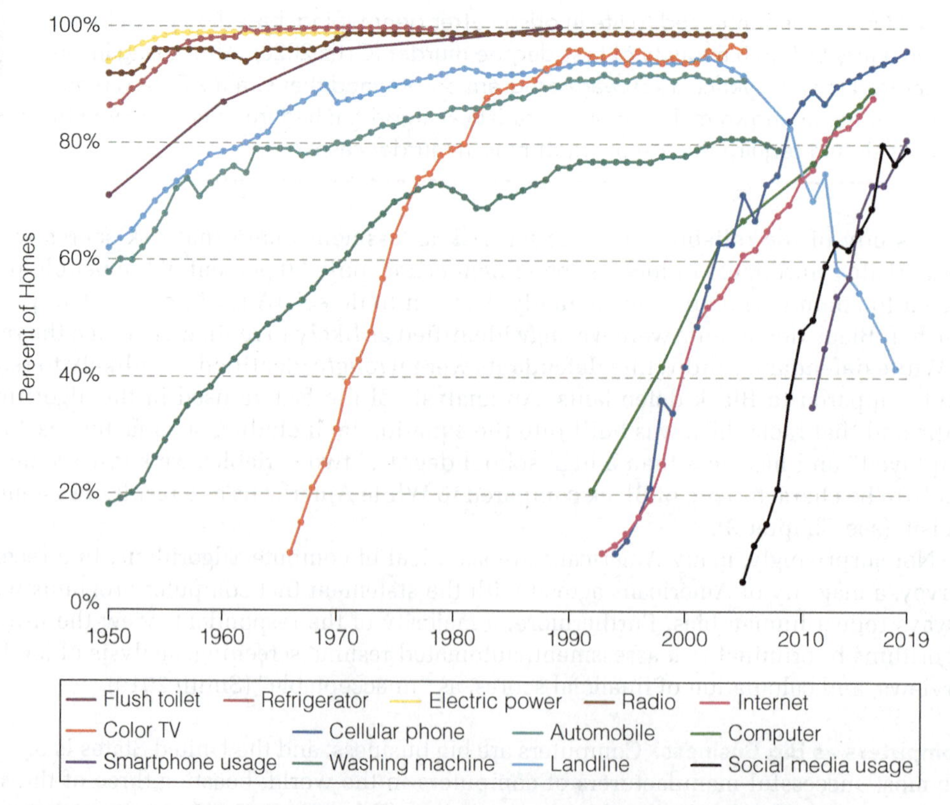

Figure 14.4 Adoption of Various Technologies in the United States, 1950–2019*

*The line stops when 100 percent adoption occurs. For example, 100 percent of households had a flush toilet(s) by 1989.

SOURCE: World in Data 2020.

The **Internet** is an international infrastructure—a network of networks—that distributes knowledge and information. In developed countries, 86 percent of the population uses the Internet compared to just 47 percent in developing countries, and 19 percent in less developed countries with males having greater access than females in each region. The global distribution of Internet users by age reveals that the majority of Internet users, 32 percent, are between the ages of 25 and 34, followed by 35- to 44-year-olds (19 percent) and 18- to 24-year-olds (18 percent). Those 65 and older are the least likely to use the Internet (Statista 2020c, 2020d).

The most recent advancement in Internet technology is the **Internet of Things (IoT)**, in which nearly every device with an on-off switch can be connected to the Internet. The vast array of network and information sharing made possible by the IoT has the potential to radically transform social relations. IoT devices, often called smart devices, have sensors (i.e., cameras) that can conduct complex data analysis similar to AI. For example, the sensors on a smart refrigerator can take inventory of what is inside through image recognition analysis and then alert you when you are running low on an item (Congressional Research Service [CRS] 2020a). IoT is being used in industry, globally the largest share of IoT projects, followed by smart cities, smart energy, and connected cars (Kumar, Tiwari, and Zymbler 2019).

AP Images/PriestmanGoode

Designers imagined what an aircraft might look like for post-pandemic travel. The redesigned cabins have "full divider screens, staggered seats, and the removal of in-flight entertainment systems." Airplane cabins also have dirt trap-free services and special colored paint designed to calm travelers (MSN News 2020, p. 1).

The number of IoT devices is dramatically increasing, growing from 9.9 billion devices in 2019 to an estimated 21.5 billion devices in 2025 (CRS 2020a). The IoT is made possible by wireless Internet access and the growth of cellular service areas. The expansion of mobile cellular service "has driven broad societal and economic transformations, [and] has changed not only the way people communicate, but also the way they plan their daily lives, organize themselves socially, and access educational, health, business, and employment opportunities" (International Telecommunication Union [ITU] 2016, p. 157). With the growth of 5G cellular networks and technologies, an even larger number of smart devices will be able to be connected simultaneously (CRS 2020a).

E-Commerce. **E-commerce** entails commercial transactions that take place over the Internet. Globally, China and the United States are primarily responsible for the recent increase in e-commerce sales (Cramer-Flood 2020). In the United States in 2019, consumers spent over $600 billion on U.S. products, a nearly 15 percent increase from the previous year (Young 2020). Amazon accounted for more than one-third of all online sales in the United States. Further, the percentage of online sales to total sales also continues to grow with e-commerce sales representing 16 percent of total retail sales in 2019 and likely a much larger proportion in 2020 as a result of the pandemic and store closures.

In the United States, more than 96 percent of Americans over the age of 18 shop online monthly. The most common device to use to shop online are mobile devices, and Gen Xers, those between the ages of 35 and 54, are the most likely to shop online, followed closely by baby boomers, consumers between the ages of 55 and 74. Women are more likely to be online shoppers than men and to shop online more often (Hwong 2018). The largest e-commerce retail purchases are books, followed by clothing; computer and computer-related equipment, including cell phones; computer software; and drugs, health aides, and beauty aids (U.S. Census Bureau 2020). Motivations to purchase products online include convenience, the ability to compare prices, and less expensive products in the few states that don't have Internet sales tax.

Health and Digital Medicine. Many Americans turn to the Internet to address health issues and concerns, and this was particularly true during the COVID-19 pandemic (see

Internet An international information infrastructure available through universities, research institutes, government agencies, libraries, and businesses.

Internet of Things (IoT) The potential connectivity of all electronic devices; widespread information sharing between people and the objects of everyday life.

e-commerce The buying and selling of goods and services over the Internet.

also Chapter 2). According to the Pew Research Center, 80 percent of Internet users have looked up health topics online, most frequently a specific disease or condition or a particular treatment or procedure (Weaver 2020). Even in countries with emerging economies, the majority of smartphone users, when seeking information, were most likely to be looking for information about health and well-being (Silver and Haung 2019).

More advanced technologies are also being used in the fight against COVID-19. For example, "China actively leveraged digital technologies such as artificial intelligence (AI), big data, cloud computing, blockchain, and 5G, which have effectively improved the efficiency of the country's efforts in epidemic monitoring, virus tracking, prevention, control and treatment, and resource allocation" (World Economic Forum 2020a, p. 1). Although predating the pandemic, South Korea uses AI to monitor the health and well-being of its elderly population (Tong-Hyung 2020). An Alexa-like speaker is placed in the homes of senior citizens and remote caretakers monitor search habits, listen for trigger words (e.g., lonely), and administer quizzes to assess memory and cognitive abilities.

Lastly, there is evidence that technology can help mediate the soaring cost of health care. High school students in Illinois used a 3-D printer that had been donated to the school to build a prosthetic hand for a 9-year-old girl who was born without fingers, for a total cost of $5 (Thornhill 2014). Other important sources of cost-effectiveness include automated data collection, electronic claims processing, reducing medical errors through more effective diagnostic and treatment interventions, remote patient monitoring, telemedicine, and the use of health care wearables.

News and Information. The Internet, perhaps more than any other technology, is the foundation of the information society. Whether reading an online book, streaming a documentary, mapping directions, taking an online course, traveling the world through virtual reality, or accessing Wikipedia, the Internet provides millions of users with 24/7 access and, with it, instant answers to questions previously requiring a trip to the library.

There is concern, however, that the very way in which the "Google generation" reads, thinks, and approaches problems has been altered by the new technology. Maps on mobile devices change the way we experience the world. Physical barriers and the risks of "getting lost" no longer limit our movements, but the immersive experience of the user-centered map restricts our sense of space and place (McMullan 2014). Journalist Nicholas Carr argues in his 2010 book *The Shallows: What the Internet Is Doing to Our Brains* that the use of the Internet promotes a chronic state of distraction that we are often unaware of:

> Our focus on a medium's content can blind us to the deep effects. We're too busy being dazzled or disturbed by the programming to notice what's going on inside our heads. In the end, we come to pretend that the technology itself doesn't matter. It's how we use it that matters, we tell ourselves. The implication, comforting in its hubris, is that we're in control. The technology is a tool, inert until we pick it up and inert again once we set it aside. (p. 3)

Finally, there is evidence that the accuracy of news and information is platform dependent. For example, analysis by the American News Pathways project indicates that people "who rely on social media for news are less likely to get the facts right about the coronavirus and politics and [are] more likely to hear some unproven claims" (Mitchell et al. 2020). Demographically, those who are most likely to use social media as a news source are disproportionately young, female, non-White, and have lower education and income levels.

Politics and e-Government. Social media sites like Twitter and Facebook have been central to political events around the world. Arab Spring protests throughout the Middle East in 2011 were coordinated through social media sites, often in spite of government efforts to restrict access, and #BLACKLIVESMATTER, YouTube videos of police violence, and support from Google, Amazon, and Apple have helped propel the issues surrounding systemic racism into the national spotlight (see Chapter 9).

Social media also played an important role in the 2020 U.S. elections. First, political candidates used Facebook, Twitter, and other social media accounts to make announcements and to share information with their followers. Second, social media was used as a venue to discuss the election with friends and family and as a source of political news. Third, because the algorithm that determines newsfeeds on Facebook, for example, favor content that you have "liked," Facebook has "become an echo chamber" (Sanders 2016, p. 1), further polarizing users along party lines. Further, unlike previous presidents, former President Trump used Twitter as his primary means of communicating with the press and his constituents. Lastly, as a result of the COVID-19 pandemic, the 2020 Democratic and Republican national conventions were held virtually, in part or in whole.

Social Media. In 2019, when the Pew Research Center ranked the top ten technology-related trends that shaped the decade, the growth and use of social media were ranked number one. Just 5 percent of U.S. adults used social media in 2005; today, 72 percent of Americans use social media (Auxier, Anderson, and Kumar 2019). Social media (e.g., Facebook, WhatsApp, Twitter) comprise a sector of the Internet called networking sites or membership communities. Membership communities have changed in three significant ways in recent years: Increasing numbers of people visit membership communities, people are spending substantially more time interacting in these communities, and older people are participating in membership communities at higher levels than ever before. In 2019, one-third of the world's population and two-thirds of Internet users used social media platforms (Ortiz-Ospina 2019).

Facebook is the largest social media platform in the world with over 2.4 billion users. Seventy percent of U.S. adults use Facebook, and no other platform comes close to the number of users with the exception of YouTube. Although social media sites surpass print newspapers as a source of news (Shearer 2018), over half of adult Facebook users report they do not understand how the platform's news feeds work, and three-quarters report not knowing that Facebook collects information about the user that, in turn, is provided to advertisers (Gramlich 2019).

In an examination of social media use over a three-year period, the most common reason given for using social media was to alleviate boredom, find information, and/or connect with others (Stockdale and Coyne 2020). Surprisingly, although today more people find their partners online than through family or friends (Rosenfeld, Thomas, and Hausen 2019), the use of social media to connect with others does not increase as users move from adolescence to adulthood.

@Kim Kardashian

I love that I can connect directly with you through instagram and Facebook, but I can't sit by and stay silent while these platforms continue to allow the spreading of hate, propaganda and misinformation - created by groups to sow division and split America apart

-Kim Kardashian West

Science and Biotechnology

Although recent computer innovations and the establishment of the Internet have led to significant cultural and structural changes, science and its resulting biotechnologies have produced not only dramatic changes but also hotly contested issues with public policy implications. In this section, we look at some of the issues raised by developments in genetics, food and biotechnology, nanotechnology, and cloning and stem cells.

What *do you* think?

Sometime in 2021, the eggs of millions of genetically modified mosquitoes will be placed throughout the Florida Keys (Milius 2020). The hope is that the specially bred male mosquito, after mating with a female mosquito, will result in female offspring that cannot survive as a result of her "father's" no-daughter genetic makeup, thereby reducing subsequent numbers of mosquitoes and the diseases they carry. Do you think you would want genetically modified mosquitoes released in your neighborhood?

Genetics. Molecular biology has led to a greater understanding of the genetic material found in all cells—DNA (deoxyribonucleic acid)—and with it the ability for **genetic testing**. Genetic testing "involves examining your DNA, the chemical database that carries instructions for your body's functions," and using that information to identify abnormalities or alterations that may lead to disease or illness (Mayo Clinic 2020, p. 1).

genetic testing Examination of DNA in order to identify abnormalities or alterations that may lead to disease or illness.

Genetic testing can be used in diagnosing the disease, determining if you are the carrier of the genetic disorder, and prenatal and newborn genetic screening.

A growing body of research indicates that human characteristics once thought to be largely social or psychological in nature are, at least in part, genetically induced. These include behavioral and psychological traits, such as personality characteristics, addiction, depression, autism, schizophrenia and anorexia, as well as physical illnesses such as sickle cell disease, breast cancer, Alzheimer's, cystic fibrosis, Down syndrome, Parkinson's disease, colon, skin, and prostate cancer, and hemophilia (National Human Genome Research Institute 2017).

The U.S. Human Genome Project (HGP), a 13-year effort to decode human DNA, was completed in 2003. Conclusion of the project has transformed medicine, particularly by revealing how the physical and the social interact:

> All diseases have a genetic component whether inherited or resulting from the body's response to environmental stresses like viruses or toxins. The successes of the HGP have ... enabled researchers to pinpoint errors in genes—the smallest units of heredity—that cause or contribute to disease. (HGP 2007, p. 1)

The hope is that, if a defective or missing gene can be identified, possibly a healthy duplicate can be acquired and transplanted into the affected cell. This is known as **gene replacement therapy**. Alternatively, gene editing, sometimes called genome editing, entails changing the DNA of an organism in order to correct a genetic mutation. Both gene replacement therapy and gene editing "offer great promise in the treatment of genetic diseases and are the result of decades of research in the laboratory, in animal models, and in clinical trials with humans" (Explorer 2020, p. 1).

gene replacement therapy The transplantation of a healthy gene to replace a defective or missing gene.

genetic engineering The manipulation of an organism's genes in such a way that the natural outcome is altered.

genetically modified food Food that has been produced by transferring genes from one plant into the genetic code of another plant.

Food and Biotechnology. **Genetic engineering** is the ability to manipulate the genes of an organism in such a way that the natural outcome is altered. A majority of Americans, 57 percent, believe genetically engineering animals in order to grow organs or tissues that could be used for humans needing a transplant is morally acceptable. The main reasons for objecting to the procedure were concerns for animal suffering and well-being and the disruption of the natural order of life (Strauss 2018).

Similarly, **genetically modified food**, also known as genetically engineered food, and genetically modified organisms involve this process of DNA recombination—scientists transferring genes from one plant into the genetic code of another plant. Figure 14.5 shows the opinions of a sample of U.S. adults concerning the negative and positive effects of genetically modified food.

Figure 14.5 Percentage of U.S. Adults Who Say It Is ________ That Genetically Modified Foods Will ________, 2018
SOURCE: Funk, Kennedy, and Hefferon 2018.

The vast majority of soybeans, corn, sugar beets, canola, and cotton grown in the United States are genetically engineered (Food and Drug Administration [FDA] 2020b). Because these ingredients are widely used in processed foods, nearly all packaged food sold in the United States and Canada contain genetically modified organisms (GMOs) (Food and Drug Administration [FDA] 2020b). When a sample of U.S. adults were asked whether genetically modified foods are worse, better, or neither better nor worse, 49 percent of Americans responded worse and 44 percent said neither worse nor better (Funk, Kennedy, and Hefferon 2018).

Effective January 1, 2020, the National Bioengineered [BE] Food Disclosure Standard went into effect. The legislation "requires that food manufacturers, importers, and other entities that label food for retail sale disclose information about BE food and BE food ingredients" (*Federal Register* 2018, p. 1). However, the Center for Food Safety (CFS) does not believe the regulation is strong enough. Consistent with other opponents of GMO, they argue that the new law, which allows food producers to use bar codes (QR codes) (i.e., consumers need a scanner) to indicate that the product is bioengineered, places too great of a burden on the consumer and is discriminatory against low-income shoppers who may not have the necessary technology (CFS 2018).

USDA

The U.S. Department of Agriculture (USDA) written regulation allows companies to label genetically modified organism food as bioengineered using pro-technology symbols such as the ones pictured here. The "smiley face" and "BE" for bioengineered are unlikely to convey that the product contains genetically modified organisms.

The U.S. Department of Agriculture (USDA) written regulation also allows companies to label genetically modified organism food as *bioengineered*, a term unfamiliar to consumers, rather than GMO and to use pro-technology symbols (see accompanying photograph) to indicated GMO products. Lastly, the CFS is concerned that the regulation will not include future forms of genetically engineered food, including food bioengineered with CRISPR, a relatively new, faster, and more precise gene editing technique (CFS 2018; National Institutes of Health 2020a).

Supporters of GM foods and biotechnology companies argue that labels are unnecessary given their research safety record. They also argue that biotechnology has the potential to alleviate hunger and malnutrition, as it enables farmers to produce higher-yield crops (Ben-Shahar and Schneider 2016). However, critics argue that the world already produces enough food for all people to have a healthy diet. If food were distributed equally, every person would be able to consume the daily recommended number of calories. The fundamental causes of hunger, these critics argue, is not lack of technology but rather poverty and unequal access to food and land.

Critics also argue that GMO practices can have wide-ranging effects on human health and the environment. Biotechnology companies claim that crops that are genetically designed to repel insects negate the need for pesticides and thus reduce chemical poisoning of land, water, animals, foods, and farmworkers. However, critics are concerned that insect populations can build up resistance to GM plants with insect-repelling traits, which would necessitate increased rather than decreased use of pesticides. The inability of humans to control chain reactions in ecosystems also raise concerns. Critics argue that the introduction of GMOs into the wild could weaken the genetic pools of wild populations, leading to their collapse and affecting the broader ecosystem (Jensen 2019).

Human health concerns include possible toxicity, carcinogenicity, food intolerance, antibiotic resistance buildup, decreased nutritional value, and food allergens to GM foods—all examples of **technology-induced diseases**. In 2015, the World Health Organization classified glyphosate, the key ingredient in Monsanto's weed killer Roundup, as a possible cause of cancer in humans. Nonetheless, in 2019, the Environmental Protection Agency said that the herbicide was not carcinogenic. However, after several successful lawsuits against the parent company, Bayer Pharmaceutical agreed to pay more than $10 billion to tens of thousands of Roundup users who had contracted cancer (Cohen 2020).

@Non GMOProject

Glyphosate-based #herbicides are the most widely used herbicides in the world—and Roundup Ready #GMO crops make a significant impact. But #glyphosate is also used on public areas, school campuses and parks: These student activists are challenging that.

-Non-GMO Project

technology-induced diseases Diseases that result from the use of technological devices, products, and/or chemicals.

Nanotechnology. **Nanotechnology** refers to the manipulation of materials and the creation of structures and systems at the scale of atoms and molecules. In science, the word *nano* means "one billionth," a particle unimaginably small. For example, there are 25,400,000 nanometers in an inch, and a sheet of newspaper is 100,000 nanometers thick. Nanotechnology can be used across a variety of sciences including chemistry, biology, physics, and engineering (National Nanotechnology Initiative [NNI] 2019a).

Nanotechnology is of particular value to the food industry because "it can be used in all areas [of the industry] such as farming, food packaging, food processing, among others, and even plays a role in the prevention of microbial contamination," which enhances food safety (Kaur and Singh 2020, p. 15). However, critics argue that adding nanomaterial to food may come with a high price for human health and the environment. In particular, nanofoods use particles of silver, titanium dioxide, zinc, and zinc oxide that have been shown to be highly toxic to animals in laboratory testing. Because of the small size of these particles, there is greater risk that these particles will access and alter human cells.

Nonetheless, nanotechnology also has many benefits. Because of its small size, materials made at the nanoscale are stronger and lighter than other materials. There are presently over 4,000 nanotechnology products on the market, including "smart fabrics" used in clothes that monitor the wearer's health, nanoscale film on eyeglasses, computers, and windshields that make them anti-reflective and self-cleaning, and nano-structured solar panels that are thinner and more efficient in converting sunlight to electricity, fulfilling the promise of affordable solar power (Nano Database 2020; NNI 2019b).

Cloning, Therapeutic Cloning, and Stem Cells. In July 1996, scientist Ian Wilmut of Scotland successfully cloned an adult sheep named Dolly. To date, cattle, goats, mice, pigs, cats, rabbits, and horses have also been cloned. In 2018, the first successful cloning of a primate took place through somatic cell nuclear transfer, a method by which the nucleus of a donor cell is placed into a fertilized egg in which its own chromosomes have been removed (Maron 2018). The result was genetically identical twin monkeys. The United States, as well as many other countries, has strict regulations on conducting research on cloning primates due to ethical concerns and the fear that it would someday lead to human cloning.

Human cloning, however, has potential medical value by allowing people to reserve therapeutic cells to treat diseases and by allowing an alternative reproductive route for infertile couples or those who are at risk of transmitting a genetic disease to their offspring. Arguments against cloning are largely based on moral and ethical considerations. Critics of human cloning suggest that, whether used for medical therapeutic purposes or as a means of reproduction, human cloning is a threat to human dignity. Cloned humans would be deprived of their individuality, and as Kahn (1997) points out, "[C]reating human life for the sole purpose of preparing therapeutic material would clearly not be for the dignity of the life created" (p. 119). In a 2018 survey, although 40 percent of Americans responded that cloning of animals was morally acceptable, just 16 percent of respondents said that cloning humans was morally acceptable (Reinhart 2018).

There are three different types of cloning. *Genetic cloning*, which is widely accepted and routinely used in laboratory work, copies genes or segments of DNA. This type of cloning is carefully regulated and not considered to have ethical implications. Reproductive and therapeutic cloning are less commonly used and raise ethical concerns. *Reproductive cloning* copies the DNA of whole animals to produce an exact genetic copy. Dolly the sheep and the twin monkeys in China are examples of reproductive cloning. This technology has the potential to address infertility issues for sterile couples and allow those who carry deleterious genes to avoid passing genetic illnesses to their offspring. However, this technology raises concerns about human dignity and individual freedom.

Therapeutic cloning is the most controversial of the three types of cloning because it uses stem cells from unused human embryos. This form of cloning has the potential to treat a wide range of illnesses and diseases; however, it requires the destruction of human embryos in order to harvest stem cells (Murnaghan 2020). **Stem cells** can produce any type of cell in the human body and thus can be "modeled into replacement parts for

nanotechnology The manipulation of materials and creation of structures and systems at the scale of atoms and molecules.

stem cells Undifferentiated cells that can produce any type of cell in the human body.

people suffering from spinal cord injuries or regenerative diseases, including Parkinson's and diabetes" (Eilperin and Weiss 2003, p. A6).

Unfortunately, companies such as U.S. Stem Cell offer desperate patients the hope of a cure through stem cell injection. The Food and Drug Administration (FDA) is suing U.S. Stem Cell for "openly violating the law and endangering patients" (McGinley and Wan 2019). Further, in 2020, after reviewing the website and Facebook pages of Dynamic Stem Cell Therapy, which, among other claims, markets "cellular products for [the] treatment or prevention of Coronavirus Disease 2019 (COVID-19)," the FDA issued a letter advising the company of apparent regulatory violations (FDA 2020c).

Despite what appears to be a universal race to the future and the indisputable benefits of scientific discoveries such as the workings of DNA and the technology of AI and stem cells, some people are concerned about the duality of science and technology. Science and the resulting technological innovations are often life-assisting and life-giving; they are also potentially destructive and life-threatening. The same scientific knowledge that led to the discovery of nuclear fission, for example, led to the development of both nuclear power plants and the potential for nuclear destruction. Thus we now turn our attention to the social problems associated with science and technology.

Societal Consequences of Science and Technology

Scientific discoveries and technological innovations have implications for all social actors and social groups. As such, they also have consequences for society as a whole. Overall, nearly three-quarters of Americans believe that science has had a mostly positive impact on society, although there are demographic differences (Thigpen and Funk 2019) (see Figure 14.6). Although the majority of Americans believe that the benefits of scientific research outweigh the potential for harmful results, there is no denying that science and the resulting technologies have some negative consequences.

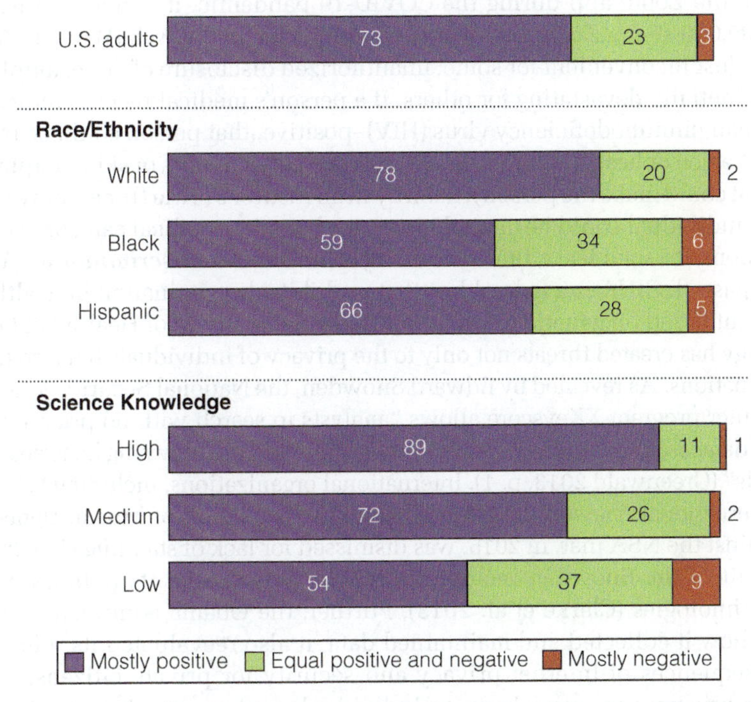

Figure 14.6 Percentage of U.S. Adults Who Say Science Has Had a(n) ____________ Effect on Society, 2019

SOURCE: Thigpen and Funk 2019.

Social Relationships, Social Media, and Social Interaction

Technology affects social relationships and the nature of social interaction. The development of telephones has led to fewer visits with friends and relatives; with the advent of DVRs, cable television, and video streaming, the number of places where social life occurs (e.g., movie theaters) has declined. The average white-collar worker checks his or her e-mail 74 times a day and reports high levels of anxiety when not at work about the need to respond to e-mails. The potential health risks of "email addiction" have led some companies to install "holiday mode" to prevent workers from receiving e-mails outside their work hours (Roberts 2014). As technology increases, social relationships and human interaction are transformed.

Students who attend schools where cell phone use is banned showed significantly higher performance on exams than those who attended schools where phones were allowed. This difference was most dramatic for underachieving students, indicating that low-achieving students suffer more from technological distraction than high-achieving students (Beland and Murphy 2015). Additionally, Rotondi, Stanca, and Tomasuolo (2017) investigated the relationship between smartphone use, life satisfaction, and the quality of face-to-face interactions with friends. The researchers conclude that the "intrusiveness of the smartphone, arising from its powerful connecting capabilities together with small size portability, reduces the quality of face-to-face social interactions ... and [thus] their positive impact on well-being" (Rotondi, Stanca, and Tomasuolo 2017, p. 24).

Loss of Privacy and Security

Through computers, individuals can obtain access to someone's phone bills, tax returns, medical reports, credit histories, bank account balances, and driving records. In 2019, nearly one-quarter of U.S. adults reported fraudulent charges on a debit or credit card, and 8 percent reported that someone took over their social media or e-mail accounts without their permission (Auxier et al. 2019). Further, nearly all Americans are affected by commercial cybersecurity breaches. In 2018, 500 million former guests of Marriott's Starwood properties had their names, e-mail addresses, phone and passport numbers, date of birth, and other personal information copied and encrypted by a hacker (Youn 2019). Due to the popularity of the Zoom app during the COVID-19 pandemic, it is not surprising that in 2020 over 500,000 stolen Zoom passwords were for sale on the web (Winder 2020).

Although just inconvenient for some, unauthorized disclosure of, for example, medical records is potentially devastating for others. If a person's medical records indicate that he or she is human immunodeficiency virus (HIV)–positive, that person could be in danger of losing his or her job or health benefits. If DNA testing of hair, blood, or skin samples reveals a condition that could make the person a liability in an insurer's (Title I) or employer's opinion (Title II), the individual could be denied insurance benefits, medical care, or even employment. In response to such fears, the ***Genetic Information Nondiscrimination Act of 2008*** (GINA) was passed. GINA is a federal law that prohibits discrimination in health coverage or employment based on genetic information (National Institute of Health 2020b).

Technology has created threats not only to the privacy of individuals but also to the security of entire nations. As revealed by Edward Snowden, the National Security Administration (NSA) computer program XKeyscore allows "analysts to search with no prior authorization through vast databases containing e-mails, online chats, and the browsing histories of millions of individuals" (Greenwald 2013, p. 1). International organizations, including Wikipedia and Amnesty International, viewed the NSA program as a violation of human rights and filed a lawsuit against the NSA that, in 2015, was dismissed for lack of standing (Thielman 2015).

As a result of the Snowden incident, the NSA reevaluated its policies on communications technologies (Clarke et al. 2013). Further, the Obama administration not only reevaluated how it collected and maintained data; it also reevaluated its role in addressing the consequences of Internet privacy and security for private citizens, stating that "breaches of privacy can cause harm to individuals and groups. It is a role of government to prevent such harm where possible, and to facilitate means of redress when harm occurs" (President's Council of Advisors on Science and Technology 2014, p. 1).

Genetic Information Nondiscrimination Act of 2008 A federal law that prohibits discrimination in health coverage or employment based on genetic information.

Walmart has filed for and received a patent for sensors that could be installed at checkout counters and throughout stores. The sensors would record interactions between customers and Walmart employees and could then be used as a measure of workplace performance (Levin 2018). Said Walmart officials, "[W]e file patents frequently but that doesn't mean the patents will actually be implemented" (p. 1). Would you take a job where there was workplace surveillance technology?

What *do you* think?

Given the foregoing, it is not surprising that, in 2019, two-thirds of Americans reported being concerned about how personal information collected by companies and the government was being used. Eighty percent reported they had very little to no control over data collected by companies or the government, the majority believing that their personal information today is less secure than in the past. Lastly, a sizable majority of Americans believe that some, most, or all of what they do online or on their cell phones is tracked by companies (91 percent) or the government (77 percent) (Auxier et al. 2019) (see Figure 14.7).

Cyberattacks and data fraud/theft are ranked alongside climate, environmental, and other natural disasters, biodiversity loss, water crises, and global governance failures as the highest-likelihood risks facing global security over the next ten years (World Economic Forum 2020b). In 2018, the European Union adopted the General Data Protection Regulation (GDPR), which "identifies legitimate bases for data processing and sets out common rules for data retention, storage limitations, and record-keeping" (CRS 2020b, p. 1). Since its implementation, violations of GDPR have resulted in heavy fines for over 273 companies including Equifax, Google, and Facebook. The only comparable law in the United States was passed by California in 2018, although other states are considering similar legislation.

AP Image/Gene J. Puskar

An autonomous (i.e., self-driving) locomotion truck parked in downtown Pittsburgh, Pennsylvania. The truck is the result of a partnership between a trucking company and a logistics company that have come together to move cargo over 400 miles, between Oregon and Idaho.

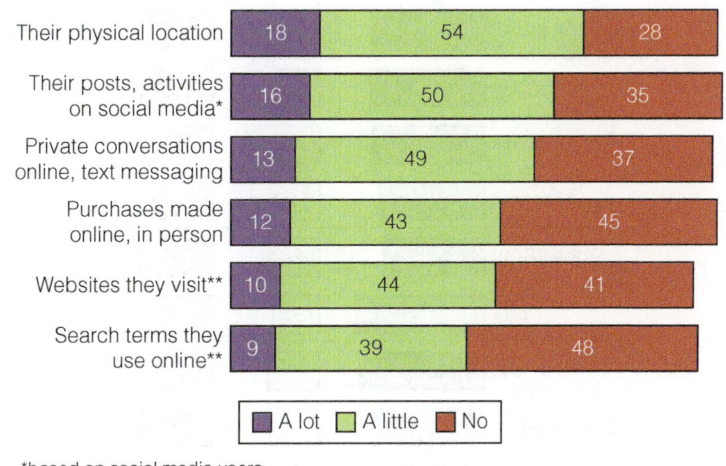

Figure 14.7 Percentage of Americans Who Say They Feel ___________ Control over Who Can Access the Following Types of Their Information, 2019
SOURCE: Auxier, Anderson, and Kumar 2019.

Unemployment and Underemployment

Some technologies replace human workers. Technology and innovations such as artificial intelligence (AI), autonomous robots, and the Internet of Things have led to predictions of mass unemployment. Indeed, a report by the McKinsey Global Institute (MGI 2017) predicts that across 46 countries, up to one-third of workers could be displaced by 2030 with workers in advanced economies feeling the greatest brunt of automation compared to workers in developing or less developed countries. Further, the report suggests that, even if there is full employment in 2030, a large proportion of the global workforce, as many as 75 million to 375 million workers, will need to transition to different occupations (MGI 2017).

In advanced economies such as the United States, there will be a demand for new skills and for increased educational requirements to meet the needs of a technology-driven society. The use of technology (e.g., robotics) to replace human labor in its initial phases replaced physical labor, thus displacing manufacturing jobs. The second phase, characterized by the use of computers and all its attendant software, impacted middle management occupations that rely on routine tasks. Newer forms of technology, however, especially AI with its focus on language, reasoning skills, and learning, are likely to displace those in higher-wage occupations. Occupations that traditionally require bachelor's degrees or higher (e.g., financial manager) are the most likely to have some *exposure* to artificial intelligence, i.e., are the most likely to be impacted by it, compared to occupations that require less than a college degree (e.g., bus driver). Demographically, the diffusion of AI into virtually every occupation is likely to disproportionately affect men, middle-age workers, and Whites and Asian Americans (Muro, Whiton, and Maxim 2019).

Americans are not oblivious to the threats technology presents. Nearly half of a sample of U.S. workers reported that in 2050 the average working person will have less job security than in 2020 (Parker, Morin, and Horowitz 2019). Figure 14.8 graphically portrays the results of a 2018 survey when respondents were asked their opinion about the impact of automation and new technologies on the workplace.

Employment challenges also arise due to the lack of highly skilled American workers in **STEM**, i.e., science, technology, engineering, and math occupations. These jobs have historically been filled by foreign workers who enter the United States on H-1B visas. However, in 2019, President Trump signed an executive order making it harder for federal agencies to hire workers on H-1B visas, an escalation of a previous policy that temporarily banned foreign workers in technology industries (McGregor 2020). There has

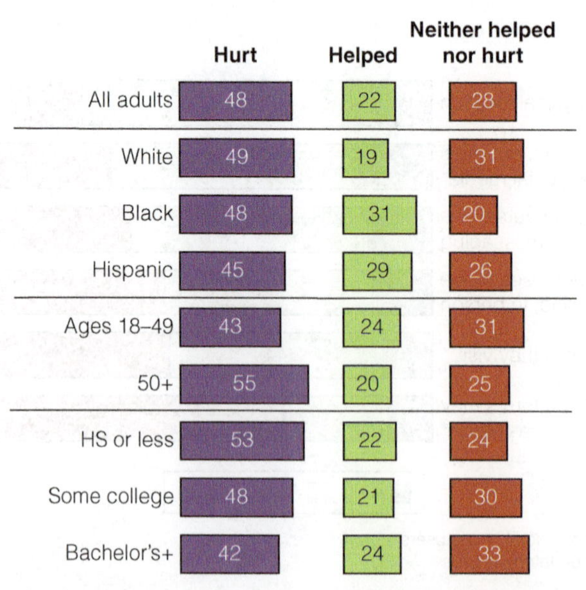

Figure 14.8 Percentage Saying That Automation of Jobs through New Technology in the Workplace Has Mostly ____________ American Workers, 2018
SOURCE: Parker, Morin, and Horowitz 2019.

STEM Science, technology, engineering, and mathematics.

also been a decrease in the number of international students attending American universities (NSF 2020), and the former administration's immigration policies made it difficult for foreign scientists to work in the United States, in contrast to China's open-door policy (Council on Foreign Relations 2018). President Biden's plan to modernize the immigration system (see Chapter 9) includes increasing the number of visas for highly skilled workers and issuing green cards to foreign graduates of U.S. PhD programs in science, technology, engineering, and mathematics (Biden 2020c).

The Digital Divide

One of the most significant social problems associated with science and technology is the increased division between social groups and regions of the world. In 2019, nearly half of the global population were not Internet users, though geographic differences and demographic differences based on region of the world exist. For example, the Internet gender gap globally is 17 percent with males having higher rates of Internet use than females. However, the Internet gender gap in the Americas is near zero. Nonetheless, in just 8 percent of the world's countries does Internet use by women exceed that by men, and gender parity in Internet use is found in just one-quarter of the world's countries (ITU 2019).

Globally, the digital divide reflects the economic and social conditions of a country. In general, wealthier and more educated countries have more technology than poorer countries with a less educated populace. Much of the gap can be explained by a lack of access to devices and/or Internet connections at home or at work. A report by the United Nations Educational, Scientific and Cultural Organization (UNESCO) (UNESCO 2019) notes that meaningful connectivity requires broadband that is "available, accessible, relevant, and affordable" (p. 1). Yet, globally, the percentage of households connected to the Internet via broadband increased by just over 1 percent in the preceding year.

Similarly, in the United States, the wealthier an individual is, the more likely she or he is to use technology and to have broadband access at home. Of individuals with annual incomes of over $100,000, 95 percent or more of the respondents reported having a smartphone, a desktop or laptop computer, and home broadband (Anderson and Kumar 2019). In contrast, of U.S. adults with incomes of $30,000 or less, 71 percent reported having a smartphone, 54 percent a desktop or laptop computer, and just over half had access to broadband at home. Not surprisingly, given the relationship between race and income, Whites compared to Blacks and Hispanics were more likely to have access to each of the three technologies (Perrin and Turner 2019).

Technology has amplified inequality."

–BARACK OBAMA, 44TH U.S. PRESIDENT

The three technology categories are linearly related to the independent variable: income. With each increase in an individual's income category, the percentage of individuals who own a smartphone, who own a desktop or laptop computer, or who have home broadband access to the Internet increases (Anderson and Kumar 2019). Furthermore, smartphone dependence, i.e., an individual's access to the Internet is limited to their smartphone, contributes to the **participation gap** (see Table 14.1). Young adults, racial and ethnic minorities, and low-income Americans are more likely to access the Internet via mobile devices and, when online, to use Twitter, play games, participate in social networking, and watch video games when compared to their advantaged, White counterparts (McCollum 2011; Tsetsi and Rains 2017). Further, Internet users who are smartphone dependent and/or without broadband access were at a disadvantage as they transitioned back to work and school during the COVID-19 pandemic (Holpuch 2020) (see Chapter 7 and Chapter 8).

participation gap The tendency for racial and ethnic minorities to participate in information and communication technologies (e.g., using smartphones to access the Internet rather than a computer) that place them in a disadvantaged position (e.g., difficulty in researching a term paper on a smartphone).

Smart Technologies. In addition to the *digital* divide, newer technologies such as autonomous vehicles, smart robots, and artificial intelligence are also likely to have a detrimental impact on certain demographic groups. Some technologies, such as industrial robots, have disproportionately led to unemployment among factory workers. A report by Oxford Economics, a forecasting company, predicts the loss of 20 million manufacturing jobs by 2030—8.5 percent of the global manufacturing labor force—as a result of automation (Oxford Economics 2020). The researchers also predict that vulnerability varies regionally, with poorer countries and manufacturing-dependent states such as North Carolina, Oregon, Indiana, Texas, and Louisiana bearing the highest burden.

There are also serious concerns about bias in artificial intelligence (AI). In the places where AI is being created, often elite university laboratories and technology companies, "AI systems are designed, [and] the cost of bias, harassment, and discrimination [is] borne ... by gender minorities, people of color, and other underrepresented groups" (West, Whitaker, and Crawford 2019, p. 7). Facial recognition technology is a particularly relevant example. Facial recognition technology is used to withdraw money, enter buildings, unlock cell phones, and identify suspects.

What *do you* think?

There has been concern across U.S. universities over the use of facial recognition technology. For example, UCLA administrators proposed using facial recognition for campus security purposes but changed their position after student protests were led by the technology rights group Fight for the Future (Paul 2020). Presently, ten schools have confirmed using facial recognition, many others will not say one way or the other whether they are using it, and approximately 50 schools have committed to not using facial recognition. Would you attend a college or university that uses facial recognition technology?

No facial recognition technology, however, is 100 percent accurate, but the errors they make are not random. An MIT study that focused on gender identification used 1,270 images of men and women from three African countries and three European countries with skin colors classified as either darker or lighter, as the benchmark data set. Three commercial face recognition companies, IBM, Microsoft, and Face ++, were evaluated. Overall, accuracy of identifying the correct gender was relatively good, averaging about ten errors for every 100 identifications, and each company performed better in identifying males than females. However, once skin color was introduced as a variable, i.e., the facial recognition software was tasked with differentiating between darker males and darker females and between lighter males and lighter females, the error rate more than tripled for two of the three companies and doubled for the third (Buolamwini and Gebru 2018). As one of the co-authors concludes:

> The deeper we dig, the more remnants of bias we will find in our technology. We cannot afford to look away this time, because the stakes are simply too high. We risk losing the gains made with the civil rights movement and women's movement under the false assumption of machine neutrality. Automated systems are not inherently neutral. They reflect the priorities, preferences, and prejudices—the coded gaze—of those who have the power to mold artificial intelligence. (Buolamwini 2020, p. 1)

Problems of Mental and Physical Health

In a review of the literature, Twenge (2020) notes that, beginning around 2012, there were sharp increases in the rates of unhappiness, life dissatisfaction, loneliness, anxiety, depression, self-harm, and suicide among adolescents and young adults in Canada, the United Kingdom, and the United States. The growing professional consensus is that these indicators of the decline in mental health are associated with increases in the use of smartphones and social media that occurred at approximately the same time.

phubbing A combination of the words *phone* and *snubbing*, which occurs when people are constantly looking at their phone while you are talking to them.

One possible explanation for the relationship between technology use and lowered mental health is the reduction of time spent in face-to-face social interaction. A second hypothesis argues that even when there is face-to-face social interaction, the presence of digital media contributes to it being an unpleasant experiences as a result of, for example, **phubbing**, a combination of the words *phone* and *snubbing*, which occurs when someone is constantly looking at their phone while you are talking to them. Additional explanations for the relationship between technology use and reduced mental health focus on the incivility of online environments, cyberbullying, and online access to self-harm information (Twenge 2020).

There is also concern over technology addiction. According to Robert Lustig, an emeritus professor of pediatric endocrinology and author of the book, *The Hacking of the American Mind: The Science behind the Corporate Takeover of Our Bodies and Brains*, "Technology, like all other 'rewards' can over-release dopamine, overexcite and kill neurons, leading to addiction ... [Technology is] not a drug, but it might as well be. It works the same way. ... It has the same results" (Lustig 2018).

Despite such expert opinion and the World Health Organization's decision to classify gaming addiction as a type of "addictive behavior disorder," there are those who argue technology addiction does not exist (Ferguson 2018). Technology is not a drug, and its mechanisms do not act in the same way as pharmacological substances. While pleasurable experiences, in general, release dopamine, a "feel-good" neurotransmitter that signals the brain, playing video games releases just 175 percent more dopamine above the baseline compared to methamphetamine that releases 1,300 percent more dopamine. Further, critics argue, fewer than 3 percent of adolescent gamers develop the problem behaviors associated with most addictions such as neglecting homework or skipping school, and research does not support the contention that those labeled as addicted to technology have more psychological problems or health issues.

Whether or not technology addiction exists is an academic issue. However, there is fact-based evidence that documents the negative consequences of technology use on health and fitness (Mustafaoglu et al. 2018). Whether it's television, computers, the Internet, smartphones, video games, or digital toys, digital technology is associated with a sedentary lifestyle, a lack of physical activity and exercise, obesity, and musculoskeletal problems. Use of digital technology is also associated with less sleep and sleep disturbances.

Malicious Use of the Internet

The Internet may be used for malicious purposes, including but not limited to crime in the dark web (see Chapter 4), malware and hacking, electronic aggression and cyberbullying, disinformation, deepfakes and conspiracy theories, and election tampering.

The Deep and Dark Web. Under the surface of the Internet lie two additional networks that cannot be reached through conventional search engines such as Google, Yahoo, or Internet Explorer. First, there is the **deep web**, which consists of nonindexed pages that can only be accessed through passwords, encryption, or specialty software. It is usually used for legal purposes that require anonymity. Alternatively, the **dark web** requires Tor Project software, which allows users to "defend [themselves] against tracking and surveillance [and] circumvent censorship" (Tor Project 2020). The dark web is a subset of the deep web and is often used for criminal activities such as buying and selling of illegal merchandise using cryptocurrency, which shields the purchaser's identity (Sheils 2020). In 2014, the FBI shut down and arrested the leader of Silk Road, a dark website notorious for the black market sales of illegal drugs and weapons using bitcoin currency (Weiser 2015). Subsequently, similar illegal marketplaces emerged to take its place.

8kun, formerly 8chan, is a dark website for anonymous user-created message boards and has been linked to several mass murderers, including Patrick Wood Crusius, the alleged 2019 El Paso, Texas shooter, who posted a right-wing manifesto on a message board before killing 26 people (Romo 2019). Ribeiro et al. (2020) investigated whether the process of right-wing radicalization begins in one or more of three online communities: the intellectual dark web (I.D.W.) (e.g., uncensored debates of controversial topics), the Alt-light (e.g., civic nationalists), or the more extreme Alt-right (white supremacist nationalists). A total of 349 channels, 330,925 YouTube videos, and over 72 million viewer comments representing nearly 6 million users were analyzed. The authors conclude that, "consistently, users who consume Alt-light or I.D.W. content in a given year go on to become a significant faction of the Alt-right user base in the following year" (p. 138).

deep web Consists of nonindexed pages that can only be accessed through passwords, encryption, or specialty software; usually used for legal purposes that require anonymity.

dark web Requires Tor Project software to access, which prevents tracking, surveillance, and censorship; usually used for illegal purposes.

Malware and Hacking. **Malware** is a general term that involves any spyware, crimeware, worms, viruses, and adware that is installed on a computer without the owner's knowledge. **Hacking** is unauthorized access to a computer to obtain data for illicit purposes. Although often financially motivated (see Chapter 4), hacking for sociopolitical reasons has increased in recent years. In 2016, Russians hacked the Democratic National Committee headquarters and used Wikileaks to share the stolen e-mails (Goel and Lichtblau 2017). In 2018, the Department of Homeland Security confirmed that Russian hackers had also accessed several states' voter registration rolls prior to the 2016 election. The U.S. Department of Justice, as a result, indicted 13 Russians and three companies (Center for Strategic and International Studies 2020).

Combining the words *hacking* and *activism*, members of the hacktivist group Anonymous has reemerged in support of the 2020 Black Lives Matter protests after several years of silence. They are accused of temporarily shutting down the Minneapolis Police Department's website after the killing of George Floyd, and there are suspicions that the group accessed a database of e-mail addresses and passwords from the police department's computer system. In 2014, after the police shooting of Michael Brown in Ferguson, Missouri, members of Anonymous threatened to retaliate against the city if any of the protesters were harmed. Subsequently, they claimed responsibility for disabling the city's website (Molloy and Tidy 2020).

Electronic Aggression and Cyberbullying. Electronic aggression is defined as any kind of aggression that takes place with the use of technology. For example, **cyberbullying** refers to the use of electronic communication (e.g., websites, e-mail, instant messaging, or text messaging) to send or post negative or hurtful messages or images about an individual or a group (Kharfen 2006). Cyberbullying differs from traditional bullying in several significant ways, including the potential for a larger audience, anonymity, the inability to respond directly and immediately to the bully, reduced levels of adult or peer supervision, and the possibility of prolonged and/or coordinated activity (Sticca and Perren 2013; Chatzakou et al. 2019).

A majority of teenagers report experiencing some form of cyberbullying or harassment with name-calling and spreading false rumors being the two most common forms of abusive behavior (Anderson 2018). Further, a quarter of teenagers say that they have been sent explicit images that they did not request, and 7 percent report that someone shared explicit images of them with others without their permission. Although boys and girls have experienced abusive online behaviors equally, girls are more likely to have been victimized by more than one type of bullying.

In an investigation of online abusive behaviors, Chatzakou et al. (2019) analyzed 1.2 million Twitter users and 2.1 million tweets. Among other conclusions, the researchers observed that bullies post more frequently than nonbullies, and their attacks are often linked to sensitive issues such as feminism, religion, and politics. They use aggressive and in some cases insulting language and are less popular than nonbullies as measured by, among other things, the number of friends and followers they have. Bullies are also more likely to have their accounts suspended by Twitter compared to random Twitter users (Chatzakou et al. 2019).

hacking Unauthorized access to a computer to obtain data for illicit purposes.

cyberbullying Use of electronic communication to send or post negative or hurtful messages or images about an individual or a group.

disinformation Misinformation that is intentionally created and distributed in order to mislead, sow conflict, or attain a specific end.

deepfake videos Digitally created with artificial intelligence, videos that can realistically portray a person or persons saying or doing things that they have never said or done.

Disinformation, Deepfakes, and Conspiracy Theories. Misinformation and disinformation are both inaccurate. In the first case, the inaccuracy is unintentional and the result of being misinformed. **Disinformation**, on the other hand, is misinformation that is intentionally created and distributed in order to mislead, sow conflict, or attain a specific end, i.e., those who spread disinformation have an agenda. The United States, as well as Europe, has been victimized by disinformation campaigns often from state actors such as Russia, which was likely the "first among major powers to … intentionally spread … inaccurate information designed to influence societies" (Polyakova and Fried 2019, p. 2). Disinformation campaigns are likely to grow as other authoritarian societies such as North Korea, China, and Iran have a vested interest in undermining democratic nations.

Technological innovation in artificial intelligence has led to **deepfake videos**, digitally created with artificial intelligence, that can realistically portray a person or persons saying or doing things that they have never said or done. At the beginning of 2019, 7,964 deepfake videos were online; nine months later, there were over 15,000 (Toews 2020). Although technology is being developed to help distinguish between deepfakes and authentic videos, deepfake videos and the technology used to make them are growing at a faster rate than the ability to distinguish between the two (Galston 2020).

In addition to disinformation and deepfake videos, the Internet facilitates a process, used broadly here, called information cascading. Given the volume of information online, it's impossible to verify it all. Over 6,000 tweets are sent every second and 300 million new photographs are posted on Facebook every day (Nemr and Gangware 2019). Nonetheless, information is often shared with others who, in turn, do the same, and with every like, user comment, and thumbs up, it moves from YouTube to Twitter to Facebook to TikTok, gaining credibility along the way. A video of the Speaker of the House, slowed down to make it appear that she's inebriated, is viewed 2 million times (Wulfsohn 2019); conspiracy theorist Alex Jones, with an audience of 1.5 million, claims that the Sandy Hook killings never took place (Massey and Robinson 2019); and COVID-19 disinformation, much of it originating with the former President, continues to cascade on social media "spreading faster than the virus itself" (Ebrahimji 2020, p. 1).

What do you think?

Following the 2021 attack on the U.S. Capitol, several social media sites banned the term "stop the steal" as well as terminating the former president's media accounts. As of February, 2021, however, a new report found that pages tied to extremist groups (e.g., QAnon) remained on Facebook (Ortutay 2021). Section 230 of the Communications Decency Act provides technology companies with immunity from lawsuits that result from what people post on their sites. Do you think this provision should be repealed, and if so, should technology companies be held responsible for the attack on the Capitol?

President Trump, through Twitter, perpetuated many conspiracy theories having tweeted about them over 1,700 times to his more than 66 million followers. After researching the president's tweets, Shear et al. (2019) observe: "Twitter is the broadcast network for Mr. Trump's parallel political reality—the "alternative facts" he has used to spread conspiracy theories, fake information and extremist content" (p. 3). His first week in office, he falsely claimed that millions of people voted illegally in the 2016 election, contending that is why he lost the popular vote. After the 2020 election, the former president tweeted, in all capital letters, "I won this election—by a lot" and claimed, among other things, that the election was rigged by the FBI and the U.S. Department of Justice (McEvoy 2020).

QAnon An international conspiracy group that has been labeled a domestic terrorist organization by the FBI.

In 2020, the former president also embraced **QAnon**, an international meta-conspiracy theory, an online cult that has been called a domestic terrorism threat by the FBI (LaFrance 2020). Initially an online fringe group, QAnon has gone mainstream with, for example, the Texas Republican Party adopting the slogan, "We Are the Storm," instantly recognized by "QAnon adherents, signaling what they claim is a coming conflagration between [former] President Trump and what they allege, falsely, is a cabal of Satan-worshiping pedophile Democrats who seek to dominate America and the world" (Rosenberg and Haberman 2020, p. 1).

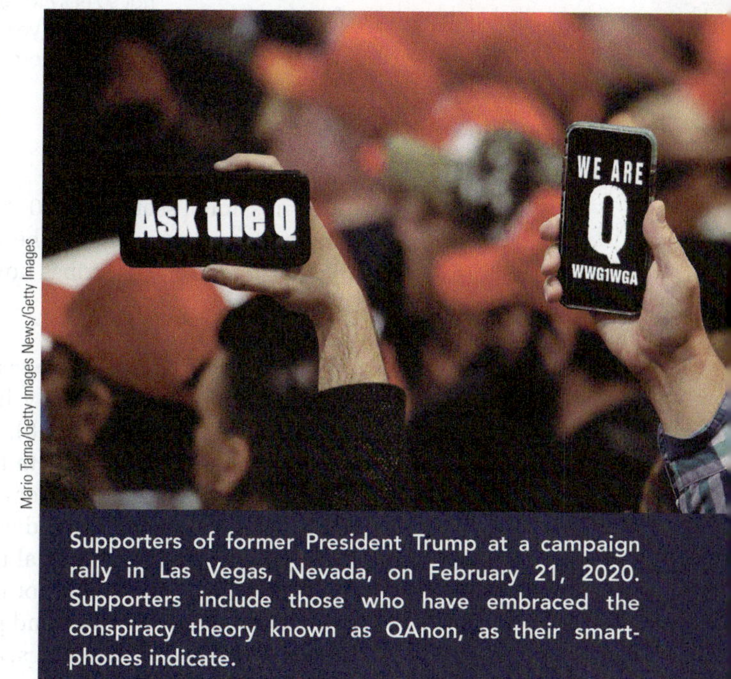

Mario Tama/Getty Images News/Getty Images

Supporters of former President Trump at a campaign rally in Las Vegas, Nevada, on February 21, 2020. Supporters include those who have embraced the conspiracy theory known as QAnon, as their smartphones indicate.

Although few in the Republican or Democratic party support the conspiracy, the former president retweeted posts from QAnon followers over 200 times, and refused to disavow them during a pre-election town hall event. Further, 24 QAnon supporters, 22 of whom were Republicans, won primaries to run for Congress and two won

Down the Rabbit Hole

QAnon is an international conspiracy theory that falsely claims that Donald Trump is leading a holy war against a deep-state cabal of Satan worshiping child molesters composed of Democratic politicians and Hollywood elites. Although some observers predicted President Trump's defeat would signal the end of QAnon, concerns over the integrity of the election promulgated by the former administration have been interpreted as evidence of the validity of QAnon (Tiffany 2020). Following are excerpts from family members and friends of QAnon members who have turned to the Internet for social support, posting their stories online.

I think this movement is very attractive to people who have basically been told their whole life that they will succeed as long as they work hard enough. But now they're finding out it isn't true and instead of blaming the real issues, they've come up with all this Satanic pedophile nonsense that puffs them up into these heroic crusaders.

~

My mom called me "pure evil" said I was a demon, that she and my dad had "failed" as parents, that I am not a Christian (I actually am ...) and that all Democrats were killing babies to drink their blood. I just don't even know what to think honestly. As a new mom, I couldn't imagine ever saying such vile things to my child. I feel like going non-contact is honestly the best thing for my well-being.

~

So sorry to hear what you've been through. I'm in Australia and it's even found its way down here. ... The thing that I always consider first is that you can't really confront these people. ... I've got a Qanon friend that I've tried to rescue but *anything* you say is instantly discredited either because your source is corrupted (because all of the fact-checking websites and news agencies are somehow beholden to the cabal or Nazis or something), or you've been 'brainwashed' by the MSM [main stream media].

~

My husband has gone so far down the qanon rabbit hole I'm almost positive there's no way to get him back. ... [H]e's always been into conspiracy theories ... but ... since covid it's now unliveable, he's convinced covid is fake (I'm a frontline health worker so that's been difficult☺) and with all his furlough time on his hands he's ended up deep in qanon. He's gone from the kindest, chilled man to constant anger and major depression. I'm at a loss, not so easy to walk away either, been together 20 years, married 14, 2 kids under 10, mortgage. It's at a point now where I can't see a way back, it makes me so sad but there's no way we can continue to live like this, kids come first.

~

One of my best friends for over a decade sent me some stupid right-wing meme and my dumbass of course fell for the bait. ... Before I knew it she was sending messages in all caps about joe Biden drinking the blood of children. I tried to explain how fundamentally absurd that was but all it managed to do was infuriate her. ... I tried pointing out how she was exhibiting all the hallmarks of someone indoctrinated into a cult but of it fell on deaf ears. I was accused of condescension and being tone deaf to conflicting ideas. ... I'm pretty sure she won't be speaking to me for a very long time. If anyone here has any advice on how they successfully navigated this mine field with a Qanon loved one I would really appreciate the input.

~

I'm glad I've found this. I've got a few friends ... who seem to have been dragged into this whole conspiracy theory world to the point of losing touch with reality. I didn't know that there was a name for it (ok maybe I'm really dumb) but the QAnon tag really encapsulates all

Congressional seats. In 2020, a resolution condemning QAnon and criminal activity by its supporters was passed in the U.S. House of Representatives (Davis 2020). This chapter's *The Human Side* describes how belief in the conspiracy theory is destroying families and friendships.

Politics and Election Tampering. In 2014, Cambridge Analytica, a British political consulting firm, surreptitiously harvested personal data from as many as 87 million Facebook accounts (Meyer 2018). The firm, linked to a former advisor to President Trump, Steve Bannon, used the data to construct "psycho-photographic" voter profiles in order to target social media users with personalized political advertisements (Rosenberg, Confessore, and Caldwallader 2018).

The accusations of digital intervention in the 2016 election (see Chapter 4) did not end with Cambridge Analytica or the hacking of the Democratic National Committee. Legions of autonomous (i.e., bots) and paid social media users (i.e., trolls) "used targeted advertisements, falsified news articles, and social media amplification tools to polarize Americans" (Kreps 2020, p. 2). For example, over 3,000 online political advertisements appeared on the stuff I see in this friend. What concerns me is that he's a smart guy, he researches in detail but seemingly only in places that support his theories. Its toxic. How do you reason with someone who is completely unreasonable?

~

I feel like someone I am close with is turning into a qanon follower. Keeps telling me in January 2021 all banks and credit unions will be dissolved and a mandatory digital currency enforced. We lose the right to use cash and the government will have full access to every inch of our income and 401ks and more.

~

I have totally lost a brother to this virtual cult. Our email correspondence took a nasty turn at an early point, and has become totally poisonous. We are not speaking to each other anymore, and the anger—mostly on his part—is utterly venomous. … I'm 76 and he is 73. Our mom is 101. When she goes, he will be basically friendless in the world. … I think that the best you can do with a QAnon nut job is to leave them alone and hope that they will somehow crawl out of the rabbit hole on their own. I'm not holding out much hope. It's honestly a family tragedy.

~

Personally, I'm scared for your mental well-being, when you're watching videos that make you pissed at the world for 5+ hours a day, every day. Being pissed for five hours a day can't be good for you mentally. And don't tell mom I told you this, but idk how long she'll be able to withstand being called delusional or blue pilled. I ask you, for the sake of the relationships you have for your family and your own mental well-being, to chill out on the Q-Anon stuff a bit. … (I love my dad and he does a lot of good for my family.)

~

I only recently learned about QAnon and now I realize that this is where all of his ideas about Covid being a hoax, Bill Gates being the antichrist and Hillary Clinton trafficking children are coming from. We had a rough few months at the start of the pandemic because we were constantly arguing. He would send me these long articles filled with medical nonsense about Covid. … While we were quarantined, it was nonstop ranting about these bizarre ideas. I couldn't handle it anymore and told him we need to find separate places to live.

~

I have a few friends who are way too deep into it. The unwavering devotion to Trump, the Democratic cabal of child molesters—everything is a conspiracy: Trump getting Covid was an assassination attempt, Biden has dementia, the world will "literally end" if the Democrats win, Antifa are coming for our children, climate change is a hoax and China controls the weather with machines so we will rely on them for energy. In the end, I just feel bad for them—but recently the pro-civil war and "taking matters into our own hands" vibe is beginning to scare me. It's not even about two separate realities anymore, it's about physically attacking those who disagree with you.

~

I broke up with my QAnon boyfriend. I moved out and I'm not ever looking back! I just want to say to people who are currently in a relationship with someone in this cult. You can get out. There's hope! I felt helpless because I lived with him. I felt sorry for him because he was usually very nice and caring towards me but then he was saying messed up things and emotionally abusing me when I would bring up my hurt feelings. He would say the abuse was all in my head and I was putting up barriers. I wouldn't open my mind to Q so I was the one who was close minded and ignorant. I finally got my copy of the lease and left. I haven't stopped smiling since.

~

My aunt who was ultra QAnon shot herself earlier today, she left a note saying she was terrified the cabal was coming for her and her kids because of Trump's loss.

SOURCE: Reddit QAnon Casualties.

Facebook between June 2015 and May 2017 that were linked to 470 fake accounts created by the Russian "troll farm," Internet Research Agency (IRA) (Shane and Goel 2017). By identifying what sites a user visited and correlating that information with online behaviors and demographic information, the IRA identified and reached groups most likely to react to socially charged content, i.e., racial, ethnic, or religious-related material (Kreps 2020).

Despite the former president's characterization of the Russian intervention into the 2016 presidential election as a "hoax," the Mueller Report concluded that there had been social media "information warfare" that favored then-candidate Trump (American Constitution Society 2020). Further, in 2020, the *bipartisan* Senate Intelligence Committee released a report that, among other things, concluded that the Russians continued to spread disinformation about their role in the 2016 election through the beginning of 2020 and that, given the prior Russian successes, Americans must remain vigilant to preserve the integrity of elections (Herb, Cohen, and Polantz 2020).

Social Media's Response. Countries, candidates, and campaigns have often interfered with the business of politics, both domestic and foreign. Malware and hacking,

@marwilliamson

Mark Zuckerberg, I don't know what universe you're living in but people are using Facebook to facilitate a fascist takeover of the United States. No, this is not just "everyone having a chance to have their voices heard." Your insistence on neutrality amounts to complicity.

-Marianne Williamson

disinformation campaigns, deepfake videos, artificial intelligence, and predictive algorithms make interference easier to do, harder to detect, and increasingly dangerous. In fact, the Senate Intelligence Committee concluded that "malicious actors will continue to weaponize information and develop increasingly sophisticated tools for personalizing, targeting, and scaling up the content" (Kreps 2020, p. 1).

That said, social media sites, some more than others, have responded to fake accounts, micro-targeted advertising, and conspiracy theories. In 2017, Twitter permanently deactivated 2,752 Russian trolls accounts associated with the IRA (Rocheleau 2017), conspiracy theorist Alex Jones was banned from Facebook (Schwartz 2019), and, in 2018, the media giant was fined $50 billion for its role in the Cambridge Analytica scandal (Wong 2019). Furthermore, Twitter banned 7,000 and gave only limited access to 150,000 QAnon-related accounts in 2020, and Facebook and YouTube moved to eliminate QAnon content from their platforms (Collins and Zadrozny 2020; Zadrozny and Collins 2020). In the same year, Facebook, Twitter, and YouTube moved to limit posts by former President Trump for containing misleading information about COVID-19 (Culliford 2020) and, in 2021, banned the former president for his alleged role in the U.S. Capitol riots. As Trump defenders fled to right-wing sites like Parler, Apple, Google, and Amazon withdrew their support for the "alt-tech" social network (Nicas and Alba 2021).

Despite Facebook's and other social media's efforts to prevent meddling in the 2020 elections, there is evidence that it occurred (Chappell 2020) (see Figure 14.9). The Director of National Intelligence released a statement just prior to the 2020 election that three nations, China, Russia, and Iran, were already engaging in overt and covert measures to interfere with the election. While Russia had a preference for the reelection of President Trump and China favored former Vice President Biden, Iran's goal was primarily to undermine U.S. democratic institutions (Press Release 2020).

Social science research has documented just how problematic such and other types of interference are. Researchers at MIT analyzed 126,000 unbroken Twitter chains (i.e., cascades) of news stories tweeted by 3 million users for a period of more than ten years. The results are troubling. False stories were 70 percent more likely to be retweeted than true stories and to reach more people quicker. Although false stories on all subjects traveled faster into Twittersphere than true stories, "false political news traveled deeper and more broadly, reached more people, and was more viral than any other category of false information" (Vosoughi, Roy, and Aral 2018, p. 49).

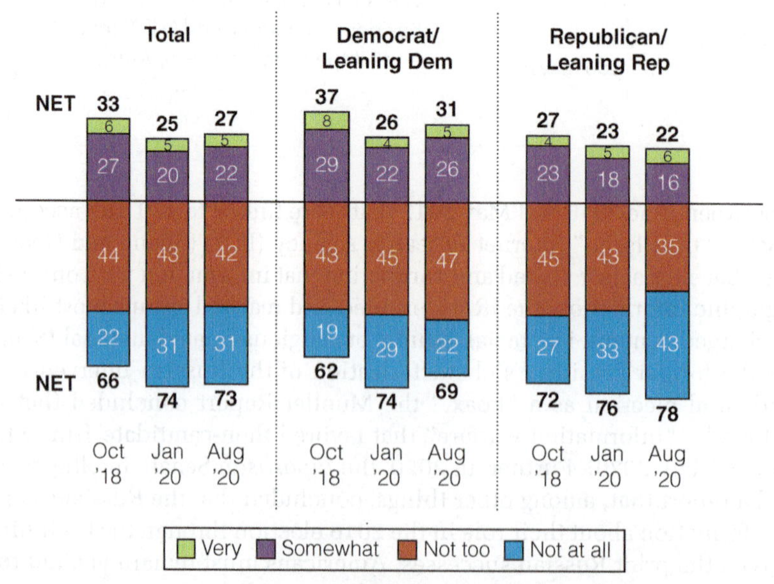

Figure 14.9 Percentage Who Say They Are__________ Confident in Technology Companies like Facebook, Twitter, and Google to Prevent Misuse of Their Platforms to Influence the Election, 2020
SOURCE: Green 2020.

The Challenge to Traditional Values and Beliefs

A statement from the Office of Science and Technology Policy, the chief technology office of the United States, notes that as a country, we do not have to sacrifice our values in the interests of emerging technologies and scientific innovations (Kratsios 2020). Yet science and technology often challenge traditionally held values and beliefs (Bugeja 2018). For example, one of the most dearly held values in the United States is *equality*. Yet artificial intelligence and automation "in the absence of mitigating policies ... are likely to exacerbate inequality and leave more Americans behind" (CFR 2018, p. 3). Figure 14.10 displays responses to a survey of U.S. adults asked about values and emerging science and technologies.

The increasing use of computers in every facet of social life and the resulting security breaches threaten the traditional values of *privacy* and *independence*. Predictive algorithms used to make sentencing decisions challenge our notions of *fairness* and *justice*. Disinformation campaigns and conspiracy theories contribute to the ongoing and public questioning of *facts*, *science*, and *honesty*. Cloning causes us to wonder about the traditional notions of *family* and *individuality*. And the American core values of *freedom*, *one person one vote*, and *democracy* are called into question as the legitimacy of elections is challenged as a result of foreign intervention. Even "*seeing is believing*" is threatened as deepfake videos digitally impersonate newsmakers, celebrities, and world leaders (O'Sullivan 2020). Toffler (1970) coined the term **future shock**

Genetic engineering

Changing a baby's genetic characteristics to reduce the risk of a serious illness that could occur over their lifetime is ...

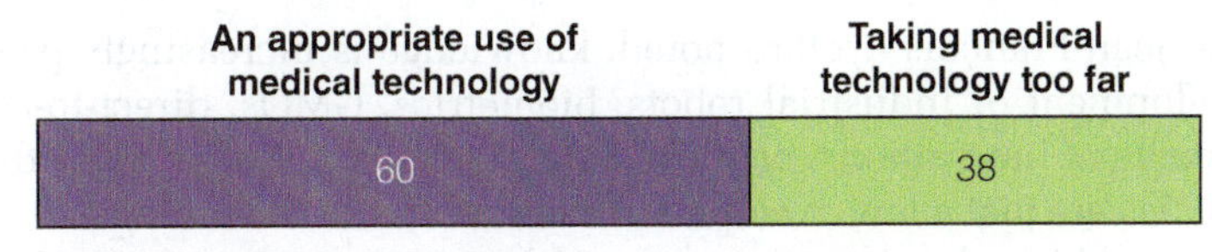

Genetic engineering of animals to grow organs/tissues for humans needing a transplant is ...

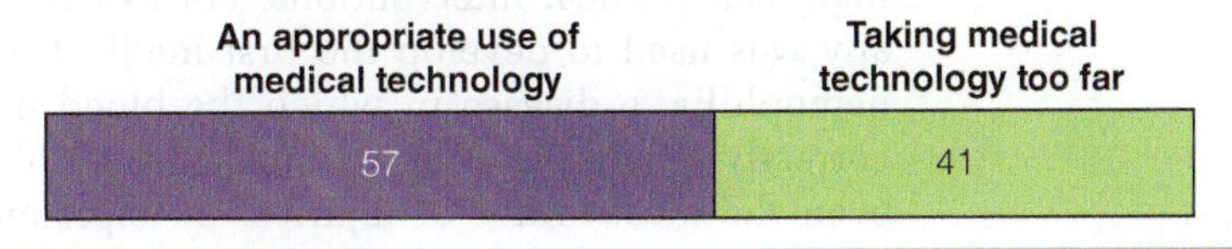

Human enhancement

They are ____ about the possibility of a brain chip implant for a much improved ability to concentrate and process information ...

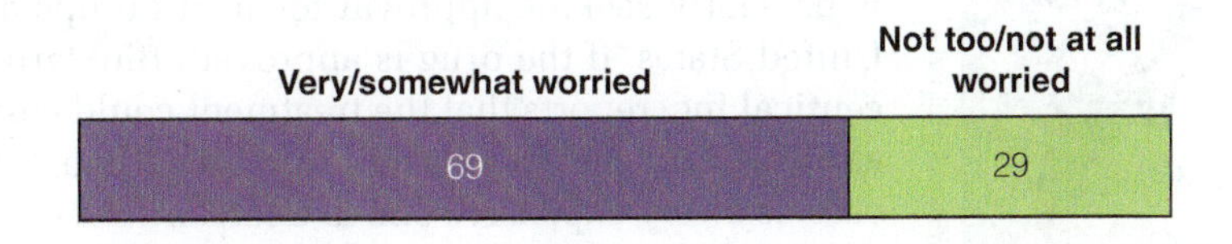

Automation

The automation of jobs through new technology in the workplace has ____ American workers ...

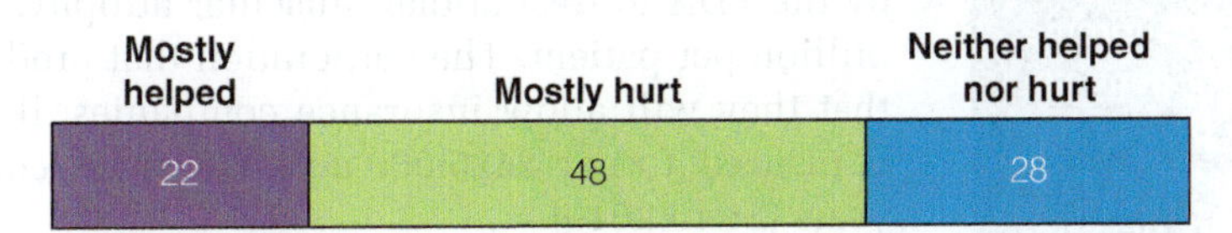

Figure 14.10 Percentage of U.S. Adults Who Say Each of the Following Concerning Emerging Science and Technology, 2019

SOURCE: Funk 2020.

future shock The state of confusion resulting from rapid scientific and technological changes that unravel our traditional values and beliefs.

to describe the confusion resulting from rapid scientific and technological changes that unravel our traditional values and beliefs.

What *do you* think?

There is no limit to the harm deepfakes can cause (Chesney and Citron 2019). Imagine being able to make anyone look and sound as though they have done anything someone might want them to—had an affair, embezzled corporate funds, abused a child, made a pornographic video, yelled racial epithets, or threatened to invade a foreign country or even overthrow a government. Do you think deep fake videos should be outlawed?

Strategies for Action: Controlling Science and Technology

As technology increases, so does the need for social responsibility. Nuclear power, artificial intelligence, genetic engineering, cloning, and computer surveillance all increase the need for social responsibility, creating both "new possibilities for social action as well as new problems that have to be dealt with" (Mesthene 1993, p. 85). In the following sections, we address various aspects of the public debate, including science, ethics and the law, the role of corporate America, and government policy.

Technology and Corporate America

As philosopher Jean-François Lyotard noted, knowledge is increasingly produced to be sold. The development of industrial robots, biometrics, GMOs, direct-to-consumer genetic testing, medical innovations, artificial intelligence, predictive algorithms, and autonomous vehicles are just a few examples of market-driven technologies.

Profit-motivated biotechnology creates a unique concern—fear that only the wealthy will have access to life-saving technologies such as genetic testing, cloned organs, and high-tech medical interventions. For example, gene therapy was used to develop the first medical treatment for hemophilia, a disease in which the blood does not clot correctly leading to internal and external bleeding with even the most minor of injuries. In experimental trials, participants, rather than having daily injections, received gene therapy which, as one recipient commented, was "life-changing" (Stein 2020).

AP Images/Tom Williams/CQ Roll Call

Patents on genes by technology companies led to protests across the United States. In 2013, the U.S. Supreme Court found that a corporation cannot own a gene sequence that, if tested, might reveal that a person has a serious disease. However, a 2019 congressional proposal is set to overturn the Supreme Court decision and "ease other restrictions on patenting software and biomedical inventions" (Servick 2019).

The company that developed the gene therapy protocol is presently seeking approval for it in Europe and in the United States. If the drug is approved, BioMarin Pharmaceutical Inc. reports that the treatment could cost as much as $3 million per patient. Hemophilia, although a rare disease, is a very expensive one to treat, and the executives of the company justify the cost arguing that since it is a one-time treatment, over the course of a hemophiliac's life, the gene therapy might save the individual millions of dollars (Stein 2020). Similarly, Zolgensma, a drug just approved by the FDA to treat spinal muscular atrophy, costs $2.13 million per patient. The corporation that produces it said that they will allow insurance companies, if the patient is insured, to pay $425,000 a year over the course of five years (Stein 2019).

The commercialization of technology causes several other concerns, including issues of quality control and the tendency for discoveries to remain closely guarded secrets rather than collaborative efforts. In 2020, Chinese and Iranian hackers targeted American universities and corporations to steal intellectual property in the race to find a vaccine for COVID-19 (Lubold and Volz 2020). Further, corporations contribute the largest share of funding for both experimental and applied R&D, 85 percent and 54 percent, respectively (NSF 2020). The dominance of such corporate involvement has made government control more difficult because researchers depend less and less on federal funding.

Finally, in 2020, the CEOs of the largest technology companies in the United States testified before the U.S. House of Representatives antitrust subcommittee. The subcommittee was investigating whether the technology companies were violating antitrust laws by engaging in such illegal practices as limiting competition by purchasing competitors. For example, Mark Zuckerberg's Facebook owns Oculus VR, WhatsApp, and Instagram. Jeff Bezos defended Amazon's near online market monopoly by stating that, compared to such retail giants as Walmart, Amazon was relatively small. He then seemingly contradicted himself by saying that "just like the world needs small companies, it also needs large ones [like Amazon]" (quoted in Holmes and Epstein 2020, p. 1).

> "Innovate in ways that stoke the economy. Because innovations in science and technology are the engines of 21st century economies."
>
> **—NEIL DEGRASSE TYSON, ASTROPHYSICIST, AUTHOR**

Net Neutrality. Concern over accessibility to broadband connectivity has led to a debate over net neutrality. **Net neutrality** advocates hold that Internet users should be able to visit any website and access any content without Internet service providers (ISPs) (e.g., cable or telephone companies) acting as gatekeepers by controlling, for example, the speed of downloads. Why would an ISP do that? Hypothetically, if Internet service provider company X signs an agreement with search engine Y, then it's in the best interest of Internet service provider X to slow down all other search engines' performances so that you will switch to search engine Y. Internet service providers argue that Internet users, be they individuals or corporations, who use more than their "fair share" of the Internet should pay more. Why should you pay the same monthly fee as your neighbor who nightly downloads full-length movie files? Others fear any government regulation of the Internet and/or prefer a strictly market model.

In 2015, the Federal Communications Commission (FCC) invoked its right to regulate net neutrality under Title II of the *Federal Communications Act,* which gives the agency authority to regulate utilities such as telephones and power companies. Former FCC chairman Tom Wheeler argued that the "Internet is simply too important to be left without rules or a referee in the field. … The Internet has replaced the function of the telephone and the post office" (Risen 2015, p. 1). The FCC's Open Internet Rules are premised on three principles: *No Blocking* (providers can't block access to otherwise legal content, services, and devices); *No Throttling* (providers can't impede the Internet traffic of otherwise legal content, services, and devices); and *No Paid Prioritization* (providers can't favor some Internet traffic over others in exchange for any other considerations, also known as "no fast lanes") (FCC 2015).

However, in 2018 the Republican-led FCC voted to repeal the Open Internet Rules and, with it, net neutrality. The ruling was appealed, and although a 2019 federal court decision favored the former Trump administration, it also held that the FCC could not block state or local net neutrality regulations. Seven states have already passed net neutrality legislation, and 20 others, the District of Columbia, and Puerto Rico, have introduced net neutrality laws into their 2020 legislative sessions (Morton 2020). At the heart of the net neutrality debate, which continues, are the questions how should Internet service providers be regulated, and by whom?

Science, Ethics, and the Law

Science and its resulting technologies alter the *culture* of society through the challenging of traditional values. Public debate and ethical controversies, however, have led to *structural* alterations in society as the legal system responds to calls for action. For example, most states have genetic privacy laws designed to protect the integrity of a patient's genetic data. However, the Electronic Privacy Information Center (EPIC), a public research organization in Washington, D.C., warns that most genetic privacy laws do not provide sufficient legal protection against misuse of genetic data (EPIC 2020). Further, the issue is

net neutrality A principle that Internet users should be able to visit any website and access any content without Internet service providers' interference.

made more complicated by home DNA-tests that disclose to the company the "biological building blocks of what makes you *you*" (Federal Trade Commission 2017, p. 1).

Are such regulations necessary? In a society characterized by rapid technological change and scientific innovation, many would say yes. Genetic engineering and cloning, for example, are two of the most hotly debated technologies in recent years. Although bioethicists and the public vehemently debate the potential costs of these techniques, from fear of harm to the resulting individual to the more general concern with "playing God," single-nucleotide polymorphism (SNP, pronounced "snip") profiling has already been implemented (*Economist* 2019). SNP allows parents to compare the DNA of embryos prior to in vitro fertilization and select the one that is genetically superior, however that term is defined.

Presently limited to selecting an embryo based on the likelihood of medical outcomes, in the not too distant future:

> For those willing to undergo IVF, and with the money to pay for it, it may also be possible to SNP-profile an embryo and thus foretell its future. As well as disease risk, height and intelligence, SNP-profiling might eventually be capable of predicting (albeit imperfectly, for environment also plays a role) thing as diverse as television-viewing habits, likelihood of being bullied at school and probability of getting divorced. (p. 2)

Should the choices that we make as a society depend on what we can do or what we should do? Whereas scientists and the agencies and corporations that fund them often determine what we *can* do, who should determine what we *should* do? Although such decisions are likely to have a strong legal component—that is, they must be consistent with the rule of law and the constitutional right of scientific inquiry—legality or the lack thereof often fails to answer the question, "What should be done?" **Heritable genome editing** occurs when DNA in sperm, eggs, or embryos that could be inherited from a parent to a child is changed, resulting in genetically modified children. In 2019, an international group of scientists called for a moratorium on the use of heritable genome editing (Lander et al. 2019).

Science, Technology, and Social Control

Science and technology raise many public policy issues. Policy decisions, for example, address concerns about the safety of automated vehicles, the privacy of e-mails, the legitimacy of elections, and the ethics of genetic engineering. In creating science and technology, have we created a monster that has begun to control us rather than the reverse? What controls, if any, should be placed on science and technology? And are such controls consistent with existing law? Consider stream-ripping using easily available software to illegally download music (the question of intellectual property rights and copyright infringement); laws limiting social media posts thought to be inaccurate (free speech issues); and machine learning from readily available online data posted by unsuspecting consumers (Fourth Amendment privacy issues).

In 2020, the Attorney General of the United States finalized an order instructing departments, including the Department of Homeland Security, to "collect DNA samples from individuals who are arrested, facing charges, or convicted, and from non–United States persons who are detained under the authority of the United States" (Department of Justice 2020). Scientists have expressed concern over the policy, arguing that it is discriminatory, violates privacy rights, and puts immigrants at risk of genetic discrimination (Wessel 2020).

Concern over the use of science and technology, as well as who controls it and who is controlled by it, is not unique to the United States. For several years, China has been collecting DNA samples from millions of boys and men ostensibly to be used in a DNA database for forensic investigations. However, unlike other countries, DNA samples are being taken from citizens who have not been detained for allegedly committing a crime or even suspected of one. The Chinese database in terms of collection or use is not governed by any laws, and human rights activists fear the data will be used in nefarious

heritable genome editing
When DNA in sperm, eggs, or embryos that could be inherited from a parent to a child is changed.

ways. The data could be used to locate those who are critical of the government or have violated China's previous one-child policy (Cyranoski 2020).

The use of predictive algorithms in child welfare cases is another case in point. Predictive algorithms are only as accurate as the data used to train them, and if there are underlying biases in the data, the prediction will reflect that. As Glaberson notes (2019), "[I]n the child welfare context, a long history of over-surveillance and over-policing of poor communities and communities of color means that those communities are disproportionately represented in any child welfare or criminal justice data set" (p. 345). Thus predictive analytics trained to assess the risk of, for example, child mistreatment, are more likely to systematically ignore White families while honing in on poor Black and/or Hispanic families.

What do you think?

Big data sets used in creating predictive algorithms "replicate and amplify the biases and discrimination inherently in society" (Valentine 2019, p. 383). The Risk Classification Algorithm (RCA), used by U.S. Immigration and Customs Enforcement (ICE), was recently "edited" to make it more punitive, undermining, to some observers, the "potential for a more humane detention system ... and the growing dismay over the mistreatment of immigrants" (Koulish and Calvo 2019, p. 13). Do you think predictive algorithms should be used in making immigrant detention decisions?

Science, Technology, and Government Policy

In 2019, President Trump signed an executive order reinstating the President's Council of Advisors on Science and Technology (PCAST), a council that had not met since January 2017 when President Obama held his administration's final meeting of PCAST (Science Policy News 2019). The charge was narrow, no reports were to be written, and when they met in February of 2020, COVID-19 was never discussed (Karlawish 2020).

PCAST is administered by the Office of Science and Technology Policy (OSTP) headed by Biden appointee Eric Lander, a trained mathematician turned molecular biologist (Mervis and Kaiser 2021). Dr. Lander was instrumental in the completion of the Human Genome Project. President Biden has charged Lander and his committee with making recommendations regarding five science and technology questions: How can the administration "(1) combat public health threats, (2) mitigate the impact of climate change, (3) keep the country a world leader in innovation, (4) use science to improve social equity, and (5) strengthen the U.S. research enterprise" (p. 1). In a further sign of President Biden's commitment to science-based policies and his acknowledgment of the importance of the social sciences, President Biden appointed Alondra Nelson, a sociologist, to a newly created position, director of science and society at OSTP. Dr. Nelson is an award-winning researcher and a faculty member at the Institute for Advanced Study at Princeton University.

Matt Wuerker/The Cartoonist Group

Congress or regulatory agencies are responsible for controlling technology, prohibiting some (e.g., assisted-suicide devices) and requiring others (e.g., seat belts). A good example is the 2020 Senate Bill 1558–*Artificial Intelligence Initiative Act*. The Act requires certain federal activities related to artificial intelligence (AI), including but not limited to the (1) establishment of an OSTP National AI

Advisory Committee, (2) establishment of an National AI Coordination Office that will provide technical and administrative support to that committee, (3) establishment of collaborative efforts by the National Institute of Science and Technology, (4) implementation of research and education programs on artificial intelligence and engineering by the NSF, and, finally, (5) establishment of grants to be awarded by the NSF to five Multidisciplinary Centers for Artificial Intelligence Research and Education (S. 1558).

The government also studies and makes recommendations on the use of science and technology through several boards and initiatives such as the National Science and Technology Council, the Office of Science and Technology Policy, and the U.S. National Nano-Technology Initiative. These agencies advise the president on matters of science and technology, including research and development, implementation, national policy, and coordination of different initiatives.

Finally, there are ongoing concerns about dissemination of disinformation. In 2019, the U.S. State Department produced a report entitled *Weapons of Mass Distraction: Foreign State Sponsored Disinformation in the Digital Age* (Nemr and Gangware 2019). In placing the role of disinformation in a social context, the authors state:

> The proliferation of social media platforms has democratized the dissemination and consumption of information, thereby eroding traditional media hierarchies and undercutting claims of authority. The environment, therefore, is ripe for exploitation by bad actors. Today, states and individuals can easily spread disinformation at lightning speed and with potentially serious impact. (p. 2)

Fighting disinformation campaigns, whether domestic or foreign, is difficult. First, there are *technology gaps.* Scientific and technological innovation outpaces the development of tools and techniques used to detect their malicious intent. Simply put, someone can spread disinformation faster than someone can fact-check it. There is also the *enormity of the task.* Even if the technology existed that could identify deepfake videos, and social media sites were required by law to assess their legitimacy, thousands of hours of videos are uploaded to social media platforms every hour and the number of platforms continues to grow. Further, as the report notes, "the private and encrypted nature of these apps prevent the platforms from publicly flagging content as false, widely disseminating corrections, or removing the objectionable content from message groups" (Nemr and Gangware 2019, p. 42).

Social scientists need to conduct research on the impact of disinformation campaigns on consumers' beliefs and behaviors (see this chapter's *World in Quarantine: Infodemic*). Studies must concentrate on whether disinformation that is embraced reflects preexisting beliefs, results in attitudinal changes, and/or behavioral shifts. Social psychological factors that make some people more vulnerable to disinformation than others such as the need to belong, the inability to process vast amounts of new information seeking rather cognitive closure, and selective bias as a result of existing polarization must also be researched (Nemr and Gangware 2019).

Understanding Science and Technology

What are we to understand about science and technology from this chapter? As structural functionalists argue, science and technology evolve as a social process and are a natural part of the evolution of society. As society's needs change, scientific discoveries and technological innovations emerge to meet these needs, thereby serving the functions of the whole. Consistent with conflict theory, however, science and technology also meet the needs of select groups and are characterized by political components. As early as 1993, former MIT professor Langdon Winner noted that the structure of science and technology conveys political messages, including "power is centralized," "there are barriers between social classes," "the world is hierarchically structured," and "good things are distributed unequally" (p. 288).

The scientific discoveries and technological innovations that society embraces are socially determined, and thus so too are the social problems they create. The democratization of information once thought to be the result of the Internet "has not led ... to greater understanding and growing peace, but instead seems to be fostering social divisions,

THE WORLD in quarantine

The Other Virus That Kills

It is estimated that 500 million people, one-third of the world's population, were infected by the H1N1 virus in the 1918 global Spanish Flu pandemic (Centers for Disease Control and Prevention [CDC] 2018). In the United States, it is believed to have started in the military, as World War I was under way although, unlike COVID-19, death rates were highest among the young and old, those under 5 years old and over 65 years old, along with 20- to 40-year-olds who were otherwise in good health. In the end, 50 million people died worldwide, 675,000 in the United States.

Like the COVID-19 pandemic, the H1N1 virus was not well understood; there was no vaccine and no antibiotics, and secondary infections were often the cause of death. And, as in 2020, the virus was controlled by "isolation [social distancing], quarantine [shelter-in-place], good personal hygiene [wash your hands], use of disinfectants [clean everything], and limitations of public gatherings [no indoor gatherings over ten people]" (CDC 2020, p. 1). However, unlike the more recent pandemic, patients had no access to essential medical technology such as diagnostic tests, ventilators, and respirators.

Nor did they have the kind of technology that allowed social interaction between distant friends and relatives. Whether it's hardware (e.g., smartphones, computers), software (e.g., Zoom), or the interactive tools of the Internet (e.g., social media), there is little doubt that today, technology makes living through a pandemic easier than in 1918. For example, during the COVID-19 pandemic, 32 percent of U.S. adults attended social gatherings online, 20 percent livestreamed a concert or play, and 18 percent participated in an online fitness class or some other kind of home workout (Spinelli 2020).

That said, misinformation and disinformation (MDI) about the origins, causes, and consequences of the 1918 H1N1 pandemic never went viral. Although it is impossible to know the human costs of COVID-19-related MDI, there is evidence of at least some MDI-related deaths (cf. Magagnoli et al. 2020). The use of hydroxychloroquine in patients with COVID-19, as suggested by former President Trump, in combination with routine care resulted in twice the number of deaths than standard care alone (Marchione 2020). Further, his remarks about injecting cleaning solutions took place on April 23, 2020, and between that date and the end of the month, eight days later, the American Association of Poison Control Center's reports that incidences of accidental poisonings from household disinfectants increased 121 percent compared to the same time period in 2019 (Kluger 2020).

Just a few weeks into what was thought to be the beginning of the pandemic, nearly half of Americans reported encountering false COVID-19 news (Jurkowitz and Mitchell 2020). The most common category of what survey respondents believed was inaccurate information concerned the severity of the virus, either exaggerated or underestimated. Despite the majority of Americans being skeptical of virus information, nearly 30 percent of the respondents reported believing that COVID-19 had been created in a laboratory rather than occurring naturally (Schaeffer 2020). Within a few months, the number of U.S. adults who believed that the risk of COVID-19 was exaggerated had grown from nearly three in ten to four in ten Americans (Mitchell et al. 2020).

Although Republicans and Democrats were equally likely to have heard the rumor that COVID-19 was intentionally created, a higher percentage of Republicans than Democrats believed it. Moreover, Americans who got their COVID-19 news from social media were more likely to have watched the online video *Plandemic* and to believe that the information in it was true (Mitchell et al. 2020). The 26-minute video claims that "a shadowy cabal of elites was using the virus and a potential vaccine to profit and gain power" (Frenkel, Decker, and Alba 2020, p. 1). Within a week of it being posted online, over 8 million social media users had viewed it on YouTube, Instagram, Facebook, and Twitter.

Pulido et al. (2020) analyzed 1,000 tweets from February 6, 2020 and February 7, 2020 that contained the key word *coronavirus* and had the highest number of retweets. The researchers concluded that, although false information was more likely to be tweeted than science-based evidence or fact-checking tweets, false information was less likely to be retweeted. Similarly, in an effort to identify and understand MDI in order to combat it, Ahmed et al. (2020) analyzed the content of a sample of tweets from a seven-day period in which the hashtag #5G coronavirus was trending in the United Kingdom. The conspiracy theory that linked 5G network capabilities with the spread of COVID-19 led to such fear of the new technology that over 20 5G network towers were set on fire. Analysis indicated that, among other things, there was "a dedicated individual Twitter account set up to spread the conspiracy theory [that] formed a cluster in the network with 408 other Twitter users. This account ... had managed to send a total of 303 tweets during this specific time period before it was closed down by Twitter" (p. 6).

The authors also note that there was no "influencer" or authority figure who could have combated the spread of the false information. Misinformation and disinformation, much like the virus itself, spread from person to person in social networks like "digital wildfires" (World Economic Forum [WEF] 2018) or, as the World Health Organization has called it, an *infodemic* (WHO 2020). The World Health Organization has been working closely with more than 50 social media and technology companies including TikTok, Google, WhatsApp, YouTube, and Facebook to stop the spread of health MDI (see Chapter 2). The goal is to ensure that "science-based health messages from the organization or other official sources appear first when people search for information related to COVID-19" (p. 1). Even the matching app Tinder has information from WHO on its webpage reminding users of the importance of social distancing, even on a date.

distrust, conspiracy theories and post factual politics" (Bridle 2018, p. 1). Artificial intelligence contributes to the spread of false information by enabling its mass production, and citizens are victimized by predictive algorithms assumed to be computationally neutral. Technology must be regulated, but given, for example, the increased use of predictive algorithms in the very sector charged with regulating them, how do governments, at every level, institute control *over* them if they are governed *by* them (Kuziemski and Misuraca 2020)? How do we gain control over technology that is both intelligent and unpredictable, technology that even its creators may not understand (Bridle 2018)?

One possible method is *public education* (Kreps 2020). For example, Google runs a digital safety program that teaches the basics of artificial intelligence, how to identify false information, and when to question the authenticity of sources and websites. (See Figure 14.11.) The digital literacy campaign teaches very simple techniques of distinguishing between likely true and likely false information by looking for grammatical and factual errors. If an article cites the source as *The New York City Times*, it is likely not credible—there is *The New York Times*, but there is no *The New York City Times*—the kind of error a foreign government might make.

In addition to public education, whether privately or publicly sponsored, *technology* itself can help mediate the problems created by technological innovations. Scientists are already creating methods to identify deepfake videos, increase the safety of industrial robots and autonomous vehicles, protect big data on the Internet, enhance e-mail security, and mitigate the harmful effects of nanoparticles. Technology can also be used to address foreign intervention campaigns intended to deceive. An AI computer model named Grover, built by the Allen Institute for Artificial Intelligence, achieved 92 percent accuracy when distinguishing between human-written and machine-written news articles (Allen Institute 2019).

> "Humankind has the science and technology to destroy itself or to provide prosperity to all. But while science offers us these opportunities, science will not make that choice for us. Only the moral power of a world acting as a community can."
>
> **–MARGARET BECKETT, MATHEMATICS AND STATISTICS, BOSTON UNIVERSITY**

Some would argue that the solution to the problems created by technology cannot be solved solely by creating more technology, which inevitably will create more problems. Nation-states must institute regulatory guidelines. Such guidelines should not limit public and private sector innovation and must be transparent. All stakeholders should have a "place at the table" in order to maximize public trust and confidence (see Figure 14.10) (Vought 2019). Further, regulations must take a "risk-based approach ... to determine which risks are acceptable and which risks present the possibility of unacceptable harm, or harm that has expected costs greater than expected benefits" (p. 4). Finally, regulatory guidelines need to be flexible, reflecting the rapidly changing technological landscape, and fair, given the potential discriminatory nature of some science and technology.

The former administration has already addressed some of the problems associated with technologies. The 2019 *National Defense Authorization Act* prioritizes measures that counter false information, and the *Honest Ads Act*, if passed, would require that all political advertising, whether in traditional or digital media, be labeled as such. Although it has not passed either House in Congress, both Twitter and Facebook have announced that they are already implementing many of the safeguards the *Honest Ads Act* requires (Kreps 2020).

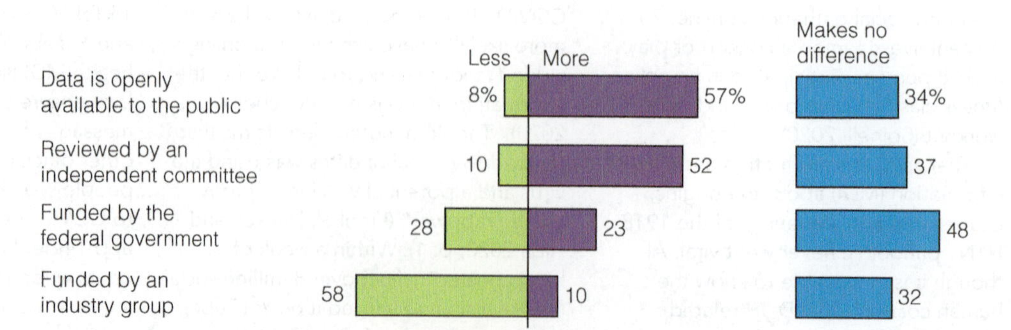

Figure 14.11 Percent of U.S. Adults Who Say When They Hear Each of the Following, They Trust Scientific Research Findings ________, 2019

SOURCE: Funk et al. 2020.

Given the global impact of technology, many multinational organizations, such as the Organization for Economic Co-operation Development (OECD), have established international AI principles, principles that can be generalized to all technologies (OECD 2020).

- Technology should benefit people and the planet.
- Technology should respect the rule of law, human rights, democratic values, and diversity.
- Technology should be transparent so that people understand the outcomes and challenges.
- Technology should be secure and safe and the potential risk continually assessed and managed.
- And the individuals or organizations that create technology should be held accountable for them. (p. 1)

These guidelines were established by a group of experts representing 20 governments, business and labor leaders, and professionals from academic and scientific communities. While they may not be legally binding, they serve as a benchmark for future governmental policies.

What do you think?

In 2017, President Trump signed legislation that removed prohibitions that protected consumers from Internet service providers (ISP) (e.g., Verizon) "collecting, storing, sharing and selling certain types" of information, including a person's "browsing history, usage history, and location details" (Fung 2017, p. 1). The rationale was that ISPs should be able to compete in the online advertising market with, for example, Google and Facebook, which do not need consent to collect user data. Do you think President Biden should reinstate the protections?

As with social problems in general, it is easy to feel despair when thinking of the challenge's science and technology create either directly (e.g., industrial robots lead to a loss of manufacturing jobs) or indirectly (e.g., sequencing of the genome facilitated genetic engineering, which led to the ability to create genetically modified organisms and, with it, genetically modified foods that may contribute to health concerns). In either case, there are negative consequences. But we should not forget that science and technology have also made our lives better, combating diseases, generating cleaner energy, increasing food production, facilitating global travel and communication, and expanding entertainment venues. As we move forward into the 21st century, one of the great challenges of civilization will be how to reorder society in a way that exploits the technological bonanza while preventing it from running roughshod over the checks and balances so delicately constructed in the simpler pre-digital years.

Chapter Review

- **What are the three types of technology?**
The three types of technology, escalating in sophistication, are mechanization, automation, and cybernation. Mechanization is the use of tools to accomplish tasks previously done by hand. Automation involves the use of self-operating machines, and cybernation is the use of machines to control machines.

- **What are some of the Internet global trends?**
In 2020, the Internet had over 4.8 billion users with 62 percent of the world's population online. Of all Internet users, the highest proportion of users come from Asia (50.3 percent), followed by Europe (15.9 percent), Africa (11.5 percent), Latin America and the Caribbean (10.1 percent), North America (7.6 percent), the Middle East (3.9 percent), and Oceania/Australia (0.6 percent) (Internet World Statistics 2020).

- **Why may the United States be "losing its edge" in scientific and technological innovations?**
The decline of U.S. supremacy in science and technology is likely to be the result of five interacting social forces. First, the federal government has been scaling back its investment in research and development. Second, corporations often focus on short-term earnings rather than new innovations. Third, developing countries, such as China

and India, are expanding their scientific knowledge and technological capabilities at a faster rate than the United States. Fourth, science and math education in U.S. schools lags behind many countries in terms of both quality and quantity. Finally, as documented in the book *Unscientific America* (Mooney and Kirshenbaum 2009), there is a disconnect between American society and the principles of science.

- **Why is the former administration called the anti-science administration?**
According to a survey conducted of agency scientists in 2020, two-thirds reported that "influence of political appointees" and "absence of leadership with needed scientific expertise" were barriers to evidence-based research (p. 9). Further, when the 2020 results were compared to data collected in 2010 and 2015 for the three agencies for which comparisons were possible, there was a greater level of inappropriate political influence under the Trump administration than under previous administrations.

- **According to Kuhn, what is the scientific process?**
Kuhn (1973) describes the process of scientific discovery as occurring in three steps. First are assumptions about a particular phenomenon. Next, because unanswered questions always remain about a given topic, science works to start filling in the gaps. Then, when new information suggests that the initial assumptions were incorrect, a new set of assumptions or framework emerges to replace the old one. It then becomes the dominant belief or paradigm until it is questioned and the process repeated.

- **How has robotics changed the workplace?**
Industrial robots have led to increased efficiency in the manufacturing process, and software robotics has led to attended and unattended automation that can perform routine and complex computer tasks, respectively, in the office and other environments.

- **How are algorithms used in the decision-making process?**
Algorithms are mathematical instructions that tell a computer how to make decisions, recommend products, or assess risks. Algorithms are often used by social media to tailor advertising and news to the online behaviors of the user. Algorithms are not always neutral and may, albeit unintentionally, lead to automated discrimination as a function of biased theoretical and mathematical models.

- **What is the Human Genome Project?**
The U.S. Human Genome Project was an effort to decode human DNA. The 13-year-old project is now complete, allowing scientists to "transform medicine" through early diagnosis and treatment as well as possibly preventing disease through gene therapy. Gene replacement therapy entails identifying a defective or missing gene and then replacing it with a healthy duplicate that is transplanted to the affected area.

- **What are the different types of cloning?**
There are three different types of cloning. Genetic cloning, which is widely accepted and routinely used in laboratory work, copies genes or segments of DNA. Reproductive cloning copies the DNA of whole animals to produce an exact genetic copy. Therapeutic cloning is the most controversial of the three types of cloning because it uses stem cells from unused human embryos; however, it has the potential to treat a wide range of illnesses and diseases.

- **What is the digital divide?**
The digital divide is the tendency for technology to be most accessible to the wealthiest and most educated. For example, some fear that there will be "genetic stratification," whereby the benefits of genetic testing, gene replacement therapy, and other genetic enhancements will be available to only the richest segments of society.

- **What are some of the problems associated with malicious use of the Internet?**
The Internet may be used for malicious purposes, including but not limited, to crime in the dark web, malware and hacking, electronic aggression and cyberbullying, disinformation, deepfake videos, conspiracy theories, and election tampering.

- **How does technology challenge traditional values and beliefs?**
The increasing use of computers in every facet of social life and the resulting security breaches threaten the traditional values of *privacy* and *independence*. Predictive algorithms used to make sentencing decisions challenge our notions of *fairness* and *justice*. Disinformation campaigns and conspiracy theories contribute to the ongoing and public questioning of *facts*, *science*, and *honesty*. Cloning causes us to wonder about the traditional notions of *family* and *individuality*. And the American core values of *freedom*, *one person/one vote*, and *democracy* are called into question as the legitimacy of elections is challenged by foreign intervention.

- **What is meant by the commercialization of technology?**
The commercialization of technology refers to profit-motivated technological innovations. For example, the development of GMOs, the commodification of women as egg donors, direct-to-consumer genetic testing, and the harvesting of regenerated organ tissues are all examples of market-driven technologies.

Test Yourself

1. Which of the following technologies is associated with industrialization?
   a. Mechanization
   b. Cybernation
   c. Hibernation
   d. Automation
2. The proposed 2021 budget, submitted by former President Trump in 2020, advocates reducing the budgets of almost all science-related agencies.
   a. True
   b. False
3. The U.S. government, as part of the technological revolution, spends more money on research and development than do educational institutions and corporations combined.
   a. True
   b. False
4. Which theory argues that technology is often used as a means of social control?
   a. Structural functionalism
   b. Social disorganization
   c. Conflict theory
   d. Symbolic interactionism
5. Autonomous vehicles use adversarial images to recognize 3-D objects.
   a. True
   b. False
6. The ability to manipulate the genes of an organism to alter the natural outcome is called
   a. gene therapy.
   b. gene splicing.
   c. genetic engineering.
   d. genetic testing.
7. Genetically modified foods have been documented as harmless to humans by the Food and Drug Administration.
   a. True
   b. False
8. An algorithm is a mathematical instruction.
   a. True
   b. False
9. Which of the following statements about technology is true?
   a. Industrial robots primarily displace white-collar workers.
   b. Autonomous vehicles have been approved for mass production and use.
   c. Internet penetration has leveled off across the world and is now fairly equal across all countries.
   d. Predictive algorithms can be used as a means of social control.
10. Methods to fight online disinformation campaigns include
    a. public education.
    b. technology.
    c. governmental regulatory guidelines.
    d. all of the above.

Answers: 1. D; 2. A; 3. B; 4. C; 5. B; 6. C; 7. B; 8. A; 9. D; 10. D.

Key Terms

algorithm 557
artificial intelligence (AI) 555
automated discrimination 559
automation 542
cultural lag 550
cyberbullying 574
cybernation 542
dark web 573
deep web 573
deepfake videos 574
disinformation 574
e-commerce 561
future shock 579
gene replacement therapy 564
genetic engineering 564
Genetic Information Nondiscrimination Act of 2008 568
genetic testing 563
genetically modified food 564
hacking 574
heritable genome editing 582
Internet 561
Internet of Things (IoT) 561
malware 574
mechanization 542
nanotechnology 566
net neutrality 581
participation gap 571
penetration rate 543
phubbing 572
QAnon 575
science 542
STEM 570
stem cells 566
technological dualism 542
technological fix 545
technology 542
technology-induced diseases 565
telepresence 556

Roberto Schmidt/AFP/Getty Images

> "Every gun that is made, every warship launched, every rocket fired signifies, in the final sense, a theft from those who hunger and are not fed, those who are cold and not clothed."
>
> **GENERAL DWIGHT D. EISENHOWER**
> Former U.S. president and military leader

15

Conflict, War, and Terrorism

Chapter Outline

Learning Objectives

After studying this chapter, you will be able to ...

1. Identify trends in global conflict over the past century.
2. Identify the causes and possible solutions for war from each sociological perspective.
3. Explain how war can be an outcome of additional social problems.
4. Identify additional social problems that occur as a consequence of war.
5. Argue for a policy change that would help create a more peaceful world.
6. Identify the pros and cons of various anti-terrorism strategies.

Human rights activist Nadia Murad speaks with Yazidi refugees at a camp in Idomeni, Greece. Along with activist Denis Mukwege of Rwanda, Murad was awarded the Nobel Peace Prize in 2018 for their work with survivors of wartime rape. In 2019, they worked together to launch the Global Survivors Fund to provide reparations to survivors of wartime rape and slavery. Human trafficking and rape have been historically used as weapons of war, and disproportionately impact women, children, and minority groups.

Joseph Galanakis/NurPhoto/Getty Images

NADIA MURAD WAS 19 YEARS OLD when she was sold into slavery. ISIS (the Islamic State in Iraq and Syria) militants attacked her village in northern Iraq in August 2014, killing her mother and six brothers, and kidnapping Nadia and her two sisters to be sold as sex slaves (Murad 2018). Nadia describes the chaos of the slave market as "like the scene of an explosion" where girls screamed, vomited, and threw themselves across their friends and sisters to try to protect them. "'Calm down!' the militants kept shouting at us. 'Be quiet!' But their orders only made us scream louder. If it was inevitable that a militant would take me, I wouldn't make it easy for him" (p. 1). Nadia's experience is part of a long history of rape and human trafficking used as weapons of war. ISIS had targeted Nadia's community because they were part of the Yazidi ethnic minority group. Capturing young women and selling them into sex slavery was one part of a larger strategy of genocide designed to eliminate Yazidis from Iraq (p. 1). In 2015, Nadia escaped and was smuggled into Germany, where she began to work as a human rights activist. Central to her activism is telling and retelling the story of her trauma. "It never gets easier to tell your story," she says. "Each time you speak it, you relive it. My story, told honestly and matter-of-factly, is the best weapon I have against terrorism" (p. 1).

War is one of the great paradoxes of human history. It both protects and annihilates. It creates and defends nations but may also destroy them. **War**, the most violent form of conflict, refers to organized armed violence aimed at a social group in pursuit of an objective. Wars have existed throughout human history and continue in the contemporary world. Whether war is just or unjust, defensive or offensive, it involves the most horrendous atrocities known to humankind. This is especially true in the 21st century, when nearly all wars are fought in populated areas rather than on remote battlefields, having deadly consequences for civilians. Thus, war is not only a social problem in and of itself but also contributes to a host of other social problems—death, disease, and disability, crime and immorality, psychological terror, loss of economic resources, and environmental devastation. In this chapter, we discuss each of these issues within the context of conflict, war, and terrorism, the most threatening of all social problems.

The Global Context: Conflict in a Changing World

As societies have evolved and changed throughout history, the nature of war has also changed. Before industrialization and the sophisticated technology that resulted, war occurred primarily between neighboring groups on a relatively small scale. In the modern world, war can be waged between nations that are separated by thousands of miles as well as between neighboring nations. Increasingly, war is a phenomenon internal to states, involving fighting between the government and rebel groups or among rival contenders for state power. Indeed, Figure 15.1 documents that wars between states—that

war Organized armed violence aimed at a social group in pursuit of an objective.

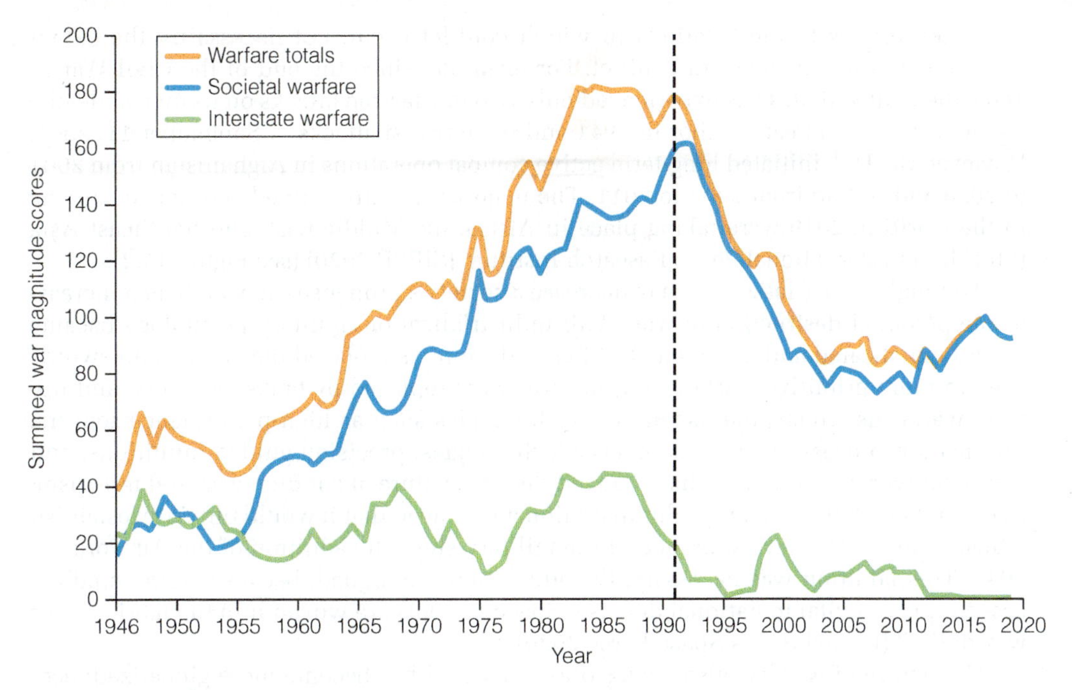

Figure 15.1 Global Trends in Armed Conflict, 1948–2019*
SOURCE: Center for Systemic Peace 2020.

is, interstate wars—recently made up the smallest percentage of armed conflicts. In the following sections, we examine how war has changed our social world and how our changing social world has affected the nature of war in the industrial and postindustrial information age.

War and Social Change

The very act that now threatens modern civilization—war—is largely responsible for creating the advanced civilization in which we live. Before large political states existed, people lived in small groups and villages. War broke the barriers of autonomy between local groups and permitted small villages to be incorporated into larger political units known as chiefdoms. Centuries of warfare between chiefdoms culminated in the development of the state. The **state** is "an apparatus of power, a set of institutions—the central government, the armed forces, the regulatory and police agencies—whose most important functions involve the use of force, the control of territory, and the maintenance of internal order" (Porter 1994, pp. 5–6). Social historian Charles Tilly famously said, "War makes states, and states make war" (1992).

> And once the state emerged, the gates were flung open to enormous cultural advances, advances undreamed of during—and impossible under—a regimen of small autonomous villages. ... Only in large political units ... was it possible for great advances to be made in the arts and sciences, in economics and technology, and indeed in every field of culture central to the great industrial civilizations of the world. (Carneiro 1994, pp. 14–15)

Industrialization and technology could not have developed in the small social groups that existed before military action consolidated them into larger states. Thus war contributed indirectly to the industrialization and technological sophistication that characterize the modern world. Industrialization, in turn, has had two major influences on war. Cohen (1986) calculated the number of wars fought per decade in industrial and preindustrial nations and concluded, "As societies become more industrialized, their proneness to warfare decreases" (p. 265). It is important to note, however, that this conclusion

state The organization of the central government and government agencies such as the military, police, and regulatory agencies.

primarily applies to the locations in which conflict occur, not necessarily the extent of a country's involvement in conflict. For example, since the end of the Civil War in 1865, the United States has experienced only two militarized attacks on its own soil—the Japanese attack on Pearl Harbor in 1941 and the terrorist attacks of September 11, 2001. However, the U.S. initiated long-term active combat operations in Afghanistan from 2001 to 2020 and in Iraq from 2003 to 2011. The majority of active armed conflicts occurring in the world in 2019 were taking place in Africa, the Middle East, and Southeast Asia (Stockholm International Peace Research Institute [SIPRI] 2020) (see Figure 15.2).

Although industrialization may decrease a society's propensity to war, it also increases the potential destruction of war. With industrialization, military technology became more sophisticated and more lethal. Rifles and cannons replaced the arrows and swords used in more primitive warfare and, in turn, were replaced by tanks, bombers, and nuclear warheads. Today, the use of new technologies such as high-performance sensors, information processors, directed energy technologies, precision-guided munitions, and computer worms and viruses has changed the very nature of conflict, war, and terrorism. In December 2019, the Trump administration announced that it would officially establish a Space Force—the newest branch of the military since the addition of the Air Force in 1947. This addition was necessary, the administration argued, because "unfettered access to space is vital to national defense. Space systems are woven into the fabric of our way of life" (United States Space Force 2020).

NATO North Atlantic Treaty Organization, founded in 1949, is a military alliance among 30 North American and European countries with the purpose of promoting the freedom and security of its member nations.

The nature of war has also changed as the world has become more globalized; economic stability and national security across nations are more intertwined than ever before. The extent to which the United States should take a leading role in providing for global security has been a major topic of public debate for at least a century (see this chapter's *Self and Society* feature). States have long used military alliances to promote their own national security interests, but these alliances come with trade-offs. One of the most important alliances that has shaped the United States in recent history is the North American Treaty Organization (NATO). **NATO** was established in 1949 as a military alliance between the United States, Canada, and several Western European nations

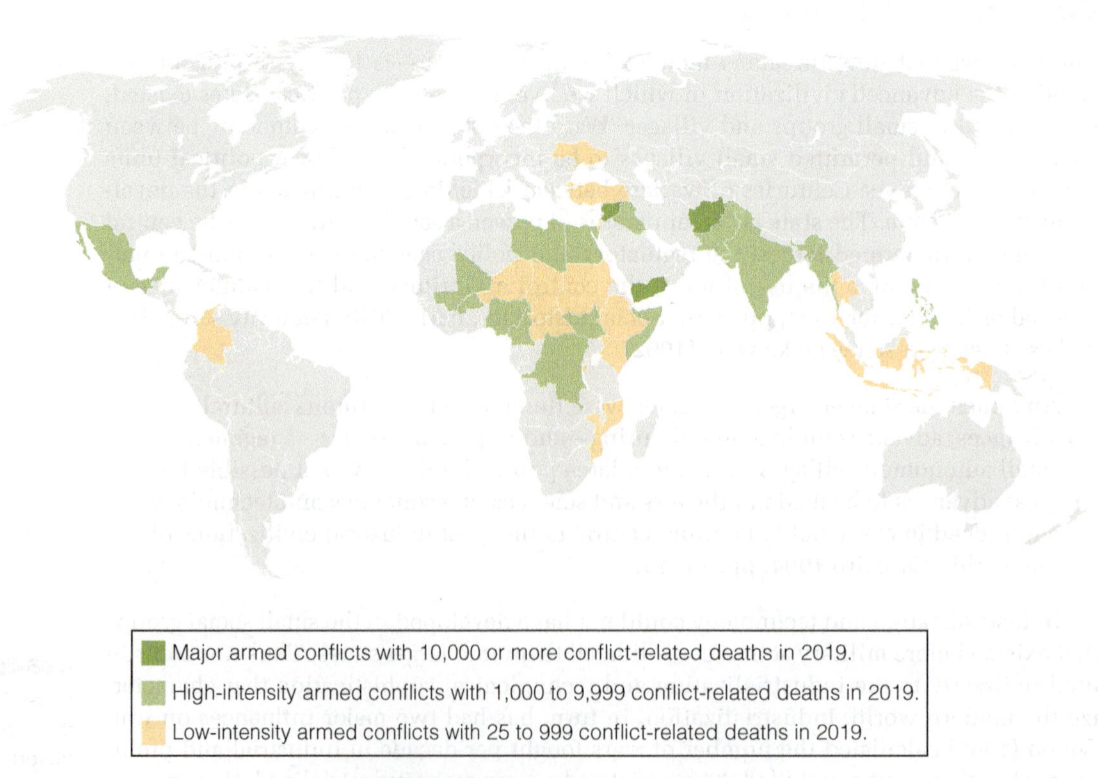

Figure 15.2 Countries with Active Armed Conflicts in 2019
SOURCE: Stockholm Institute for Peace Initiatives 2020.

to promote their mutual security against the growing aggression of the Soviet Union. This treaty marked the official beginning of the **Cold War**, the state of political tension, economic competition, and military rivalry that existed between the United States and the former Soviet Union for nearly 50 years. Since the end of the Cold War, NATO has grown to 30 member nations and continues to operate as a military and economic alliance with the purpose of promoting the freedom and security of its member nations.

Jesco Denzel /Bundesregierung/Handout/Getty Images News/Getty Images

Military alliances can help reduce the likelihood of conflict but are at times sources of international tension themselves. President Trump and German Chancellor Angela Merkel clashed at the NATO summit in July 2018. Trump publically criticized Merkel, claiming that Germany was not contributing enough to NATO's common defense fund. A few months later, Merkel hinted that America's role in the alliance was waning when she called for the formation of a European Army, saying in a speech that "the days where we can unconditionally rely on others are gone" (Glasser 2018).

NATO membership has many benefits. Not only do alliances create institutional pathways for diplomatic conflict resolution within its membership, thereby reducing the risk of war, member nations also collaborate to share military intelligence, equipment resources, and operational support. This multinational collaboration has become especially important in the current era of global terrorism and threats to cyber security. However, there are also economic and political costs to being part of an alliance. Member nations may be obligated to contribute manpower to the military operations of other member nations, which can result in political unpopularity at home. For example, U.K. Prime Minister Tony Blair's support of U.S. operations in Iraq strongly contributed to his plummeting popularity and withdrawal from political life (Naughtie 2016). NATO membership also comes with economic obligations. Members are expected to spend at least 2 percent of their GDP on national defense and are required to contribute to the NATO common fund based on a cost-sharing formula that is periodically renegotiated at the annual NATO summit (NATO 2020). Shortly after taking office in 2016, former President Trump began campaigning for a renegotiation of the cost-sharing formula, arguing that the United States was unfairly burdened with supporting the costs of the alliance (Béraud-Sudreau and Childs 2018).

Former President Trump campaigned on a platform of "America First," suggesting that his administration would renegotiate or end treaties and alliances that did not put U.S. interests above those of its allies. Critics of the America First policy—including President Joe Biden—argue that international cooperation is essential to solving problems—such as a global pandemic—and reducing the likelihood of armed conflict (Baer 2020). In an increasingly globalized world, do you think it is better for a country to look out for its own interests first or to seek cooperation with international allies?

What *do you* think?

The Economics of Military Spending

The increasing sophistication of military technology has commanded a large share of resources; world military expenditures in 2019 totaled $1.92 trillion or $249 per global citizen (SIPRI 2020). This is the highest level of military expenditure since 1988, and an increase after a brief period of reduced spending from 2011 to 2014. The United States has the highest military expenditures of any country in the world, a trend that began as a result of the competition between the United States and the Soviet Union during the Cold War. In 2019, the United States spent $732 billion on its military, which accounted for 38 percent of the world's military spending and is more than the combined military expenditures of the next nine highest-spending nations: China ($261 billion), India ($71.1 billion),

Cold War The state of military tension and political rivalry that existed between the United States and the former Soviet Union from the 1950s through the late 1980s.

The United States and Global Security

What role do you think the United States should have in Global Security?

Place a checkmark next to the response that best represents your answer for each question about the role of the United States in global security. When complete, compare your responses to those from representative samples of U.S. adults surveyed in 2019 (N = 1,503) (Pew Research Center 2019).

- Put a checkmark next to the statement you agree with more:
  - A: In foreign policy, the United States should take into account the interests of its allies even if it means making compromises with them. ___
  - B: In foreign policy, the United States should follow its *own* national interests even when its allies disagree. ___
  - Don't Know/No Answer ___
- Put a checkmark next to the statement you agree with more:
  - A: It's best for the future of our country to be active in world affairs. ___
  - B: We should pay less attention to problems overseas and concentrate on problems here at home. ___
  - Don't Know/No Answer ___
- All in all, would you say being a member of NATO is:
  - Good for the United States? ___
  - Bad for the United States? ___
  - Don't Know/No Answer ___
- Would you say the NATO Alliance is:
  - More important to the United States? ___
  - More important to NATO member countries? ___
  - About as important to the United States? as other NATO member countries? ___
  - Don't Know/No Answer ___
- Put a checkmark next to the statement you agree with more:
  - U.S. efforts to solve problems around the world usually end up making things worse. ___
  - Problems in the world would be even worse without U.S. involvement. ___
  - Don't Know/No Answer ___

Survey Results

Percent who say:	Total	Republican/Lean Republican	Democrat/Lean Democrat
In foreign policy, the United States should take into account the interests of its allies even if it means making compromises with them.	54	35	69
In foreign policy, the United States should follow its *own* national interests even when its allies disagree.	40	N/A	N/A
It's best for the future of our country to be active in world affairs.	44	37	51
We should pay less attention to problems overseas and concentrate on problems here at home.	49	N/A	N/A
Being a member of NATO is good for the United States.	77	71	82
Being a member of NATO is bad for the United States.	15	18	11
The NATO Alliance is more important to the United States than to other NATO member countries.	15	11	18
The NATO Alliance is more important to other NATO member countries than to the United States.	34	34	49
The NATO Alliance is about as important to the United States as other NATO member countries.	42	47	25
U.S. efforts to solve problems around the world usually end up making things worse.	29	N/A	N/A
Problems in the world would be even worse without U.S. involvement.	64	76	56

Russia ($65.1 billion), Saudi Arabia ($61.9 billion), France ($50.1 billion), Germany ($49.3 billion), the United Kingdom ($48.7 billion), and Japan ($47.6 billion).

Weapons Sales. The U.S. government not only spends more money than other countries on its own military and defense but also sells military equipment to other countries, either directly or by helping U.S. companies sell weapons abroad. Although the purchasing countries may use these weapons to defend themselves from hostile attack, foreign military sales may pose a threat to the United States by arming potential antagonists. For example, the United States provided weapons to the Taliban to fight against a Soviet invasion in the 1980s. Years after the Soviets left Afghanistan, rebels continued to fight for control of the country. Using weapons supplied by the United States, the Taliban took over much of Afghanistan and sheltered al-Qaeda and Osama bin Laden—also a former recipient of U.S. support—as they planned the attacks on September 11 (Bergen 2002; Rashid 2000).

The United States regularly transfers arms to countries in active conflict and is the global leader in arms transfers. Between 2007 and 2017, the last period for which the government has released data, 79 percent of the world arms trade was supplied by the United States, with 10 percent supplied by the European Union, 5 percent by Russia, and less than 2 percent by China (U.S. Department of State 2019). A 2005 report titled *U.S. Weapons at War: Promoting Freedom or Fueling Conflict?* concluded that, far "from serving as a force for security and stability, U.S. weapons sales frequently serve to empower unstable, undemocratic regimes to the detriment of U.S. and global security" (Berrigan and Hartung 2005). For example, the United States provided approximately $1.3 billion to purchase weapons for the Iraq army to support their fight against terrorist insurgents. Many of those weapons were subsequently sold by Iraqi military personnel on the black market and purchased by ISIS militants in Syria (Kirkpatrick 2014).

The Costs of War. Historically, wars have been associated with economic growth and technological innovation. During World War II, for example, a surge in government spending and investment in public works projects led to spikes in employment rates and gross domestic product (GDP) that had been at historically low levels during the Great Depression of the 1930s. At the same time, the government raised taxes to fund the war, and U.S. consumption declined as Americans at home sacrificed personal comforts as part of the war effort. The wars in Iraq and Afghanistan, however, represented the first time in modern history that the United States has gone to war without seeing an increase in GDP. It is also the first time that the government has chosen to fund a war through increased deficits rather than increased taxes (Institute for Economics and Peace [IEP] 2013).

The costs of war are difficult to estimate because it is nearly impossible to disentangle direct and indirect costs, as well as the costs associated with the loss of human life, productivity, and infrastructure. The IEP estimates that in 2017, the most conservative estimate of the direct and indirect costs associated with violence containment to the world economy was $14.8 trillion, or approximately $1,988 for each person on earth (IEP 2018). However, these costs are not evenly distributed. The economic cost of violence is 19 times higher in the ten most affected countries than in the ten least affected countries, and accounted for 45 percent of GDP in the ten more affected countries. In Syria and Afghanistan, the economic cost of violence as a percentage of GDP was 68 and 63 percent, respectively (IEP 2018).

Although violence containment efforts include non-war-related expenses, such as the costs of incarceration and violent crime, it is military expenditures that comprise the majority of global spending on violence. The direct cost of war and terrorism to the world economy was approximately $2.9 trillion in 2018 (IEP 2018). However, direct military expenditures routinely underestimate the economic cost of health care, retirement benefits, and other veterans' services that are not included in the defense budget. In 2019, these costs represented an additional $202 billion in spending on top of the $686 billion defense budget (National Center for Veterans Analysis and Statistics [NCVAS] 2020).

In addition to the disruption to social and economic development, wars also take a tremendous toll in human life within the countries where they are fought. The war in Iraq is estimated to have killed nearly 300,000 people from 2003 to 2013, with approximately 200,000 of these civilian deaths (Database 2020). While civilians are sometimes directly targeted as a strategy of war, civilian deaths are often considered "collateral," meaning they occurred as a result of being in close proximity to the violence of war but were not the intended target of a military operation. The impact of this loss of life in societies impacted by war is incalculable.

What do you think?

Near the end of his term, former President Trump announced a plan to withdraw the vast majority of troops from Iraq, Afghanistan, and Somalia—locations that are crucial to U.S. counterterrorism efforts. Trump and his allies argued that it was his last chance to fulfill a campaign promise to end the longest and most costly wars in American history. However, the move was sharply criticized by many military leaders as premature and likely to lead to a power vacuum that would only embolden terrorist groups such as al-Qaeda and ISIS (Schmitt et al. 2020). Decisions about withdrawing from a war involve weight a complex balance of costs: financial costs, but also the cost to human life both within the United States and to the lives of civilians in the countries where U.S. troops are present. Do you think an abrupt withdrawal from the wars in Iraq and Afghanistan is worth the potential long-term risks?

In addition to the direct costs of war, indirect costs occur when wars cause the destruction of critical infrastructure, the disruption in routine medical care, and increased vulnerability to disease transmission and malnutrition. The burden of these costs is primarily borne by "populations of concern," which include refugees, asylum seekers, and other people displaced by war (United Nations High Commissioner for Refugees 2020a). Approximately one-third of the 70.8 million people currently displaced by war live in refugee camps, where crowded living conditions, lack of access to medical treatment, and the prevalence of underlying conditions such as malnutrition increase the vulnerability of these populations to the COVID-19 pandemic (Volkin 2020).

Sociological Theories of War

Sociological perspectives can help us understand various aspects of war. In this section, we describe how structural functionalism, conflict theory, and symbolic interactionism can be applied to the study of war.

Structural-Functionalist Perspective

Structural functionalism focuses on the functions that war serves and suggests that war would not exist unless it had positive outcomes for society. We have already noted that war has served to consolidate small autonomous social groups into larger political states. An estimated 600,000 autonomous political units existed in the world at about 1000 BCE. Today, that number has dwindled to fewer than 200.

Another major function of war is that it produces social cohesion and unity among societal members by giving them a "common cause" and a common enemy. Unless a war is extremely unpopular, military conflict promotes economic and political cooperation. Internal domestic conflicts between political parties, minority groups, and special interest groups often dissolve as they unite to fight the common enemy. During World War II, U.S. citizens worked together as a nation to defeat Germany and Japan. The **rally round the flag effect** refers to the increase in political cohesion that occurs in times of war and international crisis. For example, after the September 11, 2001 attacks, President George W. Bush's approval rating soared to over 90 percent. One study of 51 democratic countries found that voter turnout significantly increased in the months following terrorist attacks (Robbins et al. 2013).

rally round the flag effect Phenomenon of increased political cohesion and civic engagement that commonly occurs during times of war and international crisis.

In the short term, war may also increase employment and stimulate the economy. The increased production needed to fight World War II helped pull the United States out of the Great Depression. War can also have the opposite effect, however. In a 2005 restructuring of the military, the Pentagon, seeking a "meaner, leaner fighting machine," recommended shutting down or reconfiguring nearly 180 military installations, "ranging from tiny Army reserve centers to sprawling Air Force bases that have been the economic anchors of their communities for generations" (Schmitt 2005), at a cost of thousands of civilian jobs.

AP Images/Rick Bowmer

War can encourage social reform by building social solidarity, leading people to put pressure on their government to improve social and political conditions. Here, marine veteran Todd Winn holds a sign at a Black Lives Matter protest against police brutality in May 2020. Winn said he protested because "I served with men whose skin was a different color than mine, who were the finest men that I've ever known, help to learn that really we're all the same ... [I took an oath to] Support and defend the constitution. And there's no qualification on that. It's not until this time, or only for these people" (Vaughen 2020).

Wars also function to inspire scientific and technological developments that are useful to civilians. For example, innovations in battlefield surgery during World War II and the Korean War resulted in instruments and procedures that later became common practice in civilian hospital emergency wards (Zoroya 2006). Research on laser-based defense systems led to laser surgery, research in nuclear fission facilitated the development of nuclear power, and the Internet evolved from a U.S. Department of Defense research project. Today, **dual-use technologies**, a term referring to defense-funded innovations that also have commercial and civilian applications, are quite common. DARPA (Defense Advanced Research Projects Agency) frequently partners with private and industry and public universities to develop technological innovations that have both military and civilian applications. For example, in the early 2000s, DARPA partnered with Apple to create cognitive computing systems that would make on-the-ground military decision making more efficient. The technology they developed is known as the Personalized Assistant that Learns (PAL) within the military and as Siri in the civilian world (DARPA 2020).

War also serves to encourage social reform. After a major war, members of society have a sense of shared sacrifice and a desire to heal wounds and rebuild normal patterns of life. They put political pressure on the state to care for war victims, improve social and political conditions, and reward those who have sacrificed lives, family members, and property in battle. As Porter (1994) explained, "Since ... the lower economic strata usually contribute more of their blood in battle than the wealthier classes, war often gives impetus to social welfare reforms" (p. 19). For example, much of today's social welfare system for impoverished mothers and children grew out of the benefits programs enacted for the widows and wives of deceased and injured soldiers of the U.S. Civil War (Skocpol 1992).

When members of groups who are discriminated against in society are perceived as sacrificing for the benefit of the dominant group, military policy typically precedes the expansion of equal rights laws in the society at large. Thus, the bravery and sacrifice of Black troops in segregated units during World War II led to the racial integration of the military in the early 1950s, more than a decade before the passage of civil rights legislation in U.S. law. Prior to the 2015 Supreme Court ruling that allowed for federal recognition of same-sex marriages, the military formally ended its "don't ask, don't tell" policy to allow LGBT service members to openly serve and live with their partners in 2011, and in 2013 it extended benefits to the same-sex spouses of service members (Hicks 2013; see Chapter 11).

Finally, the U.S. military has historically provided an alternative for the advancement of poor or disadvantaged groups that otherwise face discrimination or limited opportunities in the formal economy. The military's specialized training, tuition assistance programs for college education, and preferential hiring practices improve the prospects of veterans to find a decent job or career after their service (U.S. Department of Veterans Affairs 2020).

dual-use technologies
Defense-funded technological innovations with commercial and civilian use.

Jacob Silberberg/Getty Images News/Getty Images

Conflict theorists draw attention to the ways in which war benefits elites and is tied to the economic system. Antiwar demonstrations were common occurrences in the 1960s and 1970s, as students demonstrated against the war in Vietnam (1959–1975). They erupted again in opposition to the Gulf War (1990–1991), the war in Iraq (2003–2011), and the war in Afghanistan (2001–2019). More recently, there have been public demonstrations against U.S. intervention in the Syrian Civil War and use of military force against Iran.

Conflict Perspective

Conflict theorists emphasize that the roots of war are often antagonisms that emerge whenever two or more ethnic groups (e.g., Bosnians and Serbs), countries (United States and Vietnam), or regions within countries (the U.S. North and South) struggle for control of resources or have different political, economic, or religious ideologies. In addition, conflict theory suggests that war benefits the corporate, military, and political elites.

Corporate elites benefit because war often results in the victor taking control of the raw materials of the losing nations, thereby creating a bigger supply of raw materials for its own industries. Furthermore, Pentagon contracts often guarantee a profit to the developing corporations. Even if the project's cost exceeds initial estimates, called a cost overrun, the corporation still receives the agreed-on profit. In the late 1950s, President Dwight D. Eisenhower referred to this close association between the military and the defense industry as the **military-industrial complex**. For example, even as the U.S. government was deciding on the war in Iraq, "many former Republican officials and political associates of the Bush administration [were] associated with the Carlyle Group, an equity investment firm with billions of dollars in military and aerospace assets" (Knickerbocker 2002, p. 2).

War benefits the political elite by giving government officials more power. Porter (1994) observed that "throughout modern history, war has been the level by which ... governments have imposed increasingly larger tax burdens on increasingly broader segments of society, thus enabling ever-higher levels of spending to be sustained, even in peacetime" (p. 14). Political leaders who lead their country to a military victory also benefit from the hero status conferred on them.

The military elite also benefit because war and the preparations for it provide prestige and employment for military officials. Private military and security companies commonly employ retired high-ranking officers into executive leadership positions. Many seats on the boards of directors of security corporations such as DynCorp International, Mission Essential, MPRI, and Kroll Security International, just to name a few, are filled by retired generals and admirals (Spearin 2014). There is, therefore, a strong link between the for-profit private security industry and the highest levels of military leadership.

military-industrial complex A term first used by Dwight D. Eisenhower to connote the close association between the military and defense industries.

Military leaders argue that private military and security contractors (PMSCs) increase military efficiency and are more cost effective. Because PMSCs are not embedded within the formal military structure, they can be deployed with greater flexibility and speed than formal military units. By outsourcing some operational procedures to private corporations, the military also saves money because they are not necessarily responsible for long-term personnel costs such as disability and retirement benefits. Forty-four percent of PMSC personnel have formerly served in the military (Butler, Stephens, and Swed 2019).

Scholars and observers of PMSCs have raised many concerns about accountability, profiteering, and conflicts of interest among these corporations. Ethically, PSMCs raise questions about the meaning of national security in democratic societies. **Civil–military relations** refers to the nature of the relationship between society as a whole and the armed forces that are responsible for protecting its security. One foundation of civil–military relations in a democratic society is that the military must be subordinate to civilian authority. This is important to ensure that the use of force wielded by the military operates by the consent of and in the service of the people of a nation (Janowitz 1960). For example, in the United States, the Commander and Chief of the armed forces is not a military commander but rather the President because he is an elected public official. Critics argue that because PMSCs are not formally embedded within the military structure, they therefore operate outside of civilian authority and may not necessarily act by the consent of and in service to the American people (Moskos 2000).

PMSCs also raise legal and humanitarian concerns. PMSCs are for-profit businesses and, as such, do not operate under the same norms and regulations that govern the actions of national militaries. The private security firm Blackwater Worldwide came under intense scrutiny after a group of their personnel opened fire on a crowd in Baghdad while guarding a U.S. diplomatic convoy in 2007, resulting in the deaths of 17 Iraqi civilians. Four Blackwater contractors were ultimately convicted of murder or manslaughter and sentenced to lengthy prison terms (Hsu and Martin 2015).

Proponents of PMSCs argue that their use helps increase military efficiency and effectiveness by reducing costs and bureaucratic inflexibility. Critics argue that their use undermines the legitimacy of democratic governments in which national security should be considered a public good. Do you think the U.S. military should outsource some of its security operations to private corporations?

What *do you* think?

The Blackwater incident was just one high-profile example of the ethical and legal complexities associated with the use of PMSCs. In an effort to address the "legal black hole" surrounding international human rights laws and PMSCs, the Swiss government and the United Nations have worked to create international mechanisms for the standardization, monitoring, and enforcement of human rights regulations across PMSCs, first through the 2008 Montreaux Document and then through the 2010 International Code of Conduct Association for Private Security Service Providers (ICoCA) (Gasser and Malzacher 2020). As of 2020, seven countries (including the United States) and 95 private security companies have signed on to the ICoCA (ICoCA 2020).

Despite the legal and ethical concerns surrounding their use, the U.S. military has increasingly relied on PSMCs in overseas operations over the past two decades. In the war in Iraq, the ratio of PMSCs to military personnel deployed was at least 1:1, and in Afghanistan it was 3:1 (Butler, Stephens, and Swed 2019).

Feminist Theories. Feminists, as many other analysts, note the strong association between war and gender. Countries with low levels of gender equality are more likely to experience both civil and interstate war (Caprioli 2005). Furthermore, as conflict theorists also note, active combat has historically been carried out by men. Nature-based arguments about gender—that is, that men are innately aggressive or violent and women inherently peaceful—are not generally supported by social science research and do not adequately explain why men are more likely to kill than women. Feminists emphasize the social construction of aggressive masculine identities and their manipulation by elites as important reasons for the association between masculinity and militarized violence (Alexander and Hawkesworth 2008).

Although some feminists view women's participation in the military as a matter of equal rights, others see war as an extension of patriarchy and the subordination of women in male-dominated societies. Ironically, because protection of women is perceived as a

civil-military relations The relationship between society as a whole and the armed forces that are responsible for protecting its security.

feature of the masculine identity, feminists also point out that war and other conflicts are often justified using "the language of feminism" (Viner 2002). Presidents George W. Bush and Barack Obama used protection of women as partial justification for launching military attacks against the Taliban and ISIS, respectively.

Although women have historically been excluded from combat, the unconventional nature of the wars in Iraq and Afghanistan exposed thousands of women and men in occupations defined by the military as noncombat roles into harm's way. Particularly in Iraq, the use of improvised explosive devices (IEDs) turned roads into the front lines of the war, posing new dangers to traditionally noncombat roles such as truck drivers. According to the Department of Defense, as of June 2020, 173 women have been killed and 1,065 injured in overseas combat operations since the start of the war in Afghanistan in 2001 (Defense Casualty Analysis System [DCAS] 2020).

In both Iraq and Afghanistan, female linguists, intelligence specialists, and military police were needed in combat units to speak with and search Muslim women who traditionally avoid contact with men outside of their families. The realities of nonconventional warfare and the operational needs of the military put increasing pressure on the military to revise its policy that excluded women from combat roles. Beginning in 2016, the Department of Defense formally rescinded the female combat exclusion policy and began requiring each branch of service is to develop "gender-neutral" standards for each occupation; however, gender neutrality can be interpreted differently by different branches.

One option is a policy of "gender norming" in which men and woman are measured by equality of effort, rather than equality of output, with the result of men and women having different minimum standards for measures such as push-ups. This approach is generally preferred in specializations in which physical ability is not of paramount importance but there may be a shortage of qualified applicants, for example, linguists and data analysts. Past research has found that gender-norming policies result in greater numbers of women being represented in traditionally male dominated occupations; however, these women are typically perceived as incompetent in physically demanding roles, resulting in the reinforcement of gender differences among personnel (Kleykamp and Clever 2015).

A second option is a policy of occupationally specific standards validation, in which each job is assessed for the physical and mental standards necessary to accomplish its requirements, and each individual is assessed on his or her ability to meet those objectives. This approach tends to result in greater gender equality as women are perceived of as highly capable and equally contributing to mission objectives. However, women are less likely than men to meet highly demanding physical standards, resulting in a lower representation of women in physically demanding occupations. This approach is generally preferred in specializations where physical qualifications are of paramount importance. The Navy SEALS, a highly selective Special Forces combat unit that is notorious for its physically demanding training program, opted to open its officer training program to women without changing its standards. In 2019, a woman completed the grueling officer training program for the first time (Hodge Seck 2019).

Despite the many advancements women have achieved in the military, many barriers remain. Women make up approximately 15 percent of the military, but only 7 percent of generals and admirals. Because combat experience is a major factor in military promotion decisions, the integration of women into combat roles may help, over time, to bring more women into the highest leadership positions. The lack of women in high levels of leadership and the male-dominated culture of the military have been blamed for the high rate of sexual assault among women in the military, which is about twice the rate among civilian women (Portero 2013).

Equal service opportunities for transgender personnel represent the next front in the struggle for greater gender inclusion in military service. Prior to 2016, transgender people were explicitly barred from military service. From 2016 to 2019, under an Obama-era policy, transgender individuals were permitted to serve openly and access gender transition–related medical care while in service. Beginning in April 2019, however, the Trump administration reversed that policy, barring people with a diagnosis of gender dysphoria from joining the military and requiring those with this diagnosis to be discharged from service (although there are some exceptions to these rules, see Garamone 2019). Within his first week of taking office, President Biden reversed this ban through executive action. However, without a law being passed through Congress, a ban on transgender service could be implemented again in a subsequent administration.

© Carlos Bongioanni/Stars And Stripes

Transgender personnel testify before Congress about the Trump administration's efforts to block transgender people from serving in the military. Countering the administration's argument that transgender inclusion would negatively impact unit readiness and morale, Army Captain Jennifer Peace (pictured second from the left) testified: "The real impact to readiness is discharging transgender Soldiers. Losing Commanders, Intelligence Officers, medics, pilots, platoon sergeants and other critical skills. ... Excluding a portion of the population when we cannot meet recruiting goals—that impacts readiness. And watching as those who you serve with and respect are told they are no longer welcome in service—that impacts morale" (U.S. House of Representatives 2019).

Symbolic Interactionist Perspective

The symbolic interactionist perspective focuses on how meanings and definitions influence attitudes and behaviors regarding conflict and war. The development of attitudes and behaviors that support war begins in childhood. American children learn to revere and celebrate the Revolutionary War, which created our nation. Movies romanticize war, children play war games with toy weapons, and various video and computer games glorify heroes conquering villains.

Symbolic interactionism helps to explain how military recruits and civilians develop a mind-set for war by defining war and its consequences as acceptable and necessary. The word *war* has achieved a positive connotation through its use in various popular public policies—the war on drugs, the war on poverty, and the war on crime. Positive labels and favorable definitions of military personnel facilitate military recruitment and public support of armed forces. However, because different groups have different experiences in society, not all Americans are equally receptive to messages about the benefits or necessity of war. For example, one year after the United States invaded Iraq, 76 percent of Black Americans said the war was a mistake, compared with only 42 percent of White Americans (Carroll 2004).

Many government and military officials convince the public that the way to ensure world peace is to be prepared for war. Governments may use propaganda and appeals to patriotism to generate support for war efforts and to motivate individuals to join armed forces. Both proponents of war and peace employ the language of patriotism—for example, "support the troops" or "peace is patriotic"—to frame their political arguments about war (Salladay 2003; Woehrle et al. 2008).

To legitimize war, the act of killing in war is not regarded as "murder." Deaths that result from war are referred to as "casualties." Bombing military and civilian targets appears more acceptable when nuclear missiles are "peacekeepers" that are equipped with multiple "peace heads." Killing the enemy is more acceptable when derogatory and dehumanizing labels such as *Gook*, *Jap*, *Chink*, *Kraut*, and *Haji* convey the attitude that the enemy is less than human.

Such labels and their accompanying images are socially constructed, often by the media, and are presented to the public. Social constructionists, like symbolic interactionists in general, emphasize the social aspects of "knowing." Cultural historian John Dower (1986) analyzed propaganda images used by both the American and Japanese governments during World War II, revealing that both sides drew on racist stereotypes about each other to construct dehumanized images of the enemy. Japanese propaganda presented Americans as brutish, ape-like imperialists, while American propaganda presented the Japanese as devious, vermin-like predators.

Not only do media images contribute to political attitudes about war and social attitudes about the enemy, but these images also have a powerful effect on a media consumer's

@Caitlyn_Jenner

There are 15,000 patriotic transgender Americans in the US military fighting for all of us. What happened to your promise to fight for them? twitter.com/realDonaldTrump...

-Caitlyn Jenner

mental and physical health. Psychologists measured study participants' media consumption in the weeks following the 9/11 attacks in 2001 and the invasion of Iraq in 2003, and followed up with their health records annually for the next decade. They found that individuals who watched four or more hours of news coverage per day in the weeks following each event had a significantly higher likelihood than people who had less media exposure of posttraumatic stress symptoms and other health ailments years later (Silver et al. 2013).

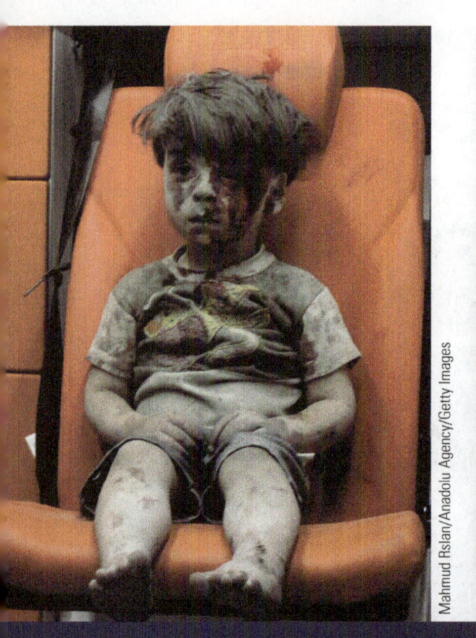

Mahmud Hslan/Anadolu Agency/Getty Images

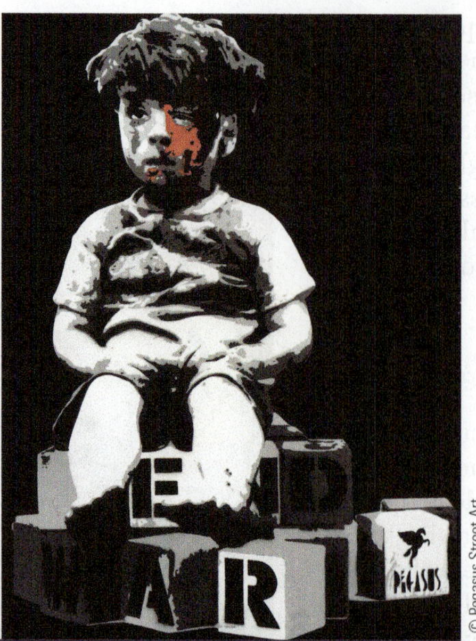

© Pegasus Street Art

Symbolic interactionists focus on how perceptions about war can be shaped by media images. The haunting photograph (left) of five-year-old Omar Daqneesh awaiting treatment in an ambulance after his family's home had been bombed by Syrian government forces appeared on the cover of the *New York Times* in August 2016. The image quickly became the center of a rallying cry for stronger U.S. intervention against the Assad regime in Syria. The image was also used as antiwar street art (right).

Causes of War

The causes of war are numerous and complex. Most wars involve more than one cause. The immediate cause of a war may be a border dispute, for example, but religious tensions that have existed between the two combatant countries for decades may also contribute to the war. The following sections review various causes of war.

Conflict over Land and Other Natural Resources

Nations often go to war in an attempt to acquire or maintain control over natural resources, such as land, water, and oil. Michael Klare, author of *Resource Wars: The New Landscape of Global Conflict* (2001), predicted that wars would increasingly be fought over resources as supplies of the most needed resources diminish. Disputed borders have been common motives for war. Conflicts are most likely to arise when borders are physically easy to cross and are not clearly delineated by natural boundaries, such as major rivers, oceans, or mountain ranges.

In the modern era, oil has been a major resource at the center of many conflicts. Not only do the oil-rich countries in the Middle East present a tempting target in themselves, but war in the region can also threaten other nations that are dependent on Middle Eastern oil. Thus, when Iraq seized Kuwait and threatened the supply of oil from the Persian Gulf, the United States and many other nations reacted militarily in the Gulf War. In the digital era, rare earth minerals needed to make cell phones, computers, and other advanced technologies have come into increasing global demand, leading to fears that control over these resources could be the next major source of conflict. In the Democratic Republic of the Congo, warring militias use the sale of rare earth minerals to Western nations to fund the ongoing civil war, and often use child slaves to work in the mines. The role of these "**conflict minerals**" in perpetuating civil war and human rights abuses associated with the mining prompted the technology firm Intel to announce that beginning in 2016, all of its products would be manufactured using conflict-free minerals (Intel 2015).

conflict minerals Minerals used in advanced technologies (e.g., cell phones) sold to Western nations by warring groups to fund ongoing civil wars; child slaves often work in the mines, leading to human rights abuses.

Water is another valuable resource that has led to wars. Unlike other resources, water is universally required for survival. At various times, the empires of Egypt, Mesopotamia, India, and China all went to war over irrigation rights. In 1998, five years after Eritrea gained independence from Ethiopia, forces clashed over control of the port city Assab and, with it, access to the Red Sea. In a document prepared for the Center for Strategic and International Studies, Starr and Stoll (1989) warned that soon "water, not oil, will be the dominant resource issue of the Middle East" (p. 1).

Despite such predictions, tensions in the Middle East have erupted into fighting repeatedly in recent years—but not over water. Beginning in 2011, pro-democracy and anti-corruption protests across the Middle East, collectively known as the Arab Spring, led to the outbreak of major civil wars in Egypt, Yemen, Libya, and Syria. In 2015, ongoing tensions in Yemen sparked by Arab Spring protests erupted into a full-scale civil war. Religious, cultural, and political differences (also see "Racial, Ethnic, and Religious Hostilities") between the Sunni-led government in the north and the Shiite rebels seeking greater political autonomy in the south have contributed to the conflict.

Mohammed Huwais/AFP/Getty Images

The causes of war are often overlapping and complex. This militia fighter loyal to the Houthi movement in Yemen joins with protestors chanting against Saudi-led airstrikes that have devastated the capital city of Sanaa. Houthis—a predominantly Shiite ethnic minority group in Yemen—began actively rebelling against the Sunni-dominated government during the Arab Spring uprisings in 2011. The Houthi movement has claimed their motivations for war include anti-imperialist ideologies directed against the United States and Saudi Arabia, anti-Semitic ethnic hatreds against Israel, religious fundamentalism of the Zaidi sect of Shia Islam, and a long history of political repression and economic deprivation at the hands of the Sunni majority government.

Conflict over Values and Ideologies

Many countries initiate war not over resources but over beliefs. World War II was largely a war over differing political ideologies: democracy versus fascism. The Cold War involved the clash of opposing economic ideologies: capitalism versus communism. Conflicts over values or ideologies are not easily resolved. They are less likely to end in compromise or negotiation because they are fueled by people's convictions.

If ideological differences can contribute to war, do ideological similarities discourage war? The answer seems to be yes; in general, countries with similar ideologies are less likely to engage in war with each other than countries with differing ideological values (Dixon 1994). Referred to as the **democratic peace theory**, research has shown that democratic nations are particularly disinclined to wage war against one another (Brown et al. 1996; Rasler and Thompson 2005).

democratic peace theory
A prevalent theory in international relations suggesting that the ideological similarities between democratic nations make it unlikely that such countries will go to war against each other.

Racial, Ethnic, and Religious Hostilities

Racial, ethnic, and religious groups vary in their cultural beliefs, values, and traditions. Thus, conflicts between racial, ethnic, and religious groups often stem from conflicting values and ideologies. Such hostilities are also fueled by competition over land and other scarce natural and economic resources. Gioseffi (1993) noted that "experts agree that the depleted world economy, wasted on war efforts, is in great measure the reason for renewed ethnic and religious strife. 'Haves' fight with 'have-nots' for the smaller piece of the pie that must go around" (p. xviii).

Racial, ethnic, and religious hostilities are sometimes perpetuated by a wealthy minority to divert attention away from their exploitations and to maintain their own position of power. Such **constructivist explanations** of ethnic conflict—those that emphasize the role of leaders of ethnic groups in stirring up intercommunal hostility—differ sharply from **primordial explanations**, or those that emphasize the existence of "ancient hatreds" rooted in deep psychological or cultural differences between ethnic groups. For example, most people are familiar with the primordial explanation of the conflict between Israel and Palestine that emphasizes ancient antagonisms between Jewish and Muslim people over the ownership of what both groups believe are their holy sites. Constructivists, however, point out historical evidence that these groups actually coexisted peacefully and shared access to these sites for many centuries. It was only after British and French authorities began dividing the land between the groups in

constructivist explanations
Those explanations that emphasize the role of leaders of ethnic groups in stirring up hatred toward others external to one's group.

primordial explanations
Those explanations that emphasize the existence of "ancient hatreds" rooted in deep psychological or cultural differences between ethnic groups, often involving a history of grievance and victimization, real or imagined, by the enemy group.

the 1920s, and in many cases promising the same land to both groups simultaneously, that the tensions that characterize the current conflict arose (Kaufman and Hassassian 2009).

As described by Paul (1998), sociologist Daniel Chirot argues that the worldwide increase in ethnic hostilities is a consequence of "retribalization"—that is, the tendency for groups, lost in a globalized culture, to seek solace in the "extended family of an ethnic group" (p. 56). Chirot identified five levels of ethnic conflict: (1) multiethnic societies without serious conflict (e.g., Switzerland); (2) multiethnic societies with controlled conflict (e.g., the United States and Canada); (3) societies with ethnic conflict that has been resolved (e.g., South Africa); (4) societies with serious ethnic conflict leading to warfare (e.g., Sri Lanka); and (5) societies with genocidal ethnic conflict, including "ethnic cleansing" (e.g., Darfur).

Religious differences as a source of conflict have recently come to the forefront. An Islamic jihad, or holy war, has been blamed for the September 11 attacks on the World Trade Center and Pentagon as well as for bombings in Saudi Arabia, Spain, Great Britain, France, Istanbul, Iraq, Sweden, and the United States. Some claim that Islamic beliefs in and of themselves have led to recent conflicts (Feder 2003). Others contend that religious fanatics, not the religion itself, are responsible for violent confrontations and emphasize that misunderstandings between cultural groups can further fuel these tensions. For example, most Americans understand the term *jihad* as radical Muslims have used it to justify violent conflict as a holy war, but many moderate Muslims point out that the more conventional understanding of the term is as a faith-based internal struggle to achieve a life of peace (Bonner 2006).

Conflicts between different sects of the same religion can also lead to long-lasting and devastating wars. Ongoing civil wars in Syria and Yemen, which have led to enormous humanitarian crises, both have roots in ethnic disputes between Sunni and Shiite populations. The conflict between these two sects has deep historical roots, and is the source of tension between Sunni and Shiite populations in Iraq, Iran, and Libya as well (Arango and Barnard 2013). Wars over differing religious beliefs have led to some of the worst episodes of bloodshed in history, in part, because some religions lend themselves to martyrdom—the idea that dying for one's beliefs leads to eternal salvation. For example, Islamic leader Osama bin Laden claimed that unjust U.S.–Middle East policies are responsible for "dividing the whole world into two sides—the side of believers and the side of infidels" (Williams 2003, p. 18). While this primordial explanation may be useful in accounting for the intractable nature of these conflicts, it does not tell the whole story. Within these countries, conflicts are also tied to power dynamics and majority–minority relations that are not unique to ethnic or religious divisions. In both Yemen and Syria, the Arab Spring movement's push for greater democratization combined with growing calls among disadvantaged groups for greater political representation and economic opportunity. Tensions flared when these protests were met with repression, leading to war. Ted Robert Gurr's (2015) theory of **relative deprivation**—that conflict is more likely when the gap between what a population expects and what they have access to in their society widens—has been strongly supported as a major cause of the wars that grew out of the Arab Spring uprisings (Korotayev and Shishkina 2020).

relative deprivation Theory that conflict is more likely when there is a meaningful gap between what a group within a population expects (for example, economic opportunity) and what they have access to within their society.

security dilemma A characteristic of the international state system that gives rise to unstable relations between states; as State A secures its borders and interests, its behavior may decrease the security of other states and cause them to engage in behavior that decreases A's security.

Defense against Hostile Attacks

The threat or fear of being attacked may cause leaders of a country to declare war on the nation that poses the threat. This is an example of what experts in international relations refer to as the **security dilemma**:

> The basic premise of the security dilemma is that as one state [nation] takes measures to increase its security. ... Another state might take similar, reactive measures to make up for the shift in the balance of power. If states perceive the actions of other actors to be offensive in nature, the change in the balance of power is perceived as detrimental to their own security. This creates a cycle in which both states will continually take measures, such as increasing military strength ... to increase their security. In turn, tensions between the two states can escalate into conflict. (Prasad 2015, p. 1)

Such situations may lead to war inadvertently. The threat may come from a foreign country or from a group within the country. After Germany invaded Poland in 1939, Britain and France declared war on Germany out of fear that they would be Germany's next victims. Germany attacked Russia in World War I, in part out of fear that Russia had entered the arms race and would use its weapons against Germany. Japan bombed Pearl Harbor, hoping to avoid a later confrontation with the U.S. Pacific fleet, which posed a threat to the Japanese military.

In 2001, a U.S.-led coalition bombed Afghanistan in response to the September 11 terrorist attacks. Moreover, in March 2003, the United States, Great Britain, and a loosely coupled "coalition of the willing" invaded Iraq in response to perceived threats of weapons of mass destruction and the reported failure of Saddam Hussein to cooperate with United Nations' weapons inspectors. Yet, in 2005, a presidential commission concluded that the attack on Iraq was based on faulty intelligence and that, in fact, "America's spy agencies were 'dead wrong' in most of their judgments about Iraq's weapons of mass destruction" (Shrader 2005, p. 1). Although the Obama administration withdrew all combat operations from Iraq in 2011, the political upheaval caused by the United States–led war contributed to the ongoing civil war in Iraq.

Revolutions and Civil Wars

Revolutions and civil wars involve citizens warring against their own government and often result in significant political, economic, and social change. The difference between a revolution and a civil war is not always easy to determine. Scholars generally agree that revolutions involve sweeping changes that fundamentally alter the distribution of power in society (Skocpol 1994). The American Revolution resulted from colonists revolting against British control. Eventually, they succeeded and established a republic where none existed before. The Russian Revolution involved a revolt against a corrupt, autocratic, and out-of-touch ruler, Czar Nicholas II. Among other changes, the revolution led to wide-scale seizure of land by peasants who formerly were economically dependent on large landowners. North and South Korea, North and South Sudan, Algeria, and Zimbabwe are just a few examples of modern countries that were formed as a result of revolution or civil war.

Civil wars may result in a different government or a new set of leaders but do not necessarily lead to such large-scale social change. Because the distinction between a revolution and a civil war depends on the outcome of the struggle, it may take many years after the fighting before observers agree on how to classify it. Revolutions and civil wars are more likely to occur when a government is weak or divided, when it is not responsive to the concerns and demands of its citizens, and when strong leaders are willing to mount opposition to the government (Barkan and Snowden 2001; Renner 2000). Recently, the civil wars in Syria and Yemen have emerged as the deadliest in recent decades and have contributed to a humanitarian crisis on a scale not seen since World War II.

@Chris MurphyCT

Soleimani was an enemy of the United States. That's not a question.

The question is this - as reports suggest, did America just assassinate, without any congressional authorization, the second most powerful person in Iran, knowingly setting off a potential massive regional war?

-Chris Murphy

What *do you* think?

The tense relationship between the United States and Iran goes back decades. The Iranian government claims that the United States has unfairly meddled in their democratic elections and control of their own natural resources and has hypocritically denied Iran the opportunity to develop itself as a nuclear power. The United States argues that Iran violates the human rights of its citizens, has sponsored terrorist groups, and has violated nuclear treaty agreements (Kaur et al. 2020). In January 2020, former President Trump authorized a military strike on Iranian bases in Iraq and Syria, killing 25 Iranian soldiers, including the top Iranian general, Qasem Solemani. Trump claimed the strike and assassination of Solemani was justifiable retaliation for Iran's support of terrorist activities that had led to the deaths of U.S. soldiers. Critics argued that the move was too provocative and dramatically increased the risk of nuclear war. Under what conditions do you think the United States is justified in using military force against foreign governments?

Life after War

Some 2.7 million Americans have served in the wars in Iraq and Afghanistan since 2001. Here, four veterans share how their experiences in war have affected them.

I don't feel good about it. It will bother me for the rest of my life and honestly I'm happy about that. I'm embarrassed by how little I knew. I was a kid from Texas. I flew on a plane for the first time when I went to boot camp. I had this visceral desire to seek vengeance for 9/11, and I believed our government when they told me that there was a connection to Iraq. So I fully supported the war. I thought we were bringing bad people to justice. I didn't understand the nuances. I didn't know anything about the people of Iraq, or the culture, or the country. And I feel ashamed about that. I'm getting my graduate degree in Middle Eastern history right now. And the more I learn, the worse I feel. I got so much personal benefit from being in the military. In many ways it was the greatest thing I've ever done. But I could have gotten those personal benefits from other means. I was in Iraq for one year. And the trauma of that year will impact me for the rest of my life. But for the people of Iraq, it's been ten years. And they're still being traumatized on a daily basis.

I don't think it's possible to be a medic in a conflict zone and not have something stay with you. … I have the hardest time forgetting this little girl. She was brought to our post one day. Two men ran toward us carrying a bundle of blankets. And they're yelling in Pashtu. And at first all I can see are these bloody blankets, but then I peel them back, and there's this little girl inside. She stepped on a landmine while playing. … And everything is seething. And I can smell the flesh. And she's screaming. But I'm trained to drown it out. I'm trained so well that I almost don't hear the screaming. I focus on our interventions. Stop the bleeding. Apply tourniquets. Administer the IV. … And she lived. And I was fine throughout the whole thing. I was just like a robot. I'd been trained for chaotic situations. But they don't train you for the aftermath. … In Afghanistan I spent so much time imagining what it would be like when I came home. I built up this perfect world. I imagined eating a big cheeseburger. And taking the longest shower. And meeting up with all my friends. … But after those visits, I was pretty much by myself. So I sat in my room and I started thinking. I'd been so busy in Afghanistan. There was always a job to do. But now it was quiet. So I thought about all the things that I'd kept at bay. I thought about the little girl that I saved. And what her life is like now. And I wondered if she's still alive. And if she is still alive, does she even want to be?

We were built to think alike. Everything is so standardized in the military that you can function without thinking. If I ever needed night vision goggles, somebody could throw me their bag, and I'd know exactly what pocket to find them in. We were like cogs in a wheel. And that may sound like a bad thing—but it's not. Nobody wants to think of themselves as cogs in a wheel, but humans love structure. … The tendency is just accelerated in the military. And it feels good. It feels good to know your place. It feels good to wear the same uniform. It feels good to know exactly what you're contributing to the mission, to the team, and to the country as a whole. Your value is so clear. But then you come home and the lines are blurred. It's hard to discover that value again. You're not sure where you fit. You're not sure how you connect to other people. You're not sure how to make a difference. It can be very isolating. And all you want to do is be back on a team.

The military is mostly filled with people who genuinely desire to do the right thing. … These people grew up as boy scouts and girl scouts. The whole reason they volunteered was because they wanted to do the right thing. But the right thing is never clear in war. If you shoot too early, an innocent person gets killed. If you shoot too late, you lose a buddy. So a lot of our injuries are moral ones. Most of us come home feeling like we did something wrong. Or we didn't give enough. Or that our friends gave too much. My best friend in the Marines was a guy named Ronnie Winchester. He was the nicest guy you can imagine. My 22nd birthday was during our officer training course. None of us had slept. We were all starving. We were only getting one ration per day. But Ronnie wanted to give me a memorable birthday. So he put a candle in his brownie and gave it to me. That's how nice of a guy he was. Ronnie ended up getting killed in Iraq. And if a guy like Ronnie got killed, you can't help but wonder why you deserve to be alive. Ronnie was 25 years old when he died. He is always going to be 25 years old. I have a wife and kids now. I get to grow old. But Ronnie Winchester is always going to be 25.

SOURCE: Humans of New York 2016.

Nationalism

Some countries engage in war in an effort to maintain or restore their national pride. For example, Scheff (1994) argued that "Hitler's rise to power was laid by the treatment Germany received at the end of World War I at the hands of the victors" (p. 121). Excluded from the League of Nations, punished by the Treaty of Versailles, and ostracized by the world community, Germany turned to **nationalism**—a sense of identity that emphasizes loyalty and devotion to the interests of one's own country, even to the exclusion or detriment of the interests of other communities or identities—as a reaction to material and symbolic exclusion. There is a well-documented association between feelings of exclusionary nationalism—that is, feelings that one's nation is superior to others—and support for an aggressive foreign policy approach (Bonikowski and DiMaggio 2016).

In 2016, Trump campaigned and was elected on the basis of nationalistic claims like "America First" and "Make America Great Again." Although nationalism can be highly polarizing, it has also proved to be a politically effective strategy by sharpening group identity boundaries and serving as the foundation on which to build coalitions of political support. There is some evidence, however, that nationalism can have a backlash effect that weakens political support in the long term. In 2020, Gallup polling found that the percentage of people saying they were "proud to be an American" had reached all-time low (41 percent). Polling also showed a sharp divide between Republicans (67 percent) and Democrats (24 percent) who expressed pride in their American identity (Brenan 2020; see Figure 15.3). One explanation for this recent decline in national pride is fear that nationalism may be contributing to an increase in violent social divisions. Shortly after a self-proclaimed white nationalist opened fire at an El Paso Walmart in August 2019, targeting Hispanic shoppers, polls indicated a sharp increase in the number of American voters who saw white nationalism as a threat (Klar 2019). A majority of Americans also said that when president, Trump had done "too little" to distance himself from white nationalist extremist groups (Pew Research Center 2019b).

The election of Joe Biden signals a change to the America First approach. The initial nominations offered by the Biden administration to top-level foreign policy positions were largely reminiscent of the Obama-era foreign policy team, one that was largely focused on diplomacy and international cooperation. While many analysts expressed relief at a return to "normalcy" in international relations, others expressed concern that this approach neglects the fundamental changes to the international order that have been ongoing since the Obama administration (Crowley 2020). Is nationalism here to stay, or will global cooperation once again take center stage in international relations?

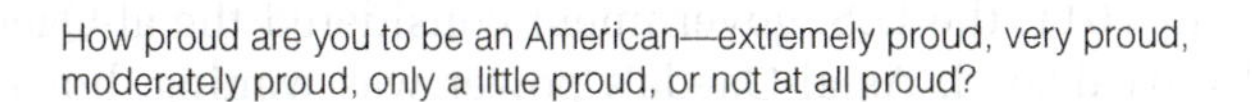

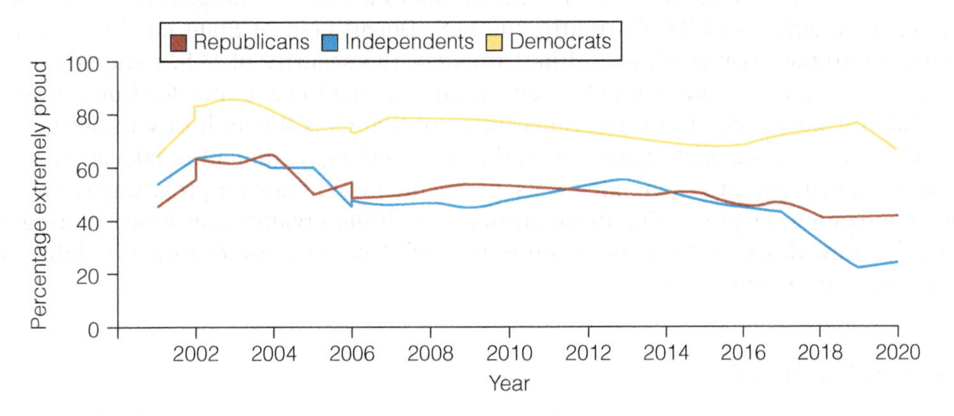

Figure 15.3 Feelings of Pride in Being American among Republicans and Democrats, 2001–2020
SOURCE: Gallup 2020.

nationalism A sense of identity that emphasizes loyalty and devotion to the interests of one's own country, even to the exclusion or detriment of the interests of other communities or identities.

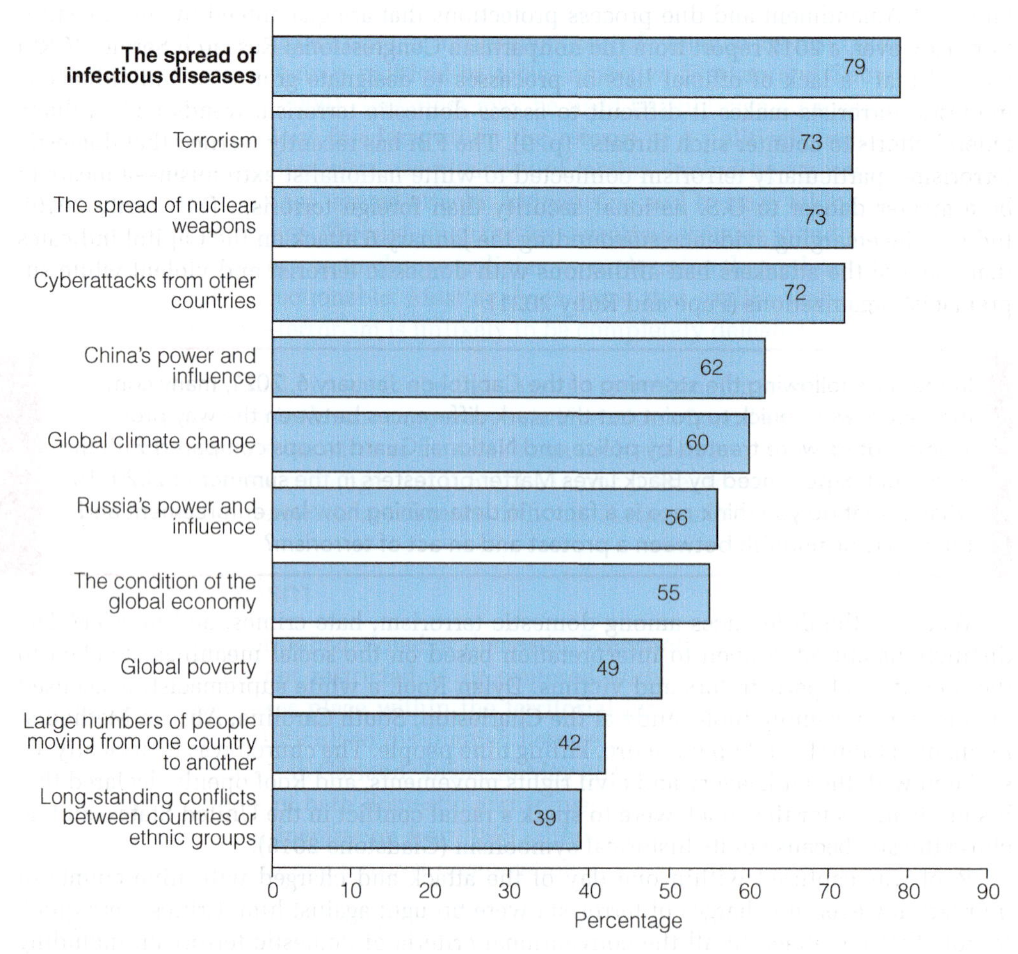

Figure 15.4 Public's Ranking of Major Threats to the United States, 2020
SOURCE: Pew 2020.

suggests that "understanding the mind-set" of a terrorist can help in the fight against terrorism. The RVE process is very complex, Borum argues, involving social relationship networks, grievances, perceived rewards, experiences with incarceration, and psychosocial vulnerabilities (see Figure 15.5).

Borum points out that the vast majority of people with militant extremist beliefs don't engage in violent action, and that counterviolent extremist (CVE) efforts need to focus on more than the "battle of ideas." Rather, CVE efforts should take a multifaceted approach to understanding and attempting to mitigate the grievances, psychosocial vulnerabilities, and social network influences that interact to produce violent extremism.

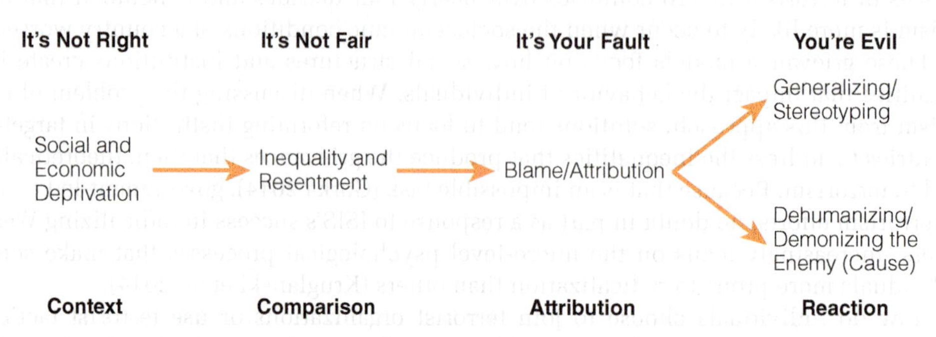

Figure 15.5 The Process of Ideological Development
SOURCE: Borum 2003.

America's Response to Terrorism

A government can use both defensive and offensive strategies to fight terrorism. Defensive strategies include using metal detectors and X-ray machines at airports and strengthening security at potential targets, such as embassies and military command posts. The Department of Homeland Security (DHS) coordinates such defensive tactics for the U.S. government. As of 2021, DHS has nearly a quarter of a million employees with a budget of $49.8 billion. The objective of the DHS is to "ensure a homeland that is safe, secure, and resilient against terrorism and other hazards" (U.S. Department of Homeland Security 2020).

Two incidents have raised questions about the U.S. government's response to terrorism. In 2010, Private Chelsea (Bradley) Manning was arrested for leaking videos and more than 700,000 pages of documents related to the conduct of the U.S. military in Iraq and Afghanistan to the website WikiLeaks. Manning was convicted in 2013 of 17 of the 22 charges that were brought against her, although she was cleared of the most serious charge, that of "aiding the enemy" (Pilkington 2013). Similarly, Edward Snowden, one of the more than 850,000 civilian contractors with top-secret clearance, leaked information about the National Security Agency's (NSA) domestic surveillance program. Snowden fled to Hong Kong and ultimately Russia where he was granted asylum. In 2016, a House intelligence report condemned Snowden saying that he was not a whistleblower and that the clear majority of documents he took were "defense secrets" that had nothing to do with privacy issues (Associated Press 2016).

Snowden's leak of the NSA surveillance program sparked a public debate about the balance between the rights of American citizens to privacy and the government's obligation to ensure public safety. In 2013, a federal judge ruled that the bulk data collection program was unconstitutional; and, in 2015, after lengthy debate, Congress voted not to renew the portion of the 2001 *PATRIOT Act* that allowed for the surveillance program. Although many continue to argue that Snowden is a traitor, others suggest that his efforts in addressing an injustice should be commended and that he should be granted a full pardon. According to one commentator:

> After 9/11 ... the trade-off between security and liberty tipped too far in the direction of intrusion and authoritarianism. Historians will record that Snowden's leaks helped, at least somewhat, to right the balance. At great risk to himself, he stood up to the immensely powerful system for which he worked, and cried foul. (Cassidy 2015, p. 1)

@Snowden

If you ever wonder where we're at on the dystopia scale, consider that it's normal to believe the government is spying on you, and crazy to believe that they're not.

-Edward Snowden

Offensive strategies include retaliatory raids, such as the U.S. bombing of terrorist facilities in Afghanistan, group infiltration, and preemptive strikes. The American public largely supports these offensive strategies; however, many policy analysts are critical about their effectiveness given the high costs of counterterrorism tactics. One study of the costs and benefits of counterterrorism spending found that approximately 300 terrorist attacks on the scale of the Boston Marathon bombing would need to occur in the United States each year to justify the annual cost of counterterrorism (Mueller and Stewert 2014).

Guantánamo Detention Center. Among the most controversial responses to the war on terrorism is the indefinite detention of "enemy combatants" at a military prison and interrogation camp in Guantánamo, Cuba. This is the primary detention center for the Taliban or their allies captured in Afghanistan, as well as suspected terrorists, including al-Qaeda members from other regions. Since 2002, a total of 779 detainees have been held at "Gitmo"; of these, 40 still remain in detention, of whom 26 have not been charged with a crime (American Civil Liberties Union [ACLU] 2018). The Bush administration argued that because these detainees were not members of a state's army, they were not covered by the **Geneva Conventions**, the principal international treaties governing the laws of war and in particular the treatment of prisoners of war and civilians during wartime. In 2006, the U.S. Supreme Court rejected this argument, ruling that the detainees were subject to minimal protections under the Conventions. Furthermore, in 2008, the Court ruled that the Constitution guarantees the right of detainees to challenge their detention in a federal court (Greenburg and de Vogue 2008).

Geneva Conventions A set of international treaties that govern the behavior of states during wartime, including the treatment of prisoners of war.

> “I have trodden a very long road in search of justice since I was taken from my wife and infant son in Karachi [Pakistan] in 2002 . . . . I told my American interrogators the truth: I was a taxi driver, a victim of mistaken identity . . . . I held out more hope for the inquest conducted by the US Senate, which began in 2009. Surely, I thought, they would learn what had happened to me, and then I would be released. No such luck. Although I appear many times in the executive summary of the report, I remain imprisoned. The inquest is now the basis of a movie starring Adam Driver. Back home in Karachi, my son is 18 years old.”
>
> **–AHMED RABBANI (DEC 5 2019), DETAINEE CURRENTLY HELD IN GUANTANAMO BAY PRISON**

The treatment of detainees at Guantánamo and at secret detention centers (so-called black sites) around the world has ignited public debate about interrogation techniques used on suspected terrorists. In 2014, a Senate committee released a report on the CIA's Detention and Interrogation Program (Senate Select Committee on Intelligence 2014). In part, the authors of the report concluded:

- The techniques used by CIA interrogators (e.g., waterboarding, isolation, sleep deprivation, beatings) constitute torture under international law.
- The CIA misrepresented the brutality and the effectiveness of the interrogation techniques to gain approval for and continuation of the program.
- The CIA intentionally impeded oversight of the program, including providing false information to the president, vice president, and the National Security Council.
- The CIA coordinated the release of classified and inaccurate information to the media concerning the effectiveness of the enhanced interrogation program.
- The CIA wrongfully detained people, including an “intellectually challenged” man whose detention was used solely as leverage to get a family member to provide information (p. 12).

Public opinion on this topic is mixed. In response to a question on whether the use of torture is ever acceptable, 49 percent of American adults said that they believed torture is never acceptable and 48 percent believed that torture is acceptable under some circumstances (Tyson 2017). A substantial majority of Republicans or those leaning Republican (71 percent) said that torture could be justified under some circumstances, compared with 31 percent of Democrats or those leaning Democratic.

Despite the understandable concerns that Americans have about terrorist attacks in general, it is important to remember that death as a result of an act of terrorism is an extraordinarily rare occurrence worldwide, especially in American society. For example, Americans are more than twice as likely to be killed by an animal on U.S. soil than killed by terrorist; Americans are also twice as likely to be killed by a native-born terrorist than a foreign-born terrorist (Nowratseh 2018). Nonetheless, fears about terrorist attacks continue to be widespread. Figure 15.4 displays the percentage of American voters who perceive of a variety of foreign policy issues, including terrorism and Russia's power and influence, as a threat (Pew Research Center 2020).

Current trends indicate that the most serious terrorist threat currently facing the United States comes from **white supremacist organizations**: extremist groups, including those aligning with neo-Nazi, alt-right, and the Ku Klux Klan, which share a belief in the inherent superiority of White people and an ideological commitment to using political violence in order to promote a racially segregated society, with only White people in a position of power. Leaders from white supremacist groups including the Proud Boys, the Boogaloos, and the Three Percenters were notably at the head of the crowd that stormed the Capitol during the riots of January 6, 2021 (Tavernise and Rosenberg 2021). The Anti-Defamation League (ADL) has recorded a substantial increase in domestic terrorist activity in recent years (see Figure 15.6). Of all the acts of domestic terrorism from 2010 to 2019, 78 percent were committed by white supremacist extremists (ADL 2020). The ADL also notes that 2019 had the highest number of incidents of white supremacist propaganda ever recorded in the organization's history. This trend has been exacerbated by social media and foreign interference in spreading disinformation (see Chapter 14). Two groups in particular—the Base and Atomwaffen Division (AWD)—have been the focus of an intensification of counterterrorism efforts by the FBI and local law enforcement agencies in recent years (Lewis et al. 2020).

white supremacist organizations Extremist groups, including those aligning with neo-Nazi, alt-right, and the Ku Klux Klan, that share a belief in the inherent superiority of White people and an ideological commitment to using political violence in order to promote a racially segregated society, with only White people in a position of power.

These groups blur the lines between domestic and transnational terrorism, a pattern that is facilitated by social media and the Internet. For example, the founder of the neo-Nazi organization The Base, Rinaldo Nazzaro, is an American citizen who is reported to be based in Russia. Localized cells of The Base operate autonomously, driven by shared ideological goals that are disseminated through social media postings. These cells have been arrested for planned and actual acts of violence against individual members of racial minority groups, synagogues, and schools in non-White

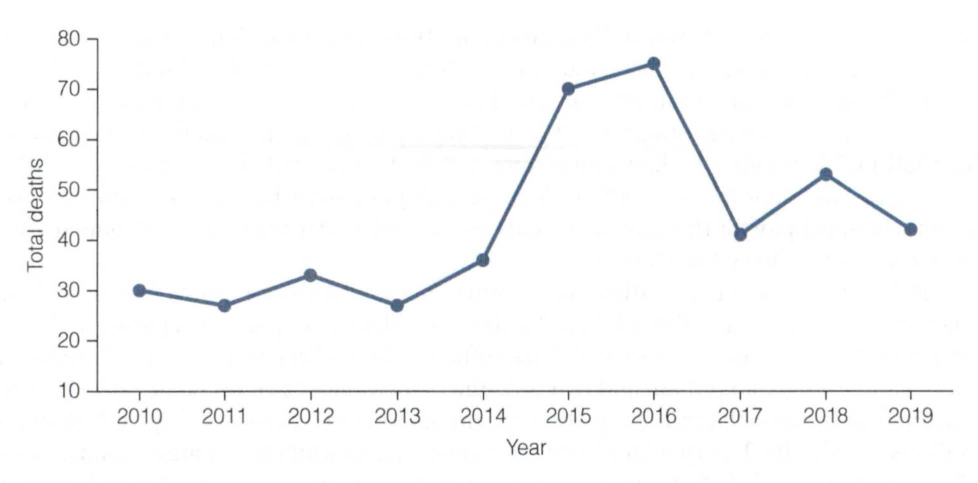

Figure 15.6 Domestic Extremist-Related Killings in the United States by Year (2010–2019)
SOURCE: ADL 2020.

communities. Another neo-Nazi organization, Atomwaffen Division (AWD), originated in Florida and has foreign branches in Germany, the United Kingdom, and the Ukraine.

Social Problems Associated with Conflict, War, and Terrorism

Social problems associated with conflict, war, and terrorism include death and disability; rape, forced prostitution, and displacement of women and children; social-psychological costs; diversion of economic resources; and destruction of the environment.

Death and Disability

Many American lives have been lost in wars, including 116,516 in World War I, 405,399 in World War II, 36,574 in Korea, and 58,220 in Vietnam (CRS 2019). As of September 2019, 7,007 U.S. military personnel have been killed in the Global War on Terror.

Many civilians and enemy combatants also die or are injured in war. Despite more than 200,000 Iraqi civilian deaths (Database 2020), many Americans are unaware of this tremendous loss of life. In a program the Pentagon developed, American reporters were "embedded" into U.S. military units to provide "journalists with a detailed understanding of military culture and life on the frontlines" (Lindner 2009, p. 21). One of the by-products of this program, however, was that 90 percent of the stories by embedded journalists were written from the perspective of American soldiers, focusing "on the horrors facing the troops, rather than upon the thousands of Iraqis who died" (p. 45).

The impact of war and terrorism extends far beyond those who are killed. Many of those who survive war incur disabling injuries or contract diseases. For example, in South Sudan alone, over 4,200 people have been killed or wounded by land mines following the end of civil war with northern Sudan in January 2005

Chip Somodevilla/Getty Images News/Getty Images

Violent clashes between white supremacist groups and antiracist protestors in Charlottesville, Virginia, in August 2017 ignited public attention to the issue of white supremacist terrorism within the United States. Six months before these white supremacist groups converged on Charlottesville for the Unite the Right rally, Elizabeth Neumann, Assistant Secretary of Threat Prevention and Security Policy for the Department of Homeland Security, testified before the Congressional committee on Intelligence and Counterterrorism about the growing threat from white supremacist organizations in the United States: "[I]t feels like we are at the doorstep of another 9/11. ...[W]e can see it building and we don't quite know how to stop it" (quoted in Lewis and Hughes 2020).

@real DonaldTrump

The case of Major Mathew Golsteyn is now under review at the White House. Mathew is a highly decorated Green Beret who is being tried for killing a Taliban bombmaker. We train our boys to be killing machines, then prosecute them when they kill! @PeteHegseth

-Donald J. Trump

(United National Mine Action Office 2011). In 1997, the Mine Ban Treaty, which requires that governments destroy stockpiles within 4 years and clear land mine fields within 10 years, became international law. To date, 162 countries have signed the agreement; 35 countries remain, including China, India, Israel, Russia, and the United States (International Campaign to Ban Landmines 2017). War-related deaths and disabilities also deplete the labor force, create orphans and single-parent families, and burden taxpayers who must pay for the care of orphans and disabled war veterans (see Chapter 2 for a discussion of military health care).

The killing of unarmed civilians undermines the credibility of armed forces, making their goals more difficult to defend. In 2019, then-President Trump issued a pardon to Major Matthew L. Golsteyn, an Army Special Forces officer, who had been convicted of the murder of an unarmed Afghan civilian, and overrode the demotions of two other officers who had been charged through the military justice system of using excessive force against civilians (Philipps 2019). The Trump administration argued that clearing the charges against these officers demonstrated dedication to America's war fighters who face unique challenges in unconventional war zones. However, many military justice insiders and Department of Defense leaders expressed concern that the move would undermine the legitimacy of U.S. Armed Forces and create new resistance against American military presence in global security. Excessive use of force by American personnel has repeatedly been used by militant groups to gain support among the civilian population and to justify the use of terrorism.

Rape, Forced Prostitution, and the Displacement of Women and Children

Despite international prohibitions against wartime rape, and "humiliating and degrading treatment, enforced prostitution and any form of indecent assault" (International Committee of the Red Cross 2015), such behaviors have routinely occurred throughout history. Before and during World War II, Japanese officials forced between 100,000 and 200,000 women and teenage girls into prostitution as military "comfort women." These women were forced to have sex with dozens of soldiers every day in "comfort stations." Many of the women died as a result of untreated sexually transmitted diseases, harsh punishment, or indiscriminate acts of torture.

Milos Bicanski/Getty Images News/Getty Images

An Afghan father plays with his son in a makeshift refugee camp on the island of Lesvos in Greece. Millions of refugees around the world have found themselves trapped in similar camps near national borders as governments enact greater restrictions on migrant movement. In February 2020, Greek police fired tear gas at refugees trying to cross the border from Turkey as they attempted to flee violence from Syria, Afghanistan, and Iraq and seek refuge in Europe.

Since 1998, Congolese government forces have fought Ugandan and Rwandan rebels. Women have paid a high price for this civil war, in which gang rape is "so violent, so systematic, so common ... that thousands of women are suffering from vaginal fistula, leaving them unable to control bodily functions and enduring ostracism and the threat of debilitating health problems" (Wax 2003, p. 1). Though much less common than violence against women, aid workers also see increasing incidents of rape and sexual violence against men as "yet another way for armed groups to humiliate and demoralize Congolese communities into submission" (Gettleman 2009, p. 1).

Feminist analyses of wartime rape emphasize that the practice reflects not only a military strategy but also ethnic and gender dominance. For example, Refugees International, a humanitarian aid group, reports that rape is "a systematic weapon of ethnic cleansing" against Darfuris and is "linked to the destruction of their communities" (Boustany 2007, p. 9). Under Darfur's traditional law, prosecution of rapists is nearly impossible: Four male witnesses are required to accuse a rapist in court, and single women risk severe corporal punishment for having sex outside marriage.

social problems
RESEARCH
UP CLOSE

The Transmission of Trauma in Refugee Families

The United Nations High Commission for Refugees (UNHCR) reports that the number of people who are currently forcibly displaced by war, persecution, and humans rights violations exceeds 70 million—the highest number since World War II.

The trauma of living through war can impact families and communities long after the violence ends. This is especially true for refugee families, whose trauma is compounded by long-term disruption in access to basic resources and the culture shock of settling into new communities.

Researchers Nin Thorup Dalgaard and Edith Montgomery wanted to better understand how the trauma experienced by refugee parents affected their children, especially young children who did not themselves have a direct experience of war trauma.

Findings from this research can help organizations who work with refugee families to provide better long-term support for the children of refugees.

Sample and Methods

The researchers recruited 30 refugee families from Iraq, Iran, Lebanon, Palestine, Syria, and Afghanistan who had been granted temporary asylum in Denmark, who had at least one parent referred for treatment for PTSD symptoms and who had at least one child between the ages of four and nine with no history of direct trauma exposure.

The researchers used a semistructured interview approach. They interviewed parents in their preferred language (Farsi, Arabic, English, or Danish) and asked them to discuss how their own trauma history and current symptoms of PTSD affected their parenting ability, their children, and the family as a whole. They also asked questions about how the parents perceived their children's psychosocial adjustment and the family's communication strategies, relationships with extended family and social networks in exile, and their feelings about maintaining connections with their extended families and cultures of origin.

To measure the psychosocial adjustment of the children, the researchers used the Strengths and Difficulties Questionnaire (SDQ), a commonly used psychological screening tool.

The parental interviews were analyzed using qualitative coding software which allow researchers to apply theoretical themes to the interview responses and analyze patterns in family functioning such as family cohesion, flexibility, roles, coping, stressor pileup, and problem-solving skills/conflict level. These qualitative results were then matched with the results of the SDQ to determine how parental trauma impacts family functioning and children's psychosocial adjustment.

Findings and Conclusion

Nearly two-thirds of the parents interviewed reported that their family functioning was negatively impacted by feelings of guilt, personal incompetence, and social isolation. This may lead to a transmission of trauma to children through what the researchers describe as a "self-fulfilling prophecy. … These parents seem to give up before even trying and to react to children's needs with withdrawal and self-blame" (p. 298). As a result, these children are at a higher risk of long-term maladjustment and inability to develop healthy coping strategies to stress. The researchers recommend that individual psychotherapy with traumatized refugee parents should focus on enhancing parental perceptions of agency and coping skills to counter the effects of their symptoms on their parenting ability.

The transmission of trauma can also occur through family role reversal or role ambiguity. The researchers note that "it appears as if many parents are so absorbed by their own grief that they fail to notice how their children are adopting a care-giving behavior toward them" (p. 299). This pattern is consistent with other prominent research findings that show children can develop symptoms of trauma by taking on a caregiver role within their families at a young age.

However, the strongest predictor of intergenerational trauma transmission between parents and children occurred when parents experienced stress pileup: a combination of worries about residency permits, finances, health, unstable housing, and concerns about the well-being of family in the country of origin. The researchers emphasize that in addition to individual psychotherapy techniques, the most direct way to reduce the transmission of trauma from refugee parents to their children is through a national immigration policy that eases the stresses and uncertainties of refugee resettlement.

ISIS has used rape and human trafficking as a strategy of ethnic cleansing against the Yazidi minority in Iraq and Syria (see the story of Nadia Murad in this chapter's opening vignette).

Globally, more than 70 million people are currently forcibly displaced from their country or from regions of their homeland (UNHCR 2020a). In 2019, 57 percent of all refugees came from three countries: Syria, Afghanistan, and South Sudan. Refugee women and female children are particularly vulnerable to sexual abuse and exploitation by locals, members of security forces, border guards, or other refugees. Wars are also dangerous for the very young. The Syrian Network for Human Rights (SNHR 2020) estimates that since the start of the Syrian Civil War in 2011, at least 29,017 children have been killed in the conflict. An estimated one in three children currently living in war-afflicted countries do not have access to schooling (UNICEF 2019). An estimated 77 percent of children born in refugee camps did not receive birth certificates, a pattern which creates long-term barriers to accessing legal rights and protections (UNHCR 2020b).

What do you think?

Large numbers of immigrants and refugees from predominantly Muslim nations in North Africa and the Middle East seeking safety in Europe and the United States have led to an increase anti-immigrant sentiments and policies. Former President Trump famously pronounced a "Muslim ban" in 2017 (National Immigration Law Center 2019), while fears that admitting refugees to the UK might increase the risk of terrorism have been linked to the 2016 "Brexit" vote that saw the UK cut ties with the European Union (Goodman and Narang 2019). Those who support increasing refugee access to asylum point out that policies targeting religious and ethnic minorities are reminiscent of policies enacted against the Jews just prior to the Holocaust. President Biden has promised to dramatically increase the number of refugees permitted to enter the United States (Biden-Harris 2020). Do you think enacting policies that limit refugees from seeking asylum in Europe and the United States is an appropriate response to the threat of terrorism?

Social-Psychological Costs

Terrorism, war, and living under the threat of war disrupt social-psychological well-being and family functioning (see this chapter's *Social Problems Research Up Close*). For example, Myers-Brown et al. (2000) report that Yugoslavian children suffered from depression, anxiety, and fear as a response to conflicts in that region, emotional responses not unlike those Americans experienced after the events of 9/11 (National Association of School Psychologists 2003). Furthermore, a study of children in postwar Sierra Leone found that over 70 percent of boys and girls whose parents had been killed were at "serious risk" of suicide (Morgan and Behrendt 2009). A study of 8- to 12-year-old Lebanese children found that children exposed to multiple events of war-related violence became increasingly desensitized to violence; as the number of war-related events children experienced increased, they became more likely to accept violence as normal and to imitate violent acts (Tarabah et al. 2015).

Guerrilla warfare is particularly costly in terms of its psychological toll on soldiers. In Iraq, soldiers were repeatedly traumatized as "guerrilla insurgents attack[ed] with impunity," and death was as likely to come from "hand grenades thrown by children, [as] earth-rattling bombs in suicide trucks, or snipers hidden in bombed-out buildings" (Waters 2005, p. 1). During the last half of the Iraq War, from 2008 to 2013, the suicide rate for U.S. soldiers surpassed that of the rate for civilians for the first time since the Vietnam War (Haiken 2013). When reservists are included in the totals with active-duty personnel, more military members committed suicide than died in combat in Iraq and Afghanistan (Donnelly 2011).

guerrilla warfare Warfare in which organized groups oppose domestic or foreign governments and their military forces; often involves small groups of individuals who use camouflage and underground tunnels to hide until they are ready to execute a surprise attack.

posttraumatic stress disorder (PTSD) A set of symptoms that may result from any traumatic experience, including crime victimization, war, natural disasters, or abuses.

Military personnel who engage in combat and civilians who are victimized by war may experience a form of psychological distress known as **posttraumatic stress disorder (PTSD)**, a clinical term referring to a set of symptoms that can result from any traumatic experience, including crime victimization, rape, or war. Symptoms of PTSD include sleep disturbances, recurring nightmares, flashbacks, and poor concentration (National Center for Posttraumatic Stress Disorder 2007). For example, Canadian Lieutenant General Roméo Dallaire, head of the United Nations peacekeeping mission in Rwanda, witnessed horrific acts of genocide. For years after his return, he continued to have images of "being in a valley at sunset, waist deep in bodies, covered in blood" (quoted in Rosenberg 2000, p. 14). PTSD is also associated with other personal problems, such as alcoholism, family violence, divorce, and suicide. According to one PTSD survivor:

> Post-traumatic stress disorder (PTSD) isn't something you just get over. You don't go back to being who you were. It's more like a snow globe. War shakes you up, and suddenly all those pieces of your life—muscles, bones, thoughts, beliefs, relationships, even your dreams—are floating in the air out of your grip. They'll come down. I'm here to tell you that, with hard work, you'll recover. But they'll never come down where they once were. You're a changed person after combat. Not better or worse, just different. (Montalván 2011, p. 50)

The rate of PTSD among soldiers is difficult to measure for several reasons. First, there is a lag, often of several years, between the time of exposure to trauma and the manifestation of symptoms. Second, many are reluctant to seek help because of the social stigma associated with being labeled mentally ill (Mittal et al. 2013). Among those who report PTSD or major depression symptoms, only about half seek treatment (Recovering Warrior Task Force 2014). The Department of Veterans Affairs estimates PTSD has afflicted 30 percent of Vietnam veterans, 12 percent of Gulf War veterans, 11 percent of Afghan war veterans, and 20 percent of Iraqi war veterans (U.S. Department of Veterans Affairs 2018). Evidence indicates that suicide prevention efforts by the Department of Veterans Affairs (VA) do work. A long-term study of suicide trends among veterans found that suicide rates among male veterans who utilized VA Health Services decreased by 30 percent over a ten-year period. Among male veterans who did not use VA Health Services, the suicide rate increased by 60 percent (Kemp 2014).

Diversion of Economic Resources

As discussed earlier, maintaining the military and engaging in warfare requires enormous financial capital and human support. However, the decision to spend, for example, $80 million for a single F-35 Joint Strike fighter jet, an amount that would pay to hire approximately 2,000 new public school teachers for a year, is a choice made by political leaders (Insinna 2019; National Education Association 2019). The U.S. military plans to purchase approximately 150 F-35 Joint Strike fighter jets per year, for an estimated annual cost of $12 billion. Allocating billions of dollars to fighter jets while schools, bridges, and other infrastructure deteriorate is a political choice.

The annual Department of Defense budget has increased from $580 billion in 2016 to $705 billion in 2021 (U.S. Department of Defense 2020). If just 1 percent of the defense budget were reallocated to social and human needs, it could (1) pay half a million households in the United States $1,200 a month for one year, (2) pay for over 200,000 scholarships for university students, (3) pay for nearly 200,000 children to attend Head Start for four years, (4) pay nearly 90,000 elementary school teachers for one year, and (5) provide over 11 million households with solar electricity for one year (National Priorities Project 2019). The requested budget for the Department of Defense for 2021 represents approximately half of all federal discretionary spending, meaning it is nearly as much as the combined total budgets for all the other federal departments and agencies combined (Office of Management and Budget 2020). (See Figure 15.7.)

Destruction of the Environment

The environmental damage that occurs during war devastates human populations long after war ends. Combatants may intentionally destroy or exploit the environment as a military strategy, a practice known as **ecocide**. For example, during the Vietnam War, the U.S. military dropped millions of gallons of herbicides, including the chemical Agent Orange, over the forests of the Mekong Delta in order to prevent the Viet Cong guerillas from finding safe haven in the dense forests (World Watch Institute 2015). The use of conflict minerals to fund guerilla groups in many African nations results in mining practices that cause harm to the local communities through forced labor, the loss of economic production, and damage to local water supplies.

ecocide The intentional destruction of the environment for military strategy.

Poaching of elephants and rhinos for the sale of ivory to fund militant groups has led to the near elimination of these species in many African nations (Geneva Graduate

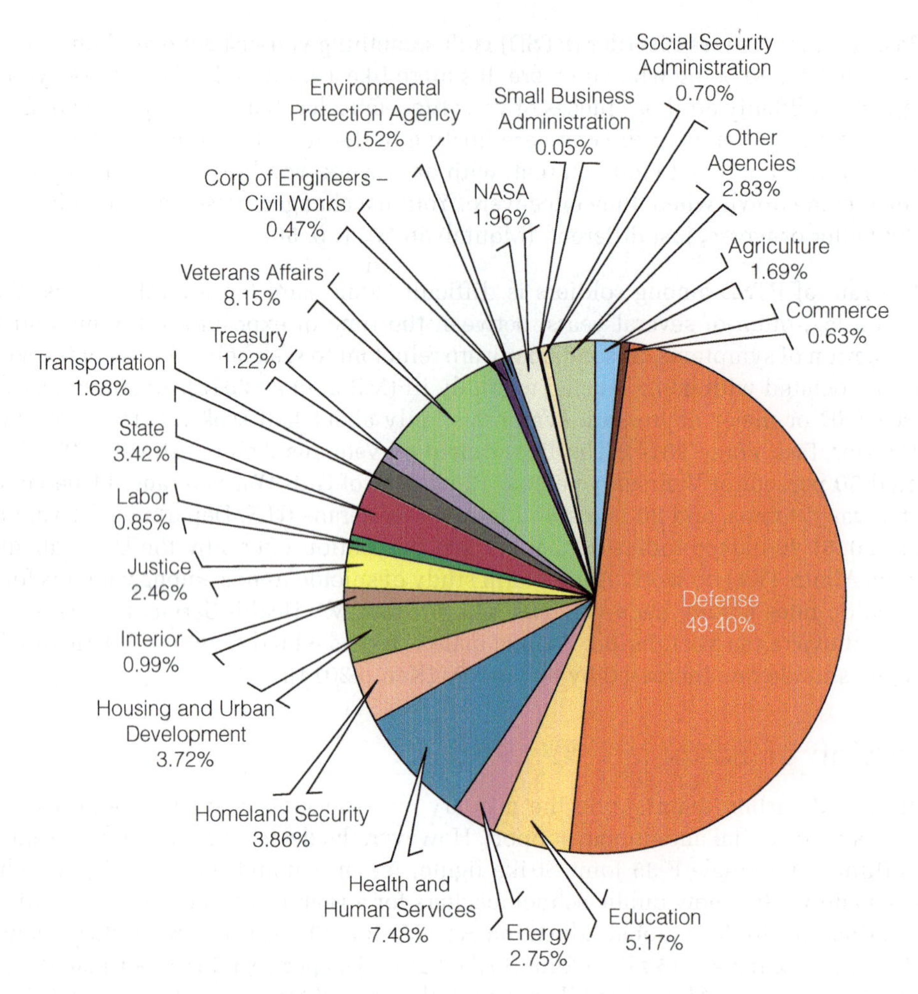

Figure 15.7 Federal Discretionary Budget Allocations, 2021 (proposed)
SOURCE: [Federal Discretionary Budget Allocations, 2021]. (n.d).

Institute for International Studies 2015). Sometimes the environmental impact of war is less deliberate. The military's disposal of garbage in burn pits and the contamination of water supplies from depleted uranium in ammunitions has contributed to the poor health conditions and high rates of cancer and birth defects in Afghanistan and Iraq (Watson Institute 2015).

In addition to being affected by war, the environment can also shape how wars are fought. Combat in many parts of the world is a seasonal activity; combatants in tropical climates must stop fighting during the rainy season, in cold climates during the winter, or in agricultural societies during planting and harvesting seasons. These patterns have helped establish some predictability in the tempo and tactics of violence, particularly for communities affected by ongoing civil war. Climate change, however, affects temperature, agriculture, sea levels, and rainfall, thus disrupting local communities' ability to anticipate and plan for fighting seasons (see Chapter 14; Geneva Graduate Institute for International Studies 2015).

nuclear winter The predicted result of a thermonuclear war whereby thick clouds of radioactive dust and particles would block out vital sunlight, lower temperature in the Northern Hemisphere, and lead to the death of most living things on Earth.

The ultimate environmental catastrophe facing the planet is thermonuclear war. Aside from the immediate human casualties, poisoned air, poisoned crops, and radioactive rain, many scientists agree that the dust storms and concentrations of particles created by a massive exchange of nuclear weapons would block vital sunlight and lower temperatures in the Northern Hemisphere, creating a **nuclear winter**. In the event of large-scale nuclear war, most living things on Earth would die.

The fear of nuclear war has greatly contributed to the military and arms buildup, which, ironically, also causes environmental destruction even in times of peace. For example, in practicing military maneuvers, the armed forces demolish natural vegetation, disturb wildlife habitats, erode soil, silt up streams, and cause flooding. Bombs exploded during peacetime leak radiation into the atmosphere and groundwater. From 1945 to 1990, 1,908 bombs were tested—that is, exploded—at more than 35 sites around the world leading to groundwater contamination. Further, over 70 years after the end of World War II, more than 2,000 tons of unexploded munitions are uncovered, for example, in Germany every year (Higginbotham 2016).

Finally, although arms control and disarmament treaties of the last decade have called for the disposal of huge stockpiles of weapons, no completely safe means of disposing of weapons and ammunition exist. Many activist groups have called for placing weapons in storage until safe disposal methods are found. Unfortunately, the longer the weapons are stored, the more they deteriorate, increasing the likelihood of dangerous leakage.

As part of the 1997 Chemical Weapons Convention, the United States agreed to destroy its entire stockpile of chemical weapons within 25 years. By the 2012 deadline, the United States had destroyed 90 percent of its stockpile, mostly mustard gas and nerve agents that were acquired during the Cold War. Destruction of the remaining 10 percent was delayed by local communities who opposed their destruction for fear of the release of harmful vapors. In February 2015, the U.S. Army announced it would begin destroying the remaining weapons using a new and safer technique of chemical neutralization. The last chemical weapons are expected to be destroyed by 2023, at a cost of $11 billion (Elliot 2015).

AP Images/Michel Lipchitz

Oil smoke from the 650 burning oil wells left in the wake of the Gulf War contains soot, sulfur dioxide, and nitrogen oxides, the major components of acid rain, along with a variety of toxic and potentially carcinogenic chemicals and heavy metals.

Strategies in Action: In Search of Global Peace

Various strategies and policies are aimed at creating and maintaining global peace. These include the redistribution of economic resources, the creation of a world government, peacekeeping activities of the United Nations, mediation and arbitration, and arms control.

Redistribution of Economic Resources

Inequality in economic resources contributes to conflict and war because the increasing disparity in wealth and resources between rich and poor nations fuels hostilities and resentment. Therefore, any measures that result in a more equal distribution of economic resources are likely to prevent conflict. John J. Shanahan (1995), retired U.S. Navy vice admiral and former director of the Center for Defense Information, suggested that wealthy nations can help reduce the social and economic roots of conflict by providing economic assistance to poorer countries. Nevertheless, U.S. military expenditures for national defense far outweigh U.S. economic assistance to foreign countries (see Figure 15.7).

As discussed in Chapter 12, strategies that reduce population growth are likely to result in higher levels of economic well-being. Funke (1994) explained that "rapidly increasing populations in poorer countries will lead to environmental overload and resource depletion in the next century, which will most likely result in political upheaval and violence as well as mass starvation" (p. 326). Although achieving worldwide economic well-being is important for minimizing global conflict, it is important that economic development does not occur at the expense of the environment.

Finally, former United Nations secretary general Kofi Annan, in an address to the United Nations, observed that it is not poverty per se that leads to conflict but rather the "inequality among domestic social groups" (Deen 2000). Referencing a research report completed by the Tokyo-based United Nations University, Annan argued that "inequality ... based on ethnicity, religion, national identity, or economic class ... tends to be reflected in unequal access to political power that too often forecloses paths to peaceful change" (Deen 2000).

The United Nations

Founded in 1945 after the devastation of World War II, the United Nations (UN) today includes 193 member states and is the principal organ of world governance. In its early years, the UN's main mission was the elimination of war from society. In fact, the UN charter begins, "We the people of the United Nations—determined to save succeeding generations from the scourge of war." During the past 70 years, the UN has developed major institutions and initiatives in support of international law, economic development, human rights, education, health, and other forms of social progress. The Security Council is the most powerful branch of the UN. Comprised of 15 member states, it has the power to impose economic sanctions against states that violate international law. It can also use force, when necessary, to restore international peace and security.

The UN has engaged in 71 peacekeeping operations since 1948, 13 of which were ongoing in 2020, involving more than 110,000 UN personnel (United Nations 2020a).

The UN has recently come under heavy criticism. First, in recent missions, developing nations have supplied more than 75 percent of the troops while developed countries—the United States, Japan, and Europe—have contributed 85 percent of the finances. As one UN official commented, "You can't have a situation where some nations contribute blood and others only money" (quoted by Vesely 2001, p. 8). Second, a review of UN peacekeeping operations noted several failed missions, including an intervention in Somalia in which 44 U.S. marines were killed (Lamont 2001). Third, as typified by the debate over the disarming of Iraq, the UN cannot take sides but must wait for a consensus of its members that, if not forthcoming, undermines the strength of the organization (Goure 2003). Even if a consensus emerges, without a standing army, the UN relies on troop and equipment contributions from member states, and there can be significant delays and logistical problems in assembling a force for intervention.

The consequences of delays can be staggering. Under the Genocide Convention of 1948, the UN is obligated to prevent instances of genocide, defined as "acts committed with the intent to destroy, in whole or in part, a national, racial, ethnical, or religious group" (United Nations 1948). In 2000, the UN Security Council formally acknowledged its failure to prevent the 1994 genocide in Rwanda (BBC 2000). After the death of ten Belgian soldiers in the days leading up to the genocide, and without a consensus for action among the members, the Security Council ignored warnings from the mission's commander about impending disaster and withdrew its 2,500 peacekeepers.

Finally, the concept of the UN is that its members represent individual nations, not a region or the world. And because nations tend to act in their own best economic and security interests, UN actions performed in the name of world peace may be motivated by nations acting in their own interests.

As a result of such criticisms, outgoing UN secretary general Kofi Annan called on the member states of the UN to approve the most far-reaching changes in the 60-year history of the organization (Lederer 2005). One of the most controversial recommendations concerns the composition of the Security Council, the most important decision-making body of the organization. Annan's recommendation that the 15 members of the Security Council—a body dominated by the United States, Great Britain, France, Russia, and China—be changed to include a more representative number of nations could, if approved, shift the global balance of power. However, the ability of the UN to promote global cooperation within a global trend of rising nationalism is uncertain. In his speech before the UN General Assembly in September 2019, former President Trump declared, "The future does not belong to globalists. The future belongs to patriots" (White House 2019). Conversely, President Biden declared "America is back" when he announced the foreign policy team for his incoming administration; "these public servants will restore America globally, its global leadership ... ready to lead the world, not retreat from it" (Jaffe, Lee, and Madahni 2020).

One major criticism of the effectiveness of UN peacekeeping operations is that wealthy nations contribute more funding and have greater influence over how operations are run, while poorer nations contribute more personnel and take on greater risks of enforcing the operations. As of May 2020, the United States contributed just 32 personnel to peacekeeping operations but nearly 28 percent of the funding, while Ethiopia contributed 6,646 personnel but less than 1 percent of the funding (United Nations 2020b). What do you think should be done with respect to the UN to ensure a fairer distribution of the risks and control of peacekeeping operations?

What *do you* think?

Mediation and Arbitration

Most conflicts are resolved through nonviolent means. Mediation and arbitration are just two of the nonviolent strategies used to resolve conflicts and to stop or prevent war. In **mediation**, a neutral third party intervenes and facilitates negotiation between representatives or leaders of conflicting groups. Good mediators do not impose solutions but rather help disputing parties generate options for resolving the conflict (Conflict Research Consortium 2003). Ideally, a mediated resolution to a conflict meets at least some of the concerns and interests of each party to the conflict. In other words, mediation attempts to find "win-win" solutions in which each side is satisfied with the solution. One of the most successful examples of mediation is the peace agreement facilitated by U.S. Senator George Mitchell that led to the Good Friday Agreement in 1998, ending nearly a half century of violent conflict in Northern Ireland.

mediation A neutral third party intervenes and facilitates negotiation between representatives or leaders of conflicting groups.

arbitration Dispute settlement in which a neutral third party listens to evidence and arguments presented by conflicting groups and arrives at a decision that the parties have agreed in advance to accept.

There are trade-offs to using mediation as a tool to resolve armed conflict. Some studies have found that while mediation can reduce violence in the short term, it can create long-term obstacles to sustainable peace, particularly once the involvement of the mediators is removed (Werner and Yuen 2005; Beardsley 2011). Other studies have found that peace is more likely to last when conflicts are resolved not through peacekeeping or mediation efforts, but when there is a decisive military victory (Toft 2010). Still, there is strong evidence that mediation saves lives, even if it doesn't lead to lasting peace between warring parties. Beardsley et al. (2019) studied all African civil wars from 1989 to 2008 and found that when mediation was combined with peacekeeping operations, there was an overall reduction in battle-related fatalities compared to conflicts that used only one or neither of these strategies.

Arbitration also involves a neutral third party who listens to evidence and arguments presented by conflicting groups. Unlike mediation, however, the neutral third party in arbitration arrives at a decision that the two conflicting parties agree in advance to accept. For instance, the Permanent Court of Arbitration—an intergovernmental organization based in The Hague since 1899—arbitrates disputes about territory, treaty compliance, human rights, commerce, and investment among any of its 121 member states who signed and ratified either of its two founding legal conventions. Recent cases include a dispute between the Netherlands and Russia over the Russian boarding of a Dutch ship, and between India and Italy over an Indian oil tanker flying an Italian flag (Permanent Court of Arbitration 2017).

Georges Gobet/AFP/Getty Images

A child soldier in Liberia points his gun at a cameraman while carting a teddy bear on his back. Although reliable figures are hard to obtain, the UN estimates that about 300,000 child soldiers are fighting in wars worldwide.

Arms Control and Disarmament

In the 1960s, the United States and the Soviet Union led the world in a nuclear arms race, with each competing to build a larger and more destructive arsenal of nuclear weapons than its adversary. If either superpower were to initiate a full-scale war, the retaliatory powers of the other nation would result in the destruction of both nations as well as much of the planet. Thus, the principle of **mutually assured destruction (MAD)** that developed from nuclear weapons capabilities transformed war from a win-lose proposition to a lose-lose scenario. If both sides would lose in a war, the theory suggested, neither side would initiate war. At its peak year in 1966, the U.S. stockpile of nuclear weapons included more than 32,000 warheads and bombs.

As their arsenals continued to grow at an astronomical cost, both sides recognized the necessity for nuclear arms control, including the reduction of defense spending, weapons production and deployment, and armed forces. Throughout the Cold War and even today, much of the behavior of the United States and the now former Soviet Union has been governed by major arms control initiatives, including the following:

- The Limited Test Ban Treaty that prohibited testing of nuclear weapons in the atmosphere, underwater, and in outer space
- The Strategic Arms Limitation Treaties (SALT I and II) that limited the development of nuclear missiles and defensive antiballistic missiles
- The Strategic Arms Reduction Treaties (START I and II) that significantly reduced the number of nuclear missiles, warheads, and bombs
- The Strategic Offensive Reduction Treaty (SORT) that required that the United States and the Russian Federation each reduce their number of strategic nuclear warheads to between 1,700 and 2,200 by 2012. This goal was met, with the United States currently holding 1,950 strategic warheads and the Russian Federation holding 1,800 (Center for Arms Control and Non-Proliferation 2013)
- The New Strategic Arms Reduction Treaty (new START) that further reduces the number of deployed strategic nuclear warheads to 1,550 for each country and significantly reduces the number of strategic nuclear missile launchers allowed (Atomic Archive 2020)

> "I will be blunt. Key components of the international arms control architecture are collapsing. The continued use of chemical weapons with impunity is driving new proliferation. Thousands of civilian lives continue to be lost because of illicit small arms and the use in urban areas of explosive weapons designed for open battlefields."
>
> **—UN SECRETARY-GENERAL ANTÓNIO GUTERRES' REMARKS TO THE CONFERENCE ON DISARMAMENT, FEBRUARY 2019**

With the end of the Cold War came the growing realization that, even as Russia and the United States greatly reduced their arsenals, other countries were poised to acquire nuclear weapons or expand their existing arsenals. Thus, the focus on arms control shifted toward the prevention of the spread of nuclear technology to nonnuclear states. Great strides have been made since the end of the Cold War in global disarmament efforts; however, these gains are currently at risk as rising nationalism around the world reduces the individual countries' trust and willingness to participate in systems of global cooperation.

mutually assured destruction (MAD) A Cold War doctrine referring to the capacity of two nuclear states to destroy each other, thus reducing the risk that either state will initiate war.

nuclear nonproliferation Efforts to prevent the spread of nuclear weapons, or the materials and technology necessary for the production of nuclear weapons.

Nuclear Nonproliferation Treaty. The **Nuclear Nonproliferation** Treaty (NPT), signed in 1970, was the first treaty governing the spread of nuclear weapons technology from the original nuclear weapons states (i.e., the United States, the Soviet Union, the United Kingdom, France, and China) to nonnuclear countries. The NPT was renewed in 2000 and is subject to review every five years. Currently, 187 countries have adopted it. The NPT holds that countries without nuclear weapons will not try to get them; in exchange, the countries with nuclear weapons agree that they will not provide nuclear weapons to countries that do not have them. Signatory states without nuclear weapons also agree to allow the International Atomic Energy Agency to verify compliance with the treaty through on-site inspections (Atomic Archive 2020). Only South Sudan, India, Israel, and Pakistan have not signed the agreement, although the last three are known to possess a nuclear arsenal. Furthermore, many experts suspect that Iran and Syria—both signatories to the NPT—are developing nuclear weapons programs. Both countries claim that their nuclear reactors are for peaceful purposes (e.g., domestic power consumption), not to develop a nuclear weapons arsenal. However, the International Atomic Energy Agency (IAEA) has repeatedly raised concerns that Iran may be violating the terms of the NPT (Arms Control Association 2018).

North Korea has been subject to harsh sanctions by world powers for decades in response to their human rights violations, aggressive weapons development, and efforts to develop nuclear warheads. Despite this global ostracism, former President Trump met with North Korean leader Kim Jong-Un at three high profile summits during 2018–2019. The Trump administration argued that the talks were a first step in encouraging North Korea to denuclearize, while critics argued that the summits simply played into the egos of both leaders and empowered North Korea to continue its aggressive actions. Do you think the Biden administration should try to bargain with North Korea to halt its weapons development or continue using sanctions to pressure the regime to change course?

What *do you* think?

As states that want to obtain nuclear weapons are quick to point out, nuclear states that advocate for nonproliferation possess well over 25,000 weapons, a huge reduction in the world's arsenal from Cold War days but still a massive potential threat to the earth. The New Strategic Arms Reduction Treaty, signed by the United States and Russia in 2009, has led to a reduction in nuclear arsenals by at least 25 percent. Still, the United States and Russia hold, by a wide margin, the largest nuclear arsenals in the world. "Do as I say, not as I do" is a weak bargaining position. This is the argument posited by Iran, which for over a decade has been subject to harsh economic sanctions spearheaded by the United States and its allies in an effort to prevent it from developing a nuclear weapons program.

The governments of North Korea and Iran, for example, have argued that they have the same right as every other sovereign nation to develop nuclear weapons for self-defense as well as energy production. Detractors have argued that both North Korea and Iran pose a substantial threat to their neighbors and to United States' interests in the region, especially U.S. allies like South Korea and Israel.

In the summer of 2015, after months of negotiations and amid substantial controversy and opposition, the Obama administration reached a historic treaty with Iran, in which the United States agreed to alleviate the crippling economic sanctions against Iran in exchange for Iran reducing its nuclear program and permitting extensive international oversight. The agreement allowed Iran to continue to develop nuclear technology, but only for energy, not for weapons. Conservative hard-liners in both countries opposed the agreement, each arguing that the other side could not be trusted. In particular, the agreement strained relations between the United States and its strongest Middle Eastern ally, Israel, which argued that Iran poses a direct threat to its existence.

Handout/Dong-A Ilbo/Getty Images News/Getty Images

Former President Trump met with North Korean leader Kim Jong-Un three times during 2018–2019. The Trump administration argued that developing a stronger relationship with the infamous dictator is an important step in encouraging North Korea to halt its nuclear weapons development. Former UN Secretary General Ban Ki-Moon publicly criticized these meetings, saying that such high-profile attention has given Kim Jong-Un exactly want he wants, saying that Kim "seems to have succeeded in acquiring de facto nuclear state status" (Haynes 2020).

The intense political challenges to the agreement within both countries highlight the difficulty of achieving successful strategies of global peace when domestic and international interests conflict. In May 2018, President Trump withdrew from the Iran nuclear deal, arguing that the agreement was not strong enough to inhibit Iran's aggression in the region. Despite the withdrawal of the United States, Iran has said that it will continue to abide by the terms of the deal with the remaining European signatories: France, Britain, and Germany (Landler 2018).

The Problem of Small Arms

Although the devastation caused by even one nuclear war could affect millions, the easy availability of

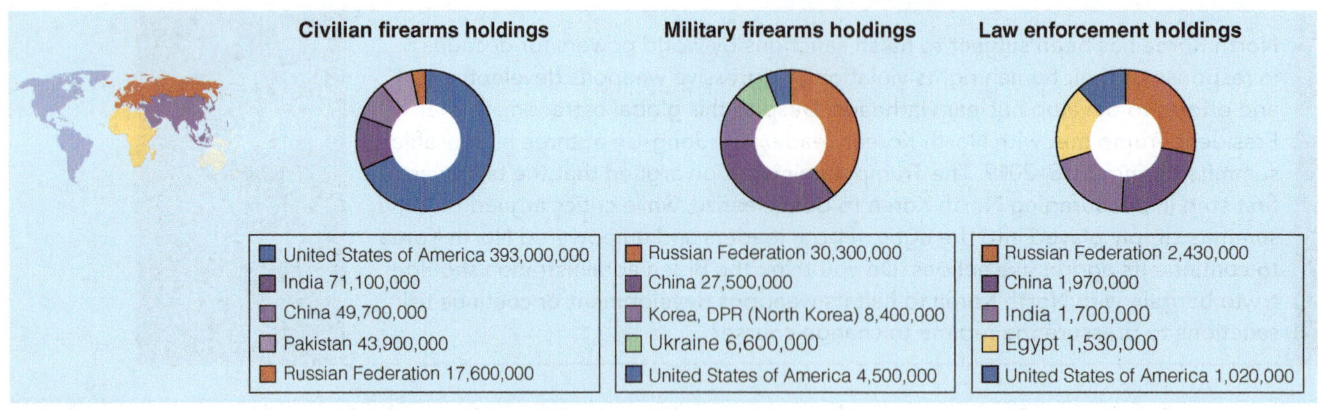

Note: Infographic produced for the release of the updated Global Firearms Holdings database in June 2018

Figure 15.8 Countries with the Largest Firearms Holdings, by Sector
SOURCE: Small Arms Survey (2018).

conventional weapons fuels many active wars around the world. Small arms and light weapons (also called conventional weapons) include handguns, submachine guns and automatic weapons, grenades, mortars, land mines, and light missiles. It is estimated that there the total number of firearms in circulation around the world increased from 875 million in 2006 to more than 1 billion in 2017 (Geneva Graduate Institute for International Studies 2018). The United States has the largest share total share of firearms, with over 393 million held by civilians, 4.5 million held by the military, and over 1 million held by law enforcement (see Figure 15.8).

Unlike control of **weapons of mass destruction** such as chemical and biological weapons, controlling the flow of small arms—especially firearms—is not easy because they have many legitimate uses by military, by law enforcement officials, and for recreational or sporting activities. Small arms are easy to afford, use, conceal, and transport illegally. Small arms trade is also lucrative. According to official records, in 2016, the most recent year for which data are available, the international small arms trade was worth at least $6.5 billion. The top three exporters of small arms and light weapons were the United States, Italy, and Brazil, which collectively were responsible for nearly 40 percent of all small arms exports (Pavesi 2019). The United States alone accounted for 18 percent of all small arms exports. In terms of moving conventional weapons, Iran, North Korea, and Saudi Arabia were the least transparent exporters.

The State Department's Office of Weapons Removal and Abatement administers a program that supports the destruction of small arms and light weapons. Between 1993 and 2019, the United States spent $2.7 billion destroying conventional weapons around the world (Office of Weapons Removal and Abatement 2020). If not destroyed, the availability of these weapons might fuel terrorist groups and undermine efforts to promote peace after wars have formally concluded.

Understanding Conflict, War, and Terrorism

As we come to the close of this chapter, how might we have an informed understanding of conflict, war, and terrorism? Each of the three theoretical positions discussed in this chapter reflects the realities of global conflict. As structural functionalists argue, war offers societal benefits—social cohesion, economic prosperity, scientific and technological developments, and social change. Furthermore, as conflict theorists contend, wars often occur for economic reasons because corporate elites and political leaders benefit from the spoils of war—land and water resources and raw materials. The symbolic interactionist perspective emphasizes the role that meanings, labels, and definitions play in creating conflict and contributing to acts of war.

weapons of mass destruction (WMD) Chemical, biological, and nuclear weapons that have the capacity to kill large numbers of people indiscriminately.

The September 11 attacks on the World Trade Center and the Pentagon and the aftermath—the battle against terrorism, the wars in Iraq and Afghanistan, the Arab Spring, and the civil war in Syria—changed the world Americans live in. For some theorists, these events were inevitable. Political scientist Samuel P. Huntington argued that such conflict represents a "clash of civilizations." In *The Clash of Civilizations and the Remaking of World Order* (1996), Huntington argued that, in the new world order,

> the most pervasive, important and dangerous conflicts will not be between social classes, rich and poor, or economically defined groups, but between people belonging to different cultural entities. ... The most dangerous cultural conflicts are those along the fault lines between civilizations ... the line separating peoples of Western Christianity, on the one hand, from Muslim and Orthodox peoples on the other. (p. 28)

Some trends seem to support Huntington's view. Polls show an increase in anti-Muslim sentiment in Europe (Massoumi 2020) and a rise in religious-based hate crimes in the United States (Beirich 2019) over the past decade.

The clash of civilizations perspective has been vehemently criticized, however, by many scholars who see this view as divisive, overly simplistic, and historically inaccurate. In particular, the clash of civilizations perspective minimizes the vast majority of citizens in predominantly Muslim nations who are religiously and politically moderate, and who share common concerns with Western nations over the dangers of religious extremism and terrorism. For example, surveys of Muslims in Middle Eastern countries have found that the majority were concerned about Islamic extremism in their country; had an unfavorable view of ISIS, al-Qaeda, the Taliban, and Boko Haram; and believed that suicide bombings were never justified (Poushter 2015; Pew Research Center 2013).

David Brooks summarizes this critique by suggesting that Huntington committed a "Fundamental Attribution Error. That is, he ascribed to traits qualities that are actually determined by context" (2011, p. 1). Huntington suggested that Arab societies are intrinsically opposed to democracy and not nationalistic, but the Arab Spring revolutions highlighted how certain political regimes can effectively, although never permanently, suppress national patriotism and the intrinsic human desire for liberty. Brooks also suggests that Huntington fundamentally misunderstood the nature of culture for, despite our intrinsic and cultural differences, we are all alike: "Huntington minimized the power of universal political values and exaggerated the influence of distinct cultural values ... underneath cultural differences there are universal aspirations for dignity, for political systems that listen to, respond to and respect the will of the people" (p. 1).

Ultimately, we are all members of one community—Earth—and have a vested interest in staying alive and protecting the resources of our environment for our own and future generations. World leaders have traditionally followed the advice of philosopher Carl von Clausewitz: "If you want peace, prepare for war." Thus, nations have sought to protect themselves by maintaining large military forces and massive weapons systems. These strategies are associated with serious costs, particularly in hard economic times. In diverting resources away from social problems, defense spending undermines a society's ability to improve the overall security and well-being of its citizens. Conversely, defense-spending cutbacks could potentially free up resources for other social agendas including lowering taxes, reducing the national debt, addressing environmental concerns, eradicating hunger and poverty, improving health care, housing and transportation, and upgrading educational services. Therein lies the promise of a "peace dividend." The hope is that future dialogue on the problems of war and terrorism will redefine national and international security to encompass social, economic, and environmental well-being.

Chapter Review

- **What is the relationship between war and industrialization?**
War indirectly affects industrialization and technological sophistication because military research and development advances civilian-used technologies. Industrialization, in turn, has had two major influences on war: The more industrialized a country is, the lower the rate of conflict; and if conflict occurs, the higher the rate of destruction.

- **What are the latest trends in armed conflicts?**
Since World War II, wars between two or more states make up the smallest percentage of armed conflicts. In the contemporary era, the majority of armed conflicts have occurred between groups in a single nation-state, who compete for the power to control the resources of the state or to break away and form their own state.

- **In general, how do feminists view war?**
Feminists are quick to note that wars are part of the patriarchy of society. Although women and children may be used to justify a conflict (e.g., improving women's lives by removing the repressive Taliban in Afghanistan), the basic principles of male dominance and control are realized through war. Feminists also emphasize the social construction of aggressive masculine identities and their manipulation by elites as important reasons for the association between masculinity and militarized violence.

- **What are some of the causes of war?**
The causes of war are numerous and complex. Most wars involve more than one cause. Relative deprivation theory focuses on how inequality can drive conflict. Values and ideologies, racial, ethnic and religious hostilities, and competition over land and resources can overlap with existing patterns of inequality and lead to armed conflict. In recent years, nationalism has emerged as a major factor driving identity-based conflicts.

- **What is terrorism, and what are the different types of terrorism?**
Terrorism is the premeditated use, or threatened use, of violence by an individual or group to gain a political or social objective. Terrorism can be either transnational or domestic. Transnational terrorism occurs when acts of terrorism are not restricted to or centered within one country. Domestic terrorism takes place within the territorial jurisdiction of one nation, such as the 1995 truck bombing of a nine-story federal office building in Oklahoma City. Increasingly, the globalization of social media technologies makes it difficult to distinguish between domestic and transnational terrorism.

- **What are some of the macro-level "roots" of terrorism?**
Many scholars have focused on macro-structural factors, such as global economic shifts, foreign occupation, repression of minorities, poverty, and weak governance as creating the underlying conditions that produce terrorism.

- **How has the United States responded to the threat of terrorism?**
The United States has used both defensive and offensive strategies to fight terrorism. Defensive strategies include using metal detectors and X-ray machines at airports and strengthening security at potential targets, such as embassies and military command posts. The Department of Homeland Security coordinates such defensive tactics. Offensive strategies include retaliatory raids such as the U.S. bombing of terrorist facilities in Afghanistan, group infiltration, and preemptive strikes. U.S. law makes it easier to designate, track, and counter the efforts of foreign terrorist organizations compared to domestic terrorism organizations.

- **What is meant by "diversion of economic resources"?**
Worldwide, the billions of dollars used on defense could be channeled into social programs dealing with, for example, education, health, and poverty. Thus, defense monies are economic resources diverted from other needy projects.

- **What are some of the criticisms of the United Nations?**
First, in recent missions, developing nations have supplied more than 75 percent of the troops. Second, several recent UN peacekeeping operations have failed. Third, the UN cannot take sides but must wait for a consensus of its members that, if not forthcoming, undermines the strength of the organization. Fourth, the concept of the UN is that its members represent individual nations, not a region or the world. Because nations tend to act in their own best economic and security interests, UN actions performed in the name of world peace may be motivated by nations acting in their own interests. Finally, the Security Council limits power to a small number of states.

- **What problems do small arms pose?**
Even after a conflict ends, these weapons circulate in society, making crime worse or falling into the hands of terrorists. Trade in small arms is legal because they have many legitimate uses—for example, by the military, police, and hunters. Because they are small and simple to handle, these weapons are easily concealed and transported, making it difficult to control them.

Test Yourself

1. War between states is still the most common form of warfare.
   a. True
   b. False
2. The rise of the modern state is most directly a result of
   a. industrialization and the creation of national markets.
   b. innovations in communications technology.
   c. the development of armies to control territory.
   d. the development of police to control a population.
3. Which sociological theory would be most likely to emphasize that war primarily benefits political and military elites through the development of the military-industrial complex?
   a. Structural-functionalist
   b. Symbolic interactionist
   c. Conflict
   d. Feminism
4. Which sociological theory would be most likely to emphasize examples of the positive impacts that war has had on society, for example, increasing social solidarity and encouraging the expansion of civil rights to minority groups?
   a. Structural-functionalist
   b. Symbolic interactionist
   c. Conflict
   d. Feminism
5. On the whole, conflicts over values and ideologies are more difficult to resolve than those over material resources.
   a. True
   b. False
6. Which of the following factors is a likely cause of revolutions or civil wars?
   a. A weak or failed state
   b. An authoritarian government that ignores major demands from citizens
   c. The availability of strong opposition leaders
   d. All of these
7. Primordial explanations of ethnic conflict suggest
   a. that ethnic leaders instigate hostilities to serve their own interests.
   b. that people become hostile when they blame their frustration with economic hardship on competing ethnic groups.
   c. that ancient hatreds compel ethnic groups to continue fighting.
   d. none of these.
8. Recent trends indicate that nationalism is declining around the world and more countries favor global cooperation over national self-interest.
   a. True
   b. False
9. Why does the federal government not follow the same process in designating domestic and foreign extremist groups as terrorist organizations?
   a. It is easier to track domestic terrorist groups compared to foreign ones.
   b. Constitutional protections limit the government's ability to prosecute and track political extremists within the United States.
   c. Domestic terrorism is not a significant issue in the United States.
   d. The FBI lacks sufficient funding to track domestic terrorist organizations.
10. The consequences of war and the military on the environment
    a. are prevalent only during wartime.
    b. are negligible.
    c. have been mostly reduced by technological innovations.
    d. persist in peacetime or for many years after a war is over.

Answers: 1. B; 2. C; 3. C; 4. A; 5. A; 6. D; 7. C; 8. B; 9. B; 10. D.

Key Terms

arbitration 623
civil–military relations 601
Cold War 595
conflict minerals 604
constructivist explanations 605
democratic peace theory 605
domestic terrorism 610
dual-use technologies 599
ecocide 619
Geneva Conventions 613
guerrilla warfare 618
mediation 623
military-industrial complex 600
mutually assured destruction (MAD) 624
nationalism 609
NATO 594
nuclear nonproliferation 624
nuclear winter 620
posttraumatic stress disorder (PTSD) 618
primordial explanations 605
rally round the flag effect 598
relative deprivation 606
security dilemma 606
state 592
Terrorism 610
Transnational terrorism 610
War 592
weapons of mass destruction (WMD) 626
white supremacist organizations 614

Epilogue

Today, there is a crisis—a crisis of faith: faith in political leadership, faith in the American dream, faith in equality and freedom, the very essence of democracy. Is American democracy in danger? Many Americans think so. After the attack on the U.S. Capitol on January 6, 2021, when a sample of U.S. adults were asked how they would describe democracy and the rule of law in America today, less than a third responded that it was "somewhat secure" while 71 percent responded that it was "threatened" (Salvanto et al. 2021).

Mettler and Lieberman (2020) identify four threats to democracy which, for the first time in American history, are copresent. The first, ***political partisanship***, was researched by sociologists Arlie Hochschild (2016), who was concerned about "the increasingly hostile split in our nation between two political camps." So she moved to Louisiana where she conducted interviews, took field notes, and asked questions, trying to understand the divisiveness by exploring the *feelings* of one-half of the equation—those on the right side of the political spectrum. After researching the topic for five years, Hochschild concluded that the "deep story" of White, older, Christian, American males, and, to a lesser extent, females was that they *feel* angry, frustrated, and marginalized by (1) demographic changes (i.e., the growth of minority and immigrant populations), (2) left-leaning Northerners imposing their liberal values on rural and Southern cultures (e.g., same-sex marriages), (3) government handouts (e.g., welfare), (4) federal regulations that lead to job losses (e.g., EPA regulations), and (5) falling or flat wages, a result, in part, of globalization.

The people she talked to felt like, after all, they had worked hard, played by the rules and, metaphorically, had been standing in line for the American dream for a long time. Now they were in the middle of the line when, suddenly, the line seemed to have stopped moving (Hochschild 2016, p. 137).

> Look! You see people *cutting in line ahead of you!* You're following the rules. They aren't. As they cut in, it feels like you are being moved back. How can they just do that? Who are they? Some are black. Through affirmative action plans, pushed by the federal government, they are being given preference for places in colleges and universities, apprenticeships, jobs, welfare payments, and free lunches… . Women, immigrants, refugees … where will it end? Your money is running through a liberal sympathy sieve you don't control or agree with. … It's not fair.

Although this is just one "deep story"—and surely there are others—Hochschild's analysis reveals the power of *identity politics* wherein race, religion, national origin, LGBTQ status, and gender in the communities she researched brought about an incalcitrant belief in the conservative agenda, Fox News, and a world divided into "we" versus "they."

And left-leaning Americans? They too have a "deep story" summed up by journalist Lee Drutman (2017). They feel betrayed by the election of a president whose "xenophobic, racist rhetoric stands in opposition to the true American vision of tolerance. It's an affront to our nation of immigrants, a country in which equality is written into our founding documents. Any Republican who supports or voted for him is guilty by association" (p. 1).

Political partisanship is destructive, leading to fractured relationships, distrust of government, legislative gridlock, and incivility. And that's not all. More than half of

Americans believe that politically motivated violence will escalate in the coming years (Bowden 2021), and a new report on domestic terrorism predicts that violence from white supremacist groups, those angered by what they believe to be a fraudulent outcome of the 2020 presidential election, and federal overreach in response to the COVID-19 pandemic will increase in 2021 (Shinkman 2021).

As the excerpt from Hochschild's book reveals and the domestic terrorism report warns, perceptions that "others" unfairly "got ahead," whether in an election or life, fuel partisanship. It is no coincidence that an analysis of the demographic composition of political parties in the United States reveals stark differences. Results of a 2020 survey indicate that Republican and Republican-leaning respondents were disproportionately White evangelical, noncollege educated, rural and Southern males who attended church at least once a week. Alternatively, Democratic and Democratic-leaning respondents were disproportionately White, college-educated women, Black women, Hispanic Catholics, the religiously unaffiliated, and urban Northeasterners (Pew Research Center 2020). Not surprisingly, Republicans and Democrats have very different perspectives on the issues of inclusivity and diversity.

Taken as a whole, Democrats support the Equal Rights Amendment (ERA), the George Floyd Justice in Policing Act, and the Equality Act, while, in general, Republicans do not. For example, while not opposing equality for women per se, some Republicans fear that passage of the ERA would strengthen abortion rights and allow transgender women to seek refuge in a battered women's shelter (Cornwell 2021). Similarly, some conservative groups, primarily opposed to transgender rights, argue that passage of the Equality Act would lead to "a backdoor way to ... let boys win in girls' sports, pressure children toward experimental cross-sex hormones—and silence all of us who disagree" (January 2021, p. 1).

Thus, there is a fundamental difference between the two parties in terms of who should have what rights or, as Mettler and Lieberman (2020) refer to it, ***conflict over who belongs***, the second threat to democracy. Nowhere is that clearer than in the debate over voter suppression laws and immigration. Presently, 43 states "are debating more than 250 voting restriction bills that all pursue the same fundamental strategy: giving one party an advantage in elections by making it harder for communities of color to cast their ballots" (Frosh, Racine, and Eisen 2021, p. 1). Alternatively, many people have very real concerns over the security of the voting process and advocate tightening regulations as a means of insuring free and fair elections.

Of late, President Biden has been accused by Republicans of creating a new border crisis, the result of campaign promises to ease immigration restrictions and proposed legislation, the *U.S. Citizenship Act of 2021*, that provides a pathway to citizenship (Antle 2021). Yet, in contrast to Republican objections, the majority of Americans support a law that would grant "Dreamers," i.e., immigrants who came to the United States as children, permanent legal status (Krogstad 2020).

Alternatively, in researching Trump's popularity, Young (2016) concludes that support "for Trump is, indeed, a function of people's degree of nativist (e.g., anti-immigrant) belief: the more nativist in orientation the person is, the more likely they are to support Trump. The same relationship, in turn, exists between identifying as a Republican and Nativism: the more nativist, the more likely to identify as a Republican" (p. 8).

Trump took advantage of *existing* fault lines, campaigning on a platform intended to unite all with unheard or unanswered grievances, the "we," not the "they" of America. In a survey, nearly one-third of Americans responded that someone could not truly be American if they weren't "Christian," and an equal number responded that someone must be "born in the United States" to be a true American (Stokes 2017). Moreover, Americans who most often responded that open borders could lead to losing our national identity were demographic clones of Hochschild's interviewees, waiting in line for the American dream—White, older men, with less than a college degree (Brockaway and Doherty 2019).

Over the course of his campaigns, the former president made it clear that immigrants are responsible for most of our most pressing social problems, echoing each of the following at some point in his many rallies: "[T]hey bring crime; they import poverty; they spread disease; they don't assimilate; they corrupt our politics; they steal our jobs; they cause our taxes to increase; they're a security risk; their religion is incompatible with American values; they can never be 'true Americans'" (Anbinder 2019, p. 1).

A nativist's ideology embraces the exclusion of all "others"—the *they*—Black or Hispanic, gay or transgender, Jew or Muslim, woman or immigrant. Nativism is increasing in the United States and is clearly an afront to democracy for it implies a preference for a tiered system of citizenship (Kleinfeld and Dickas 2020). As second-class citizens (1) *they* are not entitled to participate equally in their democracy, (2) *they* should have less weight given to their concerns than other citizens, (3) *they* have less legitimate policy concerns, (4) *they* deserve inferior public services, and/or (5) *they* do not have the right to be treated the same as other citizens.

Economic inequality, the third threat to democracy, has been growing for decades and has only increased as a result of the pandemic. In 2019, economist Elise Gould from the Economic Policy Institute testified before the U.S. House Ways and Means Committee that poor wage performance "is at the root of the large rise in overall income inequality" (Gould 2019, p. 2). As a result of wage suppression, between 2000 and 2018, household incomes grew just 0.3 percent and the number of middle-income households continued to shrink (Horowitz, Igielnik, and Kochhar 2020).

One year into the pandemic, three-quarters of lower-income respondents rated their finances negatively (Horowitz, Brown, and Minkin 2021), and in the last half of 2020, the poverty rate increased by 2.4 percent, the largest increase since the 1960s. Eight million people were added to the ranks of the poor, and nearly 30 million people lived in households where there was not enough food to eat (Scigliuzzo 2021). Moreover, since the beginning of the pandemic, more than 70 million people have applied for unemployment benefits—approximately 40 percent of the labor force (Kelly 2020).

At the other end of the continuum, the net worth of American billionaires increased by $1.1 trillion between April 2020 and January 2021, their collective wealth increasing by nearly 40 percent (Beer 2021). On March 16, 2020, five days after COVID-19 was declared a pandemic, the stock market saw its worst day in history, plunging to historic lows. Less than three months later, it rebounded (Woods 2020). The top one-fifth of Americans, owning nearly 100 percent of all privately held stocks in the United States (Scigliuzzo 2021), saw a windfall and could then reinvest their profits in the bullish market. Corporate-friendly economic policies (e.g., tax laws), each intended to prop up the struggling pandemic economy, were passed (Woods 2020) but little "trickled down" to lower-income households. Income inequality in the United States is higher than that in any other G7 country in the world, including the United Kingdom, Italy, Japan, Canada, Germany, and France (Schaeffer 2020).

The final threat to American democracy is the ***expansion of the executive powers*** of the federal government. Franklin Roosevelt, a Democrat, expanded the power of the presidency in pursuit of the New Deal during the Great Depression, and Richard Nixon, a Republican, sought power "in the service of his own political interests" in the Watergate scandal (Mettler and Lieberman 2020). But in both cases, the system responded to reign in the executive branch, in the first case through other branches of the government, and, in the second, through impeachment. It should also be noted that during these administrations, the expanding power of the executive branch was the only threat to democracy present at the time.

Mettler and Lieberman (2020) argue that by the end of the 20th century, all four threats "escalated and converged," and, by the 2010s, a "perfect storm" had been created. Enter Donald Trump, a businessman and former entertainer, with no prior political experience:

> ready to ride the storm to political victory, harnessing its fury in his quest for power, with no concern for what might be destroyed in its path. As all four threats reached high velocity and in combination generated even greater momentum, the embattled political system showed a profound lack of capacity to rein in the president. When all four threats crest in tandem, in turns out, a president and his partisan allies who control one or both chambers of Congress can threaten basic principles of American democracy in plain sight and get away with it. (p. 212)

The alleged transgressions of President Trump are too numerous to discuss here, although they are summarized in a book entitled *After Trump: Reconstructing the Presidency* (Bauer and Goldsmith 2020a). The authors acknowledge, however, that the former president simply exploited *existing* "dangerous excesses of authority and dangerous weaknesses in accountability" and that he was not the first president to do so (Bauer and Goldsmith 2020b, p. 2). Nonetheless, he is the first president to seemingly encourage citizens to violently contest the results of a lawful election and to be impeached twice.

Do you think American democracy is in danger, and, if you do, what can you do about it? What can you do about any of the social problems discussed in this book? There is always that sense that I, an individual, can't do much about the world's problems. But you can, and you must, because if you don't, who will? Regardless of being left- or right-leaning, male or female, young or old, Black or White, gay or straight, Jew, gentile, or Muslim, as trite as it sounds, the future of this country—and now more than ever—the world, is in your hands.

The social problems of today are the cumulative result of structural and cultural forces over time. Today's problems are not necessarily more or less serious than those of generations ago—they are different and, perhaps, more diverse as a result of the increased complexity of social life. They FEEL insurmountable—political partisanship, conflict over who belongs, economic inequality, and expanding executive powers. Even President Biden in his inaugural address acknowledged the fragility of democracy—but he also described its resilience. He continues (Biden 2021):

> Few periods in our nation's history have been more challenging or more difficult than the one we're in now. ... A once in a century virus... . Millions of jobs ... lost. Hundreds of thousands of businesses closed. ... A cry for racial justice some 400 years in the making ... . A cry from the planet itself. ... And now, a rise in political extremism, white supremacy, domestic terrorism ... . To overcome these challenges—to restore the soul and to secure the future of America— ... requires that most elusive things in a democracy. Unity....

Unity. Coming together as Americans working for a common good. Can you, as an individual, make a difference? Rosa Parks did. Rosa Parks was a Black seamstress in Montgomery, Alabama, in the 1950s. Like almost everything else in the South in the 1950s, public transportation was racially segregated. On December 1, 1955, Rosa Parks was on her way home from work when the "white section" of the bus she was riding became full. The bus driver told Black passengers in the first row of the "black section" to relinquish their seats to the standing White passengers. Rosa Parks refused. She was arrested and put in jail, but her treatment so outraged the Black community that a boycott of the bus system was organized by a new minister in town—Martin Luther King Jr. The Montgomery bus boycott was a success. Just 11 months later, in November 1956, the U.S. Supreme Court ruled that racial segregation of public facilities was unconstitutional. Rosa Parks had begun a process that in time would echo her actions—the civil rights movement, the March on Washington, the 1963 *Equal Pay Act*, the 1964 *Civil Rights Act*, the 1965 *Voting Rights Act*, regulations against discrimination in housing, affirmative action, and Black Lives Matter.

Was social change accomplished? In 1957, just 18 percent of Black Americans 25 years old and older had completed high school or college, compared with 43 percent of White Americans. By 2019, 88 percent of Black Americans 25 years old and older had completed high school or college, compared with 90 percent of White Americans (U.S. Census Bureau 2019). While issues of prejudice and discrimination and all their attendant problems have not been eliminated, who among us would want to return to the "good old days" of the 1950s in Montgomery, Alabama?

Although only a fraction of the readers of this text will occupy social roles that directly influence social policy, one need not be a politician or member of a social reform group to make a difference. We, the authors of this text, challenge you, the reader, to make decisions and take actions, individually or collectively, to make the world a more humane, just, and peaceful place for all. If you feel that your own actions cannot alleviate problems that are bigger than you are, consider the perspective of former President Obama: "Change will not come if we wait for some other person or some other time. We are the ones we've been waiting for. We are the change that we seek."

So where should you begin? Where Rosa Parks and others like her began—with a simple act of courage, commitment, and faith.

Appendix

Methods of Data Analysis

Description, Correlation, Causation, Reliability and Validity, and Ethical Guidelines in Social Problems Research

There are three levels of data analysis: description, correlation, and causation. Data analysis also involves assessing reliability and validity.

Description

Qualitative research involves verbal descriptions of social phenomena. Having a homeless and single pregnant teenager describe her situation is an example of qualitative research.

Quantitative research often involves numerical descriptions of social phenomena. Quantitative descriptive analysis may involve computing the following: (1) means (averages), (2) frequencies, (3) mode (the most frequently occurring observation in the data), (4) median (the middle point in the data; half of the data points are above the median and half are below it), and (5) range (the highest and lowest values in a set of data).

Correlation

Researchers are often interested in the relationship between variables. *Correlation* refers to a relationship between or among two or more variables. The following are examples of correlational research questions: What is the relationship between poverty and educational achievement? What is the relationship between race and crime victimization? What is the relationship between religious affiliation and divorce?

If there is a correlation or relationship between two variables, then a change in one variable is associated with a change in the other variable. When both variables change in the same direction, the correlation is positive. For example, in general, the more sexual partners a person has, the greater the risk of contracting a sexually transmissible disease (STD). As variable A (number of sexual partners) increases, variable B (chance of contracting an STD) also increases. Similarly, as the number of sexual partners decreases, the chance of contracting an STD decreases. Notice that in both cases, the variables change in the same direction, suggesting a positive correlation (see Figure A.1).

When two variables change in opposite directions, the correlation is negative. For example, there is a negative correlation between condom use and contracting STDs. In other words, as condom use increases, the chance of contracting an STD decreases (see Figure A.2).

The relationship between two variables may also be curvilinear, which means that it varies in both the same and opposite directions. For example, suppose a researcher finds that after drinking one alcoholic beverage, research participants are more prone to violent behavior. After two drinks, violent behavior is even more likely, and this trend continues for three and four drinks. So far, the correlation between alcohol consumption and violent behavior is positive. After the research participants have five alcoholic drinks, however, they become less prone to violent behavior. After six and seven drinks, the likelihood of engaging in violent behavior decreases further. Now the correlation between alcohol consumption and violent behavior is negative. Because the correlation changed from positive to negative, we say that the correlation is curvilinear (the correlation may also change from negative to positive) (see Figure A.3).

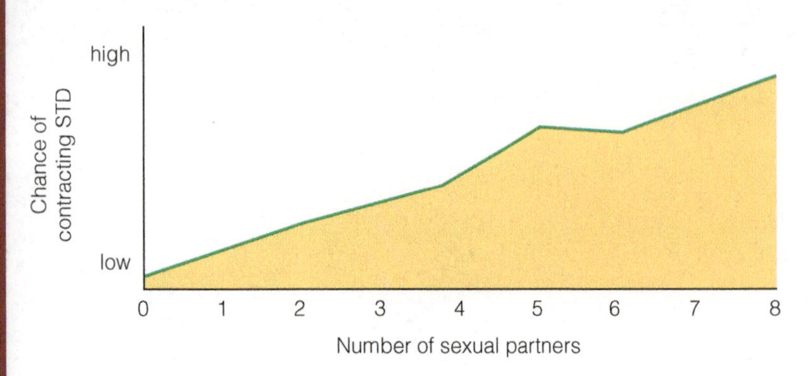

FIGURE A.1

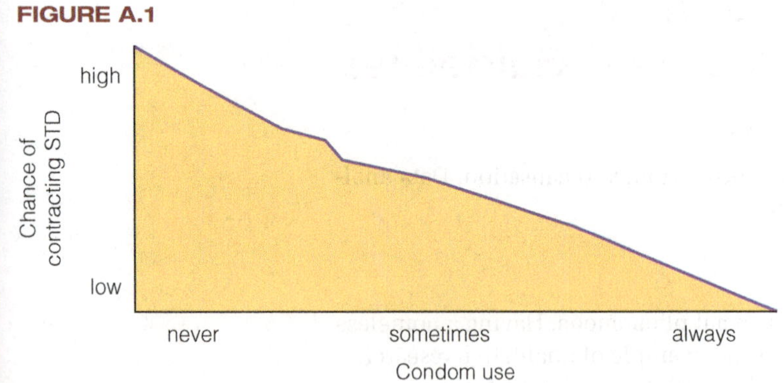

FIGURE A.2

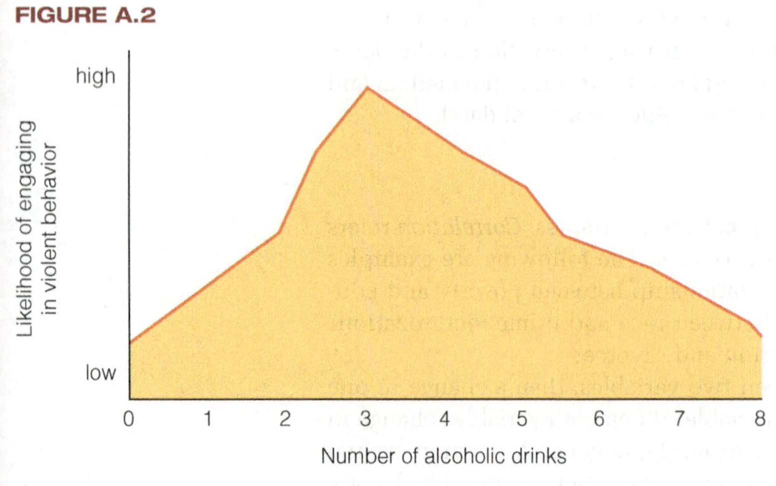

FIGURE A.3

A fourth type of correlation is called a spurious correlation. Such a correlation exists when two variables appear to be related, but the apparent relationship occurs only because each variable is related to a third variable. When the third variable is controlled through a statistical method in which the variable is held constant, the apparent relationship between the first two variables disappears. For example, Blacks have a lower average life expectancy than Whites do. Thus, race and life expectancy appear to be related. This apparent correlation exists, however, because both race and life expectancy are related to socioeconomic status. Because Blacks are more likely than Whites to be impoverished, they are less likely to have adequate nutrition and medical care.

Causation

If the data analysis reveals that two variables are correlated, we know only that a change in one variable is associated with a change in another variable. We cannot assume, however, that a change in one variable causes a change in the other variable unless our data collection and analysis are specifically designed to assess causation. The research method that best assesses causality is the experimental method (discussed in Chapter 1).

To demonstrate causality, three conditions must be met. First, the data analysis must demonstrate that variable A is correlated with variable B. Second, the data analysis must demonstrate that the observed correlation is not spurious. Third, the analysis must demonstrate that the presumed cause (variable A) occurs or changes before the presumed effect (variable B). In other words, the cause must precede the effect.

It is extremely difficult to establish causality in social science research. Therefore, much social research is descriptive or correlative rather than causative. Nevertheless, many people make the mistake of interpreting a correlation as a statement of causation. As you read correlative research findings, remember the following adage: "Correlation does not equal causation."

Reliability and Validity

Assessing reliability and validity is an important aspect of data analysis. *Reliability* refers to the consistency of the measuring instrument or technique; that is, the degree to which the way information is obtained produces the same results if repeated. Measures of reliability are made on scales and indexes (such as those in the *Self and Society* features in this text) and on information-gathering techniques, such as the survey methods described in Chapter 1.

Various statistical methods are used to determine reliability. A frequently used method is called the *test-retest method*. The researcher gathers data on the same sample of people twice (usually one or two weeks apart) using a particular instrument or method and then correlates the results. To the degree that the results of the two tests are the same (or highly correlated), the instrument or method is considered reliable.

Measures that are perfectly reliable may be absolutely useless unless they also have a high validity. *Validity* refers to the extent to which an instrument or device measures what it intends to measure. For example, police officers administer Breathalyzer tests to determine the level of alcohol in a person's system. The Breathalyzer is a valid test for measuring alcohol consumption.

Validity measures are important in research that uses scales or indexes as measuring instruments. Validity measures are also important in assessing the accuracy of self-report data that are obtained in survey research. For example, survey research on high-risk sexual behaviors associated with the spread of HIV relies heavily on self-reported data on topics such as number of sexual partners, types of sexual activities, and condom use. Yet how valid are these data? Do survey respondents underreport the number of their sexual partners? Do people who say they use a condom every time they engage in intercourse really use a condom every time? Because of the difficulties in validating self-reports of number of sexual partners and condom use, we may not be able to answer these questions.

Ethical Guidelines in Social Problems Research

Social scientists are responsible for following ethical standards designed to protect the dignity and welfare of people who participate in research. These ethical guidelines include the following:

1. *Freedom from coercion to participate.* Research participants have the right to decline to participate in a research study or to discontinue participation at any time during the study. For example, professors who are conducting research using college students should not require their students to participate in their research.
2. *Informed consent.* Researchers are required to inform potential participants of any aspect of the research that might influence a subject's willingness to participate. After informing potential participants about the nature of the research, researchers typically ask participants to sign a consent form indicating that the participants are informed about the research and agree to participate in it.
3. *Deception and debriefing.* Sometimes the researcher must disguise the purpose of the research to obtain valid data. Researchers may deceive participants as to the purpose or nature of a study only if there is no other way to study the problem. When deceit is used, participants should be informed of this deception (debriefed) as soon as possible. Participants should be given a complete and honest description of the study and why deception was necessary.
4. *Protection from harm.* Researchers must protect participants from any physical and psychological harm that might result from participating in a research study. This is both a moral and a legal obligation. It would not be ethical, for example, for a researcher studying drinking and driving behavior to observe an intoxicated individual leaving a bar, getting into the driver's seat of a car, and driving away.

   Researchers are also obligated to respect the privacy rights of research participants. If anonymity is promised, it should be kept. Anonymity is maintained in mail surveys by identifying questionnaires with a number coding system rather than with the participants' names. When such anonymity is not possible, as is the case with face-to-face interviews, researchers should tell participants that the information they provide will be treated as confidential. Although interviews may be summarized and excerpts quoted in published material, the identity of the individual participants is not revealed. If a research participant experiences either physical or psychological harm as a result of participation in a research study, the researcher is ethically obligated to provide remediation for the harm.
5. *Reporting of research.* Ethical guidelines also govern the reporting of research results. Researchers must make research reports freely available to the public. In these reports, a researcher should fully describe all evidence obtained in the study, regardless of whether the evidence supports the researcher's hypothesis. The raw data collected by the researcher should be made available to other researchers who might request it for purposes of analysis. Finally, published research reports should include a description of the sponsorship of the research study, its purpose, and all sources of financial support.

Glossary

3-D printing A revolutionary manufacturing technology that involves downloading a digital file containing a design for a product. A printer reads the file and then shoots out the product (made of specialized plastic or other raw materials) through a heated nozzle.

abortion The intentional termination of a pregnancy.

absolute poverty The lack of resources necessary for well-being—most importantly, food and water, but also housing, sanitation, education, and health care.

abusive head trauma A form of inflicted brain injury resulting from violent shaking or blunt impact; a leading cause of death in children under 1 year (*see also* shaken baby syndrome).

acculturation The process of adopting the culture of a group different from the one in which a person was originally raised.

achieved status A status that society assigns to an individual on the basis of factors over which the individual has some control (e.g., high school graduate).

acid rain The mixture of precipitation with air pollutants, such as sulfur dioxide and nitrogen oxide.

acquaintance rape Rape committed by someone known to the victim.

adaptive discrimination Discrimination that is based on the prejudice of others.

affirmative action A broad range of policies and practices in the workplace and educational institutions to promote equal opportunity as well as diversity.

Affordable Care Act (ACA) Health care reform legislation that President Obama signed into law in 2010, with the goal of expanding health insurance coverage to more Americans. Also known as the *Patient Protection and Affordable Care Act*, or Obamacare.

ageism Negative stereotyping, prejudice, and discrimination based on a person's or group's perceived chronological age.

ageism by invisibility The underrepresentation of older adults in advertising and educational materials.

algorithm A program that instructs a computer to follow a set of calculations in order to accomplish a task; i.e., it is a mathematical instruction.

alienation A sense of powerlessness and meaninglessness in people's lives.

alternative certification programs Programs whereby college graduates with degrees in fields other than education can become certified if they have "life experience" in industry, the military, or other relevant jobs.

androgyny Having both traditionally defined feminine and masculine characteristics.

anomie A state of normlessness in which norms and values are weak or unclear.

antimiscegenation laws Laws banning interracial marriage until 1967, when the Supreme Court (in *Loving v. Virginia*) declared these laws unconstitutional.

arbitration Dispute settlement in which a neutral third party listens to evidence and arguments presented by conflicting groups and arrives at a decision that the parties have agreed in advance to accept.

artificial intelligence (AI) The science of making a machine smart enough to perform tasks it hasn't been programmed to perform.

arranged marriage A type of marriage in which the bride and groom are selected by individuals other than the couple themselves, typically by family members such as parents. The bride and groom enter into the arrangement consensually.

ascribed status A status that society assigns to an individual on the basis of factors over which the individual has no control (e.g., race).

assimilation The process by which formerly distinct and separate groups merge and become integrated as one.

attributional gender bias The practice of explaining the same behavior(s) of females and males using different explanations.

automated discrimination The bias that may result, intentionally or unintentionally, as a function of theoretical and mathematical models used by AI algorithms.

automation The replacement of human labor with machinery and equipment.

aversive racism A subtle form of prejudice that involves feelings of discomfort, uneasiness, disgust, fear, and pro-White attitudes.

Baby Boomers The generation of Americans born between 1946 and 1964, a period of high birthrates.

behavior-based safety programs A strategy used by business management that attributes health and safety problems in the workplace to workers' behavior, rather than to work processes and conditions.

beliefs Definitions and explanations about what is assumed to be true.

bigamy The criminal offense in the United States of marrying one person while still legally married to another.

bilingual education In the United States, teaching children in both English and their non-English native language.

binge drinking As defined by the U.S. Department of Health and Human Services, drinking five or more drinks on the same occasion on at least one day in the past 30 days prior to the National Survey on Drug Use and Health.

biodiversity The diversity of living organisms on Earth.

bioinvasion The intentional or accidental introduction of plant, animal, insect, and other species in regions where they are not native.

biomass Material derived from plants and animals, such as dung, wood, crop residues, and charcoal; used as fuel.

biometric surveillance Surveillance used to identify a specific person through the imaging of their distinct physical characteristics.

biphobia Prejudice toward bisexual individuals.

birth rate The number of live births per 1,000 people in the population per calendar year.

bisexuality The emotional, cognitive, and sexual attraction to members of both sexes.

black lung disease A condition resulting from exposure to coal dust, which leaves the lungs of coal miners scarred, shriveled, and black, making it difficult for them to breathe and causing debilitating coughing.

blended families Also known as *stepfamilies*, families involving children from a previous relationship.

Bostock v. Clayton County, Georgia A U.S. Supreme Court decision that held that discrimination based on sexual orientation or gender identity violates Title VII of the 1964 *Civil Rights Act*.

boy code A set of societal expectations that discourages males from expressing emotion, weakness, or vulnerability, or asking for help.

breeding ground hypothesis A hypothesis that argues that incarceration serves to increase criminal behavior through the transmission of criminal skills, techniques, and motivations.

bullying Bullying occurs when "a physically stronger of socially more prominent person abuses his/her power to threaten, demean, or belittle another" (Juvonen and Graham 2014, p. 161).

Internet of Things (IoT) The potential connectivity of all electronic devices; widespread information sharing between people and the objects of everyday life.

Internet piracy Illegally downloading or distributing copyrighted material (e.g., music, games, software).

Interpol The largest international police organization in the world.

intimate partner violence (IPV) Actual or threatened violent crimes committed against individuals by their current or former spouses, cohabiting partners, boyfriends, or girlfriends.

Islamic State in Iraq and Syria (ISIS) Notorious for its brutal tactics, a transnational terrorist organization whose goal is to establish an Islamic-led state in the Levant, a region that includes Egypt, Iraq, Israel, Jordan, Lebanon, Palestine, Syria, and Turkey.

Islamophobia Anti-Muslim and anti-Islam bias.

job burnout Prolonged job stress that can cause or contribute to high blood pressure, ulcers, headaches, anxiety, depression, and other health problems.

Kyoto Protocol The first international agreement to place legally binding limits on greenhouse gas emissions from developed countries.

labor unions Worker advocacy organizations that developed to protect workers and represent them at negotiations between management and labor.

larceny Larceny is simple theft; it does not entail force or the use of force, or breaking and entering.

latent functions Consequences that are unintended and often hidden.

learning health systems The result of electronic records whereby physicians can look across patient populations and identify successful treatments or detect harmful interactions.

least developed countries The poorest countries of the world.

legalization Making prohibited behaviors legal; for example, legalizing drug use or prostitution.

legitimacy hypothesis A model that predicts that the legalization of same-sex marriages signifies approval of LGBT behaviors and therefore should reduce LGBT hate-motivated crimes.

lesbian A term referring to women who are emotionally, cognitively, and sexually attracted to women.

LGBT, LGBTQ, and LGBTQI Terms used to refer collectively to lesbian, gay, bisexual, transgender, questioning or "queer," and/or intersexed individuals.

life expectancy The average number of years that individuals born in a given year can expect to live.

light pollution Artificial lighting that is annoying, unnecessary, and/or harmful to life forms on Earth.

living wage laws Laws that require state or municipal contractors, recipients of public subsidies or tax breaks, or, in some cases, all businesses to pay employees wages that are significantly above the federal minimum, enabling families to live above the poverty line.

long-term unemployment Unemployment that lasts for 27 weeks or more.

malware A general term that includes any spyware, viruses, and adware that is installed on an owner's computer without their knowledge.

managed care Any medical insurance plan that controls costs through monitoring and controlling the decisions of health care providers.

manifest functions Consequences that are intended and commonly recognized.

marital decline perspective A view of the current state of marriage that includes the beliefs that (1) personal happiness has become more important than marital commitment and family obligations, and (2) the decline in lifelong marriage and the increase in single-parent families have contributed to a variety of social problems.

marital resiliency perspective A view of the current state of marriage that includes the beliefs that (1) marriage continues to be valued by many in society (as evidenced by same-sex marriages and remarriages) and (2) marriage is valued as a symbol of a successful personal life rather than for the tangible resources it provides.

masculine overcompensation thesis The thesis that men have a tendency to act out in an exaggerated male role when believing their masculinity is threatened.

master status The status that is considered the most significant in a person's social identity.

maternal mortality Deaths that result from complications associated with pregnancy, childbirth, and unsafe abortion.

Matthew Shepard and James Byrd, Jr. Hate Crimes Prevention Act (HCPA) This law expands the original 1969 federal hate crimes law to cover hate crimes based on actual or perceived sexual orientation, gender, gender identity, and disability.

McDonaldization The process by which principles of the fast-food industry (efficiency, calculability, predictability, and control through technology) are being applied to more sectors of society, particularly the workplace.

means-tested programs Assistance programs that have eligibility requirements based on income and/or assets.

mechanization Dominant in an agricultural society, the use of tools to accomplish tasks previously done by hand.

mediation A neutral third party intervenes and facilitates negotiation between representatives or leaders of conflicting groups.

Medicaid A public health insurance program, jointly funded by the federal and state governments, that provides health insurance coverage for the poor who meet eligibility requirements.

medical debt Debt that results when people cannot afford to pay their medical bills.

medicalization Defining or labeling behaviors and conditions as medical problems.

medical tourism A global industry that involves traveling, primarily across international borders, for the purpose of obtaining medical care.

Medicare A federally funded program that provides health insurance benefits to the elderly, disabled, and those with advanced kidney disease.

mental health Psychological, emotional, and social well-being.

mental illness Refers collectively to all mental disorders, which are characterized by sustained patterns of abnormal thinking, mood (emotion), or behaviors that are accompanied by significant distress and/or impairment in daily functioning.

meritocracy A social system in which individuals get ahead and earn rewards based on their individual efforts and abilities.

meta-analysis Meta-analysis combines the results of several studies addressing a research question; i.e., it is the analysis of analyses.

microcredit programs The provision of loans to people who are generally excluded from traditional credit services because of their low socioeconomic status.

migration The movement of people from one country or region to another; includes immigration and emigration.

Military Health System (MHS) The federal entity that provides medical care in military hospitals and clinics, and in combat zones and at bases overseas and on ships, and that provides health insurance known as Tricare to active duty service members, military retirees, their eligible family members, and their survivors.

military-industrial complex A term first used by Dwight D. Eisenhower to connote the close association between the military and defense industries.

minority group A category of people who have unequal access to positions of power, prestige, and wealth in a society and who tend to be targets of prejudice and discrimination.

modern racism A subtle form of racism that involves the belief that serious discrimination in the United States no longer exists, that any continuing racial inequality is the fault of minority group members, and that the demands for affirmative action for minorities are unfair and unjustified.

monogamy Marriage between two partners; the only legal form of marriage in the United States; also refers to the restriction of sexual behavior to between two partners

mortality Death.

mortality rate The number of deaths per 1,000 people in the population per calendar year; sometimes referred to as the death rate.

motherhood penalty The tendency for women with children, particularly young children, to be disadvantaged in hiring, wages, and the like compared to women without children.

mountaintop removal A process of coal mining in which the tops of mountains are dynamited and removed to access coal seams below.

multicultural education Education that includes all racial and ethnic groups in the school curriculum, thereby promoting awareness and appreciation for cultural diversity.

Multidimensional Poverty Index A measure of serious deprivation in the dimensions of health, education, and living standards that combines the number of deprived and the intensity of their deprivation.

multiple chemical sensitivity (MCS) Also known as "environmental illness," a condition whereby individuals experience adverse reactions when exposed to low levels of chemicals found in everyday substances.

mutually assured destruction (MAD) A Cold War doctrine referring to the capacity of two nuclear states to destroy each other, thus reducing the risk that either state will initiate war.

misogyny Hatred of women.

nanotechnology The manipulation of materials and creation of structures and systems at the scale of atoms and molecules.

nationalism A sense of identity that emphasizes loyalty and devotion to the interests of one's own country, even to the exclusion or detriment of the interests of other communities or identities.

nativist extremist groups Organizations that not only advocate restrictive immigration policy but also encourage their members to use vigilante tactics to confront or harass suspected undocumented immigrants.

NATO North Atlantic Treaty Organization, founded in 1949, is a military alliance among 30 North American and European countries with the purpose of promoting the freedom and security of its member nations.

naturalized citizens Immigrants who apply for and meet the requirements for U.S. citizenship.

neglect A form of abuse involving the failure to provide adequate attention, supervision, nutrition, hygiene, health care, and a safe and clean living environment for a minor child or a dependent elderly individual.

neonatal abstinence syndrome (NAS) A condition in which a child, at birth, goes through withdrawal as a consequence of maternal drug use.

net neutrality A principle that holds that Internet users should be able to visit any website and access any content without Internet service provider interference.

no-fault divorce A divorce that is granted based on the claim that there are irreconcilable differences within a marriage (as opposed to one spouse being legally at fault for the marital breakup).

norms Socially defined rules of behavior, including folkways, mores, and laws.

"Not in My Backyard" Opposition by local residents to a proposed new development in their community. Also known as NIMBY.

nuclear nonproliferation Efforts to prevent the spread of nuclear weapons, or the materials and technology necessary for the production of nuclear weapons.

nuclear winter The predicted result of a thermonuclear war whereby thick clouds of radioactive dust and particles would block out vital sunlight, lower temperature in the northern hemisphere, and lead to the death of most living things on Earth.

Obergefell v. Hodges The 2015 U.S. Supreme Court decision that legalized same-sex marriage in the United States.

objective element of a social problem Awareness of social conditions through one's own life experiences, through the media, and through education.

occupational sex segregation The concentration of women in certain occupations and men in other occupations.

Occupy Wall Street A protest movement that began in 2011, and is concerned with economic inequality, greed, corruption, and the influence of corporations on government.

offshoring The relocation of jobs to other countries.

online activism Activism that entails the use of social media to embrace a particular social or political issue.

oppression The use of power to create inequality and limit access to resources, which impedes the physical and/or emotional well-being of individuals or groups of people.

organized crime Criminal activity conducted by members of a hierarchically arranged structure devoted primarily to making money through illegal means.

outsource See outsourcing.

outsourcing A practice in which a business subcontracts with a third party to provide business services.

overt discrimination Discrimination that occurs because of an individual's own prejudicial attitudes.

overt differential law enforcement Criminal justice actors treating one person differently than another because of that person's characteristics, for example, race.

pandemic A worldwide disease outbreak.

pansexual Anyone whose sexual orientation is not limited by their partner's birth sex, gender, or gender identity, the prefix "pan" meaning "all."

parental alienation The intentional efforts of one parent to turn a child against the other parent and essentially destroy any positive relationship a child has with the other parent.

parent trigger laws State legislation that allows parents to intervene in their children's education and schools.

Paris Climate Accord An international agreement which calls for the countries of the world to work together to reduce greenhouse gas emissions to ward off what could become extinction-level global warming and climate change.

parity In health care, a concept requiring equality between mental health care insurance coverage and other health care coverage.

parole Release from prison, for a specific time period and subject to certain conditions, before an inmate's sentence is finished.

participation gap The tendency for racial and ethnic minorities to participate in information and communication technologies (e.g., using smartphones to access the Internet rather than a computer) that place them in a disadvantaged position (e.g., difficult to research a term paper on a smartphone).

patriarchy A male-dominated system in which men have primary power over major decision making.

penetration rate The percentage of people who have access to and use the Internet in a particular area.

perceived obsolescence The perception that a product is obsolete; used as a marketing tool to convince consumers to replace certain items even though the items are still functional.

physical violence When a person hurts or tries to hurt a partner by hitting, kicking, or using another type of physical force.

pink-collar jobs Jobs that offer few benefits, often have low prestige, and are disproportionately held by women.

planned obsolescence The manufacturing of products that are intended to become inoperative or outdated in a fairly short period of time.

phubbing A combination of the words *phone* and *snubbing*, which occurs when people are constantly looking at their phone while you are talking to them.

pluralism A state in which racial and ethnic groups maintain their distinctness but respect each other and have equal access to social resources.

plutocracy A country governed by the wealthy.

polarization hypothesis A model that argues that the legalization of same-sex marriages solidifies respective group member's beliefs—i.e., those who approve of same-sex marriage and those who disapprove of same-sex marriage—and creates a greater gulf between the two.

policy drift The failure to update labor laws to reflect changes in the broader society and economy.

political alienation A rejection of or estrangement from the political system accompanied by a sense of powerlessness in influencing government.

political partisanship When supporters of a political party are entrenched in their party's policies, with little to no motivation to compromise with opposing political views.

polyamory Multiple intimate sexual and/or loving relationships with the knowledge and consent of all partners involved.

polyandry The concurrent marriage of one woman to two or more men.

polydrug abuse A substance abuse pattern that occurs when a user becomes dependent on two or more drugs simultaneously, typically having one dominant drug addiction and then a secondary addiction to a drug that helps to counteract the negative effects of the primary drug.

polygamy A form of marriage in which one person may have two or more spouses.

polygyny A form of marriage in which one husband has more than one wife.

population growth rate The number of people added to the population in a year (e.g., births – deaths).

population momentum Continued population growth as a result of past high fertility rates that have resulted in a large number of young women who are currently entering their childbearing years.

population pyramid A graph of a country's or region's population that represents the number of males and females in each age group with a bar.

populist movements Emphasize "the people" rather than the "government elite" and their political parties, tend to be conservative, right to far-right leaning, anti-immigrant, nationalistic, and anti-globalist.

positivity rate Percentage of positive tests out of every hundred tests conducted.

postmodernism A worldview that questions the validity of rational thinking and the scientific enterprise.

posttraumatic stress disorder (PTSD) A set of symptoms that may result from any traumatic experience, including crime victimization, war, natural disasters, or abuse.

preexisting conditions Illnesses or injuries that occurred before a person begins coverage under a new health insurance plan.

prejudice Negative attitudes and feelings toward or about an entire category of people.

primary deviance Deviant behavior committed before a person is caught and labeled an offender.

primary groups Usually small numbers of individuals characterized by intimate and informal interaction.

primordial explanations Those explanations that emphasize the existence of "ancient hatreds" rooted in deep psychological or cultural differences between ethnic groups, often involving a history of grievance and victimization, real or imagined, by the enemy group.

privileged When a group has a special advantage or benefits as a result of cultural, economic, societal, legal, and political factors.

probation The conditional release of an offender who, for a specific time period and subject to certain conditions, remains under court supervision in the community.

progressive taxes Taxes in which the tax rate increases as income increases, so that those who have higher incomes are taxed at higher rates.

pronatalism A cultural value that promotes having children.

psychological aggression/abuse The use of verbal and nonverbal communication (e.g., yelling, belittling, insulting, etc.) with the intent to harm another person mentally or emotionally and/or to exert control over another person.

psychotherapeutic drug The nonmedical use of any prescription pain reliever, stimulant, sedative, or tranquilizer.

public housing Federally subsidized housing that is owned and operated by local public housing authorities (PHAs).

purchasing power parity (PPP) An economic metric used to standardize differences in currency values and standards of living in order to make international comparisons in levels of wealth and poverty; $1 PPP is equivalent to the purchasing value of $1 in the United States.

QAnon An international conspiracy group that has been labeled a domestic terrorist organization by the FBI.

qualified immunity A legal principle that protects police officers from lawsuits if, at the time of their alleged misconduct, they did not know their behavior was unlawful.

queer Although originally a pejorative term for gay men and women, the term has been reclaimed by the LGBTQ community to mean anyone who is not heterosexual or cisgender.

race A category of people who are perceived to share distinct physical characteristics that are deemed socially significant.

racial microaggressions Brief and commonplace daily verbal, behavioral, or environmental indignities, whether intentional or unintentional, that communicate hostile, derogatory, or negative racial slights and insults toward the target person or group.

racial profiling The law enforcement practice of targeting suspects on the basis of race.

racism The belief that race accounts for differences in human character and ability and that a particular race is superior to others.

radical environmental movement A grassroots movement of individuals and groups that employs unconventional and often illegal means of protecting wildlife or the environment.

rally round the flag effect Phenomenon of increased political cohesion and civic engagement that commonly occurs during times of war and international crisis.

ransomware A form of malware intrusion in which a criminal holds an individual's or company's computer "hostage."

recession A significant decline in economic activity spread across the economy and lasting for at least six months.

recidivism A return to criminal behavior by a former inmate, most often measured by rearrest, reconviction, or reincarceration.

refined divorce rate The number of divorces per 1,000 married women.

registered partnerships Federally recognized relationships that convey most but not all the rights of marriage.

rehabilitation A criminal justice philosophy that argues that recidivism can be reduced by changing the criminal through such programs as substance abuse counseling, job training, education, and so on.

rehabilitative alimony Alimony that is paid to an ex-spouse for a specified length of time to allow the recipient time to find a job or to complete education or job training.

relative deprivation Theory that conflict is more likely when there is a meaningful gap between what a group within a population expects (for example, economic opportunity) and what they have access to within their society.

relative poverty The lack of material and economic resources compared with some other population.

religious freedom laws Laws that protect business owners who discriminate against customers (e.g., gay men and women) based on religious grounds.

Religious Freedom Restoration Act of 1993 (RFRA) An act stating that the "government shall not substantially burden a person's exercise of religion" (RFRA 1993).

replacement-level fertility The level of fertility at which a population exactly replaces itself from one generation to the next; currently, the number is 2.1 births per woman.

restorative justice A philosophy primarily concerned with reconciling conflict between the victim, the offender, and the community.

roles The set of rights, obligations, and expectations associated with a status.

safety gender gap The difference between women's expressed rate of fear and men's expressed rate of fear.

sample A portion of the population, selected to be representative so that the information from the sample can be generalized to a larger population.

sanctions Social consequences for conforming to or violating norms.

sanctuary city A jurisdiction that has implemented rules to limit local law enforcement from cooperating with federal immigration authorities in order to prevent the detention or deportation of undocumented residents.

sandwich generation A generation of people who care for their aging parents while also taking care of their own children.

school-to-prison pipeline The established relationship between severe disciplinary practices, increased rates of dropping out of school, lowered academic achievement, and court or juvenile detention involvement.

school vouchers State-funded "scholarships" paid directly to parents that can be used to send qualifying public school students (e.g., low-income students) to private schools.

science The process of discovering, explaining, and predicting natural or social phenomena.

scientific racism The belief, which dominated racial thinking in the biological and social sciences in the 19th and early 20th centuries, that human groups could be ranked into hierarchies on the basis of observable characteristics, such as nose width and skin color.

secondary groups Involving small or large numbers of individuals, groups that are task-oriented and are characterized by impersonal and formal interaction.

second shift The household work and child care that employed parents (usually women) do when they return home from their jobs.

Section 8 housing A housing assistance program in which federal rent subsidies are provided either to tenants (in the form of certificates and vouchers) or to private landlords.

security dilemma A characteristic of the international state system that gives rise to unstable relations between states; as State A secures its borders and interests, its behavior may decrease the security of other states and cause them to engage in behavior that decreases A's security.

segregation The physical separation of two groups in residence, workplace, and social functions.

self-fulfilling prophecy A concept referring to the tendency for people to act in a manner consistent with the expectations of others.

Semantic Web Sometimes called Web 3.0, a version of the Internet in which pages not only contain information but also describe the interrelationship between pages; sometimes called smart media.

serial monogamy A succession of marriages in which a person has more than one spouse over a lifetime but is legally married to only one person at a time.

sex A person's biological classification as male or female.

sexism The belief that innate psychological, behavioral, and/or intellectual differences exist between women and men and that these differences connote the superiority of one group and the inferiority of the other.

sexual harassment In reference to workplace harassment, when an employer requires sexual favors in exchange for a promotion, salary increase, or any other employee benefit and/ or the existence of a hostile environment that unreasonably interferes with job performance.

sexual orientation A person's emotional and sexual attractions, relationships, self-identity, and behavior.

sexual orientation change efforts (SOCE) Collectively refers to reparative, conversion, and reorientation therapies, according to the APA.

sexual violence Forcing or attempting to force a partner to take part in a sex act, sexual touching, or a nonphysical sexual event (e.g., sexting) when the partner does not or cannot consent.

shaken baby syndrome A form of potentially fatal brain damage resulting from violently shaking a baby (*see also* abusive head trauma).

single-payer health care A health care system in which a single tax-financed public insurance program replaces private insurance companies.

slavery Any work that is performed under the threat of punishment and is undertaken involuntarily. Also known as forced labor.

slums Concentrated areas of poverty and poor housing in urban areas.

snowball sampling A technique in which one participant in a study recommends others who might be interested in participating.

social bond The bond between individuals and the social order that constrains some individuals from violating social norms (i.e., committing crime).

social group Two or more people who have a common identity, interact, and form a social relationship.

social movement An organized group of individuals with a common purpose to either promote or resist social change through collective action.

social problem A social condition that a segment of society views as harmful to members of society and in need of remedy.

social reproduction A process through which the social class structure is repeated from one generation to the next.

Social Security Also called Old Age, Survivors, Disability, and Health Insurance, a federal program that protects against loss of income due to retirement, disability, or death.

socialism An economic system characterized by state ownership of the means of production and distribution of goods and services.

social mobility The likelihood and extent to which persons will change their socioeconomic status over the course of their lives and across generations.

social stratification Systems of social inequality by which a society divides people into groups with unequal access to wealth, material and social resources, and power.

socioeconomic status (social class) A person's position in society based on the level of educational attainment, occupation, and income of that person or that person's household.

sociological imagination The ability to see the connections between our personal lives and the social world in which we live.

solar farms Large solar arrays owned by utility companies to contribute to their electricity supply, which is then sold to consumers. Also known as solar parks.

stalking A pattern of repeated, unwanted attention and contact by a partner that causes fear or concern for one's own safety or the safety of someone close to the victim.

state The organization of the central government and government agencies such as the military, police, and regulatory agencies.

status A position that a person occupies within a social group.

STEM An acronym for science, technology, engineering, and mathematics.

stem cells Undifferentiated cells that can produce any type of cell in the human body.

stereotypes Exaggerations or generalizations about the characteristics and behavior of a particular group.

stereotype threat The tendency of minorities and women to perform poorly on high-stakes tests because of the anxiety created by the fear that a negative performance will validate societal stereotypes about one's member group.

stigma A discrediting label that affects an individual's self-concept and disqualifies that person from full social acceptance.

Stonewall Uprising In 1969, patrons of this now historic Greenwich Village gay bar fought back against police brutality; often thought to be the beginning of the gay rights movement.

strip mining A process of coal mining by which large machines are used to remove topsoil and layers of rock to expose coal less than 200 feet underground.

structural sexism The ways in which the organization of society, and specifically its institutions, subordinate individuals and groups based on their sex classification.

structure The way society is organized including institutions, social groups, statuses, and roles.

structured choice Choices that are limited by the structure of society.

substance use disorder A medical diagnosis used when recurrent use of alcohol and/or drugs causes clinically significant health problems, disabilities, and inability to meet major responsibilities at work, school, or home.

subjective element of a social problem The belief that a particular social condition is harmful to society, or to a segment of society, and that it should and can be changed.

subprime mortgages High-interest or adjustable-rate mortgages that require little money down and are issued to borrowers with poor credit ratings or limited credit history.

sundown towns Communities that are purposely "all-White" and that have used various means to deliberately keep racial and ethnic minorities out.

Supplemental Nutrition Assistance Program (SNAP) The largest U.S. food assistance program.

supply reduction One of two strategies in the U.S. war on drugs (the other is demand reduction), supply reduction concentrates on reducing the supply of drugs available on the streets through international efforts, interdiction, and domestic law enforcement.

survey research A research method that involves eliciting information from respondents through questions.

sustainable development The balancing point at which all human beings can live healthy, equitable, and peaceful lives without degrading our natural environment.

Sustainable Development Goals (SDGs) A set of 17 goals that comprise an international agenda for reducing poverty and economic inequality and improving lives.

sweatshops Work environments that are characterized by less-than-minimum wage pay, excessively long hours of work (often without overtime pay), unsafe or inhumane working conditions, abusive treatment of workers by employers, and/or the lack of worker organizations aimed to negotiate better working conditions.

symbol Something that represents something else.

synthetic drugs A category of drugs that are "designed" in laboratories rather than naturally occurring in plant material.

tar sands Large, naturally occurring deposits of sand, clay, water, and a dense form of petroleum that looks like tar.

tar sands oil Oil that results from converting tar sands into liquid fuel. It is known as the world's dirtiest oil because producing it requires energy and generates high levels of greenhouse gases, and also leaves behind large amounts of toxic waste.

technological dualism The tendency for technology to have both positive and negative consequences.

technological fix The use of scientific principles and technology to solve social problems.

technology Activities that apply the principles of science and mechanics to the solutions of a specific problem.

technology-induced diseases Diseases that result from the use of technological devices, products, and/or chemicals.

teen dating violence Intimate partner violence that occurs before the age of 18.

telepresence A general term for a group of technologies that allow a person to feel that they are present or look like they are present in a remote location.

Temporary Assistance for Needy Families (TANF) A federal cash welfare program that involves work requirements and a five-year lifetime limit.

terrorism The premeditated use or threatened use of violence by an individual or group to gain a political objective.

theory A set of interrelated propositions or principles designed to answer a question or explain a particular phenomenon.

therapeutic cloning Use of stem cells to produce body cells that can be used to grow needed organs or tissues; regenerative cloning.

therapeutic communities Organizations in which approximately 35 to 500 individuals reside for up to 15 months to abstain from drugs, develop marketable skills, and receive counseling.

total fertility rates The average lifetime number of births per woman in a population.

toxic masculinity An extreme form of aggression, violence, and misogyny, socially induced and culturally specific.

transgender individual A person whose sense of gender identity is inconsistent with their birth (sometimes called chromosomal) sex (male or female).

transnational corporations Also known as multinational corporations, corporations that have their home base in one country and branches, or affiliates, in other countries.

transnational crime Criminal activity that occurs across one or more national borders.

transnational terrorism Terrorism that occurs when a terrorist act in one country involves victims, targets, institutions, governments, or citizens of another country.

TRAP laws Laws designed to restrict access to abortion through targeted restrictions on abortion providers.

under-5 mortality rate The number of deaths of children under age 5 per 1,000 live births.

underemployment Unemployed workers as well as (1) those working part-time but who wish to work full-time, (2) those who want to work but have been discouraged from searching by their lack of success, and (3) others who are neither working nor seeking work but who want and are available to work and have looked for employment in the last year. Also refers to the employment of workers with high skills and/or educational attainment working in low-skill or low-wage jobs.

undernourishment Having a caloric intake that is insufficient to meet minimum energy requirements.

unemployment To be currently without employment, actively seeking employment, and available for employment, according to U.S. measures of unemployment.

union density The percentage of workers who belong to unions.

universal health care system A system of health care, typically financed by the government, that ensures health care coverage for all citizens.

value-added measurement (VAM) VAM is the use of student achievement data to assess teacher effectiveness.

values Social agreements about what is considered good and bad, right and wrong, desirable and undesirable.

variable Any measurable event, characteristic, or property that varies or is subject to change.

Veterans Health Administration (VHA) A system of hospitals, clinics, counseling centers, and long-term care facilities that provides care to military veterans.

victimless crimes Illegal activities that have no complaining participant(s) and are often thought of as crimes against morality, such as prostitution.

visual ageism Underrepresentation of older adults in the media or representation in prejudiced or stereotypical ways.

vivisection The practice of cutting into or otherwise harming living, nonhuman animals for the purpose of scientific research.

vulnerable employment Employment that is characterized by informal working arrangements, little job security, few benefits, and little recourse in the face of an unreasonable demand.

wage theft Occurs when employers "steal" workers' wages by requiring them to work off the clock or refusing to pay them for overtime.

war Organized armed violence aimed at a social group in pursuit of an objective.

war on drugs A public policy approach to the illicit drug trade in the United States, initially implemented by the Nixon administration in the 1970s, which focused on the widespread prohibition and criminalization of drug use and distribution.

wealth The total assets of an individual or household minus liabilities.

wealthfare Laws and policies that benefit the rich.

weapons of mass destruction (WMD) Chemical, biological, and nuclear weapons that have the capacity to kill large numbers of people indiscriminately.

white-collar crime Includes *occupational crime*, in which individuals commit crimes in the course of their employment; *corporate crime*, in which corporations violate the law in the interest of maximizing profit; and *political crime*, in which government actors, by their actions or inactions, commit a crime in their own self-interest to the detriment of the state.

white supremacist organizations Extremist groups, including those aligning with neo-Nazi, alt-right, and the Ku Klux Klan, that share a belief in the inherent superiority of White people and an ideological commitment to using political violence in order to promote a racially segregated society, with only White people in a position of power.

worker cooperatives Democratic business organizations controlled by their members, who actively participate in setting their policies and making decisions. Also known as workers' self-directed enterprises.

workers' self-directed enterprises See worker cooperatives.

working poor Individuals who spend at least 27 weeks per year in the labor force (working or looking for work) but whose income falls below the official poverty level.

work/life conflict The day-to-day struggle to simultaneously meet the demands of work and other life responsibilities and goals, including family, education, exercise, and recreation.

World System Theory A theory arguing that wealth and power are divided unequally among the core countries, periphery countries, and semi-periphery countries in the world.

References*

Chapter 1

Adnane, Khalid. 2019 (April). "Trump and Globalization: The Delusion after the Fascination!" *Proceedings of the 6th International Scientific Conference on Social Sciences and Arts* 1(1):453–460. Sofia, Bulgaria: Bulgaria Academy of Sciences.

Ananyev, Maxim, Michael Poyker, and Yuan Tian. 2020 (October 20). "Exposure to Fox News Hindered Social Distancing." *The Centre for Economic Policy Research*. Available at voxeu.org.

Anderson, Stuart. 2020 (November 8). "The Biden Immigration Policy: New Hope for Immigrants and Businesses." *Forbes*. Available at www.forbes.com.

Bail, Christopher A., Lisa P. Argyle, Taylor W. Brown, John P. Bumpus, Haohan Chen, M. B. Fallin Hunzaker, Jaemin Lee, Marcus Mann, Friedolin Merhout, and Alexander Volfovsky. 2018 (September). "Exposure to Opposing Views on Social Media Can Increase Political Polarization." The National Academy of Sciences. *Proceedings of the National Academy of Sciences* 115(37):9216–9221.

Barroso, Amanda, and Rachel Minkin. 2020 (June 24). "Recent Protests Attendees Are More Racially and Ethnically Diverse, Younger Than Americans Overall." Pew Research Center. Available at www.pewresearch.org.

Baum, Sandy, and Adam Looney. 2020 (October 9). "Who Owes the Most in Student Loans: New Data from the Fed." Brookings Institute. Available at www.brookings.edu.

BBC (British Broadcasting Company). 2020. "U.S. 2020 Election: The Economy under Trump in Six Charts." BBC News. Available at www.bbc .com.

Biden, Joe. 2020 (November 7). "Speech by the President-Elect." Wilmington, Delaware.

Blankenhorn, David. 2018 (May 16). "The Top 14 Causes of Political Polarization." Available at www.the-american-interest.com.

Blumer, Herbert. 1971. "Social Problems as Collective Behavior." *Social Problems* 8(3):298–306.

Brooks, Arthur C. 2019. *Love Your Enemies: How Decent People Can Save America from Our Culture of Contempt*. New York: HarperCollins.

Calvillo, Dustin P., Bryan J. Ross, Ryan J. B. Garcia, Thomas J. Smelter, and Abraham M. Rutchick. 2020 (November 1). "Political Ideology Predicts Perceptions of the Threat of COVID-19 (and Susceptibility to Fake News about It)." *Social Psychological and Personality Science* 11(8):1119–1128.

Carothers, Thomas, and Andrew O'Donohue. 2019 (August 30). *Democracies Divided: The Global Challenge of Political Polarization*. Washington, DC: Brookings Institute Press.

Centers for Disease Control and Prevention (CDC). 2020a. "Motor Vehicle Safety: Distracted Driving." *Distracted Driver Fact Sheet*. Available at www.cdc.gov.

Centers for Disease Control and Prevention (CDC). 2020b (September 11). "Coronavirus Disease 2019 (Covid-19): Older Adults." Available at www.cdc.gov.

Chicago Tribune Wire. 2019 (October 14). "Timeline: Boeing 737 Max Jetliner Crashes and Aftermath." *Chicago Tribune*. Available at www.chicagotribune.com.

Deane, Claudia, and John Gramlich. 2020 (November 6). "2020 Election Reveals Two Broad Voting Coalitions Fundamentally at Odds." Pew Research Center. Available at www.pewresearch.org.

Deese, Kaelan. 2020 (November 10). "Poll: 70 Percent of Republicans Don't Believe the Election Was Free and Fair." *The Hill*. Available at thehill.com.

Dunn, Amina. 2020 (July 14). "As the U.S. Copes with Multiple Crises, Partisans Disagree Sharply on Severity of Problems Facing the Nation." Pew Research Center. Available at www.pewresearch.org.

Economic Recovery. 2020. "The Biden-Harris Jobs and Economic Recovery Plan for Working Families." Available at buildbackbetter.com.

Ember, Sydney. 2020 (November 8). "Where Does Joe Biden Stand on Major Policies." *New York Times*. Available at www.nytimes.com.

Farrell, Dan, and James C. Petersen. 2010. "The Growth of Internet Research Methods and the Reluctant Sociologist." *Sociological Inquiry* 80(1):114–125.

Firozi, Helena. 2019 (March 28). "The Health 202: Trump Administration Undermines Anti-Opiate Efforts by Opposing Obamacare." *Washington Post*. Available at www .washingtonpost.com.

Fleming, Zachary. 2003. "The Thrill of It All." In *In Their Own Words*, ed. Paul Cromwell, 99–107. Los Angeles: Roxbury.

Fox News. 2020 (November 10). "Presidential Election Results." Available at www.foxnews .com.

Gallup Poll. 2020a (October). "Most Important Problem." Available at news.gallup.com.

Gallup Poll. 2020b (September 13). "Satisfaction with United States." Available at news.gallup. com.

Gallup Poll. 2020c. "Confidence in Institutions." Available at news.gallup.com.

Gee, Emily. 2020 (September 25). "Less Coverage and Higher Costs: The Trump Administration's Care Legacy." *Center for American Progress*. Available at www.americanprogress.org.

Goodman, Peter S. 2019 (June 19). "Globalization Is Moving Past the U.S. and Its Vision of World Order." *New York Times*. Available at www .nytimes.com.

Gramlich, John. 2020 (April 8). "Five Facts about Fox News." Pew Research Center. Available at www.pewresearch.org.

Harvard Kennedy School. 2020 (April 23). "Harvard Youth Poll." *Institute of Politics*. Available at iop.harvard.edu.

Heltzel, Gordon, and Kristin Laurin. 2020 (August). "Polarization in America: Two Possible Futures." *Current Opinion in Behavioral Sciences* 34:179–184.

Herd, D. 2011. "Voices from the Field: The Social Construction of Alcohol Problems in Inner-City Communities." *Contemporary Drug Problems* 38(1):7–39.

Holpuch, Amanda, Jessica Glenza, Kari Paul, Sam Levin, Julian Borger, Dominic Rushe, and Tom McCarthy. 2020 (November 10). "The Tasks Joe Biden Faces: From Racial Justice to Restoring Faith in Science." *Guardian*. Available at www .theguardian.com.

Horowitz, Juliana Menasce, Ruth Igielnik, and Rakesh Kochhar. 2020 (January 9). "Trends in Income and Wealth Inequality." Pew Research Center. Available at www .pewsocialtrends.org,

Imbert, Fred, and Eustance Huang. 2020 (February 24). "Dow Plunges 1,000 Points on Coronavirus Fears, 3.5% Global Drop Is Worst in Two Years." CNBC News. Available at www .cnbc.com.

Indiana State Government. 2013. "General Information." Department of Toxicology. Available at www.in.gov/isdt/2340.htm.

*The authors and Cengage acknowledge that some of the Internet sources may have become unstable; that is, they are no longer working links to the intended reference. In that case, the reader may want to access the article through the search engine or archives of the homepage cited (e.g., fbi.gov, cbsnews.com).

Internet Live. 2020 (November 15). "Internet Users in the World." Available at www.internetlivestats.com.

Ipsos. 2020 (September 25). "'What Worries the World?' COVID-19 Is the Biggest Concern for the Six Successive Month." *Global Advisor.* Available at www.ipsos.com.

Isaac, Mike, and Kellan Browning. 2020 (November 11). "Fact Checked on Facebook and Twitter, Conservatives Switch Their Apps." *New York Times.* Available at www.nytimes.com.

Jorgensen, Frederik, Alexander Bor, Maria Fly Lindholt, and Michael Bang Peterson. 2020 (August 15). "Lockdown Evaluations during the First Wave of the COVID-19 Pandemic." *The Hope Project.* Aarhus University. Available at interactingminds.au.dk.

Jurkowitz, Mark. 2020 (July 9). "Younger Adults Differ from Older Ones in Perception of News about COVID-19, George Floyd Protests." Pew Research Center. Available at www.pewresearch.org.

Jurkowitz, Mark, and Amy Mitchell. 2020 (February). "A Sore Subject: Almost Half of Americans Have Stopped Talking Politics with Someone." Pew Research Center. Available at www.journalism.org.

Katsambekis, Giorgos, and Yannis Stavrakakis (eds). 2020. "Populism and the Pandemic: A Collaborative Report." *Populism Research Group.* Loughborough University.

Kerr, Emma. 2020 (September 15). "See 10 Years of Average Total Student Loan Debt." *U.S. News.* Available at www.usnews.com.

Kmec, Julie A. 2003. "Minority Job Concentration and Wages." *Social Problems* 50:38–59.

Krupnikov, Yanna, and John Barry Ryan. 2020 (October 20). "The Real Divide Is Between Political Junkies and Everyone Else." *New York Times.* Available at www.nytimes.com.

Kuhn, Paul. 2020. *The Hardhat Riot: Nixon, New York City, and the Dawn of the White Working-Class Revolution.* New York: Oxford University Press.

Lacdan, Joseph. 2020 (September 24). "For Massachusetts Soldier, Path to Military Service Was a Spiritual One." *Army News Service.* Available at www.army.mil.

Lantry, Lauren. 2020 (October 21). "Parents of 545 Kids Separated at Border Still Haven't Been Found: ACLU." ABC News. Available at abcnews.go.com.

Lewis, Verlan. 2019 (September 16). "The Problem of Donald Trump and the Static Spectrum Fallacy." *Party Politics* XX:1–14.

Luntz, Frank. 2018 (October 26). "No Wonder America Is Divided. We Can't Even Agree on What Our Values Mean." *Time Magazine.* Available at time.com.

Lush, Tamara. 2020 (June 16). "Poll: Americans Are the Unhappiest They've Been in 50 Years." Associated Press. Available at apnews.com.

Manchester, Julia. 2020 (October 5). "Biden Says He Should Not Have Called Trump a Clown in First Debate." *The Hill.* Available at thehill.com.

Mansbridge, Jane. 2016 (March 11). "Three Reasons Political Polarization Is Here to Stay." *Washington Post.* Available at www.washingtonpost.com.

Mason, Liliana. 2018 (March 21). "Ideologues without Issues: The Polarizing Consequences of Ideological Identities." *Public Opinion Quarterly* 82:866–887.

McCorkindale, Tina. 2020. "2020 IPR Disinformation and Society Report." *Institute for Public Relations.* Available at instituteforpr.org.

McGreal, Chris. 2020 (May 4). "How the Kent State Massacre Marked the Start of America's Polarization." *Guardian.* Available at www.theguardian.com.

McNamara, Britney. 2017. "8 Activists from around the World Share Their Best Advice for Enacting Change." *Teen Vogue.* Available at www.teenvogue.com.

Merton, Robert K. 1968. *Social Theory and Social Structure.* New York: Free Press.

Mills, C. Wright. 1959. *The Sociological Imagination.* London: Oxford University Press.

Mishel, Lawrence, and Jori Kandra. 2020 (August 18). "CEO Compensation Surge 14% in 2019 to $21.3 Million." Economic Policy Institute. Available at www.epi.org.

Misra, Jordan. 2019 (April 23). "Voter Turnout Rates among All Voting Age and Major Racial and Ethnic Groups Were Higher Than in 2014." U.S. Census Bureau. Available at www.census.gov.

Najle, Maxine, and Robert P. Jones. 2019 (February 19). "American Democracy in Crisis: The Fate of Pluralism in a Divided Nation." *Public Religion Research Institute.* Available at www.prri.org.

Norris, Louise. 2020 (October 20). "Affordable Health Insurance: Your Buying Guide." *Health Insurance.* Available at www.healthinsurance.org.

Ordonez, Franco. 2020 (November 10). "Biden Tells World Leaders 'It's Not America Alone' Anymore." National Public Radio. Available at www.npr.org.

Palacios, Wilson R., and Melissa E. Fenwick. 2003. "'E' Is for Ecstasy." In *In Their Own Words,* ed. Paul Cromwell, 277–283. Los Angeles: Roxbury.

Parker, Kim, Rich Morin, and Juliana Menasce Horowitz. 2019 (March 21). "Looking to the Future, Public Sees an America in Decline on Many Fronts." Pew Research Center. Available at www.pewsocialtrends.org.

Posen, Adam S. 2018 (March). "The Post-American World Economy." *Foreign Affairs* 97(2):28–38.

Poushter, Jacob, and J. J. Moncus. 2020 (September 23). "How People in 14 Countries View the Rest of the World in 2020." Pew Research Center. Available at www.pewresearch.org.

Public Religion Research Institute. 2020 (October 19). "Dueling Realities: Amid Multiple Crises, Trump and Biden Supporters See Different Priorities and Futures for the Nation." Available at www.prri.org.

Reiman, Jeffrey, and Paul Leighton. 2020. *The Rich Get Richer and the Poor Get Prison,* 12th ed. New York: Routledge.

Ruzza, Carlo and Rosa Sanchez Salgado. 2020 (August 1). "The Populist Turn in EU Politics and the Intermediary Role of Civil Society Organizations." *European Politics and Society* 21:1–15.

Schaeffer, Katherine. 2020 (February 7). "Six Facts about Economic Inequality in the United States." Pew Research Center. Available at www.pewresearch.org.

Serkez, Yaryna. 2020 (October 26). "Who Inspired the Trump Campaign Playbook?" *New York Times.* Available at www.nytimes.com.

Shear, Michael D., and Lisa Friedman. 2020 (November 8). "Biden's Plan for Day One in the White House." *New York Times.* Available at www.nytimes.com.

Sherman, Natalie. 2020 (November 7). "Five Questions for Joe Biden on the Economy." BBC News. Available at www.bbc.com.

Shin, Claire. 2019 (May 12). "Boeing: A Harrowing Symptom of Corporate Greed." *The Spectator.* Available at www.stuyspec.com.

Silver, Laura, Shannon Schumacher, and Mara Mordecai. 2020 (October 5). "In U.S. and UK, Globalization Leaves Some Feeling 'Left Behind' or 'Swept Up.'" Pew Research Center. Available at www.pewresearch.org.

Simi, Pete, and Robert Futrell. 2009. "Negotiating White Power." *Social Problems* 56(1):98–110.

Simmons-Duffin, Selena. 2020 (June 12). "Transgender Health Protections Reversed by Trump Administration." National Public Radio. Available at www.npr.org.

Stolzenberg, E. B., M. C. Aragon, E. Romo, V. Couch, D. McLennan, M. K. Eagan, and N. Kang. 2020. *The American Freshman: National Norms Fall 2019.* Los Angeles: Higher Education Research Institute, UCLA.

Tanenhaus, Sam. 2016 (October 17). "Rise of the Reactionary." *New Yorker.* Available at www.newyorker.com.

The Editors. 2020 (October 8)." Dying in a Leadership Vacuum." *New England Journal of Medicine* 383(15):1479–1480.

Thomas, W. I. 1931/1966. "The Relation of Research to the Social Process." In *W. I. Thomas on Social Organization and Social Personality,* ed. Morris Janowitz, 289–305. Chicago: University of Chicago Press.

U.S. Bureau of Labor Statistics. 2020. "Civilian Unemployment Rate, Seasonally Adjusted." *Graphics for Economic News Releases.* Available at www.bls.gov.

Vieten, Ulrike. 2020 (September). "The New Normal and Pandemic Populism: The COVID-19 Crisis and Anti-Hygienic Mobilisation of the Far-Right." *Social Sciences* 9(9):1–14.

Volfovsky, Alexander. 2018 (September). "Exposure to Opposing Views on Social Media Can Increase Political Polarization." *Proceedings of the National Academy of Sciences* 115(37):9216–9221.

Weir, Sara, and Constance Faulkner. 2004. *Voices of a New Generation: A Feminist Anthology.* Boston: Pearson Education.

Wilson, John. 1983. *Social Theory.* Englewood Cliffs, NJ: Prentice Hall.

Wise, Alana. 2020 (August 28). "Calling Protesters 'Thugs' and Biden Extreme, Trump's Bombastic Campaign Returns." National Public Radio. Available at www.npr.org.

Chapter 2

Adler, Loren, Sobin Lee, Kathleen Hannick, and Erin Duffy. 2019. *Provider Charges Relative to Medicare Rates, 2012–2017.* USC–Brookings Schaeffer Initiative for Health Policy. Washington, DC: Brookings Institution.

Alberga, Hannah, Josie Kao, Claire Porter Robbins, et al. 2020 (August 7). "Masks to Combat COVID-19: Which Countries Are Embracing, Requiring or Rejecting Them? A Global Guide." The Globe and Mail–Canada. Available at www.theglobeandmail.com.

American College Health Association American College Health Association (ACHA). 2019. *American College Health Association National College Health Assessment II: Reference Groups Executive Summary Spring 2015 and 2019.* Hanover, MD: American College Health Association.

American Psychiatric Association. 2017. "Mental Health Disparities: Diverse Populations Fact Sheet." Available at www.psychiatry.org.

American Psychological Association. 2019 (May). *Americans Becoming More Open about Mental Health.* Available at www.apa.org.

Arias, Elizabeth, and Jiaquan Xu. 2020 (November 17). "United States Life Tables, 2018." *National Vital Statistics Reports* 69(12). National Center for Health Statistics. Available at www.cdc.gov.

Baker, Peter, Shari L. Dwarkin, Sengfah Tong, Ian Banks, Time Shand, and Gavin Yamen. 2014. "The Men's Health Gap: Men Must Be Included in the Global Health Equity Agenda." *Bulletin of the World Health Organization* 92:618–670.

Barker, Kristin. 2002. "Self-Help Literature and the Making of an Illness Identity: The Case of Fibromyalgia Syndrome (FMS)." *Social Problems* 49(3):279–300.

Bauer, Lauren, Kristen E. Brady, Wendy Edelberg, and Jimmy O'Donnell. 2020 (September 17). *Ten Facts about COVID-19 and the US Economy.* Washington, DC: Brookings Institute. Available at www.brookings.edu.

Bauman, B. L., J. Y. Ko, S. Cox, D. V. D'Angelo, L. Warner, S. Folger, H. D. Tevendale, K. C. Coy, L. Harrison, and W. D. Barfield. 2020 (May 15). "*Vital Signs:* Postpartum Depressive Symptoms and Provider Discussions about Perinatal Depression—United States, 2018." *Morbidity and Mortality Weekly Report (MMWR).* Available at www.cdc.gov.

Beladi H, Chi-Chur Chao, Mong Shan Ee, and Daniel. Hollas. 2019. "Does Medical Tourism Promote Economic Growth? A Cross-Country Analysis." *Journal of Travel Research* 58(1):121–135.

Berchick, Edward R., Emily Hood, and Jessica C. Barnett. 2018. "Health Insurance Coverage in the United States, 2017." *Current Population Reports.* U.S. Census Bureau.

Biden, Joe. 2020. *Biden-Harris Transition.* www.Buildbackbetter.gov.

Bollyky, Thomas J., and Jennifer B. Nuzzo. 2020 (October 1). "Trump's 'Early' Travel Bans Weren't Early, Weren't Bans, and Didn't Work." *Washington Post.* Available at www.washingtonpost.com.

Brulliard, Karin. 2020 (November 11). "At Dinner Parties and Game Nights, Casual Family Life Is Fueling the Coronavirus Surge as Daily Cases Exceed 150,000." *Washington Post.* Available at www.washingtonpost.com.

Buncombe, Andrew. 2020 (March 27). "'She Was Not Alone': Nurse Uses Video Chat to Let Daughter Say Farewell to Mother Dying of Coronavirus." *The Independent.* Available at www.independent.co.uk.

Bunis, Dena. 2019 (August 21). *High Prescription Drug Prices Lead Many Consumers to Ignore Doctor's Orders.* AARP. Available at www.aarp.org.

Bureau of Consular Affairs. 2020 (June 29). *Presidential Proclamations on Novel Coronavirus.* U.S. Department of State. Available at www.travel.state.gov.

Bureau of Transportation Statistics. 2020 (March 19). *2019 Traffic Data for U.S. Airlines and Foreign Airlines U.S. Flights—Final, Full Year.* U.S. Department of Transportation. Available at www.bts.dot.gov.

Cabarkapa, S., S. E. Nadjidai, J. Murgier, and C. H. Ng. 2020. "The Psychological Impact of COVID-19 and Other Viral Epidemics on Frontline Healthcare Workers and Ways to Address It: A Rapid Systematic Review." *Brain, Behavior, & Immunity—Health* 8:100144.

Caffrey, Mary. 2017. *Physicians Far Less Likely to Take New Medicaid Patients, CDC Finds.* AJMC. Available at www.ajmc.com.

Carter, Sierra. 2020 (December 7). "Racism Literally Ages Black Americans Faster, according to Our 25-Year Study." *Guardian.* Available at www.theguardian.com.

CDC (Centers for Disease Control and Prevention). 2019 (September 5). *Racial and Ethnic Disparities Continue in Pregnancy-Related Deaths.* Press Release. Available at www.cdc.gov.

CDC (Centers for Disease Control and Prevention). 2020 (June 5). "Evidence for the Limited Spread of COVID-19 within the United States, January–February." *Morbidity and Mortality Weekly Report (MMWR)* 69(22):680–684.

Chuck, Elizabeth. 2020. "A Mother Couldn't Wait to Welcome Her Second Child. Covid-19 Killed Her before She Got to Hold the Baby." NBC News. Available at www.nbcnews.com.

Cleo. 2020 (July 13). *June Member Survey: The Burden Is Real—Immediate Action Needed to Support and Retain Working Parents.* Available at www.hicleo.com.

Cockerham, William. 2019. "Medicine and Health." In *The Wiley Companion to Sociology,* ed. George Ritzer and Wendy W. Murphy, 250–266. Hoboken, NJ: Wiley & Sons.

Cook, Won K., William C. Kerr, Katherine J. Karriker-Jaffe, Libo Li, Camilla K. Lui, and Thomas K. Greenfield. 2020 (January). "Racial/Ethnic Variations in Clustered Risk Behaviors in the U.S." *American Journal of Preventive Medicine* 58(1):e21–e29.

Correia, Tiago. 2017 (September 17). "Revisiting Medicalization: A Critique of Assumptions of What Counts as Medical Knowledge." *Frontiers in Sociology.* Available at www.frontiersin.org.

COVID Tracking Project. 2020 (December 8). *US Historical Data.* Available at www.covidtracking.com.

Dalen, James E., and Joseph S. Alpert. 2018. "Medical Tourists: Incoming and Outgoing." *American Journal of Medicine* 132(1):9–10.

de Gispert, Jaime Gonzalez. 2015. "Hispanic Paradox: Why Immigrants Have a High Life Expectancy." BBC Mundo. Available at www.bbc.com.

Deliso, Meredith. 2021 (January 28). "All of Biden's Executive Orders and Notable Actions, So Far." ABC News. Available at www.abcnews.go.com.

Emanuel, Ezekial J. 2019 (March 23). "Big Pharma's Go-To Defense of Soaring Drug Prices Doesn't Add Up." *Atlantic.* Available at www.theatlantic.com.

Eng, Renee. 2020 (September 28). "Woman Chases California Dream as More People Leave Golden State during Pandemic." *Spectrum News1.* Available at www.spectrumnews1.com.

Evers-Hillstrom, Karl. 2020 (January 25). "Lobbying Spending in 2019 Nears All-Time High as Health Sector Smashes Records." Common Dreams. Available at www.commondreams.org.

Finch, W. Holmes, and Maria E. Hernández Finch. 2020. "Poverty and COVID-19: Rates of Incidence and Deaths in the United States during the First 10 Weeks of the Pandemic." *Frontiers in Sociology* 5(47).

Garfield, Rachel, Kendal Orgera, and Anthony Damico. 2020. *The Coverage Gap: Uninsured Poor Adults in States That Do Not Expand Medicaid.* Kaiser Family Foundation. Available at www.kff.org.

Garrett, Bowen, and Anuj Gangopadhyaya. 2016. *Who Gained Health Insurance Coverage Under the ACA, and Where Do They Live?* Available at www.urban.org.

Gee, Emily. 2020 (September 25). "Less Coverage and Higher Costs: The Trump's Administration's Health Care Legacy." Center for American Progress. Available at www.americanprogress.org.

Gerdeman, Dina. 2020 (October 29). "The COVID Gender Gap: Why Fewer Women Are Dying." *Working Knowledge.* Harvard Business School. Available at www.hbswk.hbs.edu.

Gjelten, Tom. 2020 (June 1). "Peaceful Protesters Tear-gassed to Clear Way for Trump Church Photo Op." National Public Radio. Available at www.npr.org.

Global Fund. 2020. *About the Global Fund.* Available at www.theglobalfund.org.

Goldstein, Ellen, James Topitzes, Julie Miller-Cribbs, and Roger L. Brown. 2020 (October 12). "Influence of Race/Ethnicity and Income on the Link between Adverse Childhood Experiences and Child Flourishing." *Pediatric Research.* Available at https://doi.org/10.1038/s41390-020-01188-6.

Goldstein, Michael S. 1999. "The Origins of the Health Movement." In *Health, Illness, and Healing: Society, Social Context, and Self,* ed. Kathy Charmaz and Debora A. Paterniti, 31–41. Los Angeles: Roxbury.

Gracy, Delaney, Anupa Fabian, Corey Hannah Basch, Maria Scigliano, Sarah A. MacLean, Rachel K. MacKenzie, and Irwin Redlener. 2018 (January 17). "Missed Opportunities: Do States Require Screening of Children for Health Conditions That Interfere with Learning." *PLOS One* 13(1). Available at www.journals.plos.org.

Gramlich, John. 2020 (September 21). "Americans Give the US Low Marks for Its Handling of Covid-19, and So Do People in Other Countries." *FactTank*. Pew Research Center. Available at www.perwresearch.org.

Grandview Research. 2019. *U.S. Medical Tourism Market Size, Share & Trends Analysis Report by Type (Inbound, Outbound), Competitive Landscape, and Segment Forecasts, 2019–2026*. Available at www.grandviewresearch.com.

Green, Emilee. 2020. *Mental Illness and Violence: Is There a Link?* Illinois Criminal Justice Information Authority. Available at www.icjia.illinois.gov.

Haileamlak, Abraham. 2018. "Maternal and Newborn Mortality—Still the Greatest Disparity between Low-Income and High-Income Countries." *Ethiopian Journal of Health Sciences* 28(4):368.

Hamel, Liz, Ashley Kirzinger, Cailey Munana, Lunna Lopes, Audrey Kearney, and Mollyann Brodie. 2020a. *5 Charts about Public Opinion on the Affordable Care Act and the Supreme Court*. Kaiser Family Foundation. Available at www.kff.org.

Hamel, Liz, Ashley Kirzinger, Cailey Munana, and Mollyann Brodie. 2020b (December 15). *KFF COVID-19 Vaccine Monitor: December 2020*. Available at www.kff.org.

Hargraves, John, and Aaron Bloschichak. 2019 (December 17). "International Comparisons of Health Care Prices from the 2017 iFHP Survey." Washington, DC: Health Care Cost Institute Blog.

Hasell, Joe. 2020. *Testing Early, Testing Late: Four Countries' Approaches to COVID-19 Testing Compared*. Our World in Data. Available at www.ourworldindata.org.

Hathaway, Bill. 2020 (February 20). "Want to Live Longer? Stay in School, Study Suggests." *YaleNews*. Available at www.news.yale.edu.

HealthCare.gov. 2020. *The Children's Health Insurance Program*. Available at www.healthcare.gov.

Health Care Legacy. Center for American Progress. Available at www.americanprogress.org.

Healthline. 2019. "Concerned about Getting Rx Drugs from Canada? Here's What to Know." Available at www.healthline.com.

Healthy Minds Network. 2020. "The Impact of COVID-19 on College Student Well-Being." The Healthy Minds Network. Available at www.healthymindsnetwork.org.

Himmelstein, D. U., and S. Woolhandler. 2014. "High Administrative Costs: The Authors Reply." *Health Affairs* 33(1):2081.

Himmelstein, David U., Robert M. Lawless, Deborah Thorne, Pamela Foohey, and Steffie Woolhandler. 2019 (March). "Medical Bankruptcy: Still Common Despite the Affordable Care Act." *American Journal of Public Health* 109(3):431–433.

Hoffman, Kelly M., Sophie Trawalter, Jordan R. Axt, and M. Normal Oliver. 2016 (April 4). "Racial Bias in Pain Assessment and Treatment Recommendations, and False Beliefs about Biological Differences between Blacks and Whites." *Proceedings of the National Academy of Sciences of the United States of America*. Available at www.ncbi.nlm.nih.gov.

Holmes, Oliver. 2020 (October 2). "What Donald Trump Has Said about COVID-19—A Recap." *Guardian*. Available at www.guardian.com.

Holmes, Kristen, Curt Devine, and Jeremy Herb. 2020 (March 4). "Pence: 'Any American Can Be Tested' for Coronavirus." CNN Politics. Available at www.cnn.com.

Hong, Young-Rock, Juan M. Hincapie-Castillo, and Zhigang Xie. 2020. "Socioeconomic and Demographic Characteristics of US Adults Who Purchase Prescription Drugs from Other Countries." *JAMA Network Open* 3(6).

Horsley, Jamie P. 2020 (August 19). "Let's End the Covid-19 Blame Game: Reconsidering China's Role in the Pandemic." Brookings Institute. Available at www.brookings.edu.

Homan, Patricia. 2019. "Structural Sexism and Health in the United States: A New Perspective on Health, Inequality and the Gender System." *American Sociological Review* 84(3):486–516.

Huang, Pien, and Audrey Carlsen. 2021 (January 29). "How Is the COVID-19 Vaccination Campaign Going in Your State?" National Public Radio. Available at www.npr.org.

Hyland, Pat. 2019. "U.S. Medical Tourism—Steady Growth." *Tourism Review*. Available at www.tourism-review.com.

Indian Health Service. 2020. "About Indian Health Service." Available at www.ihs.gov.

Institute for Health Metrics and Evaluation. 2020 (December 4). *Covid-19 Results Briefing: The United States of America*. Available at www.healthdata.org.

Jacobs, H. E. 2016. "Pharmaceutical Greed and Its Consequences." *Connecticut Medicine* 80(5):315–316.

Jaffe, D. J. 2017. *Insane Consequences: How the Mental Health Industry Fails the Mentally Ill*. Buffalo, NY: Prometheus.

Jargon, Julie. 2020 (May 5). It's Grandparents to the Rescue for Stressed Working-from-Home Parents." *Wall Street Journal*. Available at www.wsj.com.

Jenkins, Bonnie. 2020 (March 27). "Now Is the Time to Revisit the Global Health Security Agenda." Brookings. Available at www.brookings.edu.

Johns Hopkins University. 2020a. *New Cases of COVID-10 in World Countries*. Johns Hopkins University Coronavirus Resource Center. Available at www.coronavirus.jhu.gov.

John Hopkins University. 2020b (December). *Mortality Analysis*. John Hopkins Coronavirus Resource Center. Available at coronavirus.jhu.edu.

Johnson, Lauren M. 2020 (September 21). "A North Carolina Couple Together for over 50 Years Died Minutes Apart of Covid-19 While Holding Hands." CNN. Available at www.cnn.com.

Jones, Bradley. 2020. "Increasing Share of Americans Favor a Single Government Program to Provide Health Care Coverage." *FactTank*. Pew Research Center. Available at www.pewresearch.org.

Kaiser Family Foundation. 2020. *Status of State Medicaid Expansion Decisions: Interactive Map*. Available at www.kff.org.

Kaiser Health News. 2020. *Lost on the Frontline*. Available at www.khn.org.

Kamal, Rabah, Giorlando Ramirez, and Cynthia Cox. 2020 (December 23). "How Does Health Spending in the U.S. Compare to Other Countries?" *Health System Tracker*. Available at www.healthsystemtracker.org.

Karalis, Katie. 2020 (August 21). "'No More Masks': Hundreds Attend Anti-Mask Mandate Rally in St. George." ABC4 Utah. Available at www.abc4.com.

Karpman, M., and S. Zuckerman. 2020 (November 6). *ACA Offers Protection as the Covid-19 Pandemic Erodes Employer Health Coverage*. Urban Institute. Available at www.rwjf.org.

Keisler-Starkey, Katherine, and Lisa N. Bunch. 2020 (September 15). "Health Insurance Coverage in the United States: 2019." U.S. Census Bureau. Available at www.census.gov.

Kelland, Kate. 2020 (November 29). "Malaria Death Toll to Exceed COVID-19's in Sub-Saharan Africa: WHO." Reuters. Available at www.reuters.com.

Kemp, Blakelee R., and Jennifer Karas Montez. 2020 (January 23). "Why Does the Importance of Education for Health Differ across the United States?" *Socius*. Available at journals.sagepub.com.

King, Robert. 2019. "Indian Health Service Swept Up in Efforts to Boost Transparency, Oppose Medicare for All." *Modern Healthcare*. Available at www.modernhealthcare.com.

Kirzinger, Ashley, Lunna Lopes, Bryan Wu, and Mollyann Brodie. 2019 (March 1). "KFF Health Tracking Poll—February 2019: Prescription Drugs." Kaiser Family Foundation. Available at www.kff.org.

Kochhar, Rakesh. 2020 (October 22). "Fewer Mothers and Fathers in US Are Working Due to COVID-19 Downturn: Those at Work Have Cut Hours." *FactTank*. Available at www.pewresearch.org.

Krauth, Dan. 2020 (August 31). "Pandemic Exodus: Moving Companies Turn Customers away as People Leave Tri-State in Record Numbers." ABC7. Available at www.abc7ny.com.

Kurtzman, Laura. 2020 (January 15). "Single-Payer Systems Likely to Save Money in US, Analysis Finds." University of California. Available at www.ucsf.edu.

Lagasse, Jeff. 2020 (October 6). "U.S. Insulin Prices 8 Times Higher Than in Other Countries." *Healthcare Finance*. Available at www.healthcarefinancenews.com.

Lemire, Jonathan. 2020 (July 11). "Trump Wears Mask in Public for First Time in Pandemic." Associated Press. Available at www.apnews.com.

Lipson S. K., E. G. Lattie, and D. Eisenberg. 2019. "Increased Rates of Mental Health Service Utilization by U.S. College Students: 10-Year Population-Level Trends (2007–2017)." *Psychiatric Services* 70(1):60–63.

Lyon, Edward. 2019 (February). "Imprisoning America's Mentally Ill." *Prison Legal News*. Available at www.prisonlegalnews.com.

Ma, Feinyang, Shixin Xu, Zhaoxin Tang, Zekun Li, and Lu Zhang. 2020 (published online). "Use of Antimicrobials in Food Animals and Impact of Transmission of Antimicrobial Resistance of Humans." *Biosafety and Health*. Available online at www.sciencedirect.com.

Make It OK. 2020. *Stop Mental Illness Stigma.* Available at www.makeitok.org.

Marcus, Mary Brophy. 2021 (January 20). "Ensuring Everyone in the World Gets a COVID Vaccine." *Voices of Duke Global Health Institute.* Available at www.globalhealth.duke .edu.

Markovitz, Gayle. 2020 (March 6). "Why Are Women More Depressed Than Men?" World Economic Forum. Available at www.weforum .org.

Martucci, Jessica. 2018 (December). "Beyond the Nature/Medicine Divide in Maternity Care." *AMA Journal of Ethics.* Available at www .journalofethics.ama-assn.org.

Mayo Clinic. 2020. *Herd Immunity and COVID-19 (Coronavirus): What You Need to Know.* Available at www.mayoclinic.com.

"Medicare Costs at a Glance." Undated. *Medicare.gov.* Available at Medicare.gov.

Mitchell, Ted, and Suzanne Ortega. 2019 (October 29). "Mental Health Challenges Require Urgent Response." *Inside Higher Ed.* Available at www.insidehighered.com.

Montez, Jennifer Karas, Mark D. Hayward, and Anna Zajacova. 2019 (March 11). "Educational Disparities in Adult Health: U.S. States as Institutional Actors on the Association." *Socius.* Available at journals.sagepub.com.

Morse, Susan. 2017 (October 9). "Suburban Poor, Uninsured Turn to Emergency Rooms for Care." *Healthcare Finance.* Available at www .healthcarefinancenews.com.

National Alliance on Mental Illness (NAMI). 2020a. *Mental Health Conditions.* Available at www.nami.org.

National Alliance on Mental Illness (NAMI). 2020b. *Know the Warning Signs.* Available at www.nami.org.

National Center for Health Statistics. 2016. *Health, United States, 2015.* Hyattsville, MD: U.S. Government Printing Office.

National Center for Health Statistics. 2019. *Health, United States, 2018.* Centers for Disease Control. Available at www.cdc.gov.

National Institute of Mental Health. 2020. *Mental Illness.* National Institute of Mental Health: Author. Available at www.nimh.nih.gov.

North, Anna. 2020 (July 22). "Why Masks Are (Still) Politicized in America." *Vox.* Available at www.vox.com.

Novotni, Michele. 2020. "No Judgement. No Guilt. Just ADHD Support and Understanding." *ADDitude.* Available at www.additutdemag .com.

Nunn, Ryan, Jana Parsons, and Jay Shambaugh. 2020. *A Dozen Facts about the Economics of the US Healthcare System.* Brookings Institute. Available at www.brookings.edu.

Nuzzo, Jennifer, and Tom Inglesby. 2018 (December). "U.S. Global Health Security Investments Improve Capacities for Infectious Disease Emergencies." *Health Security* 16(S-1). Available at www.lieberpub.com.

Obama, Barak. 2016 (November 4). *Executive Order—Advancing the Global Health Security Agenda to Achieve a World Safe and Secure from Infectious Disease Threats.* Office of the President of the United States.

O'Donnell, Ginger. 2019 (February 12). "College Students Form Their Own Advocacy Groups to Shape Mental Health Education on Campus." *Insight into Diversity.* Available at www .insightindiversity.com.

Office of Minority Health. 2020. *Profile: Hispanic/Latino Americans.* U.S. Department of Health and Human Services. Available at www.minorityhealth.hhs.gov.

Ortagus, Morgan. 2020 (September 3). *Update on U.S. Withdrawal from the World Health Organization.* U.S. Department of State. Available at www.state.gov.

Osorio, Victoria, Richard Prisinzano, and Mariko Paulson. 2020 (July 21). "The Increasing Mortality Gap by Education: Differences by Race and Gender." *Penn Wharton Blog.* University of Pennsylvania. Available at www .budgetmodel.wharton.upenn.edu.

Our World in Data. 2020. *Coronavirus (COVID-19) Testing.* Available at www .ourworldindata.org.

Owermohle, Sarah. 2020 (July 7). "Drug Prices Steadily Rise amid Pandemic, Data Shows." *Politico.* Available at www.politico.com.

Pear, Robert. 2019 (February 23). "Health Care and Insurance Industries Mobilize to Kill 'Medicare for All'." *New York Times.* Available at www.nytimes.com.

Peckham, Hannah, Nina M. de Gruijter, Charles Raine, Anna Radziszewski, Coziana Ciutin, Lucy R. Wedderburn, Elizabeth C. Rosser, Kate Webb, and Claire T. Deakin. 2020 (November 9). "Male Sex Identified by Global COVID-19 Meta-Analysis as a Risk Factor for Death and ITU Admission." *Nature Communications* 11(6317).

Penn, Madeline, Saurabha Bhatnagar, SreyRam Kuy, Steven Lieberman, Shereef Elnahal, Carolyn Clancy, and David Shulkin. 2019 (January). "Comparison of Wait Times for New Patients between the Private Sector and United States Department of Veterans Affairs Medical Centers." *JAMA Network Open* 2(1). Available at https://jamanetwork.com.

Pew Research Center. 2020 (June). *Republicans, Democrats Move Even Further Apart in Coronavirus Concerns.* Pew Research Center. Available at www.pewresearch.org.

Quinn, Melissa, and Margaret Brennan. 2021 (January 25). "Birx Says There Was No 'Full-Time Team' Working on COVID Response in Trump White House." *Face the Nation.* Available at CBSNews.com.

Rajkumar, S. Vincent. 2020. "The High Cost of Insulin in the United States: An Urgent Call to Action." *Mayo Clinic Proceedings* 95(1):22–28.

Reardon, Sara. 2021 (January 29). "The Most Worrying Mutations in Five Emerging Coronavirus Variants." *Scientific American.* Available at www.scientificamerican.com.

Reeves, Richard V., and Tiffany N. Ford. 2020 (May 15). *COVID-19 Much More Fatal for Men, Especially Taking Age into Account.* Brookings Institute. Available at wwww.brookings.edu.

Ritchie, Hannah. 2019. "If We Can Make Maternal Death as Rare as They Are in the Healthiest Countries We Can Save Almost 300,000 Mothers Each Year." *Our World in Data.* Available at www.ourworldindata.org.

Robinson, Carole. 2020 (October 26). "Mask Mandate Protesters Rally on Franklin Square." *Williamson Herald.* Available at www .williamsonherald.com.

Rosenbaum, Lisa. 2020. "Tribal Truce—How Can We Bridge the Partisan Divide and Conquer Covid?" *New England Journal of Medicine* 383:1682–1685.

Rosenberg, Kenneth Paul. 2019. *Bedlam: An Intimate Journey into America's Mental Health Crisis.* New York: Avery.

Rosenthal, Brian M., Joseph Goldstein, Sharon Otterman, and Sheri Fink. 2020 (November 27). "Why Surviving the Virus Might Come down to Which Hospital Admits You." *New York Times.* Available at www.nytimes.com.

Roser, Max, Hannah Ritchie, and Bernadeta Dadonaite. 2019. *Child and Infant Mortality.* Our World in Data. Available at www .ourworldindata.org.

Roser, Max, Hannah Ritchie, and Esteban Ortiz-Ospina. 2019. *World Population Growth.* Our World in Data. Available at www.worldindata .org.

Rossen, Lauren, Amy Branum, Farida Ahmad, Paul Sutton, and Robert Anderson. 2020 (October 23). "Excess Deaths Associated with Covid-19, by Age and Race and Ethnicity—United States, January 26–October 3, 2020." *Morbidity and Mortality Report.* Available at www.cdc.gov.

Rosewicz, Barb, Justin Theal, and Katy Ascanio. 2020 (January 9). "After K–12 Education, Medicaid Is the States' Biggest Expense." Pew Charitable Trust. Available at www.pewtrusts.org.

Ruff, Paul, Sana Al-Sukhun, Charmaine Blanchard, and Lawrence N. Shulman. 2016. "Access to Cancer Therapeutics in Low- and Middle-Income Countries." *American Society of Clinical Oncology Educational Book* 36:58–65.

Saad, Lydia. 2019 (December 19). "More Americans Delaying Medical Treatment Due to Cost." Gallup. Available at www.news.gallup.com.

Sasson, Isaac, and Mark D. Hayward. 2019. "Association between Educational Attainment and Causes of Death among White and Black US Adults, 2010–2017." *Journal of the American Medical Association (JAMA)* 322(8):756–763.

Schneider, Eric C. 2020 (July 23). "Failing the Test—The Tragic Gap Undermining the U.S. Pandemic Response." *New England Journal of Medicine* 383:299–302.

Schwarz, Benyamin, and Jacquelyn J. Benson. 2018. "The 'Medicalized Death': Dying in the Hospital." *Journal of Housing for the Elderly* 32(3–4):379–430.

Sepkowitz, Kent. 2020 (December 10). "Rudy Giuliani's Pricey COVID 'Cocktail'." CNN. Available at www.cnn.com.

Silva, Christianna. 2020 (September 27). "Food Insecurity in the U.S. by the Numbers." National Public Radio. Available at www.npr .org.

Slate, Risdon N., Jacqueline K. Buffington-Vollum, and W. Wesley Johnson. 2013. *The Criminalization of Mental Illness: Crisis and Opportunity for the Justice System,* 2nd ed. Durham, NC: Carolina Academic Press.

Smith, Patrick, and Dawn Liu. 2020 (February 7). "China to Investigate Death of Doctor Who Blew the Whistle on Coronavirus." NBC News. Available at www.nbcnews.com.

Snell, Kelsey. 2020 (May 15). "Here's How Much Congress Has Approved for Coronavirus Relief So Far and What It's For." National Public Radio. Available at www.npr.org.

Sommer, Lauren. 2020 (April 24). "Why the Warning That Coronavirus Was on the Move in US Cities Came So Late." National Public Radio. Available at www.npr.org.

Stein, Jeff, and Yasmeen Abutaleb. 2019 (December 20). "Congress Showers Health Care Industry with Multibillion-Dollar Victory after Wagging Finger at It for Much of 2019." *Washington Post*. Available at www.washingtonpost.com.

Stevens, Lance, and Lawrence Mallory. 2019. "U.S. Seniors Pay Billions, Yet Many Cannot Afford Healthcare." *Gallup Blog*. Available at www.news.gallup.com.

Substance Abuse and Mental Health Services Administration (SAMHSA). 2019. *Key Substance Use and Mental Health Indicators in the United States: Results from the 2018 National Survey on Drug Use and Health* (HHS Publication No. PEP19-5068, NSDUH Series H-54). Rockville, MD: Center for Behavioral Health Statistics and Quality, Substance Abuse and Mental Health Services Administration. Available from https://www.samhsa.gov/data.

Summers, Jennifer, Hao-Yuan Cheng, Hsien-Ho Lin, Lucy Telfar Barnard, Amanda Kvalsig, Nick Wildon, and Michael G. Baker. 2020. "Potential Lessons from the Taiwan and New Zealand Health Responses to the COVID-19 Pandemic." *Lancet Western Pacific* 4. Available at www.thelancet.com.

Szasz, Thomas. 1961/1970. *The Myth of Mental Illness: Foundations of a Theory of Personal Conduct*. New York: Harper & Row.

Taylor, Jamila. 2019 (December 19). *Racism, Inequality, and Health Care for African Americans*. New York: Century Foundation. Available at www.tcf.org.

Taylor, Verta. 1995. "Self-Labeling and Women's Mental Health: Postpartum Illness and the Reconstruction of Motherhood." *Sociological Focus* 28(1):23–47.

Templin, T., T. C. O. Hashiguchi, B. Thomson, J. Dieleman, and E. Bendavid. 2019. "The Overweight and Obesity Transition from the Wealthy to the Poor in Low- and Middle-Income Countries: A Survey of Household Data from 103 Countries." *PLoS Medicine* 16(11):e1002969. Available at https://doi.org/10.1371/journal.pmed.1002968.

Thames, April D., Michael R. Irwin, Elizabeth C. Breen, and S. W. Cole. 2019 (August). "Experienced Discrimination and Racial Differences in Leukocyte Gene Expression." *Psychoneuroendocrinology*. Available at www.sciencedirect.com.

Tolbert, Jennifer, Kendal Orgera, and Anthony Damico. 2020. *Key Facts about the Uninsured Population*. Kaiser Family Foundation. Available at www.kff.org.

Torres, Stacy. 2020 (August 25). "My Sister Doesn't Have COVID-19, but I'm Scared She Won't Make It out of 2020." *USA Today*. Available at www.usatoday.com.

Unchained at Last. 2020. *Forced and Child Marriage*. Available at www.unchainedatlast.org.

United Nations. 2020 (October). *Policy Brief: COVID-19 and Universal Health Coverage*. Available at unsdg.un.org.

University of Pennsylvania. 2020. *I CARE*. Available at www.caps.wellness.upenn.edu.

U.S. Department of Health and Human Services. 2020. *What Is the U.S. Opioid Epidemic?* Available at www.hhs.gov.

U.S. Government Accountability Office. 2017 (November). *Drug Industry: Profits, Research and Development Spending, and Merger and Acquisition Deals*. Report GAO-18-40. Available at www.gao.gov.

Vandebrouck, Laurent. 2020 (March 13). "What the U.S. Gets Right about Healthcare." *MedCity News*. Available at www.medcitynews.com.

Wan, William. 2020 (November 23). "For Months, He Helped His Son to Keep Suicidal Thoughts at Bay. Then Came the Pandemic." *Washington Post*. Available at www.washingtonpost.com.

Water.org. 2020. *About Us*. Available at www.water.org.

Wen, Leana S., and Nakisa B. Sadeghi. 2020 (July 20). "Addressing Racial Health Disparities in the COVID-19 Pandemic: Immediate and Long-Term Policy Solutions." Health Affairs Blog. *Health Affairs*. Available at www.healthaffairs.org.

Weyandt, Lisa L., Danielle R. Oster, Bergljiot Gyda Gudmundsdottir, George J. DuPaul, and Arthur D. Anastopoulos. 2017 (February). "Neuropsychological Functioning in College Students with and without ADHD." *Neuropsychology* 31(2):160–172.

Williams, David R. 2018 (December). "Stress and the Mental Health of Populations of Color: Advancing Our Understanding of Race-Related Stressors." *Journal of Health and Social Behavior* 59(4):466–485.

Woolf, Steven, Derek Chapman, Roy Sabo, Daniel Weinberger, Latoya Hill, and DaShanda D. H. Taylor. 2020. "Excess Deaths from COVID-19 and Other Causes, March–July 2020." *JAMA* 324(15):1562–1564. Available at www.jamanetwork.com.

World Bank. 2020a. "Life Expectancy at Birth, Total (Years)." Available at data.worldbank.org.

World Bank. 2020b. *Mortality Rate, Infant (per 1000 Live Births)*. Available at www.data.worldbank.org.

World Health Organization (WHO). 1946. *Constitution of the World Health Organization*. New York: World Health Organization Interim Commission.

World Health Organization (WHO). 2019a. *1 in 3 People Globally Do Not Have Access to Safe Drinking Water—UNICEF, WHO*. Available at www.who.int.

World Health Organization (WHO). 2019b. *Maternal Mortality*. Available at www.who.int.

World Health Organization (WHO). 2020a. *Timeline: WHO's COVID-19 Response*. Available at www.who.int.

World Health Organization (WHO). 2020b. *WHO Coronavirus Disease (COVID-19) Dashboard*. Available at www.covid19.who.int.

World Health Organization (WHO). 2020c. *Top Ten Causes of Death*. Available at www.who.int.

World Health Organization (WHO). 2020d. *World Health Statistics*. Available at www.who.int.

World Health Organization (WHO). 2020e. *COVAX: Working for Global Equitable Access to Covid-19 Vaccines*. Available at www.who.int.

World Health Organization (WHO). 2020f (April 1). *Obesity and Overweight*. Available at www.who.int.

World Health Organization (WHO). 2020g. "Violence against Women." *Fact Sheet*. Available at www.who.int.

World Health Organization (WHO). 2020h. "Gender and Women's Mental Health." Available at www.who.int.

Yamey, Gavin, and Clare Wenham. 2020 (July 1). "The US and the UK Were the Two Best Prepared Nations to Tackle a Pandemic—What Went Wrong? *Time*. Available at www.time.com.

Yong, Ed. 2020 (November 13). "No One Is Listening to Us." *Atlantic*. Available at www.theatlantic.com.

Zoellner, Danielle. 2020 (October 29). "Coronavirus: Majority of Americans Want a National Mask Mandate as Fauci Backs One for the First Time." *The Independent*. Available at www.independent.co.uk.

Chapter 3

Abadinsky, Howard. 2013. *Drug Use and Abuse: A Comprehensive Introduction*. Belmont, CA: Wadsworth.

Above the Influence. 2020. *What It Means to Live above the Influence*. Available at www.abovetheinfluence.com.

Alabama Code. 2015. "Chemical Endangerment of a Child." *Alabama Code 1975, §26-15-3.2*. Available at www.judicial.alabama.gov.

Alexander, Michelle. 2010. *The New Jim Crow: Mass Incarceration in the Age of Colorblindness*. New York: New York Press.

Armatas, C., A. Heinzerling, and J. A. Wilken. 2020. "Notes from the Field: E-Cigarette, or Vaping, Product Use—Associated Lung Injury Cases during the COVID-19 Response." *Morbidity and Mortality Weekly Report* (69):801–802.

Azofeifa, A., B. D. Rexach-Guzmán, A. N. Hagemeyer, R. A. Rudd, and E. K. Sauber-Schatz. 2019. "Driving under the Influence of Marijuana and Illicit Drugs among Persons Aged ≥16 Years—United States, 2018." *Morbidity and Mortality Weekly Report* (68):1153–1157.

Balsa, A. I., J. F. Homer, and M. T. French. 2009. "The Health Effects of Parental Problem Drinking on Adult Children." *Journal of Mental Health Policy and Economics* 12(2):55–66.

Baum, Dan. 2016 (April). "Legalize It All." *Harper's Magazine*. Available at www.harpers.org.

Beckett, Katherine, and Marco Brydolf-Horwitz. 2020. "A Kinder, Gentler Drug War? Race, Drugs, and Punishment in 21st Century America." *Punishment & Society*: 1–25. Available at https://journals.sagepub.com/doi/abs/10.1177/1462474520925145.

Behrendt, S., H.-U. Wittchen, M. Höfler, R. Lieb, and K. Beesdo. 2009. "Transitions from First Substance Use to Substance Use Disorders in Adolescence: Is Early Onset Associated with a Rapid Escalation?" *Drug and Alcohol Dependence* 99:68–78.

Berends, Lynda, Jason Ferris, and Anne-Marie Laslett. 2014. "On the Nature of Harms Reported by Those Identifying a Problem Drinker in the Family, an Exploratory Study." *Journal of Family Violence* 29:197–204.

Bergeron, Ryan. 2018 (April 23). "Woman Lost 2 Sons in One Night to Opioids: Fighting the Crisis Is Now Her Life's Work." CNN. Available at www.cnn.com.

Berman, Michael L. 2019 (June). "Using Opioid Settlement Proceeds for Public Health: Lessons from the Tobacco Experience." *Kansas Law Review.* 57:1029–1059.

Biden-Harris. 2020a. "The Biden Plan to End the Opioid Crisis." Available at www.joebiden.com.

Biden-Harris. 2020b. "The Biden Plan for Strengthening America's Commitment to Justice." Available at www.joebiden.com.

Bilefsky, Dan. 2018 (October 17). "Legalizing Recreational Marijuana, Canada Begins a National Experiment." *New York Times.* Available at www.nytimes.com.

Blomeyer, Dorothea, Chris M. Friemel, Arlette F. Buchmann, Tobias Banaschewski, Manfred Laucht, and Miriam Schneider. 2013. "Impact of Pubertal Stage at First Drink on Adult Drinking Behavior." *Alcoholism, Clinical and Experimental Research* 37(10):1804–1811.

British Medical Association. 2019 (March 13). "British Medical Association Report on Drugs of Dependence." Available at www.bma.org.

Bronson, Jennifer, and E. Ann Carson. 2019 (April). *Prisoners in 2017.* Washington, DC: U.S. Department of Justice, Bureau of Justice Statistics, NCJ 252156. Available at www.bjs.gov.

Campaign for Tobacco Free Kids. 2020. "The Rise of Cigars and Cigar-Smoking Harms Fact Sheet." Available at www.tobaccofreekids.org.

Carrigan, Matthew A., Oleg Uryasev, Carol B. Frye, Blair L. Eckman, Candace R. Myers, Thomas D. Hurley, and Stephen A. Benner. 2014 (December 1). "Hominids Adapted to Metabolize Ethanol Long before Human-Directed Fermentation." *Proceedings of the National Academy of Sciences (PNAS) of the United States of America* 12(2):458–463.

Centers for Disease Control and Prevention (CDC). 2014. "Smokefree Policies Reduce Smoking." Available at www.cdc.gov.

Centers for Disease Control and Prevention (CDC). 2015 (October 15). "Excessive Alcohol Use Continues to Be a Drain on American Economy." Available at www.cdc.gov.

Centers for Disease Control and Prevention (CDC). 2018. "Table 20. Use of Selected Substances in the Past Month among Persons Aged 12 Years and over, by Age, Sex, Race, and Hispanic Origin." National Center for Health Statistics. Available at www.cdc.gov.

Center for Disease Control and Prevention (CDC). 2019. "Fact Sheet: Alcohol Use and Your Health." Available at www.cdc.gov.

Centers for Disease Control and Prevention (CDC). 2020a (May 18). "Economic Trends in Tobacco." Available at www.cdc.gov.

Center for Disease Control and Prevention (CDC). 2020b (May 7). "Fact about FASDs." *Fetal Alcohol Spectrum Disorders (FASDs).* Available at www.cdc.gov.

Center for Disease Control and Prevention (CDC). 2020c (April 28). "Health Effects of Cigarette Smoking." *Smoking and Tobacco Use.* Available at www.cdc.gov.

Centers for Disease Control and Prevention (CDC). 2020d (March 19). "Opioid Overdose: Understanding the Epidemic." Available at www.cdc.gov.

Centers for Disease Control and Prevention (CDC). 2020e. "Current Cigarette Smoking among Adults in the United States." Available at www.cdc.gov.

Chouvy, Piere-Arnaud. 2013. "A Typology of Unintended Consequences of Drug Crop Reduction." *Journal of Drug Issues* 43(2):216–230.

Christensen, Jen, and Jacque Wilson. 2014 (January 20). "Is Marijuana as Safe as—or Safer Than—Alcohol?" CNN. Available at www.cnn.com.

Christie, Nina C., Eustace Hsu, Carol Iskiwitch, Ravi Iyer, Jesse Graham, Barry Schwartz, and John R. Monterosso. 2019. "The Moral Foundations of Needle Exchange Attitudes." *Social Cognition* 37(3):229–246.

Collier, K. M., S. M. Coyne, E. E. Rasmussen, A. J. Hawkins, L. M. Padilla-Walker, S. E. Erickson, and M. K. Memmott-Elison. 2016. "Does Parental Mediation of Media Influence Child Outcomes? A Meta-Analysis on Media Time, Aggression, Substance Use, and Sexual Behavior." *Developmental Psychology* 52(5):798–812.

Copes, Heigh, Andy Hochstetler, and J. Patrick Williams. 2008. "'We Weren't Like No Regular Dope Fiends': Negotiating Hustler and Crackhead Identities." *Social Problems* 55(2):254–270.

Count the Costs. 2019. "The Seven Costs." Available at www.countthecosts.org.

Crowley, D. Max, Damon E. Jones, Donna L. Coffman, and Mark T. Greenberg. 2014. "Can We Build an Efficient Response to the Prescription Drug Abuse Epidemic? Assessing the Cost Effectiveness of Universal Prevention in the PROSPER Trial." *Preventative Medicine* 62:71–77.

Daniller, Andrew. 2019 (November 14). "Two-Thirds of Americans Support Marijuana Legalization." Pew Research Center. Available at www.pewresearch.org.

DARE 2018 (September 27). "The New D.A.R.E.: Schools Aim to Stop Next Generation of Opioid Crisis." Available at www.dare.org.

Degenhardt, Louisa, Wai-Tat Chiu, Nancy Sampson, et al. 2008. "Toward a Global View of Alcohol, Tobacco, Cannabis, and Cocaine Use: Findings from the WHO World Mental Health Surveys." *PLoS Medicine* 5(1):1053–1077.

DeJong, William, and Jason Blanchette. 2014. "Case Closed: Research Evidence on the Positive Public Health Impact of the 21 Minimum Legal Drinking Age in the United States." *Journal of Studies on Alcohol and Drugs* 75:108–115.

Dembosky, April. 2019 (June 24). "Meth in the Morning, Heroin at Night: Inside the Seesaw Struggle of Dual Addiction." *Kaiser Health News.* Available at www.khn.org.

Demko, Paul. 2018 (May 24). "Opioid Court Fights Risk of Repeating Tobacco's Failures." *Politico.* Available at www.politico.com.

Dodes, Lance, and Zachary Dodes. 2014. *The Sober Truth: Debunking the Bad Science behind 12-Step Programs and the Rehab Industry.* Boston: Beacon.

Downey, P. Mitchell, and John K. Roman. 2014 (June). "Cost-Benefit Analysis: A Guide for Drug Courts and Other Criminal Justice Programs." *Research in Brief.* Washington, DC: U.S. Department of Justice, Office of Justice Programs, National Institute of Justice.

Drug Enforcement Administration (DEA). 2010. "Fiction: Drug Production Does Not Damage the Environment." Just Think Twice: Facts and Fiction. Available at www.justthinktwice.com.

Drug Enforcement Administration (DEA). 2014 (May). *The Dangers and Consequences of Marijuana Abuse.* Washington, DC: U.S. Department of Justice. Available at www.dea.gov.

Drug Enforcement Administration (DEA). 2019. "National Drug Threat Assessment." December. Available at www.dea.gov.

Drug Enforcement Administration (DEA). 2020. "Drug Scheduling." Available at www.dea.gov.

Drug Policy Alliance (DPA). 2018. *"Fact Sheet: Women, Prison, and the Drug War."* Available at www.drugpolicy.org/.

Drug Policy Alliance (DPA). 2020. "Drug War Statistics." Available at www.drugpolicy.org/.

Dumas, Tara M., Jordan P. Davis, and Wendy E. Ellis. 2017. "Is It Good to Be Bad? A Longitudinal Analysis of Adolescent Popularity Motivations." *Youth & Society* 51(5):659–679.

European Monitoring Centre for Drugs and Drug Addiction (EMCDDA). 2014. *Perspectives on Drugs: The EU Drugs Strategy (2013–2020).* Available at www.emcdda.europa.eu.

European Monitoring Centre for Drugs and Drug Addiction (EMCDDA). 2019. *European Drug Report 2019: Trends and Developments, Publications Office of the European Union.* Available at www.emcdda.europa.eu.

Feagin, Joe R., and C. B. Feagin. 1994. *Social Problems.* Englewood Cliffs, NJ: Prentice Hall.

Filbey, Francesca, Sina Asian, Vince D. Calhoun, Jeffrey S. Spence, Eswar Damaraju, Arvind Caprihan, and Judith Segal. 2014. "Long-Term Effects of Marijuana Use on the Brain." *Proceedings of the National Academy of Science* 111(47):16913–6918.

Flanagin, Jake. 2014 (March 25). "The Surprising Failures of 12 Steps." *The Atlantic.* Available at www.theatlantic.com.

Food and Drug Administration (FDA). 2014. "Deeming-Extending Authorities to Additional Tobacco Products." Available at www.fda.gov.

Food and Drug Administration (FDA). 2020. "Cigarette Health Warnings." Available at www.fda.gov.

Fosco, G. M., and Feinberg, M. E. 2018. Interparental Conflict and Long-Term Adolescent Substance Use Trajectories: The Role of Adolescent Threat Appraisals. *Journal of Family Psychology* 32(2):175–185.

Food and Drug Administration (FDA). 2020. "The Real Cost Campaign." Available at www.fda.gov.

Freeman, Dan, Merrie Brucks, and Melanie Wallendorf. 2005. "Young Children's Understanding of Cigarette Smoking." *Addiction* 100(10):1537–1545.

Friedersdorf, Conor. 2011 (June 15). "The War on Drugs Turns 40." *Atlantic.* Available at www .theatlantic.com.

Friedman-Rudovsky, Jean. 2009 (May 25). "Red Bull's New Cola: A Kick from Cocaine?" *Time/CNN.* Available at www.time.com.

Fryer, Ronald G, Paul S. Heaton, Stephen D. Levitt, and Kevin M. Murphy. 2014. "Measuring Crack Cocaine and Its Impact." *Economic Inquiry* 51(3):1651–1681.

Fuller, Thomas. 2020 (November 7). "Oregon Decriminalizes Small Amounts of Heroin and Cocaine; Four States Legalize Marijuana." *New York Times.* Available at www.nytimes.com.

Generations United. 2018. *Raising the Children of the Opioid Epidemic: Solutions and Support for Grandfamilies.* Available at www.gu.org.

Gilbert, R., C. S. Widom, K. Browne, D. Fergusson, E. Webb, and S. Janson. 2009. "Burden and Consequence of Child Maltreatment in High-Income Countries." *Lancet* 73(9657):68–81.

Gilligan, Conor, Emmanuel Kuntsche, and Gerhard Gmel, "Adolescent Drinking Patterns across Countries: Associations with Alcohol Policies." *Alcohol and Alcoholism* 47(6):732–737.

Gonzales, R., Mooney, L., & Rawson, R. A. 2010. "The Methamphetamine Problem in the United States." *Annual Review of Public Health* (31):385–398.

Government Accounting Office (GAO). 2013 (January). "State Approaches Taken to Control Access to Key Methamphetamine Ingredient Show Varied Impact on Domestic Drug Labs." GAO-13-204. Available at www.gao.gov.

Gramlich, John. 2020 (May 6). "Black Imprisonment Rate in the U.S. Has Fallen by a Third since 2006." Pew Research Center. Available at www.pewresearch.org.

Grucza, Richard A. Mike Vuolo, Melissa J. Krauss, Andrew D. Plunk, Arpana Agrawal, Frank J. Chaloupka, and Laura J. Bierut. 2018. "Cannabis Decriminalization: A Study of Recent Policy Changes in Five States." *International Journal of Drug Policy* 59(September):67–75.

Gusfield, Joseph. 1963. *Symbolic Crusade: Status Politics and the American Temperance Movement.* Urbana: University of Illinois Press.

Haight, S. C., J. Y. Ko, V. T. Tong, B. K. Bohm, and W. M. Callaghan. 2018. "Opioid Use Disorder Documented at Delivery Hospitalization—United States, 1999–2014." *Morbidity and Mortality Weekly Report* 67(31):845–849. Available at www.cdc.gov.

Hakim, Danny. 2019 (September 13). "New York Uncovers $1 Billion in Sackler Family Wire Transfers." *New York Times.* Available at www .nytimes.com.

Hall, Wayne, Daniel Stjepanovic, Jonathon Caulkins, Michael Lynksey, Janni Leung, Gabrielle Campbell, and Louisa Degenhardt. 2019. "Public Health Implications of Legalizing the Production and Sale of Cannabis for Medical and Recreational Use." *Lancet* 10208(26):1580–1590.

Hallett, Cynthia. 2019 (May 20). "Secondhand Marijuana Smoke Is Not Just a Growing Nuisance, It's Dangerous." NBC News. Available at www.nbcnews.com.

Halper, Evan. 2014 (December 16). "Congress Quietly Ends Federal Government's Ban on Medical Marijuana." *Los Angeles Times.* Available at latimes.com.

Hammond, D., J. L. Reid, P. Driezen, J. F. Thrasher, P. C. Gupta, N. Nargis, Q. Li, J. Yuan, C. Boudreau, G. T. Fong, K. M. Cummings, and R. Borland. 2019. "Are the Same Health Warnings Effective across Different Countries? An Experimental Study in Seven Countries." *Nicotine and Tobacco Research* 21(7): 887–895.

Hanson, David J. 1997. "History of Alcohol and Drinking around the World." *Alcohol Problems and Solutions.* Available at www.2.potsdam .edu/alcohol.

Hanson, David J. 2013. "World Alcohol and Drinking History Timeline." *Alcohol Problems and Solutions.* Available at www.2.potsdam .edu/alcohol.

Harm Reduction International (HRI). 2019. *The Death Penalty for Drug Offences: Global Overview 2018.* Available at www.hri.global.

Harvard Health. 2020. *Alcohol Withdrawal.* Available at www.health.harvard.edu.

Hedegaard, Holly, Arialdi M. Miniño, and Margaret Warner. 2020 (January). "Drug Overdose Deaths in the United States, 1999–2018." *NCHS Data Brief,* no. 356. Hyattsville, MD: National Center for Health Statistics.

Henrisken, L., N. C. Schleicher, A. L. Dauphinee, and S. P. Fortmann. 2011. "Targeted Advertising, Promotion, and Price for Menthol Cigarettes in California High School Neighborhoods." *Nicotine and Tobacco Research* 14(1):116–121.

Hingson, Ralph W., Timothy Heeren, and Michael R. Winter. 2006. "Age at Drinking Onset and Alcohol Dependence." *Archives of Pediatrics & Adolescent Medicine* 160:739–746.

Hitchman, Sara C., and Geoffrey T. Fong. 2011. "Gender Empowerment and Female-to-Male Smoking Prevalence Ratios." *Bulletin of the World Health Organization* 89:195–202.

Hoffman, Jan. 2019 (August 30). "Johnson & Johnson Ordered to Pay $572 Million in Landmark Opioid Trial." *New York Times.* Available at www.nytimes.com.

Hoffman, Jan. 2020 (March 13). "Opioid Settlement Offer Provokes Clash between States and Cities." *New York Times.* Available at www.nytimes.com.

Hollersen, Wiebke. 2013 (March 27). "'This Is Working': Portugal, 12 Years after Decriminalizing Drugs." *Spiegel International.* Available at www.spiegel.de/international.

Human Rights Watch (HRW). 2020. "Philippines' 'War on Drugs.'" Available at www.hrw.org.

Jargin, Sergei V. 2012. "Social Aspects of Alcohol Consumption in Russia." *South African Medical Journal* 102(9):719.

Journal of Urgent Care Medicine. 2018 (November 1). "Narcan or Narcan't: An Ethical Dilemma for a Modern Scourge." *JUCM Letters to the Editor.* Available at www.jucm.com.

Kaplan, Sheila. 2020 (February 12). "Teens Find a Big Loophole in the New Flavored Vaping Ban." *New York Times.* Available at www .nytimes.com.

Kelleghan, Annemarie R., Adam M. Leventhal, Tess Boley Cruz, et al. 2020. "Digital Media Use and Subsequent Cannabis and Tobacco Product Use Initiation among Adolescents." *Drug and Alcohol Dependence* 212(July):108–117.

Kelly, J. F., and B. B. Hoeppner. 2013. "Does Alcoholics Anonymous Work Differently for Men and Women? A Moderated Multiple-Mediation Analysis in a Large Clinical Sample." *Drug and Alcohol Dependence* 130(1–3):186–193.

Klassen, Mark, and Brandon P. Anthony. 2019. "The Effects of Recreational Cannabis Legalization on Forest Management and Conservation Efforts in US National Forests in the Pacific Northwest." *Ecological Economics* 162:39–48.

Knopf, Alison. 2016. "U.S. Teens Drink and Smoke Less, but Use Illicit Drugs More, Than Teens in Europe." *Child and Adolescent Behavior Letter* 32(11):4–5.

Korf, D. J. 2019. *Cannabis Regulation in Europe: Country Report Netherlands.* Amsterdam: Transnational Institute.

Lilly, Jessica. 2018 (January 19). "The Opioid Epidemic in Kermit, W.Va." *West Virginia Public Broadcasting.* Available at www .wvpublic.org.

Lipari, Rachel N., and Struther L. Van Horn. 2017 (August 24). "Children Living with Substance Dependent Parent, 2009–2014." The CBHSQ Report. Rockville, MD: Center for Behavioral Health Statistics and Quality, Substance Abuse and Mental Health Services Administration. Available at www.samhsa.gov.

Lize, Steven E., Aidyn L. Iachini, Weizhou Tang, Joshua Tucker, Kristen D. Seay, Stephanie Clone, Dana DeHart, and Teri Brown. 2016. "A Meta-Analysis of the Effectiveness of Interactive Middle School Cannabis Prevention Programs." *Prevention Science* 18:50–60.

Lee, Yon, and M. Abdel-Ghany. 2004. "American Youth Consumption of Licit and Illicit Substances." *International Journal of Consumer Studies* 28(5):454–465.

Loukas, Alexandra, Ellen M. Paddock, Xiaoyin Li, Melissa B. Harrell, Keryn E. Pasch, and Cheryl L. Perry. 2019. "Electronic Nicotine Delivery Systems Marketing and Initiation among Youth and Young Adults." *Pediatrics* 144(3):e20183601.

MacMillan, Carrie. 2020 (June 4). "Drinking More Than Usual during the COVID-19 Pandemic?" *Yale Medicine.* Available at www .yalemedicine.org.

Malinowska-Sempruch, Kasia, and Olga Rychkova. 2016 (September 28). "The Impact of Drug Policy on Women." Open Society Foundations. Available at www .opensocietyfoundations.org.

Margolis, Robert D., and Joan E. Zweben. 2011. *Treating Patients with Alcohol and Other Drug Problems: An Integrated Approach.* Chapter 3: "Models and Theories of Addiction." Washington, DC: American Psychological Association.

Marine-Street, Natalie. 2012 (April 26). "Stanford Researchers' Cigarette Ad Collection Reveals How Big Tobacco Targets Women and Adolescent Girls." *Gender News.* Available at gender.stanford.edu.

May, Philip A., Amy Baete, Jaymi Russo, et al. 2014. "Prevalence and Characteristics of Fetal Alcohol Spectrum Disorders." *Pediatrics* 134(5):855–866.

McCabe, Sean Esteban, Philip Veliz, and John E. Schulenberg. 2018. "How Collegiate Fraternity and Sorority Involvement Relates to Substance Use during Young Adulthood and Substance Use Disorders in Early Midlife: A National Longitudinal Study." *Journal of Adolescent Health* 62(3):35–43.

McGreal, Chris. 2020 (July 9). "'Opioid Overdoses Are Skyrocketing': As Covid-19 Sweeps across US an Old Epidemic Returns." *Guardian.* Available at www.theguardian.com.

McSweeney, Kendra. 2015. "The Impact of Drug Policy on the Environment." *Open Society Foundations.* Available at www .opensocietyfoundations.org.

Mears, Bill. 2012 (November 26). "Tobacco Companies Ordered to Publicly Admit Deception on Smoking Dangers." CNN. Available at www.cnn.com.

Medicine Abuse Project (MAP). 2014. *The Medicine Abuse Project.* Partnership for Drug-Free Kids. Available at www .medicineabuseproject.org.

Medicine Abuse Project (MAP). 2019. *Annual Report.* Partnership for Drug-Free Kids. Available at www.drugfree.org.

Memedovich, K. A., L. E. Dowsett, E. Spackman, T. Noseworthy, and F. Clement. (2018). "The Adverse Health Effects and Harms Related to Marijuana Use: An Overview Review." *CMAJ Open* 6(3):E339–E346.

Merolla, David. 2008 (March). "The War on Drugs and the Gender Gap in Arrests: A Critical Perspective." *Critical Sociology* 34:255–270.

Metropolitan Police Department. 2015. *The Facts on DC Marijuana Laws.* Available at www .mpdc.dc.gov.

Monitoring the Future (MTF). 2019. *Key Findings on Adolescent Drug Use: 2018 Overview.* Ann Arbor: Institute for Social Research, University of Michigan.

Myers, Matthew. 2011 (April 25). "FDA Acts to Protect Public Health by Extending Authority over Tobacco Products, Including E-Cigarettes." Available at www.tobaccofreekids.org.

Nadelmann, Ethan. 2019. "Paradigms for U.S. Drug Policy." *Brown Journal of World Affairs* 25(2):137–144.

National Academies of Sciences, Engineering, and Medicine (NASEM). 2017. *The Health Effects of Cannabis and Cannabinoids: The Current State of Evidence and Recommendations for Research.* Washington, DC: National Academies Press.

National Academies of Sciences, Engineering, and Medicine (NASEM). 2018. "Concluding Observations." In *Public Health Consequences of E-Cigarettes,* ed. D. L. Eaton, L. Y. Kwan, and K. Stratton, 657–658. Washington (DC): National Academies Press. Available at https:// www.ncbi.nlm.nih.gov/books/NBK507161/.

National Association for the Advancement of Colored People. 2020. *Criminal Justice* Fact Sheet. Available at naacp.org.

National Center on Addiction and Substance Abuse (CASA). 2019. "National Survey of American Attitudes on Substance Abuse: Teens." Available at www.casacolumbia.org.

National Conference of State Legislators (NCSL). 2020 (March 10). *State Medical Marijuana Laws.* Available at www.ncsl.org.

National Highway Traffic Safety Administration. 2019. *Drunk Driving.* Available at www.ntysa .gov.

National Institute on Alcohol Abuse and Alcoholism (NIAAA). 2013. "The Genetics of Alcoholism." *Alcohol Alert* 84:1–6.

National Institute of Alcohol Abuse and Alcoholism (NIAAA). 2020 (January). *Underage Drinking.* Available at www.niaaa .nih.gov.

National Institute on Drug Abuse (NIDA). 2018a. *Marijuana.* Bethesda, MD: US Department of Health and Human Services, National Institutes of Health. Available at www.drugabuse.com.

National Institute on Drug Abuse (NIDA). 2018b. *Principles of Drug Addiction Treatment: A Research-Based Guide, 3rd ed.* Available at drugabuse.gov.

National Survey on Drug Use and Health (NSDUH). 2019. "Key Substance Use and Mental Health Indicators in the United States: Results from the 2018 National Survey on Drug Use and Health." HHS Publication No. PEP19-5068, NSDUH Series H-54. Rockville, MD: Center for Behavioral Health Statistics and Quality, Substance Abuse and Mental Health Services Administration. Available at www .samhsa.gov.

National Institute on Drug Abuse (NIDA). 2019. *Methamphetamine Research Report.* October. Available at www.drugabuse.org.

National Institute on Drug Abuse (NIDA). 2020a. "Monitoring the Future 2012 Survey Results." Available at www.drugabuse.gov.

National Institute of Drug Abuse (NIDA). 2020b. "Table 5-5b: Trends in Annual Prevalence of Use of Various Drugs in Grades 8, 10, and 12." *Monitoring the Future: National Survey Results on Drug Use, 1975–2019.* Volume 1. Available at www.monitoringthefuture.org.

National Institute on Drug Abuse (NIDA). 2020c (June 16). *Criminal Justice Drug Facts.* Available at www.drugabuse.gov.

Nelson, Steven. 2014 (December 3). "House Leaders Rush to Defend E-Cigarettes from Possible FDA Ban." *U.S. News & World Report.* Available at www.usnews.com.

Nesi, Jacqueline, W. Andrew Rothenberg, Andrea M. Hussong, and Kristina M. Jackson. 2017. "Friends' Alcohol-Related Social Networking Site Activity Predicts Escalations in Adolescent Drinking: Mediation by Peer Norms." *Journal of Adolescent Health* 60(6):641–647.

Netherland, Julie, and Helena B. Hansen. 2016. "The War on Drugs That Wasn't: Wasted Whiteness, 'Dirty Doctors,' and Race in Media Coverage of Prescription Opioid Misuse." *Culture, Medicine, and Psychiatry* 40(4):664–686.

Nordqvist, Christian. 2012 (August 28). "Teen Cannabis Use Linked to Lower IQ." *Medical News Today.* Available at www .medicalnewstoday.com.

Office of National Drug Control Policy (ONDCP). 2006. *Methamphetamine.* Available at www./ whitehousedrugpolicy.gov/drugfact methamphetamine.

O'Neill Hayes, Tara, and Margaret Barnhorst. 2020 (June 30). "Incarceration and Poverty in the United States." *American Action Forum.* Available at www.americanactionforum.org.

Paul, Jesse. 2019 (June 12). "Where Does Colorado's Marijuana Tax Money Go? The State Made a Flow Chart to Answer the $1 Billion Question." *Colorado Sun.* Available at www .coloradosun.com.

Peralta, Robert L., Jennifer L. Steele, Stacey Nofziger, and Michael Rickles. 2010. "The Impact of Gender on Binge Drinking Behavior among U.S. College Students Attending a Midwestern University: An Analysis of Two Gender Measures." *Feminist Criminology* 10(5):355–379.

Peters, Jeremy W. 2009 (March 5). "Albany Takes Step to Repeal '70s-Era Drug Laws." *New York Times.* Available at www.nytimes.com.

Pew Research Center. 2014 (April 2). "America's New Drug Policy Landscape." Pew Research Center. Available at www.people-press.org.

Pierre, Joseph M. 2011. "Cannabis, Synthetic Cannabinoids, and Psychosis Risk: What the Evidence Says." *Current Psychiatry* 10(9): 49–58.

Primack, Brian A., James E. Bost, Stephanie R. Land, and Michael J. Fine. 2007. "Volume of Tobacco Advertising in African American Markets: Systematic Review and Meta-Analysis." *Public Health Reports* 122(5):607–615.

PSB Research. 2019. *Cannabis Culture Poll: Methodology.* Available at https://www .buzzfeednews.com/article/dominicholden /marijuana-poll-420-american-use-legalization.

Public Health and Tobacco Policy Center. 2020. *FDA Graphic Warnings.* Available at www .tobaccopolicycenter.org.

Quinones, Sam. 2019 (May). "Physicians Get Addicted Too." *The Atlantic.* Available at www.theatlantic.com.

Rabinowitz, Mikaela, and Arthur Lurigio. 2009 (August 8–11). "A Century of Losing Battles: The Costly and Ill-Advised War on Drugs." Conference Paper. American Sociological Association. San Francisco.

Remer, Cathy. 2019 (October 9). "All Our Hearts: Sharing Stories of Love, Grief and Hope from the Opioid Crisis." *Seven Days.* Available at www.sevendaysvt.com.

Ribisl, K. M., H. D'Angelo, A. L. Feld, N. C. Schleicher, S. D. Golden, D. A. Luke, and L. Henriksen. 2017. "Disparities in Tobacco Marketing and Product Availability at the Point of Sale: Results of a National Study." *Preventive Medicine* 105:381–388.

Rodriguez, Sebastian. 2018 (June 19). "'Narco-Deforestation' May Boost Disaster Risks in Central America." Reuters. Available at www.reuters.com.

Rorabaugh, W. J. 1979. *The Alcoholic Republic: An American Tradition.* New York: Oxford University Press.

Rothstein, Mark A. 2017. "The Opioid Crisis and the Need for Compassion in Pain Management." *American Journal of Public Health* 107(8):1253–1254.

Rusch, Emily. 2014 (April 11). "Marijuana-Infused Neighbor Conflicts: Ways to Clear the Air." *Denver Post.* Available at www.denverpost.com.

Saad, Lydia. 2019. "Liquor Ties Wine as Second-Favorite Adult Beverage in U.S." Available at www.gallup.com.

Saloner, B., and B. LeCook. 2013 (January 7). "Blacks and Hispanics Are Less Likely Than Whites to Complete Addiction Treatment, Largely due to Socioeconomic Factors." *Health Affairs.* Available at www.rwjf.org.

Schmidt, Lorna. 2013 (June 18). "Tobacco Company Marketing to Kids." Campaign for Tobacco Free Kids. Available at www.tobaccofreekids.org.

Schmidt, Lorna. 2014 (April 23). "Tobacco Industry Targeting of Women and Girls." Campaign for Tobacco Free Kids. Available at www.tobaccofreekids.org.

Sentencing Project. 2018a. "Trends in U.S. Corrections." Fact Sheet. Available at sentencingproject.org.

Sentencing Project. 2018b (June 22). Fact Sheet: *Trends in U.S. Corrections.* Available at www.thesentencingproject.org.

Sepulveda, Kristin, and Sarah Catherine Williams. 2019 (February 26). "One in Three Children Entered Foster Care in 2017 Because of Parental Drug Abuse." *Child Trends.* Available at www.childtrends.org.

Sesnie, Steven E., Beth Tellman, David Wrathall, Kendra McSweeney, Erik Nielsen, Karina Benessaiah, Ophelia Wang, and Luis Rey. 2017. "A Spatio-Temporal Analysis of Forest Loss Related to Cocaine Trafficking in Central America." *Environmental Research Letters* 12(5):054015.

Sifferlin, Alexandra. 2013 (June 26). "FDA Approves New Cigarettes in First Use of New Regulatory Power over Tobacco." *Time.* Available at healthland.time.com.

Smith, Chad L., Gregory Hooks, and Michael Lengefeld. 2014. "The War on Drugs in Columbia: The Environment, the Treadmill of Destruction and Risk Transfer Militarism." *Journal of World Systems Research* 20(2):185–206.

Soneji, Samir, Jessica L. Barrington-Trimis, and Thomas A. Wills. 2017. "Association between Initial Use of e-Cigarettes and Subsequent Cigarette Smoking among Adolescents and Young Adults." *Pediatrics* 171(8):788–797.

Substance Abuse and Mental Health Services Administration (SAMSHA). 2018. *Opioid Overdose Prevention Toolkit.* Available at www.samhsa.com.

Substance Abuse and Mental Health Services Administration (SAMSHA). 2020a. *Mental Health and Substance Use Disorders.* Available at www.samhsa.gov.

Substance Abuse and Mental Health Services Administration (SAMHSA). 2020b. *Laws and Regulations.* Available at www.samhsa.gov.

Surgeon General US. 2016. *Facing Addiction in America: The Surgeon General's Report on Alcohol, Drugs, and Health.* Available at https://addiction.surgeongeneral.gov/.

Taifia, Nkechi. 2006 (May). "The 'Crack/Powder Disparity': Can the International Race Convention Provide a Basis for Relief?" American Constitution Society for Law and Policy white paper. Available at acslaw.org.

Taylor, Matthew. 2014 (August 10). "Health Warnings on Alcohol Bottles Should Be Compulsory—MPs." *Guardian.* Available at www.theguardian.com.

Thio, Alex. 2007. *Deviant Behavior.* Boston: Allyn and Bacon.

Thomas, Gerald, Ginny Gonneau, Nancy Poole, and Jaclyn Cook. 2014. "The Effectiveness of Alcohol Warning Labels in the Prevention of Fetal Alcohol Spectrum Disorder: A Brief Review." *International Journal of Alcohol and Drug Research* 3(1):91–103.

Tobacco Free Kids. 2019 (December). *The Global Cigarette Industry.* Available at www.tobaccofreekids.org.

Todd, Tamar. 2018. "The Benefits of Marijuana Legalization and Regulation." *Berkeley Journal of Criminal Law* 23(1):100–114.

Tsubasa Field, Andy. 2018 (October 18). "What Happened after 2 Colleges Banned Hard Liquor at Fraternities." *Chronicle of Higher Education.* Available at www.chronicle.com.

U.S. Department of Justice. 2019 (December). *National Drug Threat Assessment 2019.* Available at www.justice.gov.

Vlaev, I., D. King, A. Darzi, and P. Dolan. 2019. "Changing Health Behaviors Using Financial Incentives: A Review from Behavioral Economics." *BMC Public Health* 19(1):1059.

Walker, Elizabeth Reisinger, and Benjamin G. Druss. 2017. "Cumulative Burden of Comorbid Mental Disorders, Substance Use Disorders, Chronic Medical Conditions, and Poverty on Health among Adults in the U.S.A." *Psychology, Health, and Medicine* 22(6)727–735.

Wechsler, William, and Toben F. Nelson. 2008. "What We Have Learned from the Harvard School of Public Health College Alcohol Study: Focusing Attention on College Student Alcohol Consumption and the Environmental Conditions That Promote It." *Journal of Alcohol Studies* (July):1–9.

Weiner, Katie. 2019 (May 19). "Overpoliced, Underrepresented: Racial Inequality and Cannabis Capitalism." *Harvard Political Review.* Available at www.harvardpolitics.com.

Weiss, Carol H., Erin Murphy-Graham, Anthony Petrosino, and Allison G. Gandhi. 2017. "The Fairy Godmother—and Her Warts: Making the Dream of Evidence-Based Policy Come True." *American Journal of Evaluation* 29(1):29–47.

West, Stephen L., and Keri K. O'Neal. 2004. "Project D.A.R.E. Outcome Effectiveness Revisited." *American Journal of Public Health* 94(6):1027–1029.

White House. 2020 (February). *FY 2021 National Drug Control Budget.* Available at www.whitehouse.gov.

White House. 2020 (February). *FY 2021 National Drug Control Budget.* Available at www.whitehouse.gov.

Willing, Richard. 2014 (May 13). "Lawsuits Target Alcohol Industry." *USA Today.* Available www.usatoday30.usatoday.com.

Winkelman, T. N. A., L. K. Admon, L. Jennings, N. D. Shippee, C. R. Richardson, and G. Bart. 2018. "Evaluation of Amphetamine-Related Hospitalizations and Associated Clinical Outcomes and Costs in the United States." *Journal of the American Medical Association Network Open* 1(6):e183758.

Witters, Weldon, Peter Venturelli, and Glen Hanson. 1992. *Drugs and Society*, 3rd ed. Boston: Jones & Bartlett.

World Drug Report (WDR). 2019. *World Drug Report 2019.* New York: United Nations, United Nations Office on Drugs and Crime (UNODC).

World Drug Report (WDR). 2020. *World Drug Report 2020.* New York: United Nations, United Nations Office on Drugs and Crime (UNODC).

World Health Organization. 2016. "Prevalence of Tobacco Smoking." Available at http://gamapserver.who.int/gho/interactive_charts/tobacco/use/atlas.html.

World Health Organization (WHO). 2018 (September). "Global Status Report on Alcohol and Health." Available at www.who.int.

World Health Organization (WHO). 2020a (May 11). "WHO Statement: Tobacco Use and COVID-19." Available at www.who.int.

World Health Organization (WHO). 2020b (May). "Tobacco: Fact Sheet." Available at www.who.int.

WVU Today. 2018 (September 27). *WVU Bans Dissociating Fraternities for 10 Years.* Available at www.wvutoday.edu.

Zimmerman, Stephanie. 2020 (April 1). "Fairness Is an Issue in Clearing Low-Level Marijuana Convictions." *American Bar Association Journal.* Available at www.abajournal.com.

Chapter 4

18 U.S. Code §2384 "Seditious Conspiracy." *Legal Information Institute.* Available at www.law.cornell.edu.

Abt, Richard, and Ernesto Lopez. 2020. "COVID-19 and Homicide: Final Report to Arnold Ventures." Arnold Ventures. Available at craftmediabucket.s3.amazonaws.com.

Agnew, Robert. 2013. "When Criminal Coping Is Likely: An Extension of General Strain Theory." *Deviant Behavior* 34(8):653–670.

Ahmed, Hauwa. 2019 (August 3). "How Private Prisons Are Profiting Under the Trump Administration." Center for American Progress. Available at www.americanprogress.org.

Allyn, Bobby. 2020 (March 25). "In Surprise Move, Christchurch Shooting Suspect Pleads Guilty to 51 Counts of Murder." National Public Radio. Available at www.npr.org.

American Civil Liberties Union (ACLU). 2020. "The Tale of Two Countries: Racially Targeted Arrests in the Era of Marijuana Reform." *ACLU Research Report.* Available at www.aclu.org.

Andrew, Scotty. 2020 (July 5). "This Town of 170,000 Replaced Some Cops with Medics and Mental Health Workers. It's Worked for over 30 Years." CNN News. Available at www.cnn.com.

Ankel, Sophia. 2020 (June 24). "30 Days That Shook America: Since the Death of George Floyd, the Black Lives Matter Movement Has Already Changed the Country." *Business Insider.* Available at www.businessinsider.com.

Ascani, Nathaniel. 2012 (Spring). "Labeling Theory and the Effects of Sanctioning on a Staff Delinquent Peer Association: A New Approach to Sentencing Juveniles." *Perspectives*: 80–84.

Asher, Jeff, and Ben Horwitz. 2020 (July 13). "It's Been 'Such a Weird Year.' That's Also Reflected in Crime Statistics." *New York Times.* Available at www.nytimes.com.

Associated Press–National Opinion Research Center (AP-NORC). 2020. "Widespread Desire for Policing and Criminal Justice Reform." Available at apnorc.org.

Associated Press. 2020 (September 30). "Court Approves $800 Million Las Vegas Shooting Settlement." *Wall Street Journal.* Available at www.wsj.com.

Balko, Radley. 2014. *Rise of the Warrior Cop: The Militarization of American Police Forces.* New York: Public Affairs.

Bates, Josiah. 2020 (November 13). "As 'Defund the Police' Splits the Democratic Caucus, Criminal Justice Activists Are Already Wary of Biden Administration." *Time.* Available at time.com.

Biden, Joe. 2020a. "The Biden Plan for Strengthening America's Commitment to Justice." Available at www.joebiden.com.

Biden, Joe. 2020b. "The Biden Plan to End Our Gun Violence Epidemic." Available at www.joebiden.com.

Boudette, Neil D. 2020 (June 10). "Former U.A.W. President Gary Jones Pleads Guilty." *New York Times.* Available at www.nytimes.com.

Boyle, Louise. 2015 (April 27). "Cleveland 'House of Horrors' Survivors Detail Their Decade of Abuse and How Ariel Castro Raped Them up to 5 Times a Day—The Only Time They Were Released from Their Chains." *Daily Mail.* Available at www.dailymail.co.uk.

Brenan, Megan. 2020 (June 23). "Record Low 54% in U.S. Say Death Penalty Morally Acceptable." Gallup Poll. Available at news.gallup.com.

Brueck, Hilary, and Shana Lebowitz. 2019 (August 5). "The Men behind the US's Deadliest Mass Shootings Have Domestic Violence—Not Mental Illness—in Common." *Insider.* Available at static.insider.com.

Buchanan, Larry, Quoctrung Bui, and Jugal K. Patel. 2020 (July 3). "Black Lives Matter May Be the Largest Movement in U.S. History." *New York Times.* Available at www.nytimes.com.

Byrne, James, and Gary Marx. 2011. "Technological Innovations in Crime Preventions and Policing: A Review of the Research on Implementation and Impact." *Journal of Police Studies* 20(3):17–38.

Byrne, James M., April Pattavina, and Faye S. Taxman. 2015. "International Trends in Prison Upsizing and Downsizing: In Search of Evidence of a Global Rehabilitation Revolution." *Victims and Offenders* 10:420–451.

Call, Corey. 2019. "Serial Entertainment: A Content Analysis of 35 Years of Serial Murder in Film." *Homicide Studies* 23(4):362–380.

Campbell, Josh, Sarah Sidner, and Eric Levinson. 2020 (June 4). "All Four Minneapolis Police Officers in George Floyd's Killing Now Face Charges." CNN. Available at www.cnn.com.

Carson, Ann E. 2020 (April). "Prisoners in 2018." U.S. Department of Justice. Bureau of Justice Statistics. Available at www.bjs.gov.

Carter, Mike. 2020 (July 29). "Seattle Police Defend Response to Weekend Black Lives Matter Protests, Deny Violations of Court's Use-of-Force Limits." *Seattle Times.* Available at www.seattletimes.com.

CATO Institute (CATO). 2020 (July 16). "Poll: 63% of Americans Favor Eliminating Qualified Immunity for Police." Available at https://www.cato.org/publications/survey-reports/poll-63-americans-favor-eliminating-qualified-immunity-police.

Chang, Justin H., and Brad J. Bushman. 2019 (May 31). "The Effect of Exposure to Gun Violence in Video Games on Children's Dangerous Behavior with Real Guns." *Pediatrics JAMA Open Network* 2(5):e194319.

Cho, Seo-Young, Axel Dreher, and Eric Neumayer. 2013. "Does Legalized Prostitution Increase Human Trafficking?" *World Development* 41:67–82.

Community Oriented Policing Services (COPS). 2020. "About the COPS Office." Available at cops.usdoj.gov.

Congressional Research Service. 2017 (June 7). "Discretionary Budget Authority by Subfunction: An Overview." Available at www.everycrsreport.com.

Conklin, John E. 2007. *Criminology,* 9th ed. Boston: Allyn and Bacon.

Corley, Cheryl. 2021 (January 6). "Massive 1-Year Rise in Homicide Rates Collided with the Pandemic in 2020." National Public Radio. Available at www.npr.org

Cramer, Maria. 2020 (May 12). "Sentenced for Three Strikes, Then Freed. Now Comes a Pushback." *New York Times.* Available at www.nytimes.com.

Cruz, Jose M., Jonathan D. Rosen, Luis Enrique Amaya, and Yulia Vorobyeva. 2017. *The New Face of Street Gangs: The Gang Phenomenon in El Salvador.* Miami: The Kimberly Green Latin American and Caribbean Center Jack D. Gordon Institute for Public Policy, Florida International University.

Culver, Jordan. 2020 (June 30). "San Quentin State Prison Is 'Deep Area of Focus and Concern' in California: Nearly One-Third of Inmates Have Coronavirus." *USA Today.* Available at www.usatoday.com.

D'Alessio, David, and Lisa Stolzenberg. 2002. "A Multilevel Analysis of the Relationship between Labor Surplus and Pretrial Incarceration." *Social Problems* 49:178–193.

Davis, Elizabeth, Anthony Whyde, and Lynn Langston. 2018 (October). "Contacts between Police and the Public, 2015." *Special Report.* U.S. Department of Justice. Bureau of Justice Statistics. Available at www.bjs.gov.

Death Penalty Information Center (DPIC). 2020. "Executions around the World." Available at deathpenaltyintheroad.org.

Debies-Carl, Jeffrey S. 2013. "Are the Kids All Right? A Critique and Agenda for Taking Youth Cultures Seriously." *Social Science Information* 52(1):110–133.

Desilver, Drew, Michael Lipka, and Dalia Fahmy. 2020 (June 3). "10 Things We Know about Race and Policing in the U.S." Pew Research Center. Available at www.pewresearch.org.

Dickman, Cassie. 2020 (June 29). "Joseph James D'Angelo Pleads Guilty to 13 Murders Tagged to California's 'Golden State Killer'." *USA Today.* Available at amp. usatoday.com.

Dixon, Emily. 2019 (July 26). "Armed Robbers Steal at Least $30 Million of Gold and Precious Metals in São Paulo Airport Heist." CNN World. Available at www.cnn.com.

DPIC. 2021. "The Death Penalty in 2020: Year End Report." *Death Penalty Information Center.* Available at reports.deathpenaltyinfo.org.

Drakulich, Kevin M. 2013. "Strangers, Neighbors, and Race: A Content Model of Stereotypes and Racial Anxieties about Crime." *Race and Justice* 2(4):322–355.

Dudley, Stephen. 2019. *MS 13 in the Americas: How the World's Most Notorious Gang Defies Logic, Resists Destruction.* Washington, DC: Office of Justice Programs, U.S. Department of Justice.

Dunlop, R. G. 2020 (January 29). "How These Jail Officials Profit from Selling E-Cigarettes to Inmates." ProPublica in Partnership with the Kentucky Center for Investigative Reporting. Available at www.propubluca.org.

Edwards, Frank, Hedwig Lee, and Michael Esposito. 2019 (August 20). "Risk of Being Killed by Police Use of Force in the United States by Age, Race-Ethnicity, and Sex." *Proceedings of the National Academy of Science* 115:16793–16798.

Edwards, Rebecca. 2020 (July 14). "The 10 Safest Metro Cities in America for 2020." *Safewise.* Available at www.safewise.com.

Elder, Brian, Laura Bennett, Shannon Gong, Felicity Rose, and Zoe Townes. 2018. "Every Second: The Impact of the Incarceration Crisis on American Families." Available at everysecond.fwd.us.

Erikson, Kai T. 1966. *Wayward Puritans.* New York: Wiley.

Europol. 2020a. "Activities and Services." Available at www.europol.europa.eu.

Europol. 2020b. "Crime Areas." Available at www.europol.europa.eu.

Fandos, Nicholas, and Katie Benner. 2021 (January 19). "Justice Department Ends Stock Trade Inquiry into Richard Burr without Charges." *New York Times.* Available at www.nytimes.com.

Federal Bureau of Investigation (FBI). 2019a. *2018 Crime in the United States.* Uniform Crime Report. Available at ucr.fbi.gov.

Federal Bureau of Investigation (FBI). 2019b (August 6). "Operation Independence Day." Available at www.fbi.gov.

Federal Bureau of Investigation (FBI). 2019c. "White-Collar Crime." *What We Investigate.* Available at www.fbi.gov.

Federal Bureau of Investigation (FBI). 2020a. *2019 Crime in the United States.* Uniform Crime Report. Available at ucr.fbi.gov.

Federal Bureau of Investigation (FBI). 2020b. *What We Investigate.* Available at www.fbi.gov.

Federal Register. 2019 (November 19). "Annual Determination of Average Cost of Incarceration Fee." Bureau of the Prisons. Available at www.federalregister.gov.

Federal Trade Commission. 2020. *Consumer Sentinel Network Report.* 2019. Available at www.ftc.gov.

Fight Crime. 2019. "From Risk to Opportunity: Afterschool Programs Keep Kids Safe When Juvenile Crime Peaks." *Fight Crime: Invest in Kids.* Available at www.strongnation.org.

Frazin, Rachel. 2020 (July 27). "Report: Oil and Gas Companies Have Extensive Ties to Police Groups." *The Hill.* Available at thehill.com.

Gallup Poll. 2020a. "Crime." *Gallup Historical Trends.* Available at news.gallup.com.

Gallup Poll. 2020b. "Guns." *In Depth Topics A to Z.* Available at news.gallup.com.

Gaston, Shytierra. 2019. "Producing Race Disparities: A Study of Drug Arrests across Place and Race." *Criminology* 57(3):424–451.

Gatto, James G. 2019 (June 4). "Federal Court 'Discards' DOJ Interpretation of Wire Act." *National Law Review* X(207). Available at www.natlawreview.com.

Gault-Sherman, Martha. 2012. "It's a Two-Way Street: The Bi-directional Relationship between Parenting and Delinquency." *Journal of Youth and Adolescence* 41(2):121–145.

Geller, Eric. 2020 (February 10). "U.S. Charges Chinese Military Hackers with Massive Equifax Breach." *Politico.* Available at www.politico.com.

Giffords Law Center. 2018 (September). "The Las Vegas Massacre's $600 Million Financial Toll." Available at lawcenter.giffords.org.

Glanton, Dahleen. 2019 (February 26). "In a City with Tens of Thousands of Surveillance Cameras, Who's Watching Whom?" *Chicago Tribune.* Available at chicagotribune.com.

Global Information Security Survey. 2020. "Is Cybersecurity about More Than Protection?" Available at assets.ey.com.

Goldenberg, Anna, Deviney Rattigan, Michael Dalton, John P. Gaughan, Scott J. Thomson, Kyle Remick, Christopher Butts, and Joshua P. Hazelton. 2019 (December). "Use of ShotSpotter Detection Technology Decreases Prehospital Time for Patients Sustaining Gunshot Wounds." *Journal of Trauma and Acute Care Surgery* 87(6):1253–1259.

Gotoff, Daniel, and Celinda Lake. 2020. "Voters Want Criminal Justice Reform. Are Politicians Listening?" *The Marshal Project.* Available at www.themarshalproject.org.

Grawert, Ames. 2020 (June 23). "What Is the First Step Act—And What's Happening with It?" The Brennan Center for Justice. Available at brennancenter.org.

Gramlich, John. 2019 (October 17). "5 Facts about Crime in the U.S." Pew Research Center. Available at www.pewresearch.org.

Gramlich, John, and Katherine Schaeffer. 2019. "7 Facts about Guns in the US." *FactTank.* Available at www.pewresearch.org.

Grzeszczak, Jocelyn. 2020 (September 30). "50 Portland Officers to Be Deputized as Federal Law Enforcement Ahead of Proud Boys Rally." *Newsweek.* Available at newsweek.com

Greenwood, Chris. 2015 (January 5). "Crime Victims Don't Trust the Police and Thousands of Offenses Are Going Unrecorded Because People Feel 'Nothing Will Be Done,' Says Federation Chairman." *Daily Mail.* Available at www.dailymail.co.uk.

Haak, Deborah M. 2018 (December 9). "Canada's Laws Designed to Deter Prostitution, Not Keep Sex Workers Safe." The Conversation. Available at theconversation.com.

Hada, Messia, Laura Smith-Spark, and Nadine Schmidt. 2018 (December 5). "90 Held in Anti-Mafia Raids across Europe." CNN World News. Available at edition.cnn.com.

Hamel, Liz, Audra Kearney, Ashley Kirzinger, Luna Lopes, Cailey Munana, and Mollyann Brody. 2020 (June 26). "Racial Disparities and Protests." Available at www.kff.org.

Hayes, Tara O'Neill. 2020 (July 16). "The Economic Costs of U.S. Criminal Justice System." American Action Forum. Available at www.americanactionforum.org.

Heron, Melonie. 2019 (June 24). "Deaths: Leading Causes for 2017." *National Vital Statistics Report* 68(6):1–75.

Hirschi, Travis. 1969. *Causes of Delinquency.* Berkeley: University of California Press.

Hosenball, Mark. 2020 (July 21). "U.S. Homeland Security Confirms Three Units Sent Paramilitary Officers to Portland." Reuters News Service. Available at www.reuters.com.

H. R. 7120. 2020. George Floyd Justice in Policing Act of 2020. Available at www.congress.gov.

Innocence Project, The. 2020. "We Advocate for Reforms That Help Prevent and Address Wrongful Convictions." Available at www.innocenceproject.org.

Insurance Institute for Highway Safety. 2019. "Dodge Cars with Big Engines Top HLDI's List of Most Stolen Vehicles." Insurance Institute for Highway Safety. Available at www.iih.org.

Interpol. 2020. "Interpol's Crime Programs." Available at www.interpol.int.

International Center for the Prevention of Crime (ICPC). 2020. "Mission." Available at icpc-cipc.org.

International Labour Office (ILO). 2017. "Frequently Asked Questions: 2016 Global Estimates of Modern Slavery." Alliance 8.7. Available at www.ilo.org.

Internet Crime Complaint Center (ICCC). 2019a. "How Does the IC3 Define Internet Crime?" Federal Bureau of Investigation. Available at the www.ic3.gov.

Internet Computer Complaint Center (ICCC). 2019b (October 2). "PSA: High-Impact Ransomware Attacks Threaten U.S. Businesses and Organizations." Federal Bureau of Investigation. Available at www.ic3.gov.

Internet Crime Complaint Center (ICCC). 2019c (December 12). "PSA: Child Predators Use Online Gaming to Contact Children." *Federal Bureau of Investigation.* Available at the www.ic3.gov.

Internet Crime Complaint Center (ICCC). 2019d. *2018 Internet Crime Report.* Available at pdf.ic3.gov.

Internet Computer Complaint Center (ICCC). 2020 (June 11). "PSA: Implementation of Fraudulent COVID-19 Shipping and Insurance Fees by Criminal Actors." Federal Bureau of Investigation. Available at www.ic3.gov.

Isidore, Chris. 2014 (November 20). "Takata Airbag Victims Looked Like They Had Been Shot or Stabbed." CNN Money. Available at money.cnn.com.

Johnson, Calvin, and William P. Quigley. 2019. "An Analysis of the Economic Cost of Maintaining a Capital Punishment System in the Pelican State." Loyola University New Orleans College of Law. Available at law.loyno.edu.

Jones, Kay, and Scottie Andrew. 2020 (August 8). "A Man Who Was Sentenced to Life in Prison for Selling $30 of Marijuana Will Be Freed." CNN News. Available at www.cnn.com.

Kaeble, Danielle, and Mariel Alper. 2020 (August). *Probation and Parole in the United States, 2017–2018.* U.S. Department of Justice, Bureau of Justice Statistics.

Kageyama, Yuri. 2017 (February 12). "Takata Expecting Red Ink from US Airbag Recall Fine." KPIC News. *Associated Press.* Available at kpic.com.

Karp, Aaron. 2018 (June). "Estimating Global Civilian-Held Firearms Numbers." *Small Arms Survey.* Available at www.smallarmssurvey.org.

Katz, Rebecca. 2012. "Environmental Pollution: Corporate Crime and Cancer Mortality." *Contemporary Justice Review: Issues in Criminal, Social, and Restorative Justice* 15(1):97–125.

Khan, Roomy. 2018 (February 22). "White-Collar Crimes—Motivations and Triggers." *Forbes.* Available at www.forbes.com.

Kincade, Brian. 2018 (March 21). "The Economics of the American Prison System." *Smart Assets.* Available at smartasset.com.

Krisher, Tom. 2020 (January 8). "Takata Recalls 10 million More Airbags from 14 Automakers including Subaru, Ford, GM, and Toyota." *USA Today.* Available at www.usatoday.com.

Kubrin, Charis E. 2005. "Gangstas, Thugs, and Hustlas: Identity and the Code of the Street in Rap Music." *Social Problems* 52(3):360–378.

Kubrin, Charis, and Ronald Weitzer. 2003. "Retaliatory Homicide: Concentrated Disadvantage and Neighborhood Culture." *Social Problems* 50:157–180.

LaFree, Gary. 2018 (January 31). "American Attitudes Are Disconnected from Reality on Crime Trends." *The Hill.* Available at thehill.com.

Landergan, Katherine. 2020 (June 12). "The City That Really Did Abolish the Police." *Politico.* Available at www.politico.com.

Langton, Lynn, and Jennifer Truman. 2014 (September). *Special Report: Social-Emotional Impact of Violent Crime.* NCJ247076. Washington, DC: U.S. Department of Justice, Office of Justice Programs. Available at www.bjs.gov.

Lawson Jr., Edward. 2018 (July 2). "Trends: Police Mobilization and the Use of Lethal Force." *Political Research Quarterly* 72(1):177–189.

Lee, Ella, and Emma Oxnevad. 2020 (July 6). "'Just a Bright, Bright Light': Remembering Gary Tinder." Available at depauliaonline.com.

Leigh, David, James Ball, Juliette Garside, and David Pegg. 2015 (February 8). "HSBC Files Show How Swiss Bank Helped Clients Dodge Taxes and Hide Millions." *Guardian*. Available at www.theguardian.com.

Leonhardt, Megan. 2019 (December 17). "The Five Biggest Data Hacks of 2019." CNBC News. Available at www.cnbc.com.

Leverentz, Andrea. 2012. "Narratives of Crime and Criminals: How Places Socially Construct the Crime Problem." *Sociological Forum* 27(2):348–371.

Lewis, Nicole, and Beatrix Lockwood. 2019 (December 17). "The Hidden Cost of Incarceration." *The Marshall Project*. Available at www.themarshallproject.org.

Lieblich, Eliav, and Adam Shinar. 2018. "The Case against Police Militarization." *Michigan Journal of Race and Law* 23(1/2):105–153.

Loeber, Rolf, David P. Farrington, Alison E. Hipwell, Stephanie D. Stepp, Dustin Pardini, and Lia Ahonen. 2015. "Constancy and Change in the Prevalence and Frequency of Offending When Based on Longitudinal Self-Reports or Official Records: Comparisons by Gender, Race and Crime Type." *Journal of Developmental and Life-Course Criminology* 1:150–168.

Lynch, Robert, and Kavya Vaghul. 2015 (December 2). *The Benefits and Cost of Investing in Early Childhood Education*. Available at equitablegrowth.org.

MAD DADS. 2019. "About Us." Available at www.maddads.com.

Mahtani, Shibani. 2018 (May 12). "'Nothing but You and the Cows and the Sirens'—Crime Tests Sheriffs Who Police Small Towns." *Wall Street Journal*. Available at www.wsj.com.

Malamuth, Neil M., Gert Martin Hald, and Mary Koss. 2012. "Pornography, Individual Differences in Risk of Men's Acceptance of Violence against Women in a Representative Sample." *Sex Roles* 66(7/8):427–439.

Malibon, Eleanor, Lisa Carson, and Sophie Yates. 2018. "What Can Policymakers Learn from Feminist Strategies to Combine Contextualized Evidence with Advocacy?" *Palgrave Communications* 4(104). Available at www .nature.com.

Mallicoat, Stacy L. 2020. *Women, Gender, and Crime*, 3rd ed. Thousand Oaks, CA: Sage Publications.

Mauer, Marc. 2018 (November 5). "Long-Term Sentences: Time to Reconsider the Scale of Punishment." *Sentencing Project*. Available at www.sentencingproject.org.

May, Channing. 2017 (March). "Transnational Crime in the Developing World." *Global Financial Integrity*. Available at secureservercdn.net.

McCarthy, Justin. 2019 (November 13). "52% Described Problem of Crime in the U.S. as Serious." *Gallup Poll*. Available at news.gallup. com.

McKinley, Jesse. 2019 (June 2). "Several U.S. States Considering Legislation to Decriminalize Prostitution." *Independent*. Available at www .independent.co.uk.

Mentch, Lucas. 2020 (January 29). "On Racial Disparities in Recent Fatal Police Shootings." *Journal of Statistics in Public Policy* 7(1):9–18.

Merton, Robert. 1957. *Social Theory and Social Structure*. Glencoe, IL: Free Press.

Michaels, Dave. 2019 (December 10). "HSBC Swiss Unit to Pay $192 Million to Settle U.S. Tax Case." *Wall Street Journal*. Available at www.wsj.com.

Midgette, Gregory, Steven Davenport, Jonathan P. Caulkins, and Beau Kilmer. 2019. "What America's Users Spend on Illegal Drugs, 2006–2016." Santa Monica, CA: RAND Corporation.

Migration Portal. 2020. "Recent Trends." Available at migrationdataportal.org.

Mitchell, Ojmarrh, Joshua C. Cochran, Daniel P Mears, and William D. Bales. 2017. "The Effectiveness of Prison for Reducing Drug Offender Recidivism: A Regressive Discontinuity Analysis." *Journal of Experimental Criminology* 13:1–27.

Mittendorf, Brian. 2020. "Why New York Is Suing NRA: 4 Questions Answered." The Conversation. Available at theconversation. com.

Mokhiber, Russell. 2007 (June 15). "Twenty Things You Should Know about Corporate Crime." Alternet. Available at www.alternet .org.

Morgan, Rachel E., and Barbara A. Oudekerk. 2019 (September). *Criminal Victimization, 2018*. U.S. Department of Justice. NCJ 253043. Available at www.bjs.gov.

Morgan, Rachel E., and Jennifer L. Truman. 2020 (September). *Criminal Victimization, 2019*. U.S. Department of Justice. NCJ 255113. Available at www.bjs.gov.

Morrison, Aaron, and Kat Stafford. 2020. "AP-NORC Poll: Support for Racial Injustice Protests Declines." ABC News. Available at abcnews.go.com.

Mouilso, Emily R., and Karen S. Calhoun. 2013. "The Role of Rape Myth Acceptance and Psychopathy in Sexual Assault Perpetration." *Journal of Aggression, Maltreatment and Trauma* 22(2):159–174.

Mugambi, Jouet. 2019. "Mass Incarceration Paradigm Shift? Convergence in an Age of Divergence." *Journal of Criminal Law and Criminology* 109(4):703–768.

Mummolo, Jonathan. 2018. "Militarization Fails to Enhance Public Safety or Reduce Crime but May Harm Police Reputation." *Proceedings of the National Academy of Sciences* 115(37):9181–9186.

Murphy, Shelley, and Milton J. Valencia. 2013 (June 17). "Confessed Murderer Ties Bulger to 6 Killings." *Boston Globe*. Available at www .bostonglobe.com.

Narcotta, Jack. 2018 (April 17). "Smart Home Surveillance Camera Market Movers Are Coming into View." *Strategy Analytics*. Available at www.strategyanalytics.com.

National Association of Town Watch. 2020a. "Membership." Available at www.natw.org.

National Association of Town Watch. 2020b. "About." Available at www.natw.org.

National Center for Missing and Exploited Children (NCMEC). 2020. "Is a Child Being Exploited Online?" Available at www .missingkids.org.

National Research Council. 1994. *Violence in Urban America: Mobilizing a Response*. Washington, DC: National Academy Press.

National Rifle Association (NRA). 2020. "Tireless Defenders of Your Second Amendment Rights." Available at https://home.nra.org.

Naylor, Brian. 2020 (August 30), "Fact Check: Trump's and Biden's Records on Criminal Justice." National Public Radio. Available at www.npr.org.

Nix, Justin, Bradley A. Campbell, Edward H. Byers, and Jeffrey P. Allport. 2017. "A Birds Eye View of Civilians Killed by Police in 2015." *Criminology and Public Policy* 16(1):309–341.

Nolan, Tom. 2020 (June 2). "Militarization Has Fostered a Policing Culture That Sets up Protesters as 'The Enemy'." The Conversation. Available at theconversation.com.

Office of Victims of Crime. 2017. "Urban and Rural Victimization." National Center for Victims of Crime. Available at ovc.ojp.gov.

Organization for Economic Cooperation and Development. 2020. *OECD Better Life Index*. Available at stats.oecd.org.

Parker, Kim, Juliana Horwitz, Ruth Igielnik, Baxter Oliphant, and Anna Brown. 2017 (June 22). *America's Complex Relationship with Guns*. Available at www.pewresearch.org.

Pasko, Lisa. 2017 (January 26). "Beyond Confinement: The Regulation of Girl Offenders' Bodies, Sexual Choices, and Behavior." *Women and Criminal Justice* 27:1:4–20.

Pasley, James. 2019 (July 25). "20 Staggering Facts about Human Trafficking in the U.S." *Business Insider*. Available at www .businessinsider.com.

Perry Preschool Project. 2020. "Who We Are, Our Research, Our Practice, Our Reach." Available at highscope.org.

Pertossi, Mayra. 2000 (September 27). "Analysis: Argentine Crime Rate Soars." Available at news.excite.com.

Peterson, Cora, Sarah Degue, Curtis Florence, and Colby N. Lokey. 2017. "Lifetime Economic Burden of Rape among U.S. Adults." *American Journal of Preventive Medicine* 52(6):691–701.

Phillips, Morgan. 2020 (June 17). "'Justice Act': What's in the Senate Republican Police Reform Bill." Fox News. Available at www.foxnews.com.

Phillips, Scott, and Justin Marceau. 2020 (July 30). "Whom the State Kills." *The Harvard Civil Rights—Civil Liberties Law Review* 55(2):1–69.

Picheta, Rob, and Henrik Pettersson. 2020 (June 8). "American Police Shoot, Kill and Imprison More People Than Other Developed Countries. Here's the Data." CNN US. Available at www .cnn.com.

Pickett, Justin T., Ted Chiricos, Kristin M. Golden, and Marc Gertz. 2012. "Reconsidering the Relationship between Perceived Neighborhood Racial Composition in Whites' Perceptions of Victimization Risk: Do Racial Stereotypes Matter?" *Criminology* 50(1):14–186.

Polaris Project. 2019. "2018 Statistics from the National Human Trafficking Hotline." Available at polarisproject.org.

PredPol. 2020. "About—Overview." Available at www.predpol.com.

Presidential Investigation Education Project. 2019. "Key Findings of the Mueller Report." Collaboration between the American Constitution Society and Citizens for Responsibility and Ethics in Washington. Available at www.acslaw.org.

Pridemore, William Alex, and Sang-Weon Kim. 2007. "Socioeconomic Change and Homicide in a Transitional Society." *Sociological Quarterly* 48:229–251.

Public Document. 2016. "Victim Impact Statement of Chanel Miller." Available at www .documentcloud.org.

Rahim, Zamira, and Rob Picheta. 2020 (June 1). "Thousands around the World Protest George Floyd's Death in Global Display of Solidarity." CNN World News. Available at www.cnn.com.

RAND Corporation. 2020 (June). "Gun Policy in America." Available at www.rand.org.

Rape, Abuse and Incest National Network (RAINN). 2017. "Perpetrators of Sexual Violence: Statistics." Available at www.rainn .org.

Rathke, Lisa. 2020 (February 6). "Bill in Vermont Would Decriminalize Prostitution." ABC News. Available at abcews.go.com.

Reiman, Jeffrey, and Paul Leighton. 2020. *The Rich Get Richer and the Poor Get Prison*, 12th ed. New York: Routledge.

Reuters. 2020 (June 4). "Bernie Madoff Fails in Bid for Compassionate Relief from Prison." Available at www.theguardian.com.

Roeder, Oliver, Lauren-Brooke Eisen, and Julia Bowling. 2015. *What Caused the Crime Decline?* New York: New York University School of Law, Brennan Center for Justice. Available at www.brennancenter.org.

Romo, Vanessa. 2020 (June 16). "PG&E Pleads Guilty on 2018 California Campfire: 'Our Equipment Started That Fire.'" National Public Radio. Available at www.npr.org.

Rostad, Whitney L., Daniel Grittins-Stone, Charlie Huntington, Christie J. Rizzo, Deborah Pearlman, and Lindsay Orchowski. 2019 (October). "The Association between Exposure to Violent Pornography and Teen Dating Violence in Grade 10 High School Students." *Archives of Sexual Behavior* 48(7):2137–2147.

Sanchez, Ray. 2020 (June 14). "Police Reforms Quickly Take Hold across America. It's Only Just Getting Started." CNN News. Available at www.cnn.com.

Sanchez, Ryan. 2020 (July 23). "Black Lives Matter Protests across America Continue Nearly 2 Months after George Floyd's Death." CNN News. Available at www.cnn.com.

Saunders, Debra J. 2020 (June 16). "Trump Signs Executive Order on Police Reform." *Review Journal*. Available at www.reviewjournal .com.

Sawyer, Wendy, and Peter Wagner. 2020 (March 24). "Mass Incarceration: The Whole Pie 2020." *Prison Policy Initiative*. Available at www .prisonpolicy.org.

Schweinhart, Lawrence J. 2007. "Crime Prevention by the High/Scope Perry Preschool Program." *Victims and Offenders* 2:141–160.

Sebastian, Dave. 2021 (January 12). "Solar Winds Discloses Earlier Evidence of Hack." *Wall Street Journal*. Available at www.wsj.com.

Seelye, Katherine Q. 2013 (August 12). "Bulger Guilty of Gangland Crimes, Including Murder." *New York Times*. Available at www.nytimes.com.

Sentencing Project. 2019. "Trends in U.S. Corrections." The Sentencing Project. Available at www.sentencingproject.org.

Sentencing Project. 2020. "Criminal Justice Facts: The United States Is the World Leader in Incarceration." Available at www .sentencingproject.org.

Sexton, Scott. 2020 (June 17). "NC Senator Burr Is Left Alone in the Crosshairs of Investigations over Stock Sales ahead of Economic Crash." *Winston-Salem Journal*. Available at journalnow.com.

Shaw, Adam. 2020 (July 11). "Romney Accuses Trump of 'Historic Corruption' after Roger Stone Commutation." Fox News. Available at www.foxnews.com.

Sherman, Lawrence. 2003. "Reasons for Emotions." *Criminology* 42:1–37.

Siegel, Larry. 2006. *Criminology*, 9th ed. Belmont, CA: Wadsworth.

Smyth, Julie. 2012. "Dual Punishment: Incarcerated Mothers and Their Children." *Columbia Social Work Review* 3:33–45.

Smith-Schoenwalder, Cecelia. 2020 (June 16). "Trump Signs Police Reform Executive Order." *U.S. News*. Available at www.usnews.com.

Sprunt, Barbara. 2020 (May 29). "The History behind "When the Looting Starts, the Shooting Starts'." National Public Radio. Available at www.npr.org.

Statista. 2019. "Crimes Americans Worry about Most in 2019." Available at www.statista.com.

Streeter, Shea. 2019 (June 7). "Lethal Force in Black and White: Assessing Racial Disparities in the Circumstances of Police Killings." *The Journal of Politics* 81(3):1124–1132.

Sundt, Jody, and Breanna Boppre. 2020 (July 15). "Did Oregon's Tough Mandatory Sentence Law 'Measure 11' Improve Public Safety? New Evidence about an Old Debate from a Multiple-Design, Experimental Strategy." *Justice Quarterly* 37(5):1–22.

Sun-Times Wire. 2020 (June 22). "104 Shot, 15 Fatally, over Father's Day Weekend in Chicago." *Chicago Sun-Times*. Available at chicago.suntimes.com.

Sutherland, Edwin H. 1939. *Criminology*. Philadelphia: Lippincott.

Sweeten, Gary, Alex R. Piquero, and Laurence Steinberg. 2013. "Age and the Explanation of Crime, Revisited." *Journal of Youth and Adolescence* 42:921–938.

Tabuchi, Hiroko. 2014 (November 7). "Takata Saw and Hid Risk in Airbags 2004, Former Workers Say." *New York Times*. Available at www.nytimes.com.

Taylor, Derrick Bryson. 2020 (July 10). "George Floyd Protest: A Timeline." *New York Times*. Available at www.nytimes.com.

Thakkar, Danny. 2019 (December 6). "Global Restrictions on Facial Recognition Paid Way to New Biometric Surveillance." *Bayometric*. Available at www.bayometric.com.

Thompson, Mark, and Evan Perez. 2015 (June 30). "BNP Paribas to Pay Nearly $9 Billion Penalty." CNN Money. Available at money.cnn. com.

Travis, Jeremy, Bruce Western, and Steve Redburn (eds.). 2014. *The Growth of Incarceration in the United States: Exploring Causes and Consequences*. Committee on Law and Justice, Division of Behavioral and Social Sciences and Education. National Research Council. Washington, DC: National Academies Press.

United Nations (UN). 2016 (May). "World Crime Trends and Emerging Issues and Responses in the Field of Crime Prevention and Criminal Justice." Vienna: Commission on Crime Prevention and Criminal Justice. Available at www.unodc.org.

United Nations Crime Trend Survey (UNCTS). 2019. *DATAUNODC*. Available at dataunodc. un.org.

United Nations Office on Drugs and Crimes (UNODC). 2019. *Global Study on Homicide*. Available at unodc.org.

U.S. Commission on Civil Rights. 2018. *Police Use of Force: An Examination of Modern Policing Practices*. Available at www.usccr.gov.

United States Department of Justice (USDOJ). 2015 (June 2). "International Organized Crime." Available at ww.justice.gov.

United States Department of Justice (USDOJ). 2020a. "Human Trafficking." Available at https://www.justice.gov.

United States Department of Justice (USDOJ). 2020b. "Child Sex Trafficking." Available at www.justice.gov.

Velez, Maria B., Christopher J. Lyons, and Wayne A. Santoro. 2015. "The Political Context of the Percent Black-Neighborhood Violence Link: A Multilevel Analysis." *Social Problems* 62(1):93–119.

Victim Statements. 2009. *U.S. v. Bernard L. Madoff*. 2009. U.S. Department of Justice. Available at www.pbs.org.

Violence Policy Center. 2019 (January 23). "States with Weak Gun Laws and Higher Gun Ownership Lead the Nation in Gun Deaths, New Data for 2017." Available at vpc.org.

Walmsley, Roy. 2019. *World Prison Population List*, 12th ed. Institute for Criminal Policy Research. Available at www.prisonstudies.org.

Warner, Tara D., and Courtney R. Thrash. 2020. "A Matter of Degree? Fear, Anxiety, and Protective Gun Ownership in the United States." *Social Science Quarterly* 101(1):285–308.

Watson, Julie. 2020 (March 17). "Ex-California Congressman Duncan Hunter Sentenced to 11 Months in Prison." NBC San Diego. Available at www.nbcsandiego.com.

Welsh-Huggins, Andrew. 2013 (July 13). "Hundreds of New Charges Filed in U.S. Kidnap Case." Associated Press. Available at apnews.com.

Weiss, Brendan, Ellen Cranley, and Grace Panetta. 2020 (February 20). "Here's Everyone Who Has Been Charged, Convicted, and Sentenced in the Russia Probe so Far." *Business Insider*. Available at www.businessinsider.com.

Williams, Emma J., Amy Beardmore, and Adam N. Joinson. 2017. "Individual Differences in Susceptibility to Online Influence: A Theoretical Review." *Computers in Human Behavior* 72:412–421.

Williams, Linda. 1984. "The Classic Rape: When Do Victims Report?" *Social Problems* 31:459–467.

Winslow, Robert W., and Sheldon Zhang. 2008. *Criminology: A Global Perspective.* Englewood Cliffs, NJ: Prentice Hall.

Yang, Allie, and Joseph Diaz. 2020 (January 2). "Michelle Knight's Triumph over 11-Year Captor Ariel Castro: 'He Doesn't Define Who I Am.'" ABC News. Available at abcnews.go.com.

Zahn, Margaret A., Robert Agnew, Diana Fishbein, Shari Miller, Donna-Marie Winn, Gayle Dakoff, Candace Kruttschnitt, Peggy Giordano, Denise C. Gottfredson, Allison A. Payne, Barry C. Feld, and Meda Chesney-Lind. 2010 (April). "Girls Study Group: Causes and Correlates of Girls' Delinquency." U.S. Department of Justice. Office of Juvenile Justice and Delinquency Prevention. Available at www.ncjrs.gov.

Zurik, Lee, and Cody Lillich. 2020 (January 21). "Former State Senator Wesley Bishop Pleads Guilty to Federal Charge, Resigns from SUNO." Fox 8 News. Available at www.fox8live.com.

Chapter 5

American Civil Liberties Union (ACLU). 2020. *Domestic Violence and Homelessness.* Available at www.aclu.org.

Ahrons, C. 2004. *We're Still Family: What Grown Children Have to Say about Their Parents' Divorce.* New York: HarperCollins.

Allendorf, Keera. 2013 (April). "Schemas of Marital Change: From Arranged Marriages to Eloping for Love." *Journal of Marriage and Family*: 453–464.

Allred, C. (2019). "Divorce Rate in the U.S.: Geographic Variation, 2018." *Family Profiles*, FP-19-23. Bowling Green, OH: National Center for Family & Marriage Research. https://doi.org/10.25035/ncfmr/fp-18-23.

Amato, Paul. 2003. "The Consequences of Divorce for Adults and Children." In *Family in Transition*, 12th ed., ed. Arlene S. Skolnick and Jerome H. Skolnick, 190–213. Boston: Allyn and Bacon.

Amato, Paul. 2004. "Tension between Institutional and Individual Views of Marriage." *Journal of Marriage and Family* 66:959–965.

Amato, P. R., A. Booth, D. R. Johnson, and S. J. Rogers. 2007. *Alone Together: How Marriage in America Is Changing.* Cambridge, MA: Harvard University Press.

Amato, P. R., and J. Cheadle. 2005. "The Long Reach of Divorce: Divorce and Child Well-Being across Three Generations." *Journal of Marriage and the Family* 67:191–206.

Amato, Paul R., and Patterson, Sarah E. 2016. "The Intergenerational Transmission of Union Instability in Early Adulthood." *Journal of Marriage and Family* 79(3):723–738.

Amundsen, Amy J., and Mike Kelly. 2014. "Trends in Alimony Modifications." Chicago: American Bar Association. Available at www.americanbar.org.

Anderson, Kristin L. 2013 (April). "Why Do We Fail to Ask 'Why' about Gender and Intimate Partner Violence?" *Journal of Marriage and Family* 75:314–318.

Associated Press. 2020 (June 29). *China Forces Birth Control on Uighurs to Suppress Population.* Available at www.apnews.com.

Astone, Nan Marie, Andrew Karas, and Allison Stolte. 2016 (December). *Father's Time with Children: Income and Residential Differences.* Urban Institute. Available at https://www.urban.org/.

Baker, Dean. 2020. *This Is What Minimum Wage Would Be If It Kept Pace with Productivity.* Center for Economic and Policy Research. Available at www.cepr.net.

BBC News. 2020 (September 21). "Abortion: How Do Trump and Biden's Policies Compared?" British Broadcasting Company. Available at www.bbc.com.

Beckmeyer, Jonathon J., Marilyn Coleman, and Lawrence H. Ganong. 2014. "Postdivorce Coparenting Typologies and Children's Adjustment." *Family Relations* 63:526–537.

Bernet, William, and Amy J. L. Baker. 2013 (October). "Parental Alienation, DSM-5, and ICD-11: Response to Critics." *Journal of the American Academy of Psychiatric Law* 41(1):98–104.

Biden, Joe. 2020a. "The Biden Plan for Strengthening America's Commitment to Justice." Available at www.joebiden.com.

Biden, Joe. 2020b. "The Biden Agenda for Women." Available at www.joebiden.com.

Biden, Joe. 2020c. "Caregiving." Available at www.joebiden.com.

Bogacz, Francois, Thierry Pun, and Olga M. Klimecki. 2020 (October 21). "Improved Conflict Resolution in Romantic Couples in Mediation Compared to Negotiation." *Humanities and Social Sciences Communications* 7(131). Available at https://doi.org/10.1057/s41599-020-00622-8.

Boghani, Priyanka. 2020 (September 8). "Children in Poverty by the Numbers." *Frontline.* Available at www.pbs.org.

Brewster, Melanie E. 2016. "Lesbian Women and Household Labor Division: A Systemic Review of Scholarly Research from 2000 to 2015." *Journal of Lesbian Studies* 20(1):47–69.

Brinig, Margaret F., and Marsha Garrison. 2018. "Getting Blood from Stones: Results and Policy Implications of an Empirical Investigation of Child Support Practice in St. Joseph County, Indiana Paternity Actions." *Family Court Review* 56(4):521–543.

Buchwald, Elizabeth. 2020 (November 18). "Joe Biden Wants a $15 Minimum Wage—Here's What's Standing in His Way. *Wall Street Journal.* Available at www.marketwatch.com.

Bullock, Penn. 2016 (October 8). "Transcript: Donald Trump's Taped Comments about Women." *New York Times.* Available at www.nytimes.com.

Bureau of Labor Statistics. 2016. "American Time Use Survey—2015 Results." News Release. Available at www.bls.gov.

Bureau of Labor Statistics. 2019 (April 21). *Employment Characteristics of Families Summary.* Available at www.bls.gov.

Butrica, Barbara A., and Karen E. Smith. 2012. "The Retirement Prospects of Divorced Women." *Social Security Bulletin* 72(1). Available at www.ssa.gov.

Carrington, Victoria. 2002. *New Times: New Families.* Dordrecht, Netherlands: Kluwer Academic.

Catalano, Shannon. 2012. *Intimate Partner Violence, 1993–2010.* Bureau of Justice Statistics. Available at www.ojp.usdoj.gov.

Cavanagh, Shannon E., and Paula Fomby. 2019. Family Instability in the Lives of American Children. *Annual Review of Sociology* 45(1):493–513.

CBS News. 2017 (January 24). *Trump Expands Anti-Abortion Ban to All U.S. Global Health Aid.* Available at www.cbsnews.com.

Center for Reproductive Rights. 2020a (May 6). "Supreme Court Hears Case Challenging Trump-Pence Denial of Birth Control Coverage." Available at www.reproductiverights.org.

Center for Reproductive Rights. 2020b. *The World's Abortion Laws.* Available at www.reproductiverights.org.

Centers for Disease Control and Prevention (CDC). 2014. *Understanding Child Maltreatment.* Fact Sheet. Available at www.cdc.gov.

Centers for Disease Control and Prevention (CDC). 2020a. *Out of School Time.* Available at www.cdc.gov.

Centers for Disease Control and Prevention (CDC). 2020b. *Preventing Intimate Partner Violence.* Available at https://www.cdc.gov/.

Cherlin, Andrew J. 2009. *The Marriage-Go-Round: The State of Marriage and Family in America Today.* New York: Knopf.

Cherlin, Andrew. 2020. "Degrees of Change: An Assessment of the Deinstitutionalization of Marriage Thesis." *Journal of Marriage and Family* 82:62–80.

Child Welfare Information Gateway. 2020 (March). "Child Abuse and Neglect Fatalities 2018: Statistics and Interventions." *Numbers and Trends.* Available at www.childwelfare.gov.

Children's Bureau. 2020. *Trends in Foster Care and Adoption 2009–2018.* U.S. Department of Health and Human Services. Available at www.acf.hhs.gov/cb/.

Cohen, Philip. 2019. "The Coming Divorce Decline." *Socius* 5:1–6.

Cohn, Jonathan. 2014 (September 14). "Five Things We Can Do to Reduce Domestic Violence." *New Republic.* Available at www.newrepublic.com.

Coontz, Stephanie. 2004. "The World Historical Transformation of Marriage." *Journal of Marriage and Family* 66(4):974–979.

Coontz, Stephanie. 2016. *The Way We Never Were: American Families and the Nostalgia Trap.* New York: Basic Books.

Cui, M., K. Ueno, M. Gordon, and F. D. Fincham. 2013 (April). "The Continuation of Intimate Partner Violence from Adolescence to Young Adulthood." *Journal of Marriage and Family* 75:300–313.

Daniel, Elycia. 2005. "Sexual Abuse of Males." In *Sexual Assault: The Victims, the Perpetrators, and the Criminal Justice System*, ed. Frances P. Reddington and Betsy Wright Kreisel, 133–140. Durham, NC: Carolina Academic Press.

Davis, John W. 2014 (July 15). "Rosemary Pate Remembered as Lawmaker Works on 'Parent Abuse' Bill." News 13. Available at www.mynews.com.

Decuzzi, A., D. Knox, and M. Zusman. 2004 (April 17). "The Effect of Parental Divorce on Relationships with Parents and Romantic Partners of College Students." Roundtable Discussion, Southern Sociological Society, Atlanta.

Dennison, Renee Peltz. 2017 (April 24). "Do Half of All Marriages End in Divorce?" *Psychology Today*. Available at www.psychologytoday.com.

Dennison, Renee Peltz, and S. Koerner. 2008. "A Look at Hopes and Worries about Marriage: The Views of Adolescents Following a Parental Divorce." *Journal of Divorce & Remarriage* 48:91–107.

Dholakia, Utpal. 2015 (November 24). "Why Are So Many Indian Arranged Marriages So Successful?" *Psychology Today*. Available at www.psychologytoday.com.

Dokko, Jane, Geng Li, and Jessica Hayes. 2015. "Credit Scores and Committed Relationships." Finance and Economics Discussion Series 2015-081. Washington, DC: Board of Governors of the Federal Reserve System. Available at www.federalreserve.gov.

Donovan, Megan K. 2017. "The Looming Threat to Sex Education: A Resurgence of Federal Funding for Abstinence-Only Programs?" *$ex Education*. Volume 20. Guttmacher Institute. Available at www.guttmacher.org.

Edin, Kathryn. 2000. "What Do Low-Income Single Mothers Say about Marriage?" *Social Problems* 47(1):112–133.

Emery, Robert E. 2014 (September 6). "How Divorced Parents Lost Their Rights." *New York Times*. Available at www.nytimes.com.

Emery, Robert E., David Sbarra, and Tara Grover. 2005. "Divorce Mediation: Research and Reflections." *Family Court Review* 43(1):22–37.

Ferraro, K. J. 2006. *Neither Angels nor Demons: Women, Crime, and Victimization*. Boston: Northeastern University Press.

Fincham, F., M. Cui, M. Gordon, and K. Ueno. 2013 (April). "What Comes before Why: Specifying the Phenomenon of Intimate Partner Violence." *Journal of Marriage and Family* 75:319–324.

Finer L. B., and Zolna, M. R. 2016. "Declines in Unintended Pregnancy in the United States, 2008–2011." *New England Journal of Medicine* 374(9):843–852.

Fingerhut, Hannah. 2017 (January 3). "About Seven in 10 Americans Opposed Overturning *Roe v Wade*." Available at www.pewresearch.org.

Finkelhor, David, Heather Turner, Brittany Kaye Wormuth, Jennifer Vanderminden, and Sherry Hamby. 2019. "Corporal Punishment: Current Rates from a National Survey." *Journal of Child and Family Studies*. Published online May 2019.

Font, Sarah A., Lawrence M. Berger, Maria Cancian, and Jennifer L. Noyes. 2018. "Permanency and the Educational and Economic Attainment of Former Foster Children in Early Adulthood." *American Sociological Review* 83(4):716–743.

FosterMore. 2020. *What Is Foster Care?* Available at www.fostermore.org.

Francis, David. 2000 (January). "Poverty and Mistreatment of Children Go Hand in Hand." *National Bureau of Economic Research Digest*. Available at www.nber.org.

Frost, Jennifer J., Adam Sonfield, Mia R. Zolna, and Lawrence B. Finer. 2014. "Return on Investment: A Fuller Assessment of the Benefits and Cost Savings of the US Publicly Funded Family Planning Program." *Milbank Quarterly* 92(4):696–749.

Frost, Jennifer J., L. F. Frohwirtt, and Mia R. Zolna. 2016. *Contraceptive Needs and Services, 2014 Update*. New York: Guttmacher Institute. Available at www.guttmacher.org.

Gadalla, Tahany M. 2009. "Impact of Marital Dissolution on Men's and Women's Income: A Longitudinal Study." *Journal of Divorce & Remarriage* 50(1):55–65.

Gager, Constance T., Scott T. Yabiku, and Miriam R. Linver. 2016. "Conflict or Divorce? Does Parental Conflict and/or Divorce Increase the Likelihood of Adult Children's Cohabiting and Marital Dissolution?" *Marriage and Family Review* 52(3):243–261.

Galvin, Gaby. 2020 (April 29). "U.S. Marriage Rate Hits Historic Low." *U.S. News & World Report*. Available at www.usnews.com.

Gallup. 2020. *Abortion*. Available at https://news.gallup.com/poll/1576/abortion.aspx.

Gartrell, Nanette K., Henny M. W. Bos, and Naomi G. Goldberg. 2010 (published online November 6). "Adolescents of the U.S. National Longitudinal Lesbian Family Study: Sexual Orientation, Sexual Behavior, and Sexual Risk Exposure." *Archives of Sexual Behavior*.

Geiger, A. W., and Gretchen Livingston. 2019 (February 13). "8 Facts about Love and Marriage in America." *FactTank*. Available at www.pewresearch.org.

Generations United. 2014. *The State of Grandfamilies in America, 2014*. Available at www.gu.org.

Glasmeier, Amy K. 2021. "Living Wage Calculator." Massachusetts Institute of Technology. Available at www.livingwage.mit.edu.

Global Initiative to End All Corporal Punishment. 2020. *Countdown to Universal Prohibition*. Available at www.endcorporalpunishment.org.

Goldberg, Abbie. E., AuliAnna Z. Smith, and Maureen Perry-Jenkins. 2012. "The Division of Labor in Lesbian, Gay, and Heterosexual New Adoptive Parents." *Journal of Marriage and Family* 74(4):812–828.

Grall, Timothy. 2020. *Custodial Mothers and Fathers and Their Child Support: 2015. Current Population Reports, P60-262*. U.S. Census Bureau. Available at www.census.gov.

Greenwood, Joleen Loucks. 2014. "Effects of a Mid- to Late-Life Parental Divorce on Adult Children." *Journal of Divorce & Remarriage* 55(7):539–556.

Gromoske, Andrea N., and Kathryn Maguire-Jack. 2012. "Transactional and Cascading Relations between Early Spanking and Children's Social-Emotional Development." *Journal of Marriage and Family* 74:1054–1068.

Gustafsson, Hanna C., and Martha J. Cox. 2012 (October). "Relations among Intimate Partner Violence, Maternal Depressive Symptoms, and Maternal Parenting Behaviors." *Journal of Marriage and Family* 74:1005–1020.

Guttmacher Institute. 2001 *Can More Progress Be Made? Teenage Sexual and Reproductive Behavior in Developed Countries*. Available at www.guttmacher.org.

Guttmacher Institute. 2019a. "Induced Abortion in the United States." Available at www.guttmacher.org.

Guttmacher Institute. 2019b (December 10). "State Policy Trends 2019: A Wave of Abortion Bans, but Some States Are Fighting Back." Available at www.guttmacher.org.

Guttmacher Institute. 2020a. "An Overview of Abortion Laws." Available at www.guttmacher.org.

Guttmacher Institute. 2020b. "Adding It Up: Investing in Sexual and Reproductive Health, 2019." Available at www.guttmacher.org.

Gutman, Matt, and Emily Shapiro 2019 (April 19). "Turpin Children Speak Out as Parents Are Sentenced in Torture Case: 'I'm Taking My Life Back.'" ABC News. Available at news.go.com.

Hailes, Helen, Rongqin Yu, Andrea Danese, and Seena Fazel. 2019. "Long-Term Outcomes of Childhood Sexual Abuse: An Umbrella Review." *Lancet Psychiatry* 6:830–839.

Halpern-Meekin, Sarah, Wendy D. Manning, Peggy C. Giordina, and Monica A. Longmore. 2013 (February). "Relationship Churning, Physical Violence, and Verbal Abuse in Young Adult Relationships." *Journal of Marriage and Family*: 2–12.

Harris, Gardiner. 2015 (April 24). "Websites in India Put a Bit of Choice into Arranged Marriages." *New York Times*. Available at www.nytimes.com.

Harrison, Mette Ivie. 2017. "Do Mormons Still Practice Polygamy? HuffPost. Available at www.huffpost.com.

Hawkins, Alan J., and Betsy VanDenBerghe. 2014. "Facilitating Forever. The National Marriage Project." Available at www.nationalmarriageproject.org.

Hitler, Adolf. 1925. *Mein Kampf*. Volume One. Munich: Eher-Verlag.

Hochschild, Arlie Russell. 1989. *The Second Shift: Working Parents and the Revolution at Home*. New York: Viking.

Horowitz, Juliana Menasce, Nikki Graf, and Gretchen Livingston. 2019 (November 6). *Marriage and Cohabitation in the U.S.* Available at www.pewresearch.org.

Human Rights Campaign. 2020. *Marriage Equality around the World*. Available at www.hrc.org.

Human Rights Watch. 2019. "Saudi Arabia: 10 Reasons Why Women Flee." Available at www.hrw.org.

Jalovaara, M. 2003. "The Joint Effects of Marriage Partners' Socioeconomic Positions on the Risk of Divorce." *Demography* 40:67–81.

James, Sarah, Louis Donnelly, Jeanne Brooks-Gunn, and Sara McLanahan. 2018. "Links between Childhood Exposure to Violent Contexts and Risky Adolescent Health Behaviors." *Journal of Adolescent Health* 63:94–101.

James, Susan Donaldson. 2014 (September 12). "'Shaken Baby Syndrome': Those Who Survive Are Horribly Afflicted." NBC News. Available at www.nbcnews.com.

Jasinski, J. L., L. M. Williams, and J. Siegel. 2000. "Childhood Physical and Sexual Abuse as Risk Factors for Heavy Drinking among African-American Women: A Prospective Study." *Child Abuse and Neglect* 24:1061–1071.

Jekielek, Susan M. 1998. "Parental Conflict, Marital Disruption, and Children's Emotional Well-Being." *Social Forces* 76:905–935.

Jindia, Shilpa. 2020 (June 30). "Belly of the Beast: California's Dark History of Forced Sterilizations." *Guardian.* Available at www .guardian.com.

Johnson, Corey. 2013 (November 8). "California Was Sterilizing Its Female Prisoners as Late as 2010." *Guardian.* Available at www .theguardian.com.

Johnson, Michael P. 2001. "Patriarchal Terrorism and Common Couple Violence: Two Forms of Violence against Women." In *Men and Masculinity: A Text Reader*, ed. T. F. Cohen, 248–260. Belmont, CA: Wadsworth.

Johnson, Michael P., and Kathleen Ferraro. 2003. "Research on Domestic Violence in the 1990s: Making Distinctions." In *Family in Transition*, 12th ed., ed. A. S. Skolnick and J. H. Skolnick, 493–514. Boston: Allyn and Bacon.

Joyce, Tina, and Martin R. Huecker. 2020. "Pediatric Abusive Head Trauma." Available at www.ncbi.nlm.nih.gov.

Kaiser Family Foundation. 2020 (June 8). "The Availability and Use of Medication Abortion." Available at www.kff.org.

Kalmijn, Matthijs, and Christiaan W. S. Monden. 2006. "Are the Negative Effects of Divorce on Well-Being Dependent on Marital Quality?" *Journal of Marriage and the Family* 68:1197–1213.

Kamp Dush, Claire M. 2013 (February). "Marital and Cohabitation Dissolution and Parental Depressive Symptoms in Fragile Families." *Journal of Marriage and Family* 75:91–109.

Kelly, Joan B., and Michael P. Johnson. 2008. "Differentiation among Types of Intimate Partner Violence: Research Update and Implications for Interventions." *Family Court Review* 46(3):476–499.

Khan, Roxanne, and Paul Rogers. 2015. "The Normalization of Sibling Violence: Does Gender and Personal Experience of Violence Influence Perceptions of Physical Assault against Siblings?" *Journal of Interpersonal Violence* 30(3):437–458.

Kim, Hyunil, and Brett Drake. 2018. "Child Maltreatment Risk as a Function of Poverty and Race/Ethnicity in the USA." *International Journal of Epidemiology* 47(3):780-787.

Kitzmann, K. M., N. K. Gaylord, A. R. Holt, and E. D. Kenny. 2003. "Child Witnesses to Domestic Violence: A Meta-Analytic Review." *Journal of Clinical and Consulting Psychology* 71:339–352.

Koball, Heather, Randy Capps, Krista Perreira, et al. 2015. "Health and Social Service Needs of U.S.-Citizen Children with Detained or Deported Immigrant Parents." Urban Institute/ Migration Policy Institute. Available at www .urban.org.

Ko, Lisa. 2016. "Unwanted Sterilization and Eugenics Programs in the United States." Available at www.pbs.org.

Koch, Wendy. 2009. "Fees Cut Down Private Adoptions." *USA Today*, p. 1A.

Kramer, Stephanie. 2019 (December 12). "U.S. Has World's Highest Rate of Children Living in Single-Parent Households." *FactTank.* Available at www.pewresearch.org.

Kuo, J. C., and R. K. Raley. 2016 (April 27). "Diverging Patterns of Union Transition among Cohabitors by Race/Ethnicity and Education: Trends and Marital Intentions in the United States." *Demography* 53(4):921–935.

Kurtzleben, Danielle. 2020 (June 28). "How Coronavirus Could Widen the Gender Wage Gap." National Public Radio. Available at www.npr.org.

Lacey, K. K., D. G. Saunders, and L. Zhang. 2011. "A Comparison of Women of Color and Non-Hispanic White Women on Factors Related to Leaving a Violent Relationship." *Journal of Interpersonal Violence* 26:1036–1055.

Lamidi, Esther O., Wendy D. Manning, and Susan L. Brown. 2019. "Change in the Stability of First Premarital Cohabitation among Women in the United States, 1983–2013." *Demography* 56(2):427–450.

Lancer, Darlene. 2020 (February 3). "Sibling Bullying and Abuse: Hidden Epidemic." *Psychology Today.* Available at www .psychologytoday.com.

Levtov, R., N. van der Gaag, M. Greene, M. Kaufman, and G. Barker. 2015. "State of the World's Fathers: Executive Summary," *A MenCare Advocacy Publication.* Available at www.sowf.men-care.org.

Lewis, Jamie M., and Rose M. Kreider. 2015 (March). *Remarriage in the United States.* United States Census Bureau. Available at www.census.gov.

Lin, I-Fen, Susan L. Brown, and Cassandra Jean Cupka. 2017. "A National Portrait of Stepfamilies in Later Life." *The Journals of Gerontology: Series B* 73(6):1043–1054.

Lindberg, Laura D., Isaac Madow-Zimet, and Heather Boomstra. 2016. "Changes in Adolescents' Receipt of Sex Education, 2006–2013." *Journal of Adolescent Health* 58(6):621–627.

Lines, Gregory. 2016. "Polymmigration: Immigration Implications and Possibilities Post *Brown v. Buhman.*" *Arizona Law Review* 58:477–510.

Livingston, Gretchen. 2018 (April 27). "About One-Third of U.S. Children Are Living with an Unmarried Parent." *FactTank.* Available at https://www.pewresearch.org. /fact-tank/2018/04/27/about-one-third-of-u-s -children-are-living-with-an-unmarried-parent/.

Long, Michelle, Amrutha Ramaswamy, and Alina Salganicoff. 2020 (October 15). "The 2020 Presidential Election: Implications for Women's Health." Kaiser Family Foundation. Available at https://www.kff.org.

Lovering, Tom. 2019 (May 21). "Moving the Needle on Child Support Compliance." Available at www.govwebworks.com.

Luan, Livia. 2018 (May 2). "Profiting from Enforcement: The Role of Private Prisons in U.S. Immigration Detention." *Migration Information Source.* Available at www .migrationpolicy.org.

Luscombe, Belinda. 2018 (November 26). "The Divorce Rate Is Dropping. That May Actually Not Be Good News." *Time.* Available at www .time.com.

Martin, A., B. E. Hamilton, M. J. K. Osterman, A. K. Driscoll, and T. J. Matthews. 2017 (January 5). "Births: Final Data for 2015." *National Vital Statistics Reports* 66(1). Available at www.cdc .gov/nchs.

Mathews, T. J., and Brady Hamilton. 2016. "Mean Age of Mothers Is on the Rise: United States, 2000–2014." Data Brief No. 232. National Center for Health Statistics. Available at www .cdc.gov.

McLanahan, Sara, and Wade Jacobsen. 2015. "Diverging Destinies Revisited." In *Families in an Era of Increasing Inequality.* National Symposium on Family Issues, vol. 5, ed. P. Amato, A, Booth, S. McHale, and J. Van Hook. New York: Springer.

Melli, Marygold S., and Patricia R. Brown. 2008. "Exploring a New Family Form—The Shared Time Family." *International Journal of Law, Policy and the Family* 231(2):231–269.

MenCare. 2020. "About MenCare." Available at www.mencare.org.

Meyer, Daniel R., Maria Cancian, and Steven T. Cook. 2017. "The Growth in Shared Custody in the United States: Patterns and Implications." *Family Court Review* 55(4):500–512.

Meyer, Daniel R., Maria Cancian, and Melody K. Waring. 2019. "Use of Child Support Enforcement Actions and Their Relationship to Payments." *Children and Youth Services Review* 108:1–11.

Mohr, Holbrook, and Garance Burke. 2014 (December 18). "786 Abused Kids Died in Plain View of Authorities: Report." *Huffington Post.* Available at www.huffingtonpost.com.

Morgan, Rachel E., and Jennifer L. Truman. 2020 (September). *Criminal Victimization, 2019.* U.S. Department of Justice. NCJ 255113. Available at www.bjs.gov.

Mortelmans, Dimitri. 2020. "Economic Consequences of Divorce: A Review." In *Parental Life Courses after Separation and Divorce in Europe.* Life Course Research and Social Policies, vol. 12, ed. M. Kreyenfeld and H. Trappe. New York: Springer.

Murray, Christine E., Alison Crowe, and Paulina Flasch. 2015. "Turning Points: Critical Incidents Prompting Survivors to Begin the Process of Terminating Abusive Relationships" *The Family Journal* 23(3):228-238.

National Center for Education Statistics. 2019. "Early Childcare and Education Arrangements." *Status and Trends in the Education of Racial and Ethnic Groups.* Available at www.nces.ed.gov.

National Center for Injury Prevention and Control. 2014a. *Child Maltreatment: Facts at a Glance.* Available at www.cdc.gov.

National Center for Injury Prevention and Control. 2014b. *Understanding Intimate Partner Violence.* Available at www.cdc.gov.

National Center for Health Statistics. 2017. *United States Life Tables, 2017, Table A.* Available at www.cdc.gov/nchs.

National Center for Health Statistics. 2019. *Births in the United States, 2018.* Center for Disease Control and Prevention. Available at www.cdc .gov/nchs.

National Center for Health Statistics. 2020. *Unmarried Childbearing.* Available at www .cdc.gov/nchs/fastats.

National Council on Aging. 2021. "Elder Abuse Facts." Available at www.ncoa.org.

National Network to End Domestic Violence. 2016. *Domestic Violence Counts 2015.* Available at www.nnedv.org.

National Public Radio (NPR). 2019a (April 17). "After Texas Abortion Ban, Clinics in Other Southwest States See Influx of Patients." Available at www.npr.org.

National Public Radio (NPR). 2019b (May 7). "'No Visible Bruises' Upends Stereotypes of Abuse, Sheds Light on Domestic Violence." Available at www.npr.org.

National Public Radio. 2020 (April 10). "Legal Fights Heat Up in Texas over Ban on Abortions amid Coronavirus." *Morning Edition.* Available at www.npr.org.

Nelson, B. S., and K. S. Wampler. 2000. "Systemic Effects of Trauma in Clinic Couples: An Exploratory Study of Secondary Trauma Resulting from Childhood Abuse." *Journal of Marriage and Family Counseling* 26:171–184.

Nelson, Nicholas, and Chuck Webber. 2020 (September). "Supreme Court Decides *Little Sisters of the Poor Saints Peter and Paul Home v. Pennsylvania* and *Trump v. Pennsylvania.*" *National Law Review* X(331):1–3.

Neuman, Scott. 2020 (April 6). "Global Lockdowns Resulting in 'Horrifying Surge' in Domestic Violence, U.N. Warns." National Public Radio. Available at www.npr.org.

Nielsen, Linda. 2014. "Shared Physical Custody: Summary of 40 Studies on Outcomes for Children." *Journal of Divorce & Remarriage* 55(8):613–635.

Nielsen, Linda. 2015. "Shared Physical Custody: Does It Benefit Most Children?" *Journal of the American Academy of Matrimonial Lawyers* 28(1):79–138.

Ogolsky, Brian G., James K. Monk, and Renee P. Dennison. 2014. "The Role of Couple Discrepancies in Cognitive and Behavioral Egalitarianism in Marital Quality." *Sex Roles* 70(7–8):329–342.

Parker, K., and W. Wang. 2013 (March 14). "Modern Parenthood." Pew Research Center. Available at www.pewsocialtrends.org.

Parker, Marcie R., Edward Bergmark, Mark Attridge, and Jude Miller-Burke. 2000. "Domestic Violence and Its Effect on Children." *National Council on Family Relations Report* 45(4):F6–F7.

Parker, Kim, and Renee Stepler. 2017 (September 14). "As U.S. Marriage Rate Hovers at 50%, Education Gap in Marital Status Widens." *FactTank.* Available at www.pewresearch.org.

Pasley, Kay, and Carmelle Minton. 2001. "Generative Fathering after Divorce and Remarriage: beyond the 'Disappearing Dad.'" In *Men and Masculinity: A Text Reader*, ed. T. F. Cohen, 239–248. Belmont CA: Wadsworth.

Payne, Krista. K. 2019. "Young Adults in the Parental Home, 2007 & 2018." *Family Profiles, FP-19-04.* Bowling Green, OH: National Center for Family & Marriage Research. Available at www.bgsu.edu/ncfmr.

Pew Research Center. 2015. *Parenting in America.* Available at https://www .pewsocialtrends.org/2015/12/17/1-the -american-family-today/.

Pew Research Center. 2016. *One-in-Five U.S. Adults Were Raised in Interfaith Homes.* Available at www.pewresearchcenter.org.

Phillips, Allie. 2014. *Understanding the Link between Violence to Animals and People.* National District Attorneys Association. Available at www.ndaa.org.

Pinsker, Joe. 2019 (May 30). "How Successful Are the Marriages of People with Divorced Parents?" *Atlantic.* Available at www .theatlantic.com.

Population Reference Bureau. 2011. *The World's Women and Girls 2011 Data Sheet.* Available at www.prb.org.

Population Reference Bureau. 2019. *2019 Family Planning Data Sheet.* Available at www.prg.org.

Provine, Doris M., and Roxanne L. Doty. 2011. "The Criminalization of Immigrants as a Racial Project." *Journal of Contemporary Criminal Justice* 27(3):261–277.

Reese, Laura Schwab, Erin O. Heiden, Kimberly Q. Kim, and Jingzhen Yang. 2014. "Evaluation of Period of PURPLE Crying, an Abusive Head Trauma Prevention Program." *Journal of Obstetric, Gynecologic, & Neonatal Nursing* 43(6):752–761.

Reiss, Fraidy. 2019. "Let's End Child Marriage in the U.S." *Unchained at Last.* Available at www .unchainedatlast.org.

Rico, Brittany, Rose M. Kreider, and Lydia Anderson. 2018 (July 9). *Growth in Interracial and Interethnic Married Couple Households.* Available at www.census.gov.

Rollè, Luca, Giulia Giardina, Angela M. Caldarera, Eva Gerino, and Piera Brustia. 2018. "When Intimate Partner Violence Meets Same Sex Couples: A Review of Same Sex Partner Violence." *Frontiers in Psychology* 9:1–13.

Rough, Bonnie J. 2018 (August 27). "How the Dutch Do Sex Ed." *Atlantic.* Available at www .theatlantic.com.

Russell, D. E. 1990. *Rape in Marriage.* Bloomington: Indiana University Press.

Salvatore, Jessica E., Sara Larsson Lonn, Jan Sundquist, Kristina Sundquist, and Kenneth S. Kendler. 2018. "Genetics, the Rearing Environment, and the Intergenerational Transmission of Divorce: A Swedish National Adoption Study." *Psychological Science* 29(3):370–378.

Santich, Kate. 2014 (February 24). "Parent Abuse Would Be a Crime under Proposed Law." *Orlando Sentinel.* Available at www .articlesorland-osentinel.com.

Sawhill, Isabel V. 2014. *Generation Unbound: Drifting into Sex and Parenthood without Marriage.* Washington, DC: Brookings Institution Press.

Sawhill, Isabel V., and Katherine Guyot. 2019. *Preventing Unplanned Pregnancy: Lessons from the States.* Washington, DC: Brookings Institution Press.

Schneider, D., K. Harknett, and S. McLanahan. 2016. "Intimate Partner Violence in the Great Recession." *Demography* 53:471–505.

Schochet, Leila. 2019 (March 28). *The Childcare Crisis Is Keeping Women out of the Workforce.* Center for American Progress. Available at www.americanprogress.org.

Schueths, April. 2018. "Not Really Single: The Deportation to Welfare Pathway for U.S. Citizen Mothers in Mixed-Status Marriage." *Critical Sociology* 45(7–8):1075–1092.

Shapiro, Sarah, and Catherine Brown. 2018 (May 9). "Sex Education Standards across the States." *Center for American Progress.* Available at cdn.americanprogress.org.

Sheff, Elisabeth. 2014. *The Polyamorists Next Door: Inside Multiple-Partner Relationships and Families.* Lanham, MD: Rowman & Littlefield.

Simiao, L., A. Levick, A. Eichman, and J. Chang. 2015. "Women's Perspectives on the Context of Violence and Role of Police in Their Intimate Partner Arrest Experiences." *Journal of Interpersonal Violence* 30(3):400–419.

Skinner, Jessica A., and Robin M. Kowalski. 2013. "Profiles of Sibling Bullying." *Journal of Interpersonal Violence* 28(8):1726–1736.

Smith, Tom W., Michael Davern, Jeremy Freese, and Stephen L. Morgan. 2019. *General Social Surveys, 1972–2018: Cumulative Codebook.* Principal Investigator, Tom W. Smith; Coprincipal Investigators, Michael Davern, Jeremy Freese, and Stephen L. Morgan. Chicago: NORC at the University of Chicago.

Snyder, Rachel. 2019. *No Visible Bruises: What We Don't Know about Domestic Violence Can Kills Us.* New York: Bloomsbury Publishing.

Sonfield, A., K. Hasstedt, and R. B. Gold. 2014. *Moving Forward: Family Planning in the Era of Health Reform.* New York: Guttmacher Institute.

Starkweather, Katherine E., and Raymond Hames. 2012. "A Survey of Non-Classical Polyandry." *Human Nature* 23:149–172.

Stepfamily Foundation. 2020. *Stepfamily Statistics.* Available at www.stepfamily.org.

Stepler, Renee. 2017. "Led by Baby Boomers, Divorce Rates Climb for America's 50+ Population." *FactTank*, Pew Research Center. Available at www.pewresearch.org.

Straus, Murray. 2000. "Corporal Punishment and Primary Prevention of Physical Abuse." *Child Abuse and Neglect* 24:1109–1114.

Stykes, Bart. 2015. "Marital Stability Following a 1st Marital Birth (FP-15-11)." National Center for Family & Marriage Research. Available at www.bgsu.edu/ncfmr.

Swan, S. C., L. J. Gambone, J. E. Caldwell, T. P. Sullivan, and D. L. Snow. 2008. "A Review of Research on Women's Use of Violence with Male Intimate Partners." *Violence and Victims* 23:301–315.

Sweeney, M. M. 2010. "Remarriage and Stepfamilies: Strategic Sites for Family Scholarship in the 21st Century." *Journal of Marriage and the Family* 72:667–684.

Swift, Art. 2016 (June 8). "Birth Control, Divorce Top List of Morally Acceptable Issues." Gallup Poll. Available at www.gallup.com.

Swiss, Liam, and Celine Le Bourdais. 2009. "Father–Child Contact after Separation: The Influence of Living Arrangements." *Journal of Family Issues* 30(5):623–652.

Tanner, Lindsey. 2019 (September 16). "Many U.S. Women Say 1st Sexual Experience Was Forced in Teens." Associated Press. Available at www.apnews.com.

Taub, Amanda, and Jane Bradley. 2020 (June 2). "As Domestic Abuse Rises, U.K. Failings Leave Victims in Peril." *New York Times.* Available at www.nytimes.com.

Thomas, Deja. 2020 (April 10). "As Family Structures Change in U.S., a Growing Share of Americans Say It Makes No Difference." *FactTank.* Pew Research Center. Available at https://www.pewresearch.org/.

Trail, Thomas E., and Benjamin R. Karney. 2012 (June). "What's (Not) Wrong with Low-Income Marriages." *Journal of Marriage and Family*: 413–427.

Truman, Jennifer L., and Rachel E. Morgan. 2016. *Criminal Victimization, 2015.* Bureau of Justice Statistics. Available at www.bjs.gov.

Umberson, D., K. L. Anderson, K. Williams, and M. D. Chen. 2003. "Relationship Dynamics, Emotion State, and Domestic Violence: A Stress and Masculine Perspective." *Journal of Marriage and the Family* 65:233–247.

UNICEF. 2020a (June). "Attitudes and Social Norms on Violence." Available at www.data.unicef.org.

UNICEF. 2020b. "Child Marriage around the World." Available at www.unicef.org.

Unchained at Last. 2017. "Child Marriage—Shocking Statistics." Available at www.unchainedatlast.org.

United Nations. 2020. "UN Chief Calls for Domestic Violence 'Ceasefire' Amid "Horrifying Global Surge.'" Sustainable Development Goals. Available at www.un.org/.

U.S. Census Bureau. 2016. "*America's Families and Living Arrangements: 2016.*" Available at www.census.gov.

U.S. Census Bureau. 2018 (November 14). "2018 Families and Living Arrangements Tables." Available at https://www.census.gov/library/visualizations/2018/comm/married.html.

U.S. Census Bureau. 2018. "People Are Waiting to Get Married." Available at www.census.gov.

U.S. Census Bureau. 2019. "Married Couple Families with Wives' Earnings Greater Than Husband's Earnings." Table F-22. Available at www.census.gov.

U.S. Department of Health and Human Services. 2017. *Child Maltreatment 2015.* Available at www.acf.hhs.gov.

De Vaus, David, Matthew Gray, Lixia Qu, and David Stanton. 2017. "The Economic Consequences of Divorce in Sex OECD Countries." *Australian Journal of Social Issues* 52(2):180–199.

Vinopal, Lauren. 2019 (June). "How Divorce Makes Men with Joint Custody Better Fathers." *Fatherly.* Available at www.fatherly.com.

Walker, Alexis J. 2001. "Refracted Knowledge: Viewing Families through the Prism of Social Science." In *Understanding Families into the New Millennium: A Decade in Review*, ed. Robert M. Milardo, 52–65. Minneapolis, MN: National Council on Family Relations.

Wallerstein, Judith S. 2003. "Children of Divorce: A Society in Search of Policy." In *All Our Families*, 2nd ed., ed. Mary Ann Mason, Arlene Skolnick, and Stephen D. Sugarman, 66–95. New York: Oxford University Press.

Wang, Wendy, and Kim Parker. 2014. "Record Share of Americans Have Never Married: As Values, Economics and Gender Patterns Change." Pew Research Center. Available at www.pewsocialtrends.org.

Wang, Wendy, and Paul Taylor. 2011 (March 19). "For Millennials, Parenthood Trumps Marriage." Pew Research Center. Available at www.pewsocialtrends.org.

Weese, Karen. 2018 (May 1). "Almost Half of Pregnancies in the U.S. Are Unplanned. There's a Surprisingly Easy Way to Change That." *Washington Post.* Available at www.washingtonpost.com.

Wiemers, Emily E., Judith A. Seltzer, Robert F. Schoeni, V. Joseph Hotz, and Suzanne M. Bianchi. 2019. "Stepfamily Structure and Transfers between Generations in U.S. Families." *Demography* 56:229–260.

Whiffen, V. E., J. M. Thompson, and J. A. Aube. 2000. "Mediators of the Link between Childhood Sexual Abuse and Adult Depressive Symptoms." *Journal of Interpersonal Violence* 15:1100–1120.

Wight, Vanessa, Suzanne M. Bianchi, and Bijou R. Hunt. 2013. "Explaining Racial/Ethnic Variation in Partnered Women's and Men's Housework: Does One Size Fit All?" *Journal of Family Issues* 34(3):394–427.

Wildsmith, Elizabeth. Jennifer Manlove, and Elizabeth Cook. 2018. *Dramatic Increase in the Proportion of Births Outside of Marriage in the United States from 1990 to 2016.* Child Trends. Available at www.childtrends.org.

Williams, K., and A. Dunne-Bryant. 2006. "Divorce and Adult Psychological Well-Being: Clarifying the Role of Gender and Child Age." *Journal of Marriage and the Family* 68:1178–1196.

World Health Organization (WHO). 2014a. "Child Maltreatment. Fact Sheet No. 150." Available at www.who.int.

World Health Organization (WHO). 2014b. "Violence against Women. Fact Sheet No. 239." Available at www.who.int.

World Health Organization (WHO). 2019 (October 25). *High Rates of Unintended Pregnancies Linked to Gaps in Family Planning Services: New WHO Study.* Available at www.who.int.

World Health Organization (WHO). 2020. "Adolescent Pregnancy." Available at www.who.int.

Wulfhorst, Ellen. 2020 (April 1). *One in Four Women Is Not Free to Say No to Sex, U.N. Research Finds.* Reuters News Agency. Available at www.reuters.com.

YaleGlobal Online. 2017 (March 16). *Out of Wedlock Births Rise Worldwide.* www.yaleglobal.yale.edu.

Zenz, Adrian. 2020. *New Research on Xinjiang Uncovers Evidence of Birth Prevention & Mass Female Sterilization.* Available at https://twtext.com/article/1277491005659295744.

Chapter 6

Aarøe, Lene, and Michael Bang Petersen. 2014. "Crowding out Culture: Scandinavians and Americans Agree on Social Welfare in the Face of Deservingness Cues." *The Journal of Politics* 76(3):684–697.

Administration for Children and Families. 2002. *Early Head Start Benefits Children and Families.* U.S. Department of Health and Human Services. Available at www.acf.hhs.gov.

Alex-Assensoh, Yvette. 1995. "Myths about Race and the Underclass." *Urban Affairs Review* 31:3–19.

Bajak, Frank. 2010 (February 27). "Chile–Haiti Earthquake Comparison: Chile Was More Prepared." *Huffington Post.* Available at www.huffingtonpost.com.

Baker, Bruce. 2020 (June 25). "The Unequal State of Public Education in the United States." *The Century Foundation.* Available at www.tcf.org.

Bhutta, Neil, Andrew C. Chang, Lisa J. Dettling, and Joanne W. Hsu. 2020 (September 28). "Disparities in Wealth by Race and Ethnicity in the 2019 Survey of Consumer Finances," *FEDS Notes.* Washington, DC: Board of Governors of the Federal Reserve System. Available at https://doi.org/10.17016/2380-7172.2797.

Bickel, G., M. Nord, C. Price, W. Hamilton, and J. Cook. 2000. *United States Department of Agriculture Guide to Measuring Household Food Security.* Alexandria, VA: U.S. Department of Agriculture, Food and Nutrition Service.

Biden-Harris. 2020. "A Tale of Two Tax Polices: Trump Rewards Wealth, Biden Rewards Work." Available at www.joebiden.com.

Bishaw, Alemayehu, and Brian Glassman. 2016. "Poverty: 2014 and 2015." *American Community Briefs.* U.S. Census Bureau. Available at www.census.gov.

Blankenhorn, David, William Galston, Jonathan Rauch, and Barbara Dafoe Whitehead. 2015 (March/April/May). "Can Gay Wedlock Break Political Gridlock?" *Washington Monthly.* Available at www.washingtonmonthly.com.

Blumenthal, Susan. 2012 (March 12). "Debunking Myths about Food Stamps." SNAP to Health. Available at www.snaptohealth.org.

Bolen, Ed. 2015 (January 5). "Approximately 1 Million Unemployed Childless Adults Will Lose SNAP Benefits in 2016 as State Waivers Expire." Center on Budget and Policy Priorities. Available at www.cbpp.org.

Bureau of Labor Statistics (BLS). 2020 (July). "A Profile of the Working Poor, 2018." *BLS Reports*. Available at www.bls.gov.

Buszkiewicz, James H., Heather D. Hill, and Jennifer J. Otten. 2020. "State Minimum Wage Rates and Health in Working-Age Adults Using the National Health Interview Survey." *American Journal of Epidemiology*. Available at https://doi.org/10.1093/aje/kwaa018.

Callahan, David, and J. Mijin Cha. 2013. "Stacked Deck: How the Dominance of Politics by the Affluent and Business Undermines Economic Mobility." Demos. Available at www.demos.org.

Carnevale, Anthony P., Nicole Smith, and Jeff Strohl. 2013 (June). "Recovery: Job Growth and Education Requirements through 2020." Center on Education and the Workforce, Georgetown Public Policy Institute. Available at www.cew.georgetown.edu.

Center on Budget and Policy Priorities (CBPP). 2014. "Chart Book: TANF at 18." Available at www.cbpp.org.

Center on Budget and Policy Priorities. 2015 (January 16). "ChartBook: The Earned Income Tax Credit and Child Tax Credit." Available at www.cbpp.org.

Center on Budget and Policy Priorities. 2020 (August 20). "Chart Book: Temporary Assistance for Needy Families." Available at www.cbpp.org.

Chappell, Bill. 2020 (February 6). "Mississippi's Ex-Welfare Director, 5 Others, Arrested over 'Massive' Fraud." National Public Radio. Available at www.npr.org.

Child Care Aware of America. 2019. *Parents and the High Cost of Child Care: 2014 Report*. Available at www.usa.childcareaware.org.

Coleman-Jensen, Alisha, Matthew P. Rabbitt, Christian A. Gregory, and Anita Singh. 2020. *Household Food Security in the United States in 2019*. USDA Economic Research Service. Available at www.ers.usda.gov.

Congressional Research Service (CRS). 2020. *The Temporary Assistance for Needy Families (TANF) Block Grant: Responses to Frequently Asked Questions*. Report RL32760. Available at www.fas.org.

Cooper, David, and Teresa Kroger. 2017 (May 10). "Employers Steal Billions from Workers' Paychecks Each Year." Economic Policy Institute. Available at www.epi.org.

Credit Suisse Research Institute. 2016. *Global Wealth Report 2016*. Available at www.publications.credit-suisse.com.

Daly, Martin. 2017. *Killing the Competition: Economic Inequality and Homicide*. Oxford and New York: Routledge, 2017.

Davis, Kingsley, and Wilbert Moore. 1945. "Some Principles of Stratification." *American Sociological Review* 10:242–249.

Dolan, Kerry A., and Luisa Kroll. 2019 (April 8). "In Defense of Kylie Jenner: Are Any of the World's Billionaires Entirely Self-Made?" *Forbes*. Available at www.forbes.com.

Dvorak, Petula. 2009 (February 5). "Increase Seen in Attacks on Homeless." *Washington Post*, p. DZ01.

Easley, Jason. 2015 (January 21). "Bernie Sanders Files a New Constitutional Amendment to Overturn Citizens United." *Politicususa*. Available at www.politicususa.org.

Economic Policy Institute (EPI). 2018 (March 18). *Family Budget Calculator*. Available at www.epi.org.

Epstein, William M. 2004. "Cleavage in American Attitudes toward Social Welfare." *Journal of Sociology and Social Welfare* 31(4):177–201.

Feeding America. 2020 (October). "The Impact of COVID-19 on Food Insecurity."Available at www.feedingamerica.com

Food and Agriculture Organization. 2020. *The State of Food Insecurity in the World 2015*. Available at www.fao.org.

Forbes. 2020. "The World's Billionaires List: The Richest in 2020." Available at www.forbes.com.

Friedman, Zack. 2020 (February 3). "Student Loan Debt Statistics in 2020: A Record $1.6 Trillion." *Forbes*. Available at www.forbes.com.

Fusaro, Vincent A., Helen G. Levy, and H. Luke Shaefer. 2018. "Racial and Ethnic Disparities in the Lifetime Prevalence of Homelessness in the United States." *Demography* 55(6):2119–2128.

Gallup News Service. 2018 (January). "Americans' Views on Economic Mobility and Economic Inequality in the U.S. (Trends)." Available at www.news.gallup.com.

Gans, Herbert. 1972. "The Positive Functions of Poverty." *American Journal of Sociology* 78:275–289.

Giannarelli, Linda, Laura Wheaton, and Gregory Acs. 2020 (July). "2020 Poverty Projections." *Urban Institute*. Available at www.urban.org.

Gilens, Martin, and Benjamin I. Page. 2014. "Testing Theories of American Politics: Elites, Interest Group, and Average Citizens." *Perspectives on Politics* 12(3):564–581.

Gilman, Michele. 2020 (February 17). "Column: How Algorithms Intended to Root out Welfare Fraud Often Punish the Poor." *PBS News Hour*. Available at www.pbs.org.

Giovanni, Thomas, and Roopal Patel. 2013. *Gideon at 50: Three Reforms to Revive the Right to Counsel*. Brennan Center for Justice. Available at www.brennancenter.org.

Golden, Olivia. 2013 (May 9). "Poverty in America: How We Can Help Families." Urban Institute. Available at www.urban.org.

Gould, Elise. 2020 (February 20). "State of Working America Wages 2019." Economic Policy Institute. Available at www.epi.org.

Graham, Carol. 2015 (February 19). "The High Costs of Being Poor in America: Stress, Pain, and Worry." *Social Mobility Memos*. Brookings Institution. Available at www.brookings.edu.

Gravelle, Jane G., and Donald J. Marples. 2019 (May 22). "The Economic Effects of the 2017 Tax Revision: Preliminary Observations." *Congressional Research Service*. Report R45736. Available at www.crs.gov.

Green, Autumn R. 2013. "Patchwork: Poor Women's Stories of Resewing the Shredded Safety Net." *Affilia* 28:51–64.

Grunwald, Michael. 2006 (August 27). "The Housing Crisis Goes Suburban." *Washington Post*. Available at www.washingtonpost.com.

Hardoon, Deborah. 2017. *An Economy for the 99%*. Nairobi: Oxfam International. Available at www.Oxfam.org.

Harell, Allison, Stuart Soroka, and Adam Mahon. 2008 (September). "Is Welfare a Dirty Word? Canadian Public Opinion on Social Assistance Policies." *Options Politiques*: 53–56.

Hickel, Jason, and Giorgos Kallis. "2020 Is Green Growth Possible?" *New Political Economy* 25(4):469–486.

Higher Ed Not Debt. 2018 (September 17). "In Their Own Words: Student Debt Stories." *Center for American Progress*. Available at www.higherednotdebt.org.

Human Rights Watch. 2020. *World Report 2015*. Available at www.hrw.org.

Huizar, Laura. 2019 (April 9). "Testimony on Wage Theft before U.S. House Subcommittee on Labor, Health, and Human Services." *National Employment Law Project*. Available at www.nelp.org.

Institute for Economics & Peace. 2020. *Pillars of Peace*. Available at economicsandpeace.org.

Iqbal, Imrana, and Charles Pierson. 2017. "A North–South Struggle: Political and Economic Obstacles to Sustainable Development." *Sustainable Development Law & Policy* 16(2):16–47.

Irvine, Leslie. 2013. "Animals as Lifechangers and Lifesavers: Pets in the Redemption Narratives of Homeless People." *Journal of Contemporary Ethnography* 42(1):3–36.

Irving, Shelley K., and Tracy A. Loveless. 2015 (May). "Dynamics of Economic Well-Being and Participation in Government Programs 2009–2012: Who Gets Assistance?" *U.S. Census Bureau* Report P70-14. Available at www.census.gov.

Jaffe, Greg. 2020 (June 6). "The Pandemic Hit and This Car Became Home to a Family of Four. Now They're Fighting to Get Out." *Washington Post*. Available at www.thewashingtonpost.com.

Katz, Michael. 2001. *The Price of Citizenship*. New York: Henry Holt and Company.

Katz, Michael B. 2013. *The Undeserving Poor: America's Enduring Confrontation with Poverty: Fully Updated and Revised*. New York: Oxford University Press.

Kertesz, Stephan G., L. Erika, Sally K. Holmes, Aerin J. DeRussy, Carol Van Deusen Lukas, and David E. Pollio. 2017. "Housing First on a Large Scale: Fidelity Strengths and Challenges in the VA's HUD-VASH Program." *Psychological Services* 14(2):118–128. Available at https://doi.org/10.1037/ser0000123.

Kessler, Glenn. 2016 (March 3). "Trump's False Claim He Built His Empire with a 'Small Loan' from His Father." *Washington Post*. Available at www.washingtonpost.com.

Kraut, Karen, Scott Klinger, and Chuck Collins. 2000. *Choosing the High Road: Businesses That Pay a Living Wage and Prosper*. Boston: United for a Fair Economy.

Ku, Leighton, and Brian Bruen. 2013 (February 19). "The Use of Public Assistance Benefits by Citizens and Non-Citizen Immigrants in the United States." CATO Working Paper. Washington, DC: Cato Institute.

Lanier, Paul, Katie Maguire-Jack, Tova Walsh, Brett Drake, and Grace Hubel. 2014 (September). "Race and Ethnic Differences in Early Childhood Maltreatment in the United States." *Journal of Developmental & Behavioral Pediatrics 35*(7):419–426.

Larcker, David F., Nicholas E. Donatiello, and Brian Tayan. 2016 (February). "Americans and CEO Pay: 2016 Public Perception Survey on CEO Compensation." Corporate Governance Research Initiative. Stanford Rock Center for Corporate Governance. Available at www.gsb.stanford.edu.

Leigh, J. Paul. 2013 (March 6). "Raising the Minimum Wage Could Improve Public Health." Economic Policy Institute Blog. Available at www.epi.org.

Lipton, Eric. 2020 (September 6). "How Trump Draws on Campaign Funds to Pay Legal Bills." *New York Times*. Available at www.nytimes.com.

Low Income Housing Authority. 2019. "In More Than 70 Cities, It's Illegal to Feed the Homeless." Available at www.lowincome.org.

Luker, Kristin. 1996. *Dubious Conceptions: The Politics of Teenage Pregnancy*. Cambridge, MA: Harvard University Press.

Lustgarten, Abrahm. 2020 (September 15). "How Climate Migration Will Reshape America." *New York Times Magazine*. Available at www.nytimes.com.

Ly, Angela, and Eric Latimer. 2015. "Housing First Impact on Costs and Associated Cost Offsets: A Review of the Literature." *Canadian Journal of Psychiatry* 60(11):475–487.

Maher, Will. 2018. "Poverty Fact Sheet: Suburban Poverty." Institute for Research on Poverty, No. 14. Available at www.irp.wisc.edu.

Massey, D. S. 1991. "American Apartheid: Segregation and the Making of the American Underclass." *American Journal of Sociology* 96:329–357.

McDonagh, Thomas. 2013. *Unfair, Unsustainable, and under the Radar: How Corporations Use Global Investment Rules to Undermine a Sustainable Future*. Democracy Center. Available at www.democracyctr.org.

McEwen, Craig A., and Bruce S. McEwen. 2017. "Social Structure, Adversity, Toxic Stress, and Intergenerational Poverty: An Early Childhood Model." *Annual Review of Sociology* 43:445–472.

McNamee, Stephen J., and Robert K. Miller Jr. 2009. *The Meritocracy Myth*, 2nd ed. Lanham, MD: Rowman & Littlefield.

Mishel, Lawrence, and Julie Wolfe. 2019 (August 14). "CEO Compensation Has Grown 940 Percent since 1978." Economic Policy Institute. Available at www.epi.org.

Mishel, Lawrence, and Jori Kandra. 2020 (August 18). "CEO Compensation Surged 14 Percent in 2019 to $21.3 Million." Economic Policy Institute. Available at www.epi.org.

National Alliance to End Homelessness. 2016 (April 20). "Housing First." Available at www.endhomelessness.org.

National Alliance to End Homelessness. 2020. *The State of Homelessness in America 2016*. Washington, DC: National Alliance to End Homelessness.

National Center for Education Statistics (NCES). 2020. "Tuition Costs of Colleges and Universities." Fast Facts. Available at www.nces.ed.gov.

National Center for Healthy Housing (NCHH). 2020. "Substandard Housing." Available at www.nchh.org.

National Coalition for the Homeless. 2016. "Vulnerable to Hate: A Survey of Hate Crimes and Violence Committed against Homeless People in 2013." Available at www.nationalhomeless.org.

National Diaper Bank Network. n.d. "What Is Diaper Need?" Available at www.nationaldia-perbanknetwork.org.

National Low Income Housing Coalition. 2020. "The Problem." Available at www.nlihc.org.

National Research Council and Institute of Medicine. 2013. *U.S. Health in International Perspective: Shorter Lives, Poorer Health*. Washington, DC: National Academies Press.

"Nipped in the Bud." 2015 (June 6–12). *The Economist*, pp. 24–25.

Olszewski-Kubilius, Paula, and Susan Corwith. 2018. "Poverty, Academic Achievement, and Giftedness: A Literature Review." *Gifted Child Quarterly* 62(1):37–55.

Organisation of Economic Co-operation and Development (OECD). 2020. "Poverty Rate (indicator)." doi: 10.1787/0fe1315d-en.

Oxfam. 2017 (January). *An Economy for the 99%*. Oxfam Briefing Paper. Available at www.oxfamamerica.org.

Oxfam. 2020 (January 20). *Time to Care*. Available at www.oxfam.org.

Pew Research Center. 2020 (August 13). *Important Issues in the 2020 Election*. Available at www.pewresearch.org.

Piketty, Thomas. 2014. *Capital in the Twenty-First Century*. Cambridge, MA: Harvard University Press.

Poor People's Campaign. 2020 (June 9). *Costs of Poverty Fact Sheet*. Available at www.poorpeoplescampaign.org.

Rank, Mark R., and Thomas A. Hirschl. 2020. "The Likelihood of Experiencing Relative Poverty over the Life Course." *PLoS One* 10(7):e0133513.

Reeves, Richard R., and Joanna Venator. 2015 (February). "Sex, Contraception, or Abortion? Explaining Class Gaps in Unintended Childbearing." Brookings Center on Children and Family. Available at www.brookings.edu.

Roseland, Mark, and Lena Soots. 2007. "Strengthening Local Economies." In *2007 State of the World*, ed. Linda Starke, 152–169. New York: Norton.

Saeed, Gohar, and Farzand Ali Jan. 2016. "Microcredit and Its Significance in Sustainable Development and Poverty Alleviation: Evidence from Asia, Africa, Latin America, and Europe." *Dialogue*, 11(3):334–345.

Schneebaum, Alyssa, and M. V. Lee Badgett. 2019. "Poverty in US Lesbian and Gay Couple Households." *Feminist Economics* 25(1):1–30.

Schnurer, Eric. 2013 (August 15). "Just How Wrong Is the Conventional Wisdom about Government Fraud?" *Atlantic*. Available at www.theatlantic.com.

Semega, Jessica, Melissa Kollar, Emily A. Shrider, and John F. Creamer. 2020 (September). "Income and Poverty in the United States: 2019." Current Population Reports P60-270. Washington, DC: U.S. Government Printing Office.

Sommeiller, Estelle, and Mark Price. 2015 (January 26). "The Increasingly Unequal States of America." Economic Policy Institute. Available at www.epi.org.

Talberth, John, Daphne Wysham, and Karen Dolan. 2013. "Closing the Inequality Divide: A Strategy for Fostering Genuine Progress in Maryland." Center for Sustainable Economy and Institute for Policy Studies. Available at www.sustainable-economy.org.

Turner, Margery Austin, Susan J. Popkin, G. Thomas Kingsley, and Deborah Kaye. 2005 (April). *Distressed Public Housing: What It Costs to Do Nothing*. Urban Institute. Available at www.urban.org.

UN-Habitat. 2018. *State of the World's Cities 2010/2011*. Available at www.unhabitat.org.

United Nations. 2020. *Inequality Matters: Report on the World Social Situation 2013*. New York: United Nations.

United Nations Development Programme (UNDP). 2010. *Human Development Report 2010*. Available at hdr.undp.org.

United Nations Development Programme (UNDP). 2020. *Charting Pathways out of Multidimensional Poverty*. Available at www.hdr.undp.org.

Urban Institute. 2017 (February). "Nine Charts about Wealth Inequality in America." Available at www.datatools.urban.org.

U.S. Census Bureau. 2016. "American Community Survey." Available at https://www.census.gov/acs/www/data/data-tables-and-tools/data-profiles/2016/.

U.S. Census Bureau. 2019. "*2013 American Community Survey*." Available at www.census.gov.

U.S. Census Bureau. 2020 (January 21). "Funding for Nutrition Benefits Program Informed by Census Statistics." *America Counts*. Available at www.census.gov.

U.S. Census Bureau. 2021. "Poverty Thresholds for 2020." Available at www.census.gov.

USDA Food and Nutrition Service. 2020. "*Supplemental Nutrition Assistance Program Participation and Costs*." Available at www.fns.usda.gov.

Weeks, Daniel. 2014 (January 10). "Why Are the Poor and Minorities Less Likely to Vote?" *The Atlantic*. Available at www.theatlantic.com.

WHO/UNICEF Joint Monitoring Programme for Water Supply and Sanitation. 2019. "Data Tables." Available at www.wssinfo.org.

Wilkinson, Richard, and Kate Pickett. 2009. *The Spirit Level: Why More Equal Societies Almost Always Do Better*. London: Penguin Books.

Wilkinson, Richard, and Kate Pickett. 2020 *The Inner Level: How More Equal Societies Reduce Stress, Restore Sanity and Improve Everyone's Well-Being*. London: Penguin Books.

Wilson, William J. 1987. *The Truly Disadvantaged: The Inner City, the Underclass, and Public Policy.* Chicago: University of Chicago Press.

Wilson, William J. 1996. *When Work Disappears: The World of the New Urban Poor.* New York: Knopf.

World Bank. 2005. *Global Monitoring Report 2005.* Available at www.worldbank.org.

World Bank. 2016. *Poverty and Shared Prosperity 2016: Taking on Inequality.* Washington, DC: World Bank.

World Bank. 2020a. "World Development Indicators." Available at wdi.worldbank.org.

World Bank. 2020b. *Global Economic Prospects, June 2020.* Washington, DC: World Bank. doi: 10.1596/978-1-4648-1553-9.

World Bank Group, International Monetary Fund. 2015. *Global Monitoring Report 2014/2015: Ending Poverty and Sharing Prosperity.* Washington, DC: World Bank Group.

World Economic Forum. 2020 (January). *The Global Social Mobility Report 2020.* Available at www.weforum.org.

World Health Organization (WHO). 2020 (September 8). "Children: Improving Survival and Well-Being." Available at www.who.int.

Wright, Erik Olin, and Joel Rogers. 2015. *American Society: How It Really Works.* New York: Norton.

York, Erica. 2020. "Summary of the Latest Federal Income Tax Data, 2020 Update." Tax Foundation. Available at www.taxfoundation .org.

Zedlewski, Sheila R. 2003. *Work and Barriers to Work among Welfare Recipients in 2002.* Washington, DC: Urban Institute. Available at www.urban.org.

Zucman, Gabriel. 2019 (January). "Global Wealth Inequality." NBER Working Paper No. 25462. National Bureau of Economic Research. Available at www.nber.org.

Chapter 7

AFL-CIO. 2020. *Death on the Job: The Toll of Neglect.* Available at www.aflcio.org.

American Society of Civil Engineers (ASCE). 2020. "Infrastructure Report Card." Available at www.infrastrcutureportcard.org.

Arnold, Taylor J., Thomas A. Arcury, Joanne C. Sandberg, Sara A. Quandt, Jennifer W. Talton, Dana C. Mora, Gregory D. Kearney, Haiying Chen, Melinda F. Wiggins, and Stephanie S. Daniel. 2020 (April). "Heat-Related Illness among Latinx Child Farmworkers in North Carolina: A Mixed-Methods Study." *New Solutions: A Journal of Environmental and Occupational Health Policy.* doi: 1048291120920571.

Austin, Colin. 2002. "The Struggle for Health in Times of Plenty." In *The Human Cost of Food: Farmworkers' Lives, Labor, and Advocacy,* ed. C. D. Thompson Jr. and M. F. Wiggins, 198–217. Austin: University of Texas Press.

Barsamian, David. 2012. "Capitalism and Its Discontents: Richard Wolff on What Went Wrong." *The Sun* 434:4–13.

Bauer-Wolf, Jeremy. 2019 (January 17). "Survey: Employers Want 'Soft Skills' from Graduates." *Inside Higher Ed.* Available at www .insidehighered.com.

Biden-Harris. 2020a. "Climate: 10 Million Clean Energy Jobs." Joe Biden for President. Available at www.joebiden.com.

Biden-Harris. 2020b. "The Biden Plan for Strengthening Worker Organizing, Collective Bargaining, and Unions." Joe Biden for President. Available at www.joebiden.com.

Brand, Jennie E. 2015. "The Far-Reaching Impact of Job Loss and Unemployment." *Annual Review of Sociology* 41:359–375.

Brenan, Megan. 2020 (September 3). "At 65%, Approval of Labor Unions in U.S. Remains High." *Gallup News.* Available at www.news .gallup.com.

Brookings Institute. 2020 (November 2). "Tracking Deregulation in the Trump Era." Available at www.brookings.edu.

Butterworth, P., L. S. Leach, L. Strazdins, S. C. Olesen, B. Rodgers, and D. H. Broom. 2011. "The Psychosocial Quality of Work Determines Whether Employment Has Benefits for Mental Health: Results from a Longitudinal National Household Survey." *Occupational and Environmental Medicine.* Advance online publication. doi:10.1136/oem.2010.059030.

Carnevale, Anthony P., Nicole Smith, and Jeff Strohl. 2014. *Recovery: Job Growth and Education Requirements through 2020.* Washington, DC: Georgetown Public Policy Institute. Available at www.cew.georgetown .edu.

Carnevale, Anthony P., and Nicole Smith. 2018. "Balancing Work and Learning: Implications for Low-Income Students." Washington, DC: Center on Education and the Workforce at Georgetown University. Available at www.cew .georgetown.edu.

Cockburn, Andrew. 2003 (September). "21st Century Slaves." *National Geographic,* pp. 2–11, 18–24.

Corley, Danielle, Sunny Frothingham, and Kate Bahn. 2017 (January 5). "Paid Sick Days and Paid Family and Medical Leave Are Not Job Killers." Center for American Progress. Available at www.americanprogress.org.

Cox, Daniel A., and Samuel J. Abrams. 2020 (July). "The Parents Are Not All Right." Washington, DC: American Enterprise Institute. Available at www.aei.org.

Davis, Kingsley, and Wilbert Moore. 1945. "Some Principles of Stratification." *American Sociological Review* 10:242–249.

Dill, Janette, Rebecca J. Erickson, and James M. Diefendorff. 2016. "Motivation in Caring Labor: Implications for the Well-Being and Employment Outcomes of Nurses." *Social Science & Medicine* 167:99–106.

Donovan, Sarah A. 2019 (May 29). "Paid Family Leave in the United States." *Congressional Research Service.* R44835. Available at www .fas.org.

Dreyfuss, Joel. 2019 (November 9). "How 3-D Printing Is Transforming the $12 Trillion Manufacturing Industry and Fueling the 4th Industrial Revolution." CNBC. Available at www.cnbc.com.

Ebeling, Richard M. 2009 (February 12). "Capitalism the Solution, Not Cause of the Current Economic Crisis." Great Barrington, MA: American Institute for Economic Research. Available at www.aier.org.

Economist, The. 2019 (July 18). "America Is the Only Rich Country without a Law on Paid Leave for New Parents." *The Economist.* Available at www.economist.com.

England, Paula. 2005. "Emerging Theories of Care Work." *Annual Review of Sociology* (31):381–399.

Fair to Wear Foundation (FWF). 2018. *The Face of Child Labour: Stories from Asia's Garment Sector.* Available at www.api.fairwear.org.

Federal Bureau of Prisons. 2020. "Work Programs." Available at www.bop.gov.

Federal Reserve. 2020 (May 21). "Report on the Economic Well-Being of Households in 2019." *Board of Governors of the Federal Reserve System.* Available at www.federalreserve.gov.

Flavin, Patrick, and Gregory Shufeldt. 2014 (October 27). "Labor Union Membership and Life Satisfaction in the United States." Working Paper, Baylor University. Available at www .blogs.baylor.edu.

Fortson, Kenneth, Dana Rotz, Paul Burkander, Annalisa Mastri, Petre Schochet, Linda Rosenberg, Sheena McConnell, and Ronald D'Amico. 2017 (May). "Providing Public Workforce Services to Job Seekers: 30-Month Impact Findings on the WIA Adult and Dislocated Worker Programs." Washington, DC: Mathematica Policy Research.

Frederick, James, and Nancy Lessin. 2000. "Blame the Worker: The Rise of Behavior-Based Safety Programs." *Multinational Monitor* 21(11). Available at www.essential.org/monitor.

Fry, Richard, Jeffry S. Passel, and D'Vera Cohn. 2020 (September 4). "A Majority of Young Adults in the U.S. Live with Their Parents for the First Time since the Great Depression." Washington, DC: Pew Research Center. Available at www.pewresearch.org.

Gallup News. 2020. "Work and Workplace." Available at www.news.gallup.com.

Galvin, Daniel J., and Jacob S. Hacker. 2020. "The Political Effects of Policy Drift: Policy Stalemate and American Political Development." *Studies in American Political Development,* 32(2):216–238.

Gardner, Matthew, Lorena Roque, and Steve Wamhoff. 2019 (December 16). "Corporate Tax Avoidance in the First Year of the Trump Tax Law." Washington, DC: Institute on Taxation and Economic Policy. Available at www.itep .org.

Global Slavery Index. 2018. Available at www .globalslaveryindex.org.

Goh, Joel, Jeffrey Pfeffer, and Stefanos A. Zenios. 2019. "Reducing the Health Toll from US Workplace Stress." *Behavioral Science & Policy* 5(1):iv–13.

Gould, Elise, Zane Mokhiber, and Julia Wolfe. 2019 (May 14). "Class of 2019: College Edition." Washington, DC: Economic Policy Institute. Available at www.epi.org.

Hebson, Gail, Jill Rubery, and Damian Grimshaw. 2015. "Rethinking Job Satisfaction in Care Work: Looking beyond the Care Debates." *Work, Employment and Society* 29(2):314–330.

Horowitz, Juliana Menasce. 2019 (September 12). "Despite Challenges at Home and Work, Most Working Moms and Dads Say Being Employed Is What's Best for Them." Washington, DC: Pew Research Center. Available at www.pewresearch.org.

Human Rights Watch. 2019 (December 18). "Fashion's Next Trend: Accelerating Supply Chain Transparency in the Apparel and Footwear Industry." New York: Human Rights Watch. Available at www.hrw.org.

Ignatius, David. 2020 (December 10). "Biden Is Picking a Cabinet Built for Comfort. What He Needs Is Vision." *Washington Post*. Available at www.washingtonpost.com.

Institute for Research on Poverty. 2019 (July). "Helping the Hard-to-Employ Transition to Employment." *Institute for Research on Poverty*. Policy brief no. 41-2019. Available at www.irp.wisc.edu.

Inter-Agency Coordination Group against Trafficking in Persons (ICAT). 2020. "Non-Punishment of Victims of Trafficking." United Nations Office on Drugs and Crime, Issue Brief no. 8. Available at www.unodc.org.

International Labour Organization (ILO). 2017. *Global Estimates of Child Labour: Results and Trends, 2012–2016*. Geneva: International Labour Organization.

International Labour Office (ILO). 2020a. *World Employment Social Outlook: Trends 2020*. Geneva: International Labour Organization. Available at www.ilo.org.

International Labour Office (ILO). 2020b (May 27). *ILO Monitor: COVID-19 and the World of Work*, 4th ed. Geneva: International Labour Organization. Available at www.ilo.org.

International Labour Office (ILO). 2020c (September 23). *ILO Monitor: COVID-19 and the World of Work*, 6th ed. Geneva: International Labour Organization. Available at www.ilo.org.

International Trade Union Confederation. 2020. *IUTC Global Rights Index*. Available at www.ituc-csi.org.

Jensen, Derrick. 2002 (June). "The Disenchanted Kingdom: George Ritzer on the Disappearance of Authentic Culture." *The Sun*, pp. 38–53.

Jones, Janelle, and Heidi Shierholz. 2018 (July 10). "Right-to-Work Is Wrong for Missouri." Washington, DC: Economic Policy Institute. Available at www.epi.org.

Khan, Kiran Shafiq, Mohammed A. Mamun, Mark D. Griffiths, and Irfan Ullah. 2020 (July). "The Mental Health Impact of the COVID-19 Pandemic across Different Cohorts." *International Journal of Mental Health and Addiction*: 1–7.

Kitroeff, Natalie. 2019 (December 16). "Fashion Nova's Secret: Underpaid Workers in Los Angeles Factories." *New York Times*. Available at www.nytimes.com.

Korten, David. 2015 (March 9). "Do Corporations Really Need More Rights? Why Fast Track for the TPP Is a Bad Idea." *Yes! Magazine*. Available at www.yesmagazine.org.

Lenski, Gerard, and J. Lenski. 1987. *Human Societies: An Introduction to Macrosociology*, 5th ed. New York: McGraw-Hill.

Lopez, Alexa. 2020 (April 28). "Poll: Voters Overwhelmingly Support More Investment in Water Infrastructure." *Infrastructure Report Card*. Available at www.infrastructurereportcard.org.

MacEnulty, Pat. 2005 (September). "An Offer They Can't Refuse: John Perkins on His Former Life as an Economic Hit Man." *The Sun* 357:4–13.

Matos, Kenneth, Ellen Galinsky, and James T. Bond. 2017. *National Study of Employers*. Alexandria, VA: Society for Human Resource Management. Available at www.shrm.org.

Maye, Adewale. 2019 (May). "No-Vacation Nation, Revised." Washington, DC: Center for Economic Policy and Research. Available at www.cepr.net.

McNicholas, Celine, Margaret Poydock, Julia Wolfe, Ben Zipperer, Gordon Lafer, and Lola Loustanunau. 2019 (December 11). "Unlawful." Washington, DC: Economic Policy Institute. Available at www.epi.org.

Miller, Claire Cain. 2015 (January 30). "The Economic Benefits of Paid Parental Leave." *New York Times*. Available at www.nytimes.com.

National Partnership for Women and Families. 2020a (October). "The Family and Medical Insurance Leave (FAMILY) Act." Fact Sheet. Available at www.nationalpartnership.org.

National Partnership for Women and Families. 2020b (October). "The Healthy Families Act." Fact Sheet. Available at www.nationalpartnership.org.

National Public Radio (NPR). 2020 (June 29). "The Undocumented Workforce." Available at www.npr.org.

National Right to Work Legal Defense Foundation. 2020. "Right to Work Frequently-Asked Questions." Available at www.nrtw.org.

Newport, Frank. 2020 (March 6). "Americans' Views of Socialism, Capitalism Are Little Changed." Gallup Organization. Available at www.gallup.com.

Nonkes, Mark. 2015 (June 10). "A Look at Child Labor inside a Garment Factory in Bangladesh." *World Vision*. Available at www.worldvision.org.

Organisation for Economic Co-operation and Development (OECD). 2019. *Negotiating Our Way Up: Collective Bargaining in a Changing World of Work*. Paris: OECD Publishing. Available at https://doi.org/10.1787/1fd2da34-en.

Office of the United States Trade Representative. 2020. *United States–Mexico–Canada Trade Fact Sheet Modernizing NAFTA into a 21st Century Trade Agreement*. Available at www.ustr.gov.

Ornstein, Daniel, and Jordan B. Glassberg. 2019 (January 29). "More Countries Consider Implementing a 'Right to Disconnect.'" *National Law Review*. Available at www.natlawreview.org.

Parenti, Michael. 2007 (February 16). "Mystery: How Wealth Creates Poverty in the World." Common Dreams NewsCenter. Available at www.commondreams.org.

Paul, Mark, William Darity Jr., and Darrick Hamilton. 2018 (March 9). "The Federal Job Guarantee—A Policy to Achieve Full Employment." Washington, DC: Center on Budget and Policy Priorities. Available at www.cbpp.org.

Perkins, John. 2004. *Confessions of an Economic Hit Man*. San Francisco: Berrett-Koehler.

Pew Research Center. 2019 (October 7). "In Their Own Words: Behind Americans' Views of 'Socialism' and 'Capitalism.'" Available at www.pewresearch.org.

Psacharopoulos, George, Harry Patrinos, Victoria Collis, and Emiliana Vegas. 2020 (April 29). "The COVID-19 Cost of School Closures." Washington, DC: Brookings Institute. Available at www.brookings.edu.

Rampell, Ed. 2013 (April 16). "An Interview with Richard Wolff." Counterpunch. Available at www.counterpunch.org.

Ritzer, George. 1995. *The McDonaldization of Society: An Investigation into the Changing Character of Contemporary Social Life*. Thousand Oaks, CA: Pine Forge Press.

Rözer, Jesper J., Bas Hofstra, Matthew E. Brashears, and Beate Volker. 2020 (October). "Does Unemployment Lead to Isolation? The Consequences of Unemployment for Social Networks." *Social Networks* 63:100–111.

Rose, Joel, and Marisa Peñaloza. 2020 (July 9). "'We Were Treated Worse Than Animals': Disaster Recovery Workers Confront COVID-19." National Public Radio. Available at www.npr.org.

Saad, Lydia. 2019 (November 25). "Socialism as Popular as Capitalism among Young Adults in U.S." *Gallup*. Available at www.news.gallup.com.

Santhanam, Laura. 2020 (July 23). "'This Is Not Working.' Parents Juggling Jobs and Child Care under COVID-19 See No Good Solutions." *PBS NewsHour*. Available at www.pbs.org.

Schwandt, Hannes, and von Wachter. 2019. "Unlucky Cohorts: Estimating the Long-Term Effects of Entering the Labor Market in a Recession in Large Cross-Sectional Data Sets." *Journal of Labor Economics* 37(S1):161–198.

Scott, Robert E., and David Ratner. 2005 (July 20). *NAFTA's Cautionary Tale*. Economic Policy Institute Briefing Paper 214. Available at www.epi.org.

Shahidi, Faraz Vahid, Carles Muntaner, Ketan Shankardass, Carlos Quiñonez, and Arjumand Siddiqi. 2019. "The Effect of Unemployment Benefits on Health: A Propensity Score Analysis." *Social Science & Medicine* 226:198–206.

Sherk, James. 2011 (November 9). "Right to Work Increases Jobs and Choices." *Heritage Foundation*. Available at www.heritage.org.

Shipler, David K. 2005. *The Working Poor*. New York: Vintage Books.

Skinner, E. Benjamin. 2008. *A Crime So Monstrous: Face-to-Face with Modern-Day Slavery*. New York: Free Press.

Society for Human Resource Management (SHRM). 2019. "Family-Friendly and Wellness." *SHRM Employee Benefits 2019*. Available at www.shrm.org.

Society for Human Resource Management (SHRM). 2020. "How the Pandemic Is Challenging and Changing Employers." *SHRM COVID-19 Research*. Available at www.shrm.org.

Strully, Kate W. 2009. "Job Loss and Health in the U.S. Labor Market." *Demography* 46(2):221–247.

SweatFree Communities. n.d. "Adopted Policies." Available at www.sweatfree.org.

Taub, Amanda. 2020 (September 26). "Pandemic Will 'Take Our Women 10 Years Back' in the Workplace." *New York Times*. Available at www.nytimes.com.

Thompson, Alex, and Theodoric Meyer. 2021 (January 1). "Janet Yellen Made Millions in Wall Street, Corporate Speeches." *Politico*. Available at www.politco.com.

Uchitelle, Louis. 2006. *The Disposable American: Layoffs and Their Consequences*. New York: Knopf.

United Students Against Sweatshops (USAS). 2020. "USAS November 5, 2020 Statement." Available at www.usas.org.

U.S. Bureau of Labor Statistics (BLS). 2019. "Union Members Summary." Available at www.bls.gov.

U.S. Bureau of Labor Statistics (BLS). 2020a (May 8). "Civilian Unemployment Rate." *The Employment Situation*. Available at www.bls.gov.

U.S. Bureau of Labor Statistics (BLS). 2020b. "Injuries, Illnesses, and Fatalities." *Census of Fatal Occupational Injuries, 2011–2018*. Available at www.data.bls.gov.

U.S. Department of Labor. 2018. "Findings from the National Agricultural Workers Survey (NAWS) 2015–2016." Research Report No. 13. Available at www.dol.gov.

VanHeuvelen, Tom. 2020. "The Right to Work, Power Resources, and Economic Inequality." *American Journal of Sociology* 125(5):1255–1302.

Vesoulis, Abby. 2020 (October 17). "'If We Had a Panic Button, We'd be Hitting It.' Women Are Exiting the Labor Force en Masse—And That's Bad for Everyone." *Time*. Available at www.time.com.

Washington Post. 2020. "Where Democrats Stand: Foreign Policy." n.d. Available at www.washingtonpost.com.

Wolff, Richard D. 2013a. "Alternatives to Capitalism." *Critical Sociology* 39(4):487–490.

Wolff, Richard D. 2013b (June 13). "Capitalism, Democracy, and Elections." *Democracy at Work* (blog). Available at www.democracyatwork.info.

World Economic Forum. 2020. *Global Risks Report 2020*, 15th ed. Geneva, Switzerland: World Economic Forum. Available at www.weforum.org.

World Policy Center. 2020. "Is Paid Leave Available to Mothers and Fathers of Infants?" Available at www.worldpolicycenter.org.

Worth Rises. 2020. *The Prison Industry: Mapping Private Sector Players*. Available at www.worthrises.org.

Wright, Erik Olin, and Joel Rogers. 2015. *American Society: How It Really Works*. New York: Norton.

Chapter 8

Abel, Jaison R., and Richard Deitz. 2019 (June 5). "Despite Rising Costs, College Is Still a Good Investment." Liberty Street Economics. Available at www.libertystreeteconomics.newyorkfed.org.

ADDitude. 2020. "Misunderstood." Available at www.additudemag.com.

Agnafors, Sara, Mimmi Barmark, and Gunilla Sydsjo. 2020 (August 19). "Mental Health and Academic Performance: A Study on Selection and Causation Effects from Childhood to Early Adulthood." *Social Psychiatry and Psychiatric Epidemiology*. Available at https://doi.org/10.1007/s00127-020-01934-5.

Allegretto, Sylvia, and Lawrence Mishel. 2020 (September 17). "Teacher Pay Penalty Dips but Persists in 2019." Economic Policy Institute. Available at www.epi.org.

Alexander, Carl, Doris Entwisle, and Linda Olson. 2014. *The Long Shadow: Family Background, Disadvantaged Urban Youth, and the Transition to Adulthood*. New York: Russell Sage Foundation.

American Statistical Association (ASA). 2014 (April 8). "ASA Statement on Using Value-Added Models for Educational Assessment." Available at www.amstat.org.

Amurao, Carla. 2013 "Fact Sheet: How Bad Is the School-to-Prison Pipeline?" PBS. Available at www.pbs.org.

ASCE (American Society of Civil Engineers). 2017. *Infrastructure Report Card*. Available at www.Infrastructurereportcard.org.

Atwell, Matthew W., Robert Balfanz, Eleanor Manspile, Vaughan Byrnes, and John M. Bridgeland. 2020. "Building a Grad Nation: Progress and Challenges in Raising High School Graduation Rates." America's Promise Alliance. Available at www.americaspromise.org.

Autor, David H., David N. Figlio, Krzysztof Karbownik, Jeffrey Roth, and Melanie Wasserman. 2016. *School Quality and the Gender Gap in Educational Achievement*. Working Paper 21908. National Bureau of Economic Research. Available at www.nber.org.

Ball, Annahita, and Candra Skrzypek. 2020 (July). "School Social Work and Educational Justice Movement: Snapshot of Practice." *Children and Schools* 42(3):179–186. Available at https://academic.oup.com/cs/article/42/3/179/5896146.

Ballantine, Jeanne H., Floyd M. Hammack, and Jenny Stuber. 2017. *The Sociology of Education: A Systematic Analysis*, 7th ed. New York: Routledge.

Barr, A., and C. Gibbs. 2017. "Breaking the Cycle? The Intergenerational Effects of Head Start." Research Paper Institute for Research on Poverty: University of Michigan.

Barrington, Kate. 2019 (March 20). "New Study Confirms That Private Schools Are No Better Than Public Schools." *Public School Review*. Available at https://www.publicschoolreview.com.

Bassok, Daphna, Jenna E. Finch, RaeHynck Lee, Sean F. Reardon, and Jane Waldfogel. 2016. "Socioeconomic Gaps in Early Childhood Experiences: 1998–2010." *AERA Open* 2(3):1–22.

Bateman, Nicole. 2020. *Working Parents Are Key to Covid-19 Recovery*. Brookings Institute. Available at www.brookings.edu.

Beard, Aaron. 2014 (October 23). "Fake Classes, Inflated Grade: Massive UNC Scandal Included Athletes over 2 Decades." *Star Tribune*. Available at www.startribune.com.

Berman, Jesse D., Meredith C. McCormack, Kirstin A. Koehler, Faith Connolly, Dorothy Clemons-Erby, Meghan F. Davis, Christine Gummerson, Philip J. Leaf, Theresa D. Jones, and Frank C. Curriero. 2018 (June). "School Environmental Conditions and Links to Academic Performance and Absenteeism in Urban, Mid-Atlantic Public Schools." *International Journal of Hygiene and Environmental Health* 221(5):800–808.

Berndtson, Dave. 2017 (February 8). "San Francisco Becomes First City to Offer Free Community College Tuition to All Residents." *The Rundown*. Available at pbs.org.

Biden, Joe. 2020. "The Biden Plan for Educators, Students, and Our Future." Available at www.joebiden.com/education.

Body, Dyvonne. 2019 (March 19). "Worse Off Than When They Enrolled: The Consequence of For-Profit Colleges for People of Color." *Aspen Institute*d. Available at www.aspeninstitute.org.

Bolick, Kristina N., and Beth A. Rogowsky. 2016 (June). "Ability Grouping Is on the Rise, but Should It Be?" *Journal of Education and Human Development* 5(2):40–51.

Broughman, Stephen P., and Nancy L. Swaim. 2016 (November). *Characteristics of Private Schools in the United States: Results from the 2013–14 Private School Universe Survey*. Available at nces.ed.gov.

Brown, Catherine, Ulrich Boser, and Perpetual Baffour. 2016. "Workin' 9 to 5." Center for American Progress. Available at www.americanprogress.org.

Brown, Catherine, Ulrich Boser, Scott Sargrad, and Max Marchitello. 2017 (January). "Implementing the Every Student Succeeds Act: Toward a Coherent, Aligned Assessment System." Available at www.american progress.org.

Bureau of Labor Statistics (BLS). 2019. "Earnings and Employment Rates by Educational Attainment, 2019." U.S. Department of Labor. Available at www.bls.org.

Bureau of Labor Statistics (BLS). 2020. *Occupational Outlook Handbook*. U.S. Department of Labor. Available at www.bls.gov.

Burke, Lilah. 2020 (October 14). "Alternatives to Austerity." Inside Higher Ed. Available at www.insidehighered.com.

Bustamante, Jaleesa. 2019 (September 23). "High School Dropout Rate." EducationData.org. Available at www.educationdata.org.

Cantor, David, Bonnie Fisher, Susan Chibnall, Shauna Harps, Reanne Townsend, Gail Thomas, Hyunshik Lee, Vanessa Kranz, Randy Herbison, and Kristin Madden. 2020 (January 17). "Report on the AAU Campus Climate Survey on Sexual Assault and Misconduct." Association of American Universities. Available at www.aau.edu.

Carlson, Deven, Elizabeth Bell, Matthew A. Lenard, Joshua M. Cowen, and Andrew McEachin. 2019. "Socioeconomic-Based School Assignment Policy and Racial Segregation Levels: Evidence from the Wake County Public School System." *American Educational Research Journal* 57(1):258–304.

Carlson, Deven D., and Joshua M. Cowan. 2015. "School Choice and Student Neighborhoods: Evidence from the Milwaukee Parental Choice Program." *Educational Policy Analysis Archives* 23(60):1–27.

Carnevale, Anthony P., and Jeff Strohl. 2013 (July). *Separate and Unequal: How Higher Education Reinforces the Intergenerational Reproduction of White Racial Privilege.* Georgetown Public Policy Institute, Center on Education and the Workforce, Georgetown University.

Catt, Andrew D., Paul DiPerna, Martin F. Lueken, Michael Q. McShane, and Michael Shaw. 2020. *The 123s of School Choice: What the Research Says about Private School Choice Programs in America.* Available at https://www.edchoice.org.

Centers for Disease Prevention and Control. (CDC). 2020 (December 28). "Children and Young People's Social, Emotional, and Mental Health." Available at www.cdc.gov.

Chen, Grace. 2019 (November 18). "When Teachers Are Graded: The Controversy of Teacher Ratings." *Public School Review.* Available at www.publicschoolreview.com.

Chen, Grace. 2020 (February 28). "Why Single-Sex Public Schools Are Growing in Popularity." *Public School Review.* Available at www.publicschoolreview.com.

Cheryan, Sapna, Sianna A. Ziegler, Victoria C. Plaut, and Andrew N. Meltzoff. 2014. "Designing Classrooms to Maximize Student Achievement." *Policy Insights from the Behavioral and Brain Sciences* 1(1):4–12.

Claims Conference. 2020. *First-Ever 50 State Survey on Holocaust Knowledge of American Millennials and Gen Z Reveals Shocking Results.* Claims Conference. Available at www.claimscom.org.

Clays, Madeline. 2020 (July 29). "Why U.S. Public School Teachers Are Leaving the Profession for Good." *Tough Nickel.* Available at www.toughnickel.com.

Coleman, James S., J. E. Campbell, L. Hobson, J. McPartland, A. Mood, F. Weinfield, and R. York. 1966. *Equality of Educational Opportunity.* Washington, DC: U.S. Government Printing Office.

Collins, Randall. 1979. *The Credential Society: An Historical Sociology of Education and Stratification.* New York: Columbia University Press.

Corbett, Christianne, Catherine Hill, and Andresse St. Rose. 2008. *Where the Girls Are.* Washington, DC: American Association of University Women.

Cowen, Joshua M., David J. Fleming, John F. Witte, Patrick J. Wolf, and Brian Kisida. 2013. "School Vouchers and Student Attainment: Evidence from a State-Mandated Study of Milwaukee's Parental Choice Program." *Policy Studies Journal* 44(1):147–168.

Craft, Lucy. 2020 (September 5). "Schools in Japan Are Back in Session amid Coronavirus Pandemic." CBS News. Available at www.cbsnews.com.

Deliso, Meredith. 2021 (January 28). "All of Biden's Executive Orders and Notable Actions, So Far." ABC News. Available at www.abcnews.go.com.

Digital Promise. 2020. "Suddenly Online: A National Survey of Undergraduates during the COVID-19 Pandemic." Available at digitalpromise.org.

DiPerna, Paul, Andrew D. Catt, and Michael Shaw. 2020. "2019 Schooling in America." EdChoice. Available at https://www.edchoice.org.

Dockterman, Eilana. 2015 (March 5). "The Hunting Ground Reignites the Debate over Campus Rape." *Time.* Available at time.com.

Douglas-Gabriel, Danielle. 2020 (October 20). "Judge Rejects Settlement over Stalled Student Debt Relief Claims, Blames DeVos for Harming Borrowers." *Washington Post.* Available at www.washingtonpost.com.

Dunietz, Galit Levi, Amilcar Natos-Moreno, Dianne C. Singer, Matthew M. Davis, Louise M. O'Brien, and Ronald D. Chervin. 2017. "Later School Start Times: What Informs Parent Support or Opposition?" *Journal of Clinical Sleep Medicine* 13(7):889–897.

Dunster, Gideon, Luciano de la Iglesia, Miriam Ben-Hamo, Claire Nave, Jason G. Fleischer, Satchidananda Panda, and Horatio O. de la Iglesia. 2018. "Sleepmore in Seattle: Later School Start Times Are Associated with More Sleep and Better Performance in High School Students." *Science Advances.* Available at www.advances.sciencemag.org.

EdChoice 2020. "School Choice: School Choice in America." Available at www.edchoice.org.

Education Week. 2017. "Single-Gender Public Schools in 5 Charts." Available at www.edweek.org.

Ertas, Nevbahar, and Christine H. Roch. 2014. "Charter Schools, Equity, and Student Enrollments: The Role of For-Profit Educational Management Organizations." *Education and Urban Society* 46(5):548–579.

Esposito, Frank. 2020. "NY Teacher Retirements Jump 121% in August amid Covid Pandemic." ABC News. Available at www.abc7ny.com.

Ewert, Stephanie. 2013 (January). *The Decline in Private School Enrollment.* Working Paper FY 12-117. Social, Economic, and Housing Statistics Division.

Flaherty, Colleen. 2019 (September 30). "Winning Tenure, in Court. *Inside Higher Ed.* Available at www.insidehighered.com.

Flannery, Mary Ellen. 2020 (May 22). "With Pandemic, Privatization Advocates Spell a Big Opportunity." *NEA News.* Available at www.nea.org.

Flannery, Mary Ellen. 2018. "Back to School without a Qualified Teacher." *NEA News.* Available at www.nea.org.

Fletcher, Robert S. 1943. *History of Oberlin College to the Civil War.* Oberlin, OH: Oberlin College Press.

Flexner, Eleanor. 1972. *Century of Struggle: The Women's Rights Movement in the United States.* New York: Atheneum.

Flood, Alison. 2018. "Growing Up in a House Full of Books Is Major Boost to Literacy and Numeracy, a Study Finds." *Guardian.* Available at www.theguardian.org.

Friedlander, Brett. 2014 (October 22). "Timeline of Events in UNC Athletic/Academic Scandal." *Star News.* Available at acc.blogs.starnewsonline.com.

Garcia, Emma. 2017. *Poor Black Children Are Much More Likely to Attend High-Poverty Schools Than Poor White Children.* Economic Policy Institute. Available at www.epi.org.

Garcia, Emma, and Elaine Weiss. 2019a (April 16). "U.S. Schools Struggle to Hire and Retain Teachers." *Economic Policy Institute.* Available at www.epi.org.

Garcia, Emma, and Elaine Weiss. 2019b (May 30). "Challenging Working Environment ('School Climates'), Especially in High Poverty Schools, Play a Role in the Teacher Shortage." Economic Policy Institute. Available at www.epi.org.

Garcia, Emma. 2020a. *Schools Are Still Segregated, and Black Children Are Paying a Price.* Economic Policy Institute. Available at www.epi.org.

Garcia, Emma. 2020b (May 7). "The Pandemic Sparked More Appreciation for Teachers, but Will It Give Them a Voice in Education and Their Working Conditions?" Economic Policy Institute. Available at www.epi.org.

Gentrup, Sarah, Georg Lorenz, Cornelia Kristen, and Irena Kogan. 2020. "Self-Fulfilling Prophecies in the Classroom: Teacher Expectations, Teacher Feedback, and Student Achievement." *Learning and Instruction* 66.

Golann, Joanne W. 2015. "The Paradox of Success at a No-Excuses School." *Sociology of Education* 88(2):103–119.

Goldenberg, Claude. 2008 (Summer). "Teaching English Language Learners." *American Educator:* 8–11, 14–19, 22–23, 42–44.

Goldstein, Dana. 2019 (December 6). "After Ten Years of Hopes and Setbacks, What Happened to the Common Core?" *New York Times.* Available at www.nytimes.com.

Government Accountability Office. 2020 (June). K–12 Education School Districts Frequently Identified Multiple Building Systems Needing Updates or Replacement." *GAO Highlights,* GAO-20-494. Available at www.gao.gov.

Greenberg, Jon. 2020 (July 20). "Fact Check: Does Joe Biden Want to End School Choice?" *Austin American Statesman.* Available at www.statesman.com.

Greene, Peter. 2020 (January 30). "Common Core Is Dead. Long Live Common Core." *Forbes.* Available at https://www.forbes.com.

Greenhouse, Linda. 2007 (June 29). "Supreme Court Votes to Limit the Use of Race in Integration Plans." *New York Times.* Available at www.nytimes.com.

Guinness, Emma. 2020 (February 26). "Teacher Resigns in Front of School Board and Sends Powerful Message to Her Students." *VT.* Available at www.vt.vo.

Haberman, Martin. 1991. "The Pedagogy of Poverty versus Good Teaching." *Phi Delta Kappan* 73(4):290–294. Available at www.det .nsw.edu.au.

Harris, Douglas N., Lihan Liu, Daniel Oliver, Cathy Balfe, Sara Slaughter, and Nicholas Mattei. 2020. *How America's Schools Responded to the COVID Crisis.* Technical Report. Education Research Alliance for New Orleans. Available at www .educationresearchalliancenola.org.

Harwell, Drew. 2020 (November 12). "Cheating Detection Companies Made Millions during the Pandemic. Now the Students Are Fighting Back." *Washington Post.* Available at https:// www.washingtonpost.com/technology.

Hay, Andrew, and Brendan O'Brien. 2020 (July 18). "In Arizona, School Reopening Sparks Protest Movement." Reuters. Available at www .reuters.com.

Head Start. 2019. "Head Start Program Facts Fiscal Year 2019." Available at eclkc.ohs.acf. hhs.gov.

Henderson, Michael B., Paul A. Peterson, and Martin R. West. 2015. "No Comment Opinion on the Common Core." *Education Next* 15(1):8–12.

Hess, Abigail. 2020 (March 26). "How Coronavirus Dramatically Changed College for over 14 Million Students." CNBC. Available at www.cnbc.com.

Higgins-Dunn, Noah, and Christina Farr. 2020 (August 19). "Trump Pushes Universities to Continue Reopening amid a String of Campus Coronavirus Outbreaks." CNBC. Available at www.cnbc.com.

Hightower, Amy M. 2013 (January 4). "States Show Spotty Progress on Education Gauges." Available at www.edweek.org.

Holcombe, Madeline. 2020. "Parents in a County with One of Georgia's Highest Coronavirus Rates Are Protesting to Get Children back in School." CNN. Available at www.cnn.com.

Hope, Elan C., Alexandria B. Skoog, and Robert J. Jagers. 2015. "'It'll Never Be the White Kids, It'll Always Be Us': Black High School Students Evolving Critical Analysis of Racial Discrimination and Inequity in Schools." *Journal of Adolescent Research* 30(1):83–112.

Hopkinson, Natalie. 2011 (January 3). "The McEducation of the Negro." *The Root.* Available at www.theroot.com.

Horowitz, Juliana Menasce. 2019 (May 8). *Americans See Advantages and Challenges in Country's Growing Racial and Ethnic Diversity.* Pew Research Center. Available at www .pewsocialtrends.org.

H.R. 2639. 2019. *Strength in Diversity Act of 2020.* 116th Congress (2019–2020). Available at www.congress.gov.

Issa, Natalie. 2019 (June 19). "U.S. Average Student Loan Debt Statistics in 2019." Credit. com. Available at www.credit.com.

Ives, Robert. 2020 (September 8). "New Research in Academic Misconduct Interventions." International Center for Academic Integrity. Available at https://www.academicintegrity .org.

Johnson, Annysa. 2019 (October 16). "Cost of Wisconsin Voucher Programs Nears $350 Million as Enrollment Surges." *Milwaukee Sentinel Journal.* Available at https://www .jsonline.com.

Juvonen, Jaana, and Sandra Graham. 2014. "Bullying in Schools: The Power of Bullies and the Plight of Victims." *Annual Review of Psychology* 65:159–185.

Kahlenberg, Richard D., Haley Potter, and Kimberly Quick. 2019. *A Bold Agenda for School Integration.* New York: Century Foundation. Available at www.tcf.org.

Kam, Katherine. 2020 (August 20). "Asian American Students Face Bullying over COVID." *WebMD Health News.* Available at www.webmd.com.

Kamenetz, Anya. 2015. *The Test: Why Our Schools Are Obsessed with Standardized Testing—But You Don't Have to Be.* Philadelphia: Public Affairs Publishing.

Kamenetz, Anya. 2020 (September 17). "'I'm Only One Person': Teachers Feel Torn between Their Students and Their Own Kids." National Public Radio. Available at www.npr.org.

Kamenentz, Anna, and Elissa Nadworny. 2020 (November 10). "What a Biden Presidency Could Mean for Education." National Public Radio. Available at www.npr.org.

Kerr, Emma. 2020 (April 8). "Covid 19 Closed Doors. Will Students Get a Refund?" *U.S. News & World Report.* Available at www.usnews.com.

Kids Count. 2020. *2020 Kids Count Data Book.* Annie E. Casey Foundation. Available at www .aecf.org.

Kingkade, Tyler. 2020 (November 12). "Biden Wants to Scrap Betsy DeVos' Rules on Sexual Assault in Schools. It Won't Be Easy." NBC News. Available at www.nbcnews.com.

King, LaGarrett J. 2017. "The Status of Black History in U.S. Schools and Society." *Social Education* 81(1):14–18.

Kinsella, Elize. 2020 (May 16). "International Students in Hardship Due to Coronavirus a 'Looming Humanitarian Crisis', Advocates Say." ABC News Australia. Available at www .abc.net.au.

Knoff, Howie. 2019. *The Impact of Inequitable School Funding: Solutions for Struggling Schools without the Money to Fully Help Struggling Students.* American Consortium for Equity in Education. Available at ace-ed.org.

Koenig, Rebecca. 2019. "Colleges Re-Enforce Inequality Rather Than Social Mobility, New Book Argues." *Ed Surge.* Available at www .edsurge.com.

Kohn, Alfie. 2011. "How Education Reform Traps Poor Children." *Education Week* 30(29):32–33.

Kozol, Jonathan. 1991. *Savage Inequalities: Children in America's Schools.* New York: Crown.

Kraft, Matthew A., and Nicole S. Simon. 2020. *School Organizational Practices and the Challenges of Remote Teaching during a Pandemic.* Albert Shanker Institute. Available at www.shankerinstitute.org.

Lafavor, Theresa, Sara E. Langworthy, Schevita Persaud, and Amanda W. Kalstabakken. 2020. "The Relationship between Parent and Teacher Perceptions and the Academic Success of Homeless Youth." *Child & Youth Care Forum* 49:449–468.

Leath, Seanna, Channing Mathews, Aysa Harrison, and Tabbye Chavous. 2019. "Racial Identity, Racial Discrimination, and Classroom Engagement Outcomes among Black Girls and Boys in Predominately Black and Predominately White School Districts." *American Educational Research Journal* 56(4):1318–1352.

Li, Zhonglu, and Zeqi Qiu. 2018. "How Does Family Background Affect Children's Educational Achievement? Evidence from Contemporary China." *The Journal of Chinese Sociology* 5(13).

Lickona, Thomas, and Matthew Davidson. 2005. *A Report to the Nation: Smart and Good High Schools.* Available at www.cortland.edu.

Lloyd, Sterling C., and Alex Harwin. 2019 (September 3). "In a National Ranking of School Systems, a New State Is on Top." *Education Week.* Available at www.edweek.org.

Locquiano, Jed, and Bob Ives. 2020. "Preliminary Findings from a Pilot Intervention to Address Academic Misconduct among First-Year College Students." *Education Research: Theory and Practice* 31(1):31–45.

London, Rebecca A. 2019. "The Right to Play: Eliminating the Opportunity Gap in Elementary School Recess." *Phi Delta Kappan.* Available at www.kappanonline.org.

Longwell-Grice, Rob, Nicole Zervas Adsitt, Kathleen Mullins, and William Serrata. 2016. "The First Ones: Three Studies on First-Generation College Students." *NACADA Journal* 36(2):34–46.

Losen, Daniel J., and Paul Martinez. 2020 (October 11). "Lost Opportunities: How Disparate School Discipline Continues to Drive Differences in the Opportunity to Learn." The Civil Rights Project. Available at www.civilrightsproject.ucla.edu.

Loveless, Tom. 2013. "The Resurgence of Ability Grouping in Persistence of Tracking." *Brown Center Report on American Education.* Available at www.brookings.edu.

Lu, Adrienne. 2013 (July 1). "Parents Revolt against Failing Schools." Pew Charitable Trust. Available at www.pewstates.org.

Lubienski, Christopher, Janelle T. Scott, John Rogers, and Kevin G. Welner. 2012 (September 5). "Missing the Target? The Parent Trigger as a Strategy for Parental Engagement and School Reform." National Education Policy Center, School of Education, University of Colorado at Boulder.

Mani, Deept, and Stefan Trines. 2018. "Education in South Korea." *World Education News and Reviews.* Available at www.wenr.wes.org.

Mantel, Barbara. 2020. "Higher Education in the COVID Era." *CQ Researcher* 30(32). Available at www.library.cqpress.com.

Mares, Marie-Louise, and Zhongdang Pan. 2013. "Effects of *Sesame Street:* A Meta-Analysis of Children's Learning in 15 Countries." *Journal of Applied Developmental Psychology* 34(3):140–151.

Maxwell, Leslie A. 2013 (January 9). "Head Start Gains Found to Wash out by Third Grade." *Education Week.* Available at www.edweek .org.

McDonnell, Sanford. 2009 (October 3). "America's Crisis of Character—and What to Do about It." *Education Week.* Available at www.edweek.org.

Means, B., and Neisler, J., with Langer Research Associates. (2020). *Suddenly Online: A National Survey of Undergraduates during the COVID-19 Pandemic.* San Mateo, CA: Digital Promise.

Meinck, Sabine, and Falk Brese. 2019. "Trends in Gender Gaps: Using 20 Years of Evidence from TIMSS." *Large-Scale Assessments in Education* 7(8).

Merton, Robert K. 1968. *Social Theory and Social Structure.* New York: Free Press.

Mettler, Suzanne. 2014. *Degrees of Inequality: How the Politics of Higher Education Sabotaged the American Dream.* New York: Basic Books.

Moody, Josh. 2019 (February 15). "The U.S. Department of Education Lists More Than 4000 Degree-Granting Academic Institutions." *U.S. News.* Available at www.thatusnews.com.

Muller, Chandra, and Katherine Schiller. 2000. "Leveling the Playing Field?" *Sociology of Education* 73:196–218.

Nadworney, Elissa, and Marco A. Treviño. 2020 (September 30). "Lies, Money, and Cheating: The Deeper Story of the College Admissions Scandal." National Public Radio. Available at www.npr.org.

National Assessment of Educational Progress (NAEP). 2019. "See How US Students Performed in Reading at Grade 12." *The Nation's Report Card.* Available at www .nationsreportcard.gov.

National Assessment of Educational Progress (NAEP). 2020. *NAEP Report Card: Reading.* Available at www.nationsreportcard.gov.

National Association of Colleges and Employers. 2019. *Job Outlook 2019.* Available at www .naceweb.org.

National Center for Education Statistics (NCES). 2016. *Digest of Education Statistics.* Available at nces.ed.gov.

National Center for Education Statistics (NCES). 2018. "Fast Facts." Available at nces.ed.gov.

National Center for Education Statistics (NCES). 2019. *Digest of Education Statistics.* Available at nces.ed.gov.

National Center for Education Statistics (NCES). 2020a. "Racial /Ethnic Enrollment in Public Schools." *The Condition of Education.* Available at nces.ed.gov.

National Center for Education Statistics (NCES). 2020b. "English Language Learners in Public Schools." *The Condition of Education.* Available at nces.ed.gov.

National Center for Education Statistics (NCES). 2020c. "Characteristics of Elementary and Secondary Schools." *The Condition of Education.* Available at nces.ed.gov.

National Center for Education Statistics (NCES). 2020d. "Back to School Statistics." *Fast Facts.* Available at nces.ed.gov.

National Center for Fair and Open Testing (NCFOT). 2017 (January). "Just Say No to Standardized Tests." Available at www.fairtest .org.

National Center for Fair and Open Testing (NCFOT). 2020 (December). "National Call to Suspend High Stakes Testing in Spring 2021." Available at www.fairtest.org.

National Public Radio. 2013 (June 2). "Why Some Schools Want to Expel Suspensions." Available at www.npr.org.

National Public Radio. 2016 (April 18). "Why America's Schools Have a Money Problem." *Morning Edition.* Available at www.npr.org.

National Strategic Planning and Analysis Research Center (NSPARC). 2017. "The Impact of National Board Certified Teachers on the Literacy Outcomes of Mississippi Kindergartners and Third Graders." Available at http://www.nbpts.org/wp-content/uploads /NBCT_MS_Report.pdf.

Network for Public Education (NPE). 2020. "About the Network for Public Education." Available at networkforpubiceducation.org.

New, Jake. 2016. "Fraud and the Final Four." *Inside Higher Education* (April 1). Available at www.insidehighered.com.

New York Equality Coalition. 2018. *Within Our Reach.* Available at www.equityinedny.edtrust .org.

Nguyen, Sophie. 2018. "Americans Think Highly of Community College." *ACCT Now.* Available at www.perspectives.acct.org.

Obama, Barack H. 2016 (October 17). "Remarks by the President on Education." White House. Available at https://obamawhitehouse .archives.gov/the-press-office/2016/10/17 /remarks-president-education.

Organisation for Economic Co-operation and Development (OECD). 2020. *Education at a Glance 2020.* Available at www.oecd.org.

Owens, Ann. 2018. "Income Segregation between School Districts and Inequality in Students' Achievement." *Sociology of Education* 91(1):1–27.

Owens, Jayanti, and Sara S. McLanahan. 2020 (June). "Unpacking the Drivers of Racial Disparities in School Suspension and Expulsion." *Social Forces* 98(4):1548–1577.

Parent Revolution. 2017. "About: Company Overview." Facebook. Available at www .facebook.com.

PDK International. 2020 (September). "52nd Annual PDK Poll of the Public's Attitudes toward the Public Schools: Public School Priorities in a Political Year." Available at www.pdkpoll.org.

Pew Research Center. 2019 (January 24). "Public's 2019 Priorities: Economy, Health Care, Education and Security All Near the Top of the List." Available at www.pewresearch.org.

Pianta, Robert C., and Arya Ansari. 2018 (July 9). "Does Attendance in Private Schools Predict Student Outcomes at Age 15? Evidence from a Longitudinal Study." *Educational Researcher* 47(7):419–434.

Poliakoff, Anne Rogers. 2006 (January). "Closing the Gap: An Overview." *ASCD InfoBrief* 44:1–10. Alexandria, VA: Association for Supervision and Curricular Development.

Pomrenze, Yon, and Bianna Golodryga. 2020 (December 9). "Night School Comes to the Rescue for Some Kindergartners and Their Parents." CNN. Available at www.cnn.com.

Power in Partnerships. 2020. "Building Connections at the Intersections of Racial Justice and LGBTQ Movements to End the School-to-Prison Pipeline." Available at https://b.3cdn.net/advancement.

Pramuk, Jacob. 2021 (February 12). "House Advances $1,400 Payments, Unemployment Boost as Part of COVID Relief Plan." CNBC. Available at www.cnbc.com.

Prothero, Arianna. 2018 (August 9). "What Are Charter Schools?" *Education Week.* Available at https://www.edweek.org.

Quinn, David M. 2015. "Kindergarten, Black–White Test Score Gaps: Re-Examining the Roles of Social and Economic Status and School Quality with New Data." *Sociology of Education* 88(2):120–139.

Quinton, Sophie. 2020 (September 14). "Coronavirus, Trump Chill International Enrollment at U.S. Colleges." *Stateline.* Pew Trusts. Available at www.pewtrusts.org.

Quirk, Abby. 2020 (July 28). "Mental Health Support for Students of Color during and after the Coronavirus Pandemic." Center for American Progress. Available at www .americanprogress.org.

Ramey, Garey, and Valerie A. Ramey. 2010 (Spring). "The Rug Rat Race." *Brookings Paper on Economic Activity*: 129–176.

Ramstetter, Katherine, and Robert Murray. 2017. "Time to Play: Recognizing the Benefits of Recess." *American Educator.* American Federation of Teachers. Available at www.aft .org.

Ravitch, Diane. 2010. *The Death and Life of the Great American School System: How Testing and Choice Are Undermining Education.* New York: Basic Books.

Ravitch, Diane. 2020 (December 12). "Teresa Thayer Snyder: What Shall We Do about the Children after the Pandemic." Available at https://dianeravitch.net/2020/12/12/teresa -thayer-snyder-what-shall-we-do-about-the -children-after-the-pandemic/.

Rawls, Kristin. 2015 (January 21). "Who Is Profiting from Charters? The Big Bucks behind Charter School Secrecy, Financial Scandal and Corruption." Available at www .alternet.org.

Reardon, Sean F. 2013a (April 27). "No Rich Child Left Behind." *New York Times.* Available at www.nytimes.com.

Reardon, Sean F. 2013b (May). "The Widening Income Gap." *Educational Leadership* 70(8):10–16.

Redden, Elizabeth. 2020 (March 24). "Stranded Abroad." *Inside Higher Ed.* Available at www .insidehighered.com.

Reilly, Katie. 2017 (October 23). "Is Recess Important for Kids or a Waste of Time? Here's What the Research Says." *Time.* Available at www.time.com.

Reilly, Katie. 2020a (August 26). "This Is What It's like to Be a Teacher during the Coronavirus Pandemic." *Time.* Available at www.time.com.

Reilly, Katie. 2020b (July 8). "With No End in Sight to the Coronavirus, Some Teachers Are Retiring Rather Than Go back to School." *Time.* Available at www.time.com.

Romo, Vanessa, and Tom Bowman. 2020 (December 21). "More Than 70 West Point Cadets Accused of Cheating in Academic Scandal." National Public Radio. Available at www.npr.org.

Rosales, John. 2018 (April 4). "The Racist Beginnings of Standardized Testing." *NEA Today.* Available at www.nea.org.

Rosen, Jill. 2014 (June 2). "Study: Children's Life Trajectories Largely Determined by Family They Are Born Into." *Johns Hopkins News Network.* Available at hub.jhu.edu.

Rosenthal, Robert, and Lenore Jacobson. 1968. *Pygmalion in the Classroom: Teacher Expectations and Pupils' Intellectual Development.* New York: Holt, Rinehart, and Winston.

Rosewicz, Barb, and Mike Maciag. 2020 (November 10). "Nearly All States Suffer Declines in Education Jobs." Pew Trust. Available at www.pewtrust.org.

Rothstein, Richard, Helen F. Ladd, Diane Ravitch, Eva L. Baker, Paul E. Barton, Linda Darling-Hammond, Edward Haertel, Robert L. Linn, Richard J. Shavelson, and Lorrie A. Shepard. 2010 (August 29). "Problems with the Use of Student Test Scores to Evaluate Teachers." Educational Policy Institute. Available at www.epi.org.

Ruszkowski, Christopher N. 2020 (February 24). "American Families Strongly Support School Choice. Educators Should Listen to Them." *USA Today.* Available at https://www.usatoday.com.

S. 2784. 2019. *Family Friendly Schools Act.* 116th Congress (2019–2020). Available at www.congress.gov.

Schanzenbach, Diane Whitmore, and Lauren Bauer. 2016. *The Long-Term Impact of the Head Start Program.* Brookings Institute. Available at www.brookings.edu.

Schleicher, Andreas. 2020. *The Impact of Covid-19 on Education: Insights from Education at a Glance 2020.* OECD. Available at www.oecd.org.

Schott Report. 2015. *Black Lives Matter: The 2012 Schott 50 State Report on Public Education and Black Males.* Schott Foundation for Public Education. Available at blackboysreport.org.

Segal, Tom. 2013 (March 26). "The Impact of Investing in Education." *Education Week.* Available at blogs.edweek.org.

Shah, Nirvi. 2013 (January). "Discipline Policy Shift with Views on What Works." *Education Week* 32(16):12.

Shamsian, Jacob, and Kelly McLaughlin. 2020 (September 2). "Here's the Full List of People Charged in the College Admissions Cheating Scandal, and Who Has Pleaded Guilty So Far." *Insider.* Available at https://www.insider.com.

Sison, Mark Allen. 2020 (July 18). "Did You Know Classrooms in the Philippines Are the Most Crowded in Asia?" *iOrbit News Online.* Available at www.iorbitnews.com.

Soares, Joseph A. 2020 (June 22). *Dismantling White Supremacy Includes Ending Racist Tests Like the SAT and ACT.* Teachers College Press Blog. Available at www.tcpress.com.

Spector, Carrie. 2019 (September 23). "School Poverty—Not Racial Composition—Limits Educational Opportunity, According to New Research at Stanford." *Stanford News.* Available at www.stanfordnews.edu.

Stanton, Zack. 2019 (March 27). "PBS Chief: 'I Wish I Knew' Why Trump Wants to Defund Us." *Politico Women Rule.* Available at www.politico.com.

Startz, Dick. 2019a. "Do Teachers Work Long Hours?" *Brown Center Chalkboard.* Brookings Institute. Available at www.brookings.edu.

Startz, Dick. 2019b. *Community College 'Free-for-All': Why Making Tuition Free Could Be Complicated.* Brookings Institute. Available at www.brookings.edu.

Stephens, Jason M., and David B. Wangaard. 2013. "Using the Epidemic of Academic Dishonesty as an Opportunity for Character Education: A Three-Year Mixed Methods Study (with Mixed Results)." *Peabody Journal of Education* 88(2):159–179.

Stoet, Gijsbert, and David C. Geary. 2015. "Sex Differences in Academic Achievement Are Not Related to Political, Economic, or Social Equality." *Intelligence* 48:137–151.

Strauss, Valerie. 2017 (September 14). "Teachers in the U.S. Paid Far Less Than Similarly Educated Professionals, Report Finds." *Washington Post.* Available at www.washingtonpost.com.

Strauss, Valerie. 2020a (August 14). "Betsy DeVos's Controversial New Rule on Campus Sexual Assault Goes into Effect." *Washington Post.* Available at www.washingtonpost.com.

Strauss, Valerie. 2020b (August 4). "With Coronavirus Cases Reported at Some Reopened Schools, Protesters Take to the Streets with Fake Coffins." *Washington Post.* Available at www.washingtonpost.com.

Strauss, Valerie. 2020c (June 21). "It Looks Like the Beginning of the End of America's Obsession with Student Standardized Tests." *Washington Post.* Available at www.washingtonpost.com.

Study International. 2019. "Private and Charter Schools in the US: What's the Difference?" Available at www.studyinternational.com.

Tepper, Taylor. 2017 (January 24). "President Trump Wants to Kill These 17 Federal Agencies and Programs: Here's What They Actually Cost (and Do)." *Time.* Available at www.time.com.

The Century Foundation. 2020 (July). *Closing America's Education Funding Gaps.* Available at www.tcfr.org.

Thomson, Sue. 2018. "Achievement at School and Socioeconomic Background: An Educational Perspective." *NPJ Science of Learning* 3(5).

Tracy, Marc. 2017 (October 13). "N.C.A.A.: North Carolina Will Not Be Punished for Academic Scandal." *New York Times.* Available at www.nytimes.com.

Tully, Sarah. 2016 (March 18). "New Parent-Trigger Bills Fail to Gain Traction in States This Year." *Education Week.* Available at https://www.edweek.org.

United Nations. 2020. "Literacy Teaching and Learning in the COVID-19 Crisis and Beyond." *International Literacy Day.* Available at www.un.org/en/observances/literacy-day.

United Nations Educational, Scientific and Cultural Organization (UNESCO). 2016. *Third Global Report on Adult Learning and Education.* Available at www.uil.unesco.org.

United Nations Educational, Scientific and Cultural Organization (UNESCO). 2017 (September). *More Than One-Half of Children and Adolescents Are Not Learning Worldwide.* Fact Sheet 46. Available at www.uis.unesco.org.

United Nations Educational, Scientific and Cultural Organization (UNESCO). 2019a (September). *New Methodology Shows That 258 Million Children, Adolescents, and Youth Are out of School.* Fact Sheet 56. Available at www.uis.unesco.org.

United Nations Educational, Scientific and Cultural Organization (UNESCO). 2019b. *Her Education, Our Future: UNESCO Fast-Tracking Girls' and Women's Education.* Available at www.uis.unesco.org.

UNICEF. 2020. "Girls' Education: Gender Equality in Education Benefits Every Child." Available at www.unicef.org.

U.S. Census Bureau. 2019. *Educational Attainment in the United States: 2019.* Detailed Tables, Table 2. Available at www.census.gov.

U.S. Bureau of Labor Statistics. 2020. *Unemployment Rates and Earnings by Educational Attainment, 2019.* Available at www.bls.gov.

U.S. Department of Education. 2020 (October 2). "Secretary DeVos Announces More Than 131 Million in New Funding to Create and Expand High-Quality Public Charter Schools." *Press Release.* Available at www.ed.gov.

U.S. Department of Education (Center for Education Statistics). 2020. *The Condition of Education 2020 (*NCES 2020-144). *Status Dropout Rates.* Available at www.nces.ed.gov.

U.S. Holocaust Memorial Museum. 2020. *Where Holocaust Education Is Required in the US.* Available at www.ushmm.org.

Venator, Joanna, and Richard V. Reeves. 2015 (March 18). "Building the Soft Skills for Success." Brookings Institute. Available at www.brookings.edu.

Vogels, Emily A. 2020. "59% of U.S. Parents with Lower Incomes Say Their Child May Face Digital Obstacles in Schoolwork." *FactTank.* Pew Research Center. Available at www.pewresearch.org.

Wang, Karla. 2019. "Teacher Turnover: Why It's Problematic and How Administration Can Address It." *Science of Learning Blog.* Available at www.scilearn.com/teacher-turnover/.

Wang, Ke, Yongqiu Chen, Jizhi Zhang, and Barbara Oudekerk. 2020. "Indicators of School Crime and Safety." National Center for Education Statistics. Available at https://nces.ed.gov/programs/crimeindicators/.

Warren, M. R. 2014. "Transforming Public Education: The Need for an Educational Justice Movement." *New England Journal of Public Policy* 26(1):1–16.

Watanabe Teresa. 2019 (December 10). "UC Violates Civil Rights of Disadvantaged Students by Requiring SAT for Admission, Lawsuits Say." *Los Angeles Times.* Available at www.latimes.com.

Weissert, Will. 2020 (December 8). "Biden Vows to Reopen Most Schools after 1st 100 Days on Job." Available at www.apnews.com.

West, Martin R. 2018 (August 1). "Privatization in American Education: Rhetoric versus Facts." *Education Next* 18(4). Available at www.educationnext.org.

Will, Madeline, Catherine Gewertz, and Sarah Schwartz. 2020. "Did Covid-19 Really Drive Teachers to Quit?" *Education Week*. Available at www.edweek.org.

Wisconsin Department of Public Instruction. 2020. "Private School Choice Programs." Available at https://dpi.wi.gov.

World Bank. 2019. Ending Learning Poverty: What Will It Take? Washington, DC: World Bank.

World Population Review. 2020a. "Per Pupil Spending by State 2020." Available at worldpopulation.com.

World Population Review. 2020b. "Common Core States 2020." Available at https://worldpopulationreview.com/state-rankings/common-core-states.

Chapter 9

Abramowitz, Alan, and Jennifer McCoy. 2019. "United States: Racial Resentment, Negative Partisanship, and Polarization in Trump's America." *The ANNALS of the American Academy of Political and Social Science* 68(1):137–156.

Alexander, Michelle. 2010. *The New Jim Crow: Mass Incarceration in the Age of Colorblindness*. New York: New Press.

Allport, Gordon Willard, Kenneth Clark, and Thomas Pettigrew. 1954. *The Nature of Prejudice*. Reading, MA: Addison-Wesley.

American Civil Liberties Union. 2014 (July 9). "United States' Compliance with the International Convention on the Elimination of All Forms of Racial Discrimination." Available at www.aclu.org.

American Council on Education and American Association of University Professors. 2000. *Does Diversity Make a Difference? Three Research Studies on Diversity in College Classrooms*. Washington, DC: American Council on Education and American Association of University Professors.

American Medical Association (AMA). 2020 (June 8). "Statement of the American Medical Association to the U.S. House of Representatives Committee on Ways and Means Re: The Disproportionate Impact of COVID-19 on Communities of Color." *Division of Legislative Counsel*. Available at www.aha.org.

Anti-Defamation League (ADL). 2019. *Pyramid of Hate*. Available at www.adl.org/education.

Apfelbaum, Evan. 2011. "Prof. Evan Apfelbaum: A Blind Pursuit of Racial Colorblindness—Research Has Implications for How Companies Manage Multicultural Teams." MIT *Sloan Experts*, MIT Sloan Management blog. Available at mitsloanexperts.mit.edu.

Armario, Christine. 2011 (May 7). "Feds: All Kids, Legal or Not, Deserve K–12 Education." *Chron*. Available at www.chron.com.

Associated Press. 2020 (August 14). "Trump Admits He's Blocking Postal Cash to Stop Mail-In Votes." *New York Times*. Available at www.nytimes.com.

Ayala, Elaine, and Ellen Huet. 2013 (February 4). "Hispanic May Be a Race on 2020 Census." *San Francisco Chronicle*. Available at www.sfgate.com.

Back, Les. 2002. "Aryans Reading Adorno: Cyber-Culture and Twenty-First Century Racism." *Ethnic and Racial Studies* 25(4):628–651.

Balaban, Samantha. 2019 (February 9). "Listeners Share Stories of Racism at School." *Weekend Edition Saturday* (NPR).

Balko, Radley. 2009 (July 6). "The El Paso Miracle." Reasononline. Available at www.reason.com.

Batalova, Jeanne, and Jie Zong. 2016 (November 11). "Language Diversity and English Proficiency in the United States." Migration Policy Institute. Available at www.migrationpolicy.org.

Batalova, Jeanne, Brittney Blizzard, and Jessica Bolter. 2020 (February 14). "Frequently Requested Statistics on Immigrants and Immigration in the United States." *Migration Information Source*. Available at www.migrationpolicy.org.

Bauer, Mary, and Sarah Reynolds. 2009. *Under Siege: Life for Low-Income Latinos in the South*. Montgomery, AL: Southern Poverty Law Center.

Beirich, Heidi. 2013 (Spring). "The Year in Nativism." *Intelligence Report* (149). Available at www.splcenter.org.

Bernstein, Hamutal, Dulce Gonzalez, Michael Karpman, and Stephen Zuckerman. 2020. "Amid Confusion over the Public Charge Rule, Immigrant Families Continued Avoiding Public Benefits in 2019." Washington, DC: Urban Institute.

Bertrand, Marianne, and Sendhil Mullainathan. 2004. "Are Emily and Greg More Employable Than Lakisha and Jamal? A Field Experiment on Labor Market Discrimination." *American Economic Review* 94(4):991–1013.

Biden-Harris. 2020a. "The Biden Plan for Securing Our Values as a Nation of Immigrants." Available at www.joebiden.com.

Bliuc, Ana-Maria, Nicholas Faulkner, Andrew Jakubowicz, and Craig McGarty. 2018. "Online Networks of Racial Hate: A Systematic Review of 10 Years of Research on Cyber-Racism." *Computers in Human Behavior* 87:75–86.

Bonilla-Silva, Eduardo. 2012. "The Invisible Weight of Whiteness: The Racial Grammar of Everyday Life in Contemporary America." *Ethnic and Racial Studies* 35(2):173–194.

Bonilla-Silva, Eduardo. 2013. *Racism without Racists: Color-Blind Racism and the Persistence of Racial Inequality*, 4th ed. Lanham, MD: Rowan and Littlefield.

Brace, C. Loring. 2005. *"Race" Is a Four-Letter Word*. New York: Oxford University Press.

Bradner, Eric. 2015 (August 18). "Huckabee: MLK Would Be 'Appalled' by Black Lives Matter Movement." CNN. Available at www.cnn.com.

Bernstein, Hamutal, Dulce Gonzalez, Michael Karpman, and Stephen Zuckerman. 2020. "Amid Confusion over the Public Charge Rule, Immigrant Families Continued Avoiding Public Benefits in 2019." Washington, DC: Urban Institute.

Brown, Anna, and Eileen Patten. 2014 (April 29). "Statistical Portrait of the Foreign-Born Population in the United States, 2012." Pew Research Center. Available at www.pewhispanic.org.

Brown, Anna, and Renee Stepler. 2016 (April 19). "Statistical Portrait of the Foreign-Born Population in the United States." Pew Research Center. Available at www.pewhispanic.org.

Brown University Steering Committee on Slavery and Justice. 2007. *Slavery and Justice*. Providence, RI: Brown University.

Bureau of Labor Statistics. 2015. "Median Weekly Earnings of Workers 25 Years and over, by Educational Attainment, Race, and Hispanic Origin, 2014." *Economic Daily*. Available at www.bls.gov.

"Campus Racial Incidents." 2020. *The Journal of Blacks in Higher Education*. Available at www.jbhe.com.

Center on Extremism. 2020 (February). "White Supremacists Double Down on Propaganda in 2019." *Anti-Defamation League*. Available at www.adl.org.

Chen, M. Keith, Kareem Haggag, Devin G. Pope, and Ryne Rohla. 2019. *Racial Disparities in Voting Wait Times: Evidence from Smartphone Data*. No. w26487. Cambridge, MA: National Bureau of Economic Research.

Chin, Mark J., David M. Quinn, Tasminda K. Dhaliwal, and Virginia S. Lovison. 2020 (July). "Bias in the Air: A Nationwide Exploration of Teachers' Implicit Racial Attitudes, Aggregate Bias, and Student Outcomes." *Educational Researcher*. doi: 0013189X20937240.

CNN. 2009 (June 18). "Senate Approves Resolution Apology for Slavery." Available at www.cnn.com.

Colby, Sandra L., and Jennifer M. Ortman. 2015 (March). "Projections of the Size and Composition of the U.S. Population: 2014 to 2060." *Current Population Reports* P25-1143. U.S. Census Bureau. Available at www.census.gov.

Colford, Paul. 2013 (April 2). "'Illegal Immigrant' No More." *The Definitive Source*, AP Blog. Available at blog.ap.org.

Conley, Dalton. 1999. *Being Black, Living in the Red: Race, Wealth, and Social Policy in America*. Berkeley: University of California Press.

Cooke, Kristina, and Ted Hesson. 2020 (February 20). "What Are 'Sanctuary' Cities and Why Is Trump Targeting Them?" Reuters. Available at www.reuters.com.

Cooper, David. 2015 (May 13). "The Policy Failures Exposed by the *New York Times'* Nail Salon Investigation." Economic Policy Institute. Available at www.epi.org.

Croll, Paul R. 2013. "Explanations for Racial Disadvantage and Racial Advantage: Beliefs about Both Sides of Inequality in America." *Ethnic and Racial Studies* 36(1):47–74.

Current Population Survey (CPS). 2020. "Educational Attainment—People 25 Years Old and over, by Total Money Earnings in 2018, Work Experience in 2018, Age, Race, Hispanic Origin and Sex." *Annual Social and Economic (ASEC) Supplement*, Persona Income Tables. Available at www.census.gov.

Daniels, Jessie. 2018. "The Algorithmic Rise of the 'Alt-Right.'" *Contexts* 17(1):60–65.

Daugherty, Owen. 2019 (May 14). "Richard Spencer: 'Charlottesville Wouldn't Have Happened without Trump.'" *The Hill*. Available at www.thehill.com.

Das, Aniruddha. 2013. "How Does Race Get "under the Skin"? Inflammation, Weathering, and Metabolic Problems in Late Life." *Social Science & Medicine* 77:75–83.

Department of Education Civil Rights Data Collection (CRDC). 2020. "2015–2016 State and National Estimations." Office of Civil Rights. Washington, DC. Available at www.ocrdata.ed.gov.

Department of Justice. 2020 (February 6). "Texas Man Charged with Federal Hate Crimes and Firearm Offenses Related to August 3, 2019, Mass-Shooting in El Paso." *Office of the United States Attorneys*. Available at www.justice.gov.

Dudziak, Mary. 2000. *Cold War Civil Rights: Race and the Image of American Democracy*. Princeton, NJ: Princeton University Press.

Dwyer, Devin. 2011 (January 5). "Opponents of Illegal Immigration Target Birthright Citizenship." ABC News. Available at www.abcnews.go.com.

Dwyer, Devin. 2020 (July 30). "Reparations for Slavery: Is Asheville a National Model?" ABC News. Available at www.abcnews.go.

Edo, Anthony. 2019. "The Impact of Immigration on the Labor Market." *Journal of Economic Surveys* 33(3):922–948.

Edwards, Griffin Sims, and Stephen Rushin. 2018 (January 14). "The Effect of President Trump's Election on Hate Crimes." *Social Science Research Network*. Available at SSRN 3102652.

Equal Employment Opportunity Commission (EEOC). 2011 (June 22). "A. C. Widenhouse Sued by EEOC for Racial Harassment." Press Release. Available at www.eeoc.gov.

Equal Employment Opportunity Commission (EEOC). 2020. "Race-Based Charges (Charges Filed with EEOC) FY 1997–FY 2019." Available at www.eeoc.gov.

Esposito, John L. 2011 (May 25). "Getting It Right about Islam and American Muslims." Huffington Post. Available at www.huffingtonpost.com.

FBI. 2016. *2015 Hate Crime Statistics*. Available at www.ucr.fbi.gov.

Federal Bureau of Investigation (FBI). 2014. *Hate Crime Statistics 2013*. Available at www.fbi.gov.

Fix, Rebecca L. 2020 (July). "Justice Is Not Blind: A Preliminary Evaluation of an Implicit Bias Training for Justice Professionals." *Race and Social Problems*: 1–13.

Frankenberg, Erica, Jongyeon Ee, Jennifer B. Ayscue, and Gary Orfield. 2019 (May 10). "Harming Our Common Future: America's Segregated Schools 65 Years after Brown." UCLA Civil Rights Project. Available at www.civilrightsproject.ucla.edu.

Frey, William H. 2020 (March 23). "Even as Metropolitan Areas Diversify, White Americans Still Live in Mostly White Neighborhoods." Brookings Institute. Available at www.brookings.edu.

Frieden, Bonnie. 2013 (April 3). "'I Don't See Race': The Pitfalls of the Colorblind Mindset." *Washington University Political Review*. Available at www.wupr.org.

Fry, Richard. 2009. "Sharp Growth in Suburban Minority Enrollment Yields Modest Gains in School Diversity." Pew Hispanic Center. Available at www.pewhispanic.org.

Fuchs, Lawrence H. 1990. *The American Kaleidoscope: Race, Ethnicity, and the Civic Culture*. Hanover, NH: Wesleyan University Press.

Gaertner, Samuel L., and John F. Dovidio. 2000. *Reducing Intergroup Bias: The Common In-Group Identity Model*. Philadelphia: Taylor & Francis.

Gallup Organization. 2020a. "Race Relations." Available at www.gallup.com.

Gallup Organization. 2020b. "Immigration." Available at www.news.gallup.com.

Gallup Organization. 2020c (July 15). "Experiences with Microaggressions, by Racial Group." Gallup Center on Black Voices. Available at www.news.gallup.com.

Gantt Shafer, Jessica. 2017. "Donald Trump's 'Political Incorrectness': Neoliberalism as Frontstage Racism on Social Media." *Social Media+Society* 3(3). doi: 2056305117733226.

Garcia, Emma. 2020 (February 12). "Schools Are Still Segregated, and Black Children Are Paying a Price." Economic Policy Institute. Available at www.epi.org.

Gates, Jr. Henry Lewis. 2013. "The Truth behind '40 Acres and a Mule.'" PBS. Available at

Gayle, Damien. 2015 (May 11). "Michelle Obama: I Was 'Knocked Back' by Race Perceptions." *Guardian*. Available at www.theguardian.com.

Godoy, Maria, and Daniel Wood. 2020 (May 30). "What Do Coronavirus Racial Disparities Look like State by State?" National Public Radio. Available at www.npr.org.

Gonzalez-Barrera, Ana. 2019 (July 2). "Hispanics with Darker Skin Are More Likely to Experience Discrimination Than Those with Lighter Skin." Pew Research Center. Available at www.pewresearch.org.

Hannon, Lance. 2015. "White Colorism." *Social Currents* 2(1):13–21.

Higginbotham, Elizabeth, and Margaret L. Andersen. 2012. "The Social Construction of Race and Ethnicity." In *Race and Ethnicity in Society*, 3rd ed., ed. E. Higginbotham and M. L. Anderson, 3–6. Belmont, CA: Wadsworth, Cengage Learning.

Hodgkinson, Harold L. 1995 (October). "What Should We Call People? Race, Class, and the Census for 2000." *Phi Delta Kappan*, pp. 173–179.

hooks, bell. 2000. *Where We Stand: Class Matters*. New York: Routledge.

Horowitz, Juliana Menasce. 2019 (May 8). "Americans See Advantages and Challenges in Country's Growing Racial and Ethnic Diversity." Pew Research Center. Available at www.pewsocialtrends.org.

Humes, Karen R., Nicholas A. Jones, and Roberto R. Ramirez. 2011 (March). "Overview of Race and Hispanic Origin: 2010." *2010 Census Briefs*. Available at www.census.gov.

Humphreys, Debra. 1999. "Diversity and the College Curriculum: How Colleges and Universities Are Preparing Students for a Changing World." *Diversity-Web*. Available at www.inform.umd.edu.

Institute on Taxation and Economic Policy. 2017 (March 1). "Undocumented Immigrants' State and Local Tax Contributions." Available at www.itep.org.

Isensee, Laura. 2015 (October 15). "Why Calling Slaves 'Workers' Is More Than an Editing Error." National Public Radio. Available at www.npr.org.

"Ivy League Legacy Admissions." 2020 (May 12). Available at www.ivycoach.com.

Jones, Nicholas A. 2015. "Update on the U.S. Census Bureau's Race and Ethnic Research for the 2020 Census." *Survey News* 3(5). Available at www.census.gov.

Kahlenberg, Richard D. 2013 (August 7). "How to Fight Growing Economic and Racial Segregation in Higher Education." *Chronicle of Higher Education*. Available at www.chronicle.com.

Kamarck, Elaine, and Christine Stenglein. 2019 (November 12). "How Many Undocumented Immigrants Are in the United States and Who Are They?" Brookings Institute. Available at www.brookings.edu.

Kamin, Debra. 2020 (August 25). "Black Homeowners Face Discrimination in Appraisals." *New York Times*. Available at www.nytimes.com.

Kaur, Harmeet. 2020 (February 16). "In Just 1 Week, 3 States Considered Bills to Ban Discrimination Based on Hair Texture or Style." CNN. Available at www.cnn.com.

King, Joyce E. 2000 (Fall). "A Moral Choice." *Teaching Tolerance* 18:14–15.

Kozol, Jonathan. 1991. *Savage Inequalities: Children in America's Schools*. New York: Crown.

Krogstad, Jens Manuel, and Jeffrey S. Passel. 2014 (November 14). "Obama's Expected Immigration Action: How Many Would Be Affected." Pew Research Center. Available at www.pewresearch.org.

Kwok, Irene, and Yuzhou Wang. 2013. "Locate the Hate: Detecting Tweets against Blacks." Proceedings of the Twenty-Seventh AAAI Conference on Artificial Intelligence. Available at www.aaai.org.

Langton, Lynn. 2017. "Hate Crime Victimization, 2004–2015." *Bureau of Justice Statistics*. Report NCJ 250653. Available at www.bjs.gov.

Levenson, Michael. 2020 (April 4). "11 Days after Fuming about a Coughing Passenger, a Bus Driver Died from the Coronavirus." *New York Times*. Available at www.nytimes.com.

Levin, Jack, and Jack McDevitt. 1995 (August). "Landmark Study Reveals Hate Crimes Vary Significantly by Offender Motivation." *Klanwatch Intelligence Report*, pp. 7–9.

Livingston, Gretchen, and Anna Brown. 2017 (May 18). "Intermarriage in the U.S. 50 Years after *Loving v. Virginia*." Pew Research Center. Available at www.pewsocialtrends.org.

Loewen, James. 2006. *Sundown Towns*. New York: Touchstone.

Long, Mark C., and Nicole A. Bateman. 2020. "Long-Run Changes in Underrepresentation after Affirmative Action Bans in Public Universities." *Educational Evaluation and Policy Analysis* 42(2):188–207.

Ly, Laura. 2013 (March 4). "Oberlin Cancels Classes to Address Racial Incidents." CNN. Available at www.cnn.com.

Marger, Martin N. 2012. *Race & Ethnic Relations: American and Global Perspectives*, 9th ed. Belmont, CA: Wadsworth, Cengage Learning.

Maril, Robert Lee. 2004. *Patrolling Chaos: The U.S. Border Patrol in Deep South Texas*. Lubbock: Texas Tech University Press.

Maril, Robert Lee. 2011. *The Fence: National Security, Public Safety, and Illegal Immigration along the U.S.–Mexico Border*. Lubbock: Texas Tech University Press.

Matthews, Cate. 2014 (September 2). "He Dropped One Letter in His Name While Applying for Jobs, and the Responses Rolled In." *Huffington Post*. Available at www.huffingtonpost.com.

McGarrity, Michael C. 2019 (May 8). "Confronting the Rise of Domestic Terrorism in the Homeland." Statement before the House Homeland Security Committee. Washington, DC. Available at www.fbi.gov.

McIntosh, Peggy. 1990 (Winter). "White Privilege: Unpacking the Invisible Knapsack." *Independent School* 49(2):31–35.

McNicholas, Celine, and Margaret Poydock. 2020 (May 19). "Who Are Essential Workers?" Economic Policy Institute. Available at www.epi.org.

Migration Policy Institute. 2020. "Profile of the Unauthorized Population: United States." *Migration Policy Institute Data Hub*. Available at www.migrationpolicy.org.

Miller, Cassie. 2020 (June 5). "The 'Boogaloo' Started as a Racist Meme." SPLC Hatewatch. Available at www.splcenter.org.

Mooney, Chris. 2014 (December 1). "The Science of Why Cops Shoot Young Black Men." *Mother Jones*. Available at www.motherjones.com.

Mukhopadhyay, Carol C., Rosemary Henze, and Yolanda T. Moses. 2007. *How Real Is Race?* Lanham, MD: Rowman & Littlefield Education.

Nadal, Kevin L., Yinglee Wong, Katie E. Griffin, Kristin Davidoff, and Julie Sriken. 2014. "The Adverse Impact of Racial Microaggressions on College Students' Self-Esteem." *Journal of College Student Development* 55(5):461–474.

Nagle, Mary Kathryn. 2015 (June 22). "Take Andrew Jackson off the $20 Bill." MSNBC. Available at www.msnbc.com.

National Academies of Science, Engineering, and Medicine. 2017. "The Economic and Fiscal Consequences of Immigration." Washington, DC: National Academies Press.

National Conference of State Legislatures (NCSL). 2013 (June). "Affirmative Action: An Overview." Available at www.ncsl.org.

Nixon, Ron. 2018 (March 22). "What Border Agents Say They Want (It's Not a Wall)." *New York Times*. Available at www.nytimes.com.

Nowrasteh, Alex. 2015 (July 14). "Immigration and Crime: What the Research Says." Cato Institute. Available at www.cato.org.

Onyejiaka, Tiffany. 2017 (August 22). "Hollywood's Colorism Problem Can't Be Ignored Any Longer." *Teen Vogue*. Available at www.teenvogue.com.

Orrenius, Pia, and Madeline Zavodny. 2019. "Do Immigrants Threaten US Public Safety?" *Journal on Migration and Human Security* 7(3):52–61.

Ossorio, Pilar, and Troy Duster. 2005. "Race and Genetics." *American Psychologist* 60(1):115–128.

Pager, Devah. 2003. "The Mark of a Criminal Record." *American Journal of Sociology* 108(5):937–975.

Parry, Marc. 2020 (August 7). "In California, Ethnic Studies Could Soon Be Required by Law. A Former Professor Is behind It." *The Chronicle of Higher Education*. Available at www.chronicle.com.

Passel, Jeffrey S., and D'Vera Cohn. 2011. "Unauthorized Immigrant Population: National and State Trends." Pew Hispanic Research Center. Available at www.pewhispanic.org.

Passel, Jeffrey S., and D'Vera Cohn. 2016 (November 17). "Children of Unauthorized Immigrants Represent Rising Share of K–12 Students." Pew Research Center. Available at www.pewresearch.org.

Passel, Jeffrey S., and D'Vera Cohn. 2018 (June 12). "Unauthorized Immigrant Workforce Is Smaller, but with More Women." Pew Research Center. Available at www.pewresearch.org.

Passel, Jeffrey S., and D'Vera Cohn. 2019 (June 12). "Mexicans Decline to Less Than Half the U.S. Unauthorized Immigrant Population for the First Time." Pew Research Center. Available at www.pewresearch.org.

Passel, Jeffrey S., and Paul Taylor. 2009. "Who's Hispanic?" Pew Hispanic Research Center. Available at www.pewhispanic.org.

Pew Research Center. 2010 (January 12). "Blacks Upbeat about Black Progress, Prospects." Available at www.pewsocialtrends.org.

Pew Research Center. 2015 (January 15). "Unauthorized Immigrants: Who They Are and What the Public Thinks." Available at www.pewresearch.org.

Picca, Leslie, and Joe R. Feagin. 2007. *Two-Faced Racism*. New York: Routledge.

Pierce, Sarah, and Jessica Bolter. 2020 (July). "Dismantling and Reconstructing the U.S. Immigration System: A Catalog of Changes under the Trump Presidency." *Migration Policy Institute*. Available at www.migrationpolicy.org.

Pollin, Robert. 2011. "Economic Prospects: Can We Stop Blaming Immigrants?" *New Labor Forum* 20(1):86–89.

Quillian, Lincoln, Devah Pager, Ole Hexel, and Arnfinn H. Midtbøen. 2017. "Meta-Analysis of Field Experiments Shows No Change in Racial Discrimination in Hiring over Time." *Proceedings of the National Academy of Sciences* 114(41):10870–10875.

Quillian, Lincoln, John J. Lee, and Brandon Honoré. 2020. "Racial Discrimination in the US Housing and Mortgage Lending Markets: A Quantitative Review of Trends, 1976–2016." *Race and Social Problems* 12(1):13–28.

Radford, Jynnah. 2019 (June 17). "Key Findings about U.S. Immigrants." Pew Research Center. Available at www.pewresearch.org.

Reflective Democracy Campaign. 2020. "National Representation Index: Texas." Who Leads US? Available at www.wholeads.us.

Rico, Brittany, Rose M. Kreider, and Lydia Anderson. 2018 (July 9). "Growth in Interracial and Interethnic Married-Couple Households." *Race, Ethnicity, and Marriage in the United States*. Available at www.census.gov.

Romo, Vanessa. 2020 (July 15). "Asheville, N.C., Approves Steps toward Reparations for Black Residents." National Public Radio. Available at www.npr.org.

Schiller, Bradley R. 2004. *The Economics of Poverty and Discrimination*, 9th ed. Upper Saddle River, NJ: Pearson Education.

Schuman, Howard, and Maria Krysan. 1999. "A Historical Note on Whites' Beliefs about Racial Inequality." *American Sociological Review* 64:847–855.

Shah, Khushbu. 2019 (July 25). "They Look White but Say They're Black: A Tiny Town in Ohio Wrestles with Race." *Guardian*. Available at www.theguardian.com.

Shapiro, Ilya, and James Knight. 2020 (June 29). "An Assessment of Minority Voting Rights Access in the United States." CATO Institute. Available at www.cato.org.

Shipler, David K. 1998. "Subtle vs. Overt Racism." *Washington Spectator* 24(6):1–3.

Sidanius, Jim, Shana Levin, Colette Van Laar, and David O. Sears. 2010. *The Diversity Challenge: Social Identity and Intergroup Relations on the College Campus*. New York: Russell Sage Foundation.

Sinclair, Stacey, Andreana C. Kenrick, and Drew S. Jacoby-Senghor. 2014 (October). "Whites' Interpersonal Interactions Shape, and Are Shaped by, Implicit Prejudice." *Policy Insights from the Behavioral and Brain Sciences* 1(1):81–87.

Southern Poverty Law Center (SPLC). 2020. *Hate Map*. Available at www.splc.org.

Southern Poverty Law Center (SPLC). 2015 (Spring). "iTunes Dumps Hate Music but Spotify and Amazon Still Selling It." *Intelligence Report*, No. 157. Available at www.splcenter.org.

Starck, Jordan G., Travis Riddle, Stacey Sinclair, and Natasha Warikoo. 2020 (January). "Teachers Are People Too: Examining the Racial Bias of Teachers Compared to Other American Adults." *Educational Researcher*. doi: 0013189X20912758.

Stepler, Renee, and Anna Brown. 2016 (April 19). "Statistical Portrait of Hispanics in the United States." Pew Research Center. Available at www.pewhispanic.org.

Sue, D. W., C. M. Capodilup, G. C. Torino, M. Bucceri, A. M. B. Holder, K. L. Nadal, et al. 2007. "Racial Microaggressions in Everyday Life: Implications for Clinical Practice." *American Psychology* 62(4):271–286.

Swift, Art. 2017 (March 15). "Americans' Worries about Race Relations at Record High." Gallup Organization. Available at www.gallup.com.

Tanneeru, Manav. 2007 (May 11). "Asian-Americans' Diverse Voices Share Similar Stories." CNN.com. Available at www.cnn.com.

Tello, Monique. 2017 (January 16). "Racism and Discrimination in Health Care: Providers and Patients." *Harvard Health Publishing.* Available at www.health.harvard.edu.

Tolbert, Caroline J., and John A. Grummel. 2003. "Revisiting the Racial Threat Hypothesis: White Voter Support for California's Proposition 209." *State Politics and Policy Quarterly* 3(2):183–202, 215–216.

Turn It Down. 2009 (May 3). "Social Networking: A Place for Hate?" Available at turnitdown.newcomm.org.

Tsu, Naomi. 2020. "What Is a Sanctuary City Anyway?" *Teaching Tolerance.* Available at www.tolerance.org.

Urban Dictionary. n.d. Available at www.urbandictionary.com.

U.S. Census Bureau. 2011 (July 3). "*2010 Census Data.*" Generated by C. Schacht. Available at 2010.census.gov/2010census/data/index.php.

U.S. Census Bureau. 2016. "2011–2015 American Community Survey 5-Year Estimates." Available at www.census.gov.

U.S. Census Bureau. 2019. "Quick Facts: Texas." Available at www.census.gov.

U.S. Census Bureau. 2020. "Questions Asked: Hispanic Origin." Available at www.2020census.gov.

U.S. Census Bureau. 2020a. "Table B01003." *American Community Survey,* 2011–2015 ACS 5-Year Estimates. Available at data.census.gov.

U.S. Census Bureau. 2020b. "Table 1.1 Population by Sex, Age, Nativity, and U.S. Citizenship Status: 2019." *Current Population Survey.* Available at www.census.gov.

U.S. Census Bureau. 2020c. "Table B05006 Place of Birth for the Foreign-Born Population in the United States." *American Community Survey,* 2018 ACES 1-Year Estimates. Available at www.census.gov.

U.S. Citizenship and Immigration Services. 2020. *A Guide to Naturalization.* Available at www.uscis.gov.

U.S. Customs and Border Protection. 2020 (February). "Snapshot: A Summary of CBP Facts and Figures." Available at www.cbp.gov.

U.S. Department of Labor. 2002. *Facts on Executive Order 11246 Affirmative Action.* Available at www.dol.gov.

Vespa, Jonathon, David M. Armstrong, and Laren Medina. 2020 (February). "Demographic Turning Points for the United States: Population Projections 2020 to 2060." U.S. Census Bureau. Report no. P25-1144. Available at www.census.gov.

Vestal, Christine. 2020 (June 15). "Racism Is a Public Health Crisis, Say Cities and Counties." Stateline. Available at www.pewtrusts.org.

Vinson, Liz. 2020 (August 14). "Treated Like Slaves: SPLC Sues Farm Labor Contractor for Underpaying H-2A Guest Workers and Threatening to Deport Them for Speaking Out." Southern Poverty Law Center. Available at www.splcenter.org.

Wang, Yu-Wei, M. Meghan Davidson, Oksana F. Yakushko, Holly Bielstein Savoy, Jeffrey A. Tan, and Joseph K. Bleier. 2003. "The Scale of Ethnocultural Empathy: Development, Validation, and Reliability." *Journal of Counseling Psychology* 50(2):221.

Washington, John. 2020 (February 25). "Family Separations at the Border Constitute Torture, New Report Claims." *The Intercept.* Available www.theintercept.com.

Weeden, L. Darnell. 2018. "Unreasonably Restrictive Voter Photo Identification Requirements Are Unequal Economic Barriers to Equal Access to the Right to Vote." *Southern California Review of Law and Social Justice* 27:73–86.

Williams, Eddie N., and Milton D. Morris. 1993. "Racism and Our Future." In *Race in America: The Struggle for Equality,* ed. Herbert Hill and James E. Jones Jr, 417–424. Madison: University of Wisconsin Press.

Wilson, Meagan Meuchel. 2014. "Hate Crime Victimization, 2004–2012—Statistical Tables." Bureau of Justice Statistics. Available at www.bjs.gov.

Wines, Michael. 2019 (July 2). "2020 Census Won't Have Citizenship Question as Trump Administration Drops Effort." *New York Times.* Available at www.nyt.com.

Winfrey, Oprah. 2009 (October). "Oprah Talks to Jay-Z." *O, The Oprah Magazine.* Available at www.oprah.com.

Winter, Greg. 2003 (January 21). "Schools Resegregate, Study Finds." *New York Times.* Available at www.nytimes.com.

Wise, Tim. 2009. *Between Barack and a Hard Place: Racism and White Denial in the Age of Obama.* San Francisco: City Light Books.

Women's Health Editors. 2020 (June 5). "'Overheard While Black' Instagram Account Shows How Causal Racism and Micro-Aggressions Happen Every Day." *Women's Health.* Available at www.womenshealthmag.com.

Wood, Graeme. 2020 (May 27). "What's Behind the COVID-19 Racial Disparity?" *The Atlantic.* Available at www.theatlantic.com.

Yeung, Jeffrey G., Lisa B. Spanierman, and Jocelyn Landrum-Brown. 2013. "'Being White in a Multicultural Society': Critical Whiteness Pedagogy in a Dialogue Course." *Journal of Diversity in Higher Education* 6(1):17–32.

Yong, Ed. 2020 (August 4). "How the Pandemic Defeated America." *Atlantic.* Available at www.theatlantic.com.

Chapter 10

Abernathy, Michael. 2003. *Male Bashing on TV.* Tolerance in the News. Available at www.tolerance.org.

Aley, Melinda, and Lindsay Hahn. 2020 (February 6). "The Powerful Male Hero: A Content Analysis of Gender Representation in Posters for Children's Animated Movies." Sex Roles. Available at www.researchgate.net.

Almed, Shumaila, and Juliana Abdul Wahab. 2014. "Animation and Socialization Process: Gender Role Portrayal in Cartoon Network." *Asian Social Science* 10(3):44–53.

American Association of University Women (AAUW). 2020a. "Fast Facts: Occupational Segregation." Available at www.aauw.org.

American Association of University Women (AAUW). 2020b. "Gender Pay Gap by State." Available at www.aauw.org.

Andersen, Margaret L. 1997. *Thinking about Women,* 4th ed. New York: Macmillan.

Anderson, Monica, and Skye Toor. 2018 (October 11). "How Social Media Users Have Discussed Sexual Harassment Since #MeToo Went Viral." Pew Research Center. Available at www.pewresearch.org.

Andrew, Scotty. 2020 (February 7). "More Than 200 Medical Professionals Contend Bills Trying to Restrict Transgender Kids from Getting Gender Reassignment Treatments." CNN Health. Available at www.cnn.com.

Anti-Defamation League (ADL). 2018. *When Women Are the Enemy: The Intersection of Misogyny and White Supremacy.* Center on Extremism. Available at www.adl.org.

Aron, Nina Renata. 2019 (March 8). "What Does Misogyny Look Like?" *New York Times.* Available at www.nytimes.com.

Arrindell, W. A., Sonja Van Well, Annemarie M. Kolk, Dick P. H. Barelds, Tian P. S. Oei, Pui Yi Lau, and the Cultural Clinical Psychology Study Group. 2013. "Higher Levels of Masculine Gender Role Stress in Masculine Than in Feminine Nations: A 13-Nations Study." *Cross-Cultural Research* 47(1):51–67.

Atwater, Leanne E., Allison M. Tringale, Rachel E. Sturm, Scott N. Taylor, and Phillip W. Braddy. 2019 (October–December). "Looking Ahead: How What We Know about Sexual Harassment Now Informs Us of the Future." *Organizational Dynamics* 48(4).

Averbuch, Maya. 2020 (March 9). "'We'll Strike': Thousands of Mexican Women Strike to Protest Femicide." *Guardian.* Available at amp.guardian.com.

Badgett, M. B. Lee, Soon Kyu Choi, and Blanco D. M. Wilson. 2019 (October). *LGBT Poverty in the United States: A Study of Differences between Sexual Orientation and Gender Identity Groups.* Williams Institute. Available at williamsinstitute.law.ucla.ed.

Barnes, Robert. 2016 (June 23). "Supreme Court Upholds University of Texas Affirmative-Action Admissions." *Washington Post.* Available at www.washingtonpost.com.

Bavel, Jan Van, Christine R. Schwartz, and Albert Esteve. 2018. "The Reversal of the Gender Gap in Education and Its Consequences for Family Life." *Annual Review of Sociology* 44:341–360.

Begley, Sharon. 2000 (November 6). "The Stereotype Trap." *Newsweek,* pp. 66–68.

Beitsch, Rebecca. 2018 (July 31). "#MeToo Has Changed Our Culture. Now It's Changing Our Laws." Pew Research Center. Available at www.pewtrusts.org.

Benham, Nicholas, Maya Desai, Madeleine Freeman, Tori Kutzner, and Karuna Srivastav. 2019. "Single-Sex Education." *Georgetown Journal of Gender and the Law* 20(2):509.

Bertrand, Marianne, Claudia Goldin, and Lawrence F. Katz. 2009 (January). "Dynamics of the Gender Gap for Young Professionals in the Financial and Corporate Sectors." Working Paper. Available at www.economics.harvard.edu.

Biana, Hazel T. 2020. "Extending Bell Hooks' Feminist Theory." *Journal of International Women's Studies* 21(1):13–29.

Biden, Joe. 2020. "The Biden Agenda for Women." Available at www.joebiden.com.

Blad, Evie. 2021 (January 28). "School Sports a Fresh Front in State Battles over Transgender Students' Rights." *Education Week.* Available at www.edweek.org.

Blau, Francine D., and Lawrence M. Kahn. 2013 (January). *Female Labor Supply: Why Is the US Falling Behind?* NBER Working Paper No.18702. National Bureau of Economic Research. Available at www.nber.org.

Bly, Robert. 1990. *Iron John: A Book about Men.* Boston: Addison-Wesley.

Boorstein, Michelle. 2014 (September 3). "US Evangelicals Headed for Showdown over Gender Roles." *Washington Post.* Available at www.washingtonpost.com.

Bower, Tim. 2019. (September–October). "The #MeToo Backlash." *Harvard Business Review.* Available at hbr.org.

Brenan, Megan. 2020 (January 29). "Women Still Handle Main Household Tasks in U.S." Available at news.gallup.com.

Brooks, John. 2018 (May 2). "The New Generation Overthrows Gender." *Your Health.* National Public Radio. Available at www.npr .org.

Brown, Elspeth H. 2019. *Work! A Queer History of Modeling.* Durham, NC: Duke University Press.

Brown, Jasmine, Gloria Riviera, and Shannon K. Crawford. 2019 (July 31). "Child Brides in the U.S. Share Stories of Exploitation, Becoming a Wife: 'I knew I was 11. I knew he was 20.'" ABC News. Available at abcnews.go.com.

Bruce, Adrienne, N., Alexis Battista, Michael W. Plankey, Lynt B. Johnson, and M. Blair Marshall. 2015 (February). "Perceptions of Gender-Based Discrimination during Surgical Training and Practice." *Medical Education Online* 20:1–10.

Bundhun, Rebecca. 2017 (February 10). "Dowries and Death Continue Apace in India." *The National.* Available at www.thenational.com.

Bureau of Labor Statistics (BLS). 2013. "Characteristics of Minimum Wage Workers: 2012." Labor Force Statistics from the Current Population Survey. Available at www.bls.gov.

Bureau of Labor Statistics (BLS). 2017. "Employed Persons by Detailed Occupation, Sex, Race, and Hispanic or Latino Ethnicity." Labor Force Statistics from the Current Population Survey. Available at www.bls.gov.

Bureau of Labor Statistics (BLS). 2020. "Employed Persons by Detailed Occupation, Sex, Race, and Hispanic or Latino Ethnicity." Labor Force Statistics from the Current Population Survey. Available at www.bls.gov.

Butler, Kiera. 2020 (March/April). "The Discredited Science behind the Rise of Single-Sex Public Schools." *Mother Jones.* Available at www.motherjones.com.

Carlana, Michela. 2019 (March 2). "Implicit Stereotypes: Evidence from Teachers' Gender Bias." *Quarterly Journal of Economics*: 1163–1224.

Carlsen, Audrey Maya Salam, Claire Cain Miller, Denise Lu, Ash Ngu, Jugal K. Patel, and Zach Wichter. 2018 (October 29). "#MeToo Brought Down 201 Powerful Men. Nearly Half of Their Replacements Are Women." *New York Times.* Available at www.nytimes.com.

Castro-Peraza, Maria Elisa, Jesús Manuel García-Acosta, Naira Delgado, Ana María Perdomo-Hernández, Maria Inmaculada Sosa-Alvarez, Rosa Llabrés-Solé, and Nieves Doria Lorenzo-Rocha. 2019. "Gender Identity: The Human Right of Depathologization." *Environmental Research and Public Health* 16(6):978.

Catalyst Research. 2019 (June 5). "Women in the Workforce—United States: Quick Take." Available at www.catalyst.org.

Cecco, Leyland. 2019 (September 27). "Toronto Man Attacked, Suspect Says He Was Radicalized Online by 'Incels.'" *Guardian.* Available at www.theguardian.com.

Center for American Women in Politics (CAWP). 2020. "Facts." Available at www.cawp.rutgers.edu.

Centers for Disease Control and Prevention (CDC). 2020 (April). *Increase in Suicide Mortality in the United States, 1999–2018.* NCHS Data Brief No. 362. Available at www.cdc.gov.

Charlesworth, Tessa P. S., and Mahzarin R. Banaji. 2019. "Gender in Science, Technology, Engineering, and Mathematics: Issues, Causes, Solutions." *The Journal of Neuroscience* 39(37):7228–7243.

Christie, Brett. 2019 (January 31). "Democrats Reintroduce Paycheck Fairness Act." *Workspan Daily.* Available at www.worldatwork.org.

Cimpian, Joseph R., Sarah T. Lubienski, Jennifer D. Timmer, Martha B. Makowski, and Emily K. Miller. 2016. "Have Gender Gaps in Math Closed? Achievement, Teacher Perceptions, and Learning Behaviors across to ECLS-K Cohorts." *American Educational Research Association Journal* 2(4):1–19. Available at journals.sagepub.com.

Coalition to End Violence against Women and Girls Globally. 2017. "IVAWA Bill Overview 2017." Available at www .futureswithoutviolence.org.

Coffey, Clare, Patricia Espinoza Revollo, Rowan Harvey, Max Lawson, Anam Parvez Butt, Kim Piaget, Diana Sarosi, Julie Thekkudan. 2020 (January). *Time to Care: Unpaid and Underpaid Care Work in the Global Inequality Crisis.* Available at oxfamilibrary.openrepository.com.

Cohen, Philip. 2004. "The Gender Division of Labor: 'Keeping House' and Occupational Sex Segregations in the United States." *Gender and Society* 18(2):239–252.

Cohen, Theodore. 2001. *Men and Masculinity.* Belmont, CA: Wadsworth.

Collins, Patricia H. 2015. "No Guarantees: Symposium on Black Feminist Thought." *Ethnic and Racial Studies* 38(13):2349–2354.

Congress. 2020a. "Pregnant Workers Fairness Act." Available at www.congress.gov.

Congress. 2020b. "Raise the Wage Act." Available at www.congress.gov.

Cook, Carolyn. 2009 (April 12). "ERA Would End Women's Second-Class Citizenship: Only Three More States Are Needed to Declare Gender Bias Unconstitutional." *Philadelphia Inquirer.* Available at www.philly.com.

Corbett, Christianne, and Catherine Hill. 2012 (October 24). *Graduating to a Pay Gap: The Earnings of Women and Men One Year after College Graduation.* American Association of University Women. Available at www.aauw .org.

Correll, Shelly J., Stephen Benard, and In Paik. 2007. "Getting a Job: Is There a Motherhood Penalty?" *American Journal of Sociology* 112(5):1297–1338.

Cox, Josie. 2020 (April 18). "Coronavirus and the Gender Pay Gap: An Excuse to Avoid Uncomfortable Facts." *Forbes.* Available at www.forbes.com.

Crary, David. 2019 (January 14). "Women Strive for Larger Roles in Male-Dominated Religions." Associated Press. Available at apnews.com.

Christie, Brett. 2019 (January 31). "Democrats Reintroduce Paycheck Fairness Act." *Workspan Daily.* Available at www.worldatwork.org.

Demberger, Brittany M., and Joanna R. Pepin. 2020. "Gender Flexibility, but Not Equality: Young Adults Division of Labor Preferences." *Sociological Science* 7:36–56.

Diaz, Jaclyn. 2019 (July 31). "States Look to Remedy Pay Gap as Federal Legislation Stalls." Bloomberg News. Available at bloomberglaw. com.

Drydakis, Nick. 2019. Trans People, Transitioning, Mental Health, Life and Job Satisfaction." 2019 (October). *IZA Institute of Labor Economics.* Discussion Paper Series. No. 12695 Available at ftp.iza.org.

Dunatchik, Allison, and Berkay Ozcan. 2019 (July). "Reducing Mommy Penalties with Daddy Quotas." *Social Policy Working Paper.* Available at www.lse.ac.uk.

Dunn, Marianne G., Aaron B. Rochlen, and Karen M. O'Brien. 2013. "Employee, Mother, and Partner: An Exploratory Investigation of Working Women with Stay-at-Home Fathers." *Journal of Career Development* 40(1):3–22.

Dwyer, Colin. 2020 (March 11). "Harvey Weinstein Sentenced to 23 Years in Prison for Rape and Sexual Abuse." National Public Radio. Available at www.npr.org.

Einwohner, Rachel L., and Elle Rochford. 2019. "After the March: Using Instagram to Perform and Sustain the Women's March." *Sociological Forum* 34(S1):1090–1111.

Eisner, Manuel, and Lana Ghuneim. 2013. "Honor Killing Attitudes amongst Adolescents in Amman, Jordan." *Aggressive Behavior* 39(5):405–417.

Elsesser, Kim. 2019 (April 1). "The Gender Pay Gap in the Career Choice Myth." *Forbes.* Available at www.forbes.com.

Equal Employment Opportunity Commission (EEOC). 2020a "Sex-Based Charges (Charges Filed with EEOC) FY 1997–FY 2019." Available at www.eeoc.gov.

Equal Employment Opportunity Commission. 2020b. "Sexual Harassment." Available at www.eeoc.gov.

Espinoza, Penelope, Ann B. Areas du Luz Fontes, and Clarissa J. Arms-Chavez. 2014. "Attributional Gender Bias: Teacher's Ability and Effort Explanations for Student's Math Performance." *Social Psychology Education* 17:105–126.

European Social Survey. 2013 (July). *Attitudes and Behaviors in a Changing Europe.* Available at www.esrc.ac.uk.

Fisher, Dana R., Don M. Danielle, and Rashawn Ray. 2017. "Intersection Takes It to the Streets: Mobilizing across Diverse Interests of the Women's March." *Science Advances* 3:1–8. Available at advancessciencemag.org.

Flaherty, Colleen. 2020 (April 21). "No Room of One's Own: Early Journal Submission Data Suggest COVID-19 Is Tanking Women's Research Productivity." *Inside Higher Education.* Available at www.insidehighered.com.

Freeman, Daniel, and Jason Freeman. 2013. *Stressed Sex: Uncovering the Truth about Men, Women, and Mental Health.* Oxford: Oxford University Press.

Friedan, Betty. 1963. *The Feminine Mystique.* New York: Norton Press.

Funnell, Nina. 2020 (May 8). "Where the Legal System Silences Women." *Washington Post.* Available at ww.washingtonpost.com.

Gallagher, Sally K. 2004. "The Marginalization of Evangelical Feminism." *Sociology of Religion* 65:215–237.

Garber, Megan. 2020 (March 5). "America Punished Elizabeth Warren for Her Competence." *Atlantic.* Available at www.theatlantic.com.

Gersen, Suk Jeannie. 2020 (May 16). "How Concerning Are the Trump's Administration's New Title IX Regulations?" *New Yorker.* Available at www.newyorker.com.

Gill, Aisha K. (2019). "Social and Cultural Implications of 'Honor'-Based Violence." In *International Human Rights of Women,* ed. Niamh Reilly, 356–380. International Singapore: Human Rights, Springer.

Girls' Attitudes Survey. 2019. "Executive Summary." Available at www.girlguiding.org.uk/.

Global Partnership. 2019 (February). *Breaking down Barriers to Girls' Education.* Global Partnership for Education. Available at www.globalpartnership.org.

Godin, Melissa. 2020 (May 20). "Canadian Team Charged with Terrorism over Attack Allegedly Motivated by 'Incel Movement.'" *Time Magazine.* Available at time.com.

Goffman, Erving. 1963. *Stigma.* Englewood Cliffs, NJ: Prentice Hall.

Goodkind, Sara, Lisa Schelbe, Andrea A. Joseph, Daphne E. Beers, and Stephanie L. Pinsky. 2013. "Providing New Opportunities or Reinforcing Old Stereotypes? Perceptions and Experiences of Single-Sex Public Education." *Children and Youth Services Review* 35:1174–1181.

Goodstein, Laurie. 2009 (July 2). "U.S. Nuns Facing Vatican Scrutiny." *New York Times.* Available at www.nytimes.com.

Goodstein, Laurie. 2012 (April 18). "Vatican Reprimands a Group of U.S. Nuns and Plans Changes." *New York Times.* Available at www.nytimes.com.

Graff, Nikki, Anna Brown, and Eileen Patton. 2019 (March 22). "The Narrowing, but Persistent, Gender Gap in Pay." Pew Research Center. Available at www.pewresearch.org.

Greijdanus, Hedy, Carlos A. de Matos Fernandes, Felicity Turner-Zwinkels, Ali Honari, Carla A. Roos, Hannes Rosenbusch, and Tom Postmes. 2020. "The Psychology of Online Activism and Social Movements: Relations between Online and Off-Line Collective Action." *Current Opinions in Psychology* 21(3):49–54.

Green, Susan. 2018 (October 16). "Women Candidates: How to Win the Second Time Around." Institute for Women's Policy Research. Available at iwpr.org.

Guy, Mary Ellen, and Meredith A. Newman. 2004. "Women's Jobs, Men's Jobs: Sex Segregation and Emotional Labor." *Public Administration Review* 64:289–299.

Hamel, Liz, Jamie Firth, and Mollyann Brodie. 2014 (December 11). "Kaiser Family Foundation/New York Times/CBS News Non-Employed Poll." Available at kff.org.

Hamilton, Mykol C., David Anderson, Michelle Broaddus, and Kate Young. 2006. "Gender Stereotyping and Under-Representation of Female Characters in 200 Popular Children's Picture Books: A Twenty-First Century Update." *Sex Roles* 55:757–765.

Harvard Men's Health Watch (HMHW). 2019 (August 26). "Mars vs. Venus: The Gender Gap in Health." *Harvard Medical School Newsletter.* Available at www.health.harvard.edu.

Hass, Ann P., Philip L. Rodgers, and Jody L. Herman. 2014. "Suicide Attempts among Transgender and Gender-Non-Conforming Adults: Findings of the National Transgender Discrimination Survey." Williams Institute. Available at williamsinstitute.law.ucla.edu.

Head, Sarah K., Sally Zweimueller, Claudia Marchena, and Elliott Hoel. 2014. *Women's Lives and Challenges: Equality and Empowerment since 2000.* United States Agency for International Development. Rockville, MD: ICF International.

Hegewisch, Ariane, and Zohal Barsi. 2020 (March 24). "The Gender Wage Gap by Occupation, 2019." Institute for Women's Policy Research Fact Sheet. Available at iwpr.org.

Heldman, Caroline, Shrikanth Narayanan, Rebecca Cooper, et al. 2020. *See Jane 2020 TV Report: Historic Screen Time & Speaking Time for Female Characters.* Geena Davis Institute for Gender in Media. Available at seejane.org.

Heilman, Brian, María Rosario Castro Bernardini, and Kimberly Pfeifer. 2020. *Caring Under COVID-19: How the Pandemic Is—and Is Not—Changing Unpaid Care and Domestic Work Responsibilities in the United States.* Boston: Oxfam, and Washington, DC: Promundo–US.

Hendrickson, Christine, and Nolan R. Theurer. 2020 (March 4). "What Is Past Is Prologue: The Ninth Circuit Again Rules That Prior Salary Cannot Justify Pay Differences." News and Insights. Available at www.seyfarth.com.

Hendrix, Steve. 2019 (June 7). "He Always Hated Women, Then He Decided to Kill Them." *Washington Post.* Available at www.washingtonpost.com.

Hersch, Joni. 2013. "Opting Out among Women with Elite Education." Vanderbilt Law and Economics Research Paper No. 13-05. Nashville, TN: Vanderbilt University.

Heyder, Anke, and Ursula Kessels. 2013. "Is School Feminine? Implicit Gender Stereotyping of School as a Predictor of Academic Achievement." *Sex Roles* 69:605–617.

Hochschild, Arlie. 1989. *The Second Shift.* London: Penguin.

Hofferth, Sandra L., and Francis Goldscheider. 2017. "Reflections on the Future of the Second Half of the Gender Revolution." Data Points. *Newsletter of the Population Association of America.* Available at populationassociation.org.

hooks, bell. 1984. *Feminist Theory from Margin to Center.* Cambridge, MA: South End Press.

Hvistendahl, Mara. 2011. *Unnatural Selection: Choosing Boys over Girls, and the Consequences of a World Full of Men.* Philadelphia: Public Affairs.

Indictment. 2019. "*United States of America v. Jeffrey Epstein, Defendant.*" United States District Court, Southern District of New York. Available at www.justice.gov.

Institute for Women's Policy Research (IWPR). 2020 (March). "The Gender Wage Gap by Occupation 2019 and by Race and Ethnicity." Available at www.iwpr.org.

International Labour Organization (ILO). 2014 (March 10). "Trafficking in Human Beings: A Severe Form of Violence against Women and Girls, and a Flagrant Violation of Human Rights." *Speeches and Statements.* Available at www.ilo.org.

International Labour Organization (ILO). 2015. *Global Wage Report 2014/15.* Available at www.ilo.org.

International Labour Organization (ILO). 2019. *A Quantum Leap for Gender Equality.* Available at www.ilo.org.

International Violence against Women Act. 2013 (June). "Issue Brief: The International Violence against Women Act." Available www.amnestyusa.org.

Ishikawa, Yumi. 2020 (May 8). "High Heels and Discrimination in Everyday Life." *Washington Post.* Available at www.washingtonpost.com.

Jackson, Janna. 2010. "'Dangerous Presumptions': How Single-Sex Schooling Reifies False Notions of Sex, Gender, and Sexuality." *Gender and Education* 22(2):227–238.

Janjuha-Jivraj, Shaheena. 2020 (October 11). "Why the Traits of Female Leadership Are Better Geared for the Global Pandemic." *Forbes.* Available at www.forbes.com.

Jitha, T. J. 2013. "Mediating Production, Re-Powering Patriarchy: The Case of Micro Credit." *India Journal of Gender Studies* 20(2):253–278.

Johns, Michelle M., Richard Lowry, Jack Andrzejewski, Lisa C. Barrios, Zewditu Demissie, Timothy McManus, Catherine N. Rasberry, Leah Robin, and J. Michael Underwood. 2019 (January 25). "Transgender Identity in the Experiences of Violence, Victimization, Substance Use, Suicide Risk and Sexual Risk Behaviors among High School Students—19 States and Large Urban School District, 2017." *Morbidity and Mortality Weekly Report* 68(3):67–71.

Johnson, Andrea, Kathryn Menefee, and Ramya Sekaran. 2019 (July). *Progress in Advancing Me Too Workplace Reforms in #20 States by 2020.* National Women's Law Center. Available at www.nwlc.org.

Kahn, Chris. 2020 (March 27). "U.S. Men Less Likely to Heed Health Warnings as Coronavirus Death Toll Mounts: Reuters Poll." Available at www.reuters.com.

Kaufman, Michelle. 2019 (September 4). "How Big Is the Gender Pay Gap in Sports? It's Much Bigger Than You Think. Here Is Proof." *Miami Herald.* Available at www.miamiherald.com.

Keplinger, Ksenia, Stephanie K. Johnson, Jessica F. Kirk, and Liza Y. Barnes. 2019 (September). "Women at Work: Changes in Sexual Harassment between September 2016 and September 2018." *Plos ONE* 14(7):1–20.

Kimmel, Michael. 2011 (Winter). "Gay Bashing Is about Masculinity." *Voice Male.* Available at www.voicemalemagazine.org.

Kimmel, Michael. 2012 (December 19). "Masculinity, Mental Illness and Guns: A Lethal Equation?" *CNN News.* Available at www.cnn.com.

Klein, Sue, Jennifer Lee, Page McKinsey, and Charmaine Archer. 2014 (December). *Identifying U.S. K–12 Public Schools with Delivered Sex Segregation.* Arlington, VA: Feminist Majority Foundation. Available at www.feminist.org/education.

Kristof, Nicholas. 2016 (January 31). "Her Father Shot Her in the Head as an 'Honor Killing.'" *New York Times.* Available at www.nytimes .com.

Kurtzleben, Danielle. 2020 (February 13). "House Votes to Provide Equal Rights Amendment, Removing Ratification Deadline." National Public Radio. Available at www.npr.org.

Lacorte, Valerie, and Jeff Hayes. 2019 (November 5). "Women's Median Earnings as a Percent of Men's, 1985–2018, with Projections for Pay Equity, by Race/Ethnicity." Institute for Women's Policy Research. Available at iwpare. org.

Lauzen, Martha M. 2020. "It's a Man (Celluloid) World: Portrayals of Female Characters in the Top Grossing Films of 2019." Women in TV and Films. Available at womenintvfilm.sdsu. edu.

Leopold, Thomas, and Jan Skopek. 2014. "Gender and the Division of Labor in Older Couples: How European Grandparents Share Market Work and Childcare." *Social Forces* 93(1):63–91.

Lepkowska, Dorothy. 2008 (December 16). "Playing Fair?" *Guardian.* Available at www .guardian.co.uk.

Leveille, Vania, and Lenora Lapidus. 2019 (April 10). "The BE HEARD Act Will Overhaul Workplace Harassment Laws." Available at www.aclu.org.

Levin, Diane E., and Jean Kilbourne. 2009. *So Sexy So Soon: The New Sexualized Childhood and What Parents Can Do to Protect Their Kids.* New York: Random House.

Levin, Sam. 2020 (November 19). "'The Fight Doesn't Stop Here': What LGBTQ+ Advocates Want from a Biden Presidency." *Guardian.* Available at www.theguardian.com.

Levenson, Eric. 2018 (April 27). "The Long, Winding Path to Bill Cosby's Guilty Verdict." CNN News. Available at www.cnn.news.

Levenson, Michael. 2020 (January 30). "U.S.A. Gymnastics Offers $215 Million to Larry Nassar Victims." *New York Times.* Available at www .nytimes.com.

Lillis, Mike. 2020 (July 21). "Ocasio-Cortez Accosted by GOP Lawmaker over Remarks: 'That Kind of Confrontation Hasn't Ever Happened to Me.'" *The Hill.* Available at thehill.com.

Lopez-Claros, Augusto, and Saadia Zahidi. 2005. *Women's Empowerment: Measuring the Global Gender Gap.* Cologny, Switzerland: World Economic Forum.

Magliozzi, Devon, Aliya Saperstein, and Laurel Westbrook. 2016. "Scaling Up: Representing Gender Diversity in Social Surveys." *Socius* 2:1–11.

Maresca, Thomas. 2020 (February 7). "Two Years Into #MeToo in South Korea, Change Is Slow to Come." *United Press International.* Available at www.upi.com.

Mascaro, Jennifer S., Kelly E. Rentscher, Patrick D. Hackett, Matthias R. Mehl, and James K. Rilling. 2017. "Child Gender Influences Paternal Behavior, Language, and Brain Function." *Behavioral Neuroscience* 131(3):262–273.

McKenzie, Sheena. 2019 (June 18). "They Wanted a Son So Much They Made Their Daughter Live as a Boy." CNN News. Available at www .cnn.news.

Medeiros, Mike, Benjamin Forest, and Patrick Öhberg. 2020. "The Case for Non-Binary Gender Questions in Surveys." *PS Political Science and Politics* 53(1):128–135.

Mehmood, Isha. 2009 (January 29). "Lilly Ledbetter Fair Pay Act Becomes Law." Available at www.civilrights.org.

Mendes, Kaitlynn, Jessica Ringrose, and Jessalynn Keller. 2018. "#MeToo and the Promise and Pitfalls of Challenging Rape Culture through Digital Feminist Activism." *European Journal of Women's Studies* 25(2):236–246.

Messner, Michael A., and Jeffrey Montez de Oca. 2005. "The Male Consumer as Loser: Beer and Liquor Ads in Mega Sports Media Events." *Signs* 30:1879–1909.

Miller, Claire Cain. 2020 (February 11). "Young Men Embrace Gender Equality, but They Still Don't Vacuum." *New York Times.* Available at www.nytimes.com.

Miller, Anna Medaris. 2020 (June 15). "An Estimated 15,000 People Rallied for Black Trans Lives in New York City." *Business Insider.* Available at www.insider.com.

Moen, Phyllis, and Yan Yu. 2000. "Effective Work/Life Strategies: Working Couples, Working Conditions, Gender, and Life Quality." *Social Problems* 47:291–326.

Morin, Rich, and Paul Taylor. 2008 (September 15). *Revisiting the Mommy Wars: Politics, Gender and Parenthood.* Available at www .pewsocialtrends.org.

National Center for Educational Statistics (NCES). 2020a. *Digest of Education Statistics.* U.S. Department of Education. Available at nces.ed.gov.

National Center for Educational Statistics (NCES). 2020b. *The Condition of Education.* U.S. Department of Education. Available at nces.ed.gov.

National Coalition for Men. 2020. "Philosophy." Available at ncfm.org.

National Coalition for Women and Girls in Education (NCWEG). 2017. *Title IX Advancing Opportunity through Equity in Education.* Washington, DC: National Coalition for Women and Girls in Education.

National Organization for Men against Sexism (NOMAS). 2020. "44 Years of NOMAS." Available at nomas.org.

National Organization for Women (NOW). 2020. "Who We Are." Available at now.org.

National Public Radio. 2011. "Two Spirits: A Map of Gender—Diverse Cultures." Public Broadcasting System. Available at www.pbs .org.

National Women's Law Center (NWLC). 2019a (September). "The Wage Gap: The Who, How, Why, and What to Do." Fact Sheet. Available at www.nwlc.org.

National Women's Law Center (NWLC). 2019b (March 14). "Voters Priorities for the New Congress." Available at www.nwlc.org.

North, Anna. 2019 (October 4). "7 Positive Changes That Have Come from the #MeToo Movement." *Vox Media.* Available at www .vox.com.

Parker, Kim, Juliana Menasce Horowitz, and Renée Stepler. 2017 (December 5). "On Gender Differences, No Consensus on Nature vs. Nurture." Pew Research Center. Available at www.pewsocialtrends.org.

Pedulla, David S., and Sarah Thebaud. 2015. "Can We Finish the Revolution? Gender, Work–Family Ideals, and Institutional Constraint." *American Sociological Review* 80(1):116–139.

Pew Research Center. 2019. "About One in Five Americans Say They Personally Know Someone Who Prefers a Pronoun Other Than 'He' Or 'She.'" Available at www.pewresearch .org.

Platt, Jonathan, Seth Prins, Lisa Bates, and Catherine Keves. 2016. "Unequal Depression for Equal Work? How the Wage Gap Explains Gender Disparities in Mood Disorders." *Social Science and Medicine* 149:1–8.

Political Parity. 2018 (June). *Path to Parity: How Women Run and Win.* Ed. Katherine Kidd. Available at www.politicalparity.org.

Poushter, Jacob, and Janell Fetterole. 2019 (April 22). "The Changing World: Global Views on Diversity, Gender Equality, Family Life and the Importance of Religion." Pew Research Center. Available at www.pewresearch.org.

Pugh, William. 2019 (June 19). "Circumcision Season." *The Sun.* Available at www.sun.co.uk.

Pollack, William. 2000. *Real Boys' Voices.* New York: Random House.

Priscott, Emily. 2018 (March 1). "Women's Liberation: What Today's #MeToo Skeptics Can Learn from Their 1970s Brothers." *The Conversation.* Available at theconversation. com.

Punshon, G., K. Maclaine, P. Trevatt, M. Radford, O. Shanley, and A. Leary. 2019. "Nursing Pay by Gender Distribution in the UK—Does the Glass Escalator Still Exist?" *International Journal of Nursing Studies*: 93:21–29.

Rehel, Erin. M. 2014. "When Dad Stays Home Too: Paternity Leave, Gender, and Parenting." *Gender and Society* 28(1):110–132.

Roman Catholic Women Priests (RCWP). 2020. "About Us." Available at www.romancatholicwomenpriests.org.

Reskin, Barbara, and Debra McBrier. 2000. "Why Not Ascription? Organizations' Employment of Male and Female Managers." *American Sociological Review* 65:210–233.

Ridgeway, Sicilia L. 2011. *Framed by Gender: How Gender Inequality Persists in the Modern World*. New York: Oxford University Press.

Rosenblatt, Joel, and Robert Burnson. 2020 (May 1). "Oracle Women Score Major Win in Court Battle over Equal Pay." Bloomberg. Available at www.bloomberg.com.

Runyan, Anne Sisson. 2018 (November–December). "What Is Intersectionality and Why Is It So Important?" *American Association of University Professors*. Available at www.aaup.org.

Sadker, David, and Karen Zittleman. 2009. *Still Failing at Fairness: How Gender Bias Cheats Boys and Girls in Schools*. New York: Simon & Schuster.

Salter, Michael. 2019 (February 27). "The Problem with a Fight against Toxic Masculinity." *Atlantic*. Available at www.theatlantic.com.

Sanchez, Diana T., and Jennifer Crocker. 2005. "How Investment in Gender Ideals Affects Well-Being: The Role of External Contingencies of Self-Worth." *Psychology of Women Quarterly* 29:63–77.

Sanger-Katz, Margo, and Erica L. Green. 2020 (June 15). "Supreme Court Expansion of Transgender Rights Undercuts Trump Restrictions." *New York Times*. Available at www.nytimes.com.

Sayman, Donna M. 2007. "The Elimination of Sexism and Stereotyping in Occupational Education." *Journal of Men's Studies* 15(1):19–30.

Schneider, Daniel. 2012. "Gender Deviance and Household Work: The Role of Occupation." *American Journal of Sociology* 117(4):1029–1072.

Schneider, Gregory S., and Laurel Vozzella. 2020 (January 27). "Virginia Finalizes Passage of Equal Rights Amendment, Setting the Stage for Legal Fight." *Washington Post*. Available at www.washintonpost.com.

Schulte, Brigid, and Haley Swenson. 2020 (June 17). "An Unexpected Upside to Lockdown: Men Have Discovered Housework." *Guardian*. Available at www.theguardian.com.

"See Jane." 2016. "PSA." Geena Davis Institute on Gender in Media. Available at seejane.org.

Selby, Danielle. 2018 (July 20). "20 Women in India Die Every Day due to Dowry Deaths." *Global Citizen*. Available at www.globalcitizen.org.

Shaw, Susan M. 2019 (March 6). "Sexism Has Long Been Part of the Culture of Southern Baptists." The Conversation. Available at thcconversation.com.

Siller, Heidi, Bettina Dickinger-Neuwirth, Nikola Komlenac, and Margarethe Hochleitner. 2019. "The Importance of Equal Treatment: Medical Students' Opinions on Affirmative Action, Equal Treatment and Discrimination." *Health Care for Women International* 40(1):47–65.

Simister, John. 2013. "Is Men's Share of House Work Reduced by Gender Deviance Neutralization? Evidence from Seven Countries." *Journal of Comparative Family Studies* 44(3):311–325.

Simpson, Ruth. 2005. "Men in Non-traditional Occupations: Career Entry, Career Orientation, and Experience of Role Strain." *Gender Work and Organization* 12(4):363–380.

Sloan, Colleen A., Danielle S. Berke, and Amos Zeichner. 2015. "Bias-Motivation against Men: Gender Expressions and Sexual Orientation as Risk Factors for Victimization." *Sex Roles* 72:140–149.

Smith, Paige. 2019 (March 27). "House Passes Legislation Intended to Curb Pay Bias." Bloomberg News. Available at news.bloomberglaw.com.

Snyder, Karrie Ann, and Adam Isaiah Green. 2008. "Revisiting the Glass Escalator: The Case of Gender Segregation in a Female Dominated Occupation." *Social Problems* 55(2):271–299.

Stebbins, Samuel, and Thomas C. Frohlich. 2020. (November 6). "The Poverty Rates for Every Group in the US: From Age and Sex to Citizenship Status." *USA Today*. Available at www.usatoday.com.

Strain, Michael R. 2013. "Single-Sex Class and Student Outcomes: Evidence from North Carolina." *Economics of Education Review* 36:73–87.

Supermajority. 2019 (August 19). "Gender Equality, the Status of Women and the 2020 Elections." *Results from a Supermajority/PerryUndem National Survey*. Available at int.nyt.com.

Taylor, Katie. 2019 (July 31). *Child Marriage: Facts from Around the World*. World Vision Advocacy. Available at www.worldvisionadvocacy.org.

Tenenbaum, Harriet R. 2009. "You'd Be Good at That: Gender Patterns in Parent–Child Talk about Courses." *Social Development* 18(2):447–463.

Thomas, Deja. 2019 (July 26). "In 2018, Two Thirds of Democratic Women Hoped to See a Woman President in Their Lifetime." Pew Research Center. Available at www.pewresearch.org.

Tin, Alexander. 2018 (August 22). "More Millennial Women Are Feminists though Overall Enthusiasm for the Term Remains Low." CBS News. Available at www.cbsnews.com.

Transcript. 2020 (July 23). "Rep. Alexandria Ocasio-Cortez (AOC) House Floor Speech on Yoho Remarks." Rev Transcripts. Available at rev.com.

Tucker, Jasmine, and Julie Vogtman. 2020 (April). "When Hard Work Is Not Enough: Women in Low-Paid Jobs." *National Women's Law Center*. Available at nwlc.org.

Uggen, C., and A. Blackstone. 2004. "Sexual Harassment as a Gendered Expression of Power." *American Sociological Review* 69:64–92.

Ungar-Sargon, Batya. 2013. "Orthodox Yeshiva Set to Ordain Three Women. Just Don't Call Them 'Rabbi.'" *Tablet Magazine*. Available at www.tabletmag.com.

United Nations (UN). 2019. "Facts and Figures: Leadership and Political Participation." UN Women. Available at www.unwomen.org.

United Nations (UN). 2014. *The Millennium Development Goals Report 2014*. Available at www.un.org.

United Nations Development Fund for Women. 2007. "Harmful Traditional Practices." Available at www.unifem.org.

United Nations Development Program (UNDP). 2020. *Tackling Social Norms: A Game Changer for Gender Inequalities*. Available at hdr.undp.org.

United Nations Educational, Scientific and Cultural Organisation (UNESCO). 2014. "Statistics on Literacy." Available at www.uis.unesco.org.

United Nations Girls Education Initiative (UNGEI). 2019. *Global Education Monitoring Report: Building Bridges for Gender Equality*. Available at www.ungei.org.

United Nations Educational, Scientific and Cultural Organization (UNESCO). 2016 (July). *Leaving No One Behind: How Far on the Way to Universal Primary and Secondary Education*. Policy Paper 27/Fact Sheet 37. Available at www.uis.unesco.org.

United Nations Population Fund. 2020. *The Cost of the Transformative Results UNFPA Is Committed to Achieving by 2030*. Chapter 5: Cost of Ending Gender Based Violence, pp. 34–39. Available at at www.unfpa.org.

United Nations Women (UNW). 2018 (July). "Facts and Figures: Economic Empowerment." Available at www.unwomen.org.

U.S. Department of State. 2012 (May 30). "The U.S. Response to Global Maternal Mortality: Saving Mothers, Giving Life." Available at www.state.gov.

Vandello, Joseph A., Jennifer K. Bosson, Dov Cohen, Rochelle M. Burnaford, and Jonathan R. Weaver. 2008. "Precarious Manhood." *Journal of Personality and Social Psychology* 95(6):1325–1339.

Vossemer, Jonas, and Stefanie Heyne. 2019. "Unemployment and Housework in Couples: Task Specific Differences and Dynamics over Time." *Journal of Marriage and Family* 81(5):1074–1090.

Walker, Shaun. 2020 (April 26). "Hungary Prepares to End Legal Recognition of Trans People." *Guardian*. Available at www.theguardian.com.

Walkington, Lori. "How Far Have We Really Come? Black Women Faculty and Graduate Students' Experiences in Higher Education." *Humboldt Journal of Social Relations* 1(39):51–65.

Wang, Yaqiu. 2020 (May 8). "#MeToo in the Land of Censorship." *Washington Post*. Available at www.washingtonpost.com.

Ward, Monique L., and Jennifer Stephens Aubrey. 2017. *Watching Gender: How Stereotypes in Movies and on TV Impact Kids Development.* San Francisco: Common Sense. Available at www.commonsensemedia.org.

Warner, Ann, Kirsten Stoebenau, and Allison M. Glinski. 2014. *More Power to Her: How Empowering Girls Can End Child Marriage.* International Center for Research on Women. Available at www.icrw.org.

Warren, Elizabeth. 2019. "The Schedules That Work after 2019." Available at www.warren .senate.gov.

Weeks, Linton. 2011 (June 23). "The End of Gender?" National Public Radio. Available at www.npr.org.

White, Alan, and Karl Witty. 2009. "Men's Under-Use of Health Services—Finding Alternative Approaches." *Journal of Men's Health* 6(2):95–97.

Willer, Robb, Christabel L. Rogalin, Bridget Conlon, and Michael T. Wojnowicz. 2013. "Overdoing Gender: A Test of the Masculine Overcompensation Thesis." *American Journal of Sociology* 118(4):980–1022.

Williams, Jamillah, Lisa Singh, and Naomi Mezey. 2019. "#MeToo as Catalyst: A Glimpse into 21st Century Activism." *University of Chicago Legal Forum* 2019(22).

Williams, Joan. 2000. *Unbending Gender: Why Family and Work Conflict and What to Do About It.* Oxford: Oxford University Press.

Williamson, Terrion. 2020. "'Sellin' Your Body': Contextualizing Racialized Gender Violence and Illicit Sexual Practices." *Signs* 45(3):524–528.

Wirtz, Andrea L., Tonya C. Poteat, Manat Malik, and Nancy Glass. 2020 (April 1). "Gender-Based Violence against Transgender People in the United States: A Call for Research and Programming." *Trauma, Violence and Abuse* 21(2):227–241.

Women's March. 2020. "Unity Principles." Available at womensmarch.com.

Wood, Wendy, and Alice H. Eagly. 2002. "A Cross-Cultural Analysis of the Behavior of Women and Men: Implications for the Origins of Sex Differences." *Psychological Bulletin* 128(5):699–727.

World Bank. 2020 (March 1). "Labor Force Participation Rate, Female (Percent of Female Population Ages 15+)—United States." Available at data.worldbank.org.

World Economic Forum (WEF). 2020. *Global Gender Gap Report 2020.* Available at www3 .weforum.org.

World Health Organization (WHO). 2009. "Ten Facts about Women's Health." Available at www.who.int.

World Health Organization (WHO). 2011. "Global Sector Strategy on HIV/AIDS 2011–2015." Available at www.who.int.

World Health Organization (WHO). 2018 (May 8). "Household Air Pollution and Health: Key Facts." Available at www.who.int.

World Health Organization (WHO). 2019 (September 19). "Maternal Mortality: Key Facts." Available at www.who.int.

World Health Organization (WHO). 2020 (February 3). "Female Genital Mutilation: Key Facts." Available at www.who.int.

Yavorsky, Jill E., and Janette Dill. 2020 (January). "Unemployment in Men's Entrance into Female Dominated Jobs." *Social Science Research* 85:1–20.

Zakrzewski, Paul. 2005 (June 19). "Daddy, What Did You Do in the Men's Movement?" *Boston Globe.* Available at www.bostonglobe.com.

Chapter 11

Abraham, Eyal, Talma Hendler, Irit Shapira-Lichter, Yaniv Kanat-Maymon, Oma Zagoory-Sharon, and Ruth Feldman. 2014. "Father's Brain Is Sensitive to Childcare Experiences." *Proceedings of the National Academy of Sciences of the United States of America: Current Issues* 111(27):9792–9797.

Adamczyk, Amy. 2017. *Cross National Public Opinion about Homosexuality: Examining Attitudes across the Globe.* Oakland: University of California Press.

African News. 2019 (October). "Uganda Government Denies Report of Reintroduction of Anti-Gay Bill." Available at www .africannews.com.

Ahlgrim, Kelly. 2020 (June 8). "47 of the Most Groundbreaking LGBTQ Characters and Relationships on TV." *Insider.* Available at www.insider.com.

Allen, Karma. 2020 (January 16). "Florida Republicans Submit Four Last-Minute Anti-LGBTQ Bills ahead of 2020 Legislative Deadline." ABC News. Available at abcnews. go.com.

Allport, G. W. 1954. *The Nature of Prejudice.* Cambridge, MA: Addison-Wesley.

American College Health Association. 2020. "Undergraduate Student Reference Group: Executive Summary, Fall 2019." Available at www.acha.org.

American Psychoanalytic Association. 2019 (June 21). "American Psychoanalytic Association Issues Overdue Apology to LGBT Community." News Release. Available at apsa.org.

American Sociological Association (ASA). 2015 (March). "Brief of *Amicus Curiae* American Sociological Association in Support of Petitioners." Available at www.asanet.org.

Associated Press. 2020a (May 5). "With Split over Gay Marriage Delayed, United Methodists Face a Year in Limbo." Available at www.nbcnews .com.

Associated Press. 2020b (February 27). "South Carolina Faces Lawsuit over LGBTQ+ Sex Education Ban." Available at www.guardian .com.

Association of American Universities (AAU). 2019. "AAU Campus Climate Survey." Available at www.aau.edu.

Aviles, Gwen, and Janelle Griffith. 2019 (April 23). "Deputy on Leave for Homophobic Comments over Teen's Suicide." NBC News. Available at www.nbcnews.com.

Badgett, M. V., Soon Kyu Choi Lee, and Bianca D. M. Wilson. 2019 (October). "LGBT Poverty in the United States: A Study of Differences Between Sexual Orientation and Gender Identity Groups." *UCLA School of Law Williams Institute.* Available at williamsinstitute.law.ucla.edu.

Baughey-Gill, Sarah. 2011. "When Gay Was Not OK with the APA: A Historical Overview of Homosexuality and Its Status as a Disorder." *Occam* 1(2):5–16.

Berg, Kristin, and Moiz Syed. 2019 (November 22). "Under Trump, LGBTQ Progress Is Being Reversed in Plain Sight." *ProPublica: Journalism in the Public Interest.* Available at projects.propublica.org.

Bhalla, Nita. 2019 (October 11). "International Donors 'Stand with' LGBT+ Ugandans over 'Kill the Gays' Bill." Available at www.reuters .com.

Biden, Joe. 2020. "The Biden Plan to Advance LGBTQ+ Equality in America and around the World." Available at www.joebiden.com.

Boertien, D., and D. Vignoli. 2019. "Legalizing Same-Sex Marriage Matters for the Subjective Well-Being of Individuals in Same-Sex Unions." *Demography* 56:2109–2121.

Bostock v. Clayton County. 2020. Available at www.supremecourt.gov.

Brady, Jeff. 2020 (June 26). "Five Years after Same Sex Marriage Decision, Equality Fight Continues." National Public Radio. Available at www.npr.org.

Brenan, Megan. 2020 (June 23). "Record Low 54% in U.S. Say Death Penalty Morally Acceptable." Gallup Poll. Available at news. gallup.com.

Brown, Michael J., and Ernesto Henriquez. 2008. "Socio-Demographic Predictors of Attitudes towards Gays and Lesbians." *Individual Differences Research* 6:193–202.

Burton, Candace W., Jung-Ah Lee, Anders Waalen, and Lisa M. Gibbs. 2019 (December 19). "'Things Are Different Now But': Older LGBT Adults' Experiences and Unmet Needs in Healthcare." *Journal of Transcultural Nursing.* Available at pubmed.ncbi.nlm.nih.gov.

Burton, Neel. 2015 (September 18). "When Homosexuality Stopped Being a Mental Disorder." *Psychology Today.* Available at www.psychologytoday.com.

Byrnes, Hristina, John Harrington, and Grant Suneson. 2020 (June 19). "Supreme Court Decision aside, Some States Are Better—and Some Are Worse—for LGBTQ Community." *USA Today.* Available at www.usatoday.com.

Carroll, A. 2016 (May). *State Sponsored Homophobia: A World Survey of Sexual Orientation Laws: Criminalization, Protection and Recognition.* Geneva: International Lesbian, Gay, Bisexual, Trans and Intersex Association.

Caruso, Kevin. n.d. "Suicide Note of a Gay Teen." Suicide Survivors Forum. Available at www .suicide.org.

Center for Disease Control and Prevention (CDC). 2018 (June 15). "Morbidity and Mortality Weekly Report." Available at www.cdc.gov.

Charlesworth, Tessa P. S., and Mahzarin Banaji. 2019 (February). "Patterns of Implicit and Explicit Attitudes: Long-Term Change and Stability from 2007 to 2016." *Psychological Science* 30(2):174–192.

Chibbaro, Lou. 2020 (February 25). "Maryland House Votes to Repeal Sodomy Laws."

Washington Blade. Available at www .washingtonblade.com.

Child Welfare League of America. 2020. "U.S. Supreme Court to Take up Philadelphia Case on Foster Placements." Available at www.cwla .org.

Choi, Soon Kyu, Bianca D. M. Wilson, Jama Shelton, and Gary Gates. 2015. *Serving Our Youth 2015: The Needs and Experiences of Lesbian, Gay, Bisexual, Transgender, and Questioning Youth Experiencing Homelessness.* Los Angeles: Williams Institute. Available at www.williamsinstitute.law.ucla.edu.

Choi, Soon Kyu, and Ilan H. Meyer. 2016 (August). *LGBT Aging: A Review of Research Findings, Needs, and Policy Implications.* Los Angeles: Williams Institute. Available at williamsinstitute.law.ucla.edu.

Ciarlante, Mitru, and Kim Fountain. 2010. "Why It Matters: Rethinking Victim Assistance for Lesbian, Gay, Bisexual, Transgender, and Queer Victims of Hate Violence and Intimate Partner Violence." National Center for Victims of Crime and the New York City Anti-Violence Project. Available at www.avp.org.

Clooney, George. 2019 (March 28). "George Clooney: Boycott Sultan of Brunei's Hotels over Cruel Anti-Gay Laws." *Deadline.* Available at deadline.com.

Conron, Kerith J., and Shoshana K. Goldberg. 2020 (April). "LGBT People in the U.S. Not Protected by State Nondiscrimination Statutes." Los Angeles: UCLA School of Law Williams Institute. Available at williamsinstitute.law.ucls.edu.

Conron, Kerith, and Kathryn O'Neill. 2020 (February). "Prohibiting Gender-Affirming Medical Care for Youth." Fact Sheet. UCLA School of Law Williams Institute. Available at Williams Institute law.ucla.edu.

Cook, Tony, Tom LoBianco, and Doug Stranglin. 2015 (April 2). "Indiana Gov. Signs Amended Religious Freedom Law." *USA Today.* Available at www.usatoday.com.

Cornell University. The Public Policy Research Portal. 2017. "What Does the Scholarly Research Say about the Effect of Gender Transition on Transgender Well-Being?" Available at whatweknow.inequality.cornell. edu.

Crump, James. 2020 (June 13). "Republican National Committee Votes to Keep 2016 Platform That Calls for Ban on Same-Sex Marriage." *Independent.* Available at www .independent.co.uk.

D'Angelo, Chris. 2019 (December 27). "Interior Department Cut 'Sexual Orientation' Anti-Discrimination Guidelines." *Huffington Post.* Available at www.huffpost.com.

DeCarlo, Aubrey. 2014. "The Relationship between Traditional Gender Roles and Negative Attitudes towards Lesbians and Gay Men in Greek-Affiliated and Independent Male College Students." Available at preserve.lib.lehigh.edu.

Dias, Elizabeth. 2019 (April 4). "Mormon Church to Allow Children of L.G.B.T. Parents to Be Baptized." *New York Times.* Available at www .nytimes.com.

Dierckx, Myrte, Petra Meier, and Joz Motmans. 2017. "'Beyond the Box': A Comprehensive Study of Sexist, Homophobic, and Transphobic Attitudes among the Belgian Population." *Journal of Diversity and Gender Studies* 4(1):5–34.

Doan-Minh, Sarah. 2019 (Winter). "Corrective Rape an Extreme Manifestation of Discrimination and the State's Complicity in Sexual Violence." *Hastings Women's Law Journal* 30(1):167–196.

Donahue, David. 2019 (Winter). "Queering California's K–12 History Curriculum." *Journal of Interdisciplinary Perspectives and Scholarship* 1(6). Available at repository.usfca.edu.

Dunigan, Jonece Star. 2019 (May 15). "How Alabama's Current Sex Ed Law Hurts LGBTQ Youth." *Alabama.* Available at www.al.com.

Durkheim, Emile. 1993 [1938]. "The Normal and the Pathological." In *Social Deviance,* ed. Henry N. Pontell, 33–63. Englewood Cliffs, NJ: Prentice Hall. (Originally published in *The Rules of Sociological Method.*)

Equality and Justice Alliance (EJA). 2020 (May). Hate Crimes against LGBT Community in the Commonwealth: A Situational Analysis." *Human Dignity Trust.* Available at sro.sussex.ac.uk.

Elliott, Marc N., David E. Kanouse, Q. Burkhart, Gary A. Abel, Georgios Lyratzopoulos, Megan K. Beckett, Mark A. Schuster, and Martin Roland. 2015. "Sexual Minorities in England Have Poorer Health and Worse Health Care Experiences: A National Survey." *Journal of General Internal Medicine* 30(1):9–16.

Everly, Benjamin A., and Joshua Schwarz. 2015. "Predictors of the Adoption of LGBT-Friendly HR Policies." *Human Resource Management* 54(2):367–384.

"FAIR Act." (2013). Available at www .faireducationact.com.

Fadel, Leila. 2019 (April 26). "Activists and Suicide Prevention Groups Seek Bans on Conversion Therapy for Minors." National Public Radio. Available at www.npr.org.

Falomir-Pichastor, Juan Manual, Carmen Martinez, and Consuelo Paterna. 2010. "Gender Role's Attitude, Perceived Similarity, and Sexual Prejudice against Gay Men." *Spanish Journal of Psychology* 13(2):841–848.

Family Acceptance Project. 2020. "Research." Available at familyproject.sfu.edu.

Federal Bureau of Investigation (FBI). 2019. "2018 Hate Crime Statistics." Available at ucr. fbi.gov.

Felter, Claire, and Danielle Renwick. 2020 (June 23). "Same-Sex Marriage: Global Comparisons." Council on Foreign Relations. Available at www.cfr.org.

Fisher, Max. 2013 (June 27). "From Colonialism to 'Kill the Gays': The Surprisingly Recent Roots of Homophobia in Africa." *Washington Post.* Available at www.washingtonpost.com.

Flores, Andrew R. 2019 (October). "Social Acceptance of LGBT People in 174 Countries: 1980 to 2017." UCLA School of Law Williams Institute. Available at williamsinstitute.law. ucla.edu.

Freedom to Marry. 2020 (May). "The Freedom to Marry Internationally." Available at www .freedomtomarry.org.

Fredriksen-Goldsen K. I., and R. Espinoza. 2014. "Transforming Public Policies to Achieve Health Equity for LGBT Older Adults." *Generations 38(4):97–106.*

Gallagher, Sophie. 2020 (March 2). "Most Anglicans Now Support Same-Sex Marriage Despite Official Church of England Stance." *The Independent.* Available at www .independent.co.uk.

Gallup Poll. 2020. "Gay and Lesbian Rights." *Gallup Historical Trends: In Depth Topics A to Z.* Available at www.gallup.com.

Ganna· Andrea, Karin J. H. Verweij, Michel G. Nivard, Robert Maier, Robbee Wedow, Alexander S. Busch, Abdel Abdellaoui, Shengru Guo, J. Fah Sathirapongsasuti, 23andMe Research Team, Paul Lichtenstein, Sebastian Lundström, Niklas Långström, Adam Auton, Kathleen Mullan Harris, Gary W. Beecham, Eden R. Martin, Alan R. Sanders, John R. B. Perry, Benjamin M. Neale, and Brendan P. Zietsch. 2019. "Large-Scale GWAS Reveals Insights into the Genetic Architecture of Same-Sex Sexual Behavior." *Science* 365:6456.

Gates, Gary J. 2014 (October). "LGB Families and Relationships: Analysis of the 2013 National Health Interview Survey." Executive Summary. Available at williamsinstitute.law.ucla.edu.

Gates, Gary J. 2016 (December 14). "Life Evaluations of LGBT Americans Decline after Election." Available at news.gallup.com.

Gates, Gary J. 2017. "Assessing LGBT Health: Demographics of Coming Out." Available at transformingcareconference.com.

Gay, Lesbian, and Straight Education Network (GLSEN). 2020. "Our Work." Available at www .glsen.org.

Gera, Vanessa, and Dorothee Thiesing. 2020. "Poland's LGBTQ Community Feels Fear and Anger after Election." *Public Broadcasting System.* Available at www.pbs.org.

Giambrone, Andrew. 2015 (May 26). "Equality in Marriage May Not Bring Equality in Adoption." *Atlantic.* Available at www.theatlantic.com.

GLAAD Media Institute. 2015. *Studio Responsibility Index.* Available at www.glaad .org.

GLAAD Media Institute. 2020a. "Accelerating Acceptance 2019: Executive Summary." Available at www.glaad.org.

GLAAD. 2020b. *Studio Responsibility Index.* Available at www.glaad.org.

Goldbach, Jeremy T., Emily E. Tanner-Smith, Meredith Bagwell, and Shannon Dunlap. 2014. "Minority Stress and Substance Use in Sexual Minority Adolescents: A Meta-Analysis." *Prevention Science* 15:350–363.

Goldberg, Abie E., Genny Beemyn, and JuliAnna Z. Smith. 2019. "What Is Needed, What Is Valued: Trans Students' Perspectives on Trans-Inclusive Policies and Practices in Higher Education." *Journal of Educational and Psychological Consultation* 29(1):27–67.

Goldberg, Shoshana K., and Kerith J. Conron. 2018 (July). "How Many Same-Sex Couples in the U.S. Are Raising Children?" *UCLA School of Law Williams Institute.* Available at williamsinstitute.law.ucla.edu.

Goldsen, Jayn. 2017 (June 21). "Are LGBT Americans Actually Reaping the Benefits of Marriage?" The Conversation. Available at www.theconversation.com.

Goldsen, K. 2018. "Shifting Social Context in the Lives of LGBTQ Older Adults." *Public Policy & Aging Report* 28(1):24–28.

Gomillion, Sarah C., and Traci A. Guiliano. 2015. "The Influence of Media on Gay, Lesbian and Bisexual Identity." *Journal of Homosexuality* 58(3):330–354.

Gonzales, Gilbert, and Kyle A. Gavulic. 2020 (June). "The Equality Act Is Needed to Advance Health Equity for Lesbian, Gay, Bisexual, and Transgender Populations." *American Journal of Public Health* 110(6):801–802.

Gonzalez, Ivet. 2013 (June 4). "Gay Parents in Cuba Demand Legal Right to Adopt." *Global Issues*. Available at www.globalissues.org.

Greenberg, Daniel, Maxine Najle, Oyindamola Bola, and Robert P. Jones. 2019 (March 26). "Part One: Nondiscrimination Protections for LGBT People." *Fifty Years after Stonewall: Widespread Support for LGBT Issues—Findings from the American Values Atlas 2018*. Public Religion Research Institute. Available at www.prri.org.

Guadalupe, Krishna L., and Doman Lum. 2005. *Multidimensional Contextual Practice: Diversity and Transcendence*. Belmont, CA: Thomson Brooks/Cole.

Guillory, Sean. 2013 (September 26). "Repression and Gay Rights in Russia." *The Nation*. Available at www.thenation.com.

Gupta, Vanita. 2020 (January 28). "Bipartisan 'Justice for Victims of Hate Crime Act' Will Strengthen Key Federal Law." News Release. Leadership Conference on Civil and Human Rights. Available at www.civilrights.org.

Hafeez, Hudaisa, Muhammad Zeshan, Muhammad A. Tahir, Nusrat Jahan, and Sadiq Naveed. 2017 (April). "Healthcare Disparities among Lesbian, Gay, Bisexual, and Transgender Youth: A Literature Review." *Cureus* 9(4). Available at www.ncbi.nlm.nih.giv.

Haider-Markel, Donald P., and Mark R. Joslyn. 2008. "Beliefs about the Origins of Homosexuality and Support for Gay Rights: An Empirical Test of Attribution Theory." *Public Opinion Quarterly* 72:291–310.

Haines, Kari, M. C. Reyn Boyer, Casey Giovanazzi, and M. Paz Galupo. 2018. "'Not a Real Family': Microaggressions Directed toward LGBTQ Families." *Journal of Homosexuality* 65(9):1138–1151.

Harper, Gary W., Nadine Jernewall, and Maria C. Zea. 2004. "Giving Voice to Emerging Science and Theory for Lesbian, Gay, and Bisexual People of Color." *Cultural Diversity and Ethnic Minority Psychology* 10:187–199.

Heck, N. C., N. A. Livingston, A. Flentje, K. Oost, B. T. Stewart, and B. N. Cochran. 2014. "Reducing Risk for Illicit Drug Use and Prescription Drug Misuse: High School Gay–Straight Alliances and Lesbian, Gay, Bisexual, and Transgender Youth." *Addictive Behaviors* 39(4):824–828.

Herek, Gregory M. 2004. "Beyond 'Homophobia': Thinking about Sexual Prejudice and Stigma in the Twenty-First Century." *Sexuality Research and Social Policy: A Journal of the NSRC* 1:6–24.

Herman, Jody L., Taylor N. T. Brown, and Ann P. Haas. 2019 (September). "Suicidal Thoughts and Attempts among Transgender Adults." *UCLA Law School Williams Institute*. Available at williamsinstitute.law.ucla.edu.

Higa, Darrel, Marilyn J. Hoppe, Taryn Lindhorst, Shaw Mincer, Blair Beadnell, Diane M. Morrion, Elizabeth A. Wells, Avry Todd, and Sarah Mountz. 2014. "Negative and Positive Factors Associated with the Well-Being of Lesbian, Gay, Bisexual, Transgender, Queer, and Questioning (LGBTQ) Youth." *Youth and Society* 46(5):663–687.

Hines, Judith M. 2014. *Internalized Heterosexism, Outness, Relationship Satisfaction, and Violence in Lesbian Relationships*. University of Illinois, Chicago. Available at indigo.uic.edu/articles.

Hoffarth, M. R., Hodson, G., and Molnar, D. S. 2018. "When and Why Is Religious Attendance Associated with Antigay Bias and Gay Rights Opposition? A Justification-Suppression Model Approach." *Journal of Personality and Social Psychology 115*(3):526–563.

H.R. 1450. *Do No Harm Act*. Available at www.congress.gov.

H.R. 5374. "Student Non-Discrimination Act of 2018." Available at www.congress.gov.

Human Rights Campaign (HRC). 2020a. "The Lies and Dangers of Efforts to Change Sexual Orientation or Gender Identity." Available at www.hrc.org.

Human Rights Campaign. 2020b. "The Lives and Livelihoods of Many in the LGBTQ Community Are at Risk amidst Covid-19 Crisis." Available at assets2.hrc.org.

Human Rights Campaign. 2020c. "State Laws and Policies." Available at www.hrc.org.

Hutt, Rosamond. 2018 (June). "This Is the State of LGBTI Rights around the World in 2018." *World Economic Forum*. Available at wwwweforum.org.

IGLA. 2016. "State-Sponsored Homophobia Report: 2016." Available at ilga.org.

ILGA-Europe. 2020. "ILGA—Europe Rainbow Map and Index 2020 Reveals That Once-Leading Countries in Europe Are Falling behind in Their Commitments to Equality for LGBTI People." Available at www.ilga-europe.org.

IGLA-RIWI. 2018 (October). *Minorities Report 2017: Attitudes to Sexual and Gender Minorities around the World*. Available at ilga.org.

It Gets Better (IGB). 2020. "About." Available at itgetsbetter.org.

Jabson, Jennifer M., Grant W. Farmer, and Deborah J. Bowen. 2014. "Stress Mediates the Relationship between Sexual Orientation and Behavioral Risk Disparities." *Biomedical Center Public Health* 14:401–407.

Jarpe-Ratner, Elizabeth. 2020 (May). "How Can We Make LGBTQ+ Inclusive Sex Education Programs Truly Inclusive? A Case Study of Chicago Public Schools Policy and Curriculum." *Sex Education* 20(3):283–299.

Johns, Michelle M., V. Paul Poteat, Stacy S. Horn, and Joseph Kosciw. 2019. "Strengthening Our Schools to Promote Resilience and Health among LGBTQ Youth: Emerging Evidence and Research Priorities from the State of LGBTQ Youth Health and Well-Being Symposium." *LGBT Health* 6(4):146–155.

Jones, Jeffrey M. 2017 (22). "In U.S., 10.2% of LGBT Adults Now Married to Same-Sex Spouse." Gallup Poll. Available at news.gallup.com.

Jones, Jeffrey M. 2021 (February 24). "LGBT Identification Rises to 5.6% in Latest U.S. Estimate." Gallup Poll. Available at news.gallup.com.

Jones, Robert J., Daniel Cox, and Juhem Navarro-Rivera. 2014 (February 26). *A Shifting Landscape*. Available at publicreligion.org.

Kaiser Family Foundation (KFF). 2020. "Poll: Large Majorities, Including Republicans, Oppose Discrimination against Lesbian, Gay, Bisexual and Transgender People by Employers and Healthcare Providers." Available at www.kff.org.

Kane, Melinda D. 2013. "LGBT Religious Activism: Predicting State Variations in the Number of Metropolitan Community Churches, 1774–2000." *Sociological Forum* 28(1):135–158.

Kaplan, Sheila. 2020 (June 13). "Healthcare Advocates Push Back against Trump's Erasure of Transgender Rights." *New York Times*. Available at www.nytimes.com.

Kates, Jennifer, Usha Ranji, Adara Beamesderfer, Alina Salganicoff, and Lindsay Dawson. 2018 (May 3). "Health and Access to Care and Coverage for Lesbian, Gay, Bisexual, and Transgender (LGBT) Individuals in the U.S." Kaiser Family Foundation. Available at www.kff.org.

Kaufman, Alexander C. 2015 (April 1). "More Than 70 Tech Execs Sign Historic Statement against LGBT Discrimination." *Huffington Post*. Available at www.huffingtonpost.com.

Kiekens, Wouter, Chaim la Roi, Henry M. W. Bos, Tina Kretschmer, Diana D. van Bergen, and René Veenstra. 2020 (January 29). "Explaining Health Disparities in Heterosexual and LGB Adolescents by Integrating the Minority Stress and Psychological Mediation Frameworks: Findings from the TRAILS Study." *Journal of Youth and Adolescence*. Available at link.springer.com.

Kimmel, Michael. 2011 (Winter). "Gay Bashing Is about Masculinity." *Voice Male*. Available at www.voicemalemagazine.org.

Knauer, Nancy J. 2019. "Implications of *Obergefell* for Same-Sex Marriage, Divorce, and Parental Rights." In *LGBTQ Divorce and Relationship Dissolution*, ed. Abby E. Goldberg and Adam P. Romero, Chapter 1, 7–30. New York: Oxford University Press.

Kosciw, Joseph G., Emily A. Greytak, Adrian E. Zongrone, Caitlin M. Clark, and Nhan L. Truong. 2018. *The 2017 National School Climate Survey: The Experiences of Lesbian, Gay, Bisexual, Transgender, and Queer Youth in Our Nation's Schools*. New York: GLSEN.

Kuhr, Elizabeth. 2020 (April 24). "1 in 5 Russians Want Gays and Lesbians 'Eliminated,' Survey Finds." Available at www.nbcnews.com.

Kurtzleben, Danielle. 2021 (February 24). "House Passes the Equality Act: Here's What It Would Do." National Public Radio. Available at www.npr.org.

Kutner, Jenny. 2015 (April 30). "Transgender Teen Leelah Alcorn's Death Ruled a Suicide—Mother Threw away Handwritten Note." *Salon*. Available at www.salon.com.

Justia. 2020. "Title VII of the Civil Rights Act." Available at www.justia.com.

LeBlanc, Paul, and Dan Merica. 2020 (March 1). "'Pete Got Me to Believe in Myself Again': Chasten Buttigieg Honors His Husband as Candidate Exits Race." CNN News. Available at www.cnn.com.

Leone, Hannah. 2019 (September 3). "New Law Requires Illinois Schools Teach Contributions of Gay, Transgender People: 'It Is Past Time Children Know the Names of LGBT+ Pioneers'" *Chicago Tribune*. Available at www .chicagotribune.com.

Levin, Dan. 2019 (July 23). "North Carolina Reaches a Settlement on 'Bathroom Bill.'" *New York Times*. Available at www.nytimes.com.

Lewis, Sophie. 2019 (May 29). "WHO Reclassifies 'Gender Identity Disorder.'" CBS News. Available at www.cbsews.com.

Lipka, Michael. 2014 (October 16). "Young U.S. Catholics Overwhelmingly Accepting of Homosexuality." Fact Tank. Available at www .pewresearch.org.

Liptak, Adam. 2020 (June 16). "Civil Rights Law Protects Gay and Transgender Workers, Supreme Court Rules." *New York Times*. Available at www.nytimes.com.

London, Emily, and Maggie Siddiqi. 2019 (April 11). "Religious Liberty Should Do No Harm." Center for American Progress. Available at www.americanprogress.org.

Madžarević, Goran, and Maria T. Soto-Sanfiel. 2018 (September). "Positive Representation of Gay Characters in Movies for Reducing Homophobia." *Sexuality and Culture* 22(3):909–930.

Mallory, Christie, Amira Hassenbush, and Brad Sears. 2015. "Harassment by Law Enforcement Officers in the LGBT Community." Williams Institute. Available at williamsinstitute.law. ucla.edu.

Mallory, Christy, and Brad Sears. 2020 (May). "The Economic Impact of Marriage Equality Five Years after *Obergefell v. Hodges*." Williams Institute. Available at williamsinstitute.law. ucla.edu.

Masci, David, Anna Brown, and Joslyn Kylie. 2019 (June 24). "5 Facts about Same-Sex Marriage." *Fact Tank*. Available at www .pewresearch.org.

McCarthy, Justin. 2014 (May 28). "Americans' Views on Origins of Homosexuality Remain Split." Available at www.gallup.com.

McDermott, Nathan. 2014 (March 21). "The Myth of Gay Affluence." *Atlantic*. Available at www .theatlantic.com.

McLaughlin, Brian, and Nathian S. Rodriguez. 2017. "Identifying with a Stereotype: Divergent Effects of Exposure to Homosexual Television Characters." *Journal of Homosexuality* 46(9):1196–1213.

Mendos, Lucas Ramon (ed.). 2019 (December). *State-Sponsored Homophobia 2019: Global Legislation Overview Update*. Geneva: ILGA World.

Metropolitan Community Church (MCC). 2020a. "About." Available at mccchurch.org.

Metropolitan Community Church (MCC). 2020b. "Denomination Core Values." Available at mccchurch.org.

Miller, Brian. 2018 (November 16). "The Age of RFRA." *Forbes*. Available at www.forbes.com.

Mithika, Stephanie. 2019 (July 25). "How Should the U.S. Approach LGBT Rights in Africa?" Atlantic Council. Available at www .atlanticcouncil.org.

Moreau, Julie. 2020 (June 23). "Supreme Court's LGBTQ Ruling Could Have 'Broad Implications,' Legal Experts Say." NBC News. Available at www.nbc.com.

Moreau, Julie. 2018. "Rights Group Sues Government after Gay Widower Denied Spousal Benefits." NBC News. Available at www.nbcnews.com.

Movement Advancement Project (MAP). 2020a "Mapping LGBTQ Equality: 2010 to 2020." Available at www.lgbtmap.org.

Movement Advancement Project (MAP). 2020b. "Religious Exemption Laws: Marriage Solemnization." Available at www.lgbtmap.org.

Movement Advance Project (MAP). 2020c. "Hate Crime Laws." Available at www.lgbtmap.org.

National Center for Transgender Equality. 2020. "Issues: Health and HIV." Available at transequality.org.

National Coalition of Anti-Violence Programs (NCAVP). 2018. "Lesbian, Gay, Bisexual, Transgender, Queer and HIV-Infected Hate and Intimate Partner Violence in 2017." *A Report from the National Coalition of Antiviolence Programs*. Available at avp.org.

National Conference of State Legislatures (NCLS). 2020. "State Partisan Composition." Available at www.ncls.org.

Nicholson, Rebecca. 2020 (June 20). "Billy Porter: "I've Lived as a Black Gay Man for 50 Years in America. Nothing Shocks Me." *Guardian*. Available at theguardian.com.

Obama, Barack. 2015 (April 8). "Petition Response: On Conversion Therapy." Available at www.obamawhitehouse.archives.gov.

Obergefell v. Hodges. 2015. Available at www .supremecourt.gov.

Ofosu, Eugene K., Michelle K. Chambers, Jacqueline M. Chen, and Eric Hehman. 2019. "Same-Sex Marriage Legalization Associated with Reduced Implicit and Explicit Antigay Bias." *Proceedings of the National Academy of Sciences* 116(18):8846–8851.

Perper, Rosie. 2020 (May 27). "The 29 Countries around the World Where Same-Sex Marriage Is Legal." *Business Inside*. Available at www .businessinsider.

Pew. 2013a (June 4). "The Global Divide on Homosexuality." Available at www.pewglobal .org.

Pew. 2013b (March 20). "Growing Support for Same-Sex Marriage: Changed Minds and Changing Demographics." Available at www .people-press.org.

Pew. 2019a. "Attitudes toward Same-Sex Marriage: Religious Attendance." Available at www.pewforum.org.

Pew. 2019b. "Attitudes toward Same-Sex Marriage: Party and Ideology." Available at www.pewforum.org.

Pew Research Center. 2020 (June 24). "Americans Are Increasingly Accepting Homosexuality in Society." Available at www.pewresearch.org.

Pew Research Center. 2019a (December 17). "In a Politically Polarized Era, Sharp Divides in Both Partisan Coalitions." Available at www .pewresearch.org.

Pew Research Center. 2019b (December 17). "Gender, Family and Marriage, Same-Sex Marriage and Religion." Available at www .pewresearch.org.

Phillips, Katherine W. 2014 (September 16). "How Diversity Makes Us Smarter." *Scientific American*. Available at www .scientificamerican.com.

Position Paper. 2015 (April). "Position Statements on Parenting of Children by Lesbian, Gay, Bisexual, and Transgender Adults." *Child Welfare League of America*. Available at www.cwla.org.

Postsecondary National Policy Institute. (2020). "Fact Sheet: LGBTQ Students in Higher Education." Available at pnpi.org.

Poushter, Jacob, and Nicholas Kent. 2020 (June 25). "The Global Divide on Homosexuality Persists." Available at www.pewresearch.org.

Price, Jammie, and Michael G. Dalecki. 1998. "The Social Basis of Homophobia: An Empirical Illustration." *Sociological Spectrum* 18:143–159.

Public Religion Research Institute (PRRI). 2020 (April 14). "Broad Support for LGBT Rights across All 50 States: Findings from the 2019 American Values Atlas." Available at www .prri.org.

Rainbow Europe. 2019 (May 13). "10th Rainbow Europe: Confirmed Stagnation and Regression on LGBTI Equality Calls for Immediate Action." *ILGA—Europe*. Available at www .ilga-europe.org.

Rankin, Susan, Jason C. Garvey, and Antonio Duran. 2020. "A Retrospective of LGBT Issues on U.S. College Campuses: 1990–2020." *International Sociology* 34(4):435–454.

Redmond-Palmer, Bill. 2020. "Same Sex Marriage Question Included in 2020 Census." *Baltimore Outloud*. Available at baltimore.com.

Religious Freedom Restoration Act of 1993 (RFRA). 1993. Available at www.govinfo.gov.

Reyna, Christine, Geoffrey Wetherell, Caitlyn Yantis, and Mark J. Brandt. 2014. "Attributions for Sexual Orientation vs. Stereotypes: How Beliefs about Value Violations Account for Attribution Effects on Anti-Gay Discrimination." *Journal of Applied Social Psychology* 4(4):289–302.

Rosenwald, Michael S. 2015 (April 6). "How Jim Obergefell Became the Face of the Supreme Court Gay Marriage Case." *Washington Post*. Available at www.washingtonpost.com.

Rosenberg, Alyssa. 2018 (January 25). "In Three Years, LGBT Americans Have Gone from Triumph to Backlash." *Washington Post*. Available at www.washingtonpost.com.

Ross, Brian, and Brian Epstein. 2017 (March 6). "Gay Conversion Therapy Advocates Heartened by Republican Electoral Victories." *ABC News*. Available atabcnews.go.com.

S. 2525. 2018 (March 8). "First Amendment Defense Act." Available at www.congress.gov.

Sanger-Katz, Margo, and Noah Weiland. 2020 (June 15). "Trump Administration Erases Transgender Civil Rights Protections in Health Care." *New York Times.* Available at www .nytimes.com.

Sasnett Sherri. 2015. "Are the Kids All Right? A Qualitative Study of Adults with Gay and Lesbian Parents." *Journal of Contemporary Ethnography* 44(2):196–222.

Scheitle, Christopher P., and Julia Kay Wolf. 2017. "The Religious Origins and Destinations of Individuals Identifying as a Sexual Minority." *Sexuality and Culture* 21:719–740.

Schiappa, E., P. B. Gregg, and D. E. Hewes. 2005. "The Parasocial Contact Hypothesis." *Communication Monographs* 72(1):92–115.

Schlatter, Evelyn, and Robert Steinback. 2014 (Update). "10 Anti-Gay Myths Debunked." *Intelligence Report* 140(Winter).

S. 1791. 2019. "Every Child Deserves a Family Act." Available at www.congress.gov.

Services and Advocacy for GLBT Elders (SAGE). 2015. "Economic Security." *SAGE.* Available at sageusa.org.

Services and Advocacy for GLBT Elders (SAGE). 2019 (March). "Aging in the LGBT Community." *SAGE.* Available at www.sageusa .org.

Shackelford, Todd K., and Avi Besser. 2007. "Predicting Attitudes toward Homosexuality: Insights from Personality Psychology." *Individual Differences Research* 5:106–114.

Sharp, Matt. 2019 (July 4). "'Disagreement Is Not Discrimination': Do No Harm Act Is a Dishonest Act to Reject Religion." *USA Today.* Available at www.usatoday.com.

Simon, Scott. 2018 (January 20). "Kicked Out of the Air Force for Being Gay, Helen Grace James Wins Honorable Discharge." Available at www .npr.org.

Singman, Brooke. 2021 (January 25). "Biden Reverses Trump Ban on Transgender Individuals Serving in the Military." Fox News. Available at www.foxnews.com.

Slaatten, Hilda, and Leena Gabrys, L. (2014). "Gay-Related Name-Calling as a Response to the Violation of Gender Norms." *Journal of Men's Studies* 22(1):28–33.

Snapp, Shannon D., Hilary Burdge, Adela C. Licona, Raymond L. Moody, and Stephen T. Russell. 2015. "Students' Perspectives on LGBTQ-Inclusive Curriculum." *Equity & Excellence in Education* 48(2): 249–265.

Socarides, Richard. 2015 (April 27). "Corporate America's Evolution on LGBT Rights." *New Yorker.* Available at www. newyorker.com.

Spence, Sarah, Charles C. Helwig, and Nicole Cosentino (2018). "Children's Judgments and Reasoning about Same-Sex Romantic Relationships." *Child Development* 89(3):988–1003.

Strauss, Valerie. 2020 (January 30). "Two Major Banks Pull Support from Florida School Voucher Programs because of Anti-LGBTQ Policies." *Washington Post.* Available at www .washingtonpost.com.

Substance Abuse and Mental Health Services Administration (SAMHSA). 2020. "2018 National Survey on Drug Use and Health: Lesbian, Gay, and Bisexual (LGB) Adults." National Survey on Drug Use and Health. U.S. Department of Health and Human Services. Available at www.samhsa.gov.

Sumerau, J. Edward, and Ryan T. Cragun. 2014. "'Why Would Our Heavenly Father Do That to Anyone?': Oppressive Othering through Sexual Classification Schemes in the Church of Jesus Christ of Latter-Day Saints." *Symbolic Interactionism* 37(3):331–352.

Summers, Bryce B. 2010. "Factor Structure and Validity of the Lesbian, Gay, and Bisexual Knowledge and Attitude Scale for Heterosexuals (LGB-KASH)." ProQuest Dissertations and Theses (UMI No. 3425038). Available at http://hdl.handle.net/10657/163.

Swenson, Kyle. 2018. "In 1955 She Was Kicked out of the Airport for Being a Lesbian. At 90, She's Fighting Back." *Stars and Stripes.* Available at www.stripes.com.

Szymanski, Dawn M., Susan Kashubeck-West, and Jill Meyer. 2008. "Internalized Heterosexism: Measurement, Psychosocial Correlates, and Research Directions." *Counseling Psychologist,* 36:525–574.

Tensley, Brandon. 2019 (October 8). "Will the Equality Act Reframe How We Talk about LGBTQ Americans?" CNN News. Available at www.cnn.com.

Title VII of the Civil Rights Act of 1964. 2020. *Justia.* Available at www.justia.com.

Tobias, Sarah, and Sean Cahill. 2003. "School Lunches, the Wright Brothers and Gay Families." National Gay and Lesbian Task Force. Available at www thetaskforce.org.

Toobin, Jeffrey. 2015 (June 30). "Why Gay Marriage Victory Anthem Was 'Star-Spangled Banner.'" *CNN News.* Available at www.cnn .com.

Trevor Project. 2020. *National Survey on LGBTQ Youth Mental Health 2020.* Available at www .thetrevorproject.org.

Trotter, J. K. 2013 (March 27). "'Skim Milk Marriage Is the New 'Broccoli.'" *Atlantic.* Available at www.theatlantic.com.

UBS. 2018. "How Planning for LGBT Retirement Differs." Available at www.ubs.com.

United Nations (UN). 2018. *Protection against Violence and Discrimination Based on Sexual Orientation and Gender Identity, Mandate of the U.N. Independent Expert: Report on Gender Identity.* Available at www.un.org.

United Nations (UN). 2020 (May 1). "Practices of So-Called 'Conversion Therapy.'" *Report of the Independent Expert on Protection against Violence and Discrimination Based on Sexual Orientation and Gender Identity.* Human Rights Council. Available at undocs.org.

Ura, Alexa. 2017 (December 5). "Are Gay Spouses Entitled to Marriage Benefits? U.S. Supreme Court Refuses Case That Would Clarify." *Governing.* Available at www.governing.com.

U.S. Census. 2019 (November 19). "U.S. Census Bureau Releases CPS Estimates of Same-Sex Households." Available at www.census.gov.

U.S. Court of Appeals, District 3. 2010 (September 22). *Florida Department of Children and Families v. In re Matter of Adoption of: X.X.G. and N.R.G.* No. 3D08-3044. Available at www.3dca.flcourts.org.

U.S. v. Windsor. 2013. U.S. Supreme Court. Available at www.supremecourt.gov.

Valencia, Zehra, Breyon Williams, and Robert Pettis. 2019 (March 26). "Pride and Prejudice: Same-Sex Marriage Legalization Announcements and LGBT Hate-Motivated Crimes." *Public Economics*: 1–46. Available at papers.ssrn.com.

Venkatraman, Sakshi. 2020 (July 2). "Election Voters Back Referendum Banning Same-Sex Marriage." NBC News. Available at www .nbcnews.com.

Victory Institute. 2020. "LGBT Elected Officials." Available at victoryinstitute.org.

Watson, J. Ryan, Minjeong Park, Ashley B. Taylor, Jessica R. Fish, Heather L. Corliss, Marla P. Eisenberg, and Elizabeth M. Saewyc. 2020. "Associations between Community-Level LGBTQ Supportive Factors and Substance Use among Sexual Minority Adolescents." *LGBT Health* 7(2):82–89.

West, Keon, and Noel M. Cowell. 2015. "Predictors of Prejudice against Lesbians and Gay Men in Jamaica." *Journal of Sex Research* 52(3):296–305.

Westcott, Ben. 2018 (September 22). "The Homophobic Legacy of the British Empire." CNN News. Available at www.cnn.com.

Wilson, Bianca D. M., Soon Kyu Choi, Gary W. Harper, Marguerita Lightfoot, Stephen Russell, and Ilan H. Meyer. 2020 (May). "Homelessness among LGBT Adults in the U.S." UCLA School of Law Williams Institute. Available at williamsinstitute.law.ucla.edu.

Winberg, Carter, Todd Coleman, Michael R. Woodford, Raymond M. McKie, Robb Travers, and Kristen A. Renn. 2019. "Hearing 'That's So Gay' and 'No Homo' on Campus and Substance Use among Sexual Minority College Students." *Journal of Homosexuality* 66(10):1472–1494.

Woodford, Michael R., Michael L. Howell, Perry Silverschanz, and Lotus Yu. 2012. "'That's So Gay!': Examining the Covariates of Hearing This Expression among Gay, Lesbian, and Bisexual Students." *Journal of American College Health* 60(6):429–434.

Yochim, Dayana. 2020 (June 22). "Pride Month: 12 Key Numbers Highlighting the Economic Status, Challenges That LGBTQ People Face." NBC News. Available at www.nbcnews.com.

Chapter 12

AARP. 2020. *What Is the Maximum Amount of Income That Is Subject to FICA Taxes?* March 25, 2020. Available at www.aarp.org.

AARP and National Alliance for Caregiving. 2020 (May). *Caregiving in the United States 2020.* Washington, DC: AARP. Available at www .caregiving.org.

American Geriatrics Society. 2020. *Why Geriatrics?* American Geriatrics Society. Available at www.americangeriatric.org.

American Psychological Association. 2020. *Older Adults: Health and Age-Related Changes.* Available at www.apa.org.

Andriano, Liliana, and Christian W. S. Monden. 2020. "The Causal Effect of Maternal

Education on Child Mortality: Evidence from a Quasi-Experiment in Malawi and Uganda." *Demography* 56:1765–1790.

Aronson, Louise. 2020 (March 28). "Ageism Is Making the Pandemic Worse." *Atlantic.* Available at www.theatlantic.com.

Ataullahjan, Anushka, Zubia Mumtaz, and Helen Valliantos. 2019. "Family Planning, Islam, and Sin: Understanding of Moral Actions in Khyber Pakhtunkhwa, Pakistan." *Social Science & Medicine* 230:49–56.

Bearak Jonathan, Anna Popinchalk, Bela Ganatra, Ann-Beth Mollar, Őzge Tunçalp, and Cynthia Beavin. 2020. "Unintended Pregnancy and Abortion by Income, Region, and the Legal Status of Abortion: Estimates from a Comprehensive Model for 1990–2019." *Lancet Global Health* 8(9):1–10.

Behjati-Ardakani, Zohreh, Mohammed Mehdi Akhondi, Homa Mahmoodzadeh, and Seyed Hasan Hosseni. 2016. "An Evaluation of the Historical Importance of Fertility and Its Reflection in Ancient Mythology." *Journal of Reproduction & Infertility* 17(1):2–9.

Biden, Joe. 2020. "Plan for Securing Our Values as a Nation of Immigrants." Available at www.joebiden.com.

Bill & Melinda Gates Foundation. 2020. *What We Do: Maternal, Newborn & Child Health Discovery & Tools.* Available at www.gatesfoundation.org

Blackstone, Amy, 2019. *Childfree by Choice: The Movement Redefining Family and Creating a New Age of Independence.* New York: Dutton.

Bloom, David, and Matthew J. McKenna. 2015. *Population, Labour Force, and Unemployment: Implications for the Creation of (Decent) Jobs.* UNDP Human Development Report Office. Available at www.hdr.undp.org.

Brenan, Megan. 2019 (June 18). *More Nonretired Americans Expect Comfortable Retirement.* Gallup. Available at www.news.gallup.com.

Brent, Harry. 2020. "Girl, 5, Writes Adorable Letter to 93-Year-Old Neighbour to Check in on Him during Lockdown." *Irish Post.* Available at www.irishpost.com.

Breytspraak, Linda, and Lynn Badura. 2015. "Facts on Aging Quiz" (revised; based on Palmore [1977; 1981]). Available at http://info.umkc.edu/aging/quiz/.

British Broadcasting Corporation. 2020a (July 7). *Coronavirus: Trump Moving to Pull US Out of World Health Organization.* Available at www.bbc.com.

British Broadcasting Corporation. 2020b (January 15). *How Do Countries Fight Falling Birth Rates?* Available at www.bbc.com.

Brandon, Emily. 2020 (July 7). "What Is the Maximum Possible Social Security Benefits in 2020? *US News & World Report.* Available at www.money.usnews.com.

Caldwell, John C., Bruce K. Caldwell, Pat Caldwell, Peter F. McDonald, and Thomas Schindlmayr. 2006. *Demographic Transition Theory.* Dordrecht, Netherlands: Springer.

Carlsson, Magnus, and Stefan Eriksson. 2019. "Age Discrimination in Hiring Decisions: Evidence from a Field Experiment in the Labor Market." *Labour Economics* 59:173–183.

Carroll, Emily Ripley. 2018. *To Reproduce or Not to Reproduce? Recontextualizing Pronatalism in Light of Climate Change.* Emory University. Unpublished Thesis.

Center for Budget and Policy Initiatives. 2020 (August 13). *Policy Basics: Top Ten Facts about Social Security.* Available at www.cbpp.org.

Center for Reproductive Rights. 2020. *The World's Abortion Laws.* Available at www.reproductiverights.org.

Center for Retirement Research. 2020 (July). "Summary of Companies That Suspended Their 401(K) Match." *Covid-19 Crisis: Economic Data.* Center for Retirement Research: Boston College. Available at www.crr.bc.edu.

Centers for Disease Control. 2021 (February 19). "Table 1: Deaths involving Coronavirus Disease 2019, Pneumonia, and Influenza Reported to NCHS by Sex and Age Group, United States, from 1/1/2020 to 2/13/2021." *Weekly Updates by Select Demographic and Geographic Characteristics: Provisional Death Counts for Coronavirus Disease 2019 (COVID-19).* Available at www.cdc.gov.

Chamie, Joseph, and Barry Mirkin. 2020. *More Countries Want More Babies.* Population Connection. Available at www.populationconnection.org.

Chase-Dunn, Christopher, and Thomas D. Hall. 2018. *Rise and Demise: Comparing World Systems.* New York: Routledge.

Cooke, L. P., and Baxter, J. 2010. "'Families' in International Context: Comparing Institutional Effects across Western Societies." *Journal of Marriage and Family* 72(3):516–536.

Culhane, Dennis, Dan Treglia, Thomas Byrne et al. 2019. *The Emerging Crisis of Aged Homelessness: Could Housing Solutions Be Funded by Avoidance of Excess Shelter, Hospital, and Nursing Home Costs?* Actionable Intelligence for Social Policy: University of Pennsylvania. Available at www.aisp.upenn.edu.

Dérer, Patrícia. 2019 (March 21). *The Iranian Miracle: The Most Effective Family Planning Program in History?* The Overpopulation Project. Available at www.overpopulation-project.com.

Duffin, Erin. 2019 (August). *Percentage of Childless Women, by Age U.S. 2018.* Statista. Available at www.statista.com.

Duflo, Esther, Pascaline Dupas, and Michael Kremer. 2015. "Education, HIV, and Early Fertility: Experimental Evidence from Kenya." *American Economic Review* 105(9):2757–2797.

Edwards, Kathryn A., Anna Turner, and Alexander Hertel-Fernandez. 2012. *A Young Person's Guide to Social Security.* Economic Policy Institute. Available at www.epi.org.

Engelman, Robert. 2011 (July 18). "The World at 7 Billion: Can We Stop Growing Now?" *Yale Environment 360.* Available at e360.yale.edu.

Ergungor, O. Emre. 2017 (October 12). *When States Default: Lessons from Law and History.* Federal Reserve Bank of Cleveland. Available at www.clevelandfed.org.

Fedor, Theresa Marie, Hans-Peter Kohler, and James M. McMahon. 2016. "Changing Attitudes and Beliefs towards a Woman's Right to Protect against HIV Risk in Malawi." *Culture, Health, and Sexuality* 18(4):435–452.

Flaherty, Colleen. 2020 (May 21). "Colleges Lower the Boom on Retirement Plans." *Inside Higher Ed.* Available at www.insidehighered.com.

Frejka, Tomas. 2017. "Childlessness in the United States." In *Childlessness in Europe: Contexts, Causes and Consequences,* ed. Michael Kreyenfield and Dirk Konietzka, 159–179. Switzerland: Springer.

Gerontological Society of America. 2018. *Longevity Economics: Leveraging the Advantages of an Aging Society.* Gerontological Society of America. Available at www.geron.org.

Gleckman, Howard. 2020 (November 10). "What Will Biden Do for Seniors?" *Forbes.* Available at www.forbes.com.

Godoy, Maria. 2020 (November 9). "Biden Said He Walked Back Trump's Walkout. Can All the Damage Be Undone?" National Public Radio. Available at www.npr.com.

Gonzalez-Barrera, Ana. 2015 (November 19). *More Mexicans Leaving Than Coming to the U.S.* Pew Research Center. Available at www.pewresearch.org.

Gosselin, Peter. 2018 (December 28). "If You're over 50, Chances Are the Decision to Leave Your Job Won't Be Yours." *ProPublica.* Available at www.features.propublica.org.

Gosselin, Peter, and Ariana Tobin. 2018 (March 22). "Cutting 'Old Heads' at IBM." *ProPublica.* Available at www.features.propublica.org.

Grandjean, Barbara, and Chad Grell. 2019. "Why No Mandatory Retirement Age Exists for Physicians: Important Lessons for Employers." *Missouri Medicine* 116(5):357–360.

Guttmacher Institute. 2020 (April). *Contraceptive Use in the United States.* Available at www.guttmacher.org.

Hassell, Joe. 2018 (April 3). "Does Population Growth Lead to Hunger and Famine?" *Our World in Data.* Available at www.ourworldindata.org.

Harris, Amy. 2017 (April 25). "How Much of the Earth's Land Is Farmable?" *Sciencing.* Available at www.sciencing.com.

Hopps, Markay, Laura Iadeluca, Margaret McDonald, and Geoffrey T. Makinson. 2017. "The Burden of Family Caregiving in the United States: Work Productivity, Health Care Resource Utilization, and Mental Health among Employed Adults." *Journal of Multidisciplinary Healthcare* 10:437–444.

Howe, Neil. 2019 (March 29). "Nations Labor to Raise Their Birthrates." *Forbes.* Available at www.forbes.com.

Intriago, Joy. 2020. "Healthcare in Sweden: A Model of Elderly Care." SeniorsMatter.com. Available at www.seniorsmatter.com.

Jefferson, Robin Seaton. 2017 (July 30). "You May Not Be as Old as You Think You Are: Researchers Are Proving 60 Really Is the New 50." *Forbes.* Available at www.forbes.com.

Johns Hopkins University. 2021 (February 28). *Covid-19 Dashboard.* Baltimore: Center for System Science and Engineering (CCSE), Johns Hopkins University.

Jones, Rachel K., and Jenna Jerman. 2017 (November). "Population Group Abortion Rates and Lifetime Incidence of Abortion: United States, 2008–2014." *American Journal of Public Health*. Available at https://pubmed.ncbi.nlm.nih.gov/29048970/.

Kabagenyi, Allen, Alice Reid, James Ntozi, and Lynn Atuyambe. 2016. "Socio-Cultural Inhibitors to Use of Modern Contraceptive Techniques in Rural Uganda: A Qualitative Study. *Pan African Medical Journal* 25(78).

Kaiser Family Foundation. 2019 (July 30). "The U.S. Government and International Family Planning & Reproductive Health Efforts." *Global Health Policy*. Available at www.kff.org.

Kamp, Jon, and Anna Wilde Mathews. 2020 (November 25). "COVID-19 Deaths Top 100,000 in U.S. Long-Term Care Facilities." *Wall Street Journal*. Available at www.wsj.com.

Karpf, Anne. 2014 (February 22). "Older Models: The Women in Their 60s, 70s, and 80s Who Are Shaking up Fashion." *Guardian*. Available at www.theguardian.com

Kearney, Melissa S., and Phillip B. Levine. 2020 (June 15). *Half a Million Fewer Children? The Coming Covid Baby Bust*. Brookings Institute. Available at www.brookings.edu.

Khazan, Olga. 2018 (October 11). "When Abortion Is Illegal, Women Rarely Die. But They Still Suffer." *Atlantic*. Available at www.theatlantic.com.

Khazan, Olga. 2020 (July 6). "The U.S. Is Repeating Its Deadliest Pandemic Mistake." *Atlantic*. Available at www.theatlantic.com.

Kiliç, Azer. 2017 (April 28). "Abortion Politics and the New Pro-Natalism in Turkey. *The Progressive Post*. Available at www.progressivepost.eu.

Kita, Joe. 2019 (December 30). *Workplace Age Discrimination Still Flourishes in America*. Available at www.aarp.org.

Knapp, Anthony. 2019 (December 30). *Net Migration between the U.S. and Abroad Added 595,000 to National Population between 2018 and 2019*. U.S Census. Available at www.census.gov.

Kolitz, Daniel. 2019 (May 20). *Is the World Really Overpopulated?* Available at www.earther.gizmodo.com.

Kornadt, Anna E., and Klaus Rothermund. 2010. "Constructs of Aging: Assessing Evaluative Age Stereotypes in Different Life Domains." *Educational Gerontology* 36(6).

Koyangi, Ai, Jordan E. DeVylder, Brendon Stubbs, André F. Carvalho, Nicola Veronese, Josep M. Haro, and Ziggi I. Santini. 2018. "Depression, Sleep Problems, and Perceived Stress among Informal Caregivers in 58 Low-, Middle-, and High-Income Countries: A Cross-sectional Analysis of Community-Based Surveys." *Journal of Psychiatric Research* 96:115–123.

Lee, Shannon. 2020. *Understanding Food Insecurity & Its Impact on Learning*. Affordable Colleges Online. Available at www.affordablecollegesonline.org.

Lev, Sagit, Susanne Wurm, and Liat Ayalon. 2018. "Origins of Ageism at the Individual Level." In *Contemporary Perspectives on Ageism*, ed. Liat Ayalon and Clemens Tesch-Romer, 51–72. Cham, Switzerland: Springer.

Lichtenstein, Bronwen. 2020 (July 28). "From 'Coffin Dodger' to 'Boomer Remover': Outbreaks of Ageism in Three Countries with Divergent Approaches to Coronavirus Control." *The Journals of Gerontology* XX(20):1–2.

Lindberg, Laura D., Alicia VanderVusse, Jennifer Mueller, and Marielle Kirstein. 2020 (June). *Early Impacts of the Covid-19 Pandemic: Findings from the 2020 Guttmacher Survey of Reproductive Health Experiences*. Guttmacher Institute. Available at www.guttmacher.org.

Loos, Eugène, and Loedana Ivan. 2018. "Visual Ageism in the Media." In *Contemporary Perspectives on Ageism*, ed. Liat Ayalon and Clemens Tesch-Romer, 163–176. Cham, Switzerland: Springer.

Luna, Taryn. 2020 (April 29). "Criticism Grows over Gov. Gavin Newsome's Management of the Coronavirus Crisis." *Los Angeles Times*. Available at www.latimes.com.

McQuillan, J., Greil, A. L., Shreffler, K. M., Wonch-Hill, P. A., Gentzler, K. C., and Hathcoat, J. D. 2012. "Does the Reason Matter? Variations in Childlessness Concerns among U.S. Women." *Journal of Marriage and Family* 74:1166–1181.

Morris, Allie, and Robert T. Garrett. 2020 (March 24). "Texas Lt. Gov. Dan Patrick Spurns Shelter in Place, Urges Return to Work, Suggests Grandparents Should Sacrifice." *Dallas Morning News*. Available at www.dallasnews.com.

Mullin, Emily. 2016 (November 11). "Why We Still Don't Have Birth Control Drugs for Men." *MIT Technology Review*.

Naso, Pedro, Bruno Lanz, and Tim Swanson. 2020. "The Return of Malthus? Resource Constraints in the Era of Declining Population Growth." *European Economic Review* 128:1–22.

National Committee to Preserve Social Security & Medicare. 2018 (February 1). "Women's Social Security Benefits." *Social Security Policy Papers*. Available at www.ncpssm.org.

Nelson, Todd D. 2011. "Ageism: The Strange Case of Prejudice against the Older You." In *Disability and Aging Discrimination*, ed. R. L. Wiener and S. L. Willborn, 37. New York: Springer Science + Business Media.

Neto, F. T. L., P. V. Bach, R. J. L. Lyra, J. C. Borges Jr, G. T. D. S. Maia, L. C. N. Araujo, and S. V. C. Lima. 2019. "Gods Associated with Male Fertility and Virility." *Andrology* 7(3):267–272.

Nuemark, David, Ian Burn, and Patrick Button. 2017 (February 27). "Age Discrimination and Hiring of Older Workers." *FRBSF Economic Letter*. Available at www.frbsf.org.

Neumark, David, Ian Burn, Patrick Button, and Nanneh Chehras. 2019. "Do State Laws Protecting Older Workers from Discrimination Reduce Age Discrimination in Hiring? Evidence from a Field Experiment." *The Journal of Law and Economics* 62(2):373–402.

Ní Bhrolcháin, M., and E. Beaujouan. 2012. "Fertility Postponement Is Largely due to Rising Educational Enrolment." *Population Studies: A Journal of Demography* 66(3): 311–327.

Notkin, Melanie. 2013 (August 1). "The Truth about the Childless Life." Huffington Post. Available at www.huffingtonpost.com.

O'Brien, Sarah. 2019 (February 23). "Social Security Expansion Bill Poised to Gain Traction in Congress, Targeting Those Who Earn over $400,000." CNBC. Available at www.cnbc.com.

Officer, Alana, Mira Leonie Schneiders, Diane Wu, Paul Nash, Jotheeswaran Amuthavalli Thiyagarajan, and John R. Beard. 2016. "Valuing Old People: Time for a Global Campaign to Combat Ageism." *Bulletin of the World Health Organization* 94:710–710A. Available at www.who.int.

Our World in Data. 2020. *Children Born per Woman Map*. Our World in Data. Available at www.ourworldindata.org.

Organization for Economic Co-operation and Development (OECD). 2020. *Investing in Women and Girls*. Available at www.oec.org.

Parker, Kim, Rich Morin, and Juliana Menasce Horowitz. 2019 (March 21). "Retirement, Social Security, and Long-Term Care." *Looking to the Future, Public Sees an America in Decline on Many Fronts*. Pew Research Center. Available at www.pewsocialtrends.org.

Peri-Rotem, N. 2016. "Religion and Fertility in Western Europe: Trends across Cohorts in Britain, France and the Netherlands." *European Journal of Population* 32:231–265.

Pinsker, Joe. 2020 (July 23). "We're Talking about More than Half a Million People Missing from the U.S. Population." *Atlantic*. Available at www.theatlantic.com.

Population Reference Bureau. 2015. "Effects of the Great Recession on Older American's Health and Well-Being." *Today's Research on Aging*, 32. Available at www.prb.org.

Population Reference Bureau. 2019. *2019 Family Planning Data Sheet*. Available at www.prg.org.

Population Reference Bureau. 2020. *2020 World Population Data Sheet*. Washington, DC: Population Reference Bureau. Available at www.prb.org.

Pritchard, David, and Francis Pakes, eds. 2014. *Riot, Unrest and Protest on the Global Stage*. London: Palgrave Macmillan.

Putin, Valdimir. 2019 (May 30). *Presentation of the Order of Parental Glory*. Kremlin. Available at www.en.kremlin.ru.

Radford, Jynnah. 2019 (June 17). "Key Findings about U.S. Immigrants." *FactTank*. Pew Research Center. Available at www.pewresearch.org.

Ramos, Juan. 2017 (November 24). "Push and Pull Factors of Migration." *Science Trends*. Available at www.sciencetrends.com.

Rau, Jordan. 2020 (June 11). *Nursing Homes Run Short of Covid-19 Protective Gear as Federal Response Falters*. National Public Radio. Available at www.npr.org.

Riffkin, Rebecca. 2015 (April 29). "Americans Settling on Older Retirement Age." Gallup Organization. Available at www.gallup.com.

Ritchie, Hannah, and Max Roser. 2019. "Age Structure." *Our World in Data*. University of Oxford. Available at www.ourworldindata.org.

Rogers, Katie, and Apoorva Mandavilli. 2020 (October 22). "Trump Administration Signals Formal Withdrawal from WHO." *New York Times*. Available at www.nytimes.com.

Romig, Kathleen. 2020a (February 20). "Social Security Lifts More Americans above Poverty Than Any Other Program." *Policy Futures.* Center on Budget and Policy Priorities. Available at www.cbpp.org.

Romig, Kathleen. 2020b (May 13). "What the 2020 Trustees' Report Shows about Social Security. *Policy Futures.* Center on Budget and Policy Priorities. Available at www.cbpp.org.

Roser, Max. 2017. "Fertility Rate." *Our World in Data.* University of Oxford. Available at www .ourworldindata.org.

Roser, Max. 2019. "Future Population Growth." *Our World in Data.* University of Oxford. Available at www.ourworldindata.org.

Roser, Max, and Esteban Ortiz-Ospina. 2020. "Global Education." *Our World in Data* University of Oxford. Available at www .ourworldindata.org.

Roser, Max, Esteban Ortiz-Ospina, and Hannah Ritchie. 2019. "Life Expectancy." *Our World in Data.* University of Oxford. Available at www .ourworldindata.org.

Roser, Max, Hannah Ritchie, and Bernadeta Dadonaite. 2019. "Child and Infant Mortality." *Our World in Data.* University of Oxford. Available at www.ourworldindata.org.

Roser, Max, Hannah Ritchie, and Esteban Ortiz-Ospina. 2019. "World Population Growth." *Our World in Data.* University of Oxford. Available at www.ourworldindata.org.

Rossman, Sean. 2017 (April 12). "Americans Are Spending More Than Ever on Plastic Surgery." *USA Today.* Available at www.usatoday.com.

Rueckert, Phineas. 2019. *10 Barriers to Education That Children Living in Poverty Face.* Available at www.globalcitizen.org.

Rutledge, Matthew S., and Geoffrey T. Sanzenbacher. 2019 (January). *What Financial Risks Do Retirees Face in Late Life?* Center for Retirement Research at Boston College. Available at https://tinyurl.com/y3s9346k.

Ryerson, William N. 2011 (March 11). "Family Planning: Looking beyond Access." *Science* 331:1265.

Sanzenbacher, Geoffrey T., Anthony Webb, Candace M. Cosgrove, and Natalia S. Orlava. 2017. *Rising Inequality in Life Expectancy by Socioeconomic Status.* Working Paper #2017-2. Center for retirement Research. Boston College. Available at www.crr.bc.edu.

Sayei, Nadja. 2017 (June 30). "Some Celebrities Care More about Helping the World Than Camera Time." *Vice.* Available at www.vice .com.

Schlein, Lisa. 2016. *Study: Negative Attitudes toward Older People Shorten Their Lives.* Available at www.voanews.com.

Sedgh, Gilda, Susheela Sing, and Rubina Hussain. 2014. "Intended and Unintended Pregnancies Worldwide in 2012 and Recent Trends." *Studies in Family Planning* 45(3):301–314.

Shain, Michelle. 2019. "Beyond Belief: How Membership in Congregations Affects the Fertility of U.S. Mormons and Jews." *Review of Religious Research* 61:201–219.

Singh, Susheela, Lisa Remez, Gilda Sedgh, Lorraine Kwok, and Tsuyoshi Onda. 2018 (March). *Abortion Worldwide 2017: Uneven Progress and Uneven Access.* Guttmacher Institute. Available at www.guttmacher.org.

Sobotka, Tomaáš, Anna Matysiak, and Zuzanna Brzozowska. 2019 (May). *Policy Responses to Low Fertility: How Effective Are They?* Working Paper No. 1. Population & Development Branch: UNFPA. Available at www.unfpa.org.

Social Security Administration. 2020a. *Starting Your Retirement Benefits Early.* Available at www.ssa.gov.

Social Security Administration. 2020b. *Income Taxes and Your Social Security Benefit.* Available at www.ssa.gov.

Social Security Board of Trustees. 2020. *The 2020 Annual Report of the Board of Trustees of the Federal Old Age and Survivors Insurance and Federal Disability Insurance Trust Funds.* Available at www.ssa.gov.

Sohngen, Tess. 2017 (July 26). *Nigeria Has the Highest Number of Out-of-School Children in the World.* Available at www.globalcitizen.org.

Somin, Ilya. 2020 (June 28). "The Danger of America's Coronavirus Immigration Bans." *Atlantic.* Available at www.theatlantic.com.

Sotamayar, Kristal. 2020. *The One-Child Policy Legacy on Women and Relationships in China.* Independent Lens.

Span, Paula. 2020 (November 27). "Biden's Plan for Seniors Is Not Just a Plan for Seniors." *New York Times.* Available at www.nytimes.com.

Starrs, Ann M. 2017 (February 4). "The Trump Global Gag Rule: An Attack on US Family Planning and Global Health Aid." *Lancet* 389:10068:485–486.

Stein, Kara. 2018 (October 16). *The New American Dream: Retirement Security.* U.S. Securities and Exchange Commission. Available at www.sec.gov.

Steger, Isabella. 2016 (November 29). "It's a Myth That China Has 30 Million 'Missing Girls' because of the One-Child Policy, a New Study Says." *Quartz.* Available at www.qz.com.

Steptoe, Andrew, Angus Deaton, and Arthur A. Stone. 2015. "Subjective Wellbeing, Health, and Aging. *Lancet* 385(9968):640–648.

Swanson, Ana. 2015 (October 30). "Why Many Families in China Won't Want More Than One Kid Even If They Can Have Them." *Washington Post.* Available at www.washingtonpost.com.

Sword, Doug. 2020 (February 28). "Retirees' Worst Nightmare: Federal Backing of Pension Funds at Risk." *Roll Call.* Available at www .rollcall.com.

The Economist. 2020 (April 18). "Hollywood's Depictions of the Elderly Are Tired Clichés." Available at www.economist.com.

Toossi, Mitra, and Elka Torpey. 2017 (May). "Older Workers: Labor Force Trends and Career Options." *Career Outlook.* U.S. Bureau of Labor Statistics. Available at www.bls.gov.

United Nations. 2013. *World Population Prospects: The 2012 Revision.* Available at www.un.org/esa/population/publications /publications.htm.

United Nations. 2018 (June 20). *Progress towards the Sustainable Development Goals, Report of the Secretary-General.* Available at https://digitallibrary.un.org /record/1627573?ln=en.

United Nations. 2019a. *World Population Prospects: The 2019 Revision.* New York: United Nations.

United Nations. 2019b (April 1). *World Economic Situation and Prospects: April 2019 Briefing No 125.* United Nations Department of Economic and Social Affairs. Available at www.un.org.

United Nations. 2020a. *World Population Ageing, 2019.* New York: United Nations. Available at www.un.org.

United Nations. 2020b. *The State of Food Security and Nutrition in the World, 2020.* New York: United Nations. Available at www.un.org.

United Nations Global Compact. 2020. *Who We Are.* Available at www.unglobalcompact.org.

U.S. Census Bureau. 2017 (May 10). "Voting Rates by Age." Available at www.census.gov.

VanDerhei, Jack. 2019 (March 7). *Retirement Savings Shortfalls: Evidence from EBRI's 2019 Retirement Security Projection Model.* Employee Benefit Research Institute. Available at www.ebri.org.

Van De Water, Paul N., and Kathleen Romig. 2020. "Social Security Benefits Are Modest." *Policy Futures.* Center on Budget and Policy Priorities. Available at www.cbpp.org.

Walker, Danielle. 2018 (November 30). "Kentucky AG Urges Lawmakers to Expand Gambling to Boost Pension Funds." *Pensions & Investments.* Available at www.pionline.com.

Wallerstein, I. 1974. "The Rise and Future Demise of the World Capitalist System: Concepts for Comparative Analysis." *Comparative Studies in Society and History* 16:387–415.

Weeks, John R. 2015. *Population: An Introduction to Concepts and Issues*, 12th ed. Belmont, CA: Wadsworth, Cengage Learning.

Weiland, Katherine. 2005. *Breeding Insecurity: Global Security Implications of Rapid Population Growth.* Washington, DC: Population Institute.

Weinberger, Daniel M., Jenny Chen, and Ted Cohen. 2020 (July 1). "Estimation of Excess Deaths Associated with the Covid-19 Pandemic in the United States, March to May 2020." *JAMA.* Available at www.jamanetwork.com.

Whorisky, Peter, Debbie Cenziper, Will Englund, and Joel Jacobs. 2020 (June 4). "Hundreds of Nursing Homes Ran Short of Staff, Protective Gear as More Than 30,000 Residents Died during Pandemic." *Washington Post.* Available at www.washingtonpost.com.

Woodward, Aylin. 2020 (September 20). "Far-Right Conspiracy Theorists Say 94% of U.S. COVID-19 Deaths Don't Count Because Those Americans Had Underlying Conditions. That's Bogus." *Business Insider.* Available at www .businessinsider.com.

World Bank. 2018. *Women, Business and the Law 2018.* Washington, DC. Available at wbl. worldbank.org/.

World Health Organization. 2019a. *Contraception: Evidence Brief.* Available at www.who.int.

World Health Organization. 2019b (June 26). *Preventing Unsafe Abortion.* Available at www .who.int.

YaleHRJ. 2018 (February 22). "Female Infanticide in China." *Yale Human Rights Journal*. Available at www.yhrj.org.

Zaidi, Batool, and S. Philip Morgan. 2017. "The Second Demographic Transition Theory: A Review and Appraisal." *Annual Review of Sociology* 43:473–492.

Zak, Danilo. 2020 (July 14). *Immigration-Related Executive Actions during the COVID-19 Pandemic*. National Immigration Forum. Available at www.immigrationforum.org.

Chapter 13

Actman, Jani. 2019. "Poaching Animals, Explained." *National Geographic*. Available at www.nationalgeographic.com.

Air Quality. 2020. "Air Pollution." Available at www.air-quality.org.uk.

Alt, Kimberly. 2020 (January 15). "LED vs. CFL vs. Incandescent vs. Fluorescent: Which Shines Cleanest?" *Earthfriends*. Available at www .earthfriends.com.

Alternative Fuels Data Center. 2020a. "Ethanol Fuel Basics." U.S. Department of Energy Office of Energy Efficiency and Renewable Energy. Available at www.afdc.energy.gov.

Alternative Fuels Data Center. 2020b. "Ethanol Laws and Incentives." U.S. Department of Energy Office of Energy Efficiency and Renewable Energy. Available at www.afdc .energy.gov.

American Civil Liberties Union (ACLU). 2020. "Second Attempt to Legislate Peaceful Protest in South Dakota Unnecessary." www.aclu.org.

American Fisheries Society. 2020. "Policy Statement on Acid Precipitation." Available at www.fisheries.org.

American Lung Association. 2020a. "Radon Basics." Available at www.lung.org.

American Lung Association. 2020b. "Indoor Air Pollutants and Health." Available at www.lung .org.

American Lung Association 2020c. "Toxic Air Pollutants." Available at www.lung.org.

American Lung Association 2020d. "State of the Air." Available at www.stateoftheair.org.

American Wind Energy Association. 2020. "Wind 101." Available at www.awea.org.

Appalachian Voices. 2020. *Ecological Impacts of Mountaintop Removal*. Available at www. appvoices.org.

Archbishop's Council. 2015 (June 16). "Archbishop of Canterbury Join Faith Leaders in Call for Urgent Action to Tackle Climate Change." Available at www.churchofengland .org.

Asaff, Beth. 2020. "Advantages and Disadvantages of Nuclear Energy." Love to Know. Available at www .greenlivinglovetoknow.com.

Atlanta Journal Constitution. 2017 (July 14). "Jimmy Carter Leases His Land to Solar Power Much of Plains." *Atlanta Journal Constitution*. Available at www.ajc.com.

Bagher, Askari Mohammed, Mirzaei Vahid, Mirhabibi Mohsen, and Dehghani Parvin. 2015. "Hydroelectric Advantages and Disadvantages." *American Journal of Energy Science* 2(2):17–20.

Bailey, Pete. 2019. "What Are the Benefits of Slower Population Growth?" *PopEd Blog*. Population Education. Available at www .populationeducation.org.

Bale, Rachael. 2019 (December 26). "How Many Species Haven't We Found Yet?" *National Geographic Newsletters*. Available at www .nationalgeographic.com.

Bamberger, Michelle, and Robert Oswald. 2014. *The Real Cost of Fracking*. Boston: Beacon Press.

Banerjee, Neela. 2017. "How Big Oil Lost Control of Its Climate Misinformation Machine." *Inside Climate News*. Available at www .insideclimatenews.org.

Belcher, Oliver, Patrick Bigger, Ben Neimark, and Cara Kennelly. 2019. "Hidden Carbon Costs of the 'Everywhere War': Logistics, Geopolitical Ecology, and the Carbon Boot-Print of the US Military." *Transactions of the Institute of British Geographers* 45:65–80.

Berardelli, Jeff. 2020 (October 5). "How Joe Biden's Climate Plan Compares to the Green New Deal." CBS News. Available at CBSNews. com.

Berkes, Howard. 2012. "As Mine Protections Fail, Black Lung Cases Surge." National Public Radio. Available at www.npr.org.

Biden, Joe. 2020. "The Biden Plan for a Clean Energy Revolution and Environmental Justice." Available at www.joebiden.com.

Bigger, P., and B. Neimark. 2017. "Weaponizing Nature: The Geopolitical Ecology of the US Navy's Biofuel Program." *Political Geography* 60:13–22.

Bogard, Paul. 2013 (August 19). "Bringing back the Night: A Fight against Light Pollution." *Yale Environment 360*. Available at e360.yale. edu.

Boggs, Amelia. 2015. "Mountaintop Removal: Is It Worth the Costs of Life and Limb in Appalachia?" *ESSAI* 13(11).

Bongaarts, John, and Brian C. O'Neill. 2018. "Global Warming Policy: Is Population Left out in the Cold?" *Science* 361(6403):650–652.

Borunda, Alejandra. 2020 (July 9). "Keystone XL Stalls—Again—along with Other Pipeline Projects." *National Geographic*. Available at www.nationalgeorgraphic.com.

Bowe, Rebecca. 2020. "Defending America's Climate Forest." Earth Justice. Available at www.earthjustice.org.

Boyle, Rebecca. 2017 (February 27). "American Kids Are about to Get Even Dumber When It Comes to Climate Science." *Mother Jones*. Available at www.motherjones.com.

Brekke, Dan. 2019. "In Remembrance: The Names of Those Lost in Camp Fire." *The California Report*. Available at www.kqed.org.

Brody, David. 2018. "Unraveling the 'Weaponization' of the EPA Is Top Priority for Scott Pruitt." CBN News. Available at www1 .cbn.com.

Brown, Alleen. 2019. "The Green Scare: How a Movement That Never Killed Anyone Became the FBI's No. 1 Domestic Terrorism Threat." *The Intercept*. Available at www.theintercept .com.

Bureau of Reclamation. 2020. "Grand Coulee Dam Statistics and Facts." U.S. Department of the Interior. Available at www.usbr.gov.

Burke, Michael. 2019 (April 3). "Trump Claims Wind Turbine 'Noise Causes Cancer.'" *The Hill*. Available at www.thehill.com.

Canu, Will H., John Paul Jameson, Ellen H. Steele, and Michael Denslow. 2017. "Mountaintop Removal Coal Mining and Emergent Cases of Psychological Disorder in Kentucky." *Community Mental Health Journal* 53:802–810.

Carrington, Damian, Niki Kommenda, Pablo Gutierrez, and Cath Levett. 2018 (June 27). "One Football Pitch of Forest Lost Every Second in 2017, Data Reveals." *Guardian*. Available at www.theguardian.com.

Carroll, Linda. 2015 (February 20). "Eco-Drones Aid Researchers in Fight to Save the Environment." NBC News. Available at www .NBCNews.com.

Ceballos, Garardo, Paul R. Ehrlich, and Peter H. Raven. 2020. "Vertebrates on the Brink as Indicators of Biological Annihilation and the Sixth Mass Extinction." *Proceedings of the National Academy of Sciences* 24:13596–13602.

Center for Disaster Philanthropy. 2020 (October 23). "2020 North American Wildfire Season." Available at www.disasterphilanthropy.org.

Centers for Disease Control. 2020. "Pesticides." Available at www.ephtracking.cdc.gov.

Cerulli, Tovar. 2014 (March 14). "A Caretaker and a Killer: How Hunters Can Save the Wilderness." *Atlantic*. Available at www .theatlantic.com.

ChargePoint. 2020. "EVs Charge ahead on College Campuses with 35% Increase in Charging Stations." Available at www.chargepoint.com.

Chen, Grace. 2016 (June 22). "Should Global Warming Be Taught in Public Schools?" *Public School Review*. Available at www.publicschool review.com.

Cheng, Lin, Jeffrey J. Opperman, David Tickner, Robert Speed, Qiaoyu Guo, and Daqing Chen. 2018. "Managing the Three Gorges Dam to Implement Environmental Flows in the Yangtze River." *Frontiers in Environmental Science* 6(64).

Cho, Renee. 2018. "What Can We Do about the Growing E-Waste Problem?" *State of the Planet*. Earth Institute, Columbia University. Available at www.blogs.ei.columbia.edu.

Clemmitt, Marcia. 2011. "Nuclear Power." *CQ Researcher* 21(22):505–528.

CNN Editorial Board. 2020a (August 23). "Flint Water Crisis Fast Facts." Available at www .cnn.com.

CNN Editorial Research. 2020b (April 8). "Kyoto Protocol Fast Facts." Available at www.cnn .com.

Cohen, Robert A., Edward L. Petsonk, Cecile Rose, Byron Young, Michael Regier, Asif Najmuddin, Jerrold L. Abraham, Andrew Churg, and Francis H. Y. Green. 2016. "Lung Pathology in U.S. Coal Workers with Rapidly Progressive Pneumoconiosis Implicates Silica and Silicates." *American Journal of Respiratory & Critical Care Medicine* 193(6):673–680.

Coleman, Clayton, and Emma Dietz. 2019. "Fact Sheet: Fossil Fuel Subsidies: A Closer Look at Tax Break and Societal Costs. Environmental and Energy Study Institute." Available at www.eesi.org.

Connor, Steve. 2015 (April 7). "Breakthrough in Hydrogen-Powered Cars May Spell End for Petrol Stations." *The Independent.* Available at www.independent.co.uk.

Cooper, Evlondo. 2020. "Responding to Public Pressure, Media Scrutiny, and Reality, Presidential Debate Moderator Chris Wallace Actually Asks about Climate Change." Media Matters for America. Available at www.mediamatters.org.

Cortes-Ramirez, Javier, Suchithra Naish, Peter D. Sly, and Paul Jagals. 2018. "Mortality and Morbidity in Populations in the Vicinity of Coal Mining: A Systematic Review." *BMC Public Health* 18(1):721–738.

Cox, Kimberly H., and Joseph S. Takahashi. 2019. "Circadian Clock Genes and the Transcriptional Architecture of the Clock Mechanism." *Journal of Molecular Endocrinology* 63(4):R93–R102.

Daley, Jim. 2020 (November 4). "U.S. Exits Paris Climate Accord after Trump Stalls Global Warming Action for Four Years." *Scientific American.* Available at www.scientificamerican.com.

d'Estries, Michael. 2018. "The Environmental Groups That Dominate Facebook." *Treehugger.* Available at www.treehugger.com.

"Decoding the Labels." 2015. *Guide to Healthy Cleaning.* Environmental Working Group. Available at www.ewg.org/guides/cleaners/content/decoding_labels.

Denchak, Melissa. 2018 (November 8). "Flint Water Crisis: Everything You Need to Know." National Resources Defense Council (NRDC). Available at www.nrdc.org.

Dennis, Brady, and Chris Mooney. 2017 (March 9). "On Climate Change, Scott Pruitt Causes an Uproar—And Contradicts the EPA's Own Website." *Washington Post.* Available at www.washingtonpost.com.

Devlin, Hannah. 2017. "Thousands of Pollution Deaths Worldwide Linked to Western Consumers—Study." *Guardian.* Available at www.theguardian.com.

Dicker, Rachel. 2016 (May 20). "Portland Public Schools Ban Education Materials Denying Climate Change." *US News & World Report.* Available at www.usnews.com.

Dormido, Hannah. 2019 (August 6). "These Countries Are the Most at Risk from the Water Crisis." Bloomberg. Available at www.bloomberg.com.

Doshkin, Fedor A. 2016. "Whose Backyard and What's at Issue? Spatial and Ideological Dynamics of Local Opposition to Hydraulic Fracturing in New York, 2010–2013." *American Sociological Review* 81(5):921–948.

Drake, Nadia. 2019. "Our Nights Are Getting Brighter, and Earth Is Paying the Price." *National Geographic.* Available at www.nationalgeographic.com.

Drugmand, Dana. 2020 (May 1). "EPA Has Been Captured by Fossil Fuel Interests, Democratic Senators Tell Courts." *Desmog.* Available at www.desmog,com.

Earth Justice. 2020. "What Is Mountaintop Removal Mining?" Available at www.earthjustice.org.

Earth Law Center. 2017. "Dams + Climate Change = Bad News." Available at www.earthlawcenter.org.

Earth Overshoot Day. 2020. "How the Date of Earth Overshoot Day 2020 Was Calculated." Available at www.overshootday.org.

Earth Works. 2013. "Cycle of Fracking Denial." Available at www.earthworks.org.

Eilperin, Juliet, and Steven Mufson. 2020. "In Rare Bipartisan Climate Agreement, Senators Forge Plan to Slash Use of Potent Greenhouse Gases." *Washington Post.* Available at www.washingtonpost.com.

Elder, Rodney. 2019. "Creating New Markets in the Lifecycle of Connected Things. *Equinix.* Available at www.blog.equinix.com.

Energy Sector Management Assistance Program. 2016. *Greenhouse Gases from Geothermal Power Production.* Technical Report 009/16. World Bank. Available at www.worldbank.org.

Environmental Defense Fund. 2020a. "Our Work." Available at www.edf.org.

Environmental Defense Fund. 2020b. "Cutting Carbon and Growing the Economy: A Decade of Cap-and-Trade Success in California." Available at www.edf.org.

Environmental Protection Agency. 2016. "Hydraulic Fracturing for Oil and Gas: Impacts from the Hydraulic Fracturing Water Cycle on Drinking Water Resources in the United States (Final Report)." EPA/600/R-16/236F. Available at www.cfpub.epa.gov.

Environmental Protection Agency (EPA). 2017 (March 28). "EPA to Review the Clean Power Plan under President Trump's Executive Order." News Release. Available at www.epa.gov.

Environmental Protection Agency (EPA). 2020a. "Regulations for Lead Emissions from Aircraft." Available at www.epa.gov.

Environmental Protection Agency (EPA). 2020b. "National Priorities List (NPL)." Available at www.epa.gov.

Environmental Protection Agency (EPA). 2020c. "Facts and Figures about Materials, Waste and Recycling." Available at www.epa.gov.

Environmental Working Group (EWG). 2020 (March 25). "EWG's 2020 Shopper's Guide to Pesticides in Produce." Available at www.ewg.org.

Fagan, Moira, and Christine Huang. 2019. "A Look at How People around the World View Climate Change." *FactTank.* Pew Research Center. Available at www.pewresearch.org.

Fellows, Andrew. 2019. *Gaia, Psyche, and Deep Ecology: Navigating Climate Change in the Anthropocene.* Oxford and New York: Routledge.

Ferris, Ann, Richard Garbaccio, Alex Marten, and Ann Wolverton. 2017. *The Impacts of Environmental Regulation on the U.S. Economy.* Working Paper 17-01. National Center for Environmental Economics, U.S. Environmental Protection Agency. Available at www.epa.gov.

Fields, Spencer. 2019. "100% Renewable." *EnergySage.* Available at www.news.energysage.com.

Food and Agriculture Organization. 2018. *The State of the World's Forests 2018—Forest Pathways to Sustainable Development.* Rome: United Nations. Available at www.fao.org.

Food & Water Watch. 2020. "Local Resolutions against Fracking." Available at www.foodandwaterwatch.org.

Forum on Religion and Ecology at Yale. n.d. Available at www.fore.yale.edu.

Foster, Russell G. 2020. "Sleep, Circadian Rhythms and Health." *Interface Focus* 10(3).

Fox, Alex. 2019 "Do Wind Turbines Really Slaughter Birds?" *The Hill.* Available at www.thehill.com.

Fridays for Future. 2020. "What We Do." Available at www.fridaysforfuture.org.

Funk, Cary, Alec Tyson, Brian Kennedy, and Courtney Johnson. 2020. "Science and Scientists Held in High Esteem across Global Publics." Pew Research Center. Available at www.pewresearch.org.

Gallup Organization. 2020. *Environment.* Available at www.gallup.org.

Geiger, Sonja Maria, Mattis Geiger, and Oliver Wilhelm. 2019. "Environment-Specific vs. General Knowledge and Their Role in Pro-Environmental Behavior." *Frontiers in Psychology* 10(1).

Georgeson, Lucien, and Mark Maslin. 2019. "Estimating the Scale of the US Green Economy within the Global Context." *Palgrave Communications* 5:121.

Gibbens, Sarah. 2019 (October 10). "Fast Food Increases Exposure to a 'Forever Chemical' Called PFAS." *National Geographic.* Available at www.nationalgraphic.com.

Global Footprint Network. 2020a. "Is Your Country Running an Ecological Deficit?" Oakland, CA: Global Footprint Network. Available at www.footprintnetwork.org.

Global Footprint Network. 2020b. *Our Past and Our Future.* Oakland, CA: Global Footprint Network. Available at www.footprintnetwork.org.

Gorski, Irena, and Brian S. Schwartz. 2019. "Environmental Health Concerns from Unconventional Gas Development." *Oxford Research Encyclopedias: Global Public Health.* Available at www.oxfordre.com.

Greene, Caitlyn, and Patrick Charles McGinley. 2019. "Yielding to the Necessities of a Great Public Industry: Denial and Concealment of the Harmful Health Effects of Coal Mining." *William and Mary Environmental Law and Policy Review* 43:689–757.

Gross, Liza. 2019 (March 21). "More Than 90 Percent of Americans Have Pesticides or Their Byproducts in Their Bodies." *The Nation.* Available at www.thenation.com.

Gross, Samantha. 2020. "The United States Can Take Climate Change Seriously While Leading the World in Oil and Gas Production." Brookings Institute. Available at www.brookings.edu.

Hansen, Johnni. 2017. "Night Shift Work and the Risk of Breast Cancer." *Current Environmental Health Reports* 4(3):325–339.

Hardin, Sally, and Claire Moser. 2019 (January 28). *Climate Deniers in the 116th Congress.* Center for American Progress Action Fund. Available at www.americanprogressaction.org.

Harvard University. 2020. "Green Revolving Fund." Available at www.green.harvard.edu.

Harvey, Chelsea. 2020 (January 22). "Closing the Ozone Hole Helped Slow Arctic Warming." *Scientific American.* Available at www .scientificamerican.com.

Hayes, Adam. 2020. "How to Profit from Solar Energy." *Investopedia.* Available at www .investopedia.com.

Helm, Sabrina, Joyce Serido, Sun Young Ahn, Victoria Ligon, and Soyeon Shim. 2019. "Materialist Values, Financial and Pro-Environmental Behaviors, and Well-Being." *Young Consumers* 20(4):264–284.

Ho, Bruce. 2020. "The Regional Greenhouse Gas Initiative Is a Model for the Nation." Natural Resources Defense Council. Available at www .nrdc.org.

Holden, Emily. 2019 (April 30). "The Media Is Failing on Climate Change—Here's How They Can Do Better ahead of 2020." *Guardian.* Available at www.theguardian.com.

Holden, Emily. 2020 (February 12). "Trump Turns back the Clock by Luring Drilling Companies to Pristine Lands." *Guardian.* Available at www .theguardian.com.

Holzman, David. 2011. "Mountaintop Removal Mining: Digging into Community Health Concerns." *Environmental Health Perspectives* 119.

Horowitz, Cara. 2020 (November 10). "How California Is an Example for Biden's Climate Change Fight." *Capradio.* Available at www .capradio.org.

Huber, Chris. 2018. "2017 Hurricane Harvey: Facts, FAQs, and How to Help." *World Vision.* Available at www.worldvision.org.

Hussain, Murtaza. 2019 (September 15). "War on the World." *The Intercept.* Available at www .theintercept.com.

Hwang, Jeongeun, Hyunjin Bae, Seunghyun Choi, Hayn Yi, Beomseok Ko, and Namkug Kim. 2020. "Impact of Air Pollution on Breast Cancer Incidence and Mortality: A Nationwide Analysis in South Korea." *Scientific Reports* 10:5392.

IHA Sustainability Ltd. 2020. "Hydropower Sustainability Tools." Available at www .hydrosustainability.org.

Inglis, Jeff, and John Rumpler. 2015. *Fracking Failures: Oil and Gas Industry Environmental Violations in Pennsylvania and What They Mean for the United States.* Raleigh: Environment North Carolina Research Policy Center. Available at www .environmentnorthcarolinacenter.org.

Intergovernmental Panel on Climate Change. 2018. "*Global Warming of 1.5C.*" United Nations Environmental Programme. Available at www.ipcc.

Intergovernmental Panel on Climate Change. 2020. "History of the IPCC." United Nations Environmental Programme. Available at www .ipcc.

International Energy Agency. 2020a. "CO_2 Emissions from Fuel Combustion: Overview." Available at www.iea.org.

International Energy Agency. 2020b. "Global Energy Review 2019." Available at www.iea .org.

International Hydropower Association. 2020. "2020 Hydropower Status Report." Available at www.hydropower.org.

International Renewable Energy Agency. 2019. "Renewable Energy and Jobs—Annual Review 2019." Available at www.irena.org.

International Renewable Energy Agency. 2020. "Geothermal Energy." Available at www.irena .org.

International Union for Conservation of Nature (IUCN). 2020. "The IUCN Red List of Threatened Species, 2020." Available at www .iucnredlist.org.

Islamic Foundation for Ecology and Environmental Sciences (IFEES). 2020. "Islamic Declaration on Global Climate Change." Available at www.ifees.org.uk.

Jacobo, Julia. 2019. "Nearly 40% Decline in Honey Bee Population Last Winter 'Unsustainable', Experts Say." ABC News. Available at www.abcnews.go.com.

Jacoby, Mitch. 2020 (March 30). "As Nuclear Waste Piles Up, Scientists Seek the Best Long-Term Storage Solutions." *Chemical & Engineering News.* Available at www.cen.acs .org.

Jennewein, Madeleine. 2018 (September 5). "Looking for a Trash Can: Nuclear Waste Management in the United States." *Science in the News.* Harvard University Graduate School of Arts and Sciences. Available at www.sitn .hms.harvard.edu.

Jerolmack, Colin, and Edward T. Walker. 2018. "Please in My Backyard: Quiet Mobilization in Support of Fracking in an Appalachian Community." *American Journal of Sociology* 124:479–516.

Johnston, Jill, and Lara Cushing. 2020. "Chemical Exposures, Health, and Environmental Justice in Communities Living on the Fenceline of Industry." *Current Environmental Health Reports* 7:48–57.

Kennedy, Brian. 2020. "Most Americans Say Climate Change Affects Their Local Community, Including 70% Living near Coast." *FactTank.* Pew Research Center. Available at www.pewresearch.org.

Khoo, Michael, and Melissa Ryan. 2020. "Climate, Clicks, Capitalism, and Conspiracists." *Medium.* Available at www .medium.com.

Kiersz, Andy, and Allana Akhtar. 2019. "21 High-Paying Careers for People Who Want to Save the Planet—and Also Have Job Security." *Business Insider.* Available at www .businessinsider.com.

Kimmerer, Robin Wall. 2013. *Braiding Sweetgrass: Indigenous Wisdom, Scientific Knowledge, and the Teachings of Plants.* Minneapolis, MN: Milkweed Editions.

Kirakozian, Ankinée. 2015. "The Determinants of Household Recycling: Social Influence, Public Policies, and Environmental Preferences." *Applied Economics* 48:1481–1503.

Koman, Patricia, Veena Singla, Juleen Lam, and Tracey J. Woodruff. 2019 (August 29). "Population Susceptibility: A Vital Consideration in Chemical Risk Evaluation under the Lautenberg Toxic Substances Control Act." *PLOS Biology.*

Komatsu, Hikaru, Jeremy Rappleye, and Iveta Silova. 2019. "Culture and the Independent Self: Obstacles to Environmental Sustainability? *Anthropocene* 26:1–13.

Kotch, Alex. 2020. "Members of Congress Own up to $93 Million in Fossil Fuel Stocks." *American Prospect.* Available at www .prospect.org.

Kowarski, Ilana. 2019 (November 27). "10 Environmentally Friendly College Campuses." *US News & World Report.* Available at www .usnews.com.

Krikorian, Shant. 2020 (January 1). *Preliminary Nuclear Power Facts and Figures 2019.* International Atomic Energy Agency. Available at www.iaea.org.

Kramar, David E., Aaron Anderson, Hayley Hilfer, Karen Branden, and John J. Gutrich. 2018. "A Spatially Informed Analysis of Environmental Justice: Analyzing the Effects of Gerrymandering and the Proximity of Minority Populations to U.S. Superfund Sites." *Environmental Justice* 11(1).

Krometis, Leigh-Anne, Julia Gohlke, Korine Kolivras, Emily Satterwhite, Susan West Marmagas, and Linsey C. Marr. 2017. "Environmental Health Disparities in the Central Appalachian Region of the United States." *Reviews on Environmental Health* 32(3):253–266.

Kurtzman, Laura. 2016. "Study Finds Wide Exposure to Environmental Toxins in Cohort of Pregnant Women." *University of California San Francisco Research Blog.* Available at www .ucsf.edu.

LaReau, Jamie L. 2020 (November 23). "GM Dumps Trump in His Attempt to Bar California from Setting Emissions." *USA Today.* Available at www.usatoday.com.

Law, Samuel, and Nicholas Troja. 2019. *Hydropower Growth and Development through the Decades.* Available at www.hydropower.org.

Li, Xiaoyang, and Yue M. Zhou. 2017. "Offshore Pollution while Offshoring Production?" *Strategic Management Journal* 38(11):2310–2329.

Lindsey, Rebecca. 2020 (August 14). "Climate Change: Global Sea Level." NOAA. Available at www.climate.gov.

Lindwall, Courtney. 2019 (May 21). "Inside the Fight for Clean Water in Newark." Natural Resources Defense Council (NRDC). Available at www.nrdc.org.

Liobikiene, Genovaite, Justina Madravickaite, and Jurga Bernatoniene. 2016. "Theory of Planned Behavior Approach to Understand the Green Purchasing Behavior in the EU: A Cross-Cultural Study." *Ecological Economics* 125:38–46.

Little, Jane Braxton. 2019 (January 16). "Fukushima Residents Return despite Radiation." *Scientific American.* Available at www. scientificamerican.com.

Loceya, Kenneth J., and Jay T. Lennona. 2016. "Scaling Laws Predict Global Microbial Diversity." *Proceedings of the National Academy of Sciences* 21:5970–5975.

Loeb, Vernon, Marianne Lavelle, and Stacy Feldman. 2020 (September 1). "President Donald Trump's Climate Change Record Has Been a Boon for Oil Companies, and a Threat to the Planet." *Inside Climate News.* Available at www.insideclimatenews.org.

Malin, Stephanie A., and Kathryn Teigan DeMaster. 2016. "A Devil's Bargain: Rural Environmental Injustices and Hydraulic Fracturing on Pennsylvania's Farms." *Journal of Rural Studies* 47:278–290.

Mansel, Tim. 2019 (July 12). "The Coal Mine That Ate Hambacher Forest." BBC News. Available at www.bbc.com.

Marsh, Jacob. 2019. "Solar Farms: What Are They, and How Do You Start One?" *Energy Sage.* Available at www.news.energysage.com.

Martin, James. 2019 (March 4). "A Rare Look at the Meltdown inside Fukushima Daiichi Nuclear Power Plant." *C|net.* Available at www.cnet.com.

Maryland Department of Natural Resources. 2020. "Deer Hunting: An Effective Management Tool." Available at www.dnr.maryland.gov.

McCarthy, Peggy. 2019. "Toxic Exposure on Army Bases Spark Battle for Health Benefits." Associated Press. Available at www.apnews.com.

McKenna, Phil. 2020 (November 25). "As Special Envoy for Climate John Kerry Will Be No Stranger to International Negotiations." *Inside Climate News.* Available at insideclimatenews.org.

McLaughlin, Tim. 2018. "U.S. Clean Coal Program Fails to Deliver on Promised Smog Cuts." Reuters. Available to www.reuters.com.

McSweeney, Robert. 2019. "Explainer: Desertification and the Role of Climate Change." Carbon Brief. Available at www.carbonbrief.org.

Melia, Michael. 2020 (February 19). "Students Push Universities to Stop Investing in Fossil Fuels." Associated Press. Available at www.apnews.com.

Menz, Fredric, and Hans M. Seip. 2004. "Acidic Rain in Europe and the United States: An Update." *Environmental Science & Policy* 7(4):253–265.

Merkley, Jeff. 2020 (July 24). *Omar, Sanders, Merkley, Markey, Barragán Introduce Bill to End Corporate Handouts to the Fossil Fuel Industry.* Available at www.merkley.senate.gov.

Michigan Civil Rights Commission. 2017. "The Flint Water Crisis: Systemic Racism through the Lens of Flint." Report of the Michigan Civil Rights Commission. Available at www.michigan.gov.

Mikata, Ihab, Adam F. Benson, Thomas J. Luben, Jason D. Sacks, and Jennifer Richmond-Bryant. 2018. "Disparities in Distribution of Particulate Matter Emission Sources by Race and Poverty Status." *American Journal of Public Health* 108(4):480–485.

Milman, Oliver. 2019 (May 22). "US Cosmetics Are Full of Chemicals Banned by Europe—Why?" *Guardian.* Available at www.theguardian.com.

Min, J., and K. Min. 2018. "Outdoor Artificial Nighttime Light and Use of Hypnotic Medications in Older Adults: A Population-Based Cohort Study." *Journal of Clinical Sleep Medicine* 14(11):1903–1910.

Mok, Kimberley. 2020 (May 7). "So Much for Fish & Chips: Greenpeace List of Most Overfished Species." *Treehugger.* Available at www.treehugger.com.

Morton, Mary Caperton. 2018. "Acid Rain Triggered Deadly Chinese Landslide." *Earth.* Available at www.magazine.com.

Muehlenbachs, L., Spiller, E., & Timmins, C. 2015. "The Housing Market Impacts of Shale Gas Development." *The American Economic Review 105*(12):3633–3659.

Mui, Simon, and Amanda Levin. 2020. "Cleaning the Air: The Benefits of the Clean Air Act." Natural Resources Defense Council. Available at www.nrdc.org.

Mulvey, Kathy, and Seth Shulman. 2015. "*The Climate Deception Dossiers: Internal Fossil Fuel Industry Memos Reveal Decades of Corporate Disinformation.*" Union of Concerned Scientists. Available at www.ucsusa.org.

Nader, Ralph. 2013 (October 14). "Why Atomic Energy Stinks Worse Than You Thought." Counterpunch. Available at www.counterpunch.org

Nassar, Atef M. K., Yehia M. Salim, and Farag M. Malhat. 2016. "Assessment of Pesticide Residues in Human Blood and Effects of Occupational Exposure on Hematological and Hormonal Qualities." *Pakistan Journal of Biological Science* 19(3):95–105.

National Academies of Sciences, Engineering, and Medicine. 2016. *Spills of Diluted Bitumen from Pipelines: A Comparative Study of Environmental Fate, Effects, and Response.* Washington, DC: National Academies Press.

National Aeronautics and Space Administration (NASA). 2016 (October 25). "2016 Antarctic Ozone Hole Attains Moderate Size, Consistent with Scientific Expectations." Available at www.nasa.gov.

National Aeronautics and Space Administration (NASA). 2020. "Images of Change." Available at www.climate.nasa.gov.

NASA Earth Observatory. 2020. "Heat and Fire Scorches Siberia." Available at www.climate.nasa.gov.

NASA Goddard Institute for Space Studies. 2020. "Graphic: Global Land–Ocean Temperature Index." Available at www.climate.nasa.gov.

National Association of Convenience Stores (NACS). 2020 (March 31). "Is Hydrogen the Future of U.S. Transport Fuel?" Available at www.convenience.org.

National Conference of State Legislators. 2017. "State Restrictions on New Nuclear Power Facility Construction." Available at www.ncsl.org.

National Geographic. 2019. "Causes of Global Warming Explained." Available at www.national geographic.com.

National Institute for Occupational Safety and Health. 2018. "Lead: Information for Workers." Center for Disease Control. Available at www.cdc.gov.

National Oceanic and Atmospheric Administration (NOAA). 2017 (April 20). "Deepwater Horizon Oil Spill Settlements: Where the Money Went." Available at www.noaa.gov.

National Science Teaching Association. 2020. "About the Next Generation Science Standards." Available at www.ngss.nsta.org.

National Oceanic and Atmospheric Administration (NOAA). 2021 (January 14). "2020 was Earth's 2nd-Hottest Year, Just behind 2016." Available at NOOA.gov.

National Toxicology Program. 2016. *14th Report on Carcinogens.* Research Triangle Park, NC: U.S. Department of Health and Human Services, Public Health Service.

Natural Resources Defense Council. 2017. "Why We Must Stop the Flow of Tar Sands Oil." Available at www.nrdc.org.

NBC News. 2019 (January 25). "Supersized Solar Farms Are Being Constructed across the World (and Soon in Space)." NBC News. Available at www.youtube.com.

Nelson, Cody. 2020. "'Their Greed Is Gonna Kill Us': Indian Country Fights against More Fracking." *Guardian.* Available at www.theguardian.com.

Netflix. 2020. *The Social Dilemma.* Available at www.netflix.com

Newburger, Emma. 2020 (May 21). "More Dams Will Collapse as Aging Infrastructure Can't Keep up with Climate Change." CNBC. Available at cnbc.com.

Nicole, Wendee. 2014. "Cooking up Indoor Air Pollution." *Environmental Health Perspectives* 122(1):A27.

No Fossil Fuel Money Pledge. 2020. Available at www.nofossilfuelmoney.org.

Norman, Jim. 2017 (March 24). "Environmental Activists Put Their Beliefs into Action." Gallup Organization. Available at www.gallup.com.

Office of Energy Efficiency & Renewable Energy. 2020. "History of Wind Energy." U.S. Department of Energy. Available at www.energy.gov.

Pappas, Stephanie. 2020. "What Is Global Warming?" *Live Science.* Available at www.livescience.com.

Pariona, Amber. 2017. "Who Are the Charismatic Megafauna of the World?" *World Atlas.* Available at www.worldatlas.com.

Parker, Laura. 2019 (June 7). "The World's Plastic Pollution Crisis Explained." *National Geographic.* Available at www.nationalgeographic.com.

Pearce, Fred. 2015. "Global Extinction Rates: Why Do They Vary so Wildly?" *Yale Environment 360.* Available at www.e350.yale.edu.

Permanent People's Tribunal. 2019 (April 12). "Advisory Opinion on the Session on Human Rights, Fracking and Climate Change." Available at www.tribunalonfracking.org.

Philip, Reena Susan, Aswathi Mary Anian, and Anand Shankar M. Raja. 2020. "Planned Fashion Obsolescence in the Light of Supply Chain Uncertainty." *Academy of Strategic Management Journal* 19(1):1–17.

Plumer, Brad. 2017 (August 23). "What 'Clean Coal' Is—and Isn't." *New York Times.* Available at www.nytimes.com.

Podesta, John. 2019. "The Climate Crisis, Migration, and Refugees." Brookings Institute. Available at www.brookings.edu.

Pope Francis. 2015 (May 24). "Encyclical Letter Laudato Si' of the Holy Father Francis." *On Care for Our Common Home. Rome:* Vatican Press. Available at www.w2.vatican.va.

Pope Francis. 2020 (November 26). "Pope Francis: A Crisis Reveals What Is in Our Hearts." *New York Times.* Available at www.nytimes.com.

Population Matters. 2020. *Quotes.* Available at www.populationmatters.org.

Prata, Joana C., Ana L. P. Silva, Tony R. Walker, Armando C. Duarte, and Teresa Rocha-Santos. 2020. "Covid-19 Pandemic Repercussions on the Use and Management of Plastics." *Environmental Science & Technology* 54:7760–7765.

Public Broadcasting Service. 2019 (November 18). "More Land Affected by Keystone Pipeline Leak Than Originally Thought." *PBS News Hour.* Available at www.pbs.org.

Redford, Robert. 2015 (July 10). "Our Last Chance." Natural Resources Defense Council. Available at www.nrdconline.org.

Riemann, D., L. B. Krone, K. Wulff, and C. Nissen. 2019. "Sleep, Insomnia, and Depression." *Neuropsychopharmacology* 45:74–89.

Reinhart, R. J. 2019 (March 27). "40 Years after Three Mile Island, Americans Split on Nuclear Power." *Gallup News.* Available at www.news.gallup.com.

Reiny, Samson. 2018. "NASA Study: First Direct Proof of Ozone Hole Recovery due to Chemicals Ban." *NASA Earth Science News.* Available at www.nasa.gov.

Research and Markets. 2019. *Global Market for Hydrogen Fueling Stations, 2019.* Available at www.researchandmarkets.com.

Rhodes, Richard. 2018. "Why Nuclear Power Must Be Part of the Energy Solution." *Yale Environment 360.* Available at www.e360.yale.edu.

Ridlington, Elizabeth, Kim Norman, and Rachel Richardson. 2016. "Fracking by the Numbers: The Damage to Our Water, Land, and Climate from a Decade of Dirty Drilling." Environment America Research & Policy Center. Available at www.environmentamerica.org.

Ritchie, Hannah. 2017 (July 24). "What Was the Death Toll from Chernobyl and Fukushima?" Our World in Data. Available at ourworldindata.org.

Ritchie, Hannah. 2018 (October 16). "Global Inequalities in CO_2 Emissions." Our World in Data. Available at www.ourworldindata.org.

Ritchie, Hannah, and Max Roser. 2018. "Plastic Pollution." Our World in Data. Available at www.ourworldindata.org.

Ritchie, Hannah, and Max Roser. 2019. "Renewable Energy." Our World in Data. Available at www.ourworldindata.org.

Rosane, Olivia. 2020 (November 18). "Activists Are Alarmed as Biden Picks White House Official Who Took Fossil Fuel Money." *EcoWatch.* Available at www.ecowatch.com.

Rosenbaum, Ava. 2018. "Personal Space and American Individualism." *Brown Political Review.* Available at www.brownpoliticalreview.org.

Rossi, Sabrina, and Alessio Pitidis. 2018. "Multiple Chemical Sensitivity: Review of the State of the Art in Epidemiology, Diagnosis, and Future Perspectives." *Journal of Occupational & Environmental Medicine* 60(2):138–146.

Rott, Nathan, and Jennifer Ludden. 2020 (March 31). "Trump Administration Weakens Auto Emissions Standards." National Public Radio. Available at www.npr.org.

Roy, Eleanor Ainge. 2019 (May 16). "'One Day We'll Disappear': Tuvala's Sinking Islands." *Guardian.* Available at www.theguardian.com.

Runge, C. Ford. 2016. "The Case against More Ethanol: It's Simply Bad for the Environment." *Yale Environment 360.* Available at www.e360yale.edu.

Rust, Niki. 2017. "Religion Can Make Us More Environmentally Friendly—Or Not." BBC. Available at www.bbc.com.

Saad, Lydia. 2020 (April 4). "Preference for Environment over Economy Largest since 2000." Gallup Organization. Available at www.gallup.org.

Sanchez, Marcio Jose, and Christopher Weber. 2020 (September 7). "California Simmers While It Burns, but No Big Power Outages." Associated Press. Available at www.apnews.com.

Schienman, Ted. 2019. "The Couples Rethinking Kids because of Climate Change." *BBC Generation Project.* Available at www.bbc.com.

Schipani, Vanessa. 2017. "The Facts on Fracking Chemical Disclosure." *SciCheck.* Available at www.factcheck.org.

SciNews. 2017. "Scientists Categorize Earth as a 'Toxic Planet.'" *SciNews.* Available at www.phys.org.

Scovronick, N., França, D., Alonso, M., Almeida, C., Longo, K., Freitas, S., Rudorff, B., and Wilkinson, P. 2016. "Air Quality and Health Impacts of Future Ethanol Production and Use in São Paulo State, Brazil." *International Journal of Environmental Research and Public Health* 13(7):695.

Seltenrich, Nate. 2015. "New Link in the Food Chain? Marine Plastic Pollution and Seafood Safety." *Environmental Health Perspectives* 123(2):A35–A41.

Sisk, Richard. 2020 (September 10). "Inside the National Guard's Daring Rescue of Hundreds from a California Wildfire." *Military News.* Available at www.military.com.

Shulz, David. 2018 (February 5). "Was Flint's Deadly Legionnaires' Epidemic Caused by Low Chlorine Levels in the Water Supply?" *Science.* Available at www.sciencemag.org.

Simkins, J. D. 2019 (August 26). "Navy Quietly Ends Climate Change Task Force, Reversing Obama Initiative." *Navy Times.* Available at www.navytimes.com.

Singh, Harsh Vijay. 2019. "The Countries Most Ready for the Global Energy Transition." World Economic Forum. Available at www.weforum.org.

Smith, Abby. 2017 (April 14). "Report Finds Global GHGs Need to Peak within Decade to Meet Paris Goal." *Inside EPA/Climate.* Available at www.insideepaclimate.com.

Staudinger, Michelle D., Nancy B. Grimm, Amanda Staudt, Shawn L. Carter, F. Stuart Chapin III, Peter Kareiva, Mary Ruckelshaus, and Bruce A. Stein. 2012. *Impacts of Climate Change on Biodiversity, Ecosystems, and Ecosystem Services: Technical Input to the 2013 National Climate Assessment.* Available at assessment.globalchange.gov.

Steiner, Irena Gorski, and Brian S. Schwartz. 2019. "Environmental Health Concerns from Unconventional Gas Development." *Oxford Research Encyclopedia of Global Public Health.* Oxford: Oxford University Press.

Stern, Todd. 2018 (October). *The Paris Agreement and Its Future.* Washington, DC: Brookings Institute. Available at www.brookings.edu.

Stiglitz, Joseph. 2019 (June 4). "The Climate Crisis Is Our Third World War. It Needs a Bold Response." *Guardian.* Available at www.theguardian.org.

Stolark, Jessie. 2018. "New Research Shows the Health Benefits of Ethanol-Fueled Cookstoves." Environmental and Energy Study Institute. Available at www.eesi.org.

Stone, Peter. 2020 (August 9). "Big Oil Remembers 'Friend' Trump with Millions in Campaign Funds." *Guardian.* Available at www.theguardian.com.

Strahan, Susan E., and Anne R. Douglass. 2018. "Decline in Antarctic Ozone Depletion and Lower Stratospheric Chlorine Determined from Aura Microwave Limb Sounder Observations." *Geophysical Research Letters* 45(1):382–390.

Strobo, R. A. 2012. "The Shape of Appalachia to Come: Coal in a Transitional Economy." *Duke Forum for Law & Social Change* 4:91–114.

Su, Zhenkuan, Michelle Ho, Zhenchun Hao, Upmanu Lall, Xun Sun, Xi Chen, and Longzeng Yan, 2019. "The Impact of the Three Gorges Dam on Summer Streamflow in the Yangtze River Basin." *Hydrological Processes* 34:705–717.

Sustainability for All. 2020. *The Battle against Planned Obsolescence.* Available at www.activesustainabilityforall.com.

Taylor, Matthew, Jonathan Watts, and John Bartlett. 2019. "Climate Crisis: 6 Million People Join Latest Wave of Global Protests." *Guardian.* Available at www.theguardian.com.

Theel, Shauna, Max Greenberg, and Denise Robbins. 2013. *Study: Media Sowed Doubt in Coverage of UN Climate Report.* Media Matters for America. Available at www.mediamatters.org.

Tomassoni, Teresa. 2020. "Colombia Was the Deadliest Place on Earth for Environmental Activists. It's Gotten Worse." NBC News. Available at www.nbcnews.com.

Tonino, Leath. 2020 (December). "Our Great Reckoning: Eileen Crist on the Consequences of Human Plunder." *The Sun*, 540, pp. 4–13.

Tran, Mark, and Claire Provost. 2012. "Controversial Dam Projects—in Pictures." *Guardian.* Available at www.theguardian.com.

Turgeon, Andrew, and Elizabeth Morse. 2018. "Petroleum." *National Geographic.* Available at www.nationalgeographic.org.

Tyson, Alec, and Brian Kennedy. 2020. "Two-Thirds of Americans Think Government Should Do More on Climate." Pew Research Center. Available at www.pewresearch.org.

UCLA Sustainability. 2014. *Green Guide to Sustainable Living at UCLA*. Available at www.sustain.ucla.edu.

Union of Concerned Scientists. 2020a. "The Connection between Climate Change and Wildfires." Available at www.ucsusa.org.

Union of Concerned Scientists. 2020b. "Each Country's Share of CO_2 Emissions." Available at www.ucsusa.org.

United Nations. 2018 (November 5). "Healing of Ozone Layer Gives Hope for Climate Action: UN Report." *UN News*. Available at www.news.un.org.

United Nations. 2019 (May 6). "UN Report: Nature's Dangerous Decline 'Unprecedented'; Species Extinction Rates 'Accelerating.'" United Nations Sustainable Development Goals Blog. Available at www.un.org.

United Nations. 2020. "Take Action for the Sustainable Development Goals." Available at www.un.org.

United Nations Environment Programme (UNEP). 2019a. "Global Chemicals Outlook II: From Legacies to Innovative Solutions." Available at www.unenvironment.org.

United Nations Environment Programme (UNEP). 2019b. "How All Religious Faiths Advocate for Environmental Protection." Available at www.unenvironment.org.

United Nations Environment Programme (UNEP). 2020. "Why Do Chemicals and Waste Matter?" Available at www.unep.org.

United Nations Framework Convention on Climate Change (UNFCCC). 2020a. "About the Secretariat." Available at www.unfccc.int.

United Nations Framework Convention on Climate Change (UNFCCC). 2020b. "United Nations Climate Change Annual Report 2019." Available at www.unfccc.int.

United States Department of Agriculture. 2016. "Pesticide Data Program: Annual Summary, Calendar Year 2015." Available at www.ams.usda.gov.

U.S. Department of Energy. 2015. "Wind Vision: A New Era for Wind Power in the United States." Available at www.energy.gov.

U.S. Department of Energy. 2020. "State Laws and Incentives. Energy Efficiency and Renewable Energy." Available at www.afdc.energy.gov.

United States Energy Information Administration 2019a. "Coal Explained." Available at www.eia.gov.

United States Energy Information Administration 2019b. "Natural Gas Explained." Available at www.eia.gov.

United States Energy Information Administration 2019c. "Solar Energy and the Environment." Available at www.eia.gov.

United States Energy Information Administration. 2019d. "Geothermal Explained." Available at www.eia.gov.

United States Energy Information Administration. 2020a. "Oil and Petroleum Products Explained." Available at www.eia.gov.

United States Energy Information Administration. 2020b. "How Many Nuclear Power Plants Are in the United States, and Where Are They Located?" Available at www.eia.gov.

United States Energy Information Administration. 2020c. "Solar Explained." Available at www.eia.gov.

United States Energy Information Administration. 2020d. "Biomass Explained." Available at www.eia.gov.

U.S. Fish and Wildlife Service. 2020. "Hunters as Conservationists." Available at www.fws.gov.

U.S. Food and Drug Administration. 2019. "Advice about Eating Fish for Women Who Are or Might Become Pregnant, Breast Feeding Mothers, and Young Children." Available at www.fda.gov.

U.S. Global Change Research Program. 2018. *Fourth National Climate Assessment: Impacts Risks, and Adaptation in the United States*. Volume II. Available at www.nca2018.globalchange.gov.

U.S. House of Representatives. 2020. House Resolution 109. 116th Congress. Available at www.congress.gov/bill.

Wallach, Omri. 2020. "Mapped: The World's Nuclear Reactor Landscape." Visual Capitalist. Available at www.visualcapitalist.com.

Wamsley, Laurel, and Scott Neuman. 2020 (September 25). "Trump Administration Moves to Expand Development in Alaska's Tongass National Forest." National Public Radio. Available at www.npr.org.

Watts, Jonathan. 2018 (February 14). "Kenya's 'Erin Brokovich' Defies Harassment to Bring Anti-Pollution Case to Courts." *Guardian*. Available at www.guardian.com.

Watts, Jonathan. 2020 (July 22). "Kenya Environmental Defenders Win Landmark $12m Court Battle." *Guardian*. Available at www.guardian.com.

Weber, Hannes, and Jennifer Dabbs Sciubba. 2019. "The Effect of Population Growth on the Environment: Evidence from European Regions." *European Journal of Population* 35(2):379–402.

Welch, Craig. 2018 (September 27). "Half the World's Orcas Could Soon Disappear—Here's Why." *National Geographic*. Available at www.nationalgeographic.com.

Willis, Haisten. 2019 (February 14). "Downsizing the American Dream: The New Trend toward 'Missing Middle Housing.'" *Washington Post*. Available at www.washingtonpost.com.

Wong, Chit Ming, Hilda Tsang, Hak Kan Lai, et al. 2016. Cancer Mortality Risks from Long Term Exposure to Ambient Fine Particle. *Cancer Epidemiology, Biomarkers & Prevention* 25(5):839–845.

Woodyatt, Amy, and Yoko Wakatsuki. 2020 (October 24). "Fukushima Water Release Could Change Human DNA, Greenpeace Warns." CNN. Available at www.cnn.com.

Worland, Justin. 2015 (June 19). "The Surprising Link between Trans Fat and Deforestation." *Time*. Available at www.time.com.

World Bank. 2014 (May 28). "State & Trends Report Charts Global Growth of Carbon Pricing." Available at www.theworldbank.org.

World Health Assembly. 2015. *Health and the Environment: Addressing the Health Impact of Air Pollution*. World Health Organization. Available at www.apps.who.int.

World Health Organization. 2018a (February 1). "Climate Change and Health." Available at www.who.int.

World Health Organization. 2018b (May 8). "Household Air Pollution and Health." Available at www.who.int.

World Health Organization. 2018c (May 2). "Ambient (Outdoor) Air Pollution." Available at www.who.int.

World Health Organization. 2018d (September 12). "Cancer." Available at www.who.int.

World Health Organization. 2020. "Flooding and Communicable Diseases Fact Sheet." Available at www.who.int.

World Nuclear Association. 2017 (March 1). "World Nuclear Power Reactors & Uranium Requirements." Available at world-nuclear.org.

World Nuclear Association. 2020a. "Nuclear Power in the USA." Available at www.world-nuclear.org.

World Nuclear Association. 2020b. "Uranium Mining Overview." Available at www.world-nuclear.org.

World Nuclear Association. 2020c. "Storage and Disposal of Radioactive Waste." Available at www.world-nuclear.org.

World Nuclear Association. 2020d. "Fukushima Daiichi Accident." Available at www.world-nuclear.org.

World Nuclear News. 2021 (January 29). "Biden Nominee Confirms Opposition to Yucca Mountain." *World Nuclear News*. Available at www.world-nuclear-news.org.

World Water Assessment Program. 2015. *The United Nations World Water Development Report: Water for a Sustainable Future*. Paris: UNESCO.

World Wildlife Fund (WWF). 2018. *Living Planet Report*. Gland, Switzerland: World Wildlife Fund, Zoological Society of London, Global Footprint Network, and Water Footprint Network. Available at www.assetsworldwildlife.org.

World Wildlife Fund (WWF). 2020. *Living Planet Report*. Gland, Switzerland: World Wildlife Fund, Zoological Society of London, Global Footprint Network, and Water Footprint Network. Available at www.wwf.org.uk.org.

Wu, Xl, R. C. Nethery, M. B. Sabath, D. Braun, and F. Dominici. 2020 (November 4). "Air Pollution and COVID-19 Mortality in the United States: Strengths and Limitations of an Ecological Regression Analysis." *Science Advances* 6(45). Available at www.advancessciencemag.org.

Yang, Sue-Ming, Yi-Yuan Su, and Jennifer V. Carson. 2014. "Eco-Terrorism and the Corresponding Legislative Efforts to Intervene and Prevent Future Attacks." TSAS Working Paper # 14-04. Canadian Network for Research on Terrorism, Security, and Society. Available at www.library.tsas.ca/tsas-working-papers.

Zaelke, Durwood, Nathan Borgford-Parnell, and Stephen O. Andersen. 2016 (September 12). "Primer on HFCs." Working Paper. Institute for Governance and Sustainable Development. Available at www.igsd.org.

Zeng, Xianlai, John A. Mathews, and Jinhui Li. 2018. "Urban Mining of E-Waste Is Becoming More Cost-Effective Than Virgin Mining." *Environmental Science & Technology* 52(8):4835–4841.

Chapter 14

Ahmed Wasim, Josep Vidal-Alaball, Joseph Downing, and Francesc López Segui. 2020 (May). "COVID-19 and the 5G Conspiracy Theory: Social Network Analysis of Twitter Data." *Journal of Medical Internet Research* 22(5):1–9.

Allen Institute. 2020. "Grover: A State of the Art Defense against Neural Fake News." *Allen Institute for AI.* Available at grover.allenai.org.

Allyn, Bobby. 2020 (November 9). "How Will Tech Policy Change in the Biden White House? Here's What You Need to Know." National Public Radio. Available at www.npr.org.

American Constitution Society. 2020. "Key Findings of the Mueller Report." Available at acslaw.org.

Anandan, Tonya M. 2019 (July 22). "Cultivating Robotics and AI for Sustainable Agriculture." *Robotic Industries Association.* Available at www.robotics.org.

Anderson, Monica. 2018 (September 27). "The Majority of Teens Have Experienced Some Form of Cyberbullying." Pew Research Center. Available at www.pewresearch.org.

Anderson, Monica, and Madhumitha Kumar. 2019 (May 7). "Digital Divide Persists Even as Lower-Income Americans Make Gains in Tech Adoption." Pew Research Center. Available at www.pewresearch.org.

Anderson, Janna, and Lee Rainie. 2018 (July 3). "Fifty-Fifty Antidotes: How Digital Life Has Been Both Positive and Negative." Pew Research Center. Available at www .pewresearch.org.

Angwin, Julia, Jeff Larson, Surya Mattu, and Lauren Kirshner. 2016 (May 23). "Machine Bias." *ProPublica.* Available at www .propublica.org grade up.

Auxier, Brooke, Monica Anderson, and Madhu Kumar. 2019 (December 20). "10 Tech-Related Trends That Shaped the Decade." Pew Research Center. Available at www .pewresearch.org.

Auxier, Brooke, Lee Rainie, Monica Anderson, Andrew Perrin, Madhu Kumar, and Erika Turner. 2019 (November 15). "Americans and Privacy: Concerned, Confused and Feeling Lack Control over Their Personal Information." Pew Research Center. Available on www .pewresearch.org.

Banker, Sachin, and Salil Khetani. 2019. "Algorithm Overdependence: How the Use of Algorithmic Recommendation Systems Can Increase Risks to Consumer Well-Being." *Journal of Public Policy and Marketing* 38(4):500–515.

Baron, Sam. 2020 (November 11). "Biden's Pivot to Science Is Welcome—Trump Only Listened to Experts When It Suited Him." The Conversation. Available at theconversation. com.

Beckett, Lois. 2020 (October 16). "QAnon: A Timeline of Violence Linked to the Conspiracy Theory." *Guardian.* Available at www .theguardian.com.

Beland, Louis-Philippe, and Richard Murphy. 2015 (May). "Communication: Technology, Distraction and Student Performance." *London School and Economics Centre for Economic Performance.* Discussion Paper No. 1350. Available at www.cep.lse.ac.uk.

Bell, Daniel. 1973. *The Coming of Post-Industrial Society: A Venture in Social Forecasting.* New York: Basic Books.

Beniger, James R. 1993. "The Control Revolution." In *Technology and the Future,* ed. Albert H. Teich, 40–65. New York: St. Martin's Press.

Ben-Shahar, Omri, and Carl Schneider. 2016. "The Better Way to Regulate 'Natural' Food." *Forbes.* Available at www.forbes.com.

Berezow, Alex. 2020 (January 21). "Who Will Lead the World in Technology: U.S., Europe, or China?" *American Council on Science and Health.* Available at www.asch.org.

Biden, Joe. 2020a. "The Biden Plan to Combat Coronavirus (COVID-19) and Prepare for Future Global Health Threats." Available at www.joebiden.com.

Biden, Joe. 2020b. "The Biden Plan to Build a Modern, Sustainable Infrastructure and Equitable Clean Energy Future." Available at www.joebiden.com.

Biden, Joe. 2020c. "The Biden Plan for Securing Our Values as a Nation of Immigrants." Available at www.joebiden.com.

Blumenthal, Richard, and Tim Wu. 2018 (May 18). "What the Microsoft Antitrust Case Taught Us." *New York Times.* Available at www .nytimes.com.

Bornmann, Lutz, and Ruediger Mutz. 2015. "Growth Rates of Modern Science: A Bibliometric Analysis Based on the Number of Publications and Cited References." *Journal of the Association for Information Science and Technology.*

Bridle, James. 2018 (June 15). "Rise of the Machines: Has Technology Evolved beyond Our Control?" *Guardian.* Available at www .theguardian.com.

Brownlee, Shannon, Kalipso Chalkidou, Jenny Doust, Adam G. Elshaug, Paul Glasziou, Iona Heath, Somil Nagpal, Vikas Saini, Divya Srivastava, Kelsey Chalmers, and Deborah Korenstein. 2017. "Evidence for Overuse of Medical Services around the World." *Lancet* 390:156–168.

Buchholz, Katherine. 2020 (June 19). "These Are the Countries That Spend the Most and Least Time on Social Media." *World Economic Forum.* Available at www.weforum.org.

Bugeja, Michael. 2018. *Interpersonal Divide in the Age of the Machine.* Oxford: Oxford University Press.

Buolamwini, Joy. 2020. "Algorithmic Bias Persists." MIT Media Lab. Available at www .media.mit.edu.

Buolamwini, Joy, and Timnit Gebru, 2018. "Gender Shades: Intersectional Accuracy Disparities in Commercial Gender Classifications." *Proceedings of Machine Learning Research* 81:1–15.

Bush, Corlann G. 1993. "Women and the Assessment of Technology." In *Technology and the Future,* ed. Albert H. Teich, 192–214. New York: St. Martin's Press.

Carr, Nicholas. 2010. *The Shallows: What the Internet Is Doing to Our Brains.* New York: Norton.

Carter, Jacob, Emily Berman, Anita Desikan, Charise Johnson, and Gretchen Goldman. 2019 (January). *In the State of Science in the Trump Era: Damage Done, Lessons Learned, and a Path to Progress.* Cambridge, MA: Union of Concerned Scientists.

Casey, Kevin. 2020 (July 30). "How to Explain Robotic Process Automation (RPA) in Plain English." *The Enterprisers Project.* Available at enterprisersproject.com.

Centers for Disease Control and Prevention (CDC). 2018. "Influenza (Flu)." *History.* Available at www.cdc.gov.

Center for Food Safety. (CFS) 2018. "USDA's Proposed Rule on GMO Food Labeling Summary." Available at www .centerforfoodsafety.org.

Center for Strategic and International Studies. 2020 (September). "Significant Cyber Incidences since 2006." Available at www.csis .org.

Ceruzzi, Paul. 1993. "An Unforeseen Revolution." In *Technology and the Future,* ed. Albert H. Teich, 160–174. New York: St. Martin's Press.

Chan, Sewell. 2007 (January 23). "New Scanners for Tracking City Workers." *New York Times.* Available at www.nytimes.com.

Chang, Chun-Chih, and Thung-Hong Lin. 2020 (April). "Autocracy Login: Internet Censorship and Civil Society in the Digital Age." *Democratization* 27(5):874–895.

Chappell, Bill. 2020 (September 3). "Facebook Clamps Down on Posts, Ads That Could Undermine U.S. Presidential Election." National Public Radio. Available at www.npr.org.

Chatzakou, Despoina, Ilias Leontiadis, Jeremy Blackburn, Emiliano DeCristofaro, Gianluca Stringhini, Athena Vakali, and Nicolas Kourtellis. 2019. "Detecting Cyberbullying and Cyberaggression in Social Media." *ACM Transactions on the Web* 13(3):1–51.

Chesney, Bobby, and Danielle Citron. 2019. "Deep Fakes: A Looming Challenge for Privacy, Democracy, and National Security." *California Law Review* 107:1753–1820.

Clarke, Richard A., Michael J. Morell, Geoffrey R. Stone, Cass R. Sunstein, and Peter Swire. 2013 (December 12). "Liberty and Security in a Changing World." National Security Administration. Available at www.nsa.gov.

Cohen, Patricia. 2020 (June 24). "Roundup Maker to Pay 10 Billion to Settle Cancer Suits." *New York Times.* Available at www.nytimes.com.

Collins, Ben, and Brandy Zadrozny. 2020 (July 21)." Twitter Bans 7,000 QAnon Accounts, Limits 150,000 Others as Part of Broad Crackdown." NBC News. Available at www .nbcnews.com.

Congressional Research Service (CRS). 2020a (February 12). "The Internet of Things (IoT): An Overview." *In Focus.* Available at crsreports.congress.gov.

Congressional Research Service (CRS). 2020b (July 17). "EU Data Protection Rules and U.S. Implications." *In Focus*. Available at crsreports.congress.gov.

Cortez-Ospina, Esteban. 2019 (October 8). "Over 2.5 Billion People Use Social Media. This Is How It Has Changed the World." World Economic Forum. Available at www.weforum.org.

Council on Foreign Relations (CFR). 2018. *The Work Ahead: Machines, Skills, and U.S. Leadership in the 21st Century.* Independent Task Force Report No. 76. Available at cdn.cfr.org.

Cramer-Flood, Ethan. 2020 (June 22). *Global E-Commerce 2020 Report.* Available at www.emarketer.com.

Culliford, Elizabeth. 2020 (August 5). "Facebook, Twitter, YouTube Pull Posts over Coronavirus Misinformation." Reuters News. Available at www.reuters.com.

Cyranoski, David. 2020 (July 7). "China's Massive Efforts to Collect Its People's DNA Concerns Scientists." *Nature*. Available at www.nature.com.

Dace, Hermione. 2020 (January 9). "China's Tech Landscape: A Primer." Institute for Global Change. Available at institute.global.

Das, Santosh. 2020 (June 11). "Top 10 Computer Manufacturing Companies in the World." Electronics and You. Available at www.electronicsandyou.com.

Davenport, Christian. 2020 (March 11). "NASA Watchdog Takes Aim at Boeing's SLS Rocket; Says Backbone of Trump's Moon Mission Could Cost a Staggering $50 Billion." *Washington Post.* Available at www.washingtonpost.com.

Davis, Susan. 2020 (October 2). "House Votes to Condemn QAnon Conspiracy Theory: 'It's a Sick Cult.'" National Public Radio. Available at www.npr.org.

Department of Justice. 2020 (March 9). "DNA-Sample Collection from Immigration Detainees." *Federal Register.* Available at www.federalregister.gov.

Desikan, Anita. 2020 (August 3). "150 Attacks on Science and Counting: Trump Administration's Anti-Science Actions Hurt People and Communities Nationwide." Union of Concerned Scientists. Available at blog.ucsusa.org.

DeSilver, Drew. 2020 (March 20). "Before the Coronavirus, Telework Was an Optional Benefit, Mostly for the Affluent Few." Pew Research Center. Available at www.pewresearch.org.

Droegemeier, Kelvin K. 2019 (April 23). "America Leading the World in Science and Technology." *Infrastructure and Technology.* Available at federallabs.org/news.

Du, Jie, and Hayden Wimmer. 2019. "Hour of Code: A Study of Gender Differences in Computing." *Information Systems Educational Journal* 17(4):91–100.

Durkheim, Emile. 1973/1925. *Moral Education.* New York: Free Press.

Dutta, Soumitra, and Bruno Lanvin. 2020 (March). *The Network Readiness in the 2019: Towards a Future-Ready Society.* Available at network readinessindex.org.

Eadicicco, Lisa. 2019 (August 8). "Apple Just Got Knocked out of the Top 3 Smart Phone Makers in the World—Here's How It Stacks up against Rivals like Samsung, Huawei, and LG." *Business Insider.* Available at www.businessinsider.com.

Ebrahimji. Alisha. 2020 (September 1). "Doctors Say Coronavirus Myths on Social Media Are 'Spreading Faster Than the Virus Itself.'" CNN Business. Available at www.cnn.com.

Economist, The. 2019 (November 9). "A New Type of Genetic Profiling Promises Cleverer, Better-Looking Children." Available at amp.economist.com.

Edelman Intelligence. 2018 (February). "GE Global Innovation Barometer 2018 Full Report." Available at www.scribd.com/document/378148615.

Eilperin, Juliet, and Rick Weiss. 2003 (February 28). "House Votes to Prohibit All Human Cloning." *Washington Post.* Available at www.washingtonpost.com.

Electronic Privacy Information Center (EPIC). 2020. "State Genetic Privacy Policy." Available at epic.org.

Explorer. 2020. "Get to Know the Different Types of Gene-Based Therapies." Explore Gene Therapy. Available at www.explorergenetherapy.com.

Federal Communications Commission (FCC). 2015. "Open Internet Rules." Available at www.fcc.gov.

Federal Register. 2018 (December 21). "National Bioengineered Food Disclosure Standard." Agricultural Marketing Service. Available at www.federalregister.gov.

Federal Trade Commission. 2017 (December 12). "DNA Test Kits: Consider the Privacy Implications." Consumer Information. Available at www.consumer.ftc.gov.

Ferguson, Christopher J. 2018 (June 18). "Debunking the Six Biggest Myths about 'Technology Addiction.'" The Conversation. Available at theconversation.com.

Food and Drug Administration. 2020a. "What GMO Crops Are Grown and Sold in the United States?" Available at www.fda.gov.

Food and Drug Administration. 2020b. "Science and History of GMO's and Other Food Modification Processes" Available at www.fda.gov.

Food and Drug Administration. 2020c (April 1). "Email to the Medical Director of Dynamic Stem Cell Therapy from the Director of the Office of Compliance and Biologics Quality." Available at www.fda.gov.

Frenkel, Sheera, Ben Decker, and Davey Alba. 2020 (May 20). "Here's How the 'Plandemic' Movie and Its Falsehoods Spread Widely Online." *New York Times.* Available at www.nytimes.com.

Fung, Brian. 2017 (April 4). "Trump Has Signed Repeal of FCC's Internet Privacy Rule. Here's What Happens Next." *Los Angeles Times.* Available at www.latimes.com.

Funk, Cary. 2020 (February 12). "Key Findings about Americans' Confidence in Science and Their Views on Scientists' Role in Society." Pew Research Center. Available at www.pewresearch.org.

Funk, Cary. 2019 (September 4). "Democrats More Supportive Than Republicans of Federal Spending for Scientific Research." Available at www.pewresearch.org.

Funk, Cary, and Meg Hefferon. 2019 (August 19). "Most Americans Have Positive Image of Research Scientists, but Fewer See Them as Good Communicators." Pew Research Center. Available at www.pewresearch.org.

Funk, Cary, Brian Kennedy, and Meg Hefferon. 2018 (November 19). "Public Perspectives on Food Risks." Pew Research Center. Available at www. pewresearch.org.

Galston, William A. 2020 (January 8). "Is Seeing Still Believing? The Deepfake—Challenge to Truth in Politics." Brookings Institute. Available at www.brookings.edu.

Gelles, David, and David Yaffe-Bellany. 2019 (August 19). "Shareholder Value Is No Longer Everything, Top CEOs Say." *New York Times.* Available at www.nytimes.com.

Glaberson, Stephanie K. 2019. "Coating over the Cracks: Predictive Analytics and Child Protection." *Fordham Urban Law Journal* 46(2):307–427.

Glass, Jennifer L., and Mary C. Noonan. 2016. "Telecommuting and Earnings Trajectories among American Women and Men 1989–2008." *Social Forces* 95(1):217–250.

Goel, Vindu, and Eric Lichtblau. 2017 (March 15). "Russian Agents Were behind Yahoo Hack, U.S. Says." *New York Times.* Available at www.nytimes.com.

Goldman, Gretchen T., Jacob M. Carter, Yun Wang, and Janice M. Larson. (2020). "Perceived Losses of Scientific Integrity under the Trump Administration: A Survey of Federal Scientists." *PLoS ONE* 15(4):1–26.

Goldman, Gretchen, Genna Reed, and Jacob Carter. 2017. "Risks to Science-Based Policy under the Trump Administration." *Stetson Law Review.* Available at www.stetson.edu.

Gowen, Annie, Juliet Eilperin, Ben Guarino, and Andrew Ba Tran. 2020 (January 23). "Science Ranks Grow Thin in Trump Administration." *Washington Post.* Available at www.washingtonpost.com.

Gramlich, John. 2019 (May 16). "10 Facts about Americans and Facebook." Pew Research Center. Available at www.pewresearch.org.

Green, Ted Van. 2020 (September 9). "Few Americans Are Confident in Tech Companies to Prevent Misuse of Their Platforms in the 2020 Election." Pew Research Center. Available at www.pewresearch.org.

Greenwald, Glenn. 2013 (July 31). "XKeyscore: NSA Tool Collects 'Nearly Everything a User Does on the Internet.'" *Guardian.* Available at www.theguardian.com.

Guyot, Katherine, and Isabel V. Sawhill. 2020 (April 6). "Telecommuting Will Likely Continue Long after the Pandemic." Brookings Institute. Available at www.brookings.edu.

Hamilton, Jon. 2020a (July 20). "Researchers Hope Experimental Gene Therapy Is an Answer to a Fatal Genetic Disorder." National Public Radio. Available at npr.org.

Hamilton, Jon. 2020b (July 27). "A Boy with Muscular Dystrophy Was Headed for a Wheelchair. Then Gene Therapy Arrived." National Public Radio. Available at npr.org.

Herb, Jeremy, Marshall Cohen, and Katelyn Polantz. 2020 (August 20). "Bipartisan Senate Report Details Trump Campaign Contacts with Russia in 2016, Adding to Mueller Findings." CNN Politics. Available at www.cnn.com.

Hoffman, Jan. 2019 (September 23). "How Anti-Vaccine Sentiment Took Hold in the United States." *New York Times.* Available at www.nytimes.com.

Holmes, Erin, and Kayla Epstein. 2020 (July 29). "Lawmakers Grilled CEOs of Apple, Google, Facebook, and Amazon in a Historic House Investigation. Here's What Happened." *Business Insider.* Available at www.businessinsider.com.

Holpuch, Amanda. 2020 (April 13). "US's Digital Divide 'Is Going to Kill People' as Covid-19 Exposes Inequalities." *Guardian.* Available at www.theguardian.com.

Human Genome Project. (HGP) 2007. *Medicine and the New Genetics.* Available at www.ornl.gov.

Hutson, Matthew. 2018 (July 19). "A Turtle—or a Rifle? Hackers Easily Fool AI's Seeing the Wrong Thing." *Science.* American Association for the Advancement of Science. Available at www.sciencemag.org.

Hwong, Connie. 2018 (March 12). "Who Is the Online Shopper?" *Consumer Insights.* Available at vertoanalytics.com.

International Federation of Robotics. 2019. "World Robotics 2019 Industrial Robots." International Federation of Robotics. Available at ifr.org.

International Telecommunication Union (ITU). 2016. *Measuring the Information Society Report.* Geneva, Switzerland: International Telecommunications Union.

International Telecommunication Union (ITU). 2019. *Measuring Digital Development: Facts and Figures 2019.* Available at www.itu.int.

Internet World Statistics. 2020 (June). "World Internet Usage and Population Statistics." Available at www.internetworldstats.com.

Iyengar, Rishi. 2018 (November). "The Future of the Internet Is Indian." CNN Business. Available at edition.cnn.com.

Jan, Tracy, and Elizabeth Dwoskin. 2019 (March 28). "HUD Is Reviewing Twitter's and Google's Ad Practices as Part of Housing Discrimination." *Washington Post.* Available at www.washingtonpost.com.

Jensen, Aaron. 2019 (May 6). "World Faces Ecological Collapse unless Urgent Action Taken, Says Devastating Global Assessment." *Friends of the Earth.* Available at foe.org.

Jurkowitz, Mark, and Amy Mitchell. 2020 (April 15). "Nearly 3 in 10." Pew Research Center. Available at www.journalism.org.

Kahn, A. 1997. "Clone Mammals . . . Clone Man?" *Nature* 386:119.

Karlawish, Jason. 2020 (May 17). "The Pandemic Plan Was in Place. Trump Abandoned It—and Science—in the Face of Covid 19." *STAT News.* Available at www.statnews.com.

Kaur, Gursheen, and Karuna Singh. 2020 (June 3). "Nanotechnology in the Food Sector." In *Emerging Technologies in Food Science,* ed. Monika Thakur and V. K. Modi, 15–36. Singapore: Springer.

Kemp, Simon. 2017 (January 24). "Digital in 2017: *Global Overview.* We Are Social." Available at www.wearesocial.com.

Kennedy, Brian, and Meg Hefferon. 2019 (March 28). "What Americans Know about Science." Pew Research Center. Available at www.pewresearch.org.

Kharfen, Michael. 2006. "1 of 3 and 1 in 6 Pre-Teens Are Victims of Cyber-Bullying." Fight Crime: Invest in Kids. Available at www.fightcrime.org.

Kharpal, Arjun. 2020 (April 27). "Power Is 'Up for Grabs': Behind China's Plan to Shape the Future of Next-Generation Tech." CNBC Markets. Available at www.cnbc.com.

Kloppenburg, Sanneke, and Irma van der Ploeg. 2018 (September). "Securing Identities: Biometric Technologies and the Enactment of Human Bodily Differences." *Science as Culture* 29(1):57–76.

Kluger, Jeffrey. 2020 (May 12). "Accidental Poisonings Increased after President Trump's Disinfectant Comments." *Time.* Available at time.com.

Koulish, Robert, and Ernesto Calvo. 2019 (April 1). "The Human Factor: Algorithms, Dissenters, and Detention in Immigration Enforcement." ICLSCL Working Paper. Available atilcss.umd.edu/papers.

Kratsios, Michael J. 2020 (January 8). "AI That Reflects American Values." Office of Science and Technology Policy. Available at www.whitehouse.gov.

Kreps, Sarah. 2020 (June). "The Role of Technology in Online Misinformation." Brookings Institute. Available at www.brookings.edu.

Krishner, Tom. 2019 (March 29). "Takata Airbags Claim Another Life after Arizona Crash." ABC News. Available at abcnews.go.com.

Kuhn, Thomas. 1973. *The Structure of Scientific Revolutions.* Chicago: University of Chicago Press.

Kumar, Sachin, Prayag Tiwari, and Mikhail Zymbler. 2019 (December 1). "Internet of Things Is a Revolutionary Approach for Future Technology Enhancement: A Review." *Journal of Big Data* 6(111):1–21.

Kuziemski, Maciej, and Gianluca Misuraca. 2020 (April). "AI Governance in the Public Sector: Three Tales from the Frontiers of Automated Decision-Making in Democratic Settings." *Telecommunications Policy* 44:1–13.

Lamoureux, Mack. 2019 (July 11). "People Tell Us How QAnon Destroyed Their Relationships." *Vice Magazine.* Available at www.vice.com.

Landrum, Asheley R., and Alex Olshansky. 2019. "The Role of Conspiracy Mentality in Denial of Science and Susceptibility to Viral Deception about Science." *Politics and Life Sciences* 38(2):193–209.

LaFrance, Adrienne. 2020 (June). "The Prophecies of Q: American Conspiracy Theories Are Entering a Dangerous New Phase." *Atlantic.* Available at www.theatlantic.com.

Lander, Eric, Francoise Baylis, Feng Zhang, Emmanuelle Charpentier, Paul Berg, and specialists from seven countries. 2019 (March 14). "Adopt a Moratorium on Heritable Genome Editing." *Nature.* Available at media.nature.com.

Lee, Dave. 2019 (November 20). "Uber Self-Driving Crash 'Mostly Caused by Human Error.'" British Broadcasting Company. Available at www.bbc.com.

Lee, Wendy. 2020 (July 20). "Popular TikTok Stars to Leave Platform for Rival App over Data Privacy Concerns." *Los Angeles Times.* Available at www.latimes.com.

Levin, Sam. 2018 (July 12). "Walmart Patents Tech That Would Allow It to Eavesdrop on Cashiers." *Guardian.* Available at www.theguardian.com.

Levine, David. L. 2017 (July 26). "Nobelists, Students and Journalists Grapple with the Anti-Science Movement." *Scientific American.* Available at scientificamerican.com.

Lubold, Gordon, and Dustin Volz. 2020 (May 14). "U.S. Says Chinese, Iranian Hackers Seek to Steal Coronavirus Research." *Wall Street Journal.* Available at www.wsj.com.

Lustig, Robert. 2018 (February 7). "Truth about Tech: A Roadmap for Kids' Digital Well-Being." Conference sponsored by Common Sense and the Center for Humane Technology. Washington, DC: Kaiser Permanente. Conference video available on YouTube.

Magagnoli, Joseph, Siddharth Narendran, Felipe Pereira, Tammy Cummings, James W. Hardin, S. Scott Sutton, and Jayakrishna Ambati. 2020 (November 1). "Outcomes of Hydroxychloroquine Usage in United States Veterans Hospitalized with Covid-19." *Med Clinical Advances* 1:1–14.

Malakoff, David, 2020 (October 15). "A Biden Presidency Could Have a 'Remarkable' Impact On Science Policy—But Also Faces Hurdles." *Science.* American Association for the Advancement of Science. Available at www.sciencemag.org.

Malakoff, David, and Jeffrey Mervis. 2020 (February 10). "Trump's 2021 Budget Drowns Science Agencies in Red Ink, Again." *Science.* American Association for the Advancement of Science. Available at www.sciencemag.org.

Marchione, Marilyn. 2020 (April 21). "More Deaths, No Benefit from Malarial Drug in VA Virus Study." *AP News.* Available at apnews.com.

Maron, Dina Fine. 2018 (January 24). "First Primate Clones Produced Using the 'Dolly' Method." *Scientific American.* Available at www.scientificamerican.com.

Massey, Eli, and Nathan J. Robinson. 2019 (June 13). "The Terrifying World of Alex Jones." *Current Affairs.* Available at www.currentaffairs.org.

Mayo Clinic. 2020. "Genetic Testing." Available at www.mayoclinic.org.

McCollum, Sean. 2011. "Getting Past the 'Digital Divide.'" *Teaching Tolerance* 39. Available at www.tolerance.org.

McEvoy, Jemima. 2020 (November 29). "Trump Claims FBI and Justice Department May Have Helped Rig the Election." *Forbes*. Available at www.forbes.com.

McDermott, John. 1993. "Technology: The Opiate of the Intellectuals." In *Technology and the Future*, ed. Albert H. Teich, 89–107. New York: St. Martin's Press.

McGinley, Laurie, and William Wan. 2019 (April 3). "This Clinic's Experimental Stem Cell Treatment Blinded Patients. Years Later, the Government Is Still Trying to Stop It." *Washington Post*. Available at www.washingtonpost.com.

McGregor, Grady. 2020 (August 4). "A New Executive Order Escalates Trump's Attack on H-1B Visas for Foreign Workers." *Fortune*. Available at fortune.com.

McKinsey Global Institute (MGI). 2017 (November 28). *Jobs Lost, Jobs Gained: What the Future of Work Will Mean for Jobs, Skills, and Wages*. Available at www.mckinsey.com.

McMullan, Thomas. 2014 (December 2). "How Digital Maps Are Changing the Way We Understand Our World." *Guardian*. Available at www.theguardian.com.

Merton, Robert K. 1973. "The Normative Structure of Science." In *The Sociology of Science*, ed. Robert K. Merton, 267–280. Chicago: University of Chicago Press.

Mervis, Jeffrey, and Jocelyn Kaiser. 2021 (January 15). "Biden Appoints Geneticist Eric Lander as Science Advisor." *Science*. Available at www.sciencemap.org.

Mesthene, Emmanuel G. 1993. "The Role of Technology in Society." In *Technology and the Future*, ed. Albert H. Teich, 73–88. New York: St. Martin's Press.

Meyer, Robinson. 2018 (March 20). "The Cambridge Analytica Scandal, in Three Paragraphs." *Atlantic*. Available at www.theatlantic.com.

Miele, Francisco, and Lia Tirabeni. 2020. "Digital Technologies and Power Dynamics in the Organization: A Conceptual Review of Remote Working and Wearable Technologies at Work." *Sociology Compass* 14(6):1–13.

Milius, Susan. 2020 (August 22). "Genetically Modified Mosquitoes Have Been OK'd for a First U.S. Test Flight." *Science News*. Available at www.sciencenews.org.

Mitchell, Amy, Mark Jurkowitz, J. Baxter Oliphant, and Elisa Shearer. 2020 (June 29). "Three Months in, Many Americans See Exaggeration, Conspiracies Theories and Partisanship in COVID-19 News." Pew Research Center. Available at www.journalism.org.

Mitchell, Amy, Mark Jurkowitz, J. Baxter Oliphant, and Elisa Shearer. 2020 (July 30). "Americans Who Mainly Get Their News on Social Media Are Less Engaged, Less Knowledgeable." Pew Research Center. Available at www.journalism.org.

Molloy, David, and Joe Tidy. 2020 (June 1). "George Floyd: Anonymous Hackers Re-Emerge amid U.S. Unrest." BBC News. Available at www.bbc.com.

Mooney, Chris, and Sheril Kirshenbaum. 2009. *Unscientific America: How Scientific Literacy Threatens Our Future*. Philadelphia: Basic Books.

Morton, Heather. 2020 (March 27). "Net Neutrality 2020 Legislation." National Conference of State Legislatures. Available at www.ncsl.org.

MSN News. 2020 (November 8). "Designers Create 'Healthier Cabins' Proposed Pandemic Air Travel." Available at www.msn.mom.

Murnaghan, Ian. 2020 (July 1). "Therapeutic Cloning." Explore Stem Cells. Available at www.explorestemcells.co.uk.

Muro, Mark, Jacob Whiton, and Robert Maxim. 2019 (November). *What Jobs Are Affected by AI?* Brookings Institute. Available at www.brookings.edu.

Mustafaoglu, Rustem, Emrah Zirek, Zeynal Yasaci, and Arzu Razak Ozdincler. 2018. "The Negative Effects of Digital Technology Usage on Children's Development and Health." *Addicta* 5(2):2149–1305.

Nano Database. 2020. "Search Database." Available at nano.db.dk.

National Human Genome Research Institute. 2017 (January 18). "Specific Genetic Disorders." National Institutes of Health. Available at www.genome.gov.

National Institute of Health. 2020a (August 17). "What Are Genome Editing and CRISPR-Cas9?" Genome Research Institute. Available at pubmed.ncbi.nlm.nih.gov.

National Institute of Health. 2020b. "What Is Genetic Discrimination?" U.S. National Library of Medicine. Available at ghr.nlm.nih.gov.

National Nanotechnology Initiative (NNI). 2019a. "What Is Nanotechnology?" Available at www.nano.gov.

National Nanotechnology Initiative (NNI). 2019b. "Benefits and Applications." Available at www.nano.gov.

National Science Foundation (NSF). 2020 (January). *The State of U.S. Science and Engineering 2020*. Available at nces.nsf.gov.

Nemr Christina, and William Gangware. 2019 (March). "Weapons of Mass Distraction: Foreign State Sponsored Disinformation in the Digital Age." U.S. Department of State's Global Engagement Center. Available at www.state.gov.

Nicas, Jack, and Davey Alba. 2021 (January 13). "Amazon, Apple and Google Cut off Parler, an App That Drew Trump Supporters." *New York Times*. Available at www.nytimes.com.

Ogburn, William F. 1957. "Cultural Lag as Theory." *Sociology and Social Research* 41:167–174.

O'Keefe, Shannon Mullen. 2020 (August 7). "One in Three Americans Would Not Get Covid-19 Vaccine." Gallup Poll. Available at news.gallup.com.

Open Access. 2019 (February 5). "Why Do Some Countries Censor the Internet?" Open Access Government. Available at www.openaccessgovernment.org.

Organization for Economic Cooperation and Development (OECD). 2020. "What Are the OECD Principles on AI?" Available at www.oecd.org.

Ortez, Erik. 2020 (February 16). "A Texas Jury Found Him Guilty of Murder. A Computer Algorithm Proved His Innocence." MSN News. Available at www.msn.com.

Ortiz-Ospina, Esteban. 2019 (September 18). "The Rise of Social Media." *Our World in Data*. Available at ourworldindata.org.

Ortutay, Barbara. 2021 (March 23). "Report: Extremist Groups Thrive on Facebook Despite Bans." *Associated Press*. Available at apnews.com.

O'Sullivan, Donie. 2020 (January). "When Seeing Is No Longer Believing: Inside the Pentagon's Race against Deepfake Videos." CNN Business. Available at www.cnn.com.

Oxford Economics. 2020. "How Robots Change the World." Available at resources.oxfordeconomics.com.

Paul, Kari. 2020 (March 2). "'Ban This Technology': Students Protest U.S. Universities' Use of Facial Recognition." *Guardian*. Available at www.theguardian.com.

Parker, Kim, Rich Morin, and Juliana Menasche Horowitz. 2019 (March 21). "The Future of Work in Automated Places." Pew Research Center. Available at www.pewsocialtrends.org.

Perrin, Andrew, and Erica Turner. 2019 (August 20). "Smartphones Help Blacks, Hispanics Bridge Some—But Not All—Digital Gaps with Whites." Pew Research Center. Available at www.pewresearch.org.

Pew Research Center. 2019a (June 12). "Mobile Fact Sheet." Available at www.pewresearch.org.

Pew Research Center. 2019b (June 12). "Social Media Fact Sheet." Available at www.pewresearch.org.

Pew Research Center. 2019c (June 12). "Internet/Broadband Fact Sheet." Available at www.pewresearch.org.

Pew Research Center. 2019d. "Science Knowledge Quiz: How Much Do You Know about Science?" Available at www.pewresearch.org.

Picchi, Aimee. 2019 (April 10). "Boeing Sued by Shareholders for Fraud after Deadly 737 Crashes." CBS News. Available at www.cbsnews.com.

Plumer, Brad, and Coral Davenport. 2019 (December 28). "Science under Attack: How Trump Is Sidelining Researchers and Their Work." *New York Times*. Available at www.nytimes.com.

Polyakova, Alina, and Daniel Fried. 2019 (June). "Democratic Defense against Disinformation 2.0." Atlantic Council Policy on Intellectual Independence. Available at www.atlanticcouncil.org.

Postman, Neil. 1992. *Technopoly: The Surrender of Culture to Technology*. New York: Knopf.

President's Council of Advisors on Science and Technology. 2014 (April). *Report to the President on Big Data and Privacy: A Technological Perspective*. Available at www.obamawhitehouse.archives.gov.

Press Release. 2020 (August 7). "Statement by NCSC Director William Evanina: Election Threat Update for the American Public." Office of the Director of National Intelligence. Available at www.dni.gov.

Pulido, Christina M., Beatriz Villarrjo-Carballido, Gisela Redondo-Sama, and Aitor Gomez. 2020. "COVID-19 Infodemic: More Retweets for Science-Based Information on Coronavirus Than for False Information." *International Sociology* 35(4):377–392.

Reardon, Sarah. 2019 (April 30). "Trump's Science Advisor on Research Ethics, Immigration and Presidential Tweets." *Nature.* Available at www.nature.com.

Reddit QAnon Casualties. 2020. Available at https://www.reddit.com/r/QAnonCasualties.

Reinhart, R. J. 2018 (June 6). "Moral Acceptability of Cloning Animals Hits New High." Gallup Poll. Available at news.gallup.com.

Research Now. 2019. *The Workers Experience Survey.* Dell Technologies, Inc. Available at www.dellemc.com.

Ribeiro, Manoel Horta, Raphael Ottoni, Robert West, Virgilio A. F. Almeida, and Wagner Meira. 2020 (January 27–30). "Auditing Radicalization Pathways on YouTube." In *Proceedings of the 2020 Conference on Fairness, Accountability, and Transparency, Barcelona, Spain,* 131–141. New York: Association for Computing Machinery.

Ridgefield's Hamlet Hub. 2020 (July 27). "Ridgefield Resident Conner Curran Is First Patient in Clinical Trial, Now He Can Run!" Available at www.hamlethub.com.

Rigano, Christopher. 2019 (January). "Using Artificial Intelligence to Address Criminal Justice Needs." *National Institute of Justice Journal* 280:1–10.

Risen, Tom. 2015 (February 26). "FCC Enacts Title II Net Neutrality Rules with Partisan Vote." *US News & World Report.* Available at www.usnews.com.

Roberts, Yvonne. 2014 (August 31). "Addicted to Email? The Germans Have an Answer." *Guardian.* Available at www.theguardian.com.

Robotic Industries Association. 2020. "The Market for Service Robots Is Expanding Rapidly." Available at www.robotics.org.

Rocheleau, Matt. 2017 (November 1). "Here's the Full List of Russian-Linked Twitter Handles." *Boston Globe.* Available at www3.bostonglobe .com.

Romo, Vanessa. 2019 (October 10). "El Paso Walmart Shooting Subject Pleads Not Guilty." National Public Radio. Available at www.npr.org.

Rosenberg, Andrea A., and Kathleen Rest. 2018 (January 1). "The Trump Administration's War on Science Agencies Threatens the Nation's Health and Safety." *Scientific American.* Available at www.scientificamerican.com.

Rosenberg, Matthew, Nicholas Confessore, and Carole Caldwallader. 2018 (March 17). "How Trump Consultants Exploited the Facebook Data of Millions." *New York Times.* Available at www.nytimes.com.

Rosenfeld, Michael J., Rubin J. Thomas, and Sonia Hausen. 2019 (September 3). "Disintermediating Your Friends: How Online Dating in the United States Displaces Other Ways of Meeting." *Proceedings of the National Academy of Sciences* 116(36):17753–17758.

Rotondi, Valentina, Luca Stanca, and Miriam Tomasuolo. 2017. "Connecting Alone: Smart Phone Use, Quality of Social Interactions and Well-Being." *Journal of Economic Psychology* 63:17–26.

S. 1558-*Artificial Intelligence Initiative Act.* 116th Congress (2019–2020). Available at www.congress.gov.

Sanders, Sam. 2016 (November 7). "Social Media's Increasing Role in the 2016 Presidential Election." *National Public Radio.* Available at www.npr.org.

Satariano, Adam. 2020 (May 7). "How My Boss Monitors Me While I Work from Home." *New York Times.* Available at www.nytimes.com.

Schaeffer, Katherine. 2020 (April 8). "Nearly 3 in 10 Americans Believe COVID-19 Was Made in a Lab." Pew Research Center. Available at www.pewresearch.org.

Schwartz, Matthew S. 2019 (May 3). "Facebook Bans Alex Jones, Louis Farrakhan and Other 'Dangerous' Individuals." National Public Radio. Available at www.npr.org.

Science Policy News. 2019 (October 22). "Trump Reconstitutes the President's Council of Advisors on Science and Technology." American Institute of Physics. Available at www.aip.org.

Servick, Kelly. 2019 (June 4). "Controversial U.S. Bill Would Lift Supreme Court Ban on Patenting Human Genes." *Science.* American Association for the Advancement of Science. Available at www.sciencemag.org.

Shane, Scott, and Vindu Goel. 2017 (September 6). "Fake Russian Facebook Accounts Bought $100,000 in Political Ads." *New York Times.* Available at www.nytimes.com.

Shear, Michael D., Maggie Haberman, Nicholas Confessore, Karen Yourish, Larry Buchanan, and Keith Collins. 2019 (November 2). "How Trump Reshaped the Presidency in over 11,000 Tweets." *New York Times.* Available at www.nytimes.com.

Shearer, Elisa. 2018 (December 10). "Social Media Outpaces Print Newspapers in the U.S. as a News Source." Pew Research Center. Available at www.pewresearch.org.

Sheils, Conor. 2020 (August 5). "Enter the Deep & Dark Web If You Dare." Available at digital.com.

Silver, Laura, and Christine Haung. 2019 (August 22). "In Emerging Economies, Smart Phone and Social Media Users Have Broader Social Networks." Pew Research Center. Available at www.pewresearch.org.

Smith, Aaron. 2019 (February 13). "7 Things We've Learned about Computer Algorithms." Pew Research Center. Available at www .pewresearch.org.

Spinelli, Dan. 2020 (July 25). "The "Plandemic" Conspiracy Theorist Is Coming to a TV Near You." *Mother Jones.* Available at www .motherjones.com.

Sprunt, Barbara. 2020 (May 15). "Trump, Unveiling Space Force Flag, Helps What He Calls New 'Super-Duper Missile.'" National Public Radio. Available at www.npr.org.

Statista. 2020a (September). "Share of U.S. Households with Computer and Broadband Use in 2018." Available at www.statista.com.

Statista. 2020b (November). "Share of Household with a Computer at Home Worldwide in 2005 to 2019." Available at www.statista.com.

Statista. 2020c (January 28). "Internet Usage Rate Worldwide in 2019, by Gender and Market Maturity." Available at www.statista.com.

Statista. 2020d (July 17). "Distribution of Internet Users Worldwide as of 2019, by Age Group." Available at www.statista.com.

Statista. 2020e. "Percentage of Employed Women in Computing Related Occupations in the United States, 2007 to 2019, by Ethnicity." Available at www.statista.com.

Stein, Rob. 2020 (July 20). "Gene Therapy Shows Promise for Hemophilia, but Could Be Most Expensive U.S. Drug Ever." National Public Radio. Available at www.npr.org.

Stein, Rob. 2019 (May 24). "At $2.1 Million, A New Gene Therapy Is the Most Expensive Drug Ever." National Public Radio. Available at www.npr.org.

Sticca, Fabio, and Sonja Perren. 2013. "Is Cyberbullying Worse Than Traditional Bullying? Examining the Differential Roles of Media, Publicity, and Anonymity for the Perceived Severity of Bullying." *Journal of Youth and Adolescence* 42(5):739–750.

Stockdale, Laura A., and Sarah M. Coyne. 2020. "Bored and Online: Reasons for Using Social Media, Problematic Social Networking Site Use, and Behavioral Outcomes across the Transition from Adolescence to Emerging Adulthood." *Journal of Adolescence* 79:173–183.

Stout, Lynn A. 2012 (June). "The Problem of Corporate Purpose." *Issues in Governance Studies* 48:1–14.

Strauss, Mark. 2018 (August 20). "A Majority of Americans Support Using Biotechnology to Grow Human Organs in Animals for Transplants." Pew Research Center. Available at www.pewresearch.org.

Sullivan, Gail. 2014 (August 26). "Students Develop Nail Polish to Detect Date-Rape Drugs." *Washington Post.* Available at www .washingtonpost.com.

Thielman, Sam. 2015 (October 18). "ACLU Lawsuit against NSA Mass Surveillance Dropped by Federal Court." *Guardian.* Available at www.theguardian.com.

Thigpen, Cary Lynne, and Cary Funk. 2019 (August 27). "Most Americans Say Science Has Brought Benefits to Society and Expect More to Come." Pew Research Center. Available at www.pewresearch.org.

Thigpen, Cary Lynne, and Cary Funk. 2020 (June 25). "Younger, More Educated U.S. Adults Are More Likely to Take Part in Citizen Science Research." Pew Research Center. Available at www.pewresearch.org.

Thornhill, Ted. 2014 (May 5). "'I Can Grab My Handlebars!' Girl Born without Fingers Delighted after High School Builds Her a Prosthetic Hand for Just Five Dollars." *Daily Mail.* Available at www.dailymail.co.uk.

Tiffany, Kaitlyn. 2020 (November 5). "QAnon Is Winning." *Atlantic.* Available at www .theatlantic.com.

Toews, Rob. 2020 (May 25). "Deepfakes Are Going to Wreak Havoc on Society. We Are Not Prepared." *Forbes.* Available at www.forbes.com.

Toffler, Alvin. 1970. *Future Shock.* New York: Random House.

Tollefson, Jeff. 2020 (November 7). "Scientists Relieved as Joe Biden Wins Tight U.S. Presidential Election." *Nature.* Available at www.nature.com.

Tong-Hyung, Kim. 2020 (March 31). "In Virus Hit South Korea, AI Monitors Lonely Elders." Associated Press. Available at apnews.com.

Tor Project. 2020. "Browse Privately. Explore Freely." Available at www.torproject.org.

Trapani, Josh, and Katherine Hale. 2019 (September 4). "Higher Education in Science and Engineering." Available at ncses.nsf.gov.

Tsetsi, Eric, and Stephen A. Rains. 2017 (June 13). "Smartphone Internet Access and Use: Extending the Digital Divide and Usage Gap." *Mobile Media and Communication* 5(3):239–255.

Twenge, Jean M. 2020 (March 27). "Increases in Depression, Self-Harm, and Suicide among U.S. Adolescents after 2012 and Links to Technology Use: Possible Mechanisms." *Psychiatric Research and Clinical Practice.* Available at prcp.psychiatryonline.org.

United Nations Educational, Scientific and Cultural Organization (UNESCO). 2019. "The State of Broadband 2019." ITU/UNESCO. Available at www.itu.int.

U.S. Census Bureau. 2020 (May 21). "E-Stats 2018: Measuring the Electronic Economy." Available at www.census.gov.

U.S. National Academy of Sciences (NAS). 2020 (June). "Statement to Restore Science-Based Policy in Government." Available at scientistsforsciencebasedpolicy.org.

Valentine, Sarah. 2019. "Impoverished Algorithms: Misguided Governments, Flawed Technologies, and Social Control." *Fordham Urban Law Journal* 46(2):364–427.

Vincent, James. 2017 (November 2). "Google's AI Thinks This Turtle Looks like a Gun, Which Is a Problem." *The Verge.* Available at www .theverge.com.

Vought, Russell T. 2019 (January 7). "Guidance for Regulation of Artificial Intelligence Applications." *Memorandum for the Heads of Executive Departments and Agencies.* Available at www.whitehouse.gov.

Vosoughi, Soroush, Deb Roy, and Sinan Aral. 2018 (March 9). "The Spread of True and False News Online." *Science* 359(6380):1151.

Wachter, Robert M. 2015 (March 21). "Why Health Care Tech Is Still So Bad." *New York Times.* Available at www.nytimes.com.

Wallach, Omri. 2020 (June 6). "How Tech Giants Make Their Billions." *Visual Capitalist.* Available at www.visualcapitalist.com.

Weaver, Jane. 2020 (July 16). "More People Search for Health Information Online but Often Can't Find What They're Looking for, Study Says." NBC News. Available at www.nbcnews .com.

Weiser, Benjamin. 2015 (February 4). "Man behind Silk Road Website Is Convicted on All Counts." *New York Times.* Available at www .nytimes.com.

Wessel, Lindzi. 2020 (October 7). "Scientists Concerned over U.S. Plans to Collect DNA Data from Immigrants." *Nature.* Available at www .nature.com.

West, Sarah Myers, Meredith Whittaker, and Kate Crawford. 2019. "Discriminating Systems: Gender, Race and Power in AI." AI Now Institute. Available at ainowinstitute.org.

Winder, Davey. 2020 (April 28). "Zoom Gets Stuffed: Here's How Hackers Got Hold of 500,000 Passwords." *Forbes.* Available at www.forbes.com.

Wong, Julia Carrie. 2019 (July 12). "Facebook to Be Fined $5BN for Cambridge Analytica Privacy Violations—Reports." *Guardian.* Available www.guardian.com.

World Economic Forum (WEF). 2018. "Digital Wildfires." *Global Risks Report 2018.* Available at reports.weforum.org.

World Economic Forum (WEF). 2020a. "How the Next Generation Information Technologies Tackled Covid-19 in China." Global Agenda. Available at www.weforum.org.

World Economic Forum (WEF). 2020b (January 15). "The Global Risks Report 2020." Available at www.weforum.org.

World Health Organization. 2020 (August 25). "Immunizing the Public against Misinformation." Available at www.who.int.

World in Data. 2020. "Technology Adoption in U.S. Household, 1950 to 2019." Available at ourworldindata.org.

Wulfsohn, Joseph A. 2019 (May 23). "Manipulated Videos of Nancy Pelosi Edited to Falsely Depict Her as Drunk Spread on Social Media." Fox News. Available at www.foxnews .com.

Youn, Soo. 2019 (July 30). "The Capital One Data Breach Is Alarming but These Are the Five Worst Corporate Hacks." ABC News. Available at abcnews.go.com.

Young, Jessica. 2020 (February 19). "U.S. E-Commerce Sales Growth 14.9% in 2019." *Digital Commerce.* Available at www .digitalcommerce360.com.

Zadrozny, Brandy and Ben Collins. 2020 (October 15). "YouTube Bans QAnon, Other Conspiracy Content that Targets Individuals." NBC News. Available at nbc.news.com

Chapter 15

Alexander, Karen, and Mary E. Hawkesworth, eds. 2008. *War and Terror: Feminist Perspectives.* Chicago: University of Chicago Press.

Anti-Defamation League (ADL). 2020 (February). "Murder and Extremism in the United States in 2019." *Center on Extremism.* Available at www.adl.org.

Arango, Tim, and Anne Barnard. 2013 (June 1). "As Syrians Fight, Sectarian Strife Infects Mideast." *New York Times.* Available at www .nytimes.com.

Arms Control Association. 2018. *IAEA Investigations of Iran's Nuclear Activities.* Available at www.armscontrol.org.

Associated Press. 2016 (September 16). "Congressional Report Slams NSA Leaker Edward Snowden." Fox News. Available at www.foxnews.com.

Atomic Archive. 2020. *Arms Control Treaties.* Available at www.atomicarchive.com.

Baer, Daniel B. 2020 (March 18). "The Virus Has Exposed the Recklessness of Trump's 'America First.'" *Foreign Policy.* Available at www .foreignpolicy.com.

Barkan, Steven, and Lynne Snowden. 2001. *Collective Violence.* Boston: Allyn and Bacon.

BBC. 2000 (April 15). "UN Admits Rwanda Genocide Failure." BBC News. Available at news.bbc.co.uk.

Beardsley, Kyle. 2011. *The Mediation Dilemma.* Ithaca, NY: Cornell University Press.

Beardsley, Kyle, David E. Cunningham, and Peter B. White. 2019. "Mediation, Peacekeeping, and the Severity of Civil War." *Journal of Conflict Resolution* 63(7):1682–1709.

Beirich, Heidi. 2019 (February 20). "The Year in Hate: Rage against Change." *Intelligence Report.* Available at www.splcenter.org.

Béraud-Sudreau, Lucie, and Nick Childs. 2018. "The US and Its NATO Allies: Costs and Value." *International Institute for Strategic Studies.* Available at iiss.org.

Bergen, Peter. 2002. *Holy War, Inc.: Inside the Secret World of Osama bin Laden.* New York: Free Press.

Berrigan, Frida, and William Hartung. 2005. *U.S. Weapons at War: Promoting Freedom or Fueling Conflict?* World Policy Institute Report. Available at www.worldpolicy.org/projects /arms/reports/wawjune2005.html.

Biden-Harris. 2020. "The Biden Plan for Securing Our Values as a Nation of Immigrants." Available at www.joebiden.com

Bjorgo, Tore. 2003. *Root Causes of Terrorism.* Paper presented at the International Expert Meeting, June 9–11. Oslo: Norwegian Institute of International Affairs.

Bonikowski, Bart, and Paul DiMaggio. 2016. "Varieties of American Popular Nationalism." *American Sociological Review* 81(5):949–980.

Bonner, Michael. 2006. *Jihad in Islamic History: Doctrines and Practice.* Princeton, NJ: Princeton University Press.

Borum, Randy. 2011. "Radicalization into Violent Extremism II: A Review of Conceptual Models and Empirical Research." *Journal of Strategic Security* 4(4):37–62. Available at scholarcommons.usf.edu/jss /vol4/iss4/3.

Boustany, Nora. 2007 (July 3). "Janjaweed Using Rape as 'Integral' Weapon in Darfur, Aid Group Says." *Washington Post.* Available at www .washingtonpost.com.

Brauer, Jurgen. 2003 (Spring). "On the Economics of Terrorism." *Phi Kappa Phi Forum* 38–41.

Brenan, Megan. 2020 (June 15). "U.S. National Pride Falls to Record Low." *Gallup.* Available at www.news.gallup.com.

Brooks, David. 2011 (March 3). "Huntington's Clash Revisited." *New York Times.* Available at www.nytimes.com.

Brown, Michael E., Sean M. Lynn-Jones, and Steven E. Miller, eds. 1996. *Debating the Democratic Peace.* Cambridge, MA: MIT Press.

Butler, John Sibley, Bryan Stephens, and Ori Swed. 2019. "Who Are the Private Military and Security Contractors? A Window to a New Profession." In *The Sociology of Privatized Security,* ed. Ori Swed and Thomas Crosbie, pp. 237–258. Cham, Switzerland: Palgrave Macmillan.

Caprioli, M. 2005. "Primed for Violence: The Role of Gender Inequality in Predicting Internal Conflict." *International Studies Quarterly* 49(2):161–178.

Carneiro, Robert L. 1994. "War and Peace: Alternating Realities in Human History." In *Studying War: Anthropological Perspectives*, ed. S. P. Reyna and R. E. Downs, 3–27. Langhorne, PA: Gordon & Breach.

Carroll, Joseph. 2004 (September 14). "Iraq Support Split along Racial Lines." *Gallup*. Available at www.news.gallup.com.

Center for Arms Control and Non-Proliferation. 2013. "Fact Sheet: Global Nuclear Weapons Inventories in 2013." Available at www.armscontrolcenter.org.

Center for Systemic Peace. 2020. "Global Trends in Armed Conflict 1948–2019." Available at systemicpeace.org.

Cohen, Ronald. 1986. "War and Peace Proneness in Pre- and Post-Industrial States." In *Peace and War: Cross-Cultural Perspectives*, ed. M. L. Foster and R. A. Rubinstein, 253–267. New Brunswick, NJ: Transaction Books.

Conflict Research Consortium. 2003. *Mediation*. Available at www.colorado.edu/conflict/peace.

Congressional Research Service (CRS). 2017. *Domestic Terrorism: An Overview*. CRS Report R44921. Available at https://crsreports.congress.gov.

Congressional Research Service [CRS]. 2019. *American War and Military Operations Casualties: Lists and Statistics*. CRS Report RL32492. Available at https://crsresports.congress.gov.

Crowley, Michael. 2020 (November 23). "An Obama Restoration on Foreign Policy? Familiar Faces Could Fill Biden's Team." *New York Times*. Available at www.nytimes.com.

Database. 2020 (April). "Documented Civilian Deaths from Violence." *Iraq Body Count*. Available at iraqbodycount.org.

Dalgaard, Nina Thorup, and Edit Montgomery. 2017. "The Transgenerational Transmission of Refugee Trauma: Family Functioning and Children's Psychosocial Adjustment." *International Journal of Migration, Health, and Social Care* 13(3):289–301.

Deen, Thalif. 2000 (September 9). *Inequality Primary Cause of Wars, Says Annan*. Available at www.hartford-hwp.com/archives.

Defense Advanced Research Projects Agency (DARPA). 2020. "Personal Assistant That Learns (PAL)." Available at www.darpa.mil.

Defense Casualty Analysis System (DCAS). 2020. "Conflict Casualties." U.S. Department of Defense. Available at www.dmdc.osd.mil.

Dixon, William J. 1994. "Democracy and the Peaceful Settlement of International Conflict." *American Political Science Review* 88(1):14–32.

Donnelly, John. 2011 (January 24). "More Troops Lost to Suicide." Available at www.congress.org.

Dower, John. 1986. *War without Mercy*. New York: Pantheon Books.

Elliot, Dan. 2015 (February 4). "U.S. to Destroy Largest Remaining Chemical Weapons Cache." *USA Today*. Available at www.usatoday.com.

Feder, Don. 2003. "Islamic Beliefs Led to the Attack on America." In *The Terrorist Attack on America*, ed. Mary E. Williams, 20–23. Farmington Hills, MA: Greenhaven Press.

Freytag, Andreas, Jens J. Kruger, Daniel Meierrieks, and Friedrich Schneider. 2011. "The Origins of Terrorism: Cross-Country Estimates of Socio-Economic Determinants of Terrorism." *European Journal of Political Economy* 27(1):S5–S16.

Funke, Odelia. 1994. "National Security and the Environment." In *Environmental Policy in the 1990s: Toward a New Agenda*, 2nd ed., ed. Norman J. Vig and Michael E. Kraft, 323–345. Washington, DC: Congressional Quarterly.

Garamone, Jim. 2019 (January 22). "DOD Moves Closer to Implementing Changes to Transgender Policy." *U.S. Department of Defense News*. Available at www.defense.gov.

Gasser, Martina, and Mareva Malzacher. 2020. "Beyond Banning Mercenaries: The Use of Private Military and Security Companies under IHL." In *International Humanitarian Law and Non-State Actors*, 47–77. The Hague, Netherlands: T.M.C. Asser Press.

Geneva Graduate Institute for International Studies. 2015. *Small Arms Survey, 2015*. New York: Oxford University Press.

Geneva Graduate Institute for International Studies. 2018. *Small Arms Survey, 2018*. New York: Oxford University Press.

Gettleman, Jeffrey. 2009 (August 4). "Symbol of Unhealed Congo—Male Rape Victims." *New York Times*. Available at www.nytimes.com.

Gioseffi, Daniela. 1993. "Introduction." In *On Prejudice: A Global Perspective*, ed. Daniela Gioseffi, xi–l. New York: Anchor Books, Doubleday.

Gladstone, Rick. 2015 (June 18). "Many Ask, Why Not Call Church Shooting Terrorism?" *New York Times*. Available at www.nytimes.com.

Glasser, Susan B. 2018 (December 17). "How Trump Made War on Angela Merkel and Europe." *New Yorker*. Available at www.newyorker.com.

Goodman, Simon, and Amrita Narang. 2019. "'Sad Day for the UK": The Linking of Debates about Settling Refugee Children in the UK with Brexit on an Anti-Immigrant News Website." *European Journal of Social Psychology* 49(6):1161–1172.

Goodwin, Jeff. 2006. "A Theory of Categorical Terrorism." *Social Forces* 84(4):2027–2046.

Gorka, Katharine C. 2014. "The Flawed Science behind America's Counter-Terrorism Strategy." Council on Global Security. Available at councilonglobalsecurity.org.

Goure, Don. 2003 (March 20). *First Casualties? NATO, the U.N.* MSNBC News. Available at www.msnbc.com/news.

Greenburg, Jan Crawford, and Ariane de Vogue. 2008 (June 12). "Supreme Court: Guantanamo Detainees Have Rights in Court." ABC News. Available at abcnews.go.com.

Gurr, Ted Robert. 2015. *Political Rebellion: Causes, Outcomes and Alternatives*. New York: Routledge.

Haiken, Melanie. 2013 (February 5). "Suicide Rate among Vets and Active Duty Military Jumps—Now 22 a Day." *Forbes*. Available at www.forbes.com.

Haynes, Suyin. 2020 (June 17). "President Trump's North Korea Summits Gave Kim Jong Un 'de Facto Nuclear State Status,' Ban Ki-moon Says." *Time*. Available at www.time.com.

Hicks, Josh. 2013 (August 14). "Pentagon Extends Benefits to Same-Sex Military Spouses." *Washington Post*. Available at www.washingtonpost.com.

Higginbotham, Adam. 2016 (January). "There Are Still Thousands of Tons of Unexploded Bombs in Germany, Left over from World War II." *Smithsonian Magazine*. Available at www.smithsonianmag.com.

Hodge Seck, Hope. 2019 (December 11). "The First Woman Has Made It through SEAL Officer Training." *Military.com*. Available at www.military.com

Hsu, Spencer S., and Victoria St. Martin. 2015 (April 13). "Four Blackwater Guards Sentenced in Iraq Shootings of 31 Unarmed Civilians." *Washington Post*. Available at www.washingtonpost.com.

Humans of New York. 2016. "Invisible Wounds." Available at www.humansofnewyork.com.

Huntington, Samuel. 1996. *The Clash of Civilizations and the Remaking of World Order*. New York: Simon & Schuster.

Insinna, Valeria. 2019 (August 21). "Inside America's Dysfunctional Trillion-Dollar Fighter-Jet Program." *New York Times*. Available at www.nytimes.com.

Institute for Economics and Peace (IEP). 2013. "The Economic Consequences of War on the U.S. Economy." Available at economicsandpeace.org.

Institute for Economics and Peace (IEP). 2018. *The Economic Value of Peace*. Available at www.visionofhumanity.org.

Intel. 2015. "In Pursuit of Conflict Free Minerals." Available at www.intel.com.

International Campaign to Ban Landmines. "Treaty Status." Available at www.icbl.org.

International Code of Conduct Association (ICoCA). 2020. *Members and Observers of the ICoCA*. Available at www.icoca.ch.

International Committee of the Red Cross. (2015). "Rule 93. Rape and Other Forms of Sexual Violence." *Customary International Humanitarian Law*. Available at www.icrc.org.

Jaffe, Alexandra, Matthew Lee, and Aamer Madhani. 2020 (November 24). "'America Is Back': Biden Pushes Past Trump Era with Nominees." Associated Press. Available at www.apnews.com.

Janowitz, Morris. 1960. *The Professional Soldier: A Social and Political Portrait*. Glencoe, IL: Free Press.

Kaufman, Edward, and Manuel Hassassian. 2009. "Understanding Our Israeli–Palestinian Conflict and Searching for Its Resolution." In *Regional and Ethnic Conflicts*, ed. Judy Carter, George Irani, and Vamik D. Volkan, 90–129. Upper Saddle River, NJ: Pearson.

Kaur, Harmeet, Allen Kim, and Ivory Sherman. 2020 (January 11). "The US–Iran Conflict: A Timeline of How We Got Here." CNN. Available at www.cnn.com.

Kemp, Janet E. 2014 (January 24). "Suicide Rates in VHA Patients through 2011 with Comparisons with Other Americans and Veterans through 2010." Veterans Health Administration. Available at www.mentalhealth.va.gov.

Kirkpatrick, David D. 2014 (November 23). "Graft Hobbles Iraq's Military in Fighting ISIS." *New York Times*. Available at www.nytimes.com.

Klar, Rebecca. 2019 (August 7). "More U.S. Voters See White Nationalism as a Threat: Poll." *The Hill*. Available at www.thehill.com.

Klare, Michael. 2001. *Resource Wars: The New Landscape of Global Conflict*. New York: Metropolitan Books.

Kleykamp, Meredith, and Molly Clever. 2015. "Women in Combat: The Quest for Full Inclusion." In *Women, War, and Violence*, vol. 2, ed. Mariam Kurtz and Lester Kurtz. Santa Barbara, CA: Praeger.

Knickerbocker, Brad. 2002 (February 13). "Return of the Military-Industrial Complex?" *Christian Science Monitor*. Available at www.csmonitor .com.

Korotayev, Andrey V., and Alisa R. Shishkina. 2020. "Relative Deprivation as a Factor of Sociopolitical Destabilization: Toward a Comparative Analysis of the Arab Spring Events." *Cross-Cultural Research* 54(2-3):296–318.

Kruglanski, Arie W., Michelle J. Gelfand, Jocelyn J. Belandr, Anna Sheveland, Malkanthi Hetiarachchi, and Rohen Gunaratna. 2014. "The Psychology of Radicalization and Deradicalization: How Significance Quest Impacts Violent Extremism." *Political Sociology* 35(1):69–93.

Lamont, Beth. 2001. "The New Mandate for UN Peacekeeping." *The Humanist* 61:39–41.

Landler, Mark. 2018 (May 8). "Trump Abandons Iran Nuclear Deal He Long Scorned." *New York Times*. Available at www.nytimes.com.

Laqueur, Walter. 2006. "The Terrorism to Come." In *Annual Editions 05–06*, ed. Kurt Finsterbusch, 169–176. Dubuque, IA: McGraw-Hill/Dushkin.

Lederer, Edith. 2005 (May 20). "Annan Lays out Sweeping Changes to U.N." Associated Press. Available at www.apnews.com.

Lewis, Jon, Seamus Hughes, Oren Segal, and Ryan Greer. 2020 (April). "White Supremacist Terror: Modernizing Our Approach to Today's Threat." Program on Extremism and Anti-defamation League at George Washington University. Available at www.extremism.gwu.edu.

Lindner, Andrew M. 2009. "Among the Troops: Seeing the Iraq War through Three Journalistic Vantage Points." *Social Problems* 56(1):21–48.

Massoumi, Narxanin. 2020 (March 6). "Why Is Europe so Islamophobic?" *New York Times*. Available at www.nytimes.com.

McGarrity, Michael C. 2019 (May 8). "Confronting the Rise of Domestic Terrorism in the Homeland." Statement before the House Homeland Security Committee. Washington, DC. Available at www.fbi.gov.

Mittal, Dinesh, K. L. Drummond, D. Belvins, G. Curran, P. Corrigan, and G. Sullivan. 2013 (June). "Stigma Associated with PTSD: Perceptions of Treatment Seeking Combat Veterans." *Psychiatric Rehabilitation Journal* 36(2):86–92.

Montalván, Luis Carlos. 2011. *Until Tuesday: A Wounded Warrior and the Golden Retriever Who Saved Him*. New York: Hyperion.

Morgan, Jenny, and Alic Behrendt. 2009. *Silent Suffering: The Psychological Impact of War, HIV, and Other High-risk Situations on Girls and Boys in West and Central Africa*. Surrey, UK: Plan International. Available at plan-international.org.

Moskos, Charles. 2000. "Toward a Postmodern Military: The United States as a Paradigm." In *The Postmodern Military: Armed Forces after the Cold War*, ed. Charles C. Moskos, John Allen Williams, and David R. Segal, 14–31. New York: Oxford University Press.

Mueller, John, and Mark G. Stewart. 2014 (Summer). "Evaluating Counterterrorism Spending." *Journal of Economic Perspectives* 28(3):237–247.

Murad, Nadia. 2018 (October 6). "I Was an Isis Sex Slave. I Tell My Story because It Is the Best Weapon I Have." *Guardian*. Available at www .theguardian.com.

Myers-Brown, Karen, Kathleen Walker, and Judith A. Myers-Walls. 2000 (November 20). "Children's Reactions to International Conflict: A Cross-Cultural Analysis." Paper presented at the National Council of Family Relations, Minneapolis.

National Association of School Psychologists. 2003. *Children and Fear of War and Terrorism*. Available at www.nasponline.org.

National Center for Posttraumatic Stress Disorder. 2007. "What Is Post Traumatic Stress Disorder?" Available at www.ncptsd.va.gov.

National Center for Veterans Analysis and Statistics. 2020. "Expenditures." U.S. Department of Veterans' Affairs. Available at www.va.gov/vetdata/Expenditures.asp.

National Education Association. 2019. "2017–2018 Average Starting Teacher Salaries by State." Available at www.nea.org.

National Immigration Law Center. 2019. "Understanding Trump's Muslim Bans." Available at www.nilc.org.

National Priorities Project. 2019. "Cost of War: Trade-Offs." Available at costofwar.com.

NATO. 2020. "Funding NATO." Available at www.nato.int.

Naughtie, James. 2016 (July 9). "How Tony Blair Came to Be So Unpopular." BBC News. Available at www.bbc.com.

New York Times. 2014 (September 10). "Transcript of Obama's Remarks on the Fight against ISIS." Available at www.nytimes.com.

Nowrasteh, Alex. 2018 (March 8). "More Americans Die in Animal Attacks Than in Terrorist Attacks." *CATO Institute*. Available at www.cato.org.

Office of the Coordinator for Counterterrorism. 2020. "Foreign Terrorist Organizations." Available at www.state.gov/s/ct.

Office of Management and Budget. 2020. "*Table S-8: Funding Levels for Appropriated ("Discretionary") Programs by Agency.*" Available at www.whitehouse.gov.

Office of Weapons Removal and Abatement. 2020. *To Walk the Earth in Safety: The United States' Commitment to Humanitarian Mine Action and Conventional Weapons Destruction*. Available at www.state.gov.

Pape, Robert A., and Keven Ruby. 2021 (February 2). "The Capitol Rioters Aren't like Other Extremists." *Atlantic*. Available at www .theatlantic.com.

Paul, Annie Murphy. 1998. "Psychology's Own Peace Corps." *Psychology Today* 31:56–60.

Pavesi, Irene. 2019 (April). "*Transfers and Transparency*." Small Arms Survey. Available at www.smallarmssurvey.org.

Permanent Court of Arbitration. 2017. "Cases: Interstate Proceedings." Available at pca-cpa.org.

Pew Research Center. 2019a (April 2). "Large Majorities in Both Parties Say NATO Is Good for the U.S." Available at www.people-press.org.

Pew Research Center. 2019b (March 28). "Majority Says Trump Has Done 'Too Little' to Distance Himself from White Nationalists." Available at www.people-press.org.

Pew Research Center. 2020 (April 10). "Americans See Spread of Disease as Top International Threat, Along with Terrorism, Nuclear Weapons, Cyberattacks." Available at www.pewresearch.com.

Philipps, Dave. 2019 (November 22). "Trump Clears Three Service Members in War Crimes Cases." *New York Times*. Available at www .nytimes.com.

Piazza, James A. 2011 (March). "Poverty, Minority Economic Discrimination, and Domestic Terrorism." *Journal of Peace Research* 28(3):339–353.

Pilkington, Ed. 2013 (July 30). "Bradley Manning Verdict: Cleared of 'Aiding the Enemy' but Guilty of Other Charges." *Guardian*. Available at www.theguardian.com

Porter, Bruce D. 1994. *War and the Rise of the State: The Military Foundations of Modern Politics*. New York: Free Press.

Portero, Ashley. 2013 (January 25). "Women in Combat Units Could Help Reduce Sexual Assaults: US Joint Chiefs Chairman." *International Business Times*. Available at www.ibtimes.com.

Poushter, Jacob. 2015 (November 17). "In Nations with Significant Muslim Populations, Much Disdain for ISIS." Pew Research. Available at www.pewresearch.org.

Prasad, Hari. 2015 (May 29). "The Security Dilemma and ISIS." *International Affairs Review*. Available at www.iar-gwu.org.

Rashid, Ahmed. 2000. *Taliban: Militant Islam, Oil, and Fundamentalism in Central Asia*. New Haven, CT: Yale University Press.

Rasler, Karen, and William R. Thompson. 2005. *Puzzles of the Democratic Peace: Theory, Geopolitics, and the Transformation of World Politics*. New York: Palgrave Macmillan.

Recovering Warrior Task Force. 2014. *Annual Report*. Washington, DC: Department of Defense. Available at rwtf.defense.gov.

Renner, Michael. 2000. "Number of Wars on Upswing." In *Vital Signs: The Environmental Trends That Are Shaping Our Future*, ed. Linda Starke, 110–111. New York: Norton.

Robbins, Joseph, Lance Hunter, and Gregg R. Murray. 2013 (July). "Voters versus Terrorists: Analyzing the Effect of Terrorist Events on Voter Turnout." *Journal of Peace Research* 50(4):495–508.

Rosenberg, Tina. 2000 (October 8). "The Unbearable Memories of a U.N. Peacekeeper." *New York Times*, pp. 4, 14.

Salladay, Robert. 2003 (April 7). *Anti-War Patriots Find They Need to Reclaim Words, Symbols, Even U.S. Flag from Conservatives*. Available at www.commondreams.org.

Scheff, Thomas. 1994. *Bloody Revenge*. Boulder, CO: Westview Press.

Schmitt, Eric. 2005 (May 14). "Pentagon Seeks to Shut down Bases across Nation." *New York Times*. Available at www.nytimes.com.

Schmitt, Eric, Thomas Gibbons-Neff, Charlie Savage, and Helene Cooper. 2020 (November 16). "Trump Is Said to Be Preparing to Withdraw Troops from Afghanistan, Iraq and Somalia." *New York Times*. Available at www.nytimes.com.

Senate Select Committee on Intelligence. 2014. "Committee Study of the Central Intelligence Agency's Detention and Interrogation Program." Available at www.documentcloud.org.

Shanahan, John J. 1995. "Director's Letter." *Defense Monitor* 24(6):8.

Shrader, Katherine. 2005 (March 31). "WMD Commission Releases Scathing Report." *Washington Post*. Available at www.washingtonpost.com.

Silver, Roxane Cohen, E. Alison Holman, Judith Pizarro Andersen, et al. 2013 (September). "Mental-and Physical-Health Effects of Acute Exposure to Media Images of the September 11, 2011, Attacks and the Iraq War." *Psychological Science* 24(9):1623–1634.

Skocpol, Theda. 1992. *Protecting Soldiers and Mothers: The Political Origins of Social Policy in the United States*. Cambridge, MA: Belknap Press of Harvard University Press.

Skocpol, Theda. 1994. *Social Revolutions in the Modern World*. Cambridge: Cambridge University Press.

Spearin, Christopher. 2014 (Summer). "Special Operations Forces and Private Security Companies." *Parameters* 44(2):61–73.

Starr, J. R., and D. C. Stoll. 1989. *U.S. Foreign Policy on Water Resources in the Middle East*. Washington, DC: Center for Strategic and International Studies.

Stockholm International Peace Research Institute (SIPRI). 2019. *SIPRI Yearbook, 2019*. Available at www.sipri.org.

Stockholm International Peace Research Institute (SIPRI). 2020 (April). *Trends in World Military Expenditures*. Available at www.sipri.org.

Syrian Network for Human Rights (SNHR). 2019. *On the Universal Children's Day: At Least 29,017 Children Have Been Killed in Syria since March 2011*. Available at www.sn4hr.org.

Tarabah, Asma, Lina Kurdahi Badr, Jinan Usta, and John Doyle. 2015 (May 26). "Exposure to Children's Desensitization Attitudes in Lebanon." *Journal of Interpersonal Violence*.

Tavernise, Sabrina, and Matthew Rosenberg. 2021 (January 7). "These Are the Rioters Who Stormed the Nation's Capitol." *New York Times*. Available at www.nytimes.com.

Tilly, Charles. 1992. *Coercion, Capital and European States: AD 990–1992*. Cambridge, MA: Basil Blackwell.

Toft, Monica Duffy. 2010. "Ending Civil Wars: A Case for Rebel Victory?" *International Security* 34(4):7–36.

Tyson, Alec. 2017 (January 26). "Americans Divided in Views of Use of Torture in U.S. Anti-Terror Efforts." Pew Research Center. Available at www.pewresearch.org.

UNICEF. 2019. *UNICEF Annual Report, 2019*. Available at www.unicef.org

United National Mine Action Office. 2011 (June). "UNMAO Regional Fact Sheet." *Southern Sudan*. Available at reliefweb.int/report/sudan/unmao-regional-fact-sheet-southern-sudan-updated-june-201.

United Nations. 1948. *Convention on the Prevention and Punishment of the Crime of Genocide*. Available at www.un.org.

United Nations. 2020a. "Summary of Troops Contributing Countries by Ranking: Police, UN Military Experts on Mission, Staff Officers and Troops." Available at www.peacekeeping.un.org.

United Nations. 2020b. "How We Are Funded." Available at www.peacekeeping.un.org.

United Nations High Commissioner for Refugees (UNCHR). 2020a. "UNHCR Population Statistics." Available at www.unhcr.org.

United Nations High Commissioner for Refugees (UNCHR). 2020b. *Born in Exile*. Available at www.unchr.org.

U.S. Department of Defense. 2020. "Fiscal Year 2021 Budget Request." Available at www.comptroller.defense.gov.

U.S. Department of Homeland Security. 2020. "FY 2021 Budget-in-Brief." Available at www.dhs.gov.

U.S. Department of State. 2019. "World Military Expenditures and Arms Transfers." Available at www.state.gov.

U.S. Department of State. 2020. "To Walk the Earth in Safety." Bureau of Political-Military Affairs. Available at www.state.gov.

U.S. Department of Veterans Affairs. 2018. "PTSD: National Center for PTSDI." Available at www.ptsd.va.gov.

U.S. Department of Veterans Affairs. 2020. "Education and Training." Available at www.benefits.va.gov.

U.S. House of Representatives. 2019 (February 27). "Hearing: Transgender Service in the Military Policy." Testimony of Captain Jennifer Peace. Available at www.docs.house.gov.

United States Space Force. 2020. "Mission." Available at www.spaceforce.mil.

Vaughen, Kelly. 2020 (June 8). "Exclusive: Utah Marine Veteran Explains Why He Chose to Wear 'I Can't Breathe' Mask." KUTV. Available at www.kutv.com.

Vesely, Milan. 2001. "UN Peacekeepers: Warriors or Victims?" *African Business* 261:8–10.

Viner, Katharine. 2002 (September 21). "Feminism as Imperialism." *Guardian*. Available at www.guardian.co.uk.

Volkin, Samuel. 2020 (April 20). "How Are Refugees Affected by COVID-19?" *The Hub*. Available at hub.jhu.edu.

Waters, Rob. 2005 (March–April). "The Psychic Costs of War." *Psychotherapy Networker*, pp. 1–3.

Watson Institute. 2015 (April). "Cost of War: Environmental Costs." Available at watson.brown.edu.

Wax, Emily. 2003 (November 3). "War Horror: Rape Ruining Women's Health." *Miami Herald*. Available at www.miami.com.

Werner, Suzanna, and Amy Yuen. 2005. "Making and Keeping Peace." *International Organization* 59(2):261–292.

White House. 2019 (September 25). *Remarks by President Trump to the 74th Session of the United Nations General Assembly*. Available at www.whitehouse.gov.

Williams, Mary E., ed. 2003. *The Terrorist Attack on America*. Farmington Hills, MA: Greenhaven Press.

Woehrle, Lynne M., Patrick G. Coy, and Gregory M. Maney. 2008. *Contesting Patriotism: Culture, Power, and Strategy in the Peace Movement*. Lanham, MD: Rowman & Littlefield.

WorldWatch Institute. 2015 (June 15). "Modern Warfare Causes Unprecedented Environmental Damage." *Vital Signs*. Available at www.worldwatch.org.

Zoroya, Gregg. 2006 (March 27). "Lifesaving Knowledge, Innovation Emerge in War Clinic." *USA Today*. Available at www.usatoday.com.

Epilogue

Anbinder, Tyler. 2019 (November 7). "Trump Has Spread More Hatred of Immigrants Than Any American in History. U.S. History Is Full of Nativists. But the President Is the Most Powerful One Yet." *Washington Post*. Available at www.washingtonpost.com.

Antle III, W. James. 2021 (March 14). "Biden Immigration Policies Cause a Predictable Border Crisis. Why Didn't He Plan for It?" NBC News. Available at www.nbcnews.com.

Bauer, Bob, and Jack Goldsmith. 2020a. *After Trump: Reconstructing the Presidency*. Washington DC: Lawfare Press.

Bauer, Bob, and Jack Goldsmith. 2020b (September 15). "Why We Wrote 'After Trump'." *Lawfare Institute*. Available at www.lawfareblog.com.

Beer, Tommy. 2021 (January 26). "Report: American Billionaires Have Added More Than $1 Trillion in Wealth during Pandemic." *Forbes*. Available at www.forbes.com.

Biden, Joseph R. 2021. "Inaugural Address by President Joseph R. Biden, Jr." White House. Available at www.whitehouse.gov.

Bowden, John. 2021 (January 17). "7 in 10 Say U.S. Democracy Is Threatened: Survey." *The Hill*. Available at www.thehill.com.

Brockaway, Claire, and Carroll Doherty. 2019 (July 17). "Growing Share of Republicans Say U.S. Risks Losing Its Identity If It Is Too Open to Foreigners." Pew Research Center. Available at www.pewresearch.org.

Cornwell, Susan. 2021 (March 17). "U.S. House Passes Resolution Aimed at Advancing Equal Rights Amendment." Reuters. Available at www.reuters.com.

Drutman, Lee. 2017 (September 5). "We Need Political Parties. But Their Rabid Partisanship Could Destroy American Democracy." *Vox*. Available at www.voxnews.com.

Frosh, Brian, Karl Racine, and Norman Elsen. 2021 (March 9). "Fight Voter Suppression in the States. Let's Not Let America Regress to Jim Crow." *USA Today.* Available at www.usatoday .com.

Gould, Elise. 2019 (March 27). "'Decades of Rising Economic Inequality in the U.S.' Testimony before the U.S. House of Representatives Ways and Means Committee." Economic Policy Institute. Available at www .epi.org.

Hochschild, Arlie R. 2016. *Strangers in Their Own Land.* New York: New Press.

Horowitz, Juliana Menasce, Anna Brown, and Rachel Minkin. 2021 (March 5). "A Year into the Pandemic, Long-Term Financial Impact Weighs Heavily on Many Americans." Pew Research Center. Available at www .pewresearch.org.

Horowitz, Juliana Menasce, Ruth Igielnik, and Rakesh Kochhar. 2020 (January 9). "1. Trends in Income and Wealth Inequality." Pew Research Center. Available at www .pewresearch.org.

January, Briana. 2021 (March 21). "Anti-LGBTQ Groups Are Using Facebook Ads to Spread Misinformation about the Equality Act." *Media Matters for America.* Available at www .mediamatters.org.

Kelly, Jack. 2020 (December 23). "803,000 Americans Filed for Unemployment Last Week: 70 Million Sought Unemployment Benefits since the Pandemic." *Forbes.* Available at www.forbes.com.

Kleinfeld, Rachel, and John Dickas. 2020 (March 5). "Resisting the Call of Nativism: What U.S. Political Parties Can Learn from Other Democracies." Carnegie Endowment for International Peace. Available at carnegieendowment.org.

Krogstad, Jens Manuel. 2020 (June 17). "Americans Broadly Support Legal Status for Immigrants Brought to the U.S. Illegally as Children." Pew Research Center. Available at www.pewresearch.org.

Mettler, Suzanne, and Robert C. Lieberman. 2020. *Four Threats: The Recurring Crisis of American Democracy.* New York: St. Martin's Press.

National Center for Educational Statistics. 2019. "Indicator 27: Educational Attainment." *Status and Trends in the Education of Racial and Ethnic Groups.* Available at nces.ed.gov.

Pew Research Center. 2020 (June 2). "In Changing U.S. Electorate, Race and Education Remain Stark Dividing Lines. Gender Gap in Party Identification Remains Widest in a Quarter Century." Pew Research Center. Available at www.pewresearch.org.

Salvanto, Anthony, Kabir Khanna, Fred Backus, and Jennifer de Pinto. 2021 (January 17). "Americans See Democracy under Threat." CBS News Poll. Available at www.cbsnews .com.

Schaeffer, Katherine. 2020 (February 27). "6 Facts about Economic Inequality in the U.S." Pew Research Center. Available at www .pewresearch.org.

Scigliuzzo, Davide. 2021 (January 17). "The Rich Are Minting Money in the Pandemic like Never Before." Bloomberg Wealth. Available at www .bloomberg.com.

Shinkman, Paul D. 2021 (March 17). "DNI: Domestic Violence Extremism Poses 'Elevated Threat in 2021'." *U.S. News and World Report.* Available at www.usnews.com.

Stokes, Bruce. 2017 (February 1). "What It Takes to Truly Be 'One of Us'." Global Attitudes & Trends. Pew Research Center. Available at www.pewresearch.org.

U.S. Census Bureau. 2019. "Percent of People 25 Years and over Who Have Completed High School or College by Race, Hispanic Origin, and Sex: Selected Years 1940–2019." *Table A-2. CPS Historical Time Series Tables.* Available at www.census.gov.

Woods, Hiatt. 2020 (October 30). "How Billionaires Saw Their Net Worth Increase by Half a Trillion Dollars during the Pandemic." *Insider.* Available at www.businessinsider.com.

Young, Clifford. 2016 (September). "It's Nativism: Explaining the Drivers of Trump's Popular Support." *Ipsos Public Affairs.* Available at www.ipsos.com.

Name Index

Page numbers in *italics* denote figures, illustrations, and captions.

Subject Index

Note: Page numbers in **boldface** type denote definitions. Page numbers in *italics* denote tables, figures, illustrations, and captions.